Physics
off the
Beaten Track

A Selection of Papers
by Robert Delbourgo

Other Related Titles from World Scientific

George Placzek: A Nuclear Physicist's Odyssey
by Aleš Gottvald and Mikhail Shifman
ISBN: 978-981-3236-91-2

Facts and Mysteries in Elementary Particle Physics
Revised Edition
by Martinus Veltman
ISBN: 978-981-3237-05-6
ISBN: 978-981-3237-49-0 (pbk)

New Perspectives on Einstein's $E = mc^2$
by Young Suh Kim and Marilyn E Noz
ISBN: 978-981-3237-70-4

From Quarks to Pions: Chiral Symmetry and Confinement
by Michael Creutz
ISBN: 978-981-3229-23-5

Physics off the Beaten Track

A Selection of Papers by Robert Delbourgo

editor

Robert Delbourgo

University of Tasmania, Australia

World Scientific

NEW JERSEY • LONDON • SINGAPORE • BEIJING • SHANGHAI • HONG KONG • TAIPEI • CHENNAI • TOKYO

Published by

World Scientific Publishing Co. Pte. Ltd.
5 Toh Tuck Link, Singapore 596224
USA office: 27 Warren Street, Suite 401-402, Hackensack, NJ 07601
UK office: 57 Shelton Street, Covent Garden, London WC2H 9HE

British Library Cataloguing-in-Publication Data
A catalogue record for this book is available from the British Library.

Cover image credit:
Hema Maps Pty Ltd
Building 14, Garden City Office Park, 2404 Logan Rd, Eight Mile Plains QLD 4113, Australia
www.hemamaps.com.au
info@hemamaps.com.au

PHYSICS OFF THE BEATEN TRACK
A Selection of Papers by Robert Delbourgo

Copyright © 2019 by World Scientific Publishing Co. Pte. Ltd.

All rights reserved. This book, or parts thereof, may not be reproduced in any form or by any means, electronic or mechanical, including photocopying, recording or any information storage and retrieval system now known or to be invented, without written permission from the publisher.

For photocopying of material in this volume, please pay a copying fee through the Copyright Clearance Center, Inc., 222 Rosewood Drive, Danvers, MA 01923, USA. In this case permission to photocopy is not required from the publisher.

ISBN 978-981-3277-89-2

For any available supplementary material, please visit
https://www.worldscientific.com/worldscibooks/10.1142/11201#t=suppl

Printed in Singapore

I dedicate this book to my wife, Elizabeth. There have been periods of intensive research when I was not of this world and she has had to tolerate my immersion into the world of symbols. She has also proved to be a great sniffer of typographical errors when looking through my manuscripts and has acted as a sounding board for more felicitous English expression.

Preface

The title of this book bears some explanation. For a good chunk of my life I have resided in Tasmania after emigrating from England. It is fair to say that forty years ago Tasmania was well off the beaten track with only occasional visitors who came to see the island out of a sense of curiosity or to sample the 'clean and green' image promoted by some travel brochures. Apart from some radio and cosmic ray astronomy the physics research carried out at the University of Tasmania was also somewhat off the radar, even though its geographic location is a great advantage for lack of radio noise, allowing exploration of low frequency radio signals.

How things have changed. Tasmania is now a hub for Southern Ocean and Antarctic Studies with regular moorings of antarctic research supply vessels and regular flights to Antarctica; tourists now flock here to enjoy its fascinating scenery, sample its pristine food and its interesting history/culture; radio and optical astronomy thrive more than ever and a small niche for theoretical physics has been established. When in Europe or USA, one feels near the epicentre of particle physics research: on track for investigating topical, active areas of interest. Imperial College, under the leadership of Salam, Matthews and Kibble, as well as ICTP were at the forefront of that wagon train and though research was feverish one could spearhead active areas in one's field. That is where I gained my apprenticeship and had my skills honed, under the guidance of experts.

There is a second interpretation of the book title. When I told Salam that I intended moving to Tasmania he gave me a piece of advice: he said that the distant location might prove a handicap (present internet notwithstanding) so he suggested that unfashionable subjects off the beaten track could be worth exploring; that this was a risky path but might pay off handsomely. So the tactic that I have followed since 1976, working as I do in a regional university 'far from the madding crowd', was to not join the main wagon train — as one would inevitably fall way behind — but to explore interesting byways which others might shun; to start off and head a new wagon train; find new angles on some unfashionable topics and try to establish methods/results that withstand the passage of time. One other thing that Salam taught me was to avoid writing potboilers and move on quickly to a new topic if one was not bearing fruit. He did this regularly himself.

Thus my physics investigations have certainly been off the beaten track for a good many years and have widened considerably. At a regional Australian university one must teach all areas of physics at all levels, from freshman to honours to graduate; inevitably one's horizons broaden and so does one's outlook on physics. One can also work at a more leisurely pace and weigh up lines of research that may prove more fruitful despite funding pressures of not following the party line which have existed for a long time. That attitude is reflected in the collection of papers selected for this book. I have compiled those papers which, for the first time, present a line of research that previously went unnoticed or introduce a novel method of tackling a problem and which also has wider applicability. I also indulged myself slightly by including a handful of papers which contain technically very challenging problems because they introduced new techniques.

The book is divided into 13 distinct sections, each beginning with a list, with an appropriate background, followed by some commentary indicating the motivation for that research and pointing to its development by others who pursued the topics.[1] The very first chapter consists of a miscellany of articles that are not readily categorized and do not fall under the headings of Sections 2 to 14. These are perhaps accessible to physics undergraduates in their final years of study, whereas the articles in the other chapters will probably be of interest just to graduate students specialising in theoretical physics.

Robert Delbourgo

[1]A full publication list can be accessed via Google Scholar.

Acknowledgments

This book owes its existence to the blandishments of Professor S C Lim who persuaded me that such an undertaking would be beneficial for warding off the intrusions of old age by giving me a project to focus upon. I was skeptical at first but perhaps he is right.

A good proportion of the papers were written with colleagues and students. If there are any errors or blemishes — I hope not — blame me, not my collaborators. It has been my great privilege and joy to work with them in any event. Having said that there are a number of persons I would like to single out as they have had a substantial impact on my career development: Abdus Salam (of course), John Strathdee, Peter Jarvis, George Thompson, Ruibin Zhang, Brian Kenny, Michael Scadron, Roland Warner, Dirk Kreimer, Andrei Davydychev, Adnan Bashir and Paul Stack; they have contributed or supported me one way or another at various stages of my research.

Finally I would like to express my appreciation to Hema Maps, QLD, for allowing me to include one of their upside-down maps on the front cover.

Contents

Part A
PHYSICS MISCELLANY

A.0. Commentaries A

This set of articles is of general interest and ought to be accessible to undergraduate students who are close to completing their bachelor degrees. They contain papers which do not fall neatly under the headings of succeeding chapters and are the products of straying off the beaten track.

A.1 Collins P., Delbourgo R. and Williams R., On the elastic Schwarzschild scattering cross-section. J. Phys. A6, 161-169 (1973)
A.2 Delbourgo R., Minimal uncertainty states for the rotation and allied groups. J. Phys. A10, 1837-1846 (1977)
A.3 Delbourgo R. and Fox J.R., Maximum weight vectors possess minimal uncertainty. J. Phys. A10, L233-235 (1977)
A.4 Delbourgo R., Universal facets of chaotic processes. ASPAP News 1, 7-11 (1986)
A.5 Delbourgo R., The floating plank. Am. J. Phys. 55, 799-802 (1987)
A.6 Shabala S., Delbourgo R. and Newman I.A., Observations of bifurcation and chaos in plant physiological responses to light. Aust. J. Plant Physiol. 24, 91-96 (1997)
A.7 Daboul J. and Delbourgo R., Matrix representation of octonions and generalisations. J. Math. Phys. 40, 4134-4150 (1999)
A.8 Delbourgo R. and Lashmar D., Born reciprocity and the $1/r$ potential. Found. Phys. 38, 995-1010 (2008)
A.9 Delbourgo R., On 2D periodic hexagonal cells. AAPPS Bulletin 27, 2-7 (2017)

One of the earliest things taught in a Classical Mechanics course is how to calculate the Rutherford scattering cross-section for particles subject to a $1/r$ potential. Article [A.1] is a general relativistic version of this problem, where the gravitational source is a Schwarzschild black hole. The complication arises from geodesics which can wrap around the black hole several times and this paper accounts for those. The extension of this work to spin $1/2$ scattering is discussed in

- C. Doran and A. Lasenby, Phys. Rev. D66, 024006 (2002)
- S. Dolan, C. Doran and A. Lasenby, Phys. Rev. D74, 064005 (2006)

A properly invariant generalisation of Heisenberg's Uncertainty Principle from Weyl commutators to other Lie algebras is treated in articles [A.2] and [A.3]. They lead to the conclusion that 'maximal weight vectors possess minimal uncertainty' and this has repercussions in other areas of physics such as entanglement and state coherence. For further information see the textbooks:

- 'Coherent States, Wavelets and their Generalisations' by S.T. Ali, J.-P. Antoine, J.P. Gazeau

and U.A. Mueller (Springer)
- 'Geometry of Quantum States: An Introduction to Quantum Entanglement' by I. Bengtsson and K. Zyczkowski (Cambridge University Press)
- 'Generalized Coherent States and their Applications' by A. Peremolov (Springer-Verlag)

Feigenbaum's constants describe the scaling properties of the period-doubling approach to chaos. Article [A.4] is a semipopular account of how the scaling relations and associated constants extend to tangent N-furcations whilst article [A.6] is a botanical example/verification of the statement that 'period three implies deterministic chaos'.

Fluid Mechanics is a subject I have had the privilege of teaching. One of the earliest topics is hydrostatics and the stability of floating bodies. Determining the metacentres of those bodies can be difficult; the case of a body with a rectangular cross-section is one of the simplest examples and is surprisingly complicated, depending on how the body floats, as described in paper [A.5]. It has spurred others into examining other shapes and configurations; for instance see

- C. Rorres, The Mathematical Intelligencer 26, 32 (2004)
- R. Trahan and T. Kalmar-Nagy, J. Comp. Nonlinear Dynamics 8, 041012 (2011)

At some stage in one's mathematical life one succumbs to the fascination of quaternions and octonions, although quantum mechanics seems to just require complex numbers. Article [A.7] represents a foray into that field and describes a way of treating octonions via an unusual interpretation of matrix multiplication. Another fascination of mine has been with Born's Reciprocity Principle, even though at first sight it seems to have little relevance to the workings of Mother Nature. In article [A.8] we have studied its possible relevance at galactic scales because it might offer an alternative explanation of galactic rotation curves without invoking dark matter — at ordinary and subatomic scales this fortunately has no observable consequences. For fully relativistic treatment of this quaplectic principle see the following papers:

- P.D. Jarvis and S.O. Morgan, Found. Phys. Lett. 19, 501 (2006)
- S.G. Low., J. Phys. A40, 14 (2007)

Solid State Physics is another area where I have had to teach (and learn for myself more deeply). 2D problems are mostly posed in rectangular configurations as they are the easiest to treat. However with the prominent advent of graphene, it has become important to consider hexagonal configurations. Paper [A.9] is meant for physics teachers and students as an alternative to rectangular shapes, with emphasis on the pictorial beauty of the hexagonal symmetry.

J. Phys. A: Math., Nucl. Gen., Vol. 6, February 1973. Printed in Great Britain. © 1973

On the elastic Schwarzschild scattering cross section

P A Collins, R Delbourgo and Ruth M Williams
Physics Department, Imperial College of Science and Technology, London SW7 2BZ, UK

MS received 11 September 1972

Abstract. We evaluate the differential cross section for scattering from a Schwarzschild source, the analogue of Rutherford scattering. The classical formula is compared with the Born approximation.

1. Introduction

The problem of a relativistic system characterized by a Schwarzschild metric has by now received exhaustive treatment (Darwin 1959, 1961, Zel'dovich and Novikov 1971) and the nature of the geodesics with their characteristic consequences are fairly well known. There is one aspect however, which, although qualitatively understood, has not to our knowledge been given any quantitative scrutiny, and this is the question of the differential scattering cross section from a Schwarzschild source, the direct analogue of Rutherford scattering. This paper is devoted to presenting the numerical results of such a computation. The main features which distinguish the Schwarzschild from the Rutherford problem are (i) the existence of a critical angular momentum below which capture occurs (absorption), (ii) multispiral scatterings which contribute to a given final scattering angle and divide the impact parameters into various 'zones', (iii) an infinite differential cross section in the backward direction which is, however, integrable and (iv) a profound difference between the classical cross section and the quantum mechanical Born approximation.

In § 2 we review the main properties of the geodesics (Synge 1960) and give implicitly the formula for the elastic differential cross section. In § 3 we give asymptotic expressions in the limit of large and critical angular momentum. The numerical results which interpolate between the two limits are presented in § 4 and are compared with the Born approximation results.

2. Geodesics and scattering

The Schwarzschild solution of Einstein's equations is described by the metric

$$\mathrm{d}s^2 = \left(1-\frac{2m}{r}\right)^{-1}\mathrm{d}r^2 + r^2(\mathrm{d}\theta^2 + \sin^2\theta\,\mathrm{d}\phi^2) - \left(1-\frac{2m}{r}\right)c^2\,\mathrm{d}t^2 \qquad (1)$$

where $m = GM/c^2$ and M is the mass of the scattering centre. Hereafter, we shall assume that r, θ, ϕ are indeed the polar coordinates which describe the motion of a particle in the gravitational field and do not require redefinition by further transformation of coordinates. If the motion of the test particle is taken to be in the $\theta = \pi/2$ plane, then the timelike geodesic equation (Synge 1960) is

$$\left(\frac{\mathrm{d}u}{\mathrm{d}\phi}\right)^2 = 2mu^3 - u^2 + 2\alpha^2 mu + \alpha^2(\beta^2 - 1) \equiv 2m(u-u_1)(u-u_2)(u-u_3) \quad (2)$$

where $u = 1/r$. The constants of motion α and β occurring above can be expressed in terms of the mass μ of the scattered particle, its energy E and its angular momentum l at asymptotic radial distances:

$$\alpha = \frac{\mu c}{l}, \qquad \beta = \frac{E}{\mu c^2}. \quad (3)$$

It is sufficient for our purposes to consider the case when the roots $u_1 \leqslant u_2 \leqslant u_3$ are all real. We are interested in situations where the test particle is initially at an infinite distance from the centre, in which case two possibilities arise: (i) scattering of the particle, occurring when $u_1 \leqslant 0 < u_2 < u_3$; (ii) capture of the particle, when $u_1 \leqslant 0 < u_2 = u_3$. (Capture also takes place when u_1 is negative and u_2 and u_3 are complex conjugates with positive real part.)

Integration of equation (2) gives the trajectory equation

$$u = u_1 + (u_2 - u_1)\,\mathrm{sn}^2\,[K - \tfrac{1}{2}\phi\{2m(u_3 - u_1)\}^{1/2}] \quad (4)$$

where we have chosen $\phi = 0$ to correspond to the perihelion ($u = u_2$) so that the constant of integration is

$$K = \int_0^1 \mathrm{d}y\{(1-y^2)(1-k^2y^2)\}^{-1/2}$$

with (5)

$$k^2 = \frac{u_2 - u_1}{u_3 - u_1}.$$

In fact $4K$ is the period of the elliptic function sn and k is its modulus. The total 'angle of deviation' χ of the test particle (see figure 1) is

$$\chi = 2\phi(u = 0) - \pi \quad (6)$$

which is an implicit equation relying on the solution of equation (4).

It is worthwhile to follow the trajectories as we decrease the angular momentum of the system (ie as the impact parameter is reduced) in order to understand the nature of the scattering and thereby to compute the cross section (see figure 1). For small α (large l) there is very little deviation of the trajectory from a straight line and the perihelion is far from the scattering centre (small u_2). As α is increased so is the deviation until a stage is reached when the particle is returned to its initial line of motion; further increase of α results in multispiral motion about the centre (one, two, ... loops), until finally when α approaches the critical values α_c the particle is no longer able to escape to infinity but spirals into the centre. This is one of the features which makes numerical computations that much more difficult than in the Rutherford case.

To arrive at a formula for the differential scattering cross section we assume a steady state situation with a stream of parallel moving particles which are continuously moving

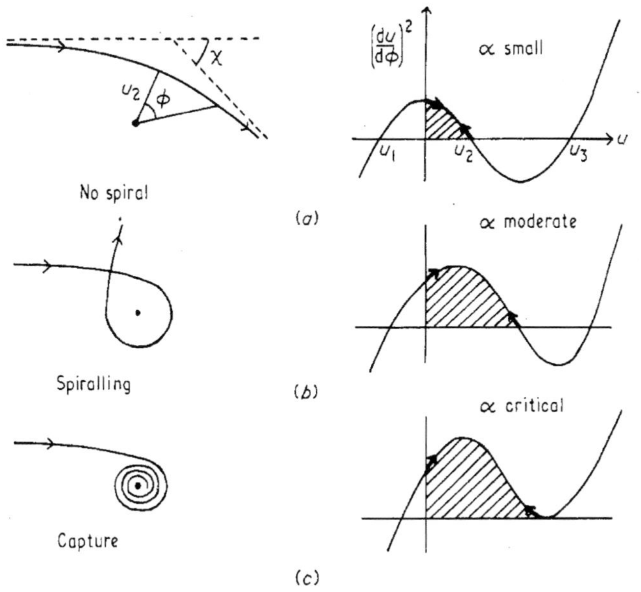

Figure 1. Trajectories and cubics in the corresponding geodesic equations for various values of α.

in from infinity, following trajectories determined by their respective impact parameters

$$b = \frac{l}{p} = \alpha^{-1}(\beta^2 - 1)^{-1/2},$$

and are scattered into the detecting apparatus. We note first of all that the measured scattering angle is $\theta = |\chi - 2n\pi|$ where n is chosen so that $0 \leqslant \theta \leqslant \pi$. The detectors collect particles which have undergone no loops, 1 loop, 2 loops, ...—in this way the impact parameters can be divided into 'Fresnel-like zones'—and theoretically we shall have to sum over all these possibilities. If there were no spiralling we would obtain the conventional formula,

$$\frac{\mathrm{d}\sigma}{\mathrm{d}\Omega} = \frac{b}{\sin\theta}\left|\frac{\mathrm{d}b}{\mathrm{d}\theta}\right| = \frac{1}{\alpha^3(\beta^2-1)\sin\theta}\left|\frac{\mathrm{d}\alpha}{\mathrm{d}\theta}\right|.$$

This has now to be replaced by

$$\frac{\mathrm{d}\sigma}{\mathrm{d}\Omega} = \frac{1}{\beta^2 - 1}\sum_n \frac{1}{\alpha_n^3}\left|\frac{\mathrm{d}\alpha_n}{\mathrm{d}\cos\theta}\right| \tag{7}$$

where α_n connotes the range of α values which result in an n spiral scattering having $\theta = \pm(\chi - 2n\pi)$. In this steady state approach the delays between particles arriving at the detectors having undergone different numbers of spirals are irrelevant.

3. Asymptotic results

We are able to obtain asymptotic expansions about two limits, $\alpha = 0$ and $\alpha = \alpha_c$. These serve to provide useful checks on numerical calculations (given in the next section) which interpolate between the two values.

By expressing u_i in terms of α, we see that small α corresponds to small k of the elliptic function since

$$k^2 = 4m\alpha(\beta^2-1)^{1/2}\{1-2m\alpha(\beta^2-1)^{1/2}+\ldots\}. \tag{8}$$

Let us therefore expand both sides of our basic formula

$$u_1(u_1-u_2)^{-1} = \mathrm{sn}^2[K-\tfrac{1}{4}(\pi+\chi)\{2m(u_3-u_1)\}^{1/2}] \tag{4}$$

in powers of k^2. This involves expansions for the roots u_i of the original cubic equation (2)

$$\begin{aligned} u_1 &= -\frac{k^2}{4m}\left(1+\frac{k^2(\beta^2-2)}{4(\beta^2-1)}+\cdots\right) \\ u_2 &= \frac{k^2}{4m}\left(1+\frac{k^2(3\beta^2-2)}{4(\beta^2-1)}+\cdots\right) \\ u_3 &= \frac{1}{2m}\left(1-\frac{k^4\beta^2}{4(\beta^2-1)}+\cdots\right). \end{aligned} \tag{9}$$

The period of the elliptic function (Magnus *et al* 1966) can likewise be expanded,

$$K = \frac{\pi}{2}\left(1+\frac{k^2}{4}+\frac{9k^4}{64}+\cdots\right)$$

and inserted in the Jacobi elliptic function expansion

$$\begin{aligned} \mathrm{sn}\,\xi = &\left(1+\frac{k^2}{16}+\frac{7k^4}{256}+\cdots\right)\sin\left(\frac{\pi\xi}{2K}\right)+\left(\frac{k^2}{16}+\frac{k^4}{32}+\cdots\right)\sin\left(\frac{3\pi\xi}{2K}\right) \\ &+\left(\frac{k^4}{256}+\cdots\right)\sin\left(\frac{5\pi\xi}{2K}\right)+\cdots \end{aligned} \tag{10}$$

to give

$$\chi = \frac{k^2(2\beta^2-1)}{2(\beta^2-1)}\left\{1+\left(1+\frac{3\pi(5\beta^2-1)}{16(2\beta^2-1)}\right)\frac{k^2}{2}+\cdots\right\}$$

or (11)

$$\theta = 2m\alpha\left(\frac{2\beta^2-1}{(\beta^2-1)^{1/2}}+\frac{3\pi}{8}m\alpha(5\beta^2-1)+\cdots\right).$$

Substituting in equation (7) we obtain the *small angle* differential cross section,

$$\frac{\mathrm{d}\sigma}{\mathrm{d}\Omega} = \left(\frac{2m(2\beta^2-1)}{\theta^2(\beta^2-1)}\right)^2+\frac{3\pi m^2(5\beta^2-1)}{4\theta^3(\beta^2-1)}+\cdots. \tag{12}$$

We see that the leading behaviour agrees with the small angle Rutherford formula in the nonrelativistic limit

$$\beta^2-1 = \frac{2T}{\mu c^2} \ll 1$$

(where T corresponds to the asymptotic kinetic energy) namely

$$\frac{\mathrm{d}\sigma}{\mathrm{d}\Omega} = \left(\frac{GM\mu}{4T}\right)^2\frac{1}{\sin^4\frac{1}{2}\theta} \tag{13}$$

as $\theta\to 0$. However, observe that discrepancies occur in next to leading order.

Near the other limit $\alpha = \alpha_c$ there is a near equality of the roots u_2 and u_3. We therefore expand about the critical value

$$\alpha_c = \left(\frac{\beta(9\beta^2-8)^{3/2}-27\beta^4+36\beta^2-8}{32m^2}\right)^{1/2} \tag{14}$$

where

$$u_{1c} = \frac{1-2(1-12m^2\alpha_c^2)^{1/2}}{6m}$$
$$u_{2c} = u_{3c} = \frac{1+(1-12m^2\alpha_c^2)^{1/2}}{6m} \tag{15}$$

and $k^2 = 1$. Thus a suitable expansion parameter is $1-k^2$ in terms of which we may write (Magnus *et al* 1966)

$$u_1 = u_{1c} + \frac{u_{2c}-u_{1c}}{4}\frac{\beta^2-1+2mu_{1c}}{\beta^2-1+2mu_{2c}}(1-k^2)^2+\ldots$$
$$u_2 = u_{2c} - \tfrac{1}{2}(u_{2c}-u_{1c})(1-k^2)+\ldots$$
$$u_3 = u_{3c} + \tfrac{1}{2}(u_{2c}-u_{1c})(1-k^2)+\ldots \tag{16}$$
$$\alpha = \alpha_c - \frac{m(u_{2c}-u_{1c})^3(1-k^2)^2}{4\alpha_c(\beta^2-1+2mu_{2c})}+\ldots$$

$$K = \sum_{n=0}^{\infty}\frac{(1/2)n(1/2)n}{(n!)^2}\{\psi(n+1)-\psi(n+\tfrac{1}{2})-\tfrac{1}{2}\ln(1-k^2)\}(1-k^2)^n$$
$$= -\tfrac{1}{2}\ln(1-k^2)\{1+\tfrac{1}{4}(1-k^2)+\ldots\}+2\ln 2+\tfrac{1}{4}(2\ln 2-1)(1-k^2)+\ldots. \tag{17}$$

Also from the exact property of the Jacobi elliptic function,

$$\mathrm{sn}(x,k) = \frac{\mathrm{i}\,\mathrm{sn}(-\mathrm{i}x,(1-k^2)^{1/2})}{\mathrm{cn}\,(-\mathrm{i}x,(1-k^2)^{1/2})}$$

we are able to use again expansions of elliptic functions of small moduli to deduce that

$$\mathrm{sn}(x,k) = \tanh x + \tfrac{1}{4}(\tanh x - x\,\mathrm{sech}^2 x)(1-k^2)+\ldots. \tag{18}$$

Equation (4) then reduces to

$$\left(\frac{u_{1c}}{u_{1c}-u_{2c}}\right)^{1/2} = \tanh[2\ln 2-\tfrac{1}{2}\ln(1-k^2)-\tfrac{1}{4}(\chi+\pi)\{2m(u_{2c}-u_{1c})\}^{1/2}]+\mathrm{O}(1-k^2)$$

giving

$$\chi = -\pi+\frac{4}{\{2m(u_{2c}-u_{1c})\}^{1/2}}\left\{-\tfrac{1}{2}\ln(1-k^2)+2\ln 2-\tanh^{-1}\left(\frac{u_{1c}}{u_{1c}-u_{2c}}\right)^{1/2}\right\}+\mathrm{O}(1-k^2)$$
$$= -\pi+\frac{1}{\{2m(u_{2c}-u_{1c})\}^{1/2}}\left\{-\ln(\alpha_c-\alpha)+8\ln 2-4\tanh^{-1}\left(\frac{u_{1c}}{u_{1c}-u_{2c}}\right)^{1/2}\right.$$
$$\left.-\ln\frac{4\alpha_c(\beta^2-1+2mu_{2c})}{m(u_{2c}-u_{1c})^3}\right\}+\mathrm{O}((\alpha_c-\alpha_n)^{1/2}) \tag{19}$$

from which we must subtract off the appropriate number of $2n\pi$ to relate to the scattering angle θ. Thus as $\alpha \to \alpha_c$ the term $\ln(\alpha_c - \alpha)$ dominates and controls the number of spirals. It follows that

$$\frac{d\chi}{d\alpha_n} = \frac{1}{\{2m(u_{2c}-u_{1c})\}^{1/2}(\alpha_c-\alpha_n)} + \ldots = \frac{\exp[\{2m(u_{2c}-u_{1c})\}^{1/2}(2n\pi \pm \theta)]}{\{2m(u_{2c}-u_{1c})\}^{1/2}} + \ldots \tag{20}$$

and therefore the ratio of successive contributions to the sum in equation (7) is essentially given by

$$\frac{\alpha_n^3}{\alpha_{n+1}^3} \exp[-\pi\{2m(u_{2c}-u_{1c})\}^{1/2}].$$

The relationship between the roots of the cubic, and the physical requirement $\beta^2 \geqslant 1$ mean that $2m(u_{2c}-u_{1c})^{1/2} \geqslant \frac{1}{2}$ and therefore the successive 'partial' differential cross sections decrease as $\exp(-\pi/\sqrt{2}) \simeq 0{\cdot}1$. Therefore the sum over spirals converges rapidly for large n and one can write

$$\frac{d\sigma}{d\Omega} = \text{zero spiral cross section} + \frac{\{2m(u_{2c}-u_{1c})\}^{1/2}}{(\beta^2-1)\sin\theta}$$
$$\times \sum_{n=1}^{\infty} \frac{\exp[-\{2m(u_{2c}-u_{1c})\}^{1/2}(2n\pi+\theta)] + \exp[-\{2m(u_{2c}-u_{1c})\}^{1/2}(2n\pi-\theta)]}{\alpha_n^3}$$
$$= \text{zero spiral cross section}$$
$$+ \frac{2\{2m(u_{2c}-u_{1c})\}^{1/2}}{(\beta^2-1)\alpha_c^3 \sin\theta} \exp[-2\pi\{2m(u_{2c}-u_{1c})\}^{1/2}] \cosh[\{2m(u_{2c}-u_{1c})\}^{1/2}\theta] + \ldots . \tag{21}$$

4. Intermediate results

For general values of angular momentum lying between the two limiting cases considered above the differential cross section has to be evaluated numerically. In order to do this we adopt the integrated form of equation (4),

$$\chi = -\pi + 2\int_0^{u_2} du\{2m(u-u_1)(u-u_2)(u-u_3)\}^{-1/2}.$$

Because the integrand has a singularity at the upper limit it is convenient to remove this by partial integration

$$\chi = -\pi + 4\left(\frac{-u_2}{2mu_1u_3}\right)^{1/2} - 2\int_0^{u_2} \left(\frac{u_2-u}{2m}\right)^{1/2} \frac{(u_1+u_3-2u)\, du}{\{(u-u_1)(u_3-u)\}^{3/2}}. \tag{22}$$

Now the integral can be evaluated by computer and in figure 2 we have plotted the differential cross section against momentum transfer t (as is more conventional in high energy physics),

$$\frac{d\sigma}{dt} = \frac{\pi}{\mu^2 c^2(\beta^2-1)} \frac{d\sigma}{d\Omega}.$$

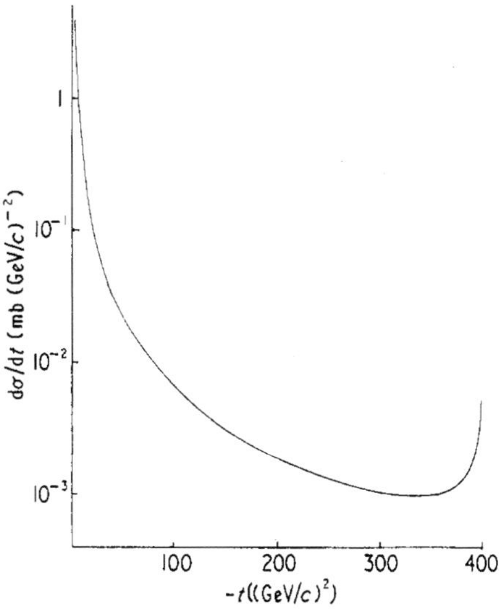

Figure 2. Differential cross section for a particle of mass 1 GeV/c^2 and momentum 10 GeV/c, with a scattering centre of mass 10^{-15} cm in gravitational units.

In our calculation we have given the test particle a typical mass of 1 GeV/c^2 and momentum 10 GeV/c; the mass of the scattering centre has been taken as 10^{-15} cm in gravitational units. The only point of note is the appearance of an infinite cross section in the backward direction which is, however, integrable. The Rutherford cross section, given by equation (13), remains finite in this direction because $db/d\chi$ vanishes there. In our calculation this does not occur since the scattering angle increases indefinitely as b decreases. The infinity in the forward direction is of course the normal one of Rutherford scattering and can be treated in the usual way.

For Coulomb scattering it is a well known fact that in a quantum mechanical treatment the Born approximation reproduces exactly the Rutherford answer and that the higher order contributions do not affect the result apart from endowing the scattering amplitude with an overall phase. It is therefore of interest to determine the Born approximation in the Schwarzschild analogue and compare it with the classical cross section just determined. Now the Green function for a scalar test particle moving in a Schwarzschild metric,

$$(\partial^2+\mu^2)\phi+\partial_\nu\{(g^{\mu\nu}-\eta^{\mu\nu})\partial_\mu\phi\} = 0$$

satisfies the equation

$$\Delta'(p) = (p^2-m^2)^{-1}\left(1+\int\frac{d^4p'}{(2\pi)^4}\,d^4x\,\exp\{i(p-p').x\}p'_\mu p_\nu(g^{\mu\nu}(x)-\eta^{\mu\nu})\Delta'(p')\right).$$

Thus the Born approximation to the scattering amplitude in the Schwarzschild geometry† reduces to

$$T(\boldsymbol{p}',\boldsymbol{p}) = \int d^3\boldsymbol{x}\,\exp\{-i(\boldsymbol{p}-\boldsymbol{p}').\boldsymbol{x}\}\left\{\frac{2m(\boldsymbol{p}^2+\mu^2)}{r-2m}+\frac{2m}{r}\left(\frac{\boldsymbol{x}.\boldsymbol{p}\boldsymbol{x}.\boldsymbol{p}'}{r^2}\right)\right\}$$

† Dr J Strathdee has evaluated a similar expression for the Born amplitude using isotropic coordinates and we are indebted to him for acquainting us with the method and results.

or

$$T(k) = 2m\pi\left(1+\frac{4E^2}{k^2}-\frac{8mE^2}{k}\{\sin(2mk)\,\mathrm{ci}(-2mk)+\cos(2mk)\,\mathrm{si}(-2mk)\}\right)$$

where $\boldsymbol{k}$ = momentum transfer, E = (conserved) energy of test particle. The corresponding cross section is depicted in figure 3 and we see that it differs considerably

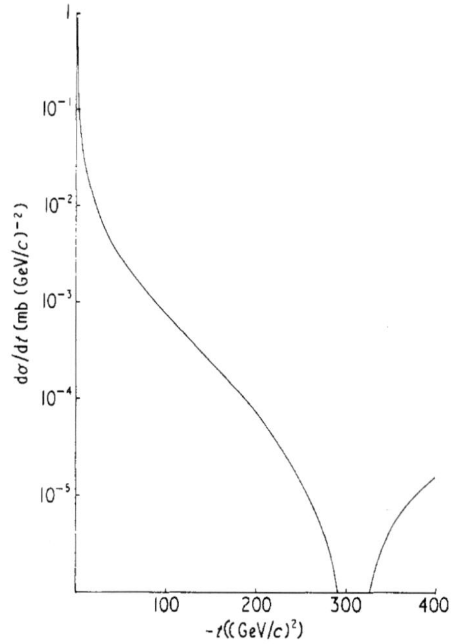

Figure 3. Differential cross section in the Born approximation for a particle of mass 1 GeV/c^2 and momentum 10 GeV/c, with a scattering centre of mass 10^{-15} cm in gravitational units.

from the classical Schwarzschild cross section. It would be rather optimistic although not inconceivable to expect that the exact Green function can reproduce the classical result. However, there are good reasons for doubting such simple minded generalizations to the quantum mechanical situation. In the first place in any realistic scattering process we should be considering the motion of a wave packet, specified by some typical dispersion and size, as it approaches the scattering centre; thus we may imagine a plane-fronted wave of large enough dimensions and characterized by appropriate width approaching the potential singularity. In the Rutherford problem one discovers a recognizable outcoming quasiradial packet emerging after some time, as one can simply ascertain by following the trajectories of a cloud of classical particles which simulate the wave packet. On the other hand, in the Schwarzschild problem this is not what happens; as a result of delays occasioned by multispiralling (due to the dispersion in the packet and the effects of near critical impact parameters) one would expect to get successions of progressively weaker outcoming radial waves, so much so that it is difficult to invisage how a viable radial wave packet can emerge and be interpreted as a particle. It may be that this phenomenon is connected with the apparent singular behaviour of the radial wavefunction at the Schwarzschild radius which suggests a redefinition of the radius r and time t to other more appropriate coordinates (Eddington 1924, Kruskal 1960).

Acknowledgments

We thank Professor Abdus Salam for suggesting this problem and for many helpful remarks. One of us (PAC) thanks the Science Research Council for a Postdoctoral Research Fellowship.

References

Darwin C 1959 *Proc. R. Soc.* A **249** 180–94
—— 1961 *Proc. R. Soc.* A **263** 39–50
Eddington A S 1924 *Nature, Lond.* **113** 192
Kruskal M D 1960 *Phys. Rev.* **119** 1743–5
Magnus W, Oberhettinger F and Soni R P 1966 *Formulae and Theorems for the Special Functions of Mathematical Physics* (Berlin: Springer-Verlag) pp 357–95
Synge J L 1960 *Relativity: the General Theory* (Amsterdam: North-Holland) pp 289–93
Zel'dovich Ya B and Novikov I D 1971 *Relativistic Astrophysics* (Chicago: University of Chicago Press) p 103

J. Phys. A: Math. Gen., Vol. 10, No. 11, 1977. Printed in Great Britain. © 1977

Minimal uncertainty states for the rotation and allied groups

R Delbourgo

Department of Physics, University of Tasmania, Hobart, Tasmania, Australia

Received 2 June 1977, in final form 5 July 1977

Abstract. We show that the measure of angular momentum uncertainty is the invariant $(\Delta J)^2 \equiv \langle \boldsymbol{J}^2 \rangle - \langle \boldsymbol{J} \rangle^2$, rather than sums of products like $\Delta J_1 \Delta J_2$, and that the critical states which minimise ΔJ are eigenstates of maximum weight of $\boldsymbol{J} \cdot \boldsymbol{n}$. We also determine the critical states for the associated groups E(2), O(2, 1), O(4) and E(3). Finally we construct coherent O(3) spin states $|z\rangle$ which are superpositions of normal spin eigenvectors and which tend to the classical limit as $|z| \to \infty$.

1. Introduction

Every basic treatise on quantum mechanics features Heisenberg's position–momentum indeterminacy principle and its deep implications for the physical measurement process. Usually, the example of the simple harmonic oscillator is quoted for the reason that the ground level is the select stationary state which minimises the uncertainty product $\Delta x \Delta p$ at $\frac{1}{2}\hbar$, in contrast to the excited states. More modern books go on to discuss coherent superpositions of energy eigenstates which retain the minimal uncertainty (Glauber 1963), and indeed an extensive literature (see Jackiw 1968, and Mathews and Eswaran 1974 for further references) has arisen over the number-phase uncertainty product appropriate to these coherent states.

So far as angular momentum $\boldsymbol{J}$ is concerned, most elementary books limit themselves to observing that only a single component $\boldsymbol{J} \cdot \boldsymbol{n}$ can be measured at once, with the resulting indeterminacy in the other two perpendicular components described by a vector model in which $\boldsymbol{J}$ rotates about the direction $\boldsymbol{n}$ of quantisation. In this paper we wish to study the indeterminacy question for $\boldsymbol{J}$ in greater depth and try to decide what is the appropriate measure for it; certainly singling out a particular product like $\Delta J_1 \, \Delta J_2$ is not enough because of the obvious bias in direction. The problem of finding the relevant uncertainty measure is not confined to O(3), of course, but applies to any group. In § 2 we discuss what is meant in general by 'quasi-classical' states with the 'least indeterminacy' before concentrating on the rotation group in § 3. Our proposal for $(\Delta J)^2 = \langle \boldsymbol{J}^2 \rangle - \langle \boldsymbol{J} \rangle^2$ as probably the correct measure of uncertainty has the virtue that ΔJ is pure scalar, and we determine the critical states which have fixed $\boldsymbol{J}^2$ and least ΔJ. (In § 6 we remove the restriction of constant j and define coherent spin states as appropriate superpositions of j levels.) Using our knowledge of O(3), quasi-classical states for O(2, 1) and E(2) are found in § 4, by continuation and contraction; and in § 5 we do the same for the associated groups O(4) and E(3).

2. Quasi-classical states

The impossibility of precisely predicting in advance the outcome of a single experiment which modifies the initial state of the system finds its natural expression in the quantum mechanical formulation of the uncertainty relations between measurements of incompatible observables. The Heisenberg uncertainty principle, $\Delta x\, \Delta p \geqslant \frac{1}{2}\hbar$, encapsulates the limitations one must face for coordinate X and momentum P measurements and sets a lower limit of precision on the accuracy of experimental determinations of these particular dynamical variables. In this case, the best one can do is to minimise the uncertainty product at $\frac{1}{2}\hbar$, because this is the closest approach to the classical situation of absolute precision; as shown in elementary textbooks this limit is reached for 'minimal states' which satisfy the equation

$$[\Delta p(X-\langle X\rangle) \pm \mathrm{i}\, \Delta x(P-\langle P\rangle)]|\psi\rangle = 0$$

the solutions of which are the well known oscillating Gaussian wave packets—these include the now familiar 'coherent states' of the oscillator (mass m, frequency ω) wherein $\Delta p = m\omega\, \Delta x$ as well.

We would like to pose a similar question for more general commutation relations, appropriate to any Lie algebra, $[F_r, F_s] = \mathrm{i}c_{rst}F_t$ where F are the generators and c are the (real) structure constants. Namely, what are the 'quasi-classical states' for which the uncertainties are minimised? Since the general problem leads us to commutators which are not c numbers, whose expectation values therefore depend on the (normalised) state $|\psi\rangle$ in question, minimisation of $\Delta F_r\, \Delta F_s$ is not guaranteed to equal $\frac{1}{2}|c_{rst}\langle F_t\rangle|$ unless the system happens to be in an eigenstate of the commutator: in that circumstance Heisenberg's derivation of the principle is correct and gives the 'minimal equation'

$$[\Delta F_s(F_r - \langle F_r\rangle) \pm \mathrm{i}\, \Delta F_r(F_s - \langle F_s\rangle)]|\psi\rangle = 0. \qquad \text{(I)}$$

Otherwise, we are obliged to resort to Jackiw's analytic method (1968) which gives instead the 'critical equation'

$$\left[\left(\frac{F_r - \langle F_r\rangle}{\Delta F_r}\right)^2 + \left(\frac{F_s - \langle F_s\rangle}{\Delta F_s}\right)^2 - 2\right]|\psi\rangle = 0 \qquad \text{(II)}$$

and to look for normalisable solutions of (II) which minimise $\Delta F_r\, \Delta F_s$. On the other hand, it should be pointed out that if one chooses to minimise the uncertainty product *ratio* $\Delta F_r\, \Delta F_s/|c_{rst}\langle F_t\rangle|$ then Heisenberg's direct method (I) does, in fact, apply.

This preamble leads us to the question of what exactly we have to minimise when we are dealing with several Lie algebra generators, and how we are to sharpen the definition of a quasi-classical state. Two problems pose themselves: (i) should one minimise just the products $\Delta F_r\, \Delta F_s$ or perhaps $\Delta F_r\, \Delta F_s/|c_{rst}\langle F_t\rangle|$ or linear combinations thereof, or even multiple products $\Delta F_r\, \Delta F_s\, \Delta F_t \ldots$? (ii) Having decided on the choice (i), how many operators should one take into account? Or, in other words, how large the Lie algebra? The rational answer to the second question is that one must include all the observables of physics (which comprise a maximally commuting set for resolving any degeneracy) and perform the chosen minimisation on the smallest Lie algebra comprising these observables. In this way we avoid considering operators which are not measurable, whose expectation values have no direct physical content. In this paper we shall have little to say about the grand case (ii), but will concentrate on the choice (i) to be made when faced with a particular Lie algebra. More specifically, we will focus our attention on the quasi-classical states for the rotation group O(3) and allied groups. Having sorted these out, some extensions to more involved cases suggest themselves.

3. A case study: O(3)

Let us record a few simple facts about angular momentum, which, if not always described in the standard texts, are trivially deduced. Let $|jm\rangle_{\boldsymbol{n}}$ be a normalised eigenstate of the angular momentum (in units of $\hbar$) directed along unit vector $\boldsymbol{n}$,

$$\boldsymbol{J}\,.\,\boldsymbol{n}|jm\rangle_{\boldsymbol{n}} = m|jm\rangle_{\boldsymbol{n}}, \qquad \boldsymbol{J}^2|jm\rangle_{\boldsymbol{n}} = j(j+1)|jm\rangle_{\boldsymbol{n}}. \tag{1}$$

Then, for such a state,

$$\langle \boldsymbol{J}\rangle = m\boldsymbol{n}, \qquad (\Delta J_1)^2 = \tfrac{1}{2}(1-n_1^2)[j(j+1)-m^2],$$

etc, and

$$(\Delta \boldsymbol{J})^2 = \langle \boldsymbol{J}^2\rangle - \langle \boldsymbol{J}\rangle^2 = \langle \boldsymbol{J}^2\rangle - \langle \boldsymbol{J}\,.\,\boldsymbol{n}\rangle^2 = j(j+1)-m^2. \tag{2}$$

The case $n_3 = 1$, $n_1 = n_2 = 0$ is the most familiar, when

$$\begin{aligned} &\langle J_1\rangle = \langle J_2\rangle = 0, \qquad \langle J_3\rangle = m, \\ &\Delta J_1 = \Delta J_2 = [\tfrac{1}{2}j(j+1) - \tfrac{1}{2}m^2]^{1/2}, \qquad \Delta J_3 = 0 \end{aligned} \tag{3}$$

and the product $\Delta J_1\,\Delta J_2$ is minimised at $\frac{1}{2}j$ by taking $|m| = j$; the other uncertainty products $\Delta J_2\,\Delta J_3$ and $\Delta J_3\,\Delta J_1$ of course being zero. This illustrates the fact that for a compact Lie group like O(3) where the ΔF are bounded it is quite easy to arrange for a number of indeterminacy products to vanish identically by simply diagonalising as many generators as simultaneously possible (the Cartan sub-algebra defining the rank).

If we plot the $\Delta \boldsymbol{J}$ in three dimensions we recognise the products $\Delta J_1 \Delta J_2$, etc, as uncertainty areas. Several natural measures of total uncertainty come to mind, namely

$$\text{the 'uncertainty volume'}, \ \Delta^2 J \equiv \Delta J_1\,\Delta J_2\,\Delta J_3 \tag{4a}$$

$$\text{the 'uncertainty surface'}, \ \Delta^2 J \equiv [(\Delta J_1\,\Delta J_2)^2 + (\Delta J_2\,\Delta J_3)^2 + (\Delta J_3\,\Delta J_1)^2]^{1/2} \tag{4b}$$

$$\text{the 'uncertainty radius'}, \ \Delta J \equiv [(\Delta J_1)^2 + (\Delta J_2)^2 + (\Delta J_3)^2]^{1/2}. \tag{4c}$$

We note that† only the latter is a true scalar. $\Delta^3 J$ becomes a rotational invariant in the infinitesimal limit only, while the components $\Delta J_i\,\Delta J_j$ in $\Delta^2 J$ can be regarded as defining a surface vector. Since it is also true that

$$\Delta J_1\,\Delta J_2 \geqslant \tfrac{1}{2}|\langle J_3\rangle| \tag{5}$$

etc, more exotic measures of uncertainties can be contemplated, involving weightings by expectation values, for instance

$$\delta^3 J \equiv (\Delta J_1\,\Delta J_2\,\Delta J_3)^2 / |\langle J_1\rangle\langle J_2\rangle\langle J_3\rangle| \tag{6a}$$

$$\delta^2 J \equiv [(\Delta J_1\,\Delta J_2)^2 / |\langle J_3\rangle|^2 + \text{cyclic}]^{1/2}, \tag{6b}$$

† From the inequalities between geometric and arithmetic means, note also that $(\Delta^2 J) \geqslant 3^{1/2}(\Delta^3 J)^{2/3}$ and $(\Delta J)^3 \geqslant 3^{3/2}\Delta^3 J$.

and so on. Because we can analyse the case of angular momentum in some detail, fortunately we are able to decide which uncertainty measure is the most appropriate and obtain some clues as to how to proceed with more complicated groups.

It is immediately apparent that minimising the uncertainty volume gives precious little information; thus $\Delta^3 J \equiv 0$ when we are dealing with eigenstates of *any* individual component J_i. Unfortunately too, as a measure of indeterminacy it depends on one's choice of axes. For instance, if we select $\boldsymbol{n} = (1, 1, 1)/\sqrt{3}$ and $m = j$, then according to (2),

$$\Delta J_1 = \Delta J_2 = \Delta J_3 = (j/3)^{1/2}$$

giving $\Delta^3 J = (j/3)^{3/2}$ for this choice of quantisation direction. On the other hand, referred to axes parallel and perpendicular to $\boldsymbol{n}$, $(\Delta^3 J)_{\boldsymbol{n}}$ is zero. For these reasons we shall reject measure (4*a*) as totally unsatisfactory.

Turning to (4*b*), we know from the usual direct derivation of the relations (5) that

$$\Delta^2 J \geqslant \tfrac{1}{2}[\langle J_3\rangle^2 + \langle J_2\rangle^2 + \langle J_1\rangle^2]^{1/2} = \tfrac{1}{2}[j(j+1) - (\Delta J)^2]^{1/2}. \tag{7}$$

In fact, we can remove the equality sign from (7) since the lower bound can only be attained for states which satisfy

$$c_1(J_1 - \langle J_1\rangle)|\psi\rangle = c_2(J_2 - \langle J_2\rangle)|\psi\rangle = c_3(J_3 - \langle J_3\rangle)|\psi\rangle$$

where c_2/c_1, c_3/c_2 and c_1/c_3 are all imaginary, which is clearly impossible. (The case when one of the uncertainties, say ΔJ_3, is zero, causing some c_i to vanish, has to be examined separately; here the answer is already known, namely $\Delta^2 J$ is indeed minimised at $\frac{1}{2}j$ for $|\psi\rangle = |jj\rangle$. Then the equality sign in (7) does indeed apply.) Because there is a strict inequality in (7) for all $\Delta J_i \neq 0$, this implies a failure of the direct method, and to minimise $\Delta^2 J$ we must resort to Jackiw's analytic method which supplies a weaker condition on the wavefunction. Here one finds the critical equation

$$\{(J_1 - \langle J_1\rangle)^2[(\Delta J_2)^2 + (\Delta J_3)^2] + \text{cyclic terms} - 2(\Delta^2 J)^2\}|\psi\rangle = 0 \tag{8}$$

which can be cast in the form

$$[\nu_1^2(J_1^2 - \langle J_1\rangle^2) + \text{cyclic} - (\Delta^2 J)^2(\Delta J)^{-2}]|\psi\rangle = 0 \tag{9}$$

where the components of the unit vector $\boldsymbol{\nu}$ are given by

$$\nu_1^2 = [(\Delta J_2)^2 + (\Delta J_3)^2]/2(\Delta J)^2 \qquad \text{etc.} \tag{10}$$

For reasons of symmetry $\Delta^2 J$ attains its physical extremum† when $\boldsymbol{\nu} = (1, 1, 1)/\sqrt{3}$ with all ΔJ_i equal. As shown in the appendix the solution of (9) with the least $\Delta^2 J$ is none other than $|jj\rangle_{\boldsymbol{\nu}}$ and yields $\Delta^2 J = j$. On the other hand, we clearly do better for the uncertainty surface by taking one uncertainty to vanish (say ΔJ_3) since then $\Delta^2 J$ is as small as $\frac{1}{2}j$. As it is unreasonable to say that an eigenstate of J_3 or J_1 or J_2 has a smaller uncertainty than an eigenstate of $\boldsymbol{J}\cdot\boldsymbol{n}$ we conclude that $\Delta^2 J$ is not a good measure of indeterminacy either, primarily because of its lack of rotational invariance.

The last simple case to examine is the uncertainty radius (4*c*). Here the critical states are given by

$$(\boldsymbol{J} - \langle \boldsymbol{J}\rangle)^2|\psi\rangle = (\Delta J)^2|\psi\rangle. \tag{11}$$

† The hypothetical extremum $\Delta^2 J = 0$ cannot occur physically unless $j = 0$. Other possibilities like $\boldsymbol{\nu} = (1, 1, 0)/\sqrt{2}$ can be ruled out as they imply that $\Delta J_1 = \Delta J_2 = 0$, an impossibility. The case $\Delta J_3 = 0$, $\Delta J_1 = \Delta J_2 \neq 0$ with $\boldsymbol{\nu} = (1, 1, \sqrt{2})/2$ has already been considered.

In the appendix we prove that the solution of (11) is the naive answer $|\psi\rangle = |jj\rangle_{\boldsymbol{n}}$ which yields the minimal $\Delta J = j^{1/2}$. This answer has the virtue that it does not depend on the orientation of axes, and we therefore suggest that the radius ΔJ is much the best measure of uncertainty. It can be argued that what we are proposing for angular momentum contradicts the standard procedure for position–momentum uncertainty. However, it must be pointed out that the Weyl group of Heisenberg commutators is different from O(3) and is, moreover, invariant under the scaling $X \to \lambda X$, $P \to P/\lambda$—hence the area $\Delta x \Delta p$ is the natural measure of uncertainty, not $(\Delta x)^2 + (\Delta p)^2$. For $\boldsymbol{J}$ there is no such scaling invariance, rather the relative weights of $\boldsymbol{J}$ components are absolutely fixed via the Casimir operator $\boldsymbol{J}^2$. And that is why we are advancing ΔJ as the relevant measure, rather than $\Delta^2 J$ or $\Delta^3 J$.

The more sophisticated measures of uncertainty (6*a*) and (6*b*) would require an analysis in their own right were it not for their lack of rotational symmetry. Without examining them in detail we note that for $|jm\rangle_{\boldsymbol{n}}$ states considered in (1) and (2),

$$\delta^3 J = [\tfrac{1}{2}j(j+1) - \tfrac{1}{2}m^2]^3[(1-n_1^2)(1-n_2^2)(1-n_3^2)/n_1n_2n_3|m|^3]$$

$$\delta^2 J = [\tfrac{1}{2}j(j+1) - \tfrac{1}{2}m^2][(1-n_1^2)(1-n_2^2)/n_3^2 + \text{cyclic}]^{1/2}$$

become infinite when quantisation along conventional axes (x or y or z) is made, and they achieve their minimal values of $1/3\sqrt{3}$ and $j\sqrt{3}$ respectively along axis $\boldsymbol{n} = (\pm 1, \pm 1, \pm 1)/\sqrt{3}$ when $m = j$. This suffices to demonstrate their uselessness although one could no doubt arrive at the same conclusion by investigating the critical states of $\delta^3 J$ and $\delta^2 J$.

In a nutshell, the final result of this section is that the states of minimal uncertainty are $|jj\rangle_{\boldsymbol{n}}$ and possess the uncertainty radius $\Delta J = j^{1/2}$. Of course there is no uncertainty in $\boldsymbol{J}^2$.

4. E(2) and O(2, 1)

We can approach the Euclidean group by a contraction of O(3); by putting $P_1 = cJ_1$, $P_2 = cJ_2$ and taking the limit $c \to 0$ in the resulting commutators

$$[P_1, P_2] = \mathrm{i}c^2 J_3, \qquad [P_2, J_3] = \mathrm{i}P_1, \qquad [J_3, P_1] = \mathrm{i}P_2 \tag{12}$$

one arrives at the E(2) Lie algebra. The Casimir

$$\boldsymbol{J}^2 = (P_1^2 + P_2^2)/c^2 + J_3^2 \to \infty$$

is this limit, so by fixing $p = jc$ and finite, p^2 remains as the eigenvalue of the E(2) Casimir $\vec{P}^2$. The relevant uncertainty here is ΔP and since, for the critical angular momentum states $|jj\rangle_{\boldsymbol{n}}$ already found,

$$(\Delta P_1)^2 + (\Delta P_2)^2 = c^2[(\Delta \boldsymbol{J})^2 - (\Delta J_3)^2] = \tfrac{1}{2}c^2 j(1+n_3^2) = \tfrac{1}{2}pc(1+n_3^2),$$

we see that as $c \to 0$, $\Delta p \to 0$ for all $\boldsymbol{n}$. However, $(\Delta J_3)^2 = \frac{1}{2}(1-n_3^2)p/c \to \infty$ unless $n_3 = 1$. These minimal states of E(2) are none other than the eigenstates $|\psi\rangle$ of the translation group, but because they are not strictly normalisable we must analyse them more carefully.

Form the normalised wave packet

$$|f_{\vec{k}}\rangle = \int \mathrm{d}\vec{k}\, f(\vec{k})|\vec{k}\rangle, \qquad \int |f(\vec{k})|^2\, \mathrm{d}\vec{k} = 1.$$

Therefore

$$(\Delta\vec{P})^2 = \int \vec{k}^2 |f(\vec{k})|^2 \,d\vec{k} - \left(\int \vec{k}|f(\vec{k})|^2 \,d\vec{k}\right)^2.$$

In the limit that f becomes a distribution: $|f|^2 \to \delta(\vec{p}-\vec{k})$, we strictly get $\Delta p \to 0$, but the problematic quantity is now $\langle J_3\rangle = -\frac{1}{2}\int f^*(\mathrm{i}\partial f/\partial\phi)\,dk^2\,d\phi$. In order to ensure Hermiticity of J_3, we ought more properly to form periodic wave packets, for example

$$f(\phi) \propto [\exp(-\tfrac{1}{2}\cos^2\tfrac{1}{2}\phi/\sigma)](\sin\tfrac{1}{2}\phi)^{1/2}, \qquad 0<\phi<2\pi$$

with

$$\left.\begin{array}{l}\langle\cos\tfrac{1}{2}\phi\rangle = 0,\ \langle\cos^2\tfrac{1}{2}\phi\rangle \simeq \sigma/2 \\ \langle\cos\phi\rangle \simeq -1+\sigma,\ \langle\sin\phi\rangle = 0\end{array}\right\}\text{as } \sigma\to 0$$

in which case $\Delta J_3 = 0$ trivially. Minimal wave packets of this type correspond to the contracted angular momentum states $|jj\rangle_z$ quantised along the third direction. The other familiar E(2) eigenstates,

$$J_3|p^2m\rangle = m|p^2m\rangle, \qquad (P_1 \pm \mathrm{i}P_2)|p^2m\rangle = (\surd p^2)|p^2m\pm 1\rangle$$

have the variances $\Delta p = p$ and $\Delta J_3 = 0$ and are not minimal states unless $p=0$. They correspond to taking m finite in $|jm\rangle_z$ while letting $j = p/c \to \infty$.

Turning next to the continuation from O(3) to (2, 1), we put $J_1 = \mathrm{i}K_1$, $J_2 = \mathrm{i}K_2$ to arrive at

$$[K_1, K_2] = -\mathrm{i}J_3, \qquad [J_3, K_1] = \mathrm{i}K_2, \qquad [J_3, K_2] = -\mathrm{i}K_1 \tag{13}$$

wherein $\vec{K}$ and J_3 are now the Hermitian generators. The O(2, 1) unitary representations are characterised in the same way as angular momentum

$$\begin{aligned}(J_3^2 - K_1^2 - K_2^2)|jm\rangle &= j(j+1)|jm\rangle \\ J_3|jm\rangle &= m|jm\rangle\end{aligned} \tag{14}$$

but the range of j and m values is different. There are:

(a) the discrete series $D^{\pm}$ where

$$m = \pm(j+1), \pm(j+2), \ldots \qquad j = -\tfrac{1}{2}, 0, \tfrac{1}{2}, \ldots; \tag{15a}$$

(b) the principal series D^{P} where

$$m = \text{integer or half-integer} \qquad j+\tfrac{1}{2} = \text{pure imaginary} = \mathrm{i}\rho; \tag{15b}$$

(c) the complementary series D^{c} where

$$m = \text{integer} \quad \text{and} \quad -1<j<0. \tag{15c}$$

Continuous cases (b) and (c) have a negative Casimir and the case $j=-1$ gives a trivial representation. Having learnt the lesson of O(3), we suggest that the quasi-classical states here are the ones which minimise

$$\Delta K \equiv [(\Delta K_1)^2 + (\Delta K_2)^2 - (\Delta J_3)^2]^{1/2}. \tag{16}$$

If we stick to the standard states (14) for the moment, this means that, since $\Delta K = [m^2 - j(j+1)]^{1/2}$, for ($a$) $D^{\pm}$, ΔK minimises at $(j+1)^{1/2}$ when $m = \pm(j+1)$; while for (b) and (c) $D^{\mathrm{P,c}}$, ΔK minimises at $(\frac{1}{4}+\rho^2)^{1/2}$ or $[-j(j+1)]^{1/2}$ when $m=0$. The O(2, 1) invariance of ΔK ensures that the same conclusions are true for any other direction of quantisation determined by a 'unit vector' $\boldsymbol{n}=(n_1, n_2, n_3)$; $n_3^2-n_1^2-n_2^2=1$. The formulae

$$\begin{aligned}(\Delta K_{1,2})^2&=\tfrac{1}{2}(1+n_{1,2}^2)[m^2-j(j+1)]\\(\Delta J_3)^2&=\tfrac{1}{2}(n_3^2-1)[m^2-j(j+1)]\end{aligned}\tag{17}$$

are the appropriate continuations of (2).

5. O(4) and E(3)

O(4) being locally isomorphic to O(3)×O(3), we can immediately apply the results of § 3 to it. In the usual way, form two vectors $\boldsymbol{J}=(J_{23}, J_{31}, J_{12})$, $\boldsymbol{K}=(J_{14}, J_{24}, J_{34})$ out of the O(4) generators $J_{\mu\nu}$, and in the ensuing commutators,

$$\boldsymbol{J}\times\boldsymbol{J}=\boldsymbol{K}\times\boldsymbol{K}=\mathrm{i}\boldsymbol{J}, \qquad \boldsymbol{J}\times\boldsymbol{K}=\mathrm{i}\boldsymbol{K}\tag{18}$$

take independent linear combinations $\boldsymbol{J}(\pm)=\frac{1}{2}(\boldsymbol{J}\pm\boldsymbol{K})$ to obtain

$$\boldsymbol{J}(\pm)\times\boldsymbol{J}(\pm)=\mathrm{i}\boldsymbol{J}(\pm), \qquad \boldsymbol{J}(\pm)\times\boldsymbol{J}(\mp)=0.\tag{19}$$

The O(4) states are, in effect, labelled by the $\mathrm{O}_\pm(3)$ Casimirs $j_\pm(j_\pm+1)$ and written $|j_+m_+, j_-m_-\rangle_{\boldsymbol{n}(+)\boldsymbol{n}(-)}$ when quantised along axes $\boldsymbol{n}(+)$ and $\boldsymbol{n}(-)$. The O(4) Casimirs themselves are given by

$$\begin{aligned}\Sigma^2&=\tfrac{1}{4}J_{\mu\nu}J_{\mu\nu}=\tfrac{1}{2}(\boldsymbol{J}^2+\boldsymbol{K}^2)=\boldsymbol{J}(+)^2+\boldsymbol{J}(-)^2=j_+(j_++1)+j_-(j_-+1)\\\Pi^2&=|\tfrac{1}{8}\epsilon_{\mu\nu\kappa\lambda}J_{\mu\nu}J_{\kappa\lambda}|=|\boldsymbol{J}\cdot\boldsymbol{K}|=|\boldsymbol{J}(+)^2-\boldsymbol{J}(-)^2|=|j_+(j_++1)-j_-(j_-+1)|.\end{aligned}\tag{20}$$

These are the invariant dispersions we are suggesting should be minimised in trying to define a quasi-classical state. The smallest dispersions occur for $m_\pm=j_\pm$ whereupon

$$\begin{aligned}(\Delta\Sigma)^2&\equiv\tfrac{1}{2}[(\Delta\boldsymbol{J})^2+(\Delta\boldsymbol{K})^2]=\langle\tfrac{1}{4}J_{\mu\nu}J_{\mu\nu}\rangle-\tfrac{1}{4}\langle J_{\mu\nu}\rangle\langle J_{\mu\nu}\rangle\\&=j_++j_-=-1+(\tfrac{1}{2}+\Sigma^2+\Pi^2)^{1/2}+(\tfrac{1}{2}+\Sigma^2-\Pi^2)^{1/2}\end{aligned}\tag{21}$$

$$\begin{aligned}(\Delta\Pi)^2&\equiv|\langle\boldsymbol{J}\cdot\boldsymbol{K}\rangle-\langle\boldsymbol{J}\rangle\cdot\langle\boldsymbol{K}\rangle|=\tfrac{1}{8}\epsilon_{\mu\nu\kappa\lambda}|\langle J_{\mu\nu}J_{\kappa\lambda}\rangle-\langle J_{\mu\nu}\rangle\langle J_{\kappa\lambda}\rangle|\\&=|j_+-j_-|=|(\tfrac{1}{2}+\Sigma^2+\Pi^2)^{1/2}-(\tfrac{1}{2}+\Sigma^2-\Pi^2)^{1/2}|.\end{aligned}\tag{22}$$

In fact, these critical states are eigenfunctions of $\frac{1}{2}J_{\mu\nu}n_{\mu\nu}$ and of $\frac{1}{4}\epsilon_{\mu\nu\kappa\lambda}J_{\mu\nu}n_{\kappa\lambda}$ where the 'unit tensor' $n_{\mu\nu}$ is composed of $\boldsymbol{n}_+$ and $\boldsymbol{n}_-$ similarly to the way $J_{\mu\nu}$ is broken up into $\boldsymbol{J}(+)$ and $\boldsymbol{J}(-)$. If we do not place any further restrictions on Π and Σ that is all there is to say. However, if we require further a vanishing pseudo-scalar Casimir Π, then $j_+=j_-(=j$ say) and $\Delta\Sigma=(2j)^{1/2}$ is the minimal scalar uncertainty.

The contraction to E(3) is fairly straightforward. Put $\boldsymbol{P}=c\boldsymbol{K}$ as in § 4 and let $c\to0$ with $\boldsymbol{J}$ held finite, but $|\boldsymbol{K}|\to\infty$ somehow. Because $\boldsymbol{J}=\boldsymbol{J}(+)+\boldsymbol{J}(-)$, this means we must let j_+ and $j_-\to\infty$ with $j_+-j_-=\lambda$ held fixed. In this limit, the E(3) Casimirs $\boldsymbol{P}^2$ and $\boldsymbol{J}\cdot\boldsymbol{P}$ arise from O(4) Casimirs $\Sigma^2\to\frac{1}{2}\boldsymbol{P}^2/c^2$, $\Pi^2\to\boldsymbol{J}\cdot\boldsymbol{P}/c$, and thus, putting $p=2cj_+$,

$$\begin{aligned}\boldsymbol{P}^2&\to2c^2[j_+(j_++1)+j_-(j_-+1)]\to p^2\\\boldsymbol{P}\cdot\boldsymbol{J}&\to c\Pi^2\to c[j_+(j_++1)-j_-(j_-+1)]\to\lambda p.\end{aligned}\tag{23}$$

The E(3) variances tend, as $c \to 0$, to

$$\Delta p = c[(\Delta \boldsymbol{K})^2]^{1/2} \to 2c\ \Delta\Sigma = 2c(2j_+)^{1/2} \to 0$$
$$p\Delta\lambda \equiv |\langle \boldsymbol{P}.\boldsymbol{J}\rangle - \langle \boldsymbol{P}\rangle.\langle \boldsymbol{J}\rangle| = c(\Delta\Pi)^2 \to c|\lambda| \to 0 \tag{24}$$

i.e. the minimal states $|p\lambda\rangle$, eigenstates of $\boldsymbol{P}^2$ and $\boldsymbol{P}.\boldsymbol{J}$, can have vanishing dispersion after strongly localised wave packets are constructed in the manner of § 4. We note that the angular momentum dispersion becomes infinite

$$\begin{aligned}(\Delta J)^2 &= \langle \boldsymbol{J}^2\rangle - \langle \boldsymbol{J}\rangle.\langle \boldsymbol{J}\rangle \\ &= j_+(j_+ + 1) + j_-(j_- + 1) + 2j_+j_-\boldsymbol{n}_+.\boldsymbol{n}_- - (\boldsymbol{n}_+j_+ + \boldsymbol{n}_-j_-)^2 \\ &= (j_+ + j_-) \to \infty\end{aligned}$$

and this seems to be the price we must pay to be certain of the helicity.

6. Coherent spin states

A great deal is known about coherent states for the harmonic oscillator, but far less about coherent states for angular momentum. (See Bacry *et al* (1976) for a recent review.) In making an analogy between J_3 and the number operator, and between J_- and the annihilation operator, Radcliffe (1971) was able to construct a coherent spin basis and demonstrate its usefulness in the context of statistical mechanics for ferromagnetic systems. (Subsequently Kolodziejczyk and Ryter (1974) showed that a state of minimal uncertainty $(\Delta J_1)(\Delta J_2)$ was only possible for Radcliffe's ground state, i.e. $|J_3| = j$, as we proved more directly in § 3.) His formulation fixed with certainty the Casimirs $j_1, j_2, \ldots$ of a series of angular momenta. Owing to the finite degeneracy of spin vectors having no counterpart in the oscillator we would advocate the alternative course of equating j, *not* J_3, with the number operator of the oscillator and then constructing a coherent basis as an infinite superposition of different spin-j states. We can minimise ΔJ by superposing $|jj\rangle_{\boldsymbol{n}}$ states, so one rather obvious possibility is to form the linear combination

$$|\alpha\rangle = \sum_j \alpha^j[\exp(-\tfrac{1}{2}|\alpha|^2)]|jj\rangle/(j!)^{1/2} \tag{25}$$

for Bose systems say, from which it is readily established that

$$\langle \boldsymbol{J}^2\rangle = |\alpha|^2(|\alpha|^2 + 2), \qquad \langle \boldsymbol{J}\rangle^2 = |\alpha|^2(|\alpha|^2 + 1), \qquad \Delta J = |\alpha|. \tag{26}$$

A more complete procedure, which includes integer and half-integer spins, is to construct creators $a^\dagger$ and annihilators a of spin-$\tfrac{1}{2}$

$$[a_\alpha, a_\beta^\dagger] = \delta_{\alpha\beta}; \qquad \alpha, \beta = \uparrow, \downarrow \tag{27}$$

in terms of which $\boldsymbol{J} = \tfrac{1}{2}a^\dagger\boldsymbol{\sigma}a$. The usual angular momentum states are created as follows:

$$|jm\rangle = \frac{(a_\uparrow^\dagger)^{j+m}(a_\downarrow^\dagger)^{j-m}}{[(j+m)!(j-m)!]^{1/2}}|0\rangle. \tag{28}$$

We can build up coherent spin states, labelled by a two spinor $z=(z_\uparrow, z_\downarrow)$ as follows:

$$|z\rangle=\sum_{jm}\frac{(z_\uparrow)^{j+m}(z_\downarrow)^{j-m}}{[(j+m)!(j-m)!]^{1/2}}|jm\rangle\exp(-\tfrac{1}{2}|z_\uparrow|^2-\tfrac{1}{2}|z_\downarrow|^2)=\mathrm{e}^{a^\dagger z}|0\rangle\mathrm{e}^{-\frac{1}{2}z^\dagger z}. \tag{29}$$

The exponential factors $\mathrm{e}^{-|z|^2/2}$ are included to simplify the 'orthogonality' and 'completeness' relations:

$$\begin{aligned}1&=\int|z\rangle\,\mathrm{d}^2z_\uparrow\,\mathrm{d}^2z_\downarrow\langle z|/\pi^2\\ \langle z'|z\rangle&=\mathrm{e}^{z'^\dagger z}\,\mathrm{e}^{-(z^\dagger z+z'^\dagger z')/2}.\end{aligned} \tag{30}$$

These new states incorporate the simple proposal (25) if we set $z_\downarrow=0$ and sum over half-integer as well as integer j. They may also be related to Radcliffe's states,

$$|\mu\rangle=\sum_m\frac{\mu^{j-m}|jm\rangle}{(1+|\mu|^2)^j}\left(\frac{(2j)!}{(j+m)!(j-m)!}\right)^{1/2} \tag{31}$$

if in (29) we put $z_\uparrow=1$, $z_\downarrow=\mu$ and do not sum over j:

$$|z_\uparrow=1, z_\downarrow=\mu\rangle\sim\frac{\mathrm{e}^{-(1+|\mu|^2)/2}}{(1+|\mu|^2)^j[(2j)!]^{1/2}}|\mu\rangle. \tag{32}$$

Because they are eigenstates of the annihilation operator, it is relatively easy to work out the dispersion of $\boldsymbol{J}^2$. One finds

$$\begin{aligned}\langle z|\boldsymbol{J}|z\rangle&=\tfrac{1}{2}z^\dagger\boldsymbol{\sigma}z\\ \langle z|\boldsymbol{J}^2|z\rangle&=\tfrac{1}{4}z^\dagger z(z^\dagger z+3)\end{aligned}$$

therefore

$$(\Delta J)^2/\langle\boldsymbol{J}^2\rangle=(1+\tfrac{1}{3}z^\dagger z)^{-1} \tag{33}$$

so as $|z|\to\infty$ the classical limit is reached. Such states could conceivably be useful in describing highly excited molecular rotation levels. Having said that, it is not immediately clear, however, what is the analogue of the oscillator phase operator ϕ, though it must surely be connected to the a and $a^\dagger$ via the structure

$$\cos\phi\sim a^\dagger+a,\qquad \sin\phi\sim a^\dagger-a.$$

We emphasise that all recent studies of coherent spin states (Belissard and Holtz 1974, Hioe 1974, Peremolov 1972, Onofri 1975) base themselves on Radcliffe's original suggestion and confine themselves to fixed j. Thus they differ in an essential way from our proposal (29).

7. Generalisations

We have lowered our sights to O(3) and closely related groups in this paper to deduce that quasi-classical states are those which minimise the uncertainty invariant ΔJ; the vectors turn out to be eigenstates of highest weight of $\boldsymbol{J}.\boldsymbol{n}$. For other Lie algebras a natural generalisation is to look for minima of the quadratic

$$(\Delta F)^2\equiv c_{rst}c_{rsu}(\langle F_tF_u\rangle-\langle F_t\rangle\langle F_u\rangle)$$

and higher-order Casimirs. We anticipate again that the quasi-classical vectors have highest weight. For Lie algebras which are semi-direct products of an Abelian algebra and a compact algebra, our experience with E(2) and E(3) suggests that the minimal states are eigenstates of the Abelian sub-algebra and any other Casimirs. Finally, if one abandons the restriction that Casimirs themselves be precisely known, the method of § 6 suggests a route by which one can construct appropriate coherent states different from those of Peremolov (1972).

Appendix

We wish to solve the equation

$$(\boldsymbol{J}-\langle\boldsymbol{J}\rangle)^2|\psi\rangle=(\Delta J)^2|\psi\rangle$$

wherein $\Delta J_1=\Delta J_2=\Delta J_3$ and therefore $\Delta^2 J=3^{-1/2}(\Delta J)^2$. We look for solutions where $|\psi\rangle$ is the usual eigenstate of $\boldsymbol{J}^2$. The equation then reduces to

$$2\boldsymbol{J}\cdot\langle\boldsymbol{J}\rangle|\psi\rangle=[j(j+1)+\langle\boldsymbol{J}\rangle^2-(\Delta J)^2]|\psi\rangle.$$

Thus $|\psi\rangle$ is in an eigenstate of $\boldsymbol{J}\cdot\boldsymbol{n}$ where $\boldsymbol{n}$ is a unit vector parallel to $\langle\boldsymbol{J}\rangle$ and therefore

$$\boldsymbol{J}\cdot\langle\boldsymbol{J}\rangle|\psi\rangle=m|\langle\boldsymbol{J}\rangle||\psi\rangle. \tag{A.1}$$

The equality of uncertainties entails that $\langle\boldsymbol{J}\rangle=m(\pm1,\pm1,\pm1)/\sqrt{3}$ and correspondingly $(\Delta J_1)^2=(\Delta J_2)^2=(\Delta J_3)^2=\frac{1}{3}[j(j+1)-m^2]$ giving $\Delta^2 J=[j(j+1)-m^2]$. The critical state with the least $\Delta^2 J$ then corresponds to the selection $m=j$.

If we abandon the constraint $\Delta J_1=\Delta J_2=\Delta J_3$, the solution to (A.1) is simply that $\boldsymbol{J}\cdot\boldsymbol{n}|\psi\rangle=m|\psi\rangle$ with $\langle\boldsymbol{J}\rangle=m\boldsymbol{n}$, namely the case considered at the beginning of § 3.

References

Bacry H, Grossman A and Zak A 1976 *Lecture Notes in Physics* vol. 50 (Berlin: Springer)
Belissard J and Holtz R 1974 *J. Math. Phys.* **15** 1275
Glauber R J 1963 *Phys. Rev.* **131** 2766
Hioe F T 1974 *J. Math. Phys.* **15** 1174
Jackiw R 1968 *J. Math. Phys.* **9** 339
Kolodziejczyk L and Ryter A 1974 *J. Phys. A: Math., Nucl. Gen.* **7** 213
Mathews P M and Eswaran K 1974 *Nuovo Cim.* B **19** 99
Onofri E 1975 *J. Math. Phys.* **16** 1087
Peremolov A M 1972 *Commun. Math. Phys.* **26** 222
Radcliffe J M 1971 *J. Phys. A: Gen Phys.* **4** 313

J. Phys. A: Math. Gen., Vol. 10, No. 12, 1977. Printed in Great Britain. © 1977

LETTER TO THE EDITOR

Maximum weight vectors possess minimal uncertainty

R Delbourgo and J R Fox

Department of Physics, University of Tasmania, Hobart, Tasmania, 7001, Australia

Received 6 July 1977

Abstract. An appropriate uncertainty measure for a compact Lie group is the invariant dispersion

$$(\Delta F)^2 = \langle g^{rs}(F_r - \langle F_r\rangle)(F_s - \langle F_s\rangle)\rangle.$$

We prove that it is minimised for maximum weight vectors (of greatest length in the weight space), and those unitarily related to them.

In a recent paper (Delbourgo 1977) we examined the problem of how to define quasiclassical states (characterised by a least uncertainty) for O(3) and closely associated groups. Our investigation showed that the most natural measure of indeterminancy, the invariant dispersion,

$$(\Delta J)^2 \equiv \langle \boldsymbol{J}^2\rangle - \langle \boldsymbol{J}\rangle \cdot \langle \boldsymbol{J}\rangle,$$

was minimised for maximum weight angular momentum states

$$\boldsymbol{J}\cdot\boldsymbol{n}|jj\rangle_n = j|jj\rangle_n,$$

and we guessed that the result carried over to arbitrary compact Lie groups; namely, the maximum weight vectors corresponded most closely to the classical situation of absolute precision. In this letter we would like to outline a simple proof of this conjecture.

Let F_s denote the set of generators of a compact Lie algebra obeying the commutation rules

$$[F_r, F_s] = \mathrm{i}\, C_{rs}{}^t F_t.$$

The positive-definite Cartan metric

$$g_{rs} = \tfrac{1}{2} C_{rp}{}^q C_{qs}{}^p,$$

and its inverse g^{rs}, can be used to construct the quadratic Casimir $F^2 \equiv g^{rs}F_rF_s$. Following our O(3) analysis we contend that the most appropriate measure of quantum indeterminancy is the invariant dispersion (variance)

$$(\Delta F)^2 \equiv \langle(\Delta\hat{F})^2\rangle \equiv \langle g^{rs}(F_r - \langle F_r\rangle)(F_s - \langle F_s\rangle)\rangle = \langle g^{rs}(F_rF_s - \langle F_r\rangle\langle F_s\rangle)\rangle. \qquad (1)$$

We will now show that eigenvectors of the Casimirs $F^2, F^3, \ldots$ have least ΔF when they are of maximum weight (i.e. have weight vectors of maximum length).

First we choose a canonical basis whereupon $g^{rs} \propto \delta^{rs}$, and make the conventional split into Cartan operators H_k (defining the rank) and $E_{\pm\alpha}$ (changing the weights). As $(\Delta\hat{F})^2$ is a positive-definite operator, it has a lowest eigenvalue, and in that lowest eigenstate the lowest expectation value is attained. The minimum dispersion states are therefore among the eigenstates of $(\Delta\hat{F})^2$, satisfying

$$(F_sF_s - 2F_s\langle F_s\rangle_\psi + \langle F_s\rangle_\psi\langle F_s\rangle_\psi)|\psi\rangle \propto |\psi\rangle.$$

But $|\psi\rangle$ is taken to be already an eigenvector of F_sF_s (among other possible Casimirs). Therefore a necessary condition for minimal dispersion is

$$F_s\langle F_s\rangle_\psi|\psi\rangle \propto |\psi\rangle. \tag{2}$$

To find solutions of (2), first suppose that $|\psi\rangle$ is an eigenvector $|\boldsymbol{h}\rangle$ of the (rank) r commuting generators H_k which form a basis for the Cartan sub-algebra. Then

$$\begin{aligned} H_k|\boldsymbol{h}\rangle &= h_k|\boldsymbol{h}\rangle, \qquad k=1,\ldots,r, \\ \langle\boldsymbol{h}|E_\alpha|\boldsymbol{h}\rangle &= 0, \qquad \text{all roots } \alpha, \end{aligned} \tag{3}$$

because of the step-up and -down action of the E_α. For generators F_s chosen so that r of them are the H_k (the others being Hermitian linear combinations of the E_α),

$$F_s\langle F_s\rangle_h|\boldsymbol{h}\rangle = H_k\langle H_k\rangle_h|\boldsymbol{h}\rangle = h_kh_k|\boldsymbol{h}\rangle$$

showing that any state $|\boldsymbol{h}\rangle$ solves (2), and ΔF is minimised when $\langle F_s\rangle\langle F_s\rangle$ is maximised, i.e. for greatest $|\boldsymbol{h}|^2 \equiv h_kh_k$; namely for *vectors of maximum weight.*

The second part of the proof consists in finding the most general solutions of (2). For this we note that it is always possible to find a group transformation such that F_sn_s, with $\boldsymbol{n}$ a unit vector, is transformed into an element H_ka_k in the Cartan sub-algebra, where $\boldsymbol{a}$ is a unit vector in r-dimensional space, i.e.

$$U(\boldsymbol{n})F_sn_sU^{-1}(\boldsymbol{n}) = H_ka_k(\boldsymbol{n})$$

where $U(\boldsymbol{n})$ is an operator† lying in the coset space G/H. The vectors $|\boldsymbol{h}\rangle$ are eigenvectors of this expression, so

$$F_sn_sU^{-1}(\boldsymbol{n})|\boldsymbol{h}\rangle = h_ka_k(\boldsymbol{n})U^{-1}(\boldsymbol{n})|\boldsymbol{h}\rangle$$

or

$$\frac{F_s\langle F_s\rangle_\psi}{(\langle F_t\rangle_\psi\langle F_t\rangle_\psi)^{1/2}}|\psi\rangle = h_ka_k(\boldsymbol{n})|\psi\rangle \tag{4}$$

if we set $U^{-1}(\boldsymbol{n})|\boldsymbol{h}\rangle = |\psi\rangle$ and choose $n_s = \langle F_s\rangle_\psi/(\langle F_t\rangle_\psi\langle F_t\rangle_\psi)^{1/2}$. For this choice of $\boldsymbol{n}$, $a_k = h_k/|\boldsymbol{h}|$, as can be seen by taking the inner product of (4) with $|\psi\rangle$, yielding

$$(\langle F_s\rangle_\psi\langle F_s\rangle_\psi)^{1/2} = h_ka_k.$$

But the left-hand side as an invariant under group transformations and is equal to $h_kh_k/|\boldsymbol{h}|$ when $U(\boldsymbol{n})=1$, i.e. $|\psi\rangle = |\boldsymbol{h}\rangle$, so that $a_k = h_k/|\boldsymbol{h}|$. Therefore $U^{-1}|\boldsymbol{h}\rangle$ solves (2) with eigenvalue h_kh_k. These vectors are none other than Perelomov's coherent states (Perelomov 1972).

Finally, to minimise ΔF, we need to maximise $\langle F_s\rangle\langle F_s\rangle$, and this is achieved by taking $|\boldsymbol{h}\rangle$ to be a maximum weight state. The foregoing shows that any vector $U|\boldsymbol{h}\rangle$ unitarily related to a maximum weight state equally well minimises ΔF. This is, of course, to be expected, as the choice of the sub-algebra basis $(H_1,\ldots,H_r)$ is not

† For instance, if we are dealing with O(N) and the rotation operators $J_{\mu\nu}$, $UJ_{\mu\nu}n^{\mu\nu}U^{-1} = J_{12}a^{12} + J_{34}a^{34} + \ldots$ and U is an orthogonal transformation which rotates the special tensor a into n. Similarly, if G is the group SU(N), $UF^b_an^a_bU^{-1} = H^1_1a^1_1 + H^2_2a^2_2 + \ldots$ where U is a unitary transformation.

unique and the maximum weight eigenstates of any unitarily equivalent basis (the states $U|\boldsymbol{h}\rangle$) should also minimise ΔF.

For illustration take SU(3) and representations $\Phi^{(b_1 \ldots b_q)}_{(a_1 \ldots a_p)}$ labelled by the pair (p, q). In this case the least

$$\Delta F \equiv (\tfrac{1}{2}\langle F^a_b F^b_a\rangle - \tfrac{1}{2}\langle F^a_b\rangle\langle F^b_a\rangle)^{1/2}$$

equals $(p+q)^{1/2}$ and occurs, of course, for highest weight vectors like $\Phi^{(2\ldots 2)}_{(1\ldots 1)}$. The value of $\langle F^b_a\rangle\langle F^c_b\rangle\langle F^a_c\rangle$ is thereby completely determined and there is no point in considering uncertainties related to the cubic Casimir.

References

Delbourgo R 1977 *J. Phys. A: Math. Gen.* **10** 1837–46
Peremolov A M 1972 *Commun. Math. Phys.* **26** 222–34

Universal Facets of Chaotic Processes

It was found that the rate of approach to turbulence does not really depend on the details of the nonlinear equations used in the model, which implies that there exist universal characteristics in many apparently disparate chaotic systems.

R. Delbourgo

In teaching science and tracing its development it is customary and politic to concentrate on those phenomena that are smooth or simply periodic because they are amenable to standard linear analysis. And it cannot be denied that such is the way in which laws of nature are discovered: from the uniform, controlled response of systems under preselected changes of experimental parameters. Unfortunately, that is hardly ever the way that nature chooses to operate. More often than not systems will behave in a chaotic fashion because of the existing natural conditions. Witness the erratic climate or the swirling of water in the oceans or the sudden solar flare-ups.

Until about 10 years ago the subject of turbulent behaviour was in a state of chaos. The most popular scenario at that time was probably Landau's which viewed turbulence as a succession of instabilities arising through variations in the parameters, giving rise to (harmonics of) an ever-increasing number of instability mode frequencies. Today we know a lot more about turbulent motion and the scenarios tend to be much simpler in conception. In particular, I would like to relate Feigenbaum's scenario, which envisages the passage to chaos as a succession of bifurcations or frequency-halvings. This view of the development of turbulence has now received impressive experimental support from many different branches of science.

Such a breakthrough in our understanding of erratic motion has come through the realisation that very simple dynamical systems can exhibit very complicated behaviour for certain ranges of parameters. A crucial aspect is the occurrence of nonlinearity, allowing the present state of the system to ensue from a multiplicity of past states. Because many physical processes are modelled by equations with intrinsic nonlinearity (e.g. the Navier-Stokes equation for viscous fluids) even when the behaviour is steady, our newly gained insights into chaos restore our faith in the ability of these equations to describe turbulent motion as well. Furthermore, Feigenbaum's discovery that the rate of approach to turbulence does not really depend on the details of the nonlinear equations means that we can expect universal characteristics in many apparently disparate systems, which is indeed the case!

Let me begin by sketching out a simple model that leads to the period-doubling route to chaos. (Later on I will mention the extension of the idea to more complicated systems and will elaborate upon the experimental confirmation of the universality concept). For those of you interested in following up the subject matter, I have provided a number of popular science magazines and easy-to-read review articles at the end from which more basic references may be gleaned (in particular, the reprint selections by Cvitanovic and Hao).

The basic example with the potential to display the main features of erratic behaviour is the Julia model,

$$x_{n+1} = F(x_n) = Lx_n(1 - x_n) \tag{1}$$

One can choose to regard this nonlinear relation as a model for a bacterial population confined to some container carrying nutrients as well; x_n denotes the proportion of the container occupied by the bacteria on day n and L parametrises the rate of growth of the population for small x. (L is constrained to be less than 4 so as to bound x by 1). The factor $(1-x)$ in eq. (1) is present to limit the colony to the container size and it crudely takes into account the competition between the bacteria for food. The question then arises, "How does the population x vary from day to day?" A superficial guess would have the population settling down to a steady state in due time. However, in general, that turns out to be very far from the truth — x can fluctuate wildly for certain ranges of the parameter L. Stated more precisely, a small change in the initial value of x can cause the subsequent value of x to diverge rapidly away from the initial state; in other words, there is *extreme* sensitivity to initial conditions. The root cause of erraticism is the nonlinearity of the model, so let us trace this effect in a bit more detail.

If the x-value were to settle down to a single number, it could only be when $x = F(x) = Lx(1-x)$, at the so-called "fixed point of the map". We can determine this value by looking for the intersection of the equation $y = x$ with the $y = F(x)$. When $L < 1$ there is just one solution, $x = 0$; the population dies out in time since the reproduction rate L is too small. For $L > 1$ there is one intersection point, at $x = x^* = 1 - 1/L$ and it would appear that this be the steady situation. Before asserting this categorically we have to check that the final state is a *stable* one. It is easy to prove that stability requires, $|F'(x^*)| < 1$, implying, for our particular model,

$$|L(1 - 2x^*)| = |2 - L| < 1$$

or that $L < 3$. We may readily test the correctness of this analysis by tracking the behaviour of x when $L = 2.5$, say. In Figure 1 we indicate how convergence to the single stable x^* takes place, no matter what the initial value x.

R. Delbourgo is professor in theoretical physics at the University of Tasmania, Australia.

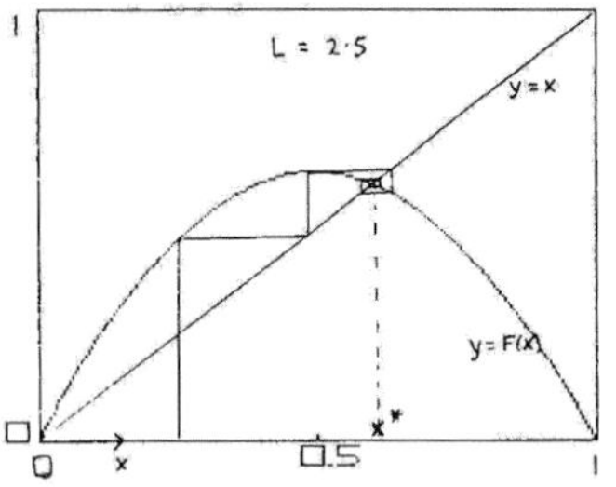

Figure 1. Mapping $x \to Lx(1-x)$ for $L = 2.5$ showing convergence of x-values to the fixed point $x^* = 0.6$.

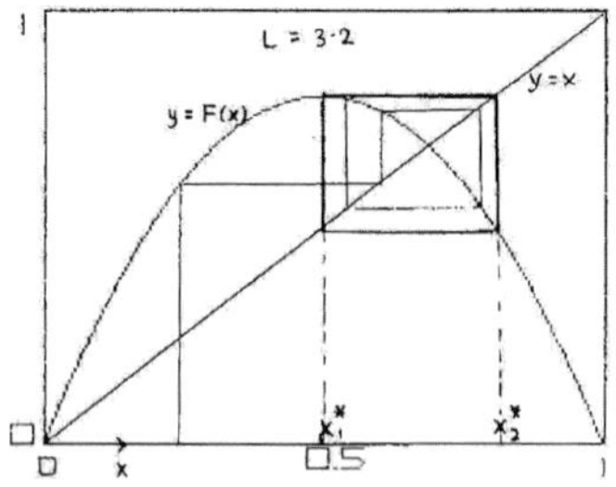

Figure 2a. Mapping $x \to Lx(1-x)$ for $L = 3.2$ showing period-doubling and the occurrence of two fixed points.

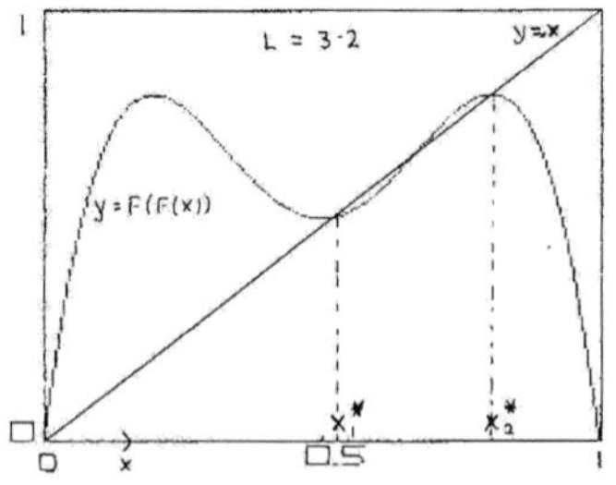

Figure 2b. Iteration of the Julia mapping for $L = 3.2$ showing the stability of the two fixed points.

Beyond $L = 3$ the analysis indicates instability and the practical consequence is (at first) a bifurcation to two neighbouring stable fixed points x_1^*, x_2^* such that

$$x_1^* = F(x_2^*) \quad , \quad x_2^* = F(x_1^*) \tag{2}$$

or

$$x_i^* = F(F(x_i^*)) \quad ; \quad i = 1 \text{ or } 2$$

In these circumstances history repeats itself every second day, signifying a period-doubling, or *frequency-halving* phenomenon, in contrast to the more familiar phenomenon of harmonics (frequency doubling, etc.). We speak of a 2-cycle being born by "pitchfork bifurcation". This is exhibited in Figures 2a and 2b. When the parameter L is cranked up, further instabilities occur and the fixed points continue to bifurcate. Eventually when L attains the critical value

$$L_{2^\infty} = 3.56994567...$$

the period is effectively infinite and chaos is fully developed. What this means is that it is virtually impossible to predict the difference between final x-values which start out nearly equal.

It is very easy to demonstrate these properties of the mapping even on the most primitive personal computers. Figure 3a is a printout of a graphic display of the fixed points of the scheme as L is varied from 2 to 4. The computer was programmed to display the stable x-values only after sufficiently many iterations were carried out to allow the system to settle down. The bifurcation phenomenon is plainly visible even with the pixel limitations on resolution.

The same figure also exhibits the occurrence of "windows of stability" beyond bifurcation chaos, of which the 3-cycle is the most prominent since it is the widest. Such a cycle is born at $L = 3.828427$ by "tangent bifurcation" in the manner sketched in Figures 4a, 4b and 4c. Just prior to reaching the critical value of L there is the phenomenon of "intermittency" whereby the x-values spend most of their time in the vicinity of the three quasistable x^*: the system makes a valiant attempt to settle down. In Figure 3a one may also discern the further bifurcation of the 3-cycle (by the same pitchfork mechanism as before) as the parameter L is increased, until chaos re-emerges in a triple band. I would like you to notice that the approach to chaos on this side is a microcosm of what happened previously, before L_{2^∞} was reached from below.

If you examine Figure 3 even more carefully you should be able to make out the existence of other narrower windows associated with other cycles; for instance the four cycle to the right at $L = 3.960$ and perhaps the three 5-cycles at $L = 3.738$, 3.905, 3.990. In fact there are an infinite number of such windows (though some are excessively narrow) in the interval to the right of pitchfork chaos. Equally, there are an infinite number of places which are truly chaotic: an analogy can be made between the rational numbers and the denumerable cycles of x, and the irrational numbers and the turbulent motions of x.

When several cycles of the same period arise, it is possible to distinguish between them by the manner in which the various limit points x^* are visited. For example, 4-cycles can be produced in one of two ways: either by two successive pitchfork bifurcations or by tangent bifurcation. If we arrange the x^* in increasing magnitude, then the sequence of points is visited as shown in Figures 5a and 5b. (These are the only possibilities because the rightmost limit point must move to the leftmost one). One can categorise the various 5-cycles in much the same way, and so on for other cycles. In fact, Metropolis, Stein and Stein gave the complete group-theoretic catalo-

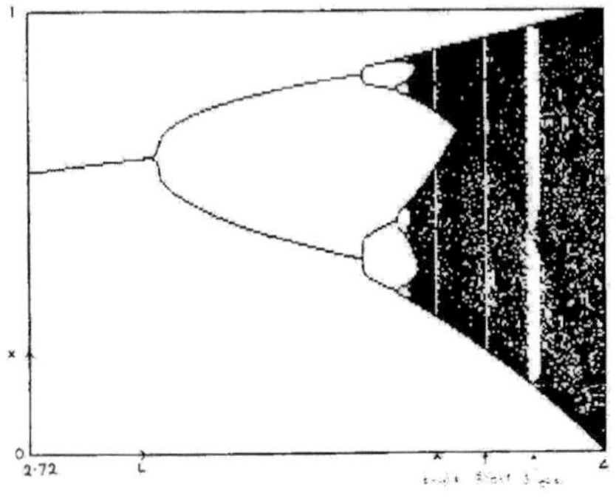

Figure 3a. The fixed points of the map $x \to Lx(1-x)$ for L in the range 2 to 4. Note the bifurcations, chaotic regions and windows of stability.

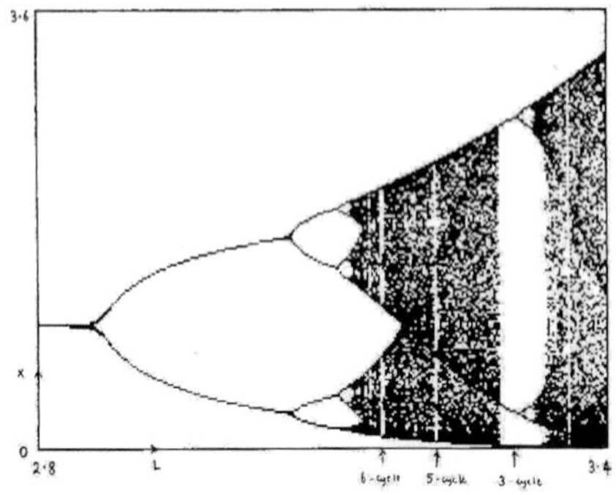

Figure 3b. The fixed points of the map $x \to x\, exp\, (L(1-x))$ for L in the range 1 to 3.56.

gue of the cycle sequences many years ago in a remarkable paper.

The most striking thing about their mathematical analysis is that the *sequence of cycles does not depend on the precise form of the mapping*. The only requirement on $F(x)$ is that it be continous, differentiable and have one maximum over the relevant range of x, as does the Julia map. This result is termed "*structural universality*". Perhaps if you see the sequence of cycles for another popular model (see Figure 3b)

$$x_{n+1} = F(x_n) = x_n \exp\left[L(1 - x_n)\right] \qquad (3)$$

you may be convinced about the truth of this (qualitative) statement. This is the first facet of universality to be stressed.

The second facet was discovered by Feigenbaum almost accidentally while carrying out computations on a programmable calculator. In trying to locate the accumulation points of various cycle sets he noticed that convergence to these points was geometrical. In the context of pitchfork bifurcations he found that L-spacings between adjoining doublings were contracted successively by the same factor d. Numerical work showed that the spacing ratios tended to

$$d = \lim_{n\to\infty} (L_{2^{n+2}} - L_{2^{n+1}})/(L_{2^{n+1}} - L_{2^n}) = 4.6692\ldots \qquad (4)$$

and, more surprisingly, the convergence rate did not depend on the specific form of the map (beyond the requirement that F be quadratic at the maximum). Not only this ratio but the ratio between successive x-spacings in the forks themselves was also insensitive to the details of the nonlinear relation:

$$a = \lim_{n\to\infty} (Dx^*_{n+1})/(Dx^*_n) = -2.5029\ldots \qquad (5)$$

These two (quantitative) discoveries go under the name of "*metric universality*". The next figure (Figure 6) depicts these scaling features in more graphic detail: the various L separations are drawn logarithmically to make the repetitive branching more obvious, while x is appropriately scaled.

From Feigenbaum's research we therefore learn that two universal constants a and d describe the approach to turbulance. The actual values of a and d stated in eq. (4) and (5) just depend on the parabolic maximum of the nonlinear transformation. If the nature of the maximum is changed then so are the values of a and d. For instance, a quartic maximum associated with the map

$$F(x) = 1 - Mx^4$$

yields the numbers

$$d = 7.2851\ldots, \qquad a = -1.6903\ldots$$

but, otherwise, the universality concepts survive in their entirety.

Recently, Kenny, Hart and I have tried to search for other universal features which

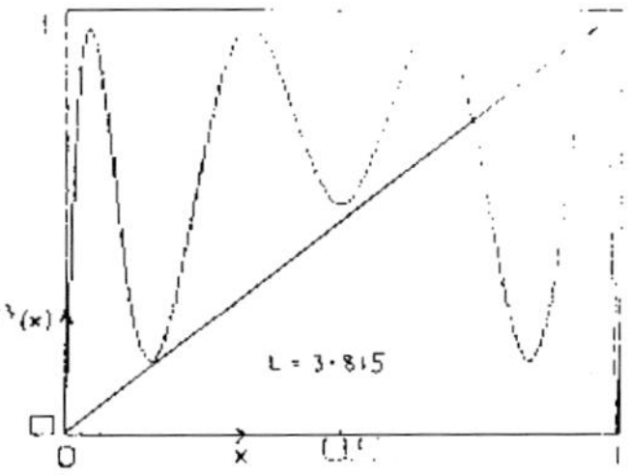

Figure 4a. Triple iterate of the Julia map at $L = 3.815$, indicating intermittency.

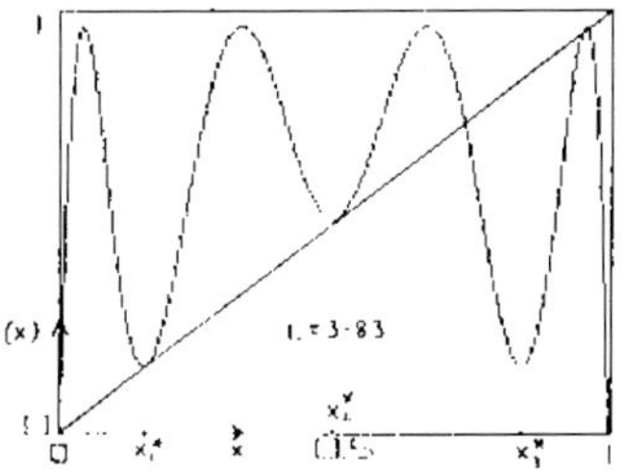

Figure 4b. Triple iterate of the Julia map at $L = 3.83$, showing birth of the 3-cycle.

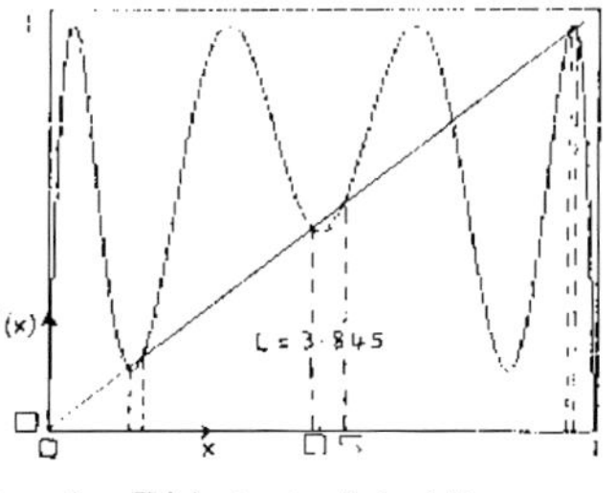

Figure 4c. Triple iterate of the Julia map at $L = 3.845$, showing pitchfork bifurcation to a 6-cycle.

applies to tangent bifurcations in the region beyond pitchfork chaos (the accumulation point of period doublings)†. We have looked at sequences of period triplings, quadruplings, etc. (these technically go under the name of 'kneading sequences') of particular types. We have demonstrated that here too one may extract universal characteristics (map-independent properties) that only depend on the particular sequence studied. For example the period triplings or "trifurcatons" have associated the constants

$$d_3 = 55.26\ldots, \qquad a_3 = -9.277\ldots$$

while the "quadrifurcations" (to be distinguished from double pitchforking) lead to the values

$$d_4 = 981.6\ldots, \qquad a_4 = -38.82\ldots;$$

and so on for other multiplication factor N sets.

---

†Phys. Rev. **A31**, 514 (1985).

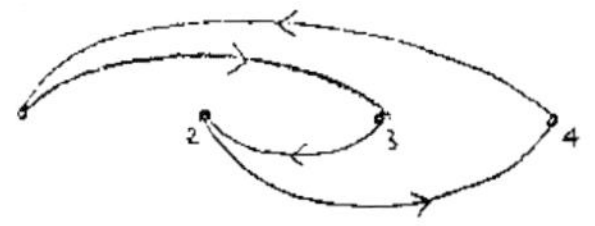

Figure 5a. Order of cycling for the fixed points x^* of the 4-cycle arising by successful pitchforks.

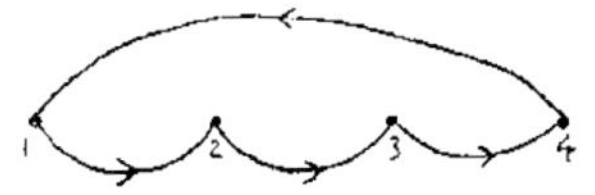

Figure 5b. Order of cycling for the fixed points x^* of the 4-cycle arising by tangent bifurcation.

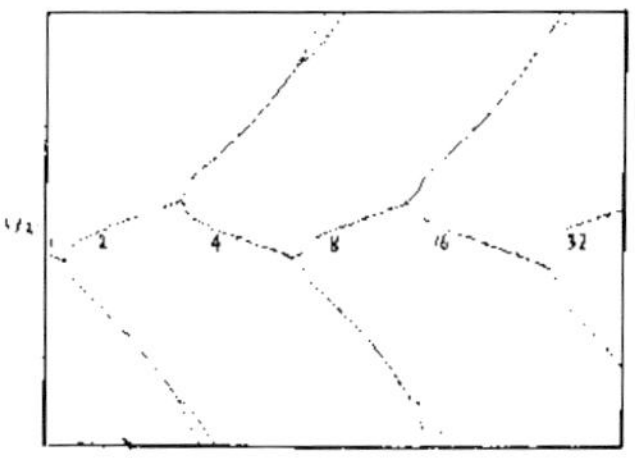

Figure 6. Plot of $x - 1/2$ (scaled by a^N) against the logarithm of $(L_\infty - L)$ which shrinks as d^{-N}.

It is quite remarkable that if one plots ln d against ln a one discovers a linear relation to an excellent approximation. Specifically, the fit

$$\ln d = 2 \ln a + \ln(2/3)$$

seems to cover practically all the sequences and improves with increasing period factor N. Table 1 and Figure 7 tell the story about tangent bifurcations. Indeed the prediction that

$$3d = 2a^2 \qquad (6)$$

was made by Eckmann, Epstein and Wittwer on analytical grounds, but only for the rightmost tangent bifurcations and only in the asymptotic limit of N. Our work has shown that the relation between the universal constants has a much wider validity.

The reason why such universal features should exist at all was first exposed by Feigenbaum. He examined the pitchfork sequence and saw that each period doubling was influenced more and more strongly by the nature of the mapping near the maximum of F (quadratic in most models) and the corresponding iterates

$$(F)^n(x) = \underbrace{F(F(\ldots(F(x)))}_{n-\text{fold}}$$

For sufficiently large n only a region in x, shrunken by the scale factor $(-a)^n$ governed the duplication characteristics of the system. Specifically, the scaled down function

$$g(x) = \lim_{n\to\infty} (-a)^n [f]^{2^n} (L_\infty, x/(-a)^n) \qquad (7)$$

Table 1. Universal constants and accumulation points for Julia Map.

Cycle	L	a	d
2	3.569945671871..	2.5029..	4.6692..
3	3.854077963591..	9.277..	55.26..
4b	3.961556587172..	38.82..	981.6..
5a	3.743666637542..	20.13..	255.5..
5b	3.906698486337..	45.80..	1287.1..
5c	3.990344147459..	160.0..	16931
6a	3.633007276957..	20.93..	218.4..
6c	3.937689018821..	115.0..	8508
6d	3.977813426669..	207.6..	28020
6e	3.997587860857..	645.0..	279125
7a	3.702578617486..	49.17..	1446
7b	3.774752160872..	58.63..	2254
7c	3.886170825806..	131.4..	10170
7d	3.922249877845..	191.5..	22840
7e	3.951069246131..	230.1..	35306
7f	3.968997497961..	317.1..	63629
7g	3.984755539188..	503.9..	167161
7h	3.994540535630..	858.2..	487353
7i	3.999397355799..	2603	4511991
8b	3.662694216049..	66.39..	2305
8c	3.800982575491..	88.12..	5830
8d	3.870605069125..	190.4..	19679
8e	3.899517548758..	205.1..	26486
9b	3.687339675888..	112.5..	7918
10a	3.606403696015..	47.96..	1111
10b	3.647301729914..	98.86..	4523
11a	3.681735129699..	352.2..	70063

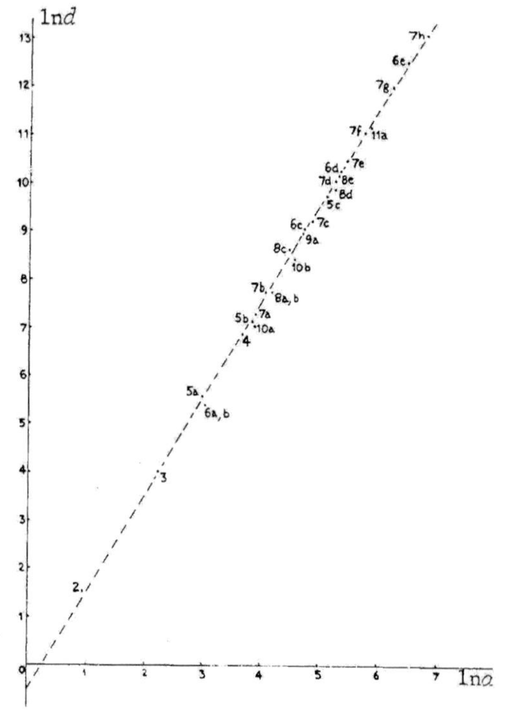

Figure 7. Ln d versus Ln a, indicating the near validity of the relation $3d = 2a^2$.

was itself universal and obeyed the renormalization group equation

$$g(x) = -ag(g(-x/a)) \quad (8)$$

in much the same guise that critical phenomena today are studied in other fields of physics. Similar kinds of equation apply to other multiplication periods; for instance the tripling case has its own

$$G(x) = \lim_{n\to\infty}(-a_3)^M[F]^{3^n}(L_{3\infty}, x/(-a_3)^n)$$

which satisfies

$$G(x) = -a_3G(G(G(x/-a_3))) \quad (9)$$

Having given the background to the universality idea, it is time now to look at the experimental repercussions and see whether the concept has any relevance for natural phenomena. The period-doubling feature itself has been observed in a very large variety of processes (see Table 2), but I would like to concentrate on one of these processes, Rayleigh-Benard convection, since it is relatively easy to grasp.

The experiment itself is conceptually trivial. One detects with great delicacy the temperature x at a point in a fluid (liquid helium in fact) held between two surfaces whose temperature difference L is under experimental control. For L small the measured x settles down to a single fixed value. As L is increased beyond a critical value, convective rolls start to form. Initially there is just one roll and x fluctuates between two values with some period. A further increase in L results in more bifurcation and period doubling of x; until, at a critical L, turbulence is complete and x is essentially indeterminate. In practice it is quite hard to detect more than 4 doublings because the precision of temperature control has to increase by a factor of about 5 at every stage and the precision of x-measurement must get finer by a factor of 3. Obviously this becomes impossibly hard, and anyway the inherent noise tends to wash things out, not to mention the delays entailed in waiting for transients to fade away. However the numbers discovered so far agree uncannily with the theoretical predictions in this and very many other physical, chemical and biological experiments. What is more the Fourier amplitudes at the various frequency halvings agree reasonably well with Feigenbaum's predictions for them (Figure 8).

You must be wondering how it is that we can get so far with one-dimensional mappings. The reason is that most complex systems involve energy dissipation or degradation and damp out in the end to leave one operative degree of freedom—though it should be emphasised that the contraction to one dimension can be very complicated and entail stretching and folding effects in the many-dimensional phase space. The Henon-Heiles coupled equations

$$x_{n+1} = 1 - ax_n^2 + y_n$$

$$y_{n+1} = bx_n$$

Table 2. Experiments revealing period-doubling route to chaos.

Investigation	Doublings	d	a
Chemical reactions	3	?	?
Water flow	4	4.3 ± .8	?
Helium flow	4	3.5 ± 1.5	?
Mercury flow	4	4.4 ± .1	?
Diode oscillations	5	4.3 ± .1	2.4 ± .1
Transistors	4	4.7 ± .3	?
Josephson junctions	3	4.5 ± .3	2.7 ± .2
Laser feedback	3	4.3 ± .3	?
Acoustic (He)	3	4.8 ± .6	?
Brusselator (Computer)	7	4.6 ± .2	?
Nerve stimulation	?	?	?
Cardiac dysrythmia	?	?	?
Sociology (Wars)	??	?	?

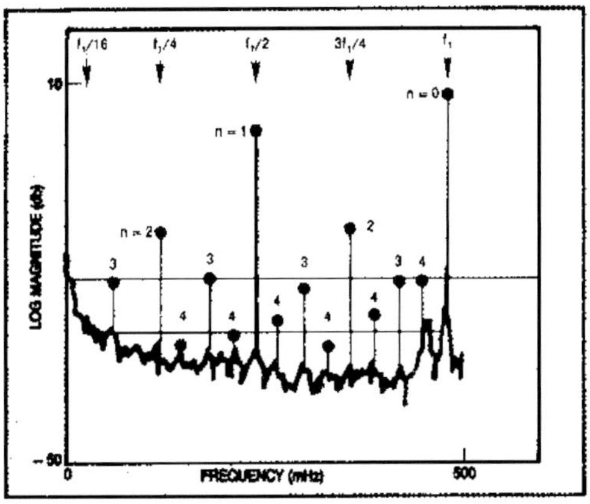

Figure 8. Power spectrum in Rayleigh-Bernard convective flow near the critical value of the parameter. Vertical lines are the theoretical predictions for particular frequency-halvings.

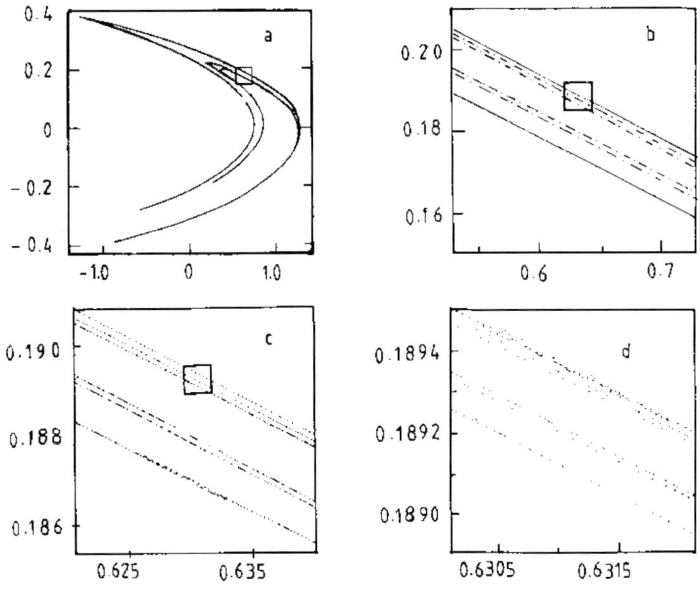

Figure 9. The Henon-Heiles system with parameter values $a = 1.4$, $b = .3$, showing self-similarity at various levels of magnification.

have been much analysed and they show how strange but self-similar is the situation in two dimensions. See Figure 9.

For conservative systems friction will not reduce the degrees of freedom so one has to study nonlinear iterations in more than one variable. An analogue of period doubling has been shown to arise here too. Elliptic fixed points (on surfaces with constant energy in phase space) will, at certain parameter values, turn hyperbolic and give rise to pairs of nearby elliptic fixed points. The process repeats itself until chaos pervades phase space with a cloud of points. It is worth pointing out that the universal scaling constants are not the same as those occurring in dissipative systems. Other applications have been to complex variable maps (Fatou's generalisation of the Julia map to $z = x + iy$) and to circle maps where x corresponds to an angle and the mapping takes the form

$$x_{n+1} = x_n + M - L\sin(2\pi x_n) \qquad (10)$$

with $L > 2\pi$ to ensure noninvertibility. Such circle maps are of considerable interest in turbulent cylindrical motion and in nonlinear oscillators that are driven externally (with a controllable frequency). Self-similarity is again in evidence.

Another direction that the research has taken is towards the quantum generalisation. The chaotic features are associated with matrix representatives of the observables that are truly random and of much the same size everywhere so that they cannot be 'simplified' by unitary transformation to another basis. Evidently the research into this subject of turbulence is still in its infancy but already it shows great promise of at last unravelling this most knotty of natural phenomena. I hope that I have stimulated your interest in this fascinating area of science and that you will find the references useful if you wish to pursue the subject.

References

Scientific semipopular magazines:

R.M. May, Nature **261**, 459 (1976)

G.B. Lubkin, Physics Today, March (1981)

D.R. Hofstadter, Scientific American, November (1981)

L.P. Kadanoff, Physics Today, December (1983)

Review Articles and Books:

M.J. Feigenbaum, Los Alamos Science 1, 4 (1980)

J.P. Eckmann, Reviews of Modern Physics **53**, 643 (1981)

E. Ott, Reviews of Modern Physics **53**, 655 (1981)

B. Hu, Physics Reports **91**, 233 (1982)

P. Cvitanovic, *Universality in Chaos* (Adam Hilger, Bristol, 1984).

Hao Bai-Lin, *Chaos* (World Scientific, Singapore, 1984).

Heinz Georg Schuster, *Deterministic Chaos: An Introduction* (Physik-Verlag, Weinheim, 1984).

The floating plank

R. Delbourgo
Physics Department, University of Tasmania, Hobart, Australia 7005

(Received 17 July 1986; accepted for publication 20 December 1986)

The stable floating configuration of a long plank of rectangular cross section depends on the relative density of the plank to the fluid and on the ratio of the sides. The complete solution of this metacentric problem is given.

I. INTRODUCTION

Courses in continuum mechanics include a component of hydrostatics as a preliminary to the more difficult task of understanding fluid mechanics. Although hydrostatics is considered to be an easy topic, there is one part that certainly is not, namely the determination of equilibrium configuration of floating bodies. This is discussed in some detail in Sommerfeld's classic text.[1] From our point of view the relevant points are that (i) when a body floats in equilibrium, the weight and buoyancy forces are equal and opposite and pass through the same point, either the center of gravity G or the center of buoyancy C; (ii) this equilibrium may be stable or not as the body undergoes a tilt through an angle θ and the center of buoyancy moves to a new position C_θ, such that the magnitudes of weight and buoyancy forces remain equal; (iii) for small θ the metacenter M lies at the center of curvature of the buoyancy arc CC_θ; (iv) if the equilibrium is to be stable, the couple on the body must act in such a way that it should restore equilibrium and this corresponds to a metacenter which lies above the center of gravity. Thus the problem is one of finding the metacentric point M of the buoyancy forces as the body undergoes a virtual rotation and this depends critically on the geometry of the system. In fact, the design of keels of boats, for instance, is very much an exercise of discovering the shape which maximizes the metacentric height for large rotations of the boat. This is no simple matter and requires as much flair as it does science.

A seemingly trivial example of stability concerns a two-dimensional rectangular body floating in a fluid; in particular this can be modeled by a long plank having a rectangular cross section of sides a, b, where we may disregard the long side if it lies nearly horizontally. The analysis of this example is surprisingly complicated and I often set this as a challenging problem to final year undergraduates. The book on hydrostatics by Lamb[2] and the text by Ramsey[3] treat parts of this problem and mention that Huygens devoted much effort into studying the various possibilities, with a final solution having to await Korteweg. However, the ideas are rather scattered and I thought it would be worthwhile to provide the complete analysis in this Journal in order to expose the richness of this type of problem. I make no claim to originality; indeed there is little one needs to know, beyond first-year university mathematics and physics, in order to follow the arguments.

It is clear that the stable configuration can only depend on the aspect ratio $r = b/a$ and on the density of the plank relative to that of the fluid s (<1). Without loss of generality, we may choose $r \leqslant 1$, by swapping the two sides if necessary. At the same time we can take $s \leqslant \frac{1}{2}$ due to the following duality argument.[2,3]

Let G be the location of the center of mass of the entire body, C the center of mass of the immersed part (i.e., the center of buoyancy), and C' the center of mass of the exposed part. See Fig. 1(a). Obviously, G lies on the line between C and C'. If V denotes the immersed volume and V' the exposed volume, Archimedes' principle tells us that

$$s(V+V') = V \text{ or } sV' = (1-s)V. \tag{1}$$

Thus an interchange $s \leftrightarrow (1-s)$ corresponds to $V \leftrightarrow V'$. Hence to each equilibrium configuration there is a dual one where $s \to 1-s$ and the body is rotated through 180°: see Fig. 1(b). In fact one can also prove that duality preserves stability.[3] For this reason I shall restrict the discussion to $s \leqslant \frac{1}{2}$. (It is easy enough to make the appropriate substitutions to cover the case where $\frac{1}{2} < s < 1$.)

The problem can be categorized into six possible configurations according as to whether (1) the shorter side is immersed symmetrically as in Fig. 2(a), (2) the longer side is immersed symmetrically as in Fig. 2(b), (3) the shorter side is immersed asymmetrically as in Fig. 2(c),

 © 1987 American Association of Physics Teachers

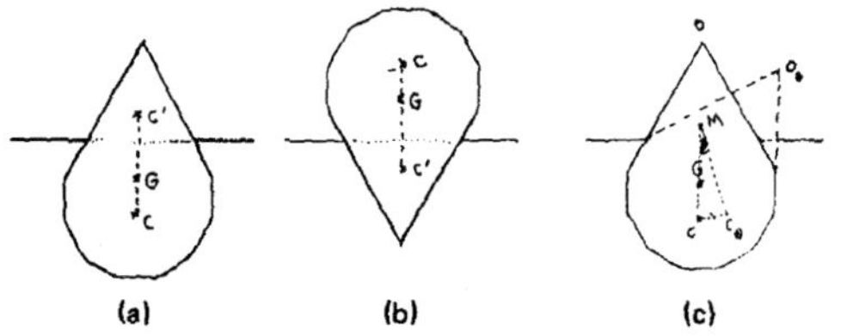

Fig. 1. (a) A stable float when $s > \frac{1}{2}$. (b) The dual float corresponding to $s \to 1 - s$. (c) Relative position of the metacenter M, center of mass G, and center of buoyancy C, for stable flotation. C moves to C_θ and O to O_θ as the body is tilted through θ (dashed shape). M is the center of curvature of the buoyancy arc CC_θ for small θ.

(4) the longer side is immersed asymmetrically as in Fig. 2(d), (5) only one corner is immersed "quasisymmetrically" as in Fig. 2(e), (6) only one corner is immersed "asymmetrically" as in Fig. 2(f). In the last two cases, the definition of "quasisymmetric" refers to a central root of a quartic that corresponds to a symmetrically floating square plank when $r = 1$; it retains that significance even for $r < 1$. I shall analyze cases (1) and (3) in Sec. II; cases (2) and (4) are simply derived from these by inverting r. Cases (5) and (6) are more difficult and are left to Sec. III. Evidently, there is a critical value of the specific gravity s for which cases (2) and (4) merge into case (6) and this will be mentioned at the relevant place in the analysis. Section IV contains the conclusions, comprising Fig. 3, in which all the various possibilities are neatly graphed. A glance at that diagram will reveal which of the configurations the plank will settle down to.

In order to determine the metacenter M, as the body undergoes a rotation θ (with the same immersed volume), it proves more convenient to virtually rotate the fluid level rather than the body. For small θ, M is located at the center of curvature of the buoyancy arc and has to lie vertically above G if the equilibrium is to be stable—see Fig. 1(c). Since, for small virtual θ, MC is a constant and $GC < MC$, GC has to be a minimum as a function of θ, to guarantee stability. This is the rule which gives all the results; the rest is algebra and some calculus.

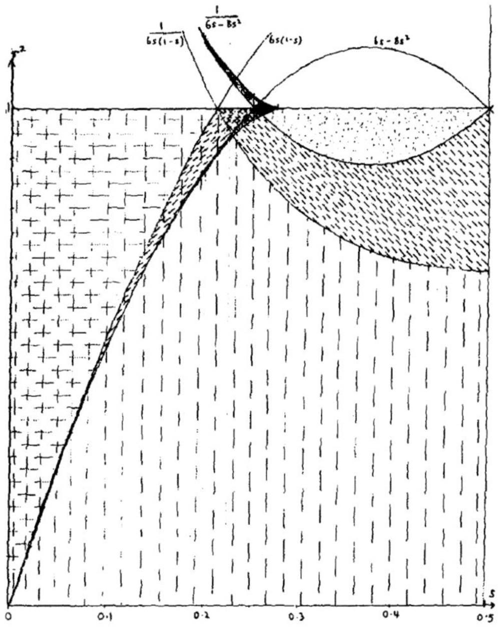

Fig. 3. The various regions of stability marked on an r^2-s plot. Configurations: (1) —shading, (2) | shading, (3) / shading, (4) \ shading, (5) dotted shading, and (6) solid shading.

II. CONFIGURATIONS WITH TWO CORNERS IMMERSED

Examine Fig. 4 which corresponds to a general inclined configuration (it includes the symmetric choice $\theta = 0$ as a special case). We will require GC to be vertical for the θ in question. Using the Cartesian axes as drawn, and remembering $b = ra$, the center of gravity is at

$$G = \tfrac{1}{2}a(1,r)$$

and the center of buoyancy (the center of mass of the immersed trapezium) is readily calculated to lie at

$$C = \tfrac{1}{2}a(s + r^2 \tan\theta/12s,\ r + r^2 \tan\theta/6s).$$

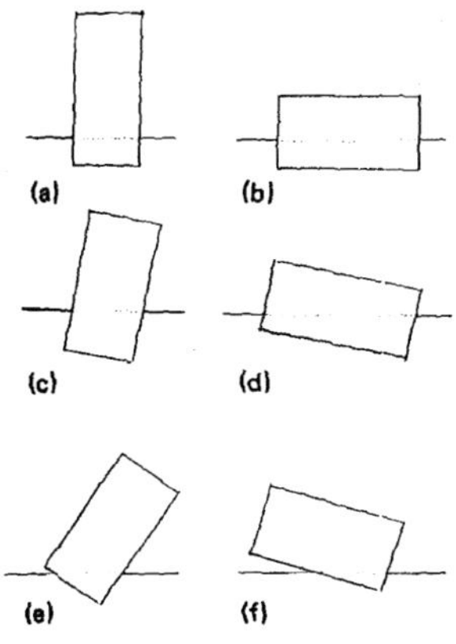

Fig. 2. Cases (a)–(f) correspond to configurations (1)–(6) stated in the Introduction.

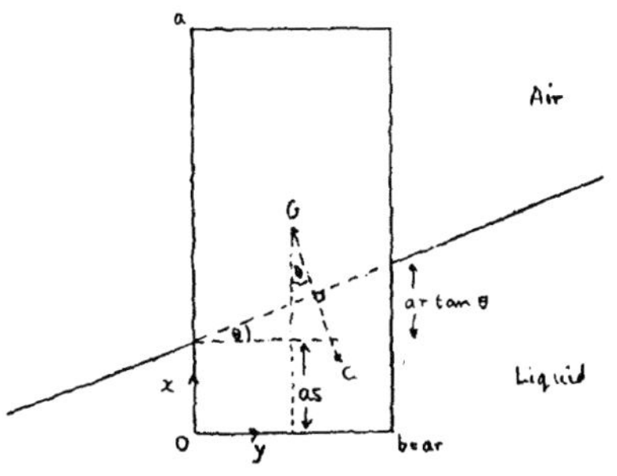

Fig. 4. Hydrostatic configuration with two corners immersed. Here GC is normal to the surface, which is drawn at inclination θ for convenience.

Hence we can find the coordinate vector **CG**. Since CG is to be vertical one needs $y_{CG} = -x_{CG} \tan\theta$. We thereby obtain the equilibrium condition,

$$t[s(s-1) + r^2/6 + r^2t^2/12] = 0, \tag{2}$$

where $t = \tan\theta$. At the same time we have

$$4(CG)^2/a^2 = (s - 1 + r^2t^2/12s)^2 + (r^2t/6s)^2, \tag{3}$$

which has to be a minimum for stability. The vanishing of the first derivative of Eq. (3) just coincides with Eq. (2) and the requirement of positive second derivative gives

$$s(s-1) + r^2/6 + r^2t^2/4 \geqslant 0. \tag{4}$$

There are two equilibrium solutions of Eq. (2), namely the symmetrical float $t = 0$, stable [according to (4)] if $r^2 \geqslant 6s(1-s)$ and asymmetrical float $t^2 = 12s(1-s)/r^2 - 2$, stable if $r^2 \leqslant 6s(1-s)$. Thus the curve $r^2 = 6s(1-s)$ demarcates the boundary between a symmetric and an asymmetric equilibrium stance. This is not quite the full story. Because both corners are immersed by assumption, there is an upper limit to θ, given by $\tan\theta = 2s/r$ (at which stage the trapezuim in Fig. 4 reduces to a triangle with one vertex at the origin of coordinates). This critical asymmetric case corresponds to $r^2 = 6s - 8s^2$.

The conclusion of this section is that case 1: symmetric, stable occurs when $6s(1-s) \leqslant r^2$ and is otherwise unstable, case 3: asymmetric, stable occurs when $6s - 8s^2 \leqslant r^2 \leqslant 6s(1-s)$ and is otherwise unstable. When r attains the lower limit here, one corner is on the verge of penetrating the liquid surface.

Obviously the configurations where the sides a and b are swapped are simply obtained by the substitution $r \to 1/r$ above. That is case 2: symmetric, stable occurs when $r^2 \leqslant 1/6s(1-s)$, whereas case 4: asymmetric, stable arises for $1/6s(1-s) \leqslant r^2 \leqslant 1/(6s - 8s)^2$; here the critical stance arises where r reaches the upper limit. These regions are sketched out in Fig. 3. Note the overlap between them corresponding to the simultaneous existence of equilibrium configurations.

III. CONFIGURATIONS WITH ONE CORNER IMMERSED

Refer to Fig. 5 for this case. Here a' and b' are the lengths of the immersed sides and Archimedes' principle gives $sab = a'b'/2$. Since $\tan\theta = a'/b'$ we deduce

$$a'^2 = 2sabt \text{ and } b'^2 = 2sab/t, \text{ where } t = \tan\theta.$$

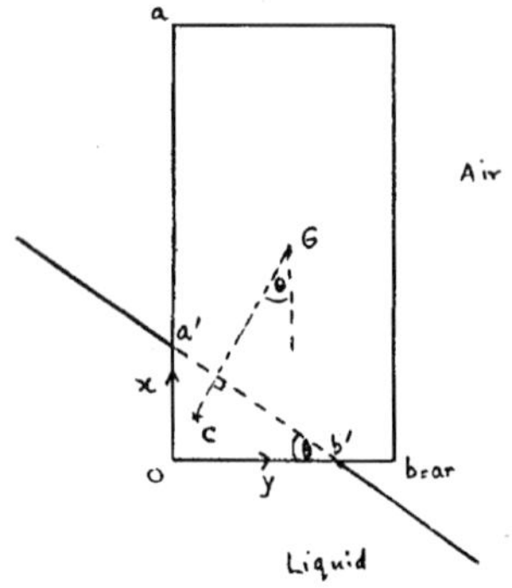

Fig. 5. Hydrostatic configuration with only one corner immersed.

Because we are assuming that only the origin O is submerged, this means that $a' \leqslant a$ and $b' \leqslant b$. Thus we are restricted to θ values bounded via

$$2s/r \leqslant t \leqslant 1/2sr. \tag{5}$$

Now C is at $(a'/3, b'/3)$ and G is at $(a/2, b/2)$ relative to the axes. Since GC is normal to the liquid surface (equilibrium), one needs

$$t = a'/b' = y_{GC}/x_{GC} = (b - 2b'/3)/(a - 2a'/3)$$

or

$$t = [3b - (8sab/t)^{1/2}]/[3a - (8sab/t)^{1/2}]$$
$$= [r - q(r/t)^{1/2}]/[1 - q(rt)^{1/2}], \tag{6}$$

where $q^2 = 8s/9$. $q < \frac{2}{3}$ if r is less than $\frac{1}{2}$. Observe that Eq. (6) is a quartic in $t^{1/2}$, namely

$$t^2 - t^{3/2}r^{-1/2}/q + t^{1/2}r^{1/2}/q - 1 = 0 \tag{6'}$$

and is not going to yield its secrets readily. Note also that the distance CG is given by

$$4(CG)^2 = (a - 2a'/3)^2 + (2 - 2b'/3)^2$$
$$= a^2[(1 - q(rt)^{1/2} + (r - q(r/t)^{1/2}].$$

Its first derivative vanishes automatically when Eq. (6) is obeyed. Its second derivative with respect to $t^{1/2}$ has to be positive to ensure stability, implying that

$$t^2 - 2t^{1/2}r^{1/2}/q + 3 \geqslant 0. \tag{7}$$

Rather than tackle the combination of (5), (6'), and (7) at once, let me first treat the case of the square plank, $r = 1$, as it helps to set the scene for the case of general r. When $r = 1$ the restrictions reduce to

Equilibrium: $t^2 - t^{3/2}/q + t^{1/2}/q + 1 = 0,$ (6s)

Stability: $t^2 - 2t^{1/2}/q + 3 \geqslant 0,$ (7s)

Configuration: $3q/2 \leqslant t^{1/2} \leqslant 2/3q.$ (8s)

Fortunately (6s) factorizes nicely and provides the three roots

$$t = 1, \quad t = t_{\pm} = [1/2q \pm (1/4q^2 - 1)^{1/2}]^2 \tag{9s}$$

(Note that $t_+ = 1/t_-$ indicates equal asymmetric inclinations about the symmetric configuration $t = 1$.) We see that if $t = 1$ holds, stability requires that $q \geqslant \frac{1}{2}$, whereupon $t_{\pm}$ go complex, thereby excluding the asymmetric solutions. On the other hand if $t = t_+$ or t_-, reality necessitates $q \leqslant \frac{1}{2}$ and the stability condition is obeyed. As far as (8s) is concerned, $t = 1$ requires $q \leqslant \frac{2}{3}$ while the roots $t_{\pm}$ require $q^2 \geqslant \frac{2}{9}$. Putting all this together, one can assert that for a square plank the symmetric state is the stable one for $\frac{1}{2} \leqslant q \leqslant \frac{2}{3}$ or $\frac{9}{32} \leqslant s \leqslant \frac{1}{2}$ while the asymmetric state applies for $2^{1/2}/3 \leqslant q \leqslant \frac{1}{2}$ or $\frac{1}{4} \leqslant s \leqslant \frac{9}{32}$. This is marked in Fig. 3. The demarcation point between these two situations, $s = \frac{9}{32}$, corresponds to the case of repeated roots of the quartic [an inflection point of Eq. (6s)].

Let us now return to the case of general r. Recognizing that a "central" root will arise—shifted from $t = 1$— as well as "asymmetric" roots $t_{\pm}$, we shall simply trace out the boundary curve between them by finding out the condition that leads to repeated roots, namely the vanishing derivative of (6'). This corresponds to

$$4t^{2/3} - 3t/qr^{1/2} + r^{1/2}/q = 0. \tag{9}$$

Together, Eqs. (6') and (9) specify the separation line between quasisymmetric and asymmetric configurations and describe some curve in the $r-s$ plane. The curve is best represented in parametric form, with t itself acting as the parameter:

$$s = 9/(6t^2 + 20 + 6/t^{-2}),$$
$$r = t(t^2+3)/(3t^2+1). \qquad (10)$$

This includes the special point, $r=1$, $t=1$, $s=\frac{9}{32}$ and is sketched in Fig. 3. The constraint (5) is obeyed in the lenticular region where the curve is drawn.

The critical case where $a' = a$ corresponds to the upper limit $t_+ = 1/2sr$ and it is easy to check that this gives $r^2 = 1/(6s - 8s^2)$, in agreement with the discussion in Sec. II. Likewise the other critical case has $b' = b$ or $t_- = 2s/r$, in conformity with the boundary curve $r^2 = 6s - 8s^2$ associated with the other corner becoming immersed.

IV. CONCLUSIONS

The six possible configurations of the floating plank mentioned in the Introduction have been carefully studied and the results appear in Fig. 3. The reader can easily inspect from the diagram which of the floating states of the plank is feasible given the specific gravity s and the aspect ratio $r = a/b$. Although the plotted region is limited to $r \leqslant 1$ and $s \leqslant \frac{1}{2}$, it is quite simple to discover the floating configuration for $r > 1$ by simply redefining the sides a, b, and when $s > \frac{1}{2}$, by applying duality (rotation of the body through 180° about the surface).

It is perhaps surprising that such a simple excercise as this turns out to be so complicated in practice. The example only serves to underline how subtle metacentric problems can be in general.[1] I have assumed throughout this article that the plank is long and that the other dimension can be neglected. With a shorter block this is clearly a false premise. Ramsey[3] provides a brief discussion of the floating cube but his treatment is far from exhaustive and does not apply to the general rectangular block anyway. The interested reader is invited to generalize my presentation to that situation.

[1]A. Sommerfeld, *Mechanics of Deformable Bodies, Vol. II* (Academic, London, 1967).
[2]H. Lamb, *Statics* (Cambridge U. P., London, 1921).
[3]A.S. Ramsey, *Hydrostatics* (Cambridge U.P., London, 1936).

Research Note

Aust. J. Plant Physiol., 1997, **24**, 91–96

Observations of Bifurcation and Chaos in Plant Physiological Responses to Light

Sergey Shabala[AB], Robert Delbourgo[A] and Ian Newman[A]

[A]*Physics Department, University of Tasmania, GPO Box 252C, Hobart, Tas 7001, Australia.*
[B]*Corresponding author; email: Sergey.Shabala@phys.utas.edu.au*

Abstract. Although some theoretical predictions have been made, no experimental evidence of chaotic behaviour in plant physiological responses has been reported. Here we present observations of period-doubling and tripling in higher plants. For leaf bioelectric and temperature responses of maize, tomato, and burweed plants to rhythmical light, two different routes to chaos were found experimentally. One was via successive period-doubling and the other via the formation of intermittently chaotic oscillations from a subharmonic synchronisation. Because these effects appeared in intact plants, under conditions close to those found in nature, they may have wide significance, including for plant phylogenesis.

Introduction

Biological systems, from the molecular level to that of communities, are governed by non-linear mechanisms. It is not surprising that such complex non-linear systems possess non-linear dynamics leading to 'strange' behaviour, such as bifurcations and chaos. We must expect to see chaos as often as we see cycles or steadiness (May 1989). However, among 7157 titles in a recent bibliographic search related to chaotic dynamics in complex systems (Zhang 1990), only 236 directly concern living systems. Most of these are theoretical papers, analysing complex behaviour of ecological, biochemical, immunological, or neurophysiological models (Lawton *et al.* 1975; Markus and Hess 1984; May 1989; Pool 1989). Experimental papers are rare. What is the reason for the dearth of experimental evidence for deterministic chaos in plant behaviour? Is chaotic behaviour a characteristic physiological response of plant control systems, or is it just an engaging theoretical concept? Most experimental ecologists and biologists have ignored chaos, and seem to regard unstable patterns as curiosities or annoyances (Pool 1989). The problem is compounded by the difficulties of filtering information from superimposed environmental noise and by the short lengths of data series.

The largest number of experimental observations was accumulated in neurophysiology and medicine. Examples include the period-doubling bifurcation in embryonic chick heart cell activity (Guevara *et al.* 1981; Glass *et al.* 1984), 'Devil's staircase' in experimental data from the intact human heart (Glass 1987), and complexity in brain activity. Other model systems studied include excitable biological membranes. The periodically forced Hodgkin-Huxley oscillator has been shown to exhibit periodic (harmonic or subharmonic synchronisation) and non-periodic (quasi-periodic or chaotic) motions (Aihara *et al.* 1984). The motions were determined by the amplitude and the frequency of the stimulating current. For plants, there are only a few examples of chaotic behaviour for the unicellular alga *Nitella* after periodic stimulation by applied electric current (Hayashi *et al.* 1982, 1983). We have failed to uncover any experimental report of bifurcations or chaotic regimes in higher plant physiological behaviour.

However, conditions for chaos are clearly present in plants. Oscillations in plant photosynthesis (Ogawa 1982), transpiration and stomatal conductance (Reich 1984; Siebke and Weis 1995), growth and movement (Ensminger and Lipson 1992; Antkowiak and Engelmann 1995), root ion exchange (Shabala *et al.* 1997) and other parameters have been reported. They occur as a result of feedback loops in physiological control systems. Hence, deterministic chaos might be present in plant physiological responses. Apart from a few purely theoretical papers which have predicted the possibility of Hopf bifurcation in stomatal responses (Rand *et al.* 1981), this issue of chaos has not arisen in plant physiological literature. Here we present the first experimental observations of period-doubling and tripling in higher plants for leaf bioelectric and temperature responses to rhythmical light. These effects appeared in intact plants, under conditions close to those found in nature, and they may be of great adaptive significance in plant phylogenesis.

Materials and Methods

Plant leaf bioelectric and temperature responses were measured under variable light conditions for different lines and hybrids of maize (*Zea mays* L.), soybean (*Glycine hispida* Max.), tomato (*Lycopersicum esculentum* Mill.), burweed (*Xanthium strumarium* L.) and *Tradescantia virginiana* L. Plants were grown from seeds in laboratory conditions as described by Shabala (1996) and measured at ages from 10 days to 2 months. Specific details concerning the plant material used are given in the figure captions.

10.1071/PP96075 0310-7841/97/010091

Leaf–air temperature differences were recorded using differential copper–constantan thermocouples (Shabala 1996): measurement of temperature functions as an indirect measure of leaf transpiration. For bioelectric measurements, standard electrophysiological techniques for extracellular bioelectric measurements were used (Shabala *et al.* 1993). Briefly, non-polarising Ag/AgCl electrodes were connected to the leaf surface via an electrolyte bridge (BSM solution: 0.1 mM KCl + 0.1 mM $CaCl_2$, pH 6). Stability of electric contact was ensured by using a clip made from Perspex and gently pressing a wet cotton wick to the leaf surface. The other end of the wick was immersed in BSM solution alongside the Ag/AgCl electrode. This electrode comprised a plastic tube about 1 mm diameter filled with 1 M KCl in 3% agar, with impaled chlorided silver wire. Up to six separate electrodes were used simultaneously in experiments. Amplified bioelectric potentials were recorded either with a standard chart recorder or via a computer data acquisition system.

Irradiance variation was achieved by light on/off 'square wave' cycles of duration from 1 to 60 min. A fibre optic light source was normally used, incorporating a water filter to prevent leaf temperature changing with light. Adjustment of the irradiance was achieved by regulating the aperture of the light source. The whole leaf was illuminated during measurements.

Calculations of the amplitude characteristics of each cycle shown in Fig. 2*A* were done using the method of equivalent sine functions (Shabala 1989), and plotted against irradiance *I* (Fig. 2*B*). Briefly, for each light/dark cycle, the amplitude of the sine function having the same graphical area as the periodically changing bioelectric potential was calculated. These calculations were done using the following equation for amplitude:

$$A = 1 / T_0 \int_0^{T_0} \sqrt{f^2(t)}dt,$$

where f (t) is the value of the measured leaf bioelectric potential at time t, and T_0 is the period of the light/dark cycle. For the normal bioelectric responses, 8–9 successive values of A were averaged. For the period-doubling cases, two different mean amplitudes were calculated using odd and even amplitude values (A_1, A_3, A_5, ... and A_2, A_4, A_6, ... respectively). Four different mean amplitudes were similarly calculated for the responses with $T = 4T_0$. Means ± standard error for the Fig. 2*A* data, from one individual plant, are given in Fig. 2*B*.

Results and Discussion

All plants studied possess chaotic physiological responses without any apparent correlation with plant age or genetic specificity. In this communication, we present only a small part of the experimental evidence for the chaotic physiological responses which were typical for all plants studied. Being subjected to light/dark variations with periods of the order of minutes, plants exhibited rhythmical changes of leaf bioelctric potential (Fig. 1). The leaf bioelectric response was normally at the frequency of the forcing light pacemaker (Fig.1*A*). The first 40–50 min show transients caused by the change of the average level of irradiance. After that, oscillations are seen with nearly constant amplitude, modulated only by a slow 24-h circadian rhythm (data not shown). However, for some values of light intensities and light/dark cycle durations, the leaf biolectric responses can exhibit 'strange' features. An example is shown in Fig. 1*B*. Starting at 80 min, the leaf develops a clear double-period response to the rhythmical light although no external conditions have changed. This bifurcation degenerates into erratic behaviour after 130 min. After 170 min the oscillatory responses again follow the forcing light rhythm.

Experiments and theory have identified some distinguishing routes to chaos (Berge *et al.* 1984). One of the principal routes is a period-doubling cascade. Another possibility is intermittency, in which long periods of relative regularity are abruptly and irregularly interrupted by bursts of quite different activity. Both of these routes may be inferred from Fig. 1*B*. Yet another route, for forced cyclic systems, is the mode-locking phenomenon over a small range of parameters (Berge *et al.* 1984).

Measured bioelectric responses arise from light-induced modulation of membrane ion transporters (Marre *et al.* 1989; Okazaki *et al.* 1994). Many different cells are involved in the integrated response measured by the extracellular bioelectric technique (Shabala *et al.* 1993). They contain various light receptors linked to a variety of membrane-located control systems, each of them being governed by non-linear mechanisms. So the question arises: what is the external variable controlling the period-doubling observed in our experiments?

Most experimentally observed chaos in biological membranes occurs under artificial circumstances, with artificial electrical or chemical stimulation (Hayashi *et al.* 1982, 1983; Aihara *et al.* 1984) which we do not expect under natural conditions. For this study, we investigated the possibility that chaotic regimes may be driven by the irradiance. Fig. 2*A* shows a sequence of bioelectric responses of one maize leaf to increasing irradiance (I) in the physiological range typically found within the plant canopy (Knapp and Smith 1990; Pearcy 1990; Cardon *et al.* 1994). Normal bioelectric responses, with the same frequency as the rhythmical light, were observed for $I = 5$ (not shown), 10 and 15 W m^{-2}. At increased irradiance two period-doubling bifurcations were observed ($T = 2T_0$ for $I = 20$ W m^{-2} and $T = 4\ T_0$ for $I = 25$ W m^{-2}). According to the theory, the width of successive bifurcations towards chaos becomes very narrow (Glass 1987). Feigenbaum (1978) showed that, for systems which exhibit a sequence of bifurcations, the ratio of the parameter values in which successive cycles appear approaches a universal value $\delta = 4.6692...$ Because the irradiance was changed in discrete steps $\Delta I = 5$ W m^{-2} , the subsequent chain of period-doubling (periods $8T_0$, 16 T_0, ...) and the transition to chaos was missed. Nevertheless,

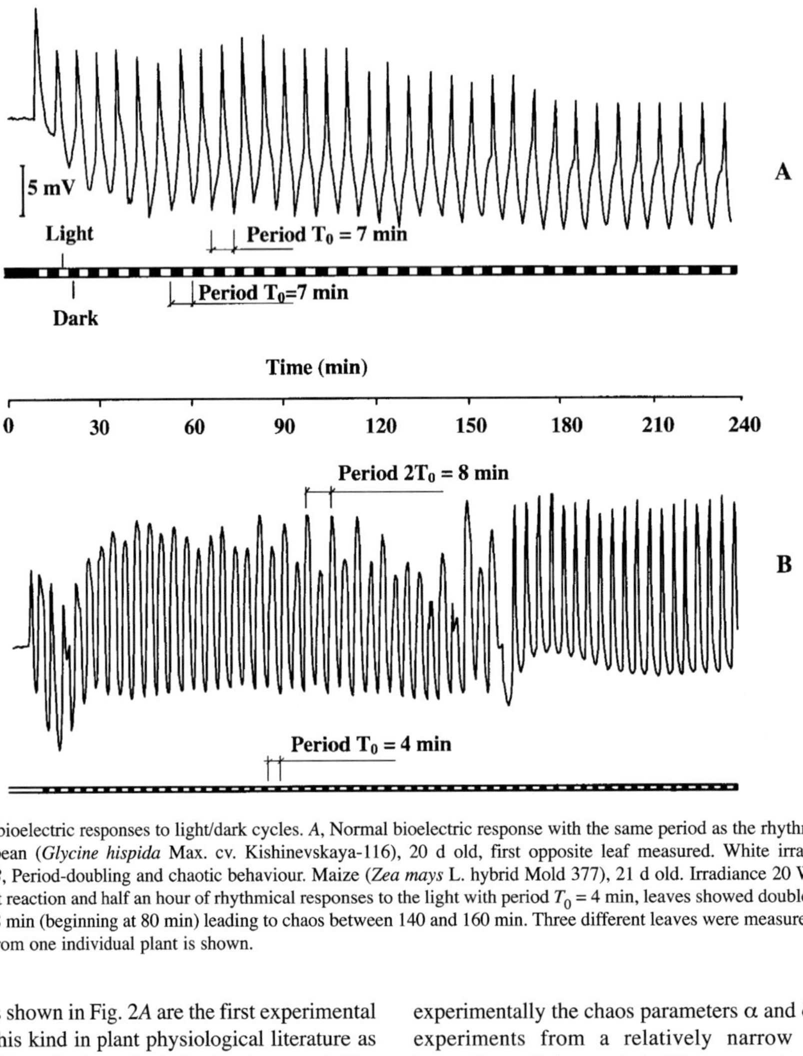

Fig. 1. Plant leaf bioelectric responses to light/dark cycles. *A*, Normal bioelectric response with the same period as the rhythmical light changes (T_0 = 7 min). Soybean (*Glycine hispida* Max. cv. Kishinevskaya-116), 20 d old, first opposite leaf measured. White irradiance 15 W m^{-2}; temperature 24°C. *B*, Period-doubling and chaotic behaviour. Maize (*Zea mays* L. hybrid Mold 377), 21 d old. Irradiance 20 W m^{-2}, temperature 20°C. After transient reaction and half an hour of rhythmical responses to the light with period T_0 = 4 min, leaves showed double-period bioelectric responses with T = 8 min (beginning at 80 min) leading to chaos between 140 and 160 min. Three different leaves were measured at the same time. Typical behaviour from one individual plant is shown.

the two first steps shown in Fig. 2*A* are the first experimental observations of this kind in plant physiological literature as far as we know. When the intensity is further increased, Fig. 2*A* shows a return from chaos via double period at 35 W m^{-2} to single period response at 40 W m^{-2}, a phenomenon known as 'period bubbling' (Berge *et al.* 1984).

As mentioned above, the Feigenbaum number is δ = 4.6692... For the data summarised in Fig. 2*B*, this ratio is about 4. Because of the coarseness of the irradiance steps, it may be not significantly different from the Feigenbaum number. It is also possible that the effect of light on the leaf bioelectric response was mediated by some intermediate parameters with non-linear dependence upon *I*. Penetrating light tends to change its characteristics within the leaf (Vogelmann *et al.* 1989), affecting different light receptors associated with non-linear ion channel conductances (Marre *et al.* 1989; Shabala 1989). In order to evaluate experimentally the chaos parameters α and δ, more accurate experiments from a relatively narrow range of light intensities will be needed. (The parameter δ describes the shrinkage rate of bifurcation fork lengths, and α describes the shrinkege rate of fork width in the period-doubling cascade).

Another kind of chaotic behaviour was observed in leaf temperature responses. Rhythmically changing irradiance causes oscillatory stomatal opening and closing, producing periodical variation of leaf–air temperature difference (LATD). Under normal conditions this variation of LATD had the same period as the rhythm of light cycles T_0 (Fig. 3*A*). In a theoretical paper, Rand *et al.* (1981) showed that a hydraulic model of stomata can exhibit a diversity of dynamic behaviour, including Hopf (Berge *et al.* 1984) and other bifurcations. That the system in our experiments must have passed through a chaotic region is supported by the

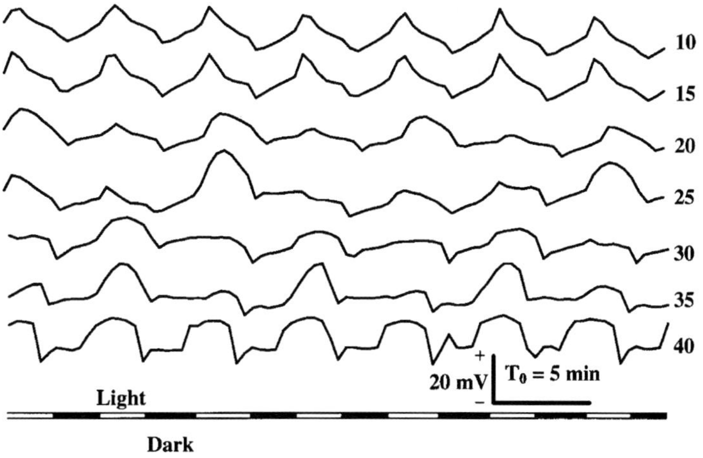

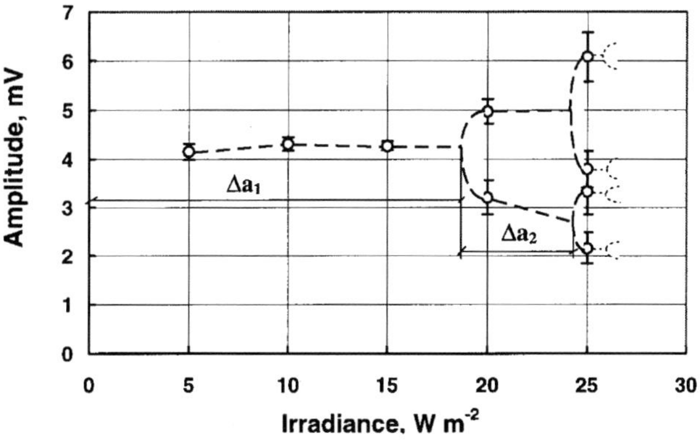

Fig. 2. A sequence of period-doubling in leaf bioelectric responses to light-dark cycles. *A*, Fragments of records for different irradiances ranging from 10 to 40 W m^{-2}. Maize (*Zea mays* L. hybrid Mold 377), 21 d old. The third true leaf was measured. Temperature 22°C. A discrete increase of irradiance $\Delta I = 5$ W m^{-2} was made every hour. Bioelectric responses with the same period $T_0 = 5$ min were observed for $I = 5$ (not shown), 10,15, and 40 W m^{-2}. Double period $T = 10$ min is apparent for $I = 20$, 30, and 35 W m^{-2}. Bioelectric responses with a further doubling $T = 20$ min can be seen for $I = 25$ W m^{-2}. Three different plants were used in the experiments; all exhibited the same features but the data shown relate to one individual plant. *B*, Bifurcational diagram illustrating the period-doubling sequence of leaf bioelectric responses to light/dark cycles. Amplitude of each cycle shown in *A* has been calculated using the method of equivalent sine functions and plotted against irradiance. Δa_i represents the range of irradiances where a stable cycle of period $i \times T_0$ is present. Then the ratio $\Delta a_1/\Delta a_2 \cong 4$ is a roughly estimated value of the Feigenbaum number δ.

pronounced period-tripling of light-induced leaf temperature responses (Fig. 3*B*). Period-tripling is one of the main signals of non-linearity and it spells the existence of chaos (Li and Yorke 1975) for some range of forcing parameters. It can represent a stable stage within an intermittent regime (Aihara *et al.* 1984), and we have now shown that it arises in plant physiological responses.

Conclusions

The observations reported here appear to be the first experimental evidence of bifurcations and chaotic behaviour in physiological responses of higher plants. Four issues arise from these observations.

• First, the behaviour occured under experimental conditions frequently found in nature. Light is a widely

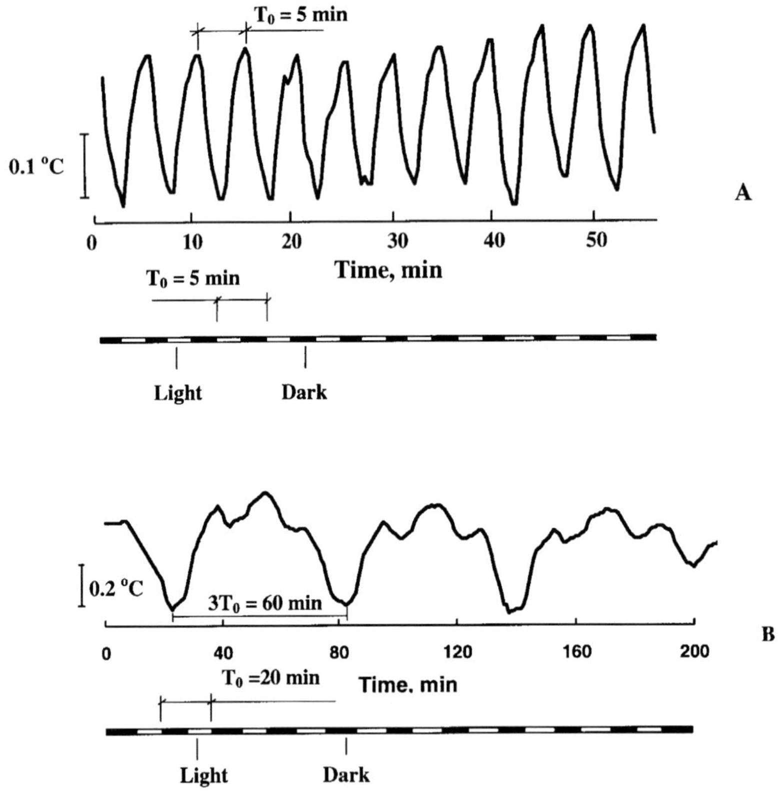

Fig. 3. Bifurcation in leaf temperature responses to light/dark cycles. *A*, Normal responses with period the same as the light/dark variations (T_0 = 5 min). Tomato (*Lycopersicum esculentum* Mill. cv. Utro), 40 d old. Measurements from the second true leaf. Irradiance 15 W m^{-2}. Temperature 22°C. *B*, Bifurcation in leaf temperature responses to rhythmical light for the same plant as in *A*. Period of light/dark cycles T_0 = 20 min. As well as showing 20-min responses, leaf temperature oscillated with three times the period T = 60 min.

fluctuating ubiquitous natural stimulus. Light fluctuations of 5–15 min duration are typical of cloud movements and have been reported earlier (Knapp and Smith 1990; Cardon *et al.* 1994). Also light sunflecks of around 30 W m^{-2} are typical for understorey plants (Pearcy 1990). Discrete Fourier analysis of the natural light fluctuations during a typical autumn day in Tasmania showed the presence of three significant components with periods about 14, 6, and 3 min in the irradiance range from 20 to 60 W m^{-2} (data not shown). Hence the bifurcational responses reported here were occurring in near-natural conditions. Perhaps chaos in plant behaviour is more prevalent than plant physiologists have imagined?

• Second, this kind of dynamic behaviour was recorded for a variety of plants: maize (a cultivated species, C_4 photosynthesis, monocotyledonous), tomato (cultivated, C_3 photosynthesis, dicotyledonous), and burweed (weed, C_3 photosynthesis, dicotyledonous). Thus chaotic behaviour is likely to occur in a wide range of plants in natural conditions.

• Third, these 'strange' responses were found in such different physiological systems as membrane transport and thermoregulation. Hence chaos may also be found in other control systems in plants if it is actively sought.

• Finally, what is the physiological significance of these phenomena? Because of the lack of experimental observation, this question had not arisen before in plant physiology. In medicine the opinion exists that many medical problems may arise from physiological changes causing normally rhythmic processes to show erratic or chaotic fluctuations (May 1989) which are usually regarded as harmful (Glass 1987) . It is hard to believe however, for plants in the natural range of fluctuating environmental conditions, that these bifurcational responses represent abnormal or damaging behaviour, since they have 'survived' after millions of years of evolution under natural conditions. Has this sort of dynamic behaviour provided flexibility in responses of plants to a changing environment (Markus and Hess 1984)? The investigation of this question in relation to plant breeding and speciation may be a priority for future research.

Acknowledgments

We are very grateful to Dr S.N. Maslobrod, Institute of Genetics, Moldavian Academy of Sciences, for his keen interest and constructive suggestions which stimulated this research.

References

Aihara, K., Matsumoto, G., and Ikegaya, Y. (1984). Periodic and non-periodic responses of a periodically forced Hodgkin-Huxley oscillator. *Journal of Theoretical Biology* **109**, 249–269.

Antkowiak, B., and Engelmann, W. (1995). Oscillations of apoplasmic K^+ and H^+ activities in *Desmodium motorium* (Houtt.) Merrill pulvini in relation to the membrane potential of motor cells and leaflet movements. *Planta* **196**, 350–356.

Berge, F., Pomean, Y., and Vidal, C.H. (1984). 'Order within Chaos.' (John Wiley and Sons: New York.)

Cardon, Z.G., Berry, J.A., and Woodrow, I.E. (1994). Dependence of the extent and direction of average stomatal response in *Zea mays* L. and *Phaseolus vulgaris* L. on the frequency of fluctuations in environmental stimuli. *Plant Physiology* **105**, 1007–1013.

Ensminger, P.A., and Lipson, E.D. (1992). Growth rate fluctuations in *Phycomyces* sporangiophores. *Plant Physiology* **99**, 1376–1380.

Feigenbaum, M.J. (1978). Quantitative universality for a class of nonlinear transformations. *Journal of Statistical Physics* **19**, 25–52.

Glass, L., Guevara, M.R., Belair, J., and Shrier, A. (1984). Global bifurcations of a periodically forced biological oscillator. *Physical Review A* **29**, 1348–1357.

Glass, L. (1987). Complex cardiac rhythms. *Nature* **330**, 695–696.

Guevera, M.R., Glass, L., and Shrier, A. (1981). Phase locking, period-doubling bifurcations, and irregular dynamics in periodically stimulated cardiac cells. *Science* **214**, 1350–1353.

Hayashi, H., Nakao, M., and Hirakawa, K. (1982). Chaos in the self-sustained oscillation of an excitable biological membrane under sinusoidal stimulation. *Physics Letters* **88A**, 265–266.

Hayashi, H., Nakao, M., and Hirakawa, K. (1983). Entrained, harmonic, quasiperiodic and chaotic responses of the self-sustained oscillation of *Nitella* to sinusoidal stimulation. *Journal of the Physical Society of Japan* **52**, 344–351.

Knapp, A.K., and Smith, W.K. (1990). Contrasting stomatal responses to variable sunlight in two subalpine herbs. *American Journal of Botany* **77**, 226–231.

Lawton, J.H., Hassel, M.P., and Beddington, J.R. (1975). Dynamic complexity in predator–prey models framed in difference equations. *Nature* **255**, 58–60.

Li, T.Y., and Yorke, J.A. (1975). Period three implies chaos. *American Mathematical Monthly* **82**, 985–992.

Markus, M., and Hess, B. (1984). Transition between oscillatory modes in a glycolytic model system. *Proceedings of the National Academy of Sciences USA* **81**, 4394–4398.

Marre, M.T., Albergoni, F.G., Moroni, A., and Marre, E. (1989). Light-induced activation of electrogenic H^+ extrusion and K^+ uptake in *Elodea densa* depends on photosynthesis and is mediated by the plasma membrane H^+ ATPase. *Journal of Experimental Botany* **40**, 343–352.

May, R. (1989). The chaotic rhythms of life. *New Scientist* **124** (1691), 21–25.

Ogawa, T. (1982). Simple oscillations in photosynthesis of higher plants. *Biochimica et Biophysica Acta* **681**, 103–109.

Okazaki, Y., Tazawa, M., and Iwasaki, N. (1994). Light-induced changes in cytosolic pH in leaf cells of *Egeria densa*: measurements with pH-sensitive microelectrodes. *Plant and Cell Physiology* **35**, 943–950.

Pearcy, R.W. (1990). Sunflecks and photosynthesis in plant canopies. *Annual Review of Plant Physiology and Plant Molecular Biology* **41**, 421–453.

Pool, R. (1989). Ecologists flirt with chaos. *Science* **243**, 310–313.

Rand, R.H., Upadhyaya, S.K., Cooke J.R., and Stori, D.W. (1981). Hopf bifurcation in a stomatal oscillator. *Journal of Mathematical Biology* **12**, 1–11.

Reich, P.B. (1984). Oscillations in stomatal conductance of hybrid poplar leaves in the light and dark. *Physiologia Plantarum* **61**, 541-548.

Siebke, K., and Weis, E. (1995). Assimilation images of leaves of *Glechoma hederacea*: analysis of non-synchronous stomata related oscillations. *Planta* **196**, 155–165.

Shabala, S.N. (1989). Light-induced bioelectric oscillations and their relation to other physiological processes. Ph.D. Thesis. Institute of Experimental Botany, Minsk.

Shabala, S.N. (1996). Leaf temperature kinetics measure plant adaptation to extreme high temperatures. *Australian Journal of Plant Physiology* **23**, 445–452.

Shabala, S.N., Martynenko, A.I., Maslobrod, S.N., and Sadovoy, A.F. (1993). Topography of distribution of bioelectrical potentials of the leaf surface in statics and dynamics. *Physiology & Biochemistry of Cultivated Plants (Russia)* **25**, 500–505.

Shabala, S.N., Newman, I.A., and Morris, J. (1997). Oscillations in H^+ and Ca^{2+} ion fluxes around the elongation region of corn roots and effects of external pH. *Plant Physiology* **113** (in press).

Vogelmann, T.C., Bornman, J.F., and Josserand, S. (1989). Photosynthetic light gradients and spectral regime within leaves of *Medicago sativa*. *Philosophical Transactions of the Royal Society of London* **B 323**, 411–421.

Zhang, S. (1990). 'Bibliography on Chaos.' (World Scientific: Singapore.)

Manuscript received 8 July 1996, accepted 6 November 1996

JOURNAL OF MATHEMATICAL PHYSICS VOLUME 40, NUMBER 8 AUGUST 1999

Matrix representation of octonions and generalizations

Jamil Daboul[a] and Robert Delbourgo[b]
School of Mathematics and Physics, University of Tasmania, Hobart, Australia

(Received 31 December 1998; accepted for publication 25 March 1999)

We define a special matrix multiplication among a special subset of $2N\times 2N$ matrices, and study the resulting (nonassociative) algebras and their subalgebras. We derive the conditions under which these algebras become alternative nonassociative, and when they become associative. In particular, these algebras yield special matrix representations of octonions and complex numbers; they naturally lead to the Cayley–Dickson doubling process. Our matrix representation of octonions also yields elegant insights into Dirac's equation for a free particle. A few other results and remarks arise as byproducts. © *1999 American Institute of Physics.* [S0022-2488(99)03108-4]

I. INTRODUCTION

Complex numbers and functions have played a pivotal role in physics for three centuries. On the other hand, their generalization to other Hurwitz algebras does not seem to have fired the interest of physicists to the same extent, because there is still no *compelling* application of them. Thus, despite the fascination of quaternions and octonions for over a century, it is fair to say that they still await universal acceptance. This is not to say that there have not been valiant attempts to find appropriate uses for them. One can point to their possible impact on quantum mechanics and Hilbert space,[1] relativity and the conformal group,[2] field theory and functional integrals,[3] internal symmetries in particle physics,[4] color field theories,[5] and formulations of wave equations.[6]

In all these cases, there is nothing that stands out and commands our attention; rather, the attempts to describe relativistic physics in terms of quaternions and octonions look rather contrived if not forced, especially for the case of octonions. In this paper we describe a generalization of octonions that allows for Lie algebras beyond the obvious SU(2) structure that is connected with quaternions. We do not presume that they will lead to new physics, but we do think they will at least provide a new avenue for investigation.

Since octonions are not associative, they cannot be represented by matrices with the usual multiplication rules. In this paper, we give representations of octonions and other nonassociative algebras by special matrices, which are endowed with very special multiplication rules; these rules can be regarded as an adaptation and generalization of Zorn's multiplication rule.[7] These matrix representations suggest generalizations of octonions to other nonassociative algebras, which, in turn, lead one almost automatically to a construction of new algebras from old ones, with double the number of elements; we have called these "double algebras." Closer inspection reveals that our procedure can be made to correspond to the Cayley–Dickson construction method,[8] except that in our case the procedure seems rather natural, once one accepts the multiplication rule, whereas the Cayley–Dickson rule looks *ad hoc* at first sight.

II. DEFINITIONS, NOTATIONS, AND A REVIEW OF THE OCTONION ALGEBRA $\mathcal{O}$

The Cayley or the octonion algebra $\mathcal{O}$ is an eight-dimensional nonassociative algebra, which is defined in terms of the basis elements e_μ $(\mu=0,1,2,\ldots,7)$ and their multiplication table. e_0

[a]Permanent address: Department of Physics, Ben Gurion University of the Negev, 84105 Beer Sheva, Israel; electronic mail: daboul@bgumail.bgu.ac.il
[b]Electronic mail: Bob.Delbourgo@utas.edu.au

0022-2488/99/40(8)/4134/17/$15.00 © 1999 American Institute of Physics

stands for the unit element. We can efficiently summarize the table by introducing the following notation [in general, we shall use Greek indices $(\mu,\nu,...)$ to include the 0 and latin indices $(i,j,k,...)$ when we exclude the 0]:

$$\hat{e}_k \equiv e_{4+k}, \quad \text{for } k=1,2,3. \tag{1}$$

The multiplication rules among the basis elements of octonions e_μ can be expressed in the form

$$-e_4 e_i = e_i e_4 = \hat{e}_i, \quad e_4 \hat{e}_i = -\hat{e}_i e_4 = e_i, \quad e_4 e_4 = -e_0, \tag{2}$$

$$e_i e_j = -\delta_{ij} e_0 + \epsilon_{ijk} e_k, \tag{3}$$

$$\hat{e}_i \hat{e}_j = -\delta_{ij} e_0 - \epsilon_{ijk} e_k \quad (i,j,k=1,2,3), \tag{4}$$

$$-\hat{e}_j e_i = e_i \hat{e}_j = -\delta_{ij} e_4 - \epsilon_{ijk} \hat{e}_k. \tag{5}$$

We can formally summarize the rules above by

$$e_\mu e_\nu = g_{\mu\nu} e_0 + \sum_{k=1}^{7} \gamma^k_{\mu\nu} e_k, \quad g_{\mu\nu} := \mathrm{diag}(1,-1,...,-1), \quad \gamma^k_{ij} = -\gamma^k_{ji}, \tag{6}$$

where $\mu,\nu=0,1,...,7$, and $i,j,k=1,...,7$. The multiplication properties are sometimes displayed graphically by a circle surrounded by a triangle, but we shall not bother to exhibit that.

The multiplication law (3) shows that the first four elements form a closed *associative* subalgebra of $\mathcal{O}$, which is known as the *quaternion algebra*,

$$\mathcal{Q} \equiv \langle e_0, e_1, e_2, e_3 \rangle_{\mathrm{R}}, \tag{7}$$

while the other rules (2), (4), and (5) show that $\mathcal{O}$ can be graded as follows:

$$\mathcal{O} = \mathcal{Q} \oplus \hat{\mathcal{Q}}, \quad \text{where } \hat{\mathcal{Q}} := e_4 \mathcal{Q}. \tag{8}$$

$\mathcal{O}$ is a nonassociative algebra. Now a measure of the nonassociativity in any algebra $\mathcal{A}$ is provided by the *associator*, which is defined for any three elements, as follows:

$$(x,y,z) := (xy)z - x(yz), \quad \text{for } x,y,z \in \mathcal{A}. \tag{9}$$

In particular, the associators for the octonion basis are

$$(e_i, e_j, e_k) = 2\epsilon_{ijkl} e_l, \tag{10}$$

where ϵ_{ijkl} are *totally* antisymmetric[9] and equal to unity for the following seven combinations:[5]

$$1247, \quad 1265, \quad 2345, \quad 2376, \quad 3146, \quad 3157, \quad \text{and} \quad 4576. \tag{11}$$

The quaternionic subalgebra $\mathcal{Q}$. It is very well known that the quaternions form an associative subalgebra $\mathcal{Q}$, which can be represented by the Pauli matrices:

$$e_0 \rightarrow \sigma_0 = 1, \quad \text{and} \quad e_j \rightarrow -i\sigma_j \quad (j=1,2,3), \tag{12}$$

where, as usual,

$$\sigma_0 = \begin{pmatrix} 1 & 0 \\ 0 & 1 \end{pmatrix}, \quad \sigma_1 = \begin{pmatrix} 0 & 1 \\ 1 & 0 \end{pmatrix}, \quad \sigma_2 = \begin{pmatrix} 0 & -i \\ i & 0 \end{pmatrix}, \quad \sigma_3 = \begin{pmatrix} 1 & 0 \\ 0 & -1 \end{pmatrix}. \tag{13}$$

It is trivial to check that the above map is an isomorphism:

$$e_i e_j \Leftrightarrow -\sigma_i \sigma_j = -(\delta_{ij} + i\epsilon_{ijk}\sigma_k) \Leftrightarrow -\delta_{ij} + \epsilon_{ijk} e_k . \tag{14}$$

III. NONASSOCIATIVE MULTIPLICATION

In contrast to $\mathcal{Q}$, *the Cayley algebra* $\mathcal{O}$ *cannot be represented by matrices with the usual multiplication rules*, because $\mathcal{O}$ is not associative. However, as we demonstrate below, it is possible to represent octonions by matrices, provided one defines a special multiplication rule among them.

A. Zorn's representation of octonions

Zorn[7] gave a representation of the octonions[8] in terms of 2×2 matrices M, whose diagonal elements are scalars and whose off-diagonal elements are three-dimensional vectors:

$$\mathcal{O} \ni x \rightarrow \begin{pmatrix} \alpha & \mathbf{a} \\ \mathbf{b} & \beta \end{pmatrix}, \tag{15}$$

and invoked a peculiar multiplication rule for these matrices.[7] With a slight modification of the rule adopted by Humphreys,[10] p. 105, our rule is

$$\begin{pmatrix} \alpha & \mathbf{a} \\ \mathbf{b} & \beta \end{pmatrix} * \begin{pmatrix} \alpha' & \mathbf{a}' \\ \mathbf{b}' & \beta' \end{pmatrix} = \begin{pmatrix} \alpha\alpha' + \mathbf{a}\cdot\mathbf{b}' & \alpha\mathbf{a}' + \beta'\mathbf{a} - \mathbf{b}\times\mathbf{b}' \\ \alpha'\mathbf{b} + \beta\mathbf{b}' + \mathbf{a}\times\mathbf{a}' & \beta\beta' + \mathbf{b}\cdot\mathbf{a}' \end{pmatrix}. \tag{16}$$

We propose to adapt this multiplication law to octonions and also replace the necessary three-dimensional basis vectors $\hat{v}_k$ by Pauli matrices σ_k $(k=1,2,3)$, so that the octonions can be represented by the following ordinary 4×4 matrices:

$$\begin{aligned} e_0 \Leftrightarrow \Omega_0 &\equiv \begin{pmatrix} 1 & 0 \\ 0 & 1 \end{pmatrix}, \quad e_k \Leftrightarrow \Omega_k \equiv \begin{pmatrix} 0 & -\sigma_k \\ \sigma_k & 0 \end{pmatrix} \quad (k=1,2,3), \\ e_4 \Leftrightarrow \Omega_4 &\equiv \begin{pmatrix} i & 0 \\ 0 & -1 \end{pmatrix}, \quad \hat{e}_k \Leftrightarrow \hat{\Omega}_k \equiv \begin{pmatrix} 0 & i\sigma_k \\ i\sigma_k & 0 \end{pmatrix}. \end{aligned} \tag{17}$$

[Note the equality of Ω_k $(k=1,2,3)$ to the Dirac matrices γ_k, and Ω_4 to $i\gamma_0$ in the Pauli–Dirac representation.] It can be shown by explicit multiplication, that the above map (17) becomes an isomorphism, provided we define the modified product rule, which we denote by ♥:

$$\begin{pmatrix} \alpha & A \\ B & \beta \end{pmatrix} \heartsuit \begin{pmatrix} \alpha' & A' \\ B' & \beta' \end{pmatrix} = \begin{pmatrix} \alpha\alpha' + \frac{1}{2}\mathrm{Tr}(AB') & \alpha A' + \beta' A + \dfrac{i}{2}[B,B'] \\ \alpha' B + \beta B' - \dfrac{i}{2}[A,A'] & \beta\beta' + \frac{1}{2}\mathrm{Tr}(BA') \end{pmatrix}, \tag{18}$$

where $[A,B] \equiv AB - BA$ is the commutator of A and B. Of course, $A = \mathbf{a}\cdot\boldsymbol{\sigma}$ and $B = \mathbf{b}\cdot\boldsymbol{\sigma}$ are traceless: $\mathrm{Tr}\,A = \mathrm{Tr}\,B = 0$.

In particular, the above multiplication rule yields the following relations:

$$\begin{aligned} \begin{pmatrix} 0 & \eta\sigma_i \\ \xi\sigma_i & 0 \end{pmatrix} \heartsuit \begin{pmatrix} 0 & \eta'\sigma_j \\ \xi'\sigma_j & 0 \end{pmatrix} &= \begin{pmatrix} \eta\xi'\delta_{ij} & \xi\xi'\dfrac{i}{2}[\sigma_i,\sigma_j] \\ -\eta\eta'\dfrac{i}{2}[\sigma_i,\sigma_j] & \xi\eta'\delta_{ij} \end{pmatrix} \\ &= \delta_{ij}\begin{pmatrix} \eta\xi' & 0 \\ 0 & \xi\eta' \end{pmatrix} + \epsilon_{ijk}\begin{pmatrix} 0 & -\xi\xi'\sigma_k \\ \eta\eta'\sigma_k & 0 \end{pmatrix}, \end{aligned} \tag{19}$$

which are helpful for checking the multiplication rules (2)–(5), by substituting the appropriate coefficients, η and ξ.

B. The standard conjugate of octonions

Usually, octonions are studied over the field of real numbers $\mathbb{R}$,

$$x=\sum_{\mu=0}^{7} x_\mu e_\mu \equiv x_0+\mathbf{x}, \quad \text{for } x_\mu \in \mathbb{R}, \tag{20}$$

although later we will find it interesting to deal with their complex extension. The standard *conjugate* $\bar{x}$ of an octonion over $\mathbb{R}$ is defined by

$$\bar{x}:=x_0 e_0-\sum_{i=1}^{7} x_i e_i \equiv x_0-\mathbf{x}. \tag{21}$$

The reason for this definition is that the product of $\bar{x}$ with x yields a positive definite norm:

$$n(x)=x\bar{x}=\bar{x}x=\sum_{\mu=0}^{7} x_\mu^2 \geqslant 0. \tag{22}$$

Moreover, this norm obeys the decomposition law,

$$n(xy)=n(x)n(y). \tag{23}$$

However, with complex octonions [real $x \rightarrow$complex z in (20)], we shall still formally define the *conjugate* $\bar{z}$ of z, to be

$$\bar{z}:=z_0 e_0-\sum_{i=1}^{7} z_i e_i, \quad \text{for } z_\mu \in \mathrm{C}. \tag{24}$$

It follows that the product $z\bar{z}$ is again proportional to unity:

$$n(z)=z\bar{z}=\bar{z}z=\sum_{\mu=0}^{7} z_\mu^2 \in \mathrm{C}, \tag{25}$$

but $n(z)$ ceases to be real, in general; therefore $n(z)$ should simply be regarded as a scalar function, but not a norm.

It is interesting to calculate $n(z)$ by using the matrix representation (26): First, we note that if z is mapped into the matrix Z, then $\bar{z}$ will be mapped into $\bar{\mathrm{Z}}$, as follows:

$$z \rightarrow Z \equiv \sum_{\mu=0}^{7} z_\mu \Omega_\mu = \begin{pmatrix} \alpha & A \\ B & \beta \end{pmatrix}, \quad \bar{z} \rightarrow \bar{Z} \equiv \begin{pmatrix} \beta & -A \\ -B & \alpha \end{pmatrix}, \tag{26}$$

where $A=\mathbf{a}\cdot\boldsymbol{\sigma}$ and $B=\mathbf{b}\cdot\boldsymbol{\sigma}$, with

$$\alpha=z_0+iz_4, \quad \beta=z_0-iz_4, \quad a_k=-z_k+iz_{4+k}, \quad b_k=z_k+iz_{4+k} \quad (k=1,2,3). \tag{27}$$

Second,

$$z\bar{z} \leftrightarrow Z \heartsuit \bar{Z} = \begin{pmatrix} \alpha & A \\ B & \beta \end{pmatrix} \heartsuit \begin{pmatrix} \beta & -A \\ -B & \alpha \end{pmatrix} = \left(\alpha\beta - \frac{1}{2}\mathrm{Tr}(AB) \right) \begin{pmatrix} 1 & 0 \\ 0 & 1 \end{pmatrix} = n(z) I_{4\times 4}. \tag{28}$$

Therefore, we reproduce the expression (25), as expected:

$$n(z):=\frac{1}{4}\mathrm{Tr}(Z\heartsuit \bar{Z})=\alpha\beta-\frac{1}{2}\mathrm{Tr}(AB)=\alpha\beta-\mathbf{a}\cdot\mathbf{b}=\sum_{\mu=0}^{7} z_\mu^2. \tag{29}$$

C. Hermitian conjugate of octonions

Since σ_i are Hermitian matrices, *all our representation matrices* Ω_k *are anti-Hermitian, with the exception of the identity* Ω_0 (which is Hermitian, of course):

$$\Omega_k^\dagger=-\Omega_k, \quad k=1,2,...,7. \tag{30}$$

This fact enables us to prove that the following "hermiticity" property also holds for the ♥ products:

$$(\Omega_\mu \heartsuit \Omega_\nu)^\dagger=\Omega_\nu^\dagger \heartsuit \Omega_\mu^\dagger, \quad \text{for } \mu,\nu=0,1,...,7. \tag{31}$$

First, we note that this equality holds trivially for $(\Omega_0 \heartsuit \Omega_\mu)^\dagger=\Omega_\mu^\dagger=\Omega_\mu^\dagger \heartsuit \Omega_0^\dagger$. Second, we prove (31) for $j,k\neq 0$ by using (6) and noting that γ_{ij}^k are real and antisymmetric in j, k, so that

$$(\Omega_j \heartsuit \Omega_k)^\dagger=-\delta_{jk}\Omega_0+\sum_{i=1}^{7}\gamma_{jk}^i\Omega_i^\dagger=-\delta_{kj}\Omega_0+\sum_{i=1}^{7}\gamma_{kj}^i\Omega_i=\Omega_k \heartsuit \Omega_j=\Omega_k^\dagger \heartsuit \Omega_j^\dagger, \quad j,k=1,...,7. \tag{32}$$

The conjugation property (31) of Ω_μ suggests the following formal definition for the *Hermitian conjugate* of the octonionic basis:

$$e_0^\dagger=e_0, \quad e_j^\dagger=-e_j \quad (j=1,2,...,7), \tag{33}$$

whereupon the "*number operators*" become equal to the identity element:

$$N_\mu:=e_\mu^\dagger e_\mu=e_\mu e_\mu^\dagger=e_0=1 \quad (\text{no summation}) \quad (\mu=0,1,...,7). \tag{34}$$

We can now define the *Hermitian conjugate* of the complex octonions z in a natural way, by

$$z^\dagger:=\sum_{\mu=0}^{7}\bar{z}_\mu e_\mu^\dagger=\bar{z}_0 e_0-\sum_{i=1}^{7}\bar{z}_i e_i\equiv\bar{z}_0-\bar{\mathbf{z}}, \quad \text{where } z_i\in\mathbb{C}. \tag{35}$$

We then calculate

$$\begin{aligned} zz^\dagger &\equiv(z_0+\mathbf{z})(\bar{z}_0-\bar{\mathbf{z}})=|z_0|^2+(\bar{z}_0\mathbf{z}-z_0\bar{\mathbf{z}})-\mathbf{z}\bar{\mathbf{z}} \\ &=\sum_{\mu=0}^{7}|z_\mu|^2-\sum_{k=1}^{7}(z_0\bar{z}_k-z_k\bar{z}_0)e_k+\sum_{1\leq i<j}^{7}(z_i\bar{z}_j-z_j\bar{z}_i)e_ie_j \\ &=N(z)+2i\sum_{k=i}^{7}\mathfrak{I}\left(\bar{z}_0 z_k+\sum_{1\leq i<j\leq 7}z_i\bar{z}_j\gamma_{ij}^k\right)e_k, \end{aligned} \tag{36}$$

where

$$N(z)=\sum_{\mu=0}^{7}|z_\mu|^2. \tag{37}$$

J. Math. Phys., Vol. 40, No. 8, August 1999 Matrix representation of octonions and . . . 4139

The definition $N(z)$ is perfectly reasonable for a norm, although the decomposition law (23) is *not* satisfied. We see that the ''space components'' $(zz^\dagger)_i$ of $zz^\dagger$ are pure imaginary. To understand why this is expected on general grounds, it is useful to introduce the concept of a *Hermitian octonion:* $y^\dagger = y$, which signifies that

$$\bar{y}_0 = y_0, \quad \bar{y}_i = -y_i \quad (i=1,\ldots,7), \tag{38}$$

so that y_0 must be real and all the space components must be pure imaginary.

Since $zz^\dagger$ is Hermitian by (31), we see that its space components can only be pure imaginary. If we wish to get rid of these components and retain only the zero component, we must add the standard conjugate. Thus, we may define the *Hermitian norm* by

$$N(z) = (zz^\dagger + \overline{zz^\dagger})/2. \tag{39}$$

Hence, if z is mapped into Z, then $z^\dagger$ will be mapped into $Z^\dagger$, which is obtained by the standard Hermitian conjugation of the matrix Z.

One of the main insights gained by using the matrix representation is when we calculate the Hermitian norm. If

$$z \to Z = z^\mu \Omega_\mu = \begin{pmatrix} \alpha & A \\ B & \beta \end{pmatrix}, \quad \text{then } z^\dagger \to Z^\dagger = \begin{pmatrix} \bar{\alpha} & B^\dagger \\ A^\dagger & \bar{\beta} \end{pmatrix}. \tag{40}$$

The product

$$Z \heartsuit Z^\dagger = \begin{pmatrix} \alpha & A \\ B & \beta \end{pmatrix} \heartsuit \begin{pmatrix} \bar{\alpha} & B^\dagger \\ A^\dagger & \bar{\beta} \end{pmatrix} = \begin{pmatrix} \alpha\bar{\alpha} + \frac{1}{2}\mathrm{Tr}(AA^\dagger) & \alpha\beta^\dagger + \bar{\beta}A + \frac{i}{2}[B, A^\dagger] \\ \bar{\alpha}\beta + \beta A^\dagger - \frac{i}{2}[A, B^\dagger] & \beta\bar{\beta} + \frac{1}{2}\mathrm{Tr}(BB^\dagger) \end{pmatrix}. \tag{41}$$

The zero component of $zz^\dagger$ is proportional to the trace of $Z \heartsuit Z^\dagger$, so that the new *Hermitian norm* can be expressed in terms of the representation matrices Z as follows:

$$\begin{aligned} N(z) &= \tfrac{1}{4}\mathrm{Tr}(Z \heartsuit Z^\dagger) = \tfrac{1}{2}(|\alpha|^2 + |\beta|^2) + \tfrac{1}{4}(\mathrm{Tr}(AA^\dagger) + \mathrm{Tr}(BB^\dagger)) \\ &= \frac{1}{2}\left(|\alpha|^2 + |\beta|^2 + \sum_{k=1}^{3} (|a_k|^2 + |b_k|^2) \right) = \sum_{\mu=0}^{7} |z_\mu|^2, \end{aligned} \tag{42}$$

in accordance with (37). For real z_μ we get $\beta \to \bar{\alpha}$, $B \to -A^\dagger$ in (26). Therefore, $AB \to -AA^\dagger = -\mathbf{a}\cdot\boldsymbol{\sigma}\bar{\mathbf{a}}\cdot\boldsymbol{\sigma} = -\mathbf{a}\cdot\bar{\mathbf{a}}$. Thus, the formally defined scalar reduces to a conventional norm:

$$N(z) = \alpha\beta - A\cdot B = |\alpha|^2 + \mathbf{a}\cdot\bar{\mathbf{a}} \to |\alpha|^2 + \sum_{k=1}^{3} |a_k|^2 = \sum_{\mu=0}^{7} x_\mu^2 \equiv n(z) \geqslant 0. \tag{43}$$

D. Nonassociative algebras from Lie algebras

The main advantage of our matrix representation over the Zorn vector representation is that our multiplication rule can be generalized to *any number n of dimensions*, whereas the Zorn rule is restricted, since it is defined in terms of vector product $\mathbf{a}\times\mathbf{b}$, which only applies to 3-vectors!

In particular, given *any* representation of an n-dimensional Lie algebra $\mathbf{g}$ in terms of Hermitian $N\times N$ matrices $\lambda_k (k=1,2,\ldots,n)$, we can then define $2n+2$ different $2N$-dimensional matrices,

$$e_0 \Leftrightarrow \Omega_0 \equiv \begin{pmatrix} 1 & 0 \\ 0 & 1 \end{pmatrix}, \quad e_k \Leftrightarrow \Omega_k \equiv \begin{pmatrix} 0 & -\lambda_k \\ \lambda_k & 0 \end{pmatrix} \quad (k=1,\ldots,n),$$

$$\hat{e}_0 \Leftrightarrow \Omega_{n+2} \equiv \begin{pmatrix} i & 0 \\ 0 & -i \end{pmatrix}, \quad \hat{e}_k \Leftrightarrow \hat{\Omega}_k \equiv \begin{pmatrix} 0 & i\lambda_k \\ i\lambda_k & 0 \end{pmatrix}. \tag{44}$$

If we multiply these matrices using the ♥ rule, we end up with a closed algebra, which we shall call the *double algebra* g^D, with the following product rules for their basis elements ($\hat{\Omega}_0 \equiv \Omega_{n+2}$):

$$-\hat{\Omega}_0\Omega_k = \Omega_k\hat{\Omega}_0 = \hat{\Omega}_k, \quad \hat{\Omega}_0\hat{\Omega}_k = -\hat{\Omega}_k\hat{\Omega}_0 = \Omega_k, \quad \hat{\Omega}_0\hat{\Omega}_0 = -\Omega_0,$$

$$\Omega_i\Omega_j = -\delta_{ij}\Omega_0 + f_{ijk}\Omega_k, \quad \hat{\Omega}_i\hat{\Omega}_j = -\delta_{ij}\Omega_0 - f_{ijk}\Omega_k, \tag{45}$$

$$-\hat{\Omega}_j\Omega_i = \Omega_i\hat{\Omega}_j = -\delta_{ij}\hat{\Omega}_0 - f_{ij}\hat{\Omega}_k.$$

Above, the f_{ijk} are the structure constants of the Lie algebra **g**, defined as usual by

$$[L_i, L_j] = i f_{ijk} L_k. \tag{46}$$

The matrices (44) can be regarded as the ♥- matrix representation of the following (nonassociative) abstract algebra:

$$-\hat{e}_0 e_k = e_k \hat{e}_0 = \hat{e}_k, \quad \hat{e}_0 \hat{e}_k = -\hat{e}_k \hat{e}_0 = e_k, \quad \hat{e}_0 \hat{e}_0 = -e_0, \quad \text{where } \hat{e}_0 \equiv e_{n+2}, \tag{47}$$

$$e_i e_j = -\delta_{ij} e_0 + f_{ijk} e_k, \tag{48}$$

$$\hat{e}_i \hat{e}_j = -\delta_{ij} e_0 - f_{ijk} e_k, \tag{49}$$

$$-\hat{e}_j e_i = e_i \hat{e}_j = -\delta_{ij}\hat{e}_0 - f_{ijk}\hat{e}_k. \tag{50}$$

These rules (47)–(50) can all be summarized by ($\mu,\nu=0,1,2,\ldots,2n+2$),

$$e_\mu e_\nu = g_{\mu\nu} e_0 + \sum_{k=1}^{2n+2} \gamma^k_{\mu\nu} e_k, \quad g_{\mu\nu} := \text{diag}(1,-1,\ldots,-1), \quad \gamma^k_{ij} = -\gamma^k_{ji}. \tag{51}$$

We note from (44) that the e_μ, $\mu=0,1,\ldots,n$ correspond to a *subalgebra* $\mathbf{g}_+$ of $\mathbf{g}^D$. The rules (47) show that the double algebra $\mathbf{g}^D$ is obtained from $\mathbf{g}_+$ simply by adding a new element, called $\hat{e}_0$, and defining the other $\hat{e}_k$. This Lie algebra example then automatically leads us to a more general doubling procedure, which can be applied to *any algebra* and not just to those constructed from Lie algebras. In fact, this doubling idea is exactly the procedure that is known as the *Cayley–Dickson process*, as we shall see below.

IV. DEFORMED MULTIPLICATION AND THE $\mathcal{A}^♥$ ALGEBRA

Begin with the following subset of $2N\times 2N$-matrices:

$$\mathcal{A} := \left\{ \begin{pmatrix} \alpha & A \\ B & \beta \end{pmatrix} \middle| A, B \in M_{N\times N} \right\}, \tag{52}$$

where the $N\times N$ matrices α and β in the first and fourth quadrants are proportional to unit matrices.

Among these matrices we may define a more general multiplication rule than that given in (18).[11] We shall still denote it by ♥ since it only introduces two complex *deformation parameters* λ_0 and λ (their values will be restricted as we impose further conditions on the subalgebras):

$$X \heartsuit X' \equiv \begin{pmatrix} \alpha & A \\ B & \beta \end{pmatrix} \heartsuit \begin{pmatrix} \alpha' & A' \\ B' & \beta' \end{pmatrix} := \begin{pmatrix} \alpha\alpha' + \lambda_0 A \cdot B' & \alpha A' + \beta' A - \lambda[B, B'] \\ \alpha' B + \beta B' + \lambda[A, A'] & \beta\beta' + \lambda_0 B \cdot A' \end{pmatrix}. \tag{53}$$

As before, $[A,B] \equiv AB - BA$ denotes the commutator, but $A \cdot B$ may now be chosen to be any suitable *bilinear map* into an appropriate field F. For example, one might define $A \cdot B$ by $A \cdot B \equiv \mathrm{Tr}(AB)/N$, or if A and B belong to a Lie algebra, then one could take $A \cdot B$ to be the adjoint trace: $A \cdot B := \mathrm{Tr}(\mathrm{ad}\, A \;\mathrm{ad}\, B)$, where ad denotes the adjoint representation.[12]

When $\lambda = 0$ and $\lambda_0 = 1$ the multiplication rule (53) looks *almost* like the usual one for matrices. However, it still yields nonassociativity, since we are replacing matrix products, such as AB, by $A \cdot B$ times the unit matrix. But in any case, it is evident that with the ♥ product the set $\mathcal{A}$ becomes a closed algebra, which we denote by $\mathcal{A}^\heartsuit$.[13]

A. Complex numbers from real

Before continuing, let us consider the simplest example of the above matrices, namely, the case $N = 1$. In this circumstance, the matrices A and B become simple *commuting* numbers, a and b. If we specialize further, and choose $\beta = \alpha$ and $b = -a$ to be real, we end up with two-parameter matrices. Their products are

$$X \heartsuit X' \equiv \begin{pmatrix} \alpha & a \\ -a & \alpha \end{pmatrix} \heartsuit \begin{pmatrix} \alpha' & a' \\ -a' & \alpha' \end{pmatrix} := \begin{pmatrix} \alpha\alpha' - \lambda_0 a a' & \alpha a' + \alpha' a \\ \alpha' a + \alpha a' & \alpha\alpha' - \lambda_0 a a' \end{pmatrix}, \tag{54}$$

and this is nothing but the multiplication rule of two complex numbers z and z', provided that we set $\lambda_0 = 1$ and identify α and a with the real and imaginary parts of z. Thus, a *subalgebra* of $\mathcal{A}^\heartsuit$ for $N = \lambda_0 = 1$ becomes isomorphic to the complex numbers $\mathbb{C}$:

$$\mathcal{A}^\heartsuit \ni X = \begin{pmatrix} \alpha & a \\ -a & \alpha \end{pmatrix} \Leftrightarrow z \equiv \alpha + ia \in \mathbb{C}. \tag{55}$$

B. Simple and Hermitian conjugates

The attractive feature of the generalization (53) is that most results and definitions needed for octonions apply almost automatically to $\mathcal{A}^\heartsuit$. For example, for every element $X \in \mathcal{A}^\heartsuit$ we can define a conjugate element $\bar{X}$, as follows:

$$\bar{X} = \overline{\begin{pmatrix} \alpha & A \\ B & \beta \end{pmatrix}} := \begin{pmatrix} \beta & -A \\ -B & \alpha \end{pmatrix}. \tag{56}$$

By substituting $A' \to -A$, $B' \to -B$, $\alpha' \to \beta$, and $\beta' \to \alpha$ in (53), we get immediately

$$X \heartsuit \bar{X} = \begin{pmatrix} \alpha & A \\ B & \beta \end{pmatrix} \heartsuit \begin{pmatrix} \beta & -A \\ -B & \alpha \end{pmatrix} = (\alpha\beta - \lambda_0 A \cdot B) \begin{pmatrix} 1 & 0 \\ 0 & 1 \end{pmatrix} \equiv n(X) I_{2N \times 2N}, \tag{57}$$

where $n(X) \in \mathbb{C}$. In the meantime, we should again look upon $n(X)$ simply as a scalar function, defined by the map $\mathcal{A}^\heartsuit \to \mathbb{C}$ in (58). Later, we shall study the conditions on $\mathcal{A}^\heartsuit$ under which $n(X)$ becomes a norm.

V. SUBALGEBRAS OF $\mathcal{A}^\heartsuit$

The algebra $\mathcal{A}^\heartsuit$ has several interesting subalgebras.

(i) An obvious subalgebra is the one obtained by choosing both matrices A and B to be traceless:

$$\mathcal{A}_0^{\heartsuit}:=\left\{\begin{pmatrix}\alpha & A\\ B & \beta\end{pmatrix}\middle|\,\mathrm{Tr}\,A=\mathrm{Tr}\,B=0\right\}. \tag{58}$$

(ii) This subalgebra has, in turn, another subalgebra $\mathcal{A}_A^{\heartsuit}\subset\mathcal{A}_0^{\heartsuit}$, in which A and B become antisymmetric matrices.

(iii) A third subalgebra, which we denote by $\mathcal{A}_+^{\heartsuit}$, is obtained by choosing $\beta=\alpha$ and $B=-A$:

$$\mathcal{A}_+^{\heartsuit}:=\left\{\begin{pmatrix}\alpha & A\\ -A & \alpha\end{pmatrix}\right\}. \tag{59}$$

It is easily verified that products of such matrices stay in the same class:

$$\begin{aligned}X\heartsuit X'&=\begin{pmatrix}\alpha & A\\ -A & \alpha\end{pmatrix}\heartsuit\begin{pmatrix}\alpha' & A'\\ -A' & \alpha'\end{pmatrix}\\ &=\begin{pmatrix}\alpha\alpha'-\lambda_0 A\cdot A' & \alpha A'+\alpha' A-\lambda[A,A']\\ -\alpha A'-\alpha' A+\lambda[A,A'] & \alpha\alpha'-\lambda_0 A\cdot A'\end{pmatrix}\in\mathcal{A}_+^{\heartsuit}.\end{aligned} \tag{60}$$

Moreover, $\mathcal{A}_+^{\heartsuit}$ has an interesting property.

Proposition 1: The subalgebra $\mathcal{A}_+^{\heartsuit}$ is flexible for all matrices A.

To put this result into perspective, we note that all Abelian or anticommutative algebras are flexible; thus, if $yx=\pm xy$, then $x(yx)=\pm(yx)x=(xy)x$, so that $(x,y,x)=0$. Therefore, it is of interest to show that $\mathcal{A}_+^{\heartsuit}$, which is neither Abelian nor anticommutative, is also flexible.

Proof: We shall prove the above assertion by explicit multiplication. However, to simplify the calculations we first note that the multiples of unity added to each element do not affect the associators:

$$(X+\alpha 1,Y+\beta 1,Z+\gamma 1)=(X,Y,Z), \tag{61}$$

where 1 is the identity matrix. This follows immediately from the linearity of associators:

$$(X+\alpha 1,Y,Z)=(X,Y,Z)+\alpha(1,Y,Z)=(X,Y,Z). \tag{62}$$

The property (61) is helpful for calculating associators of the subalgebra $\mathcal{A}_+^{\heartsuit}$, since we can set the α's equal to zero, when calculating the associators.

We now calculate explicitly the associator (X_1,X_2,X_3) for general matrices from $\mathcal{A}_+^{\heartsuit}$, but using only those with $\alpha_i=0$, i.e.,

$$X_i=\begin{pmatrix}0 & A_i\\ -A_i & 0\end{pmatrix}\in\mathcal{A}_+^{\heartsuit},\quad \text{for } i=1,2,3. \tag{63}$$

We get

$$(X_1,X_2,X_3)\equiv(X_1X_2)X_3-X_1(X_2X_3)=\begin{pmatrix}p & P\\ -P & p\end{pmatrix}-\begin{pmatrix}q & Q\\ -Q & q\end{pmatrix}=\begin{pmatrix}p-q & P-Q\\ Q-P & p-q\end{pmatrix}, \tag{64}$$

where

$$p=\lambda\lambda_0([A_1,A_2]\cdot A_3), \tag{65}$$

$$P=-\lambda_0(A_1\cdot A_2)A_3+\lambda^2[[A_1,A_2],A_3] \tag{66}$$

and

$$q=\lambda\lambda_0(A_1\cdot[A_2,A_3]), \tag{67}$$

$$Q=-\lambda_0(A_2\cdot A_3)A_1+\lambda^2[A_1,[A_2,A_3]]. \tag{68}$$

Therefore, the elements of the associator (X_1,X_2,X_3) are

$$p-q=\lambda\lambda_0([A_1,A_2]\cdot A_3-A_1\cdot[A_2,A_3])=0, \tag{69}$$

$$\begin{aligned}P-Q&=-\lambda_0((A_1\cdot A_2)A_3-(A_2\cdot A_3)A_1)+\lambda^2([[A_1,A_2],A_3]-[A_1,[A_2,A_3]])\\&=-\lambda_0((A_1\cdot A_2)A_3-(A_2\cdot A_3)A_1)+\lambda^2[A_2,[A_3,A_1]].\end{aligned} \tag{70}$$

In other words, *the associator (X,Y,X) vanishes identically, for any* λ, λ_0, $A_1=A_3$ and A_2,

$$(X,Y,X)=0, \quad \text{for } X,Y\in\mathcal{A}_+^\heartsuit. \tag{71}$$

□

(iv) As a fourth subalgebra, let **g** be a given Lie algebra of dimension n, and let $V_{\mathbf{g}}$ be the algebra spanned by the representation matrices of **g**. Then, we can define a subalgebra of $\mathcal{A}^\heartsuit$ via

$$\mathbf{g}^D:=\left\{\begin{pmatrix}\alpha & A\\ B & \beta\end{pmatrix}\middle| A,B\in V_{\mathbf{g}}\right\}. \tag{72}$$

Clearly the off-diagonal elements, such as $\alpha A'+\beta' A+\lambda[B,B']$, of the products, $X\heartsuit X'$ belong to $V_{\mathbf{g}}$. Hence, $\mathbf{g}^D$ are subalgebras of $\mathcal{A}_0$. Moreover, half of $\mathbf{g}^D$, obtained by the intersection of $\mathbf{g}^D$ with $\mathcal{A}_+^\heartsuit$, will be a subalgebra of $\mathbf{g}^D$:

$$\mathbf{g}_0^D:=\left\{\begin{pmatrix}\alpha & A\\ -A & \alpha\end{pmatrix}\middle| A\in V_{\mathbf{g}}\right\}\subset\mathbf{g}^D\subset\mathcal{A}_+^\heartsuit. \tag{73}$$

The commutators of the elements of $\mathbf{g}_0^D$ constitute a Lie algebra, which is isomorphic to the original algebra **g**.

VI. GRADING OF $\mathcal{A}^\heartsuit$

Proposition 2: The algebra $\mathcal{A}^\heartsuit$ can be graded, as follows:

$$\mathcal{A}^\heartsuit=\mathcal{A}_+^\heartsuit\oplus\mathcal{A}_-^\heartsuit=\mathcal{A}_+^\heartsuit\oplus K\mathcal{A}_+^\heartsuit=\mathcal{A}_+^\heartsuit\oplus K\heartsuit\mathcal{A}_+^\heartsuit, \tag{74}$$

where the "grading matrix" is

$$K\equiv\begin{pmatrix}1 & 0\\ 0 & -1\end{pmatrix}. \tag{75}$$

Observe that $K\heartsuit X=KX$ for any $X\in A^\heartsuit$. Also, of course,

$$\mathcal{A}_\eta^\heartsuit\heartsuit\mathcal{A}_{\eta'}^\heartsuit\subseteq\mathcal{A}_{\eta\eta'}^\heartsuit. \tag{76}$$

Proof: Every matrix $X\in\mathcal{A}^\heartsuit$ can be decomposed, as follows:

$$X\equiv\begin{pmatrix}\alpha & A\\ B & \beta\end{pmatrix}=\begin{pmatrix}\alpha_+ & A_+\\ -A_+ & \alpha_+\end{pmatrix}+\begin{pmatrix}\alpha_- & A_-\\ A_- & -\alpha_-\end{pmatrix}\equiv X_++\hat{X}_- \tag{77}$$

$$\equiv X_{+}+KX_{-}, \tag{78}$$

where

$$\alpha_{\pm}\equiv\frac{1}{2}(\alpha\pm\beta),\quad A_{\mp}\equiv\frac{1}{2}(A\pm B),\quad K\equiv\begin{pmatrix}1 & 0\\ 0 & -1\end{pmatrix}. \tag{79}$$

The first set of matrices (with $\beta=\alpha$ and $B=-A$) constitutes the subalgebra $\mathcal{A}_{+}^{\heartsuit}$, which we defined earlier in (59). The second set of matrices (with $\beta=-\alpha$ and $B=A$) will be called $\mathcal{A}_{-}^{\heartsuit}$. Since $\mathcal{A}_{+}^{\heartsuit}$ is a subalgebra of $\mathcal{A}^{\heartsuit}$, clearly $\mathcal{A}_{+}^{\heartsuit}\mathcal{A}_{+}^{\heartsuit}=\mathcal{A}_{+}^{\heartsuit}$. The rest of the inclusion relations (76), namely,

$$\hat{X}\heartsuit\hat{X}'\in\mathcal{A}_{+}^{\heartsuit},\quad X\heartsuit\hat{X}'\in\mathcal{A}_{-}^{\heartsuit},\quad \hat{X}\heartsuit X'\in\mathcal{A}_{-}^{\heartsuit}, \tag{80}$$

follow immediately from the equalities (85)–(87), which we shall prove below. □

Proposition 3: The following equalities hold for any $X,X'\in\mathcal{A}_{+}^{\heartsuit}$:

$$KXK=\bar{X}, \tag{81}$$

$$(KX)\heartsuit(KX')=X'\heartsuit\bar{X}, \tag{82}$$

$$X\heartsuit(KX')=K(\bar{X}\heartsuit X'), \tag{83}$$

$$(KX)\heartsuit X'=K(X'\heartsuit X). \tag{84}$$

Proof: The proof follows simply by explicit matrix multiplication, using (60):

$$KX\heartsuit KX'=\begin{pmatrix}\alpha & A\\ A & -\alpha\end{pmatrix}\heartsuit\begin{pmatrix}\alpha' & A'\\ A' & -\alpha'\end{pmatrix}=\begin{pmatrix}\alpha\alpha'+\lambda_0 A\cdot A' & \alpha A'-\alpha' A-\lambda[A,A']\\ -\alpha A'+\alpha' A+\lambda[A,A'] & \alpha\alpha'+\lambda_0 A\cdot A'\end{pmatrix}$$

$$=\begin{pmatrix}\alpha' & A'\\ -A' & \alpha'\end{pmatrix}\heartsuit\begin{pmatrix}\alpha & -A\\ A & \alpha\end{pmatrix}\equiv X'\heartsuit\bar{X}\in\mathcal{A}_{+}^{\heartsuit}, \tag{85}$$

$$X\heartsuit(KX')=\begin{pmatrix}\alpha & A\\ -A & \alpha\end{pmatrix}\heartsuit\begin{pmatrix}\alpha' & A'\\ A' & -\alpha'\end{pmatrix}=\begin{pmatrix}\alpha\alpha'+\lambda_0 A\cdot A' & \alpha A'-\alpha' A+\lambda[A,A']\\ \alpha A'-\alpha' A+\lambda[A,A'] & -\alpha\alpha'-\lambda_0 A\cdot A'\end{pmatrix}$$

$$=\begin{pmatrix}1 & 0\\ 0 & -1\end{pmatrix}\left\{\begin{pmatrix}\alpha & -A\\ A & \alpha\end{pmatrix}\heartsuit\begin{pmatrix}\alpha' & A'\\ -A' & \alpha'\end{pmatrix}\right\}=K(\bar{X}\heartsuit X')\equiv\widehat{\bar{X}\heartsuit X'}\in\mathcal{A}_{-}^{\heartsuit}, \tag{86}$$

$$KX\heartsuit X'=\begin{pmatrix}\alpha & A\\ A & -\alpha\end{pmatrix}\heartsuit\begin{pmatrix}\alpha' & A'\\ -A' & \alpha'\end{pmatrix}=\begin{pmatrix}\alpha\alpha'-\lambda_0 A\cdot A' & \alpha A'-\alpha' A+\lambda[A,A']\\ \alpha A'-\alpha' A+\lambda[A,A'] & -\alpha\alpha'-\lambda_0 A\cdot A'\end{pmatrix}$$

$$=\begin{pmatrix}1 & 0\\ 0 & -1\end{pmatrix}\left\{\begin{pmatrix}\alpha' & A'\\ -A' & \alpha'\end{pmatrix}\heartsuit\begin{pmatrix}\alpha & A\\ -A & \alpha\end{pmatrix}\right\}=K(X'\heartsuit X)\in\mathcal{A}_{-}^{\heartsuit}. \tag{87}$$

A. Matrix representation of the Cayley–Dickson process

If we multiply the grading matrix K by a real or complex scalar ν, and let $\mu \equiv \nu^2$, we get

$$\nu_{\text{op}} := \nu K, \quad \nu_{\text{op}} \nu_{\text{op}} := \nu^2 1 = \mu 1. \tag{88}$$

Therefore, using the relations (81)–(84), we get the multiplication rule

$$(X_1 + \nu_{\text{op}} X_2) \heartsuit (X_3 + \nu_{\text{op}} X_4) = (X_1 \heartsuit X_3 + \mu \overline{X_4 \heartsuit X_2}) + \nu_{\text{op}}(\bar{X}_1 \heartsuit X_4 + X_3 \heartsuit X_2), \quad \forall X_i \in \mathcal{A}_+^{\heartsuit}. \tag{89}$$

This is exactly the multiplication rule given by Cayley and Dickson, where one starts with an abstract algebra $\mathcal{B}$ and defines an abstract operator ν_{op}, and essentially *postulates* the following multiplication rule:[8]

$$(b_1 + \nu_{\text{op}} b_2)(b_3 + \nu_{\text{op}} b_4) = (b_1 b_3 + \mu b_4 \bar{b}_2) + \nu_{\text{op}}(\bar{b}_1 b_4 + b_3 b_2), \quad b_i \in \mathcal{B}, \tag{90}$$

where $\bar{b}_i \in \mathcal{B}$ is the conjugate of b_i, and $\nu_{\text{op}} \notin \mathcal{B}$, such that $\nu_{\text{op}}^2 = \mu \cdot 1$.

Observe that the $\heartsuit$ multiplication rule provides an explicit matrix representation of the Cayley–Dickson process,[8] provided that the original algebra $\mathcal{B}$ can be represented by $\mathcal{A}_+^{\heartsuit}$.

B. Composition algebras from 2×2 matrices

One may wonder what happens if we allow the rudimentary 2×2 matrices to contain arbitrary complex elements. Since

$$X \heartsuit X' \equiv \begin{pmatrix} \alpha & a \\ b & \beta \end{pmatrix} \heartsuit \begin{pmatrix} \alpha' & a' \\ b' & \beta' \end{pmatrix} := \begin{pmatrix} \alpha\alpha' + \lambda_0 a b' & \alpha a' + \beta' a \\ \alpha' b + \beta b' & \beta b' + \lambda_0 b a' \end{pmatrix}, \tag{91}$$

when $X' = \bar{X}$ this product yields a "norm,"

$$X * \bar{X} = n(X) \begin{pmatrix} 1 & 0 \\ 0 & 1 \end{pmatrix}, \quad \text{where } n(x) = \alpha\beta - \lambda_0 ab. \tag{92}$$

It is easy to check, by explicit multiplication, that the following identity holds for any $\lambda_0 \in \mathbb{C}$:

$$\begin{aligned} n(x)n(x') &= (\alpha\beta - \lambda_0 ab)(\alpha'\beta' - \lambda_0 a'b') \\ &= (\alpha\alpha' + \lambda_0 ab')(\beta\beta' + \lambda_0 ba') - \lambda_0(\alpha a' + \beta' a)(\alpha' b + \beta b') = n(xx'), \end{aligned} \tag{93}$$

which informs us that the standard norm (92) for $N=1$ obeys the composition law.

Clearly, the norm (92) is degenerate for any $\lambda_0 \in \mathbb{C}$, if we allow X to be any 2×2 matrix. [For example, simply choosing $\beta = a = b = 1$ and $\alpha = \lambda_0$ will yield an $x \neq 0$ with $n(x) = \alpha\beta - \lambda_0 ab = 0$.] The question then arises, when is the norm $n(x)$ in (92) nondegenerate? We can certainly guarantee that $n(x)$ is nondegenerate, if we restrict X to have the following special form:

$$X = \begin{pmatrix} z & w \\ -\bar{w} & \bar{z} \end{pmatrix}, \quad \text{and} \quad \Re\lambda_0 > 0, \tag{94}$$

whence

$$n(x) = |z|^2 + \lambda_0 |w|^2 \neq 0, \quad \text{for } \Re\lambda_0 \neq 0. \tag{95}$$

[Of course, there exist a few equivalent variations of the conditions (94). For instance, we can replace $-\bar{w}$ by $\bar{w}$, but demand that $\Re\lambda_0 < 0$.]

Anyhow, this means that we are dealing with a division algebra, which must therefore be one of the four possibilities. Because $X \in M_{2\times 2}$, we may expand it in terms of Pauli matrices, getting

$$X=\begin{pmatrix} z & w \\ -\bar{w} & \bar{z} \end{pmatrix}=z_1\begin{pmatrix} 1 & 0 \\ 0 & 1 \end{pmatrix}+iw_2\begin{pmatrix} 0 & 1 \\ 1 & 0 \end{pmatrix}+iw_1\begin{pmatrix} 0 & -i \\ i & 0 \end{pmatrix}+iz_2\begin{pmatrix} 1 & 0 \\ 0 & -1 \end{pmatrix}$$

$$\equiv x_0\sigma_0 - i\mathbf{x}\cdot\boldsymbol{\sigma} \leftrightarrow x_0 e_0 + \sum_{k=1}^{3} x_k e_k, \quad \text{where } w_i, z_i \in \mathbb{R}. \tag{96}$$

Since $e_0 \to \sigma_0$ and $e_j \to -i\sigma_j$ $j=1,2,3$ are known representations of the quaternions, we conclude that this is the quaternion algebra over the *real* field $\mathbb{R}$, as expected. Indeed, the matrix (96) is the usual representation of $\mathcal{Q}$ in terms of standard matrices. Later on we shall describe another representation by nonstandard matrices.

VII. CONDITIONS ON DEFORMATION PARAMETERS

Previously we showed that $\mathcal{A}_+^{\heartsuit}$ is flexible. We now ask under what conditions $\mathcal{A}_+^{\heartsuit}$ can become alternative.

For this, we must have $(X_1, X_1, X_3)=0$. By setting $X_2 = X_1$ and noting Eq. (70), we get the condition

$$[A_1,[A_3,A_1]]=\frac{\lambda^2}{\lambda_0}=((A_1\cdot A_1)A_3-(A_1\cdot A_3)A_1). \tag{97}$$

This condition can be satisfied if $\lambda^2=\lambda_0/4$ and $A_i=\mathbf{a}_i\cdot\boldsymbol{\sigma}$: Indeed, for such A_i, we get

$$\begin{aligned}[A_2,[A_3,A_1]]=[\mathbf{a}_2\cdot\boldsymbol{\sigma},2i[\mathbf{a}_3\times\mathbf{a}_1]\cdot\boldsymbol{\sigma}]&=-4[\mathbf{a}_2\times[\mathbf{a}_3\times\mathbf{a}_1]]\cdot\boldsymbol{\sigma}\\ &=4((\mathbf{a}_2\cdot\mathbf{a}_3)\mathbf{a}_1-(\mathbf{a}_2\cdot\mathbf{a}_1)\mathbf{a}_3)\cdot\boldsymbol{\sigma}\\ &=4((A_1\cdot A_2)A_3-(A_2\cdot A_3)A_1),\end{aligned} \tag{98}$$

where we used $(A\cdot B)=\frac{1}{2}\mathrm{Tr}(AB)$. Hence, we can have $\lambda=\pm\frac{1}{2}\sqrt{\lambda_0}$. For the special choice $\lambda_0=-1$, we get $\lambda=\pm 1/2$.

This sign ambiguity is the origin of the nonuniqueness of the ♥ product.[14]

VIII. SUMMARY

Our algebras provide concrete the matrix representation of a big class of nonassociative algebras. They may suggest new constructions in the future. One such possibility, which leads to notions of triality, is described in Appendix A, but anyhow the formulation (18) and its deformations (53) permit generalizations that have an obvious affinity to higher symmetry groups, rather than the simple case of SU(2). We also believe that our treatment of hermiticity and norm for the complex case is reasonable; we illustrate their utility with reference to the Dirac equation in Appendix B.

ACKNOWLEDGMENTS

We are extremely grateful to Dr. G. Joshi for supplying us with a comprehensive list of references in this vast topic and for pointing us in the right direction, and to Dr. B. Gardner for useful comments.

APPENDIX A: THE ◇ PRODUCT

In this appendix we try another type of product, which we denote by ◇, where the commutators $[B,B']$ in the ♥ product are now replaced by the standard matrix products BB'.

Let us first consider the simplest case, $N=1$, where the matrices A and B become scalars, so that we shall first deal with 2×2 matrices:

$$X\equiv\begin{pmatrix}\alpha & a\\ b & \beta\end{pmatrix}. \tag{A1}$$

We define the new matrix product, as follows:

$$X\diamond X'\equiv\begin{pmatrix}\alpha & a\\ b & \beta\end{pmatrix}\diamond\begin{pmatrix}\alpha' & a'\\ b' & \beta'\end{pmatrix}:=\begin{pmatrix}\alpha\alpha'+\lambda_0 ab & \alpha a'+\beta' a+\lambda bb'\\ \alpha' b+\beta b'+\eta\lambda aa' & \beta\beta'+\lambda_0 ba'\end{pmatrix}, \tag{A2}$$

where η, λ, λ_0 are arbitrary complex numbers. We now ask the question, whether for such a product, we can define for every X a conjugate $\bar{X}$, such that $X\diamond\bar{X}=n(X)\cdot 1$, where $n(X)$ is some quadratic form of X, i.e., $n(sX)=s^2 n(X)$.

Let us try the following ansatz:

$$\bar{X}\equiv\begin{pmatrix}\beta+\gamma & -a\\ -b & \alpha+\delta\end{pmatrix}. \tag{A3}$$

We want to determine γ and δ, and derive conditions on α, β, a,b, by demanding that $X\diamond\bar{X}\propto 1$:

$$X\diamond\bar{X}=\begin{pmatrix}\alpha & a\\ b & \beta\end{pmatrix}\diamond\begin{pmatrix}\beta+\gamma & -a\\ -b & \alpha+\delta\end{pmatrix}=\begin{pmatrix}\alpha(\beta+\gamma)-\lambda_0 ab & \delta a-\lambda b^2\\ \gamma b-\eta\lambda a^2 & \beta(\alpha+\delta)-\lambda_0 ab\end{pmatrix}$$

$$=n(X)\begin{pmatrix}1 & 0\\ 0 & 1\end{pmatrix},\quad \text{where } n(X)=\alpha(\beta+\gamma)-\lambda_0 ab. \tag{A4}$$

This condition is obeyed, if

$$\delta a-\lambda b^2=0,\quad \gamma b-\eta\lambda a^2=0,\quad \text{and}\quad \alpha\gamma=\beta\delta. \tag{A5}$$

We cannot satisfy these conditions for general a and b. (For example, if $a=0$ and $b\neq 0$, then δ would be infinite.) Thus, for $\eta\neq 0$ we must assume that either both a and b are zero or both are unequal to zero. But for $\eta=0$ we must demand $b=0$. With these restrictions, we get

$$\delta(X)=\lambda\frac{b^2}{a},\quad \gamma(X)=\eta\lambda\frac{a^2}{b},\quad \text{and}\quad \frac{\beta}{\alpha}=\frac{\gamma}{\delta}=\eta\frac{a^3}{b^3}. \tag{A6}$$

An additional condition can be obtained by demanding that the adjoint operation is an involution, so that $\bar{\bar{X}}=X$, whence

$$\bar{\bar{X}}=\begin{pmatrix} \alpha+\delta(X)+\gamma(\bar{X}) & a \\ b & \beta+\gamma(X)+\delta(\bar{X}) \end{pmatrix}$$

$$=\begin{pmatrix} \alpha+\lambda\left(\frac{b^2}{a}-\eta\frac{a^2}{b}\right) & a \\ b & \beta+\lambda\left(\eta\frac{a^2}{b}+\frac{b^2}{-a}\right) \end{pmatrix}$$

$$=\begin{pmatrix} \alpha & a \\ b & \beta \end{pmatrix}=X. \tag{A7}$$

Thus, we get the new condition

$$b^3=\eta a^3, \quad \text{or} \quad b=\xi a, \quad \text{where } \xi=\eta^{1/3}. \tag{A8}$$

Note that for each η we have three cubic roots ξ. Substituting (A8) into the third equality in (A6), we get

$$\alpha=\beta \quad \text{and} \quad \gamma=\delta. \tag{A9}$$

Hence, a can be any complex number as long as $b=\xi a$. Finally, by noting all the above conditions, we get for every $\eta\in\mathrm{C}$ three sets of 2×2 matrices, which are closed algebras under the product $\diamond$:

$$X(\xi)=\left\{\begin{pmatrix} \alpha & a \\ \xi a & \alpha \end{pmatrix}\right\}, \quad \xi=\eta^{1/3}\in\mathrm{C}. \tag{A10}$$

For these matrices, the adjoint and the corresponding quadratic form are

$$\bar{X}\equiv\begin{pmatrix} \alpha+\xi^2\lambda\alpha & -a \\ -\xi a & \alpha+\xi^2\lambda a \end{pmatrix}, \tag{A11}$$

and

$$n(X)=\alpha(\beta+\gamma)-\lambda_0 ab=\alpha^2+\eta\lambda\alpha\frac{a^2}{b}-\lambda_0 ab=\alpha^2+\xi^2\lambda\alpha a-\xi\lambda_0 a^2. \tag{A12}$$

Note that $n(X)$ is quadratic in X, i.e., $N(sX)=s^2N(X)$, $s\in\mathrm{C}$.

For the special case $b=-a$, or $\xi=-1$, we obtain a known quadratic form,[11]

$$n(X)=\alpha^2+\lambda\alpha a+\lambda_0 a^2. \tag{A13}$$

Proceeding to larger matrices, let

$$X\diamond X'\equiv\begin{pmatrix} \alpha & A \\ B & \beta \end{pmatrix}\diamond\begin{pmatrix} \alpha' & A' \\ B' & \beta' \end{pmatrix}:=\begin{pmatrix} \alpha\alpha'+\lambda_0 A\cdot B' & \alpha A'+\beta' A+\lambda BB' \\ \alpha' B+\beta B'+\eta\lambda AA' & \beta\beta'+\lambda_0 B\cdot A' \end{pmatrix}, \quad \eta\in\mathrm{C}. \tag{A14}$$

One can readily check that such matrices yield closed algebras with respect to the above $\diamond$ product. By restricting B to be ξA, we get the following subalgebra:

$$X(\xi)=\left\{\begin{pmatrix} \alpha & A \\ \xi A & \alpha \end{pmatrix}\right\}, \quad \xi=\eta^{1/3}\in\mathrm{C}. \tag{A15}$$

We can check, using (A14), that products of two such matrices yield a matrix of the same type:

$$X \diamond X' \equiv \begin{pmatrix} \alpha & A \\ \xi A & \alpha \end{pmatrix} \diamond \begin{pmatrix} \alpha' & A' \\ \xi A' & \alpha' \end{pmatrix} = \begin{pmatrix} \alpha\alpha' + \xi\lambda_0 A \cdot A' & \alpha A' + \alpha' A + \xi^2 \lambda A A' \\ \xi(\alpha' A + \alpha A') + \eta\lambda A A' & \alpha\alpha' + \xi\lambda_0 A \cdot A' \end{pmatrix}. \tag{A16}$$

However, if we replace the *scalar a* in Eqs. (A11) and (A12) by a *matrix A*, we do not get an adjoint nor a bilinear form, since the appropriate items do not stay scalar, as they should.

Finally, we note that if we replace the simple products AA' in (A16) by anticommutators $\{A,A'\}/2$, and if we make the scalar products $A \cdot A'$ symmetric, i.e.,

$$\begin{pmatrix} \alpha & A \\ \xi A & \alpha \end{pmatrix} \diamond_S \begin{pmatrix} \alpha' & A' \\ \xi A' & \alpha' \end{pmatrix} := \begin{pmatrix} \alpha\alpha' + \xi\lambda_0 A \cdot A' & \alpha A' + \alpha' A + \xi^2 \lambda \{A,A'\}/2 \\ \xi(\alpha' A + \alpha A') + \eta\lambda\{A,A'\}/2 & \alpha\alpha' + \xi\lambda_0 A \cdot A' \end{pmatrix},$$

then the product becomes Abelian. Therefore the new algebra will automatically become *flexible*. One can similarly symmetrize the more general product (112) by defining, $X \diamond_S X' := (X \diamond X' + X' \diamond X)/2$, and also get a flexible algebra.

APPENDIX B: APPLICATION TO THE DIRAC EQUATION

In momentum space, the free Dirac equation reads as

$$P\Psi \equiv (p_0 - \mathbf{p}\cdot\boldsymbol{\alpha} - m\beta)\Psi = \begin{pmatrix} p_0 - m & -\mathbf{p}\cdot\boldsymbol{\sigma} \\ -\mathbf{p}\cdot\boldsymbol{\sigma} & p_0 + m \end{pmatrix} \Psi = 0, \tag{B1}$$

and it can be rewritten in terms of octonions as follows:

$$p\psi \equiv (p_0 + i\mathbf{p}\cdot\hat{\mathbf{e}} + ime_4)\psi = 0, \tag{B2}$$

where we have used the correspondence

$$\alpha_k = \begin{pmatrix} 0 & \sigma_k \\ \sigma_k & 0 \end{pmatrix} = -i\hat{\Omega}_k \Leftrightarrow -i\hat{e}_k, \quad \beta = \begin{pmatrix} 1 & 0 \\ 0 & -1 \end{pmatrix} = -i\Omega_4 \Leftrightarrow -ie_4. \tag{B3}$$

Notice that we cannot write the Dirac equation in terms of quaternions alone, since we require five different basis elements: $(e_0, e_4, \hat{e}_k;\ k=1,2,3)$ or $(e_0, e_4, e_k;\ k=1,2,3)$.

The octonion p is nicely Hermitian $(p = p^\dagger)$ and has zero norm:

$$p\bar{p} = n(p) = p_0^2 - \mathbf{p}^2 - m^2 = 0. \tag{B4}$$

Therefore the solution ψ of the *Dirac octonionic equation* (B2) is elegantly given by the standard conjugate of p, namely,

$$\psi = \bar{p} \equiv p_0 - i\mathbf{p}\cdot\hat{\mathbf{e}} - ime_4 \Leftrightarrow \Psi = \bar{P} = \begin{pmatrix} p_0 + m & \mathbf{p}\cdot\boldsymbol{\sigma} \\ \mathbf{p}\cdot\boldsymbol{\sigma} & p_0 - m \end{pmatrix}. \tag{B5}$$

The first and second columns of Ψ are proportional to the positive energy solutions $u^1(p)$ and $u^2(p)$, while the third and fourth columns yield the negative energy solutions $v^1(p)$ and $v^2(p)$, if we replace m by $-m$, because

$$(p_0 - \mathbf{p}\cdot\boldsymbol{\alpha} - m\beta)u^i(p) = 0, \quad (p_0 - \mathbf{p}\cdot\boldsymbol{\alpha} + m\beta)v^i(p) = 0 \quad (i = 1,2). \tag{B6}$$

Therefore the *physical* and *normalizable* solutions can be expressed as

4150 J. Math. Phys., Vol. 40, No. 8, August 1999 J. Daboul and R. Delbourgo

$$\Psi_p \equiv (u^1|u^2|v^1|v^2) = \frac{1}{\sqrt{2m(p_0+m)}}\begin{pmatrix} p_0+m & \mathbf{p}\cdot\boldsymbol{\sigma} \\ \mathbf{p}\cdot\boldsymbol{\sigma} & p_0+m \end{pmatrix} \Leftrightarrow \frac{p_0+m-i\mathbf{p}\cdot\hat{\mathbf{e}}}{\sqrt{2m(p_0+m)}}, \tag{B7}$$

ensuring that

$$\bar{u}^i(p)u^j(p) \equiv u^{i\dagger}(p)\beta u^j(p) = \delta_{ij} \quad \text{and} \quad \bar{v}^i(p)v^j(p) \equiv v^{i\dagger}(p)\beta v^j(p) = -\delta_{ij} \quad (i,j=1,2). \tag{B8}$$

[1]D. Finkelstein, J. M. Jauch, S. Shiminovitch, and D. Speiser, J. Math. Phys. **3**, 207 (1962); **4**, 136 (1963); **4**, 788 (1963); G. G. Emch, Helv. Phys. Acta **36**, 739 (1963); L. P. Horwitz and L. C. Biedenharn, *ibid.* **38**, 385 (1965); Ann. Phys. (N.Y.) **157**, 432 (1984); M. Gunaydin, C. Piron, and H. Ruegg, Commun. Math. Phys. **61**, 69 (1978); C. G. Nash and G. C. Joshi, J. Math. Phys. **28**, 2883 (1987); **28**, 2886 (1987); S. de Leo and P. Rotelli, Prog. Theor. Phys. **92**, 917 (1994).

[2]F. Gursey, *Proceedings of "Symmetries in Physics (1600–1980)"* Sant Feliu de Guixols, 1992; S. de Leo, J. Math. Phys. **37**, 2955 (1996). These articles contain reviews of this topic.

[3]F. Gursey, *Quaternion Methods in Field Theory* John Hopkins Workshop, QCD 161, J55, 1980; K. Morita, Prog. Theor. Phys. **67**, 1860 (1982); S. L. Adler, Phys. Rev. Lett. **55**, 783 (1985); Commun. Math. Phys. **104**, 611 (1986); C. Nash and G. C. Joshi, J. Math. Phys. **28**, 463 (1987); Int. J. Theor. Phys. **27**, 409 (1988); S. L. Adler, *Quaternionic Quantum Mechanics and Quantum Fields* (Oxford University Press, Oxford, 1995).

[4]F. Gursey, "Algebraic methods and quark structure," in *Proceedings of the Kyoto Symposium*, January 1975; F. Gursey, "Octonionic structure in particle physics," *Proceedings of ICGTMP78*, Berlin, 1979, p. 509; T. Kugo and P. Townsend, Nucl. Phys. B **221**, 357 (1983); A. Sudbery and J. Phys. A **12**, 939 (1984); P. S. Howe and P. C. West, Int. J. Mod. Phys. A **7**, 6639 (1992); S. de Leo, Int. J. Theor. Phys. **35**, 1821 (1996); F. Gursey and C. H. Tze, *On the Role of Division, Jordan and Related Algebras in Particle Physics* (World Scientific, Singapore, 1996).

[5]F. Gursey, "Color, quarks and octonions," John Hopkins Workshop, January, 1974; S. L. Adler, Phys. Rev. D **21**, 2903 (1980); S. Catto and F. Gursey, Nuovo Cimento A **99**, 685 (1988); S. de Leo and P. Rotelli, J. Phys. G **22**, 1137 (1996).

[6]A. J. Davies and G. C. Joshi, J. Math. Phys. **27**, 3036 (1986); P. Rotelli, Mod. Phys. Lett. A **4**, 933 (1989); S. de Leo and P. Rotelli, *ibid.* **11**, 357 (1996); Int. J. Theor. Phys. **37**, 1511 (1998).

[7]M. Zorn, Abh. Mat. Sem. Hamburg **9**, 395 (1933); R. D. Schafer, Am. J. Math. **76**, 435 (1954); G. Moreno, Bol. Soc. Mat. Mexicana **4**, 13 (1998).

[8]R. D. Schafer, *An Introduction to Non-Associative Algebras* (Academic, New York, 1966); H. P. Peterson, Abh. Mat. Sem. Hamburg **9**, 35 (1971); S. Okubo, *Introduction to Octonions and Other Non-Associative Algebras in Physics* (Cambridge University Press, Cambridge, 1995).

[9]The complete antisymmetry of ϵ_{ijkl} in its first three indices is an immediate consequence of the fact that $\mathcal{O}$ is an alternative algebra, such algebras being defined by the two "*alternative conditions*." $x^2y=x(xy)$, and $(yx)x=yx^2$, $\forall x,y\in\mathcal{A}$. These conditions can be expressed in terms of associators as $(x,x,y)=(y,x,x)=0$, $\forall x,y\in\mathcal{A}$. They enable one to prove that the associator (x,y,z) of an alternative algebra is totally antisymmetric in its three arguments. The proof is based on the "*linearization technique*," which consists of replacing one or more of the arguments, say x, by $x+\lambda w$. By these means we get $(x+\lambda w,x+\lambda w,y)=(x,x,y)+\lambda^2(w,w,y)+\lambda(w,x,y)+\lambda(x,w,y)=\lambda((w,x,y)+(x,w,y))=0$, so that $(w,x,y)=-(x,w,y)$. One can similarly prove the antisymmetry among the other arguments. In particular, the total antisymmetry yields $(x,y,x)=0$, or $(z,y,x)+(x,y,z)=0$, where the second equality follows the first, again by linearization. Algebras that obey these last conditions are called *flexible*. These conditions are much less restrictive than the alternative conditions. In fact, most of the interesting nonassociative algebras are flexible. For example, both the Lie and Jordan algebras are flexible, the *Jordan algebras* $\mathcal{T}$ being defined by the (Jordan) identities: $(x^2,y,x)=(x,y,x^2)=0$, $\forall x,y\in\mathcal{T}$. See N. Jacobson, *Structure and Representations of Jordan Algebras*, (American Mathematical Society, 1968). We shall find another class of flexible algebras below.

[10]J. E. Humphreys, *Introduction to Lie Algebras and Representation Theory* (Springer-Verlag New York, 1972).

[11]K. A. Zhevlakov, A. M. Slin'ko, I. P. Shestakov, and A. I. Shirshov, *Rings That Are Nearly Associative* (Academic, New York, 1982).

[12]Note, that although $X_i\cdot X_j=-2\delta_{ij}$ $(i,j=1,2,3)$ for $X_i\in so(3)$, the diagonal elements of $X_i\cdot X_j$ can have different signs, as in $so(2,1)$, where $(X_i\cdot X_j)=\mathrm{diag}(-2,-2,2)$.

[13]Actually we can generalize $\mathcal{A}^\heartsuit$ further, by replacing the product of the complex numbers $\alpha\beta$ with a bilinear product $\alpha\beta\leftrightarrow\alpha\cdot\beta=\alpha_+^2+\lambda_1\alpha_-^2$, where $\alpha_\pm=(\alpha\pm\beta)/\sqrt{2}$. We have refrained from doing so in the text.

[14]Thus, if we reverse the sign in front of the σ in (17), we only need to change the sign of the commutators in (18) to fix the multiplication rule.

Found Phys (2008) 38: 995–1010
DOI 10.1007/s10701-008-9247-8

Born Reciprocity and the 1/r Potential

R. Delbourgo · D. Lashmar

Received: 21 April 2008 / Accepted: 2 October 2008 / Published online: 11 October 2008
© Springer Science+Business Media, LLC 2008

Abstract Many structures in nature are invariant under the transformation pair, $(\mathbf{p}, \mathbf{r}) \to (b\,\mathbf{r}, -\mathbf{p}/b)$, where b is some scale factor. Born's reciprocity hypothesis affirms that this invariance extends to the entire Hamiltonian and equations of motion. We investigate this idea for atomic physics and galactic motion, where one is basically dealing with a $1/r$ potential and the observations are very accurate, so as to determine the scale $b \equiv m\Omega$. We find that an $\Omega \sim 1.5 \times 10^{-15}\ \mathrm{s}^{-1}$ has essentially no effect on atomic physics but might possibly offer an explanation for galactic rotation, without invoking dark matter.

Keywords Born reciprocity · Atomic physics · Galactic rotation · Dark matter

1 Born's Reciprocity Principle

One cannot help but be struck by the way that numerous structures in physics look the same under the simultaneous substitution between momentum $\mathbf{p}$ and position $\mathbf{r}$,

$$\mathbf{p} \to b\,\mathbf{r}, \quad \mathbf{r} \to -\mathbf{p}/b, \tag{1}$$

where b is a scale with the dimensions of M/T. This applies to the classical Poisson brackets $\{r_i, p_j\} = \delta_{ij}$, the quantum commutator brackets $[r_i, p_j] = i\hbar\delta_{ij}$, and the form of the Hamiltonian equations (classical or quantum), $\dot{r}_i = \partial H/\partial p_i$, $\dot{p}_i = -\partial H/\partial r_i$ and of the angular momenta, $L_{ij} = r_i p_j - r_j p_i$. It leads to the concept

R. Delbourgo (✉) · D. Lashmar
School of Mathematics and Physics, University of Tasmania, P.O. Box 252-21, Hobart, Australia 7001
e-mail: bob.delbourgo@utas.edu.au

D. Lashmar
e-mail: dlashmar@utas.edu.au

of phase-space, Fourier transforms and uncertainty relations. The conjugacy between space and momentum is extensible to energy and time, this being required for special relativity. However these observations do not presume that Hamiltonians are invariant under transformation (1).

Born's reciprocity principle [1–3] goes one stage further and assumes that all physical equations of motion are *invariant* and not just covariant under such conjugacy transformations, so that

$$H(\mathbf{r},\mathbf{p}) = H(-\mathbf{p}/b, b\,\mathbf{r}). \tag{2}$$

At first sight this seems a patently absurd idea for anything but oscillators and even there it seems quite silly because it leads to fixed frequency for all vibrations, determined by the value of b. For these reasons and for its failure in accounting for the observed particle masses [1] the principle has naturally fallen into disrepute and has never been taken seriously by physicists. There are also some fundamental philosophical objections to the idea, which will be mentioned later. In spite of these very valid criticisms, we wish to explore the principle and see if we can determine a non-zero value of b (which is probably tiny indeed and tied to cosmic scales). The incentive/reason why we wish to entertain the chance that (2) may be valid arises also from vibrations. The point is that any oscillator contains some measure of anharmonicity; how this is manifested depends on the physical context, but we can be sure that the linear force and its associated potential V cannot increase without end. For instance we might suppose that in reality, for mass m, the true potential is $V(r) \simeq (m\omega^2 r^2/2)\exp(-cr^2)$, where c is a small parameter which sets the distance at which anharmonicity kicks in, and ω is the natural rotational frequency for displacements that are not excessive. If we attempt to make H reciprocity invariant, we may arrive at

$$2H = (p^2 + b^2r^2)/m + m\omega^2[r^2\exp(-cr^2) + (p^2/b^2)\exp(-cp^2/b^2)].$$

It follows that if b is miniscule on ordinary momentum scales, the dangerous last term is minute, as is the correction to the kinetic energy, so the standard picture prevails for small or moderate r. This example indicates we have no right to be so dismissive of Born's principle.

If we view the reciprocity substitution as the transformation

$$\begin{pmatrix}\mathbf{p}\\ b\,\mathbf{r}\end{pmatrix} \to \begin{pmatrix}0 & 1\\ -1 & 0\end{pmatrix}\begin{pmatrix}\mathbf{p}\\ b\,\mathbf{r}\end{pmatrix},$$

we can think of another reciprocity transformation which also leaves most of our physical structures intact, namely

$$\begin{pmatrix}\mathbf{p}\\ b\,\mathbf{r}\end{pmatrix} \to \begin{pmatrix}0 & -i\\ -i & 0\end{pmatrix}\begin{pmatrix}\mathbf{p}\\ b\,\mathbf{r}\end{pmatrix}, \quad \begin{pmatrix}H\\ bt\end{pmatrix} \to \begin{pmatrix}i & 0\\ 0 & -i\end{pmatrix}\begin{pmatrix}H\\ bt\end{pmatrix}, \quad m \to im.$$

In this paper we will consider a non-relativistic potential which is fully established on the large-scale (Newtonian gravity) and on the small scale (atomic physics), namely $1/r$; it has its roots in graviton and photon exchange and has been thoroughly studied over the centuries! We shall explore how Born's principle affects it. The immediate question is how to make $V(r) \propto 1/r$ compatible with (1). The substitution, $1/r \to 1/\sqrt{r^2 + p^2/b^2}$ can be rejected outright as it would enormously enhance the velocity dependence for small b, in fact ridiculously so. However a more reasonable alternative is $1/r \to (1/r + b/p)$, since the last term fades out[1] as $b \to 0$. There may exist other choices for making V compatible with Born's principle, but they are probably less natural than the proposal:

$$H(\mathbf{r}, \mathbf{p}) = \frac{\mathbf{p}^2 + b^2\mathbf{r}^2}{2m} - \alpha\left(\frac{1}{r} + \frac{b}{p}\right), \tag{3}$$

where α signifies the interaction strength ($Ze^2/4\pi\epsilon_0$ for a hydrogenic atom or GMm for gravity). At this stage we will take the sign of b positive even though it hails from $1/\sqrt{\mathbf{p}^2}$ and we are not entirely sure about the sign of the root because the underlying dynamics (and associated field) is unclear. The choice of sign will become firmer in Sect. 4 but it is fully consonant with the second form of reciprocity substitution. In short, for what it's worth, (3) will be our object of study.[2] Since r stands for the relative distance between the test body and the centre of influence, it makes more sense to reinterpret $b = m\Omega$ where m is the reduced mass and regard Ω as the truly universal constant. Only by doing so will we ensure that the modified kinetic energy assumes the same form no matter how coordinates are chosen:

$$p_1^2/m_1 + p_2^2/m_2 + m_1\Omega^2 r_1^2 + m_2\Omega^2 r_2^2 = P^2/M + p^2/m + M\Omega^2 R^2 + m\Omega^2 r^2,$$

and that the motion of the test body is geometrically determined by the gravitational field in which it finds itself; otherwise the b/p reciprocity term will affect test bodies differently and be at variance with the equivalence principle. In the equation above R is the centre of mass location, $M = m_1 + m_2$ is the total mass and P is the total momentum. Because of P, R dependence this means that H is not translation invariant under $\mathbf{r}_i \to \mathbf{r}_i + \mathbf{a}$:

$$H \to H + (2\mathbf{a}.\mathbf{R} + a^2)M\Omega^2/2$$

but this is no different from the usual Coulomb Hamiltonian transforming under translations or boosts; thus H is only *covariant* under those transformations. Therefore this represents a philosophical problem for relativity (even Galilean) and we shall worry about it later. In what follows we shall be working in the centre of mass frame, $\mathbf{R} = \mathbf{P} = 0$ as has been implicitly assumed in (3).

The paper is set out as follows. Section 2 discusses the classical problem and trajectories as b is varied. Not unexpectedly we find that the distorted orbits precess around the force centre and we determine the rate for small b; this is much like general

[1]If a Yukawa potential $\exp(-\mu r)/r$ is forcibly modified to obey reciprocity, the additional term is $b\exp(-\mu p/b)/p$ and becomes negligible for small b; thus nuclear physics is unlikely to be affected.

[2]Such a Hamiltonian in turn spawns a strange-looking Lagrangian when it is expressed in terms of position and velocity. If we think of (3) in a relativistic context we are working in the static approximation in the Coulomb gauge.

relativistic corrections to Newtonian gravity. Section 3 deals with the quantum version; we find the change in energy levels to first order in b by the variational method and perturbation theory—which happen to agree with one another. In this way we set limits on the value of b so as not to disturb experimental atomic results, namely $b \leq 10^{-26}$ kg/s, a rather weak conclusion. More stringent limits come by looking at galactic scales, where the $1/v$ term can influence rotation rates profoundly. Section 4 contains our investigations of (3) for galaxies (possessing a supermassive black hole at the centre); there we find that the $1/v$ term can yield a velocity which at first increases linearly from the centre and then steadies out to a constant value, before rising again due to the effect of the harmonic b^2r acceleration. Our conclusions and continued worries with Born's principle end the paper in Sect. 5.

2 Classical Motion

The equations of motion arising from the non-relativistic expression (3) are

$$\dot{\mathbf{r}} = \mathbf{p}(1/m + b\alpha/p^3), \qquad \dot{\mathbf{p}} = -\mathbf{r}(b^2/m + \alpha/r^3), \tag{4}$$

with $H = E$, the conserved "energy". This means there is a cubic relation between momentum and speed, force and displacement. For positive b the speed/momentum can never vanish, and if $\alpha > 0$ (an attractive interaction) the force cannot disappear either. Minimising both quantities leads to $p_m = (2b\alpha m)^{1/3}$, $r_m = (2\alpha m/b^2)^{2/3}$, whereupon $v_m = 3p_m/2m$. For the purpose of the ensuing analysis we shall take b to be a small and positive quantity; many of the results can be continued to negative b without endangering the steps and we shall anyway be expressing the (small) precessions of the orbit to first order in b where the sign is somewhat irrelevant.

It is quite difficult to solve these equations in general but it helps to use rotational invariance of (4) and the conserved angular momentum $\ell = |\mathbf{r} \times \mathbf{p}|$ to simplify the Hamiltonian in the usual way:

$$H = \frac{1}{2m}\left[p_r^2 + \frac{\ell^2}{r^2} + b^2r^2\right] - \alpha\left[\frac{1}{r} + \frac{b}{\sqrt{p_r^2 + \ell^2/r^2}}\right] = E; \quad p_r \equiv \mathbf{p}\cdot\hat{\mathbf{r}}. \tag{5}$$

(Because of reciprocity we could equally well have used $r_p \equiv \mathbf{r}\cdot\hat{\mathbf{p}}$ in place of p_r and p instead of br. However the form (5) is more familiar and this is the framework we shall adopt.)[3] In consequence the radial equations are

$$\dot{r} = \frac{p_r}{m}\left[1 + \frac{\alpha b m}{(p_r^2 + \ell^2/r^2)^{3/2}}\right], \qquad \dot{p}_r = \frac{\ell^2}{mr^3} - \frac{b^2 r}{m} - \frac{\alpha}{r^2} + \frac{\alpha b \ell^2}{(r^2p_r^2 + \ell^2)^{3/2}}. \tag{6}$$

[3]Having introduced angular momentum and noting that ℓ is reciprocity invariant, one may be inclined to approach (3) differently and only interchange r with p_r/b. This is not quite in the spirit of Born's coordinate-independent suggestion and although it nicely removes the square roots from (5), it also means that the kinetic energy piece $p_r^2 + \ell^2/r^2$ has to be supplemented by an extra term ℓ^2b^2/p_r^2; in turn this means that deviations from circular orbits are mandatory no matter what the nature of the potential energy, because $p_r \to 0$ will lead to infinite E.

For determining the trajectory in the orbital plane, remember that the rate of change of azimuthal angle is $\dot{\varphi} = \ell/mr^2$, so the trajectory equation is obtained by integrating

$$\frac{d\varphi}{dr} = \frac{\ell}{r^2 p_r[1 + \alpha bm/(\ell^2/r^2 + p_r^2)^{3/2}]}, \tag{7}$$

in which p_r has to be eliminated in terms of E via (5). This is a hard problem for general b so we shall first turn to Mathematica for some elucidation of the motion and orbits.

By looking at the E-contours in phase space $(r - p_r)$ one finds that for $\ell \neq 0$ the trajectories at lower energy are bounded with $p_r = \dot{r} = 0$ at perigee or apogee. Absolute minima of E arise when $\partial E/\partial r = 0$ or where

$$-\frac{\ell^2}{mr^3} + \frac{b^2 r}{m} + \frac{\alpha}{r^2} - \frac{\alpha b}{\ell} = 0.$$

This cubic equation in r is readily solved and it has three real roots in the region of interest,

$$r_\pm = \frac{1}{2b}\left[\frac{\alpha m}{\ell} \pm \sqrt{\frac{\alpha^2 m^2}{\ell^2} - 4bl}\right], \qquad r_0 = \sqrt{\frac{\ell}{b}}, \tag{8}$$

with corresponding extremal energy values

$$E_\pm = -\frac{1}{2m}\left(\frac{\alpha^2 m^2}{\ell^2} + 2\ell b\right), \qquad E_0 = b\ell - 2\alpha\sqrt{\frac{b}{\ell}}. \tag{9}$$

E_0 is an unstable maximum, while $E_\pm$ are equal value stable minima, and this is best revealed by plotting $E(r, p_r = 0)$ against r in Fig. 1. The phase space portrait is drawn in Fig. 2 and the corresponding trajectory is depicted in Fig. 3 over a timespan of two seconds. In all diagrams we have assumed unit mass and taken exaggerated values, $\ell = 1, \alpha = 10, b = 1$ in order to emphasize the misshapen and precessing orbit, as this is all we wish to depict.

It is apparent that for $b \neq 0$ orbits become distorted (sometimes very pronouncedly) from Keplerian ellipses and precession occurs. It is of interest to work out the precessional rate for model (3) and small b when the distortions/changes are tiny too. To do so we make the standard change of variable $u = 1/r$ and expand (5) and (3) to first order in b. Thus

$$\frac{d\varphi}{du} = \frac{\ell}{p_r}\left[1 + \frac{\alpha bm}{(\ell^2 u^2 + p_r^2)^{3/2}}\right]^{-1} \simeq \frac{\ell}{p_r}\left[1 - \frac{\alpha bm}{(\ell^2 u^2 + p_r^2)^{3/2}}\right], \tag{10}$$

and

$$p^3 - 2m(E + \alpha u)p - 2m\alpha b + b^2 u^{-2} = 0; \quad p = \sqrt{p_r^2 + \ell^2 u^2}.$$

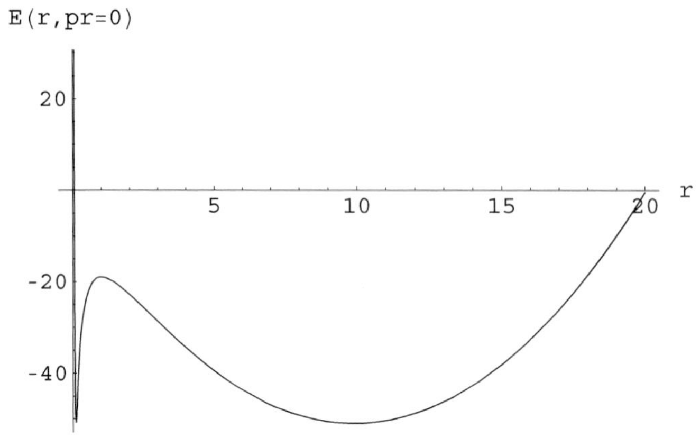

Fig. 1 Dependence of energy on radius when the radial velocity vanishes

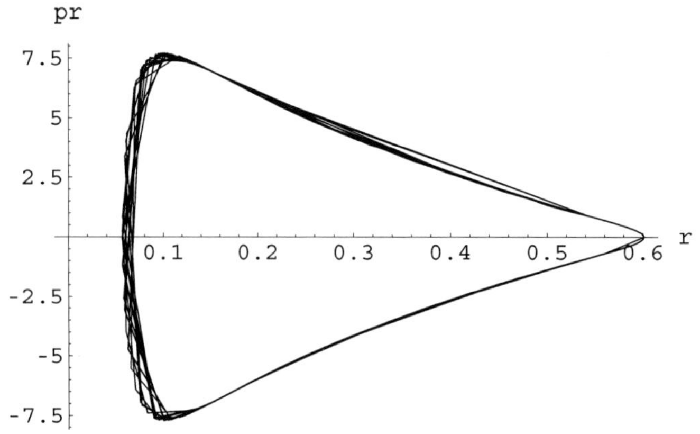

Fig. 2 Phase portrait when $E = -21$ in SI units, from $t = 0$ to 2 s

The root we need is $p_r \simeq \sqrt{2m(E + \alpha u) - \ell^2 u^2}$, so expanding around that the trajectory equation (10) simplifies to

$$\begin{aligned} \frac{d\varphi}{du} &\simeq \frac{\ell}{\sqrt{2m(E+\alpha u) - \ell^2 u^2}}\left[1 - \frac{\alpha b m}{[2m(E+\alpha u)]^{3/2}}\right], \\ \frac{d\varphi}{du} &= \frac{1}{\sqrt{(u-u_1)(u_2-u)}}\left[1 - \frac{\alpha b m \ell^3}{[(u_1+u_2)u - u_1 u_2]^{3/2}}\right], \end{aligned} \tag{11}$$

where $u_1 = 1/r_1, \varphi_1 = 0$ at apogee and $u_2 = 1/r_2, \varphi_2 = \pi/2$ at perigee—the turning points in r. The first term on the right of (11) with $b = 0$ produces the usual elliptical orbit answer

$$2u = (u_1 + u_2) + (u_1 - u_2)\cos(2\varphi)$$

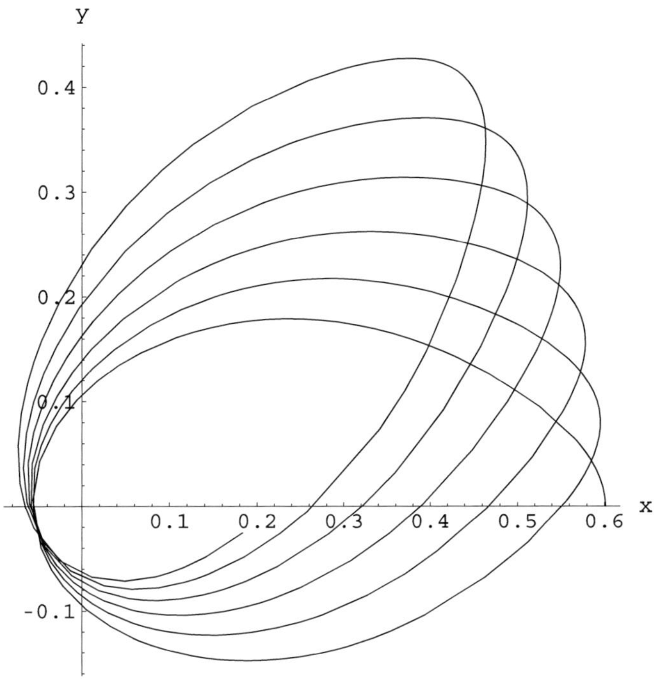

Fig. 3 Trajectory when $E = -21\text{J}$ over a 2 s interval

and the integration of the second term with small nonzero b is connected with the precession. Over an orbit the additional change in azimuth is

$$\begin{aligned} -\Delta\varphi_b &= 2\alpha bm\ell^3 \int_{u_1}^{u_2} \frac{du}{\sqrt{(u-u_1)(u_2-u)}} \frac{1}{[(u_1+u_2)u - u_1u_2]^{3/2}} \\ &= 2\alpha bm\ell^3 \int_0^{\pi} \frac{d\phi}{(\frac{1}{2}[(u_1^2+u_2^2)+(u_1^2-u_2^2)\cos\phi])^{3/2}} \\ &= \frac{4\alpha bm\ell^3}{u_1^2u_2} E(1-u_1^2/u_2^2) = 4\alpha bm\ell^3 r_1^2 r_2 E(1-r_2^2/r_1^2). \end{aligned} \tag{12}$$

Here $E(k)$ is the complete elliptic integral, which for small argument behaves as $E(k) = (\pi/2)(1-k/4-\cdots)$ as $k \to 0$. Hence for small eccentricity ($r_1 \simeq r_2 \simeq a$), the change in azimuth per revolution is

$$\Delta\varphi_b \simeq -4\pi\alpha bm\ell^3 a^3 = -4\pi ba^2(ma\alpha)^{5/2}, \tag{13}$$

to a good approximation, where a is the semimajor axis. In the following section we shall set limits on b (for electrons at least) such that accurate atomic physics experiments are not substantially disturbed. For now it is interesting to compare (13) with the result from general relativity,

$$\Delta\varphi_{\rm GR} \simeq 3\pi MG\left(\frac{1}{r_1}+\frac{1}{r_2}\right) \simeq \frac{6\pi\alpha}{ma},$$

for small eccentricity. This is *very* different from the reciprocity result (13).

3 Quantum Mechanical Considerations

We will now be dealing with the operator version of (3) for hydrogenic atoms when $\alpha = Ze^2/4\pi\epsilon_0$. and

$$H = \frac{P_r^2 + b^2R^2 + L^2/R^2}{2m} - \alpha\left[\frac{1}{R} + \frac{b}{\sqrt{P_r^2 + L^2/R^2}}\right]. \tag{14}$$

It is evident that we are dealing with a tricky problem due to the last term, connected with $1/|P|$, even for eigenfunctions of angular momentum $Y_{\ell m}(\theta,\phi)$ so $L^2 \to \ell(\ell+1)\hbar^2$ in the Schrödinger equation.

Any serious attempt to try to solve (14) needs an interpretation of P_r^{-1} when $\ell = 0$. As $P_r R(r) \to (-i\hbar/r)\partial(rR(r))/\partial r$, when acting on the radial part $R(r)$ of the wave function, a reasonable definition of the inverse is the indefinite integral:

$$P_r^{-1}R(r) \to -\frac{i}{\hbar r}\int_r^\infty dr' r' R(r') \tag{15}$$

for solutions where $rR(r) \to 0$ as $r \to \infty$. One readily checks that $P_r P_r^{-1}\psi = P_r^{-1}P_r\psi = \psi$ for normalizable wave functions. As well, $P_r^{-1}.\exp(ikr)/r = (1/\hbar k).\exp(ikr)/r$ for outgoing waves, giving extra credence to the interpretation (15). Unitarity of the theory is ensured if Hamiltonian is hermitian and its eigenvalues are real; in the ordinary Coulomb problem (even with the addition of a harmonic potential) this requires P_r to be hermitian which in turn is guaranteed if $r\psi(r) \to 0$ as $r \to 0$. Born-extended to (3), the Hamiltonian also needs the reciprocal operator R_p to be hermitian, so in addition one must impose the condition $p\phi(p) \to 0$ as $p \to 0$, on the momentum space wave function, in accordance with reciprocity. This is the very least we must do to conserve probability.

In the event we have not succeeded in obtaining a complete solution of $u(r) = r\psi(r)$ of type $P(r)\exp(-\kappa r - br^2/2\hbar)$ to the equation for the radial wavefunction

$$-\frac{\hbar^2}{2m}\frac{d^3u}{dr^3} + \frac{d}{dr}\left(\frac{b^2r^2u}{2m} - \frac{\alpha u}{r}\right) - \frac{i\alpha bu}{\hbar} = E\frac{du}{dr},$$

because of the troublesome last term on the lefthand side. No matter. One can still obtain a sensible estimate for the change in energy levels to $O(b)$ either by using perturbation theory or making a simple variational calculation. Since the effect is greatest for the ground state, we will analyse the displacement for the lowest energy level both ways—whose answers fortunately agree.

To apply perturbation theory, take the unperturbed wave function in coordinate and momentum space:

$$\psi(r) = \frac{e^{-r/a}}{\sqrt{\pi a^3}}, \quad \phi(p) = \frac{8\sqrt{\pi a^3}}{(p^2a^2/\hbar^2+1)^2}; \quad a = \frac{\hbar^2}{m\alpha} = \text{Bohr radius} \tag{16}$$

and use both to work out the expectation value,

$$\Delta E = \langle b^2R^2/2m - \alpha b/P\rangle = \frac{2\pi b^2}{m}\int_0^\infty (r\psi(r))^2\,dr - \frac{4\pi\alpha b}{h^3}\int_0^\infty p\phi^2(p)\,dp$$
$$= 3b^2a^2/2m - 16\alpha ba/3\pi\hbar \simeq -16b\hbar/3\pi m, \tag{17}$$

to first order in b.

With the variational method adopt a trial wavefunction just like (16), except that a is no longer identified with the Bohr radius. Then one gets (as a function of a),

$$E(a) = \frac{\hbar^2}{2ma^2} + \frac{3b^2a^2}{2m} - \frac{\alpha}{a} - \frac{16\alpha ba}{3\hbar\pi}. \tag{18}$$

Minimising the energy, a cubic equation for the trial radius a is found:

$$\frac{\hbar^2}{ma^3} - \frac{\alpha}{a^2} + \frac{16\alpha b}{3\hbar\pi} = 0, \tag{19}$$

whose solution to order b is $a \simeq \hbar^2/m\alpha + 16b\hbar^5/3m^3\alpha^3$. Hence the optimal energy is obtained as

$$E = -m\alpha^2/2\hbar^2 - 16b\hbar/3\pi m + \cdots,$$

thereby agreeing with (17).

Taking the H-atom as the archetypical case we must make sure that atomic values are hardly touched by the reciprocity term $\alpha b/P$. The Rydberg energy of 13.6... eV $\sim 10^{-18}$ J is known to nine places of decimals and because $\Delta E \sim b\hbar/m_e \sim 10^{-4}b$ J we shall require this to be $\leq 10^{-30}$ J, just to be on the safe side. This sets a limit $b < 10^{-26}$ kg/s. This may seem miniscule but if we translate it into a universal angular frequency via $\Omega \equiv b/m_e$, we obtain a value $\Omega \leq 10^4$ Hz that is not so small. Even with this bound on Ω, we can determine the effect of the reciprocity term on the fine structure of hydrogen (which is normally of the order μeV). The b/p perturbation results in a ^{2}S-^{2}P level separation

$$\langle b\alpha/P\rangle_{20} - \langle b\alpha/P\rangle_{21} = 64b\hbar/15\pi m = 64\Omega\hbar/15\pi$$

which is at most of order 10^{-12} eV. As we shall see presently, galactic considerations produce Ω which are very much tinier or periods which are *many* orders of magnitude greater; therefore we can state with some confidence that the reciprocity modification with small enough b has essentially no effect on the accuracy of atomic calculations.

4 Reciprocity and Rotation of the Galaxy

We now treat the gravitational case, based on classical Newtonian dynamics and, presently, its reciprocity variant (3). It is a really important subject as the existence of possible local dark matter owes a great deal to this study. We wish to investigate whether Born's reciprocity modification (3) has anything of consequence to say on the topic of dark matter and, in particular, if the observed tangential velocity profile is consistent with the *visible* matter distribution, without invoking local dark matter.

In order to make any headway, observations of the motion/distribution of (visible) matter have to be taken into account [4–6]. It has been established that the Milky Way contains a spinning supermassive black hole at centre with a mass of at least 4×10^6 suns or about 10^{37} kg. Visible mass is about 10^{11} suns and on the basis of perceived rotation rates it is inferred that about ten times that mass is hidden in dark matter, as far out as we can see—*assuming ordinary Newtonian* gravity. The rotation speed of the Milky Way [7, 8] seems to increase linearly with radius over a small distance, of about 0.3 kpc $\sim 10^{19}$ m, peaks, oscillates a bit and settles down to some 230 km/s out to 5×10^{20} m from the black hole at the centre. We shall take this as a crude description of our galaxy, neglecting spiral arm structure and the effect of some barring of mass in the middle on the motion.

Since we are presuming that the rotation is mainly tangential, if we consider a typical star such as the sun, we may take $p_r \simeq 0$ to a good approximation. So we just need to sum over all other masses M and their relative speeds v to obtain an average value for $\langle \alpha(1/r + b/p)\rangle = \langle GmM(1/r + \Omega/v)\rangle$. Here we have adopted a positive b-sign, with $b \equiv m\Omega > 0$, as before, to fit in with the observations to come; the opposite sign gives nothing but grief.

We cannot perform such averaging without some idea of the mass density distribution $\rho(\mathbf{r})$. *Unmodified* Newtonian gravity indicates that the matter distribution is roughly *spherical* and $\rho(r) \sim 1/r^2$ as far out as one can see, to ensure that the enclosed mass increases linearly with radius and reproduce a constant tangential star speed. Of course the visible mass distribution flatly contradicts this (and hence the need for dark matter): visible material is mostly disk-shaped, with a concentration near the galaxy centre. This is what we shall model by taking a cylindrical Gaussian distribution,

$$\rho(\varrho, z) \simeq \kappa^2 \delta \mathcal{M} e^{-\kappa^2(\varrho^2+\delta^2 z^2)}/\varrho\pi^2, \tag{20}$$

to which we shall add a contribution from the central bulge including a black hole. (Here ϱ is the distance from the central axis and z is the distance from the galactic plane.) The integral over the visible mass density produces a finite galactic mass $\mathcal{M}$; the parameter κ specifies the size of the galaxy disk while δ is the ratio of disk diameter to disk thickness which varies from about 10 near the centre to 50 at the outer reaches of the disk; so we will take an average value $\delta^2 \simeq 1000$ presently.

Firstly we derive the potential energy in the plane of the disk at location $(\varrho, z = 0)$ for the cylindrical mass approximation above.

$$V(\varrho) = -\frac{Gm\mathcal{M}\kappa^2\delta}{\pi^2} \iiint d\phi dz' d\varrho' \frac{e^{-\kappa^2(\varrho'^2+\delta^2 z'^2)}}{\sqrt{\varrho^2 + \varrho'^2 - 2\varrho\varrho' \cos\phi + z'^2}}$$

Found Phys (2008) 38: 995–1010 1005

$$\simeq -\frac{2Gm\mathcal{M}\kappa^2\delta}{\pi}\iint dz'd\varrho'\,\frac{e^{-\kappa^2(\varrho'^2+\delta^2 z'^2)}}{\sqrt{\varrho^2+\varrho'^2+z'^2}},\quad \text{taking } \langle\cos\phi\rangle = 0$$

$$= -\frac{Gm\mathcal{M}\kappa}{\sqrt{\pi}}\int_0^\infty \frac{e^{-u\kappa^2\varrho^2}\,du}{\sqrt{u(1+u)(1+u/\delta^2)}}, \tag{21}$$

where we have used $2\int_0^\infty e^{-\xi^2 X}d\xi = \sqrt{\pi/X}$. For large ϱ note that $V(\varrho) \simeq -Gm\mathcal{M}/\varrho$ as expected for a finite-sized source, while for small distances $V(\varrho) \simeq 2Gm\mathcal{M}\kappa^2\delta\varrho$ vanishes as $\varrho \to 0$. To (21) we will shortly add a contribution from the central bulge, approximated by a mass $c\mathcal{M}$ placed at the middle.

The average over relative velocity v is a lot cruder and relies on data amassed over many years by astronomers as well a number of numerical approximations. The galactic disk [4, 5] exerts a strong attraction towards the plane on stars which stray from the disk and this results in a velocity dispersion $\Delta v_z \sim 30$ km/s. As we are supposing that the majority of stars are turning at the same orbital speed v, the only relevant quantity is the angle ϕ between the azimuths of the two bodies and the velocity dispersion along the z-axis, $|\mathbf{v}'-\mathbf{v}| = \sqrt{(2v\sin(\phi/2))^2 + 4\Delta v_z^2}$. Therefore the sum simplifies to an integration over azimuth:

$$\int_0^{2\pi}\frac{d\phi}{2\pi|\mathbf{v}'-\mathbf{v}|} \simeq \frac{1}{4\pi v}\int_0^{2\pi}\frac{d\phi}{\sqrt{\sin^2\phi/2+\epsilon^2}} = \frac{K(\frac{1}{1+\epsilon^2})}{\pi v\sqrt{1+\epsilon^2}} \equiv \frac{L}{v},$$

where K is the elliptic integral of the first kind and $\epsilon \equiv \Delta v_z/v \sim 0.15$ roughly [4, 5]. (We use L as a parameter to fit the data in due course.) So without much compunction we shall simply take $\langle\Omega/|\mathbf{v}-\mathbf{v}'|\rangle \sim L\Omega/v$ as a fair approximation; after all we are only striving to get an idea of the size of the reciprocity constant Ω here.

Finally there is the matter of the halo contribution. Unlike mainstream ideas inferring dark matter, we rely on observations of *visible* matter (population II stars, white or brown dwarfs, star clusters, etc.) to put a bound of about $c_H \simeq 1\%$ on the amount of halo mass $c_H\mathcal{M}$. Furthermore we shall suppose, like everyone else, that the halo velocity distribution is largely thermal [6]. In order to model these effects we will neglect the small oblateness of the halo and use a radial halo density, $\rho_H(r) \simeq c_H\kappa_H\mathcal{M}/2\pi^2r^2(1+\kappa_H^2r^2)$, multiplied by a normalized velocity distribution $\rho_H(v) = (\beta/\pi)^{3/2}e^{-\beta v^2}$. This fixes the mass of the halo to be $c_H\mathcal{M}$ and the halo velocity dispersion [6] to be $(\Delta v_H)^2 = 3/2\beta$, while the halo size is determined by $1/\kappa_H$. It allows us to work out the contribution to the gravitational potential energy due to the halo:

$$V_H(r) = -Gm\left[\frac{1}{r}\int_0^r \rho_H(r')4\pi r'^2dr' + \int_r^\infty \rho_H(r')4\pi r'dr'\right]$$

$$= -\frac{Gmc_H\kappa_H\mathcal{M}}{\pi}\left[\frac{2}{\kappa_H r}\arctan(\kappa_H r) + \ln\left(1+\frac{1}{\kappa_H^2 r^2}\right)\right], \tag{22}$$

as well as the reciprocity halo contribution:

$$\left\langle \frac{1}{v} \right\rangle_H = \frac{1}{v} \int_0^v \rho_H(v') 4\pi v'^2 dv' + \int_v^\infty \rho_H(v') 4\pi v' dv' = \mathrm{erf}(\sqrt{\beta} v)/v. \quad (23)$$

Putting all this together, and including reciprocity terms, we obtain the total energy dependence on angular momentum $\ell = m\varrho v$ (conserved by axial symmetry) and distance from axis ϱ for a test mass m:

$$\begin{aligned} E(\varrho, \ell) &= \frac{\ell^2}{2m\varrho^2} + \frac{1}{2} m\Omega^2 \varrho^2 - (L+c) \frac{Gm^2 \mathcal{M}\Omega\varrho}{\ell} \\ &\quad - Gm\mathcal{M} \left(\frac{c}{\varrho} + \frac{\kappa\delta}{\sqrt{\pi}} \int_0^\infty \frac{\mathrm{e}^{-u\kappa^2\varrho^2}\, du}{\sqrt{u(1+u)(\delta^2+u)}} \right) \\ &\quad - \frac{Gm^2 \mathcal{M} c_H \Omega\varrho}{\ell} \mathrm{erf}(\sqrt{\beta}\ell) m\varrho \\ &\quad - \frac{Gm\mathcal{M} c_H \kappa_H}{\pi} \left[\frac{2}{\kappa_H r} \arctan(\kappa_H r) + \ln\left(1 + \frac{1}{\kappa_H^2 r^2}\right) \right]. \end{aligned}$$

Since we are presuming that the velocity is largely tangential, $\dot{p_\varrho} = 0 = \partial E/\partial\varrho$, which leads to the force equation:

$$\begin{aligned} 0 &= -\frac{mv^2}{\varrho} + m\Omega^2\varrho - (L+c) Gm\mathcal{M} \frac{\Omega}{v\varrho} \\ &\quad + Gm\mathcal{M} \left(\frac{c}{\varrho^2} + \frac{2\kappa^3\varrho}{\sqrt{\pi}} \int_0^\infty \sqrt{\frac{u}{(1+u)(1+u/\delta^2)}} \mathrm{e}^{-u\kappa^2\varrho^2}\, du \right) \\ &\quad + \frac{2Gm\mathcal{M} c_H}{\pi\varrho^2} \arctan(\kappa_H \varrho) + \frac{Gm\mathcal{M}\Omega c_H}{\varrho} \left[2\sqrt{\frac{\beta}{\pi}} \mathrm{e}^{-\beta v^2} - \frac{\mathrm{erf}(\sqrt{\beta} v)}{v} \right]. \quad (24) \end{aligned}$$

We can simplify the look of this equation by rescaling to dimensionless variables. Let $\mathcal{V} = v/\sqrt{G\mathcal{M}\kappa}$, $\mathcal{R} = \kappa\varrho$, $\omega = \Omega/\sqrt{G\mathcal{M}\kappa^3}$, $\mathcal{B} = \beta\kappa G\mathcal{M}$. The tangential velocity profile then reduces to solving an equation for $\mathcal{V}$ as a function of $\mathcal{R}$:

$$\begin{aligned} &\mathcal{V}^2 + \frac{\omega(L+c)}{\mathcal{V}} + c_H\omega \left[\frac{\mathrm{erf}(\sqrt{\mathcal{B}}\mathcal{V})}{\mathcal{V}} - 2\sqrt{\frac{\mathcal{B}}{\pi}} \mathrm{e}^{-\mathcal{B}\mathcal{V}^2} \right] \\ &= \omega^2\mathcal{R}^2 + \left[\frac{c}{\mathcal{R}} + \frac{2\mathcal{R}^2}{\sqrt{\pi}} \int_0^\infty \frac{\sqrt{u}\, \mathrm{e}^{-u\mathcal{R}^2}\, du}{\sqrt{(1+u)(1+u/\delta^2)}} \right] + \frac{2c_H}{\pi\mathcal{R}} \arctan\left(\frac{\kappa_H \mathcal{R}}{\kappa} \right). \end{aligned}$$

(25)

The chosen sign for Ω is highly significant in determining the behaviour of the velocity for small $\mathcal{R}$, viz. $\mathcal{V} \to (L+c)\mathcal{R}\omega/c$, as required by the data. (Had we reversed the sign of ω or b we would not have been able to fit observations even remotely.)

A numerical solution of (23) is possible once a few measured values are input. The galaxy disk is 50000 light-years or more in radius, i.e. about 5×10^{20} m, so let us set the scale $\kappa \sim 0.7 \times 10^{-20}$ m^{-1} or so. Taking the visible galactic mass to be roughly $\mathcal{M} \sim 2 \times 10^{41}$ kg, one estimates that $\kappa^3 G\mathcal{M} \sim 5 \times 10^{-30}/\text{s}^2$. Also the rotation speed outside the innermost part of the galaxy equals about 230 km/s and v grows to this value over about 0.3 kpc or 10^{19} m, telling us that $\Omega(L+c)/c \sim 2.5 \times 10^{-14}$ s^{-1}. In attempting to fit the observed velocity profile, use values $c \simeq 0.05$, as the proportion of galactic mass concentrated in the central region (a bit on the low side), and $c_H = 0.01$ as the ratio of halo mass to disk mass. For simplicity take the halo radius to equal the disk radius or $\kappa = \kappa_H$; although this is an underestimate it makes little difference to the numerical results because c_H is quite small anyhow. A similar comment attaches to our chosen $\mathcal{B}$-value of about 13; see [6]. Finally set $L \simeq 0.9$ and $\Omega \simeq 1.4 \times 10^{-15}$ Hz, implying $\omega \simeq 0.17$. None of these inputs is at all absurd.

The profile equation (25) now determines $\mathcal{V}$ as a function of $\mathcal{R}$ and the numerical results are plotted in Fig. 4. In trying to fit the data it is important to mention that there are two roots for the cubic in $\mathcal{V}$. In the very inner region we use the smaller root, corresponding to the linear relation $v \simeq r\Omega(L+c)/c$, associated with the b/p Born term, and beyond about $r = 10^{19}$ m we adopt the larger root. Although the Fig. 4 graph is rather high below 2 kpc ($\simeq 6 \times 10^{19}$ m)z and shows a drop off to steadying speed that is a bit faster than what is observed, the general shape of the curve is moderately satisfactory. Moreover the flattish part of the curve has $v \sim 200$ km/s, which is the correct magnitude; its value depends on the galactic mass and the way it is distributed. (Beyond the flat part v is expected to rise slowly linearly because the competition between the gravitational attraction due to the galactic mass distribution and the harmonic Born term is lost.) However it must be admitted that our model and calculation have many rough edges and need a lot more refinement before they can be judged a success in any way; it is even conceivable that dressing the model may spoil it rather than enhance it.

While we are on the subject of the Born attractive deceleration in our local vicinity, one might wonder whether it has anything useful to add to the Pioneer anomaly

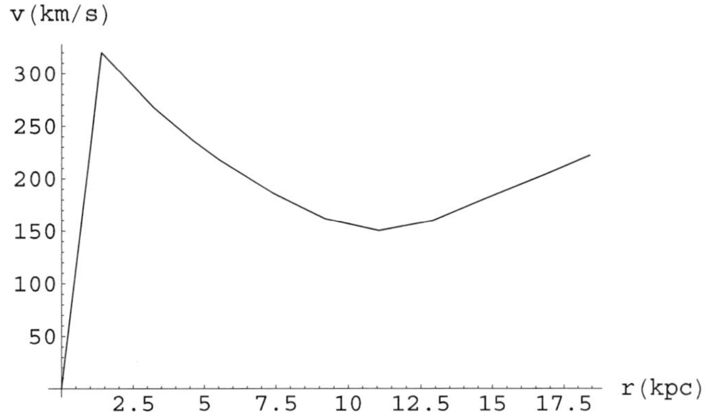

Fig. 4 Predicted tangential velocity curve with radius

controversy. The answer is no: $\Omega^2 r$ is far too small to account for the observed [9] deceleration of 10^{-9} m/s^2 as seen in Doppler and ranging data; even at distances of 10^{13} m it is seven or eight orders of magnitude too small, given our estimate of Ω. Probably an explanation for the anomaly will be found in more conventional physics which has so far escaped everyone's notice.

5 Conclusions and Criticisms

The above fit has some good features and some bad ones; on the positive side it is able in principle to produce an orbital velocity curve that matches the data near the centre (linear rise) and at moderate distances (almost constant speed), without calling upon a dark matter component—Born's reciprocity principle has after all modified Newtonian dynamics in a radical way. But it must be confessed that the detailed fit above is far from convincing. This negative feature may be due to the crudeness of our visible mass and velocity distributions. It would be more realistic to model properly matter in the inner galactic bulge, introduce barring plus spiral arms and generally deal more accurately with the velocity distribution when working out the gravitational potential and its reciprocity counterpart. This is clearly fertile ground for future research. Meanwhile we can probably be content with our *estimate* of $\Omega \simeq 1.5 \times 10^{-15}$ Hz for Born's hypothesis and not much more; it is truly a galactic scale and corresponds to a period of 1.3×10^8 years, which may be connected with the rotation rate of the galaxy.

Since reciprocity might do away with dark matter by providing a cosmic attractive force other than gravity, this raises a number of questions. How does the harmonic term affect the bending of light? Until someone succeeds in extending Born's hypothesis in a convincing way to massless particle interactions (which entails redoing the whole of general relativity in a reciprocally invariant manner) one cannot make precise quantitative assessments. However we can venture to suppose that the attractive force between photons and matter will be increased somewhat by the harmonic term involving b and the effect will likely become more noticeable at larger impact parameters; this is something to be welcomed because it means that weak lensing of galaxy superclusters (the other bit of observational evidence for dark matter) will show up without the need for dark material. Even so one might well doubt that reciprocity can explain the observed mass distributions and lensing effects seen in cluster mergers [10]; this needs to be researched more fully once light-bending and galaxy distortion profiles are worked out properly for reciprocally-invariant interactions.

A second question is the impact of harmonic interactions on the expanding universe. If dark matter is excluded then the value of the cosmological constant, which affects the Hubble rate, must be suitably increased; in doing so it is vital to include the reciprocal terms which can, through the universal harmonic force, potentially cause a *deceleration* of the expansion, in addition to what is expected from gravity via the *visible* energy density. This too needs further research and calls for a fully relativistic formulation, different from Born's attempts, which incorporates the high energies in the early history of the universe and the high speeds seen at the outer regions of the visible cosmos. Only then can we be more certain as to what effect there will be on the resultant cosmic acceleration.

Serious concerns remain. The scheme destroys translational invariance because of the way that momentum and position are tied, so a violation of Galilean relativity is to be expected, never mind Lorentzian relativity. Choice of origin is another issue. We have picked the galactic centre as the obvious place; however one might fret about effects of harmonic acceleration ($\Omega^2\mathbf{r}$) which can get quite large from outer regions of the universe, but is less significant within a galaxy. Fortunately it appears that on cosmic scales the galaxies are uniformly distributed in every direction all around so one may presume that such attractions will cancel out overall on an individual galaxy. And because galaxies are receding away from us at Hubble rates the relative velocity term Ω/v diminishes and balances out as one goes outward. So it would seem, superficially at least, that one could choose any other galaxy and take the origin at its centre to reproduce its own galactic rotation without worrying unduly about other galaxies (except for the decelerating effect on cosmic expansion of course).

Fundamental theoretical criticisms can nevertheless be levelled at (3). It is explicitly non-relativistic and should at least be made to conform with special relativity, To carry out that program sensibly one would need to study the quaplectic group [11–13] and augment the electromagnetic or gravitational field (photon/graviton exchange) by their reciprocity analogues or some other contributions. This signifies overhauling the whole of standard field theory and the task seems rather difficult, if not vague, at this stage. In the end it may turn out that Born reciprocity is unable to fit the data properly or mesh in with our familiar relativistic field theory concepts. That would be a pity as it would prevent an extension of the idea to relativistic cosmology. Even if it works in limited fashion it would have to be seen as one of a panoply of modified Newtonian descriptions (MOND) of gravity [14–18] over large distances. Taking a skeptical point of view, Born's idea will probably be found wanting, dark matter will be required and the problem of seeing it by non-gravitational methods will remain with us for a good while. But before the death knell is finally sounded on the subject of reciprocity, a proper relativistic version must be sought and specialist galactic modellers need to investigate its ramifications comprehensively.

Acknowledgement We would like to thank Professor John Dickey and Dr. Andrew Cole for enlightening discussions on galactic structure and for pointing us in the right direction.

References

1. Born, M.: Rev. Mod. Phys. **21**, 463 (1949)
2. Born, M.: Nature **163**, 207 (1949)
3. Green, H.S.: Nature **163**, 208 (1949)
4. Olling, R.P., Merrifield, M.R.: Mon. Not. R. Astron. Soc. **326**, 164 (2001)
5. Korchagin, V.I., et al.: Astrophys. J. **126**, 2896 (2003)
6. Battaglia, G., et al.: Mon. Not. R. Astron. Soc. **370**, 1055 (2005)
7. Clemens, D.P.: Astrophys. J. **205**, 422 (1985)
8. McClure-Griffiths, N.M., Dickey, J.M.: Astrophys. J. **671**, 427 (2007)
9. Anderson, J.D., et al.: Phys. Rev. Lett. **81**, 2858 (1998)
10. Clowe, D., et al.: Astrophys. J. Lett. **648**, 109 (2006)
11. Low, S.G.: J. Phys. A **35**, 5711 (2002)
12. Jarvis, P.D., Morgan, S.O.: Found. Phys. Lett. **19**, 501 (2006)

13. Govaerts, J., Jarvis, P.D., Morgan, S.O., Low, S.G.: J. Phys. A **40**, 12095 (2007)
14. Milgrom, M.: Astrophys. J. **270**, 365 (1983)
15. Milgrom, M.: Astrophys. J. **302**, 617 (1986)
16. Milgrom, M.: Ann. Phys. **229**, 384 (1994)
17. Bekenstein, J.D.: Contemp. Phys. **47**, 387 (2006)
18. Bekenstein, J., Magueijo, J.: Phys. Rev. D **73**, 103513 (2006)

On 2D Periodic Hexagonal Cells

R. DELBOURGO
SCHOOL OF MATHEMATICS AND PHYSICS, UNIVERSITY OF TASMANIA

ABSTRACT

Graphene, the new wondrous material, is a perfect example of a two-dimensional hexagonal crystal unlike any other. Here we exhibit some of the characteristic directional features associated with hexagonal cells, emphasising the sixfold symmetry. We depict the X-ray, vibrational and electronic band structures to be expected in such systems via 2 dimensional contour plots.

INTRODUCTION

It is little wonder that graphene has captured the interest of physicists or materials scientists and no surprise that it was the centrepiece of the 2010 Nobel Physics Prize. The reason for the great excitement lies in is its remarkable transport properties and amazing strength plus the fact that technology has advanced to the stage where graphene can be produced in relatively large sheets, on a variety of substrates. A very bright future is predicted for it. Thus it possesses very low resistivity, high opacity and can furnish the photonics, spintronics and electronic industry with a low cost alternative to silicon.

There are numerous review articles [1, 2, 3, 4, 5] on this subject and a veritable avalanche of research papers has appeared, describing or exploring different properties. These concern its electronic characteristics, its vibrational properties, its edge effects, the Dirac-like properties of quasiparticles at the corners of Brillouin zones [6, 7, 8, 9], anomalous Hall effects, etc. Our aim in this paper is very modest: we wish to illustrate its beautiful directional properties in a manner that might benefit teachers of condensed matter physics: the 2D hexagonal symmetry is somewhat unfamiliar and can serve as a nice extension of the standard square symmetry analysis [10].

The paper is set out as follows. Section 2 contains the basic notation and its consequences for glancing X-rays. In Section 3 we discuss the oscillation modes within a hexagon, clamped at its edges. These represent the energy levels of a particle held in the infinite hexagonal well; the results are no doubt familiar to the graphene experts but are probably not widely known to the usual quantum mechanics practitioners, who often resort to rectangular boundary conditions. We also consider the modes corresponding to antinodes at the edges because these are subsequently needed for discussing the Kronig-Penney model in Section 5. Section 4 is devoted to idealised vibrational modes (longitudinal and transverse) of the atoms themselves, assuming nearest neighbour interactions. Here, as in Section 3, our aim is to illustrate the hexagonal symmetry of the motions. In the last Section we describe the electronic band structure, using the approximation of delta-function walls between adjoining cells rather than the tight-binding approximation used in most expositions. In order to keep the analysis simple, nowhere do we study edge effects (armchair, zigzag, corrugations [11]) or multilayer graphene; for such elaboration we refer the reader to research reviews. We hope that this paper will serve as a useful extension of the usual quantum mechanical problems encountered by college students and highlight the beautiful hexagonal symmetry of the principal properties of graphene.

BASES AND X-RAY FEATURES

Let us set down our notation which will be used in subsequent sections. The basic hexagonal cell (with side length ℓ) is drawn in Figure 1 and our coordinate system is centred there. The fundamental cell vectors are

$$\mathbf{a} = (3,-\sqrt{3})\ell/2, \ \mathbf{b}=(3,\sqrt{3})\ell/2. \qquad (1)$$

Any two nearby corner points such as carbon atoms at U and V can be taken as centres of X-ray scattering so that all lattice points are generated by translation:

$$\mathbf{u}_{rs}=\mathbf{u}_{00}+r\mathbf{a}+s\mathbf{b} = (3r+3s-1,\sqrt{3}(s-r-1))\ell/2$$
$$\mathbf{v}_{rs}=\mathbf{v}_{00}+r\mathbf{a}+s\mathbf{b} = (3r+3s+1,\sqrt{3}(s-r-1))\ell/2, \qquad (2)$$

where r and s are integers. The basic reciprocal lattice, drawn in Figure 2, is nothing more than a scaled version of the original lattice, rotated by 30° with side length $4\pi/3\sqrt{3}\ell$. Its fundamental vectors are

$$\mathbf{A} = (1,-\sqrt{3})\, 2\pi/3\ell, \ \mathbf{B}=(1,\sqrt{3})\, 2\pi/3\ell. \qquad (3)$$

Then, according to the Von Laue conditions for constructive interference, whenever the change of wave number is (R and S are integers)

$$\Delta\mathbf{k} = R\mathbf{A}+S\mathbf{B} = (R+S,(S-R)\sqrt{3})\, 2\pi/3\ell, \qquad (4)$$

X-rays directed *along* the graphene plane will of course be diffracted by the lattice. However because there are *two* atoms decorating each unit cell there is a modulating factor $1+\exp i[(R+S)\, 2\pi/3] \propto 2\cos[\pi(R+S)/3]$ apart from an irrelevant phase; this means that diffraction will be greatest when $R+S$ is exactly divisible by 3 but that there are no missing orders of diffraction.

THE HEXAGONAL WELL

The literature on the equilateral triangular drum is old indeed; it goes back to Lamé [12] and Pockels [13]. A very nice series of reviews on this topic has been written by McCartin [14], discussing the completeness, orthogonality and multiplicity of the solutions, subject to Dirichlet and Neumann conditions at the edge of the triangle. It is not hard to adapt the solutions to the case of the hexagon and we shall do so shortly. (As a matter of fact the solutions of the equilateral triangle are embedded in a one-sixth section of the hexagon.) Here is an outline of the main steps.

First of all introduce triangular coordinates, u, v, w centred at the origin with $u+v+w=0$, as illustrated in Figure 3. The maximum values of these coordinates within the hexagon is the radius of the incircle $r=\ell\sqrt{3}/2$ and we see that in terms of the normal Cartesian coordinates,

$$u = -y, \quad v-w=\sqrt{3}x. \qquad (5)$$

The second step is to note that the Schrödinger or Helmholtz equation for a free particle of energy E in the cell, namely

$$\left[\frac{\partial^2}{\partial x^2}+\frac{\partial^2}{\partial y^2}+k^2\right]\psi(x,y)=0; \quad k^2=\frac{2ME}{\hbar^2},$$

may be transcribed into triangular coordinates:

$$\left[\frac{\partial^2}{\partial u^2}+3\frac{\partial^2}{\partial(v-w)^2}+k^2\right]\psi(x,y)=0, \qquad (6)$$

cleverly allowing for cyclic sums of separable solutions of the form,

$$\begin{aligned}\psi(u,v,w) &= [\cos(k_1u)+C\sin(k_1u)] \\ &\quad [\cos(k_2-k_3)\frac{(v-w)}{3}+D\sin(k_2-k_3)\frac{(v-w)}{3}] \\ &+[\cos(k_2u)+C\sin(k_2u)]\,[\cos(k_3-k_1)\frac{(v-w)}{3}+D\sin(k_3-k_1)\frac{(v-w)}{3}] \\ &+[\cos(k_3u)+C\sin(k_3u)]\,[\cos(k_1-k_2)\frac{(v-w)}{3}+D\sin(k_1-k_2)\frac{(v-w)}{3}] \qquad (7)\end{aligned}$$

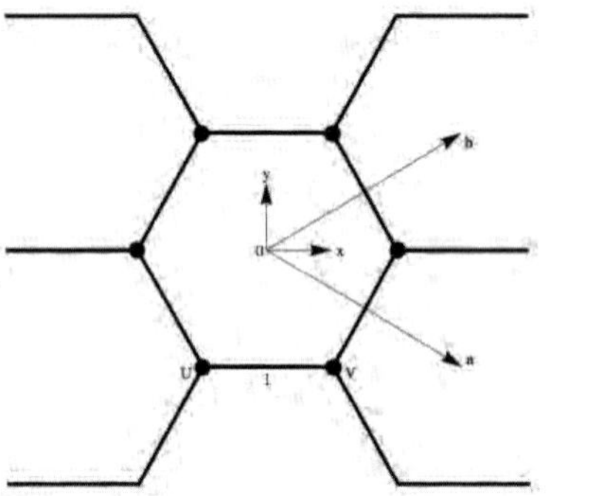

Fig. 1: Basic cell vectors. Any two adjoining C atoms such as U and V can be taken as the basic cell components.

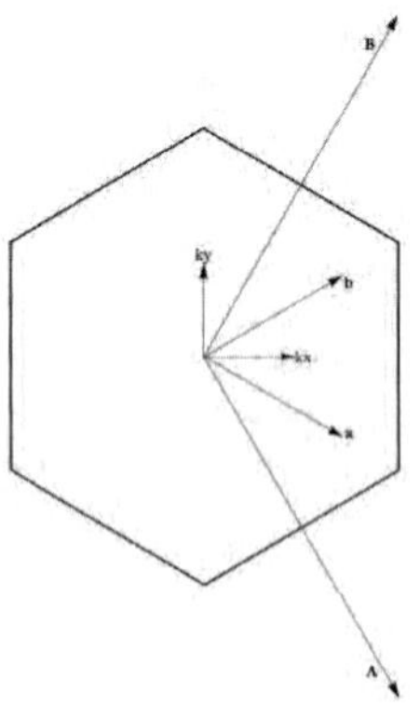

Fig. 2: Brillouin zone and its basis vectors **A and B**.

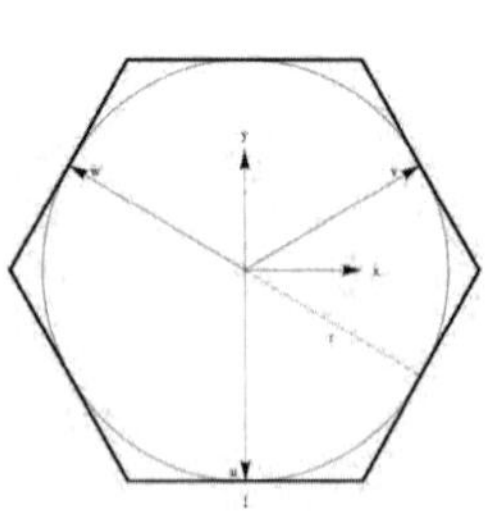

Fig. 3: Triangular coordinates for describing modes of oscillation.

where $k_1+k_2+k_3 = 0$ and $k^2 = 2(k_1^2+k_2^2+k_3^2)/3 = 4(k_1^2+k_2^2+k_1k_2)/3$. The third step is to pick out symmetric or antisymmetric solutions under reflection about a vertical diagonal; this selects out sin or cos functions in fact. Typically one finds [14] the (Dirichlet) quantized solutions $k_i = \pi n_i/r = 2\pi n_i/\sqrt{3}\ell$ – the case of interest for this section:

$$\psi_{D\,sym/anti} = \sin\frac{\pi n_1 u}{r}.\cos/\sin\frac{\pi(n_2-n_3)(v-w)}{3r} + \sin\frac{\pi n_2 u}{r}.\cos/\sin\frac{\pi(n_3-n_1)(v-w)}{3r} + \sin\frac{\pi n_3 u}{r}.\cos/\sin\frac{\pi(n_1-n_2)(v-w)}{3r}, \quad (8)$$

where n_i are non-zero integers subject to $n_1+n_2+n_3 = 0$. It suffices to take $n_1 \geq n_2 > 0$ to obtain distinct modes (in the antisymmetric configuration n_1 and n_2 need to be different for a nonvanishing answer). Hence the quantized energy levels are given by

$$k^2_{n_1,n_2,n_3} = 8\pi^2(n_1^2+n_2^2+n_3^2)/9\ell^2 = 16\pi^2(n_1^2+n_2^2+n_1n_2)/9\ell^2. \quad (9)$$

While we are discussing this topic let us note [14] the Neumann solutions (effectively antinodes along the hexagon edges) as we will require them later:

$$\psi_{N\,sym/anti} = \cos\frac{\pi n_1 u}{r}.\cos/\sin\frac{\pi(n_2-n_3)(v-w)}{3r} + \cos\frac{\pi n_2 u}{r}.\cos/\sin\frac{\pi(n_3-n_1)(v-w)}{3r} + \cos\frac{\pi n_3 u}{r}.\cos/\sin\frac{\pi(n_1-n_2)(v-w)}{3r}, \quad (10)$$

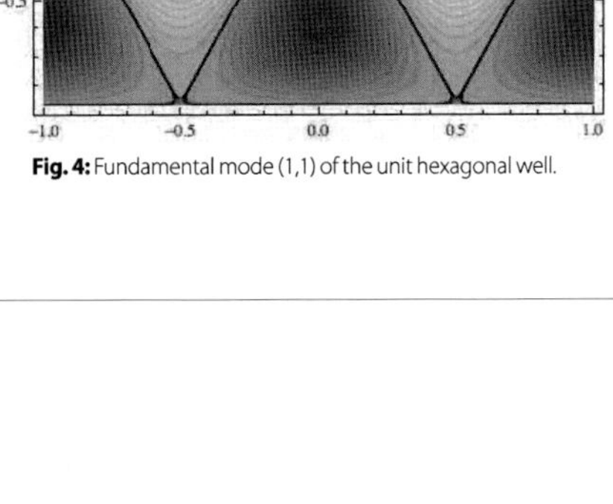

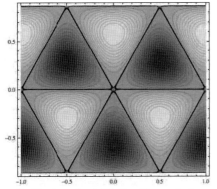

Fig. 4: Fundamental mode (1,1) of the unit hexagonal well.

We have illustrated some of the lowest few modes in Figures 4 to 6 in the form of contour plots rather than 3D plots as they are easier to comprehend (the lightest shaded ares are the maxima while the darkest correspond to minima). They show the lovely hexagonal features expected of the vibrations. The fundamental Neumann mode is depicted in Figure 7 even though it is not required for the present purposes. [It should be noted that triangular cuts within the hexagon describe the modes for an equilateral triangle.] McCartin has analysed the orthogonality and completeness properties of solutions (8) and (10) so we refer the reader to his review article.

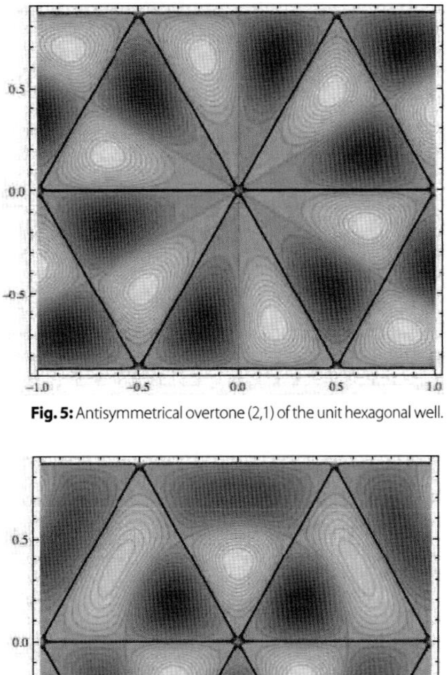

Fig. 5: Antisymmetrical overtone (2,1) of the unit hexagonal well.

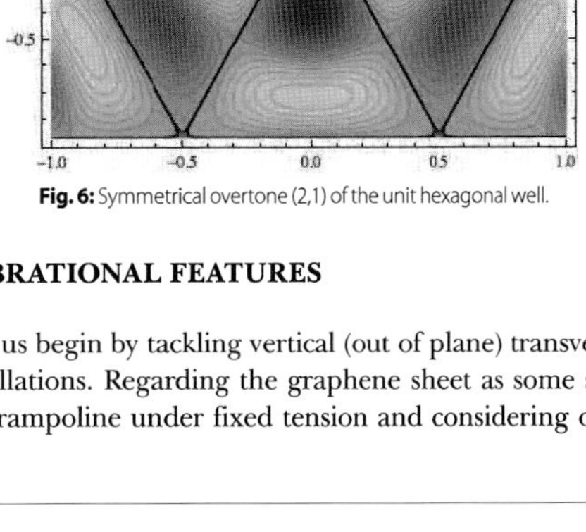

Fig. 6: Symmetrical overtone (2,1) of the unit hexagonal well.

VIBRATIONAL FEATURES

Let us begin by tackling vertical (out of plane) transverse oscillations. Regarding the graphene sheet as some sort of trampoline under fixed tension and considering only

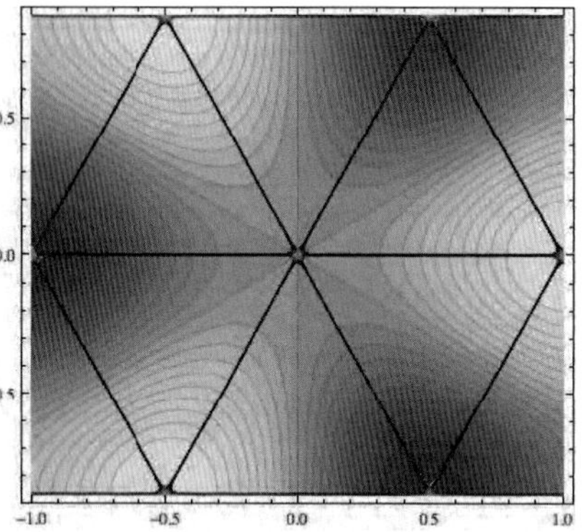

Fig. 7: Basic Neumann mode (1,0) of the unit hexagonal well.

nearest neighbour interactions, the downward acceleration on a C atom located at $\mathbf{u}_{rs}$ is determined by the adjoining vertical displacements relative to the atoms at $\mathbf{v}_{rs}$, $\mathbf{v}_{rs\text{-}1}$, $\mathbf{v}_{r\text{-}1s}$. Hence vibrations are described by the simple harmonic equation,

$$d^2U_{rs}/dt^2 = -\Omega^2(3U_{rs}-V_{rs}-V_{rs\text{-}1}-V_{r\text{-}1s}), \qquad (11)$$

where Ω is a characteristic frequency associated with the "tension" of the links and how strongly or loosely the C atoms adhere to the substrate. Similarly for the twin atom located at $\mathbf{v}_{rs}$ one obtains

$$d^2V_{rs}/dt^2 = -\Omega^2(3V_{rs}-U_{rs}-U_{rs+1}-U_{r+1s}). \qquad (12)$$

Looking for plane wave solutions [$\mathbf{k}=(k_x, k_y)$ below], at the positions summarized by (2),

$$U_{rs} = U\exp i[(\mathbf{k}.\mathbf{u}_{rs}) - \omega t], \;\; V_{rs} = V\exp i[(\mathbf{k}.\mathbf{v}_{rs}) - \omega t]$$

one arrives at two coupled equations for the amplitudes of the cell atoms at U and V:

$$\omega^2 U = \Omega^2[3U - Ve^{ik_x\ell} - 2Ve^{-ik_x\ell/2}\cos(k_y\ell\sqrt{3}/2)]$$

$$\omega^2 V = \Omega^2[3V - Ue^{-ik_x\ell} - 2Ue^{-ik_x\ell/2}\cos(k_y\ell\sqrt{3}/2)] \qquad (13)$$

producing the eigenvalue equation

$$(\omega^2/3\Omega^2-1)^2 = (1+4c_y^2+4c_xc_y)/9;$$
$$c_x \equiv \cos(3k_x\ell/2),\; c_y \equiv \cos(\sqrt{3}k_y\ell/2)$$

or dispersion relation between frequency and wavenumber for transverse vertical acoustic (−) and optical (+) modes:

$$\omega_\pm = \Omega[3 \pm \sqrt{1+4c_xc_y+4c_y^2}\,. \qquad (14)$$

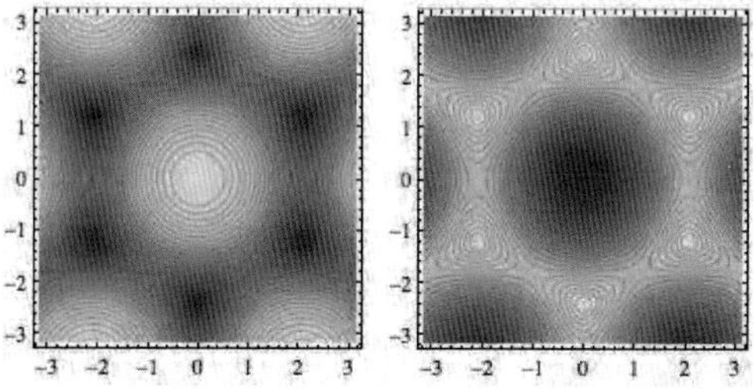

Fig. 8: ω-k dispersion relation for acoustic and optical transverse modes (scaled by Ω) within the fundamental Brillouin zone.

This dispersion relation for the two modes is shown in Figure 8. In the long wavelength limit,

$$\omega_-^2/\Omega^2 \simeq 3\ell^2(k_x^2+k_y^2)/4, \text{ so } v_{acoustic} = 2\Omega/\sqrt{3}\ell, \qquad (15)$$

whereas $\omega_+ \simeq \sqrt{6}\Omega$ for optical vibrations. The acoustic and optical modes remain separated when the wave vector $\mathbf{k}$ is directed along x even at the edge of the Brillouin zone, but they merge when $\mathbf{k}$ is directed along y and a corner (a so-called Dirac point) of the zone is approached. It is the instabilities in the vertical motion that apparently tend to make graphene sheets ripple or even curl up into tubes.

Next we shall study the oscillations of the C atoms within the graphene plane. This time we are dealing with two-dimensional vector displacements $\mathbf{U}=(U_x,U_y)$, $\mathbf{V}=(V_x,V_y)$ for the atoms located at positions determined by (2), rather than the vertical displacements considered previously. For small vibrations we assume that the restoring force is proportional to the extension in the direction of the link; so if one defines three unit vectors $\mathbf{e}_{1,2,3}$ along each of the links, the total restoring force of the atom at U summed over the three links is

$$[(\mathbf{V}_{rs}-\mathbf{U}_{rs}).\mathbf{e}_1]\mathbf{e}_1+[\mathbf{V}_{r\text{-}1s}-\mathbf{U}_{rs}).\mathbf{e}_2]\mathbf{e}_2+[\mathbf{V}_{rs\text{-}1}-\mathbf{U}_{rs}).\mathbf{e}_3]\mathbf{e}_3$$

apart from an overall factor. Resolving the $\mathbf{e}_i$ into Cartesian components, we get the equation pair for the U atom (ϖ is the horizontal characteristic frequency),

$$\ddot{U}_{xrs}/\varpi^2 = -3U_{xrs}/2 + V_{xrs} + (V_{xr\text{-}1s}+V_{xrs\text{-}1})/4 + \sqrt{3}(V_{yrs\text{-}1}-V_{yr\text{-}1s})/4 \qquad (16)$$

$$\ddot{U}_{yrs}/\varpi^2 = -3U_{yrs}/2 + 3(V_{yr\text{-}1s}+V_{yrs\text{-}1})/4 + \sqrt{3}(V_{xrs\text{-}1}-V_{xr\text{-}1s})/4. \qquad (17)$$

A similar equation applies to the vector displacement $\mathbf{V}$ of the V atom alongside:

$$\ddot{V}_{xrs}/\varpi^2 = -3V_{xrs}/2 + U_{xrs} + (U_{xr+1s}+U_{xrs+1})/4 + \sqrt{3}(U_{yrs+1}-U_{yr+1s})/4 \qquad (18)$$

$$\ddot{V}_{yrs}/\varpi^2 = -3V_{yrs}/2 + 3(V_{yr+1s}+V_{yrs+1})/4 + \sqrt{3}(U_{xrs+1}-U_{xr+1s})/4. \qquad (19)$$

For vibrations of circular frequency ω and wave number **k**, one arrives at the secular equation,

$$\begin{pmatrix} W & 0 & 1+c_y e^{-i\chi}/2 & -i\sqrt{3}s_y e^{-i\chi}/2 \\ 0 & W & -i\sqrt{3}s_y e^{-i\chi} & 3c_y e^{-i\chi}/2 \\ 1+c_y e^{i\chi}/2 & i\sqrt{3}s_y e^{i\chi}/2 & W & 0 \\ i\sqrt{3}s_y e^{i\chi}/2 & 3c_y e^{i\chi}/2 & 0 & W \end{pmatrix} \begin{pmatrix} U_x \\ U_y \\ V_x e^{ik_x\ell} \\ V_y e^{ik_x\ell} \end{pmatrix} = 0, \quad (20)$$

where $W \equiv \omega^2/4\varpi^2-3/2$, $\chi \equiv 3k_x\ell/2$, $c_y \equiv \cos(\sqrt{3}k_x\ell/2)$, $s_y \equiv \sin(\sqrt{3}k_y\ell/2)$ and ϖ stands for a characteristic frequency of planar oscillations, which is not necessarily equal to the value Ω for vertical oscillations because of the the substrate's influence.

This leads to the set of roots (dispersion relations):

$$\omega/\varpi = 0,\ 2\sqrt{3},\ (3-\sqrt{1+4c_xc_y+4c_y^2}),\ (3+\sqrt{1+4c_xc_y+4c_y^2}). \quad (21)$$

The first two roots may be ignored (the first is static displacement and the second is dispersionless). However, the third and fourth planar roots are strikingly similar to the vertical motion roots (14) (acoustic and optical), except that the oscillation frequency is not necessarily the same.

KRONIG-PENNEY ELECTRON BAND MODEL

We now turn to a model of the electronic energy bands which is a generalization of the one-dimensional Kronig-Penney scheme – instead of the tight-binding approximation which is more popular. To that end, we envisage a series of δ-function potential walls lying along the carbon links as that is where the electron clouds are at their most prominent. Any stray negative charge carrier would encounter these repulsive barriers, except of course at the cell corners where they would feel the attraction from the C nuclei. Therefore as our model we may envisage loose electrons as moving freely within the cells until they hit a cell edge, where the wave function discontinuity across the cell boundary is proportional to the wave function there (which is itself continuous). This should mimic the periodic properties and symmetry features of more realistic models.

Consider then the following parametrizations of the symmetric wave function Ψ_0 for the cell round the origin, drawn in Figure 1, and the wave function Ψ_1 for the cell below it (centred at $x=0$, $y=-\sqrt{3}\ell$). One must allow for both Dirichlet *and* Neumann boundary conditions – the analogues of $\sin(kx)$ and $\cos(kx)$ that are included in the one-dimensional analysis. Here we do not consider antisymmetric solutions, since the model merely serves as an example to illustrate the hexagonal features of the bands – the true nature of the bands can in fact only be revealed through experiments.

$$\begin{aligned} \Psi_j = {} & [A_j \sin(k_1u)+B_j\cos(k_1u)].\cos(k_1+2k_2)(v-w)/3+ \\ & [A_j\sin(k_2u)+B_j\cos(k_2u)].\cos(2k_1+k_2)(v-w)/3- \\ & [A_j\sin((k_1+k_2)u)-B_j\cos((k_1+k_2)u)].\cos(k_1-k_2)(v-w)/3, \end{aligned} \quad (22)$$

for cells $j = 0$ or 1. There are the (dis)continuity conditions which we can apply at $v = w$ or $x = 0$,

$$\begin{aligned} \Psi_1|_{u=r} - \Psi_0|_{u=r} &= 0, \\ (\partial\Psi_1/\partial u)|_{u=r} - (\partial\Psi_0/\partial u)|_{u=r} &= (V/r)\,\Psi_0|_{u=r}, \end{aligned} \quad (23)$$

as well as the Bloch periodicity conditions,

$$\begin{aligned} \Psi_1|_{u=r} &= e^{2iqr}.\ \Psi_0|_{u=-r} \\ (\partial\Psi_1/\partial u)|_{u=r} &= e^{2iqr}.\ (\partial\Psi_0/\partial u)|_{u=-r}, \end{aligned} \quad (24)$$

where q is the (real) quasi-momentum and V corresponds to the size of the δ function barrier. We also remind the reader that the electronic energy is given by $E = (2\hbar^2/3m^2)\,[k_1^2+k_2^2+k_1k_2]$.

The equations (23) and (24) determine the allowed energy bands from the reality of q via the conditions' combination:

$$\begin{aligned} & (1/2+\cos(2qr))\,[(\alpha+\beta)\cos(\alpha-\beta)-\beta\cos(2\alpha+\beta)-\alpha\cos(\alpha+2\beta)+ \\ & \alpha(\cos\alpha-\cos 2\alpha)+\beta(\cos\beta-\cos 2\beta)-(\alpha+\beta)[\cos(\alpha+\beta)-\cos(2\alpha+2\beta)] \\ & +V[2\sin\alpha-\sin 2\alpha+2\sin\beta-\sin 2\beta-2\sin(\alpha+\beta)+\sin(2\alpha+2\beta)]=0, \end{aligned}$$

where $\alpha \equiv k_1r$, $\beta \equiv k_2r$. With k_1 and k_2 inclined at 120° with respect to one another and $-1 \le \cos(qr) \le 1$, one may plot the allowed regions of k_i space for the energy bands. These are drawn in Figures 9 and 10, where we see that if we follow the k_x axis, the bands get narrower as V increases [15]. Indeed as $V \to \infty$ the ratio $B/A \to 0$ in equation (22) and the k values become quantized as in the infinite hexagonal well. More generally the energy bands lie between concentric circles at low energies because of the nature of the expression $k_1^2+k_2^2+k_3^2$ with triangular axes. For the tight binding model we refer the reader to reference [10].

CONCLUSIONS

Our aim in this paper has not been to provide an exhaustive analysis of all properties of graphene, not even in summarized form. Plentiful reviews exist which do

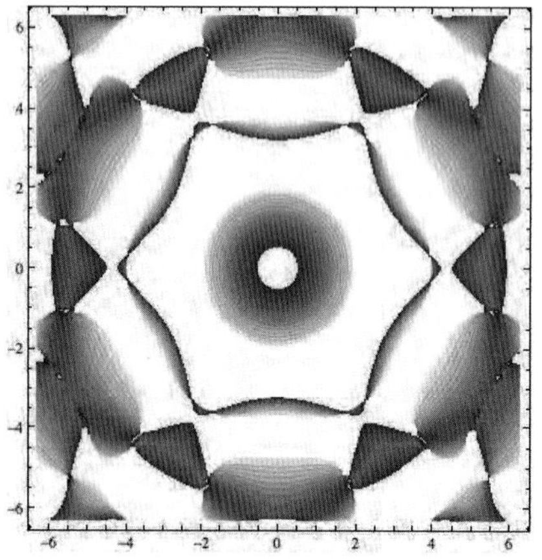

Fig. 9: Allowed band (shaded region) for V=1.

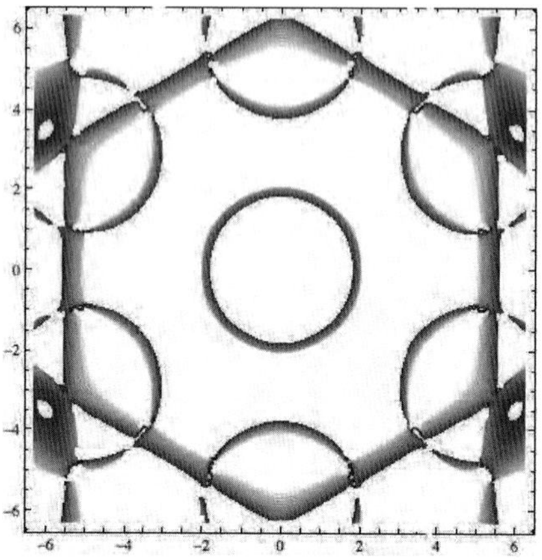

Fig. 10: Allowed band (shaded region) for V=10.

just that. Rather our purpose has been to highlight the *symmetry* features to expect from this remarkable material both in the electronic and vibrational characteristics, in the idealized situation that one has an infinite lattice of it. In that way, the directional properties of graphene come much more into evidence. In practice of course we expect the symmetry to be ruined by boundary effects, though the departure is probably significant only near the zigzag and armchair edges.

Graphene is a perfect vehicle for teaching two-dimensional aspects of solid state physics and makes for a very useful variation on the standard theme of square lattices. University teachers can use the material to illustrate many aspects of condensed matter and enlarge the students' horizons.

Acknowledgments: The use of Mathematica has greatly eased the task of producing the various diagrams.

References

[1] A.K. Geim and K.S. Novoselov, Nature Materials 6, 183 (2007).
[2] A.K. Geim and P. Kim, Scientific American, 90, April (2008).
[3] A.K. Geim, Science 324, 1530 (2009).
[4] A.H. Castro-Neto, F. Guinea, N.M.R. Peres, K.S. Novoselov and A.K. Geim, Rev.Mod. Phys. 81, 109 (2009).
[5] M.J. Allen, V.C. Tung and R.B. Kaner, Chem. Rev. 110, 132 (2010).
[6] J.W. McClure, Phys Rev. 108, 612 (1957); ibid Phys. Rev. 112, 715 (1958).
[7] J.C. Charlier, P.C. Eklund, J. Zhu and A.C. Ferrari, Top. Appl. Phys. 111 673 (2008).
[8] K.S. Novoselov et al, Nature 438, 197 (2005).
[9] G. Murguia, A. Raya, A. Sanchez and E. Reyes, Am. J. Phys. 78,700 (2010).
[10] P.R. Wallace, Phys. Rev. 71, 622 (1947).
[11] A. Matulis and F.M. Peeters, Am. J. Phys. 77, 595 (2009).
[12] G. Lamé, J. de l'Ecole Poly. 22, 194 (1833).
[13] F. Pockels, "Uber die partielle Differentialgleichung $\Delta u+k^2u=0$", B.G. Teubner, Leipzig (1891).
[14] B.J. McCartin, Amer. Conf. on Applied Mathematics (MATH '08) Harvard 195, March (2008), and references therein.
[15] The result above may be contrasted with the well-known 1D result: $\cos q\ell = \cos k\ell + V(\sin k\ell)/k\ell$.

Robert Delbourgo obtained his PhD in 1963 at Imperial College under the supervision of Prof Abdus Salam. The Nobel Prize was awarded to Salam in 1979 and he and Robert have coauthored 30 research publications. Robert held various appointments at the University of Wisconsin, the International Centre for Theoretical Physics at Trieste and the Weizmann Institute before being appointed to a lectureship At Imperial College in 1966, followed by a readership. In 1976 he was awarded a DSc by the University of London and accepted a Chair of Physics at the University of Tasmania where he has also served as Dean of the Faculty of Science and of Graduate Studies. After election to fellowship of the Australian Academy of Science, Robert served as Chair of the National Committee for Physics. He has published over 250 scientific papers and supervised over 30 higher degree students. In 1989 Professor Delbourgo was awarded the Thomas Ranken Lyle Medal (AAS), following the the Walter Boas Medal (AIP) in 1988. Finally he gained the prestigious Harrie Massey medal and prize (AIP/IOP) in 2002 for his contributions to quantized gauge-field theories and their symmetry properties.

Part B
CHAOS THEORY

B.0. Commentaries B

A hot subject in the early 1980's was the realisation that perfectly deterministic nonlinear systems can lead to chaotic behaviour, following the classic meteorological work of Lorenz (J. Atmos. Sc. 20, 130 (1963)) and his strange attractor. This sensitivity to initial conditions means that predictability of even simple systems can sometimes be placed in jeopardy. In particular the period-doubling route to this state of confusion is a sure sign of chaotic motion and this has been observed in many systems. Pioneering work in this area has been carried out by Eckmann (Rev. Mod. Phys. 53, 643 (1981)) and experimentally by Libchaber (J. de Phys., Coll. C3, 41 (1980)) confirming the renormalisation equations discovered by Feigenbaum. The following articles mainly concern tangent bifurcations in such systems beyond period doubling and how they lead to relations between the two Feigenbaum constants α and δ describing the approach to turbulence from stable cycles.

B.1 Delbourgo R. and Kenny B.G., Universal features of tangent bifurcation. Aust. J. Phys. 38, 1-22 (1985)
B.2 Delbourgo R. and Kenny B.G., Universality relations. Phys. Rev. A33, 3292-3302 (1986)
B.3 Delbourgo R., Hughes P. and Kenny B.G., Islands of stability and complex universality relations. J. Math. Phys. 28, 60-63 (1987)

In the first paper [B.1] we noticed that other period multiplications, besides doubling, have renormalisation group constants δ_N, α_N of their own associated with their N-cycle structures and can be encapsulated by functional equations too. Paper [B.2] determined the dependence of these constants on N, which becomes linear in N asymptotically; by eliminating N one may thereby establish a connection between δ_N and α_N. *This is a very general trick for relating previously unsuspected connections between fundamental constants — seek their dependence upon an auxiliary variable and then eliminate that variable.* In fact in the paper (Aust. J. Phys. 39, 189-201 (1986)) we have empirical/numerical evidence to show that the relation between the Feigenbaum constants is given by $\delta_N = [(z-1)/(z-\frac{1}{2})]\alpha_N^2$ for cycles undergoing N-multiplication when the maximum of the mapping order is z. Article [B.3] extends this body of work into the complex plane.

Readers who are interested in the development of this subject could not do better than consult the classic papers by May, Feigenbaum, Cvitanovic, Procaccia and others:

- R.M. May, Nature 261, 459 (1976)
- M.J. Feigenbaum, J. Stat. Phys. 19, 25 (1978)
- M.J. Feigenbaum, J. Stat. Phys. 21 669 (1979)
- M.J. Feigenbaum, Physica. D7, 16-39 (1983)

- P. Grassberger and I. Procaccia, Phys. Rev. Lett. 50, 346 (1983)
- P. Cvitanovic, Acta Phys. Polon. 65, 203 (1984)
- D. Auerbach, P. Cvitanovic, J.-P. Eckmann, G. Gunaratne and I. Procaccia, Phys. Rev. Lett 58, 2387 (1987)
- P. Cvitanovic and J. Myrheim, Comm. Math. Phys. 121, 225 (1989)

Aust. J. Phys., 1985, **38**, 1–22

Universal Features of Tangent Bifurcation

R. Delbourgo[A] *and B. G. Kenny*[B]

[A] Physics Department, University of Tasmania,
Hobart, Tas. 7001.
[B] Physics Department, University of Western Australia,
Nedlands, W.A. 6009.

Abstract

We exhibit certain universal characteristics of limit cycles pertaining to one-dimensional maps in the 'chaotic' region beyond the point of accumulation connected with period doubling. Universal, Feigenbaum-type numbers emerge for different sequences, such as triplication. More significantly we have established the existence of *different* classes of universal functions which satisfy the *same* renormalization group equations, with the *same* parameters, as the appropriate accumulation point is reached.

1. Introduction

Considerable progress has been achieved during the last few years in our understanding of turbulent or chaotic behaviour in natural processes (Ruelle and Takens 1971; Ott 1981; Eckmann 1981; Hu 1982). Much of the insight has come from a study of one-dimensional nonlinear mappings, both in qualitative and quantitative terms (May 1976; Collet and Eckmann 1980). Experimental evidence from diverse scientific fields ranging from physics through chemistry to biology has accumulated, which provides substantial support to the scenario based upon the period-doubling route to chaos, not only in the regime before the onset of chaos, but in the regime beyond where turbulence has developed. However, in that chaotic domain there exist certain windows of stability connected with low period cycle structures and in their vicinity one may observe the phenomenon of intermittent periodicity (Manneville and Pomeau 1980; Hirsch *et al.* 1982; Hu and Rudnick 1982).

Most of the theoretical studies (Feigenbaum 1983) have been focussed on the neighbourhood of the accumulation point of the first pitchfork bifurcation sequence and many of the characteristic universal properties have been thoroughly investigated there. In this paper we wish to highlight a number of universal properties that lie beyond this region and pertain to tangent bifurcations. These properties are partly implied in the paper by Derrida *et al.* (1979) which described the self-similarity of chaotic bands and cycles in that regime, but they are not widely known. We will exhibit what we believe are several new features associated with windows of stability to the right of the onset of chaos. Apart from demonstrating the occurrence of universal numbers connected with period multiplications in a 'forward' and reverse' sense (see Sections 2–4), we also show that the solution of the corresponding renormalization group equations near the accumulation points is by no means unique,

0004-9506/85/010001$02.00

despite the scaling parameters being the same. This provides the necessary graphic support to McCarthy's (1983) mathematical analysis which also proposed a multiplicity of such solutions.

We have attempted to make this paper self-contained by providing all the numerical and other evidence needed. Sometimes we have not been able to avoid covering familiar ground; still, we believe that the various tables and figures will be of real value to the expert and nonexpert alike by exposing, at a glance, all the numerical details* about the attainment of the various limits for two typical mappings. In Section 2 we summarize the well-known properties of pitchfork sequences, and in Section 3 we show that another 'reverse' period-doubling sequence in the chaotic region is governed by the same universal constants, subject to one important proviso: namely, the occurrence of families of solutions of the (duplication) functional equation. Sections 3 and 4 generalize the work to period triplings; again we demonstrate the existence of many solutions to the (triplication) functional equation by examining the reverse sequences of functions as one approaches the period-tripling accumulation point. We conclude in Section 5 with a number of comments about fractional universal functions by tracking other function sequences.

2. Windows of Stability

It has long been established (Metropolis *et al.* 1973) that for smooth maps of the real axis onto itself of the type

$$x \to F(\lambda, x); \qquad a \leqslant x \leqslant b,$$

where F has a unique maximum $x = X$ in the interval $[a, b]$ and λ is constrained to lie in some specified range, there is a universal sequence of limit cycles. This sequence is independent of the detailed form of the mapping F beyond the conditions stated. For example, it applies to the typical maps

$$\text{(A)} \qquad x \to \lambda x(1-x), \quad 0 < x < 1, \; 1 < \lambda < 4 \qquad \text{with } X = \tfrac{1}{2}; \tag{1a}$$

$$\text{(B)} \qquad x \to x e^{\lambda(1-x)}, \quad 0 < x < \infty, \; 0 < \lambda < \infty \qquad \text{with } X = 1/\lambda. \tag{1b}$$

(In fact, for these two examples F possesses a quadratic maximum and we will largely be restricting our attention to this class of functions.) In the chaotic region, beyond some critical value of λ (see equations 2 below), there exist infinitely many parameter values characterized by stable limit cycles of finite order; 'windows of stability', as May (1976) has phrased the regions in their vicinity. The order in which these windows succeed one another is independent of the map (even the character of the F maximum) within the constraints. This is the content of *structural universality*.

As they appear in order, the low period cycles up to 8 are listed in Table 1 for the reader's convenience. There we specify the 'superstable' λ-parameter values for which $x = X$ is one of the fixed points of the cycle for maps A and B. In what follows we shall denote by λ_{2^n} that value of λ for which the 2^n cycle is superstable. As the cycle period increases, so does the multiplicity of cycle *structures* as has been fully documented by Metropolis *et al.* (1973). We have followed May (1976) in Table 1 by

* All our computations were carried out on TRS-80 microcomputers to double precision.

appending a lower case letter to distinguish between different cycles of the same order, although this labelling is not of much value except for the low order cycles: already at period 9 there occur 28 different cycles and not enough letters in the alphabet to accommodate them.

Table 1. Superstable parameter values (up to 8 cycles) for mappings A and B

Cycle	Map A	Map B	Cycle	Map A	Map B
2	3·23606798	2·25643121	8h	3·94421350	3·48286345
4a	3·49856170	2·59351893	7e	3·95103216	3·50943386
8a	3·55464086	2·67100426	4b	3·96027013	3·59011302
6a	3·62755753	2·77263994	8i	3·96093370	3·60907717
8b	3·66219250	2·81656251	7f	3·96897686	3·70138725
7a	3·70176915	2·85991838	8j	3·97372426	3·73428947
5a	3·73891491	2·91759985	6d	3·97776642	3·77387587
7b	3·77421419	2·98514113	8k	3·98140895	3·80592983
8c	3·80077094	3·03277660	7g	3·98474762	3·82392739
3	3·83187406	3·11670045	8l	3·98774550	3·85101848
6b	3·84456879	3·17360416	5c	3·99026705	3·92280940
8d	3·87054098	3·25777911	8m	3·99251952	4·02352830
7c	3·88604588	3·29362781	7h	3·99453781	4·07007407
8e	3·89946895	3·33449413	8n	3·99621960	4·10314846
5b	3·90570647	3·36398510	6e	3·99758312	4·18096812
8f	3·91204662	3·39276769	8o	3·99864115	4·30421131
7d	3·92219340	3·41870460	7i	3·99939706	4·39226269
8g	3·93047300	3·43427458	8p	3·99984936	4·57119266
6c	3·93753644	3·45595376			

The first truly chaotic place is the accumulation point of the 2^n cycles

$$\lambda_{2^\infty} = \lim_{n\to\infty} \lambda_{2^n} = 3\cdot569945671\,, \qquad \text{map A;} \tag{2a}$$

$$= 2\cdot692368853\,, \qquad \text{map B;} \tag{2b}$$

associated with 'pitchfork' or 'forward' bifurcations, and the passage to it from smaller parameter values λ is known as the 'period-doubling route to chaos'. Feigenbaum (1978, 1979) noticed that when the chaotic point was approached from below, the ratios of the relative differences between successive λ_{2^n}

$$\lim_{n\to\infty} \frac{D\lambda_{2^n}}{D\lambda_{2^{n+1}}} \left(\equiv \frac{\lambda_{2^n} - \lambda_{2^{n-1}}}{\lambda_{2^{n+1}} - \lambda_{2^n}} \right) = \delta\,, \tag{3a}$$

as well as the ratios of the relative spacings,

$$\lim_{n\to\infty} \frac{Dx_{2^n}}{Dx_{2^{n+1}}} \left(\equiv \frac{x^*_{2^n} - X}{x^*_{2^{n+1}} - X} \right) = -\alpha\,, \tag{3b}$$

between the central fixed point X and the nearest fixed point

$$x^*_{2^n} = [F]^{2^{n-1}}(\lambda_{2^n}, X)\,, \tag{4}$$

tended geometrically to two universal constants δ and α respectively, independently of mapping details (see Table 2).† This is termed *metric universality*. When the functions possess a quadratic maximum as in (1a) or (1b) the universal constants turn out to be

$$\delta = 4\cdot 6692\ldots, \qquad \alpha = 2\cdot 5029\ldots. \tag{5}$$

To simplify the notation it proves useful to shift origin and rescale x by a factor a,

$$x \to X + ax,$$

whereupon

$$F(x) \to f(x),$$

with an a-dependent normalization $f(0)$. Often a is chosen so that $f(0) = 1$. For instance, with our two mappings, we have

$$\text{(A)} \quad f(x) = -1/2a + \lambda(1/4a - ax^2); \qquad a = \tfrac{1}{4}\lambda - \tfrac{1}{2} \;\text{ ensures }\; f(0) = 1; \tag{6a}$$

$$\text{(B)} \quad f(x) = -1/\lambda a + (1/\lambda a + x)e^{\lambda - 1 - a\lambda x}; \qquad a = (e^{\lambda-1} - 1)/\lambda \;\text{ fixes }\; f(0) = 1. \tag{6b}$$

In this way the spacing between the centremost fixed points may be reinterpreted as

$$Dx_{2^n} = [f]^{2^{n-1}}(\lambda_{2^n}, 0).$$

As noted, there exist windows of stability to the right of λ_{2^∞} and it is on these windows that we wish to exclusively focus attention. In any given window of period k, it is well known (Feigenbaum 1978, 1979) that if one studies harmonics of period $k\cdot 2^n$ which arise by pitchfork bifurcation, the sequences of

$$Dx_{k\cdot 2^n} = x^*_{k\cdot 2^n} - X = [f]^{c\cdot 2^{n-1}}(\lambda_{k\cdot 2^n}, 0), \tag{7a}$$

$$D\lambda_{k\cdot 2^n} = \lambda_{k\cdot 2^n} - \lambda_{k\cdot 2^{n-1}}, \tag{7b}$$

are again characterized by the same universal constants α and δ (see Table 3):

$$\lim_{n\to\infty} \frac{Dx_{k\cdot 2^n}}{Dx_{k\cdot 2^{n+1}}} = -\alpha, \qquad \lim_{n\to\infty} \frac{D\lambda_{k\cdot 2^n}}{D\lambda_{k\cdot 2^{n+1}}} = \delta. \tag{8a, b}$$

In (7a) the integer c is determined by the precise details of the k-cycle structure. For the 3-cycle we have $c = 2$.

The most noticeable window in the chaotic region, because it is the widest, is connected with the 3-cycle; that cycle is born (May 1976; Collet and Eckmann (1980)

† A few notational points: Feigenbaum (1978, 1979) used d_n in place of our Dx_{2^n}, and λ_n instead of our λ_{2^n}. Our more explicit formulae are necessary later. Also, N-iterates of F, as in (4), will be written as $[F]^N$ rather than as F^N to avoid subsequent confusion when a host of universal functions are introduced.

Table 2*a*. Forward bifurcations $1\cdot 2^n$ for mappings A and B

Cycle	Map A		Map B	
	λ	Dx	λ	Dx
2	3·236067978	−0·310016994	2·256431209	−1·113644594
4a	3·498561699	0·116401770	2·593518933	0·200505017
8a	3·554640863	−0·045975211	2·671004264	−0·107148265
16a	3·566667380	0·018326176	2·687782643	0·037739867
32a	3·569243532	−0·007318431	2·691386189	−0·015822652
64a	3·569795294	0·002923675	2·692158376	0·006197856
128a	3·569913465	−0·001168087	2·692323776	−0·002495613
256a	3·569938774	0·000466690	2·692359200	0·000993953
512a	3·569944195	−0·000186459	2·692366787	−0·000397622

Table 2*b*. Ratios of successive $D\lambda$ and Dx for mappings A and B

Cycle	Map A		Map B	
	$R\lambda$	Rx	$R\lambda$	Rx
4a	4·681	−2·663	4·350	−5·554
8a	4·663	−2·532	4·618	−1·871
16a	4·668	−2·509	4·656	−2·839
32a	4·669	−2·504	4·667	−2·385
64a	4·669	−2·503	4·669	−2·553
128a	4·669	−2·503	4·669	−2·484
256a	4·669	−2·503	4·669	−2·511
512a	—	−2·503	—	−2·500

Table 3*a*. Forward bifurcations $3\cdot 2^n$ for mappings A and B

Cycle	Map A		Map B	
	λ	Dx	λ	Dx
3	3·831874055	−0·457968514	3·116700451	−2·343057612
6b	3·844568792	0·027235706	3·173604163	0·081983089
12b	3·848344657	−0·011051342	3·190739426	−0·037885476
24b	3·849198054	0·004430880	3·194602580	0·014375982
48b	3·849383110	−0·001771810	3·195440367	−0·005874092
96b	3·849422845	0·000708039	3·195620245	0·002326936
192b	3·849431360	−0·000282899	3·195658791	−0·000932966
384b	3·849433184	0·000113019	3·195667047	0·000372242
768b	3·849433575	−0·000045159	3·195668815	−0·000148815

Table 3*b*. Ratios of successive $D\lambda$ and Dx for mappings A and B

Cycle	Map A		Map B	
	$R\lambda$	Rx	$R\lambda$	Rx
6b	3·362	−16·815	3·321	−28·580
12b	4·419	−2·464	4·436	−2·164
24b	4·617	−2·494	4·611	−2·635
48b	4·657	−2·501	4·658	−2·447
96b	4·667	−2·502	4·667	−2·524
192b	4·669	−2·503	4·669	−2·494
384b	4·669	−2·503	4·669	−2·506
768b	—	−2·503	—	−2·501

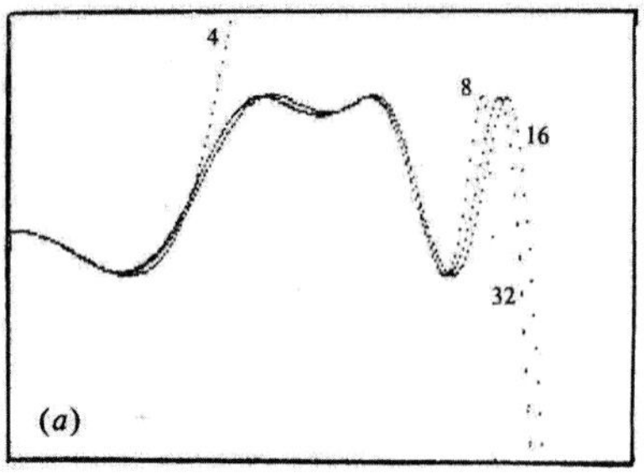

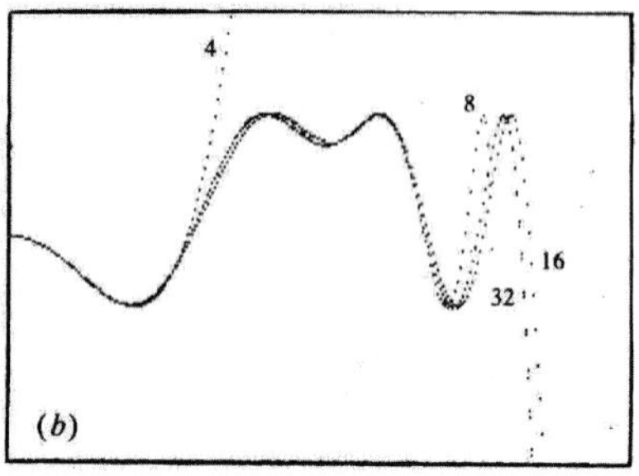

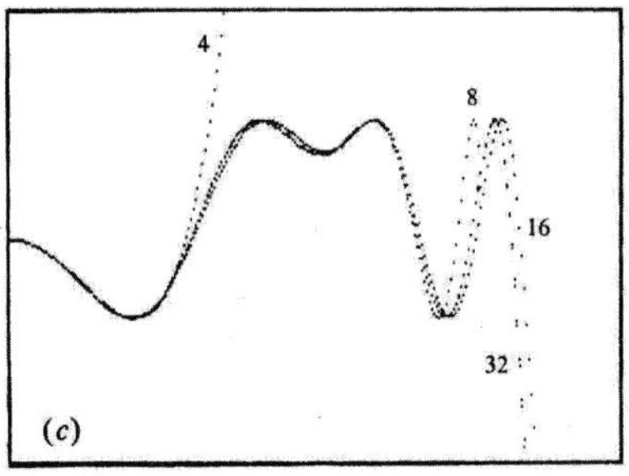

Fig. 1. Sequence of functions tending to
(*a*) $g_0(x)$,
(*b*) $g_1(x)$,
(*c*) $g(x)$,
with the integers denoting the orders of iteration.

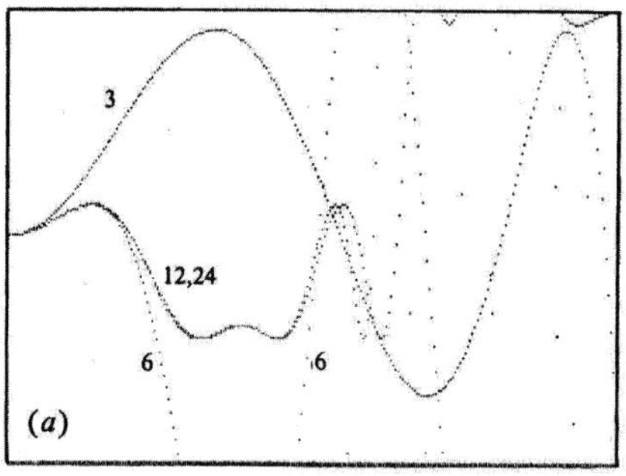

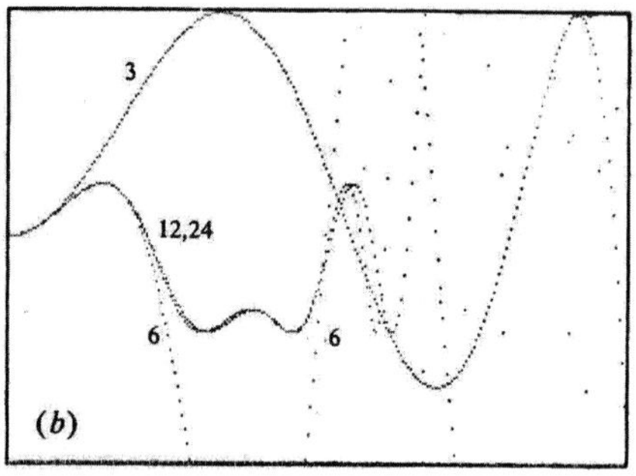

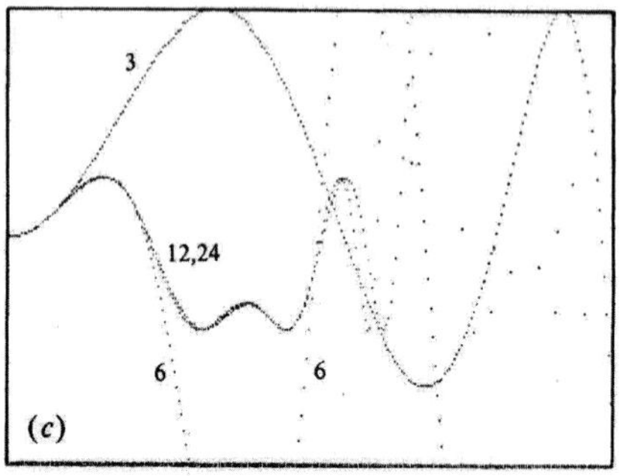

Fig. 2. Forward sequence of functions tending to (scaled)
(*a*) $g_0(x)$,
(*b*) $g_1(x)$,
(*c*) $g(x)$,
for iterations $3 \cdot 2^n$.

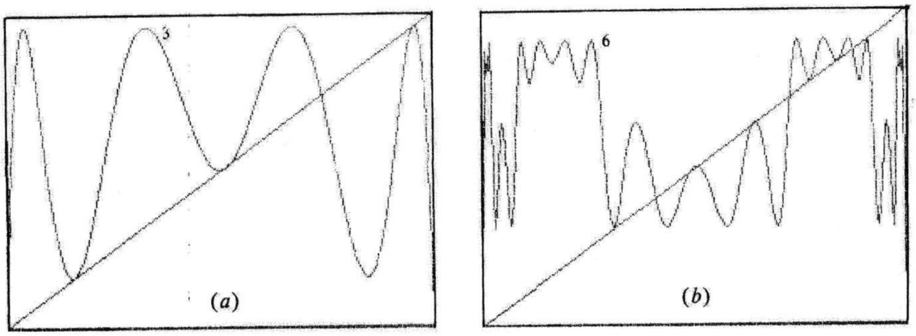

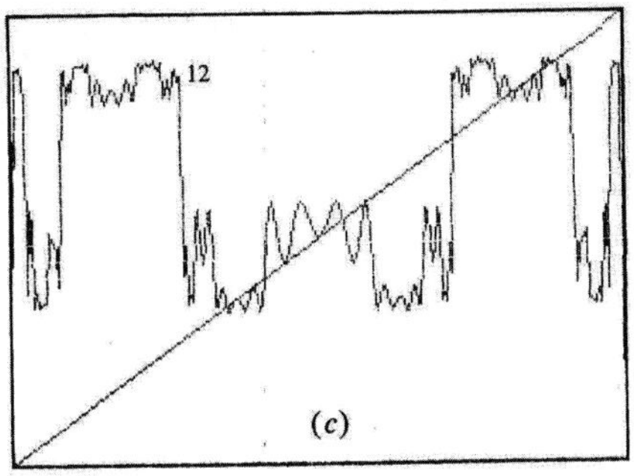

Fig. 3. The iterate
(a) $[F]^3$,
(b) $[F]^6$,
(c) $[F]^{12}$,
evaluated at λ_3, $\lambda_{2\cdot 3}$ and $\lambda_{2^2\cdot 3}$ respectively.

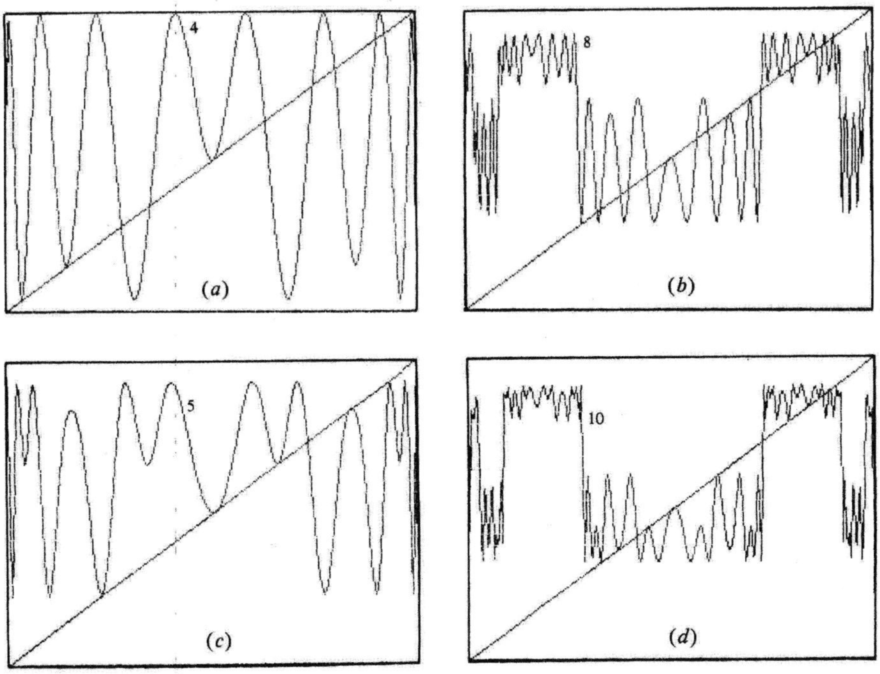

Fig. 4. Iterate (a) $[F]^4$, (b) $[F]^8$, (c) $[F]^5$ and (d) $[F]^{10}$ evaluated at λ_4, $\lambda_{2\cdot 4}$, λ_{5a} and $\lambda_{2\cdot 5a}$ respectively.

by 'tangent bifurcation' in the region to the right of λ_{2^∞}. [Just below this window there is an almost stable triplication pattern giving rise to the 'intermittency' phenomenon (see Manneville and Pomeau 1980; Hirsch *et al.* 1982; Hu and Rudnick 1982).] For this particular case we shall term the tangent bifurcation a 'trifurcation', recognizing it as a mathematical misdemeanour—thus period tripling $N \to 3N$ does not happen as λ varies continuously. Likewise, there are possibilities of fourfold, fivefold period multiplications as we keep on increasing λ.

Our investigations are primarily concerned with the sequences $k\cdot 2^n$ and $k\cdot 3^n$ in the forward and backward sense (see Section 3), and we have discovered that for such period doublings and triplings properties analogous to Feigenbaum's metric universality prevail. Specifically we have studied these sequences for the two popular maps A and B. It should be clear that any conclusions we draw from both mappings almost certainly apply to other maps with the same general characteristics; namely, one-dimensional non-invertible maps with a unique quadratic maximum.

By comparing the shapes of iterated maps $[f]^{2^n}$ in the central region, using computer techniques, Feigenbaum (1978, 1979) was able to demonstrate the existence of a universal function

$$g_1(x) = \lim_{n\to\infty} (-\alpha)^n [f]^{2^n} (\lambda_{2^{n+1}}, x/(-\alpha)^n).$$

He also showed that one could define a whole sequence of functions

$$g_r(x) = \lim_{n\to\infty} (-\alpha)^n [f]^{2^n} (\lambda_{2^{n+r}}, x/(-\alpha)^n), \tag{9}$$

satisfying

$$g_{r-1}(x) = -\alpha g_r(g_r(-x/\alpha)). \tag{10}$$

Feigenbaum then conjectured that this sequence of functions converged to a unique limit

$$g(x) = \lim_{r\to\infty} g_r(x) = \lim_{n\to\infty} (-\alpha)^n [f]^{2^n} (\lambda_{2^\infty}, x/(-\alpha)^n), \tag{11}$$

which satisfied the fixed point Feigenbaum–Cvitanovic relation

$$g(x) = -\alpha g(g(-x/\alpha)). \tag{12}$$

[The existence of g in certain cases was in fact proved by Collet *et al.* (1980) and Lanford (1982), who also proved existence and uniqueness for the mappings $x \to 1-\mu x^{1+\varepsilon}$, with ε small.] The scale of g is arbitrary and is set through the normalization condition $g(0) = 1$. In an effort to make the present paper self-contained, as well as for later comparison with other universal functions, we give g_0, g_1 and g in Fig. 1.

Associated with the cycles k born by tangent bifurcation is a cascade of harmonics $k\cdot 2^n$ emerging by subsequent period doubling (forward bifurcation). One may again abstract the same universal function $g_1(x)$—and indeed the entire sequence $g_r(x)$ culminating in $g(x)$—by approaching the accumulation point $\lambda_{k\cdot 2^\infty}$. This is shown

Table 4*a*. Backward bifurcations $2^n \cdot 3$ for mappings A and B

Cycle	Map A		Map B	
	λ	Dx	λ	Dx
3	3·831874056	0·345710203	3·116700451	−2·343057611
6a	3·627557530	−0·140860795	2·772639937	0·266331112
12a	3·582229836	0·056600411	2·709628054	−0·189584964
24a	3·572577293	−0·022642507	2·696053119	0·061965036
48a	3·570509238	0·009049220	2·693157769	−0·026946032
96a	3·570066370	−0·003615723	2·692537798	0·010418610
192a	3·569971522	0·001444641	2·692405037	−0·004218240
384a	3·569951208	−0·000577183	2·692376604	0·001676493
768a	3·569946858	0·000230607	2·692370514	−0·000671228

Table 4*b*. Ratios of successive $D\lambda$ and Dx for mappings A and B

Cycle	Map A		Map B	
	$R\lambda$	Rx	$R\lambda$	Rx
6a	4·508	−2·454	5·460	−8·798
12a	4·696	−2·498	4·642	−1·405
24a	4·667	−2·500	4·689	−3·060
48a	4·700	−2·502	4·670	−2·300
96a	4·669	−2·503	4·670	−2·586
192a	4·669	−2·503	4·669	−2·470
384a	4·669	−2·503	4·669	−2·516
768a	—	−2·503	—	−2·498

Table 5. Backward bifurcations for mapping A

Cycle	λ	Dx	$R\lambda$	Rx
		(*a*) $2^n \cdot 4$		
4	3·960270127	0·351775477	—	—
8b	3·662192504	−0·146589904	4·095	−2·400
16b	3·589399844	0·059085959	4·763	−2·481
32b	3·574118089	−0·023662325	4·660	−2·497
64b	3·570839054	0·009458540	—	−2·502
		(*b*) $2^n \cdot 5$a		
5a	3·738914930	−0·158342067	—	—
10	3·605385838	0·064326385	4·797	−2·462
20	3·577549811	−0·025819307	4·658	−2·491
40	3·571573647	0·010325161	4·671	−2·501
80	3·570294339	−0·004126144	—	−2·502
		(*c*) $2^n \cdot 5$b		
5b	3·905706470	0·180442003	—	—
10	3·647048802	−0·076515812	4·257	−2·358
20	3·586281315	0·030775652	4·735	−2·486
40	3·573447578	−0·012325970	4·663	−2·497
80	3·570695539	0·004926708	—	−2·502
		(*d*) $2^n \cdot 5$c		
5c	3·990267047	0·353121938	—	—
10	3·673008246	−0·148321143	3·894	−2·381
20	3·591544528	0·059811996	4·802	−2·480
40	3·574581219	−0·023962174	4·656	−2·496
80	3·570938128	0·009578873	—	−2·502

in Fig. 2 for the particular case of $3 \cdot 2^n$ cycles and is fairly well understood; thus (the existence of g for $\varepsilon = 1$ has been proved by Campanino and Epstein 1981)

$$g_1(\mu x) = \mu \lim_{n\to\infty} (-\alpha)^n [f]^{k\cdot 2^n} (\lambda_{k\cdot 2^{n+1}}, x/(-\alpha)^n),$$

with the magnification μ being the only k-dependent ingredient. An appreciation of the scale μ may be gained by comparing Figs 1 and 2.

3. Reverse Bifurcations

In this paper we wish to draw attention to quite distinct limiting sequences in which 2^n multiples of the k cycle occur to the *left* of the basic k cycle, i.e. they are *not* harmonics of that cycle but are instead born by tangent bifurcation. For any $k > 2$ these sequences also approach λ_{2^∞} but in the *reverse* order

$$2^\infty \cdot k \leftarrow \ldots \leftarrow 4 \cdot k \leftarrow 2 \cdot k \leftarrow k.$$

We call this the 'reverse' or 'backward' bifurcation sequence (Feigenbaum 1980; Kopylov and Sivac 1982). Tables 4 and 5 provide the superstable λ values for various low order cycles, mainly for map A. If one examines the $2^n \cdot 3$ order iterates of F (Fig. 3) one can pick out a copy, reduced in scale, of the basic 3 cycle in the vicinity of X. Fig. 4 shows that a similar pattern prevails for other cycles. Moreover one can establish numerically, beyond reasonable doubt (see Tables 4 and 5), that for this backward sequence the usual Feigenbaum constants arise:

$$\lim_{n\to\infty} R\lambda_{2^n\cdot k} \;(\equiv D\lambda_{2^n\cdot k}/D\lambda_{2^{n+1}\cdot k}) = \delta = 4\cdot 6692\ldots, \tag{13a}$$

$$\lim_{n\to\infty} Rx_{2^n\cdot k} \;(\equiv Dx_{2^{n-1}\cdot k}/Dx_{2^n\cdot k}) = -\alpha = -2\cdot 5029\ldots. \tag{13b}$$

This suggests that a universal function of the type g_0 or g_1 may exist for *each* of the 'reverse bifurcation' sequences connected with a particular k cycle. Indeed, we see strong indications of this in Figs 3 and 4 by observing that a copy of the fundamental $[f]^k$ occurs in the vicinity of the central fixed point for the various iterates as the period doubles up in the reverse order.

By appropriate computational procedures (enlarging and inverting the central region at each stage) we have established that the limiting function

$$g_0^k(x) = \lim_{n\to\infty} (-\alpha)^n [f]^{2^n\cdot k} (\lambda_{2^n\cdot k}, x/(-\alpha)^n)$$

exists and is *distinct for every cycle.* This is a totally new phenomenon and is quite different from what happens when the pitchfork sequence is studied.* It is in fact possible to define a whole sequence of functions

$$g_r^k(x) = \lim_{r\to\infty} (-\alpha)^n [f]^{2^n\cdot k} (\lambda_{2^{n+r}\cdot k}, x/(-\alpha)^n), \tag{14}$$

* We are careful to distinguish between the forward superstable values $\lambda_{k\cdot 2^n}$ and the backward superstable values $\lambda_{2^n\cdot k}$ in what follows. Of course, as far as functional iterates are concerned, there is no difference between $[f]^{2^n\cdot k}(\lambda, x)$ and $[f]^{k\cdot 2^n}(\lambda, x)$ at the *same* λ-parameter value.

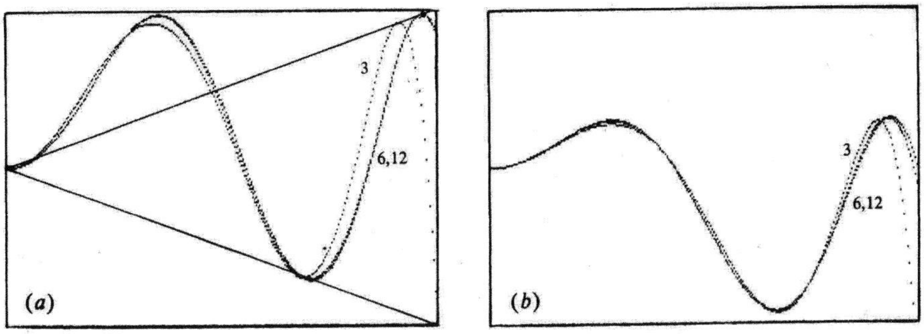

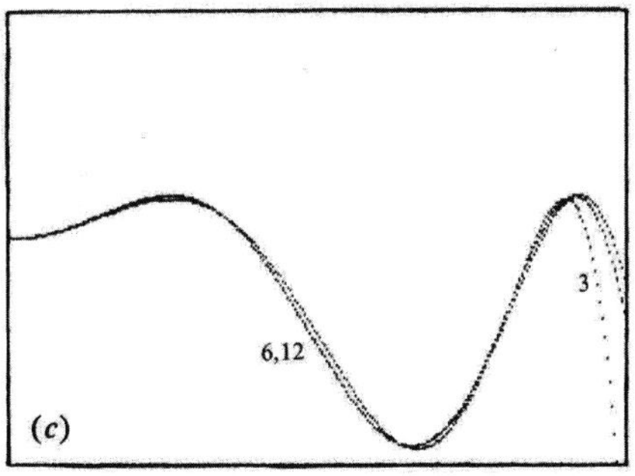

Fig. 5. Reverse sequence of functions tending to
(*a*) g_0^3,
(*b*) g_1^3,
(*c*) g^3,
for $2^n \cdot 3$ iterates. In (*a*) the fixed points nearest the origin occur where the $y = \pm x$ lines intersect the extrema.

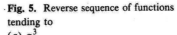

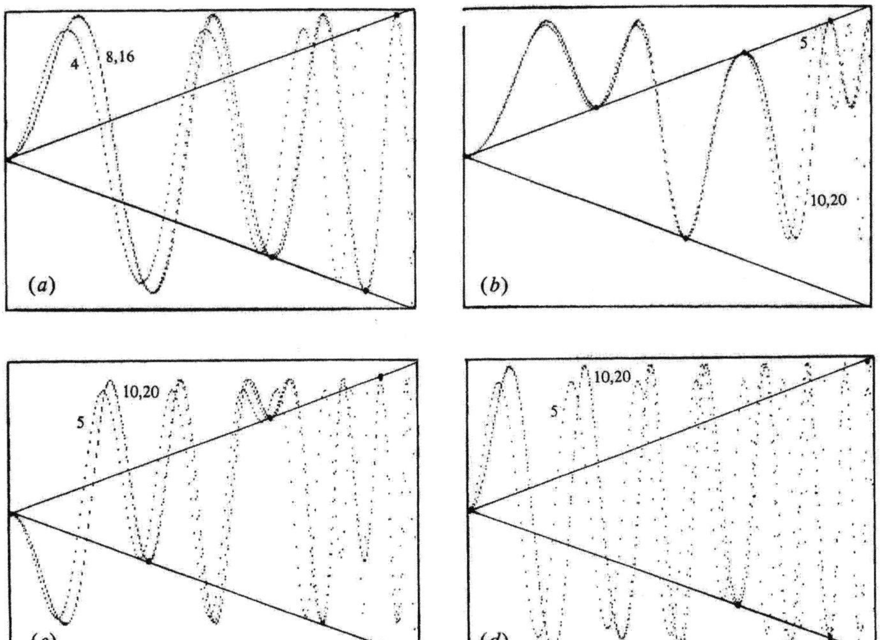

Fig. 6. Reverse sequence of functions tending to (*a*) g_0^4, (*b*) g_0^{5a}, (*c*) g_0^{5b} and (*d*) g_0^{5c} for $2^n \cdot 4$, $2^n \cdot 5a$, $2^n \cdot 5b$ and $2^n \cdot 5c$ iterates respectively. The fixed points are displayed similarly to Fig. 5*a*.

such that

$$g_{r-1}^k(x) = -\alpha g_r^k(g_r^k(-x/\alpha)). \tag{15}$$

Figs 5*a* and 5*b* evidently point to convergence toward a limit function

$$g^k(x) = \lim_{r\to\infty} g_r^k(x),$$

and it is clear that $g^k(x)$ obeys the standard fixed point equation

$$g^k(x) = -\alpha g^k(g^k(-x/\alpha)), \tag{16}$$

where again we have the freedom to set the scale through $g^k(0) = 1$.

Thus it appears that we have an *infinite* class of universal functions (McCarthy 1983) satisfying the standard fixed point equation. In order to distinguish between these functions we may utilize the characteristic structures of the universal functions g_0^k. It is obvious that even when k is specified, there is a variety of functions associated with the classification of Metropolis *et al.* (1973). This is illustrated by the four distinct universal functions corresponding to the 4, 5a, 5b and 5c cycles as shown in Figs 6*a*–6*d*.

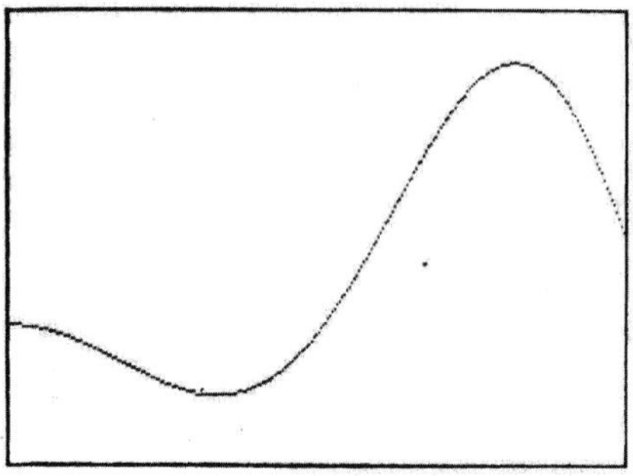

Fig. 7. Third iterate of the standard Feigenbaum universal function $g(x)$. (Compare this with Fig. 5*c*.)

Note, however, that from equation (14), if we go to the limit $r \to \infty$, then an alternative definition is

$$g^k(x) = \lim_{n\to\infty} (-\alpha)^n [f]^{2^n \cdot k}(\lambda_{2^\infty}, x/(-\alpha)^n), \tag{17}$$

assuming as always that the orders of limits can be reliably interchanged. This indicates that the *different* function sequences for fixed k all converge to the *same* limit g^k which depends *only* on the order k of the cycle and *not* on its structure! Further, since the standard universal function is defined by equation (11), we infer that

$$g^k(\mu x) = \mu[g]^k(x). \tag{18}$$

(This receives numerical support in Fig. 7 for the case $k = 3$.) Certainly when $g(g(x)) = -g(-\alpha x)/\alpha$, it is straightforward to verify that

$$[g]^k[g]^k = -[g]^k(-\alpha x)/\alpha. \tag{19}$$

It is worth observing, though, that for every k we are allowed independently to set the scale* by $g^k(0) = 1$, which means that $g^k(x)$ cannot simply be the kth iterate of the standard function $g(x)$ scaled to $g(0) = 1$; indeed, (18) is only correct up to a scaling μ. In any event, each of these g^k satisfies the familiar fixed point relation (16), indicating that an *infinite* number of solutions to that renormalization group equation exists.

4. Triplications

We now turn to a systematic study of cycle sequences of the type $k \cdot 3^n$, where the basic cycle has period k. We call these triplications of the k cycle and they correspond to a particular type of tangent bifurcation. First, we shall distinguish between two distinct sequences which we denote by $k \cdot 3^n$ and $3^n \cdot k$ associated with forward and backward (or reverse) triplications. The forward sequence arises to the right of every k cycle and converges to an accumulation point which depends on k and its cycle structure—in many ways it is analogous to the pitchfork sequence $k \cdot 2^n$. (As a special instance the 3 cycle spawns the sequence 3^n.) Such cycles may be identified by studying the three bands associated with the chaotic region to the right of the pitchfork accumulation point $\lambda_{3 \cdot 2^\infty}$. As far as $k = 3$ is concerned, there is the distinct point of accumulation

$$\lambda_{3^\infty} = 3 \cdot 854077963591, \quad \text{map A};$$

$$= 3 \cdot 216164774983, \quad \text{map B}.$$

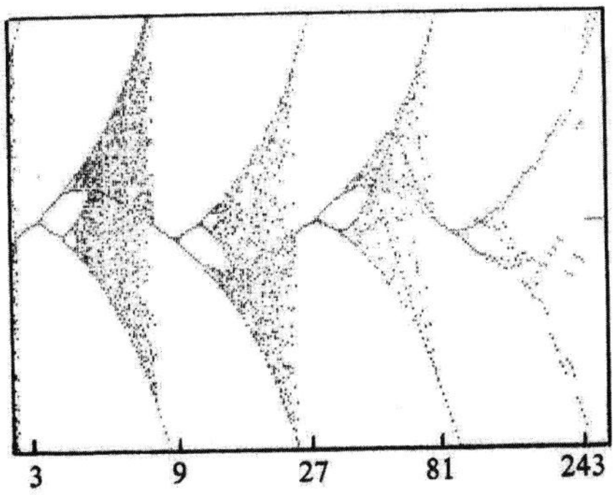

Fig. 8. Repeating triplication pattern as λ_{3^∞} is approached (see text). The density of points along the vertical axis thins out since the total number of iterations is fixed as λ varies along the horizontal axis.

The backward or reverse sequence, which we have denoted by $3^n \cdot k$, is characterized by the fact that it always converges to λ_{3^∞}, irrespective of k; indeed, this sequence is very similar to the reverse bifurcations discussed in the previous section. However, whether these triplings converge to λ_{3^∞} from the left or right depends upon whether or not the basic k cycle lies to the left or right of the 3 cycle. Thus $3^n \cdot 5a$ converges from the left—so the terminology 'backward' is rather a misnomer for it— whereas $3^n \cdot 5b$ and $3^n \cdot 5c$ converge from the right. This triplication pattern is exhibited in Fig. 8 for $x(\lambda - \lambda_{3^\infty})^{-\ln A/\ln \Delta}$ against $\log(\lambda - \lambda_{3^\infty})$ with the constants Δ and A in (21a) and (21b) already anticipated.

* One must be careful not to confuse $g^k(0)$ with the related quantity $Dx_{2^n \cdot k} = [f]^{2^n \cdot C}(\lambda_{2^n \cdot k}, 0)$, where C is the number of iterations needed to bring x to X for $n = 0$.

Table 6*a*. Forward trifurcations $1\cdot 3^n$ for mappings A and B

Cycle	Map A λ	Map A Dx	Map B λ	Map B Dx
3	3·831874055283	0·3457102029	3·116700451066	−2·343405761
9	3·853675276839	−0·0356580611	3·214601114697	0·129432449
27	3·854070677510	0·0038524177	3·216136205149	−0·015927717
81	3·854077831706	−0·0004151813	3·216164258172	0·001690728
243	3·854077961203	0·0000447527	3·216164765631	−0·000182565

Table 6*b*. Ratios of successive $D\lambda$ and Dx for mappings A and B

Cycle	Map A $R\lambda$	Map A Rx	Map B $R\lambda$	Map B Rx
9	55·13	−9·695	63·78	−18·105
27	55·27	−9·256	54·72	−8·126
81	55·25	−9·279	55·28	−9·421
243	—	−9·277	—	−9·261

Table 7. Forward trifurcations for mapping A

Cycle	λ	Dx	$R\lambda$	Rx
		(*a*) $4\cdot 3^n$		
4	3·9602701272212	0·3517754767	—	—
12	3·9614314419566	−0·0084164470	56·00	−41·796
36	3·9614521815369	0·0008811366	54·99	−9·552
108	3·9614525586728	−0·0000951532	55·26	−9·260
324	3·9614525654981	0·0000102551	—	−9·279
		(*b*) $5a\cdot 3^n$		
5a	3·738914912970	−0·1583420673	—	—
15	3·744016873483	0·0232593167	58·05	−6·808
45	3·744104768920	−0·0024341389	54·93	−9·556
135	3·744106369092	0·0002628481	55·27	−9·261
405	3·744106398046	−0·0000283256	—	−9·280
		(*c*) $5b\cdot 3^n$		
5b	3·905706469831	0·1804420034	—	—
15	3·906641328957	−0·0097690137	56·51	−18·471
45	3·906657872652	0·0010263518	54·96	−9·518
135	3·906658173687	−0·0001108003	55·26	−9·263
405	3·906658179135	0·0000119443	—	−9·276
		(*d*) $5c\cdot 3^n$		
5c	3·990267046974	0·3531219383	—	—
15	3·990335169048	−0·0020549297	56·45	−17·184
45	3·990336375733	0·0002136359	54·94	−9·619
135	3·990336397698	−0·0000230798	55·19	−9·256
405	3·990336398096	0·0000024874	—	−9·279

In Tables 6–8 the numerical results for both trifurcating sequences are presented, chiefly for map A, in a similar way to the tabulation for bifurcations. For the forward sequence we define the relevant superstable values by $\lambda_{k\cdot 3^n}$ and for the backward sequence by $\lambda_{3^n\cdot k}$. As pointed out, we have

$$\lim_{n\to\infty} \lambda_{k\cdot 3^n} \equiv \lambda_{k\cdot 3^\infty} \neq \lambda_{3^\infty}, \quad \text{except for } k = 3; \tag{20a}$$

$$\lim_{n\to\infty} \lambda_{3^n\cdot k} = \lambda_{3^\infty}. \tag{20b}$$

Table 8. Backward trifurcations for mapping A

Cycle	λ	Dx	$R\lambda$	Rx
		(*a*) $3^n \cdot 4$		
4	3·960270127221	0·3517754767	—	—
12	3·855993729675	−0·0374971999	55·43	−9·381
36	3·854112685005	0·0040525443	55·17	−9·253
108	3·854078592024	−0·0004367589	55·25	−9·279
324	3·854077974966	0·0000470753	—	−9·278
		(*b*) $3^n \cdot 5$a		
5a	3·738914912971	0·158342067319		
15	3·852099410353	−0·015978208764	58·21	−9·910
45	3·854042076559	0·001725644420	55·13	−9·259
135	3·854077314123	−0·000185974316	55·26	−9·279
405	3·854077951835	0·000020045340	55·25	−9·278
		(*c*) $3^n \cdot 5$b		
5b	3·905706469831	0·1804420034	—	—
15	3·854991046674	−0·0190731288	56·57	−9·461
45	3·854094524136	0·0020604249	55·13	−9·257
135	3·854078263307	−0·0002220630	55·25	−9·279
405	3·854077969016	0·0000239337	—	−9·278
		(*d*) $3^n \cdot 5$c		
5c	3·990267046974	0·3531219383	—	—
15	3·856587276680	−0·0379534384	54·26	−9·304
45	3·854123393250	0·0041020168	55·23	−9·252
135	3·854078785902	−0·0004420946	55·25	−9·279
405	3·854077978475	0·0000476519	—	−9·278

With *both* sequences we find that there is a scaling law determining the relative window sizes, analogous to (8b),

$$\lim_{n\to\infty} R\lambda_{k\cdot 3^n} \left(\equiv \frac{D\lambda_{k\cdot 3^n}}{D\lambda_{k\cdot 3^{n+1}}} = \frac{\lambda_{k\cdot 3^n} - \lambda_{k\cdot 3^{n-1}}}{\lambda_{k\cdot 3^{n+1}} - \lambda_{k\cdot 3^n}}\right)$$

$$= \lim_{n\to\infty} R\lambda_{3^n\cdot k} \left(\equiv \frac{D\lambda_{3^n\cdot k}}{D\lambda_{3^{n+1}\cdot k}} = \frac{\lambda_{3^n\cdot k} - \lambda_{3^{n-1}\cdot k}}{\lambda_{3^{n+1}\cdot k} - \lambda_{3^n\cdot k}}\right) = \quad = 55\cdot 26, \tag{21a}$$

with a universal constant Δ that is map-independent, apart from the quadratic maximum requirement. As well, there is a second scaling law determining the trident sizes, analogous to (8a),

$$\lim_{n\to\infty} Rx_{k\cdot 3^n} \left(\equiv \frac{Dx_{k\cdot 3^{n-1}}}{Dx_{k\cdot 3^n}} = \frac{x^*_{k\cdot 3^{n-1}} - X}{x^*_{k\cdot 3^n} - X}\right)$$

$$= \lim_{n\to\infty} Rx_{3^n\cdot k} \left(\equiv \frac{Dx_{3^{n-1}\cdot k}}{Dx_{3^n\cdot k}} = \frac{x^*_{3^{n-1}\cdot k} - X}{x^*_{3^n\cdot k} - X}\right) = -A = -9\cdot 277, \tag{21b}$$

governed by another universal constant A.†

There is a striking resemblance between pitchfork bifurcations and the forward triplications—both converge to separate points of accumulation $\lambda_{k\cdot 2^\infty}$ and $\lambda_{k\cdot 3^\infty}$

† After determining values for Δ and A, we realized that Derrida *et al.* (1979) had determined them for the first class of sequences. However, it is not entirely obvious that their work extends to the second class or that the numbers are truly universal (map-independent).

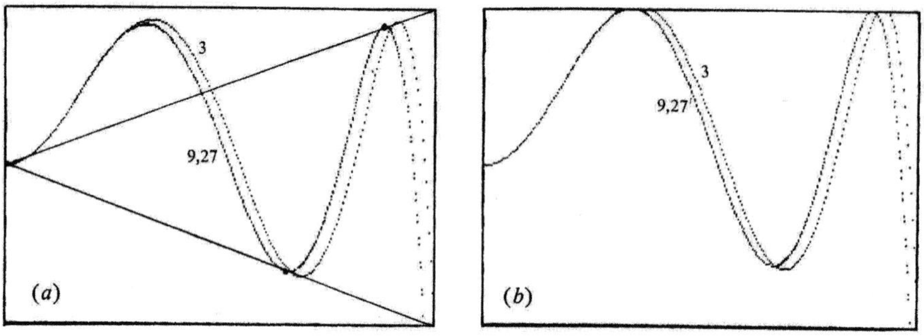

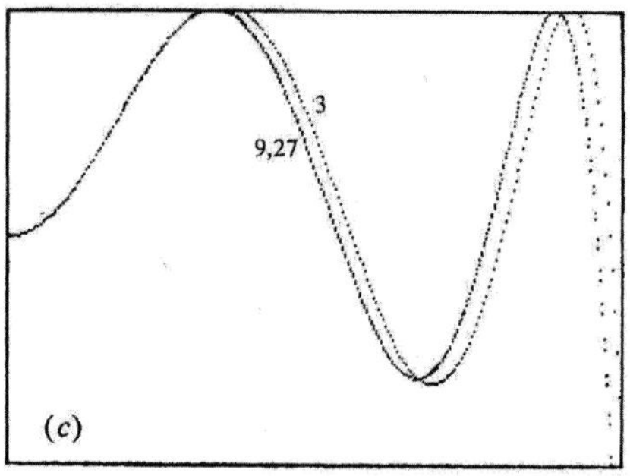

Fig. 9. Forward triplication sequence of functions tending to
(*a*) G_0,
(*b*) G_1,
(*c*) G.
The order of iteration is indicated by the integers. In (*a*) the fixed points nearest the origin occur where the $y = \pm x$ lines intersect the extrema.

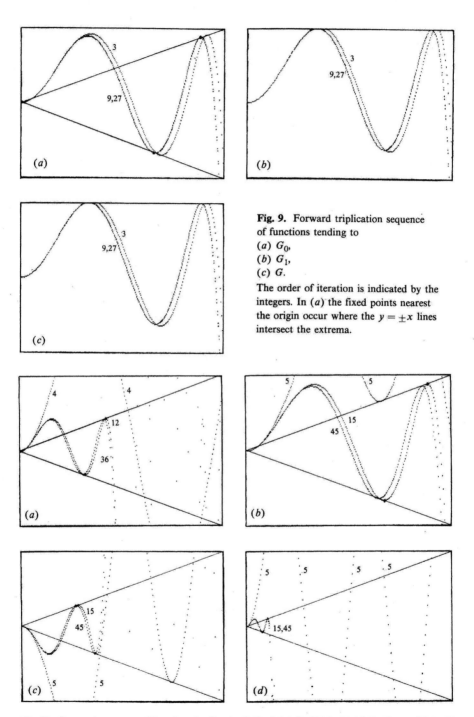

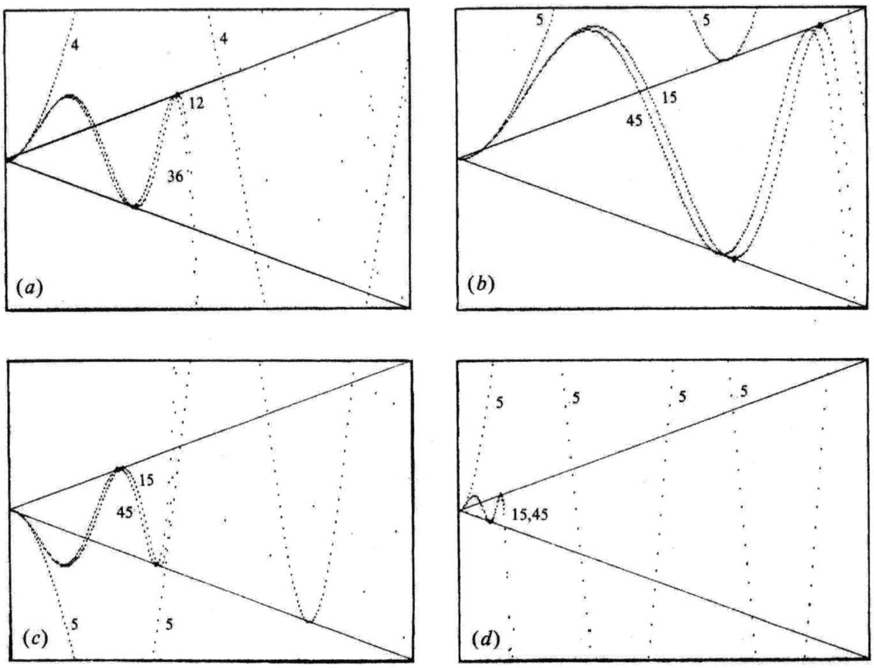

Fig. 10. Forward sequence of functions tending to G_0 for (*a*) $4 \cdot 3^n$, (*b*) $5a \cdot 3^n$, (*c*) $5b \cdot 3^n$ and (*d*) $5c \cdot 3^n$ iterates. Observe how (*a*) is a scaled version of Fig. 9*a*. The fixed points in each case are displayed similarly to Fig. 9*a*.

respectively—as well as between reverse bifurcations and trifurcations—both converge to the same point of accumulation $\lambda_{2\infty}$ and $\lambda_{3\infty}$ respectively. These analogies suggest that we should pursue the idea of universal trifurcation functions as we have already done for both kinds of cycle doublings.

5. Universal Triplication Functions

We begin by focussing on the analogue of the pitchfork sequence, namely forward period tripling $k\cdot 3^n$. By standard computational techniques we have shown that the limiting function

$$G_0(x) = \lim_{n\to\infty} (-A)^n[f]^{3^n}(\lambda_{3^n}, x/(-A)^n)$$

exists; it is depicted in Fig. 9*a*. Of course one can also define a series of functions via

$$G_r(x) = \lim_{n\to\infty} (-A)^n[f]^{3^n}(\lambda_{3^{n+r}}, x/(-A)^n), \tag{22}$$

whereupon

$$G_{r-1}(x) = -AG_r(G_r(G_r(-x/A))). \tag{23}$$

Computations (see Figs 9*b* and 9*c*) suppport the expectation that this sequence G_r converges to a limit function

$$G(x) = \lim_{r\to\infty} G_r(x) = \lim_{n\to\infty} (-A)^n[f]^{3^n}(\lambda_{3^\infty}, x/(-A)^n), \tag{24}$$

such that

$$G(x) = -AG(G(G(-x/A))). \tag{25}$$

Again we may fix the scale through $G(0) = 1$. (Actually more is necessary, as shown in the following section.) Equation (25) is in direct analogy to the Feigenbaum–Cvitanovic equation (12).

A study of the first sequence $k\cdot 3^n$ produces the same universal function (see Fig. 10). Thus, if

$$G_0(\mu x) = \mu \lim_{n\to\infty} (-A)^n[f]^{k\cdot 3^n}(\lambda_{k\cdot 3^n}, x/(-A)^n),$$

we find that $G(x)$ emerges (again up to some magnification) when we go to the accumulation point directly:

$$G(\mu x) = \mu \lim_{n\to\infty} (-A)^n[f]^{k\cdot 3^n}(\lambda_{k\cdot 3^\infty}, x/(-A)^n). \tag{26}$$

This is shown for the case $k = 4$ in Fig. 11.

However, a study of the second reverse class of sequences $3^n\cdot k$ leads to *different* universal functions which we denote by G_0^k:

$$G_0^k(x) = \lim_{n\to\infty} (-A)^n[f]^{3^n\cdot k}(\lambda_{3^n\cdot k}, x/(-A)^n).$$

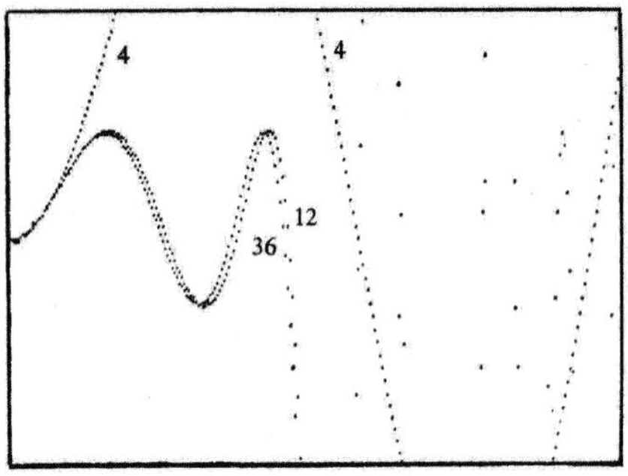

Fig. 11. Forward triplicating sequence $4 \cdot 3^n$ tending to scaled G.

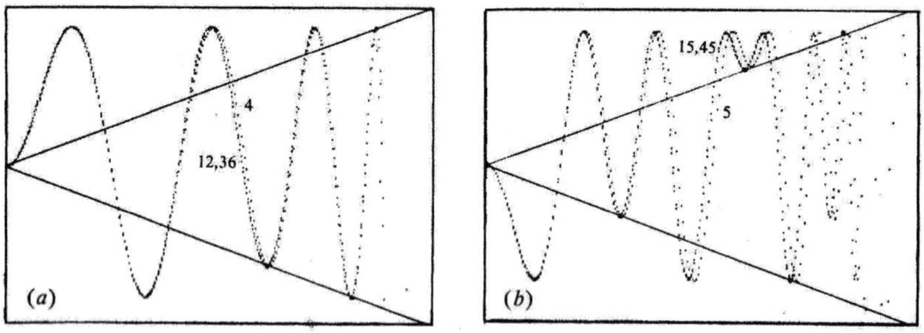

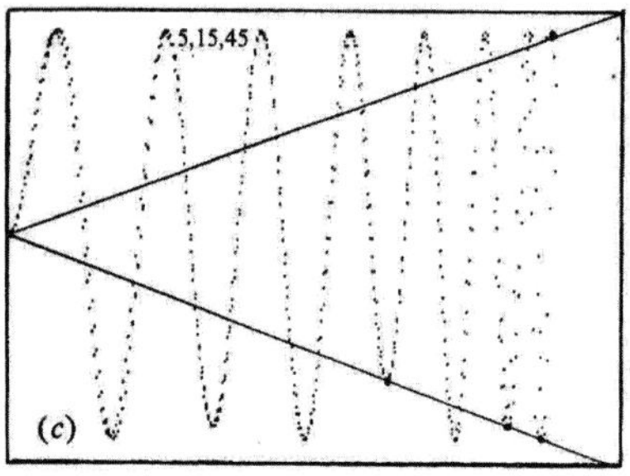

Fig. 12. Reverse sequence of functions
(*a*) $3^n \cdot 4$,
(*b*) $3^n \cdot 5\mathrm{b}$,
(*c*) $3^n \cdot 5\mathrm{c}$,
tending to G_0^4, $G_0^{5\mathrm{b}}$ and $G_0^{5\mathrm{c}}$ respectively.
In each part the fixed points
are displayed similarly to Fig. 9*a*.

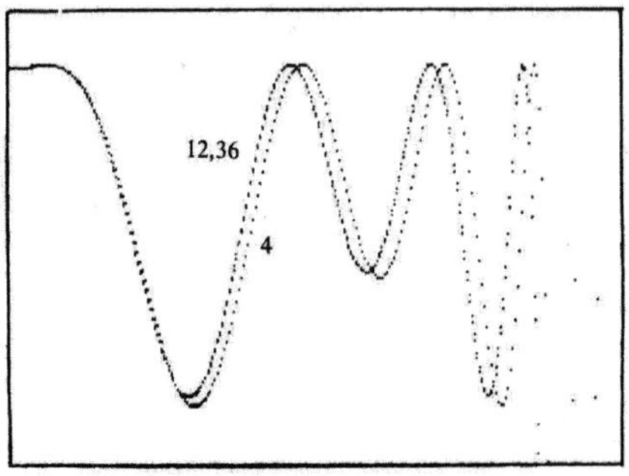

Fig. 13. Reverse sequences G^4 associated with $3^n \cdot 4$ cycles. We know this to be the fourth iterate of G, up to a scaling.

Figs 12*a*–12*c* show how this happens for $k = 4$, 5*b* and 5*c* respectively. Again one may define a sequence of functions

$$G_r^k(x) = \lim_{n\to\infty} (-A)^n [f]^{3^n \cdot k}(\lambda_{3^{n+r} \cdot k}, x/(-A)^n), \tag{27}$$

satisfying

$$G_{r-1}^k(x) = -AG_r^k(G_r^k(G_r^k(-x/A))). \tag{28}$$

There is graphic support (see Fig. 13) for the assumption that this sequence converges to a limit function

$$G^k(x) = \lim_{r\to\infty} G_r^k(x),$$

obeying the triplication equation

$$G^k(x) = -AG^k(G^k(G^k(-x/A))). \tag{29}$$

Alternatively, if we allow r to tend to infinity in (27) we see that formally

$$G^k(x) = \lim_{n\to\infty} (-A)^n [f]^{3^n \cdot k}(\lambda_{3^\infty}, x/(-A)^n). \tag{30}$$

Once more we come across an *infinite* class of functions obeying the same fixed point triplication equation (25). It would seem from (30) that one can identify

$$G^k(\mu x) = \mu[G]^k(x) \tag{31}$$

up to a magnification μ. Nonetheless, it is important to realize that, as was the case for $g^k(x)$ in Section 3, it is admissible to set the normalization $G^k(0) = 1$ for each k *a priori*. It is certainly true that when $G(x)$ obeys (25) so does $[G]^k(x)$.

The examination of these universal functions and sequences reinforces the parallel between the 2^n and 3^n harmonics. We have little doubt that these considerations apply to other kinds of period multiplications; for instance, the fourfold sequences associated with the 4-cycle window born by tangent bifurcation.

6. Fractional Universal Functions

The fact that the sequence

$$(-\alpha)^n f^{k\cdot 2^n}(\lambda_{k\cdot 2^\infty}, x/(-\alpha)^n)$$

converges to $g(x)$ as $n \to \infty$ suggests that the related sequence

$$g_n^{1/k}(x) \equiv (-\alpha)^n f^{2^n}(\lambda_{k\cdot 2^\infty}, x/(-\alpha)^n)$$

may converge in the limit $n \to \infty$ to some 'fractional' universal function which we denote by $g^{1/k}(x)$. Even if $g^{1/k}(x)$ is not unique, the kth iterate clearly is unique and it would be very surprising if no sort of convergence occurred as $n \to \infty$. If $g^{1/k}$ exists, obviously

$$[g^{1/k}]^k = g(x), \qquad g^{1/k}(g^{1/k}(x)) = -g^{1/k}(-\alpha x)/\alpha. \tag{32a, b}$$

When $k = 4$ there are solid grounds for anticipating such convergence because, *up to a scaling*, we know that

$$g(x) = \lim_{n\to\infty} (-\alpha)^n f^{4\cdot 2^n}(\lambda_{4\cdot 2^\infty}, x/(-\alpha)^n),$$

and therefore

$$\begin{aligned} g^{1/4}(x) &= \lim_{n\to\infty} (-\alpha)^n f^{2^n}(\lambda_{4\cdot 2^\infty}, x/(-\alpha)^n) \\ &= \lim_{n\to\infty} (-\alpha)^{n+2} f^{4\cdot 2^n}(\lambda_{4\cdot 2^\infty}, x/(-\alpha)^{n+2}) \\ &= \alpha^2 g(x/\alpha^2) \end{aligned} \tag{33}$$

evidently exists. Direct computation (see Fig. 14) bears out this assertion. Contrary to what is commonly believed (Feigenbaum 1983), this argument shows that *there are other points of accumulation besides* $\lambda = \lambda_{2^\infty}$ at which

$$\lim_{n\to\infty} (-\alpha)^n f^{2^n}(\lambda, x/(-\alpha)^n)$$

exists. More generally, we would expect convergence of this function sequence for all $k = 2^m$ (m integer > 1) *provided we fix* $\lambda = \lambda_{2^m\cdot 2^\infty}$, in which case the limit function is of the type

$$g^{2^{-m}} = (-\alpha)^m g(x/(-\alpha)^m), \tag{34}$$

and satisfies (32).

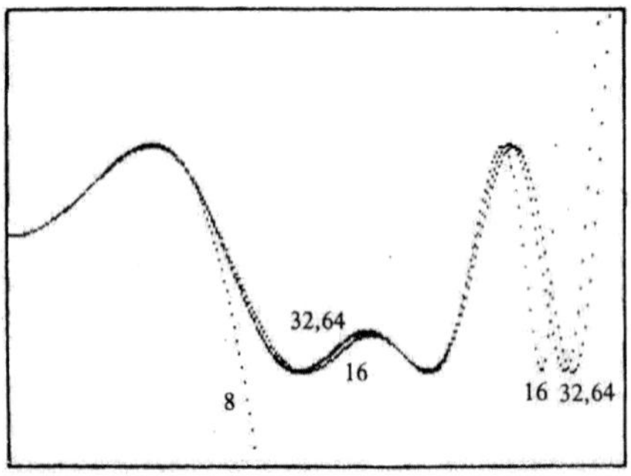

Fig. 14. Sequence of functions leading to $g^{1/4}$ at the accumulation point $\lambda_{4\cdot 2^\infty} = 3{\cdot}96119824$.

The idea can be further generalized in a straightforward way to universal functions associated with period tripling. Thus, there must exist fractional functions such that

$$[G^{1/k}]^k = G, \qquad G^{1/k}(G^{1/k}(G^{1/k}(x))) = -G^{1/k}(-Ax)/A, \tag{35a, b}$$

for $k = 3^m$ and $m > 1$. For instance, when $m = 2$ we must have

$$\begin{aligned} G^{1/9}(x) &= A^2G(x/A^2) \\ &= \lim_{n\to\infty} (-A)^n f^{3^n}(\lambda_{9\cdot 3^\infty}, x/(-A)^n), \end{aligned} \tag{36}$$

where $\lambda_{9\cdot 3^\infty}$ is a point of accumulation ($\neq \lambda_{3^\infty}$) associated with one of the many 9 cycles.

We conclude with some remarks on the normalization of the universal functions. The basic $G_0(x)$ has three superstable fixed points near the origin, one of which may be taken to be $x = 0$; see Fig. 9*a*. The function $G_1(x)$ is, of course, related to $G_0(x)$ through

$$G_0(x) = -AG_1(G_1(G_1(-x/A))).$$

Now, in the case of pitchfork bifurcation, the curve associated with $g_1(x)$ supports a circulation square such that

$$g_1(0) = 1, \qquad g_1(1) = 0.$$

Alternatively we may regard $x = 0, 1$ as fixed points of the iterate $[g_1]^2$, i.e. $[g_1]^2(0) = 0$ and $[g_1]^2(1) = 1$. However, for period tripling the curve associated with $G_1(x)$ supports a circulation polygon, corresponding to three superstable fixed points of $[G_1]^3(x)$, which we may take to be 1, 0 and γ_1; having set the scale with the first two fixed points, γ_1 is some new universal constant† lying between 0 and -1. Here we have

$$G_1(0) = 1, \qquad G_1(1) = \gamma_1, \qquad G_1(\gamma_1) = 0, \tag{37a, b, c}$$

thereby setting the scale for the first function in the sequence G_r. The limiting function G as $r \to \infty$ must then satisfy

$$G(0) = 1, \qquad G(1) = \gamma, \qquad G(\gamma) = -1/A, \tag{38a, b, c}$$

in contrast to (37), because of the fixed point relation (29). The need for three boundary or normalization conditions (38) for G is dictated by G being a solution of a period-tripling functional equation. While the origin of the third boundary condition is clear, it is not obvious to us at present whether or not it is a truly independent condition.

Note added in proof: A good approximation to $G(x)$ in lowest order is given by

$$G(x) = 1 - \mu_{3\infty} x^2.$$

For curves with a quadratic maximum μ is related to λ via equation (6a), so the value $\lambda_{3\infty} = 3\cdot 8541$ determines $\mu_{3\infty} = 1\cdot 7864$. To this order we find that equations (38) become approximately

$$G(0) = 1, \quad G(1) = \gamma = 1 - \mu_{3\infty}, \quad G(\gamma) = 1 - \mu_{3\infty}\gamma^2 = -1/A,$$

† In relation to this, and by examining our various universal curves which exhibit nontrivial local fixed point structure (Figs 5 and 6), we see that in every case one can choose the centremost fixed point to be $x^* = 0$ and (by scaling the horizontal axis) the rightmost fixed point to be $x^* = 1$. The values x^* of all other fixed points are then determined to lie between -1 and 1, at universal locations.

which illustrates that the second boundary condition is *not* independent. In numerical terms this gives

$$\gamma = -0{\cdot}7864, \qquad A = 9{\cdot}534,$$

yielding a value for A within 3% of the experimental value. Higher order (in x^2) approximations to $G(x)$ shift the above numerical value of γ by less than 1%, while A agrees with the experimental value to rather better than 1% already at second order.

References

Campanino, M., and Epstein, H. (1981). *Commun. Math. Phys.* **79**, 261.

Collet, P., and Eckmann, J. P. (1980). 'Iterated maps on the interval as dynamical systems'. *In* 'Progress in Physics', Vol. 1 (Birkhauser: Boston).

Collet, P., Eckmann, J. P., and Lanford, O. E. (1980). *Commun. Math. Phys.* **76**, 211.

Derrida, B., Gervois, A., and Pomeau, M. (1979). *J. Phys.* A **12**, 269.

Eckmann, J. P. (1981). *Rev. Mod. Phys.* **53**, 643.

Feigenbaum, M. J. (1978). *J. Stat. Phys.* **19**, 25.

Feigenbaum, M. J. (1979). *J. Stat. Phys.* **21**, 669.

Feigenbaum, M. J. (1980). *Commun. Math. Phys.* **77**, 65.

Feigenbaum, M. J. (1983). *Physica* D **7**, 16.

Hirsch, J., Nauenberg, M., and Scalapino, D. J. (1982). *Phys. Lett.* A **87**, 391.

Hu, B. (1982). *Phys. Rep.* **91**, 233.

Hu, B., and Rudnick, J. (1982). *Phys. Rev. Lett.* **48**, 1645.

Kopylov, C. F., and Sivak, A. G. (1982). *Ukr. Acad. Sci.* **517**, 53.

Lanford, O. E. (1982). *Bull. Am. Phys. Soc.* **6**, 427.

McCarthy, P. J. (1983). *Commun. Math. Phys.* **91**, 431.

Manneville, P., and Pomeau, Y. (1980). *Commun. Math. Phys.* **74**, 189.

May, R. M. (1976). *Nature* **261**, 459.

Metropolis, N., Stein, M. L., and Stein, P. R. (1973). *J. Combinatorial Theory* A **15**, 25.

Ott, E. (1981). *Rev. Mod. Phys.* **53**, 655.

Ruelle, D., and Takens, F. (1971). *Commun. Math. Phys.* **20**, 167.

Manuscript received 20 September, accepted 30 November 1984

PHYSICAL REVIEW A VOLUME 33, NUMBER 5 MAY 1986

Universality relations

R. Delbourgo
Physics Department, University of Tasmania, G.P.O. Box 252C, Hobart, Tasmania, Australia 7001

B. G. Kenny
Physics Department, University of Western Australia, Nedlands, Perth, Western Australia, Australia 6009
(Received 18 November 1985)

By analyzing the functional equations governing N replication for real nonlinear maps on the (unit) interval (of order z), we have derived the asymptotic ($N \to \infty$) behaviors of the scaling constants α and δ, $\alpha(z) \sim (2z)^{(N-1)/(z-1)}$, $\delta/\alpha^2 = (z-1)/(z-\frac{1}{2})$. In addition, we have determined the limiting forms of the universal functions. The predictions are in excellent agreement with (separately compiled) numerical data over a wide range of z values and cycle structures.

I. INTRODUCTION

In recent years theoretical and experimental evidence has steadily mounted that infinite sequences of period doublings frequently occur in dissipative dynamical systems. According to the universality theory of one-dimensional maps developed by Feigenbaum,[1] variation of a single tuning parameter drives the system through a series of bifurcations characterized by scaling constants δ and α which depend only on the nature of the maximum. From the point of view of the physicist, the exciting fact is that the scaling constants associated with the (quadratic maximum) map appear to be observed and measured[2,3] in relatively complicated dynamical systems. More recently, related scaling constants corresponding to the map of a circle onto itself[4] seem to have been observed experimentally.[5]

However, there is still some degree of mystery about these scaling constants in the sense that they have only been determined by *purely numerical* techniques. So far it has not proved possible to calculate them analytically in terms of known quantities nor is it known whether or not α and δ are independent of each other. In an attempt to shed some light on these questions, we have therefore examined their behavior as a function of the parameter z characterizing the maximum of the map, realizing that the case $z=2$ is likely to be the only one of physical interest. We have further considered[6] more general period N-tuplings connected with such real function iterations again realizing that only the case $N=2$ is experimentally detectable. The motivation for our studies is that if we can gain some insight into the behavior of $\delta(N,z)$ and $\alpha(N,z)$ as N and Z are varied, then this may help in understanding the particular case $z=N=2$.

It turns out that our ambition has been fulfilled, at least for large N. We have been able to find exact asymptotic analytical forms for δ and α as a function of N and z for a large class of kneading sequences. (Our approach has something in common with $1/N$ methods applied to quantum mechanical models.) In particular, the $z=2$ case is exactly soluble in closed form. This has been recognized[7] only for the situation when the fixed points of an N-cycle are permuted cyclically by the mapping. We have succeeded in generalizing this to most kneading sequences and to arbitrary z values; in this way we can calculate the asymptotic form of δ and α to arbitrary accuracy. What is more, we can find an asymptotic relation between δ and α as stated in the abstract.

The paper is organized as follows. Section II concerns period doubling ($N=2$) and tripling ($N=3$), where we show that truncations of the functional equations describing α and δ scaling do lead to α-δ relations that are quite well obeyed, even though one is far from the small-z region where they are expected to be correct. Various clues about the properties of the universal eigenfunctions are used to derive the asymptotic behaviors of those functions and the scaling constants in Sec. III. These are tested against the numerical solutions of the functional equations, as described in Sec. IV including a discussion of the obstacles that need to be overcome as $N \to \infty$. The Appendix focuses on the case $z=2$ where exact solutions for the asymptotic δ and α are extracted. We have tabulated the most important results in various parts of the text and deliberately omitted others that have no significant bearing on our arguments. If readers are interested in "missing" cases or more accurate values they should write directly to us.

II. DOUBLING AND TRIPLING RELATIONS

In order to lay the foundation for a study of the exact analytic form of δ and α for high order cycles, we shall begin by showing how δ and α may be related for period doubling and tripling when $z-1$ is not large.

A. Period doubling

Here α is determined by the functional equation[1]

$$g(g(x)) = -g(-\alpha x)/\alpha \; . \tag{2.1}$$

Provided that z is not much larger than 1, a reasonable approximation to g is provided by

 ©1986 The American Physical Society

$$g(x)=1-a|x|^z, \tag{2.2}$$

where we have ignored terms of order $|x|^{2z}$. We later demonstrate numerically that Eq. (2.2) is a sensible approximation under the stated assumption. Inserting the truncated Taylor expansion in Eq. (2.1) and comparing coefficients of powers of $|x|$, it is readily seen that a and α are related through

$$a-1=1/\alpha, \tag{2.3}$$

$$za=\alpha^{z-1}, \tag{2.4}$$

from which one may find $\alpha=\alpha(z)$ and $a=a(z)$.

Similarly, δ is determined by the functional equation[1]

$$g'(g(x))h(x)+h(g(x))=-\delta h(-\alpha x)/\alpha. \tag{2.5}$$

Again, provided $z-1$ is small, a reasonable approximation for $h(x)$ is given by

$$h(x)=1-b|x|^z, \tag{2.6}$$

as we shall confirm numerically later. Inserting the expressions (2.2) and (2.6) for $g(x)$ and $h(x)$ into Eq. (2.5) and comparing coefficients, one readily finds that

$$za+b-1=\delta/\alpha \tag{2.7}$$

$$2zab+z(z-1)a^2=\delta b\alpha^{z-1}. \tag{2.8}$$

Together with Eqs. (2.3) and (2.4) one may find $\delta=\delta(z)$ and $b=b(z)$. Rather than do this, we eliminate δ from Eq. (2.7) and (2.8) to get

$$[2zab+z(z-1)a^2]/[za+b-1]=b\alpha^z, \tag{2.9}$$

and use Eqs. (2.3) and (2.4) to express b directly in terms of a as follows:

$$ab(za+b-1)=(a-1)[2ab+(z-1)a^2]. \tag{2.10}$$

This admits two solutions,

$$b=a-1=1/\alpha \tag{2.11}$$

and

$$b=a(1-z). \tag{2.12}$$

The second solution corresponds to the marginal eigenvalue $\delta=1$ and is associated with the class of solutions of Eq. (2.5) noted by Feigenbaum,

$$h(x)=g(x)-xg'(x)=1-a(1-z)|x|^z, \tag{2.13}$$

in this order of approximation. The other solution is the interesting one and we note that it has $b=1/\alpha$, which becomes very small as z tends to 1 because α increases without limit. If one sets $b=0$ in Eq. (2.7), we obtain the lowest-order computational result[8]

$$\delta/\alpha=za-1 \tag{2.14}$$

or

$$\delta=\alpha^z-\alpha=\alpha^z(1-1/za). \tag{2.14'}$$

However, we can do better than that. From Eqs. (2.7) and (2.11), we see that

$$\delta/\alpha=za+1/\alpha-1 \tag{2.15}$$

or

$$\delta=\alpha^z-\alpha+1 \tag{2.15'}$$

relating δ and α directly. This relation works surprisingly well as may be seen from Table I. If one inserts the exact value for α, δ is predicted correctly to within a couple of percent over a wide range of z.

One may hone this argument by going to the next order in $|x|^z$. Table II gives the solutions of (2.1) and (2.5) after truncation to order $|x|^{2z}$. Although it is not obvious from the table, the α and δ are much improved (typically by 10% or so) in relation to the first-order predictions ensuing from (2.3), (2.4), (2.7), and (2.8). Also, it is worth mentioning that we have not yet succeeded in obtaining an analytic relation between δ and α from the second-order equations analogous to (2.15), but have been obliged to resort to numerical methods in deriving the entries in Table II.

B. Period tripling

In this case the appropriate universal constant α is determined by the functional equation[9]

$$G(G(G(x)))=-G(-\alpha x)/\alpha. \tag{2.16}$$

Provided $z-1$ is small, a reasonable approximation to G is provided by

$$G(x)=1-A|x|^z, \tag{2.17}$$

where we have again ignored higher-order terms. [In this approximation it turns outs that $A=\mu_{3\infty}$ for the map $f(x)=1-\mu|x|^z$.] Substituting the truncated Taylor series equation (2.17) into Eq. (2.16) and comparing coefficients of powers of $|x|$, it is readily seen that A and α are related through

$$A(A-1)^z-1=1/\alpha, \tag{2.18}$$

$$z^2A^2(A-1)^{z-1}=\alpha^{z-1}, \tag{2.19}$$

from which one may find the dependence of A and α on z. There is agreement with the exact numerical results to

TABLE I. Duplication constants. δ is predicted from exact α via Eq. (2.15), $\delta=\alpha^z-\alpha+1$.

z	Exact α	Exact δ	Predicted δ
1.1	7.97	2.83	2.84
1.2	5.37	3.14	3.15
1.5	3.39	3.80	3.85
2	2.50	4.67	4.76
3	1.93	6.08	6.26
4	1.69	7.29	7.47
5	1.56	8.35	8.69
7	1.41	10.2	10.7
10	1.29	12.3	12.5

TABLE II. Self-consistent solutions of the truncated duplication equations to second order in $|x|^z$. The quoted coefficients arise as $g(x)=1-a|x|^z+e|x|^{2z}$, $h(x)=1-b|x|^z-f|x|^{2z}$. Note how rapidly the coefficients decrease with z. The exact α and δ are given in Table I for comparison purposes.

z	a	e	α	b	f	δ
1.1	1.132	0.006	7.981	0.123	0.001	2.838
1.2	1.203	0.018	5.395	0.180	0.003	3.151
1.5	1.354	0.061	3.415	0.278	0.006	3.864
2	1.522	0.128	2.534	0.360	0.018	4.844
3	1.729	0.220	1.966	0.436	0.068	6.624
4	1.856	0.278	1.732	0.471	0.138	8.322
5	1.939	0.314	1.600	0.493	0.203	10.00
7	2.058	0.368	1.449	0.509	0.352	13.21
10	2.151	0.402	1.335	0.525	0.496	18.01

within about 3% even for $z=1.5$. As before, the predictions improve as $z-1$ becomes smaller. Purely numerical techniques have shown us that the higher-order coefficients in the expansion of $G(x)$ are quite small for modest z.

It is straightforward to show that the δ associated with period tripling is determined by the functional equation

$$G'([G]^2(x))H(x)+H(G(x))G'(G(x)) + H([G]^2(x))=-\delta H(-\alpha x)/\alpha . \quad (2.20)$$

Now for small $z-1$ a good approximation to $H(x)$ is

$$H(x)=1-B|x|^z . \quad (2.21)$$

The coefficient B decreases very rapidly as z tends to 1 [indeed, $B=O(1/\alpha)$, as we shall see] and the higher coefficients vanish even more quickly.

Inserting the expressions (2.17) and (2.21) for $G(x)$ and $H(x)$ into Eq. (2.20) we find

$$z^2A^2(A-1)^{z-1}-zA(A-1)^{z-1}(1-B) +B(A-1)^z-1=\delta/\alpha , \quad (2.22)$$

$$z^2A^2\{3B(A-1)^{z-1}-(z-1)(A-1)^{z-2} \times[-zA^2+A(A-1)+(B-1)A]\} =B\delta\alpha^{z-1} . \quad (2.23)$$

We may solve these equations as we did for period doubling although the algebra is rather more tedious here. Again there are two solutions:

$$B=z^2A^2(A+zA-2)/\alpha^z(A-1)(A-1+zA) \quad (2.24)$$

and

$$B=A(1-z) . \quad (2.25)$$

As with period doubling, the second solution corresponds to the marginal eigenvalue $\delta=1$, again with eigenfunction

$$H(x)=G(x)-xG'(x)=1-A(1-z)|x|^z , \quad (2.13')$$

in this order of approximation. Focusing on the other solution, and after some scrutiny of Eqs. (2.24) and (2.19), we find that in this case $B=O(1/\alpha)$, as mentioned before. If one then sets $B=0$ in Eq. (2.22), one arrives at the lowest-order answer,

$$z^2A^2(A-1)^{z-1}-zA(A-1)^{z-1}-1=\delta/\alpha \quad (2.26)$$

or

$$\delta=\alpha^z-\alpha^z/zA-\alpha . \quad (2.26')$$

However, we can improve on this by not setting $B=0$. After considerable algebraic manipulation we end up with the direct relation

$$\delta=\alpha^z[1+(z-1)/(\alpha+1)z^2-1/z]+(z-1)\alpha+z+2 . \quad (2.27)$$

This connection between the universal triplication constants is surprisingly good as may be seen from Table III. To within a few percent and over a wide range of z, δ is correctly predicted from inputing the true α.

III. ANALYTIC RELATIONS FOR ASYMPTOTIC N

In an earlier paper[6] we reported the existence of an empirical relation between successive universal functions for $z=2$, namely

$$^{N+1}g(x)-1=4[^{N}g(x/2)-1] \quad (3.1)$$

when N was large. The significance of this relation is

TABLE III. Duplication constants. δ is predicted from exact α via Eq. (2.27), $\delta=\alpha^z[1+(z-1)/(\alpha+1)z^2-1/z]+(z-1)\alpha+z+2$.

z	Exact α	Exact δ	Predicted δ
1.2	635	514.5	514.4
1.5	30.1	73.1	72.4
2	9.28	55.3	54.2
3	4.37	67.0	65.9
4	3.16	86.4	85.8
5	2.60	107	109
7	2.08	154	159
10	1.73	229	236

more readily appreciated if we expand $g(x)$ in a power series,

$$^{N}g(x)=\sum_{n=0}^{\infty}{}^{N}g_n x^{2n} , \tag{3.2}$$

for we learn that the coefficients behave asymptotically as

$$\begin{aligned} &^{N}g_0=1\ ,\ \ ^{N}g_1=-2\ , \\ &^{N}g_2=c4^{1-N}\ ,\ \ ^{N}g_3=c16^{1-N}\ , \end{aligned} \tag{3.3}$$

and so on. In other words, we may expect that the terms of increasing powers of x^2 decrease very rapidly with N in a systematic way.

More generally, for arbitrary z-order maximum mappings, there is the asymptotic relation

$$^{N+1}g(x)-1=(2z)^{1/(z-1)}[^{N}g(x(2z)^{1/z(1-z)})-1] , \tag{3.4}$$

implying that in the expansion

$$^{N}g(x)=\sum_{n=0}^{\infty}{}^{N}g_n\,|\,x\,|^{zn} , \tag{3.5}$$

the coefficients behave asymptotically as

$$\begin{aligned} &^{N}g_0=1\ ,\ \ ^{N}g_1=-2\ , \\ &^{N}g_n\sim(2z)^{(1-N)(n-1)/(z-1)}\ ,\ \ n\geq 2\ . \end{aligned} \tag{3.6}$$

Once again we have a systematic suppression of higher-order terms in $|\,x\,|^z$. Furthermore, since it has been established that

$$\alpha_N\sim(2z)^{(N-1)/(z-1)} , \tag{3.7}$$

one notices that the general coefficient in the $^{N}g(x)$ expansion behaves as

$$^{N}g_n\sim\alpha_N{}^{1-n}\ ,\ \ n\geq 2\ . \tag{3.8}$$

Armed with these clues we shall set about deriving some proper analytic results about the asymptotic form of α and the first few $^{N}g_n$. Then we will extend our analysis to δ_N and the associated eigenfunction $h(x)$; in the process we will extract the asymptotic analytic connection between δ and α.

From now on we shall be dealing with the N-tupling renormalization-group functions g obeying

$$[^{N}g]^{N}(x)=-{}^{N}g(-\alpha_N x)/\alpha_N\ . \tag{3.9}$$

To save index clutter we shall delete the superscript N on g and the subscript N on α and δ; the dependence on N is to be implicitly understood hereafter. Knowing that as we go up in powers of $|\,x\,|^z$, the coefficients in the g expansion decrease dramatically with N, we shall truncate the series to at most three terms as an acceptable first approximation to g; therefore we take

$$g(x)=1+g_1\,|\,x\,|^z+g_2\,|\,x\,|^{2z} , \tag{3.10}$$

where $g_1\to -2$ as $N\to\infty$. The mth iterate of g is likewise truncated to

$$[g]^m(x)=-a_m+b_m\,|\,x\,|^z-c_m\,|\,x\,|^{2z}\ . \tag{3.11}$$

Because $[g]^{m+1}=g([g]^m)$, we have the recurrence property of the truncated coefficients,

$$\begin{aligned} &a_{m+1}\simeq 2a_m{}^z-1\ ,\ \ b_{m+1}\simeq 2za_m{}^{z-1}b_m\ , \\ &c_{m+1}\simeq z[(z-1)a_m{}^{z-2}b_m{}^2+2a_m{}^{z-1}c_m]\ . \end{aligned} \tag{3.12}$$

Here we have set $g_2\simeq 0$ for truncation purposes and $g_1\simeq -2$, an approximation which is excellent for large N.

Now the universality equation (3.9) connecting the Nth iterate of g with the starting function tells us that

$$a_N\simeq 1/\alpha\ ,\ \ b_N\simeq 2\alpha^{z-1}\ ,\ \ c_N\simeq g_2\alpha^{2z-1}\ . \tag{3.13}$$

On the other hand, from the recurrences (3.12) we have

$$b_N=2(2z)^{N-1}\prod_{j=1}^{N-1}a_j{}^{z-1}\ , \tag{3.14}$$

which immediately informs us that the universal N-plication constant depends on z and (asymptotic) N as

$$\alpha(N,z)=d(z)(2z)^{(N-1)/(z-1)}\ , \tag{3.15}$$

with

$$d(z)=\prod_{j=1}^{N-1}a_j\simeq\prod_{j=1}^{\infty}a_j\ . \tag{3.16}$$

This argument does not depend on the cycle structure (kneading sequence) although the value of the coefficient $d(z)$ does.

We may work out the asymptotic form of d by using the recurrence relation for a_n and using the asymptotic property $\alpha\to\infty$ as $N\to\infty$. Thus

$$a_{N-1}=[(1+1/\alpha)/2]^{1/z}\simeq(\tfrac{1}{2})^{1/z}\ . \tag{3.17}$$

The choice of sign connecting a_n with a_{n-1} is important here and hinges on the nature of the cycle structure, which at present is taken as the rightmost one. The asymptotic infinite product (3.16) can thus be expressed as

$$d(z)=\prod_{j=1}^{\infty}p_j\ ,\ \ p_j=[(1+p_{j-1})/2]^{1/z}\ ,\ \ p_0=0\ . \tag{3.18}$$

The infinite product is readily evaluated for $z=2$, yielding

$$d(2)=2/\pi\ . \tag{3.19}$$

This answer has been discovered by Eckmann, Epstein, and Wittwer.[7] Here, however, we see how to evaluate $d(z)$ for arbitrary z; see the Appendix for further details.

In fact, it is straightforward to evaluate $d(z)$ for cycles other than the rightmost. Let us denote the second rightmost by a prime, adding more primes as we move left across the kneading sequences. This sequence has

$$p_0'=0\ ,\ \ p_1'=-(\tfrac{1}{2})^{1/z}$$

(because of the α sign change), and

$$p_j'=[(1+p_{j-1}')/2]^{1/z}\ ,\ \ j\geq 2\ . \tag{3.18'}$$

We may again evaluate the infinite product (see Appendix) to find now

$$d'(2)=-2/3\pi\ . \tag{3.19'}$$

The same procedure works for the third rightmost cycles; this time the sequence yields

$$p_0''=0\ ,\ p_1''=-(\tfrac{1}{2})^{1/z}\ ,$$
$$p_2''=-[(1+p_1'')/2]^{1/z}\ , \tag{3.18''}$$
$$p_j''=[(1+p_{j-1}''/2]^{1/z}\ ,\ j\geq 3$$

and

$$d''(2)=2/5\pi\ . \tag{3.19''}$$

The whole process is readily continued down the kneading sequences. Numerical results for $d(z),d'(z),d''(z),\ldots$ are displayed in Table IV. Note the alternation of the sign. It is instructive to compare them against the analytical predictions (asymptotic N) for $z=2$, which are

$$d=2/\pi\ ,\ d'=-2/3\pi\ ,\ d''=2/5\pi\ ,$$
$$d'''=-2/7\pi\ ,\ \ldots\ .$$

Even for $z\neq 2$ the predicted forms for α (or d) are in excellent agreement with the data—see Sec. IV—although the precise analytic form is not so transparent and the infinite product for d can only be computed numerically.

One may also obtain an asymptotic analytical expression for g_2 in (3.10) which is helpful in deciding when numerical solutions of the functional equation (3.19) reach asymptotia. The expression which we shall derive also turns up in considering the eigenvalue equation for δ, which we discuss below. For now we look at the third relation in (3.13). By successive applications of (3.12) we find that

$$c_N\simeq g_2\alpha^{2z-1}$$
$$=z(z-1)[a_N{}^{z-2}b_N{}^2+2za_N{}^{z-2}a_{N-1}{}^{z-1}b_{N-1}{}^2$$
$$+(2z)^2(a_Na_{N-1})^{z-2}a_{N-2}{}^{z-1}b_{N-2}{}^2+\cdots]\ . \tag{3.20}$$

Then, making use of (3.12),

$$g_2=\frac{z-1}{z}\alpha^{1-2z}\left[\frac{1}{a_N{}^z}+\frac{1}{2z}\frac{1}{a_N{}^{z-1}a_{N-1}{}^z}+\frac{1}{(2z)^2}\frac{1}{a_N{}^{z-1}a_{N-1}{}^{z-1}a_{N-2}{}^z}+\cdots\right]$$
$$\equiv c(z)/\alpha^{2z-1}\ ,$$

with

$$c(z)\equiv\frac{z-1}{z}\left[\frac{1}{a_N{}^z}+\frac{1}{2za_N{}^{z-1}a_{N-1}{}^z}+\frac{1}{(2z)^2a_N{}^{z-1}a_{N-1}{}^{z-1}a_{N-2}{}^z}+\cdots\right]\ . \tag{3.21}$$

In the asymptotic limit the series for $c(z)$ extends to infinity and the form of successive terms follows immediately from the previous considerations; thus the sum depends on the cycle structure. It is readily evaluated for $z=2$ (see Appendix) and we find, going across the kneading sequences, starting from rightmost cycles, that

$$c(z)=4/\pi\ ,\ c'(z)=4/3\pi\ ,\ c''(z)=4/5\pi\ ,\ \ldots\ . \tag{3.22}$$

The numerical results for a variety of z are to be found in Table IV and we again find excellent agreement with "experimental" values in the asymptotic regime.

At this point we are ready to tackle the eigenvalue equation for δ for the case of N-tupling. This eigenvalue equation can be derived by a generalization of the technique used by Feigenbaum[1] for period doubling; it reads

$$h([g]^N)+h([g]^{N-1})g'([g]^N)+h([g]^{N-2})[g]^{2\prime}([g]^{N-1})+\cdots+h(g)[g]^{N-1\prime}([g]^2)+h(x)[g]^{N\prime}(g)=-\delta h(-\alpha x)/\alpha\ . \tag{3.23}$$

Again we may expand h as a power series in $|x|^z$ (truncated to $m+1$ terms),

$$h(x)=\sum_{n=0}^{m}h_n|x|^{zn}\ , \tag{3.24}$$

starting with $h_0=1$ as the normalization condition. One may then obtain a numerical solution of (3.23) to order $m+1$ if a solution g of the functional equation (3.9) has already been found—we shall amplify on this later. Here we focus on extracting an algebraic relation between δ and α for large N in a way similar to that in Sec. II for $N=2,3$. The results are very accurate asymptotically owing to the fact that the coefficients h_n die off very rapidly

TABLE IV. Numerically determined coefficients c and d for rightmost, next to rightmost, etc., sequences arising in Eqs. (3.15) and (3.21). Only for $z=2$ can the answers be found analytically.

z	d	$-d'$	d''	c	$-c'$	c''
1.2	0.397 556	0.086 469	0.042 302	0.495 879	0.123 933	0.140 931
1.5	0.514 739	0.136 900	0.074 042	0.917 821	0.266 137	0.228 880
2	0.636 620	0.212 207	0.127 324	1.273 24	0.424 413	0.254 648
3	0.759 931	0.329 187	0.221 182	1.568 60	0.593 029	0.205 138
4	0.821 227	0.413 814	0.296 469	1.695 51	0.677 545	0.144 841
5	0.857 706	0.477 735	0.357 220	1.765 36	0.726 829	0.093 813

with n; specifically,

$$h_n \sim 1/\alpha^n \tag{3.25}$$

signifies that we are justified in truncating the series (3.24) to lowest order as we did earlier for $N=2,3$ (though in the latter case it was done to keep the algebra simple). Our plan then is to expand both sides of (3.23) only to order $|x|^z$ and equate coefficients of x^0 and $|x|^z$ on each side.

The general term on the left-hand side of (3.23) is of the form

$$h([g]^{m-1}[g]^{n\prime}([g]^m) \text{ with } n+m=N+1 .$$

Because of our truncation, we take, from (3.11),

$$[g]^m = -a_m + b_m |x|^z + O(|x|^{2z}) ,$$

and therefore

$$h([g]^{m-1}) = 1 - h_1 a_{m-1}{}^z + z h_1 a_{m-1}{}^z b_{m-1} |x|^z . \tag{3.26}$$

Note that for almost all m (except for m near N) $a_{m-1}=1$ to an excellent approximation.

The expression (3.26) multiplies $[g]^{n\prime}([g]^m)$ which we now evaluate

$$\begin{aligned}[g]^{n\prime}([g]^m) &= g'([g]^m)\cdots g'([g]^N)\\ &\simeq (2z)^n |a_m - b_m|x|^z|^{z-1}\cdots|a_N - b_N|x|^z|^{z-1}\\ &= (2z)^n\left[\prod_{j=m}^{N} a_j{}^{z-1}\right]\left[1-|x|^z(z-1)\left[\frac{b_N}{a_N}+\cdots+\frac{b_m}{a_m}\right]\right] .\end{aligned} \tag{3.27}$$

For large N and the majority of m values, we may use (3.16) to write

$$\prod_{j=m}^{N} a_j{}^{z-1} \simeq [d(z)]^{z-1} . \tag{3.16'}$$

The coefficient of $|x|^z$ within the large square brackets in (3.27) may be recast as

$$\begin{aligned}b_N/a_N + \cdots + b_m/a_m &= \alpha^{z-1} z^{-1}(z-1)(1/a_N{}^z + 1/2z a_N{}^{z-1} a_{N-1}{}^z + \cdots)\\ &= c(z)\alpha^{z-1}\end{aligned} \tag{3.28}$$

for large N and most m. [Here, for simplicity, we have ignored the contribution $b_{m-1}/a_{m-1} + \cdots + b_1/a_1$, which is nondominant for large N but which can be included in a more careful treatment, the result of which we quote in Eq. (3.30) below.] Thus the right-hand side of (3.27) asymptotically tends to

$$\alpha^{z-1}(2z)^{n-N}[1 - c(z)\alpha^{z-1}|x|^z] .$$

Multiplying this by (3.26), we obtain the typical term in (3.23) to be

$$\begin{aligned}h([g]^{m-1})[g]^{n\prime}([g]^m) \simeq (1-h_1)\alpha^{z-1}(2z)^{n-N}&\\ + h_1\alpha^{z-1}|x|^z&\\ \times\left[1-\frac{(1-h_1)c(z)\alpha^{z-1}}{(2z)^{N-n}}\right]&.\end{aligned} \tag{3.29}$$

Returning to (3.23), summing over all values of m from 0 to N, but paying particular attention to the case where $m=0$ ($n=N$) because the term differs slightly from (3.29), and equating coefficients on each side, we arrive at the following pair of relations:

$$\alpha^{z-1}\left[\frac{2-2z}{2z-1} - \frac{h_1}{2z-1}\right] = -\frac{\delta}{\alpha} , \tag{3.30a}$$

$$(N+1)h_1\alpha^{z-1} + N(1-h_1)\frac{(2z-2)}{2z-1}\alpha^{z-1} + c(z)\alpha^{2z-2} - \frac{(1-h_1)c(z)\alpha^{2z-2}}{2z-1} = \delta h_1 \alpha^{z-1} . \tag{3.30b}$$

There are two solutions for h_1 from which δ is derived.

(i) The marginal eigenvalue case

$$h_1 = 2z-2 \text{ with } \delta = 1 . \tag{3.31}$$

This is expected since the solution $\delta=1$ is anticipated, when to this order of approximation

$$\begin{aligned}h(x) &= g(x) - x g'(x)\\ &= 1 - 2|x|^z + 2xz|x|^{z-1}\\ &= 1 + (2z-2)|x|^z .\end{aligned}$$

(ii) The eigenvalue of interest ($\delta > 1$),

$$h_1 = -c(z)/\alpha - N/\alpha^z$$

with

$$\delta = \alpha^z[1 + c(z)/2\alpha(z-1)](2z-2)/(2z-1) . \tag{3.32}$$

Remembering that $\alpha \gg 1$, a reasonable approximation for large N is

$$h_1 = -c(z)/\alpha \text{ with } \delta/\alpha^z = (z-1)/(z-\tfrac{1}{2}) . \tag{3.32'}$$

As z increases we shall see (Sec. IV) that α grows less rapidly with N than for small z. In that case the sharper prediction (3.32) for δ agrees slightly better with the numerical data than the limiting prediction (3.32'). In any event, comparison of (3.32') with (3.21) shows that, asymptotically,

$$g_2 \simeq -h_1 \simeq c(z)/\alpha \ . \tag{3.33}$$

It is worth remarking that the analytical form (3.32) for δ depends only on the cycle structure via α, which itself is determined through the coefficient $d(z)$ in (3.15). Thus the $z=2$ relation, $\delta/\alpha^2 \simeq \frac{2}{3}$ holds *irrespective of the kneading sequence,* as we have noted before.

In the next section we shall present very convincing numerical evidence to support the analytical forms just derived for α, δ, g_2, and h_1 for a wide range of z, N, and for three distinct cycle structures. This evidence has been amassed by high-precision solutions[10] to the functional eigenvalue equations (3.9) and (3.23).

IV. NUMERICAL SOLUTION OF EIGENVALUE FUNCTIONAL EQUATIONS

A. α eigenvalue

The N-tupling renormalization-group functions determining α_N through Eq. (3.9) were studied numerically in a similar fashion to that described by Feigenbaum[1] for period doubling. For large α_N, the coefficients g_n in the Taylor-series expansion of $g(x)$ in Eq. (3.5) decreased very rapidly with n. In order to reliably determine $g_2, g_3, \ldots$ via such numerical techniques, it proved necessary to enhance the importance of the higher-order terms somehow. Two techniques were employed.

(a) Enlarging the interval on which $g(x)$ was evaluated from [0,1] to, say, [0,10]. The higher-order g_n were thus successively weighted more heavily by up to factors of 10^z. Since $g_n/g_{n+1}=O(\alpha_N)$, which was frequently of order 1000 or more, a weighting of this magnitude was often needed.

(b) Rescaling $g(x)$. Conventionally, $g(x)$ is normalized to $g(0)=1$. However, given any solution $g(x)$ of (3.9), $\lambda g(x/\lambda)$ is also a solution, which is, however, normalized to λ at $x=0$. Because

$$\lambda g(x/\lambda)=\sum_n g_n |x|^{zn}\lambda^{1-zn} \ , \tag{3.5'}$$

we see that by choosing $\lambda=0.1$, say, successive g_n are enhanced by a constant factor 10^z. Such a trick proved essential for reliable computation even of g_2 for $z=1.2$ for high-order cycles because of the astronomical value of α.

We should note that, especially for $z>2$, some difficulty was encountered in achieving reasonable accuracy in our approximation for the various $g(x)$. A significant improvement was made by *not* choosing a linear spacing x_1/I $(i=1,2,\ldots,I)$ but rather choosing a spacing of the type $x_i=(i/I)^{1/z}$ $(i=1,2,\ldots,I)$. The procedure makes good sense since the function has a zth order maximum at $x=0$. A nonlinear choice of the x_i as indicated insures that $g(x)$ is evaluated at equally spaced values of the function. (A related point was made by Feigenbaum *et al.*[5] with respect to the solution for the circle map for $z=3$.) As the order of the cycle increased, particularly for large z, it became necessary to adopt double precision to avoid accumulation of rounding errors due to the enormous number of iterations involved.

The numerical solution of Eq. (3.9) proved to be relatively straightforward for rightmost cycles. Given an approximate solution $g(x)=1-2|x|^z$ and a value of α reasonably close to the anticipated value, the solution converged quite swiftly. The values thus determined were in excellent agreement with those determined previously from cycle data, at least for $z=<4$. However, as the N increases, the number of solutions to the functional equation increases exponentially fast and it is not a trivial exercise to locate solutions for cycles other than rightmost, which also have $g(x)=1-2|x|^z$ to an excellent approximation, but which have quite different α values.

In order to cope with this situation, earlier experience of cycle studies via the map

$$f(x)=1-\mu|x|^z$$

was put to good use. These previous investigations had provided points of accumulation $\mu_{n\infty}$ for say the $5a, 5b, 5c$ sequences. These gave good first-order approximations to $g(x)$ through

$$g(x)=1-\mu_{n\infty}|x|^z \ .$$

In addition, the associated values of α were known at least for some z. Armed with this information, rapid convergence to the appropriate $g(x)$ was achieved. When trying to solve Eq. (3.9) for large N values that had not previously been examined, it was necessary to make a reasonable estimate of the relevant point of accumulation using the asymptotic formula

$$(2-\mu_{n\infty})/(2-\mu_{n+1\infty})=2z \ . \tag{4.1}$$

This makes it relatively easy to solve (3.9) for say the $5a$ cycle: the lowest-order approximation to the relevant $g(x)$ yields a very close value for $g_1=-\mu_{n\infty}$. This, in turn, provides an approximate value for the g_1 coefficient for the $6c$ cycle through (4.1), whereupon a more precise value is pinpointed through numerical solution of (3.9).

We stopped increasing N it became clear that the asymptotic regime had been reached. Usually this happened around $N=10$ or more (depending on how accurate we wanted our solutions to be).

B. δ eigenvalue

Once α and g are determined, one may proceed to the solution of δ and h from Eq. (3.23). Here we again essentially follow Feigenbaum.[1] A finite-dimensional approximation to the eigenvalue problem is found by numerical techniques. Like bifurcation, a spectrum of eigenvalues is obtained, only one of which exceeds 1 and corresponds to δ. The results were in excellent agreement with those obtained by the more tedious process of studying N-furcation. An extra eigenvalue (the marginal case) close to one was also found; its appearance and the smallness of its deviation from unity served as a useful check of the goodness of fit of $g(x)$.

A further useful check on the numerical work is the asymptotic relation $g_2=-h_1$ derived in Sec. III. It gives a strong indication of how closely the current solutions approach the asymptotic regime. In Table V we exhibit the systematic approach of the solutions to asymptotia for rightmost cycles. Values of α and δ are quoted to a few

TABLE V. Exact values of δ, α, g_2, and h_1 obtained by solving the functional equations compared with the asymptotic values predicted by Eqs. (3.15), (3.21), and (3.33) for rightmost cycles and (a) $z=1.2$, (b) $z=1.5$, (c) $z=2.0$, (d) $z=3$, (e) $z=4$, and (f) $z=5$.

N	Exact δ	δ_{asy}	Exact α	α_{asy}	Exact g_2	Exact $-h_1$	Asymptotic $g_2=-h_1$
				(a)			
2	3.132	18.05	5.374	31.7	0.0175	0.1794	0.0157
3	514.5	3449	634.7	2521	4.78×10^{-4}	0.0016	1.97×10^{-4}
4	2.46×10^{5}	6.59×10^{5}	9.80×10^{4}	2.01×10^{5}	4.27×10^{-6}	8.16×10^{-6}	2.5×10^{-6}
5	7.63×10^{7}	1.26×10^{8}	1.11×10^{7}	1.6×10^{7}	6.03×10^{-8}	$5.8/\times10^{-8}$	3.10×10^{-8}
7	4.07×10^{12}	4.60×10^{12}	9.28×10^{10}	1.01×10^{11}	5.3×10^{-12}	5.8×10^{-12}	4.9×10^{-12}
9	1.63×10^{17}	1.68×10^{12}	6.30×10^{14}	6.43×10^{14}	7.9×10^{-16}	8.0×10^{-10}	7.7×10^{-16}
				(b)			
2	3.801	4.98	3.389	4.63	0.0562	0.2698	0.198
3	73.12	134.6	30.10	41.7	0.0198	0.0377	0.0220
4	2720	3635	323.4	375	0.0025	0.0033	0.0024
5	86 913	98 131	3175	3377	2.79×10^{-4}	3.09×10^{-4}	2.72×10^{-4}
7	7.02×10^{7}	7.15×10^{7}	2.71×10^{5}	2.74×10^{5}	3.38×10^{-6}	3.43×10^{-6}	3.36×10^{-6}
9	5.20×10^{10}	5.22×10^{10}	2.21×10^{7}	2.22×10^{7}	4.15×10^{-8}	4.16×10^{-8}	4.14×10^{-8}
				(c)			
2	4.669	4.32	2.503	2.543	0.105	0.326	0.5
3	55.25	69.2	9.277	10.2	0.094	0.128	0.125
4	981.6	1107	38.82	40.7	0.030	0.033	0.031
5	16 931	17 707	160	163	7.79×10^{-3}	8.00×10^{-3}	7.81×10^{-3}
7	4.51×10^{6}	4.53×10^{6}	2603	2608	4.89×10^{-4}	4.90×10^{-4}	4.90×10^{-4}
9	1.16×10^{9}	1.16×10^{9}	41 715	41 722	3.05×10^{-5}	3.05×10^{-5}	3.05×10^{-5}
				(d)			
2	6.085	5.16	1.928	1.86	0.109	0.291	0.843
3	66.82	75.8	4.364	4.56	0.254	0.270	0.344
4	967	1115	10.63	11.17	0.137	0.134	0.140
5	14 859	16 380	26.45	27.35	0.058	0.056	0.057
7	3.43×10^{6}	3.54×10^{6}	162.5	164.1	9.61×10^{-3}	9.60×10^{-3}	9.57×10^{-3}
9	7.59×10^{8}	7.64×10^{8}	982.4	984.9	1.60×10^{-3}	1.60×10^{-3}	1.59×10^{-3}
				(e)			
2	7.285	6.238	1.690	1.643	0.312	0.540	1.032
3	85.79	99.8	3.152	3.285	0.368	0.349	0.516
4	1275	1597	6.192	6.570	0.246	0.231	0.258
5	20 992	25 550	12.49	13.14	0.134	0.126	0.129
7	5.96×10^{6}	6.54×10^{6}	51.32	52.56	0.033	0.032	0.032
9	1.62×10^{9}	1.67×10^{9}	208.4	210.2	0.0081	0.0081	0.0081
				(f)			
2	8.345	7.34	1.556	1.52	0.517	0.864	1.16
3	107.2	130.5	2.600	2.71	0.428	0.377	0.651
4	1701	2320	4.520	4.82	0.358	0.309	0.366
5	30 514	41 261	8.059	8.56	0.216	0.196	0.206
7	1.09×10^{7}	1.30×10^{7}	26.16	27.1	0.068	0.065	0.065
9	3.80×10^{9}	4.13×10^{9}	84.36	85.8	0.021	0.021	0.021

places (more accurate values can be obtained upon request) for easy reference. A similar tabulation is made in Table VI for next to rightmost sequences for selected z; and the exercise is repeated for the third kneading sequence in Table VII. We have suppressed a lot more information in the interest of clarity and brevity. The interested reader can approach us directly for more details pertaining to other z,N values.

V. SUMMARY

In Sec. II we addressed the question of solving the functional eigenvalue equations for α and δ to lowest order by purely algebraic methods, both for period doubling and tripling, with z arbitrary. Apart from the marginal eigenvalue $\delta=1$ there is a solution $\delta>1$ related to α. For period doubling it is given by the simple formula (2.15),

TABLE VI. Similar to Table V but for the second rightmost set, and (a) $z=1.5$, (b) $z=2$, (c) $z=3$, and (d) $z=4$.

N	Exact δ'	$\delta'_{\rm asy}$	Exact α'	$\alpha'_{\rm asy}$	Exact g_2	Exact $-h_1$	Asymptotic $g_2=-h_1$
				(a)			
5	4984	13 460	-514.3	-898	6.98×10^{-4}	1.32×10^{-3}	2.96×10^{-4}
7	8.62×10^{6}	9.82×10^{6}	$-67\,771$	$-72\,754$	4.24×10^{-6}	4.56×10^{-6}	3.66×10^{-6}
9	7.02×10^{9}	7.15×10^{9}	-5.84×10^{6}	-5.89×10^{6}	4.68×10^{-8}	4.65×10^{-8}	4.52×10^{-8}
				(b)			
5	1287	1967	-45.8	-54.3	0.0106	0.0151	0.0078
7	4.87×10^{5}	5.04×10^{5}	-858	-869	5.01×10^{-4}	5.14×10^{-4}	4.88×10^{-4}
9	1.29×10^{8}	1.29×10^{8}	-13893	-13907	3.06×10^{-5}	3.06×10^{-5}	3.05×10^{-5}
				(c)			
5	1106	1331	-11.25	-11.85	0.042	0.050	0.050
7	2.82×10^{5}	2.88×10^{5}	-70.6	-71.1	0.0081	0.0080	0.0083
9	6.18×10^{7}	6.21×10^{7}	-426.0	-426.6	1.38×10^{-3}	1.38×10^{-3}	1.39×10^{-3}
				(d)			
5	1418	1647	-6.40	-6.62	0.0529	0.0599	0.102
7	4.03×10^{6}	4.22×10^{6}	-26.2	-26.5	2.26×10^{-2}	2.24×10^{-2}	2.56×10^{-2}
9	1.06×10^{8}	1.08×10^{8}	-105.4	-105.9	6.23×10^{-3}	6.21×10^{-3}	6.40×10^{-3}

for tripling by (2.27). Although these relations are exact as $z\to1$, they happen to have a much wider range of validity (to within 5% up to $z=10$). These gave us confidence that a first-order approach can be extended to period N-tuplings and can provide a relation between δ and α for a large range of z.

In Sec. III we tackled the solution of the renormalization-group equation and found that $\alpha(z,N)$ has the exact form (3.15) for asymptotic N. The coefficient $d(z)$ is readily calculated numerically for arbitrary z but for the case $z=2$ can be found analytically; d assumes the simple values $2/\pi, -2/3\pi, 2/5\pi$ etc., depending on the kneading sequence. The associated eigenfunction g can also be calculated asymptotically and we found that the coefficients g_2, g_3 fell off as powers of α^{-1}, thus justifying the truncation of the series expansions for large α. Given the g we went on to tackle the eigenvalue problem for δ and associated eigenfunction h. The eigenvalue δ of interest was given to leading order in terms of α by (3.32′), *irrespective of the kneading sequence.* The universal function h also has x-expansion coefficients which fell off as powers of α and we were thus able to prove (3.33).

TABLE VII. Similar to Table V but now for the third rightmost set and (a) $z=1.5$, (b) $z=2$, (c) $z=3$, and (d) $z=4$.

N	Exact δ''	$\delta''_{\rm asy}$	Exact α''	$\alpha''_{\rm asy}$	Exact g_2	Exact $-h_1$	Asymptotic $g_2=-h_1$
				(a)			
5	644.2	5354	129.2	486	$-0.000\,65$	0.0029	0.000 47
7	2.96×10^{6}	3.90×10^{6}	3.35×10^{4}	3.93×10^{4}	6.00×10^{-6}	7.44×10^{-6}	5.82×10^{-6}
9	2.74×10^{9}	2.84×10^{9}	3.12×10^{6}	3.19×10^{6}	7.24×10^{-8}	7.4×10^{-8}	7.18×10^{-8}
				(b)			
5	255.5	708	20.1	32.6	-0.0117	0.0034	0.0078
7	1.67×10^{5}	1.81×10^{5}	503.9	521.5	4.54×10^{-4}	4.96×10^{-4}	4.88×10^{-4}
9	4.61×10^{7}	4.64×10^{7}	8323	8344	3.04×10^{-5}	3.06×10^{-5}	3.05×10^{-5}
				(c)			
5	240.4	404	6.72	7.96	-0.087	-0.069	0.0258
7	85 230	87 237	47.4	47.8	2.95×10^{-3}	3.08×10^{-3}	4.29×10^{-3}
9	1.88×10^{7}	1.88×10^{7}	286.4	286.7	6.84×10^{-4}	6.84×10^{-4}	7.16×10^{-4}
11	4.07×10^{9}	4.07×10^{9}	1720	1720	1.18×10^{-4}	1.18×10^{-4}	1.19×10^{-4}
				(d)			
5	292	434	4.29	4.74	-0.207	-0.207	0.031
7	1.09×10^{5}	1.11×10^{5}	18.89	18.97	-4.7×10^{-4}	-3.6×10^{-4}	0.0076
9	2.83×10^{7}	2.84×10^{7}	75.8	75.9	0.0014	0.0014	0.0019
13	1.86×10^{12}	1.86×10^{12}	1214	1214	1.18×10^{-4}	1.17×10^{-4}	1.19×10^{-4}

The details of the numerical solution of the functional equations were sketched out in Sec. IV. Representative tables were compiled which gave excellent agreement with the earlier asymptotic predictions for large N (usually around 10). Indeed, the theoretical predictions provided a guide of how far to push the numerical work since it was not difficult to achieve an accuracy of better than 0.1% for any of the above quantities over a wide range of z and many cycle structures. The surprising thing is that the asymptotic predictions are quite accurate for relatively small N, say, for $z=2$ and rightmost cycles. Thus the investigation proved to be more fruitful than we originally anticipated. It will guide us in future work on universality for complex mappings, where we again hope to discover relations between complex δ and α.

ACKNOWLEDGMENTS

One of us (B.G.K.) wishes to thank Mr. A. M. Aitken for extensive programming assistance. The calculations were carried out in double precision (28 decimal places) on a Control Data Corporation Cyber computer.

APPENDIX

We shall make repeated use of the identity

$$\prod_{i=1}^{\infty}\cos(\theta/2^i)=(\sin\theta)/\theta\ , \tag{A1}$$

as well as the identity obtained by taking the logarithm of (A1) and differentiating both sides with respect to θ, namely

$$\sum_{i=1}^{\infty}2^{-i}\tan(\theta/2^i)=1/\theta-\cot\theta\ . \tag{A2}$$

In Sec. III we showed that asymptotically α was given by (3.15) with the coefficient $d(z)$ in (3.16) determined by the recurrences (3.18). One elegant way of solving these recurrences for $z=2$ is to define $p_0=\cos\theta$ and $p_i=\cos\theta_i$. By so doing, Eq. (3.18) can be immediately solved to

$$\theta_i=\theta/2^i\ . \tag{A3}$$

Consequently the infinite product in (3.16) reduces to

$$d(2)=\prod_{i=1}^{\infty}\cos(\theta/2^i)=(\sin\theta)/\theta\ , \tag{A4}$$

by virtue of Eq. (A1) . For the rightmost cycle, the relevant sequence is is $p_0=0$, $p_1=-1/\sqrt{2}$, etc., so that the correct choice for the initial θ is $\theta=\pi/2$. In this case

$$d(2)=[\sin(\pi/2)]/(\pi/2)=2/\pi\ , \tag{A5}$$

as has previously been observed by Eckmann *et al.*[7]

For the second rightmost cycle, the relevant sequence begins with $p_0=0$, $p_1=-1/\sqrt{2}$, $p_2=(1-1/\sqrt{2})^{1/2}$, etc. This time the correct choice for θ is $3\pi/2$, whereupon

$$d'(2)=[\sin(3\pi/2)]/(3\pi/2)=-2/3\pi\ . \tag{A6}$$

For the third rightmost cycle, the appropriate sequence begins with $p_0=0,p_1=-1/\sqrt{2},p_2=-(1-1/\sqrt{2})^{1/2},p_3=+\dots$, etc., so that the correct choice for θ is now $\theta''=5\pi/2$. Here

$$d''(2)=[\sin(5\pi/2)]/(5\pi/2)=-2/5\pi\ , \tag{A7}$$

and so on for other cycles.

In Sec. III we also showed that asymptotically g_2 and h_1 were given by formula (3.33) with the constant $c(z)$ determined via (3.21). Alternatively, expressed in terms of the p_i, we can find $c(2)$ as

$$\begin{aligned}2c(2)&=(p_1^{-2}+p_1^{-1}p_2^{-2}/4+p_1^{-1}p_2^{-1}p_3^{-2}/16\ \cdots)\\&=\left[\prod_{j=1}^{\infty}p_j\right]^{-1}\left[\left[\prod_{2}^{\infty}p_i\right]\Big/p_1+\left[\prod_{3}^{\infty}p_i\right]\Big/4p_2+\left[\prod_{4}^{\infty}p_i\right]\Big/16p_3+\cdots\right]\\&=\lim_{\theta\to\pi/2}(4/\sin\theta)[\tan(\theta/2)+2^{-1}\tan(\theta/4)+4^{-1}\tan(\theta/8)+\cdots]\\&=\lim_{\theta\to\pi/2}(4/\sin\theta)(1/\theta-\cot\theta)=8/\pi\ .\end{aligned} \tag{A8}$$

Thus $c(2)=4/\pi$. For the next two cycles one must take the limit as $\theta\to3\pi/2$ and $5\pi/2$, respectively, yielding $c'(2)=-4/3\pi$ and $c''(2)=-4/5\pi$. One may continue in a similar fashion for other kneading sequences.

By combining (A5) with (A8) one has

$$g_2=-h_1=c(z)/\alpha=c(z)/d(z)4^{n-1}=2^{3-2n}\ , \tag{A9}$$

independent of the sequence (in the asymptotic limit, be it understood). This result is peculiar to $z=2$ and is in good agreement with the numerical data. See Tables V–VII.

[1]M. J. Feigenbaum, J. Stat. Phys. **21**, 669 (1979).
[2]A. Libchaber, C. Lanouche and S. Fauve, J. Phys. (Paris) Lett. **43**, L211 (1982).
[3]V. Franceschini and C. Tebaldi, J. Phys. **21**, 707 (1979).
[4]J. Stavans, F. Heslot, and A. Libchaber, Phys. Rev. Lett. **55**, 596 (1985).
[5]M. J. Feigenbaum, L. P. Kadanoff, and S. J. Shenker, Physica **5D**, 370 (1982); D. Rand, S. Ostlund, J. Sethna, and E. Siggia,

Phys. Rev. Lett. **49**, 132 (1982); Physica **6D**, 303 (1984).
[6]R. Delbourgo, B. G. Kenny, and W. Hart, Phys. Rev. A **31**, 514 (1985); Aust. J. Phys. (to be published).
[7]J. P. Eckmann, H. Epstein, and P. Wittwer, Commun. Math. Phys. **93**, 495 (1984).
[8]This was observed by Feigenbaum, Ref. 1.
[9]R. Delbourgo and B. G. Kenny, Aust. J. Phys. **38**, 1 (1985).
[10]See also the work by A. Mo and P. C. Hemmer, Phys. Scr. **29**, 296 (1984), which treats the same iterative map though chiefly for z near 1 or z very large.

Islands of stability and complex universality relations

R. Delbourgo and P. Hughes
Department of Physics, University of Tasmania, Hobart, Australia 7005
B. G. Kenny
Department of Physics, University of Western Australia, Perth, Western Australia 6009

(Received 10 April 1986; accepted for publication 10 September 1986)

For complex mappings of the type $z \to \lambda z(1-z)$, universality constants α and δ can be defined along islands of stability lying on filamentary sequences in the complex λ plane. As the end of the filament is approached, asymptotic values $\alpha_N \sim \lambda_\infty^{N-1}$, $\delta_N/\alpha_N^2 \sim 1$ are attained, where $\mu_\infty = \lambda_\infty(\lambda_\infty - 2)/4$, is associated with the limiting form of the universal function for that sequence, $g(z) = 1 - \mu_\infty z^2$. These results are complex generalizations of the real mapping case (applying to tangent bifurcations and windows of stability) where $\mu_\infty = 2$ and $\delta/\alpha^2 \to \frac{2}{3}$ correspond to the filament running along the real axis.

I. INTRODUCTION

In many branches of physics there is a real gain in understanding by extending the analysis of real dynamical variables into the complex plane. For dynamical nonlinear maps in one variable the advantages of such a generalization are much less obvious, but this has not hindered several researchers[1-3] from pursuing these studies. In this report we wish to point out the existence of complex universality relations between the analogs of the Feigenbaum constants α and δ as they pertain to various "kneading" sequences in the complex plane, thereby generalizing earlier work of ours[4] applying to the real case.

II. THEORY

We focus as usual on the Julia–Fatou mapping

$$z \to \lambda z(1-z), \tag{1a}$$

or

$$z \to 1 - \mu z^2, \tag{1b}$$

with $\mu = \lambda(\lambda - 2)/4$ providing the equivalence between (1a) and (1b). The detailed fractal nature of these maps have been highlighted by Mandelbrot[1] and Cvitanovic and Myrheim[5] and many fascinating properties have emerged. One of the most interesting of these[2] is the fact that "islands of stability" belonging to the Mandelbrot set are all connected to the main cactus by filaments. As a special instance, the real filament (with $\text{Im}\,\lambda = \text{Im}\,\mu = 0$) will connect the windows of stability that are seen in the counterpart real problem; that particular filament ends at $\lambda_\infty = 4$ or $\mu_\infty = 2$. If one follows the kneading sequences of the same type along that real filament then one can discover that as the end is approached, the universal constants associated with N-replication attain the asymptotic values,

$$\alpha \sim 4^{N-1}, \quad \delta/\alpha^2 \to \tfrac{2}{3}.$$

Cvitanovic and Myrheim[3,5] have shown that subcacti sprouting from the main Mandelbrot cactus allow one to define complex Feigenbaum constants which characterize their universal rates of shrinkage as one follows a particular Farey sequence. They have provided extensive lists of δ and α values connected with Farey sets m/n. In this note we will be examining other sequences which represent the analog of tangent bifurcations for real nonlinear maps. These sequences correspond to isolated islands of stability associated with N-replication that lie *outside* the main Mandelbrot cactus and its many leaves. It will be recalled that there exist as many as 2^N superstable N-cycles for quadratic maximum maps *some of which* touch the main cactus (and thus belong to the Farey sets). The islands we are referring to are the ones which do *not* touch the cactus, and they are generally quite small in size. In fact some are so miniscule that they are not at all easy to locate without some finesse. In Table I we have listed all of the island locations up to $N = 6$ for the readers' convenience, while in Table II we give the main islands for $N = 7$ and 8 that are relevant to our work, in that they lie on certain filaments. Also listed are some of the computed α and δ values for those N-plications as one follows a filament (see Figs. 1 and 2), which were determined by solving the complex versions of the functional equations.

III. ASYMPTOTIC RELATIONS

The values of α and δ are important only insofar as they relate to a particular kneading sequence. (The reader can convince himself that the asterisked sets do belong to the same sequences by tracing out the limit cycles in z.) There are of course an infinite number of such sequences or filaments in the complex plane. Our point is that for each of these sequences one can find an asymptotic δ–α relationship, showing that the two constants are not really independent of one another. In order to establish the α–δ relation, we turn to the renormalization group equations for them; they apply to complex just as well as real maps. The crucial observation is that as one approaches the end of a filament the renormalization group function $\overset{N}{g}(z)$ that determines α, namely

$$[\overset{N}{g}]^{N}(z) = -\overset{N}{g}(-\alpha z)/\alpha, \tag{2}$$

tends to the limiting quadratic form ($N \to \infty$),

$$\overset{\infty}{g}(z) = 1 - \mu_\infty z^2, \tag{3}$$

where μ_∞ labels the end point of the filament. This result is readily verified by solving Eq. (2) for successively large values of N and noticing how quickly the coefficients g^n in the

 0022-2488/87/010060-04$02.50 © 1987 American Institute of Physics

TABLE I. Superstable λ-values and cycle constants. A single asterisk refers to the rightmost real filamentary sequence, a double asterisk to the main complex filament, and a triple asterisk represents a subsidiary filament sequence. A # connotes a leaf on the main cactus. There exist other complex conjugate sequences and constants as well as reflected ones (where $\lambda \to 2 - \lambda$), which have been deliberately omitted.

Cyc. No.	Re λ	Im λ	Re α	Im α	Re δ	Im δ
2*	2.236 07	0	2.503	0	4.669	0
3*	3.831 87	0	9.277	0	55.26	0
3	2.552 65	0.959 46	2.097	− 2.358	4.600	8.981
4*	3.961 56	0	38.82	0	981.6	0
4#***	1.998 77	1.061 43	1.135	− 3.260	− 0.853	18.11
4**	2.741 22	1.185 66	10.56	− 5.375	100.4	− 69.34
5	3.743 91	0	20.13	0	255.5	0
5	3.905 71	0	45.80	0	1287.1	0
5*	3.990 27	0	160	0	16 931	0
5	1.667 62	0.988 55	0.380	− 3.554	− 9.520	26.37
5	2.841 45	0.611 22	3.874	− 2.181	18.97	14.56
5***	2.041 61	1.233 69	9.554	− 9.078	114.3	− 184.7
5	2.627 10	1.212 69	− 5.85	− 17.47	281.9	54.46
5**	2.808 89	1.216 51	40.43	− 2.526	1399	253
5	3.473 87	0.307 47	5.583	22.09	205.4	287.1
6	3.627 56	0	20.9	0	218.4	0
6	3.844 57	0	115.0	0	8508	0
6	3.937 54	0	207.6	0	28 020	0
6*	3.997 58	0	645.0	0	279 130	0
6	1.485 25	0.889 66	− 0.160	− 3.626	− 20.66	0
6	3.365 02	0.203 24				
6	2.610 95	1.068 40				
6	2.624 84	1.256 12	27.65	64.54	− 1458	4208
6	2.781 41	1.232 49	− 8.274	− 69.98	− 4270	462.9
6**	2.830 89	1.217 39	124.2	41.97	8769	12142
6	2.960 83	0.676 23	19.92	− 7.643	421.8	− 188.9
6	1.673 45	1.107 62	8.284	− 11.43	91.37	− 342.8
6***	2.080 40	1.267 61	37.22	− 8.569	1772	34.60
6	1.974 85	1.239 54	− 6.824	− 23.89	− 777.4	− 5.382

expansion,

$$g(z) = \sum_{n=0}^{\infty} \overset{N}{g}_n z^{2n} = 1 + \overset{N}{g}_1 z^2 + \cdots,$$

settle down to the limiting case (3). Of course the particular case of the rightmost cycle sequence for real maps, when $\mu_\infty = 2$, is now well known.[6] In much the same way, the renormalization group function $h(z)$, which determines δ,

$$\sum_{m=0}^{\infty} h([g]^{N-m})[g]^{m\prime}([g]^{N-m+1}) = -\delta/\alpha, \qquad (2')$$

also has coefficients h_n (in a power series in z^{2n}), which die off rapidly with n. The technique[4] for deriving a relation between α and δ makes great use of these facts.

One truncates the m"th iterate of g (here we drop the N superscript),

$$[g]^m(z) = -a_m + b_m z^2 + \cdots,$$

and uses the recurrence property $[g]^{m+1} = g([g]^m)$ to derive the recurrence relations of the truncated coefficients,

$$a_{m+1} = \mu_\infty a_m^2 - 1, \quad b_{m+1} = 2\mu_\infty a_m b_m. \qquad (4)$$

These formulas become more and more accurate for large N. Referring to (2) we also readily see that

$$a_N = 1/\alpha, \quad b_N = \mu_\infty \alpha \qquad (5)$$

offer separate ways of calculating α provided that the end point μ_∞ of the filament is known. Now, from the recur-

TABLE II. Superstable λ-values and some cycle constants for selected islands of stability corresponding to $N = 7,8,9$ replication.

Cycle	Re λ	Im λ	Re α	Im α	Re δ	Im δ
7*	3.999 40	0	2603	0	4.51E6	0
7**	2.837 39	1.215 05	310.8	269.8	− 2.2E4	1.5E5
7***	2.098 49	1.271 46	97.2	27.8	6436	10555
8*	3.999 85	0	10424	0	7.24E7	0
8**	2.839 03	1.213 56	567.1	1142	− 1.1E6	8.4E5
8***	2.105 37	1.269 36	179.7	180.2	− 35140	70041
9*	3.999 9	0	41715	0	1.16E9	0
9**	2.839 3	1.212 6				
9***	2.107	1.267				

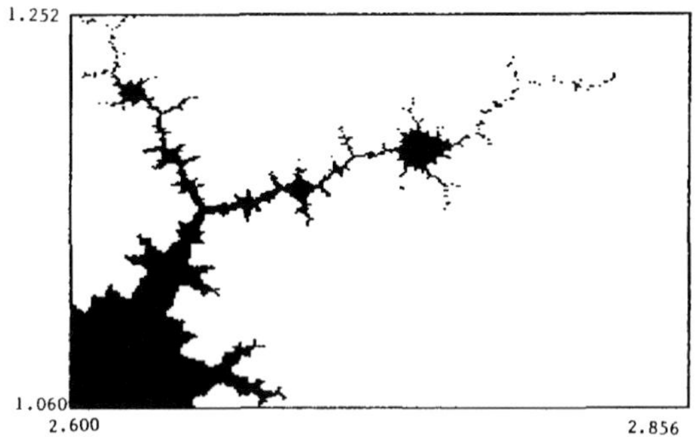

FIG. 1. Strings of islands in the range $2.600 < \mathrm{Re}\, z < 2.856$, $1.06 < \mathrm{Im}\, z < 1.252$. The ** filament extends to the right and terminates.

rence property (4), one easily derives

$$b_N = (2\mu_\infty)^{n-1}\mu_\infty \prod_{m=2}^{N-1} a_m$$

or (6)

$$\alpha = d(2\mu_\infty)^{N-1} \quad \text{with} \quad d = \prod_{m=2}^{N-1} a_m.$$

So it only remains to estimate the product d of the a-coefficients; this is possible by working downwards in m via (4), starting with (5):

$$\begin{aligned} a_{N-1} &= \pm[(1 + 1/\alpha)/\mu]^{1/2}, \\ a_{m-1} &= \pm[(1 + a_m)/\mu]^{1/2}, \quad m < N. \end{aligned} \tag{7}$$

The only *significant* ambiguity in (7) is the choice of sign for the root. This depends on the kneading sequence and completely characterizes it in fact. For instance the rightmost real filamentary sequence has the same root sign throughout.

Let us define the branch cut of $z^{1/2}$ to lie along the negative real axis (i.e., $|\arg z| < \pi$). Then for the ** complex filament sequence (see Fig. 1), the roots have to be taken as

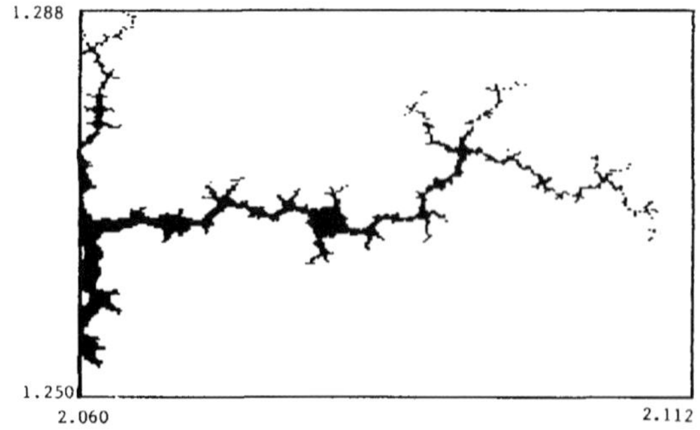

FIG. 2. Strings of islands in the range $2.060 < \mathrm{Re}\, z < 2.112$, $1.25 < \mathrm{Im}\, z < 1.288$. The *** filament branches to the right; another large filament straggles upwards.

$$\begin{aligned} &a_m \leftrightarrow +, \quad m = N-1, N-2,\ldots,3 \\ &a_2 \text{ and } a_1 \leftrightarrow -, \end{aligned} \tag{8**}$$

whereas for the *** sequence the correct choice of signs is

$$\begin{aligned} &a_m \leftrightarrow +, \quad m = N-1,\ldots,4 \\ &a_3 \,\&\, a_2 \,\&\, a_1 \leftrightarrow -. \end{aligned} \tag{8***}$$

[These choices can be verified directly by iterating (4) up in m, starting with $a_1 = -1$, which method gives no sign ambiguity.] The relevant point about this procedure is that for N asymptotic, we get to a good approximation,

$$a_{N-1} = -(1/\mu_\infty)^{1/2},\ldots, \quad a_2 = \mu_\infty - 1, \quad a_1 = -1,$$

with *most* of the coefficients a_m (of a particular root sign) settling down to solutions of the algebraic equation,

$$a = \mu_\infty a^2 - 1$$

or

$$\begin{aligned} a &= [(1 \pm (1 + 4\mu_\infty)^{1/2})/2\mu_\infty] \\ &= (\lambda_\infty/2\mu_\infty) \end{aligned}$$

or

$$(2 - \lambda_\infty)/2\mu_\infty.$$

Consequently a *crude* estimate of the asymptotic form of the product d is

$$d_{**} \simeq (\lambda/2\mu)^{N-3}\mu^{-1/2}, \quad \alpha_{**N} \simeq 4\mu^{3/2}\lambda^{N-3}, \tag{9**}$$

$$\begin{aligned} &d_{***} \simeq (1-\mu)(\lambda/2\mu)^{N-4}\mu^{-1/2}, \\ &\alpha_{***N} \simeq (2\mu)^3(1-\mu)\lambda^{N-4}, \end{aligned} \tag{9***}$$

each evaluated at the respective end points of the filaments. Clearly these estimates can be much improved numerically, but the salient prediction is that as $N \to \infty$

$$\lim \alpha_{N+1}/\alpha_N = \lambda_\infty,$$

which is well substantiated by the numerical facts; thus

$$\alpha_{**8}/\alpha_{**7} = 2.86 + 1.21i \quad (\text{cf. } \lambda_\infty = 2.84 + 1.20i),$$

$$\alpha_{***8}/\alpha_{***7} = 2.20 + 1.23i \quad (\text{cf. } \lambda_\infty = 2.11 + 1.27i),$$

although $N = 8$ is hardly an asymptotic number.

Turning next to the eigenvalue equation for δ, one finds that the coefficients h_n in the expansion,

$$h(z) = \sum_{h=0}^{\infty} h_n z^{2n} = 1 + h_1 z^2 + \cdots, \tag{10}$$

vanish very rapidly with $n (> = 1)$ for large N, ensuring that the lowest order approximation of (2') has exactly the same form as in the real case,[4] namely

$$\sum_{m=0}^{N} [g]^{m\prime}([g]^{N-m+1}(0)) \simeq -\delta/\alpha. \tag{11}$$

Since $g(z) \simeq 1 - \mu_\infty z^2$ and $g'(z) \simeq -2\mu_\infty z$, one has

$$[g]^{n\prime}([g]^{N-n+1}(0)) = (2\mu_\infty)^n \prod_{j=N-n+1}^{N} a_j.$$

Remembering the definition d of (6), and its relation to α, we may rewrite (11) as

$$\frac{\delta}{\alpha^2} = \left[1 - \frac{1}{2\mu_\infty} - \frac{1}{(2\mu_\infty)^2 a_2} - \frac{1}{(2\mu_\infty)^3 a_2 a_3} - \cdots - \frac{1}{(2\mu_\infty)^{N-1} a_2 a_3 \cdots a_{N-1}}\right] \equiv c,$$

where c is completely fixed by μ_∞ and the kneading sequence of root signs for the a_m. Thus even in the complex case, we discover that δ is proportional to α^2. The facts bear out this prediction:

$$(\delta_8 \alpha_8^{-2})_{***} = 1.083 + 0.540\,i, \quad \text{while} \quad (\delta_7/\alpha_7^{-2})_{***} = 1.081 + 0.544\,i,$$

and

$$(\delta_8 \alpha_8^{-2})_{**} = 0.837 + 0.245\,i, \quad \text{while} \quad (\delta_7 \alpha_7^{-2})_{**} = 0.835 + 0.248\,i.$$

Of course one can also estimate the coefficient c by approximating a with $(\lambda/2\mu)$ for the most part but the results are not especially reliable and we will therefore omit details at this point. There is one important point though that needs exposing. This has to do with the algebraic determination of the end point λ_∞. Consider the ** sequence. As $N \to \infty$, a_3 can be evaluated in one of two ways; either as $\mu(\mu-1)^2 - 1$ by going up in N for $N = 2$, or as a solution $(\lambda/2\mu)$ of $\mu a^2 - a = 1$. This means that

$$2\mu^2(1-\mu)^2 = \lambda + 2\mu,$$

yielding a quartic in λ with roots $\lambda = 0$, 4, $2.83929 \pm 1.21258i$, wherein we recognize the last pair as corresponding to the ** filament (see Table I). Similarly, the *** filament end point is obtained as a root of the equation

$$2\mu^2(1-\mu)^2 = -\lambda + 2\mu,$$

and occurs in the vicinity of $\lambda = 2.107 \pm 1.267i$.

Clearly all the above considerations can be extended to other filamentary sequences. In all the cases we can anticipate that

$$\alpha_N \propto \lambda_\infty^{N-1} \quad \text{and} \quad \delta \propto \alpha_N^2,$$

with the proportionality constants depending on the end value μ_∞ of the filament and the kneading sequence. These results represent the complex generalization of the δ–α relation found in the real case.[4] We envisage that the same ideas will find application in circle maps.

[1]B. B. Mandelbrot, Ann. N.Y. Acad. Sci. **357**, 249 (1980).
[2]A. Douady and B. Hubbard, C. R. Acad. Sci. Paris **294**, 123 (1982).
[3]P. Cvitanovic and J. Myrheim, Phys. Lett. A **94**, 329 (1983).
[4]R. Delbourgo, W. Hart, and B. G. Kenny, Phys. Rev. A **31**, 514 (1985); Aust. J. Phys. **39**, 189 (1986); R. Delbourgo and B. G. Kenny, "Universality relations," Phys. Rev. A **33**, 3292 (1986).
[5]P. Cvitanovic and J. Myrheim, "Complex universality," Nordita preprint, 1984.
[6]J. P. Eckmann, H. Epstein, and P. Wittwer, Commun. Math. Phys. **93**, 495 (1984).

Part C
QUANTUM ANOMALIES AND DIMENSIONAL CONTINUATION

C.0. Commentaries C

The surprising discovery of anomalies in conservation laws due to quantum effects by Adler (Phys. Rev. 177, 2426 (1969)) and by Bell & Jackiw (Nuovo Cim. A60, 47 (1969)) (specifically with respect to the axial current) had a profound effect on the development of the standard model of particle physics since it jeopardised renormalisability. It was thanks to the existence of the GIM mechanism (Phys. Rev. D2, 1285 (1970)) and the emergence of quark/lepton generations that the axial anomaly could be tamed. Georgi and Glashow (Phys. Rev. D6, 429 (1972)), as well as Bouchiat *et al.* (Nuovo Cim. A56, 1150 (1968)) showed that the standard representations arising in the model, when taken together, mercifully saved the day. This established the veracity of the renormalisation program for electroweak theory. Today second order quantum effects are being calculated with full confidence in the results, thereby sharpening the accuracy of electroweak parameters.

C.1 Delbourgo R. and Salam A., The gravitational correction to PCAC. Phys. Lett. B40, 381-382 (1972)
C.2 Akyeampong D.A. and Delbourgo R., Dimensional regularisation, abnormal amplitudes and anomalies. Nuovo Cim. A17, 578-586 (1973)
C.3 Akyeampong D.A. and Delbourgo R., Anomalies via dimensional regularisation. Nuovo Cim. A19, 219-224 (1974)
C.4 Delbourgo R. and Prasad V.B., Spinor interactions in the two-dimensional limit. Nuovo Cim. A21, 32-44
C.5 Delbourgo R., A dimensional derivation of the gravitational PCAC correction. J. Phys. A10, L237-240 (1977)
C.6 Delbourgo R. and Matsuki T., Chiral and gravitational anomalies in any dimension. J. Math. Phys. 26, 1334-1336 (1985)
C.7 Delbourgo R. and Waites A.B., Induced parity violation in odd dimensions. Aust. J. Phys. 47, 465-474 (1994)
C.8 Delbourgo R. and Roberts M.L., Relativistic phase space: Dimensional recurrences. J. Phys. A36, 1719-1727 (2003)

One way of understanding quantum anomalies is through the effects of infinite mass regulator fields, but there are other ways of comprehending them; see for instance

- K. Fujikawa, Phys. Rev. D21, 2848 (1980) and Erratum, Phys. Rev. D22, 1499 (1980)
- S.L. Adler, J.C. Collins, and A. Duncan, Phys. Rev. D15, 1712 (9177)
- L. Alvarez-Gaume and E. Witten, Nucl. Phys. B234, 269 (1984)

Except for paper [C.1] which extends the regulator method to the gravitational contribution (as was independently discovered by T. Kimura using point splitting in the article Prog. Theor. Phys, 42, 1191 (1969)) the remainder of the listed articles concern anomalies within the framework of dimensional continuation. The latter is the framework of choice for maintaining vectorial gauge invariance and was initiated by

- C.G. Bollini and J.J. Giambiagi, Nuovo Cim. B12, 20 (1972)
- G. 't Hooft and M. Veltman, Nucl. Phys. B44, 189 (1972)
- J.F. Ashmore, Lett. Nuovo Cim. 4, 289 (1972)
- G.M. Cicuta and E. Montaldi, Lett. Nuovo Cim. 4, 329 (1972)

However the occurrence of γ_5 in the axial current can turn out to be problematic in this framework. Ignoring the suggestion that the trace cyclicity should be abandoned — which I personally dislike — there are basically two ways of handling the difficulty presented by γ_5: the first by 't Hooft and Veltman maintains γ_5 as a product of just $\gamma_0\gamma_1\gamma_2\gamma_3$, whereas our work [C.2] regards the axial current as an antisymmetric tensor product of three different γ matrices and the pseudoscalar current as that of four different γ in a higher dimensional space. The consequence of either method is that the conservation of the axial current can no longer be preserved except in 4D because of the emergence of 'evanescent' terms. For the first method see

- G. 't Hooft and M. Veltman, Nucl. Phys. B44, 189 (1972)
- P. Breitenlohner and D. Maison, Comm. Math. Phys. 52, 11 (1977)

and for the second method see the further elaboration of our work above by

- T.L. Trueman, Phys. Lett. B88, 331 (1979)
- G. Bonneau, Nucl. Phys. B171, 477 (1980)
- G. Bonneau, Int. J. Mod. Phys. A5, 3831 (1990)
- S.A. Larin, Phys. Lett. B113, 303 (1993)

Article [C.3] establishes the simple rule that *when a conservation law only holds in a particular dimension but fails in other dimensions, a quantum anomaly should be expected*, and that this can lead to finite corrections in certain processes that must be cancelled out by appropriate counter-terms. Paper [C.5] closes the circle by proving that the gravitational contribution to the axial anomaly can be neatly derived by this 'evanescent' procedure. The interest in paper [C.4] lies in the section dealing with Fierz reshuffling in any dimension; there we have derived the generating function for the elements of the Fierz matrix, for the first time. In paper [C.6] we derived all anomalies in any dimension by enlarging the work of L. Alvarez-Gaume & E. Witten (Nucl. Phys. B234, 269 (1984)) involving gravity itself. Finally in articles [C.7] and [C.8] we have studied other repercussions of dimensional continuation which have general validity and may be useful to others.

Volume 40B, number 3 PHYSICS LETTERS 10 July 1972

THE GRAVITATIONAL CORRECTION TO PCAC

R. DELBOURGO and A. SALAM

Physics Department, Imperial College, London S.W. 7 2B7, U.K.

Received 26 May 1972

A simple-minded perturbation calculation to order G (the Newtonian gravitational constant) of the two-graviton coupling to a pseudoscalar (P) or axial (A) current via a fermion loop gives an anomalous contribution which can be added to the Adler term in the form $\partial_\alpha A_\alpha = 2m\,P + e^2\epsilon_{\kappa\lambda\mu\nu}F_{\kappa\lambda}F_{\mu\nu}/16\pi^2 + \epsilon_{\kappa\lambda\mu\nu}R_{\kappa\lambda\rho\sigma}R_{\mu\nu\rho\sigma}/768\pi^2$. The anomalous terms can be interpreted as arising from an infinite-mass regulating loop.

The occurrence of anomalous terms in PCAC relations in naive perturbation theory computations involving bilinear currents has prompted us to examine the question for the case when the currents involve derivatives of the fermion fields. In particular we wish to report on the one-fermion loop modifications to such PCAC relations when the stress-tensor current is considered, and especially to consider the basic if academic problem* of the two-graviton decay mode of the π^0 meson, the direct analogue of $\pi^0 \to 2\gamma$. Our result, incorporating the Adler anomaly [1], can be stated in the form

$$\partial_\alpha A_\alpha = 2m\,P + e^2\,\epsilon_{\kappa\lambda\mu\nu}F_{\mu\nu}F_{\mu\nu}/16\pi^2 + \\ + \epsilon_{\kappa\lambda\mu\nu}R_{\kappa\lambda\rho\sigma}R_{\mu\nu\rho\sigma}/768\pi^2 \qquad (1)$$

where A is the axial current, P is the pseudoscalar current and R is the Riemann curvature tensor, the gravity analogue of the Maxwell tensor F. All quantities in (1) are in the context of a naive perturbation calculation in lowest order of the relevant couplings (the one-loop graphs), and indeed the anomalous terms can be interpreted as the contribution from an infinite mass fermion regulator in the gravitational as well as the electromagnetic parts. We see in (1) the role played by the two fundamental longrange forces of nature and the parallelism between the Maxwell and Riemann tensors.

In this note we will only mention the salient features of the calculation leaving the considerable details and interpretation of our results to another publication. Because we are dealing with gravity there are two complications* which do not arise with electromagnetism: (a) many more Feynman graphs need to be taken into account when working to any given order in the Newtonian constant $G = 16\pi\kappa^2$, and (b) the resulting integrals are superficially more divergent. Nevertheless, the saving grace which renders subsequent analysis possible and makes the integrals finite is gravitational gauge invariance. In contrast to the two-photon mode of a pseudoscalar or axial current, the two graviton mode brings in three distinct Feynman graphs, as depicted in fig. 1, of which the graviton pole diagram vanishes identically.

With physical gravitons the general form of the matrix element for π^0 decay must, by gravitational gauge invariance, reduce to

$$\epsilon_{\mu\nu}^{*}(k)\epsilon_{\mu'\nu'}^{*}(k')\epsilon_{\mu\mu'\lambda\lambda'}k_\lambda k'_{\lambda'}(\eta_{\nu\nu'}kk' - k_{\nu'}k'_\nu)G_{\mathrm{P}}(k\cdot k'). \qquad (2)$$

Likewise the axial (index α) decay must be writable in the form + (3)

$$\epsilon_{\mu\nu}^{*}(k)\epsilon_{\mu'\nu'}^{*}(k')\epsilon_{\mu\mu'\lambda\lambda'}k_\lambda k'_{\lambda'}(\eta_{\nu\nu'}k\cdot k' - k_{\nu'}k'_\nu)(k+k')_\alpha G_{\mathrm{A}}(k\cdot k').$$

* This ceases to be an academic problem for the case of strong gravity and the two-f meson decay mode of a (therefore) massive pseudoscalar meson.

* A necessary complication when dealing with spinors is the use of the vierbein formalism.

\+ The basic reason for the form (3) is that the 1^+ component of the axial current cannot couple to a pair of identical helicity 2 particles; only the 0^- piece can and this enters in a derivative fashion.

Volume 40B, number 3 PHYSICS LETTERS 10 July 1972

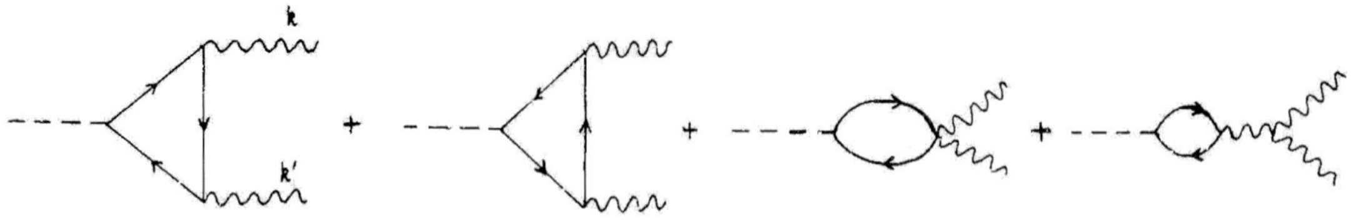

Fig. 1. One-fermion loop contributions to the two-graviton mode.

Expressions (2) and (3) can be directly compared with the matrix elements for the two-photon decay modes [2]

$$\epsilon^*_\mu(k)\,\epsilon^*_{\mu'}(k')\,\epsilon_{\mu\mu'\lambda\lambda'}k_\lambda k'_{\lambda'}F_\mathrm{P}(k\cdot k') \qquad \text{and}$$

$$\epsilon^*_\mu(k)\,\epsilon^*_{\mu'}(k')\,\epsilon_{\mu\mu'\lambda\lambda'}k_\lambda k'_{\lambda'}(k+k')_\alpha F_\mathrm{A}(k\cdot k').$$

Now a naive evaluation of the diagrams of fig. 1 yields*

$$2m\,G_\mathrm{P} = \frac{m^2\kappa^2}{8\pi^2}\int_0^1 \frac{\theta(1-x-y)xy\,\mathrm{d}x\,\mathrm{d}y}{(2k\cdot k'xy-m^2)},$$

$$G_\mathrm{A} = \frac{-\mathrm{i}\kappa^2}{8\pi^2}\int_0^1 \frac{\theta(1-x-y)x^2y^2\,\mathrm{d}x\,\mathrm{d}y}{(2k\cdot k'xy-m^2)}.$$

We observe an anomaly in that

$$2ik\cdot k'G_\mathrm{A} = 2m\,G_\mathrm{P} + \kappa^2/192\pi^2 \neq 2mG_\mathrm{P},$$

$$\text{c.f. } 2ik\cdot k'F_\mathrm{A} = 2m\,F_\mathrm{P} + e^2/2\pi^2 \neq 2mF_\mathrm{P}.$$

By recasting the structure of the effective pseudoscalar Lagrangian in terms of Christoffel symbols { }, we can rewrite the anomalous gravity contribution in terms of the Riemann tensor to this order in G,

$$\epsilon_{\mu\mu'\lambda\lambda'}(\partial_{\nu'}\partial_\lambda g_{\mu\nu}\partial_\nu\partial_{\lambda'}g_{\mu'\nu'} - \partial_\kappa\partial_\lambda g_{\mu\nu}\partial_\kappa\partial_{\lambda'}g_{\mu'\nu'}) =$$

$$= \epsilon_{\mu\lambda\mu'\lambda'}(\partial_\kappa\{\mu\nu,\lambda\}\partial_\kappa\{\mu'\nu,\lambda'\} - \partial_{\nu'}\{\mu\nu,\lambda\}\partial_\nu\{\mu'\nu',\lambda'\})$$

$$= \tfrac{1}{2}\epsilon_{\kappa\lambda\mu\nu}R_{\kappa\lambda\rho\sigma}R_{\mu\nu\rho\sigma}$$

finally giving us eq. (1). Furthermore we note that just as the e.m. anomalous term $e^2/2\pi^2 = \lim_{m\to\infty} 2mF_\mathrm{P}$ so is the gravitational anomalous term $\kappa^2/192\pi^2 = \lim_{m\to\infty} 2mG_\mathrm{P}$, and therefore all anomalous corrections can be identified with contributions from infinite mass regulating loops, which inturn allows for possible re-interpretation of the Ward-Takahashi identity. The attractive possiblity that eq. (1) can be made generally covariant is pure speculation at the present time.

We have greatly benefited from conversations with Dr. C. Isham.

References

[1] S. Adler, Phys. Rev. 177 (1969) 2426.
[2] J. Steinberger, Phys. Rev. 76 (1949) 1180;
L. Rosenberg, Phys. Rev. 129 (1963) 1786.

* In deriving the pseudoscalar result, for instance, one encounters a constant nongauge invariant contribution $\eta_{\nu\nu'}\epsilon_{\mu\mu'\lambda\lambda'}k_\lambda k'_{\lambda'}$ which must be eliminated in the same way as one does in the two-photon decay mode of a scalar meson.

IL NUOVO CIMENTO VOL. 17 A, N. 4 21 Ottobre 1973

Dimensional Regularization, Abnormal Amplitudes and Anomalies.

D. A. AKYEAMPONG (*) and R. DELBOURGO

Physics Department, Imperial College - London

(ricevuto il 27 Marzo 1973)

Summary. — Identifying axial vectors and pseudoscalars as antisymmetric 3-component and 4-component tensors in space-time of arbitrary dimensions, we discover a new PCAC law: $\partial_{[K}\bar{\psi}\Gamma_L\Gamma_M\Gamma_{N]}\psi = 2mi\bar{\psi}\Gamma_{[K}\Gamma_L\Gamma_M\Gamma_{N]}\psi - \frac{1}{5}i\bar{\psi}(i\overleftrightarrow{\partial}^J + 2eA^J)\Gamma_{[J}\Gamma_K\Gamma_L\Gamma_M\Gamma_{N]}\psi$, where the extra term on the right-hand side does not exist in 4 dimensions. However, when we do descend to 4 dimensions after dimensional regularization it is precisely the axial vector anomaly. In a parallel calculation we have also shown how to obtain the anomaly in the matrix elements of the trace of the stress tensor.

1. – Introduction.

The technique of dimensional regularization (¹) is relatively new. Its greatest success to date has been the ability to treat perturbation graphs of non-Abelian gauge group (²) models in an explicitly gauge-invariant manner. However, it has not yet proved its worth in amplitudes involving pions and axial mesons which are overall abnormal, principally because the interpretation of the pseudoscalar current has been inadequate. This paper is an attempt to remedy

(*) On leave of absence from University of Ghana, Legon.

(¹) J. ASHMORE: *Lett. Nuovo Cimento*, **4**, 289 (1972); C. BOLLINI and J. GIAMBIAGI: *Nuovo Cimento*, **12** B, 20 (1972); G. 'T HOOFT and M. VELTMAN: *Nucl. Phys.*, **44** B, 189 (1972).

(²) G. 'T HOOFT and M. VELTMAN: *Nucl. Phys.*, **50** B, 318 (1972); D. CAPPER and M. R. MEDRANO: Trieste IC/72/138; M. BROWN: *Methods for perturbation calculations in gravity*, ICTP/72/13.

this defect. Its degree of success may be judged from the very natural emergence of the axial anomaly (3).

When we are working in a space of even-n dimensions wherein generalized Γ-matrices can be defined, there exists an analogue of the four-dimensional γ_5: it is given by the product of all n Γ-matrices and characteristically it anticommutes with all the vector matrices Γ_M and yields vanishing trace unless it is multiplied by at least n different Γ_M. It is this last property which is the most damaging to the calculation of $\pi^0 \to 2\gamma$, for if we identify the pion current with this generalized γ_5, we get a zero answer for all $n>4$. (Another possibility would be to identify the pion current with the traditional product $\Gamma_0\Gamma_1\Gamma_2\Gamma_3$. This unfortunately is noncovariant and therefore unattractive.)

The point of view which we have taken is that the pion is to be regarded as a quadruply antisymmetric tensor in n dimensions, the axial mesons as a triply antisymmetric tensor. We are guaranteed to get the usual 4-dimensional characteristics this way. However what is just as important is that we are now able to evaluate abnormal amplitudes for arbitrary n and get nonzero answers. If we make this interpretation, we are stuck with a new axial current divergence relation

$$\partial_{[K} A_{LMN]} = 2mi\, P_{[KLMN]} + P'_{[KLMN]}\,,$$

where A is the axial current, P is the pseudoscalar current and

$$P' = i\bar{\psi}(i\overleftrightarrow{\partial}^J + 2eA^J)\Gamma_{[J}\Gamma_K\Gamma_L\Gamma_M\Gamma_{N]}\psi/5$$

is a new contribution, nonexistent in 4 dimensions. If we evaluate the lowest-order fermion loop contribution to the $\pi^0 \to 2\gamma$ matrix element in n dimensions, we find that the P'-term leads one precisely to the axial anomaly when we go to $n=4$. We believe therefore that our choices of abnormal vertices are justified by this result.

In Sect. **2** we have collected all the properties of the generalized Γ-matrices that are useful to the following work. Section **3** contains a discussion of the normal vertex $\sigma \to \gamma\gamma$ and we show there how dimensional regularization directly produces the anomaly (4) in the trace of the stress tensor element (5), *viz.*

$$\langle j_\mu(k)\theta^\lambda_\lambda(0)\, j_\nu(-k)\rangle = \left(k\cdot\frac{\partial}{\partial k} - 2\right)\Pi_{\mu\nu}(k) - \frac{e^2}{6\pi^2}(\eta_{\mu\nu}k^2 - k_\mu k_\nu)\,.$$

Our treatment of the axial anomaly is to be found in Sect. **4** as well as some discussion of the axial vector self-energy.

---

(3) J. SCHWINGER: *Phys. Rev.*, **82**, 664 (1951); J. BELL and R. JACKIW: *Nuovo Cimento*, **60** A, 47 (1969); S. ADLER: *Phys. Rev.*, **177**, 2426 (1969).

(4) C. CALLAN, S. COLEMAN and R. JACKIW: *Ann. of Phys.*, **59**, 42 (1970).

(5) M. CHANOWITZ and J. ELLIS: *Phys. Lett.*, **40** B, 397 (1972).

2. – Some properties of generalized Γ-matrices.

It is a well-known fact that one can generalize the γ-matrices to an n-dimensional vector space. Without going into great detail we shall content ourselves here by listing some of their more relevant properties. One begins with the Clifford algebra

$$\{\Gamma_M, \Gamma_N\} = 2\eta_{MN}, \tag{1}$$

where the indices M, N run from 0, 1, 2, ... up to $n-1$. The metric η is appropriate to the $SO_{n-1,n}$ group, *i.e.* $\eta_{00}=1$, $\eta_{0N}=0$ and $\eta_{MN}=-\delta_{MN}$ when $M\geqslant 1$. It can be used in the normal way to raise and lower indices, *e.g.* $\Gamma^M\equiv\eta^{MN}\Gamma_N$. There are some differences between even- and odd-dimensional spaces, but by and large these are not terribly significant to the later work:

i) When $n=2l$ is even, the Γ-matrices are of dimension $2^l\times 2^l$. There will be a total of $2^{2l}-1=n^2-1$ matrices obtained by multiplication and these form a complete set. Among these is the generalization of the four-dimensional γ_5 which we shall refer to as

$$\Gamma_{-1}\equiv\Gamma_0\Gamma_1\Gamma_2\ldots\Gamma_{n-1}. \tag{2}$$

This anticommutes with all the vector matrices $\{\Gamma_{-1}, \Gamma_M\}=0$ and $(\Gamma_{-1})^2=(-1)^{l+1}$. Also there exists (*) the generalization C of the charge conjugation matrix

$$C\tilde{\Gamma}_M C^{-1} = -\Gamma_M.$$

ii) When $n=2l+1$ is odd, the Γ's still have dimension $2^l\times 2^l$. The vector set $\Gamma_M^{(2l+1)}$ consists of the vector matrices $\Gamma_M^{(2l)}$ $(M=0, 1, \ldots, 2l-1)$ appropriate to the even-dimensional case as well as $\Gamma_{2l}^{(2l+1)}\equiv\Gamma_{-1}^{(2l)}$.

One can realize these matrices as direct products of Pauli matrices. In the generalized Weyl representation, with a Euclidean metric, we can write down a recurrence relation between the matrices. Thus

$$\Gamma_0^{(2l+2)} = 1\times 1\times\ldots\times 1\times\sigma_1,$$
$$\Gamma_1^{(2l+2)} = \underbrace{1\times 1\times\ldots\times\sigma_3}_{l}\times\sigma_2,$$

and

$$\Gamma_M^{(2l+2)} = \Gamma_{M-2}^{(2l)}\times\sigma_2, \qquad \text{where} \qquad 2\leqslant M\leqslant 2l+1.$$

(*) Actually this also exists in a space of $4k-1$ dimensions (k integer). We thank Prof. T. W. B. KIBBLE for demonstrating all of these facts to us.

Thus

$$\Gamma^{(2l+2)}_{-1} = 1\times 1\times \ldots \times 1\times \sigma_3 = \Gamma^{(2l+3)}_{2l+2}$$

and

$$\begin{aligned} C^{(2l)} &= \sigma_2\times 1\times \sigma_2\times 1\times \ldots \times\sigma_2 \quad \text{when } l \text{ is odd},\\ &= C^{(2l-2)}\times\sigma_3 \qquad\qquad\quad \text{when } l \text{ is even}. \end{aligned}$$

The reader may readily pass to the pseudo-Euclidean metric by inserting the necessary factors of i.

Hereafter we shall suppose that the vector space is even-dimensional $n=2l$, so that Γ_{-1} exists. Our dimensional regularization will correspond to continuation in l. Clearly the product of an odd number of Γ's has vanishing trace and for the rest

$$(3) \qquad \begin{cases} 2^{-l}\,\mathrm{Tr}\,(\Gamma_M\Gamma_N) = \eta_{MN}\,, \\ 2^{-l}\,\mathrm{Tr}\,(\Gamma_K\Gamma_L\Gamma_M\Gamma_N) = \eta_{KL}\eta_{MN} - \eta_{KM}\eta_{LN} + \eta_{KN}\eta_{LM}\,, \end{cases}$$

etc.; of especial significance is the fact that

$$\mathrm{Tr}\,\big(\Gamma_{-1}(\Gamma)^r\big) \equiv 0 \qquad \text{unless} \qquad r=2l\,.$$

In what follows we shall find it convenient to define the antisymmetric product

$$(4) \qquad \Gamma_{[M_1}\Gamma_{M_2}\ldots\Gamma_{M_r]} \equiv \sum_{M_r \text{ perm}} (-1)^p\,\Gamma_{M_1}\Gamma_{M_2}\ldots\Gamma_{M_r}\,,$$

for which

$$(5) \qquad \mathrm{Tr}\,\big(\Gamma_{[M_1}\Gamma_{M_2}\ldots\Gamma_{M_r]}(\Gamma)^s\big) = 0 \text{ unless } s=r.$$

Otherwise

$$(6) \qquad 2^{-l}\,\mathrm{Tr}\,(\Gamma_{[M_1}\Gamma_{M_2}\ldots\Gamma_{M_r]}\Gamma^{N_1}\Gamma^{N_2}\ldots\Gamma^{N_r}) = (-1)^{[\frac{1}{2}r]}\,\delta^{[N_1}_{[M_1}\ldots\delta^{N_r]}_{M_r]}\,,$$

$$(7) \qquad 2^{-l}\,\mathrm{Tr}\,(\Gamma_{[M_1}\Gamma_{M_2}\ldots\Gamma_{M_r]}\Gamma_K\Gamma^{[N_1}\ldots\Gamma^{N_r]}\Gamma_L) =$$

$$= (-1)^{[3r/2]}(-\delta^{[N_1}_K\eta_{L[M_1}\delta^{N_2}_{M_2}\ldots\delta^{N_r]}_{M_r]} - \delta^{[N_1}_L\eta_{K[M_1}\delta^{N_2}_{M_2}\ldots\delta^{N_r]}_{M_r]} + \eta_{KL}\,\delta^{[N_1}_{[M_1}\delta^{N_2}_{M_2}\ldots\delta^{N_r]}_{M_r]})\,,$$

and so on. The only other formulae we shall require later on are the multiplication rules

$$(8) \qquad \begin{cases} \Gamma_N\Gamma_{[M_1}\Gamma_{M_2}\ldots\Gamma_{M_r]}\Gamma^N = (n-2r)(-1)^r\,\Gamma_{[M_1}\Gamma_{M_2}\ldots\Gamma_{M_r]}\,, \\ [\Gamma_N, \Gamma_{[M_1}\Gamma_{M_2}\ldots\Gamma_{M_r]}] = 2\eta_{N[M_1}\Gamma_{M_2}\ldots\Gamma_{M_r]}\,, \\ \{\Gamma_N, \Gamma_{[M_1}\Gamma_{M_2}\ldots\Gamma_{M_r]}\} = 2\Gamma_{[N}\Gamma_{M_1}\ldots\Gamma_{M_r]}/(r+1)\,, \end{cases} \qquad r \text{ even},$$

and the contraction formula

$$(9) \qquad \delta^N_{[N}\,\delta^{N_1}_{M_1}\ldots\delta^{N_r}_{M_r]} = (n-r)\,\delta^{N_1}_{[M_1}\ldots\delta^{N_r}_{M_r]}\,.$$

3. – Normal vertices and CVC.

Imagine electrodynamics in 2^l dimensions described by the Lagrangian

$$\mathscr{L} = \bar{\psi}(\Gamma^J(\tfrac{1}{2}i\overleftrightarrow{\partial}_J + eA_J) - m)\psi\,.$$

From the equation of motion

$$[\Gamma\cdot(i\partial + eA) - m]\,\psi = 0$$

it follows by trivial manipulations that the vector electromagnetic current is conserved:

$$\partial^J(\bar{\psi}\Gamma_J\psi) \equiv 0\,.$$

Alternatively stated as a Ward identity,

$$[\Gamma\cdot(p+k)-m]^{-1}k\cdot\Gamma[\Gamma\cdot p - m]^{-1} = [\Gamma\cdot p - m]^{-1} - [\Gamma\cdot(p+k)-m]^{-1};$$

we expect that Green's functions involving external photons and charge zero lines should obey divergence-free conditions on each of the photon indices. This is indeed borne out by the gauge-invariant (¹) evaluation of the photon self-energy:

$$\Pi_{MN}(k) = (\eta_{MN}k^2 - k_Mk_N)\cdot\frac{2e^2}{(2\pi)^l}\int_0^1 dx\,\frac{x(1-x)\Gamma(2-l)}{[k^2x(1-x)-m^2]^{2-l}}\,, \tag{10}$$

a calculation which we shall not repeat. Instead we shall consider two other examples involving overall normal amplitudes.

The first case worth studying is the 2γ-decay of a scalar meson σ, because it has bearing on the possible anomaly of the trace of the stress tensor and because a *naive* evaluation can produce a non-gauge-invariant answer. In general the fermion loop contribution to the off-shell amplitude is

$$T_{MN}(k,k') = -2ie^2\int\frac{d^{2l}p}{(2\pi)^{2l}}\cdot\frac{\mathrm{Tr}\,[(\Gamma\cdot(p-k')+m)\Gamma_N(\Gamma\cdot p+m)\Gamma_M(\Gamma\cdot(p+k)+m)]}{[(p-k')^2-m^2][p^2-m^2][(p+k)^2-m^2]}\,.$$

If we make the external photons physical,

$$T_{MN}(k,k') \to -4ie^2\int\frac{d^{2l}p}{(2\pi)^{2l}}\frac{dx\,dy\,\theta(1-x-y)}{[p^2+2k\cdot k'xy-m^2]^3}\,m\cdot 2^l\cdot$$
$$\cdot\left(\eta_{MN}\left\{\left(\frac{2}{l}-1\right)p^2+m^2-2k\cdot k'xy\right\}+(k_Nk'_M-\eta_{MN}k\cdot k')(1-4xy)\right),$$

and we would be inclined to drop the p^2-term in passing to 4 dimensions ($l \to 2$). (Certainly this is the simple-minded way the calculation is carried out in space-time—it leads to a finite nongauge-invariant result which is unacceptable and requires regularization.) However, the rules of dimensional regularization require us to evaluate the Feynman integral for all l (away from the singularities in the l-plane) before proceeding to $l=2$. When this is done we obtain a manifestly gauge-invariant answer:

$$T_{MN}(k,k') \to (\eta_{MN}k\cdot k' - k_N k'_M)\cdot\frac{2me^2}{(2\pi)^l}\Gamma(3-l)\int_0^1 \mathrm{d}x\,\mathrm{d}y\,\frac{(1-4xy)\theta(1-x-y)}{[2k\cdot k'xy-m^2]^{3-l}}\,, \tag{11}$$

which reduces to the correctly regularized answer in 4 dimensions ([6]).

If instead we keep the photons off shell but pass to the limit $k'=-k$, we get

$$T_{MN}(k,-k) = (k_M k_N - \eta_{MN}k^2)\cdot\frac{4me^2}{(2\pi)^l}\int_0^1 \mathrm{d}x\,\frac{x(1-x)\Gamma(3-l)}{[k^2x(1-x)-m^2]^{3-l}}\,. \tag{12}$$

Significantly, this is finite as $l\to 2$.

The relevance of this result comes when we consider the photon matrix element of the trace of the energy-momentum tensor. To order e^2 one has to add two seagull diagrams to the usual vertex diagrams. The sum total gives us the formal relation

$$\langle j_M(k)\,\theta^L_L(x=0)\,j_N(k')\rangle = mT_{MN}(k,k')$$

for arbitrary l, and in the limit of zero momentum transfer we obtain

$$\langle j_M(k)\theta^L_L(0)\,j_N(-k)\rangle = mT_{MN}(k,-k) = -m\frac{\partial}{\partial m}\Pi_{MN}(k) = \tag{13}$$

$$= \left(k\cdot\frac{\partial}{\partial k}-2\right)\Pi_{MN}(k) - (\eta_{MN}k^2-k_Mk_N)\cdot\frac{4e^2}{(2\pi)^l}\int_0^1 \mathrm{d}x\,\frac{x(1-x)\Gamma(3-l)}{[k^2x(1-x)-m^2]^{2-l}}\,.$$

There is an «anomalous contribution» on the right of (13) which is finite as $l\to 2$ and equals $-e^2(\eta_{MN}k^2-k_Mk_N)/6\pi^2$, in agreement with CHANOWITZ and ELLIS ([5]).

Let us next turn to the axial vector self-energy. We shall first have to decide on what we mean by an axial vector in $2l$ dimensions. The obvious candidate (which we shall strongly argue against in the following Section) is to interpret the axial vector current as $\bar\psi i\Gamma_M\Gamma_{-1}\psi$, where Γ_{-1} is the generalization of γ_5. If we adopt this (wrong) point of view, the lowest-order axial

([6]) R. DELBOURGO and A. SALAM: *PCAC anomalies and gravitation*, Trieste, IC/72/86.

vector self-energy reads

$$(14) \qquad \Pi^{\text{axial}}_{MN}(k) =$$

$$= \frac{2(-1)^l}{(2\pi)^l} \int_0^1 \mathrm{d}x \frac{\Gamma(2-l)}{[k^2x(1-x)-m^2]^{2-l}} \{-m^2\eta_{MN} + (k^2\eta_{MN} - k_M k_N)\, x(1-x)\}\,.$$

It becomes formally gauge invariant in the limit of zero fermion mass. In any case both transverse and longitudinal components require renormalization in general as we pass to $l = 2$.

4. – Abnormal vertices and PCAC.

There is a well-known difficulty with pions in dimensional regularization. If we stick to the traditional idea of identifying the pseudoscalar current with $\bar{\psi}\Gamma_{-1}\psi$, then we are faced with a zero answer in the computation of $\pi^0 \to 2\gamma$. The basic reason is that the trace of Γ_{-1} with four Γ-matrices vanishes for arbitrary $l \geqslant 3$. The way out of this dilemma is rather to represent the pions as a quadruply antisymmetric tensor having the current $\bar{\psi}\Gamma_{[K}\Gamma_L\Gamma_M\Gamma_{N]}\psi/4!$, an identification which is correct for four dimensions and which prevents the relevant trace from vanishing. However we have to be more careful about PCAC. In the electrodynamics model considered earlier we find from the equations of motion that

$$(15) \qquad -i\partial_{[K}\bar{\psi}\Gamma_L\Gamma_M\Gamma_{N]}\psi = 2m\bar{\psi}\Gamma_{[K}\Gamma_L\Gamma_M\Gamma_{N]}\psi - \bar{\psi}(i\overleftrightarrow{\partial}^J + 2eA^J)\Gamma_{[J}\Gamma_K\Gamma_L\Gamma_M\Gamma_{N]}\psi/5\,.$$

In the same vein we will interpret the axial current as the triply antisymmetric quantity $\bar{\psi}\Gamma_{[L}\Gamma_M\Gamma_{N]}\psi$, of which the generalized curl appears on the left of (15). To lowest order in e we can restate (15) as the Ward identity

$$(16) \qquad [\Gamma\cdot(p+k)-m]^{-1} k_{[K}\Gamma_L\Gamma_M\Gamma_{N]}[\Gamma\cdot p - m]^{-1} =$$

$$= [\Gamma\cdot(p+k)-m]^{-1}(2m\Gamma_{[K}\Gamma_L\Gamma_M\Gamma_{N]} - \tfrac{1}{5}(2p+k)^J\Gamma_{[J}\Gamma_K\Gamma_L\Gamma_M\Gamma_{N]})[\Gamma\cdot p - m]^{-1} +$$
$$+ [\Gamma\cdot(p+k)-m]^{-1}\Gamma_{[K}\Gamma_L\Gamma_M\Gamma_{N]} + \Gamma_{[K}\Gamma_L\Gamma_M\Gamma_{N]}[\Gamma\cdot p - m]^{-1}\,.$$

In four dimensions, since

$$\gamma_{[\lambda}\gamma_\mu\gamma_{\nu]} = -6\varepsilon_{\lambda\mu\nu\varrho}\gamma^\varrho\gamma_5\,, \qquad \gamma_{[\varkappa}\gamma_\lambda\gamma_\mu\gamma_{\nu]} = 4!\,\gamma_5\,\varepsilon_{\varkappa\lambda\mu\nu}\,, \qquad \gamma_{[\varkappa}\gamma_\lambda\gamma_\mu\gamma_\nu\gamma_{\varrho]} = 0$$

and

$$\partial_{[\varkappa}\varepsilon_{\lambda\mu\nu]\varrho} = -\partial_\varrho\varepsilon_{\varkappa\lambda\mu\nu}\,,$$

we find that (15) and (16) formally reduce to the usual PCAC identities.

However, for arbitrary dimensions we see that PCAC is no longer quite true. Rather the « divergence of the axial vector » current is the « pseudoscalar » current plus a *new current* $\psi\overleftrightarrow{\partial}^J\Gamma_{[J}\Gamma_K\Gamma_L\Gamma_M\Gamma_{N]}\psi$ which is overall « pseudoscalar ». We shall now demonstrate that, when we pass to four dimensions, it is this new term which is responsible for the anomaly.

Consider firstly the usual pseudoscalar triangle graph for $\pi^0 \to 2\gamma$, which is now interpreted to be

$$(17)\qquad T^{[IJKL]}_{MN}(k, k') = $$

$$= \frac{2ie^2}{4!}\int \frac{d^{2l}p}{(2\pi)^{2l}} \cdot \frac{\mathrm{Tr}\,\{\Gamma^{[I}\Gamma^J\Gamma^K\Gamma^{L]}(\Gamma\cdot(p-k')+m)\Gamma_N(\Gamma\cdot p+m)\Gamma_M(\Gamma\cdot(p+k)+m)\}}{[(p-k')^2-m^2][p^2-m^2][(p+k)^2-m^2]}.$$

If we place the photons on their mass shells and use the trace relation (6), the decay amplitude reduces quite simply to

$$(18)\qquad T^{[IJKL]}_{MN}(k, k') \to \frac{1}{4!}\,\delta^{[I}_N\delta^J_M\delta^K_R\delta^{L]}_S k^R k'^S \cdot \frac{2e^2 m}{(2\pi)^l}\int_0^1 dx\,dy\,\frac{\theta(1-x-y)\Gamma(3-l)}{[2k\cdot k'xy-m^2]^{3-l}}.$$

It is manifestly gauge invariant and becomes the Steinberger amplitude when $l \to 2$. Far more interesting is the contribution from the new current

$$T'^{[IJKL]}_{MN}(k, k') = \frac{2ie^2}{4!(2\pi)^{2l}}\int d^{2l}p\{[(p-k')^2-m^2][p^2-m^2][(p+k)^2-m^2]\}^{-1}\cdot$$

$$\cdot(2p+k)_H\,\mathrm{Tr}\,\{\Gamma^{[H}\Gamma^I\Gamma^J\Gamma^K\Gamma^{L]}(\Gamma\cdot(p-k')+m)\Gamma_N(\Gamma\cdot p+m)\Gamma_M(\Gamma\cdot(p+k)+m)\} \to$$

$$\to \frac{4e^2(2/l-1)}{4!(2\pi)^l\Gamma(l)}\int \frac{p^{2l}dp^2\,dx\,dy\,\theta(1-x-y)}{[p^2+2k\cdot k'xy-m^2]^3}\,\delta^{[I}_M\delta^J_N\delta^K_R\delta^{L]}_S k^R k'^S.$$

Naively we might suppose that this contribution disappears (*i.e.* is not present) when $l=2$. Before we can arrive at this conclusion we shall have to evaluate the momentum integral. We find instead that

$$(19)\qquad T'^{[IJKL]}_{MN}(k, k') \to \frac{2e^2}{4!(2\pi)^l}\,\delta^{[I}_M\delta^J_N\delta^K_R\delta^{L]}_S k^R k'^S \int_0^1 dx\,dy\,\frac{\theta(1-x-y)\Gamma(3-l)}{[2k\cdot k'xy-m^2]^{2-l}},$$

which has the same structure as T and is none other than the famous axial anomaly when we pass to the limit $l=2$. We believe that this calculation amply confirms the interpretation we have given to axial vectors and pseudoscalars in the context of dimensional regularization. It means that one can now study interactions involving 0^- and 1^+ currents with some degree of confidence and compute overall abnormal amplitudes.

To close let us therefore return to the axial self-energy. We now assert that the correct result is given by the tensor

$$\Pi^{[LMN]}_{[IJK]}(k) = -i\int \frac{\mathrm{d}^{2l}p}{(2\pi)^{2l}}\,\frac{\mathrm{Tr}\,\{\Gamma_{[I}\Gamma_J\Gamma_{K]}(\Gamma\cdot(p+k)+m)\Gamma^{[L}\Gamma^M\Gamma^{N]}(\Gamma\cdot p+m)\}}{[(p+k)^2-m^2][p^2-m^2]}\,.$$

If we use the identities (6), (7) and (8), the answer reduces to

$$\Pi^{[LMN]}_{[IJK]} = \frac{2}{(2\pi)^l}\int_0^1 \mathrm{d}x\,\frac{\Gamma(2-l)}{[k^2x(1-x)-m^2]^{2-l}}\left(m^2\delta^{[L}_{[I}\delta^M_J\delta^{N]}_{K]}+k_{[I}k^{[L}\delta^M_J\delta^{N]}_{K]}x(1-x)\right)- \\ -\frac{1}{(2\pi)^l}\int_0^1 \mathrm{d}x\,\frac{(2l-4)\Gamma(1-l)}{[k^2x(1-x)-m^2]^{1-l}}\,\delta^{[L}_{[I}\delta^M_J\delta^{N]}_{K]}\,. \tag{20}$$

It differs from the previous result (14) in the final longitudinal term which reduces to the quadratic polynomial $(-\frac{1}{3}k^2+2m^2)\eta_{\mu\nu}$ in 4 dimensions. Thus, repercussions of modified PCAC enter already at the level of the two-point function, as noted elsewhere ([6]).

* * *

One of us (D.A.A.) would like to thank Prof. T.W.B. KIBBLE for hospitality at Imperial College and the Associate of Commonwealth Universities for the award of a Fellowship.

● RIASSUNTO (*)

Identificando i vettori assiali e gli pseudoscalari con tensori antisimmetrici a 3 e 4 componenti in uno spazio-tempo di dimensioni arbitrarie, si scopre una nuova legge di PCAC $\partial_{[K}\bar{\psi}\Gamma_L\Gamma_M\Gamma_{N]}\psi = 2mi\bar{\psi}\Gamma_{[K}\Gamma_L\Gamma_M\Gamma_{N]}\psi - \frac{1}{5}i\bar{\psi}(i\overleftrightarrow{\partial}^J+2eA^J)\Gamma_{[J}\Gamma_K\Gamma_M\Gamma_{N]}\psi$, in cui il termine addizionale a destra non esiste in 4 dimensioni. Tuttavia, quando si scende a 4 dimensioni dopo la regolarizzazione dimensionale, esso è precisamente l'anomalia vettoriale assiale. In un calcolo parallelo si è anche mostrato come ottenere l'anomalia negli elementi di matrice della traccia del tensore di sforzo.

(*) *Traduzione a cura della Redazione.*

Размерная регуляризация, анормальные амплитуды и аномальности.

Резюме (*). — Идентифицируя аксиальные векторы и псевдоскалры как антисимметричные трех-компонентные и четырех-компонентные тензоры в пространстве-времени произвольного числа измерений, мы приходим к новому закону PCAC:

$$\partial_{[K}\bar{\psi}\Gamma_L\Gamma_M\Gamma_{N]}\psi = 2mi\bar{\psi}\Gamma_{[K}\Gamma_L\Gamma_M\Gamma_{N]}\psi - \tfrac{1}{5}i\bar{\psi}(i\overleftrightarrow{\partial}^J+2eA^J)\Gamma_{[J}\Gamma_K\Gamma_L\Gamma_M\Gamma_{N]}\psi\,,$$

где дополнительный член справа не существует в четырехмерном случае. Однако когда мы спускаемся до четырех измерений после размерной регуляризации, получается в точности аксиальная векторная аномалия. При параллельном вычислении мы также показываем, как возникает аномалия в матричных элементах для шпура тензора напряжений.

(*) *Переведено редакцией.*

IL NUOVO CIMENTO VOL. 19 A, N. 2 21 Gennaio 1974

Anomalies Via Dimensional Regularization.

D. A. AKYEAMPONG (*) and R. DELBOURGO

Physics Department, Imperial College - London

(ricevuto il 9 Luglio 1973)

Summary. — Anomalous Ward identities arise when the Lagrangian in $2l$-dimensional space-time is not invariant under the symmetry transformation in question with l arbitrary. (The anomalous symmetry-breaking terms have associated a factor $2l-4$ and disappear at the classical level in four dimensions.) In particular, the axial anomaly is connected with the chiral breaking due to the fermion kinetic energy and the canonical trace anomaly is a consequence of dilatation non-invariance of any interaction.

In two recent papers (1) we showed how to extract the famous axial anomalies using the technique of dimensional regularization. We also indicated that the same approach could be used with profit to obtain the anomalies in the trace of the canonical stress tensor. In this note we would like to elaborate on the entire method of continuing the space-time dimensions so as to pinpoint the source of anomalies in general. Our contention, simply stated, is that the anomalous terms in naive four-dimensional Ward identities are the quantum loop corrections to matrix elements of operators which arise in the naive $2l$-dimensional Ward identities and which disappear classically in the limit $l \to 2$.

A) For the case of the axial current we made it reasonably clear that the source of difficulties with naive PCAC was the chiral noninvariance of the fer-

(*) On leave of absence from the University of Ghana, Legon.

(1) D. A. AKYEAMPONG and R. DELBOURGO: *Nuovo Cimento*, **17** A, 578 (1973); **18** A, 94 (1973).

mion kinetic energy when $2l$ is allowed to be arbitrary. Correspondingly the naive $2l$-dimensional PCAC relation contains the « anomalous » current $\bar{\psi}\partial^J\cdot\Gamma_{[J}\Gamma_K\Gamma_L\Gamma_M\Gamma_{N]}\psi$ and as $l\to 2$ this gives nonvanishing closed fermion loop corrections to matrix elements with at most five external meson legs; the reason is that one encounters the product of an integral which diverges as $l\to 2$, and which thereby carries a factor $\Gamma(N-l)$ with $N\leqslant 2$, multiplied by a tracing factor $(l-2)$ due to the anomalous tensor. Indeed this technique of continuation in l gives a very specific gauge-invariant regularization (²) of the formally divergent four-dimensional integrals. We illustrated the correctness of the approach by evaluating the $\langle VAA\rangle$, $\langle AAA\rangle$ and $\langle VAAA\rangle$ anomalies. What we did not make clear however, was that the procedure yields a whole host of other anomalies associated with external scalar and pseudoscalar lines, which, although self-consistent, have to be subtracted out if one wishes to recover the minimal Bardeen set (³). To third order these extra anomalies are easily computed:

(1) $\langle P'P\rangle = m(k^2/6-m^2)/\pi^2\,,$

(2) $\langle P'A_\mu\rangle = ik_\mu(k^2/6-m^2)/2\pi^2\,,$

(3) $\langle P'PS\rangle = (3k^2+3kk'+k'^2-18m^2)/6\pi^2\,,$

(4) $\langle P'A_\mu S\rangle = imk'_\mu/\pi^2\,,$

(5) $\langle P'PV_\mu\rangle = m(k-k')_\mu/3\pi^2\,,$

(6) $\langle P'A_\mu V_\nu\rangle = i[(k'_\mu k_\nu+2k_\mu k_\nu+3k_\mu k'_\nu)-\eta_{\mu\nu}(3k^2+k'^2+3kk'-6m^2)]/6\pi^2\,.$

k refers to the momentum carried by the pseudoscalar field P' having the anomalous kinetic interaction and k' refers to the momentum of the other unnatural leg of the three-point vertex. We see that the consistency conditions

$$\langle S(0)A_\mu(-k)P'(k)\rangle = (\partial/\partial m)\langle A_\mu(-k)P'(k)\rangle\,,$$
$$\langle V_\nu(0)A_\mu(-k)P'(k)\rangle = (\partial/\partial k^\nu)\langle A_\mu(-k)P'(k)\rangle\,,$$

etc. are all obeyed. Observing that $P'/(P, A, S, V)$ has the dimensions of mass, we can begin the elimination of these normal anomalies by introducing the

(²) Given a convergent integral like $g\int m\,d^4k/(k^2-m^2)$ the introduction of regulator fields modifies this to $\sum g_i\int m_i\,d^4k/(k^2-m_i^2)$ which converges if $\sum g_i m_i$, $\sum g_i m_i^3$ and $\sum g_i m_i^3\log m_i$ are all finite. In dimensional regularization, this is replaced by the unambiguous convergent integral $(l-2)\int d^{2l}k/(k^2-m^2)$. This explains how it is that our anomalies differ from those of R. BROWN, C. SHIH and B. L. YOUNG: *Phys. Rev.*, **186**, 1491 (1969) obtained by means of « universal regularization ». They agree, however, with those of P. BREITENLOHNER and H. MITTER: *Nuovo Cimento*, **10** A, 655 (1972).

(³) W. A. BARDEEN: *Phys. Rev.*, **184**, 1848 (1969).

counter-termı Lagrangians

$$\delta \mathscr{L}_1 = -m[\tfrac{1}{6}\partial^J P^{KLMN}\partial_J P'_{KLMN} - m^2 P^{KLMN} P'_{KLMN}]/4!\,,$$
$$\delta \mathscr{L}_2 = \tfrac{1}{2}[\tfrac{1}{6}\partial^J A^{KLM}\partial_J \partial^N P'_{KLMN} - m^2 A^{KLM}\partial^J P'_{JKLM}]/3!\,,$$

(defined up to coupling constant factors) which serve to cancel (1) and (2). As they stand the $\delta\mathscr{L}$ are not vector gauge invariant and we are obliged to make the substitution $\partial^J \varphi \to (\partial^J + V^J \times)\varphi$ so as to restore the U_3 gauge symmetry. In this manner we automatically cancel anomalies (5) and (6). If at the same time we interpret m in (1) and (2) as the vacuum expectation value of the scalar field S, then anomalies (3) and (4) can also be removed by making the further replacements $m \to m + S$ and

$$m\partial P \partial P' \to \tfrac{1}{2}[3\partial S \partial(PP') + 3\partial P \partial(SP') - \partial P' \partial(SP)] \text{ etc.}$$

in $\delta\mathscr{L}$, and gauging these as well. It is not unreasonable to expect that this procedure serves to cancel all the extra higher-point anomalies involving P and S (since the consistency conditions are satisfied) and that we remain with the minimal $\langle A^n V^{n'}\rangle$ anomalies.

B) For the trace of the canonical stress tensor we previously sketched how the anomaly emerged in the context of the photon propagator. We shall try to be more explicit now and spell out precisely the source of the anomaly ([4]). Start by fixing the canonical scale dimension d of the fields by requiring that the free field Lagrangian

$$\mathscr{L}_f = i\bar{\psi}\partial^J \Gamma_J \psi + \tfrac{1}{2}(\partial^J\varphi)(\partial_J\varphi) - \tfrac{1}{4}(\partial^J V^K - \partial^K V^J)(\partial_J V_K - \partial_K V_J)$$

is dilatation invariant. At this stage $d = l - 1$ for bosons and $l - \frac{1}{2}$ for fermions, and coincides with the mass dimensions of the fields. Upon introducing the mass terms $-m\bar{\psi}\psi$, $-\frac{1}{2}\mu^2\varphi^2$, the dilatation symmetry is naturally gone. Much more significant is the fact that interaction terms of any description are bound to spoil dilatation invariance in $2l$ dimensions when l is continued from 2 to arbitrary values; for although interactions like

$$\mathscr{L}_i = -h\varphi^4/4! + g\bar{\psi}\psi\varphi + e\bar{\psi}\Gamma^J\psi V_J$$

appear to be dilatation invariant in four dimensions (with the numbers e, g and h dimensionless), they are no longer so for other l-values. The same argument

---

([4]) K. WILSON: *Phys. Rev. D*, **2**, 1473 (1970); C. G. CALLAN, S. COLEMAN and R. JACKIW: *Ann. of Phys.*, **59**, 42 (1970); M. CHANOWITZ and S. ELLIS: *Phys. Lett.*, **40** B, 317 (1972).

applies to the interaction $(\bar{\psi}\gamma\psi)^2$ of the two-dimensional Thirring model. And this is the sign of an anomaly. Indeed the divergence of the dilatation current reads

$$\partial^\lambda D_\lambda = \frac{\partial\mathscr{L}}{\partial\varphi}(l-1)\varphi + \frac{\partial\mathscr{L}}{\partial\psi}\left(l-\frac{1}{2}\right)\psi + \bar{\psi}\left(l-\frac{1}{2}\right)\frac{\partial\mathscr{L}}{\partial\bar{\psi}} + \frac{\partial\mathscr{L}}{\partial V_\lambda}(l-1)V_\lambda - 2l\mathscr{L}$$

or

$$\theta^\lambda_\lambda = \mu^2\varphi^2 + m\bar{\psi}\psi + (l-2)[-h\varphi^4/12 + g\bar{\psi}\psi\varphi + e\bar{\psi}\Gamma^\lambda\psi V_\lambda].$$

It is the interaction terms multiplying $l-2$ which are responsible for the anomalous scale dimensions among other things. For instance, at zero momentum transfer, the trace identity is

$$(p\,\partial/\partial p - 2l + 2d)G(p) = \langle\{\mu^2\varphi^2 + \ldots + (l-2)e\bar{\psi}\Gamma^\lambda\psi V_\lambda\}(0)\,\xi(p)\,\bar{\xi}(-p)\rangle$$

where $G(p) = \langle\xi(p)\bar{\xi}(-p)\rangle$ is the two-point Green's function. To discard all the anomalous interaction effects would be to invite false theorems ([4,5]). Retaining them instead, we obtain polynomial higher-order perturbation corrections to the classical propagator which can be reinterpreted as anomalous additions to the scale dimension. Three examples should suffice to demonstrate the mechanism:

i) $\mathscr{L}(\varphi) = \frac{1}{2}(\partial\varphi)^2 - h\varphi^4/4!$. The propagator equation reads

$$-i(p\,\partial/\partial p - 2)\,D(p) = \langle\{(l-2)\,h\varphi^4(0)/12\}\,\varphi(p)\,\varphi(-p)\rangle$$

or

$$-\left(p\frac{\partial}{\partial p} - 2\right)D^{-1}(p) = i(l-2)\int\frac{d^{2l}k}{(2\pi)^{2l}}\frac{1}{k^2} + \frac{1}{3}i(l-2)h^2\int\exp[ipx](D(x))^3 d^{2l}x + \ldots,$$

where $D(x) = \Gamma(l-1)(-x^2+i\varepsilon)^{1-l}/4\pi^l$ is the massless causal propagator. The first integral gives zero in the limit $l\to 2$ and the second integral can be reduced to

$$\frac{\frac{1}{3}(l-2)h^2}{(4\pi^2)^2}\int_0^\infty dr\,r^{6-5l}\frac{J_{l-1}(r\sqrt{-p^2})}{(\sqrt{-p^2})^{l-1}} = \frac{\frac{1}{3}(l-2)h^2}{(4\pi^2)^2}(-p^2)^{2l-3}2^{6-5l}\frac{\Gamma(3-2l)}{\Gamma(3l-3)} \to$$
$$\to -h^2p^2/3(32\pi^2)^2\,.$$

To this order this corresponds to a change in scale from $d=1$ to $d = 1 + h^2/3(32\pi^2)^2$.

---

(5) S. Coleman and R. Jackiw: *Ann. of Phys.*, **67**, 552 (1971).

ii) $\mathscr{L}(\psi, \varphi) = \frac{1}{2}(\partial\varphi)^2 + i\bar{\psi}\Gamma\partial\psi + g\bar{\psi}\psi\varphi$. The fermion equation is

$$i(p\,\partial/\partial p - 1)\,S(p) = \langle\{(l-2)\,g\bar{\psi}\psi\varphi\}(0)\,\psi(p)\,\bar{\psi}(-p)\rangle$$

or

$$\left(p\frac{\partial}{\partial p} - 1\right)S^{-1}(p) = 2i(l-2)g^2\int\frac{d^{2l}k}{(2\pi)^{2l}}\frac{\Gamma(k+p)}{k^2(k+p)^2} + \ldots =$$

$$= 2(l-2)g^2\frac{\Gamma p}{(2\pi)^l}\int dx\frac{(1-x)\Gamma(2-l)}{[p^2x(1-x)]^{2-l}} + \ldots \to -g^2\Gamma p/4\pi^2\,,$$

and once again the net effect is to change the scale dimension to $d = \frac{3}{2} - g^2/4\pi^2$.

iii) $\mathscr{L}(\psi, V) = \bar{\psi}(i\partial\Gamma - m)\psi + e\bar{\psi}\Gamma^J\psi V_J - \frac{1}{4}(\partial_K V_J - \partial_J V_K)^2$. The photon propagator equation is

$$i(k\,\partial/\partial k - 2)\,D_{MN}(k) = \langle\{m\bar{\psi}\psi + (l-2)\,e\bar{\psi}\Gamma^J\psi V_J\}(0)\,V_M(k)\,V_N(-k)\rangle$$

or

$$(k\partial/\partial k - 2)\,D^{-1}_{MN}(k) = -[m\partial/\partial m + 2(l-2)]\Pi_{MN}(k)\,,$$

where

$$\Pi_{MN}(k) = (\eta_{MN}k^2 - k_M k_N)\frac{2e^2}{(2\pi)^l}\int dx\frac{x(1-x)\Gamma(2-l)}{[k^2x(1-x) - m^2]^{2-l}}$$

is the photon self-energy to order e^2. The anomalous contribution equals $\lim_{l\to 2} 2(l-2)\Pi_{MN}(k) = -e^2(\eta_{MN}k^2 - k_M k_N)/6\pi^2$ as we reported earlier.

It does not take much imagination to see that the repercussions of dilatation breaking will enter into all the higher-point functions and will not just correspond to a change of canonical scale. For instance, the four-point meson coupling h will receive anomalous corrections from $(l-2)\,h\varphi^4$ in higher orders which are not simply external-line propagator insertions.

In conclusion we cannot emphasize enough the advantages of continuation of space-time dimensions. The method was originally conceived to provide a gauge-invariant treatment of vector (6) and tensor (7) fields, it was later extended to deal with chiral symmetries and abnormal amplitudes (1), and has

---

(6) J. ASHMORE: *Lett. Nuovo Cimento*, **4**, 289 (1972); C. BOLLINI and J. GIAMBIAGI: *Nuovo Cimento*, **12** B, 20 (1972); G. 'T HOOFT and M. VELTMAN: *Nucl. Phys.*, **44** B, 189 (1972).

(7) D. M. CAPPER, G. LEIBBRANDT and M. RAMON-MEDRANO: Trieste preprint IC/73/26; M. BROWN: Imperial College preprint ICTP/72/13.

now been shown to generate all the anomalous Ward identities in an almost trivial fashion. In fact the technique of dimensional regularization has not failed us so far; it is appealingly simple, easy to apply and so obviously « right » that in our opinion it supersedes Pauli-Villars regularization.

* * *

One of us (D.A.A.) would like to thank Prof. T. W. B. KIBBLE for hospitality at Imperial College and the Association of Commonwealth Universities for the award of a Fellowship.

● RIASSUNTO (*)

Le identità di Ward anomale si presentano quando nello spazio-tempo a $2l$ dimensioni il lagrangiano non è invariante rispetto alle trasformazioni di simmetria in questione con l arbitrario. (I termini anomali che rompono la simmetria hanno associato un fattore $2l-4$ e scompaiono al livello classico in quattro dimensioni.) In particolare l'anomalia assiale è connessa con la rottura chirale dovuta all'energia cinetica dei fermioni e l'anomalia nella traccia canonica è conseguenza della noninvarianza di dilatazione di ogni interazione.

(*) *Traduzione a cura della Redazione.*

Аномальности через регуляризацию числа измерений.

Резюме (*). — Аномальные тождества Уорда возникают в том случае, когда Лагранжиан в $2l$-мерном пространстве-времени является неинвариантным относительно преобразования симметрии, рассматриваемого при произвольном l. (Аномальные члены, нарушающие симметрию, объединены с множителем $(2l-4)$ и исчезают на классическом уровне в случае четырех измерений.) В частности, аксиальная аномалия связывается с киральным нарушением из-за фермионной кинетической энергии. Аномалия канонического следа представляет следствие неинвариантности расширения для любого взаимодействия.

(*) *Переведено редакцией.*

IL NUOVO CIMENTO VOL. 21 A, N. 1 1 Maggio 1974

Spinor Interactions in the Two-Dimensional Limit (*).

R. DELBOURGO and V. B. PRASAD

Physics Department, Imperial College - London

(ricevuto il 3 Dicembre 1973)

Summary. — It is shown how the Thirring model with its anomalies can be treated by dimensional regularization.

1. – Introduction.

One of the important lessons we have learnt during the course of the past year in applying dimensional continuation to quantum field theories is that the limit to integer dimensions must be left to last [1]. Besides the practical advantage of allowing us to manipulate our integrals with ease till the very end, it also has the physical consequence of pinpointing the source of anomalous quantum corrections [2]. Therefore if one thing is clear, it is that we must consistently set up our field theory in *arbitrary* dimensions, which, if we wish to accommodate electromagnetism, is necessarily even ($2l$) before continuation. The actual continuation to any dimension is a trivial affair in cases where only scalar or vector particles are involved, but when spinors or abnormal-parity objects arise the generalization requires considerable care.

Our aim in this paper is to examine spinor-spinor interactions for arbitrary l and then to see what emerges for the Thirring model in the two-dimensional limit. Of course we find no surprises, so in a sense we are using the Thirring model as a testing ground for the dimensional techniques proposed to date.

(*) To speed up publication, the authors of this paper have agreed to not receive the proofs for correction.

(1) J. ASHMORE: *Lett. Nuovo Cimento*, **4**, 289 (1972); C. BOLLINI and J. GIAMBIAGI: *Nuovo Cimento*, **12** B, 20 (1972); G. 'T HOOFT and M. VELTMAN: *Nucl. Phys.*, **44** B, 189 (1972).

(2) D. A. AKYEAMPONG and R. DELBOURGO: *Nuovo Cimento*, **17** A, 578 (1973); **18** A, 94 (1973); **19** A, 141 (1974).

It is comforting to find them validated. In Sect. **2** we shall give a brief resume of the properties of the Fierz reshuffling matrix for arbitrary l. (Remarkably enough this Fierz transformation was determined by CASE (3) as early as 1955 though the motivation for this was apparently lacking.) In the following two Sections are sketched the computations which give the perturbative corrections in the Thirring model, and in the last Section we show how the axial and scaling anomalies can be extracted as l approaches 1.

2. – The Fierz matrix.

Before we can write down the Fierz transformation we shall need to be precise with the definitions of all the various Γ-matrices. Starting with the Clifford algebra

$$\{\Gamma_M, \Gamma_N\} = 2\eta_{MN} = 2\,\mathrm{diag}\,(1, -1, -1, \ldots, -1)\,, \tag{1}$$

where M and N are vector indices running from 0 to $2l-1$, we can form the raised index matrices $\Gamma^M = \eta^{MN}\Gamma_N$ etc. This fixes $\mathrm{Tr}\,(\Gamma_M\Gamma^N) = 2^l\delta^N_M$. We shall normalize the remaining set of Γ-matrices in identical fashion. First there is the unit matrix with trace 2^l, then there are the vector matrices Γ_M, then we have the « spin » matrices

$$\Gamma_{[KL]} = \tfrac{1}{2}i[\Gamma_K, \Gamma_L] = i\Gamma_K\Gamma_L\,, \qquad K<L\,,$$

the « axial » matrices

$$\Gamma_{[KLM]} = i\Gamma_K\Gamma_L\Gamma_M\,, \qquad K<L<M\,,$$

the « pseudoscalar » matrices

$$\Gamma_{[JKLM]} = \Gamma_J\Gamma_K\Gamma_L\Gamma_M\,, \qquad J<K<L<M\,,$$

and so on, normalized by

$$\mathrm{Tr}\,(\Gamma_{[KL]}\Gamma^{[MN]}) = 2^l(\delta^M_K\delta^N_L - \delta^N_K\delta^M_L) = 2^l\delta^{[MN]}_{[KL]}\,,$$

$$\mathrm{Tr}(\Gamma_{[IJK]}\Gamma^{[LMN]}) = 2^l\delta^{[LMN]}_{[IJK]}\,, \text{ etc}\,.$$

The procedure terminates with $\Gamma_{[012\ldots 2l-1]}$, the analogue of γ_5 in four dimensions. For short we shall denote all these matrices by $\Gamma_{(r)}$, where r refers to the

(3) K. M. CASE: *Phys. Rev.*, **97**, 810 (1955).

number of antisymmetric indices carried by Γ. All the above is then conveniently summarized by one formula

$$\operatorname{Tr}(\Gamma_{(r)}\Gamma^{(s)}) = \delta^{(s)}_{(r)}\,. \tag{2}$$

Note the charge conjugation property $C\Gamma_{(r)}C^{-1} = \eta_r\tilde{\Gamma}_{(r)}$, where

$$\eta_r = 1, -1, -1, 1, -1, \ldots \quad \text{for } r = 0, 1, 2, 3, 4, 5, \ldots\,.$$

The completeness of the Γ-matrices in the spinor space $(\alpha = 1, 2, \ldots, 2^l)$ means that we can write

$$\delta^\beta_\alpha\delta^\delta_\gamma = 2^{-l}\sum_r \Gamma_{(r)\alpha}{}^\delta\,\Gamma^{(r)\beta}{}_\gamma = 2^{-l}\sum_{\text{distinct } r}(\Gamma^{[M_1\ldots M_r]})^\delta_\alpha(\Gamma_{[M_1\ldots M_r]})^\beta_\gamma\,. \tag{3}$$

For general matrices F and G the reshuffling theorem follows:

$$F^\beta_\alpha G^\delta_\gamma = 2^{-l}\sum_r (F\Gamma_{(r)}G)^\delta_\alpha(\Gamma^{(r)})^\beta_\gamma\,. \tag{4}$$

Suppose now that we consider spinor-spinor interactions expanded in terms of the $2l+1$ kinematic covariants $K(r) = \Gamma_{(r)}\otimes\Gamma^{(r)}$ defined explicitly as

$$\begin{cases} K(0)^{\beta\delta}_{\alpha\gamma} = \delta^\beta_\alpha\delta^\delta_\gamma\,, \quad K(1)^{\beta\delta}_{\alpha\gamma} = (\Gamma_M)^\beta_\alpha(\Gamma^M)^\delta_\gamma\,, \quad K(2) = (\Gamma_{[MN]})^\beta_\alpha(\Gamma^{[MN]})^\delta_\gamma, \ldots, \\ K(2l) \;= (\Gamma_{[01\ldots 2l-1]})^\beta_\alpha(\Gamma^{[01\ldots 2l-1]})^\delta_\gamma\,. \end{cases} \tag{5}$$

By the reshuffling rule (4) we can equally well use the crossed kinematic covariants $\tilde{K}(r)^{\beta\delta}_{\alpha\gamma} = (\Gamma_{(r)})^\delta_\alpha(\Gamma^{(r)})^\beta_\gamma$ for the expansion. The linear relation between K and $\tilde{K}$ is

$$\tilde{K}(s)^{\beta\delta}_{\alpha\gamma} = 2^{-l}\sum_r(\Gamma_{(s)}\Gamma_{(r)}\Gamma^{(s)})^\beta_\alpha(\Gamma^{(r)})^\delta_\gamma = \sum_r C^{2l}(s, r)K(r)^{\beta\delta}_{\alpha\gamma}\,, \tag{6}$$

where $C^{2l}(s, r)$, the sr-element of the Fierz transformation in $2l$ dimensions, is fixed through the equation

$$C(s, r)\Gamma_{[M_1\ldots M_r]} = 2^{-l}\sum_{\text{distinct } s}\Gamma_{[N_1\ldots N_s]}\Gamma_{[M_1\ldots M_r]}\Gamma^{[N_1\ldots N_s]}\,. \tag{7}$$

By double reshuffling we may be sure that $C^2 = 1$ and we already know the boundary elements $C(0, r) = 2^{-l}$.

In order to determine the general element $C^{2l}(s, r)$ it is sufficient to pick a particular matrix $\Gamma_{(r)} \to \Gamma_0\Gamma_1\ldots\Gamma_{r-1}$ in (7). By summing over indices N which are and which are not included in $\Gamma_{(r)}$ one can arrive at the general formula

$$C(s,r)=2^{-l}(-1)^{sr}\sum_q(-1)^q\binom{2l-r}{s-q}\binom{r}{q}=\\ =\sum_q\frac{2^{-l}(-1)^{sr+q}(2l-r)!\,r!}{(s-q)!\,(2l-r-s+q)!\,q!\,(r-q)!}. \tag{8}$$

Hence the reflection rules

$$C(2l-s,r)=(-1)^rC(s,r)\,,\qquad C(s,2l-r)=(-1)^sC(s,r)\,. \tag{9}$$

We can construct the matrices for 2, 4 and 6 dimensions from a knowledge of

$$C(0,r)=2^{-l}\,,\quad C(1,r)=2^{-l}(-1)^r(2l-2r)\quad\text{and}\quad C(2,r)=2^{-l}[2(l-r)^2-l]\,,$$

so we might as well note them

$$C^2=\frac{1}{2}\begin{pmatrix}1&1&1\\2&0&-2\\1&-1&1\end{pmatrix},\qquad C^4=\frac{1}{4}\begin{pmatrix}1&1&1&1&1\\4&-2&0&2&-4\\6&0&-2&0&6\\4&2&0&-2&-4\\1&-1&1&-1&1\end{pmatrix},$$

$$C^6=\frac{1}{8}\begin{pmatrix}1&1&1&1&1&1&1\\6&-4&2&0&-2&4&6\\15&5&-1&-3&-1&5&15\\20&0&-4&0&4&0&-20\\15&-5&-1&3&-1&-5&15\\6&4&2&0&-2&-4&-6\\1&-1&1&-1&1&-1&1\end{pmatrix}.$$

With no prior information about the origins of (8) the square property $C^2=1$ would be far from obvious. If one wants to prove this from scratch the best avenue is first to construct a generating function for the Fierz matrix (which yields other dividends besides). One readily checks that

$$(1+z)^{2l}\left[1-w\left(\frac{1-z}{1+z}\right)\right]^{-1}\equiv 2^l\sum_{r,s}(-1)^{sr}C^{2l}(s,r)\,w^rz^s \tag{10}$$

reproduces formula (8). Hence the representation

$$2^l(-1)^{sr}s!\,C(s,r)=\left(\frac{\mathrm{d}}{\mathrm{d}z}\right)^s[\{1-z\}^r\{1+z\}^{2l-r}]\big|_{z=0}\,.$$

It becomes straightforward to verify now that

$$\sum_r C(s,r)\,C(r,t) =$$
$$= 2^{-2l}\frac{1}{s!}\left(\frac{\mathrm{d}}{\mathrm{d}z}\right)^s \left[\{1+z-(-1)^{s+t}(1-z)\}^t\{1+z+(-1)^{s+t}(1-z)\}^{2l-t}\right]\Big|_{z=0} = \delta(r,t)\,.$$

The generating function (10) leads immediately to the reflection rules (9) and furthermore provides recurrence relations such as

$$2C^{2l+2}(s,r) \quad = C^{2l}(s,r) + 2(-1)^r C^{2l}(s-1,r) + C^{2l}(s-2,r)\,,$$
$$2C^{2l+2}(s,r+2) = C^{2l}(s,r) - 2(-1)^r C^{2l}(s-1,r) + C^{2l}(s-2,r)\,,$$

which, if desired, can be used to build up matrices of larger dimension from those of smaller dimension.

3. – Generalized self-energies. Two-dimensional accidents.

In calculating higher-order corrections to four-Fermi interactions one encounters fermion loops. By Fierz transformation these can be recast in the form of meson self-energy parts

$$\Pi^{(s)}_{(r)}(x) = i\langle T[\bar\psi(x)\,\Gamma_{(r)}\,\psi(x)\,\bar\psi(0)\,\Gamma^{(s)}\,\psi(0)]\rangle\,,$$

which we now proceed to evaluate for later reference. For the moment let us be as general as possible by supposing the fermion is massive. Fixing $r \geqslant s$ for definiteness we have to evaluate the momentum-space integral

$$\Pi^{[N_1\dots N_s]}_{[M_1\dots M_r]}(k) = i\int\frac{\mathrm{d}^{2l}p}{(2\pi)^{2l}}\,\frac{\mathrm{Tr}\,\{\Gamma_{[M_1\dots M_r]}(\Gamma\cdot p+m)\,\Gamma^{[N_1\dots N_s]}(\Gamma\cdot p+\Gamma\cdot k+m)\}}{[p^2-m^2][(p+k)^2-m^2]}\,.$$

Introducing a Feynman parameter α in the usual way and utilizing (7) one arrives at

$$\Pi^{[N_1\dots N_s]}_{[M_1\dots M_r]}(k) = \frac{\Gamma(1-l)}{(2\pi)^l}\int_0^1\frac{\mathrm{d}\alpha}{[m^2-k^2\alpha(1-\alpha)]^{1-l}}\,(-1)^r\,(2l-r-1)\,\delta^{[N_1\dots N_s]}_{[M_1\dots M_r]} -$$
$$-\frac{\Gamma(2-l)}{(2\pi)^l}\int_0^1\frac{\mathrm{d}\alpha}{[m^2-k^2\alpha(1-\alpha)]^{2-l}}\,\cdot$$
$$\cdot\left(m^2\{-(-1)^r\}\,\delta^{[N_1\dots N_s]}_{[M_1\dots M_r]} - mk^{[N_1}\delta^{N_2\dots N_s]}_{[M_1\dots M_r]} - 2\alpha(1-\alpha)\,k_{[M_1}k^{[N_1}\delta^{\dots N_s]}_{\dots M_r]}\right).$$

The results for $r=1$ and $r=3$ have been given previously ([1,2]). In particular, we observe that when the fermion mass in zero Π is *diagonal* in the number of indices:

$$\Pi^{[N_1 \dots N_r]}_{[M_1 \dots M_r]}(k) = \frac{\Gamma(1-l)}{(2\pi)^l} \int_0^1 \frac{d\alpha}{[-k^2\alpha(1-\alpha)]^{1-l}} \cdot \left((-1)^r(2l-r-1)\,\delta^{[N_1 \dots N_r]}_{[M_1 \dots M_r]} + 2(l-1)\,k_{[M_1}k^{[N_1}\delta^{N_2 \dots N_r]}_{M_2 \dots M_r]}/k^2\right). \tag{11}$$

The vector self-energy complies with gauge invariance

$$\Pi^{MN}(k) = \frac{2\Gamma(2-l)}{(2\pi)^l} \int_0^1 \frac{d\alpha}{[-k^2\alpha(1-\alpha)]^{1-l}} \left(\eta^{MN} - \frac{k^M k^N}{k^2}\right) = \left(\eta^{MN} - \frac{k^M k^N}{k^2}\right) \Pi(k^2)\,, \tag{12}$$

and its x-space transform can be usefully expressed in the form

$$i\Pi^{MN}(x) = 2^l(l-1)(2l-1)^{-1}(\partial^M\partial^N - \eta^{MN}\partial^2)\,D^2(x)\,, \tag{12'}$$

corresponding to the Fourier transform

$$i\int \exp\,[-ik\cdot x]\,\Pi(k^2)\,d^{2l}k/k^2(2\pi)^{2l} = 2^l(l-1)(2l-1)^{-1}D^2(x)\,, \tag{12''}$$

where the massless causal propagator is

$$iD(x) = \Gamma(l-1)(-x^2+i\varepsilon)^{1-l}/4\pi^l\,. \tag{13}$$

In the two-dimensional limit $\Pi_{MN} \to (\eta_{MN} - k_M k_N/k^2)/\pi$ is finite! For future reference we also remark that near $l=1$

$$(l-1)\,D^2(x) \to \ln\,(-x^2+i\varepsilon)/8\pi^2 = -\,iD(x)/2\pi\,. \tag{13'}$$

So far as the Fierz transformation is concerned, inspection of C^2 reveals that a vector is reshuffled into a scalar and a tensor (which equals the pseudoscalar in two dimensions)—in fact the Fierz invariants are

$$K(0)-K(2)\,, \qquad K(0)+K(1)-K(2) \qquad \text{and} \qquad K(0)-K(1)-K(2)$$

with crossing eigenvalues 1, 1 and -1 respectively. This suggests that a Feynman diagram like Fig. 2*b*) will vanish in two dimensions; this intuitive guess is substantiated by a careful passage to $l=1$ only because the vector self-energy happens to be finite in this limit and multiplies a fading Fierz matrix factor $C(1,1)=2^{1-l}(1-l)$. We can anticipate many happy simplifications for the Thirring model therefore. In any other dimension but two the argument could not be pushed through.

There exist a number of purely spinorial identities peculiar to two dimensions which are vital for summing the $l\to 1$ perturbation series. The general identity

$$\Gamma_L\Gamma_M\Gamma_N-\Gamma_N\Gamma_M\Gamma_L=-2i\Gamma_{[LMN]} \tag{14}$$

reduces in two dimensions to

$$\gamma_\lambda\gamma_\mu\gamma_\nu=\gamma_\nu\gamma_\mu\gamma_\lambda\,, \tag{14'}$$

providing we do not close the fermion lines between which this matrix acts (or if we do so, providing the fermion loop integral converges). Letting $S(p)=(p\cdot\Gamma)^{-1}$ stand for the massless fermion propagator, another general identity is

$$S(p)\Gamma_M S(p-k)=(k_M+ik^N\Gamma_{[MN]})\big(S(p-k)-S(p)\big)/k^2-\\ -2ip^Lk^N\Gamma_{[LMN]}S(k)S(p-k)/p^2\,, \tag{15}$$

and it collapses into the two-dimensional form

$$S(p)\gamma_\mu S(p-k)=k^\nu(\eta_{\mu\nu}-\varepsilon_{\mu\nu}\gamma_5)\big(S(p-k)-S(p)\big)/k^2 \tag{15'}$$

because $\Gamma_{[\mu\nu]}\to i\varepsilon_{\mu\nu}\gamma_5\equiv i\varepsilon_{\mu\nu}\gamma_0\gamma_1$. In view of (14′), an equivalent way to express (15′) is

$$S(p)\gamma_\mu S(p-k)=S(k)\gamma_\mu S(p-k)-S(p)\gamma_\mu S(k)\,. \tag{15''}$$

Rules (14′), (15′) and (15″) are absolutely crucial in summing the perturbation graphs (4) for the Thirring model and in proving that $j=\bar\psi\gamma\psi$ behaves like a free field (5).

4. – Perturbation theory for the Thirring model.

We shall now focus on the lowest-order quantum loop corrections to the Thirring model in $2l$ dimensions:

$$\mathscr{L}=\bar\psi(i\Gamma\cdot\partial-m)\psi+\tfrac{1}{2}g(\bar\psi\Gamma_M\psi)(\bar\psi\Gamma^M\psi) \tag{16}$$

(4) K. Johnson: *Nuovo Cimento*, **20**, 773 (1961); A. H. Mueller and T. L. Trueman: *Phys. Rev. D*, **4**, 1635 (1971).
(5) H. Georgi and J. M. Rawls: *Phys. Rev. D*, **3**, 874 (1971).

in an effort to understand the miracles which make the perturbation graphs summable in the limit $l \to 1$. We will essentially be repeating the calculations of MUELLER and TRUEMAN [4] except that instead of introducing regulator fields and nonlocal interactions we shall be relying on l-continuation.

The simplest calculation, that of $\Pi_{MN}(x) = i\langle T[j_M(x) j_N(0)]\rangle$, has already been done in lowest order and is stated in (12). Higher-order corrections to $\Pi_{\mu\nu}$ amount to summing the bubble graphs because other diagrams give zero at $l = 1$ after Fierz reshuffling; it is not hard to see that the complete sum is

$$\Pi_{\mu\nu}(k) = (\eta_{\mu\nu} - k_\mu k_\nu/k^2)/(\pi + g)\,.$$

For our second task let us evaluate the second-order fermion self-energy: the two graphs depicted in Fig. 1*a*) and *b*) (the latter being zero at $l = 1$). For general l they add up to

$$\Sigma(p) = ig^2(2\pi)^{-2l}\int \mathrm{d}^{2l}k\, \Gamma_{(1)} S(p+k)[\Pi^{(1)}_{(1)}\Gamma^{(1)} + \sum_r C(1, r)\Pi^{(r)}_{(1)}\Gamma_{(r)}]\,.$$

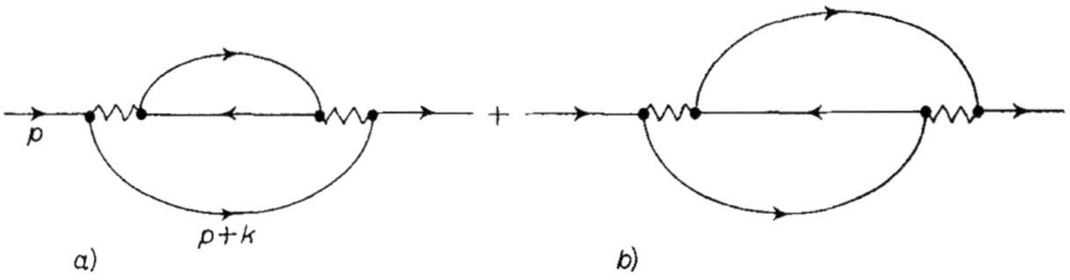

Fig. 1. – Fermion self-energy graphs in order g^2.

Since Π is diagonal and vectorial and since $C(1, 1) = 2^{1-l}(1 - l)$ we get

$$\Sigma(p) = ig^2[1 + 2^{1-l}(1 - l)]\int \frac{\mathrm{d}^{2l}k}{(2\pi)^{2l}}\, \Gamma_M \frac{(\eta^{MN} - k^M k^N/k^2)}{\Gamma\cdot(p+k)}\, \Gamma_N \Pi(k^2)\,. \tag{17}$$

Now one of the important consequences of dimensional regularization is that the total volume element $\int \mathrm{d}^{2l}k$ and the massless tadpole graphs $D(0) = \int \mathrm{d}^{2l}k/k^2$ can be consistently set equal to zero (which incidentally explains why Σ vanishes to order g). Therefore

$$-iS(p)\Sigma(p)S(p) = g^2[1 + 2^{1-l}(1 - l)](2\pi)^{-2l}\int \mathrm{d}^{2l}k\, \Pi(k^2)/k^2\, \Gamma\cdot(p+k).$$

If we put $l = 1$ at once, $\Pi(k^2) \to 1/\pi$ and we recognize the Fourier transform of $g^2 D(x) S(x)/\pi$, the free massless scalar and spinor propagators being

$$iD(x) \to \ln(-x^2 + i\varepsilon)/4\pi\,, \qquad S(x) = i\gamma\cdot\partial D(x) = \gamma\cdot x/2\pi(x^2 - i\varepsilon)\,.$$

However if we reserve $l \to 1$ till the very end and make use of (12″) we deduce that

$$(18) \qquad iS'(x) = \langle T\psi(x)\bar{\psi}(0)\rangle = \\ = iS(x)[1 + g^2\{2^l + 2(1-l)\}(l-1)D^2(x)/(2l-1) + \ldots].$$

Johnson's answer is recovered (4) by recalling the limit formula (13′) so that $iS'(x) \to iS(x)[1 - ig^2 D(x)/\pi + \ldots]$. Summing over higher-order graphs in the manner of MUELLER and TRUEMAN (which simply corresponds to dressing internal vector lines with bubbles) one ends up with

$$(19) \qquad iS'(x) = iS(x)\exp[-ig^2 D(x)/(\pi + g)].$$

Consider next the vector Green's function $\langle\psi(x)j^M(z)\bar{\psi}(y)\rangle$ which also receives two first-order contributions (Fig. 2*a*) and *b*)). Following the same line of reasoning which led to eq. (17) we obtain the momentum-space amplitude

$$J^M(p, p-k) = g[2^l + 2(1-l)]S(p)(\eta^{MN} - k^M k^N/k^2)\Gamma_N S(p-k) = \\ = ig[1 + 2^{1-l}(1-l)]\Pi(k^2)\Gamma_{[MN]}k^N[S(p-k) - S(p)]/k^2.$$

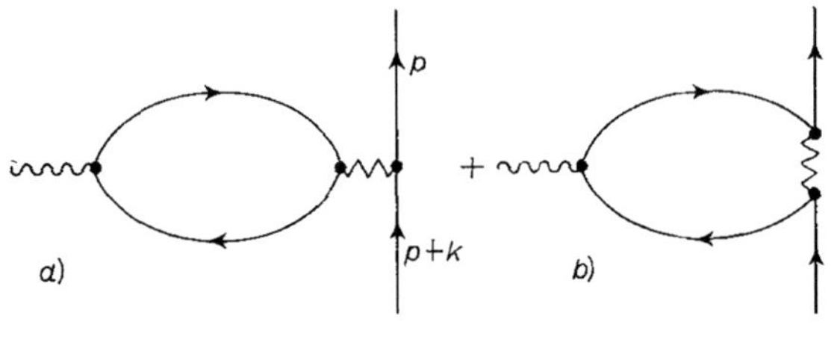

Fig. 2. – Vertex corrections in order g^2.

Taking its Fourier transform we get the x-space Green's function

$$g(2l-1)^{-1}[2^l + 2(1-l)](l-1)\Gamma_{[MN]}\partial^N[D^2(z-x) - D^2(z-y)]S(x-y),$$

and as $l \to 1$ we can recognize this as the first-order term in the complete answer

$$(20) \qquad [\eta_{\mu\nu} + \varepsilon_{\mu\nu}\gamma_5(1 + g/\pi)^{-1}]\partial^\nu[D(z-x) - D(z-y)]\cdot iS'(x-y),$$

which comes by summing the bubble graphs and dressing the fermions.

To close this Section let us sketch how one may work out the four-point Green's function $\langle T[\psi(x)\bar{\psi}(y)\psi(x')\bar{\psi}(y')]\rangle$. Up to order g^2 we have to contend with six diagrams as well as their crossed versions (Fig. 3*a*) to *f*)). The main

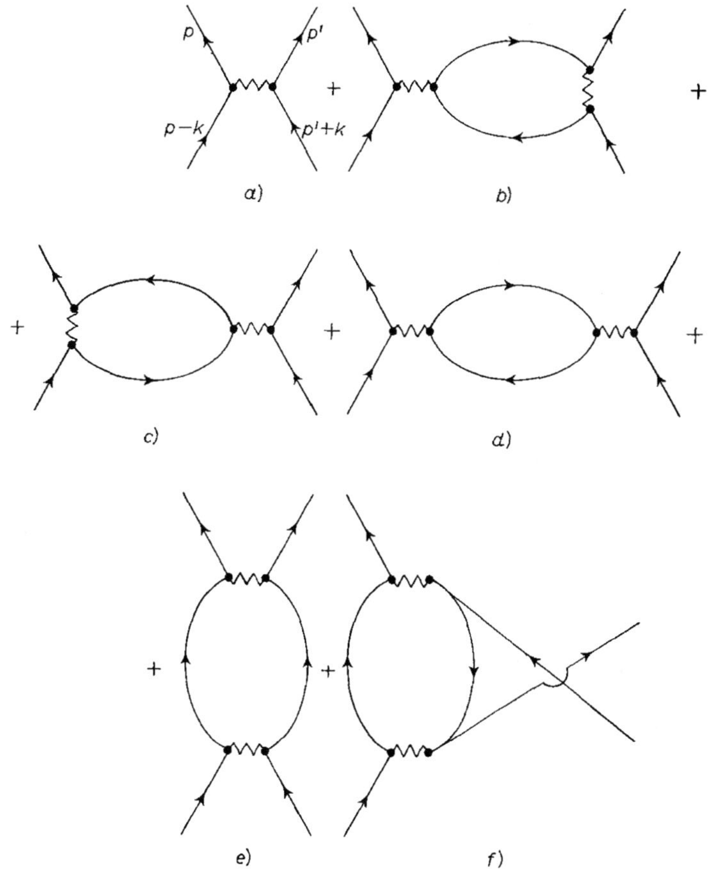

Fig. 3. – Fermion-fermion scattering in order g^2.

thing to expose is how vector exchange can be replaced by scalar and pseudo-scalar exchange through a judicious use of (15′). For instance, in lowest order

$$(21)\qquad gS(p)\gamma_\mu S(p-k)\otimes S(p')\gamma^\mu S(p'+k)=$$

$$=gk^\nu(\eta_{\mu\nu}+\varepsilon_{\mu\nu}\gamma_5)[S(p-k)-S(p)]\otimes k_\lambda(\eta^{\mu\lambda}+\varepsilon^{\mu\lambda}\gamma_5)[S(p')-S(p'+k)]/k^4=$$

$$=g[S(p-k)-S(p)](1\otimes 1+\gamma_5\otimes\gamma_5)[S(p')-S(p'+k)]/k^2\,.$$

Together with its crossed version amplitude (21) has the x-transform

$$g(1\otimes 1-\gamma_5\otimes\gamma_5)\cdot$$

$$\cdot[D(x-x')-D(x-y')-D(x'-y)+D(y-y')]S(x-y)S(x'-y')\,.$$

In order g^2 diagrams 3*b*), *c*) and *d*) have already been discussed and it is easy to see that 3*e*) and *f*) have cancelling ultraviolet characteristics as

$l \to 1$; both lead to the amputated amplitude

$$(2\pi)^{2l} T(p, p', k) =$$
$$= i\int d^{2l}q\, \Gamma_M S(q) \Gamma_N \otimes [\Gamma^M S(p+p'-q)\Gamma^N + \Gamma^N S(p'-p+k+q)\Gamma^M] =$$
$$= i\int d^{2l}q\, \Gamma_M S(q) \Gamma_N \otimes \Gamma^M [S(p+p'-q) + S(p'-p+k+q)]\Gamma^N,$$

if we drop all $\Gamma_{[MLN]}$ terms. By persistently applying (15), the Green's function can be expressed as linear combinations of integrals like

$$(1\otimes 1 - \gamma_5 \otimes \gamma_5)\int \frac{d^{2l}q}{(2\pi)^{2l}} \frac{1}{q^2(k-q)^2} [S(p-k) - S(p-q)] \otimes [S(p'+k) - S(p'+q)],$$

which are in turn recognizable as Fourier transforms of mixed terms

$$(1\otimes 1 - \gamma_5 \otimes \gamma_5) D(x-x') D(x-y') S(x-y) S(x'-y') .$$

All this is just to indicate how the general sum of eikonal graphs (4) can be performed. With dressed fermions, the complete Green's function is

$$\exp\left[ig\{1\otimes 1 - \gamma_5 \otimes \gamma_5(1+g/\pi)^{-1}\}\cdot \right.$$
$$\left.\cdot\{D(x-x') - D(x-y') - D(x'-y) + D(y-y')\}\right]\cdot iS'(x-y)\cdot iS'(x'-y'), \tag{22}$$

a result which otherwise appears rather mysterious.

5. – Anomalies.

In two dimensions one identifies γ_5 with the spin tensor $\Gamma_0\Gamma_1$ although its charge parity is negative. The correct approach to extract the two-dimensional axial anomaly (5) is to associate chirality with the set of spin transformations $\psi \to \exp[\frac{1}{2} i\theta^{KL}\Gamma_{[KL]}]\psi$ and to take the $l \to 1$ limit of the axial Ward identity. Thus, under an infinitesimal « chiral » transformation, the change in $\mathscr{L}$, expressed in the two ways

$$\delta\mathscr{L} = \tfrac{1}{2} i\delta\theta^{KL}(i\bar\psi\{\tfrac{1}{2}\Gamma\cdot\overleftrightarrow{\partial}, \Gamma_{[KL]}\}\psi + g\bar\psi\{\Gamma_M, \Gamma_{[KL]}\}\psi\bar\psi\Gamma^M\psi),$$
$$= -\tfrac{1}{4}\delta\theta^{KL}\partial^M(\bar\psi[\Gamma_M, \Gamma_{[KL]}]\psi),$$

yields the PCAC relation

$$i\partial_{[K}\bar\psi\Gamma_{L]}\psi = \tfrac{1}{2}\bar\psi\{\Gamma\cdot\overleftrightarrow{\partial}, \Gamma_{[KL]}\}\psi + g\bar\psi\{\Gamma_M, \Gamma_{[KL]}\}\psi\bar\psi\Gamma^M\psi . \tag{23}$$

The anomalous term on the right-hand side of (23) is nonexistent in two dimensions. As we have stressed before [2] it would be wrong to delete them for $l=1$ owing to fermion loop corrections. Specifically, there is a transition from vector current to axial divergence

$$\langle i\partial_{[K}\bar\psi\Gamma_{L]}\psi,\ \bar\psi\Gamma_M\psi\rangle = \tfrac{1}{2}\langle\bar\psi\{\Gamma\cdot\overleftrightarrow{\partial},\Gamma_{[KL]}\}\psi,\ \bar\psi\Gamma_M\psi\rangle\,,$$

where the $(\bar\psi\Gamma\psi)^2$ anomalous term can be disregarded because tadpole graphs are necessarily involved. In momentum space the calculation devolves to the product of an integral which diverges as $l\to1$ and a trace which disappears as $l\to1$:

$$k_{[K}\Pi_{L]M} = \int\frac{d^{2l}p}{(2\pi)^{2l}}\,\frac{\mathrm{Tr}\,[\{2p+k,\Gamma,\Gamma_{[KL]}\}\Gamma\cdot p\Gamma_M\Gamma\cdot(p+k)]}{(p+k)^2p^2} = \tag{24}$$

$$= \frac{2(1-l)}{l(2\pi)^l}(\eta_{KM}k_L-\eta_{KL}k_M)\int_0^1\frac{d\alpha\,\Gamma(1-l)}{[k^2\alpha(\alpha-1)]^{1-l}}\underset{l\to1}{\rightrightarrows}(\eta_{KM}k_L-\eta_{KL}k_M)/\pi\,.$$

If we were to introduce electromagnetism $e\bar\psi\Gamma\cdot\mathscr{A}\psi$ into the Thirring model, one would interpret (24) as an anomalous PCAC equation $\partial^\mu j_{\mu5} = -e\varepsilon^{\mu\nu}\partial_\mu\mathscr{A}_\nu/\pi$ in lowest order. Summing the higher-order bubble graphs, the effect, as usual, is to replace π by $\pi+g$. It is perhaps important to stress that no other anomalous amplitudes such as $\langle i\partial_{[K}\bar\psi\Gamma_{L]}\psi,\ \bar\psi\psi\bar\psi\Gamma_M\psi\rangle$ survive the limit $l\to1$ because the fermion loop integrals are finite and multiply zero kinematic traces.

The scaling anomaly is a similar story, and we shall show it can be consistently determined from the trace of the stress tensor

$$\theta^M_M = (l-1)g(\bar\psi\Gamma_M\psi)(\bar\psi\Gamma^M\psi) = 2(l-1)\mathscr{L}_{\mathrm{int}}\,,$$

which does *not* vanish except when $l=1$. The anomalous scale dimension of ψ can be obtained from the pair of relations

$$i\left\langle\psi(x)\int\theta^M_M(z)\,d^{2l}z\,\bar\psi(0)\right\rangle = 2(l-1)\,g\frac{\partial}{\partial g}S'(x)\,,$$

$$= \left(x\cdot\frac{\partial}{\partial x}-2d\right)S'(x)\,.$$

Having already seen that near $l=1$,

$$S'(x) = S(x)[1+2g^2(l-1)D^2(x)+\ldots]\,,$$

we can proceed to the two-dimensional limit

$$\lim_{l\to1}2(l-1)g\frac{\partial}{\partial g}S'(x) = \lim_{l\to1}8g^2(l-1)^2D^2(x)S(x) = -\frac{g^2}{2\pi^2}s(x)\,,$$

and get the anomalous scale corrections of $g^2/4\pi^2$ to this order. If we sum the higher-order bubble graphs, the total scale dimension is amended to

$$d = \frac{1}{2} + \frac{g^2/4\pi^2}{1 + g/\pi},$$

which could in fact have been immediately deduced from the nonperturbative answer (19) as did WILSON (6). Since a change of gauge alters the scale dimension without affecting S-matrix elements, MUELLER and TRUEMAN have questioned the physical significance of d.

* * *

We wish to thank Prof. GURALNIK and Dr. AKYEAMPONG for discussions.

---

(6) K. G. WILSON: *Phys. Rev. D*, **2**, 1473 (1971).

● RIASSUNTO (*)

Si mostra come si possa trattare il modello di Thirring e le sue anomalie per mezzo di regolarizzazione dimensionale.

(*) *Traduzione a cura della Redazione.*

Спинорные взаимодействия в двумерном пределе.

Резюме (*). — Показывается, как модель Тирринга с ее аномалиями может трактоваться как регуляризация многомерности.

(*) *Переведено редакцией.*

J. Phys. A: Math. Gen., Vol. 10, No. 12, 1977. Printed in Great Britain. © 1977

LETTER TO THE EDITOR

A dimensional derivation of the gravitational PCAC correction

R Delbourgo

Department of Physics, University of Tasmania, Hobart, 7001, Australia

Received 18 October 1977

Abstract. The gravitational contribution to the anomaly in the axial current divergence, i.e. $RR^*/384\pi^2$, is derived via dimensional continuation.

The contribution of the gravitational field to the axial current anomaly (Kimura 1969, Delbourgo and Salam 1972) has assumed a new importance in the context of gravitational pseudoparticles (Belavin and Burlankov 1976, Eguchi and Freund 1976, Hawking 1977) and because of its possible effect on the topology of Yang–Mills fields (Charap and Duff 1977). The derivations of the gravitational part of the anomaly have in the past been obtained on the basis of regulator fields or through a point-splitting procedure. Now that dimensional regularisation has become the method of choice for dealing with formally divergent integrals in gauge theories, it is probably worthwhile to rederive the anomaly by this technique. We do so below.

The vector field contributions to the anomaly were worked out by dimensional continuation some time ago (Akyeampong and Delbourgo 1973a, b) and the procedures used there can guide us in our determination of the gravitational contributions. The idea is simply to continue the field theory of spinors, vectors and tensors to arbitrary dimension $n=2l$, and to identify axial vectors and pseudoscalars as three-fold or four-fold antisymmetric tensors respectively. The axial anomaly is then simply given in terms of expectation values of the evanescent current

$$\bar{\psi}\{\Gamma\,.\,\overleftrightarrow{\partial},\Gamma_{[\kappa\lambda\mu\nu]}\}\psi \tag{1}$$

which vanishes in four dimensions at the classical level but can leave its imprint through fermion loop integrals which diverge in four dimensions at the quantum level; the product of the disappearing trace, proportional to $(l-2)$, and the divergence pole $(l-2)^{-1}$ leaves the finite anomaly.

Clearly we must set up general relativity for fermions in arbitrary space–time dimensions in preparation for continuation to $l=2$. This is straightforwardly accomplished in terms of a *vierbein* field L^{α}_{μ} by simply generalising the normal four-dimensional procedure. The relevant part of the Lagrangian is then

$$\mathscr{L}(\psi,L)=(\det L)^{-1}[\tfrac{1}{2}\mathrm{i}L^{\mu\alpha}(\bar{\psi}\Gamma_{\alpha}\overleftrightarrow{\partial}_{\mu}\psi)-m\bar{\psi}\psi+\tfrac{1}{8}\bar{\psi}\{\Gamma^{\alpha},\Gamma^{[\beta\gamma]}\}\psi B_{\alpha\beta\gamma}] \tag{2}$$

with

$$B_{\alpha\beta\gamma}\equiv L^{\mu}_{\alpha}L^{\nu}_{\beta}(L_{\gamma\rho}\Gamma^{\rho}_{\mu\nu}-\partial_{\mu}L_{\nu\gamma})$$

$$\Gamma^{\rho}_{\mu\nu}=\tfrac{1}{2}g^{\rho\lambda}(\partial_{\nu}g_{\lambda\mu}+\partial_{\mu}g_{\lambda\nu}-\partial_{\lambda}g_{\mu\nu})$$

$$g_{\mu\nu}=\eta^{\alpha\beta}L_{\alpha\mu}L_{\beta\nu}.$$

The $\Gamma^{[\alpha_1\ldots\alpha_r]}$ denote r-fold antisymmetric products of the basic Γ matrices associated with the flat-space limit and suitably normalised. We may define the quantum graviton field f through the split $L^\mu_\alpha = \delta^\mu_\alpha + \kappa f^\mu_\alpha$ and thereby reduce our Lagrangian to the form

$$\mathscr{L}(\psi, L) = \{1 - \kappa \operatorname{Tr} f + \tfrac{1}{2}\kappa^2[(\operatorname{Tr} f)^2 + (\operatorname{Tr} f^2)] + \ldots\}$$
$$\times[\tfrac{1}{2}\mathrm{i}\bar{\psi}\Gamma \cdot \overleftrightarrow{\partial}\psi - m\bar{\psi}\psi + \tfrac{1}{2}\mathrm{i}\kappa f^{\alpha\beta}\bar{\psi}\Gamma_\alpha\overleftrightarrow{\partial}_\beta\psi + \tfrac{1}{16}\kappa^2\bar{\psi}\Gamma^{[\alpha\beta\gamma]}\psi f_{\beta\delta}\overleftrightarrow{\partial}_\alpha f^\delta{}_\gamma + \mathrm{O}(\kappa^3)].$$

As we are going to treat gravity classically, we shall impose the mass shell conditions $\partial^\alpha f_{\alpha\beta} = f_{\gamma\gamma} = 0$ to obtain the *effective* interaction Lagrangian

$$\mathscr{L}_{\mathrm{int}}(\psi, L) = \tfrac{1}{2}\mathrm{i}\kappa f^{\alpha\beta}\bar{\psi}\Gamma_\alpha\overleftrightarrow{\partial}_\beta\psi + \tfrac{1}{2}\kappa^2 f^{2\alpha}{}_\alpha\bar{\psi}(\tfrac{1}{2}\mathrm{i}\Gamma \cdot \overleftrightarrow{\partial} - m)\psi - \tfrac{1}{16}\kappa^2 f_{\beta\delta}\overleftrightarrow{\partial}_\alpha f^\delta{}_\gamma\bar{\psi}\Gamma^{[\alpha\beta\gamma]}\psi + \mathrm{O}(\kappa^3) \quad (3)$$

and hence the *effective* Feynman rules shown in figure 1.

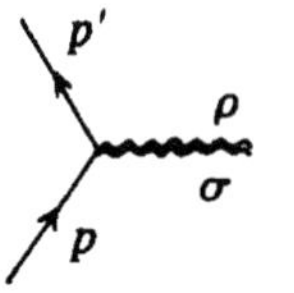

$$\tfrac{1}{4}\kappa[(p+p')_\rho\Gamma_\sigma + (p+p')_\sigma\Gamma_\rho]$$

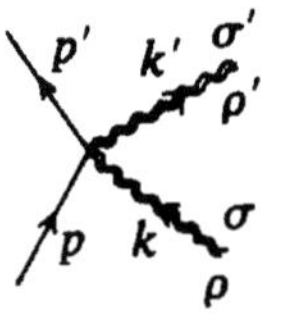

$$\tfrac{1}{2}\kappa^2(\eta_{\rho\rho'}\eta_{\sigma\sigma'} + \eta_{\rho\sigma'}\eta_{\sigma\rho'})[\tfrac{1}{2}\Gamma \cdot (p+p') - m]$$
$$+ \tfrac{1}{16}\kappa^2(\eta_{\rho\rho'}\delta^\lambda_\sigma\delta^\mu_{\sigma'} + \eta_{\sigma\sigma'}\delta^\lambda_\rho\delta^\mu_{\rho'} + \eta_{\rho\sigma'}\delta^\lambda_\sigma\delta^\mu_{\rho'} + \eta_{\sigma\rho'}\delta^\lambda_\rho\delta^\mu_{\sigma'})(k+k')^\nu\Gamma_{[\lambda\mu\nu]}$$

Figure 1. Effective spinor graviton couplings to order κ^2.

The pesuodoscalar–two-graviton amplitude, subject to gauge invariance and mass shell conditions, can be decomposed as follows:

$$F^{(\rho_1\sigma_1)(\rho_2\sigma_2)}_{[\kappa\lambda\mu\nu]}(k_1k_2k_3) = P(k_3^2)\delta^{\sigma_1}_{[\kappa}\delta^{\sigma_2}_{\lambda}k_{1\mu}k_{2\nu]}(\eta^{\rho_1\rho_2}k_1 \cdot k_2 - k_1{}^{\rho_2}k_2{}^{\rho_1}) + (\rho \leftrightarrow \sigma \text{ perms}) \quad (4)$$

where P is a scalar invariant of k_3^2 only (remember $k_1^2 = k_2^2 = 0$). It automatically satisfies $k_{1\rho_1}F^{\rho_1\cdots}_{\cdots} = k_{2\rho_2}F^{\cdot\cdot\rho_2\cdot}_{\cdots} = 0$ from the antisymmetrical structure on the right of (3). Likewise the axial–two-graviton amplitude reads

$$F^{(\rho_1\sigma_1)(\rho_2\sigma_2)}_{[\kappa\lambda\mu]} = A(k_3^2)\delta^{\rho_1\rho_2\pi\tau}_{[\kappa\lambda\mu\nu]}k_{1\pi}k_{2\tau}k_3{}^\nu(\eta^{\sigma_1\sigma_2}k_1 \cdot k_2 - k_1^{\delta_2}k_2^{\sigma_1}) + (\rho \leftrightarrow \sigma \text{ perms})$$

and satisfies the requisite constraints. The axial anomaly itself arises from the diagrams of figure 2. Now the beauty of the dimensional technique is that the individual axial A and pseudoscalar P form factors need not be separately evaluated; rather, the anomaly is immediately obtained by insertion of the evanescent vertex $\{\Gamma \cdot p, \Gamma_{[\kappa\lambda\mu\nu]}\}$ at the meson leg and extracting the appropriate 'divergent' parts of the diagrams.

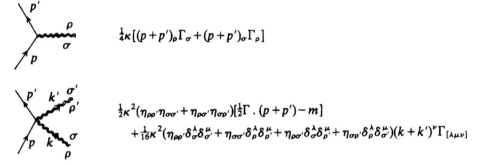

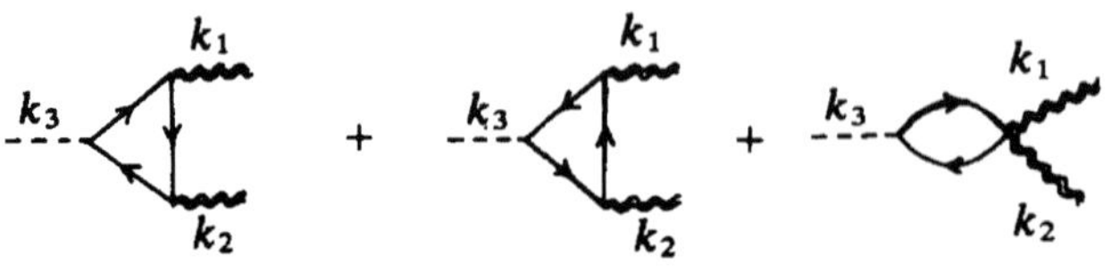

Figure 2. Gravitational contributions to the axial anomaly.

Thus we need only evaluate the loop integrals†

$$\mathrm{Tr}\int \bar{d}^{2l}p\{\Gamma.p,\Gamma_{\kappa\lambda\mu\nu}\}[\Gamma.(p+k_1)-m]^{-1}p^{\rho_1}\Gamma^{\sigma_1}(\Gamma.p-m)^{-1}p^{\rho_2}\Gamma^{\sigma_2}[\Gamma.(p-k_2)-m]^{-1}$$

and

$$\mathrm{Tr}\int \bar{d}^{2l}p\{\Gamma.p,\Gamma_{\kappa\lambda\mu\nu}\}\eta^{\sigma_1\sigma_2}\delta_\xi^{\rho_1}\delta_\eta^{\rho_2}(k_1-k_2)_\zeta[\Gamma.(p+k_1)-m]^{-1}\Gamma^{[\xi\eta\tau]}[\Gamma.(p-k_2)-m]^{-1}.$$

Only the first integral is relevant if we concentrate on the kinematic piece $\delta_{[\kappa}^{\sigma_1}\delta_\lambda^{\sigma_2}k_{1\mu}k_{2\nu]}k_2^{\rho_1}k_1^{\rho_2}$ of (4). By introducing Feynman parameters, the triangle anomaly is thereby reduced to a consideration‡ of

$$\begin{aligned}
2\int_0^1 & \mathrm{d}x\,\mathrm{d}y\,\mathrm{d}z\,\delta(1-x-y-z)\int \bar{d}^{2l}p(p^2+k_3^2xy-m^2)^{-3}\\
&\times\mathrm{Tr}\{\{\Gamma.p,\Gamma_{\kappa\lambda\mu\nu}\}[\Gamma.(p+zk_1-xk_3)+m](p+xk_2)^{\rho_1}\Gamma^{\sigma_1}\\
&\times[\Gamma.(p+xk_2-yk_1)+m](p-yk_1)^{\rho_2}\Gamma^{\sigma_2}[\Gamma.(p+yk_3-zk_2)+m]\}\\
&=2k_2^{\rho_1}k_1^{\rho_2}\int_0^1 \mathrm{d}x\,\mathrm{d}y\,\theta(1-x-y)xy\\
&\times\int \bar{d}^{2l}p(p^2+k_3^2xy-m^2)^{-3}\,\mathrm{Tr}(\{\Gamma.p,\Gamma_{\kappa\lambda\mu\nu}\}\Gamma.p\Gamma^{\sigma_1}\Gamma^{\sigma_2}\Gamma.k_1\Gamma.k_2)+\dots\\
&=2k_2^{\rho_1}k_1^{\rho_2}\int_0^1 \mathrm{d}x\,\mathrm{d}y\,\theta(1-x-y)xy\\
&\times\int \bar{d}^{2l}p\,p^2(p^2+k_3^2xy-m^2)^{-3}.2\Big(\frac{l-2}{l}\Big).\mathrm{Tr}(\Gamma_{\kappa\lambda\mu\nu}\Gamma^{\sigma_1}\Gamma^{\sigma_2}\Gamma.k_1\Gamma.k_2)+\dots\\
&=2^{l+1}(l-2)k_2^{\rho_1}k_1^{\rho_2}\delta_{[\kappa}^{\sigma_1}\delta_\lambda^{\sigma_2}k_{1\mu}k_{2\nu]}(4\pi)^{-l}\Gamma(2-l)\\
&\times\int (k_3^2xy-m^2)^{l-2}\mathrm{i}xy\theta(1-x-y)\,\mathrm{d}x\,\mathrm{d}y\\
&\xrightarrow[l\to 2]{}-\mathrm{i}\frac{k_2^{\rho_1}k_1^{\rho_2}}{12\pi^2}\delta_{[\kappa}^{\sigma_1}\delta_\lambda^{\sigma_2}k_{1\mu}k_{2\nu]}+\dots.
\end{aligned}$$

Inserting the appropriate numerical factor $\frac{1}{16}\kappa^2$ and the various index permutations, the result may be interpreted as the generally covariant anomaly

$$R_{\kappa\lambda\rho\sigma}R_{\mu\nu}{}^{\rho\sigma}\epsilon^{\kappa\lambda\mu\nu}/384\pi^2.$$

The derivation is therefore really quite easy despite the profusion of indices.

References

Akyeampong D A and Delbourgo R 1973a *Nuovo Cim.* A **17** 578
—— 1973b *Nuovo Cim.* A **18** 94
Belavin A A and Burlankov D E 1976 *Phys. Lett.* **58A** 7

† Remember that we are able to drop all $\eta_{\rho_1\sigma_1}$, $k_{1\rho_1}$, $k_{2\rho_2}$ terms etc for real gravitons.
‡ Divergent numerators of order p^4 have identically zero trace and are safely zero; we can also discard all convergent numerators of order p^0 for purposes of the anomaly.

Charap J M and Duff M 1977 *Phys. Lett.* **69B** 445
Delbourgo R and Salam A 1972 *Phys. Lett.* **40B** 381
Eguchi T and Freund P G O 1976 *Phys. Rev. Lett.* **37** 1251
Hawking S 1977 *Phys. Lett.* **60A** 81
Kimura R 1969 *Prog. Theor. Phys.* **42** 1191

Chiral and gravitational anomalies in any dimension

R. Delbourgo
Department of Physics, University of Tasmania, Hobart, Tasmania, Australia 7001
T. Matsuki
Department of Physics, University of Tasmania, Hobart, Tasmania, Australia 7001 and Department of Physics and Astronomy, Louisiana State University, Baton Rouge, Louisiana 70803

(Received 26 September 1984; accepted for publication 25 January 1985)

Gravitational contributions to the chiral anomaly in $4N$ space-time dimensions as well as the purely gravitational anomaly in $4N-2$ dimensions are expressed in terms of the Riemann–Christoffel tensor. Using this formula, we give a simple proof that if $N \geqslant 4$ there is no way to cancel the gravitational anomalies using fields of spin-$\frac{1}{2}$, -$\frac{3}{2}$, and -1.

I. INTRODUCTION

It seems that physicists and mathematicians have independently developed the same theory of the chiral anomaly.[1,2] Motivated by the unification of gravity with other gauge fields in higher dimensions, many physicists have recently calculated the chiral anomalies due to gauge fields[3] as well as the gravitational field[4] in higher dimensions. They have searched for a consistent theory in which anomalies are canceled among fields of different representations of a gauge group and/or different spin. To arrive at their results they have resorted to Feynman diagram or path integral methods, neither procedure being too difficult to apply for external spin-1 fields. In the case of a gravitational field, however, it seems that the higher the dimension the more complicated the calculation becomes.[4,5] On the other hand, mathematicians[2] have studied the subject in arbitrary dimensions from the beginning but have expressed their results in their own fashion, in a terminology slightly unfamiliar to physicists. The mathematicians' results have often preceded the physicists' derivation.

Recognizing this fact, we shall express the mathematicians' expressions for the chiral anomaly (for "spin"-$\frac{1}{2}$ and -$\frac{3}{2}$ elds) in terms of the Riemann–Christoffel tensor more familiar to physicists. The newly discovered pure gravitational anomaly[6] receives contributions from spinor fields and antisymmetric tensor fields among others and it can also be written out in terms of the curvature tensor in a similar way to the chiral case. Finally, using the explicit formulas derived by us, we can discuss a cancellation of the gravitational anomaly among fields of different spin in arbitrary dimensions.

II. THE GENERATING FUNCTION AND THE COEFFICIENTS

Let us start by describing the mathematical terminology for the chiral anomaly in a gravitational background. The index theorem[2] tells us that the contributions of one left-handed spin-$\frac{1}{2}$ and one left-handed spin-$\frac{3}{2}$ field to the chiral anomaly in $D=4N$ space-time dimensions are given, respectively, by

$$A_D^{1/2} = C_D \prod_{i=1}^{D/2} \frac{x_i/2}{\sinh(x_i/2)} \tag{1}$$

and

$$A_D^{3/2} = C_D \prod_{i=1}^{D/2} \frac{x_i/2}{\sinh(x_i/2)} \left[2\sum_{j=1}^{D/2} \cosh x_j - 1 \right], \tag{2}$$

where

$$C_D = (4\pi)^{-D/2} \tag{3}$$

and the x_i's are defined as follows. The curvature tensor can be regarded as an antisymmetric two-form matrix R,

$$(R)^{ab} = R^{ab}{}_{\mu\nu} dx^\mu \wedge dx^\nu . \tag{4}$$

This antisymmetric $D \times D$ matrix can be expressed in terms of "eigenvalues" x_i's:

$$R = \begin{pmatrix} 0 & x_1 & & & & & & \\ -x_1 & 0 & & & 0 & & \cdots & \\ & & 0 & x_2 & & & & \\ & 0 & -x_2 & 0 & & & & \\ & \vdots & & & \ddots & & & \\ & & & & & & 0 & x_{D/2} \\ & & & & & & -x_{D/2} & 0 \end{pmatrix}. \tag{5}$$

Given (1)–(5), the anomaly is written as

$$\int d^D x\, \partial_\mu (\sqrt{-g} J_5^\mu) = \int (A_D^{1/2} + A_D^{3/2}) \tag{6}$$

in the presence of one left-handed spin-$\frac{1}{2}$ and one left-handed spin-$\frac{3}{2}$ field in extended Minkowski space [i.e., one-time and $(D-1)$-space dimensions]. Here J_5^μ is a contravariant axial current composed of the spin-$\frac{1}{2}$, spin-$\frac{3}{2}$, and associated Fadeev–Popov–Nielsen ghosts.

The problem is to express (1) and (2) in terms of R in (5). Instead of treating each case separately, let us develop the problem in a more general way. We define

$$A_D = C_D \prod_{i=1}^{D/2} f(x_i), \tag{7}$$

where the function $f(x)$ is even in x and may take one of the forms (1) or (2) say. A simple computation, using (5), provides the relation between x and R:

$$\sum_{i=1}^{D/2} x_i^{2m} = \frac{1}{2} \mathrm{Tr}(iR)^{2m} . \tag{8}$$

 0022-2488/85/061334-03$02.50 © 1985 American Institute of Physics

In terms of the lhs of (8), (7) is expanded and expressed in terms of R:

$$A_D = C_D \exp[\tfrac{1}{2}\,\mathrm{Tr}\ln f(iR)] . \tag{9}$$

Suppose the function $f(x)$ possesses the series expansion

$$f(x) = 1 + \sum_{n=1}^{\infty} b_n x^{2n} . \tag{10}$$

Then, with the help of the logarithmic function series, we may write (9) as

$$A_D = C_D \sum_{l=0}^{\infty} \frac{1}{l!}\left[\mathrm{Tr}\sum_{m=1}^{\infty} -\frac{1}{2m}\left\{-\sum_{n=1}^{\infty} b_n (iR)^{2n}\right\}^m\right]^l . \tag{11}$$

Hence the quantity A_D assumes the form

$$A_D = C_D \sum_{l=0}^{\infty}\frac{1}{l!}\left[\mathrm{Tr}\sum_{m=1}^{\infty} a_m R^{2m}\right]^l , \tag{12}$$

whereupon the desired final expression which includes only D-form terms reads

$$A_D = C_D \sum_{\{n_i\}}^{n_1+2n_2+\cdots+Nn_N=N} \frac{1}{n_1!n_2!\cdots n_N!} \times (a_1\,\mathrm{Tr}\,R^2)^{n_1}(a_2\,\mathrm{Tr}\,R^4)^{n_2}\cdots(a_N\,\mathrm{Tr}\,R^{2N})^{n_N} . \tag{13}$$

Here the summation over n_i runs from zero to a certain integer such that the constraint

$$n_1 + 2n_2 + 3n_3 + \cdots + Nn_N = N$$

is obeyed.

The problem is thus reduced to discovering how to express the $\{a_m\}$ in terms of the $\{b_n\}$. This can be done by picking up only R^{2m} terms within the square brackets of (11). The answer is

$$a_m = \frac{(-1)^{m+1}}{2}\sum_{\{n_i\}}^{\Sigma jn_j=m}\frac{(n_1+n_2+\cdots+n_N-1)!}{n_1!n_2!\cdots n_N!} \times(-b_1)^{n_1}(-b_2)^{n_2}\cdots(-b_N)^{n_N} , \tag{14}$$

where one should notice that the summation is actually over a set of $\{n_i : 1 \leqslant i \leqslant m\}$. The last step to our goal consists in replacing the constraint over the $\{n_i\}$, essentially a Kronecker delta, by

$$\delta_{m,n_1+2n_2+\cdots+Nn_N} = \frac{1}{2\pi}\int_0^{2\pi} d\theta\, e^{2i\theta(n_1+2n_2+\cdots+Nn_N-m)} . \tag{15}$$

[The factor of 2 in the exponent on the rhs of (15) is chosen for later convenience.] Then (14) can be written as

$$\begin{aligned} a_m &= \frac{(-1)^{m+1}}{4\pi}\int_0^{2\pi} d\theta\, e^{-2im\theta} \times \sum_{n=1}^{\infty}\frac{(-1)^n}{n}(b_1e^{2i\theta}+b_2e^{4i\theta}+\cdots+b_Ne^{2iN\theta})^n \\ &= \frac{i(-1)^{m+1}}{4\pi}\oint dz\, z^{-2m-1} \times\sum_{n=1}^{\infty}\frac{(-1)^n}{n}(b_1z^2+b_2z^4+\cdots+b_Nz^{2N})^n . \end{aligned} \tag{16}$$

Since $m \leqslant N$, we may add terms like $z^{2N+l}(l \geqslant 1)$ within the brackets of (16) without affecting the residue. In other words the interior of the bracket can be replaced by $f(z) - 1$. This finally yields the elegant result

$$\begin{aligned} a_m &= \frac{i(-1)^{m+1}}{4\pi}\oint dz\, z^{-2m-1}\ln f(z) \\ &= \frac{i(-1)^{m+1}}{8\pi m}\oint dz\, z^{-2m}\frac{d}{dz}\ln f(z) , \end{aligned} \tag{17}$$

where the second line is obtained by partially integrating the first line and observing that $\ln f(z)/z^{2m} \to 0$ as $z \to \infty$. Summarizing, the answer is given by (13) with coefficients a_m's calculated by (17).

Now we are in a position to compute each case quickly.

(i) Spin-$\frac{1}{2}$ field: Here (1) provides the function

$$f(x) = (x/2)/\sinh(x/2) \tag{18}$$

and (17) reduces to

$$a_m^{1/2} = \frac{i(-1)^m}{8\pi m}\oint\frac{dz}{z^{2m}(e^z-1)} . \tag{19}$$

Since

$$\frac{z}{e^z-1} = \sum_{n=0}^{\infty} B_n\frac{z^n}{n!} ,$$

where B_n is the Bernoulli number, we obtain

$$a_m^{1/2} = \frac{(-1)^{m+1}}{4m(2m)!}B_{2m} . \tag{20}$$

This result agrees with the one in Ref. 5 which was derived by a complicated Feynman graph calculation.

(ii) Spin-$\frac{3}{2}$ field: This case resembles the spin-$\frac{1}{2}$ field. The difference lies in the extra factor on the right of (2), which can be manipulated with the help of (8); viz.

$$\begin{aligned} 2\sum_j \cosh x_j &= \sum_{n=0}^{\infty}\frac{1}{(2n)!}\mathrm{Tr}(iR)^{2n} \\ &= \mathrm{Tr}(e^{iR}) \end{aligned} \tag{21}$$

where the last line is obtained because $\mathrm{Tr}(R^{2n+1}) = 0$. Recalling how we derived (13), and combining (12) with (21), we arrive at

$$A_D^{3/2} = C_D\sum_{\{n_i\}}^{n_0+n_1+2n_2+\cdots+Nn_N=N}\frac{1}{(2n_0)!n_1!n_2!\cdots n_N!} \times\left[\mathrm{Tr}(-R^2)^{n_0} - \delta_{n_0,0}\right] \times(a_1^{1/2}\,\mathrm{Tr}\,R^2)^{n_1}(a_2^{1/2}\,\mathrm{Tr}\,R^4)^{n_2}\cdots(a_N^{1/2}\,\mathrm{Tr}\,R^{2N})^{n_N} . \tag{22}$$

The answer involves the spin-$\frac{1}{2}$ coefficients of (20). The first term in the square bracket in (22) represents the pure spin-$\frac{3}{2}$ part while the second term corresponds to the contribution of the fictitious particles.

Next we consider the purely gravitational anomaly[6] which arises in $(4N-2)$ dimensions in the context

$$\begin{aligned} &-\frac{1}{2}\int d^{4N-2}x\sqrt{-g}\,T^{\mu\nu}(D_\mu\epsilon_\nu + D_\nu\epsilon_\mu) \\ &= 4\pi\int(D_a\epsilon_b - D_b\epsilon_a)\frac{\delta}{\delta R^{ab}}(A_{D=4N}) . \end{aligned} \tag{23}$$

Here the lhs is the variation of the one-loop effective action for matter fields under the infinitesimal general coordinate transformation $x^\mu \to x^\mu + \epsilon^\mu$, while the rhs comes by differentiating a certain $4N$-form A_{4N} with respect to a matrix element of the two-form matrix R of (4). If the rhs does not vanish it implies that invariance under the infinitesimal coordinate transformation is broken; equivalently when one writes the lhs of (23) as

$$\int d^{4N-2}x\sqrt{-g}\,\epsilon^\nu D^\mu T_{\mu\nu}\,,$$

one infers that the induced energy-momentum tensor is not conserved. Indeed there are at least three kinds of field (spin-$\frac{1}{2}$, spin-$\frac{3}{2}$, and antisymmetric tensor) which can contribute to the rhs of (23). For spin-$\frac{1}{2}$ and spin-$\frac{3}{2}$, A_{4N} is none other than the chiral anomaly (13), explicitly worked out in (20) and (22).

The corresponding quantity for an antisymmetric tensor field in $D = 4N$ dimensions is given by

$$A^{1}_{D} = -\frac{1}{8}C_D \prod_{i=1}^{D/2} \frac{x_i}{\tanh x_i}\,. \tag{24}$$

We can just as easily apply the former procedure to express (24) in terms of R by setting

$$f(x) = x/\tanh x\,. \tag{25}$$

This time (17) becomes

$$a^{1}_{n} = \frac{i(-1)^m}{2\pi m}\oint dz\, z^{-2m}\frac{e^{2z}}{e^{4z}-1}$$
$$= \frac{(-1)^{m+1}2^{4m}}{4m(2m)!}B_{2m}\left(\frac{1}{2}\right), \tag{26}$$

where we have used the generating function for the Bernoulli polynomial,

$$\frac{ze^{xz}}{e^z-1} = \sum_{n=0}^{\infty} B_n(x)\frac{z^n}{n!}\,. \tag{27}$$

Combining (13) with (24), the antisymmetric contribution can be written as

$$A^{1}_{D} = -\frac{C_D}{8}\sum_{\{n_i\}}^{n_1+2n_2+\cdots+Nn_N=N}\frac{1}{n_1!n_2!\cdots n_N!}$$
$$\times(a^{1}_{1}\,\mathrm{Tr}\,R^2)^{n_1}(a^{1}_{2}\,\mathrm{Tr}\,R^4)^{n_2}\cdots(a^{1}_{N}\,\mathrm{Tr}\,R^{2N})^{n_N} \tag{28}$$

with the coefficients derived in (26). Incorporating (28) into the rhs of (23) we easily obtain the antisymmetric tensor contribution to the gravitational anomaly.

At last we have all the information needed to look for a cancellation of the gravitational anomaly among fields of different spin. This problem has already been addressed by the authors of Ref. 6 who analyzed the expressions up to 14 dimensions. However, we are armed with general explicit formulas for $A^{1/2}_{D}$, $A^{3/2}_{D}$, and A^{1}_{D} in arbitrary dimensions, and can tackle the problem more comprehensively. We give below the simple explicit analysis as an alternative to that of Ref. 6. Since we are seeking a nontrivial solution of the equation

$$c_1 A^{1/2}_{D} + c_2 A^{3/2}_{D} + c_3 A^{1}_{D} = 0\,, \tag{29}$$

we may set $c_3 = 1$ in general. After obtaining the solution for c_1 and c_2 as rational numbers (if it exists) we can convert the solution to integers by appropriate multiplication.

In order to reveal an inconsistency in a certain dimension we need at least *three* equations for c_1 and c_2. We choose those three equations as the ones that provide the coefficients of $(\mathrm{Tr}\,R^2)^N$, $(\mathrm{Tr}\,R^2)^{N-2}\,\mathrm{Tr}\,R^4$, and $(\mathrm{Tr}\,R^2)^{N-3}\,\mathrm{Tr}\,R^6$, which are, respectively,

$$c_1 - (20N+3)c_2 + (-8)^{N-1} = 0\,, \tag{30}$$
$$c_1 - (20N-285)c_2 - \tfrac{7}{2}(-8)^{N-1} = 0\,, \tag{31}$$
$$c_1 - (20N+435)c_2 + \tfrac{31}{4}(-8)^{N-1} = 0\,. \tag{32}$$

Amazingly, Eqs. (30)–(32) are linearly dependent and give a unique solution

$$c_1 = (20N-61)(-8)^{N-3}\,, \tag{33}$$
$$c_2 = (-8)^{N-3}\,. \tag{34}$$

This then is perfectly acceptable when $N = 1, 2$, and 3, i.e., in two, six, and ten space-time dimensions. For higher N there are further consistency conditions. It is enough to examine the coefficient of $(\mathrm{Tr}\,R^2)^{N-4}(\mathrm{Tr}\,R^4)^2$ which, if it is to vanish, necessitates that

$$c_1 - (20N-573)c_2 + \tfrac{49}{4}(-8)^{N-1} = 0\,. \tag{35}$$

There is now disagreement between (33), (34), and (35). And this proves that there is no way to cancel the gravitational anomalies among spin-$\frac{1}{2}$, spin-$\frac{3}{2}$, and antisymmetric tensor fields if $N \geqslant 4$, or in space-times of dimension $D \geqslant 14$. Of course it is not inconceivable that cancellation can be achieved by including extra fields belonging to even more exotic Lorentz group representations, if we are not deterred by the cause of renormalizability which is anyway lost.

Note added in proof: After submitting this paper, we received a preprint from Osaka by R. Endo and M. Takao who derive the gravitational anomalies for spin-$\frac{1}{2}$ and spin-$\frac{3}{2}$ via Fujikawa's path integral method. They give explicit answers up to 16 dimensions as in Ref. 6, which are particular cases of our general formulas (20), (22), and (28). Their paper is to be published in the Prog. Theor. Phys.

ACKNOWLEDGMENTS

This research was supported by the Australian Research Grants Scheme through Grant No. 81-15663 and by the U. S. Department of Energy under Contract No. DE-AS05-77ER05490.

[1]S. Adler, Phys. Rev. **177**, 2426 (1969); J. Bell and R. Jackiw, Nuovo Cimento A **60**, 47 (1969); S. Adler and W. Bardeen, Phys. Rev. **182**, 1517 (1969); W. Bardeen, Phys. Rev. **184**, 1848 (1969); T. Kimura, Prog. Theor. Phys. **42**, 1191 (1969); R. Delbourgo and A. Salam, Phys. Lett. B **40**, 381 (1972); K. Fujikawa, Phys. Rev. D **21**, 2848 (1980); N. K. Nielsen, M. T. Grisaru, H. Romer, and P. Van Nieuwenhuizen, Nucl. Phys. B **140**, 477 (1978).
[2]M. F. Atiyah and I. M. Singer, Ann. Math. **87**, 485,546 (1968); **93**, 1, 119, 139 (1971); M. F. Atiyah and G. B. Segal, Ann. Math. **87**, 531 (1968); T. Eguchi, P. B. Gilkey, and A. J. Hanson, Phys. Rep. **66**, 213 (1980).
[3]P. H. Frampton and T. W. Kephart, Phys. Rev. Lett. **50**, 1343, 1347 (1983); P. K. Townsend and G. Sierra, Nucl. Phys. B **222**, 493 (1983); T. Matsuki, Phys. Rev. D **28**, 2107 (1983); R. Delbourgo and P. D. Jarvis, J. Phys. G **10**, 591 (1984); B. Zumino, W. Yong-Shi, and A. Zee, Nucl. Phys. B **239**, 477 (1984).
[4]T. Matsuki and A. Hill, OSU preprint, 1983; R. Delbourgo and P. D. Jarvis, Phys. Lett. B **136**, 43 (1984); R. Delbourgo and J. J. Gordon, Univ. of Tasmania preprint, 1984.
[5]J. J. Gordon, Classical Quantum Gravity **1**, 673 (1984).
[6]L. Alvarez-Gaume and E. Witten, Nucl. Phys. B **234**, 269 (1984).

Aust. J. Phys., 1994, **47**, 465–74

Induced Parity Violation in Odd Dimensions

R. Delbourgo and A. B. Waites

Physics Department, University of Tasmania,
GPO Box 252C, Hobart, Tas. 7001, Australia.

Abstract

One of the interesting features about field theories in odd dimensions is the induction of parity-violating terms and well-defined *finite* topological actions via quantum loops if a fermion mass term is originally present and conversely. Aspects of this issue are illustrated for electrodynamics in 2+1 and 4+1 dimensions.

1. Introduction

There are a few curiosities associated with field theories in an odd number of space–time dimensions. The first is that the overall degree of divergence of an integral possessing an odd mass scale cannot be taken at face value, since such an integral behaves like the gamma function at half-integral argument values. This is most easily seen by considering dimensional regularisation of the typical integral as D tends to an odd integer (when r below is an integer),

$$I = -i\int \frac{\Gamma(r)\,\bar{d}^D p}{(p^2-M^2)^r} = \frac{(-1)^r\,\Gamma(r-D/2)}{(4\pi)^{D/2}\,(M^2)^{r-D/2}}\,. \tag{1}$$

A second noteworthy property is that in odd dimensions, one commonly encounters couplings which are odd powers of mass. This can be understood by considering a free-field theory in D space–time dimensions,

$$\int d^Dx\,[((\partial\phi)^2-\mu^2\phi^2)/2+\bar{\psi}(i\gamma.\partial-m)\psi-F^{\mu\nu}F_{\mu\nu}/4]\,, \tag{2}$$

where typically ϕ is a scalar, ψ is a spinor and $F_{\mu\nu}\equiv\partial_\mu A_\nu-\partial_\nu A_\mu$ is a Maxwell gauge field. The dimensionlessness of the action (in natural units) specifies the mass dimensions of the fields,

$$[\phi],[A]\sim M^{D/2-1}, \qquad [\psi]\sim M^{(D-1)/2},$$

whereupon interaction Lagrangians like

$$\int d^Dx\,[e\bar{\psi}\gamma.A\psi+f\phi^3+g\phi\bar{\psi}\psi+\lambda\phi^4+G(\bar{\psi}\psi)^2+\ldots] \tag{3}$$

0004-9506/94/040465$05.00

will have coupling constants with prescribed dimensions,

$$[e], [g] \sim M^{2-D/2}, \quad [f] \sim M^{3-D/2}, \quad [\lambda] \sim M^{4-D}, \quad [G] \sim M^{2-D}. \tag{4}$$

One then observes that in odd dimensions the couplings e, f, g have odd $\sqrt{M}$ scales. That does not matter much for electrodynamics since we meet powers of e^2 in the perturbation expansion but for f, g we can potentially encounter single powers of the coupling. In particular for three dimensions,

$$[f] \sim M^{\frac{3}{2}}, \qquad [e], [g] \sim M^{\frac{1}{2}}, \qquad [\lambda] \sim M, \qquad [G] \sim M^{-1}, \tag{5}$$

telling us that electrodynamics, chromodynamics, $\lambda\phi^4$ and Yukawa interactions become *super*-renormalisable, while Fermi interactions remain unrenormalisable. Combined with the first property, this has the effect of eliminating certain ultraviolet infinities in theories such as $\lambda\phi^4$; thus the tadpole graphs of order λ and higher are perfectly finite and the *only* infinity in that model is the self-mass of ϕ due to the three-ϕ intermediate state. For electrodynamics in 2+1 dimensions, the situation is even better—no infinities at all.

A third peculiar aspect of odd D dimensions stems from the algebra of the Dirac γ-matrices

$$\{\gamma_\mu, \gamma_\nu\} = 2\eta_{\mu\nu}\,.$$

When D is even it is well-known that the γ have size $2^{D/2} \times 2^{D/2}$ and there exists a γ_5 matrix which is the product of all the different $D\gamma$ and which anticommutes with each γ_μ; one can always arrange it to have square -1, like all the space-like γ (in our metric $+,-,-,\ldots$). It is not so well known that in one higher dimension, the size of the γ remains the same—all that happens is that the 'γ_5' matrix becomes the last element of the D Dirac matrices.

For instance, in three dimensions one can take the two-dimensional Pauli $\gamma_0 = \sigma_3$, $\gamma_1 = i\sigma_1$ and simply append $\gamma_2 =$ 'γ_5' $= i\gamma_0\gamma_1 = -i\sigma_2$ to complete the set, without altering the size of the representation. At the same time it should be noticed that one can get a non-zero trace from the product of *three* gamma-matrices, viz. $\mathrm{Tr}[\gamma_\rho\gamma_\sigma\gamma_\tau] = 2i\epsilon_{\rho\sigma\tau}$. Similarly, in five dimensions one can take the usual four-dimensional ones and just append $\gamma_5 \equiv \gamma_0\gamma_1\gamma_2\gamma_3$ as the fifth component; here as well the product of the full five γ gives a non-vanishing trace: $\mathrm{Tr}[\gamma_\mu\gamma_\nu\gamma_\rho\gamma_\sigma\gamma_\tau] = -4\epsilon_{\mu\nu\rho\sigma\tau}$. The lesson is that when D is odd, one should be careful before discarding traces of odd monomials of gamma-matrices if there are sufficiently many γ, since at least

$$\mathrm{Tr}[\gamma_{\mu_1}\gamma_{\mu_2}\cdots\gamma_{\mu_D}] = (2i)^{(D-1)/2}\epsilon_{\mu_1\mu_2\cdots\mu_D}.$$

Another property worth remembering in odd dimensions is that if one constructs suitably normalised antisymmetric products of r matrices, $\gamma_{[\mu_1\mu_2\cdots\mu_r]}$ (the total set of these from $r = 0$ to $r = D$ generates a complete set into which any $2^{[D/2]} \times 2^{[D/2]}$ matrix can be expanded), then there exists the relation

$$\gamma_{[\mu_1\mu_2\cdots\mu_r]} = i^{(D-1)/2}\epsilon_{\mu_1\mu_2\cdots\mu_D}\gamma^{[\mu_{r+1}\mu_{r+2}\cdots\mu_D]}/(D-r)!$$

This often helps in simplifying products of matrices.

Turning to discrete operations, a charge conjugation operator $\mathcal{C}$ with the transposition property

$$\mathcal{C}\gamma_{[\mu_1\mu_2\cdots\mu_r]}\mathcal{C}^{-1} = (-1)^{[(r+1)/2]}(\gamma_{[\mu_1\mu_2\cdots\mu_r]})^T \tag{6}$$

always exists in even dimensions, but cannot be defined at the odd values $D = 5, 9, 13\ldots$. This is intimately tied with the existence of topological terms in the action for the pure gauge field, as we shall see. As for parity $\mathcal{P}$, it corresponds to an inversion of all the *spatial* coordinates for even D, since that is an improper transformation. However when D is odd, it should be regarded as a reflection of all the space coordinates *except the very last one*, x_{D-1}, in order to ensure that the determinant of the transformation remains negative. It is straightforward to verify that this corresponds to the unitary change

$$\begin{aligned}\mathcal{P}\psi(x_0, x_1, ..., x_{D-2}, x_{D-1})\mathcal{P}^{-1} =& -i\eta\gamma_0\gamma_{D-1}\psi(x_0, -x_1, ..., -x_{D-2}, x_{D-1}) \\ =& \eta\gamma_1\cdots\gamma_{D-2}\psi(x_0, -x_1, ..., -x_{D-2}, x_{D-1}),\end{aligned} \tag{7}$$

where η is the intrinsic parity of the fermion field. In that regard, one can just as easily check that a mass term like $m\bar{\psi}\psi$ in the action is *not* invariant under parity for D odd.

This potential to induce other parity-violating interactions in the theory forms the subject of this paper. In Section 2 we shall examine the induction of a Chern–Simons term in 2+1 electrodynamics from a fermion mass; the converse process is considered in an Appendix. In Section 3 we generalise this to 4+1 QED and to higher D, with the converse effect also treated in the Appendix. An explanation of why this is a pure one-loop effect is also provided. Finally we discuss in Section 4 what happens when electrodynamics is purely topological (no free F^2 term in the Lagrangian) as this represents a system quite different from what we are accustomed to.

2. 3D Electrodynamics

Turning to QED in 2+1 dimensions, we are blessed with a coupling with positive mass dimension $[e^2] \sim M$, so we anticipate a finite number of ultraviolet singularities. But in fact *none exists* thanks to gauge invariance. The standard diagrams for photon polarisation, electron self-energy and vertex corrections are all perfectly finite—barring infrared problems, which actually correct themselves non-perturbatively through the dressing of the photon line, as first demonstrated by Jackiw and Templeton (1981), although some doubts about the loop expansion have been expressed by Pisarski and Rao (1985).

Straightforward evaluation of the graphs produces the one-loop results

$$\Pi_{\mu\nu}(k) = (k_\mu k_\nu - k^2\eta_{\mu\nu})\frac{e^2}{2\pi}\int\frac{\alpha(1-\alpha)d\alpha}{\sqrt{m^2-k^2\alpha(1-\alpha)}}$$

$$+ im\epsilon_{\lambda\mu\nu}k^\lambda\frac{e^2}{4\pi}\int\frac{d\alpha}{\sqrt{m^2-k^2\alpha(1-\alpha)}}, \quad (8)$$

$$S(p) = \langle T(\bar{\psi}(p)\psi(0))\rangle = \frac{e^2}{16\pi}\int\frac{dw}{w(\gamma.p-w)}\left[\frac{\xi}{w}-\frac{4m}{(w-m)^2}\right], \quad (9)$$

where ξ is a parameter that fixes a Lorentz-covariant gauge ($\xi = 0$ is the Landau gauge). The vertex corrections are obviously finite too, because '$Z_1 = Z_2$'. Higher powers of e^2 only serve to make the diagrams more convergent than they already are in lowest order, since the series expansion of a physical amplitude will take the form

$$T(k) = T_0(k)\left[1 + c_1\frac{e^2}{m} + c_2\frac{e^4}{m^2} + \dots\right]; \quad c_i = f(k/m),$$

where k signifies the external momentum variables. Broadhurst *et al.* (1993) have calculated some of these coefficients for any D.

An important aspect of (8) is the induced parity-violating Chern–Simons interaction, $\epsilon_{\lambda\mu\nu}A^\lambda F^{\mu\nu}$. It is not surprising that it should have sprouted, since we started with a fermion mass $m\bar{\psi}\psi$ term which is intrinsically $\mathcal{P}$-violating in odd dimensions; but the value of that induced photon term is finite and disappears as $m \to 0$. Note, however, that in the infrared limit it reduces to $ie^2\epsilon_{\lambda\mu\nu}k^\lambda/4\pi$ *provided that* $m \neq 0$.

There has been some argument in the literature that this Chern–Simons term may produce anomalies in some processes because it contains an ϵ tensor specific to three dimensions (in much the same way that the axial anomaly is connected with the ϵ tensor in four dimensions). This cannot be true because vacuum polarisation to order e^2 is perfectly sensible and *finite* when evaluated by any reasonable regularisation. Anomalies only arise when a *divergence* multiplies an apparent *zero*, as in the Pauli–Villars method, when a mass regulator contribution M^2 multiplies an integral of order $1/M^2$; or in a dimensional context, when a pole $1/(D-3)$ multiplies an evanescent zero $(D-3)$. But, as we have already seen, the photon self-energy diagram contains no such singularity (Delbourgo and Waites 1993). Thus an anomaly is indeed absent.

3. Electrodynamics in Higher Dimensions

A subject of debate has been whether the induced Chern–Simons term suffers from higher-order loop corrections. There is a simple proof, presented later, which says this cannot be, but before describing it, let us exhibit the induced term for any odd D dimension, as it is a 'clean' result. We begin as before with massive fermion QED. For arbitrary D the induced topological term takes the form of an n-point function,

$$C\epsilon_{\mu_1\mu_2\cdots\mu_D}A^{\mu_1}F^{\mu_2\mu_3}\ldots F^{\mu_{D-1}\mu_D}; \qquad n=(1+D)/2. \tag{10}$$

Notice that this conforms perfectly with charge conjugation: when $D=3$ and $\mathcal{C}$ is conserved, the topological term involves an even number $n=2$ of photons; when $D=5$ and $[e^2]\sim M^{-1}$, we encounter three photon lines but then $\mathcal{C}$ is no longer valid; when $D=7$, $\mathcal{C}$-invariance becomes operative again and the number of photon lines is $n=4$; and so on.

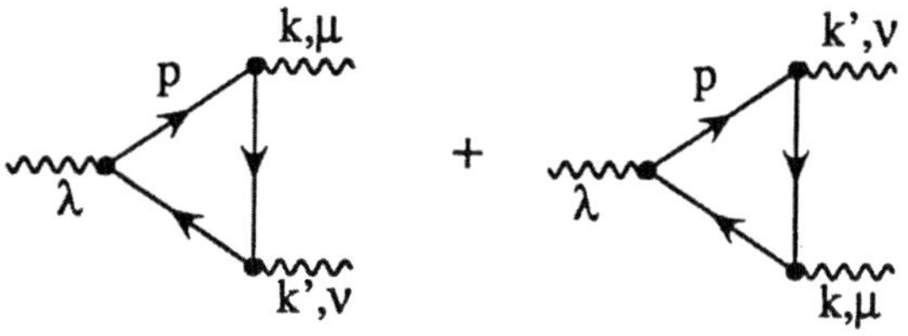

Fig. 1. One-loop induction of a Chern–Simons amplitude in five dimensions.

The result of the one-loop contribution to the topological term has already been quoted in equation (8). Looking at the next odd dimension, $D=5$, the relevant one-loop graphs are shown in Fig. 1, leading to the induced vertex

$$\Gamma_{\lambda\mu\nu}(k,k')=-2ie^3\int\bar{d}^5p\,\frac{\mathrm{Tr}[\gamma_\nu(\gamma.p+m)\,\gamma_\mu(\gamma.(p+k)+m)\,\gamma_\lambda(\gamma.(p-k')+m)]}{(p^2-m^2)\,[(p+k)^2+m^2]\,[(p-k')^2-m^2]}.$$

Introducing Feynman parameters in the usual way to combine denominators and picking out the term with five gamma-matrices in the trace, we end up with

$$\begin{aligned}\Gamma_{\lambda\mu\nu}(k,k')&=-\,16ie^3m\int\bar{d}^5p\,\frac{d\alpha\,d\beta\,d\gamma\,\delta(1-\alpha-\beta-\gamma)\epsilon_{\lambda\mu\nu\rho\sigma}\,k^\rho\,k'^\sigma}{[p^2-m^2+k^2\alpha\beta+k'^2\gamma\alpha+(k+k')^2\beta\gamma]^3}\\&=-\,\frac{e^3m\epsilon_{\lambda\mu\nu\rho\sigma}\,k^\rho\,k'^\sigma}{8\pi^2}\int_0^1\frac{d\alpha\,d\beta\,d\gamma\,\delta(1-\alpha-\beta-\gamma)}{\sqrt{m^2-k^2\alpha\beta-k'^2\gamma\alpha-(k+k')^2\beta\gamma}}.\end{aligned}\tag{11}$$

One can regard this amplitude as the five-dimensional description of the process $\pi^0\to 2\gamma$, because one of the indices (4) of the Levi–Civita tensor just corresponds to the standard pseudoscalar and the residual four indices (0 to 3) are the normal 4-vector ones. Just as with 2+1 QED, we see that the induced term in 4+1 QED vanishes with the fermion mass m. [Contrariwise, one can check that if $m=0$, but a term (10) is present from the word go, then a fermion mass term, amongst other parity-violating ones, will arise. See the Appendix.]

We are now in a position to quote the topological vertex induced for arbitrary odd D by the fermion mass term. Introduce $n=(D+1)/2$ and Feynman parameters α_i, $i=1\ldots n$, for each internal line (Fig. 2). Call the momentum flowing across

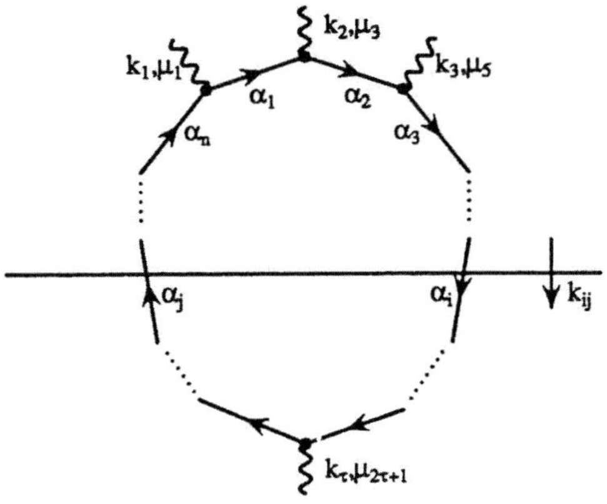

Fig. 2. One-loop induction of a Chern–Simons term in D dimensions.

each possible cutting of two lines k_{ij} if those lines have parameters α_i, α_j. The calculation then produces the result

$$\Gamma_{\mu_1\mu_2\cdots\mu_n}(k) = -\frac{me^n i^{n-1}}{2(2\pi)^{n-1}}\epsilon_{\mu_1\mu_2\cdots\mu_D} k_1^{\mu_2} k_2^{\mu_4}\dots k_n^{\mu_D}$$

$$\times \int_0^1 \prod_{k=1}^n d\alpha_k \frac{\delta\left(1-\sum_k \alpha_k\right)}{\sqrt{m^2 - \sum_{i<j=1}^n k_{ij}^2 \alpha_i \alpha_j}}. \tag{12}$$

One may readily check that this collapses to the results (8) and (11) for $D=3$ and $D=5$ respectively. It corresponds to the Chern–Simons term (10) where $C = e^n/2n!(4\pi)^{n-1}$ if one goes to the soft photon limit, always assuming $m \neq 0$.

To finish off this section let us explain why this one-loop answer (12) is all there is. In three dimensions, the Lagrangian $\epsilon_{\lambda\mu\nu}A^\lambda F^{\mu\nu}$ will change by a pure divergence under the gauge transformation, $\delta A \to \partial\chi$, so the action remains invariant for all normal field configurations that vanish at ∞. However, a fourth-order interaction like

$$\epsilon_{\lambda\mu\nu}A^\lambda F^{\mu\nu}F_{\rho\sigma}F^{\rho\sigma}$$

will *not* be invariant under the gauge change; thus it is not permitted. More generally, in odd D dimensions the interaction

$$\epsilon_{\mu_1\mu_2\cdots\mu_D}A^{\mu_1}F^{\mu_2\mu_3}\dots F^{\mu_{D-1}\mu_D}(F_{\rho\sigma}F^{\rho\sigma})^N; \qquad N \geq 1, \tag{13}$$

and ones like it, are forbidden by gauge invariance and thus cannot be produced. On the other hand a two-loop contribution to the fundamental topological term can be regarded as an integration of (13), with $N=1$, over one of the photon momenta. Since we have just concluded that (13) must be absent, we deduce that the induced topological term (10) cannot receive any two-loop (or higher-loop) quantum corrections. In this respect, it is a pristine result similar to the Adler–Bardeen theorem for the axial anomaly; nevertheless it is only of academic interest in as much as QED becomes unrenormalisable (cf. the dimensions of e^2) when $D \geq 5$, unless the space–time is compact, e.g. in some Kaluza–Klein geometries.

4. Topological QED

So far we have considered models where the initial Lagrangian contains the normal free gauge kinetic energy $F_{\mu\nu}F^{\mu\nu}$ term, and seen what transpires as a result of parity violation primarily through the fermion field. Now we shall consider what happens when the initial Lagrangian has no gauge field kinetic energy but starts off life instead with a Chern–Simons piece such as (10). In 2+1 dimensions, this still means a bilinear term in the gauge field capable of launching a propagator,

$$D_{\mu\nu} = \left[\frac{i\epsilon_{\mu\nu\lambda}k^\lambda}{\mu k^2} - \xi\frac{k_\mu k_\nu}{k^4}\right], \tag{14}$$

where we have taken account of gauge-fixing, with parameter ξ. (The very same expression can be obtained by adding a conventional kinetic term $-ZF_{\mu\nu}F^{\mu\nu}/4$ and taking the limit $Z \to 0$.) Evaluating the fermion self-energy now yields

$$\Sigma(p) = e^2 \int \bar{d}^3k\, \frac{\gamma_\mu(\gamma.p - \gamma.k)\gamma_\nu}{(p-k)^2} D^{\mu\nu}(k) = -e^2\left[\frac{\xi\gamma.p}{16\sqrt{-p^2}} + \frac{\sqrt{-p^2}}{8\mu}\right], \tag{15}$$

which contains a mass-like term where none previously existed. In the same vein, we may compute the vacuum polarisation correction to (14) and arrive at

$$\Pi_{\mu\nu}(k) = ie^2\mathrm{Tr}\int \bar{d}^3p\, \frac{\gamma_\mu\gamma.p\gamma_\nu\gamma.(p-k)}{p^2(p-k)^2} = (-k^2\eta_{\mu\nu} + k_\mu k_\nu)\frac{e^2}{8\sqrt{-k^2}}, \tag{16}$$

which has the effect of leaving $D(k) \sim 1/k$. In higher orders of perturbation theory we may expect that

$$\Sigma(p) = \gamma.p\, f\left(\frac{e^2}{\sqrt{-p^2}}, \frac{e^2}{\mu}\right) + \sqrt{-p^2}\, g\left(\frac{e^2}{\sqrt{-p^2}}, \frac{e^2}{\mu}\right),$$

$$\Pi_{\mu\nu} = (-k^2\eta_{\mu\nu} + k_\mu k_\nu)\,\pi\left(\frac{e^2}{\sqrt{-k^2}}, \frac{e^2}{\mu}\right),$$

where f, g and π are scalar functions of their arguments. It is fascinating to speculate on the full form of those functions by applying some non-perturbative method of solution.

The situation is radically different in 4+1 dimensions since the Chern–Simons term is *trilinear* in the gauge field and alone cannot engender a propagator. Rather, one must resort to quantum corrections to get something of that ilk;

the vacuum polarisation graph (initially from massless fermions) produces a hard quantum loop contribution:

$$D_{\mu\nu} = \left(-\frac{\eta_{\mu\nu}}{k^2} + \frac{k_\mu k_\nu}{k^4} \right) \frac{512\pi}{3e^2\sqrt{-k^2}} - \xi \frac{k_\mu k_\nu}{k^4}. \tag{17}$$

Taken with the trilinear gauge interaction, this can produce a vacuum polarisation effect from the gauge field itself, namely

$$\Pi_{\mu\nu} = i\left(\frac{512\pi C}{3e^2}\right)^2 \int \frac{d^5k'}{[k'^2(k-k')^2]^{\frac{3}{2}}} \epsilon_{\mu\rho\sigma\alpha\beta}\, k^\alpha\, k'^\beta\, \epsilon_\nu{}^{\rho\sigma\gamma\delta}\, k_\gamma\, k'_\delta$$

or

$$\Pi_{\mu\nu} = (\eta_{\mu\nu}\, k^2 - k_\mu k_\nu) \left(\frac{512C}{3e^2}\right)^2 \frac{\sqrt{-k^2}}{12}.$$

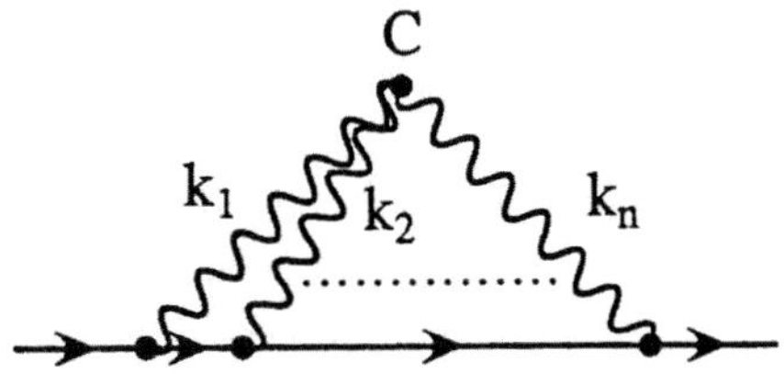

Fig. 3. Induction of a fermion mass term through a topological interaction.

Interestingly, (17) does not give birth to a mass-like fermion self-energy at the one-loop level—five gamma-matrices are needed to obtain that. This means we have to consider two-loop effects, either to order e^4 or to first order in the Chern–Simons coupling C, as sketched in Fig. 3. Quite generally in 4+1 dimensions we may anticipate that

$$\Sigma(p) = \gamma.p\, f(e^2\sqrt{-p^2}, C/e^3) + \sqrt{-p^2}\, g(e^2\sqrt{-p^2}, C/e^3),$$

$$\Pi_{\mu\nu} = (-k^2\eta_{\mu\nu} + k_\mu k_\nu)\, \pi(e^2\sqrt{-k^2}, C/e^3).$$

However, we must be on guard that higher-order contributions in e^2 and C are very likely unrenormalisable now and possibly of academic interest only. Still, our discussion does indicate the nature of such parity-violating contributions in these models and how they spring from just one source.

If we could trust some non-perturbative method for resumming the Feynman diagrams then we might be able to estimate the quantum effects associated with the dimensional couplings e^2 and C. The same of course applies with even greater force to pure Chern–Simons theory in higher odd dimensions.

Acknowledgments

We thank Dirk Kreimer for discussions. The ARC, under Grant A69231484, has kindly supported this research.

References

Jackiw, R., and Templeton, S. (1981). *Phys. Rev.* D **23**, 2291.
Pisarski, R., and Rao, S. (1985). *Phys. Rev.* D **32**, 2081.
Broadhurst, D., Fleischer, C., and Tarasov, V. (1993). *Z. Phys.* C **60**, 287.
Delbourgo, R., and Waites, A. B. (1993). *Phys. Lett.* B **300**, 241.

Appendix

Here we shall examine the converse of Sections 2 and 3, in as much as we deal with massless electrodynamics (A and ψ) but introduce the parity violation through a primary Chern–Simons term, not a fermion mass. Our treatment is to be contrasted with that in Section 4, where a kinetic term for the photon was absent. In the present circumstances,

$$\mathcal{L} = \bar{\psi}\gamma.(i\partial - eA)\psi - F_{\mu\nu}F^{\mu\nu}/4 + C\epsilon_{\mu_1\mu_2\cdots\mu_D}A^{\mu_1}F^{\mu_2\mu_3}\dots F^{\mu_{D-1}\mu_D},$$

we can be certain the gauge field will propagate at the bare level in any dimension D. It so happens that when $D = 3$ the Chern–Simons piece is also bilinear and can be incorporated with the standard F^2 term to give the initial two-point function,

$$D_{\mu\nu} = \frac{-\eta_{\mu\nu} + k_\mu k_\nu/k^2}{k^2 - \mu^2} + i\frac{\mu\epsilon_{\mu\nu\lambda}k^\lambda}{k^2(k^2-\mu^2)} - \xi\frac{k_\mu k_\nu}{k^4}.$$

Parity-violating terms then arise through quantum corrections in other Green functions.

Probably the most significant of these is in the fermion self-energy,

$$\Sigma(p) = \frac{e^2\gamma.p}{16\pi p^2}\left[\frac{p^4-\mu^4}{\mu^2 p}\ln\left(\frac{\mu+p}{\mu-p}\right) - \frac{2(p^2-\mu^2)}{\mu} + \frac{\pi p^2\sqrt{-p^2}}{\mu^2} - \tfrac{3}{2}\pi\xi\sqrt{-p^2}\right]$$
$$+ \frac{e^2}{8\mu\pi}\left[\frac{p^2-\mu^2}{p}\ln\left(\frac{\mu+p}{\mu-p}\right) - 2\mu + \pi\sqrt{-p^2}\right].$$

We should notice that in the limit of small μ, this expression reduces to

$$\Sigma(p) = \frac{e^2\gamma.p}{4\pi p^2}\left[\frac{\mu}{2} - \frac{3\pi\xi\sqrt{-p^2}}{8}\right] + \frac{e^2\mu}{8\sqrt{-p^2}}.$$

and could have been directly evaluated by regarding ϵAF as an interaction, rather than combining it with the bare photon propagator as above. (It will disappear, of course, when $\mu \to 0$ in the Landau gauge.)

Indeed this is the only sensible treatment for higher dimensions D since the Chern–Simons term is no longer bilinear. For $D = 5$ to first order in $C\epsilon AFF$, one engenders the mass term (and no kinetic term)

$$\Sigma(p) = -ie^3 C \int \bar{d}^D k\, \bar{d}^D k'\, \frac{\epsilon_{\mu\nu\lambda\alpha\beta}\, k^\alpha\, k'^\beta\, \gamma^\mu\, \gamma.(p-k-k')\gamma^\nu\, \gamma.(p-k)\gamma^\lambda}{k^2\, k'^2\, (k+k')^2\, (p-k-k')^2\, (p-k)^2}$$

$$= \frac{3\Gamma(3-D)\, p^4 e^3 C}{(16)^3 \pi^4};$$

unfortunately this is divergent as $D \to 5$, which is not too surprising. There is likewise a two-loop contribution of the same type to the photon self-energy, but this cannot add a parity-violating part to Π because such a term would violate gauge invariance for $D = 5$, as we have already explained in Section 3.

Manuscript received 29 April, accepted 25 May 1994

INSTITUTE OF PHYSICS PUBLISHING JOURNAL OF PHYSICS A: MATHEMATICAL AND GENERAL

J. Phys. A: Math. Gen. **36** (2003) 1719–1727 PII: S0305-4470(03)56180-7

Relativistic phase space: dimensional recurrences

R Delbourgo and M L Roberts

School of Mathematics and Physics, University of Tasmania, GPO Box 252-21, Hobart, Tasmania 7001, Australia

E-mail: Bob.Delbourgo@utas.edu.au and Martin.Roberts@utas.edu.au

Received 15 November 2002, in final form 17 December 2002
Published 29 January 2003
Online at stacks.iop.org/JPhysA/36/1719

Abstract

We derive recurrence relations between phase space expressions in different dimensions by confining some of the coordinates to tori or spheres of radius R and taking the limit as $R \to \infty$. These relations take the form of mass integrals, associated with extraneous momenta (relative to the lower dimension), and produce the result in the higher dimension.

PACS numbers: 11.10.Kk, 11.80.Cr, 12.90.+b

1. Establishing the recurrences

Formulations of field theories in higher dimensions are now quite commonplace, with 10 to 12 dimensions featuring prominently, especially in the context of M-theory. Mostly the coordinates which are additional to the usual four spacetime ones are either very small or the fields possess a severely damped behaviour as one moves away from the 3-brane. Thus the extra coordinates are characterized by one or more length scales R, which are generally miniscule or, if larger, only affect gravity. At the other extreme one can contemplate the R as being huge; indeed the method of 'box normalization' with a large R has a venerable pedigree and allows us to describe vacuum diagrams or compute properties per unit volume when they so depend. The same method also permits us to make sense of quantities like the volume of spacetime or $\delta^4(0)$ in the limit $R \to \infty$.

One of the primary objects of interest in these higher dimensions $D \equiv 2\ell$ is the behaviour of relativistic N-body phase space ρ_N^D since it primarily governs the magnitudes of transmutations, ignoring amplitude modulation. Based on earlier coordinate space methods [1], this behaviour has been recently studied [2] and codified [3]; a summary of how the N-body result in *fixed*, flat D space can be evaluated by means of Almgren recurrences [4]—mass integrals over smaller N but with the *same* D—was given in [2]. In this paper we establish relations between ρ having the same N, but different D, which are quite interesting in their own right. They may well have some use in the context of recent developments in string or M-theory or p-brane physics; for such theories possess a set of length scales R_i (or parameters arising in the extended metric) which serve to constrain the motion of particles to subspaces or 'branes' of lower dimension. Each theory produces its own particle spectrum whose spacing is determined by the R_i. As the limit $R_i \to \infty$ is taken, we may anticipate that the particles freely explore the higher-dimensional space, and the corresponding phase space at a given energy reduces to the relativistic phase space for the entire 'bulk'.

The strategy for deriving such recurrence relations in the flat space limit is quite simple. We just confine one or several of the D-dimensional coordinates to a torus or sphere of radius R and take the limit $R \to \infty$ at the very end, in much the same way that box normalization is handled. The act of confinement creates a series of discrete modes in the restricted coordinates and phase space must be summed over the various modes, subject to mass bounds. By choosing the topology of the extra space to be spherical or toroidal, the masses of the modes are easy to work out. (Had we been considering more complicated topologies, the mass spectra would have been much harder to calculate although we still expect the $R_i \to \infty$ limit to yield results coinciding with our choices.) Because the contributions to the masses from the confined coordinates are inversely proportional to R, the summation reduces to a mass integral in the limit of enormous R: hence the result ρ_N^D in higher D is expressible as a set of mass integrals of ρ over lower D, but with the same N; one can readily understand this as the effect of integrating over extra momentum components relative to those in the lower dimension. The forms of such connections are rather intriguing and some are not at all obvious; in fact, for larger N they are quite intricate. We know, of course, in advance that they must work out somehow; the interest is in the 'somehow', not why they do so. For smaller values of N we are able to check their validity, but verifying them for $N > 3$ is a daunting task in many cases.

In the following section we suppose that one or several of the coordinates are confined to a torus, so the recurrence relation is effectively that between ρ^D and ρ^{D-1}. Its nature is simple since there is only one extra coordinate to contend with, so we are dealing with a one-dimensional sum or integral in the continuum limit. We show how the recurrence pans out for few-body processes. The next two sections deal with the case when there are several extra coordinates (confined to a sphere) and here we encounter a multidimensional summation or integration, which is nontrivial. When $N = 2$ we demonstrate how the relations work out for any number of extra coordinates, but for $N = 3$ we have only succeeded in following through the connection when the dimension difference equals two or more, though no doubt it must apply to any N value.

2. Relations between D and $D - 1$ phase space

Let us begin by specifying our notation. Our metric is $+, -, -, -, \ldots$ with a total of D coordinates. The N-body phase space integral in *flat* space is defined by

$$\rho_N^D = \left(\prod_{i=1}^N \int \mathrm{d}\Omega_{p_i}\right)(2\pi)^D \delta^D\left(p - \sum_{i=1}^N p_i\right) \tag{1}$$

where $\mathrm{d}\Omega_{p_i} \equiv \theta(p_{i0})\delta\big(p_i^2 - m_i^2\big)\,\mathrm{d}^D p_i/(2\pi)^{D-1}$. We separate spacetime coordinates x into $(D - n)$ coordinates called $\mathbf{x}$ and extra ones labelled $\vec{y}$, n of which are independent; likewise for the conjugate momenta. (For the purpose of this section n equals 1, but we shall consider other n values later on.) The general aim of the exercise is to establish a connection between ρ_N^D and ρ_N^{D-n} and see how that works out analytically because it is nontrivial for large N or n.

The first step is to confine (periodic) y to the circumference of a circle of radius R. Thus the space is considered to have the direct product topology $M^{D-1}(x) \otimes T^1(y)$. Then Fourier expand a (real) field Φ into modes in the standard way so as to fix the normalization correctly:

$$\Phi(\mathbf{x}, y) = \sum_{k=-\infty}^{\infty} \phi_k(\mathbf{x}) \exp(iky/R)/\sqrt{2\pi R} \qquad \phi_k^*(\mathbf{x}) = \phi_{-k}(\mathbf{x}). \tag{2}$$

Being complex in general, ϕ_k^* can be regarded as the antiparticle field to ϕ_k, where k is positive say. y must run from $-\pi R$ to πR, to ensure that, upon integration over y, the free action takes its proper form,

$$S_{\text{free}} = \frac{1}{2}\int \mathrm{d}^D x[(\partial_x\Phi)^2 - m^2\Phi^2] = \sum_k \frac{1}{2}\int \mathrm{d}^{D-1}\mathbf{x}\left[(\partial_{\mathbf{x}}\phi_k^*)(\partial_{\mathbf{x}}\phi_k) - m_k^2\phi_k^*\phi_k\right] \tag{3}$$

where $m_k^2 \equiv m^2 + (k/R)^2$ corresponds to the mode k mass squared. Note that if one restricts the sum to positive k (because of repetition) the factor of $1/2$ disappears and one gets the right normalization for a complex field. A zero-mode field which is of course real and y-independent is given by $\Sigma(\mathbf{x}) = \sigma(\mathbf{x})/\sqrt{2\pi R}$.

To determine the phase space integral we consider the point interaction between a heavy field Σ (with mass m_0) decaying into N distinguishable fields which carry their own distinct quantum numbers, so Σ matches all of them. (One may also consider the case where some of the final particles are identical, but that just serves to introduce symmetry factors which must be taken into account and adds little to the discussion.) Write the effective interaction as $\mathcal{L} = \phi_1 \ldots \phi_N \Sigma$, so that integration over the 'extra' coordinate produces

$$\int \mathcal{L}(\mathbf{x}, y)\,\mathrm{d}y = \frac{\sigma}{(2\pi R)^{(N-1)/2}}\left(\prod_{i=1}^{N}\phi_{ik_i}\right)\delta_{\sum_i k_i,0}$$

with the sum taken over positive and negative k values. Thus we deduce the effective coupling of σ with the various ϕ in the lower dimension to be $g_{k_1\ldots k_N} = \delta_{\sum_i k_i,0}/(2\pi R)^{(N-1)/2}$. This means that the higher-dimensional phase space can be written in the more explicit form,

$$\rho^D_{m_0\to m_1+\cdots+m_N} = \sum_{k_i}\frac{\delta_{\sum_i k_i,0}}{(2\pi R)^{N-1}}\rho^{D-1}_{m_0\to m_{1k_1}+\cdots+m_{Nk_N}} \tag{4}$$

subject to energy–momentum conservation of course, which thus provides upper bounds on the magnitudes of the running k-values.

The second step is to take the limit as $(R, k) \to \infty$ and let $\mu_i = k_i/R$. We see that the connection (4) reduces to the continuous version,

$$\rho^D_{m_0\to m_1+\cdots+m_N} = \left(\prod_i \int \frac{\mathrm{d}\mu_i}{2\pi}\right) 2\pi\delta\left(\sum_i \mu_i\right)\rho^{D-1}_{m_0\to m_{1\mu_1}+\cdots+m_{N\mu_N}} \qquad m_{i\mu}^2 \equiv m_i^2 + \mu^2 \tag{5}$$

where again the range of μ-values is restricted by the condition $m_0 \geqslant m_{1\mu_1} + \cdots + m_{N\mu_N}$. This last form is quite readily understood as a consequence of writing the mass shell condition, $0 = p^2 - m^2 = \mathbf{p}^2 - (m^2 + \mu^2)$, where μ stands for the last momentum component p_y. The phase space integral over p_y then produces the delta function $\delta(\sum_i \mu_i)$, because the initial zero-mode has no dependence on p_y.

Equation (5) is the recurrence relation we were seeking, so now let us see how it works out for those cases which we can tackle explicitly. Start with the easiest case, $N = 2$, where we know that

$$\rho^{2\ell}_{m_0\to m_1+m_2} = \frac{\pi^{1-\ell}\Gamma(\ell-1)\left(\lambda\left(m_0^2, m_1^2, m_2^2\right)\right)^{2\ell-3}}{2^{2\ell-1}m_0^{2\ell-2}\Gamma(2\ell-2)} \tag{6}$$

involving the Källen function, $\lambda(a, b, c) \equiv \sqrt{a^2 + b^2 + c^2 - 2ab - 2bc - 2ca}$. The recurrence relation (5) reduces to the prediction that

$$\rho^{2\ell}_{m_0\to m_1+m_2} = \frac{1}{2\pi}\int \mathrm{d}\mu\, \rho^{2\ell-1}_{m_0\to m_{1\mu}+m_{2\mu}}\theta(m_0 - m_{1\mu} - m_{2\mu}) \tag{7}$$

and its verification relies upon the observation that

$$\lambda^2(m_0^2, m_{1\mu}^2, m_{2\mu}^2) = \lambda^2(m_0^2, m_1^2, m_2^2) - 4m_0^2\mu^2$$

plus the basic integral ($M^2 \equiv \lambda(m_0^2, m_1^2, m_2^2)/2m_0$ and $r \equiv 2\ell - 3$ below)

$$\int \mathrm{d}\mu\, \theta(m_0 - m_{1\mu} - m_{2\mu})\lambda^r(m_0^2, m_{1\mu}^2, m_{2\mu}^2) = \int_0^{M^2} \frac{\mathrm{d}\mu^2}{\sqrt{\mu^2}}\left(2m_0\sqrt{M^2-\mu^2}\right)^r$$

$$= \frac{\lambda^{r+1}(m_0^2, m_1^2, m_2^2)\,\Gamma\left(\frac{r}{2}+1\right)\sqrt{\pi}}{2m_0\Gamma\left(\frac{r}{2}+\frac{3}{2}\right)}.$$

The recurrence relation becomes much more interesting for the three-body case,

$$4\pi^2\rho^{2\ell}_{m_0\to m_1+m_2+m_3} = \iiint \mathrm{d}\mu_1\,\mathrm{d}\mu_2\,\mathrm{d}\mu_3\,\delta(\mu_1+\mu_2+\mu_3)\rho^{2\ell-1}_{m_0\to m_{1\mu_1}+m_{2\mu_2}+m_{3\mu_3}} \tag{8}$$

upon recalling that even- and odd-dimensional phase space behave rather differently: odd D leads to a Laurent series in the masses, while even D generally leads to elliptic functions [3–5]; equation (8) signifies that there exists an integral relation between elliptic functions and polynomials/poles. Moreover, the nature of the integral displays *explicit* symmetry in the masses which is useful. Because of the δ function constraint and the fact that the μ run over positive and negative values, the rhs of (8) can be broken up into the sum of three terms:

$$4\pi^2\rho^{2\ell}_{m_0\to m_1+m_2+m_3} = 2\int_0\int_0 \mathrm{d}\mu_1\,\mathrm{d}\mu_2\,\rho^{2\ell-1}_{m_0\to m_{1\mu_1}+m_{2\mu_2}+m_{3\mu_1+\mu_2}} + 2\text{ cyclic perms.}$$

We may quickly check the truth of relation (8) for the test case $\ell = 2$ when all particles are massless, since that limit of phase space simply yields

$$\rho^4_{m_0\to 0+0+0} = \lim_{\ell\to 2}\frac{(4\pi)^{1-2\ell}[\Gamma(\ell-1)]^3 m_0^{4\ell-6}}{2\Gamma(3\ell-3)\Gamma(2\ell-2)} = \frac{m_0^2}{256\pi^3}$$

for the lhs. On the other hand, for the rhs, substituting the 3D result [6],

$$\rho^3_{m_0\to m_1+m_2+m_3} = (m_0 - m_1 - m_2 - m_3)\theta(m_0 - m_1 - m_2 - m_3)/16\pi m_0$$

each of the three permutations produces the same answer and we obtain a perfect check of the recurrence relation. However, one learns something new in the massive case since a *new* symmetrical representation for 4D phase space emerges:

$$\rho^4_{m_0\to m_1+m_2+m_3} = \frac{1}{32\pi^3 m_0}\int_0\int_0 \mathrm{d}\mu_1\,\mathrm{d}\mu_2\left(m_0 - m_{1\mu_1} - m_{2\mu_2} - m_{3(\mu_1+\mu_2)}\right) + 2\text{ other perms.} \tag{9}$$

One expects the right-hand integral to produce elliptic functions, but the main virtue of (9) is that we get a pleasingly symmetrical sum of them which in principle ought to match an earlier form [7], albeit obtained in a different manner. Although we have not yet succeeded in establishing the precise relation with the Jacobian zeta function form, we have performed a series of *successful* numerical checks of (9), using typical mass values.

The same thing happens for larger *N*. For instance in four-body decay there arise four permutations with three of the μ_i having the same sign, opposite to the last one, plus six permutations where two pairs of μ_i have the same sign and opposite to the other pair. This would provide an elegant symmetrical way of evaluating four-body decay without resorting to Almgren's nonsymmetrical way [4] of pairing two bodies together and then summing over their pairwise mass sums.

One may of course continue in this vein and discuss spaces with the topology $M^{D-n} \otimes T^n$, but one learns very little new by this ruse because the process just yields a set of angles $\theta_j = y_j/R_j$ and a set of mode numbers k_j which collectively lead to $m_{\mathbf{k}}^2 = m^2 + \sum_j (k_j/R_j)^2$. There is little gain in taking the limit as each $R_j \to \infty$ separately because we only care for the final result where none of the radii is finite.

3. Relations between D and $D-2$ phase space

Next we shall suppose that we are dealing with the direct topology $M^{D-2} \otimes S^2$ so that the two extra angular coordinates are confined to the surface of a 2-sphere having radius R; thus $\vec{y} = R(\sin\theta\cos\varphi, \sin\theta\sin\varphi, \cos\theta)$, whereas previously y was identified with the circumference $R\theta$ rather than the radius vector. In such a situation expand the fields in spherical harmonics,

$$\Phi(\mathbf{x}, \hat{y}) = \sum_{j,k} \phi_{jk}(\mathbf{x}) Y_{jk}(\theta, \varphi)/R \tag{10}$$

before considering the limit of large R. The chosen factors ensure that the lower-dimensional field modes ϕ are properly normalized:

$$S_{\text{free}} = \frac{1}{2}\int \mathrm{d}^D x[(\partial_x \Phi)^2 - m^2\Phi^2] = \sum_{j,k} \frac{1}{2}\int \mathrm{d}^{D-2}\mathbf{x}\left[(\partial_{\mathbf{x}}\phi_{jk}^*)(\partial_{\mathbf{x}}\phi_{jk}) - m_j^2 \phi_{jk}^* \phi_{jk}\right] \tag{11}$$

where $m_j^2 = m^2 + j(j+1)/R^2$. (In equation (11) it is really meant that $\mathrm{d}^D x = \mathrm{d}^{D-2}\mathbf{x} R^2\, \mathrm{d}^2\Omega$ and $(\partial\Phi)^2 = g^{ab}\partial_a\Phi\partial_b\Phi$.) Again we note that $\phi_{jk} = \phi_{j-k}^*$, so the complex modes are found by just summing over positive k and discarding the factor of $1/2$.

The two-body recurrence relation can be verified in its entirety, since the effective coupling of a zero-mode mass m_0 field $\Sigma = \sigma/\sqrt{4\pi}R$ to two others yields a uniform amplitude, independently of the angular momentum eigenvalues j, k as we see from

$$\int \mathrm{d}^D x\, \Phi_1\Phi_2\Sigma = \sum_{j,k}\int \mathrm{d}^{D-2}\mathbf{x}\, \sigma\, \Phi_{1jk}\Phi_{2j-k}/\sqrt{4\pi}R. \tag{12}$$

The $(2j+1)$ degeneracy in k leads to

$$\rho^D_{m_0\to m_1+m_2} = \frac{1}{4\pi R^2}\sum_j (2j+1)\rho^{D-2}_{m_0\to m_{1j}+m_{2j}} \qquad m_{ij}^2 = m_i^2 + \left(\frac{j}{R}\right)^2. \tag{13}$$

In the limit $(R, j) \to \infty$, with $\mu = j/R$, this reduces to the continuum prediction,

$$\rho^D_{m_0\to m_1+m_2} = \int_0 \mathrm{d}\mu^2 \rho^{D-2}_{m_0\to m_{1\mu}+m_{2\mu}}/4\pi \tag{14}$$

which is readily verified from the explicit result (6).

The three-body case is altogether more fascinating. Here the effective interaction reduces to

$$\int \mathrm{d}^D x\, \Phi_1\Phi_2\Phi_3\Sigma = \frac{1}{\sqrt{4\pi}R^2}\sum_{j_1,k_1}\sum_{j_2,k_2}\sum_{j_3,k_3}\int \mathrm{d}^{D-2}\mathbf{x}\, \sigma\, \Phi_{1j_1k_1}\Phi_{2j_2k_2}\Phi_{3j_3k_3}$$
$$\times \int \mathrm{d}^2\Omega\, Y_{j_1k_1}Y_{j_2k_2}Y_{j_3k_3}. \tag{15}$$

To make progress, use the orthogonality property of spherical harmonics [8]

$$\int \mathrm{d}^2\Omega\, Y_{j_1k_1}Y_{j_2k_2}Y_{j_3k_3} = \sqrt{\frac{(2j_1+1)(2j_2+1)(2j_3+1)}{4\pi}}\begin{pmatrix} j_1 & j_2 & j_3 \\ 0 & 0 & 0\end{pmatrix}\begin{pmatrix} j_1 & j_2 & j_3 \\ k_1 & k_2 & k_3\end{pmatrix}.$$

This then specifies the magnitudes of the mode couplings. When evaluating the sum over modes, apply the completeness relation of C-G coefficients,

$$\sum_{k_2,k_3}\begin{pmatrix} j_1 & j_2 & j_3 \\ k_1 & k_2 & k_3 \end{pmatrix}\begin{pmatrix} j_4 & j_2 & j_3 \\ k_4 & k_2 & k_3 \end{pmatrix} = \frac{\delta_{j_1j_4}\delta_{k_1k_4}}{2j_1+1}$$

signifying

$$\sum_{k_1,k_2,k_3}\begin{pmatrix} j_1 & j_2 & j_3 \\ k_1 & k_2 & k_3 \end{pmatrix}^2 = \sum_{k_1}\frac{1}{2j_1+1} = 1.$$

Therefore, the summation over modes produces the discrete recurrence relation,

$$\rho^D_{m_0\to m_1+m_2+m_3} = \sum_{j_1,j_2,j_3}\frac{(2j_1+1)(2j_2+1)(2j_3+1)}{(4\pi R^2)^2}\begin{pmatrix} j_1 & j_2 & j_3 \\ 0 & 0 & 0 \end{pmatrix}^2 \rho^{D-2}_{m_0\to m_{1j_1}+m_{2j_2}+m_{3j_3}} \tag{16}$$

whose continuum limit is of interest. To take $R\to\infty$, first note that Wigner's 3-j symbol is [8]

$$\begin{pmatrix} j_1 & j_2 & j_3 \\ 0 & 0 & 0 \end{pmatrix} \equiv (-1)^{-(j_1+j_2+j_3)/2}\frac{\left(\frac{j_1+j_2+j_3}{2}\right)!\sqrt{(-j_1+j_2+j_3)!(j_1-j_2+j_3)!(j_1+j_2-j_3)!}}{\sqrt{(1+j_1+j_2+j_3)!}\left(\frac{-j_1+j_2+j_3}{2}\right)!\left(\frac{j_1-j_2+j_3}{2}\right)!\left(\frac{j_1+j_2-j_3}{2}\right)!}.$$

Since we are concerned with the large j limit, apply Stirling's formula,

$$\frac{a!}{(a/2)!^2} \simeq \frac{2^{a+1/2}}{\sqrt{\pi a}}$$

to the previous expression. The square of the 3-j symbol then magically simplifies to the inverse area of a triangle having sides j_1, j_2, j_3:

$$\begin{pmatrix} j_1 & j_2 & j_3 \\ 0 & 0 & 0 \end{pmatrix}^2 \simeq \frac{2\theta(\lambda_E)}{\pi\lambda_E\left(j_1^2, j_2^2, j_3^2\right)} \qquad \lambda_E^2 \equiv -\lambda^2 \tag{17}$$

which makes good sense, recalling the vector addition formula for angular momenta. One finishes with the continuum result

$$\rho^D_{m_0\to m_1+m_2+m_3} = \frac{1}{4\pi^3}\int_0\int_0\int_0 \mathrm{d}\mu_1\,\mathrm{d}\mu_2\,\mathrm{d}\mu_3\frac{\mu_1\mu_2\mu_3}{\lambda_E\left(\mu_1^2,\mu_2^2,\mu_3^2\right)}\rho^{D-2}_{m_0\to m_{1\mu_1}+m_{2\mu_2}+m_{3\mu_3}}. \tag{18}$$

This recurrence is very difficult to verify in general, especially for even D. We can however make a fist of it for 5D in the massless limit when the check collapses to the veracity of

$$\rho^5_{m_0\to 0+0+0} = \frac{1}{64\pi^4}\int_0\int_0\int_0 \mathrm{d}\mu_1\,\mathrm{d}\mu_2\,\mathrm{d}\mu_3\frac{\mu_1\mu_2\mu_3\theta(\lambda_E)}{\lambda_E\left(\mu_1^2,\mu_2^2,\mu_3^2\right)}\left(1-\frac{\mu_1+\mu_2+\mu_3}{m_0}\right). \tag{19}$$

The lhs of (19) is known to be equal to $m_0^4/53760\pi^3$. To integrate the rhs, change variables according to $\mu_1 = w/2-u$, $\mu_2 = w/2-v$ and $\mu_3 = v+u$, so,

$$\int_0\int_0\int_0 \mathrm{d}\mu_1\,\mathrm{d}\mu_2\,\mathrm{d}\mu_3\,X = 2\int_0^{m_0}\mathrm{d}w\int_0^{w/2}\mathrm{d}v\int_0^{w/2-v}\mathrm{d}u\,X$$

may be carried out with $X \propto \frac{(w/2-u)(w/2-v)(u+v)(m_0-w)}{\sqrt{uv[2w(u+v)-w^2]}}$. The result indeed reproduces the lhs. We have also verified the truth of (19) numerically.

The miraculous birth of the Euclidean Källen function λ_E in (17) and (19) can be rendered less mysterious if we look upon this as the result of integrating over the last two components of momentum $\vec{\mu}$, corresponding to a 'radial kernel'. Thus from the mathematical fact that

$$\left(\prod_{i=1}^{3}\int\frac{\mathrm{d}^2\vec{\mu}_i}{(2\pi)^2}\right)(2\pi)^2\delta^2\left(\sum_{i=1}^{3}\vec{\mu}_i\right)F(|\vec{\mu}_j|)=\left(\prod_{i=1}^{3}\int\frac{\mu_i\,\mathrm{d}\mu_i\,\mathrm{d}\theta_i}{(2\pi)^2}\right)F(\mu_j)\int\mathrm{d}^2\vec{k}\,\mathrm{e}^{\mathrm{i}\vec{k}\cdot\sum\vec{\mu}}$$

we find that this expression equals [9, 10]

$$\left(\prod_{i=1}^{3}\int\frac{\mu_i\,\mathrm{d}\mu_i}{2\pi}\right)F(\mu_j)2\pi\int_0^\infty J_0(k\mu_1)J_0(k\mu_2)J_0(k\mu_3)k\,\mathrm{d}k$$

$$=\left(\prod_{i=1}^{3}\int\frac{\mu_i\,\mathrm{d}\mu_i}{2\pi}\right)\frac{4\theta(\lambda_E)F(\mu_j)}{\lambda_E\left(\mu_1^2,\mu_2^2,\mu_3^2\right)}.$$

Moreover, the extension to N-body phase space suggests itself immediately via an N-fold kernel, associated with the integral $2\pi\int_0^\infty\left(\prod_{i=1}^N J_0(k\mu_i)\right)k\,\mathrm{d}k$ although this is not readily stated in terms of simple functions for $N>4$; something geometrical associated with the lengths μ_i is certainly involved. This kernel has to be folded over $\rho^{D-2}_{m_0\to m_{1\mu_1}+\cdots+m_{N\mu_N}}$ and integrated with respect to all $\mu_i\,\mathrm{d}\mu_i$ to establish the recurrence.

4. Relations between D and $D-n$ phase space

With the torus and 2-sphere thoroughly understood, it is natural to extend the argument to coordinates confined to an n-sphere where the topology is $M^{D-n}\otimes S^n$. Here we need to make use of hyperspherical harmonics [11] defined over n angles. Associated with them are the generalized quadratic Casimir operator with eigenvalue $j(j+n-1)$ and $n-1$ angular momentum components (generically labelled $\mathbf{k}$), producing a degeneracy of $h_{jn}=(2j+n-1)\cdot(j+n-2)!/j!(n-1)!$ Thus the free action may be normalized according to equation (11), where

$$\Phi(\mathbf{x},\hat{y})=\sum_{j,\mathbf{k}}\phi_{j,\mathbf{k}}(\mathbf{x})Y_{j,\mathbf{k}}(\hat{y})/R^{n/2}$$

and we must sum over $(n-1)$ of the $\mathbf{k}$ labels. Now the squared mass equals $m^2+j(j+n-1)/R^2$ of course, because the last term corresponds to a hyperspherical Laplacian eigenvalue [11].

Running through similar steps as before and skipping details, we arrive at the two-body recurrence relation for finite R,

$$\rho^D_{m_0\to m_1+m_2}=\sum_j h_{jn}\rho^{D-n}_{m_0\to m_{1j}+m_{2j}}\Big/\Omega_n R^n \tag{20}$$

where $\Omega_n=2\pi^{(n+1)/2}/\Gamma((n+1)/2)$ is the total solid angle corresponding to n angular coordinates. In the limit of large R and therefore large j, since $h_{jn}\simeq 2j^{n-1}/\Gamma(n)$, we obtain the continuum limit,

$$\rho^D_{m_0\to m_1+m_2}=\frac{\Gamma((n+1)/2)}{\pi^{(n+1)/2}\Gamma(n)}\int_0\mathrm{d}\mu\,\mu^{n-1}\rho^{D-n}_{m_0\to m_{1\mu}+m_{2\mu}} \tag{21}$$

and this is readily checked via the explicit answer (6). Equations (7) and (14) are particular cases of (21).

While we have succeeded in treating the two-body decay by this procedure, it clearly becomes unwieldy and probably useless for $N>2$ and larger values of n, since we would have to integrate over multiproducts of hyperspherical harmonics, which are not exactly well known, though some fancy generalizations of 3-j symbols and the like must exist. On the other hand, one can make much better progress by regarding the recurrence as the result of integrating over the last n momenta $\vec{\mu}$:

$$\rho^D_{m_0\to m_1+\cdots+m_N} = \left(\prod_{i=1}^{N}\int \frac{\mathrm{d}^n\vec{\mu}_i}{(2\pi)^n}\right)\int \mathrm{d}^n\vec{k}\,\mathrm{e}^{\mathrm{i}\vec{k}\cdot\sum_i\vec{\mu}_i}\rho^{D-n}_{m_0\to m_{1\mu_1}+\cdots+m_{N\mu_N}}. \tag{22}$$

Now in general [1],

$$\int \mathrm{d}^n\vec{\mu}\exp(\mathrm{i}\vec{k}\cdot\vec{\mu})f(\mu) = \int_0^\infty (2\pi\mu)^{n/2}J_{n/2-1}(k\mu)f(\mu)\,\mathrm{d}\mu/k^{n/2-1}$$

and $\mathrm{d}^n\vec{k} = k^{n-1}\,\mathrm{d}k\Omega_{n-1} = 2k^{n-1}\,\mathrm{d}\mu\cdot\pi^{n/2}/\Gamma(n/2)$, so (20) simplifies to

$$\rho^D_{m_0\to m_1+\cdots+m_N} = \left(\prod_{i=1}^{N}\int_0^\infty \frac{\mu_i^{n/2}\,\mathrm{d}\mu_i}{(2\pi)^{n/2}}\right)\int_0^\infty \frac{2\pi^{n/2}k^{n-1}\,\mathrm{d}k}{\Gamma(n/2)}\left(\prod_{i=1}^{N}\frac{J_{n/2-1}(k\mu_i)}{k^{n/2-1}}\right)$$
$$\times\,\rho^{D-n}_{m_0\to m_{1\mu_1}+\cdots+m_{N\mu_N}}. \tag{23}$$

As far as we are aware there is no amenable formula for the radial kernel: an integral over the product of N Bessel functions of the first kind with different arguments for all N. However, the cases $N=2$ and $N=3$ are known and for positive a, b, c, read [9]

$$\int_0^\infty J_\nu(ax)J_\nu(bx)x\,\mathrm{d}x = 2\delta(a^2-b^2)$$

$$\int_0^\infty J_\nu(ax)J_\nu(bx)J_\nu(cx)x^{1-\nu}\,\mathrm{d}x = 2\theta(\lambda_E)\lambda_E^{2\nu-1}/(8abc)^\nu\Gamma(\nu+1/2)\sqrt{\pi}.$$

For $N=4$ it is even known that [10]

$$\int_0^\infty J_0(ax)J_0(bx)J_0(cx)J_0(\mathrm{d}x)x\,\mathrm{d}x = \mathbf{K}(\sqrt{abcd}/\Delta)/\pi^2\Delta$$

where $16\Delta^2 = (a+b+c-d)(b+c+d-a)(c+d+a-b)(d+a+b-c)$ is associated with the maximal area of a (cocyclic) quadrilateral formed by the lengths a, b, c, d and $\mathbf{K}$ is the complete elliptic integral of the first kind. However, this means that we may at least write a simple closed form for the three-body recurrence relation, when n is arbitrary:

$$\rho^D_{m_0\to m_1+m_2+m_3} = \left(\prod_{i=1}^{3}\int_0^\infty \frac{\mu_i\,\mathrm{d}\mu_i}{(2\pi)^{n/2}}\right)\frac{2^{3-n/2}\theta(\lambda_E)\pi^{n/2-1}}{\lambda_E^{3-n}\left(\mu^2,\mu_2^2,\mu_3^2\right)\Gamma(n-1)}\rho^{D-n}_{m_0\to m_{1\mu_1}+m_{2\mu_2}+m_{3\mu_3}}. \tag{24}$$

An interesting case occurs when $N=n=3$, whereupon (24) reduces to

$$\rho^D_{m_0\to m_1+m_2+m_3} = \int_0^\infty\int_0^\infty\int_0^\infty \mathrm{d}\mu_1^2\,\mathrm{d}\mu_2^2\,\mathrm{d}\mu_3^2\frac{\theta(\lambda_E)}{64\pi^4}\rho^{D-3}_{m_0\to m_{1\mu_1}+m_{2\mu_2}+m_{3\mu_3}} \tag{25}$$

and because 3D phase space is so simple [6], we obtain an intriguing representation for 6D phase space on setting $D=6$. Specifically,

$$\rho^6_{m_0\to m_1+m_2+m_3} = \int_0^\infty \mathrm{d}a\int_0^\infty \mathrm{d}b\int_0^\infty \mathrm{d}c\frac{\theta(\lambda_E(a,b,c))}{64\pi^4}\frac{m_0-\sqrt{m_1^2+a}-\sqrt{m_2^2+b}-\sqrt{m_3^2+c}}{16\pi m_0}.$$

With larger N, presumably the radial kernel on the right-hand side of (23) involves all lengths μ_i and something geometrically more complicated than λ_E, connected with the closed figure $\sum_i\vec{\mu}_i = 0$. (Thus we expect it to vanish when any length exceeds the lengths of the other three, amongst other conditions. This is a very interesting topic worth future investigation.)

5. Conclusions

In this paper we have established the connection between relativistic phase space for different dimensions by two methods. They complement Almgren's connection between phase space in the same dimension, but for different numbers of decay products. Our recurrence relations have practical utility when the difference in dimensions n is odd since it leads to elegant symmetrical representations of ρ for even D involving elliptic functions.

Last but not least, one should observe that phase space is nothing but the imaginary part of a sunset Feynman diagram. Since recurrence relations between sunset diagrams differing in dimensionality by 2 have been found by other methods [12], it should be possible to rewrite our results that way, for even n at any rate. We should also point out that as a rule those relations do not connect even and odd D because they are obtained by integration by parts from scalar particle Feynman graphs.

Acknowledgments

We are pleased to acknowledge financial support from the Australian Research Council under grant number A00000780. Comments and suggestions by A I Davydychev are also much appreciated.

References

[1] Akyeampong D A and Delbourgo R 1974 *Nuovo Cimento* A **19** 141
Mendels E 1978 *Nuovo Cimento* A **45** 87
[2] Groote S, Körner J G and Pivovarov A A 1999 *Nucl. Phys.* B **542** 515
Groote S, Körner J G and Pivovarov A A 1999 *Eur. Phys. J.* C **11** 279
Groote S and Pivovarov A A 2000 *Nucl. Phys.* B **580** 459
[3] Bashir A, Delbourgo R and Roberts M L 2001 *J. Math. Phys.* **42** 5553
[4] Almgren B 1968 *Ark. för Phys.* **38** 161
[5] Bauberger S, Berends F A, Böhm M and Buza M 1995 *Nucl. Phys.* B **434** 383
[6] Rajantie A K 1996 *Nucl. Phys.* B **480** 729
[7] Davydychev A I and Delbourgo R 2002 Three-body phase space: symmetrical treatments *Proc. Australian Institute Physics Biennial Congress* (Sydney, 2002) *Preprint* hep-th/0209233 (AIPC 2002 vol 347 at press)
[8] Wigner E P 1968 *Group Theory* (London: Academic)
[9] Gervois A and Navelet H 1985 *J. Math. Phys.* **26** 633
[10] Prudnikov A P, Brychkov Yu A and Marychev O I 1986 *Integrals and Series* vol 2 (Amsterdam: Gordon and Breach) (see sections 2.12.42 and 2.12.44 in particular)
[11] Bateman H 1953 *Higher Transcendental Functions* vol 2 (New York: McGraw-Hill)
[12] Tarasov O V 1997 *Nucl. Phys.* B **502** 455
Tarasov O V 2000 *Nucl. Phys. Proc. Suppl.* **89** 237
See also equation (25) in Davydychev A I and Tausk J B 1996 *Phys. Rev.* **53** 7381

Part D
GAUGE TECHNIQUE

D.0. Commentaries D

The idea of using the Ward-Green-Takahashi identities to relate (parts of) Green functions in gauge theories, thereby gaining some leverage on renormalisation and nonperturbative behaviour, is due to Salam (Phys. Rev. 130, 1287 (1963)). The method was dubbed *the gauge technique* and it has been applied extensively in QED and QCD to discover the asymptotic behaviour of propagators and vertex functions. The way it has been manipulated depends on the approach. Following the lead of Ball & Chiu (Phys. Rev. D22, 2542 (1980)) and Cornwall (Phys. Rev. D26, 1453 (1982)), many authors have opted to work on amputated Green functions, which lead to difficult nonlinear equations often requiring numerical computer-aided calculation; they call this approach *solving the Dyson-Schwinger equations.* Instead we have taken the view that the identities are just as easily solved for the full Green functions because they lead to linear equations that are more amenable to analysis.

In any of these methods a dependence on the gauge fixing method is prominent because one is going off-shell. Furthermore, there are parts of Green functions which remain unknown by the technique — these are transverse to the gauge field momentum. In that respect the Landau-Khalatnikov transformations (Sov. Phys. JETP 2, 69 (1956)) are useful because they stipulate how the amplitudes change as the gauge fixing function is varied. Those transformations do provide information about the transverse amplitudes, as well as the longitudinal ones, but only if one has full confidence of the transverse plus longitudinal results in one particular gauge. What we do know is that the transverse part disappears in the infrared limit so that the results derived by the gauge technique become precise in that limit.

D.1 Delbourgo R. and West P., A gauge covariant approximation to quantum electrodynamics. J. Phys. A10, 1049-1056 (1977)

D.2 Delbourgo R. and West P., Infrared behaviour of a gauge covariant approximation. Phys. Lett. B72, 96-98 (1977)

D.3 Delbourgo R., Solving the gauge identities. J. Phys. A11, 2057-2071 (1978)

D.4 Delbourgo R., The gauge technique. Nuovo Cim. A49, 484-496 (1979)

D.5 Delbourgo R. and Keck B.W., On the gauge dependence of spectral functions. J. Phys. A13, 701-711 (1980)

D.6 Delbourgo R. and Zhang R., Transverse vertices in electrodynamics and the gauge technique. J. Phys. A17, 3593-3607 (1984)

The articles for scalar and vector electrodynamics (Phys. Rev. B135, 1398 (1964)) and the spinor version by Strathdee (Phys. Rev. B135, 1428 (1964)) were the first attempts to apply the gauge technique; they focussed on the amputated Green functions and were beset by nonlinearities. They depended on a full understanding of the first order perturbation theory results off-shell. Ball & Chiu (Phys. Rev. D22, 2550 (1980) and Erratum, Phys. Rev. D23, 3085 (1981)) pursued this matter and calculated all the off-shell functions to first order perturbation theory as a check, while Kizilersu, Reenders and Pennington (Phys. Rev. D52, 1242 (1995)) did the same in the fermionic case, giving the full dependence on the covariant gauge parameter. Papers [D.1] and [D.2] introduced two new tricks which served to linearise the problems of the technique: the first trick was to use the full Green functions, since the gauge identities still apply linearly, and the second trick was to adopt the spectral representation of the propagator, leading to a linear equation for the spectral function that may be solved exactly. In [D.3] it was proved that this method may be extended to all n-point Green functions. Paper [D.4] is a short review of the entire technique and how it can be applied to the axial currents too, while article [D.5] for the first time describes the dependence of the spectral function on the choice of gauge parameter.

While the technique's conclusions are correct in the infrared and ultraviolet regimes, they are faulty at intermediate momenta because they fail to take account of transverse vector contributions. In article [D.6] we attempted to remedy this problem by including transverse parts which are exact to first order perturbation theory and by solving the identity for the 4-point function rather than restricting the method to the 3-point function only. Thus they are exact to the next order in perturbation theory. In that context readers should also consult the paper by He, Khanna and Takahashi (Phys. Lett. 480, 222 (2000)) which focusses on the corresponding identity for the transverse vertex.

There is still much activity in solving Schwinger-Dyson equations in the context of QCD. This brings in the ghost particles, needed to restore unitarity, which complicates the analysis. In fact the results depend critically on the behaviour of the ghost propagator in the infrared regime. There is still some argument as to whether the gauge field acquires a mass by this mechanism. See the review and sample papers by

- C.D. Roberts and A.G. Williams, Prog. Part. Nucl. Phys. 33, 477 (1994)
- A. Bashir, A. Kizilersu, and M.R. Pennington, Phys. Rev. D57, 1242 (1998)
- J.M. Cornwall, Phys. Rev. D26, 1453 (1982)
- D.C. Curtis and M.R. Pennington, Phys. Rev. D42, 4165 (1990)
- A.C. Aguilar and A.A. Natale, J. H. E. Phys. (2004)
- A.C. Aguilar, D. Binosi, and J. Papavassiliou, Phys. Rev. D78, 025010 (2008)

in which further references may be followed up.

J. Phys. A: Math. Gen., Vol. 10, No. 6, 1977. Printed in Great Britain. © 1977

A gauge covariant approximation to quantum electrodynamics

R Delbourgo† and P West‡

† Department of Physics, University of Tasmania, Box 252C, GPO, Hobart, Tasmania, Australia 7001

‡ Ecole Normale Superieure, 24 Rue Lhomond, 75231 Paris Cedex 05, France

Received 11 February 1977

Abstract. A non-perturbative method of solving the Dyson–Schwinger equations in QED, which *preserves* the gauge identities, is considered. The starting point is determined by an integral equation for the electron propagator spectral function which is explicitly solved in the Landau gauge; this determines the Green functions in successive orders of iteration since no spurious infinities arise beyond the usual renormalizable ones.

1. Introduction

A favourite pastime of theoreticians has been to look for approximate solutions of the complete set of equations linking the Green functions in various quantum field models like QED. Most of these approximations amount to summing specific sets of perturbation graphs with the foreknowledge or hope that the selection will provide the dominant contributions in the kinematic region of interest. In gauge theories most such approximations unfortunately violate the gauge constraints among the Green functions and this makes it difficult to judge the correctness or otherwise of the solutions found. There is one approximation method however which has the virtue of preserving the Ward identities at every stage: this is Salam's gauge technique (1963); by contrast, here it becomes difficult to judge the 'order of approximation' because the iteration procedure is basically non-perturbative. In early papers the gauge technique was applied to electrodynamics of mesons and spinors. The zeroth-order Green functions were obtained by truncating the Dyson–Schwinger equations so as to satisfy two-particle unitarity and could be simply calculated by applying first-order perturbation theory to the Lehmann spectral functions (Delbourgo and Salam 1964, Strathdee 1964). It then was verified that the resulting asymptotic behaviours of the Green functions were no different in the next order of iteration, signifying that the procedure was 'asymptotically stable'. However it remained unclear what the iterated sum of the series would yield and what bearing, if any, this has on renormalization group aspects of the problem (Manoukian 1974).

In this paper we wish to return to the gauge technique for QED but without resorting to two-particle unitarity for providing the starting point. Rather we shall solve the Dyson–Schwinger equation for the spinor propagator as a proper integral equation, neglecting photon dressing in the first place since that has no important bearing on the gauge identities. We shall then use the solution to determine its influence on the photon self-energy and other Green functions such as the vertex part. We do not look for finite electrodynamics and thus get no eigenvalue equation for e^2.

2. The zeroth gauge approximation

The Ward identities between Green functions in gauge theories are now well known (Nishijima 1960, Rivers 1966) even for non-Abelian gauges (Lee 1974, Kluberg-Stern and Zuber 1975). Thus with photon legs amputated the first few identities read

$$k^\mu S(p)\Gamma_\mu(p, p-k)S(p-k) = S(p-k) - S(p) \tag{1}$$

$$k^\mu S(p')\Gamma_{\nu\mu}(p'k'; pk)S(p)$$
$$= S(p')\Gamma_\nu(p', p'+k')S(p'+k') - S(p-k')\Gamma_\nu(p-k', p)S(p) \tag{2}$$

etc, where S denotes the complete electron propagator and Γ stands for the fully amputated connected Green functions with appropriate arguments and with coupling constants factorized out. In QED the propagators S of the electron and D of the photon, and the vertex part Γ_μ, play a central role via the Dyson–Schwinger equations

$$1 = Z_\psi(\gamma . p - m + \delta m)S(p) - \mathrm{i}e^2 Z_\psi \int \bar{\mathrm{d}}^4k\, S(p)\Gamma_\mu(p, p-k)S(p-k)\gamma_\nu D^{\mu\nu}(k) \tag{3}$$

$$D^{-1}_{\mu\nu}(k) = Z_A[-k^2\eta_{\mu\nu} + k_\mu k_\nu(1-a^{-1})] + \mathrm{i}e^2 Z_\psi \operatorname{Tr} \int \bar{\mathrm{d}}^4p\gamma_\nu S(p)\Gamma_\mu(p, p-k)S(p-k) \tag{4}$$

$$\Gamma_\mu(p, p-k) = Z_\psi\gamma_\mu - \mathrm{i}e^2 Z_\psi \int \bar{\mathrm{d}}^4p' K(p, p'; p-k, p'-k)S(p')\Gamma_\mu(p', p'-k)S(p'-k) \tag{5}$$

in which we have adopted a covariant photon gauge parametrized by the bare constant a. One can make the gauge identities look more obvious by replacing (5) with

$$\Gamma_\mu(p, p-k) = Z_\psi\gamma_\mu - \mathrm{i}e^2 Z_\psi \int \bar{\mathrm{d}}^4p'\gamma_\lambda S(p')\Gamma_{\nu\mu}(p'k'; pk)D^{\lambda\nu}(k). \tag{5'}$$

Multiplication of (5′) by k^μ yields (3) immediately, and in that sense incorporates it.

In the gauge technique (Salam 1963) one seeks solutions to equations (3)–(5) consistent with the Ward–Takahashi identities (1), (2), etc. To see how these can be determined iteratively, let us begin with the Lehmann spectral representation for the spinor propagator in the form†

$$S(p) = \left(\int_{-\infty}^{-m} + \int_m^\infty\right) \frac{\rho(W)\,\mathrm{d}W}{\gamma . p - W + \mathrm{i}\epsilon(W)0} \tag{6}$$

where $\rho(W)$ is a positive definite distribution in a non-gauge theory.

† The form

$$S(p) = \int_m^\infty \frac{(\gamma p\rho_1(s) + m\rho_2(s))\,\mathrm{d}s}{p^2 - s + \mathrm{i}0}$$

with $m_0 = Z\int m\rho_2(s)\,\mathrm{d}s$, $1 = Z\int \rho_1(s)\,\mathrm{d}s$ is probably more familiar. The connection with the form (6) is provided by

$$\rho(W) = \epsilon(W)(W\rho_1(W^2) + m\rho_2(W^2)).$$

Since

$$S(p-k)-S(p)=\int \mathrm{d}W\rho(W)\frac{1}{\gamma.p-W}\gamma.k\frac{1}{\gamma.(p-k)-W}$$

the simplest possible (but by no means the unique) solution of (1) is to take

$$S(p)\Gamma_\mu^{(0)}(p,p-k)S(p-k)=\int \mathrm{d}W\rho(W)\frac{1}{\gamma.p-W}\gamma_\mu\frac{1}{\gamma.(p-k)-W}. \tag{7}$$

One can of course add to (7) any arbitrary transverse function of the type $(k^2\eta_{\mu\nu}-k_\mu k_\nu)F^\nu(p,p-k)$ which would have no effect on the gauge identities but we neglect this to begin with—a more precise reason follows shortly. Likewise a possible solution of identity (2) is provided by

$$\begin{aligned} S(p')\Gamma_{\nu\mu}^{(0)}(p'k';pk)S(p) &= \int \mathrm{d}W\rho(W)\frac{1}{\gamma.p'-W} \\ &\times\left(\gamma_\nu\frac{1}{\gamma.(p'+k')-W}\gamma_\mu+\gamma_\mu\frac{1}{\gamma.(p'-k)-W}\gamma_\nu\right)\frac{1}{\gamma.p-W} \end{aligned} \tag{8}$$

and so on. In analogy to (7) this is a weighted sum over electron mass distributions of the tree graphs. Indeed if we go to the mass shell of the charged spinor lines by picking out pole terms through the substitution $\rho(W)\to\delta(W-m)$ we get precisely the Born terms. We shall take this criterion as *defining* the zeroth gauge approximation $\Gamma^{(0)}$ of all the charged line Green functions; the photon lines are left undressed in this initial stage. Thus $\Gamma^{(0)}$ are functionals of the electron propagator S which is all we have to find, and this we can do by solving the electron line equation (3) as a true integral equation without resorting to two-particle unitarity. Using the basic $\Gamma^{(0)}$ we can then determine the photon propagator, vertex part and other connected functions in a recursive way† via

$$\begin{aligned} D^{-1(n+1)} &= Z_A k^{-2}+Z_\psi e^2\int S^{(n)}\Gamma^{(n)}S^{(n)}\gamma \\ S^{-1(n+1)} &= Z_\psi(\gamma.p-m_0)+Z_\psi e^2\int \Gamma^{(n)}S^{(n)}D^{(n)}\gamma \\ \Gamma^{(n+1)} &= Z_\psi\gamma+Z_\psi e^2\int K[\Gamma^{(n)}]S^{(n)}\Gamma^{(n)}S^{(n)} \end{aligned} \tag{9}$$

the hope being that the iterations will converge as $n\to\infty$ to the final true answers. Certainly for the iteration scheme to make sense it is necessary that higher-order multi-electron functions come out to be finite; otherwise all the advantages of conventional renormalizability would be lost and the gauge technique would become worthless. Finiteness is guaranteed when the stability criterion $\Gamma SD^{1/2}\sim 1/k^2$ is satisfied. We shall check *a posteriori* that our zeroth-order expressions do obey this asymptotic stability property and thus do not jeopardize renormalizability. Beyond this we have to look for some property of $\rho(W)$ which stands out as the iterations proceed since it is quite clear that the entire scheme is non-perturbative and we have not exactly

† Note that $S^{(1)}=S^{(0)}$ and $k.\Gamma^{(1)}=k.\Gamma^{(0)}$ in the first iteration.

expansions in e^2 to guide us about what we mean about the 'order of iteration'. Unfortunately we have not gone far enough in the iteration scheme to find out what it is except for the fact that the lowest-order spectral function (22) receives logarithmic corrections in succeeding orders.

3. The zeroth Green functions

In lowest order the equations (9) devolve to finding a solution of

$$Z_\psi^{-1}=(\gamma\,.\,p-m_0)S^{(0)}(p)-\mathrm{i}e^2\int \bar{\mathrm{d}}^4k\,S^{(0)}(p)\Gamma_\mu^{(0)}(p,p-k)S^{(0)}(p-k)D^{\mu\nu(0)}(k)$$

$$=(\gamma\,.\,p-m_0)\int\frac{\rho(W)\,\mathrm{d}W}{\gamma\,.\,p-W}-\mathrm{i}e^2\int\bar{\mathrm{d}}^4k\;\mathrm{d}W\rho(W)\frac{1}{\gamma\,.\,p-W}\gamma_\mu$$

$$\times\frac{1}{\gamma\,.\,(p-k)-W}\gamma_\nu\,.\left(-\eta^{\mu\nu}+\frac{k^\mu k^\nu}{k^2}(1-a)\right)\frac{1}{k^2}$$

$$=\int\frac{\rho(W)\,\mathrm{d}W}{\gamma\,.\,p-W}(\gamma\,.\,p-m_0+\Sigma(p,W)) \tag{10}$$

where $\Sigma(p, W)$ is obtained from lowest-order perturbation theory for an electron mass W. Thus

$$\operatorname{Im}\Sigma(p,W)=\frac{e^2(p^2-W^2)}{16\pi p^3}[a(p^2+W^2)-(a+3)pW]\theta(p^2-W^2) \tag{11}$$

yielding $\Sigma(p, W)$ via a dispersion integral. Recalling that

$$1=Z_\psi\int\rho(W)\,\mathrm{d}W \qquad \text{and} \qquad m_0=Z_\psi\int W\rho(W)\,\mathrm{d}W, \tag{12}$$

let us perform our renormalizations on (10) and remove the pole term by putting $\rho(W)=\delta(W-m)+\sigma(W)$. Then

$$\frac{\Sigma(W,m)}{W-m}+\int(W'-m+\Sigma(W,W'))\frac{\sigma(W')\,\mathrm{d}W'}{W-W'+\mathrm{i}\epsilon(W')0}=0 \tag{13}$$

with a once-subtracted

$$\Sigma(W,m)=\frac{(W-m)}{\pi}\int\frac{\operatorname{Im}\Sigma(W',m)\,\mathrm{d}W'}{(W'-m)(W-W')}. \tag{14}$$

Upon taking imaginary parts of (13) we remain with the integral equation

$$\epsilon(W)(W-m)\sigma(W)=\frac{\operatorname{Im}\Sigma(W,m)}{\pi(W-m)}+\int\mathrm{d}W'\,\sigma(W')\frac{\operatorname{Im}\Sigma(W,W')}{\pi(W-W')} \tag{15}$$

for the spectral function $\sigma(W)$. To show that exact solutions can be found let us specialize to the Landau gauge $a=0$ where $\operatorname{Im}\Sigma$ is particularly easy† and the equation

† It is also relatively simple in the Yennie gauge $a=3$ where to lowest order, $\sigma(W)=3e^2(W^2-m^2)\epsilon(W)/16\pi^2W^3$ guarantees infrared finiteness but gives an ultraviolet divergence for Z_ψ.

reduces to ($W^2 \geqslant m^2$)

$$\epsilon(W)(W-m)W^2\sigma(W)$$
$$= -\frac{3e^2}{16\pi^2}\left(m(W+m)+\int^W \mathrm{d}W'\,\sigma(W')(W+W')W'\right).$$

Using dimensionless variables

$$\omega = W/m, \qquad \xi = -3e^2/16\pi^2, \qquad s(\omega)=\epsilon(\omega)\omega^2\sigma(\omega), \tag{16}$$

the equation simplifies to

$$\frac{(\omega-1)s(\omega)}{\xi} = (\omega+1)+\left(\int_1^{\omega\epsilon(\omega)} - \int_{-\omega\epsilon(\omega)}^{-1}\right)\mathrm{d}\omega'\,s(\omega')\left(1+\frac{\omega}{\omega'}\right). \tag{17}$$

This can in turn be reduced to a pair of coupled equations by making the substitution $s(\omega)=\omega s_1(\omega^2)+s_2(\omega^2)$. Thus

$$\begin{aligned}(s_2(\omega^2)-s_1(\omega^2))/\xi &= \int_1^{\omega^2}\mathrm{d}\omega'^2\,s_2(\omega'^2)/\omega'^2+1\\ (\omega^2 s_1(\omega^2)-s_2(\omega^2))/\xi &= \int_1^{\omega^2}\mathrm{d}\omega'^2\,s_1(\omega'^2)+1\end{aligned} \tag{18}$$

Finally the pair can be reconverted into hypergeometric equations

$$\begin{aligned}&\left(Z(1-Z)\frac{\mathrm{d}^2}{\mathrm{d}Z^2}-[1-Z(3-2\xi)]\frac{\mathrm{d}}{\mathrm{d}Z}-(1-\xi)^2\right)s_1(Z)=0\\ &\left(Z(1-Z)\frac{\mathrm{d}^2}{\mathrm{d}Z^2}-2(1-\xi)Z\frac{\mathrm{d}}{\mathrm{d}Z}+\xi(1-\xi)\right)s_2(Z)=0.\end{aligned} \tag{19}$$

The appropriate solutions, satisfying the boundary conditions embodied in (18) and incorporating an infrared† cut-off μ^2 are

$$s_1(Z)=\frac{2\xi}{(Z-1)}\left(\frac{Z-1}{\mu^2/m^2}\right)^{2\xi}F(\xi,\xi;2\xi;1-Z)$$
$$s_2(Z)=\frac{2\xi Z}{(Z-1)}\left(\frac{Z-1}{\mu^2/m^2}\right)^{2\xi}F(\xi,\xi+1;2\xi;1-Z).$$

In terms of the original variables this gives

$$\sigma(W)=\epsilon(W)\theta(W^2-m^2)\frac{2\xi}{W}\left(\frac{W^2-m^2}{\mu^2}\right)^{2\xi}\frac{m^2}{W^2-m^2}\left(F\left(\xi,\xi;2\xi;1-\frac{W^2}{m^2}\right)\right.$$
$$\left.+\frac{W}{m}F\left(\xi,\xi+1;2\xi;1-\frac{W^2}{m^2}\right)\right). \tag{20}$$

† We can verify the necessity of a μ^2 at the lower limit of integration if we attempt a perturbation expansion in ξ of equations (18). It is less obviously needed in the quoted solution (20) where we might even dispense with it by dropping μ^2/m^2 altogether.

The complete zeroth-order electron propagator follows:

$$S(p)=\frac{1}{\gamma\cdot p-m}+\int\frac{\mathrm{d}W^2}{W^2}\,\frac{\gamma\cdot p\,s_1(W^2/m^2)+m s_2(W^2/m^2)}{p^2-W^2+\mathrm{i}\epsilon}$$

$$=\frac{1}{\gamma\cdot p-m}-\left(\frac{m^2}{\mu^2}\right)^{2\xi}\Gamma(1-\xi)\Gamma(1-\xi)\Gamma(1+2\xi)$$

$$\times\left[\frac{\gamma\cdot p}{p^2}\left(F\left(1-\xi,1-\xi;1;\frac{p^2}{m^2}\right)-1\right)+\frac{1-\xi}{m}F\left(1-\xi,2-\xi;2;\frac{p^2}{m^2}\right)\right] \qquad (21)$$

and the integral† for it converges comfortably since

$$\sigma(W)\sim\frac{\Gamma(1+2\xi)}{\Gamma(\xi)\Gamma(\xi)}\,\frac{m^{2-2\xi}W^{2\xi-3}}{\mu^{4\xi}}\left[1+\frac{m}{\xi W}+\mathrm{O}\left(\frac{1}{W^2}\right)\right]. \qquad (22)$$

Furthermore we may actually evaluate the electron renormalization constants in this gauge by going to asymptotic values of p in (21) or else from the formal expressions for the bare quantities:

$$Z_\psi^{-1}=1+\int\sigma(W)\,\mathrm{d}W=1+(m^2/\mu^2)^{2\xi}\Gamma(1-\xi)\Gamma(1-\xi)\Gamma(1+2\xi)\simeq 2 \qquad \text{for small } \xi$$

$$Z_\psi^{-1}m_0=m+\int W\sigma(W)\,\mathrm{d}W=m\left[1+\left(\frac{m^2}{\mu^2}\right)^{2\xi}\frac{\Gamma(1+2\xi)}{\Gamma(\xi)\Gamma(1+\xi)}\lim_{d\to 0}\frac{\Gamma(d-\xi)\Gamma(d-\xi-1)}{\Gamma(d)}\right]$$

$$=m.$$

This ultraviolet finiteness is, of course, characteristic of the Landau gauge and not expected for other values of a.

The other Green function $\Gamma^{(0)}$ is given in this zeroth order by bare photon lines and by tree graphs weighted by the just found electron mass distribution. Expressions (7) and (8) are particular examples. For convergence the important point is the asymptotic behaviour (22).

4. First-order Green functions

In the next stage of the iteration we have to evaluate $D^{(1)}$ and the transverse part of $\Gamma^{(1)}$ using the lowest-order $D^{(0)}$, $\Gamma^{(0)}$ and $S^{(0)}$ just found. There are no new infinities because the asymptotic stability condition is amply satisfied so the Z will just perform their usual

† We have used the basic integral

$$\int_0^\infty x^{c-1}(x+y)^{-d}F(a,b;c,-x)\,\mathrm{d}x$$

$$=\frac{\Gamma(a-c+d)\Gamma(b-c+d)\Gamma(c)}{\Gamma(a+b-c+d)\Gamma(d)}F(a-c+d,b-c+d;a+b-c+d;1-y)$$

in this and succeeding expressions. Note the reality of σ in (20) and the fact that $S(p)$ correctly shows a cut for $p^2\geqslant m^2$ in formula (21) whose discontinuity is of course σ.

function. Concentrate first on the photon self-energy to first order,

$$\Pi^{(1)}_{\mu\nu}(k) = \mathrm{i}e^2 Z_\psi \,\mathrm{Tr}\int \bar{\mathrm{d}}^4p\, \mathrm{d}W \rho(W)\left(\gamma_\nu \frac{1}{\gamma\,.\,p - W}\gamma_\mu \frac{1}{\gamma\,.\,(p-k)-W}\right)$$

$$= \mathrm{i}e^2 Z_\psi \int \mathrm{d}W \rho(W)(-k^2\eta_{\mu\nu} + k_\mu k_\nu)\Pi(k^2, W^2) \tag{23}$$

where it is known from QED that

$$\Pi(k^2, m^2) = \frac{e^2}{12\pi^2}\left[\ln\frac{\Lambda^2}{m^2} - \frac{11}{6} - \frac{4m^2}{k^2} + \left(1+\frac{2m^2}{k^2}\right)\left(1-\frac{4m^2}{k^2}\right)^{1/2}\right.$$

$$\left.\times \ln\left(\frac{[1-(4m^2/k^2)]^{1/2}+1}{[1-(4m^2/k^2)]^{1/2}-1}\right)\right]. \tag{24}$$

The renormalized propagator is thereby obtained as

$$D^{(1)-1}_{\mu\nu}(k) = (-k^2\eta_{\mu\nu} + k_\mu k_\nu)\left(1 + Z_\psi \int \mathrm{d}W \rho(W)(\Pi(k^2, W^2) - \Pi(0, W^2))\right) - k_\mu k_\nu Z_A/a. \tag{25}$$

Since

$$\Pi(k^2, W^2) - \Pi(0, W^2)$$

$$= \frac{e^2}{12\pi^2}\left[-\frac{5}{3} - \frac{4W^2}{k^2} + \left(1+\frac{2W^2}{k^2}\right)\left(1-\frac{4W^2}{k^2}\right)^{1/2} \ln\left(\frac{[1-(4W^2/k^2)]^{1/2}+1}{[1-(4W^2/k^2)]^{1/2}-1}\right)\right]$$

we see that (25) will not carry any transverse infinities if $\rho(W) \sim W^\epsilon$ with $\epsilon < -1$. This is visibly true in the Landau gauge where (22) is the zeroth solution, but in fact it is also true in other gauges as well because there is asymptotic stability. It may be of interest to spell out the first-order photon propagator to two extreme limits.

(i) As $k^2 \to 0$,

$$\int \mathrm{d}W \rho(W)(\Pi(k^2, W^2) - \Pi(0, W^2))$$

$$\to \frac{e^2}{60\pi^2}\int \frac{k^2\rho(W)\,\mathrm{d}W}{W^2} = \frac{e^2k^2}{60\pi^2m^2}\left[1 + \left(\frac{m^2}{\mu^2}\right)^{2\xi}\Gamma(2-\xi)\Gamma(2-\xi)\Gamma(1+2\xi)\right].$$

Therefore

$$D^{(1)-1}_{\text{transv}} \to k^2\left[1 + \frac{e^2k^2}{60\pi^2m^2}\left(\frac{1+(m^2/\mu^2)^{2\xi}\Gamma(2-\xi)\Gamma(2-\xi)\Gamma(1+2\xi)}{1+(m^2/\mu^2)^{2\xi}\Gamma(1-\xi)\Gamma(1-\xi)\Gamma(1+2\xi)}\right)\right].$$

(ii) As $k^2 \to \infty$

$$\int \mathrm{d}W \rho(W)\Pi(k^2, W^2) \to +\frac{e^2}{12\pi^2}\int \rho(W)\ln\left(-\frac{k^2}{W^2}\right)\mathrm{d}W.$$

Because $\int \mathrm{d}W \rho(W) \ln W$ is finite like Z_ψ, in the Landau gauge, there remains a logarithmic dependence on k^2 in $D^{(1)}$; this shows that Z_A^{-1} is logarithmically infinite in this next order. All in all, the first-order corrections are manageable and cannot greatly affect the propagator $S^{(2)}$ when we go to the next order of iteration.

More significant probably are the transverse parts to the vertex part $\Gamma^{(1)}$ that enter into (9) at the next level. It is conceptually simpler to deal with this equation in the form (5′) whereupon

$$\Gamma^{(1)}_{\mu}(p, p-k)Z_{\psi}^{-1} = \gamma_{\mu} - ie^2 \int \bar{d}^4p' \, \gamma_{\lambda} S^{(0)}(p')\Gamma^{(0)}_{\nu\mu}(p'k'; pk)D^{(0)\lambda\nu}(k). \tag{26}$$

Since the longitudinal part $k \,.\, \Gamma^{(1)}$ equals $k \,.\, \Gamma^{(0)}$, it is already known; and so is, of course, the $k \to 0$ limit via the differential Ward identity. We have not made a detailed study of (26) since the number of kinematic terms that can turn up lead to very complicated expressions. However we may note that if one goes to the electron mass shell, then because $\Gamma^{(0)}$ reduces to the Born term, the calculation is exactly the order α correction of the form factor in QED and cannot fail to reproduce $g-2$ in this order. However we have no reason to believe that $\Gamma^{(2)}$ includes the α^2 correction of the form factor because $\Gamma^{(1)}_{\mu\nu}$ may have no direct connection with Born graphs.

To summarize our work thus far: we have found a ***gauge covariant*** solution of the Dyson–Schwinger equation for the electron propagator and it provides the basis of a subsequent iteration scheme† to yield all the remaining Green functions. The solution has the merit of exactly satisfying gauge identity (1) and in that sense is a significant generalization of the Baker *et al* (1967) solution as well as the parallel later work that has consisted in replacing Γ by the bare vertex γ in the equation, including the work on self-consistent dynamical symmetry breaking.

References

Baker M, Johnson K and Willey R 1967 *Phys. Rev.* **163** 1699
Delbourgo R and Salam A 1964 *Phys. Rev.* **135** B1398
Kluberg-Stern K and Zuber J B 1975 *Phys. Rev.* D **12** 467
Lee B W 1974 *Phys. Lett.* **46B** 214
Manoukian E B 1974 *Ann. Phys.* **82** 248
Nishijima K 1960 *Phys. Rev.* **119** 485
Rivers R J 1966 *J. Math. Phys.* **7** 385
Salam A 1963 *Phys. Rev.* **130** 1287
Strathdee J 1964 *Phys. Rev.* **135** B1428

† Though whether the iterations converge to well defined answers is a matter of speculation.

Volume 72B, number 1 PHYSICS LETTERS 5 December 1977

INFRARED BEHAVIOUR OF A GAUGE COVARIANT APPROXIMATION

R. DELBOURGO
Department of Physics, University of Tasmania, Hobart, Tasmania, Australia

and

P. WEST[‡]
Laboratoire de Physique Théorique de l'Ecole Normale Supérieure, 24 rue Lhomond, 75231 Paris, France*

Received 28 September 1977

The infrared behaviour in scalar and fermion Q.E.D. is calculated using a gauge covariant approximation scheme to the Dyson–Schwinger equations.

A nonperturbative scheme of solving the Dyson–Schwinger equations of Q.E.D. was considered in a recent paper [1], a scheme which had the virtue of preserving the Ward–Takahashi identities at every stage, rather than to any given order in perturbation theory. In the zeroth stage of the scheme, the validity of which was not restricted to any momentum regime, considerable structure was built into the Green's functions, and the electron propagator equation was solved in the Landau gauge. However, it was assumed that the electron line contained a simple pole near its mass shell apart from a branch cut. In this letter we do not regard the structure near the mass shell as given a priori, but instead we shall determine it by the gauge scheme itself. By a simple calculation, in an arbitrary gauge, we shall derive the pure branch point characteristics of the electron propagator, coinciding with those obtained previously by several authors [2; a more complete list can be found in 3] in their resolution of the infrared problem in Q.E.D. We shall also present the improved solution for the electron propagator spectral function (valid for all momenta) in the Landau gauge.

Firstly, we define our notation and outline the method of the covariant gauge approximation first used by Salam [4]. We refer the reader to the recent paper for more details. The zeroth approximation is obtained by seeking an expression for the vertex, $\Gamma_\mu(p, p-k)$ as a functional of the fermion propagator, $S(p)$, which satisfies the Ward identity,

$$k^\mu S(p)\Gamma_\mu(p,p-k)S(p-k) = S(p-k) - S(p). \qquad (1)$$

As the Ward identity does not determine Γ_μ given S, there exists considerable arbitrariness, but an elegant and solvable starting ansatz was found to be

$$S(p)\Gamma_\mu^{(0)}(p,p-k)S(p-k) = \int \mathrm{d}\omega\, \rho(\omega)\frac{1}{\not{p}-\omega}\gamma_\mu\frac{1}{\not{p}-\not{k}-\omega} \qquad (2)$$

Here $\rho(\omega)$ is the Lehmann spectral function given by

$$S(p) = \int \rho(\omega)\,\mathrm{d}\omega/(\not{p}-\omega+\mathrm{i}\epsilon(\omega)0). \qquad (3)$$

For future use we note the relations

$$m_0 = \int \omega\,\rho(\omega)\,\mathrm{d}\omega, \qquad Z_\psi^{-1} = \int \rho(\omega)\,\mathrm{d}\omega,$$

where m_0 is the bare fermion mass and Z_ψ is the fermion wavefunction renormalization.

Eq. (2) is then used to eliminate Γ_μ in favour of S in the two Dyson–Schwinger equations;

$$D_{\mu\nu}^{-1}(k) = Z_A\,[-k^2\eta_{\mu\nu} + k_\mu k_\nu (1-a^{-1})] + \mathrm{i}e^2 Z_\psi \,\mathrm{Tr}\int \mathrm{d}^4p\,\gamma_\nu S(p)\Gamma_\mu(p,p-k)S(p-k), \qquad (4)$$

$$1 = Z_\psi(\not{p}-m_0)S(p) - \mathrm{i}e^2 Z_\psi \times \int \mathrm{d}^4k\, S(p)\Gamma_\mu(p,p-k)S(p-k)\gamma_\nu D^{\mu\nu}(k), \qquad (5)$$

and so obtain two closed coupled equations for S and

[‡] Supported by a Royal Society fellowship.
* Laboratoire propre du Centre National de la Recherche Scientifique.

Volume 72B, number 1 PHYSICS LETTERS 5 December 1977

$D_{\mu\nu}$. Because of the complexity of these equations further progress could only be made by setting (in the zeroth approximation)

$$D_{\mu\nu}(k) = k^{-2}(-\eta_{\mu\nu} + (k_\mu k_\nu/k^2)(1-a))$$

in eq. (5); that is leaving the photon lines undressed. Without assuming any mass shell behaviour the resulting equation for the imaginary part of eq. (5) is

$$\epsilon(\omega)(\omega - m)\rho(\omega)$$

$$= \int d\omega' \rho(\omega') g_m \Sigma(\omega,\omega')/\pi(\omega - \omega'), \qquad (6)$$

where $\Sigma(\omega,\omega')$ is the second order self energy Feynman graph with mass ω', namely

$$g_m \Sigma(\omega,\omega') = \frac{e^2}{16\pi^2}\frac{\omega - \omega'^2}{\omega^3}$$

$$\times\, [a(\omega^2 + \omega'^2) - (a+3)\omega\omega']\,\theta(\omega^2 - \omega'^2). \qquad (7)$$

The reader is referred to the original paper [1] for details of the higher order corrections and their renormalizability.

Making the change of variables

$$\epsilon(\omega)\omega^2\rho(\omega) = \omega s_1(\omega^2) + s_2(\omega^2) = r(\omega),$$

$$z = (\omega/m)^2, \qquad \xi^2 = e^2/16\pi^2, \qquad (8)$$

we obtain

$$z s_1(z) - s_2(z) = -3\xi^2 \int_1^z dz' s_1(z')$$

$$+ a z \xi^2 \int_1^z dz' s_1(z')/z' - s_1(z) + s_2(z)$$

$$= -3\xi^2 \int_1^z dz' s_2(z')/z' + (a\xi^2/z)\int_1^z s_2(z')\, dz'. \qquad (9)$$

To study the infrared behaviour of the theory we must examine $\rho(\omega)$ in the $\omega \to m$ limit. In this limit eqs. (6) and (9) reduce to

$$(w-1) r(w) \approx 2\xi^2(a-3)\int_1^w dw' r(w'),$$

where $w = \omega m^{-1}$, which has solution

$$r(w) = R(w-1)^{-1+2(a-3)\xi^2}. \qquad (10)$$

The propagator, near mass shell, is given by eq. (3) and is

$$S(p) = \text{constant} \cdot \frac{(\not p + m)}{(p^2 - m^2 + i0)^{1-2(a-3)\xi^2}}\,.$$

This is the simplest derivation we know of this well-known result.

An exact solution (for all z) of eq. (9) is possible in the Landau gauge, giving

$$\rho(\omega) = \frac{6\xi^2 R(\xi^2)}{(\omega^2/m^2 - 1)^{1+6\xi^2}}$$

$$\times \left[\frac{1}{\omega} F\left(-3\xi^2, -3\xi^2; -6\xi^2; 1 - \frac{\omega^2}{m^2}\right)\right.$$

$$\left. + \frac{1}{m} F\left(-3\xi^2, 1-3\xi^2; -6\xi^2; 1 - \frac{\omega^2}{m^2}\right)\right], \qquad (11)$$

where $R(0) = 1$. This solution differs from the one given in our original paper [1] since we have *not* separated out a pole term from ρ and thus have not needed to introduce infrared cut-offs in the resulting equations for σ. In eq. (11) R is an arbitrary constant (because the equation is homogeneous) normalized to give unit residue in the free field limit. As $e^2 \to 0$ $\rho(\omega) \to \delta(\omega - m)$. In any other gauge we have not been able to obtain the result as a closed form in terms of familiar transcendental functions.

We can go through a similar procedure for scalar electrodynamics [5]. Discarding seagull graphs (which are of higher order) the integral equation for the scalar spectral function again leads to a parallel result

$$\Delta(p) \approx (p^2 - m^2 + i0)^{-1+(a-3)2\xi^2}.$$

So far as axial electrodynamics is concerned, the solutions of the Ward identities, the first of which is

$$S(p)\Gamma_{\mu 5}(p, p-k) S(p-k)$$

$$= \int d\omega \rho(\omega) \frac{1}{\not p - \omega}\left[i\gamma_\mu\gamma_5 - \frac{i k_\mu 2\omega\gamma_5}{k^2}\right]\frac{1}{\not p - \not k - \omega},$$

mean that the infrared difficulties are obviated since the pole parts (k^{-2} pieces) of the Green's functions dynamically [6] generate axial meson masses. In such problems the gauge approximation is just useful for providing some clues about the high energy behaviours and does not affect the spontaneous breaking of the chiral symmetry.

These nonperturbative answers, with characteristic branchpoint behaviour, of course imply that asymptotic

Volume 72B, number 1 PHYSICS LETTERS 5 December 1977

states [7] contain not only single particle states, but also coherent soft photon states. The near mass shell behaviour of other charged particle Green's functions (like $\Gamma_\mu(p, p-k)$), are obtained by substituting $\rho(\omega)$ from eqs. (8) and (10) in the spectral weighted expressions (for example eq. (2)).

Hence, we have shown that the zeroth order covariant gauge approximation to Q.E.D. reproduces the "correct" behaviour near mass shell. As this approximation is expected to be equally applicable to all momentum regimes and was not designed to solve the infrared problem in particular, it gives us confidence to hope that it is reliable in other regimes.

It will be interesting to explore the analogous approximation for nonabelian gauge theories, in the hope that it may provide their infrared behaviour correctly; however, before one can start on the fermion propagator here one must determine self-consistently the long-range nature of the gluon propagator from its own Dyson–Schwinger equation as the dressing [8] becomes all important and strongly influences the quark lines.

One of the authors (P.C.W.) wishes to thank Professor D. Zwanziger for helpful discussions.

References

[1] R. Delbourgo and P. West, J. Phys. A10 (1977) 1049.
[2] C. Hagen, Phys. Rev. 130 (1963) 813;
T. Kibble, Phys. Rev. 173 (1968) 1527;
J.K. Storrow, Nuovo Cimento 54A (1968) 15.
[3] D. Zwanziger, Phys. Rev. D11 (1975) 3481.
[4] A. Salam, Phys. Rev. 130 (1963) 1287.
[5] R. Delbourgo, J. Phys. A10 (1977) 1369.
[6] R. Jackiw and K. Johnson, Phys. Rev. 8 (1973) 2386.
[7] T. Kibble, Phys. Rev. 174 (1968) 1882.
[8] H. Pagels, Phys. Rev. 14 (1976) 2747.

J. Phys. A: Math. Gen., Vol. 11, No. 10, 1978. Printed in Great Britain

Solving the gauge identities

R Delbourgo
Department of Physics, University of Tasmania, Hobart, Tasmania 7001, Australia

Received 5 December 1977

Abstract. We show how one may 'solve' the Ward–Takahashi identities of a gauge theory to determine the longitudinal Green functions in terms of the basic source propagators and in such a way that on-shell amplitudes reduce to their classical values. We demonstrate the method for scalar, spinor and vector sources in Lorentz covariant and non-covariant Abelian gauges; but for non-Abelian groups we work in the axial gauge in order to avoid fictitious terms which otherwise spoil the procedure.

1. Introduction

Renormalisable quantum field theories which unify the basic forces of nature are founded on an underlying gauge principle and are consequently endowed with many attractive features, not the least being calculability. The gauge invariance possessed by the action finds its expression in the Ward–Takahashi identities (or their non-Abelian counterparts) connecting the various Green functions—relations between amplitudes involving an additional gauge vector current or extra gauge vector line. These gauge identities play such a crucial role in the renormalisation programme that one tries to preserve them at all costs by the regularisation scheme needed to define quantum loop corrections (although this is sometimes not possible for chiral groups) and, as a result, one can relate the various action counterterms and finish up with overall multiplicative renormalisations.

Such renormalisable gauge identities connect the divergence of an $(n+1)$-point amplitude with an n-point amplitude; therefore, for an amplitude with only two source legs and an arbitrary number of gauge lines, one can successively move down to the source propagator by taking a sufficient number of divergences at the vector ends. Working back, one can determine a good part of the Green functions (more precisely, the longitudinal pieces) in terms of the two-point propagator. This is the essence of the gauge technique (Salam 1963, Delbourgo and Salam 1964, Strathdee 1964) and it has the virtue of being a gauge-covariant procedure, suitable of being adaptable to any other approximation method for extracting solutions of the Green functions equations. Naturally there is vast ambiguity in the determination of the amplitudes since any transverse component (orthogonal to the contracting momentum) can be acceptably added without affecting the gauge covariance. As we shall see, these ambiguities can be effectively eliminated by requiring that on the source shell the amplitudes reduce to their classical values: this then provides the starting gauge approximation of the Green functions; the field equations provide subsequent transverse corrections.

In earlier papers the gauge technique was applied to scalar and spinor sources. We briefly recapitulate the method in § 2 and show how it can be generalised to vector

0305-4770/78/0010-2057$02.00 © 1978 The Institute of Physics

particles (and higher-spin fields) in Lorentz covariant Abelian gauge theories; used in conjunction with Dyson–Schwinger equations we derive the infrared behaviour of charged spin-1 electrodynamics, paralleling the work on scalar and spinor electrodynamics (Delbourgo and West 1977a, b, Delbourgo 1977). The extension of the gauge technique to chiral groups and to (spontaneously broken) pseudovector electrodynamics is given in § 3.

Since unified gauge models centre round non-Abelian groups, the resulting Slavnov–Taylor identities in relativistic gauges (Slavnov 1972, Taylor 1971, Lee 1974) exhibit a rather complex form owing to the occurrence of fictitious particle terms; indeed, the latter render the gauge technique almost intractable since they include Bethe–Salpeter kernels of ghost-source scattering, whose spectral representations are barely known, and which can only be evaluated in a certain approximate sense†. Therefore for non-Abelian groups we prefer to stick to ghost-free axial gauges (Kummer 1961, Arnowitt and Feckler 1962, Schwinger 1963, Fradkin and Tyutin 1970) in which the Ward–Takahashi identities for the one-particle irreducible amplitudes do assume their characteristically simple form and become amenable to the gauge technique. One must, of course, pay the price of lack of relativistic invariance and we show how this can be met for electrodynamics in § 4 before going on to the Yang–Mills problem in § 5; the choice of axial gauge is vindicated by the simple structure of the full Green functions.

2. The gauge technique for electrodynamics

Let A_μ stand for the electromagnetic field, interacting with some quantised source field ϕ (and its adjoint $\bar\phi$), for which the connected vacuum generating functional W is defined through

$$\exp(\mathrm{i}W[j_\mu, j, \bar j]) = \int (\mathrm{d}A\, \mathrm{d}\phi\, \mathrm{d}\bar\phi) \exp\left(\mathrm{i}\int \mathrm{d}^4x(\mathscr{L}(A,\phi,\bar\phi) - j^\mu A_\mu - \bar j\phi - \bar\phi j - F(A))\right)$$

where $F(A)$ is some gauge fixing term—it needs no compensating in this Abelian case. The phase invariance of $\mathscr{L}$ under

$$\delta\phi = \mathrm{i}\Lambda\phi, \qquad \delta\bar\phi = -\mathrm{i}\bar\phi\Lambda, \qquad \delta A = -\partial\Lambda/e$$

results in the functional gauge identity,

$$\left[e\left(j\frac{\delta}{\delta j} - \bar j\frac{\delta}{\delta \bar j}\right) + \partial_\mu\left(\frac{\delta F}{\delta A_\mu}\right)\right]_{A_\nu = \mathrm{i}\delta/\delta j^\nu} W + \partial^\mu j = 0 \tag{1}$$

and functional current derivatives of the above equation yield the Ward–Takahashi identities between the connected Green functions

$$C(x_1, \ldots, x_n) \equiv \mathrm{i}^{n+1}\delta^n W/\delta j(x_1)\ldots\delta j(x_n).$$

For instance, in a covariant gauge specified by $F = -(\partial A)^2/2a$, one obtains the identity

$$a^{-1}\partial^\mu\,\partial^2 C_\mu(x;y,z) = e[\delta^4(x-z)C(y,x) - \delta^4(x-y)C(x,z)], \tag{2}$$

typical of the more general relation between $\partial C^{(n+1)}$ and $C^{(n)}$. One can arrive at

† Pagels (1976) has nevertheless done wonders with them by assuming $1/q^4$ behaviour of the gluon propagator and by picking out the infrared singular terms of the field equations.

similar identities for the one-particle irreducible Green functions $\Gamma^{(n)}$ defined through ($i\Delta \equiv C^{(2)} = \Gamma^{(2)-1}$),

$$C(x_1, \ldots, x_n) = \Big(\prod_j \int i\Delta(x_j, y_j)\, d^4y_j\Big)\, \Gamma(y_1, \ldots, y_n)$$

or the functional field derivatives of

$$\Gamma(A, \phi, \bar{\phi}) = W + \int (Aj + \bar{j}\phi + \bar{\phi}j).$$

For example,

$$\partial^\mu \Gamma_\mu(x; y, z) = e[\delta^4(x-z)\Gamma(y, x) - \delta^4(x-y)\Gamma(x, z)] \tag{3}$$

replaces identity (2), and so on. Extracting powers of e for each source line and taking Fourier transforms, we get the more familar versions of the identities† (Nishijima 1960, Rivers 1966),

$$\begin{aligned} k^\mu \Gamma_\mu(p, p-k) &= \Delta^{-1}(p) - \Delta^{-1}(p-k) \\ k'^\nu \Gamma_{\nu\mu}(p'k', pk) &= \Gamma_\mu(p+k, p) - \Gamma_\mu(p', p'-k) \end{aligned} \tag{4}$$

which were first derived via the canonical commutation formalism. These identities are multiplicately renormalised by the same (infinite) constant.

We aim to obtain 'solutions' of the gauge set (4) with the subsequent intention of inserting the solutions in the Green function equations. Gauge covariance is assured and not something which needs to be imposed at the end, making this the primary virtue of the gauge technique. Clearly an infinite number of possible 'solutions' can be found (Rivers 1966) all differing by transverse components that disappear upon contraction with the gauge field momentum; but this is not to deny that longitudinal components are interconnected by the identities and that equations (4) do embody considerable information about the Green functions. At this juncture it may be worth pointing out that in an axial gauge which specifies $n \,.\, A = 0$ (by the choice of gauge fixing term $F = Bn \,.\, A$, with B acting as a Lagrangian multiplier field), the identities (4) remain intact but identities like (2) are altered to orthogonality conditions

$$n^\mu C_\mu(x; y, z) = 0 \qquad \text{etc.}$$

These axial gauge identities (Delbourgo *et al* 1974, Kummer 1975) furthermore generalise very simply to non-Abelian groups, unlike the Lorentz covariant gauges where radical modifications become necessary.

One can arrive at a non-trivial class of solutions to (4) by noticing that: (i) the classical values of Γ (bare vertices) or of C (tree graphs) automatically obey the identities; (ii) successive divergences at the gauge legs bring us down to the two-point functions; and (iii) the propagators can themselves be represented as weighted spectral sums over free propagators. It follows that we can construct a gauge covariant set of quantum amplitudes by taking mass weighted sums of classical amplitudes. One could think of a more general procedure starting with a general representation of an N-point function and working up to the higher-point functions by some well defined algorithm—working down to lower-point functions is trivial—but this is extremely difficult to put into practice because spectral representations of fully off-shell amplitudes are hardly known or even guessed when $N > 3$. In any case it would be absurd

† Please note that there is no discrepancy between these identities and those quoted in previous researches. There the Γ correspond to amputated amplitudes, not one-particle irreducible parts.

to go to such lengths since the initial gauge approximation must be subject to transverse corrections entailed by the coupled Green functions equations, and therefore the starting point may as well be chosen simply. We believe we have done this by reverting in the end to the spectral form of the basic Born amplitudes with all gauge lines removed, namely the propagator, if there is a single source line. Besides, the renormalisations need only be carried out on the propagators and gauge-related vertex functions with the higher-point amplitudes then generated through the skeleton expansion.

In relativistically covariant gauges (axial gauges are considered below) where the propagators for scalar or spinor sources are rigorously known to possess the representations

$$\Delta(p)=\int \mathrm{d}W^2\rho(W)(p^2-W^2+\mathrm{i}0)^{-1}$$

or (5)

$$S(p)=\int \mathrm{d}W\rho(W)(\gamma\,.\,p-W+\mathrm{i}0\epsilon(W))^{-1},$$

the simplest gauge technique leads to the solutions

$$C(p',k_i,p)=\int \mathrm{d}W^2\rho(W^2)c(p',k_i,p|W^2)$$

or (6)

$$C(p',k_i,p)=\int \mathrm{d}W\rho(W)c(p',k_i,p|W),$$

where $c(\ldots|W)$ stand for the classical functions for a source of mass W. The first non-trivial illustration, the vertex function, explicitly reads

$$\Delta(p)\Gamma_\mu(p,p-k)\Delta(p-k)=\int \mathrm{d}W^2\rho(W^2)(p^2-W^2)^{-1}(2p-k)_\mu[(p-k)^2-W^2]$$

or (7)

$$S(p)\Gamma_\mu(p,p-k)S(p-k)=\int \mathrm{d}W\rho(W)[\gamma\,.\,p-W]^{-1}\gamma_\mu[\gamma\,.\,(p-k)-W]^{-1}$$

from which one readily sees how to write down the higher-point functions in this gauge approximation. One facet of this construction is that on-shell, where $\rho(W^2)\to\delta(W^2-m^2)$ or $\delta(W-m)$, the amplitudes become identically equal to the classical ones, and we may adopt this as the criterion which *defines* the initial gauge approximation. There only remains to find the spectral functions and this can be achieved via the Dyson–Schwinger equations as we have reported elsewhere; the procedure, which leads to an integral equation for ρ, is intrinsically non-perturbative. It also yields the infrared behaviour of the Green functions very economically (Delbourgo and West 1977b).

Electrodynamics of charged vector mesons is complicated by the occurrence of two spectral functions†

$$\Delta_{\mu\nu}(p)=\int \mathrm{d}W^2[(-\eta_{\mu\nu}+p_\mu p_\nu/W^2)\rho(W^2)-\eta_{\mu\nu}\tau(W^2)](p^2-W^2+\mathrm{i}0)^{-1} \quad (8)$$

† Actually the spinor case implicitly contains two spectral functions, the even and odd parts of $\rho(W)$, and it suggests that a closer analogy with mesons can be achieved by using spectral representations in Kemmer's β-formalism rather than (8).

but the solution here is readily understood if one or other of the weights is taken to be zero in turn. Thus $\tau=0$ corresponds to an integral over spin-1 vector mesons of mass W whose field theory is governed by

$$\mathscr{L}_\rho=-\tfrac{1}{2}(D_\mu V^+_\nu-D_\nu V^+_\mu)(D^\mu V^\nu-D^\nu V^\mu)+W^2V^+_\mu V^\mu$$

with bare vertices

$$\begin{aligned}R_{\lambda\mu\nu}(p,p-k)&=-\eta_{\mu\nu}(2p-k)_\lambda+(p-k)_\mu\eta_{\nu\lambda}+p_\nu\eta_{\mu\lambda},\\ R_{\kappa\lambda\mu\nu}&=-2\eta_{\mu\nu}\eta_{\kappa\lambda}+\eta_{\mu\kappa}\eta_{\nu\lambda}+\eta_{\nu\kappa}\eta_{\mu\lambda},\end{aligned}\tag{9}$$

whereas $\rho=0$ refers to an (unphysical) theory

$$\mathscr{L}_\tau=-D_\mu V^+_\nu D^\mu V^\nu+W^2V^+_\mu V^\mu$$

having for its bare $\Gamma^{(n)}$,

$$T_{\lambda\mu\nu}(p,p-k)=-\eta_{\mu\nu}(2p-k)_\lambda,\qquad T_{\kappa\lambda\mu\nu}=-2\eta_{\mu\nu}\eta_{\kappa\lambda}.\tag{10}$$

After these observations it comes as no surprise that the gauge identities are solved by

$$\begin{aligned}&\Delta^{\mu\mu'}(p)\Gamma_{\lambda\mu'\nu'}(p,p-k)\Delta^{\nu'\nu}(p-k)\\ &=\int\frac{dW^2}{(p^2-W^2)[(p-k)^2-W^2]}\left[\left(\eta^{\mu\mu'}-\frac{p^\mu p^{\mu'}}{W^2}\right)R_{\lambda\mu'\nu'}\right.\\ &\qquad\left.\times\left(\eta^{\nu'\nu}-\frac{(p-k)^{\nu'}(p-k)^\nu}{W^2}\right)\rho(W^2)-T_{\lambda\mu\nu}\tau(W^2)\right]\end{aligned}\tag{11}$$

as one can easily check; and similarly for the higher-point functions. In the appendix we have pursued (11) and determined the integral equations for ρ and τ that are provided by the field equations. In the infrared limit we find $\tau\to0$ and $\rho\to(W^2-m^2)^{-1-e^2(3-a)/8\pi^2}$ in complete analogy to the scalar and spinor cases, which strongly suggests spin independence of infrared behaviour.

3. Pseudovector electrodynamics

Next we consider abnormal (1^{++}) photons. For simplicity and also for aesthetic reasons we suppose that the chiral symmetry is exact at the Lagrangian level so that its eventual breaking is spontaneous or dynamical but not as the result of quantum regularisation, i.e. we introduce enough sources to cancel out offending anomalies. Being a true symmetry of the action the chiral U(1) group leads to its own set of gauge identities, quite analogous to (1). For fermion pseudovector electrodynamics where

$$\mathscr{L}_5=\bar\psi i\gamma(\partial+eA\gamma_5)\psi-\tfrac{1}{4}F_{\mu\nu}F^{\mu\nu},$$

the generalisations of (4) read

$$\begin{aligned}-ik^\mu\Gamma_{\mu5}(p.p-k)&=S^{-1}(p)\gamma_5+\gamma_5S^{-1}(p-k)\\ ik^\mu\Gamma_{\nu5\mu5}(p'k',pk)&=\Gamma_{\nu5}(p',p'+k')\gamma_5+\gamma_5\Gamma_{\nu5}(p-k',p)\end{aligned}\tag{12}$$

etc, and are trivially obeyed at the bare level (massless fermions, $\Gamma_{\mu5}=i\gamma_\mu\gamma_5$, $\Gamma^{(n)}=0$ for $n>3$). After quantum corrections however, $S(p)$ propagates with a whole spectrum of intermediate massive fermions from which it necessarily follows that $\Gamma_{\mu5}$

contains a pole; this implies dynamical chiral breakdown (Jackiw and Johnson 1973) and the generation of a pseudovector photon mass through Schwinger's mechanism (1962). Thus

$$\gamma_5 S(p-k)+S(p)\gamma_5$$
$$=\int \mathrm{d}W\rho(W)\{\gamma_5[\gamma\,.\,(p-k)-W]^{-1}+(\gamma\,.\,p-W)^{-1}\gamma_5\}$$
$$=\int \mathrm{d}W\rho(W)(\gamma\,.\,p-W)^{-1}(k\,.\,\gamma-2W)\gamma_5[\gamma\,.\,(p-k)-W]^{-1}$$

provides the solution,

$$S(p)\Gamma_{\mu5}(p,p-k)S(p-k)=\int \mathrm{d}W\rho(W)\frac{1}{\gamma\,.\,p-W}\left(\mathrm{i}\gamma_\mu\gamma_5-\frac{2\mathrm{i}Wk_\mu\gamma_5}{k^2}\right)\frac{1}{\gamma\,.\,(p-k)-W} \tag{13}$$

manifesting the $k^2\to 0$ singularity with all its consequences.

The solutions of the gauge identities for the higher-point amplitudes can be extracted from the work of Jackiw and Johnson (1973). Those authors noted that the phenomenological Lagrangian

$$\mathcal{L}_{5M}=\bar\psi\gamma\,.\,(\partial+eA)\gamma_5\psi-\tfrac{1}{4}F_{\mu\nu}F^{\mu\nu}+\tfrac{1}{2}\mu^2A^2-\mu A\,.\,\partial\phi+\tfrac{1}{2}(\partial\phi)^2-m\bar\psi\exp(2e\gamma_5\phi/\mu)\psi$$

of *massive* mesons and fermions also possesses a local chiral symmetry, it being the coupling of A to the massless ϕ which is responsible for the $1/k^2$ poles at the meson legs. By drawing all the tree graphs of this theory and suitably summing over masses, one arrives at the requisite solution. For instance, the four-point amplitude, with vector lines amputated, reads

$$G_{\nu5\mu5}(p'k',pk)$$
$$=\int \mathrm{d}W\rho(W)\frac{1}{\gamma\,.\,p'-W}\left[\left(\gamma_\nu\gamma_5-\frac{2Wk'_\nu\gamma_5}{k'^2}\right)\frac{1}{\gamma\,.\,(p+k)-W}\right.$$
$$\times\left(\gamma_\mu\gamma_5-\frac{2Wk_\mu\gamma_5}{k^2}\right)+\left(\gamma_\mu\gamma_5-\frac{2Wk_\mu\gamma_5}{k^2}\right)\frac{1}{\gamma\,.\,(p'-k)-W}\left(\gamma_\nu\gamma_5-\frac{2Wk'_\nu\gamma_5}{k'^2}\right)$$
$$\left.+\frac{4Wk_\mu k'_\nu}{k^2k'^2}\right]\frac{1}{\gamma\,.\,p-W}.$$

This initial gauge approximation can be taken as a suitable basis for a self-consistent determination of $\rho(W)$, but the evaluation is considerably harder than normal vector electrodynamics because the self-energy corrections of the pseudovector lines *cannot* be dropped in the first instance as they serve to render the meson massive and thereby alter the entire cut structure of $S(p)$. Parallel comments apply to scalar and vector sources.

4. Electrodynamics in axial gauges

We shall stop using Lorentz covariant gauges from now on. The reason is that we are placing our entire emphasis on the gauge identities, so it is crucial for us that they be as simple as possible. Non-Abelian groups in relativistic invariant groups are known to lead to highly complicated Slavnov–Taylor identities riddled with fictitious particle terms and various Bethe–Salpeter kernels; these make our task of solving the identities well nigh impossible and force us into adopting an axial gauge where the Green functions cease to be Lorentz covariant (they depend on an external four-vector n) but where the identities for $\Gamma^{(n)}$ are straightforward (Bernstein 1977, Delbourgo *et al* 1974, Kummer 1975). As an introduction to Yang–Mills theory let us first study electrodynamics.

As we mentioned, in the gauge $n\,.\,A=0$, the Ward–Takahashi identities for one-particle irreducible amplitudes are the naive ones obtained by canonical methods, i.e. the set (4). The one and only problem is that the $\Gamma^{(n)}$ are functions of n as well as the momenta; e.g. in the scalar case, $\Delta(p,n)=\mathcal{C}(p^2,p\,.\,n)$. The initial gauge approximation writes the $\Gamma^{(n)}$ as functionals of Δ, and it is therefore essential to know something about the cut structure of the propagators in axial gauges. An examination of the source self-energy past $\Pi(p,n)$ in lowest-order perturbation theory reveals quite a lot: Π is a function of p^2 and $(p\,.\,n)^2$ and when $(p\,.\,n)^2<m^2$, there is just the usual relativistic cut for $p^2\geqslant m^2$. It is a strong indication that the correct representation† of the propagator is

$$\Delta(p,n)=\int \mathrm{d}W^2\rho(W^2,p\,.\,n)(p^2-W^2+\mathrm{i}0)^{-1} \tag{15}$$

providing $(p\,.\,n)^2<W^2$ (threshold). Note how the covariant gauge parameter is replaced by the parameter $p\,.\,n/m$ in the axial gauge. With the fermion self-energy, one finds instead

$$\Sigma(p,n)=\gamma\,.\,p\Sigma_1(p^2,p\,.\,n)+m\Sigma_2(p^2,p\,.\,n)+\gamma\,.\,n\Sigma_0(p^2,p\,.\,n)$$

which in turn leads to the spectral form

$$S(p,n)=\tfrac{1}{2}\int \mathrm{d}W[(\gamma\,.\,p-W)^{-1},\rho(W,p\,.\,n)+\gamma\,.\,n\rho_0(W,p\,.\,n)] \tag{16}$$

if $(p\,.\,n)^2<m^2$. The absence of terms $[\gamma\,.\,n,\gamma\,.\,p]$ in S or Σ is a consequence of C-invariance, since the Lagrangian multiplier field B like A_μ possesses negative charge parity. We defer the vector case to the next section.

Finding 'solutions' of the Ward–Takahashi identities is complicated by the $p\,.\,n$ dependence of the spectral functions. For scalar electrodynamics we can tackle the difficulty by writing the propagator difference as

$$\begin{aligned}\Delta(p)-\Delta(p')&=\int \mathrm{d}W^2\left(\frac{\rho(W^2,p\,.\,n)}{p^2-W^2}-\frac{\rho(W^2,p'\,.\,n)}{p'^2-W^2}\right)\\&=\tfrac{1}{2}\int \mathrm{d}W^2\bigg[\left(\frac{1}{p^2-W^2}-\frac{1}{p'^2-W^2}\right)(\rho(W^2,p\,.\,n)+\rho(W^2,p'\,.\,n))\\&\quad+\left(\frac{1}{p^2-W^2}+\frac{1}{p'^2-W^2}\right)(\rho(W^2,p\,.\,n)-\rho(W^2,p'\,.\,n))\bigg].\end{aligned}$$

† To our knowledge a *rigorous* proof of (15) or (16) is lacking, but this has not prevented its use (Frenkel and Taylor 1976).

Then, by inspection, an appropriate solution of the vertex identity,

$$\Delta(p)\Gamma_\mu(p,p')\Delta(p')$$
$$=\tfrac{1}{2}\int \mathrm{d}W^2 \frac{1}{p^2-W^2}(p+p')_\mu \frac{1}{p'^2-W^2}(\rho(W^2,p.n)+\rho(W^2,p'.n))$$
$$-\tfrac{1}{2}\int \mathrm{d}W^2\left(\frac{1}{p^2-W^2}+\frac{1}{p'^2-W^2}\right)n_\mu \frac{\rho(W^2,p.n)-\rho(W^2,p'.n)}{p.n-p'.n} \qquad (17)$$

is rightly Bose symmetric. Observe also that the n_μ component disappears from the full Green function (multiplication by $D^{\lambda\mu}(p-p')$) so that the only true signal of non-covariance is the n-dependence of the averaged spectral function. The same analysis can be carried over to the higher-point functions and again it is only the mean ρ which enters the expressions.

With spinors the decomposition problem is not more difficult. Write

$$S(p)-S(p')=\tfrac{1}{4}\int \mathrm{d}W\left\{\frac{1}{\gamma.p-W}-\frac{1}{\gamma.p'-W},\rho+\gamma.n\rho_0+\rho'+\gamma.n\rho_0'\right\}$$
$$+\tfrac{1}{4}\int \mathrm{d}W\left\{\frac{1}{\gamma.p-W}+\frac{1}{\gamma.p'-W},\rho+\gamma.n\rho_0-\rho'-\gamma.n\rho_0'\right\}.$$

Then again,

$$S(p)\Gamma_\mu(p,p')S(p')$$
$$=\tfrac{1}{4}\int \mathrm{d}W\left\{\frac{1}{\gamma.p-W}\gamma_\mu\frac{1}{\gamma.p'-W},\rho+\gamma.n\rho_0+\rho'+\gamma.n\rho_0'\right\}$$
$$-\tfrac{1}{4}\int \mathrm{d}W n_\mu\left\{\frac{1}{\gamma.p-W}+\frac{1}{\gamma.p'-W},\frac{\rho+\gamma.n\rho_0-\rho-\gamma.n\rho_0'}{p.n-p'.n}\right\} \qquad (18)$$

and so on to higher-point functions, with n_μ pieces being essentially irrelevant. These gauge covariant expressions are ready for redeployment in the field equations, but that exercise lies outside the scope of this paper. Rather, we pass on to discovering solutions for the more interesting non-Abelian problem.

5. Non-Abelian theories in axial gauges

Let A_μ stand for the gauge meson fields in the regular representation of the internal symmetry and let f^{abc} be the structure constants. In axial gauges $n.A^a=0$, the connecting Green functions involving at least two vector lines,

$$C^{a_1a_2\dots}_{\mu_1\mu_2\dots}(x_1,x_2,\dots)=\delta^{\cdots}W/\delta j^{a_1}_{\mu_1}(x_1)\delta j^{a_2}_{\mu_2}(x_2)\dots,$$

including any number of source derivatives $\delta/\delta j$, $\delta/\delta K$ (where K is the current of the multiplier field B), obey

$$n^{\mu_i}C_{\cdot\mu_i\cdot}(\cdot,x_i,\cdot)=0 \qquad (19)$$

because of f-antisymmetry†. A particular case is the propagator

$$\mathrm{i}\delta^{ab}\Delta_{\mu\nu}(x-y)=\delta^2W/\delta j^a_\mu(x)\delta j^b_\nu(y); \qquad n^\mu\Delta_{\mu\nu}=0,$$

† And the basic functional gauge equation $(D_\mu j^\mu+\mathrm{i}n.D\,\delta/\delta K)W=0$.

which together with the mixed propagator

$$i\delta^{ab}\Delta_{\mu B}(x-y)=\delta^2 W/\delta j^a_\mu(x)\delta K^b(y)=i\delta^{ab}(n\,.\,\partial)^{-1}\,\partial_\mu\delta^4(x-y) \qquad (20)$$

and

$$i\Delta_{BB}(x-y)=\delta^2 W/\delta K^a(x)\delta K^b(y)=0$$

make up the full A_μ, B propagator matrix. The result (19) is not very surprising in view of the fact that $C_{\cdot\mu\cdot}(\cdot\, x\,\cdot)$ factorises as $\int \Delta_{\mu\nu}(x-y)\Gamma_{\cdot\nu\cdot}(\cdot\, y\,\cdot)\,d^4y$.

The $\Gamma^{(n)}$ functions in the pure gauge theory satisfy the canonical identities

$$\begin{aligned}
&\partial^\lambda(\Delta^{-1})^{ab}_{\lambda\mu}(x-y)=0\\
&\partial^\lambda\Gamma^{abc}_{\lambda\mu\nu}(xyz)=f^{abe}\delta^4(x-y)(\Delta^{-1})^{ce}_{\mu\nu}(x-z)+f^{ace}\delta^4(x-z)(\Delta^{-1})^{be}_{\mu\nu}(y-x)\\
&\partial^\kappa\Gamma^{abcd}_{\kappa\lambda\mu\nu}(xyzw)=f^{abe}\delta^4(x-y)\Gamma^{ecd}_{\lambda\mu\nu}(xzw)+f^{ace}\delta^4(x-z)\Gamma^{bed}_{\lambda\mu\nu}(yxw)\\
&\qquad\qquad+f^{ade}\delta^4(x-w)\Gamma^{bce}_{\lambda\mu\nu}(yxw)
\end{aligned} \qquad (21)$$

etc. And even if additional sources ϕ_i—on which the generators are represented by matrices $(T^a)^j_i$—are incorporated, the identities retain their traditionally simple form,

$$\partial^\lambda\Gamma^a_\lambda(x;y,z)=\delta^4(x-y)\Delta^{-1}(x-z)T^a-T^a\delta^4(x-z)\Delta^{-1}(y-x). \qquad (22)$$

How do we go about solving such involved relations as (21)? Many clues have been offered in previous sections, but before they can be applied one should appreciate that the multiplier field plays an important role and that the full propagator matrix Δ and its inverse Δ^{-1}, with elements $(\Delta^{-1})^{ab}_{\mu\nu}$, $\Delta^{-1}(^{ab}_{B\mu})=n_\mu\delta^{ab}$ (remaining unrenormalised like $\Delta_{B\mu}$) and $(\Delta^{-1})_{BB}=0$, figure in many places. Also recall that $\Gamma_{B\mu_1\mu_2\ldots}=0$ for one-particle irreducible amplitudes comprising at least two vectors. Let us therefore start with the spectral representation†, expected to be valid when $(p\,.\,n)^2<0$,

$$\begin{aligned}
\Delta_{\mu\nu}(p)=\int dW^2(p^2-W^2)^{-1}\Big[&\Big(-\eta_{\mu\nu}+\frac{p_\mu n_\nu+n_\mu p_\nu}{p\,.\,n}-\frac{p_\mu p_\nu}{(p\,.\,n)^2}\Big)\alpha(W^2,p\,.\,n)\\
&+\Big(-\eta_{\mu\nu}+\frac{n_\mu n_\nu}{n^2}\Big)\beta(W^2,p\,.\,n)\Big]
\end{aligned} \qquad (23)$$

and of course the unaffected parts,

$$\Delta_{B\mu}=p_\mu/(p\,.\,n), \qquad \Delta_{BB}=0. \qquad (24)$$

The bare propagator is obtained by substituting $\alpha(W^2)\to\delta(W^2)$, $\beta(W^2)\to 0$. Our objective is to write down all higher-point functions in terms of α and β so as to satisfy identities (21), and to achieve it we will have to synthesise the methods of §§ 2 and 4 in a manner which represents the *full Bose symmetry* of the amplitudes.

For the moment, forget about the n-dependence of α and β. When $\beta=0$, one is dealing with a massive vector theory with inverse propagator

$$\Delta^{-1}_{\mu\nu}(p)=(p^2-W^2)(-\eta_{\mu\nu}+p_\mu p_\nu/p^2)\equiv\mathscr{A}^{-1}_{\mu\nu}, \qquad \Delta^{-1}_{B\mu}(p)=n_\mu, \qquad \Delta^{-1}_{BB}=0. \qquad (25)$$

Since

$$\Delta^{-1}_{\mu\nu}(q)-\Delta^{-1}_{\mu\nu}(r)=p^\lambda\Lambda^0_{\lambda\mu\nu}(p,q,r)+W^2(r_\mu r_\nu/r^2-q_\mu q_\nu/q^2) \qquad (26)$$

† We have extracted the internal factor δ^{ab} out of Δ^{ab}; likewise below we remove the factor f^{abc} from Γ^{abc}, always assuming the gauge symmetry is not spontaneously broken.

where

$$\Lambda^0_{\lambda\mu\nu}(p, q, r) = -\eta_{\mu\nu}(q-r)_\lambda - \eta_{\mu\nu}(r-p)_\mu - \eta_{\lambda\mu}(p-q)_\nu \tag{27}$$

is the bare Yang–Mills vertex, an acceptable solution is

$$\begin{aligned}\Delta^{\mu\mu'}(q)\Gamma_{\lambda\mu'\nu'}(p, q, r)\Delta^{\nu'\nu}(r) &= \int \mathrm{d}W^2\alpha(W^2)\left(-\eta^{\mu\mu'} + \frac{q^{\mu'}n^{\mu} + q^{\mu}n^{\mu'}}{q\,.\,n} - \frac{q^{\mu}q^{\mu'}}{(q\,.\,n)^2}\right)\frac{1}{q^2 - W^2}\\ &\quad\times \Lambda^{(a)}_{\lambda\mu'\nu'}\left(-\eta^{\nu'\nu} + \frac{r^{\nu'}n^{\nu} + r^{\nu}n^{\nu'}}{r\,.\,n} - \frac{r^{\nu'}r^{\nu}}{(r\,.\,n)^2}\right)\frac{1}{r^2 - W^2}\end{aligned}$$

with

$$\Lambda^{(a)}_{\lambda\mu\nu} = \Lambda^0_{\lambda\mu\nu} + W^2\left[\frac{n_\lambda}{p\,.\,n}\left(\frac{r_\mu r_\nu}{r^2} - \frac{q_\mu q_\nu}{q^2}\right) + (p\lambda \leftrightarrow q\mu \leftrightarrow r\nu \text{ perms})\right]. \tag{28}$$

A similar analysis, when $\alpha = 0$ but $\beta \neq 0$, gives

$$\Delta^{\mu\mu'}(q)\Gamma_{\lambda\mu'\nu'}(pqr)\Delta^{\nu'\nu}(r) = \int \mathrm{d}W^2\beta(W^2)\frac{-\eta^{\mu\mu'} + n^{\mu}n^{\mu'}/n^2}{q^2 - W^2}\Lambda^{(b)}_{\lambda\mu'\nu'}\frac{-\eta^{\nu'\nu} + n^{\nu'}n^{\nu}/n^2}{r^2 - W^2}$$

with

$$\Lambda^{(b)}_{\lambda\mu\nu} = \Lambda^0_{\lambda\mu\nu} + n_\lambda(r_\mu r_\nu - q_\mu q_\nu)/p\,.\,n + \text{perms}. \tag{29}$$

The lack of total Bose symmetry in $\Lambda^{(a)}$ and $\Lambda^{(b)}$ has been remedied by adding supplementary terms, proportional to n_μ and n_ν, which vanish upon contraction with $\Delta(q)$ and $\Delta(r)$. Observe too that the n_λ-pieces themselves will disappear when $\Delta(p)$ is applied on (28) and (29) to obtain the full Green function, but that they cannot be dropped if we require the mixed function $\Delta^{B\lambda}(p)\Gamma_{\lambda\mu\nu}$.

A more serious obstacle is presented by the $p\,.\,n$ dependence of the spectral functions. To circumvent this, define the components,

$$\begin{aligned}\mathscr{A}^{\mu\nu}(p, W) &\equiv \left(-\eta^{\mu\nu} + \frac{p^{\mu}n^{\nu} + p^{\mu}n^{\nu}}{p\,.\,n} - \frac{p^{\mu}p^{\nu}}{(p\,.\,n)^2}\right)\frac{1}{p^2 - W^2}\\ \mathscr{B}^{\mu\nu}(p, W) &\equiv (-\eta^{\mu\nu} + n^{\mu}n^{\nu}/n^2)/(p^2 - W^2)\end{aligned} \tag{30}$$

and set $\beta = 0$ at first. Then

$$\begin{aligned}[\Delta(q)p\,.\,\Gamma\Delta(r)]_{\mu\nu} &= \tfrac{1}{2}\int \mathrm{d}W^2[\alpha(W^2, q\,.\,n) + \alpha(W^2, r\,.\,n)][\mathscr{A}(q, W)p\,.\,\Lambda^a\mathscr{A}(r, W)]_{\mu\nu}\\ &\quad + \tfrac{1}{2}\int \mathrm{d}W^2[\alpha(W^2, r\,.\,n) - \alpha(W^2, q\,.\,n)][\mathscr{A}(q, W) + \mathscr{A}(r, W)]_{\mu\nu}\end{aligned} \tag{31}$$

immediately suggests a possible factorisation of p, following the manoeuvre of § 4. Unfortunately (31) is not yet ready for the operation since the straight recipe would not provide a totally symmetric Γ. Instead we must carry on with our manipulations on (31) and write

$$[\Delta(q)p\,.\,\Gamma\Delta(r)]_{\mu\nu}$$

$$=\tfrac{1}{3}\int \mathrm{d}W^2[\alpha(W^2,q\,.\,n)+\alpha(W^2,r\,.\,n)$$

$$+\alpha(W^2,p\,.\,n)][\mathscr{A}(q,W)p\,.\,\Lambda^a\mathscr{A}(r,W)]_{\mu\nu}$$

$$+\tfrac{1}{6}\int \mathrm{d}W^2[\alpha(W^2,q\,.\,n)+\alpha(W^2,r\,.\,n)$$

$$-2\alpha(W^2,p\,.\,n)][\mathscr{A}(r,W)-\mathscr{A}(q,W)]_{\mu\nu}$$

$$+\tfrac{1}{2}\int \mathrm{d}W^2[\alpha(W^2,r\,.\,n)-\alpha(W^2,q\,.\,n)][\mathscr{A}(r,W)+\mathscr{A}(q,W)]_{\mu\nu}.$$

We deduce that

$$[\Delta(q)\Gamma\Delta(r)]_{\lambda\mu\nu}$$

$$=\tfrac{1}{3}\int \mathrm{d}W^2[\alpha(W^2,q\,.\,n)+\alpha(W^2,r\,.\,n)$$

$$+\alpha(W^2,p\,.\,n)][\mathscr{A}(q,W)\Lambda^{(a)}\mathscr{A}(r,W)]_{\lambda\mu\nu}$$

$$+\frac{n_\lambda}{6p\,.\,n}\int \mathrm{d}W^2[\alpha(W^2,q\,.\,n)+\alpha(W^2,r\,.\,n)$$

$$-2\alpha(W^2,p\,.\,n)][\mathscr{A}(r,W)-\mathscr{A}(q,W)]_{\mu\nu}$$

$$+\frac{n_\lambda}{2p\,.\,n}\int \mathrm{d}W^2[\alpha(W^2,r\,.\,n)-\alpha(W^2,q\,.\,n)][\mathscr{A}(r,W)+\mathscr{A}(q,W)]_{\mu\nu}$$

up to n_μ, n_ν terms in Γ.

Having outlined the essential steps we may reasonably quote the complete, symmetrised answer:

$$\Delta^{\lambda\lambda'}(p)\Delta^{\mu\mu'}(q)\Delta^{\nu\nu'}(r)\Gamma_{\lambda'\mu'\nu'}(pqr)$$

$$=\int \mathrm{d}W^2\mathscr{A}^{\lambda\lambda'}(p,W)\mathscr{A}^{\mu\mu'}(q,W)\mathscr{A}^{\nu\nu'}(r,W)\Lambda^{\alpha}_{\lambda'\mu'\nu'}(pqr|Wn)$$

$$+\int \mathrm{d}W^2\mathscr{B}^{\lambda\lambda'}(p,W)\mathscr{B}^{\mu\mu'}(q,W)\mathscr{B}^{\nu\nu'}(r,W)\Lambda^{\beta}_{\lambda'\mu'\nu'}(pqr|Wn) \tag{32}$$

in which $\mathscr{A}$ and $\mathscr{B}$ are given by (30) and

$$\Lambda^{\alpha}_{\lambda\mu\nu}(pqr|Wn)=\tfrac{1}{3}[\alpha(W^2,p\,.\,n)+\alpha(W^2,q\,.\,n)+\alpha(W^2,r\,.\,n)]\Lambda^{(a)}_{\lambda\mu\nu}$$

$$+\frac{n_\lambda}{6p\,.\,n}[\alpha(W^2,q\,.\,n)+\alpha(W^2,r\,.\,n)$$

$$-2\alpha(W^2,p\,.\,n)][\mathscr{A}^{-1}(q,W)-\mathscr{A}^{-1}(r,W)]_{\mu\nu}$$

$$+\frac{n_\lambda}{2p\,.\,n}[\alpha(W^2,r\,.\,n)-\alpha(W^2,q\,.\,n)][\mathscr{A}^{-1}(q,W)+\mathscr{A}^{-1}(r,W)]_{\mu\nu}$$

$$+(p\lambda\leftrightarrow q\mu\leftrightarrow r\nu\ \text{perms}) \tag{33}$$

$$\Lambda^{\beta}_{\lambda\mu\nu}(pqr|Wn)=\tfrac{1}{3}[\beta(W^2,p\,.\,n)+\beta(W^2,q\,.\,n)+\beta(W^2,r\,.\,n)]\Lambda^{(b)}_{\lambda\mu\nu}$$
$$+\frac{n_\lambda}{6p\,.\,n}[\beta(W^2,q\,.\,n)+\beta(W^2,r\,.\,n)$$
$$-2\beta(W^2,p\,.\,n)][\mathscr{B}^{-1}(r,W)-\mathscr{B}^{-1}(q,W)]_{\mu\nu}$$
$$+\frac{n_\lambda}{2p\,.\,n}[\beta(W^2,r\,.\,n)-\beta(W^2,q\,.\,n)][\mathscr{B}^{-1}(r,W)+\mathscr{B}^{-1}(q,W)]_{\mu\nu}$$
$$+(p\lambda\leftrightarrow q\mu\leftrightarrow r\nu\ \text{perms}). \tag{34}$$

As (32) stands, we can discard all n-pieces from Λ^α and Λ^β to leave the *relativistically covariant Yang–Mills vertex* Λ^0 weighted by the *average spectral function*,

$$[\Delta(p)\Delta(q)\Delta(r)\Gamma(pqr)]_{\lambda\mu\nu}$$
$$=\int \mathrm{d}W^2\tfrac{1}{3}[\alpha(W^2,p\,.\,n)+\alpha(W^2,q\,.\,n)+\alpha(W^2,r\,.\,n)][\mathscr{A}(p)\mathscr{A}(q)\mathscr{A}(r)\Lambda^0]_{\lambda\mu\nu}$$
$$+\int \mathrm{d}W^2\tfrac{1}{3}[\beta(W^2,p\,.\,n)+\beta(W^2,q\,.\,n)$$
$$+\beta(W^2,r\,.\,n)][\mathscr{B}(p)\mathscr{B}(q)\mathscr{B}(r)\Lambda^0]_{\lambda\mu\nu} \tag{35}$$

a pleasingly simple answer for the only Green function which really matters. The n-terms in Λ^α and Λ^β come into their own for the other Green functions which are intimately tied to the gauge identities:

$$\Delta^{B\lambda}(p)\Delta^{\mu\mu'}(p)\Delta^{\nu\nu'}(r)\Gamma_{\lambda\mu'\nu'}(pqr)$$
$$=(p\,.\,n)^{-1}[\Delta(q)p\,.\,\Gamma\Delta(r)]^{\mu\nu}$$
$$=(p\,.\,n)^{-1}[\Delta^{\mu\nu}(r)-\Delta^{\mu\nu}(q)]$$
$$=\int \mathrm{d}W^2(p\,.\,n)^{-1}p^\lambda\mathscr{A}^{\mu\mu'}(q,W)\mathscr{A}^{\nu\nu'}(r,W)\Lambda^{\alpha}_{\lambda\mu'\nu'}(pqr|Wn)$$
$$+\int \mathrm{d}W^2(p\,.\,n)^{-1}p^\lambda\mathscr{B}^{\mu\mu'}(q,W)\mathscr{B}^{\nu\nu'}(r,W)\Lambda^{\beta}_{\lambda\mu'\nu'}(pqr|Wn).$$

The solution (35) is tailor-made for insertion in the Dyson–Schwinger equations and a proper treatment of the (non-perturbative) infrared behaviour of the gluon propagator. However, a first investigation of the Abelian problem will be needed to set the scene.

Appendix. Infrared behaviour of charged vector meson theory

The Dyson–Schwinger equations in vector electrodynamics include the source equation

$$Z^{-1}\delta^\nu_\mu=\Delta_{\mu\lambda}(p)[-(p^2-m_0^2)\eta^{\lambda\nu}+p^\lambda p^\nu]$$
$$-\mathrm{i}e^2\int \bar{\mathrm{d}}^4kD_{\lambda\lambda'}(k)\Delta_{\mu\mu'}(p)\Gamma^{\lambda'\mu'\nu'}(p,p-k)\Delta_{\nu'\rho}(p-k)R^{\lambda\nu\rho}(p,p-k)$$
$$+\text{tadpole term}+\text{two-photon contribution} \tag{A.1}$$

wherein

$$R_{\lambda\mu\nu} = -\eta_{\mu\nu}(2p-k)_\lambda + (p-k)_\mu\eta_{\nu\lambda} + p_\nu\eta_{\mu\lambda} - M(k_\mu\eta_{\nu\lambda} - k_\nu\eta_{\mu\lambda}) \qquad (A.2)$$

is the bare vertex (9), obtained by minimal substitution, to which we have added a magnetic moment M. The luxury of magnetic interactions must be afforded if vector electrodynamics is eventually regarded as part of a Yang–Mills system, when $M=1$ becomes the norm rather than $M=0$. Similarly, in the initial gauge approximation, we will include M as part of the vertex R, since it is not determined by the gauge identity. Actually for infrared behaviour magnetic terms are innocuous as they are visibly soft, but for ultraviolet characteristics they become important.

Make the initial substitution (11) in (A.1) to obtain the integral equations for the spectral functions. Dropping the e^4, 2γ-piece, (in analogy to scalar electrodynamics) the problem reduces to

$$Z^{-1}\eta_{\mu\nu} = (\eta_{\mu\nu}p^2 - p_\mu p_\nu)\int ds\,\frac{\rho(s)+\tau(s)}{p^2-s} + m_0^2\int \frac{(-\eta_{\mu\nu}+p_\mu p_\nu/s)\rho(s) - \eta_{\mu\nu}\tau(s)}{p^2-s}\,ds$$

$$+\int \frac{ds}{p^2-s}\,[\rho(s)(-\delta^\lambda_\mu + p_\mu p^\lambda/s)\Pi^R_{\lambda\nu}(p,s) + \tau(s)\Pi^T_{\mu\nu}(p,s)] \qquad (A.3)$$

where Π^R and Π^T are self-energies in lowest-order perturbation theory for a meson of mass $\sqrt{s}$ possessing interactions (A.2) and (10) respectively at one vertex. If we decompose Π into longitudinal and transverse parts,

$$\Pi_{\mu\nu}(p,s) = (-\eta_{\mu\nu} + p_\mu p_\nu/p^2)\Pi_\perp(p^2,s) + (p_\mu p_\nu/p^2)\Pi_\parallel(p^2,s)$$

and take imaginary parts of (A.3), we remain with the coupled scalar equations†

$$(s-m^2)(\rho(s)+\tau(s))$$

$$= \frac{1}{\pi}\int_{m^2}^{s} \frac{ds}{s-s'}\,[\rho(s')\,\mathrm{Im}\,\Pi^R_\perp(s,s') - \tau(s')\,\mathrm{Im}\,\Pi^T_\perp(s,s')] \qquad (A.4)$$

$$-m^2\tau(s) = \frac{1}{\pi}\int_{m^2}^{s} \frac{ds'}{s-s'}\left[\left(\frac{s}{s'}-1\right)\rho(s)\,\mathrm{Im}\,\Pi^R_\parallel(s,s') + \tau(s')\,\mathrm{Im}\,\Pi^T_\parallel(s,s')\right].$$

The infrared behaviour concerns the limit $s\to m^2$, whereas when $s\to\infty$ we anticipate that the ultraviolet behaviour will show symptoms of non-renormalisability if developed perturbatively in powers of e^2, starting with $\rho(s)=\delta(s-m^2)$, $\tau(s)=0$. However, let us look for *non-perturbative* solutions of (A.4). The absorptive parts entering there are straightforwardly calculated to equal ($p^2 \geqslant m^2$),

$$\mathrm{Im}\,\Pi^R_\perp(p^2,m^2) = -\frac{e^2(p^4-m^4)}{96\pi m^2 p^4}[12m^2p^2 - (M^2+3M+1)(p^2-m^2)^2$$

$$+\tfrac{3}{2}(1-a)(p^2+m^2)^2]$$

† Note the sum rules $Z^{-1} = \int(\rho(s)+\tau(s))\,ds = m_0^2\int s^{-1}\rho(s)\,ds$ which follow from (18). At the level of (A.4) subtractions are made to interpret

$$\Pi(p^2,s) = \frac{p^2-s}{\pi}\int \frac{\mathrm{Im}\,\Pi(s',s)\,ds}{(s'-s)(s'-p^2)}$$

as the renormalised self-energy.

$$\operatorname{Im}\Pi_{\parallel}^{R}(p^2,m^2)=\frac{e^2(p^2-m^2)}{32\pi p^4}\{6m^2p^2+M(p^2-m^2)[(M+1)(p^2-m^2)-3(p^2+m^2)]$$
$$-\tfrac{3}{2}(1-a)m^2(p^2+m^2)\} \tag{A.5}$$
$$\operatorname{Im}\Pi_{\perp}^{T}(p^2,m^2)=\frac{e^2(p^2-m^2)}{64\pi p^4}[8p^2(p^2+m^2)+\tfrac{1}{3}(p^2-m^2)^2+(1-a)(p^2+m^2)(3p^2+m^2)]$$
$$\operatorname{Im}\Pi_{\parallel}^{T}(p^2,m^2)=\frac{e^2(p^2-m^2)}{64\pi p^4}[(p^2-m^2)^2-3(1-a)(p^4-m^4)].$$

Notice the threshold behaviour of $\operatorname{Im}\Pi_{\parallel}^{T}$ which, associated with $(p^2/m^2-1)\operatorname{Im}\Pi_{\parallel}^{R}$, means that τ vanishes relative to ρ as the mass shell is approached. Also, as expected, magnetic M-contributions disappear as $p^2\to m^2$. In fact, in the infrared limit, the relevant equation simplifies to ($a\neq 3$),

$$(p^2-m^2)\rho(p^2)\simeq\frac{1}{\pi}\int_{m^2}^{p^2}\mathrm{d}s\,\rho(s)\frac{\operatorname{Im}\Pi_{\perp}^{R}(p^2,s)}{p^2-s}\underset{p^2\to s}{\cong}\frac{e^2(a-3)}{8\pi^2}\int_{m^2}^{p^2}\mathrm{d}s\,\rho(s)$$

possessing the solution

$$\rho(p^2)\propto(p^2-m^2)^{-1+e^2(a-3)/8\pi^2} \tag{A.6}$$

and leading to

$$\Delta_{\mu\nu}(p)\underset{p^2\to m^2}{\to}(-\eta_{\mu\nu}+p_\mu p_\nu/m^2)/(p^2-m^2)^{1-e^2(a-3)/8\pi^2}. \tag{A.7}$$

The answer is identical to scalar and spinor electrodynamics, and we conjecture that such behaviour is valid for any spin field.

The ultraviolet limit is not so trivial since we must deal with the coupled pair (A.4). Even in the Fermi gauge ($a=1$) with M set equal to zero, the reduced equations look pretty formidable:

$$(z-1)(\rho(z)+\tau(z))=-\frac{e^2}{16\pi^2}\int_1^z\mathrm{d}z'\left\{\rho(z')\left(5-\frac{z'}{z}\right)+\tau(z')\left[2+\frac{2z'}{z}+\frac{1}{3}\left(1-\frac{z'}{z}\right)^2\right]\right\}$$
$$\tau(z)=-\frac{e^2}{16\pi^2}\int_1^z\mathrm{d}z'\left[\rho(z')\left(3-\frac{3z'}{z}\right)+\tau(z')\frac{1}{4}\left(1-\frac{z'}{z}\right)^2\right]$$

since the solutions cannot be simple power dependences $\rho\sim z^r$, $\tau\sim z^t$ even asymptotically.

References

Arnowitt R and Fickler S 1962 *Phys. Rev.* **127** 1821
Bernstein J 1977 *Phys. Rev.* D **15** 2273
Delbourgo R 1977 *J. Phys. A: Math. Gen.* **10** 1369
Delbourgo R and Salam A 1964 *Phys. Rev.* **135** B1398
Delbourgo R, Salam A and Strathdee J 1974 *Nuovo Cim.* A **23** 23
Delbourgo R and West P 1977a *J. Phys. A: Math. Gen.* **10** 1049
—— 1977b *Phys. Lett.* **72B** 96
Fradkin E S and Tyutin I V 1972 *Phys. Rev.* D **2** 2841
Frenkel J and Taylor J C 1976 *Nucl. Phys.* B **109** 526
Jackiw R and Johnson K 1973 *Phys. Rev.* D **8** 2386

Kummer W 1961 *Acta Phys. Aust.* **14** 149
—— 1975 *Acta Phys. Aust.* **41** 315
Lee B W 1974 *Phys. Lett.* **46B** 214
Nishijima K 1960 *Phys. Rev.* **119** 485
Pagels H 1976 *Phys. Rev.* **14** 2747
Rivers R J 1966 *J. Math. Phys.* **7** 385
Salam A 1963 *Phys. Rev.* **130** 1287
Schwinger J 1962 *Phys. Rev.* **125** 397
—— 1963 *Phys. Rev.* **130** 402
Slavnov A A 1972 *Theor. Math. Phys.* **10** 99
Strathdee J 1964 *Phys. Rev.* **135** B1428
Taylor J C 1971 *Nucl. Phys.* B **33** 436

IL NUOVO CIMENTO VOL. 49 A, N. 4 21 Febbraio 1979

The Gauge Technique (*).

R. DELBOURGO
Department of Physics, University of Tasmania - Hobart, Tasmania 7001

(ricevuto il 9 Ottobre 1978)

Summary. — The gauge technique is a procedure whereby the Ward identities of a gauge theory are exatly solved to provide the longitudinal components of the Green's functions as functionals of the pure charged-line amplitudes. In quantum electrodinamics, for instance, the two-electron multiphoton amplitudes are determined by the electron propagator, which can itself be found via the Dyson-Schwinger equations: thus the procedure is gauge covariant and self-consistent. In this manner one readily obtains the infra-red properties of the amplitudes both for relativistic and nonrelativistic (axial) gauge-fixing schemes. The transverse components of the amplitudes are generated iteratively via the skeleton expansion using the gauge-covariant vertices and propagators determined by the gauge technique. As the method is nonperturbative, the prospect of applying it to the infra-red behaviour in chromodynamics looks promising.

1. – Introduction.

In 1963, SALAM [1], with the aim of finding conditions necessary to render vector electrodynamics renormalizable, introduced a new nonperturbative procedure for gauge field theories in which the gauge identities between Green's functions are satisfied *ab initio* before further exploitation of the field theory equations. For the most primitive Ward-Takahashi identity in electrodynamics,

(*) To speed up publication, the author of this paper has agreed to not receive the proofs for correction.

(1) ABDUS SALAM: *Phys. Rev.*, **130**, 1287 (1963).

the procedure specifically provides a vertex function as a functional of the charged-line propagator and the method strongly suggests a more convergent ultraviolet behaviour (through the skeleton expansion) than is obtained from normal perturbation theory. In a subsequent paper (2) with the author, the method was dubbed « The Gauge Technique » and it was applied to scalar and vector electrodynamics with the propagator and vertex function determined via two-particle unitarity; STRATHDEE (3) treated the parallel spinor case.

Now in « solving » the gauge identities the Green's functions may be split into longitudinal and transverse parts, where the latter decouple from the identities involving lower point functions. Consequently there is an intrinsic ambiguity in the way the split is made into transverse and longitudinal parts and the wide scope afforded to exponents of the technique was discussed by RIVERS (4). In an entirely different context, MCCLURE and DRELL, as well as KROLL (5), were concerned with solving the Ward-Takahashi identities, but their motivation was to look for possible modifications of electrodynamics rather than to view the technique as a powerful new gauge-covariant tool. Apart from a relatively recent paper by MANOUKIAN (6), who attempted to get initial propagators consistent with the field equations instead of two-particle unitarity, the subject lay dormant for over a decade. The reason for the disenchantment lies in the fact that any iteration scheme which computes the transverse from the longitudinal components of the Green's functions does not seem capable of a unique formulation because the order of iteration is not immediately associated with the coupling constant. There is the added difficulty that the technical problems associated with sequential iterations become progressively unmanageable in « higher orders ».

In spite of these troubles, which to this day remain largely unresolved, interest has been reawakened in the gauge technique by the discovery of a simple ansatz (7) which satisfied the gauge identities and which provides the multiphoton amplitudes in terms of the no-photon amplitude. Furthermore, this ansatz allows one (8) to determine simply the starting longitudinal Green's functions via the Dyson-Schwinger equations. In two-dimensional electrodynamics (9) it reproduces Schwinger's original solution and in four dimensions it provides a *linear* equation for the spinor spectral function which can be

(2) ABDUS SALAM and R. DELBOURGO: *Phys. Rev.*, **135**, 1398 (1964).
(3) J. STRATHDEE: *Phys. Rev.*, **135**, 1428 (1964).
(4) R. J. RIVERS: *Journ. Math. Phys.*, **7**, 385 (1966).
(5) J. A. MCCLURE and S. D. DRELL: *Nuovo Cimento*, **37**, 1640 (1965); N. M. KROLL: *Nuovo Cimento*, **45** A, 65 (1966).
(6) E. B. MANOUKIAN: *Ann. of Phys.*, **82**, 248 (1974).
(7) R. DELBOURGO: *J. Phys. A*, **11**, 2057 (1978).
(8) R. DELBOURGO and P. WEST: *J. Phys. A*, **10**, 1049 (1977); R. DELBOURGO: *J. Phys. A*, **10**, 1369 (1977).
(9) R. DELBOURGO and T. SHEPHERD: *J. Phys.*, **46**, *L* 197 (1978).

solved explicitly. Perhaps the most encouraging of all is the way in which the infra-red behaviour is trivially obtained ([10]) without going through the tedious process ([11]) of summing subclasses of Feynman diagrams. Hence the technique is still full of promise and may provide some useful clues about the infra-red characteristics ([12]) of quantum chromodynamics. However, because it succeeds only if the gauge identities are amenable to being « solved », it becomes very necessary to work in axial gauges to avoid fictitious particle amplitudes and Bethe-Salpeter kernels. A preliminary investigation of the infra-red behaviour of electrodynamics in the axial gauge ([13]) has been undertaken in order to set the scene for the more complicated non-Abelian case. Extensions to Weyl gravity have also been promoted ([14]).

In sect. **2**, we will discuss methods for solving the gauge identities and their inherent ambiguities. For the most part, here as in later sections, we shall stick to spinor electrodynamics within a class of covariant gauges; but at appropriate places we will indicate the generalizations to scalar and vector sources, chiral enlargements ([15]), and we shall mention which changes are involved by adoption of noncovariant gauges. We show, in sect. **3**, how starting electron spectral functions can be determined by means of the Dyson-Schwinger equations in a way which corrects the gauge noncovariance of earlier attempts; especially interesting are the infra-red and ultraviolet properties of the answers. In the final section **4**, we discuss further the renormalization subtractions and the vexed problem of getting the transverse Green's functions from the coupled field equations.

2. – Solving the gauge identities.

The gauge invariance of electrodynamics is embodied in the Ward-Takahashi identities for the Green's functions. These identities relate amplitudes with N photons to divergences of amplitudes with $N+1$ photons, and for spinor

---

([10]) R. Delbourgo and P. West: *Phys. Lett.*, **72** B, 96 (1977).
([11]) A. Abrikosov: *Sov. Phys. JETP*, **3**, 71 (1956). See also D. Yennie, S. Frautschi and H. Suura: *Ann. of Phys.*, **13**, 379 (1961); C. R. Hagen: *Phys. Rev.*, **130**, 813 (1963); T. W. B. Kibble: *Phys. Rev.*, **173**, 1527 (1968); D. Zwanziger: *Phys. Rev. D*, **11**, 3481 (1975).
([12]) H. Pagels: *Phys. Rev. D*, **15**, 2991 (1975); W. Marciano and H. Pagels: *Phys. Rep.*, **36** C, 3 (1978).
([13]) R. Delbourgo and P. Phocas-Cosmetatos: *J. Phys. A*, to appear (1979).
([14]) Abdus Salam and J. Strathdee: IC/78/12, to be published.
([15]) H. Schnitzer and S. Weinberg: *Phys. Rev.*, **164**, 1878 (1967) were also interested in solving chiral gauge identities. However, they were concerned with low-energy behaviour and polynomial approximations to the amplitudes.

electrodynamics the first few identities read

(1) $$(p'-p)^\mu S(p')\Gamma_\mu(p',p)S(p) = S(p)-S(p'),$$

(2) $$k^\mu S(p')\Gamma_{\mu\nu}(p'k',pk)S(p) = $$
$$= S(p')\Gamma_\nu(p',p'+k')S(p'+k') - S(p-k')\Gamma_\nu(p-k',p)S(p),$$

etc., where S denotes the complete spinor propagator and Γ stands for the fully amputated connected Green's functions (but not one-particle irreducible) with appropriate arguments and with the coupling constant factorized out. Corresponding identities can be written down for the one-particle–irreducible amplitudes.

From the propagator representation

(3) $$S(p) = \int \mathrm{d}W\,\varrho(W)[\gamma\cdot p - W]^{-1}$$

it follows that the spectrally weighted expressions

(4) $$S(p')\Gamma^{\parallel}_\mu(p',p)S(p) = \int \mathrm{d}W\,\varrho(W)[\gamma\cdot p' - W]^{-1}\gamma_\mu[\gamma\cdot p - W]^{-1},$$

(5) $$S(p')\Gamma^{\parallel}_{\nu\mu}(p',p)S(p) = \int \mathrm{d}W\,\varrho(W)[\gamma\cdot p' - W]^{-1}\cdot$$
$$\cdot(\gamma_\nu[\gamma\cdot(p+k)-W]^{-1}\gamma_\mu + \gamma_\mu[\gamma\cdot(p'-k)-W]^{-1}\gamma_\nu)[\gamma\cdot p - W]^{-1},$$

etc., identically satisfy the gauge identities (1), (2), ... by themselves. The amplitudes certainly include the tree graphs if we substitute $\varrho(W) = \delta(W-m)$ and are capable of describing *some* measure of quantum corrections by the very appearance of a nontrivial ϱ. But it should also be obvious that they represent a far-from-complete picture of the true functions because they entirely miss out transverse components $\Gamma^{\perp}_\mu$ which are orthogonal photon momenta and never enter in the contracted identities. This indeterminateness of $\Gamma^{\perp}_\mu$ is the main criticism one can level at the gauge technique and there is no way to answer it without at some point referring to the Green's function equations which necessarily couple longitudinal to transverse Γ. Although the ansätze (4), (5), ... express the Γ as functionals of the electron propagator, they do not tell us how to find a meaningful ϱ, which at this early stage remains quite arbitrary; once again we shall need to turn to the Dyson-Schwinger equations for providing a physically relevant ϱ.

Before we consider these questions, it is worth recalling that the gauge technique was originally invoked for solving the identities between one-particle irreducible functions, for instance the basic

(1′) $$(p'-p)^\mu \Gamma_\mu(p',p) = S^{-1}(p') - S^{-1}(p).$$

If we use the inverse spectral representation

$$(3') \qquad \begin{cases} S^{-1}(p) = -\int (W-m)^2[\gamma\cdot p - W]^{-1} r(W)\,\mathrm{d}W \\ \text{with} \\ \int Wr(W)\,\mathrm{d}W = \int mr(W)\,\mathrm{d}W = m\,, \end{cases}$$

then we arrive at a seemingly alternative expression for the vertex function

$$(4') \qquad \Gamma_\mu^\parallel(p',p) = \int \frac{m-W}{\gamma\cdot p' - W}\,\gamma_\mu\,\frac{m-W}{\gamma\cdot p - W}\,r(W)\,\mathrm{d}W\,.$$

In fact, however, (4′) is completely equivalent to version (4) as one can easily check by proceeding to the mass shell of one of the spinor lines. Note also the connection between (3′) and (3)

$$\varrho(W) = |(W-m)\,S(W)|^2 r(W)\,.$$

In practice we shall find it more convenient to work with version (4) rather than (4′).

From the structure of the ansätze, it is evident that no kinematic dependence on the photon momentum has been incorporated into our starting longitudinal amplitudes. This is demonstrated most graphically by going to the electrons' mass shells, when $\bar{u}\Gamma_\mu^\parallel u \to \bar{u}\gamma_\mu u$, with no sign of an anomalous magnetic correction or even a charge form factor. These important aspects of the complete vertex (the transversal $\sigma_{\mu\nu}k^\nu$ among them) will have to emerge as corrections to the initial set (4), (5), ... when one studies the full dynamical equations of the theory. At the level of our initial solutions, all *off-shell dependence* in the Green's functions *resides in the electron lines.* One could think of modifying [2,3] the amplitudes by including transverse corrections arising from vacuum polarization

$$\Pi_{\mu\nu}(k) = ie^2\,\mathrm{Tr}\int [S(p)\gamma_\mu S(p-k)\Gamma_\nu]\,\mathrm{d}^4\bar{p}$$

and/or just adding the appropriate anomalous magnetic corrections; but that would seem to be totally *ad hoc* since there is no rationale behind these modifications without referring back to the field equations and the coupling between $\Gamma_\mu^\parallel$ and $\Gamma_\mu^\perp$. We shall come back to this matter in sect. **4**, but, without prejudicing the issue there, the first task confronting us is the determination of a physically sensible ϱ. If we can find a meaningful expression for it, the initial gauge set (4), (5) must contain a considerable amount of information even if it does not tell the whole story.

In the original papers [2,3] it was proposed to determine the spectral functions consistent with two-particle unitarity, *viz.*

$$\mathrm{Im}\, S^{-1} = |\Gamma|^2 \qquad \text{(2-particle phase space)}.$$

This amounted to using a ϱ equalling the first-order perturbation result and it gave

$$S(p) \sim \ln p^2/\gamma \cdot p\,, \qquad \Gamma_\mu \sim \gamma_\mu/(\ln p^2)^2$$

at asymptotic momenta. And for meson electrodynamics it implied that a $(\varphi^\dagger \varphi)^2$ counterterm was not necessary to the theory providing $\Gamma_\mu^\perp$ and S were used in the skeleton expansion (2,16,17). Shortly, we shall be discussing another way to get at ϱ which is closer in spirit to field theory, but before we get embroiled with technicalities let us mention a few generalizations to other sources and other gauge models.

If we limit ourselves to electrodynamics, we may also consider charged scalar and vector mesons (propagators Δ and $\Delta_{\mu\nu}$). The identities analogous to (1), (2), ... can again be solved by spectral weighting; thus from

$$\Delta(p) = \int \mathrm{d}W^2 \varrho(W^2)[p^2 - W^2]^{-1}, \tag{6}$$

$$\Delta_{\mu\nu}(p) = \int \mathrm{d}W^2 [(-\eta_{\mu\nu} + p_\mu p_\nu/W^2)\sigma(W^2) - \eta_{\mu\nu}\tau(W^2)][p^2 - W^2]^{-1}, \tag{7}$$

it is straightforward to arrive at

$$\Delta(p')\Gamma_\mu(p', p)\Delta(p) = \int \mathrm{d}W^2 \varrho(W^2)[p'^2 - W^2]^{-1}(p' + p)_\mu [p^2 - W^2]^{-1},$$

$$\Delta^{\mu\mu'}(p')\Gamma_{\lambda\mu'\nu'}(p', p)\,\Delta(p) = \int \mathrm{d}W^2 [p'^2 - W^2]^{-1}[p^2 - W^2]^{-1}\cdot$$
$$\cdot\left[\sigma(W^2)\left(\eta^{\mu\mu'} - \frac{p'^\mu p'^{\mu'}}{W^2}\right)(-\eta_{\mu'\nu'}(p' + p)_\lambda + p_{\mu'}\eta_{\nu'\lambda} + p'_{\nu'}\eta_{\mu'\lambda})\cdot\right.$$
$$\left.\cdot\left(\eta^{\nu\nu'} - \frac{p^\nu p^{\nu'}}{W^2}\right) + \eta^{\mu\nu}(p + p')_\lambda \tau(W^2)\right]$$

with correspondingly obvious expressions for multiphoton processes.

The method can also be extended (7) to axial electrodynamics in which U_1 gauge chirality is exact and the identities start with

$$-i(p' - p)^\mu \Gamma_{\mu 5}(p', p) = S^{-1}(p')\gamma_5 + \gamma_5 S^{-1}(p)\,. \tag{8}$$

---

(16) R. Delbourgo, Abdus Salam and J. Strathdee: *Nuovo Cimento*, **23** A, 237 (1974).
(17) Abdus Salam and J. Strathdee: IC/78/44, to be published.

Borrowing from the equivalent Jackiw-Johnson ([18]) field theory, the identities find their solution in expressions like

$$-iS(p)\Gamma_{\mu 5}(p, p-k)S(p-k) = \tag{9}$$

$$= \int \mathrm{d}W \varrho(W) \frac{1}{\gamma\cdot p - W}\left(\gamma_\mu\gamma_5 - \frac{2Wk_\mu\gamma_5}{k^2}\right)\frac{1}{\gamma\cdot(p-k) - W}.$$

The attendant $k^2 \to 0$ singularities signify dynamical symmetry breaking and show all the hallmarks of the Schwinger mechanism ([19]) for generating an axial-meson mass.

Nor need one stick to covariant vector gauges of the kind usually invoked:

$$D^a_{\mu\nu}(k) = [-\eta_{\mu\nu} + (1-a)k_\mu k_\nu/k^2]/k^2.$$

On the contrary, since the gauge technique is prefaced upon getting workable solutions of the gauge identities, generalization to axial gauges ($n\cdot A = 0$) is strongly indicated, for only in such gauges do the non-Abelian identities retain their simple naive form. In these gauges the bare photon propagator reads

$$D^n_{\mu\nu}(k) \equiv \mathscr{A}_{\mu\nu}(k)/k^2 = \left[-\eta_{\mu\nu} + \frac{k_\mu n_\nu + k_\nu n_\mu}{k\cdot n} - \frac{k_\mu k_\nu n^2}{(k\cdot n)^2}\right]\Big/ k^2 \tag{10}$$

and, after dressing,

$$D_{\mu\nu}(k) = \int \frac{\mathrm{d}W^2}{k^2 - W^2}[\mathscr{A}_{\mu\nu}\alpha(W^2, k\cdot n) + \mathscr{B}_{\mu\nu}\beta(W^2, k\cdot n)], \tag{11}$$

where

$$\mathscr{B}_{\mu\nu} = -\eta_{\mu\nu} + n_\mu n_\nu/n^2. \tag{12}$$

The ansätze which satisfy the gauge identities are not so trivially found now because the spectral functions for the gauge factors, as well as the sources, depend on $n\cdot p$ among other possible invariants. Despite this complication, one can determine the appropriate gauge solutions without very much trouble, the net result being that the Green's functions are still given by similar expressions except that the weighting also includes an average over the incoming lines. Specifically one finds in quantum electrodynamics that

$$S(p')\Gamma^{\parallel}_\mu(p', p)S(p) = \int \mathrm{d}W \frac{1}{2}\{\varrho(W, n\cdot\gamma, n\cdot p') + \varrho(W, n\cdot\gamma, n\cdot p)\}\cdot \tag{13}$$

$$\cdot \frac{1}{\gamma\cdot p' - W}\gamma_\mu \frac{1}{\gamma\cdot p - W}$$

([18]) R. Jackiw and K. Johnson: *Phys. Rev.*, **8**, 2386 (1973).

([19]) J. Schwinger: *Phys. Rev.*, **128**, 2425 (1962).

up to n_μ terms which will vanish anyhow upon contraction with $\varepsilon^\mu(k)$ or $D^{\mu\nu}(k)$. Analogous formulae apply to scalar electrodynamics and, as for the three-gluon vertex, the solution is

$$(14)\qquad \Delta^{\lambda\lambda'}(p)\Delta^{\mu\mu'}(q)\Delta^{\nu\nu'}(r)\Gamma_{\lambda'\mu'\nu'}(pqr) =$$
$$= \int \mathrm{d}W \frac{1}{3}\{\alpha(W^2, p\cdot n) + \alpha(W^2, q\cdot n) + \alpha(W^2, r\cdot n)\}\cdot$$
$$\cdot[\mathscr{A}(p)\mathscr{A}(q)\mathscr{A}(r)\Lambda]^{\lambda\mu\nu}/(p^2 - W^2)(q^2 - W^2)(r^2 - W^2) +$$
$$+ \int \mathrm{d}W^2 \frac{1}{3}\{\beta(W^2, p\cdot n) + \beta(W^2, q\cdot n) + \beta(W^2, r\cdot n)\}\frac{[\mathscr{B}(p)\mathscr{B}(q)\mathscr{B}(r)\Lambda]^{\lambda\mu\nu}}{(p^2 - W^2)(q^2 - W^2)(r^2 - W^2)},$$

where

$$(15)\qquad \Lambda_{\lambda\mu\nu}(pqr) = \eta_{\mu\nu}(r-q)_\lambda + \eta_{\nu\lambda}(p-r)_\mu + \eta_{\lambda\mu}(p-q)_\nu$$

is the normal Yang-Mills symmetric vertex.

We turn now to computation of the starting spectral functions which appear in all the above formulae.

3. – Self-consistent evaluation of the spectral functions.

The BCS theory of superconductivity has prompted the application of similar ideas to relativistic quantum field theory. A prominent feature of these studies is the role played by the « gap equations » as a means of self-consistently determining the source mass. In electrodynamics this self-consistency equation (up to a renormalization factor) is normally written as

$$(16)\qquad S^{-1}(p) = \gamma\cdot p - m_0 + \Sigma(p) = \gamma\cdot p - m_0 + ie^2\int \gamma_\mu S(p-k)\gamma^\mu \bar{\mathrm{d}}^4k/k^2\,.$$

The object of the exercise is then to determine the behaviour of $\Sigma(p)$ as a function of m_0, e^2 and its argument $\gamma\cdot p$. The low-energy properties describe the mass renormalization and the high-energy properties usually contain the dependence on e^2. Unfortunately the approximation (16) of the Dyson-Schwinger equation in which Γ is replaced by the bare vertex is not a gauge-covariant one and it is, therefore, very hard to judge the significance of any nonperturbative structure in e^2 which emerges. Luckily the gauge technique is able to remedy just this deficiency for it is guaranteed to be gauge covariant; what is more, it leads to a perfectly amenable linear equation to be contrasted with the nonlinear nature of (16).

The correct gap equation is the complete one

$$(17)\qquad Z_1^{-1} = (\gamma\cdot p - m_0)S(p) - ie^2\int \bar{\mathrm{d}}^4k\, S(p)\Gamma_\mu(p, p-k)S(p-k)\gamma_\nu D^{\mu\nu}(k)$$

and, if the ansatz (4) is used and photon dressing is neglected, leads to the linear equation

$$0 = \int dW \, [W - m_0 - \Sigma(p, W)] [\gamma \cdot p - W]^{-1} \varrho(W) \,, \tag{18}$$

where $\Sigma(p, W)$ is the *lowest*-order self-energy appropriate to an electron of mass W. After appropriate renormalizations (see the following section), Σ is interpreted as the subtracted dispersion integral

$$\Sigma(W, W') = \frac{W - W'}{\pi} \int \frac{\operatorname{Im} \Sigma(W'', W') \, dW''}{(W'' - W')(W'' - W)}$$

and m_0 is replaced by m. The imaginary part of (18) then yields the spectral equation

$$(W - m) \varrho(W) = \int_m^W \frac{\operatorname{Im} \Sigma(W, W')}{\pi (W - W')} \varrho(W') \, dW'' \,. \tag{19}$$

An elementary computation gives

$$\operatorname{Im} \Sigma(p, W) = \frac{e^2 (p^2 - W^2)}{16 \pi p^3} [a(p^2 + W^2) - (a + 3) p W] \, \theta(p^2 - W^2) \,,$$

in the covariant gauge class D^a, which is to be substituted in (19). It is possible to solve (19) exactly in terms of familiar transcendental functions only in the Landau gauge $a = 0$. The answer is

$$\varrho(W) = \frac{2 \varepsilon R(\varepsilon)}{(W^2/m^2 - 1)^{1 + 2\varepsilon}} \cdot \left[\frac{1}{W} F\left(-\varepsilon, -\varepsilon; -2\varepsilon; 1 - \frac{W^2}{m^2} \right) + \frac{1}{m} F\left(-\varepsilon, 1 - \varepsilon; -2\varepsilon; 1 - \frac{W^2}{m^2} \right) \right] \tag{20}$$

with $\varepsilon = 3e^2/16\pi^2$. But it is also possible to find the infra-red behaviour in any of these gauges by virtue of $\varrho(W') \to \varrho(W)$ on the r.h.s. of (19) in the limit $W \to m$. Thus the ϱ equation simplifies to

$$\frac{d}{dW} [(W - m) \varrho(w)] = \varrho(W) \lim_{W \to W' \to m} \frac{\operatorname{Im} \Sigma(W, W')}{\pi (W - W')} = \varrho(W) \frac{e^2 (a - 3)}{8 \pi^2} \,.$$

It follows that near the electron mass shell

$$S(p) \to \frac{\gamma \cdot p + m}{p^2 - m^2} \left(\frac{m^2}{p^2 - m^2} \right)^{e^2 (a - 3)/8\pi^2} \,. \tag{21}$$

When the gauge technique is applied to carry out the parallel calculations for scalar and vector mesons, the answers are much the same:

$$\Delta(p) \to \frac{1}{p^2 - m^2}\left(\frac{m^2}{p^2 - m^2}\right)^{e^2(a-3)/8\pi^2}, \tag{22}$$

$$\Delta_{\mu\nu}(p) \to \frac{-\eta_{\mu\nu} + p_\mu p_\nu/m^2}{p^2 - m^2}\left(\frac{m^2}{p^2 - m^2}\right)^{e^2(a-3)/8\pi^2}. \tag{23}$$

These infra-red behaviours support the calculations of Gorkov ([20]).

One can go a stage further and use the gauge technique for determining the infra-red characteristics in axial gauges $n\cdot A = 0$. One finds that near the mass shell all noncovariance gets absorbed into an exponent. Explicitly,

$$\begin{cases} S(p) \to \dfrac{\gamma\cdot p + m}{2(p^2 - m^2)}\left[\left(\dfrac{m^2}{p^2 - m^2}\right)^{\zeta_+} + \left(\dfrac{m^2}{p^2 - m^2}\right)^{\zeta_-}\right] + \\ \qquad\qquad + \dfrac{m\gamma\cdot n}{2(p^2 - m^2)}\left[\left(\dfrac{m^2}{p^2 - m^2}\right)^{\zeta_+} - \left(\dfrac{m^2}{p^2 - m^2}\right)^{\zeta_-}\right], \\ \Delta(p) \to \dfrac{1}{p^2 - m^2}\left(\dfrac{m^2}{p^2 - m^2}\right)^{\zeta}, \end{cases} \tag{24}$$

where

$$\zeta = -\frac{e^2}{2\pi^2}\int_0^1 du\left[1 + u^2\left(\frac{p^2 n^2}{(p\cdot n)^2} - 1\right)\right]^{-1}$$

and

$$\zeta_\pm = \frac{m\zeta}{m \pm p\cdot n}. \tag{25}$$

(A similar sort of phenomenon was found by HAGEN ([11]) in the radiation gauge.) The vector case is currently being studied and one may anticipate analogous answers. The extension to Yang-Mills theory is considerably harder and the computations remain to be done. It will be interesting to see if the calculated infra-red properties show indications of a confinement mechanism ([12]), $D(p)$ being considerably more singular than $1/p^2$, or whether it is imperative to include different topological configurations in order to get the phenomenon.

([20]) L. P. GORKOV: *Sov. Phys. JETP*, **3**, 762 (1956).

4. – Transverse corrections.

The renormalizations which take us from (17) to (19) were only sketched out in sect. **3**; they bear closer inspection. As we shall soon see, they force us to make more than a passing mention to transverse components. Let us first make the exact substitution $\Gamma = \Gamma^{\parallel} + \Gamma^{\perp}$, where $\Gamma^{\parallel}$ is found by the gauge technique and $\Gamma^{\perp}$ is the transverse remainder. Notice that $\Gamma^{\perp}$ is at least of order e^2 because $\Gamma^{\parallel}$ includes the primary γ. With definition (3) and remembrance of $Z_1^{-1} = \int \varrho(W)\,\mathrm{d}W$, we have the *exact* Dyson-Schwinger equation

$$(17') \qquad 0 = \int \frac{\mathrm{d}W \varrho(W)}{\gamma\cdot p - W}\left[W - m_0 - ie^2\int \mathrm{d}^4k\, \gamma_\mu \frac{1}{\gamma\cdot(p-k) - W}\gamma_\nu D^{\mu\nu}(k)\right] - \\ - ie^2\int \mathrm{d}^4k\, S(p)\Gamma_\mu^{\perp}(p, p-k)S(p-k)\gamma_\nu D^{\mu\nu}(k)\,.$$

There is no question but that it is possible to renormalize (17) or (17′) order by order in e^2, just like the equations for the photon propagator and the vertex part. Thus $\varrho(W)$ can be expressed as a series in e^2 *with no cut-off Λ dependence.* If we view (17′) as providing an equation for $\varrho(W)$, the Λ dependence *must* disappear after renormalizations are dutifully done.

Second, let us replace D by the bare photon propagator—all this means is that we can forego certain Z_3 subtractions. Then (17′) simplifies to

$$0 = \int \frac{W - m_0 + \Sigma(p, W)}{\gamma\cdot p - W}\varrho(W)\,\mathrm{d}W - ie^2\int \mathrm{d}^4k\, S(p)\Gamma_\mu^{\perp}(p, p-k)S(p-k)\gamma_\nu D_a^{\mu\nu}(k)$$

and its imaginary part yields

$$(19') \qquad [W - m_0 - \mathrm{Re}\,(W, W)]\varrho(W) = \int \frac{\mathrm{Im}\,\Sigma(W, W')}{\pi(W - W')}\cdot \\ \cdot\varrho(W')\,\mathrm{d}W' + \text{transverse terms}\,.$$

Now $m = m_0 - \Sigma(m, m)$ is exact to order e^2, but also receives e^4 and higher terms. These higher corrections are, of course, also contained in the transverse component $\Gamma_\mu^{\perp}$. Because

$$\Sigma(W, W) - \Sigma(m, m) \sim e^2\left(W \ln\frac{W}{\Lambda} - m\ln\frac{m}{\Lambda}\right)$$

has an intrinsic dependence on Λ, the only way that the Λ pieces can cancel in (19′) is if the transverse additions fulfil that function. It is no accident that they are of the right order in e^4 to accomplish the cancellation. It follows that

$$(W-m)\varrho(w)=\int\frac{\operatorname{Im}\Sigma(W,W')}{\pi(W-W')}\varrho(W')\,dW'+\text{renormalized transverse terms}$$

is the correct equation for the spectral function. What we have done in (19) is to drop the transverse additions to provide the initial ϱ.

We conclude with a brief discussion about how one may evaluate the transverse Green's functions. For this purpose we can lean on Dyson's skeleton expansion which gives the four- and higher-point amplitudes as functionals of the fundamental S, D and Γ, themselves connected by the Dyson-Schwinger equations:

$$S^{-1}=Z_1(\gamma\cdot p-m_0)+Z_1e^2\int\Gamma SD\gamma\,, \tag{26}$$

$$D^{-1}=Z_3k^2+Z_1e^2\int S\Gamma S\gamma\,, \tag{27}$$

$$\Gamma\;=Z_1\gamma+Z_1e^2\int\gamma SDK[\Gamma,S,D]\,. \tag{28}$$

We have basically already constructed a substantial component of the longitudinal functions by the ansätze (4), (5), etc., the functional dependence being purely on S. To work out the transverse parts (as well as further longitudinal modifications), we need to include $\Gamma^\perp_\mu$ vertices in the skeleton expansion. The $\Gamma^\perp$ can themselves be generated via eq. (28) preferably by incorporating photon dressing. The following iteration scheme, therefore, suggests itself:

$$\left\{\begin{aligned}S^{-1(n+1)}&=Z_1(\gamma\cdot p-m_0)+Z_1e^2\int\Gamma^{(n)}S^{(n)}D^{(n)}\gamma\,,\\ D^{-1(n+1)}&=Z_3k^2+Z_1e^2\int S^{(n)}\Gamma^{(n)}S^{(n)}\gamma\,,\\ \Gamma^{(n+1)}&=Z_1\gamma+Z_1e^2\int\gamma S^{(n)}D^{(n)}K[\Gamma^{(n)},S^{(n)},D^{(n)}]\,,\end{aligned}\right. \tag{29}$$

where $S^{-1(1)}=S^{-1(0)}=k\cdot\Gamma^{(1)}=k\cdot\Gamma^{(0)}$. $S^{(0)}$ is, of course, determined in zeroth approximation through (17). The reader will observe that there is nothing sacred, nor unique, about this iteration programme for working out $\Gamma^\perp_\mu$ plus successive changes to $\Gamma^\parallel_\mu$. It is intrinsically nonperturbative and e^2 does *not directly* correspond to the order of iteration. It is very likely that an improved and more systematic way of generating the quantum corrections will be formulated in the light of experience. Certainly as the scheme (29) stands, the higher-order computations seem inordinately complicated.

In this review we have largely dwelt on the positive virtues of the gauge technique. It is relatively easy to point out the flaws, and we have not refrained from so doing. Nevertheless, the ease with which certain nonperturbative results can be derived by the method is a strong point in its favour. We hope that it will stimulate further research and lead to a clearing of the more nebulous areas in this subject.

● RIASSUNTO (*)

La tecnica di gauge è una procedura con la quale le identità di Ward di una teoria di gauge sono risolte esattamente per fornire i componenti longitudinali delle funzioni di Green come funzionali delle semplici ampiezze di linea carica. Nell'elettrodinamica quantica, per esempio, le ampiezze multifotoniche a due elettroni sono determinate dal propagatore di elettroni, che può esso stesso essere trovato mediante le equazioni di Dyson-Schwinger: così la procedura è covariante di gauge e autoconsistente. In questo modo si ottengono prontamente le proprietà nell'infrarosso delle ampiezze sia per schemi relativistici che per schemi non relativistici (assiali) che fissano il gauge. I componenti trasversali delle ampiezze sono generati iterativamente mediante lo sviluppo a scheletro che usa vertici covarianti di gauge e propagatori determinati con la tecnica di gauge. Poiché il metodo è non perturbativo, la prospettiva di applicarlo al comportamento nell'infrarosso nella cromodinamica appare promettente.

(*) *Traduzione a cura della Redazione.*

Резюме не получено.

J. Phys. A: Math. Gen., **13** (1980) 701–711. Printed in Great Britain

On the gauge dependence of spectral functions†

R Delbourgo and B W Keck
Physics Department, University of Tasmania, Hobart, Tasmania, Australia

Received 28 March 1979

Abstract. An integral relation is derived between charged source spectral functions in photon gauges differing by longitudinal terms $ak_\mu k_\nu/k^4$. The a dependences of the spectral and Green functions supplied by the gauge technique automatically satisfy these integral relations.

1. Introduction

The relations between Green functions in different Lorentz covariant gauges were obtained many years ago (Landau and Khalatnikov 1956, Fradkin 1956) and subsequently rederived by functional (Zumino 1960, Bialynicki–Birula 1960) and other (Okubo 1960) methods. These relations serve to connect the charged particle Green functions for photon propagators $D_{\mu\nu}(x)$ differing by the longitudinal term $\partial_\mu\partial_\nu M$ and they were effectively used by Johnson and Zumino (1959) and Zumino (1960) to highlight the gauge dependence of the renormalisation constants as well as the infrared and ultraviolet behaviours of the electron propagator. The connection between the Green functions involves the phase factor $\exp(ie^2M(x))$, and, as such, is most easily expressed in configuration space. In this paper we wish to discuss the gauge dependence of the charged propagator spectral function ρ; since this is couched in momentum space the relationship cannot be expected to remain so simple. For the class of covariant gauges where $D_{\mu\nu}(k)$ varies by $ak_\mu k_\nu/k^4$,

$$\exp[ie^2M(x)] = (-m^2x^2)^{-e^2a/16\pi^2}$$

represents the multiplicative effect of the phase factor. We shall determine (see equations (22) and (23)) the connection between ρ in two different a gauges and then demonstrate how the explicit ρ obtained (Delbourgo and West 1977a, b, Delbourgo 1977) by means of the gauge technique (Salam 1963, Delbourgo 1979) automatically satisfy the identity. To expose this more transparently, we go on to find the corresponding x space charged particle Green functions.

A rapid derivation of the gauge relations is provided in § 2, as well as the corresponding steps for the non–abelian case and why they are not particularly fruitful. In § 3 we obtain the integral connection between the spectral functions and finally we study the repercussions for the gauge technique in § 4.

† Supported by the ARGC under grant no. B77/15249

0305-4470/80/020701 + 11$01.00 © 1980 The Institute of Physics

2. Gauge dependence of Green functions

Here we shall rapidly derive the gauge properties of amplitudes by functional methods. We shall treat the abelian case first and stick to Zumino's (1960) notation. Further on, we shall discuss the generalisation to the non-abelian case and show that it fails to provide closed form relations that are of any use.

In a gauge† $a_\mu A^\mu = A$ one begins with the vacuum functionals

$$Z_\Lambda[u, v, J] = \int [\mathrm{d}A_\mu\, \mathrm{d}\psi\, \mathrm{d}\bar{\psi}\, \mathrm{d}B]\psi(u_1)\ldots\psi(u_m)\bar{\psi}(v_1)\ldots\bar{\psi}(v_n) \times \exp\left[\mathrm{i}\int (\mathscr{L} - JA - B(\Lambda - aA))\right] \tag{1}$$

where B acts as an auxiliary multiplier field, enforcing the gauge condition. As a further extension, which for the Landau gauge amounts to an addition $\partial_\mu\partial_\nu M(x-y)$ to the photon propagator $D_{\mu\nu}(x-y)$, one may envisage

$$Z_{\Lambda,M}[u, v, J] = \int \mathrm{d}A_\mu \ldots \mathrm{d}B\, \pi\psi(u_i)\pi\bar{\psi}(v_i) \exp\left(\mathrm{i}\int (\mathscr{L} - JA - B(\Lambda - aA) - \tfrac{1}{2}BMB)\right] \tag{2}$$

and assume without loss that $\partial_\mu a^\mu = 1$.

The Λ, M dependence may be elucidated by gauge transforming the integration variables $A, \psi, \bar{\psi}$ by an amount depending on B:

$$A_\mu \to A_\mu - \partial_\mu \chi \qquad \psi \to \exp(\mathrm{i}e\chi)\psi \qquad \bar{\psi} \to \exp(-\mathrm{i}e\chi)\bar{\psi} \tag{3}$$

where

$$\chi(x) = g(x) + \int \mathrm{d}y\, h(x-y)B(y). \tag{4}$$

We obtain

$$Z_{\Lambda,M}[u, v, J] = \int [\mathrm{d}A_\mu \ldots \mathrm{d}B]\pi \exp[\mathrm{i}e(g+hB)(u_i)]\psi(u_i)\pi \exp[-\mathrm{i}e(g+hB)(v_i)]\bar{\psi}(v_i) \times \exp\left\{\mathrm{i}\int [\mathscr{L} - J(A-\partial)(g+hB)] - B(\Lambda - aA + g + hB) - \tfrac{1}{2}BMB\right\} \tag{5}$$

and so

$$Z_{\Lambda,M}[u, v, J] = \exp\left(-\mathrm{i}\int (g\partial J)\right]\pi \exp(\mathrm{i}eg(u_i))\pi \exp(-\mathrm{i}eg(v_i))Z_{\Lambda',M'}[u, v, J] \tag{6}$$

where

$$\begin{aligned} \Lambda'(x) &= \Lambda(x) - e\Sigma h(u_i - x) + e\Sigma h(v_i - x) + \int \mathrm{d}y\, h(y-x)(\partial J)(y) + g(x) \\ M'(x) &= M(x) + h(x) + h(-x) \end{aligned} \tag{7}$$

Clearly by suitably choosing g and h we can connect (Λ, M) to any (Λ', M').

In particular, with a little algebra, we find

$$Z_{\Lambda,M}[u, v, J] = \exp(\mathrm{i}\theta)Z_{\Lambda,0}[u, v, J] \tag{8}$$

where

$$\theta = -\tfrac{1}{2}(\partial J)\,.\,M(\partial J) + e[\Sigma(M\partial J)(u_i) - \Sigma(M\partial J)(v_i)] + e^2(-\tfrac{1}{2}\Sigma M(u_i - u_j) + \Sigma M(u_i - v_j) - \tfrac{1}{2}\Sigma M(v_i - v_j)). \tag{9}$$

† $(a_\mu A^\mu)(x) = \int \mathrm{d}y\, a_\mu(x-y)A^\mu(y)$

Expanding in J and suppressing Λ we find for the propagators and vertex function

$$D_{\mu\nu}(x-y;M)=D_{\mu\nu}(x-y;0)-\partial_\mu\partial_\nu M(x-y) \tag{10}$$

$$S(x-y;M)=\exp(\mathrm{i}e^2(M(x-y)-M(0))]S(x-y;0) \tag{11}$$

$$\{D_{\mu\nu}S\Gamma^\nu S\}(x,y,z;M)=\exp[\mathrm{i}e^2(M(x-y)-M(0))][\{D_{\mu\nu}S\Gamma^\nu S\}(x,y,z;0)$$
$$+\mathrm{i}eS(x-y;0)\partial^z_\mu\{M(x-z)-M(y-z)\}] \tag{12}$$

The above relations are for unrenormalised fields and expectation values. Since (Johnson and Zumino 1959)

$$Z_2(M)/Z_2(0)=\exp(-\mathrm{i}e^2M(0)) \tag{13}$$

one deduces the following relations between renormalised Green functions:

$$S(x;M)=\exp(\mathrm{i}e^2M(x))S(x;0) \tag{11'}$$

$$\{DS\Gamma S\}(x,y,z;M)=\exp(\mathrm{i}e^2M(x-y))[\{DS\Gamma S\}(x,y,z;0)$$
$$+\mathrm{i}eS(x-y;0)\partial^z(M(x-y)-M(y-z))] \tag{12'}$$

Note that since

$$e_r^2=Z_3e_u^2,\qquad M_r=Z_3^{-1}M_u,\qquad e_r(S\Gamma SD)_r=Z_2^{-1}e_u(S\Gamma SD)_u$$

the combinations e^2M and $e\Gamma SD$ are renormalisation invariant. Indeed, consistent with the transformation properties (12), it is easily verified that the renormalised Dyson–Schwinger equation

$$Z_2^{-1}\delta(x)=(\mathrm{i}\gamma\,.\,\partial-m_0)S(x)+\mathrm{i}e\gamma_\mu\{S\Gamma_\nu SD^{\mu\nu}\}(x,0,x)$$

is valid for any fixed value of M, as we know it must be.

Let us generalise from QED to QCD to see how the argument goes awry. We begin with the Faddeev–Popov modified version of (2),

$$Z_{\Lambda\mu}=\int[\mathrm{d}A_\mu\ldots\mathrm{d}B]\Delta(A)\pi\psi(u)\pi\bar\psi(v)\exp\left[\mathrm{i}\int(\mathscr{L}-JA-B(\Lambda-aA)-\tfrac12BMB)\right]. \tag{14}$$

The measure $[\mathrm{d}A]\Delta(A)$ is invariant only under certain A-dependent transformations of Slavnov (1972) type,

$$\delta A_\mu=-D_\mu\delta\chi(A)=-D_\mu(a\,.\,D)^{-1}\delta\Lambda \tag{15}$$

where D is the normal covariant derivative and $\delta\Lambda$ is A independent. (The usual proof of this invariance (Lee and Zinn-Justin 1973) is perhaps rather complicated, so we thought it worthwhile to present a more transparent and general derivation in Appendix A). As for QED, $\delta\Lambda$ can depend on B. We obtain

$$Z_{\Lambda\mu}=\int[\mathrm{d}A\ldots\mathrm{d}B]\Delta(A)\pi\exp(\mathrm{i}eT\delta\chi)\psi(u)\,.\,\pi\exp(-\mathrm{i}e\bar T\delta\chi)\bar\psi(v)$$
$$\times\exp\left[\mathrm{i}\int(\mathscr{L}-J(A-D\delta\chi)+B(aA-\Lambda+\delta\Lambda)-\tfrac12BMB)\right] \tag{16}$$

as the direct analogue of (5). In (16), T is the generator in the fermion representation, and is one of the unpleasant complications over the abelian case. For one thing, the factors $\exp(\mathrm{i}eT\delta\chi)$ and $\exp(-\mathrm{i}e\bar T\delta\chi)$ are not phases, so cannot be combined with the rest of the expression as a change in Λ. For another, they depend in a complicated way on A, so replacement of A by $i\delta/\delta J$ can only lead to relations involving infinite numbers of Green functions. In the absence of fermions the situation is hardly better due to the $\delta\chi$ term in the square brackets. The simplicity, even for infinitesimal transformations, is therefore lost in the non-abelian case, although the gauge–dependence of the scaling properties has been further investigated (Hosoya and Sato 1974, Tarasov and Vladimirov 1977).

3. Gauge dependence of spectral functions

In view of our failure with QCD we restrict ourselves to QED hereafter. Recall that when the photon propagator is varied as

$$D^{M}_{\mu\nu}(k) = D^{0}_{\mu\nu}(k) + k_\mu k_\nu M(k) \tag{10}$$

the renormalised propagators for different M values are connected by

$$S(x; M) = \exp(\mathrm{i}e^2 M(x)) S(x; 0) \tag{11'}$$

On the other hand, we know that the propagator admits the spectral decomposition

$$S(x) = \int \rho(W) S(x \mid W)\, \mathrm{d}W \tag{17}$$

where $S(x \mid W) = (\mathrm{i}\gamma . \partial + W)\Delta(x \mid W^2)$ is the free (mass W) fermion propagator. This yields a complicated relation

$$\int \rho(W; M) S(x \mid W)\, \mathrm{d}W = \exp(\mathrm{i}e^2 M(x)) \int \rho(W; 0) S(x \mid W)\, \mathrm{d}W,$$

between the corresponding spectral functions in the two gauges. It can be simplified somewhat by taking the discontinuity of the Fourier transform:

$$-\pi\rho(p; M) = \mathrm{Im} \int \mathrm{d}W \rho(W; 0) \int \mathrm{d}^4x \exp(\mathrm{i}p . x) \exp(\mathrm{i}e^2 M(x)) S(x \mid W) \tag{18}$$

This is now an integral relation between the different ρ.

To make further progress, the form of M needs to be specified. In the class of covariant gauges

$$D_{\mu\nu}(k) = [-\eta_{\mu\nu} + k_\mu k_\nu (1-a)/k^2]/k^2 \tag{19}$$

one identifies $M(k) = -a/k^4$ at a formal level. Let us use dimensional regularisation to give meaning to $\exp[\mathrm{i}e^2 M(x)]$, instead of introducing Pauli–Villars regulators as did Johnson and Zumino (1959). In $2l$ dimensions, we substitute $e^2(m^2)^{2-l}$ for e^2 to maintain the coupling constant dimensionless, m being the electron mass. Hence, in the four-dimensional limit

$$\begin{aligned} e^2 M(x) &\to -\lim_{l\to 2} e^2(m^2)^{2-l} \int \bar{d}^{2l}k \exp(-\mathrm{i}k . x)\frac{a}{k^4} \\ &= \lim_{l\to 2} \frac{e^2(m^2)^{2-l} a}{4(l-2)} \int \bar{d}^{2l}k \exp(-\mathrm{i}k . x) \frac{\partial}{\partial k^\mu}\frac{\partial}{\partial k_\mu}\frac{1}{k^2} \\ &= \lim_{l\to 2} \frac{e^2(m^2)^{2-l}}{4(2-l)} x^2 D(x) \end{aligned}$$

where $\mathrm{i}D(x)=\Gamma(l-1)(-x^2+\mathrm{i}0)^{1-l}/4\pi^l$ is the causal massless propagator for arbitrary l. Thus, up to an x independent constant factor,

$$e^2M(x)\to \mathrm{i}e^2a\ln(-m^2x^2)/16\pi^2$$

or

$$\exp(\mathrm{i}e^2M(x))=(-m^2x^2)^{-e^2a/16\pi^2}\equiv(-m^2x^2)^{-a\epsilon} \tag{20}$$

The abbreviation $\epsilon=e^2/16\pi^2$ has been used as it recurs throughout.

In order to discover the relation between $\rho(W;a)$ and $\rho(W;0)$, we shall need the transform

$$\int \mathrm{d}^4x\,\exp(\mathrm{i}px)(-m^2x^2)^{-a\epsilon}(\mathrm{i}\gamma.\partial+W)\{-\mathrm{i}WK_1[W(-x^2)^{1/2}]/4\pi^2(-x^2)^{1/2}\}$$
$$=\int_0^\infty \mathrm{d}r\frac{r^2J_1[r(-p^2)^{1/2}]}{(-p^2)^{1/2}}.(m^2r^2)^{-a\epsilon}W^2\left[\frac{K_1(Wr)}{r}+\mathrm{i}\gamma.x\frac{K_2(Wr)}{r^2}\right]$$
$$=\left(\frac{4m^2}{W^2}\right)^{-a\epsilon}\frac{\Gamma(1-a\epsilon)}{W}\left[\Gamma(2-a\epsilon)F\left(1-a\epsilon,2-a\epsilon;2;\frac{p^2}{W^2}\right)\right.$$
$$\left.+\frac{\gamma.p}{2W}\Gamma(3-a\epsilon)F\left(1-a\epsilon,3-a\epsilon;3;\frac{p^2}{W^2}\right)\right]. \tag{21}$$

Since the hypergeometric solution $F(\alpha,\beta;\gamma;z)$, regular at $z=0$, has a branch point at $z=1$ with a discontinuity given by

$$\mathrm{Im}\{\Gamma(\alpha)\Gamma(\beta)F(\alpha,\beta;\gamma;z)\}$$
$$=-\frac{\pi\Gamma(\gamma)\theta(z-1)}{\Gamma(1-\alpha-\beta+\gamma)}(z-1)^{\gamma-\alpha-\beta}F(\gamma-\alpha,\gamma-\beta;\gamma-\alpha-\beta+1;1-z)$$

there follows the fundamental integral relation

$$\rho(p;a)=\left(\int_m^p+\int_{-p}^{-m}\right)\mathrm{d}W\frac{\rho(W;0)}{W}\left(\frac{W}{2m}\right)^{2a\epsilon}\frac{(p^2/W^2-1)^{-1+2a\epsilon}}{\Gamma(2a\epsilon)}$$
$$\times[F(a\epsilon,1+a\epsilon;2a\epsilon;1-p^2/W^2)$$
$$+W^{-1}\gamma.pF(a\epsilon,2+a\epsilon;2a\epsilon;1-p^2/W^2)] \tag{22}$$

between the spinor spectral functions. The connection is visibly non-trivial; nor is it more trivial in the companion scalar case,

$$\rho(p^2,a)=\int_{m^2}^{p^2}\mathrm{d}W^2\frac{\rho(W^2;0)}{W^2}\left(\frac{W^2}{4m^2}\right)^{a\epsilon}\frac{(p^2/W^2-1)^{-1+2a\epsilon}}{\Gamma(2a\epsilon)}F(a\epsilon,1+a\epsilon;2a\epsilon;1-p^2/W^2) \tag{23}$$

after one traces out the parallel argument.

At asymptotic values the relations look more tractable. In the infrared limit, with scalars (near $p^2=m^2$), the connection reads

$$\rho(p^2;a)\simeq\int_{m^2}^{p^2}\frac{\mathrm{d}W^2}{m^2}\rho(W^2;0)\frac{2^{-2a\epsilon}}{\Gamma(2a\epsilon)}\left(\frac{p^2}{W^2}-1\right)^{-1+2a\epsilon}$$

and with spinors (near $\gamma . p = m$) it reads, very similarly,

$$\rho(p; a) \simeq \left(1 + \frac{\gamma . p}{m}\right) \int_m^p \frac{\mathrm{d}W}{m} \rho(W, 0) \frac{2^{-2a\epsilon}}{\Gamma(2a\epsilon)} \left(\frac{p^2}{W^2} - 1\right)^{-1+2a\epsilon}.$$

If we assume that there are threshold singularities

$$m^2\rho(p^2; 0) \sim (p^2/m^2 - 1)^{\zeta_0} \qquad \text{and} \qquad m\rho(p; 0) \sim (p/m - 1)^{\zeta_{1/2}}$$

in the Landau gauge ($a = 0$), we readily arrive at the general infrared behaviours,

$$\begin{aligned} &m^2\rho(p^2; a) \sim (p^2/m^2 - 1)^{\zeta_0 + 2a\epsilon} \\ &m\rho(p; a) \sim (1 + \gamma . p/m)(p^2/m^2 - 1)^{\zeta_{1/2} + 2a\epsilon}. \end{aligned} \tag{24}$$

Likewise, in the ultraviolet domain ($p^2 \gg m^2$) for integrals dominated by the upper end point and the postulated asymptotic behaviour†

$$m^2\rho(p^2; 0) \simeq (p^2/m^2)^{\eta_0} \qquad m\rho(p; 0) \simeq (p^2/m^2)^{\eta_{1/2}}$$

we arrive, via a scaling of (22) and (23), at the general ultraviolet characteristics

$$m^2\rho(p^2; a) \simeq (p^2/m^2)^{\eta_0 + a\epsilon} \qquad m\rho(p; a) \simeq (p^2/m^2)^{\eta_{1/2} + a\epsilon}. \tag{25}$$

The connecting formulae (22) and (23) do not, however, tell us what ρ is in a chosen gauge or what values the coefficients η and ζ take. To find these out one is obliged to turn to perturbation theory and the renormalisation group. In QED we know by infrared freedom that

$$\zeta_0 = \zeta_{1/2} = -1 + 6\epsilon$$

which may be substituted in (26) to give the complete infrared answers for ρ or the Green functions. At the other, high-energy extreme, if we suppose that the propagators‡ are indeed governed by a dimension η as in (25), then up to *first order in* e^2 a simple calculation based upon extraction of leading logarithms gives

$$\eta_0 = \eta_{1/2} = -1 - 3\epsilon$$

From (25) one can read off the ultraviolet behaviour in any covariant gauge a. The virtues of the Landau and Yennie gauges were emphasised long ago by Johnson and Zumino (1959).

4. The gauge technique

If we knew $\rho(W)$ for all W in a given a gauge, we could determine ρ and S in any other a gauge fairly easily. Unfortunately we do not know the full ρ (for, if we did, it would mean that we should have solved QED completely) except in some approximation. One such approximation is provided by the gauge technique which offers, at a first level, the

† Strictly, we are considering the even part of $\rho(p)$, viz. $\rho(p) + \rho(-p)$, as this supplies the dominant behaviour (Atkinson and Slim 1979) of S at asymptotic values.

‡ In so far as the propagators have the same $p \to \infty$ (and $x \to 0$) behaviour as their spectral functions, the small distance limit is more easily discussed directly in configuration space from (11) and (20).

Landau gauge spectral functions†

$$m^2\rho(W^2;0)=\frac{(W^2/m^2-1)^{-1-6\epsilon}}{2^{-6\epsilon}\Gamma(-6\epsilon)}F(-3\epsilon,1-3\epsilon;-6\epsilon;1-W^2/m^2) \quad (26)$$

$$m\rho(W;0)=\frac{\epsilon(W)(W^2/m^2-1)^{-1-6\epsilon}}{2^{-6\epsilon}\Gamma(-6\epsilon)} \times\left[\frac{m}{W}F(-3\epsilon,-3\epsilon;-6\epsilon;1-W^2/m^2) +F(-3\epsilon,1-3\epsilon;-6\epsilon;1-W^2/m^2)\right] \quad (27)$$

These can be substituted in (24) and (25) respectively to give us the ρ for $a\neq 0$. In the scalar case one is confronted by the integral

$$\int_1^x \mathrm{d}y\frac{y^{-1+a\epsilon}}{2^{(2a-6)\epsilon}}\frac{(x/y-1)^{-1+2a\epsilon}}{\Gamma(2a\epsilon)}\frac{(y-1)^{-1-6\epsilon}}{\Gamma(-6\epsilon)}F(a\epsilon,1+a\epsilon;2a\epsilon;1-x/y) \times F(-3\epsilon,1-3\epsilon;-6\epsilon;1-y)$$

wherein $x=p^2/m^2$ and $y\equiv W^2/m^2$. This is evaluated in Appendix B and the result is

$$m^2\rho(W^2;a)=\frac{(W^2/m^2-1)^{-1+(2a-6)\epsilon}}{2^{(2a-6)\epsilon}\Gamma((2a-6)\epsilon)}F((a-3)\epsilon,1+(a-3)\epsilon;(2a-6)\epsilon;1-W^2/m^2)) \quad (28)$$

which is precisely the answer supplied the gauge technique for any a. (In the spinor case the integral cannot be reduced to a simple ${}_2F_1$ function‡, and we shall not quote it.) We have thus verified that the gauge techniques ρ are entirely consistent with the gauge dependence (23) expected on general grounds; this is reassuring if not especially surprising bearing in mind the gauge covariance of the technique.

We can clear much of the mystery and avoid these contortions with hypergeometric functions if we go to x space. As an intermediate step, evaluate

$$\begin{aligned}\Delta(p;0)&=\int_{m^2}^{\infty}\mathrm{d}W^2\,\rho(W^2;0)/(p^2-W^2)\\ &=\int_0^{\infty}\frac{x^{-1-6\epsilon}(x+1-p^2/m^2)^{-1}}{2^{-6\epsilon}\Gamma(-6\epsilon)m^2}F(-3\epsilon,1-3\epsilon;-6\epsilon;-x)\\ &=-m^{-2}2^{6\epsilon}\Gamma(1+3\epsilon)\Gamma(2+3\epsilon)F(1+3\epsilon,2+3\epsilon;2;p^2/m^2).\end{aligned} \quad (29)$$

Thus

$$\begin{aligned}\mathrm{i}\Delta(x;0)&=\int_0^{\infty}\mathrm{d}q\,q^2J_1[q(-x^2)^{1/2}]\Delta((-q^2)^{1/2};0)/4\pi^2(-x^2)^{1/2}\\ &=mK_1[m(-x^2)]^{1/2}(-m^2x^2)^{3\epsilon}/4\pi^2(-x^2)^{1/2}\end{aligned} \quad (30)$$

† The normalisation factors, which are not provided by the technique, have been carefully introduced to give the correct free propagators as $\epsilon\to 0$ and to maintain their simple character in any gauge. Thus $\rho\to\delta(W^2-m^2)$ or $\delta(W-m)$ as the coupling vanishes.

‡ Rather it leads to ${}_4F_3$ functions, again in conformity with the differential equation for $\rho(W;a)$ provided by the technique.

in the Landau gauge, according to the initial gauge approximation. Hence in other gauges,

$$\mathrm{i}\Delta(x;a)=(-m^2x^2)^{(3-a)\epsilon}mK_1[m(-x^2)^{1/2}]/4\pi^2(-x^2)^{1/2}. \tag{31}$$

The Yennie gauge $a=3$ would thus have the free propagator as the starting function.

Carrying out the same steps in the spinor case, we find (see also Khare and Kumar 1978)

$$S(p;0)=-2^{6\epsilon}\Gamma^2(1+3\epsilon)[(\gamma\cdot p)^{-1}(F(1+3\epsilon,1+3\epsilon;1;p^2/m^2)-1)$$
$$+m^{-1}(1+3\epsilon)F(1+3\epsilon,2+3\epsilon;2;p^2/m^2)] \tag{32}$$

and therefore † $(r^2\equiv -x^2)$

$$\mathrm{i}S(x;0)=\frac{(m^2r^2)^{3\epsilon}m^2}{4\pi^2}\left(\frac{K_1(mr)}{r}+\frac{\mathrm{i}\gamma\cdot x}{r^2}\left\{K_2(mr)+\frac{2K_1(mr)}{mr}[\exp(-3\mathrm{i}\pi\epsilon)S_{1+6\epsilon,0}(\mathrm{i}mr)-1]\right.\right.$$
$$\left.\left.+\frac{12\mathrm{i}\epsilon}{m^2r^2}K_0(mr)\exp(-3\mathrm{i}\epsilon)S_{6\epsilon,1}(\mathrm{i}mr)\right\}\right) \tag{33}$$

where $S_{\mu,\nu}$ is the Lommel function. In the limit as $\epsilon\to 0$, because $S_{1,0}(z)=1$, one verifies that (33), like (32), tends to the free-field propagator—a useful boundary condition to test the correctness of the expressions. Multiplication of (33) by $(m^2r^2)^{-a\epsilon}$ then provides the Green function for arbitrary a. We see that, unlike the scalar case, the spinor result is non-trivial no matter what a value is chosen. The reason why the scalar propagator (31) looks so simple for $a=3$ is because the e^2 correction to the scalar spectral function happens to vanish for that gauge; this never happens in any gauge with spinors.

This work sheds considerably light on the gauge technique, by the explicit demonstration that the procedure respects the general relations (10), (11), etc and by the relatively simple nature of the configuration space propagators (31) and (33). Give the spectral functions (26) and (27) one may go on to construct the gauge-covariant Green functions (up to transverse parts) in x space by appropriate weighting:

$$\{S\Gamma_\mu S\}(xy,z)=\int \mathrm{d}W\rho(W)S(x-z\mid W)\gamma_\mu S(z-y\mid W)$$

$$\{S\Gamma_{\mu\nu}S\}(xy,zw)=\int \mathrm{d}W\rho(W)(S(x-z\mid W)\gamma_\nu S(z-w\mid W)\gamma_\mu S(w-y\mid W)$$
$$+S(x-w\mid W)\gamma_\mu S(w-z\mid W)\gamma_\nu S(z-y\mid W)).$$

† Some work is needed to pass from (32) to (33). The intermediate steps involve the use of

$$\int_0^\infty rJ_1(qr)F\left(b,b;1;\frac{-q^2}{m^2}\right)=-\Gamma^{-2}(b)G_{13}^{30}\left(\frac{1}{4}r^2m^2\middle|\begin{matrix}1\\0\;b\;b\end{matrix}\right);$$

the identifications, via the Barnes–Millin representation of G,

$$G_{13}^{30}\left(z\middle|\begin{matrix}0\\0\;b\;b\end{matrix}\right)=G_{02}^{20}(z\mid bb)=2z^bK_0(2z^{1/2})$$

$$G_{13}^{\prime 30}\left(z\middle|\begin{matrix}1\\0\;b\;b\end{matrix}\right)=-z^{-1}G_{13}^{30}\left(z\middle|\begin{matrix}0\\0\;b\;b\end{matrix}\right)$$

and the integral

$$\int z^\mu K_\nu(z)\,\mathrm{d}z=(-\mathrm{i})^\mu z[(\mu-\nu-1)K_\nu(z)S_{\mu-1,\nu+1}(\mathrm{i}z)-\mathrm{i}K_{\nu+1}(z)S_{\mu,\nu}(\mathrm{i}z)]$$

Naturally, these higher Green functions are no longer simple transcendental functions like (31) and (33), except in the limit of vanishing photon momentum, i.e. upon integration over photon location.

Appendix A

We wish to give what we believe is a fairly transparent proof of the invariance of $\int[dA]\Delta(A)$ under transformations of Slavnov type, such as (15). The presence of space–time coordinates and indices is inessential; consider just a 'manifold' with coordinates A_i (so in practice $i = \mu, \alpha, x$) and a 'Lie group' of transformations g with coordinates g_a (in practice $a = \alpha, x$) preserving the measure $[dA]$. For infinitesimal transformations we write

$$\delta A_i = D_{ia}(A) \,.\, \delta\chi_a.$$

We are interested in $\int \mu^T[dA]f(A)$ for surfaces T transverse to the group action, where μ^T is the measure on T obtained from the measure $[dA]$ and a left invariant measure on the group. For invariant f we know that the integral is independent of T. In particular, if T is given by the gauge condition $L_a(A) = \Lambda_a$, then the integral can be written as

$$\int \mu^T[dA]f(A) = \int [dA]\Delta^L(A)\delta(L(A)-\Lambda)f(A)$$

where $\Delta^L = \det F^L$ and F^L_{ab} is defined via the infinitesimal transformations of L, viz

$$\delta L = F^L \,.\, \delta\chi \qquad \text{or} \qquad F_{ab} = (\partial L_a/\partial A_i)D_{ib}.$$

Now consider A-dependent transformations of the type

$$A' = g_A(A) \qquad \text{or} \qquad A = g'_{A'}(A').$$

Given a gauge function L define L' by $L'(A') = L(A)$, and similarly $f'(A') \equiv f(A)$. Since we know that

$$\int_T \mu^T[dA]f(A) = \int_{T'} \mu^{T'}[dA']f'(A')$$

where T and T' are given by L, $L' = \Lambda$, we may write

$$\int [dA]\Delta^L(A)\delta(L(A)-\Lambda)f(A) = \int [dA']\Delta^{L'}(A')\delta(L'(A')-\Lambda)f'(A').$$

By integrating over Λ we can express this quite generally as

$$\int [dA]\Delta^L(A)f(A) = \int [dA']\Delta^{L'}(A')f(g'_{A'}(A'))$$

and changing variable on the right-hand side to A, we find

$$\Delta^L(A) = |\partial A'/\partial A|\Delta^{L'}(A').$$

This establishes the invariance of $\int[dA]\Delta^L(A)$, provided that $\Delta^{L'} = \Delta^L$, or $F^{L'} = F^L$, which in its turn is satisfied if *L' and L differ by a gauge-invariant function.*

Consider now an infinitesimal A-dependent gauge transformation $A' = A + D(A) \,.\, \delta\chi(A)$. From the definition of F^L and L, we see that $L'(A')+$

$F^L(A)\,.\,\delta\chi(A)=L(A)$. Hence if we choose

$$\delta\chi(A)=[F^L(A)]^{-1}\,.\,\delta\Lambda$$

where $\delta\Lambda$ is gauge-invariant, we guarantee the invariance of the measure. In particular for the gauge fixing term $a\,.\,A=\Lambda$, $F=(a\,.\,D)$ and we therefore require the A dependence of the transformations to be of Slavnov type,

$$\delta A_\mu=D_\mu(a\,.\,D)^{-1}\,.\,\delta\Lambda$$

Appendix B

We wish to prove that when $a'+b'-c'=b-a$, the integral

$$\Gamma(c)\Gamma(c')I\equiv\int_0^1 \mathrm{d}t\,(1-tz)^{-a}(1-t)^{c-1}t^{c'-1}F\left(a,b;c;\frac{z(1-t)}{1-tz}\right)F(a',b';c';tz)$$

equals

$$I=F(a+a',b+b';c+c';z)/\Gamma(c+c').$$

Integrals comparable with this can be found in standard texts dealing with hypergeometric functions but none exactly having the above form. We shall therefore return to first principles for the proof.

Substituting the series expansion of F,

$$I=\sum_{n,n'=0}^{\infty}\int_0^1 \mathrm{d}t\,(1-tz)^{-a-n}(1-t)^{c+n-1}t^{c'+n'-1}z^{n+n'}$$

$$\times\left(\frac{\Gamma(a+n)\Gamma(b+n)}{\Gamma(a)\Gamma(b)\Gamma(c+n)\Gamma(1+n)}\right)\left(\frac{\Gamma(a'+n')\Gamma(b'+n')}{\Gamma(a')\Gamma(b')\Gamma(c'+n')\Gamma(1+n')}\right)$$

$$=\sum_{n,n'}F(a+n,c'+n';c+c'+n+n';z)\left(\frac{z^{n+n'}}{\Gamma(c+c'+n+n')}\right)$$

$$\times\left(\frac{\Gamma(a+n)\Gamma(b+n)}{\Gamma(a)\Gamma(b)\Gamma(1+n)}\right)\left(\frac{\Gamma(a'+n')\Gamma(b'+n')}{\Gamma(a')\Gamma(b')\Gamma(1+n')}\right)$$

$$=\sum_r\sum_{n,n'}\left(\frac{z^r}{\Gamma(a)\Gamma(b)\Gamma(a')\Gamma(b')}\right)\left(\frac{\Gamma(a+r-n')\Gamma(c'+r-n)}{\Gamma(c+c'+r)\Gamma(1+r-n-n')}\right)$$

$$\times\left(\frac{\Gamma(b+n)}{\Gamma(1+n)}\right)\left(\frac{\Gamma(a'+n')\Gamma(b'+n')}{\Gamma(c'+n')\Gamma(1+n')}\right).$$

Recalling the combinatorial formula

$$\sum_n\frac{\Gamma(A+n)\Gamma(B-n)}{\Gamma(C+n)\Gamma(D-n)}=\frac{\Gamma(A+B)\Gamma(A-C+1)\Gamma(B-D+1)}{\Gamma(C+D-1)\Gamma(A+B-C-D+2)}$$

we can reduce the integral to the double sum

$$I=\sum_r\frac{z^r\Gamma(b+c'+r)}{\Gamma(a)\Gamma(a')\Gamma(b')\Gamma(c+c'+r)}\sum_{n'}\frac{\Gamma(a'+n')\Gamma(b'+n')\Gamma(a+r+n')}{\Gamma(b+c'+n')\Gamma(1+n')\Gamma(1+r-n')}.$$

Further simplification is possible because the n' summation is Saalschutzian and one can make use of the identity

$$\sum_n \frac{\Gamma(A+n)\Gamma(B+n)\Gamma(C-A+r-n-B)}{\Gamma(C+n)\Gamma(1+n)\Gamma(1+r-n)}$$

$$=\frac{\Gamma(A)\Gamma(B)\Gamma(C-A-B)\Gamma(C-A+r)\Gamma(C-B+r)}{\Gamma(C-A)\Gamma(C-B)\Gamma(C+r)\Gamma(1+r)}.$$

Recalling the condition $a'+b'-c'+a=b$, we obtain

$$I=\sum_r \frac{z^r\Gamma(a+b'+r)\Gamma(a+a'+r)}{\Gamma(a+b')\Gamma(a+a')\Gamma(c+c'+r)\Gamma(1+r)}=\frac{F(a+a',b+b';c+c';z)}{\Gamma(c+c')}$$

which was to be proved. As far as we know, this is a new integral identity.

Such an integral arises in the text in the formula succeeding equation (27). The substitution $y=1+t(x-1)$ there and the identifications

$$a\to a\epsilon, \qquad b\to 1+a\epsilon, \qquad c\to 2a\epsilon,$$

$$a'\to -3\epsilon, \qquad b'\to 1-3\epsilon, \qquad c'\to -6\epsilon$$

bring it into the desired form and lead to the result (28).

References

Atkinson D and Slim H A 1979 *Nuovo Cim.* **50** 555
Bialynicki-Birula I 1960 *Nuovo Cim.* **17** 951
Delbourgo R 1977 *J. Phys. A: Math. Gen.* **10** 1369
—— 1979 *Nuovo Cim.* A **49** 484
Delbourgo R and West P 1977a *J. Phys. A: Math. Gen.* **10** 1049
—— 1977b *Phys. Lett.* **72**B 96
Fradkin E 1956 *Sov. Phys.-JETP* **2** 361
Hosoya A and Sato A 1974 *Phys. Lett.* **48**B 36
Johnson K and Zumino B 1959 *Phys. Rev. Lett.* **3** 351
Khare A and Kumar S 1978 *Phys. Lett.* **78**B 94
Landau L and Khalatnikov I 1956 *Sov. Phys.-JETP* **2** 69
Lee B W and Zinn-Justin J 1973 *Phys. Rev.* **7**D 1049
Okubo S 1960 *Nuovo Cim.* **15** 949
Salam A 1963 *Phys. Rev.* **130** 1287
Slavnov A A 1972 *Zh. Teor. Math. Fiz.* **10** 153
Tarasov O V and Vladimirov A A 1977 *Nucl. Phys.* (*USSR*) **25** 1104
Zumino B 1960 *J. Math. Phys.* **1** 1

J. Phys. A: Math. Gen. **17** (1984) 3593–3607. Printed in Great Britain

Transverse vertices in electrodynamics and the gauge technique

R Delbourgo and R Zhang

Department of Physics, University of Tasmania, Hobart, Australia 7001

Received 18 April 1984

Abstract. The inclusion of transverse vertex corrections in the gauge technique, needed for the restoration of gauge covariance, results in a self-consistent equation for the source propagator spectral function which agrees with perturbation theory when expanded to order e^4. We have managed to solve the equation in the infrared and ultraviolet limits, for scalar and spinor sources, in an arbitrary covariant gauge. The gauge covariance at asymptopia is thus established.

1. Introduction

As a non-perturbative approach to gauge theories the gauge technique (Delbourgo and Salam 1964, Strathdee 1964) transcends the original Baker–Johnson–Willey (1964) scheme. The notable successes of the technique (Delbourgo 1979) have largely stemmed from first gauge approximation (Delbourgo and West 1977a, b) whereby transverse corrections to the Green functions are neglected. While this may be fine at low- and high-energy limits, it is not at intermediate energies (Slim 1981, Delbourgo *et al* 1981) and it becomes desirable to improve the results by going to the next gauge approximation taking account of the transverse amplitudes in the vertex (and thus incorporating the charge and magnetic form factors etc). Apart from restoring gauge covariance at all momenta, the need for transverse vertices is most pressing in two- and three-dimensional theories (Gardner 1981, Roo and Stam 1984) where the vector particle acquires a mass (Deser *et al* 1982). In fact, in the two-dimensional case the exact form of the transverse vertex can be deduced from the axial gauge identities (Delbourgo and Thompson 1982).

For four-dimensional electrodynamics there have been two notable attempts at incorporating transverse vertices into the technique. King (1983), by looking at leading logarithmic terms in perturbation theory, introduced an approximate transverse vertex in QED which was asymptotically gauge covariant and which exactly renormalised the Dyson–Schwinger equation. However, the attempt was deficient in respect of the low-energy properties: the transverse vertex did not agree even with e^2 perturbation theory and had incorrect gauge dependence. Parker (1984) largely rectified these faults in the way he introduced transverse corrections, though only for scalar electrodynamics and only in the Fermi gauge. The purpose of this paper is to extend Parker's work to the spinor case and to allow for all possible gauge parameters. In the end we shall obtain a self-consistent equation for the propagator incorporating the transverse vertex which, when expanded to order e^4 agrees *exactly* with perturbation theory.

The outline of the gauge technique in the next gauge approximation appears in § 2 and is applied in § 3, where we derive the self-consistent equation of the spinor spectral

0305-4470/84/183593 + 15$02.25 © 1984 The Institute of Physics

function. This equation is very complicated but becomes amenable in the infrared and ultraviolet limits, yielding the leading behaviours:

$$S(p) \sim (\gamma p - m)^{-1+(a-3)e^2/8\pi^2}$$

$$S(p) \sim \left(\frac{\gamma p}{(p^2)^{1-ae^2/16\pi^2}} + \frac{m}{(p^2)^{1+3e^2/16\pi^2}}\right)\left(1 - \frac{3e^2}{16\pi^2}\ln\frac{p^2}{m^2}\right).$$

Finally, we extend Parker's scalar electrodynamics work to arbitrary gauges in § 4. The purpose of appendix 1 is to substantiate the vanishing of the transverse Green function in the soft-photon limit (i.e. proving the absence of a soft-photon singularity even when electrons are off-shell); while appendix 2 contains a few details about the determination of the discontinuity in the fermion self-energy to order e^4, the kernel of our equation for the spectral function.

2. The technique to order e^4

There are three major steps in applying the technique to gauge theories:

(a) Setting up the Dyson–Schwinger (DS) equations for the Green function of interest for the gauge model in question.

(b) Solving the gauge identities (up to transverse corrections be it understood) that involve the Green function; in this connection spectral representations can often facilitate solution.

(c) Truncating the DS equation to a particular order in e so as to obtain a self-consistent equation for the amplitude in terms of which the (longitudinal) higher amplitudes are expressed.

Until recently, researchers had contented themselves with studying the two-point function and, in essence, were applying the gauge technique in its most primitive form. Thus they were led to a self-consistent, but non-perturbative, solution of the propagator which was only exact to order e^2 and which sacrificed gauge covariance (Slim 1981, Delbourgo *et al* 1981) except at the asymptopia.

In this paper we wish to include some degree of transversality into the amplitudes by extending Parker's (1984) recent work on scalar electrodynamics to the more realistic spinor case, QED. We shall ascend to the next level of the technique by studying the three-point vertex (the two-point function follows from it) and constructing an ansatz for it which is exact to order e^3 and which provides a non-perturbative solution for the fermion propagator that is exact† to order e^4. These improvements mean that we are incorporating magnetic effects in the technique for the first time and, by the very inclusion of the transverse vertices, can look forward to an amelioration of the gauge covariance properties at intermediate momentum values.

This is how the three basic tools are wielded in practice:

(a) Including sources, the QED action is

$$S = \int d^4x[-\tfrac{1}{4}F_{\mu\nu}F^{\mu\nu} + \bar{\psi}(\gamma(i\delta - eA) - m)\psi - (\delta A)^2/2a - \bar{j}\psi - \bar{\psi}j - J^\mu A_\mu] \qquad (1)$$

in a class of covarient gauges parametrised by (a). (All fields and constants in (1) are

† Apart from internal vacuum polarisation corrections which do not affect the issue of gauge covariance and can be separately accounted for.

as yet unrenormalised.) The generating functional of connected Green functions

$$W(j,\bar{j},J) = -\mathrm{i}\ln\int(\mathrm{d}A\,\mathrm{d}\psi\,\mathrm{d}\bar{\psi})\exp(\mathrm{i}S)$$

satisfies the basic fermionic (DS) equation

$$\frac{\delta W}{\delta j(p)}(\gamma p - m_0) = e\int \bar{\mathrm{d}}^4 k\left(\mathrm{i}\frac{\delta^2 W}{\delta J^\mu(k)\delta j(p-k)} - \frac{\delta W}{\delta J^\mu(k)}\frac{\delta W}{\delta j(p-k)}\right)\gamma_\mu - \bar{j}(p) \tag{2}$$

when expressed in momentum space. Additional j, J functional derivatives provide the DS equations of all fermionic amplitudes.

It is convenient to define photon amputated amplitudes $G(\equiv S\Gamma S)$ by

$$\left.\frac{\delta^{n+2}W}{\delta\bar{j}(p')\delta j(p)\delta J_{\mu_1}(-k_1)\ldots\delta J_{\mu_n}(-k_n)}\right|_{j=J=0}$$
$$= \bar{\delta}^4(p'-p-k_1-\ldots k_n)$$
$$\times(-e)^n D^{\mu_1\nu_1}(k_1)\ldots D^{\mu_n\nu_n}(k_n)G_{\nu_1\ldots\nu_n}(p',p;k_1\ldots k_n) \tag{3}$$

where p' is outgoing, p and k_i are incoming momenta. Operating on (2) with $\delta^{n+1}/\delta\bar{j}(\delta J)^n$ with increasing n and, bearing in mind that

$$\delta^{n+1}W/(\delta J)^n\delta j = \delta^{n+1}W/(\delta J)^n\delta j = \delta W/\delta J = 0$$

for vanishing sources, one arrives at the successive DS equations:

$$S(p)(\gamma p - m_0) = Z^{-1} + \mathrm{i}e^2\int \bar{\mathrm{d}}^4 k\, D^{\kappa\lambda}(k)G_\kappa(p,p-k;k)\gamma_\lambda \tag{4}$$

$$G_\mu(p',p;p'-p)(\gamma p - m_0)$$
$$= S(p')\gamma_\mu - \mathrm{i}e^2\int \bar{\mathrm{d}}^4 k\, D^{\kappa\lambda}(k)G_{\mu\kappa}(p',p-k;p'-p,k)\gamma_\lambda \tag{5}$$

$$G_{\mu\nu}(p',p;q,p'-p-q)(\gamma p - m_0)$$
$$= -G_\mu(p',p'-q;q)\gamma_\nu - G_\nu(p',p+q;p'-p-q)\gamma_\mu$$
$$-\mathrm{i}e^2\int \bar{\mathrm{d}}^4 k\, D^{\kappa\lambda}(k)G_{\mu\nu\kappa}(p',p-k;q,p'-p-q,k)\gamma_\lambda \tag{6}$$

which are now renormalised. One may, of course, derive similar equations from the adjoint of (2) in which the differential operator $(\gamma p' - m_0)$ acts on the left. For instance:

$$(\gamma p' - m_0)G_\mu(p',p;p'-p)$$
$$= \gamma_\mu S(p) - \mathrm{i}e^2\int \bar{\mathrm{d}}^4 k\, D^{\lambda\kappa}(k)\gamma_\lambda G_{\mu\kappa}(p'+k,p;p'-p,k). \tag{5a}$$

Before moving to the next phase, it is worth observing that the gauge identities (arising by contractions with k) relate one equation to the next; in effect the lower point DS equations are subsumed in the higher point equations. For example, contracting (5) with k^μ yields (4).

(b) The next step is to exploit the gauge identities so as to gain some information about the longitudinal parts of Green functions on the RHS of (4), (5) and (6). The first few identities read:

$$(p'-p)^\lambda G_\lambda(p',p;p'-p)=S(p)-S(p') \tag{7}$$

$$k^\lambda G_{\lambda\mu}(p',p;k,k')=G_\nu(p',p+k;k')-G_\nu(p'-k,p;k') \tag{8}$$

$$k^\lambda G_{\lambda\mu\nu}(p',p;k,k',k'')=G_{\mu\nu}(p',p+k;k',k'')-G_{\mu\nu}(p'-k,p;k',k'') \tag{9}$$

and in the limit as the photon momentum vanishes they lead to the differential forms of the identities:

$$G_\lambda(p,p;0)=-\partial S(p)/\partial p_\lambda \qquad \text{etc.}$$

From these relations one may abstract the longitudinal parts of the n-point G (those that survive contraction with k) in terms of the full $(n-1)$-point G. The transverse parts G^{T} remain undetermined of course, but if we insist that the longitudinal G^{L} are non-singular and obey the differential identities then we can safely disregard the G^{T} in the infrared domain†. Actually one can reconstruct a good measure of the amplitudes in terms of the propagator S by using the exact spectral representation

$$S(p)=\int \rho(W)\,\mathrm{d}W/(\gamma p-W) \tag{10}$$

and making the ansätze

$$G^{\mathrm{L}}_{\mu_1\ldots\mu_n}(p',p;k_1\ldots k_n)=\int \rho(W)\,\mathrm{d}W\,G^0_{\mu_1\ldots\mu_n}(p',p;k_1\ldots k_n|W) \tag{11}$$

where $G^0(|W)$ is the Born amplitude for a fermion of mass W. Not only are all gauge identities automatically respected and the infrared properties exact, but the lowest-order perturbation results are naturally incorporated. Put another way, the longitudinal amplitudes (11), including coupling factors, are exact to order e^n and do not affect the analytic properties anticipated from Feynman diagrammatics. We could sharpen the G ansätze if general spectral representations were available for all momenta off-shell, allowing solution of the identities at higher levels. Unfortunately, even for the three-point function, a totally general and usable representation does not exist; however, we are fortunately able to overcome this problem below.

(c) A *closed* form equation for the amplitudes is attained if one *truncates* the DS equations at some level. Heretofore almost all research has been directed at the primitive equation (4) in the first gauge approximation $G_\mu \to G^{\mathrm{L}}_\mu[S]$, $D_{\mu\nu}(k)\to D^{\mathrm{bare}}_{\mu\nu}(k)$. Although the resulting non-perturbative solutions are endowed with many attractive features (correct analyticity, asymptotics, exact infrared behaviour), they are also deficient in many respects, primarily through the neglect of G^{T} in (4). This is what we shall try to remedy here.

The idea is to move up to the next level equations (5) and (8) where the full three-point vertex is determined by the four-point amplitude. In principle (8) gives the longitudinal $G_{\mu\nu}$ in terms of the complete G_μ which can then be substituted into (5)—the second gauge approximation—so as to yield a self-consistent vertex equation‡

† And quite often in the ultraviolet too as we shall see; at least the G^{T} do not spoil the asymptotic gauge covariance of the amplitudes.

‡ These steps can probably be carried out, complicated though they certainly are. The subsequent task of solving for G_μ is even more difficult, if not intractable. Since we already know that $\mathrm{O}(e^6)$ corrections to S must come in through $G^{\mathrm{T}}_{\mu\nu}$, which modify the resulting G_μ, it makes as much sense to follow the simpler approach advocated below with no sacrifice in the degree of accuracy.

from which S can be extracted. Since transverse corrections of order e^4 are neglected by dropping $G^{\mathrm{T}}_{\mu\nu}$ it follows that the resulting Γ can only be precisely correct to order e^4. In practice then, to the same order of accuracy, it is sufficient to adopt the ansatz (11)

$$G^{\mathrm{L}}_{\mu\nu} = \int \mathrm{d}W \rho(W)(\gamma p' - W)^{-1}\{\gamma_\nu[\gamma(p+k) - W]^{-1}\gamma_\mu + \gamma_\mu[\gamma(p+k') - W]^{-1}\gamma_\nu\}(\gamma p - W)^{-1} \tag{12}$$

before substituting in (8). The resulting non-perturbative ρ (or propagator S) will then be correct to fourth order; and gauge covariance will be restored to the same order with (one may hope) a corresponding improvement in the momentum at non-asymptotic values.

This is the general strategy, which is executed in the following section.

3. The self-consistent equation for ρ

To arrive at the improved spectral equation we must substitute for $G_{\mu\nu}$ in (4) using (5) and (8). This is practicable (and correct to order e^4) when we use the longitudinal ansatz $G^{\mathrm{L}}_{\mu\nu}$, equation (12), on the right of (5). Take the difference of (5) and its adjoint

$$\gamma p' G_\mu - G_\mu \gamma p = \gamma_\mu S(p) - S(p')\gamma_\mu - \mathrm{i}e^2 \int \bar{\mathrm{d}}^4 k\, D^{\kappa\lambda}(k)[\gamma_\lambda G^{\mathrm{L}}_{\mu\kappa}(p'+k, p; p'-p, k) - G^{\mathrm{L}}_{\mu\kappa}(p', p-k; p'-p, k)\gamma_\lambda] + \mathrm{O}(e^4) \text{ terms from } G^{\mathrm{T}}.$$

Next, multiply the LHS by $\gamma p'$, the RHS by γp, take the difference and use the spectral representations for S and G^{L}. This yields the *exact* expression

$$G_\mu(p', p) = \int \mathrm{d}W \frac{\rho(W)}{\gamma p' - W}\left(\gamma_\mu + \Lambda^2_\mu(p', p|W) + \frac{(\gamma p'\gamma_\mu + \gamma_\mu \gamma p)}{p'^2 - p^2}\Sigma^2(p|W) - \Sigma^2(p'|W)\frac{(\gamma p'\gamma_\mu + \gamma_\mu\gamma p)}{p'^2 - p^2}\right)\frac{1}{\gamma p - W} + \mathrm{O}(e^4) \tag{13}$$

where

$$\Lambda^2_\mu(p', p|W) = \mathrm{i}e^2 \int \bar{\mathrm{d}}^4 k\, D^{\kappa\lambda}(k)\gamma_\kappa \frac{1}{\gamma(p'-k) - W}\gamma_\mu \frac{1}{\gamma(p-k) - W}\gamma_\lambda \tag{14}$$

$$\Sigma^2(p|W) = -\mathrm{i}e^2 \int \gamma_\kappa \frac{1}{\gamma(p-k) - W}\gamma_\lambda D^{\kappa\lambda}(k)\, \bar{\mathrm{d}}^4 k \tag{14a}$$

are the order e^2 perturbative results for the vertex and self-energy of a mass W electron. The second gauge approximation consists in dropping the finite $\mathrm{O}(e^4)$ correction on the RHS of (13). It is important to realise that, even so, (13) does include transversal corrections to G_μ; indeed if we identify

$$G^{\mathrm{L}}_\mu(p', p) = \int \mathrm{d}W \rho(W)\frac{1}{\gamma p' - W}\gamma_\mu \frac{1}{\gamma p - W} \tag{15}$$

as the longitudinal part, obeying identity (7), it is possible to prove that the remainder G^{T} in the second gauge approximation (13) is non-singular† as the photon momentum vanishes. More particularly:

$$\lim_{p\to p'} G^{\mathrm{T}}_\mu(p',p)=0.$$

(Appendix 1 gives an explicit demonstration.) Furthermore, the approximated vertex incorporates full second-order perturbation theory via the direct substitution $\rho(W)=\delta(W-m)$.

One may tidy up the appearance of (13) by using the dispersion representation

$$\Sigma(p|W)=-\frac{1}{\pi}\int\frac{\operatorname{Im}\Sigma(W'|W)\,\mathrm{d}W'}{\gamma p-W'} \tag{16}$$

whereupon

$$G_\mu(p',p)=\int \mathrm{d}W\frac{\rho(W)}{\gamma p'-W}\left(\gamma_\mu+\Lambda^2_\mu(p',p|W)+\int\frac{\mathrm{d}W'}{\pi}\frac{\operatorname{Im}\Sigma^2(W'|W)}{\gamma p'-W'}\gamma_\mu\frac{1}{\gamma p-W'}\right)$$
$$\times\frac{1}{\gamma p-W}+\mathrm{O}(e^4). \tag{13a}$$

The spectral function equation comes by inserting (13) into (4)

$$\int \mathrm{d}W\frac{\rho(W)(W-m_0)}{\gamma p-W}$$
$$=\mathrm{i}e^2\int \bar{\mathrm{d}}^4k\,D^{\kappa\lambda}(k)\int\frac{\mathrm{d}W\rho(W)}{\gamma p-W}$$
$$\times\left(\gamma_\kappa+\Lambda^2_\kappa(p,p-k|W)+\int\frac{\mathrm{d}W'}{\pi}\frac{\operatorname{Im}\Sigma^2(W'|W)}{\gamma p-W'}\gamma_\kappa\frac{1}{\gamma(p-k)-W'}\right)$$
$$\times\frac{1}{\gamma(p-k)-W}\gamma_\lambda$$
$$+G^{\mathrm{T}}_{\mu\nu}\text{ contributions of }\mathrm{O}(e^6) \tag{17}$$

and may be linearised by leaving the photon undressed.

$$D^{\kappa\lambda}(k)\to(-\eta^{\kappa\lambda}+(1-a)k^\kappa k^\lambda/k^2)/k^2,$$

which implies that the (logarithmic, ultraviolet) effects of vacuum polarisation on electron propagation are being neglected. That being so, ρ satisfies the standard equation (Delbourgo and West 1977a):

$$\int \mathrm{d}W\rho(W)[W-m_0+\Sigma(p|W)]/(\gamma p-W)=0$$
$$+\text{terms of order }e^6 \tag{17a}$$

where

$$\Sigma(p|W)=\Sigma^2(p|W)+\Sigma^4(p|W) \tag{18}$$

† In this respect we disagree with King (1983) who includes just the last two terms of (13) as part of his longitudinal vertex and then must *impose* $G^{\mathrm{T}}(p,p;0)=0$ through some 'regularisation'. In our case the vanishing of the transverse vertex is automatic.

and

$$\Sigma^4(p|W) = -\mathrm{i}e^2 \int \bar{\mathrm{d}}^4k\, D^{\kappa\lambda}(k)\Bigg(\Lambda^2_\kappa(p, p-k|W) + \int \frac{\mathrm{d}W'}{\gamma p - W'}\gamma_\kappa \times \frac{\mathrm{Im}\,\Sigma^2(W'|W)/\pi}{[\gamma(p-k)-W'][\gamma(p-k)-W]}\Bigg)\gamma_\lambda \tag{18a}$$

is associated with a purely transverse vertex. Taking the discontinuity of (17*a*) it follows that

$$\varepsilon(W)\rho(W)[W - m_0 + \Sigma(W|W)] = \int \mathrm{d}W' \frac{\rho(W')}{\pi}\frac{\mathrm{Im}\,\Sigma(W|W')}{W - W'} + \mathrm{O}(e^6). \tag{19}$$

The evaluation of Σ^2 and Σ^4 is so involved that we have relegated it to appendix 2 and have only outlined the essential details at that. Here we shall require formulae (A2.1), (A2.2), (A2.7), (A2.8)–(A2.11). Since $m_0 = \mathrm{Re}\ \Sigma(m, m)$, this still leaves a wavefunction renormalisation infinity on the LHS of (19) of order e^4 which we anticipate should cancel against an infinity of the same order on the right since ρ is renormalised. The cancellation happens in any perturbative expansion for the ρ and it happens for us as well (because we incorporate perturbation theory exactly up to order e^4, barring internal vacuum polarisation). Thus†

$$\mathrm{Im}\,\Sigma^4(W|m) \supset [\mathrm{Im}\,\Sigma^2(W|m)/(W-m)][\Sigma^2(W|W) - \Sigma^2(m|m)] \tag{20}$$

ensures that the divergent $\log \Lambda^2$ match up on each side of (19). Consequently, in the second gauge approximation, we finally have the finite renormalised equation:

$$\pi\varepsilon(W)\rho(W)(W-m)\left[1 + \frac{e^2}{16\pi^2}\left(1 + \frac{3W}{W-m}\ln\left(\frac{W^2}{m^2}\right)\right)\right] = \int \mathrm{d}W'\,\rho(W')\frac{\mathrm{Im}\,\Sigma_\mathrm{f}(W|W')}{W-W'} \tag{21}$$

where the logarithmic divergences are absent in Σ_f. Specifically, using the abbreviations $x = W'/W$, $\eta = e^2/16\pi^2$, the kernel on the RHS of (21) reads

$$\begin{aligned}\mathrm{Im}\,\Sigma_\mathrm{f}(W|W')/\pi w\eta \\ &= (1-x^2)\theta(1-x^2)[a(1+x^2) - x(a+3)] \\ &\quad\times\Bigg\{1 - 4\eta\ln x^2 + \eta(1+x) \\ &\quad\times\left[1 + \frac{2(1-4x)}{1-x^2}\ln x^2 + 2(1-4x+x^2)\ln\left(\frac{1-x^2}{x^2}\right)\right]\Bigg\} \\ &\quad + 4\eta\theta(1-x^2)[x(1+x^2+2x^3)Z_1 - (1-x-x^2)Z_2] \\ &\quad + \eta\theta(1-9x^2)[xY_1 + Y_2] - \eta(1-x^2)\theta(1-x^2)\Big\{-\tfrac{7}{2} + 24x - \tfrac{1}{2}x^2 - 2x^3\end{aligned}$$

† $\Sigma^2(W|W) = -(3e^2W/16\pi^2)[\ln(\Lambda^2/W^2) + 1]$ independently of the gauge parameter a, and our calculations indeed confirm that

$$\mathrm{Im}\,\Sigma^4(p|W) \supset -(3e^4/256\pi^2)(p+W)\{ap[1 + (W^2/p^2)] - (a+3)W\}\{p[\ln(\Lambda^2/p^2) + 1] - W[\ln(\Lambda^2/W^2) + 1]\}$$

$$+a(-3+x+2x^2+x^3)+[15x^3+a(1-4x)(1+x^3)]\ln x^2/(1-x^2)$$
$$+2[(1-a)(-1+4x-x^2-x^3+4x^4-x^5)+(-6x+8x^2)$$
$$\times(1-x^2)]\ln\left(\frac{1-x^2}{x^2}\right)+\frac{4x(1+x)}{1-x}\left[f\left(\frac{1}{1-x^2}\right)-\ln x^2\ln\left(\frac{1-x^2}{x^2}\right)\right]\Bigg\}. \quad (22)$$

In (22) we recall that f is the dilogarithmic function (see (A2.10)) and that (Z_1, Z_2) and (Y_1, Y_2) correspond to electron–two photon and two electron–positron cut contributions, respectively, expressed in dimensionless variables, integral representations for which are stated in appendix 2.

Equation (21) is obviously extremely difficult to solve. However, some real simplifications occur in the infrared limit as we might expect. Let $\omega \equiv W/m$, then as $\omega \to 1$, the LHS of the equation tends to

$$\pi\varepsilon(\omega)\rho(\omega)(\omega-1)[1+7\eta].$$

In that limit $f \to x^2-1$, $Z_1 \to 1-x$, $Z_2 \to 2(1-x)^2$, Y_1, $Y_2 \to 0$, so the kernel (22) 'miraculously' tends to

$$2\eta(1-x)(a-3)(1+7\eta).$$

As a consequence, a remarkable cancellation of the $(1+7e^2/16\pi^2)$ factor on each side leaves us with

$$(\omega-1)\rho(\omega) \approx 2\eta(a-3)\int_1^\omega d\omega'\,\rho(\omega'),$$

and the standard infrared result:

$$\rho(W) \approx (W-m)^{-1+(a-3)e^2/8\pi^2}. \quad (23)$$

The fact that this behaviour emerges properly is a useful check that our kernel (22) is right.

We may carry out a similar analysis in the ultraviolet domain $\omega \gg 1$. Here we have to be more careful with the asymptotic estimation as typical integrals

$$\int \frac{\ln(W'/W)}{W'-W}\,dW'$$

must be treated with care. Because η is so small one can extract the dominant behaviour through the ansatz (Parker 1984)

$$\rho(W) = \varepsilon(W)[W(W^2)^{k_1}(1+b\ln W^2)+m(W^2)^{k_2}(1+c\ln W^2)] \quad (24)$$

and find self-consistency in QED with

$$k_1 = -1+a\eta, \qquad k_2 = -1-3\eta, \qquad b = c = -3\eta$$

leading to the expression for $S(p)$ quoted in the introduction. It would be misleading not to mention that the logarithmic term is undoubtedly influenced by internal vacuum polarisation corrections so the coefficient b will correspondingly alter. Moreover, higher orders of gauge approximation will bring in further powers of $(\eta \ln W^2)$; we are ignorant about the coefficients of these terms and whether the leading logs can be summed.

To find the complete analytical solution of (21) is probably impossible but it should be feasible to obtain a numerical solution for $\rho(W)$ and hence for $S(p)$ in any gauge. This task remains to be done and will be reported separately. Knowledge of ρ is vital for seeing how the gauge independence of vacuum polarisation is substantiated at least to order e^4, and thus how reliable are the results on dynamically generated vector boson masses (Delbourgo *et al* 1982). Also it will provide a non-perturbative approximation for the off-shell charge and magnetic form factors when one integrates over W in equation (13), which may be testable against known experimental data.

4. Scalar electrodynamics

We round off our study by treating the parallel case of scalar electrodynamics. Although Parker (1984) did examine the problem in the Fermi gauge ($a=1$) there remain questions associated with gauge dependence that can only be answered by finding the results for arbitrary a. We sketch the computations below.

All equations up to (11), excepting (4)–(6), can be taken over by the straight replacements $\rho(w)\to\rho(W^2)$, $S(p)\to\Delta(p)$, $(\gamma p-m)\to(p^2-m^2)$, etc. In place of (4), (5) and (12) we have

$$\Delta(p)(p^2-m_0^2)=Z^{-1}+\mathrm{i}e^2\int \bar{\mathrm{d}}^4k\, D^{\kappa\lambda}(k)G_\kappa(p,p-k;k)(2p-k)_\lambda$$

$$-e^4\int \bar{\mathrm{d}}^4k\,\bar{\mathrm{d}}^4k'\,\eta_{\kappa\lambda}D^{\kappa\mu}(k')D^{\lambda\nu}(p-k-k')G_{\mu\nu}(p-k-k',p;k,k')$$

$$+\text{tadpole term.} \tag{4s}$$

$$[(p'^2-p^2)+\text{tadpole}]G_\mu(p',p)$$

$$=(p'+p)_\mu\Delta(p)-2\mathrm{i}e^2\int \bar{\mathrm{d}}^4k\,D_{\mu\lambda}(k)G^\lambda(p-k,p;-k)$$

$$-\mathrm{i}e^2\int \bar{\mathrm{d}}^4k(2p'-k)_\kappa D^{\kappa\lambda}(k)G_{\lambda\mu}(p'-k,p;-k,p'-p)$$

$$+e^4\int \bar{\mathrm{d}}^4k\,\bar{\mathrm{d}}^4k'\,D^\lambda_\rho(k')D^{\rho\kappa}(k)G_{\lambda\kappa\mu}(p'-k-k',p;-k,-k',p'-p). \tag{5s}$$

$$G^{\mathrm{L}}_{\nu\mu}=\int\frac{\mathrm{d}W^2\rho(W^2)}{(p'^2-p^2)(p^2-W^2)}\left(2\eta_{\mu\nu}-\frac{(2p'+k')_\nu(2p+k)_\mu}{(p+k)^2-W^2}-\frac{(2p'-k)_\mu(2p-k')_\nu}{(p-k')^2-W^2}\right). \tag{12s}$$

In the second gauge approximation, and exactly to order e^4 in (4s), we may use (12s) on the RHS of (5s) and drop $G_{\lambda\kappa\mu}$. This gives the analogue of (13), namely,

$$G_\mu(p',p)=\int\frac{\mathrm{d}W^2\,\rho(W^2)}{(p'^2-W^2)(p^2-W^2)}\left((p'+p)_\mu+\Lambda^2_\mu(p',p|W)\right.$$

$$\left.+\frac{\Pi^2(p|W)-\Pi^2(p'|W)}{p'^2-p^2}(p'+p)_\mu\right) \tag{13s}$$

where

$$\Lambda^2_\mu(p',p|W)=\mathrm{i}e^2\int \bar{\mathrm{d}}^4k\,D^{\kappa\lambda}(k)\bigg[\frac{(2p'+k)_\lambda(p+p'+2k)_\mu(2p+k)_\kappa}{[(p'+k)^2-W^2][(p+k)^2-W^2]}$$
$$-2\eta_{\mu\lambda}\bigg(\frac{(2p'+k)_\lambda}{(p'+k)^2-W^2}+\frac{(2p+k)_\lambda}{(p+k)^2-W^2}\bigg)\bigg] \qquad (14s)$$

and

$$\Pi^2(p|W)=-\mathrm{i}e^2\int \bar{\mathrm{d}}^4k\,D^{\kappa\lambda}(k)\frac{(2p+k)_\kappa(2p+k)_\lambda}{(p+k)^2-W^2} \qquad (14s')$$

are the order e^2 vertex and self-energy respectively for a charged scalar meson of mass W. The last two terms in $(13s)$ correspond to the transverse vertex correction (which again vanishes as $p'\to p$). If one substitutes these last equations into $(4s)$ and takes the discontinuity, there results the renormalised equation

$$\pi\rho(W^2)(W^2-m^2)\bigg(1+\eta\frac{3W^2}{W^2-m^2}\ln\frac{W^2}{m^2}\bigg)$$
$$=\int_{m^2}^{W^2}\mathrm{d}W'^2\,\rho(W'^2)\frac{\mathrm{Im}\,\Pi_{\mathrm{f}}(W^2|W'^2)}{W^2-W'^2}. \qquad (21s)$$

In arriving at $(21s)$ we have cancelled off the e^4 divergences† on each side—there is a perfect match—to leave us with the finite kernel (here $u=W'^2/W^2$).

$$\mathrm{Im}\,\Pi_{\mathrm{f}}(W^2|W'^2)/\pi\eta W^2\theta(W^2-W'^2)$$
$$=(a-3)(1-u^2)+\eta I+\theta(1-9u)\eta k-\eta(1-u)$$
$$\times\bigg[\frac{75+143u}{4}-\frac{5+6u}{2(1-u)}-2(1+14u+u^2)\ln\bigg(\frac{1-u}{u}\bigg)$$
$$+3(1-a)\bigg(1-u+\frac{(1-2u)(1+u)}{1-u}\ln u\bigg)$$
$$+\frac{(1+3u)^2}{1-u}\bigg(3f\bigg(\frac{1}{u}\bigg)-2\ln u\ln\bigg(\frac{1-u}{u}\bigg)\bigg)\bigg] \qquad (22s)$$

where

$$I=\int_u^1 \mathrm{d}u'(1-u')\{(4/u')(u'-u)[1+a+\tfrac{5}{8}(1-a)^2]$$
$$-(1/2u'^2)(3u'+3-2a)(3u'+3u-2au)\}$$

$$K=\int_0^{(1-\sqrt{u})^2}\mathrm{d}v\bigg(L+\frac{2(2+2u-v)(3+u-2v)}{v+u-1}\ln\bigg|\frac{L+v+u+1}{L-v-u+1}\bigg|\bigg)$$

† The divergent term is

$$\frac{e^4}{256\pi^3}(3-a)\bigg(1+\frac{m^2}{W^2}\bigg)\bigg(3\ln\frac{\Lambda^2}{m^2}+7\bigg)$$

if the interested reader would like to check it.

with

$$L=\left[\left(\frac{2-2u-v}{2+2u-v}\right)(1+u^2+v^2-2u-2v-2uv)\right]^{1/2}.$$

We have omitted the calculational details: suffice it to say that there are two photon–one meson cuts and a three-meson cut (in K).

If we approach the infrared region ($W^2\to m^2$ or $u\to 1$). The kernel simplifies to

$$\operatorname{Im}\Pi_{\rm f}(W^2|W'^2)/\pi(W^2-W'^2)\to 2\eta(a-3)(1+3\eta)$$

and the equation tends to

$$\rho(W^2)(W^2-m^2)\approx 2\eta(a-3)\int_{m^2}^{W^2}\mathrm{d}W'^2\,\rho(W'^2)$$

producing the standard infrared result

$$\rho(W^2)\sim(W^2-m^2)^{-1+(a-3)e^2/8\pi^2}. \tag{23s}$$

Turning to the ultraviolet domain ($W^2/m^2\to\infty$ or $u\to 0$), we achieve asymptotic self-consistency with

$$\rho(W^2)\sim(W^2)^{-1+k\eta}(1+b\ln W^2)\quad\text{if}\quad k=(a-3)\quad\text{and}\quad b=-3\eta.$$

Here, again, the logarithmic term is affected by photon dressing and we cannot even hazard a guess about what happens in higher orders of gauge approximation. (The same criticism, incidentally, applies to Parker's results in the ultraviolet limit.)

We are slightly more hopeful that (21s) will yield itself to a full solution than the corresponding spinor equation (21). At all events, armed with known asymptotic solutions, we ought to be able to do a full numerical analysis of the spectral equations and go from there to determine the induced effects on vacuum polarisation. That work lies ahead.

Appendix 1

We want to prove that

$$G^{\rm T}_\mu(p',p)=\int \mathrm{d}W\frac{\rho(W)}{\gamma p'-W}\left[\Lambda^2_\mu(p',p|W)+\frac{\gamma p'\gamma_\mu+\gamma_\mu\gamma p}{p'^2-p^2}\Sigma^2(p|W)\right.$$
$$\left.-\Sigma^2(p'|W)\frac{\gamma p'\gamma_\mu+\gamma_\mu\gamma p}{p'^2-p^2}\right]\frac{1}{\gamma p-W}$$

vanishes as $p'\to p$. Consider the expression in the square brackets and let $\delta p=p'-p$ be small. Then to order $(\delta p)^0$

$$[\;]\approx \mathrm{i}e^2\int \bar{\mathrm{d}}^4k\,D^{\kappa\lambda}(k)\left[\gamma_\kappa\frac{1}{\gamma p-\gamma k-W}\gamma_\mu\frac{1}{\gamma p-\gamma k-W}\gamma_\lambda\right.$$
$$+\left(1+\delta p^\rho\frac{\partial}{\partial p^\rho}\right)\left(\gamma_\kappa\frac{1}{\gamma p-\gamma k-W}\gamma_\lambda\right)\frac{2p_\mu+\gamma\delta p\gamma_\mu}{2p\cdot\delta p}$$
$$\left.-\frac{2p_\mu+\gamma\delta p\gamma_\mu}{2p\cdot\delta p}\left(\gamma_\kappa\frac{1}{\gamma p-\gamma k-W}\gamma_\lambda\right)\right]$$

$$= \left(\eta_{\mu\rho} - \frac{p_\mu \delta p_\rho}{p\cdot\delta p}\right)\frac{\partial\Sigma^2(p|W)}{\partial p_\rho} - \left[\Sigma^2(p|W), \frac{\gamma\delta p\gamma_\mu}{2p\cdot\delta p}\right].$$

Define $\Sigma(p|W) = AW + \gamma p B$ where A and B are invariant scalar functions of p^2 and W^2. The derivatives of A and B give zero since p_ρ terms vanish upon contraction: also A drops out of the commutator term. Hence

$$[\ ] \approx B\left(\eta_{\mu\rho} - \frac{p_\mu\delta p_\rho}{p\cdot\delta p}\right)\gamma_\rho - B\left[\gamma p, \frac{\gamma\delta p\gamma_\mu}{2p\cdot\delta p}\right] \equiv 0$$

by γ-matrix algebra. This then confirms the finite behaviour of G^{T} and the dominance of G^{L} in the infrared, explaining why the gauge technique is so successful in that regime.

Appendix 2

The calculation of Σ^2 in (14′) is so straightforward that we shall just quote the answer:

$$\mathrm{Im}\,\Sigma^2(p|W) = \frac{e^2(p^2-W^2)}{16\pi p^4}\theta(p^2-W^2)[a\gamma p(p^2+W^2) - (a+3)Wp^2] \tag{A2.1}$$

$$\Sigma^2(p|W) = \frac{e^2}{16\pi^2}\left\{[a\gamma p - (a+3)W]\ln\frac{\Lambda^2}{W^2} + \gamma p\left(1 - a + a\frac{W^2}{p^2}\right)\right.$$
$$\left. + \left(-a\gamma p\left[1 + \frac{W^2}{p^2}\right] + [3+a]W\right)\left(1 - \frac{W^2}{p^2}\right)\ln\left(1 - \frac{p^2}{W^2}\right)\right\}. \tag{A2.2}$$

In equation (19) we also require the discontinuity of Σ^4 where

$$\Sigma^4(p|W) = -\mathrm{i}e^2\int \bar{\mathrm{d}}^4k\, D^{\kappa\lambda}(k)\left[\Lambda^2_\kappa(p, p-k|W)\right.$$
$$\left. + \int\frac{\mathrm{d}W'}{\gamma p - W'}\frac{\mathrm{Im}\,\Sigma^2(W'|W)}{\pi(W-W')}\gamma_\kappa\left(\frac{1}{\gamma(p-k)-W} - \frac{1}{\gamma(p-k)-W'}\right)\right]\gamma_\lambda$$
$$= -\mathrm{i}e^2\int\bar{\mathrm{d}}^4k\, D^{\kappa\lambda}(k)\Lambda^2_\kappa(p,p-k|W)\gamma_\lambda + \int\frac{\mathrm{d}W'\,\mathrm{Im}\,\Sigma^2(W'|W)}{\pi(\gamma p - W')(W-W')}$$
$$\times[\Sigma^2(p|W) - \Sigma^2(p|W')]$$
$$\equiv I_\Lambda + I_\Sigma \tag{A2.3}$$

The hard work commences now.

Decompose I_Λ into gauge-independent and gauge-dependent parts†

$$I_\Lambda = e^4\int\bar{\mathrm{d}}^4k\,\bar{\mathrm{d}}^4l\,\gamma_\kappa(\gamma p - \gamma k - W)^{-1}\gamma_\mu[\gamma(p-k-l) - W]^{-1}\gamma_\lambda(\gamma p - \gamma l - W)^{-1}$$
$$\times[-\eta^{\kappa\lambda} + (1-a)k^\kappa k^\lambda/k^2]/k^2l^2$$
$$\equiv I^{\mathrm{i}}_\Lambda + I^{\mathrm{g}}_\Lambda \tag{A2.4}$$

where

$$I^{\mathrm{g}}_\Lambda = \mathrm{i}e^2(1-a)\int\bar{\mathrm{d}}^4k\,\gamma k(\gamma p - \gamma k - W)^{-1}[\Sigma^2_{a=0}(p-k|W) - \Sigma^2_{a=0}(p|W)] \tag{A2.5}$$

† $(1-a)ll/l^2$ disappear because Σ^4 comes from a transverse G_μ.

and

$$I^i_\Lambda = e^4 \int \bar{d}^4k\, \bar{d}^4l\, \gamma_\lambda(\gamma p - \gamma k - W)^{-1}\gamma_\mu[\gamma(p-k-l) - W]^{-1}\gamma^\lambda \times(\gamma p - \gamma l - W)^{-1}\gamma^\mu / k^2 l^2. \tag{A2.6}$$

From (A2.5) it is not very difficult to arrive at

$$\begin{aligned} \operatorname{Im} I^g_\Lambda = \frac{e^2(a-1)}{16\pi}\Bigg\{ &[\operatorname{Re}\Sigma^2_{a=0}(p|W) - \operatorname{Re}\Sigma^2_{a=0}(W|W)](1+\gamma p W^3/p^4) \\ &+\left(1-\frac{W^2}{p^2}\right)\left[\frac{\gamma p(p^2-5W^2)}{2p^2} + W - \frac{3Wp^2}{p^2-W^2}\ln\frac{p^2}{W^2}\right] \\ &+\operatorname{Im}\Sigma^2_{a=0}(p|W)\left[\ln\frac{\Lambda^2}{W^2} - 1 - \frac{\gamma p W}{p^2} - \left(1+\frac{\gamma p W^3}{p^4}\right)\ln\left|\frac{p^2-W^2}{W^2}\right|\right]\Bigg\} \end{aligned} \tag{A2.7}$$

but it is much harder to extract the imaginary part of (A2.6). We first disentangle the γ-algebra by writing

$$\begin{aligned} I^i_\Lambda = e^4 \int \bar{d}^4k\, \bar{d}^4l \{ &4W[(2p-k-l)(p-k-l) + (p-k)(p-l) - 2m^2] \\ &+ (4/\gamma p)[W^2 p(3p-2k-2l) - 2p(p-k-l)(p-k)(p-l)]\} \\ &\times\{[(p-k)^2 - W^2][(p-k-l)^2 - W^2][(p-l)^2 - W^2]k^2 l^2\}^{-1} \end{aligned}$$

and then apply the Cutkosky–Nakanishi cutting rules: the imaginary parts then appear by letting various combinations of propagators go on-shell: either two photons and one electron, one photon and one electron, or two electrons and one positron. After much arduous computation one ends up with

$$\begin{aligned} \operatorname{Im} I^i_\Lambda &= W I^{(1)}_\Lambda + \gamma p I^{(2)}_\Lambda \\ I^{(1)}_\Lambda &= X_1 + (1 + W^2/p^2)Z_1 + Z_2 + Y_1 \\ I^{(2)}_\Lambda &= X_2 + (2W^4/p^4)Z_1 - (1 - W^2/p^2)Z_2 + Y_2 \end{aligned} \tag{A2.8}$$

where

$$\begin{aligned} X_1 &= -\frac{e^4}{32\pi^3\xi}\left\{\ln\frac{\Lambda^2}{W^2} + 1 + \frac{3}{2\xi}\ln(\xi-1) + (\xi - \tfrac{1}{2})\left[f(\xi) - \ln(\xi-1)\ln\left(\frac{\xi}{\xi-1}\right)\right]\right\} \\ X_2 &= -\frac{e^4}{32\pi^3\xi}\left\{-\frac{1}{4\xi}\ln\frac{\Lambda^2}{W^2} - \frac{1}{2} + \frac{1-\xi}{\xi^2}\ln(\xi-1) + \frac{(\xi-1)^2}{\xi}\right. \\ &\qquad\left.\times\left[f(\xi) - \ln(\xi-1)\ln\left(\frac{\xi}{\xi-1}\right)\right]\right\} \\ Z_1 &= \frac{e^4}{64\pi^3}\int_0^{(p-W)^2} dq^2 \frac{\ln\Phi}{p^2 - W^2 - q^2} \\ Z_2 &= \frac{e^4}{64\pi^3 p^2}\int_0^{(p-W)^2} dq^2 \ln\Phi \end{aligned} \tag{A2.9}$$

$$Y_1 = \frac{e^4}{64\pi^3 p^2}\int_{4W^2}^{(p-W)^2} dq^2 \frac{2q^2-p^2-5W^2}{p^2+3W^2-q^2}\ln\Psi$$

$$Y_2 = \frac{e^4}{64\pi^3 p^4}\int_{4W^2}^{(p-W)^2} dq^2 \frac{p^2q^2+5W^2q^2-q^4-4W^4-2p^2W^2}{p^2+3W^2-q^2}\ln\Psi$$

in which

$$\xi \equiv p^2/(p^2-W^2)$$

$$f(\xi) \equiv \int_0^1 \frac{d\xi'}{\xi-\xi'}\ln|\xi'-1| \qquad \text{(Spence's dilogarithm function),}$$

$$\Phi \equiv \left|\frac{p^2-q^2-W^2+\Delta}{p^2-q^2-W^2-\Delta}\right|; \qquad \Delta \equiv [p^4+q^4+W^4-2p^2W^2-2q^2W^2-2p^2q^2]^{1/2}$$

$$\Psi \equiv \left|\frac{p^2+3W^2-q^2+\Delta(1-4W^2/q^2)^{1/2}}{p^2+3W^2-q^2-\Delta(1-4W^2/q^2)^{1/2}}\right|. \tag{A2.10}$$

The cuts in X and Z begin at $p^2=W^2$, those in Y at $p^2=9W^2$.

Finally we need the imaginary part of I_Σ. This is extracted from (A2.3) either through the discontinuity in the denominator multiplying $\mathrm{Im}\,\Sigma^2\times\mathrm{Re}\,\Sigma^2$ or as a principal value integral over $\mathrm{Im}\,\Sigma^2\times\mathrm{Im}\,\Sigma^2$. One eventually arrives at

$$\begin{aligned}
\mathrm{Im}\, I_\Sigma = {} & \frac{e^2}{16\pi^2}\,\mathrm{Im}\,\Sigma^2(p|W)\left\{-4\ln\frac{\Lambda^2}{W^2}+2\left(1+\frac{W}{\gamma p}\right)\right. \\
& \times\left[4W\left(\ln\frac{W^2}{p^2-W^2}+\frac{p^2}{p^2-W^2}\ln\frac{p^2}{W^2}\right)\right. \\
& \left.\left.-\gamma p\left(\frac{p^2+W^2}{p^2}\ln\frac{W^2}{p^2-W^2}+\frac{p^2}{p^2-W^2}\ln\frac{p^2}{W^2}\right)\right]\right\} \\
& +\frac{e^4}{256\pi^3}\frac{p^2-W^2}{p^2}\left(4W-\gamma p\frac{(p^2+W^2)}{p^2}\right)\left(a\ln\frac{\Lambda^2}{W^2}+1\right) \\
& +\frac{e^4}{256\pi^3}\frac{a(p^2-W^2)}{p^2}\left[W\left(\frac{2W^2}{p^2}-\frac{4p^2}{p^2-W^2}\ln\frac{p^2}{W^2}\right)\right. \\
& \left.+\gamma p\left(\frac{9p^2-7W^2}{2p^2}+\frac{p^4-4W^4}{p^2(p^2-W^2)}\ln\frac{p^2}{W^2}\right)\right] \\
& -\frac{e^4}{64\pi^3}(a+3)W\frac{(p^2-W^2)}{p^2}\left[1-\frac{W^2}{p^2-W^2}\ln\frac{p^2}{W^2}\right].
\end{aligned} \tag{A2.11}$$

References

Baker M, Johnson K and Willey R 1964 *Phys. Rev.* B **136** 1111
Delbourgo R 1979 *Nuovo Cim.* **49A** 484
Delbourgo R, Keck B and Parker C N 1981 *J. Phys. A: Math. Gen.* **14** 921
Delbourgo R, Kenny B G and Parker C N 1982 *J. Phys. G: Nucl. Phys.* **8** 1173
Delbourgo R and Salam A 1964 *Phys. Rev.* **135** B1398
Delbourgo R and Thompson G 1982 *J. Phys. G: Nucl. Phys.* **8** L185

Delbourgo R and West P 1977a *J. Phys. A: Math. Gen.* **10** 1049
—— 1977b *Phys. Lett.* **72B** 96
Deser S, Jackiw R and Templeton S 1982 *Ann. Phys., NY* **140** 372
Gardner E 1981 *J. Phys. G: Nucl. Phys.* **7** L269
King J 1983 *Phys. Rev.* D **27** 1821
Parker C N 1984 *J. Phys. A: Math. Gen.* **17** 2873
Roo H and Stam K 1984 *University of Groningen preprint*
Slim H A 1981 *Nucl. Phys.* B **177** 172
Strathdee J 1964 *Phys. Rev.* **135** B1428

Part E
NON-POLYNOMIAL LAGRANGIANS

E.0. Commentaries E

There was a period in the late sixties when non-polynomial interactions became all the rage giving some hope that this might cure the renormalisation difficulties occurring in theories that contain coupling constants with inverse mass dimensions. This was because the behaviour of the Green functions at ultraviolet momenta is very different from what emerges in polynomial interaction models. The spark for this activity was a series of papers by Efimov (Sov. Phys. JETP 17, 1417 (1963)) who considered exponential interaction terms, coinciding with nonlinearly realisations of chiral models wherein the pion field is the matrix exponent. Alas, although they occupied attention from quite a lot of researchers they did not fulfill their potential in as much it became unclear how to handle higher orders of perturbation theory. Our own interest in this subject was focussed mainly on gravity when the graviton field arises as an exponent when parametrising the metric.

E.1 Delbourgo R., Salam A. and Strathdee J., Infinities of nonlinear and Lagrangian theories. Phys. Rev. 187, 1999-2007 (1969)

E.2 Delbourgo R., Koller K. and Salam A., Renormalization of rational Lagrangians. Ann. Phys. 66, 1-19 (1971)

E.3 Ashmore J.F. and Delbourgo R., On matrix superpropagators I. J. Math. Phys. 14, 176-181 (1973)

E.4 Ashmore J.F. and Delbourgo R., On matrix superpropagators II. J. Math. Phys. 14, 569-571 (1973)

The article [E.1] concerned the effect of nonlinear pion interactions and various technical isssues which are entrained; in paper [E.2] we showed that even nonlinear rational interactions like $\phi^4/(1+\phi)$ can lead to non-renormalisable effects although they might be construed as behaving like ϕ^3 asymptotically. Papers [E.3] and [E.4] have, in technical terms, been the most demanding of my career, involving the graviton field or an SU(N) matrix field through their exponential realisations. They have never been bettered even though they are of passing interest nowadays and contain some elegant mathematics.

PHYSICAL REVIEW VOLUME 187, NUMBER 5 25 NOVEMBER 1969

Infinities of Nonlinear and Lagrangian Theories

R. DELBOURGO,* ABDUS SALAM,† AND J. STRATHDEE
International Atomic Energy Agency, International Centre for Theoretical Physics, Trieste, Italy
(Received 1 April 1969; revised manuscript received 8 July 1969)

The technique for summing perturbation contributions introduced by Efimov and Fradkin is extended and applied to nonlinear (chiral) Lagrangian theories. It is shown that the only likely infinites in these theories are those associated with self-mass and self-charge.

I. INTRODUCTION

ONE of the significant recent advances in particle theory has been the formulation of chirally invariant Lagrangian theories.[1] These theories have so far been used with reasonable success for predicting low-energy (soft-meson) amplitudes in the following way: The interaction Lagrangian—an exponential or rational function of the spin-zero meson fields φ^i—is expanded as an infinite power series in φ^i and then used to evaluate tree-diagram[2] contributions to the amplitudes. Clearly, at the next level of sophistication one is interested in the closed-loop contributions, at which stage two related problems arise:

(i) Since the Lagrangian itself is expressed as an infinite power series, $\mathcal{L}_{\rm int}=\sum_n a_n g^n \varphi^n(\partial\varphi)^2$, the *number* of perturbation diagrams in each order n increases (typically) as fast or faster than $n!$. On any reasonable estimate, the perturbation expansion must be a *divergent series.* For respectable theories like quantum electrodynamics, with Lagrangians which are polynomials in the field variables, one has always suspected[3] that the perturbation expansion provides an *asymptotic series* in $e^2/\hbar c$; here, with Lagrangians which are themselves infinite series, this behavior appears to be a virtual certainty.

(ii) Each of the terms in the expansion of the Lagrangian [terms like $\varphi^n(\partial\varphi)^2$; $n\geqslant 1$] represents a nonrenormalizable interaction in the conventional sense. The ultraviolet infinities of the perturbation expansion therefore get progressively more virulent. On the face of it, this is rather surprising, since it is well known that every nonlinear theory can be reformulated as a theory of linear group representations[4] with polynomial Lagrangians together with a certain number of constraints on the fields φ^i. Before the imposition of the constraint, the theories are renormalizable; if any nonrenormalizability occurs, it must arise through the imposition of the constraint.

In this paper, we argue that *both* difficulties (i) and (ii) stem from the same circumstance, namely, the expansion of the Lagrangian in a power series of the field variables, and that a summation,[5] or even a partial summation, of the divergent perturbation series is likely at the same time to reduce the problem of ultraviolet infinities.[6]

An advance was made towards the (partial) summation of the perturbation series arising from rational and exponential Lagrangians in a series of papers by Efimov and Fradkin[7,8] during 1963. Like all summation methods for divergent series, the problem of uniqueness of the sum remains unresolved in their technique. Efimov, however, has claimed that besides satisfying the usual analyticity requirements, the Efimov-Fradkin (EF) summation method meets the demand of *consistency with Landau-Cutkosky unitarity* at least for the self-energy and vertex functions. In this paper, we wish to apply the EF method to summing the perturbation series of nonlinear Lagrangians of the chiral variety.[9] We wish to show that the infinities in such theories appear to be no worse after summation than those encountered in conventionally renormalizable theories. Central to our discussion is the result which states that the degree of ultraviolet infinity of EF sums depends on the growth of $\mathcal{L}_{\rm int}(\varphi)$ as $\varphi\to\infty$ for nonlinear theories just as for the usual linear theories. To be more specific, the result (extended below to include derivative cou-

* Imperial College, London, England.
† On leave of absence from Imperial College, London, England.

[1] These effective Lagrangians were originated by F. Gürsey, Nuovo Cimento **16**, 230 (1960); S. Weinberg, Phys. Rev. Letters **18**, 188 (1967); J. Schwinger, Phys. Letters **24B**, 473 (1967). Two useful reviews are F. Gürsey and N. Chang, Phys. Rev. **164**, 1752 (1967); and S. Gasiorowicz and D. Geffen, Argonne National Laboratory Report No. ANL/HEP 6801 (unpublished).

[2] B. W. Lee and M. Nieh, Phys. Rev. **166**, 1507 (1968); Y. Nambu, Phys. Letters **26B**, 626 (1968); L. Prokhorov, Nuovo Cimento **57A**, 245 (1968).

[3] C. A. Hurst, Phys. Rev. **85**, 920L (1952); F. J. Dyson, *ibid.* **85**, 631 (1952).

[4] Abdus Salam and J. Strathdee, Phys. Rev. (to be published); C. Isham, Nuovo Cimento **59A**, 356 (1969).

[5] We list here some of the papers where summation of perturbation diagrams of infinite parts of such diagrams has been carried out using widely different techniques. R. Arnowitt and S. Deser, Phys. Rev. **100**, 349 (1955); G. Feinberg and A. Pais, *ibid.* **131**, 2724 (1963); T. D. Lee and C. N. Yang, *ibid.* **128**, 885 (1962); Abdus Salam, *ibid.* **130**, 1287 (1963); Abdus Salam and R. Delbourgo, *ibid.* **135**, B1398 (1964); M. Baker, K. Johnson, and R. Willey, *ibid.* **136**, B1111 (1964); T. D. Lee, Nuovo Cimento **59A**, 579 (1969).

[6] Throughout this paper we use the terms "divergent" for *series* and "ultraviolet infinite" for *integrals*.

[7] G. V. Efimov, Zh. Eksperim. i Teor. Fiz. **44**, 2107 (1963) [English transl.: Soviet Phys.—JETP **17**, 1417 (1963)].

[8] E. S. Fradkin, Nucl. Phys. **49**, 624 (1963). See also subsequent work by G. V. Efimov, Nuovo Cimento **32**, 1046 (1964); Nucl. Phys. **74**, 657 (1965).

[9] After completion of this paper we became aware of the work of H. M. Fried [Phys. Rev. **174**, 1725 (1968); New Phys. (Korean Phys. Soc.) Suppl. **7**, 23 (1968)] which suggests applying the EF methods to chiral Lagrangians, though derivative couplings were not considered.

plings so essential in nonlinear chiral Lagrangians) can be stated as follows:

(i) Assign to each scalar field $\varphi(x)$ (with the propagator $\langle T\{\varphi(x)\varphi(0)\}\rangle=\Delta(x)\approx x^{-2}$ as $x^2\to 0$) the "singularity" behavior $\varphi(x)\approx 1/\sqrt{(x^2)}\approx 1/x$ as $x\to 0$ or equivalently $\varphi\sim M$ with $M\to\infty$.

(ii) Likewise assign the behaviors

$$\partial_\mu\varphi(x)\underset{x\to 0}{\approx} 1/x^2 \quad\text{or}\quad \partial_\mu\varphi\underset{M\to\infty}{\sim} M^2;$$

$$\psi(x)\underset{x\to 0}{\approx} 1/x^{3/2} \quad\text{or}\quad \psi\underset{M\to\infty}{\sim} M^{3/2},\quad \psi=\text{spin-}\tfrac{1}{2}\text{ field};$$

$$U_\mu(x)\underset{x\to 0}{\approx} 1/x^2 \quad\text{or}\quad U_\mu\underset{m\to\infty}{\sim} M^2,\quad U=\text{spin-1 field}.$$

A theory is expected to be renormalizable, with only a few types of integrals that are ultraviolet infinite, if $\mathfrak{L}_{\text{int}}\underset{M\to\infty}{\sim} M^4$. *This criterion applies equally to integrals in conventional polynomial Lagrangians like* $\mathfrak{L}_{\text{int}}=g\varphi^4$ *or* $g\bar{\psi}\psi\varphi$, *as well as to EF sums in theories with Lagrangians like* $g\varphi^2(\partial\varphi)^2/(1+\varphi^2)$. We shall call such theories *normal*. These like $\mathfrak{L}_{\text{int}}=g\varphi^3$ or $g(\partial\varphi)^2/(1+\varphi^2)$ which behave like M^3 or M^2 or lower ($\mathfrak{L}_{\text{int}}\sim M^n$; $n<4$) will be called ***supernormal***. All theories which behave worse than φ^4, i.e., for which $\mathfrak{L}_{\text{int}}\sim M^n$, $n>4$, will be called ***abnormal***. For supernormal theories there is the attractive possibility that when $n<2$ all integrals, including those for self-mass and self-charge, may be finite.

The plan of the paper is as follows: In Sec. II we give an outline of the EF method which has two ingredients: (i) Hori's exponential representation[10] of Wick's normal-ordering theorem and (ii) the EF integral representation[7,8] of Hori's exponential operator, which essentially performs a "Borel" sum of the divergent perturbation series. The power-counting rules for estimating *over-all* ultraviolet infinities of EF sums is given in Sec. III. We consider derivative couplings in Sec. IV and formulate the rules for writing EF sums in such a manner that the ultraviolet power-counting estimate can also be stated here. Section V contains the application of these results to the nonlinear (chiral type) Lagrangians in an $SU(2)\otimes SU(2)$ symmetric theory. Since equivalence theorems, which state that on-mass-shell S-matrix elements are unaltered by contact transformations in field space, play such a critical role,[11] we devote Sec. VI to a nonrigorous discussion of the circumstances in which such transformations are permissible. Not discussed in this paper is the problem of absorbing these infinities into counter-term Lagrangians.

[10] S. Hori, Progr. Theoret. Phys. (Kyoto) **7**, 589 (1952).

[11] The type of problem which one meets is that $\mathfrak{L}_{\text{int}}(\varphi)$ may look abnormal when expressed in terms of one set of fields but normal in another formulation. Consider, for example, a theory with $\mathfrak{L}_{\text{int}}=2\varphi(1+\varphi)(\partial\varphi)^2+\varphi^3(1+\varphi)^3$, which looks hideously abnormal but can be transformed into the familiar supernormal φ^3 theory with substitution $\varphi\to\varphi(1+\varphi)$.

II. THEORIES WITH NONDERIVATIVE COUPLINGS

We summarize below the steps needed to arrive at the EF representation[7,8] of the S matrix, assuming that the interaction Lagrangian contains no field derivatives. (In Sec. IV we extend the techniques to cover situations where derivatives are encountered.) An illustrative example is presented to demonstrate the power of the EF method.

Step 1. Begin with the standard perturbation expansion of the S matrix,

$$S=\sum_N \frac{i^N}{N!}S^{(N)},$$

where

$$S^{(N)}=g^N\int d^4z_1\cdots d^4z_N T[L\{\varphi(z_1)\}\cdots L\{\varphi(z_N)\}] \quad (1)$$

and we are supposing in this section that

$$\mathfrak{L}_{\text{int}}=gL\{\varphi(x)\}, \quad (2)$$

where φ denotes a real scalar field.

The further expansion of the S matrix into normal Wick products can be compactly expressed through *Hori's functional operator*[10] as follows:

$$S^{(N)}=g^N\int d^4z_1\cdots d^4z_N \times\exp\left(\frac{1}{2}\int d^4x_1d^4x_2\Delta(x_1-x_2)\frac{\delta^2}{\delta\varphi(x_1)\delta\varphi(x_2)}\right) \times[L\{\varphi(z_1)\}\cdots L\{\varphi(z_N)\}], \quad (3)$$

where $\Delta(x_1-x_2)$ denotes the bare causal propagator for the scalar field φ. This formula can be simplified to read

$$S^{(N)}=g^N\int d^4z_1\cdots d^4z_N \exp\left(\tfrac{1}{2}\sum_{i,j=1}^{N}\Delta_{ij}\frac{\partial^2}{\partial\varphi_i\partial\varphi_j}\right) \times[L\{\varphi_1\}\cdots L\{\varphi_N\}]_{\varphi_k=\varphi^{\text{ext}}(z_k),\Delta_{kl}=\Delta(z_k-z_l)}. \quad (4)$$

Here $\varphi_k=\varphi^{\text{ext}}(z_k)$ is the wave function of any external particle which may be acting at the point z_k. One may rewrite (4) in a form where these external wave functions are exhibited separately by writing

$$S^{(N)}=\int d^4x_1\cdots d^4x_n :\varphi(x_1)\cdots\varphi(x_n): S^{(N)}(x_1,\cdots,x_n), \quad (5)$$

where the n-point function in the Nth order equals

$$S^{(N)}(x_1,\cdots,x_n)=\langle 0|\frac{\delta^n}{\delta\varphi(x_1)\cdots\delta\varphi(x_n)}S^{(N)}|0\rangle = g^N\int d^4z_1\cdots d^4z_N\sum_{m_i}\delta^{m_1}(x-z_1)\cdots\delta^{(m_N)}(x-z_N)S_{m_1\cdots m_N}(\Delta), \quad (6)$$

($\sum m_i = n$) with

$$\delta^m(x-z) = \delta(x_{i_1}-z)\delta(x_{i_2}-z)\cdots\delta(x_{i_m}-z), \quad (7)$$

and

$$S_{m_1\cdots m_N}(\Delta) \equiv \exp\left(\tfrac{1}{2}\sum_{ij}\Delta_{ij}\frac{\partial^2}{\partial\varphi_i\partial\varphi_j}\right)\left(\frac{\partial}{\partial\varphi_1}\right)^{m_1} \cdots\left(\frac{\partial}{\partial\varphi_N}\right)^{m_N}[L(\varphi_1)\cdots L(\varphi_N)]_{\varphi=0}. \quad (8)$$

The vacuum graphs are given in their entirety by

$$S = \sum_N \frac{(ig)^N}{N!}\int d^4z_1\cdots d^4z_N S_{00\cdots 0}(\Delta), \quad (9)$$

$$S_{00\cdots 0}(\Delta) = \exp\left(\tfrac{1}{2}\sum_{ij}^{N}\Delta_{ij}\frac{\partial^2}{\partial\varphi_i\partial\varphi_j}\right)L(\varphi_1)\cdots L(\varphi_N), \quad (10)$$

while the two-point (self-energy) graphs are completely described by

$$S(x_1,x_2) = \sum_{N\geqslant 2}\frac{(ig)^N}{N!}\int d^4z_1\cdots d^4z_N$$
$$\times\{\delta(x_1-z_1)\delta(x_2-z_2)S_{200\cdots}(\Delta)$$
$$+\delta(x_1-z_2)\delta(x_2-z_2)S_{020\cdots}(\Delta)+[\delta(x_1-z_1)\delta(x_2-z_2)$$
$$+\delta(x_1-z_2)\delta(x_2-z_1)]S_{110\cdots}(\Delta)\}, \quad (11)$$

with

$$S_{200\cdots}(\Delta) = \exp\left[\tfrac{1}{2}\sum\Delta\frac{\partial^2}{\partial\varphi\partial\varphi}\right]\frac{\partial^2}{\partial\varphi_1{}^2}L(\varphi_1)\cdots L(\varphi_N)\Big|_{\varphi=0}$$

$$S_{110\cdots}(\Delta) = \exp\left[\tfrac{1}{2}\sum\Delta\frac{\partial^2}{\partial\varphi\partial\varphi}\right] \times\frac{\partial^2}{\partial\varphi_1\partial\varphi_2}L(\varphi_1)\cdots L(\varphi_N)\Big|_{\varphi=0,\text{etc.}}, \quad (12)$$

which expressions are represented graphically in Figs. 1(a) and 1(b), respectively.

Step 2. Give a simple integral representation of Hori's exponential operator by making use of the EF lemma[7,8]

$$\exp\left(\Delta\frac{\partial^2}{\partial\varphi\partial\varphi'}\right)F(\varphi,\varphi')$$
$$=\frac{1}{\pi}\int d^2u\exp\left(-|u|^2+uc\frac{\partial}{\partial\varphi}+u^*c'\frac{\partial}{\partial\varphi'}\right)F(\varphi,\varphi')$$
$$=\frac{1}{\pi}\int d^2u\exp(-|u|^2)F(\varphi+uc,\,\varphi'+u^*c'), \quad (13)$$

with the parameters c and c' constrained to satisfy $cc'=\Delta$, but otherwise arbitrary. (They can be chosen to suit one's purpose. Thus $c=c'=\sqrt{\Delta}$ would correspond to the most symmetric choice, one we often make; $c=\Delta$, $c'=1$ to the most asymmetric choice. In any event, the final result cannot explicitly involve any square roots of Δ and must depend only on the product $cc'=\Delta$.) Since the final expression on the right-hand side of (13) involves as integrand the function F shifted from its value at φ, φ' to $\varphi+uc$, $\varphi'+u^*c'$, we shall call this the ***exponential-shift*** lemma.

FIG. 1. (a) Self-energy diagrams $S_{200\cdots}$. (b) Self-energy diagrams $S_{110\cdots}$.

(a) (b)

Applying the lemma to the Nth order S matrix by introducing complex variables u_{ij}, c_{ij} between every two pairs of points ij, one has the representation

$$\exp\left(\tfrac{1}{2}\sum_{ij}\Delta_{ij}\frac{\partial^2}{\partial\varphi_i\partial\varphi_j}\right)L(\varphi_1)\cdots L(\varphi_N) = \prod_{i\geqslant j}\left(\frac{1}{\pi}\int d^2u_{ij}\right)$$
$$\times\exp(-\tfrac{1}{2}\sum_{ij}|u_{ij}|^2)L(\varphi_1+\sum_k c_{1k}u_{1k})$$
$$\cdots L(\varphi_N+\sum_k c_{Nk}u_{Nk}), \quad (14)$$

with

$$c_{ij}c_{ji}=\Delta_{ij} \text{ (no summation over } ij)$$

and

$$u_{ij}=u_{ji}{}^*. \quad (15)$$

As an application of the lemma, consider all vacuum graphs of order g^N. These are given by

$$S_{00\cdots 0}(\Delta) = \prod_{ij}\left(\frac{1}{\pi}\int d^2u_{ij}\right)\exp(-\sum|u_{ij}|^2)$$
$$\times L(\sum_k c_{1k}u_{1k})\cdots L(\sum_k c_{Nk}u_{Nk}). \quad (16)$$

Likewise, the self-energy graphs of order g^N are given in terms of

$$S_{20\cdots 0}(\Delta) = \prod_{ij}\left(\frac{1}{\pi}\int d^2u_{ij}\right)\exp(-\sum|u_{ij}|^2)$$
$$\times L''(\sum_k c_{1k}u_{1k})\cdots L(\sum_k c_{Nk}u_{Nk}),$$

$$S_{110\cdots}(\Delta) = \prod_{ij}\left(\frac{1}{\pi}\int d^2u_{ij}\right)\exp(-\sum|u_{ij}|^2)$$
$$\times L'(\sum_k c_{1k}u_{1k})L'(\sum_k c_{2k}u_{2k})\cdots, \quad (17)$$

and so on. Hence $L'\equiv\partial L/\partial\varphi$, $L''=\partial^2L/\partial\varphi^2$, etc.

To see how this works in practice, take the model for which $gL(\varphi)=g\varphi^4/(1+\lambda^2\varphi^2)$. The power of the technique, which explicitly displays sums of perturbation series to each order in g, is already apparent since *all*

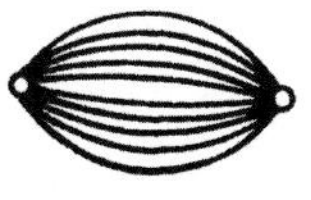

FIG. 2. Typical vacuum transition S_{00}.

orders in λ^2 are automatically taken into account by the EF expressions. Thus, to second order in g and all orders in λ^2, the vacuum contribution equals

$$g^2S_{00}(x_1,x_2)=g^2\int\frac{d^2u}{\pi}e^{-|u|^2}\frac{c^4u^4}{1+\lambda^2c^2u^2}\frac{c'^4u^{*4}}{1+\lambda^2c'^2u^{*2}},$$

where $cc'=\Delta(x_1-x_2)$. Likewise, the two relevant self-energy terms to second order in g but all orders in λ^2 are

$$g^2S_{20}=g^2\int\frac{d^2u}{\pi}e^{-|u|^2}\frac{d^2}{c^2du^2}\left(\frac{c^4u^4}{1+\lambda^2c^2u^2}\right)\frac{c'^4u^{*4}}{1+\lambda^2c'^2u^{*2}}$$

and

$$g^2S_{11}=g^2\int\frac{d^2u}{\pi}e^{-|u|^2}\frac{d}{cdu}\left(\frac{c^4u^4}{1+\lambda^2c^2u^2}\right)\frac{d}{c'du^*}\left(\frac{c'^4u^{*4}}{1+\lambda^2c'^2u^{*2}}\right).$$

The simplification of these integrals rests on the pair of relations[12]

$$\frac{1}{\pi}\int d^2u\, u^{*m}u^nf(|u|^2)=\delta_{nm}\int_0^\infty d\xi\xi^nf(\xi),$$

$$\frac{1}{\pi}\int d^2u\frac{f(|u|^2)}{(1+\alpha u^2)(1+\beta u^{*2})}=\int_0^\infty d\xi\frac{f(\xi)}{1-\alpha\beta\xi^2},$$

and derivatives thereof. Thus we find, as expected, that the integrals only involve the product $cc'=\Delta$ and not the parameters c and c' separately. Explicitly (see Figs. 2 and 3),

$$g^2S_{00}=g^2\int_0^\infty d\xi\frac{\Delta^4\xi^4e^{-\xi}}{1-\lambda^4\Delta^2\xi^2},\tag{18}$$

$$g^2S_{20}=-g^2\int_0^\infty d\xi\frac{\lambda^2\Delta^4\xi^6e^{-\xi}}{1-\lambda^4\Delta^2\xi^2},$$

$$g^2S_{11}=g^2\int_0^\infty d\xi\Delta^3\xi^4e^{-\xi}\left(\frac{2}{1-\lambda^4\Delta^2\xi^2}+\frac{2}{(1-\lambda^4\Delta^2\xi^2)^2}\right),\tag{19}$$

In particular, when we set $\lambda=0$, we recover the φ^4 perturbation-theory results, viz.,

$$S_{00}=4!\Delta^4,\quad S_{20}=S_{02}=0,\quad S_{11}=4(4!)\Delta^3.$$

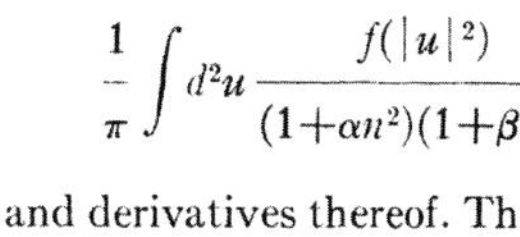

(a) (b)

FIG. 3. (a) Self-energy diagram S_{20}. (b) Self-energy diagram S_{11}.

[12] It is clear from these identities that the E-F summation method is equivalent to a Borel summation of divergent series.

We shall return to the ultraviolet properties of these integrals after we have discussed the question of infinities.

III. ULTRAVIOLET INFINITIES OF EF SUMS

Physically, we are only concerned with S-matrix elements in momentum space,[13] i.e., the Fourier transforms

$$\tilde{S}(p)=\prod\left(\int d^4x\, e^{ipx}\right)S(\Delta(x_{ij})).\tag{20}$$

On account of the causal character of the propagators $\Delta(x)$, the task of defining the x-space contours of integration in integrals like (20) is not trivial. As is well known,[14] the light-cone singularity of $\Delta(x)$ is given by the following expression:

$$\begin{aligned}4\pi i\Delta(x;\mu)&=\delta(x^2)-\frac{\mu\theta(x^2)}{2\sqrt{(x^2)}}J_1(\mu\sqrt{(x^2)})+i\mu\left[\frac{\theta(x^2)}{2\sqrt{(x^2)}}N_1(\mu\sqrt{(x^2)})\right.\\&\qquad\left.+\frac{\theta(-x^2)}{\pi\sqrt{(-x^2)}}K_1(\mu\sqrt{(-x^2)})\right]\\&=\delta(x^2)-\frac{i}{\pi x^2}-\tfrac{1}{4}\mu^2\left[\theta(x^2)-\frac{2i}{\pi}\ln(\tfrac{1}{2}\mu\sqrt{|x^2|})\right]\\&\qquad+O((\sqrt{|x^2|})\ln|x^2|).\end{aligned}\tag{21}$$

The crucial part of Efimov's work is a method of carrying out the x-space integrals, with the demonstration that one may define them so as to preserve the unitarity of the S matrix in the perturbation sense, i.e., in the expansion of $S(\Delta)$ in powers of Δ. Efimov's procedure consists in concentrating firstly on the Euclidean or Symanzik region of the external momenta.[15] For this region of p space, it must be assumed that x-space contours of integration have been rotated from the Minkowskian into the Euclidean region of x. (For the theories under consideration, the Minkowskian integrals may not be well defined.) For other regions of p-space, Efimov makes suitably defined continuations from the

[13] The integrands (18) and (19) exhibit poles on the real axis, $\xi>0$, and pose the problem of defining the correct contour of integration in the ξ plane, such that power series expansion of EF integrals coincides with the perturbation expansion and satisfies causality and unitarity requirements. We believe that this problem is bound up with the problem of defining the Fourier integral (20) away from the Symanzik region in p space. Efimov (Ref. 7) in his calculation of self-masses takes the ***principal-value*** integral in ξ space. This has been further discussed by B. W. Lee and B. Zumino, CERN Report No. TH1053, 1969 (unpublished).

[14] See, e.g., N. N. Bogolubov and D. V. Shirkov, ***Introduction to the Theory of Quantized Fields*** (Wiley-Interscience, Inc., New York, 1959).

[15] The Symanzik region is defined by the condition that the linear combinations $\sum_j\alpha_jp_j$ be spacelike for any choice of real parameters α_j; i.e., the Gram determinants of $-(p_i\cdot p_j)$ are positive for all sets of i and j.

Symanzik region. In this paper we are only concerned with the ultraviolet infinities associated with integrals (20), so for our purpose it is sufficient to remain in the Symanzik region—or, to make matters simpler, on its edge, where all external momenta p_μ are zero. Thus we examine the infinities associated with the Euclidean x-space integrals ($x_{ij}^2<0$)

$$\tilde{S}(0)=\int \Pi d^4x\, S(\Delta),$$

where the Δ *assume real values.* A naive power count of the over-all[16] infinities can be made by considering the appropriate proper diagrams and retaining the most singular parts of all the propagators Δ. Applying the lower cutoff $x^2=M^{-2}$ ($M^2\to\infty$) to all x-space integrations, it is evident that we can associate a factor M to each $\sqrt{\Delta}$ that occurs; and since what in fact determines the infinities is the powers of $L(\varphi\approx\sqrt{\Delta})$, we may easily estimate the over-all infinity to be expected by setting $\varphi=M$ in $L(\varphi)$ and letting $M\to\infty$.

Consider, therefore, an n-point function and follow the Dyson power-counting procedure.[17] Suppose that $L(\varphi=M)$ behaves as M^ν for large M. The integrand of $S_{m_1\cdots m_N}(\Delta)$ in (14) contains the term (putting $c_{ij}=c_{ii}=\sqrt{(\Delta_{ij})}$ for simplicity)

$$\Delta^{-\Sigma m_i/2}[L((\sqrt{\Delta})u)]^N\sim M^{-n+N\nu},$$

where n denotes the number of external lines and N the order of the graph (number of "vertices"). The singularity produced at $x^2=0$ ($M\to\infty$) is compensated by $4(N-1)$ integrations, four integrations being omitted because the integrand is independent of the over-all c.m. coordinates. There,

$$\int (d^4x)^{N-1}S(\Delta)\sim M^{-4(N-1)}M^{-n+N\nu}. \tag{22}$$

If the integral is to be regular in the limit $M\to\infty$, then

$$N(4-\nu)+n>4.$$

This is the same criterion which one encounters in renormalization theory of polynomial Lagrangians. In this count we have included tadpole contributions [$i=j$ terms in Eq. (14)]. If these were left out, the count would be jeopardized in a subtle manner to be discussed elsewhere.

We return to the example above to see that this naive infinity count is sensible. The self-energy contributions (19) to second order in g (but all orders in λ) read, in momentum space,

$$\tilde{S}(p)=2g^2\int S_{20}(\Delta(x))d^4x+2g^2\int S_{11}(\Delta(x))e^{ipx}d^4x. \tag{23}$$

Taking $p^2\leqslant 0$ (Euclidean region), the integrals reduce to

$$i\tilde{S}(p)=4\pi^2g^2\int_{1/M}^{\infty}dr\, r^3S_{20}\left(\frac{\mu K_1(\mu r)}{4\pi^2 r}\right)+\frac{8\pi^2g^2}{\sqrt{(-p^2)}}\times\int_{1/M}^{\infty}dr\, r^2J_1(r\sqrt{(-p^2)})S_{11}\left(\frac{\mu K_1(\mu r)}{4\pi^2 r}\right), \tag{24}$$

where we cut off the integrations at $r=M^{-1}$ in order to estimate the infinity as $x^2\to 0$. Since the ultraviolet behavior of the integral is independent of the value of p^2, we set this equal to 0:

$$i\tilde{S}(0)=4\pi^2g^2\int_{1/M}^{\infty}dr\, r^3\left[S_{20}\left(\frac{\mu K_1(\mu r)}{4\pi^2 r}\right)+S_{11}\left(\frac{\mu K_1(\mu r)}{4\pi^2 r}\right)\right].$$

As $r\to 0$,

$$\frac{\mu K_1(\mu r)}{4\pi^2 r}\to\frac{\mu^2}{8\pi^2}\ln(\tfrac{1}{2}\mu r)+\frac{1}{4\pi^2r^2},$$

so that the lethal infinities at the lower limit are obtained, using (19), as

$$\lim_{M\to\infty}4\pi^2g^2\int_{1/M}dr\, r^3\left[S_{20}\left(\frac{1}{4\pi^2r^2}\right)+S_{11}\left(\frac{1}{4\pi^2r^2}\right)\right]$$
$$=\lim_{M\to\infty}4\pi^2g^2\int_{1/M}dr\, r^3\left(\frac{-12}{\lambda^2 16\pi^4r^4}-\frac{8}{\lambda^4}\right)$$
$$\sim\frac{3g^2}{\pi^2\lambda^2}\ln M; \tag{25}$$

i.e., we meet a logarithmic infinity at most. A naive power count (up to these logarithms) would have agreed with this result since when we set $\Delta=M^2\to\infty$,

$$\int d^4x[S_{20}(\Delta)+S_{11}(\Delta)]$$
$$\sim M^{-4}[S_{20}(M^2)+S_{11}(M^2)]\to -12/\lambda^2. \tag{26}$$

An interesting feature of the result is the pole $1/\lambda^2$ of the "leading infinity" in the λ^2 plane. This is not entirely surprising in view of the fact that for $\lambda=0$, we must necessarily recover the conventional quadratic perturbation infinity.

IV. THEORIES WITH DERIVATIVE COUPLINGS

In this section we extend the summation technique to cases where $\mathcal{L}_{\text{int}}$ contains derivatives of the φ field,

$$\mathcal{L}_{\text{int}}=gL(\varphi,\partial_\mu\varphi). \tag{27}$$

It is common knowledge that for such situations the Hamiltonian contains surface-dependent terms and formula (1) for the S matrix holds only if suitable modifications are made to the definition of time-ordering

---

[16] There may, of course, be hidden infinities from subintegrations which, though they are not discussed here, form an integral part of the renormalization program. We hope to study these in future work.

[17] F. J. Dyson, Phys. Rev. **75**, 1736 (1949).

products of $\partial_\mu\varphi$, More specifically, using a theorem[18] first proved by Matthews for Lagrangians involving one time derivative, and later extended by Dyson to Lagrangians with two time derivatives, the S matrix is covariantly defined if we invert the order of differentiation and the time-ordering operation in vacuum expectation values of the following variety:

$$\langle T^*\varphi_\mu(x)\varphi(x')\rangle=\Delta_\mu(x-x')\equiv\partial_\mu\Delta(x-x'),$$
$$\langle T^*\varphi_\mu(x)\varphi_\nu(x')\rangle=\Delta_{\mu\nu}(x-x')\equiv-\partial_\mu\partial_\nu\Delta(x-x'),$$

where $\varphi_\mu\equiv\partial_\mu\varphi$, provided one leaves out all terms which involve $\delta^4(0)$ whenever it occurs. Given, then, the modified time-ordering operation T^*, we have

$$S^{(N)}=g^N\int d^4z_1\cdots d^4z_N T^*[L\{\varphi(z_1),\varphi_\mu(z_1)\}\cdots L\{\varphi(z_N),\varphi_\mu(z_N)\}]. \quad (28)$$

The Wick reduction can be carried through by extending Hori's exponential operator to include differentiation with respect to the derived fields φ_μ as follows:

$$S^{(N)}=g^N\int d^4z_1\cdots d^4z_N\exp\left(\frac{\partial}{\partial\boldsymbol{\varphi}}\boldsymbol{\Delta}\frac{\partial}{\partial\boldsymbol{\varphi}}\right)\times[L\{\varphi_1,\varphi_{\mu1}\}\cdots L\{\varphi_N,\varphi_{\mu N}\}]_{\varphi_k=\varphi^{\text{ext}}(z_k),\,\varphi_{\mu k}=\partial_\mu\varphi^{\text{ext}}(z_k)}, \quad (29)$$

where

$$\frac{\partial}{\partial\boldsymbol{\varphi}}\boldsymbol{\Delta}\frac{\partial}{\partial\boldsymbol{\varphi}}\equiv\tfrac{1}{2}\sum_{ij}\frac{\partial}{\partial\boldsymbol{\varphi}_i}\boldsymbol{\Delta}_{ij}\frac{\partial}{\partial\boldsymbol{\varphi}_j}$$
$$\equiv\tfrac{1}{2}\sum_{ij}\Delta(z_i-z_j)\frac{\partial^2}{\partial\varphi_i\partial\varphi_j}+\Delta_\mu(z_i-z_j)\frac{\partial^2}{\partial\varphi_{\mu i}\partial\varphi_j}$$
$$+\Delta_\nu(z_i-z_j)\frac{\partial^2}{\partial\varphi_i\partial\varphi_{\nu j}}+\Delta_{\mu\nu}(z_i-z_j)\frac{\partial^2}{\partial\varphi_{\mu i}\partial\varphi_{\nu j}}. \quad (30)$$

In order to give a simple integral representation of this generalized operator we must be prepared to introduce auxiliary vector variables. (This representation will be needed in its full generality only for interactions which are not polynomials in the vector variables.) To see how this is achieved, it is enough to consider a pair of points, since the extension to the whole series of points is easily performed by the method outlined in Sec. II. Since

$$\Delta_\mu(x)=2x_\mu\frac{d}{dx^2}\Delta(x^2)\equiv2x_\mu\Delta'(x^2),$$
$$\Delta_{\mu\nu}(x)=-4x_\mu x_\nu\Delta''(x^2)-2g_{\mu\nu}\Delta'(x^2), \quad (31)$$

we need to introduce at most one auxiliary vector and four auxiliary scalar complex integrations. Showing this in detail,

$$\exp\left(\frac{\partial}{\partial\boldsymbol{\varphi}}\Delta(x)\frac{\partial}{\partial\boldsymbol{\varphi}'}\right)F(\boldsymbol{\varphi};\,\boldsymbol{\varphi}')$$
$$=\exp\left(\Delta\frac{\partial^2}{\partial\varphi\partial\varphi'}+2x_\mu\Delta'\frac{\partial^2}{\partial\varphi_\mu\partial\varphi'}-2x_\nu\Delta'\frac{\partial^2}{\partial\varphi\partial\varphi_\nu'}-4x_\mu x_\nu\Delta''\frac{\partial^2}{\partial\varphi_\mu\partial\varphi_\nu}-2\Delta'\frac{\partial^2}{\partial\varphi_\mu\partial\varphi_\mu}\right)F(\varphi,\varphi_\lambda;\,\varphi',\varphi_\lambda')$$
$$=\frac{1}{\pi^8}\int d^8c_{(\mu)}d^2c\,d^2b_1d^2b_2d^2a\exp\left(-|a|^2+\alpha a\frac{\partial}{\partial\varphi}+\alpha'a^*\frac{\partial}{\partial\varphi'}\right)\exp\left(-|b_2|^2+\beta_2b_2 2x_\mu\frac{\partial}{\partial\varphi_\mu}+\beta_2'b_2^*\frac{\partial}{\partial\varphi'}\right)$$
$$\times\exp\left(-|b_1|^2+\beta_1b_1\frac{\partial}{\partial\varphi}+\beta_1'b_1^*2x_\nu\frac{\partial}{\partial\varphi_\nu'}\right)\exp\left(-|c|^2+\gamma_1c2x_\mu\frac{\partial}{\partial\varphi_\mu}+\gamma_1'c^*2x_\nu\frac{\partial}{\partial\varphi_\nu'}\right)$$
$$\times\exp\left(-|c_\lambda c_\lambda|^2+\gamma_2c_\mu\frac{\partial}{\partial\varphi_\mu}+\gamma_2'c_\nu^*\frac{\partial}{\partial\varphi_\nu'}\right)F(\varphi,\varphi_\lambda;\,\varphi',\varphi_\lambda')$$
$$=\frac{1}{\pi^8}\int d^8c_{(\mu)}d^2c\,d^2b_1d^2b_2a\exp[-(|a|^2+|b_1|^2+|b_2|^2+|c|^2+|c_\lambda c_\lambda|)]F(\varphi+\alpha a+\beta_1b_1,$$
$$\varphi_\lambda+2x_\lambda\beta_2b_2+2x_\lambda\gamma_1c+\gamma_2c_\lambda;\,\varphi'+\alpha'a^*+\beta_2'b_2^*,\,\varphi_\lambda'+2x_\lambda\beta_1'b_1^*+2x_\lambda\gamma_1'c^*+\gamma_2'c_\lambda^*), \quad (32)$$

where

$$\alpha\alpha'=\Delta,\quad\gamma_1\gamma_1'=-\Delta'',$$
$$\beta_2\beta_2'=-\beta_1\beta_1'=-\tfrac{1}{2}\gamma_2\gamma_2'=\Delta'. \quad (33)$$

The result cannot depend on the individual $\alpha,\beta,\cdots$, but only on the products $\alpha\alpha'=\Delta$, etc. For the remainder of this discussion we choose to make the quasisymmetric split

$$\alpha=\alpha'=\beta_1=\beta_2'=\surd\Delta,\quad\gamma_1=\gamma_1'=\surd(-\Delta''),$$
$$\gamma_2=\gamma_2'=\surd(-2\Delta'),\text{ and }\beta_2=-\beta_1'=\Delta'/\surd\Delta. \quad (34)$$

[18] P. T. Matthews, Phys. Rev. **75**, 1270 (1949); F. J. Dyson, *ibid.* **83**, 608 (1951).

Now because in the limit $x^2 \to 0$,

$$\Delta'(x^2) \sim 1/x^4 \text{ and } \Delta''(x^2) \sim 1/x^6, \qquad (35)$$

one can see that, consistently for all integrations over the shifted functional, we can ascribe the "singularity factors"

$$\varphi \sim M \quad \text{and} \quad \varphi_\mu \sim M^2 \qquad (36)$$

owing to the terms Δ and $(x\Delta'/\sqrt{\Delta}+x\sqrt{\Delta''})$ occurring, respectively, in the shifted arguments. Perhaps the clearest way to appreciate this conclusion is to realize that most of the auxiliary integrations are redundant and that for the simple case treated above, only one auxiliary vector and one auxiliary scalar variable suffice to make the exponential shift defined in Sec. II. Thus, write

$$\exp\left(\frac{\partial}{\partial \boldsymbol{\varphi}}\boldsymbol{\Delta}\frac{\partial}{\partial \boldsymbol{\varphi}'}\right)=\exp\left[\left(c_{\mu\lambda}\frac{\partial}{\partial\varphi_\mu}+c_\lambda\frac{\partial}{\partial\varphi}\right)\right.$$
$$\left.\times\left(c_{\lambda\nu}'\frac{\partial}{\partial\varphi_\nu'}+c_\lambda'\frac{\partial}{\partial\varphi'}\right)+cc'\frac{\partial^2}{\partial\varphi\partial\varphi'}\right], \qquad (37)$$

with

$$cc'+c_\lambda c_\lambda'=\Delta, \quad c_{\mu\lambda}c_\lambda=\Delta_\mu, \quad c_{\mu\lambda}c_{\lambda\nu}'=\Delta_{\mu\nu}. \qquad (38)$$

$$\exp\left(\frac{\partial}{\partial \boldsymbol{\varphi}}\boldsymbol{\Delta}\frac{\partial}{\partial \boldsymbol{\varphi}'}\right)F(\boldsymbol{\varphi};\boldsymbol{\varphi}')=\frac{1}{\pi^5}\int d^8u_{(\mu)}d^2u$$
$$\times\exp[-|u|^2+|u_\lambda u_\lambda|)]F(\varphi+cu+c_\mu u_\mu,\ \varphi_\lambda+c_{\lambda\mu}u_\mu;$$
$$\varphi'+c'u^*+c_\mu'u_\mu^*,\ \varphi_\lambda'+c_{\lambda\nu}u_\nu^*). \qquad (39)$$

Again the result can only depend on the products $cc'=\Delta$; if we make the symmetrical choice $c=c'$ for simplicity, then (see the Appendix) in the (Euclidean) limit $x\to 0$,

$$c\sim 1/x, \quad c_\mu\sim 1/x, \quad c_{\mu\nu}\sim 1/x^2.$$

The association (36) of the ultraviolet factors $\varphi\sim M$ and $\varphi_\mu\sim M^2$ then becomes more obvious. The following identities prove useful for the vector integrations:

$$\frac{1}{\pi^4}\int d^8u_{(\mu)}u_\kappa u_\lambda^* e^{-u_\mu u_\mu^*}=\tfrac{1}{4}g_{\kappa\lambda}, \qquad (40)$$

$$\frac{1}{\pi^4}\int d^8u_{(\mu)}u_\kappa u_\lambda f(u_\nu u_\nu^*)=0, \text{ etc.} \qquad (41)$$

Here we are concerned only with a superficial count of the over-all infinities to be expected in a given S-matrix element.

The procedure is the same as before and will not be repeated, since the result is that derivatives make no essential difference to the infinity count beyond what is expected for conventional polynomial Lagrangian theories.

V. NONLINEAR REALIZATIONS OF $SU(2)\otimes SU(2)$

The simplest practical applications of our conclusions about derivative couplings are to be found in the nonlinear realizations of chiral groups. We shall study the case of $SU(2)\otimes SU(2)$ symmetry for definiteness, because the features which emerge will apply to more complicated cases as well.

Describe the mesons of the $(\frac{1}{2},\frac{1}{2})$ representation by the field matrix

$$\mathcal{S}=\sigma+i\boldsymbol{\tau}\cdot\boldsymbol{\varphi}\Lambda(\boldsymbol{\varphi}^2), \qquad (42)$$

where the nonlinearity is introduced by imposing the constraint

$$\mathcal{S}\mathcal{S}^\dagger=1 \text{ or } \sigma^2+\boldsymbol{\varphi}^2\Lambda^2(\boldsymbol{\varphi}^2)=1. \qquad (43)$$

The choice of the function $\Lambda(\boldsymbol{\varphi}^2)$ corresponds to different parametrizations of the nonlinear coordinates [σ and $\boldsymbol{\varphi}$ are coordinates of the differential manifold (43)], and with each such choice of Λ the corresponding interpolating field φ is different.[19] (However, we shall use the same symbol in every case.)

The *unique*[20] $SU(2)\otimes SU(2)$-invariant Lagrangian which contains only two derivatives of the fields is

$$\mathcal{L}=\tfrac{1}{4}\Lambda^{-2}(0)\,\mathrm{Tr}[(\partial_\mu\mathcal{S})(\partial_\mu\mathcal{S}^\dagger)]=\tfrac{1}{4}\Lambda^{-2}(0)\,\mathrm{Tr}[\mathcal{G}_\mu\mathcal{G}_\mu], \qquad (44)$$

where we write

$$\mathcal{G}_\mu\equiv -i\mathcal{S}^\dagger\partial_\mu\mathcal{S}=\mathcal{G}_\mu^\dagger. \qquad (45)$$

If we substitute for $\mathcal{S}$ the expression (42) and eliminate σ by means of the constraint equation (43), we then find

$$\mathcal{G}_\mu=\boldsymbol{\tau}\cdot[\Lambda(\sigma\partial_\mu\boldsymbol{\varphi}-\boldsymbol{\varphi}\partial_\mu\sigma+\Lambda\boldsymbol{\varphi}\times\partial_\mu\boldsymbol{\varphi})+2\Lambda'\sigma\boldsymbol{\varphi}(\boldsymbol{\varphi}\cdot\partial_\mu\boldsymbol{\varphi})]$$
$$=\Lambda\boldsymbol{\tau}\cdot\left((1-\Lambda^2\varphi^2)^{1/2}\partial_\mu\boldsymbol{\varphi}+\frac{\boldsymbol{\varphi}(\boldsymbol{\varphi}\partial_\mu\boldsymbol{\varphi})}{(1-\Lambda^2\varphi^2)^{1/2}}\right.$$
$$\left.\times(\Lambda^2+2\Lambda^{-1}\Lambda')+\Lambda\boldsymbol{\varphi}\times\partial_\mu\boldsymbol{\varphi}\right) \qquad (46)$$

and

$$\mathcal{L}_{\text{int}}=\tfrac{1}{4}\Lambda^{-2}(0)\,\mathrm{Tr}[\mathcal{G}_\mu\mathcal{G}_\mu]-\tfrac{1}{2}(\partial_\mu\boldsymbol{\varphi})\cdot(\partial_\mu\boldsymbol{\varphi})$$
$$=\tfrac{1}{2}[\Lambda^{-2}(0)\Lambda^2-1](\partial_\mu\boldsymbol{\varphi})\cdot(\partial_\mu\boldsymbol{\varphi})$$
$$+\frac{(\boldsymbol{\varphi}\cdot\partial_\mu\boldsymbol{\varphi})(\boldsymbol{\varphi}\cdot\partial_\mu\boldsymbol{\varphi})}{2\Lambda^2(0)(1-\Lambda^2\varphi^2)}(\Lambda^4+4\Lambda\Lambda'+4\Lambda'^2\varphi^2), \qquad (47)$$

where $\Lambda'=d\Lambda/d\varphi^2$. The ensuing equations of motion can be conveniently remembered in the Sugawara form[21]

$$\partial_\mu\mathcal{G}_\mu=0, \quad \partial_\mu\mathcal{G}_\nu-\partial_\nu\mathcal{G}_\mu+i[\mathcal{G}_\mu,\mathcal{G}_\nu]=0. \qquad (48)$$

---

[19] For criteria when the S matrices are equivalent, see Sec. VI and Refs. 23 and 24.

[20] The coordinate independence of the Lagrangian on the differential manifold has been proved by S. Coleman, J. Wess, and B. Zumino, Phys. Rev. **177**, 2239 (1969); and C. Isham (Ref. 4).

[21] H. Sugawara and M. Yoshimura, Phys. Rev. **173**, 1419 (1968).

We may now inquire about the "ultraviolet behavior" of the interaction Lagrangian with a view to possible renormalizability.[22] Begin by supposing that for large $\varphi\sim M$

$$\Lambda(\varphi^2)\to\varphi^k\sim M^k,\quad \sigma\sim[1-M^{2+2k}]^{1/2},$$

$$\mathcal{J}_\mu\sim M^{k+2}\left[(1-M^{2+2k})^{1/2}+\frac{M^2(M^{2k}+M^{-2})}{(1-M^{2+2k})^{1/2}}+M^{k+1}\right],\tag{49}$$

and

$$\mathcal{L}_{\rm int}\sim(M^{2k}-1)M^4+\frac{M^6}{1-M^{2+2k}}(M^{4k}+M^{2k-2}).\tag{50}$$

Hence

for $k>0$, $\mathcal{J}_\mu\sim M^{2k+3}$ and $\mathcal{L}_{\rm int}\sim M^{2k+4}$;

for $-1<k<0$, $\mathcal{J}_\mu\sim M^{2k+3}$ and $\mathcal{L}_{\rm int}\sim M^4$;

and

for $k<-1$, $\mathcal{J}_\mu\sim M^{k+2}$ and $\mathcal{L}_{\rm int}\sim M^4$.

This shows that nonlinear realizations of chiral groups, for the *preferred meson fields*, yield normal ($k<0$) or seemingly abnormal ($k>0$) Lagrangians, *but not supernormal ones.* The reason for this is not far to seek. For $k<0$, $\mathcal{L}\sim M^{2k+4}$, so that subtracting off $\mathcal{L}_f=(\partial_\mu\varphi)^2\sim M^4$, we meet a normal situation.

The question now poses itself: Since we can pass from one set of coordinates φ to another, φ', by a point transformation

$$S=\sigma+i\boldsymbol{\tau}\cdot\boldsymbol{\varphi}\Lambda(\boldsymbol{\varphi}^2)=\sigma'+i\boldsymbol{\tau}\cdot\boldsymbol{\varphi}'\Lambda'(\boldsymbol{\varphi}'^2),\tag{51}$$

what is the significance of the abnormal parametrizations ($k>0$)? In Sec. VI we argue that the invariance of the total Lagrangian ($\mathcal{J}_\mu\mathcal{J}_\mu$) should imply that the S-matrix elements on the mass shell do not differ from one parametrization to the next, so that the theory is normal irrespective of the possibility $k>0$. We list below some special choices of parametrization.

(i) Gasiorowicz-Geffen coordinates.

$$\Lambda(\varphi^2)=\lambda,\ \text{a constant (i.e., } k=0)$$

$$\mathcal{J}_\mu=\lambda\boldsymbol{\tau}\cdot[\sigma\partial_\mu\boldsymbol{\varphi}-\boldsymbol{\varphi}\partial_\mu\sigma+\lambda\boldsymbol{\varphi}\times\partial_\mu\boldsymbol{\varphi}],$$

with $\sigma=(1-\lambda^2\varphi^2)$. Also,

$$2\mathcal{L}_{\rm int}=\lambda^2(\boldsymbol{\varphi}\cdot\partial_\mu\boldsymbol{\varphi})(\boldsymbol{\varphi}\cdot\partial_\mu\boldsymbol{\varphi})/(1-\lambda^2\varphi^2)\sim M^4.\tag{52}$$

(ii) Schwinger coordinates.

$$\Lambda(\varphi^2)=\lambda(1+\lambda^2\varphi^2)^{-1/2};\ \lambda\text{ constant (i.e., } k=-1).$$

Thus $\sigma=(1+\lambda^2\varphi^2)^{-1/2}$,

$$\mathcal{J}_\mu=\frac{\lambda\boldsymbol{\tau}}{1+\lambda^2\varphi^2}[\partial_\mu\boldsymbol{\varphi}+\lambda\boldsymbol{\varphi}\times\partial_\mu\boldsymbol{\varphi}],$$

and

$$2\mathcal{L}_{\rm int}=-\frac{\lambda^2}{1+\lambda^2\varphi^2}\left(\varphi^2(\partial_\mu\boldsymbol{\varphi})\cdot(\partial_\mu\boldsymbol{\varphi})+\frac{(\boldsymbol{\varphi}\cdot\partial_\mu\boldsymbol{\varphi})(\boldsymbol{\varphi}\cdot\partial_\mu\boldsymbol{\varphi})}{1+\lambda^2\varphi^2}\right)\sim M^4.\tag{53}$$

(iii) Weinberg coordinates.

$$\Lambda(\varphi^2)=2\lambda(1+\lambda^2\varphi^2)^{-1};\ \lambda\text{ constant (i.e., } k=-2)$$

$$\sigma=(1-\lambda^2\varphi^2)(1+\lambda^2\varphi^2)^{-1},$$

giving

$$\mathcal{J}_\mu=\frac{2\lambda\boldsymbol{\tau}}{(1+\lambda^2\varphi^2)^2}[(1-\lambda^2\varphi^2)\partial_\mu\boldsymbol{\varphi}+2\lambda\boldsymbol{\varphi}\times\partial_\mu\boldsymbol{\varphi}+2\lambda^2\boldsymbol{\varphi}(\boldsymbol{\varphi}\cdot\partial_\mu\boldsymbol{\varphi})]$$

and

$$2\mathcal{L}_{\rm int}=(\partial_\mu\boldsymbol{\varphi})\cdot(\partial_\mu\boldsymbol{\varphi})\left(\frac{1}{(1+\lambda^2\varphi^2)^2}-1\right)\sim M^4.\tag{54}$$

(iv) Harmonic coordinates. A set of coordinates which may prove useful in the vector problem is defined by the condition

$$\Lambda(1-\varphi^2\Lambda^2)^{1/2}=\lambda^2,$$

where λ is a constant. In these coordinates, which we shall call harmonic, the current operator is given by

$$\mathcal{J}_\mu=\left[\partial_\mu\boldsymbol{\varphi}+\frac{2\lambda}{1+(1-4\lambda^2\varphi^2)^{1/2}}\boldsymbol{\varphi}\times\partial_\mu\boldsymbol{\varphi}+\frac{2\lambda\varphi(\boldsymbol{\varphi}\cdot\partial_\mu\boldsymbol{\varphi})}{(1-4\lambda^2\varphi^2)^{1/2}}\right]\lambda\boldsymbol{\tau}.$$

In this form the linear term $\partial_\mu\varphi$ appears multiplied by a constant rather than by a function of φ^2.

VI. FIELD TRANSFORMATIONS

In Sec. V we assumed the correctness of the basic equivalence theorem, which states that if a local point transformation of fields is made such that the physical spectrum associated with these fields is unaltered—and therefore also the Hilbert spaces of in and out states remains the same—then the on-mass-shell (physical) S-matrix elements, computed using either the original or the transformed Lagrangians, are identical. This theorem,[23] first stated by Chisholm, Kamefuchi, O'Raifeartaigh, and Salam, has been proved to varying degrees of restrictiveness on field transformations and rigor by the above-mentioned authors and in axiomatic field theory by Borchers. It has latterly been extended by Coleman, Wess, and Zumino[20] who claim to sharpen the result to apply even to diagrams with equal numbers of closed loops. The weak point, when one comes to applying the theorem in practical cases, is the lack of criteria whereby one may judge what transformations

[22] In applying the exponential-shift lemma for making an ultraviolet count, one has to introduce isotopic labels U^i to the auxiliary variables of integration.

[23] J. S. R. Chisholm, Nucl. Phys. **26**, 469 (1961); H. J. Borchers, Nuovo Cimento **15**, 784 (1960); S. Kamefuchi, L. O'Raifeartaigh, and Abdus Salam, Nucl. Phys. **28**, 529 (1961).

leave unchanged the in and out limits of the interpolating fields. For practical purposes, the only procedure known to us is the adiabatic switching on and off of charges; this implies that a point transformation is allowed if:

(i) In the limit $g\to 0$ for a transformation like $\varphi(x)\to\varphi'(x)=a_1\varphi(x)+a_2\varphi^2(x)+\cdots$, the $a_i\to 0$, $i>1$, and $a_1\to\text{const}\neq 0$. ($a_1\neq 1$ implies a wave-function renormalization).

(ii) In the language of axiomatic field theory, all transformations $\varphi\to\varphi'$ are allowed, provided φ and φ' are mutually local operators, $[\varphi'(x),\varphi(y)]=0$, $(x-y)^2<0$ and provided $\langle 0|\varphi|p\rangle=z\langle 0|\varphi'|p\rangle$, $z\neq 0$, where $|p\rangle$ is the appropriate one-particle state.

(iii) The only known procedure for computing S-matrix elements for given Lagrangians is essentially the Dyson perturbation procedure which relies on identifying that part of the Lagrangian which depends *bilinearly* on field variables as $\mathcal{L}_f$. In this paper, when making point transformations we have separated out *all* bilinear terms; thus a term like $\mathcal{L}=(\partial_\mu\varphi)^2/(1+\varphi^2)$ will contribute $(\partial_\mu\varphi)^2$ to $\mathcal{L}_f$ and $[\varphi^2/(1+\varphi^2)](\partial_\mu\varphi)^2$ to $\mathcal{L}_{\text{int}}$.

(iv) A consequence of the split mentioned in (iii) is that in our power-counting theorem, $\mathcal{L}=(\partial_\mu\varphi)^2/(1+\varphi^2)$, does not behave supernormally like M^2 (assuming $\varphi\sim M$, $\partial\varphi\sim M^2$) but normally like $[\varphi^2/(1+\varphi^2)]\times(\partial\varphi)^2\sim M^4$. This may mean that our estimates of singularity behavior are likely to be overestimates and that a future formulation of a new computational procedure may depress our estimates of likely infinities.

(v) Regarding our discussion of nonlinear realizations of chiral groups in Sec. V, it is important to realize that the interpolating fields for two different choices of coordinates can be related to each other; thus, writing

$$S=\sigma(\varphi^2)+i\boldsymbol{\tau}\cdot\boldsymbol{\varphi}\Lambda(\varphi^2)=\sigma'(\varphi'^2)+i\boldsymbol{\tau}\cdot\boldsymbol{\varphi}'\Lambda'(\varphi'^2),\quad(51')$$

one can express $\boldsymbol{\varphi}$ fields in terms of $\boldsymbol{\varphi}'$ fields by comparing terms of the power series in the φ. We have assumed that the adiabatic limits of both φ and φ' are the same, so that the on-mass-shell S matrices are equal and so is the singularity behavior of S-matrix elements. It is well known that this result does not apply to the n-point Green's functions.

VII. CONCLUSIONS

We have shown in this paper that a simple power count of ultraviolet infinite integrals in Efimov-Fradkin sums of perturbation diagrams suggests that nonlinear meson theories may behave in the same way as polynomial Lagrangian theories so far as the infinity count is concerned.

A number of fundamental problems remain, basic to the whole approach, which are unresolved. There is the difficult problem of ***uniqueness*** of the sums, the renormalization program, and the problem of defining the contours in the auxiliary variable planes. It is important to realize that the proof of the absence of infinities in this paper has been given with all vectors x_μ Wick-rotated and Eulidean. It appears that for nonlinear theories a "Euclidean continuation postulate" must be an essential feature of the theories to render their matrix elements finite. This principle is not new. It has been suggested by Schwinger,[24] Symanzik,[25] Fradkin,[8] and others. The only thing one must guarantee is that the unitarity relation $T+T^\dagger+TT^\dagger=0$ is preserved when the continuation in external momenta to the physical region is made.

ACKNOWLEDGMENTS

We are indebted to Professor P. T. Matthews and Professor G. Feldman for their support.

APPENDIX

We give here proofs of the singular behaviors of the shifted arguments occurring in (39) for derivative-coupling theories. Our only concern is the (Euclidean) limit $x\to 0$, of Eqs. (38), where $\Delta(x)\sim 1/x^2$. To solve Eqs. (38), let

$$c_{\mu\nu}\equiv d_{\mu\nu}(x)c_1+e_{\mu\nu}(x)c_0,$$

$$c_\mu=x_\mu c_2,$$

and

$$d_{\mu\nu}(x)\equiv g_{\mu\nu}-x^{-2}x_\mu x_\nu\equiv g_{\mu\nu}-e_{\mu\nu}(x),$$

and make the symmetrical choice $c=c'$ as in the text. Since

$$\Delta_{\mu\nu}=-2\Delta' d_{\mu\nu}-(2\Delta'+4x^2\Delta'')e_{\mu\nu}$$

and

$$\Delta_\mu=2x_\mu\Delta',$$

we obtain the equations

$$c_1^2=-2\Delta',\quad c_0^2=-2(\Delta'+2x^2\Delta''),$$

$$c_0c_2=2\Delta',\quad c^2+x^2c_2^2=\Delta,$$

which are solved by

$$c_1=[-2\Delta']^{1/2}\sim 1/x^2,$$

$$c_0=[-2(\Delta'+2x^2\Delta'')]^{1/2}\sim i(\sqrt{6})/x^2,$$

$$c_2=2\Delta'/c_0\sim i\sqrt{2}/\sqrt{3}x^2,$$

and

$$c=[\Delta-x^2c_2^2]^{1/2}\sim 1/\sqrt{(3x^2)}.$$

This proves the statement that a correct estimate of the most singular behavior is given by

$$c_{\mu\nu}\sim 1/x^2,\quad c_\mu\sim 1/x,\quad\text{and}\quad c\sim 1/x.$$

[24] J. Schwinger, Proc. Natl. Acad. Sci. U. S. **44**, 956 (1958); Phys. Rev. **115**, 721 (1959).
[25] K. Symanzik, J. Math. Phys. **7**, 510 (1966).

ANNALS OF PHYSICS: **66**, 1–19 (1971)

Renormalization of Rational Lagrangians

R. DELBOURGO, K. KOLLER, AND ABDUS SALAM

Physics Department, Imperial College, London SW7

Received May 2, 1970

We show that rational Lagrangians of the type $G\phi^{\nu_0}(1 + \lambda\phi)^{-\nu_1}$ can be renormalized by introducing a finite class of infinite counter terms providing that the Dyson index $\nu_0 - \nu_1 \leqslant 3$. The form of the counter terms is explicitly exhibited. The theories become unrenormalizable when $\nu_0 - \nu_1 > 3$; we discuss, in particular, the case $\nu_0 - \nu_1 = 4$, which resembles a $g\phi^4$ theory for $\phi \to \infty$, and is nonrenormalizable, contrary to what one may have naively expected.

1. INTRODUCTION

In two earlier papers [1], the problem of infinities associated with nonpolynomial Lagrangians, in general, and rational Lagrangians, in particular, was considered. In this paper, we study the problem of absorbing these infinities into a renormalization of constants in the theory by a small number of counterterms. Our results can be stated for the typical Lagrangian $V = G\phi^{\nu_0}(1 + \lambda\phi)^{-\nu_1}[\nu_1 \geqslant 1$, ν_0 integer] and sums thereof:

(i) When $\nu_0 - \nu_1 \leqslant 1$ all S-matrix elements are completely free of infinities and the theory is superrenormalizable.

(ii) When $\nu_0 - \nu_1 = 2$ and 3, there are a finite number of distinct types of infinities in the S-matrix elements which can be absorbed into a finite class of counterterms that serve to renormalize the constants in the theory.

(iii) When $\nu_0 - \nu_1 \geqslant 4$, the theory is nonrenormalizable. This conclusion is especially interesting for the case $\nu_0 - \nu_1 = 4$; thus, although the polynomial interaction $g\phi^4$ is renormalizable by itself, we find that the rational interaction $G\phi^5/(1 + \lambda\phi)$ which possesses the same limit when $\phi \to \infty$ is not.

Any rational Lagrangian may be decomposed in the general form

$$L = \sum_{r=0}^{\nu} G_r\phi^r + \sum_s \frac{\gamma_s}{(1 + \lambda_s\phi)^{\kappa_s}}$$

© 1971 by Academic Press, Inc.

and since $(1 + \lambda\phi)^{-2} = (1 + \lambda(\partial/\partial\lambda))(1 + \lambda\phi)^{-1}$, etc., it is in practice quite sufficient to study the model

$$L = G_\nu \phi^\nu + G_{\nu-1}\phi^{\nu-1} + \cdots + G/(1 + \lambda\phi)$$

to gain insight into the infinities one may expect in more complicated examples. We shall call ν the Dyson index of the theory, defined by $\lim_{\phi\to\infty} L(\phi) = 0(\phi^\nu)$. Rephrasing our results we can say that *when* $\lambda \neq 0$ the theory is (i) superrenormalizable when $\nu \leqslant 1$, (ii) renormalizable for $\nu = 2$ or 3, but (iii) nonrenormalizable for $\nu \geqslant 4$. We prove these results in Sections 3, 4, and 5 after surveying the general techiques for computing S-matrix elements in Section 2. Of special interest are the explicit forms of the counter Lagrangians needed to absorb all infinities of the theory (as exhibited in Eq. (22)). Appendix A discusses various aspects of normally ordered Lagrangians and resulting tadpole terms. Appendix B contains a simplified representation of the momentum space superpropagator for particles of arbitrary mass. Appendix C shows that by taking the limit $\lambda^2 \to 0$ of the model $g\phi^4/(1 + \lambda\phi)$ one may recover the conventional infinities of the ϕ^4 theory.

2. Representations of S-Matrix Elements

An exposition (I) of methods for computing S-matrix elements in perturbation theory for nonpolynomial Lagrangians appears in papers I and II. Basically, they can be divided into p-space and x-space methods, both of which we shall use where convenient for isolating the infinities and their renormalization. The p-space methods have the advantage of familiarity, generality, and of giving an idea of possible infinities of subgraphs, in addition to the overall infinity of the graph. The x-space methods are better suited for writing compactly the counterterm Lagrangians. We shall give a brief survey of these methods.

Let us write the interaction Lagrangian as the series

$$V(\phi) = G \sum_{r=\nu_0}^{\infty} \frac{v(r)\,\phi^r}{r!}, \tag{1}$$

where G is the major coupling constant to which the perturbation expansion refers. To order G^N the conventional Feynman perturbation theory gives the following rules for constructing a graph (see Fig. 1) and its corresponding S-matrix element:

(a) Draw points $x_1, x_2, \ldots, x_N$ to represent the vertices.

(b) Attach m_i external lines to each point x_i and connect *each* pair of points

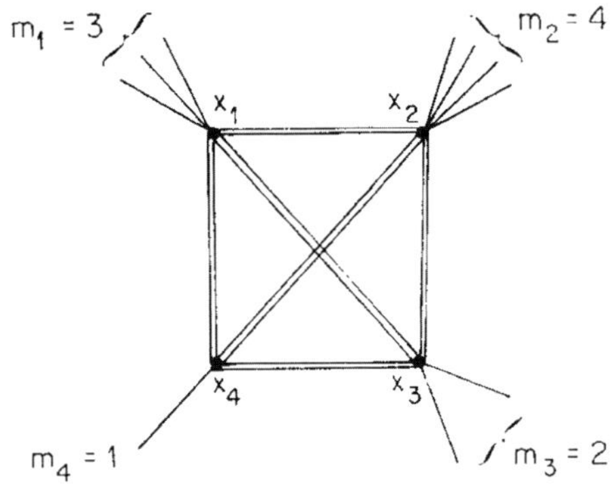

FIG. 1. An example of a general supergraph.

x_i and x_j by a cocoon of n_{ij} lines which we shall term a superline. Thus, $n_i \equiv \sum_j n_{ij}$ is the total number of internal lines emanating from vertex i.

(c) Sum over all n_{ij} with the condition that the total number of lines coming into each vertex, $m_i + n_i$ is not less than ν_0.

The corresponding S-matrix element is written as

$$S_{m_1 m_2 \cdots m_N}(x) = \frac{(iG)^N}{N!} \sum_{n_{ij}} \prod_{i>j}^{N} v\left(m_i + \sum_l n_{il}\right) \frac{[\Delta(x_i - x_j)]^{n_{ij}}}{n_{ij}!}, \tag{2}$$

subject to $m_i + \sum_j n_{ij} \geqslant \nu_0$. Correspondingly there is the more physically relevant momentum space transform

$$\tilde{S}_{m_1 m_2 \cdots m_N}(p) = \int \prod_j d^4x_j \, e^{ip_j x_j} S_{m_1 m_2 \cdots m_N}(x). \tag{3}$$

Note that in stating how diagrams are to be constructed and calculated, we have not included tadpole lines which emanate from and close onto the same vertex, so have interpreted $V(\phi)$ in (1) as already normally ordered, each term in the series being considered as $: \phi^n :$. If, however, we wish to abandon this assumption of normal ordering and allow for tadpole insertions, we can continue to use (2) but relax the condition that $i \neq j$ in the summations. We have then to "define" the singular quantity $\Delta(O)$ as the infinite integral $\int [d^4p/(2\pi)^4] \, \Delta(p)$. Appendix A contains a full treatment of normal ordering and its effect on interaction Lagrangians.

It is possible to obtain a closed form for (2) or (3), interpreted as an asymptotic series, by exploiting Laplace transform methods. Write

$$V(\phi) = \int_0^\infty \mathscr{V}(\zeta) \, e^{-\zeta\phi} \, d\zeta \tag{4}$$

defined by analytic continuation where appropriate. One may then express (2) in the form,

$$S_{m_1 m_2 \cdots m_N}(x) = \left[\prod_{k=1}^{N} \int_0^\infty d\zeta_k \, \mathscr{V}(\zeta_k) \, \zeta_k^{m_k}\right] e^{\frac{1}{2}\Sigma_{ij}\zeta_i \Delta_{ij} \zeta_j}, \tag{5}$$

where $\Delta_{ij} = \Delta(x_i - x_j)$ and

$$\mathscr{V}(\zeta) = G \sum_{r=\nu_0}^{\infty} \frac{v(r)\, \delta^{(r)}(\zeta + \epsilon)}{r!} \bigg|_{\epsilon \to 0}. \tag{6}$$

From the closed form (5) we note the following identity:

$$S_{m_1+1, m_2+1, m_3 \cdots}(x) = \frac{\partial}{\partial \Delta_{12}} S_{m_1 m_2 m_3 \cdots}(x), \tag{7}$$

which leads to important results like

$$S_{l+m, l}(x_1, x_2) = \frac{\partial^l}{\partial \Delta_{12}^l} S_{m0}(x_1, x_2), \qquad N = 2, \tag{8}$$

$$S_{200}(x_1, x_2, x_3) = \int_{-\infty}^{\Delta_{23}} d\Delta_{23} \frac{\partial^2}{\partial \Delta_{12}\, \partial \Delta_{13}} S_{000}(x_1, x_2, x_3); \qquad N = 3, \text{ etc.}$$

These relations will prove useful in studying asymptotic behaviors and providing power counting of infinities[1,2].

An alternative way of obtaining closed expressions and one which is perhaps better suited for obtaining *explicitly* high energy behaviors is the p-space method

[1] The point is that high energy behavior is determined by small x behavior, i.e., the limit as x_i approach one another; this corresponds to taking the limit $\Delta \to \infty$ which means, from (5), that the high energy behavior is governed by small ζ behavior of the Laplace transform $\mathscr{V}(\zeta)$.

[2] The momentum space expression corresponding to (5) is

$$\tilde{S}_{m_1 \cdots m_N}(p) = \left[\prod_k \int_0^\infty d\zeta_k \mathscr{V}(\zeta_k) \zeta_k^{m_k}\right] K(\zeta, p),$$

with the *interaction-independent* kernel

$$K(\zeta, p) = \left[\prod_k \int d^4 x_k \, e^{i p_k x_k}\right] \exp\left[\frac{1}{2} \sum_{ij} \zeta_i \Delta_{ij} \zeta_j\right]$$

again governed by small ζ as $p \to \infty$.

of writing $\tilde{S}(p)$ through a Watson Sommerfeld transformation on (2). This allows one to cast the series as an integral

$$S_{m_1 \cdots m_N}(x) = \frac{(iG)^N}{N!} \prod_{i>j} \int \frac{dz_{ij}}{2\pi i} \Gamma(-z_{ij})\, v\left(m_i + \sum_l z_{il}\right) [-\Delta(x_i - x_j)]^{z_{ij}}, \quad (9)$$

where one can rotate the contours in z_{ij} space to lie parallel to the imaginary axes such that $m_i + \mathrm{Re} \sum_j z_{ij} = \nu_0 - \epsilon$. Denote the Fourier transform of $[\Delta(x)]^z$ by $\Delta(k, z)$ viz.

$$[\Delta(x)]^z = \int \frac{d^4k}{(2\pi)^4} e^{-ikx} \Delta(k, z). \quad (10)$$

If we associate the internal momentum k_{ij} to each superpropagator (cocoon) joining the point x_i and x_j, the space integrations in (3) can be performed to give

$$\tilde{S}_{m_1 m_2 \cdots m_N}(p) = \frac{(iG)^N}{N!} \prod_{i>j} \int \frac{dz_{ij}}{2\pi i} \Gamma(-z_{ij})\, e^{i\pi z_{ij}} v\left(m_i + \sum_l z_{il}\right)$$

$$\cdot \int \frac{d^4k_{ij}}{(2\pi)^4} \Delta(k_{ij}, z_{ij}) \cdot (2\pi)^4\, \delta^4\left(p_i + \sum_l k_{il}\right) \quad (11)$$

an expression which resembles very closely the normal Feynman graph result (where $z_{ij} = 1$ or 0) and which makes it possible to use Dyson power counting techniques. The Fourier transform $\Delta(k, z)$ is welldefined only for $0 < \mathrm{Re}\, z < 2$ and is given outside this range by analytical continuation. A convenient approximate form for $\Delta(k, z)$ is given in Appendix B for arbitrary mass μ which becomes exact as $\mu \to 0$ or $k \to \infty$ or $z \to 0, 1$ and carries the correct threshold behaviors in momentum space. Because we shall only be interested in the high energy behavior, where $k^2/\mu^2 \to \infty$, it will be sufficient to approximate the superpropagator by the propagator for the massless case

$$iD(k, z) = \frac{\Gamma(2 - z)}{\Gamma(z)} \frac{(16\pi^2)^{1-z}}{(-k^2)^{2-z}}. \quad (12)$$

In passing, we should note that since the definition of the chronological product $\langle 0 \mid T[\phi^n(x), \phi^n(0)] \mid 0 \rangle$ is ambiguous to the extent of terms like $\sum_{r=0}^{n-2} a_r(\partial^2)^r\, \delta^4(x)$, with a_r real and finite, the Fourier transform of $\langle 0 \mid T[e^{\zeta_1\phi(x)}, e^{\zeta_2\phi(0)}] \mid 0 \rangle = e^{\zeta_1\zeta_2\Delta(x)}$ carries an ambiguity of the type $\sum_{n=0}^{\infty} a_n(\partial^2)^n\, \delta^4(x)$, where—to satisfy Jaffe localizability for the operator $e^{\zeta\phi(x)} - \sum_{n=0}^{\infty} a_n Z^n$ should be an entire function of order less than $\frac{1}{2}$. This ambiguity in the definition of the T-product would carry through for all nonpolynomial theories (whether superrenormalizable or otherwise and whether rational or not) and includes the case of rational Lagrangians through formula (15). This has been remarked upon (4) by Efimov, Lee, and Zumino and more recently by Lehmann and Pohlmeyer, Mitter, Blomer and Constantinescu. For the particular case of the exponential Lagrangian $e^{\zeta\phi}$ and for the two-point superpropagator $e^{\zeta_1\zeta_2\Delta(x)}$, Lehmann and Pohlmeyer have given a detailed discussion of this ambiguity and show that the ansatz adopted by Volkov, Arbuzov and Fillipov, and Salam and Strathdee (which amounts to setting all the $a_n = 0$) picks out what they call "minimally singular superpropagator". For the rest of this paper we shall always use the minimal singularity ansatz and neglect the type of ambiguity mentioned, even though the rigorous discussion of Lehmann and Pohlmeyer is so far available only for the two-point superpropagator in an exponential Lagrangian theory. This is because we believe, on the basis of the work in this paper, that all such ambiguous terms are finite renormalization counter Lagrangians which however contain derivatives of field operators of infinitely high order (corresponding to the derivatives occurring in $\sum^{\infty} a_n(\partial^2)^n\, \delta^4(x)$) and will inevitably give a nonlocal Lagrangian. Thus if we make an ansatz, like we have done in this paper, that if one starts with a local Lagrangian any counter Lagrangians must also be local if the theory is to be renormalizable, we shall be able to discard the ambiguities discussed above.

Returning to the consideration of high energy behavior, as far as high momentum power counts are concerned it is enough to notice that as $k^2/\mu^2 \to \infty$, expression (11) consists of products of factors like $\int dz\, \mathscr{K}(z, m)(-k^2)^{z-2}$ integrated along contours and with kernels $\mathscr{K}(z, m)$ specified by the interaction $V(\phi)$. These kernels exhibit successions of poles at values of z lying to the left of the contour and in the limit $k^2 \to -\infty$, it is the nearest pole, at $z = z_0$ say, which dominates and provides the momentum dependence $(k^2)^{z_0-2}$ wherefrom one may judge whether or not the integration over internal momenta gives finite results or not. This we shall now investigate.

3. Superrenormalizable Models

As a very simple illustration of the more general superrenormalizable case $\phi^{\nu_0}/(1 + \lambda\phi)^{\nu_1}$ with $\nu_0 - \nu_1 \leqslant 1$, take the model theory

$$V(\phi) = \frac{G\phi^2}{1 + \lambda\phi} \tag{13}$$

and apply the momentum space method for studying the ultraviolet behavior of integrals. In this model, contours lie to the left of $\sum_j z_{ij} + m_i = 2$ and the Taylor coefficients are

$$v(z) = (-\lambda)^{z-2}\, \Gamma(1 + z). \tag{14}$$

Note that because of the positions of the contours on the Re z axes, $\sum_{ij} z_{ij} + \sum_i m_i = 2F + m < 2N$, where $F = N(N-1)/2$ is the total number of internal super lines joining every pair of supervertices, and m is the total number of external lines.

To estimate the overall asymptotic p-behavior, compute all matrix elements in the Symanzik region of external momenta ($p_i \cdot p_j \leqslant 0$, etc.). As is well known, one can here perform Wick rotations and employ Euclidean internal momenta ($k_{ij}^2 < 0$). Also use the massless superpropagator (12) as we are only interested in large k^2 contributions. The integral (11) becomes

$$\tilde{S}_{m_1 m_2 \cdots m_N}(p) \approx \frac{(iG)^N}{N!} \lambda^{m-2N} \prod_{i>j} \int \frac{dz_{ij}}{2\pi} \frac{\Gamma(1-z_{ij})\,\Gamma(2-z_{ij})}{\Gamma(1+z_{ij})} \Gamma\Big(1 + m_i + \sum_l z_{il}\Big)$$

$$\cdot \int \frac{d^4 k_{ij}}{\pi^2} \left(\frac{\lambda^2}{16\pi^2}\right)^{z_{ij}} \frac{1}{(k^2)^{2-z_{ij}}} \cdot (2\pi)^4\, \delta^4\Big(p_i + \sum_l k_{il}\Big). \tag{15}$$

The number of independent internal momenta, i.e., the number of loops l, equals $F - N + 1$ for an N-th order graph with F internal lines. (Unlike Feynman diagrams in polynomial theories, $F = \frac{1}{2}N(N-1)$ is quadratically dependent on N.) Thus the overall power count of the integral gives:

$$\begin{aligned} \tilde{S} &\approx \int \Pi\, dz\, \mathscr{K}(m, z) \int (d^4k)^l\, (k^2)^{\sum_{i>j} z_{ij} - 2F} \\ &\approx \int \Pi\, dz\, \mathscr{K}(m, z) \int (d^4k)^{-N+1}\, (k^2)^{\sum z_{ij}}. \end{aligned} \tag{16}$$

In the limit $-k^2 = M^2 \to \infty$ the internal integrations provide the ultraviolet[3] factor $\lim_{M\to\infty} M^{-4N+4+2\sum_{i>j} z_i}$ which describes the overall infinite character of $\tilde{S}$ on dimensional grounds. Recalling that the nearest singularity to the left of the z contour governs the high energy behavior so that

$$\operatorname{Re} \sum_{i,j} z_{ij} < 2N - m, \tag{17}$$

we see that $\tilde{S} < M^{4-m-2N}$ and therefore ($m \geqslant 0$, $N \geqslant 2$) there is never any overall infinity in $\tilde{S}$. This argument applies to *each* subintegration and we can therefore conclude that all S-matrix elements are finite in this model theory.

[3] The quadratic dependence of F on N necessitates that the superpropagators fall faster than k^{-2} to guarantee finiteness of the integrals. To get the order of magnitude right, note that $\tilde{S} \approx \int (d^4k)^{F-N+1} (k^2)^{zF-2F}$, where we have very roughly replaced $\sum z$ by zF. Thus finiteness of $\tilde{S}$ demands $\operatorname{Re} z \leqslant 4/N$ or alternatively stated an average behavior of superpropagators like $(k^2)^{-2+4/N}$ rather than $(k^2)^{-1}$ is needed.

In more general examples like $\phi^{\nu_0}/(1+\lambda\phi)^{\nu_1}$ the z contours lie to the left of $\text{Re}\sum_l z_{il} = \nu_0 - m_i$ and at first sight it would seem that $\text{Re}\sum_{ij} z_{ij} < \nu_0 N - m$ might violate our statement of finiteness since $\tilde{S} < M^{4+(\nu_0-4)N-m}$ could possibly still be infinite. However, a careful analysis reveals that the first singularity in $\mathscr{K}(m, z)$ to the left of the contour actually occurs below

$$\text{Re}\sum_l z_{il} = \nu_0 - \nu_1 + 1 - m_i$$

because of the presence of binomial factors in the expansion of $(1+\lambda\phi)^{-\nu_1}$ with $\nu_1 > 1$ and which provide appropriate zeros. Thus, in fact,

$$\text{Re}\sum_{ij} z_{ij} < (\nu_0 - \nu_1 + 1)\, N - m$$

as regards the high energy behavior of $\tilde{S}$ (as one might have expected from the naive consideration of the Dyson index). The correct ultraviolet behavior is given by

$$\tilde{S} < M^{4+(\nu_0-\nu_1-3)N-m}, \tag{18}$$

so the theory is superrenormalizable if $\nu_0 - \nu_1 \leqslant l$. Likewise, theories which are linear combinations $\sum_i G_i\phi^{\nu_i}(1+\lambda_i\phi)^{-\kappa_i}$ with $\nu_i - \kappa_i \leqslant 1$ can be shown to be perfectly finite in all their S-matrix elements.

4. Renormalizable Models

We shall now study some simple examples where infinities do arise but can be renormalized by counterterm Lagrangians. We take our prototype superrenormalizable model and add to it a succession of polynomial terms with Dyson indices >1, viz. $G_2\phi^2 + G_3\phi^3 + \cdots$. Since $G_2\phi^2$ can be simply absorbed into a mass renormalization we shall in this section examine the first nontrivial model

$$V(\phi) + V'(\phi) \equiv \frac{G\phi^2}{1+\lambda\phi} + G'\phi^3. \tag{19}$$

In setting up a perturbation series in G and G' a term of order $G^N G'^{N'}$ consists of supergraphs from $[V(\phi)]^N$, pure ϕ^3 graphs from $[V'(\phi)]^{N'}$, and joining terms. We showed that $V(\phi)$ by itself has no infinities, and one knows that ϕ^3 theory by itself has three primitive infinities (see Fig. 2) which can be renormalized by introducing corresponding counterterms.

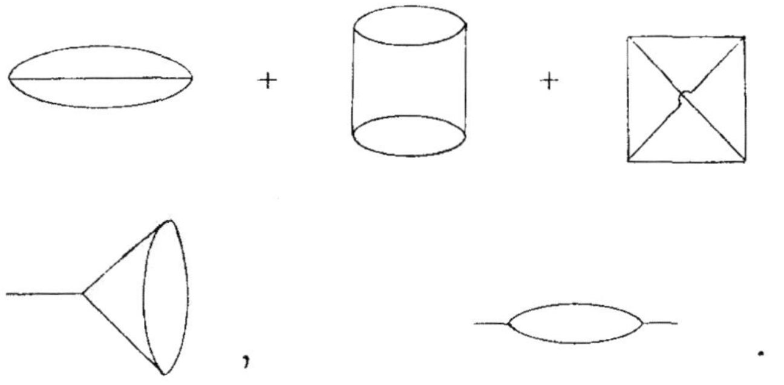

FIG. 2. The primitive infinities of ϕ^3 theory.

More specifically one has

$$A_0 \approx G'^2 c_2 + G'^4 c_0'', \qquad A_1 \approx G'^3 c_0', \qquad A_2 \approx G'^2 c_0,$$

where c_0 and c_2 are logarithmic and quadratic infinities such as

$$c_0 = i \int \Delta^2(x)\, d^4x \sim \log M, \qquad c_2 = i \int \Delta^3(x)\, d^4x \sim M^2.$$

For future reference we note that the renormalized Lagrangian of ϕ^3 theory, including counterterms, can be compactly expressed as

$$:V' + \delta V': = :\left[\exp G'\left(c_2 \frac{d^3}{d\phi^3} + c_0 \frac{d}{d\phi} + c_0' G' \frac{d^2}{d\phi^2} + c_0'' G'^2 \frac{d^3}{d\phi^3}\right)\right] V':. \quad (20)$$

We now turn to possible infinities which arise when internal lines connect V and V' interactions. (The V' infinities by themselves are assumed to be renormalized as above). To lowest order GG' one has the two infinities shown in Fig. 3 which we can term "vacuum-like" and "self-energy-like".

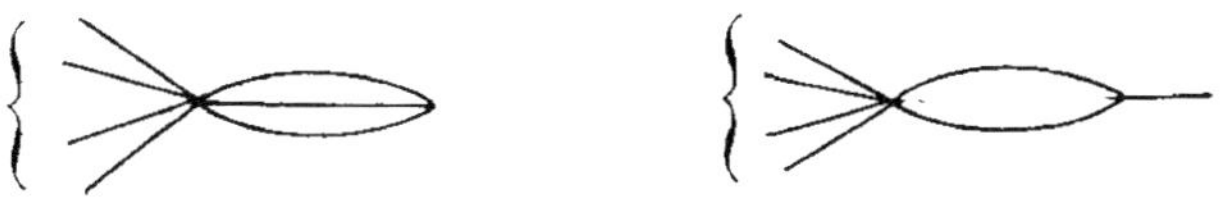

FIG. 3. Infinities to order GG' in $((G\phi^2/1 + \lambda\phi) + G'\phi^3)$ theory.

Because any number of external lines can emanate from the G-vertex we can call these two graphs primitive since they can be regarded as supervertex modifications of $V(\phi)$. The counterterm

$$\delta V = G'\left[c_0 \frac{d}{d\phi} + c_2 \frac{d^3}{d\phi^3}\right] \frac{G\phi^2}{1 + \lambda\phi}$$

can be added to the Lagrangian to cancel these infinities making *every* S-matrix element finite to order GG'.

Before we consider further infinite graphs of order $GG'^{N'}$, there is a basic repetition pattern worth recognizing as shown in Fig. 4.

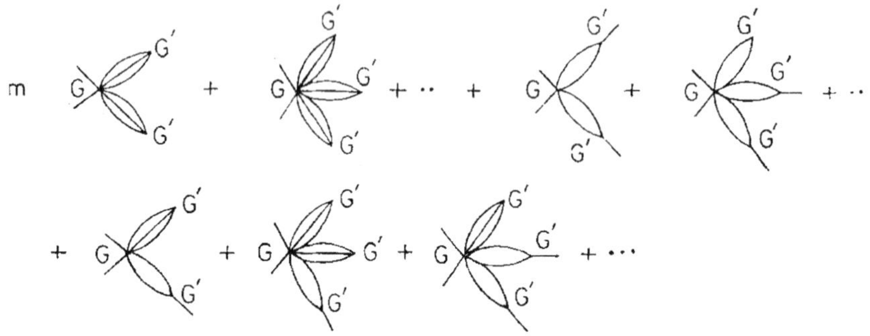

FIG. 4. Repetitions of lowest order infinities.

These repetition graphs modify the super-vertex and their effect can be elegantly expressed in the compact form:

$$\left[\exp G'\left(c_0 \frac{d}{d\phi} + c_2 \frac{d^3}{d\phi^3}\right)\right] \frac{G\phi^2}{1 + \lambda\phi}\,.$$

We need only therefore consider genuine new primitive infinities such as shown in Fig. 5 since further repetitions of them are simply given by modifying the exponential operator above. Although there are an infinite number of such graphs[4]

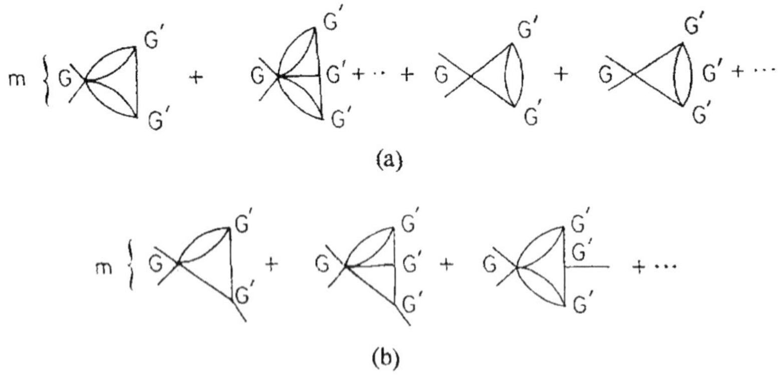

FIG. 5 (a). Higher order vacuumlike primitive infinities; (b). Higher order self-energylike primitive infinities in $((G\phi^2/1 + \lambda\phi) + G'\phi^3)$ theory.

[4] The graphs in Fig. 5 can be obtained from those of Fig. 3 by either drawing new lines out of the supervertex and attaching them through ϕ^3 vertices to the lines of Fig. 3 or by joining two lines of Fig. 3 by an extra connecting line. The first kind of attachment does not change the ultraviolet behavior; the second kind diminishes it by a factor M^2.

(similar to the infinite set of irreducible vertex parts in quantum electro-dynamics) the overall ultraviolet behaviours are still no worse than quadratic. It is this circumstance which permits us to renormalize them along conventional lines by adding further (at most quadratically) infinite counter-terms. In all we get

$$V + \delta V = \left[\exp\left(\sum_{r=0}^{\infty} C_2^{(r)} G'^{r+1} \frac{d^{r+3}}{d\phi^{r+3}} + \sum_{r=0}^{\infty} C_0^{(r)} G'^{r+1} \frac{d^{r+1}}{d\phi^{r+1}}\right)\right] V(\phi), \quad (21)$$

where $C_2^{(r)}$ denotes a sum of quadratically infinite constants $\sim M^2$ (as given by Fig. 5(a)) and $C_0^{(r)}$ a sum of logarithmically infinite constants $\sim(\log M)^\nu$ (as given by Fig. 5(b)).

We shall now show that with the counterterms $\delta V + \delta V'$ there are no further infinities in any order. Consider first graphs to order $G^2G'^N$. Characterize it as a $(n_1 + n_2 + m)$-point function of ϕ^3 theory connected by $n_1 + n_2$ lines to the super-propagator of $V(\phi)$ theory having $m_1 + m_2$ external lines as depicted in Fig. 6.

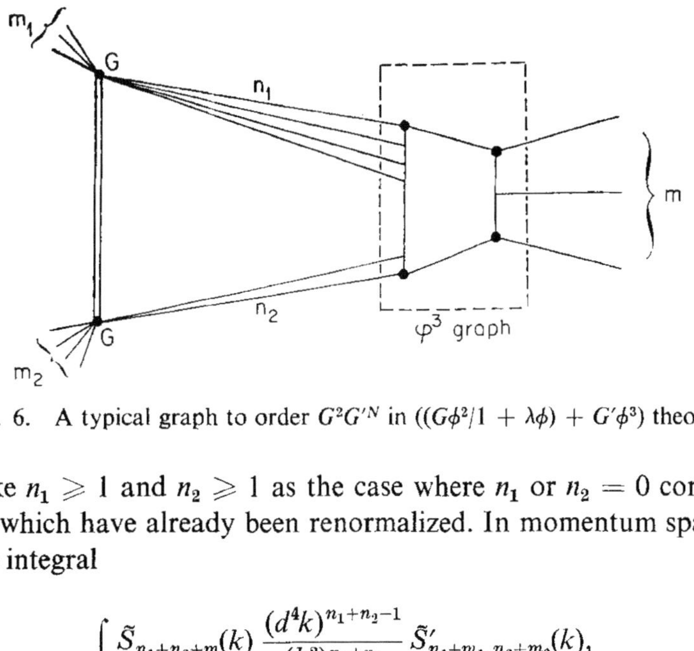

FIG. 6. A typical graph to order $G^2G'^N$ in $((G\phi^2/1 + \lambda\phi) + G'\phi^3)$ theory.

We must take $n_1 \geqslant 1$ and $n_2 \geqslant 1$ as the case where n_1 or $n_2 = 0$ corresponds to attachments which have already been renormalized. In momentum space we must deal with the integral

$$\int \tilde{S}_{n_1+n_2+m}(k) \frac{(d^4k)^{n_1+n_2-1}}{(k^2)^{n_1+n_2}} \tilde{S}'_{n_1+m_1, n_2+m_2}(k),$$

where S and S' refer to elements corresponding to V and V'. It is well known in ϕ^3 theory that $\tilde{S}_{n_1+n_2+m}(k) \sim k^{4-N-m-n_1-n_2}$ while in the supertheory according to Eq. (7), $S_{n_1+m_1,\, n_2+m_2}(\Delta) = (\partial/\partial\Delta)^{n_1+m_1} S_{0, n_2+m_2-n_1-m_1}$ if $n_1 + m_1 < n_2 + m_2$ say; or $S_{n_1+m_1, n_2+m_2}(k) \sim (k^2)^{-n_1-m_1} k^{-2}$. Therefore the ultraviolet behavior of our integral is given by

$$M^{n_1+n_2-N-m-2-2(n_1+m_1)} < M^{-n_1+n_2-N-2}.$$

But the connectedness properties of our graphs always ensure that $N - n_2 \geqslant -1$, so that in every case the integrals converge at least as rapidly as M^{-2}.

With more supervertices the situation keeps on improving owing to greater convergence of the internal integrations. Thus all graphs of order $G^2G'^N$, $G^3G'^N$,... are finite. In short the renormalized theory, described by the Lagrangian

$$L = V + \delta V + V' + \delta V'$$
$$= \left[\exp\left(\sum_{r=0}^{\infty} C_2^{(r)} G'^{r+1} \frac{d^{r+3}}{d\phi^{r+3}} + \sum_{r=0}^{\infty} C_0^{(r)} G'^{r+1} \frac{d^{r+1}}{d\phi^{r+1}}\right)\right](V + V') \tag{22}$$

has finite S-matrix elements (including self-mass and self-charge).

It is instructive to see what our modifications to the Lagrangian do to the Laplace transformed interaction $\mathscr{L}(\zeta)$. Before renormalization the infinities come essentially from $\phi \to \infty$ or $\zeta \to 0$. The introduction of counterterms to $V(\phi)$ gives the new interaction

$$\left[\exp\left(c\frac{d^2}{d\phi^2} + \cdots\right)\right] V(\phi) \qquad \text{or} \qquad \mathscr{V}(\zeta) \exp[c\zeta^2 + \cdots].$$

Thus, renormalization affects the behavior of the Lagrangian in the limit $\zeta \to 0$ so as to eliminate all perturbation infinities.

5. Unrenormalizable Models

If we add to the supernormalizable interaction $V(\phi)$ the polynomial terms $G_5\phi^5 + G_6\phi^6 + \cdots$ it is obvious that the theory cannot be renormalized since ϕ^5, ϕ^6 are already intractable by themselves. However, if we take the model theory

$$V + V'' = \frac{G\phi^2}{1 + \lambda\phi} + G''\phi^4, \tag{23}$$

it looks, on the face of it, capable of renormalization as the Dyson index ν of the theory $[L(\phi) \to \phi^\nu$ as $\phi \to \infty]$ is 4. We shall now prove that even so the theory cannot be renormalized in the standard way by introducing counterterms. Basically this is because the infinities progressively get worse in higher orders and require counterterms with an increasing number of derivatives. The theory then becomes manifestly nonlocal and no S-matrix is calculable in terms of a few basic parameters.

The demonstration follows the methods of the previous section. To order G^N *or* to order $(G'')^N$ there is evidently no problem since in one case the theory is finite or in the other case it is easily renormalized. The difficulties arise from cross terms. Adopting the most optimistic attitude, disregard vacuumlike graphs as in Fig. 7(a)

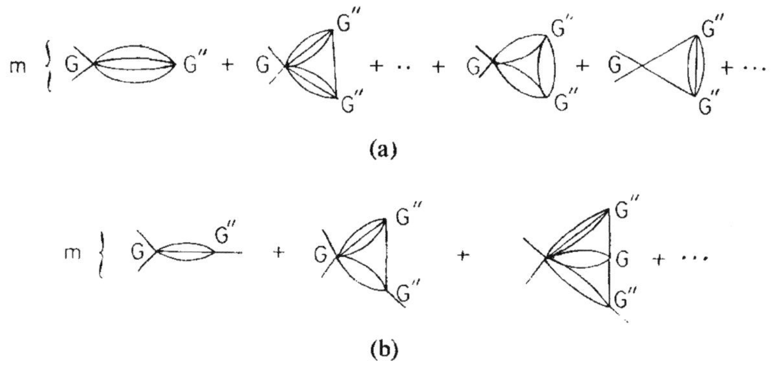

FIG. 7(a). Vacuumlike graphs to order GG''^N in $((G\phi^2/1 + \lambda\phi) + G''\phi^4)$ theory. (b). Self-energylike graphs to order GG''^N in $((G\phi^2/1 + \lambda\phi) + G''\phi^4)$ theory.

and their repetitions, since they only contribute multiplicative terms to the supervertex (although the infinities do get worse as the order of G'' is increased!). Consider the self energylike graphs of Fig. 7(b) and repetitions thereof. Unlike vacuum graphs, these are momentum dependent and the counterterms needed will necessarily involve derivatives of the field ϕ. Since their infinities go as $M^2, M^4, M^6 ...$, an arbitrarily increasing number of subtraction terms are needed with higher and higher derivatives. *This is the typical nonrenormalizable situation.* The same difficulty occurs for *all* other modifications of the supervertex, so that not only are self-energylike modifications of the supervertex infinite but also *all* the scatteringlike modifications. The same situation holds for more complicated graphs with more than one supervertex.

The problems studied in this paper are a preliminary step to the consideration of the renormalization status of chiral, Yang–Mills and Weak interaction Lagrangians all of which, by suitable field transformations, can be written (3) as a mixture of polynomial and nonpolynomial terms. These more realistic cases are currently under study.

APPENDIX A: NORMAL ORDERING

Suppose we are given an interaction Lagrangian $L(\phi)$. This can be rewritten as a normally ordered interaction : $V(\phi)$: for all polynomial L by repeated application of the identity $\phi^2 = :\phi^2: + i\Delta_+(0)$. Noting that if the approach to zero is made in the timelike direction, $i\Delta_+(0) = \Delta(0)$, where $\Delta(x)$ is the causal propagator, let us rewrite the operation as

$$\phi^2 = :\left[\exp\left(\frac{1}{2}\,\Delta(0)\,\frac{d^2}{d\phi^2}\right)\right]\phi^2: .$$

This normal ordering operation actually applies to polynomials ϕ^ν and can thereby be generalized to arbitrary interaction functions. Thus, we have the theorem that

$$U(\phi) = :V(\phi):\,,$$

where

$$V(\phi) = \left[\exp\left(\frac{1}{2}\,\Delta(0)\,\frac{d^2}{d\phi^2}\right)\right] U(\phi)$$

or

$$U(\phi) = \left[\exp\left(-\frac{1}{2}\,\Delta(0)\,\frac{d^2}{d\phi^2}\right)\right] V(\phi). \tag{A.1}$$

One can write an integral representation to compute $V(\phi)$,

$$V(\phi) = \frac{1}{\sqrt{2\pi}} \int_{-\infty}^{\infty} e^{-\frac{1}{2}u^2} U(\phi + u\Delta^{\frac{1}{2}}(0))\, du. \tag{A.2}$$

For example if

$$\begin{aligned} U(\phi) &= G\phi^\nu, \\ V(\phi) &= G(-\Delta(0))^{\frac{1}{2}\nu}\, H_\nu[\phi(-\Delta(0))^{-\frac{1}{2}}], \\ H &= \text{Hermite polynomial}; \end{aligned} \tag{A.3a}$$

$$\begin{aligned} U(\phi) &= Ge^{-\lambda\phi}, \\ V(\phi) &= Ge^{-\lambda\phi + \frac{1}{2}\lambda^2\Delta(0)}, \text{ etc.} \end{aligned} \tag{A.3b}$$

Diagrammatically the operation $\exp[\frac{1}{2}\,\Delta(0)\, d^2/d\phi^2]$ has the effect of introducing tadpole loops *at* every vertex of a Feynman graph and is analogous to the action of $\exp[\frac{1}{2}\sum_{i\neq j} \Delta_{ij}\, \partial^2/\partial\phi_i\partial\phi_j]$ which introduces propagators *between* vertices x_i and x_j of the graph. In general, it should be noted that the functional form of V may be very different from U. Only for exponential and polynomial interactions do both functions have the same character.

APPENDIX B: The Superpropagator

For a massive free particle the causal two-point function can be written as

$$\Delta(x) = \frac{\mu K_1(\mu r)}{4\pi^2 r}; \qquad r^2 = -x^2. \tag{B.1}$$

Δ is real for $x^2 < 0$ (spacelike) and is cut in the x^2 plane from 0 to ∞, the value on the physical sheet for $x^2 > 0$ (timelike) being obtained from below the cut by an $x^2 - i\epsilon$ prescription. In the limit $\mu \to 0$, $D(x) = 1/4\pi^2(r^2 + i\epsilon)$ is the massless propagator. The Fourier transforms are of course given by

$$\tilde{\Delta}(k) = i(k^2 - \mu + i\epsilon)^{-1}, \qquad \tilde{D}(k) = i(k^2 + i\epsilon)^{-1}.$$

In the text we are interested in the Fourier transform of the superpropagator $[\Delta(x)]^z$ for arbitrary complex z,

$$\Delta(k, z) \equiv \int d^4x \, e^{ikx}[\Delta(x)]^z. \tag{B.2}$$

In the *massless* case Gel'fand an Shilov [2] give the result as

$$iD(k, z) = \frac{\Gamma(2 - z)}{\Gamma(z)} \cdot \frac{(16\pi^2)^{1-z}}{(-k^2)^{2-z}} \, . \tag{B.3}$$

This can be analytically continued outside the range $0 < \operatorname{Re} z < 2$, where it is originally well-defined. In doing so it is necessary to see the prescription $k^2 \to k^2 + i\epsilon$ to obtain the correct distributions at nonpositive integer z values; for example,

$$\lim_{z \to 0} D(k, z) = (2\pi)^4 \, \delta^4(k), \text{ etc.}$$

We shall give here an *approximate* solution to the problem for the *massive* case which has the virtue of reducing to well-known functions in various limits ($z \to 0, 1$; $\mu \to 0$, $p^2 \to \infty$,...) and which also has the correct thresholds and types of discontinuities at the branch points in the p^2 plane. To this end note that $[\Delta(x)]^z$ is cut in the x^2 plane from 0 to ∞, and that

$$\begin{aligned} [K_1(\mu r)]^z &\approx (\mu r)^{-z} && \text{as } x^2 \to 0, \\ &\approx \left(\frac{\pi}{2\mu r}\right)^{\frac{1}{2}z} e^{-\mu r z} && \text{as } x^2 \to \infty. \end{aligned}$$

After Wick rotation, the integration we must perform is ($\kappa^2 = -k^2$)

$$i\Delta(k, z) = \frac{4\pi^2}{\kappa} \int_0^\infty dr \cdot r^2 J_1(\kappa r) \left[\frac{\mu r K_1(\mu r)}{4\pi^2 r^2}\right]^z, \tag{B.4}$$

which cannot be done in general except in very special circumstances ($z = 1$ or $\mu = 0$). Let us therefore replace $[K_1(\mu r)]^z$ by $K_\nu(\rho)$ multiplied into some simple polynomial of μr such that the behaviors as $r \to 0$ ($p^2 \to \infty$) and $r \to \infty$ (p^2 thresholds) are correctly reproduced. This suggests the approximation

$$[K_1(\mu r)]^z \approx \frac{z^{\frac{1}{2}(1+z)}}{\Gamma[\frac{1}{2}(1 + z)]} (2\mu r)^{\frac{1}{2}(1-z)} \, K_{\frac{1}{2}(1+z)}(\mu r z)$$

or

$$[\Delta(x)]^z \approx (8\pi^2 r^2)^{-z} \frac{(2\mu r z)^{\frac{1}{2}(1+z)}}{\Gamma[\frac{1}{2}(1+z)]} K_{\frac{1}{2}(1+z)}(\mu r z). \tag{B.5}$$

Hence (B.4) reads

$$i\Delta(k, z) \approx \frac{(8\pi^2)^{1-z}\,(2\mu z)^{\frac{1}{2}(1+z)}}{2\kappa\Gamma[\frac{1}{2}(1+z)]} \int_0^\infty dr \cdot r^{\frac{1}{2}(5-3z)} J_1(\kappa r)\, K_{\frac{1}{2}(1+z)}(\mu z r).$$

Carrying out the integration we obtain

$$i\,\Delta(k, z) \approx (16\pi^2)^{1-z}\,\Gamma(2-z)\,\Gamma[\tfrac{1}{2}(5-z)]\,\Gamma^{-1}[\tfrac{1}{2}(1+z)] \times (\mu^2 z^2)^{z-2}\, F(\tfrac{1}{2}(5-z), 2-z; 2; k^2/\mu^2 z^2). \tag{B.6}$$

This superpropagator therefore has a cut in the k^2 plane from the threshold $\mu^2 z^2$ with discontinuity of (B.6) given by

$$\operatorname{disc} \Delta(k, z) \approx \frac{(16\pi^2)^{1-z}\,(\mu^2 z^2)^{\frac{1}{2}(1+z)}\,(k^2 - \mu^2 z^2)^{\frac{1}{2}(3z-5)}}{\Gamma[\frac{1}{2}(1+z)]\,\Gamma(\frac{1}{2}[-3+3z)]} \cdot F\left(\frac{1}{2}(z-1), z; \frac{3}{2}(z-1); 1 - \frac{k^2}{\mu^2 z^2}\right). \tag{B.7}$$

To take the *high energy limit* $k^2/\mu^2 \to \infty$ we shall work in the region $\operatorname{Re} z > -1$ and then continue our results to other z values. Thus

$$\lim_{k^2/\mu^2\to\infty} i\Delta(k, z) \approx \frac{\Gamma(2-z)}{\Gamma(z)} \frac{(16\pi^2)^{1-z}}{(-k^2)^{2-z}} F\left(2-z, 1+z; \frac{1}{2}(1-z); \frac{\mu^2 z^2}{k^2}\right) \approx iD(k, z). \tag{B.8}$$

Note that the *threshold* behavior in k^2 is corectly obtained at all *positive integer* z since $\Delta(k, z)$ possesses a square-root branch cut for even z and a logarithmic branch cut for odd z according to (B.7). Lastly, we may check that as $z \to 0$ or 1 the correct distributions follow, viz.,

$$\lim_{z\to 1} \Delta(k, z) = i(k^2 - \mu^2),$$

$$\lim_{z\to 0} \Delta(k, z) = (2\pi)^4\, \delta^4(k).$$

APPENDIX C: Polynomial Models as Limits of Rational Models

As a corollary of our work we shall show that in the lim $\lambda^2 \to 0$ of $G\phi^3(1 + \lambda\phi)^{-1}$ and $g\phi^4(1 + \lambda\phi)^{-1}$ we recover the conventional renormalization counterterms of $G\phi^3$ and $g\phi^4$ theory from the counterterms of the rational models.

First, take the simpler case

$$V_3(\phi) = \frac{G\phi^3}{1 + \lambda\phi} = \frac{G}{\lambda}\left[\phi^2 - \frac{\phi^2}{1 + \lambda\phi}\right]. \tag{C.1}$$

The counterterms correspond to vacuumlike infinities c_0 of ϕ^2 theory, and if we follow the same steps as in Section 5, the final renormalized Lagrangian L_3 free of infinities[5] is

$$L_3 = V_3 + \delta V_3 = \frac{c_0 G^2}{\lambda^2} + \left[\exp\left(\frac{c_0 G}{\lambda}\frac{d^2}{d\phi^2}\right)\right]\frac{G\phi^3}{1 + \lambda\phi}\cdot \tag{C.2}$$

Now proceed to the limit $\lambda \to 0$ counting $1/\lambda^2$ as a quadratic infinity like $c_2 = i\int \Delta^3(x)\, d^4x \sim M^2$. Then we get

$$\begin{aligned} L_3 \to \frac{c_0 G^2}{\lambda^2} - \frac{6!}{3!}c_0{}^3 G^4 + \frac{5!}{2!}c_0{}^2 G^3\phi - \frac{4!}{2!}c_0 G^2\phi^2 + G\phi^3 \\ - 4!\,\frac{c_0{}^2 G^3}{\lambda} + 3!\,\frac{c_0 G^2}{\lambda}\phi, \end{aligned} \tag{C.3}$$

which exactly corresponds to the renormalized Lagrangian for $G\phi^3$ theory apart from the linearly divergent terms $G^3/\lambda + G^2\phi/\lambda$. These terms have no counterpart in momentum space integrals and can be eliminated by taking the principal value limit $\frac{1}{2}(\lim_{\lambda\to 0+} + \lim_{\lambda\to 0-})$ which averages out the singularity at $\lambda \to 0$.

This next case is more subtle. Begin with the unrenormalised Lagrangian

$$V_4 = \frac{g\phi^4}{1 + \lambda\phi} = \frac{g}{\lambda^2}\left[\lambda\phi^3 - \phi^2 + \frac{\phi^2}{1 + \lambda\phi}\right] \tag{C.4}$$

[5] Note that the exponential term just corresponds to a mass shift by G/λ in the tadpole graphs, viz,

$$\begin{aligned} \Delta\left(0, m^2 + \frac{G}{\lambda}\right) &= \int d^4p(p^2 - m^2 + G/\lambda)^{-1} \\ &= \int d^4p(p^2 - m^2)^{-1} + \frac{G}{\lambda}\int d^4p(p^2 - m^2)^{-2} + \cdots \\ &= \Delta(0, m^2) + c_0 G/\lambda + \text{finite terms}. \end{aligned}$$

and look for all infinite vertex attachments as previously. Keep the leading infinities for taking the limit $\lambda^2 \to 0$, and add counter terms

$$V_4 + \delta V_4 \approx \frac{g^2}{\lambda^2}(c_2 + c_0\varphi^2) + \\ + \left[\exp \sum_{r=0}^{\infty} \left(\frac{g}{\lambda}\right)^{r+1} \left(C_2^{(r)} \frac{d^{r+3}}{d\phi^{r+3}} + C_0^{(r)} \frac{d^{r+1}}{d\phi^{r+1}}\right]\frac{g\phi^4}{1+\lambda\phi},$$

where C_2 stands for the infinite set of quadratic and C_0 for the infinite set of logarithmic infinities. Expanding the exponential operator, the denominator $(1 + \lambda\phi)$ and eliminating linear and cubic infinities by taking the principal value limit, we obtain the series of terms $A_0 + A_2\phi^2 + A_4\phi^4$, where the most infinite contributions are

$$A_0 \sim \frac{c_2 g^2}{\lambda^2} + \frac{1}{\lambda^2}\sum_{r=1}^{\infty}(-g)^{r+1} C_2^{(r)}(r+3)!,$$

$$A_2 \sim \frac{c_0 g^2}{\lambda^2} + \sum_{r=0}^{\infty}(-g)^{r+1} C_2^{(r)} \frac{(r+5)!}{2} + \frac{1}{\lambda^2}\sum_{r=0}^{\infty}(-g)^{r+1} C_0^{(r)} \frac{(r+3)!}{2},$$

$$A_4 \sim g + \lambda^2 \sum_{r=0}^{\infty}(-g)^{r+1} C_2^{(r)} \frac{(r+7)!}{4!} + \sum_{r=0}^{\infty}(-g)^{r+1} C_0^{(r)} \frac{(r+5)!}{4!}.$$

If as $\lambda^2 \to 0$, we count C_2/λ^2 as a quartic infinity $\mathscr{C}_4 \sim M^4$, C_0/λ^2 as a quadratic infinity $\mathscr{C}_2 \sim M^2$, and $\lambda^2 C_2$ as a logarithmic infinity $\mathscr{C}_0 \sim (\log M^\nu)$, the Lagrangian we get is $\mathscr{C}_4 + \mathscr{C}_2\phi^2 + (\mathscr{C}_0 + 1)\, g\phi^4$, so that $\mathscr{C}_0$ corresponds to vacuum graphs, $\mathscr{C}_2$ can be interpreted as a self-mass term and $\mathscr{C}_4$ as a meson scattering renormalization. The reason why the wavefunction renormalization is missing is that we must be more careful when we take the limit $\lambda^2 \to 0$. The point is that self-energy graphs of Fig. 5(b) give convergent contributions $C_{-2}p^2/\lambda^2$ to the kinetic energy only for $\lambda^2 \neq 0$, where $C_{-2} \sim M^{-2}$. If we pass to $\lambda^2 \to 0$ these turn into logarithmically infinite counterterms $C_{-2}\lambda^{-2} \cdot (\partial\phi)^2 \to \mathscr{C}_0'(\partial\phi)^2$. With the identifications $\mathscr{C}_2 = Z_3\, \delta m^2$, $\mathscr{C}_0 = Z_4 - 1$, and $\mathscr{C}_0' = Z_3 - 1$, the correspondence with the renormalized $g\phi^4$ is complete.

A model which avoids the previous cubic and linear infinity difficulties associated with $\lambda \to 0$ is provided by $U_4 = g\phi^4/(1 - \lambda^2\phi^2)$. When $\lambda \neq 0$, the renormalized Lagrangian is

$$U_4 + \delta U_4 = \frac{g^2 c_0}{\lambda^4} + \left[\exp\left(\frac{g c_0}{\lambda^2}\frac{d^2}{d\phi^2}\right)\right]\frac{g\phi^4}{1 - \lambda^2\phi^2}. \tag{C.6}$$

Now the infinite irreducible set of quartic, quadratic and logarithmic infinities of ϕ^4 theory is exactly recovered by taking the limit $\lambda^2 \to 0$, as the reader may easily check from the power series expansion of (C.6).

References

1. R. Delbourgo, Abdus Salam, and J. Strathdee, Infinities of nonlinear Lagrangian theories, *Phys. Rev.* **187** (1969), 1909; Abdus Salam and J. Strathdee, "Momentum space behavior of integrals in nonpolynomial Lagrangian theories, Phys. Rev. (to appear). We shall refer to these as papers I and II in the text.
2. I. M. Gel'fand and G. E. Shilov, "Generalized Functions," Vol. I, Academic Press, New York, 1964.
3. Abdus Salam, Nonpolynomial Lagrangian theories, "Proceedings of the Coral Gables Conference on Fundamental Interactions at High Energy, Miami, Gordon & Breach, New York, 1970.
4. G. V. Efimov, *Nucl. Phys.* **74** (1965), 657; B. W. Lee and B. Zumino, *Nucl. Phys.* **B13** (1969), 671; P. K. Mitter, "On an Analytic Approach to the Regularization of Weak Interaction Singularities," Oxford preprint; H. Lehmann and K. Pohlmeyer, "On the Superpropagator of Fields with Exponential Coupling," DESY preprint; R. Blomer and F. Constantinscu, "On the Zero Mass Superpropagator," Munich preprint.
5. M. K. Volkov, *Ann. Phys. (New York)* **49** (1968), 202; B. A. Arbuzov and A. T. Fillipov, *Soviet Phys. JETP* **49** (1965), 990.

On matrix superpropagators. I

J. Ashmore and R. Delbourgo
Physics Department, Imperial College, London SW7 2BZ, England
(Received 29 July 1971)

Given the free propagator of a matrix-valued field $\phi_{\alpha\beta}$ in the form $\langle\phi_{\alpha\beta}(x), \phi_{\gamma\delta}(0)\rangle = (1/2)(\delta_{\alpha\gamma}\delta_{\beta\delta} + \delta_{\alpha\delta}\delta_{\beta\gamma} - 2c\delta_{\alpha\beta}\delta_{\gamma\delta})\Delta(x)$, we derive an integral representation for the matrix superpropagator $\langle\phi^N_{\alpha\beta}(x), \phi^N_{\gamma\delta}(0)\rangle$ for arbitrary N, and apply this to find the exponentially parametrized gravity superpropagator $\langle|-g(x)|^\omega g_{\alpha\beta}(x), |-g(0)|^\omega g_{\gamma\delta}(0)\rangle$ with $g_{\mu\nu}(x) \equiv [\exp\kappa\phi(x)]_{\mu\nu}$. Other applications are also mentioned.

1. INTRODUCTION

Nonpolynomial Lagrangians are finding increasing application in quantum field theories of elementary particle interactions. The techniques needed to cope with such complicated Lagrangians and the results that have appeared to date have been largely limited to the case when the field $\phi(x)$ is scalar[1] or a member of an $SU(2)$ multiplet,[2] though some progress has lately been achieved towards extending the methods[3] to $SU(3)$ fields. In the meantime one interesting problem which has arisen concerns the exponential parameterization of gravity[4] $g_{\mu\nu}(x) = [\exp\kappa\phi(x)]_{\mu\nu}$, and its associated superpropagator

$$\langle|-g(x)|^\omega g_{\alpha\beta}(x),\ |-g(0)|^\omega g_{\gamma\delta}(0)\rangle$$

for damping out ultraviolet infinities and providing a localizable field theory—a nontrivial generalization of the scalar field situation which is *not* amenable to any of the nonpolynomial techniques[2] presented heretofore. The main object of our paper will be to provide the answer in *closed form* in terms of the free propagator Δ and the gauge parameter c occurring in

$$\langle\phi_{\alpha\beta}(x), \phi_{\gamma\delta}(0)\rangle = \tfrac{1}{2}[\delta_{\alpha\gamma}\delta_{\beta\delta} + \delta_{\alpha\delta}\delta_{\beta\gamma} - 2c\,\delta_{\alpha\beta}\delta_{\gamma\delta}]\Delta(x), \quad (1)$$

although the new techniques which we develop find numerous other applications.

Basic to the success of the whole approach is to look for an integral representation in which an *invariant function* of the matrix field $\phi(x)$ appears to an *arbitrary power*. Only when this is found can one, by appropriate manipulation, derive an integral representation for $\langle\mathrm{Tr}[\phi^N(x)], \mathrm{Tr}[\phi^N(o)]\rangle$, with N *arbitrary*, and thence proceed to the matrix function

$$\langle\phi^N_{\alpha\beta}(x), \phi^N_{\gamma\delta}(0)\rangle \equiv K^{(N)}_{\alpha\beta\gamma\delta}\Delta^N(x)$$

as we shall show. Thus whereas the simple integral representation

$$|-g|^{-1/2} = \pi^{-2}\int d^4n\, e^{n_\alpha g_{\alpha\beta} n_\beta}$$

has been successfully used in the past[5] for dealing with the rational parameterization of gravity $g = \eta + \kappa\phi$, it is quite futile for our purpose since the (invariant) determinant $|-g|$ appears to be a fixed power of $-\frac{1}{2}$ and will not lend itself to determining matrix superpropagators

$$\langle F[\phi(x)]_{\alpha\beta}, F[\phi(0)]_{\gamma\delta}\rangle$$

for *general* matrix functions F. Combing through the literature, it seems that generalizations of integral representations of a single variable y to a $\nu\times\nu$ matrix Y involve the *determinant* $|Y|$ of the matrix. Indeed one of the simplest such generalizations is Siegel's integral[6] which provides an expression for $|Y|$ to an arbitrary power:

$$\int dX\,|X|^\mu e^{-\mathrm{Tr}(XY)} = \pi^{\nu(\nu-1)/4}\Gamma_\nu(\mu)|Y|^{-\mu-(1/2)(\nu+1)}, \quad (2)$$

where

$$\Gamma_\nu(\mu) \equiv \Gamma(\mu+1)\Gamma(\mu+\tfrac{3}{2})\cdots\Gamma[\mu+\tfrac{1}{2}(\nu+1)],$$
$$dX = d^{\nu(\nu+1)/2}x \equiv \prod_{\alpha\leqslant\beta} dx_{\alpha\beta}, \quad (3)$$

and the integration is taken over all $\frac{1}{2}\nu(\nu+1)$ elements $x_{\alpha\beta}$ which maintain the $\nu\times\nu$ matrix X *symmetric* and *positive definite*. Initially representation (2) is defined for $\mathrm{Re}\mu > -1$; but we shall later analytically continue the formula to other μ values. When $\nu = 1$, Siegel's integral will be recognized as the conventional definition of the gamma function.

In Sec. 2 we shall show how (2) and (3) provide the key for finding integral representations of other invariants and specifically $\mathrm{Tr}(X^N)$. Used in conjunction with (1) we show how this enables us to arrive at integral representations of general matrix superpropagators in Sec. 3. In Sec. 4 we reduce these integrals to Pfaffians over double variable integrals following methods which have been extensively used in statistical mechanics.[7] The relevant Pfaffians are evaluated for gravity in Sec. 5 and the exponential superpropagator explicitly spelled out. We conclude the paper with the applicability of our new method to other cases such as chiral $SU(3)$.

2. AN INTEGRAL REPRESENTATION FOR $\langle\mathrm{Tr}(\phi^N), \mathrm{Tr}(\phi^N)\rangle$

Begin with Siegel's integral (2) and make the substitution $Y = 1 + \kappa\phi$. Then take the vacuum expectation value of the product of two such integrals, remembering from (1) that

$$\langle e^{[-\mathrm{Tr}(A\phi(x))]}, e^{[-\mathrm{Tr}(B\phi(0))]}\rangle = e^{\{[\mathrm{Tr}(AB) - c\,\mathrm{Tr}A\,\mathrm{Tr}B]\Delta(x)\}}. \quad (4)$$

Thus we get

$$\begin{aligned}
&\langle|1+\kappa\phi(x)|^{-\mu-(1/2)(\nu+1)}, |1+\kappa'\phi(0)|^{-\mu'-(1/2)(\nu+1)}\rangle \\
&= \int \frac{dX\,dX'}{\pi^{\nu(\nu-1)/2}}\frac{|X|^\mu|X'|^{\mu'}}{\Gamma_\nu(\mu)\Gamma_\nu(\mu')} \times \langle e^{\{-\mathrm{Tr}(X[1+\kappa\phi(x)])\}}, e^{\{-\mathrm{Tr}(X'[1+\kappa\phi(0)])\}}\rangle \\
&= \int \frac{dX\,dX'}{\pi^{\nu(\nu-1)/2}}\frac{|X|^\mu|X'|^{\mu'}}{\Gamma_\nu(\mu)\Gamma_\nu(\mu')} \times e^{-\mathrm{Tr}X - \mathrm{Tr}X' + \kappa\kappa'\Delta[\mathrm{Tr}(XX') - c\,\mathrm{Tr}X\,\mathrm{Tr}X']} \quad (5)\\
&= \int \frac{dX}{\pi^{\nu(\nu-1)/4}}\frac{|X|^\mu}{\Gamma_\nu(\mu)} \times e^{-\mathrm{Tr}X}|1-\kappa\kappa'\Delta(X - c\,\mathrm{Tr}X)|^{-\mu'-(1/2)(\nu+1)}.
\end{aligned}$$

 Copyright © 1973 by the American Institute of Physics

Noting that

$$|1+\kappa\phi|^{-n} = \exp[-n\ \mathrm{Tr}\ \log(1+\kappa\phi)] = 1 - n\ \mathrm{Tr}\ \log(1+\kappa\phi) + o(n^2),$$

we can take μ derivatives of (5) at $-\frac{1}{2}(\nu+1)$ to get

$$\langle \mathrm{Tr}\ \log[1+\kappa\phi(x)], \mathrm{Tr}\ \log[1+\kappa'\phi(0)]\rangle = \frac{\partial}{\partial\mu}\bigg|_{(\nu+1)/2} \int \frac{dX}{\pi^{\nu(\nu-1)/4}} \frac{|X|^{\mu}}{\Gamma_\nu(\mu)} \times e^{-\mathrm{Tr}X}\ \mathrm{Tr}\ \log[1-\kappa\kappa'\Delta(X - c\ \mathrm{Tr}X)]$$

and then pick terms of order $(\kappa\kappa')^N$ to reach our desired integral representation

$$\langle \mathrm{Tr}[\phi^N(x)], \mathrm{Tr}[\phi^N(0)]\rangle = N\Delta^N \frac{\partial}{\partial\mu} I_\nu^{(c)}(\mu,N)\bigg|_{\mu=-(\nu+1)/2}, \quad (6)$$

where

$$I_\nu^{(c)}(\mu,N) \equiv \int \frac{dX}{\pi^{\nu(\nu-1)/4}} \frac{|X|^{\mu}}{\Gamma_\nu(\mu)} e^{-\mathrm{Tr}X}\ \mathrm{Tr}((X - c\ \mathrm{Tr}X)^N). \quad (7)$$

None of the nonpolynomial Lagrangian methods to be found in the literature have succeeded in deriving such a closed form expression, so in (6) and (7) we have a powerful new result. Of course as they stand the formulas are not useful because I is a $\frac{1}{2}\nu(\nu+1)$—fold integral which has to be differentiated; but in Secs. 4 and 5 we show how to carry out such calculations. For the present, however, we simply note that the $c \neq 0$ case follows by straightforward binomial manipulation from the case $c = 0$. Thus

$$\int dX\, |X|^{\mu} e^{-\mathrm{Tr}X}\ \mathrm{Tr}((X - c\ \mathrm{Tr}X)^N)$$

$$= \sum_{n=0}^{N} \binom{N}{n} \left(c\frac{\partial}{\partial b}\right)^{N-n} \int dX\, |X|^{\mu} e^{-b\,\mathrm{Tr}X}\ \mathrm{Tr}(X^n)\bigg|_{b=1}$$

$$= \sum_{n=0}^{N} \binom{N}{n} \frac{\Gamma(N+\mu\nu+\frac{1}{2}\nu(\nu+1))}{\Gamma(n+\mu\nu+\frac{1}{2}\nu(\nu+1))} (-c)^{N-n} \times \int dX\, |X|^{\mu} e^{-\mathrm{Tr}X}\ \mathrm{Tr}(X^n)$$

upon rescaling X by the factor b. Hence

$$I_\nu^{(c)}(\mu,N) = \sum_{n=0}^{\infty} \frac{\Gamma(N+1)\Gamma(N+\mu\nu+\frac{1}{2}\nu(\nu+1))(-c)^{N-n}}{\Gamma(n+1)\Gamma(n+\mu\nu+\frac{1}{2}\nu(\nu+1))\Gamma(N-n+1)} I_\nu^{(0)}(\mu,n). \quad (8)$$

3. MATRIX SUPERPROPAGATORS

We shall now prove that once we have calculated (6) so that the coefficients a_N in

$$\langle \mathrm{Tr}(\phi^N(x)), \mathrm{Tr}(\phi^N(0))\rangle \equiv N!\,\nu a_N \Delta^N(x) \quad (9)$$

are known, this is sufficient to determine all matrix superpropagators

$$\langle F_{\alpha\beta}(\phi), F_{\gamma\delta}(\phi)\rangle, \qquad F(\phi) \equiv \sum_N \frac{F_N \phi^N(x)}{N!}. \quad (10)$$

To that end we first write the general form for the N'th power

$$\langle \phi^N_{\alpha\beta}(x), \phi^N_{\gamma\delta}(0)\rangle \equiv N![\tfrac{1}{2}(\delta_{\alpha\gamma}\delta_{\beta\delta} + \delta_{\alpha\delta}\delta_{\beta\gamma})b_N - \delta_{\alpha\beta}\delta_{\gamma\delta}c_N]\Delta^N \quad (11)$$

and show that b_N and c_N are determined from the a_N. For if we trace over (11),

$$a_N = b_N - \nu c_N,$$

whereas a direct application of the Wick expansion on (9) gives

$$a_N = \tfrac{1}{2}(\nu + 1 - 2c)b_{N-1} - (1 - c\nu)c_{N-1}.$$

The pair of recurrence conditions demonstrate that

$$\tfrac{1}{2}(\nu+2)(\nu-1)b_N = \nu a_{N+1} - (1-c\nu)a_N,$$
$$\tfrac{1}{2}(\nu+2)(\nu-1)c_N = a_{N+1} - \tfrac{1}{2}(\nu+1-2c)a_N, \quad (12)$$

so the particular matrix superpropagator (11) reads

$$\langle \phi^N_{\alpha\beta}(x), \phi^N_{\gamma\delta}(0)\rangle = \frac{N!}{(\nu+2)(\nu-1)} \times \begin{pmatrix} \{\delta_{\alpha\gamma}\delta_{\beta\delta} + \delta_{\alpha\delta}\delta_{\beta\gamma} - 2\nu^{-1}\delta_{\alpha\beta}\delta_{\gamma\delta}\}\nu a_{N+1} \\ + \{(\delta_{\alpha\gamma}\delta_{\beta\delta} + \delta_{\alpha\delta}\delta_{\beta\gamma})(c\nu - 1) + \delta_{\alpha\beta}\delta_{\gamma\delta}(\nu+1-2c)\}a_N \end{pmatrix} \quad (11')$$

Then for the most general matrix superpropagator (10) it follows that

$$\langle F_{\alpha\beta}(x), F_{\gamma\delta}(0)\rangle = \sum_N \frac{F_N^2 \Delta^N}{N!} \langle \phi^N_{\alpha\beta}(x), \phi^N_{\gamma\delta}(0)\rangle = \frac{1}{(\nu+2)(\nu-1)} \begin{pmatrix} \{\delta_{\alpha\gamma}\delta_{\beta\delta} + \delta_{\alpha\delta}\delta_{\beta\gamma} - 2\nu^{-1}\delta_{\alpha\beta}\delta_{\gamma\delta}\}\nu a'(\Delta) \\ + \{(\delta_{\alpha\gamma}\delta_{\beta\delta} + \delta_{\alpha\delta}\delta_{\beta\gamma})(c\nu - 1) + \delta_{\alpha\beta}\delta_{\gamma\delta}(\nu+1-2c)\}a(\Delta) \end{pmatrix}, \quad (10')$$

where

$$a'(\Delta) = \sum_N a_{N+1} F_N^2 \Delta^N/N!,$$
$$a(\Delta) = \sum_N a_N F_N^2 \Delta^N/N! = \nu^{-1}\langle \mathrm{Tr}F, \mathrm{Tr}F\rangle. \quad (13)$$

In particular, for the interesting exponential matrix case where $F(\phi) = \exp\kappa\phi$, $F_N = \kappa^N$. Therefore

$$a(\Delta) = \sum_N a_N (\kappa^2\Delta)^N/N!,$$
$$a'(\Delta) = da(\Delta)/d(\kappa^2\Delta) \quad (13')$$

are supposedly determinate functions and have only to be substituted into (12) to provide us with the answer for

$$\langle [\exp\kappa\phi(x)]_{\alpha\beta}, [\exp\kappa\phi(0)]_{\gamma\delta}\rangle.$$

4. REDUCTION TO PFAFFIAN FORM

Our next task is to actually evaluate the integral

$$I_\nu^{(0)}(\mu,N) = \int \frac{dX}{\pi^{\nu(\nu-1)/4}} \frac{|X|^{\mu}}{\Gamma_\nu(\mu)} e^{-\mathrm{Tr}X}\ \mathrm{Tr}(X^N), \quad (7')$$

since every quantity we need stems from it. Observing that the integrand involves functions of X invariant under similarity transformations, it serves to parametrize the symmetric positive definite matrices X via the orthogonal group which diagonalizes them to $\Lambda = \mathrm{diag}(\lambda_1, \lambda_2, \ldots, \lambda_\nu)$. Thus we define

$$X(\theta, \lambda) = S(\theta)\Lambda \tilde{S}(\theta),$$

where θ are a set of $\frac{1}{2}\nu(\nu-1)$ angular parameters. The Jacobian of the transformation is

$$d^{\nu(\nu+1)/2}\, x = \prod_{j>i} |\lambda_j - \lambda_i| \prod_k d\lambda_k \cdot J(\theta)\, d^{\nu(\nu-1)/2}\, \theta,$$

and the change of variables simplifies (7) to

$$I_\nu^{(0)}(\mu, N) = \frac{\gamma_\nu}{\Gamma_\nu(\mu)} \prod_k \int_0^\infty d\lambda_k\, e^{-\lambda_k} \lambda_k^\mu \prod_{j>i} |\lambda_j - \lambda_i| \sum_{l=1} \lambda_l^N, \tag{14}$$

where γ_ν is a normalization constant (coming from the angular integration) determined by the condition that $I_\nu^{(0)}(\mu, 0) = \nu$ according to Siegel's integral. Integrals of the type (14) involve the Vandermonde determinant

$$\prod_{j>i} |\lambda_j - \lambda_i| = \prod_{j>i} \epsilon(\lambda_j - \lambda_i) \cdot \begin{vmatrix} 1 & 1 & \ldots & 1 \\ \lambda_1 & \lambda_2 & \ldots & \lambda_\nu \\ \lambda_1^2 & \lambda_2^2 & \ldots & \lambda_\nu^2 \\ \cdot & \cdot & \ldots & \cdot \\ \lambda_1^{\nu-1} & \lambda_2^{\nu-1} & \ldots & \lambda_\nu^{\nu-1} \end{vmatrix} \tag{15}$$

and occur frequently in statistical mechanics where methods have been developed[7] for simplifying them which we shall adopt. From (14) and (15), using properties of determinants, the calculation reduces to finding

$$I_\nu^{(0)}(\mu, N) = \frac{\gamma_\nu \nu!}{\Gamma_\nu(\mu)} \sum_{l=1}^{\nu} \int_0^\infty d\lambda_1 \cdots d\lambda_\nu \prod_{j>i} \epsilon_{ji} E_1(\lambda_1) E_2(\lambda_2) \cdots E_{N+l}(\lambda_l) \cdots E_\nu(\lambda_\nu)$$

with

$$E_\mu(\lambda) \equiv \lambda^{\mu+n-1} e^{-\lambda}. \tag{16}$$

Hereafter we shall suppose that ν is even (the case ν odd is treated in the Appendix). Recall now that the Pfaffian of antisymmetric $\nu \times \nu$ matrices A is defined as

$$|A| \equiv \begin{vmatrix} a_{12} & a_{13} & \cdots & a_{1\nu} \\ & a_{23} & \cdots & a_{2\nu} \\ & & \cdots & \\ & & & a_{\nu-1\,\nu} \end{vmatrix} = \frac{2}{\nu} \sum_{j>i} a_{ij} A_{ij}$$

$$= [2^{\nu/2}(\tfrac{1}{2}\nu)!]^{-1} \sum_{j\ \mathrm{perms}} (-1)^P a_{j_1 j_2} a_{j_3 j_4} \cdots a_{j_{\nu-1} j_\nu}, \tag{17}$$

where A_{ij} is the cofactor of the element a_{ij}. It so happens that

$$|\epsilon| = \prod_{j>i} \epsilon_{ji} = \prod_{j>i} \epsilon(\lambda_j - \lambda_i). \tag{18}$$

Used in conjunction with (16) I becomes a sum of double integrals in a Pfaffian expansion

$$I_\nu^{(0)}(\mu, N) = \frac{\gamma_\nu \nu!}{\Gamma_\nu(\mu)} \sum_{l=1} \begin{vmatrix} a_{12} & a_{13} & \cdot & a_{1\,l+N} & \cdot & a_{1\nu} \\ & a_{23} & \cdot & a_{2\,l+N} & \cdot & a_{2\nu} \\ & & & & \cdots & \\ & & & & & \cdot \end{vmatrix}$$

with elements

$$a_{ij} \equiv \int_0^\infty d\lambda\, d\lambda' E_i(\lambda) E_j(\lambda') \epsilon(\lambda' - \lambda).$$

Furthermore, since Pfaffians are linewise additive,

$$I_\nu^{(0)}(\mu, N) = \frac{2\gamma_\nu \nu!}{\Gamma_\nu(\mu)} \sum_{j>i} a_{ij}^{(N)} A_{ij}, \tag{19}$$

where A_{ij} is the appropriate cofactor and

$$a_{ij}^{(N)} \equiv \tfrac{1}{2}[a_{i\,N+j} + a_{N+i\,j}], \quad a_{ij}^{(0)} = a_{ij}. \tag{20}$$

Thus the normalization constant $\gamma_\nu = \Gamma_\nu(\mu)/(\nu-1)!\,|A|$.

It remains to work out the coefficients (20) which enter in the Pfaffian expansion (19). Changing variables to $\lambda + \lambda' = s$, $\lambda - \lambda' = st$, the integral representation of $a_{ij}^{(N)}$ boils down to

$$a_{ij}^{(N)} \equiv (\tfrac{1}{2})^{2\mu+i+j+N}\Gamma(2\mu + i + j + N)\alpha_{ij}^{(N)}$$

$$\alpha_{ij}^{(N)} \equiv \int_0^1 dt\,(1-t^2)^\mu[(1-t)^N + (t \leftrightarrow -t)] \times[(1-t)^{i-1}(1+t)^{j-1} - (t \leftrightarrow -t)]. \tag{21}$$

These α coefficients are embodied in the generating function

$$\left.\begin{aligned} \alpha_{ij}(z) &\equiv \sum_{N=0}^\infty \frac{\alpha_{ij}^{(N)} z^N}{N!} = 2\int_0^1 dt\,(1-t^2)^\mu e^z \cosh zt \begin{pmatrix} (1-t)^{i-1}(1+t)^{j-1} \\ -(i \leftrightarrow j) \end{pmatrix} = e^z \partial_{ij}\alpha(z) \\ \partial_{ij} &\equiv (1-\partial)^{i-1}(1+\partial)^{j-1} - (i \leftrightarrow j), \quad \partial \equiv \frac{d}{dz} \\ \alpha(z) &= 2\int_0^1 dt\,(1-t^2)^\mu \sinh zt = \frac{\sqrt{\pi}\,\Gamma(\mu+1)}{(\frac{1}{2}z)^{\mu+(1/2)}} L_{\mu+(1/2)}(z) = \sqrt{\pi} \sum_{k=0}^\infty \frac{\Gamma(\mu+1)(\frac{1}{2}z)^{2k+1}}{\Gamma(k+\mu+2)\Gamma(k+\frac{3}{2})} \end{aligned}\right\}, \tag{22}$$

where L is the modified Struve function. Thus $\alpha_{ij}(z)$ is an entire function of z. Since the only cases of physical interest are for $\nu \leqslant 6$, when the i, j values are small integers, the most practical way of discovering the $\alpha_{ij}^{(N)}$ is to work out

$$\alpha_{ij}^{(N)} = \partial^N e^z \partial_{ij}\alpha(z)\Big|_{z=0} = (1+\partial)^N \partial_{ij}\alpha(z)\Big|_{z=0}, \tag{23}$$

a method which has another advantage as we shall find. In passing let us record the lowest few N values when $\nu = 4$ for future use:

$$\alpha_{12}^{(0)} = \alpha_{12}^{(1)} = \tfrac{1}{2}\alpha_{13}^{(0)} = \tfrac{1}{2}\alpha_{13}^{(1)} = -\tfrac{4}{3},$$
$$\alpha_{23}^{(0)} = \alpha_{23}^{(1)} = \tfrac{1}{2}\alpha_{24}^{(0)} = \tfrac{1}{2}\alpha_{24}^{(1)} = -4,$$

$$\alpha_{14}^{(0)} = \alpha_{14}^{(1)} = -\tfrac{4}{3}, \quad \alpha_{34}^{(0)} = \alpha_{34}^{(1)} = 4,$$
$$\alpha_{12}^{(2)} = \tfrac{1}{2}\alpha_{13}^{(2)} = \tfrac{4}{3}, \quad \alpha_{23}^{(2)} = \tfrac{1}{2}\alpha_{24}^{(2)} = -12,$$
$$\alpha_{14}^{(2)} = \tfrac{52}{3}, \quad \alpha_{34}^{(2)} = \tfrac{20}{3}, \text{etc.}$$

With $c \neq 0$, therefore, the final Pfaffian expansion reads

$$I_\nu^{(c)}(\mu, N) = \frac{\gamma_\nu \nu!}{\Gamma_\nu(\mu)} \sum_{j>i} \sum_{n=0}^{N} \times \frac{\Gamma(N+1)\Gamma(N+\mu\nu+\frac{1}{2}\nu(\nu+1))(-c)^{N-n}}{\Gamma(n+1)\Gamma(n+\mu\nu+\frac{1}{2}\nu(\nu+1))\Gamma(N-n+1)} a_{ij}^{(n)} A_{ij}$$

to be differentiated at $\mu = -\frac{1}{2}(\nu+1)$. Since

$$\Gamma_\nu(\mu) = \pi^{\nu/4} 2^{-\mu\nu+\nu^2/4} \Gamma(2\mu+2)\Gamma(2\mu+4)\cdots\Gamma(2\mu+\nu)$$

introduces multiple zeros which are cancelled by zeros in the Pfaffian coefficients, we finally write for the superpropagator

$$\langle \mathrm{Tr}\phi^N(x), \mathrm{Tr}\phi^N(o)\rangle = N!\Delta^N \nu! \gamma_\nu \sum_{n=0}^{N} \binom{N}{n} (-c)^{N-n} \times \frac{\partial}{\partial\mu}\bigg|_{-(\nu+1)/2} \sum_{j>i} \frac{a_{ij}^{(n)} A_{ij}}{\Gamma_\nu(\mu)\Gamma(n+\mu\nu+\frac{1}{2}\nu(\nu+1))} \tag{24}$$

which is now a matter of straightforward computation as we will exhibit for the case of gravity ($\nu = 4$) below.

5. THE GRAVITY SUPERPROPAGATOR

In Lagrangian field theories which conform to the principles of general relativity, one meets interactions of the type

$$L = g^{\alpha\beta}(x)|-g(x)|^{\omega} T_{\alpha\beta},$$

where $g_{\alpha\beta}$ is the metric tensor, $T_{\alpha\beta}$ is a covariant tensor density of some fields ψ, and ω is the weight needed to render L a scalar density. The corresponding superpropagators of interest are defined by (the second order in L) expectation values

$$\langle g^{\alpha\beta}(x)|-g(x)|^{\omega}, g^{\gamma\delta}(0)|-g(0)|^{\omega}\rangle = K^{\alpha\beta\gamma\delta}(\Delta, c') \tag{25}$$

and are given functions of the flat space free graviton (h-field) propagator

$$\langle h^{\alpha\beta}(x), h^{\gamma\delta}(0)\rangle = \tfrac{1}{2}(\eta^{\alpha\gamma}\eta^{\beta\delta} + \eta^{\alpha\delta}\eta^{\beta\gamma} - 2c'\eta^{\alpha\beta}\eta^{\gamma\delta})\Delta(x) \tag{26}$$

in a particular gauge specified by c'. η is the Minkowski metric and $g = \eta + \kappa h + o(\kappa^2)$ where κ is related to the Newtonian gravitational constant.

Such a superpropagator was evaluated previously[5] for the rational parameterization $g = \eta + \kappa h$ when $l = \frac{1}{2}$ and 1. However, a localizable version of gravity[4] uses instead the exponential parameterization $g = \exp\kappa h$, and this is the example we shall concentrate on here as it is not amenable to any of the previous treatments.[2,5] Noting that

$$g^{\alpha\beta}(x)|-g(x)|^{\omega} = \{\exp\kappa[h(x) + \omega\eta\, \mathrm{Tr}h(x)]\}^{\alpha\beta},$$

we can interpret $(h + \omega\eta\,\mathrm{Tr}h)$ as a new field ϕ and by rotating to a Euclidean metric reestablish Eq. (1) with c given in terms of c' by

$$(1-4c) = (1-4c')(1+4\omega)^2.$$

All we have left to do is to evaluate the coefficients a_N appearing in (9),

$$a_N = 3!\gamma_4 \sum_{n=0}^{N} \binom{N}{n} (-c)^{N-n} \frac{\partial}{\partial\mu}\bigg|_{-5/2} \times \sum_{j>i} \frac{a_{ij}^{(n)} A_{ij}}{\Gamma_4(\mu)\Gamma(n+4\mu+10)}, \tag{27}$$

to obtain the required form (12). Now in

$$\sum_{j>i}^{4} \frac{a_{ij}^{(n)} A_{ij}}{\Gamma_4(\mu)} = \frac{2^{4\mu+4}/\pi}{\Gamma(2\mu+2)\Gamma(2\mu+4)} \times \begin{pmatrix} a_{12}^{(n)} a_{34} - a_{13}^{(n)} a_{24} + a_{14}^{(n)} a_{23} \\ + a_{34}^{(n)} a_{12} - a_{24}^{(n)} a_{13} + a_{23}^{(n)} a_{14} \end{pmatrix}$$

we encounter terms

$$\frac{a_{ij}^{(n)} a_{\bar{i}\bar{j}}}{\Gamma(2\mu+2)\Gamma(2\mu+4)} = \frac{\Gamma(2\mu+i+j+n)\Gamma(2\mu+10-i-j)}{\Gamma(2\mu+4)\Gamma(2\mu+2)2^{4\mu+10+n}} \alpha_{ij}^{(n)} \alpha_{\bar{i}\bar{j}}^{(0)}$$

which vanish at $2\mu + 5 = 0$ and are easily differentiated there; a little work using definition (21) gives

$$32\pi \frac{\partial}{\partial\mu}\bigg|_{-5/2} \sum_{j>i} \frac{a_{ij}^{(n)} A_{ij}}{\Gamma_4(\mu)\Gamma(n+4\mu+10)} = 12\delta_{n0}\alpha_{14}^{(0)}\alpha_{23}^{(0)} - 3\delta_{n1} \times (\alpha_{13}^{(1)}\alpha_{24}^{(0)} + \alpha_{12}^{(1)}\alpha_{34}^{(0)}) + \tfrac{3}{2}\delta_{n2}\alpha_{12}^{(2)}\alpha_{34}^{(0)} + 2^{-n}3[n(n+1) \times \alpha_{34}^{(n)}\alpha_{12}^{(0)} + 2n\alpha_{24}^{(n)}\alpha_{13}^{(0)} + 2\alpha_{14}^{(n)}\alpha_{23}^{(0)} + 2\alpha_{23}^{(n)}\alpha_{14}^{(0)}].$$

Inserting this into (27), we finally arrive at

$$a_N = \sum_{n=0}^{N} \binom{N}{n} (-c)^{N-n} \times \begin{pmatrix} 2\delta_{n0} - \frac{3}{2}\delta_{n1} + \frac{1}{2}\delta_{n2} \\ -2^{-n-2}\begin{Bmatrix} \frac{1}{2}n(n+1)\alpha_{34}^{(n)} + 2n\alpha_{24}^{(N)} \\ + 3\alpha_{14}^{(n)} + \alpha_{23}^{(n)} \end{Bmatrix} \end{pmatrix} \tag{28}$$

with the α_{ij} provided by Eqs. (22) and (23). Thus in terms of the generating Struve function $\alpha(z)$ of (22) having $\mu = -\frac{5}{2}$,

$$a_N = 2(-c)^N - 3N(-c)^{N-1}/2 + N(N-1)(-c)^{N-2}/8 - \tfrac{1}{4}\sum_{n=0}^{N}\binom{N}{n}(-c)^{N-n}(\tfrac{1}{2}\partial)^n e^z \times \begin{pmatrix} n(n-1)(1-\partial^2)^2 \\ + 2n(1-\partial^2)(5-\partial^2) \\ + 4(5+\partial^2) \end{pmatrix} \partial\alpha(z)\big|_{z=0}, \tag{28'}$$

and the first few coefficients are

$$a_0 = 4, \quad a_1 = 1-4c, \quad a_2 = \tfrac{5}{2} - 2c + 4c^2,$$
$$a_3 = 4 - \tfrac{15}{2}c + 3c^2 - 4c^3, \text{etc.},$$

which may also be checked by the (more tedious) direct Wick expansion.

Lastly we form the function (13′):

$$a(\Delta) = \sum_N \frac{a_N(\kappa^2\Delta)^N}{N!} = \left(2 - \frac{3(\kappa^2\Delta)}{2} + \frac{(\kappa^2\Delta)^2}{8}\right)e^{-c\kappa^2\Delta}$$

$$-\frac{1}{4}e^{-\kappa^2\Delta(c-1/2)}\begin{pmatrix}\frac{1}{4}(\kappa^2\Delta\partial)^2 e^z(1-\partial^2)^2 \\ +(\kappa^2\Delta\partial)e^z(1-\partial^2)(5-\partial^2) \\ +4e^z(5+\partial^2)\end{pmatrix}\partial\alpha(z)\Big|_0$$

$$= (2 - 3z + \tfrac{1}{2}z^2)e^{-2cz} + e^{z(1-2c)} \times \begin{pmatrix} 2+3z-z^2 \\ +\frac{1}{2}z(z+\frac{1}{2})3\pi L_0(z) \\ -\frac{1}{2}z^2\pi L_1(z)\end{pmatrix}\Bigg|_{z=\kappa^2\Delta/2} \tag{29}$$

(where L is the modified Struve function) which enters in

$$\langle g^{\alpha\beta}(x)| - g(x)|^{\omega}, g^{\gamma\delta}(0)| - g(0)|^{\omega}\rangle$$

$$= \langle[\exp\kappa\phi(x)]^{\alpha\beta}, \exp[\kappa\phi(0)]^{\gamma\delta}\rangle$$

$$= \frac{1}{18}\begin{pmatrix}\{\eta^{\alpha\gamma}\eta^{\beta\delta} + \eta^{\alpha\delta}\eta^{\beta\gamma} - \frac{1}{2}\eta^{\alpha\beta}\eta^{\gamma\delta}\}\, 4\dfrac{d}{d(\kappa^2\Delta)} + \\ \{[\eta^{\alpha\gamma}\eta^{\beta\delta} + \eta^{\alpha\delta}\eta^{\beta\gamma}](4c-1) + \eta^{\alpha\beta}\eta^{\gamma\delta}(5-2c)\}\end{pmatrix}a(\Delta). \tag{30}$$

This is the gravity superpropagator in the gauge c' of (26) where $(1-4c) = (1-4c')(1+4\omega)^2$. Note that when $\omega = -\frac{1}{2}$, $c = c'$, and furthermore in the De Donder gauge $c = \frac{1}{2}$ makes for some simplification in (29) and (30). However, (29) and (30) cannot really be simplified very much further and by their nature we see how entirely nontrivial is the result for exponential gravity. It goes without saying, however, that the superpropagator is an entire function of Δ; indeed the leading behavior as $\Delta \to \infty$ is

$$\langle g^{\alpha\beta}| - g|^{\omega}, g^{\gamma\delta}| - g|^{\omega}\rangle \sim [\eta^{\alpha\gamma}\eta^{\beta\delta} + \eta^{\alpha\delta}\eta^{\beta\gamma} + \eta^{\alpha\beta}\eta^{\gamma\delta}](\kappa^2\Delta)^{3/2}e^{\kappa^2\Delta(1-c)},$$

where as $\Delta \to 0$ we of course recover the perturbation series.

6. OTHER APPLICATIONS

We have demonstrated above how Siegel's integral (2) enables one to derive closed expressions for superpropagators of matrix fields as in (6) and how the calculation boils down to taking the derivative of a ν-dimensional Pfaffian as in (24). In practice this is not too difficult since the cases of practical interest involve $SU(\nu)$ or $SO(\nu)$ fields with $\nu \leqslant 6$, where the number of Pfaffian terms is small.[8] Thus for the example of $SU(3)$ there will be six terms in all to be differentiated, as we briefly discuss in the Appendix. For instance $SU(3)\otimes SU(3)$ with matrix interactions of the type $m\bar{\psi}^{\alpha}[\exp\gamma_5\kappa\phi]_{\alpha}^{\beta}\psi_{\beta}$, where ϕ is a nonet of pseudoscalar mesons propagating as

$$\langle\phi_{\alpha}^{\beta}(x)\phi_{\gamma}^{\delta}(o)\rangle = \delta_{\gamma}^{\beta}\delta_{\alpha}^{\delta}\Delta(x)$$

in the interaction picture, is perfectly amenable to our treatment. This is perhaps the single most important application of the new method besides the case of gravity covered in this paper, and will be the subject of a separate publication. What is not so obvious is how the new technique helps for calculating higher orders of perturbation theory which the previously used transform methods are at some pains to solve in any case.

APPENDIX

Assume ν is odd. All the manipulations done until Eq. (16) in Sec. 4 remain valid. However, as it stands (18) fails. In its place one can prove by induction that

$$\prod_{j>i}\epsilon(\lambda_j - \lambda_i) = Pf(\epsilon^+),$$

where the bordered $(\nu+1)\times(\nu+1)$ matrix ϵ^+ is defined by adding an extra column and row of ones:

$$\epsilon^+ = \left(\begin{array}{ccccc|c} & & & & & 1 \\ & & & & & 1 \\ & & & & & \cdot \\ & & & & & \cdot \\ \hline -1 & -1 & \cdot & \cdot & & 0\end{array}\right)$$

with ϵ_{ji} defined as a $\nu\times\nu$ matrix element. Doing the same symmetrization steps as in the even-dimensional case, one obtains

$$I_{\nu}^{(0)}(\mu, N) = \frac{2\gamma_{\nu}\nu!}{\Gamma_{\nu}(\mu)}\sum_{j>i}^{\nu+1} a_{ij}^{(N)}A_{ij},$$

where a_{ij} are defined by (20) for $1 \leqslant i < j \leqslant \nu$ while

$$a_{i\,\nu+1} = -a_{\nu+1\,i} = \int_0^{\infty} E_i(\lambda)\,d\lambda = \Gamma(\mu+i),$$

$$a_{i\,\nu+1}^{(N)} = \int_0^{\infty}\lambda^N E_i(\lambda)\,d\lambda = \Gamma(\mu+i+N), \quad i = 1,\ldots,\nu,$$

and trivially $a_{\nu+1\,\nu+1} = 0$. [Again by construction we have $I_{\nu}^{(0)}(\mu, o) = \nu$.] In this way we can take over formula (24) with the summation over i, j values running from 1 to $\nu+1$.

It is quite straightforward to check that for $\nu = 1$ one correctly reproduces the known result for the scalar case, viz., $a_N = (1-c)^N$. The case $\nu = 3$ includes the superpropagator for chiral $SU(3)\otimes SU(3)$, and the relevant computation devolves upon differentiating at $\mu = -2$,

$$\frac{1}{\Gamma_3(\mu)}\sum_{1\leqslant i\leqslant j<4} a_{ij}^{(N)}A_{ij} = \frac{2^{2\mu+2}}{\sqrt{\pi}\,\Gamma(\mu+1)\Gamma(2\mu+3)}\begin{pmatrix}a_{12}^{(N)}a_{34} - a_{13}^{(N)}a_{24} \\ +a_{14}^{(N)}a_{23} + a_{34}^{(N)}a_{12} \\ -a_{24}^{(N)}a_{13} + a_{23}^{(N)}a_{14}\end{pmatrix}$$

$$= \frac{(\frac{1}{2}\partial)^N e^z}{\Gamma(2\mu+3)}\begin{pmatrix}\{\Gamma(\mu+3)\Gamma(2\mu+3+N) + \Gamma(2\mu+3)\Gamma(\mu+3+N)\}\partial_{12} \\ -\{\Gamma(\mu+2)\Gamma(2\mu+4+N) + \Gamma(2\mu+4)\Gamma(\mu+2+N)\}\partial_{13} \\ +\{\Gamma(\mu+1)\Gamma(2\mu+5+N) + \Gamma(2\mu+5)\Gamma(\mu+1+N)\}\partial_{23}\end{pmatrix}\frac{\alpha(z)}{\Gamma(\mu+1)}\Bigg|_0.$$

The details and results of this example will be given elsewhere because of its particular importance in strong interaction physics.

---

[1]G. V. Efimov, Nucl. Phys. **74**, 657 (1965); H. M. Fried, Phys. Rev. **174**, 1725 (1968).

[2]R. Delbourgo, J. Math. Phys. **13**, 464 (1972); A. Hunt, K. Koller and Q. Shafi, Phys. Rev. D 3, 1327 (1971); T. Martin and J. G. Taylor (Southampton preprint).

[3]J. Charap (private communication).

[4]C. Isham, A. Salam, and J. Strathdee, Phys. letters **35B**, 585 (1971).

[5]R. Delbourgo and A. Hunt. Nuovo Cimento Lett. **4**, 1010 (1970); C. Isham, A. Salam, and J. Strathdee, Phys. Rev. D 2, 685 (1970).

[6]C. L. Siegel, Ann. Math. **36**, 527 (1935); R. Bellmann, Duke Math. J. **23**, 571 (1956); I. Olkin, Duke Math. J. **26**, 207 (1959).

[7]M. L. Mehta, Nucl. Phys. **18**, 395 (1960); N. G. de Bruijn, J. Ind. Math. Soc. **19**, 133 (1955).

[8]When $\nu=2$, there is only one such Pfaffian term and the results of Ref. 2 are easily recovered.

On matrix superpropagators. II

J. Ashmore and R. Delbourgo

Physics Department, Imperial College. London SW7 2BZ
(Received 21 September 1971)

The techniques developed in a previous paper (I) to compute the T product $\langle\phi^N{}_{\alpha\beta}(x), \phi^N{}_{\gamma\delta}(o)\rangle$ for arbitrary N are extended to cover the case when $\phi(x)$ is a Hermitian matrix-valued field in ν dimensions. We obtain a closed expression which is used to determine superpropagators like $\langle[\exp\kappa\phi(x)]_{\alpha\beta}, [\exp\kappa\phi(o)]_{\gamma\delta}\rangle$, which occur in strong interaction physics when ϕ is an SU(3) field, say.

1. INTRODUCTION

In a recent paper (I) bearing the same title,[1] a new method was developed for computing the vacuum expectation value $\langle\phi^N(x)_{\alpha\beta}, \phi^N(0)_{\gamma\delta}\rangle$ of matrix-valued symmetric fields ϕ in ν dimensions. The result for $\nu = 4$ was applied to finding the exponentially parameterized gravity superpropagator

$$\langle g_{\alpha\beta}(x), g_{\gamma\delta}(0)\rangle, \quad \text{where } g_{\alpha\beta}(x) = [\exp\kappa\phi(x)]_{\alpha\beta},$$

which occurs in localizable nonpolynomial models of quantized gravity.

In strong interaction physics the interest in nonpolynomial Lagrangians is mainly centred on nonlinear realizations of chiral $SU(\nu)$. The purpose of this paper is to show how the techniques of I are readily extended to Hermitian fields ϕ making it possible to deal with chiral $SU(3)$ matrix interactions of the type

$$m\bar{\psi}^\alpha F_\alpha^\beta(\phi)\psi_\beta, \quad F = \text{unitary function of } \gamma_5\phi,$$

which previous methods[2,3] were at great pains to tackle. We shall deduce the superpropagator $\langle F(\phi(x))_\alpha^\beta, F(\phi(0))_\gamma^\delta\rangle$ in closed form when ϕ propagates as

$$\langle\phi_\alpha^\beta(x), \phi_\gamma^\delta(0)\rangle = (\delta_\gamma^\beta\delta_\alpha^\delta - c\delta_\alpha^\beta\delta_\gamma^\delta)\Delta(x). \tag{1}$$

(Taking $c = 0$, ϕ has the interpretation of a nonet, while $c = \frac{1}{3}$ corresponds to an octet of pseudoscalar mesons.) The power of the method allows one to calculate the superpropagators for arbitrary parameterizations of the matrix group, such as the exponential or Cayley parameterizations:

$$F = e^{\gamma_5\kappa\phi} \text{ or } (1 + \tfrac{1}{2}\gamma_5\kappa\phi)(1 - \tfrac{1}{2}\gamma_5\kappa\phi)^{-1}.$$

Central to the whole approach of I was an integral representation due to Siegel giving the determinant $|Y|$ of a $\nu \times \nu$ symmetric matrix Y to an arbitrary power:

$$\left.\begin{aligned} &\int dX|X|^{\mu}e^{-\mathrm{Tr}(XY)} = \pi^{\nu(\nu-1)/4}\Gamma_\nu(\mu)|Y|^{-\mu-(\nu+1)/2},\\ &\Gamma_\nu(\mu) \equiv \Gamma(\mu+1)\Gamma(\mu+\tfrac{3}{2})\cdots\Gamma[\mu+\tfrac{1}{2}(\nu+1)]\\ &dX = \prod_{i\le j} dx_{ij}\end{aligned}\right\}, \tag{2}$$

the integration being taken over the space of all real, positive definite *symmetric* matrices. Formula (2) is therefore perfectly adequate for dealing with quantum gravity. However for Hermitian fields one needs a generalization of (2), and this reads

$$\left.\begin{aligned} &\int dZ|Z|^{\mu}e^{-\mathrm{Tr}(ZY)} = \pi^{\nu(\nu-1)/2}\Gamma_\nu^*(\mu)|Y|^{-\mu-\nu},\\ &\Gamma_\nu^*(\mu) \equiv \Gamma(\mu+1)\Gamma(\mu+2)\cdots\Gamma(\mu+\nu)\\ &dZ = \prod_{i=1} dz_{ii}\prod_{i<j} d^2z_{ij}\end{aligned}\right\}, \tag{3}$$

where the ν^2-fold integral is taken over the space of all positive definite Hermitian matrices $Z = (z_{ij})$. The proof of (3) is to be found in the Appendix. It turns out that the calculations evolving from (3) although similar to the ones evolving from (2) are in some respects simpler—for instance there is no distinction between even and odd dimensions—and this seems to be a reflexion of the phenomenon that real analysis is often harder than complex analysis.

The plan of the paper is as follows: In Sec. 2 we show how the matrix-valued superpropagator may be deduced from a knowledge of $\langle\mathrm{Tr}\phi^N, \mathrm{Tr}\phi^N\rangle$ and in Sec. 3 we indicate how this quantity may be explicitly computed by a use of representation (3). Finally in Sec. 4 we write down the superpropagators for chiral $SU(3)$ in exponential and Cayley coordinates. In places we shall be a little sketchy since most of the algebraic steps are to be found in I.

2. MATRIX SUPERPROPAGATORS

As in I we first show that the problem of arriving at the general superpropagator

$$\langle F(\phi(x))_\alpha^\beta, F(\phi(0))_\gamma^\delta\rangle, \quad F(z) = \sum_N F_N z^N/N! \tag{4}$$

is completely determined by a knowledge of the coefficients a_N in

$$\langle\mathrm{Tr}(\phi^N(x)), \mathrm{Tr}(\phi^N(0))\rangle \equiv N!\nu a_N\Delta^N(x). \tag{5}$$

For, making the ansatz,

$$\langle\phi^N{}_\alpha^\beta(x), \phi^N{}_\gamma^\delta(0)\rangle = N!(\delta_\alpha^\delta\delta_\gamma^\beta b_N - \delta_\alpha^\beta\delta_\gamma^\delta c_N)\Delta^N(x), \tag{6}$$

we obtain directly from Wick's theorem the pair of recurrence relations

$$\begin{aligned} a_N &= b_N - \nu c_N,\\ a_N &= (\nu - c)b_{N-1} - (1 - c\nu)c_{N-1}.\end{aligned} \tag{7}$$

Therefore writing the generating function of (5) as

$$\begin{aligned} a(\kappa^2\Delta) &= \nu^{-1}\langle\mathrm{Tr}\,[\exp\kappa\phi(x)], \mathrm{Tr}\,[\exp\kappa\phi(0)]\rangle\\ &= \sum_N a_N(\kappa^2\Delta)^N/N!,\end{aligned} \tag{8}$$

we find the generating function of the matrix-valued fields (6) to be

$$\langle[\exp\kappa\phi(x)]_\alpha^\beta, [\exp\kappa\phi(0)]_\lambda^\delta\rangle$$

$$= (\nu^2-1)^{-1}\begin{pmatrix}\{\nu\delta_\alpha^\delta\delta_\gamma^\beta - \delta_\alpha^\beta\delta_\gamma^\delta\}d/d(\kappa^2\Delta) +\\ \{(c\nu-1)\delta_\alpha^\delta\delta_\gamma^\beta + (\nu - c)\delta_\alpha^\beta\delta_\gamma^\delta\}\end{pmatrix} a(\kappa^2\Delta). \tag{9}$$

Hence when we come to the general case (4) it is only necessary to represent the coefficients F_N as moments, $F_N = \int t^N d\mu(t)$ to make these superpropagators integral transforms, e.g.,

$$\langle\mathrm{Tr}F(\phi), \mathrm{Tr}F(\phi')\rangle = \iint d\mu(t)d\mu(t')\nu a(tt'\Delta).$$

Copyright © 1973 by the American Institute of Physics

3. EVALUATION OF <Trϕ^N, Trϕ^N>

Take the vacuum expectation value of two integrals such as (3), making the substitution $Y = 1 + \kappa\phi$ and remembering that

$$\langle \exp[\mathrm{Tr}Z\phi(x)],\ \exp[\mathrm{Tr}Z'\phi(x)]\rangle = \exp[\mathrm{Tr}(ZZ') - c\ \mathrm{Tr}Z\ \mathrm{Tr}Z']\Delta(x)$$

from (1). Following the same steps as in I we arrive at

$$\langle \mathrm{Tr}(\phi^N(x)),\ \mathrm{Tr}(\phi^N(0))\rangle = N\Delta^N\left(\frac{\partial}{\partial\mu}\right)_{\mu=-\nu} I_\nu^{(c)}(\mu,N), \qquad (10)$$

where

$$I_\nu^{(c)}(\mu,N) = \int \frac{dZ}{\pi^{\nu(\nu-1)/2}} \frac{|Z|^\mu}{\Gamma_\nu^*(\mu)} e^{-\mathrm{Tr}Z} T_\nu((Z - c\ \mathrm{Tr}Z)^N)$$
$$= \sum_{n=0}^{\infty} \binom{N}{n}(-c)^{N-n} \frac{\Gamma(N+\mu\nu+\nu^2)}{\Gamma(n+\mu\nu+\nu^2)} I_\nu^{(0)}(\mu,n). \qquad (11)$$

To make tractable the integral $I_\nu^{(0)}(\mu,N)$, one notes that the integrand is a function only of the eigenvalues of the matrices involved which are necessarily real and positive. Changing to this set of variables, we have

$$dZ \to \gamma_\nu \prod_{k=1}^{\nu} d\lambda_k \prod_{i<j} (\lambda_j - \lambda_i)^2,$$

where γ_ν is a normalization constant [determined by $I_\nu^{(0)}(\mu,0) = \nu$] coming from the angular integrations of parameters of $SU(\nu)$ which diagonalize Z. Hence the integral (11) reduces to

$$I_\nu^{(0)}(\mu,N) = \frac{\gamma_\nu}{\Gamma_\nu^*(\mu)} \prod_k \left(\int d\lambda_k e^{-\lambda_k}\lambda_k^\mu\right) \prod_{i<j} (\lambda_j-\lambda_i)^2 \sum_{l=1}^{\nu} \lambda_l^N \qquad (12)$$

and involves the square of the Vandermonde determinant $\Pi(\lambda_j - \lambda_i)$ rather than its absolute value as we had in I; so we can apply a well-known identity[4] originally due to Lagrange for integrals involving products of determinants

$$\int_a^b d^\nu\lambda\ \det(\psi_i(\lambda_j))\ \det(\chi_i(\lambda_j)) = \nu!\ \det\int d\lambda\psi_i(\lambda)\chi_j(\lambda) \qquad (13)$$

to (12) and obtain the basic form

$$I_\nu^{(0)}(\mu,N) = \frac{\nu!\gamma_\nu}{\Gamma_\nu^*(\mu)} \sum_{ij} a_{ij}^{(N)} \frac{\partial}{\partial a_{ij}} |A| \qquad (14)$$

with

$$a_{ij}^{(N)} = \int_0^\infty d\lambda\ e^{-\lambda}\lambda^{N+\mu+i+j-2} = \Gamma(\mu+i+j+N-1)$$

and

$$a_{ij}^{(0)} = a_{ij}. \qquad (15)$$

This should be contrasted with the expression one encounters in the real symmetric case, where in place of of the determinant in (14) one meets a Pfaffian (= det$^{1/2}$) possessing much more complicated matrix elements.

4. CHIRAL $SU(3)$ PROPAGATORS

Nonlinear realizations of $SU(\nu) \otimes SU(\nu)$ bring in unitary functions $F(\phi)$ of pseudoscalar meson fields transforming under the $(\nu,\nu) \oplus (\bar\nu,\nu)$ representation, and derivatives thereof. Correspondingly one is faced with superpropagators (4) when one performs a perturbation expansion in the Lagrangian rather than the coupling constant κ. For $\nu = 2$ there exist perfectly adequate and simple transform methods[2] for dealing with matrix interactions F; but for $\nu = 3$ these transform methods already become far too difficult to apply in practice even if they are still valid in principle. It is, therefore, when $\nu \geq 3$ that our new techniques can be used to advantage. We shall show below how they work for chiral $SU(3) \otimes SU(3)$ in the exponential and Cayley parameterizations of F.

To begin we need the basic coefficients a_N of (5) which are in this case given by

$$a_N = 3!\gamma_3 \sum_n \binom{N}{n}(-c)^{N-n} \frac{\partial}{\partial\mu}\bigg|_{-3} \frac{1}{\Gamma^*(\mu)\Gamma(n+3\mu+9)} \times \sum_{ij} a_{ij}^{(n)} \frac{\partial |A|}{\partial a_{ij}}$$

with the A matrix elements (15). The differentiation is straightforwardly carried out to yield

$$a_N = \sum_{n=0}^{N} \binom{N}{n}(-c)^{N-n}[2\delta_{n0} - \delta_{n1} - \tfrac{1}{3}\delta_{n2} + \tfrac{1}{6}(n+2)(n+3)], \qquad (16)$$

so the first few coefficients are

$$a_0 = 3, \quad a_1 = 1 - 3c, \quad a_2 = 3 - 2c + 3c^2,$$
$$a_3 = 5 - 9c + 3c^2 - 3c^3, \quad \text{etc.},$$

as can be checked by direct Wick expansion. More relevant is the generating function of (8):

$$a(\zeta) = [2 - \zeta - (\zeta^2/6)]e^{-c\zeta} + [1 + \zeta + (\zeta^2/6)]e^{(1-c)\zeta}, \qquad (17)$$

where $\zeta = \kappa^2\Delta$.

For definiteness now suppose we have a nonet of pseudoscalar mesons, so that $c = 0$ in (1), and consider firstly the exponential parameterization

$$F(\phi) = e^{\gamma_5\kappa\phi} = \cos\kappa\phi + \gamma_5 \sin\kappa\phi. \qquad (18)$$

By taking even and odd parts in κ^2 in (9) we get the desired superpropagator

$$16\langle[\exp\gamma_5\kappa\phi(x)]_\alpha^\beta, [\exp\gamma_5\kappa\phi(0)]_\gamma^\delta\rangle$$
$$= 1\otimes 1\begin{pmatrix}(3\delta_\alpha^\delta\delta_\gamma^\beta - \delta_\alpha^\beta\delta_\gamma^\delta)a'(-\kappa^2\Delta) - \\ -(\delta_\alpha^\delta\delta_\gamma^\beta - 3\delta_\alpha^\beta\delta_\gamma^\delta)a(-\kappa^2\Delta)\end{pmatrix}$$
$$+ (\kappa^2 \leftrightarrow -\kappa^2),$$
$$+ \gamma_5\otimes\gamma_5\begin{pmatrix}(3\delta_\alpha^\delta\delta_\gamma^\beta - \delta_\alpha^\beta\delta_\gamma^\delta)a'(-\kappa^2\Delta) - \\ -(\delta_\alpha^\delta\delta_\gamma^\beta - 3\delta_\alpha^\beta\delta_\gamma^\delta)a(-\kappa^2\Delta)\end{pmatrix}$$
$$- (\kappa^2 \leftrightarrow -\kappa^2) \qquad (19)$$

with

$$a(\zeta) = 3 + [1 + \zeta + (\zeta^2/6)](e^\zeta - 1), \qquad (20)$$

which demonstrates that the exponential superpropagator[5] is nothing more than polynomials multiplying hyperbolic functions of Δ, as was found[2] with chiral $SU(2)$; thus it is an entire function of Δ as expected. Indeed as $\Delta \to \infty$,

$$\langle(\exp\gamma_5\kappa\phi)_\alpha^\beta, (\exp\gamma_5\kappa\phi')_\gamma^\delta\rangle \sim (1\otimes 1 - \gamma_5\otimes\gamma_5) \times (\delta_\alpha^\delta\delta_\gamma^\beta + \delta_\alpha^\beta\delta_\gamma^\delta)(\kappa^2\Delta)^2 e^{\kappa^2\Delta}, \qquad (21)$$

whereas as $\Delta \to 0$ we, of course, recover the perturbation series. Turning next to the Cayley parameterization one can exploit the result (19) by taking an integral transform

$$V = \frac{1 + \frac{1}{2}\gamma_5 \kappa \phi}{1 - \frac{1}{2}\gamma_5 \kappa \phi} = \int_0^\infty e^{-t}(2e^{\gamma_5 \kappa \phi t/2} - 1)\,dt$$

$$\therefore \langle V^{\beta}_{\alpha}, V'^{\delta}_{\gamma}\rangle = \int_0^\infty dt\, dt'\, e^{-(t+t')}\langle (2e^{\gamma_5\kappa\phi t/2} - 1)^{\beta}_{\alpha}, (2e^{\gamma_5\kappa\phi' t'/2} - 1)^{\delta}_{\gamma}\rangle$$

$$= \int_0^\infty dt\, dt'\, e^{-(t+t')}\begin{pmatrix} 4\langle (e^{\gamma_5\kappa\phi t/2})^{\beta}_{\alpha}, (e^{\gamma_5\kappa\phi t'/2})^{\delta}_{\gamma}\rangle \\ -\,3\; 1 \otimes 1\delta^{\beta}_{\alpha}\delta^{\delta}_{\gamma} \end{pmatrix}. \tag{22}$$

We shall not belabor the issue by giving the answer (22) in gory detail except to mention that, as with all rational Lagrangians, one meets at the very least incomplete functions like $\int dt\, e^{-t}(1 - \kappa^2 \Delta t)^{-1}$, and their derivatives, with their possible inherent ambiguities. By taking other transforms of (19) the reader is equipped to deal with other nonlinear realizations of chiral $SU(3)$.

APPENDIX

For completeness we give a proof of formula (3):

$$J_\nu(\mu; Y) = \int_{Z>0} dZ\, |Z|^{\mu} e^{-\mathrm{Tr}(ZY)}$$
$$= \pi^{\frac{1}{2}\nu(\nu-1)}\Gamma^*_\nu(\mu)\,|Y|^{-\mu-\nu}. \tag{3}$$

Since dZ is the invariant measure on the space of Hermitian matrices for $|Y| > 0$, we have on transforming $ZY \to Z$

$$J_\nu(\mu; Y) = |Y|^{-\mu-\nu} J_\nu(\mu; 1). \tag{A1}$$

Introduce the notation $Z_k = (z_{ij})$, $\nu \geqslant i, j \geqslant \nu - k + 1$, and $u_1 = z_{11}$ (real); $v_i = z_{1i}$ ($i = 2, 3, \ldots, \nu$). Expanding by the first row and column,

$$|Z| \equiv |Z_\nu| = [u_1 - Z^{-1}_{\nu-1}(v, \bar{v})]\,|Z_{\nu-1}|, \tag{A2}$$

where $Z^{-1}_\nu(\,,\,)$ is the (real) quadratic form obtained from the $\nu \times \nu$ matrix (Z^{-1}_ν). By hypothesis on the integration region, $|Z_\nu|$ and $|Z_{\nu-1}|$ are positive. So changing to the variable $w = u_1 - Z^{-1}_{\nu-1}(v, \bar{v})$, we obtain

$$J_\nu(\mu; 1) = \int_{Z_\nu > 0} dZ_\nu\, |Z_\nu|^{\mu} e^{-\mathrm{Tr}\, Z_\nu}$$
$$= \int_{Z_{\nu-1}>0} dZ_{\nu-1}\, |Z_{\nu-1}|^{\mu} e^{-\mathrm{Tr} Z_{\nu-1}} \int_0^\infty dw \iint_{-\infty}^{+\infty} d^2v_2\, d^2v_3 \cdots d^2v_\nu \times (w^{\mu}\, e^{-w - Z^{-1}_{\nu-1}(v,\bar{v})})$$
$$= \pi^{\nu-1}\Gamma(\mu+1)\int_{Z_{\nu-1}>0} dZ_{\nu-1}\, |Z_{\nu-1}|^{\mu}\, |Z^{-1}_{\nu-1}|^{-1} e^{-\mathrm{Tr}\, Z_{\nu-1}}$$

$$\text{or } J_\nu(\mu; 1) = \pi^{\nu-1}\Gamma(\mu+1)\, J_{\nu-1}(\mu+1; 1). \tag{A3}$$

On iterating (A3) with (A1) we obtain (3).

---

[1]J. Ashmore and R. Delbourgo, J. Math. Phys. **14,** 569 (1973).
[2]R. Delbourgo, J. Math. Phys. **13**, 464 (1972). J. Charap (private communication).
[3]C. L. Siegel, Ann. Math. **36,** 527 (1935).
[4]N. G. de Bruijn, SIAM J. Appl. Math. (Soc. Ind. Appl. Math.) **19,** 133 (1955).
[5]If one had been dealing only with the propagation of the symmetric field components ϕ, corresponding to the matrices $\lambda^0, \lambda^1, \lambda^3, \lambda^4, \lambda^6, \lambda^8$ of $SU(3)$, then the procedure of Ref. 1 would be relevant and the generating function which replaces (20) is $a(\zeta) = \zeta e^{-c\zeta}[Ei(\zeta) - 2Ei(\zeta/2) + \ln\zeta/4 + \gamma - 2] + e^{\zeta(1/2-c)}[2\zeta - 3)\sinh(\zeta/2) + (2\zeta + 5)\cosh(\zeta/2) + 4]$, a much more complicated result, although also entire in $\zeta = k^2\Delta$.

Part F
SUPERSYMMETRY

F.0. Commentaries F

Much excitement was generated in the mid 1970's by the discovery of supersymmetry (SUSY) through the work of Volkov & Akulov (Phys. Lett. B46, 109 (1973)), Wess & Zumino (Nucl. Phys. B78, 1 (1974)) and Salam & Strathdee (Phys. Lett. B51, 353 (1974)). This 'square-rooting' of the Poincaré group genuinely offered a higher symmetry generalisation that evaded the strictures of O'Raifeartaigh's (Phys. Rev. 139, B1052 (1965)) and Coleman & Mandula's (Phys. Rev. 159, 1251 (1967)) no-go theorems about higher symmetries. Furthermore the fermion-boson cancellation of quantum loop corrections served to reduce, or at least relate, the renormalisation constants. Above all, SUSY ensured that the vacuum zero-point energy was truly zero and that the Higgs mass need not be excessive.

F.1 Delbourgo R., Superfield perturbation theory and renormalisation. Nuovo Cim. A25, 646-656 (1975)

F.2 Delbourgo R. and Ramón-Medrano M., Supergauge theories and dimensional regularisation. Nucl. Phys. B110, 473-487 (1976)

The optimism in supersymmetry, as being an all encompassing symmetry of spacetime, survives in some quarters to this day; this is despite the lack of experimental evidence for the existence of superpartners of the ordinary particles, like squarks, gluinos, etc. I was myself very enthusiastic about this brilliant idea and spent the mid-seventies in exploring the repercussions and simplifications it afforded for unifying the infinities in quantum field theory through a judicious supersymmetric formalism. This is basically what paper [F.1] is all about, while paper [F.2] combines the scheme with dimensional continuation. By the 1980's disillusionment set in for me in pursuing this line of research, owing to lack of any experimental justification, and I moved on to other areas. Readers interested in this topic should also consult the following articles as well as references therein:

- D.M. Capper and G. Leibbrandt, Nucl. Phys. B85, 492 (1975)
- A. Salam and J.A. Strathdee, Nucl. Phys. B86, 142 (1975)
- M.T. Grisaru, W. Siegel and M. Rocek, Nucl. Phys. B159, 429 (1979)

IL NUOVO CIMENTO VOL. 25 A, N. 4 21 Febbraio 1975

Superfield Perturbation Theory and Renormalization (*) (**).

R. DELBOURGO

Physics Department, Imperial College - London

(ricevuto il 14 Agosto 1974)

Summary. — The perturbation theory graphs and divergences in supersymmetric Lagrangian models are studied by using superfield techniques. In super Φ^3-theory very little effort is needed to arrive at the single infinite (wave function) renormalization counter-term, while in Φ^4-theory the method indicates the counter-Lagrangians needed at the one-loop level and possibly beyond.

1. – Preliminaries.

In their analysis of the supersymmetric Φ^3-model, WESS and ZUMINO (1) discovered the remarkable fact that just a wave function factor sufficed to renormalize the theory at the one-loop level, a conclusion which was later confirmed to all orders by ILIOPOULOS and ZUMINO (2) by using the Ward identities appropriate to supergroup transformations. However elegant their work, it still involved considerable labour simply because their description brings in large numbers of component fields which have to be tediously written out every time. The powerful technique of superfields, invented by SALAM and STRATHDEE (3), which combines all the component fields into a single entity $\Phi(x, \theta)$, avoids these cumbersome features, makes short work of supertrans-

(*) To speed up publication, the author of this paper has agreed to not receive the proofs for correction.

(**) After completing this work we received a preprint by Dr. D. M. CAPPER which essentially covers the same ground, though his emphasis is on momentum space and restricted mainly to Φ^3-theory. We thank him for sending us an advance copy of his paper.

(1) J. WESS and B. ZUMINO: *Phys. Lett.*, **49** B, 52 (1974).

(2) J. ILIOPOULOS and B. ZUMINO: *Nucl. Phys.*, **76** B, 301 (1974).

(3) A. SALAM and J. STRATHDEE: *Nucl. Phys.*, **76** A, 477 (1974). See also S. FERRARA, J. WESS and B. ZUMINO: CERN preprint TH 1863.

formations, and provides new insights into supergroups. This paper is devoted to a re-examination of Φ^3-theory in the framework of superfields, and to a corresponding first look at Φ^4-theory.

In the following we shall borrow heavily from the recent article by SALAM and STRATHDEE [4], particularly from their Appendix C where the Feynman rules for superfields are spelt out. Starting from the free Lagrangian of a scalar superfield

$$\mathscr{L}_0 = \tfrac{1}{8}\nabla_\theta^2(\varphi_+\varphi_-) - \tfrac{1}{2}m\nabla_\theta(\varphi_+^2 + \varphi_-^2)\,, \tag{1}$$

they obtain the free propagators

$$\begin{cases} \Delta_{\pm\pm}(x\theta, x'\theta') = -i\langle T[\varphi_\pm(x\theta)\varphi_\pm(x'\theta')]\rangle = \\ \qquad = -\tfrac{1}{2}m(\bar\theta - \bar\theta')(\theta - \theta')_\pm \exp[\tfrac{1}{2}i\bar\theta\not\partial\theta']\Delta_c(x - x')\,, \\ \Delta_{\pm\mp}(x\theta, x'\theta') = -i\langle T[\varphi_\pm(x\theta)\varphi_\mp(x'\theta')]\rangle = \\ \qquad = \exp[\tfrac{1}{2}i\bar\theta\not\partial\theta' \pm \tfrac{1}{4}(\bar\theta - \bar\theta')\not\partial\gamma_5(\theta - \theta')]\Delta_c(x - x')\,. \end{cases} \tag{2}$$

Here $\varphi_\pm(x\theta)$ refer to complex superfields of particular chirality

$$\varphi_\pm(x\theta) = \exp[\mp\tfrac{1}{4}\bar\theta\not\partial\gamma_5\theta][A_\pm + \bar\theta\psi_\pm + \tfrac{1}{2}(\bar\theta\theta_\pm)F_\pm] \tag{3}$$

with $\psi_\pm = \frac{1}{2}(1 \pm i\gamma_5)\psi$, etc. and $\nabla_\theta = \partial^2/\partial\theta_\alpha\partial\bar\theta^\alpha$. Of course one may easily pass from (2) to the momentum-space expressions

$$\begin{cases} \Delta_{\pm\pm}(p\theta, p\theta') = -\tfrac{1}{2}m(\bar\theta - \bar\theta')(\theta - \theta')_\pm \exp[\tfrac{1}{2}\bar\theta(\gamma\cdot p)\theta']/(p^2 - m^2)\,, \\ \Delta_{\pm\mp}(p\theta, p\theta') = \exp[\tfrac{1}{2}\bar\theta(\gamma\cdot p)\theta' \pm \tfrac{1}{4}(\bar\theta - \bar\theta')i(\gamma\cdot p)\gamma_5(\theta - \theta')]/(p^2 - m^2)\,. \end{cases} \tag{4}$$

The formulae (3) conveniently summarize the propagators of all the field components

$$A_\pm = \varphi_\pm|_{\theta=0}\,, \qquad \psi_\pm = \partial\varphi_\pm/\partial\theta|_{\theta=0}\,, \qquad F_\pm = \tfrac{1}{2}\nabla_\theta\varphi_\pm|_{\theta=0}\,,$$

and one easily deduces [2] that

$$\begin{cases} \langle T(A_\pm F_\pm)\rangle = -im/(p^2 - m^2)\,, \\ \langle T(\psi_\pm\bar\psi_\pm)\rangle = \tfrac{1}{2}im(1 \pm i\gamma_5)/(p^2 - m^2)\,, \\ \langle T(A_\pm A_\mp)\rangle = i/(p^2 - m^2)\,, \\ \langle T(\psi_\pm\bar\psi_\mp)\rangle = \tfrac{1}{2}i(1 \pm i\gamma_5)(\gamma\cdot p)/(p^2 - m^2)\,, \\ \langle T(F_\pm F_\mp)\rangle = ip^2/(p^2 - m^2)\,, \end{cases} \tag{5}$$

all other propagators vanishing.

---

[4] A. SALAM and J. STRATHDEE: Trieste preprint IC/74/42.

We shall be studying theories which are given in the first place by the interaction Lagrangian

$$\mathscr{L}_1 = \nabla_\theta[V(\varphi_+) + V(\varphi_-)] \equiv \nabla_\theta V(\Phi)\,, \tag{6}$$

where $V(\Phi)$ may even be a nonpolynomial function of its argument. Then in N-th–order perturbation theory the amputated n-point Green's function (m_i lines attached to vertex $x_i\theta_i$) is given by the Hori expression

$$S_{m_1\ldots m_N}(x_1\theta_1, \ldots, x_N\theta_N) = \exp\left[\sum_{ij}\frac{1}{2}\frac{\partial}{\partial\varphi_i}\Delta_{ij}\frac{\partial}{\partial\varphi_j}\right]\prod\left(\frac{\partial}{\partial\varphi_i}\right)^{m_i} V(\Phi_i)\big|_{\theta=0} \tag{7}$$

before we act with the Laplacian ∇_θ (see Fig. 1).

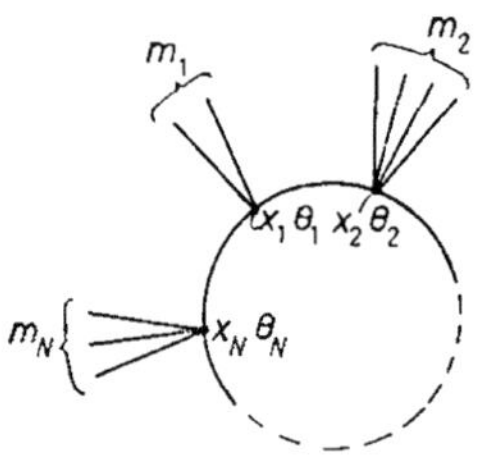

Fig. 1. – A general supergraph in N-th order.

The θ-differentiation can either be applied to the external or internal lines. Thus if all outside legs are A-lines, then the associated Green's function is

$$\nabla_{\theta_1}\ldots\nabla_{\theta_N} S_{m_1\ldots m_N}(x_1\theta_1, \ldots, x_N\theta_N)\,.$$

However if all outside legs are pure F-lines, the diagram is described by S itself. For intermediate cases, such as external ψ-lines, one θ-derivative is taken internally and another externally. In any case, since $\partial^2/\partial\theta^2$ has the dimensions of mass, we conclude that the most singular diagrams are the external A-line graphs when ∇_θ is operating at each vertex.

One important point before we particularize to Φ^3 and Φ^4 theories; namely, we need not concern ourselves with tadpole graphs. These vanish identically for $\theta = \theta'$ from (2), which can be interpreted as cancellation among field component loops. Therefore in (7), i and j really do refer to different points.

2. – One-loop graphs in Φ^3-theory.

Let us specialize now to

$$V(\Phi) = \tfrac{1}{6}g[\varphi_+^3 + \varphi_-^3]\,, \tag{8}$$

where g is a dimensionless coupling constant. Because ∇_θ annihilates combinations like $\bar\theta\partial\!\!\!/\gamma_5\theta$ or $\bar\theta\gamma_5\theta$ we can effectively replace our propagators (2) by the simpler expressions

$$(9)\qquad \begin{cases} \Delta_{\pm\pm} \to -\tfrac{1}{2}m(\bar\theta-\bar\theta')(\theta-\theta')_\pm \Delta_c\,, \\ \Delta_{\pm\mp} \to \exp[i\bar\theta\partial\!\!\!/\theta'_\mp]\,\Delta_c \end{cases}$$

in all subsequent computations. To show how simple are the calculations, consider the basic one-loop graphs, the self-energies Σ. According to Fig. 2 we can read (*) these off:

$$(10)\qquad \begin{cases} i\Sigma_{\pm\mp} = g^2\Delta^2_{\pm\mp} = g^2\exp[i\bar\theta\partial\!\!\!/\theta'_\mp]\,\Delta^2_c\,, \\ i\Sigma_{\pm\pm} = g^2\Delta^2_{\pm\pm} = 0\,. \end{cases}$$

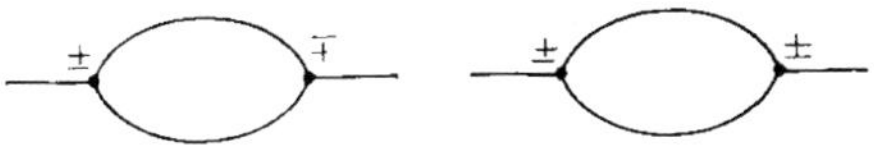

Fig. 2. – Self-energies of Φ^3-theory in second order.

In terms of the component fields, the only nonvanishing pieces are obtained as

$$i\Sigma_{A_+A_-} = -g^2\partial^2\Delta_c^2\,,\qquad i\Sigma_{\psi_+\bar\psi_-} = \tfrac{1}{2}g^2(1\pm i\gamma_5)\,i\partial\!\!\!/\Delta_c^2\,,\qquad i\Sigma_{F_+F_-} = g^2\Delta_c^2\,.$$

The significant point is that the *derivatives appear externally* in (10), so the graph has just the usual logarithmic divergence of φ^3-theory (implying that the quadratic infinities from individual field contributions cancel among themselves). This infinity can be removed by a supersymmetric counter-term

$$\tfrac{1}{8}(Z-1)\nabla^2_\theta(\varphi_+\varphi_-)\,,$$

which corresponds to wave function renormalization.

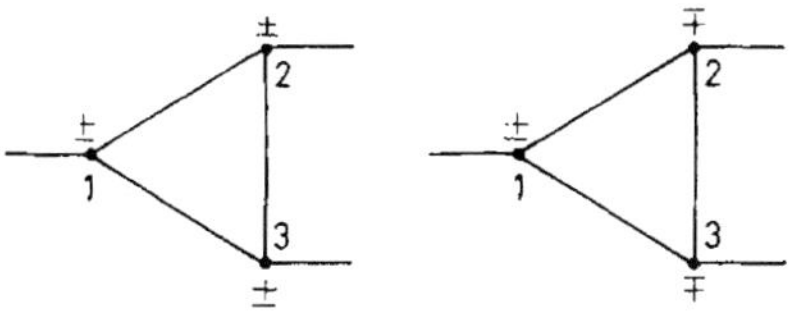

Fig. 3. – Vertex parts of Φ^3-theory in second order.

(*) To understand how formulae (9) follow from (8) note that the argument of Δ_c undergoes the complex displacement $\tfrac{1}{4}i\bar\theta\gamma_\mu(1\pm i\gamma_5)\theta'$ and also that $(\bar\theta\theta_\pm)^2=0$ according to (A.2).

With the vertex corrections of Fig. 3 we use the lemma (A.6) proved in the Appendix that

$$[(\bar\theta_1-\bar\theta_2)(\theta_1-\theta_2)_\pm][(\bar\theta_2-\bar\theta_3)(\theta_2-\theta_3)_\pm]\ldots[(\bar\theta_n-\bar\theta_1)(\theta_n-\theta_1)_\pm]=0$$

to prove that $\Gamma_{\pm\pm\pm}=0$, while

$$\Gamma_{\pm\mp\mp}=ig^3\Delta_{\pm\mp}(12)\,\Delta_{\mp\mp}(23)\,\Delta_{\pm\mp}(13)= \tag{11}$$
$$=\tfrac{1}{2}ig^3m[\exp[i\bar\theta_1\partial\!\!\!/\theta_{2\mp}]\Delta_{12}\cdot\exp[-\tfrac{1}{2}i\bar\theta_3\partial\!\!\!/\theta_{1\pm}]\Delta_{13}\cdot(\bar\theta_2-\bar\theta_3)(\theta_2-\theta_3)_\mp\Delta_{23}]\,.$$

If one picks out the nonzero-component contributions

$$\Gamma_{A_\pm A_\mp A_\mp}=img^3\Delta_{23}\partial_1^2[\Delta_{12}\Delta_{13}],$$
$$\Gamma_{\psi_\pm A_\mp\bar\psi_\mp}=\tfrac{1}{2}img^3\Delta_{23}i\partial\!\!\!/_1(1\mp i\gamma_5)[\Delta_{12}\Delta_{13}]\,,$$
$$\Gamma_{F_\pm\psi_\mp\bar\psi_\mp}=-\tfrac{1}{2}img^3\Delta_{23}(1\mp i\gamma_5)[\Delta_{12}\Delta_{13}]\,,$$

one again notices the feature that all derivetives latch on to the external legs; thus the graph is perfectly finite and the proper vertex requires no renormalization to order g^2.

For other one-loop diagrams it is not very difficult to demonstrate finiteness by means of dimensional analysis and power counting: *i*) we need only worry about the most virulent diagrams, *viz.* those with external A-lines; ii) graphs with a surplus of $\pm$ compared to $\mp$ vertices necessarily involve a chirality-preserving propagator $\Delta_{\pm\pm}$ and contribute factors of mass which serve to cut down the degree of singularity; iii) thus the worst graphs have equal numbers of $+$ and $-$ vertices which occur in alternating fashion, *i.e.* $\Gamma_{+-+-\ldots-}$. Evaluating the even n-point function $\Gamma_{A_+A_-A_+\ldots A_-}$ at zero external momentum, to pick out the possible divergence, we get

$$\Gamma_{A_+A_-\ldots A_-}=i\nabla_{\theta_1}\nabla_{\theta_2}\ldots$$
$$\ldots\nabla_{\theta_n}\int d^4k\,(k^2-m^2)^{-n}\exp[\bar\theta_2(\gamma\cdot k)(\theta_1-\theta_2)_+ +\ldots+\bar\theta_n(\gamma\cdot k)(\theta_1-\theta_{n-1})_+]=$$
$$=i\nabla_{\theta_1}\nabla_{\theta_2}\ldots\nabla_{\theta_{n-1}}\int d^4k\,(k^2)^{\frac{1}{2}n}(k^2-m^2)^{-n}\cdot(\bar\theta_1-\bar\theta_3)(\theta_1-\theta_3)\ldots(\bar\theta_{n-1}-\bar\theta_1)(\theta_{n-1}-\theta_1)=0$$

by (A.6). Hence the graph is bound to contain factors of external momentum, guaranteed to render the result finite.

3. – Higher-loop graphs in Φ^3-theory.

Let us look at some two-loop graphs contributing to the self-energy before we make the necessary generalization. Figure 4*a*) gives zero by simple dimensional counting: one encounters products of the type $(\bar{\theta}\theta m\Delta)^5$ with a surplus of θ—after acting with ∇^4 the surviving θ^2 terms mean a zero answer as $\theta \to 0$. Figure 4*b*) contribution disappears for a different reason:

$$g^4 m^2(\bar{\theta}_1 - \bar{\theta}_2)(\theta_1 - \theta_2)_+ \Delta_{12} \cdot (\bar{\theta}_2 - \bar{\theta}_3)(\theta_2 - \theta_3)_+ \Delta_{23} \cdot \\ \cdot \exp[i\bar{\theta}_2 \not{\partial} \theta_{4-}] \Delta_{24} \cdot \exp[i\bar{\theta}_1 \not{\partial} \theta_{4-}] \Delta_{14} \cdot \exp[i\bar{\theta}_3 \not{\partial} \theta_{4-}] \Delta_{34}$$

leads to

$$\Sigma_{A_+A_-} \sim \int d^4x_2\, d^4x_4\, g^4 m^2 \partial_4^2 [\Delta_{34} \Delta_{14} \Delta_{24} \Delta_{12} \Delta_{13}] = 0 \,.$$

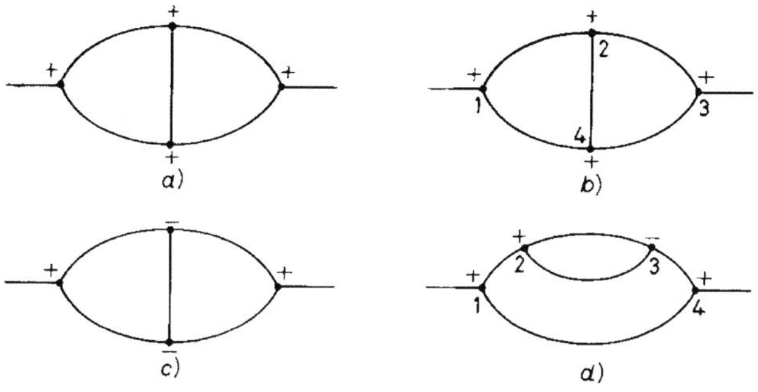

Fig. 4. – Σ_{++} self-energy in Φ^3-theory to second order.

Figure 4*c*) vanishes again by θ counting, and diagram 4*d*) gives zero for the same reason as diagram 4*b*), *viz.* the derivatives migrate to an internal point where they integrate out to zero. Thus to order g^4 we have obtained $\Sigma_{A_+A_+} = 0$. On the other hand, for $\Sigma_{A_-A_+}$, although it is true that Fig. 5*a*) and *b*) give zero, the remaining Fig. 5*c*), *d*) and *e*) are nonvanishing; thus Fig. 5*c*) reduces to

$$g^4 m^2 \int \Delta_{12} \Delta_{34} \partial_1^2 [\Delta_{23} \Delta_{14} \Delta_{24}]\, d^4x_2\, d^4x_4 \,,$$

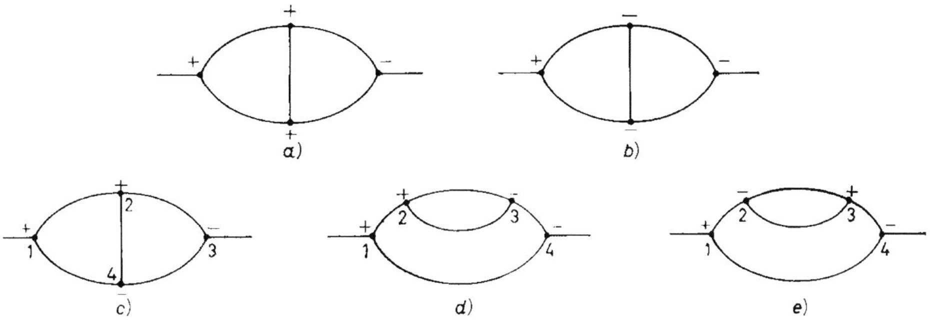

Fig. 5. – Σ_{+-} self-energy in Φ^3-theory to second order.

which is finite, Fig. 5*d*) simplifies to

$$g^4 m^2 \int \Delta_{12}\, \partial_1^2 (\Delta_{23}^2 \Delta_{34} \Delta_{14})\, d^4x_2\, d^4x_3\,,$$

which is finite once the second-order renormalization of Δ^2 has been accomplished, and finally Fig. 5*e*) reduces to

$$g^4 \partial_1^2 \int \Delta_{12}\, \partial_2^2 (\Delta_{23}^2 \Delta_{34} \Delta_{14})\, d^4x_2\, d^4x_3\,,$$

which represents a logarithmic infinity, again connected with wave function renormalization.

The g^4 analysis above suggests that $\Sigma_{A_\pm A_\pm} = 0$, and that the most singular parts (logarithmic) of $\Sigma_{A_\pm A_\mp}$ are associated with chirality-changing self-energy insertions in internal lines. The following general argument substantiates these statements: supergroup invariance of the vacuum ([4]) entails that

$$i\Delta(x\theta, x'\theta') = \langle T[\Phi(x\theta)\Phi(x'\theta')]\rangle =$$
$$= \langle T[\Phi(x - x' + \tfrac{1}{2} i\bar\theta\gamma\theta', \theta - \theta')\Phi(0)]\rangle = \exp[\tfrac{1}{2} i\bar\theta \not\partial \theta']\tilde\Delta(x - x'; \theta - \theta')\,.$$

By the same token, the self-energies must always be expressible as

$$i\Sigma(x\theta, x'\theta') = \exp[\tfrac{1}{2} i\bar\theta\not\partial\theta']\,\tilde\Sigma(x - x'; \theta - \theta')\,. \tag{12}$$

Then if one picks out θ with particular chiralities

$$i\Sigma_{\pm\pm}(x\theta, x'\theta') = \exp[i\bar\theta_\pm\not\partial\theta_\pm]\,\tilde\Sigma_{\pm\pm}(x - x'; \theta - \theta'_\pm) \to \tilde\Sigma(x - x'; \theta_\pm - \theta'_\pm)$$

and

$$i\Sigma_{\pm\mp}(x\theta, x'\theta') \to \exp[i\bar\theta_\pm\not\partial\theta'_\mp]\,\tilde\Sigma_{\pm\mp}(x - x'; \theta - \theta')\,. \tag{13}$$

If one expands $\tilde\Sigma_{AA}$ in powers of $(\theta - \theta')_\pm$ previous experience shows that at least two powers of m are involved, and, since $[m(\bar\theta - \bar\theta')(\theta - \theta')_\pm]^2 = 0$, we deduce that $\Sigma_{A_\pm A_\pm} = 0$. On the contrary, for $\Sigma_{A_\pm A_\mp}$ these mass factors do not always arise and all we can say is that $\Sigma_{A_\pm A_\mp}$ is finite, or at worst diverges logarithmically.

Likewise for the vertex function a supertransformation establishes that

$$\Gamma(x_1\theta_1, x_2\theta_2, x_3\theta_3) = \exp[\tfrac{1}{2} i(\bar\theta_1\not\partial_1\theta_3 + \bar\theta_2\not\partial_2\theta_3)]\,\tilde\Gamma(x_1 - x_3, x_2 - x_3; \theta_1 - \theta_3, \theta_2 - \theta_3)\,.$$

Again $\Gamma_{A_\pm A_\pm A_\pm}(123) = \tilde\Gamma_{A_\pm A_\pm A_\pm}(1 - 3; 2 - 3) \to 0$ by θ-expansion, symmetry and the use of lemma (A.6). On the other hand

$$\Gamma_{++-}(123) = \exp[\tfrac{1}{2}i(\bar\theta_1\bar\partial_1 + \bar\theta_2\bar\partial_2)\theta_{3-}]\tilde\Gamma_{+--}(1-3;\,2-3)$$

will either vanish or carry a factor $m(\bar\theta_1 - \bar\theta_2)(\theta_1 - \theta_2)_+$ in the θ-expansion, giving a finite answer once internal self-energy renormalizations have been performed. By similar arguments the reader may readily infer the finiteness of all higher-point irreducible vertices.

4. – One-loop graphs in Φ^4-theory.

Now let us begin afresh with the interaction

$$V(\Phi) = G(\varphi_+^4 + \varphi_-^4)/4! \tag{14}$$

though, as we shall soon discover, the theory must be fundamentally modified by renormalization terms. Before getting involved with details, note that G has the dimensions of inverse mass, so the effective coupling constant is bound to be G^2m^2 or $G^2\partial^2$ in some sense. For reasons already given, tadpole graphs can be disregarded.

Lagrangian (14) provides the self-energies to order G^2

$$\begin{cases} \Sigma_{\pm\pm} = iG^2\Delta^3_{\pm\pm} = 0\,, \\ \Sigma_{\pm\mp} = iG^2\Delta^3_{\pm\mp} = iG^2\exp[i\theta\bar\partial\theta'_\mp]\Delta_c^3\,, \end{cases} \tag{15}$$

corresponding to the component field self-energies

$$\Sigma_{A_\pm A_\mp} = iG^2\partial^2\Delta_c^3\,, \qquad \Sigma_{\psi_\pm\bar\psi_\mp} = iG^2 i\partial\!\!\!/\,\tfrac{1}{2}(1\mp i\gamma_5)\Delta_c^3\,, \qquad \Gamma_{F_\pm F_\mp} = iG^2\Delta_c^3\,. \tag{16}$$

The Δ_c^3 infinity requires two subtractions; in the ordinary φ^4-theory the subtractions have the significance of mass and wave function renormalization, but in our case one of them cannot be so interpreted (the one corresponding to fixing the second p^2-derivative of the inverse propagator). In fact the order-G^2 counter-Lagrangian

$$\delta\mathscr{L}_0 = \tfrac{1}{8}(Z-1)\nabla_\theta^2(\varphi_+\varphi_-) + Y\nabla_\theta^2(\partial\varphi_+\cdot\partial\varphi_-) \tag{17}$$

with $Z \sim 1 + G^2\Lambda^2$ and $Y \sim G^2\ln\Lambda^2$ already differs radically from the original free Lagrangian $\mathscr{L}_0$ in the number of derivatives associated with the Y term. Before worrying about the repercussions of (17) let us pursue the other one-loop corrections due to (14).

The G-vertex corrections (see Fig. 6) are either zero (*) or logarithmically infinite, as in ordinary φ^4-theory. Thus

$$(18)\qquad \begin{cases} i\Gamma_{++++} = G^2 \Delta^2_{++} = 0 \\ \text{and} \\ i\Gamma_{++--} = G^2 \Delta^2_{+-} = G^2 \exp[\bar{\theta} i \not{\partial} \theta'_-] \Delta^2_c \,, \end{cases}$$

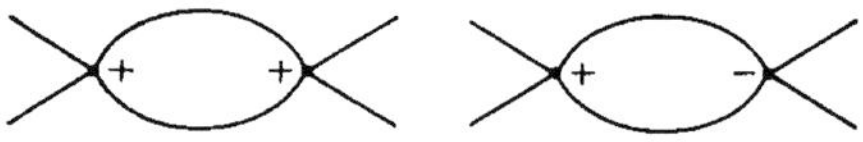

Fig. 6. – Rescattering corrections of Φ^4-theory in second order.

the derivatives again migrating to the outside lines. Now we need a new counter-Lagrangian

$$(19)\qquad \delta \mathscr{L}_1 = X \nabla^2_\theta (\varphi^2_+ \varphi^2_-)$$

with $X \sim G^2 \ln \Lambda^2$ to eliminate the infinities (18) of the scattering graphs. We remark that $\delta \mathscr{L}_1$ is inherently different from the starting interaction $\mathscr{L}_1$ despite its manifest supersymmetry.

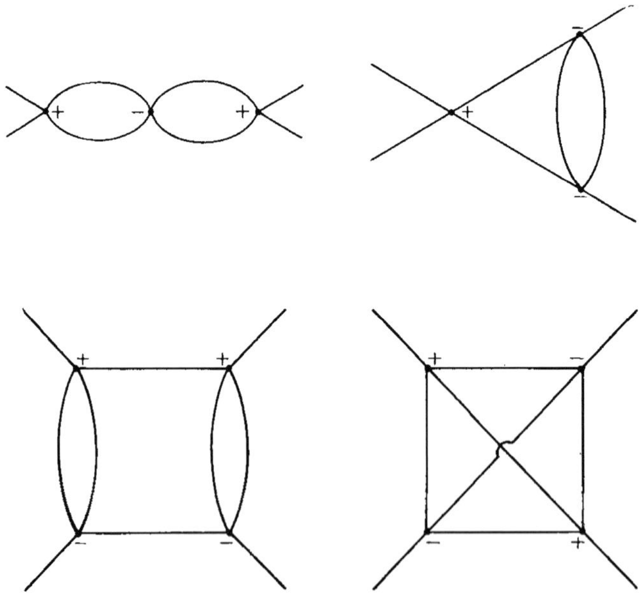

Fig. 7. – Selected higher-order graphs of Φ^4-theory.

(*) Which does not imply that Σ_{++} vanishes identically. Indeed Fig. 4c) gives a finite correction to the « mass terms » $\Sigma_{A_+F_+}$, $\Sigma_{\psi_+\bar{\psi}_+}$.

We have investigated a number of multiloop corrections provided by (14) —for example those depicted in Fig. 7—and have found that in each case the three counter-terms (17) and (19) (analogous to the three renormalizations of ordinary φ^4) are enough, per one or two loops.

Unfortunately, that is far from being the whole story. Having generated new bilinear and quadrilinear Lagrangians, these must now be studied in their own right. Indeed the situation closely resembles the state of affairs in gravity theory ([5]) where $R_{\mu\nu}R^{\mu\nu}$ and R^2 counter-Lagrangians are generated from the original Einstein Lagrangian R. In particular, the modified free Lagrangian $\mathscr{L}_0 + \delta\mathscr{L}_0$ contains quartic derivative terms which can lead to damped propagators, albeit with a possible ghost problem. Since the interactions have at most two derivatives we cannot venture to say that renormalizability is lost. Another point to be borne in mind is that, with derivative interactions present, we may expect $\delta^4(0)$ type terms as a consequence of canonical quantization which also tend to cancel off the worst effects of (19). In fact the situation is so murky at the present time that it is rather premature to announce ([6]) categorically the demise of super Φ^4 as a possible renormalizable model. Anyhow, whatever the eventual outcome of a thorough investigation of this question, we certainly feel that the superfield methods of SALAM and STRATHDEE offer the most attractive and economical way of tackling the problems, and we hope that the calculations outlined in this paper will have convinced the reader about this.

* * *

This work owes a great deal to the enlightening remarks of Prof. A. SALAM to whom we are most grateful for discussions.

APPENDIX

Here we shall list a number of useful properties of Majorana spinors which make for crucial simplifications in the text.

Let $\theta_\pm = \frac{1}{2}(1 \pm i\gamma_5)\theta$. Then

$$\theta_\pm(\bar{\theta}_\pm\theta_\pm) = 0 \,, \tag{A.1}$$

which is obvious in a two-component basis. In particular

$$(\bar{\theta}\theta_\pm)^n = 0 \qquad \text{for } n \geqslant 2 \,. \tag{A.2}$$

([5]) D. M. CAPPER, M. J. DUFF and L. HALPERN: IC/73/130 (to appear in *Phys. Rev.*); S. DESER and P. VAN NIEUWENHUIZEN: *Phys. Rev. Lett.*, **32**, 245 (1974).

([6]) W. LANG and J. WESS: Karlsruhe preprint.

With two different spinors

$$(\text{A.3})\qquad (\bar\theta(\gamma\cdot P)\theta'_\pm)(\bar\theta(\gamma\cdot Q)\theta'_\pm) = \tfrac{1}{2}(\bar\theta\theta_\mp)(\bar\theta'\theta'_\pm)P\cdot Q$$

follows by Fierz transformation. Hence from (A.1)

$$(\text{A.4})\qquad (\bar\theta(\gamma\cdot P)\theta'_\pm)(\bar\theta(\gamma\cdot Q)\theta'_\pm)(\bar\theta(\gamma\cdot R)\theta'_\pm) = 0\,.$$

Thus

$$(\text{A.5})\qquad \exp[\bar\theta(\gamma\cdot P)\theta'_\pm] = 1 + \bar\theta(\gamma\cdot P)\theta'_\pm + \tfrac{1}{4}P^2(\bar\theta\theta_\mp)(\bar\theta'\theta'_\pm)\,.$$

In the text we meet the following product of spinor differences:

$$F_n(\theta_1, \theta_2, \dots, \theta_n) \equiv (\bar\theta_1-\bar\theta_2)(\theta_1-\theta_2)_\pm\cdot(\bar\theta_2-\bar\theta_3)(\theta_2-\theta_3)_\pm \dots (\bar\theta_n-\bar\theta_1)(\theta_n-\theta_1)_\pm\,.$$

Now in view of (A.1)

$$F_n(\theta_1, \dots, \theta_n) = 2(\bar\theta_1\theta_{1\pm})(\bar\theta_2\theta_{2\pm})\dots(\bar\theta_n\theta_{n\pm}) + (-2)^n\bar\theta_1\theta_{2\pm}\bar\theta_2\theta_{3\pm}\dots\bar\theta_n\theta_{1\pm}\,.$$

A Fierz reshuffle of the last term reorganizes the product as

$$F_n(\theta_1, \dots, \theta_n) = 2(\bar\theta_1\theta_{1\pm})[(\bar\theta_2\theta_{2\pm})\dots(\bar\theta_n\theta_{n\pm}) + (-2)^{n-1}\bar\theta_2\theta_{3\pm}\dots\bar\theta_n\theta_{2\pm}] =$$
$$= (\bar\theta_1\theta_{1\pm})F_{n-1}(\theta_2, \dots, \theta_n)\,.$$

By successive iterations

$$(\text{A.6})\qquad F_n(\theta_1, \dots, \theta_n) = (\bar\theta_1\theta_{1\pm})(\bar\theta_2\theta_{2\pm})\dots(\bar\theta_{n-2}\theta_{n-2\pm})F_2(\theta_{n-1}, \theta_n) = 0\,,$$

since

$$F_2(\theta, \theta') = [(\bar\theta-\bar\theta')(\theta-\theta')_\pm]^2 = 0\,.$$

● RIASSUNTO (*)

Per mezzo delle tecniche dei supercampi si studiano i grafici della teoria perturbativa e le divergenze nei modelli lagrangiani supersimmetrici. Nella teoria del supercampo Φ^3 con poco sforzo si perviene al singolo controtermine infinito di rinormalizzazione (delle funzioni d'onda), mentre per la teoria Φ^4 il metodo indica il controlagrangiano necessario al livello di un solo cappio e possibilmente oltre.

(*) *Traduzione a cura della Redazione.*

Теория возмущений для суперполей и перенормировка.

Резюме (*). — Используя технику суперполей, исследуются графики теории возмущений и расходимости в моделях суперсимметричных Лагранжианов. В теории суперполя Φ^3 требуется незначительное усилие для получения единственного бесконечного перенормировочного контр-члена, тогда как в Φ^4 теории предложенный метод требует контр-Лагранжианов на одно-петельном уровне и, возможно, помимо этого.

(*) *Переведено редакцией.*

Nuclear Physics B110 (1976) 473–487
© North-Holland Publishing Company

SUPERGAUGE THEORIES AND DIMENSIONAL REGULARIZATION

R. DELBOURGO and M. Ramón MEDRANO *
Blackett Laboratory of Physics, Imperial College, London SW7 2BZ

Received 30 April 1976

The Becchi-Rouet-Stora identities for a non-Abelian supergauge vector theory are derived. They are explicitly verified for the two-point functions by a dimensional continuation which manifestly preserves the supersymmetry; the formulation is unusual in that the propagators become more strongly damped as the number of dimensions is increased.

1. Introduction

Formal Ward identities [1] between Green functions in non-Abelian gauge theories can be straightforwardly derived by means of the functional integral formalism. In the improved version [2] of Becchi, Rouet and Stora (BRS) the identities prove very useful for relating the renormalization counter terms [3] that are needed in such theories. The implicit assumption normally made, that a regularization procedure can be devised which maintains the gauge symmetry and thus makes sense of the identities, is confirmed by the technique of dimensional [4] continuation.

Now over the last two years, with considerable ingenuity, Abelian and non-Abelian supersymmetric gauge theories, comprising bosons and fermions, have been formulated [5] and their corresponding Ward identities have been worked out [6]. However no regularization method has yet been proposed which exactly conserves the supergauge symmetry to *every* order of quantum loops; therefore a verification of the formal supergauge relations is still lacking. In this paper we describe a regularization procedure, again founded on dimensional continuation, which is guaranteed to do just that. A novel feature of the method is the occurrence of varying numbers of derivatives in the Lagrangian, increasing with the number $2l$ of dimensions, forced on us by the requirement of supersymmetry for all l.

In sect. 2 we review the formal BRS identities among the supergauge and superghost fields, as they apply in four dimensions. Our approach differs in some important respects from previous [6] work, so as to facilitate the actual derivation of the identities and to permit straightforward generalization to arbitrary dimensions, as

* Work supported by GIFT (Grupo Interuniversitario de Fisica Teorica) Spain.

presented in sect. 3. Following that, we examine the relations between the two-point functions and associated renormalization constants, which we verify by explicit computation, this being the first manifestly supersymmetric evaluation of a one-loop diagram in Yang-Mills theory, matching the calculations for the Abelian case [7] and simpler φ^3 type theories [8].

2. Non-Abelian supergauges in four dimensions

It is well-known that in order to incorporate a local supergauge symmetry [5] one must first consider chiral superfields because of their closure property. In a vector gauge theory the chiral matter fields $\chi_\pm$ transform according to

$$\chi_\pm \to \exp[ig\Lambda_\pm]\chi_\pm\,, \qquad \Lambda_- = \Lambda_+^\dagger\,, \tag{1}$$

and the supergauge field Φ compensates for this by transforming as

$$(1+g\Phi) \to \exp[ig\Lambda_-](1+g\Phi)\exp[-ig\Lambda_+]\,, \tag{2}$$

so as to leave invariant the matter field Lagrangian

$$\mathcal{L}_\chi = \int d^2\theta_+ d^2\theta_- [\chi_+^\dagger(1+g\Phi)\chi_- + \chi_-^\dagger(1+g\Phi)^{-1}\chi_-]\,. \tag{3}$$

Here we differ from previous papers on the subject in our choice of interpolating field; thus we treat $g\Phi = [\exp(g\Psi) - 1]$ rather than Ψ as the fundamental quantity. As we shall soon see, this saves a lot of clutter in discussing the superghost terms [6] that have to be included when deciding upon a manifestly supersymmetric gauge. The imbalance in non-polynomiality between χ_+ and χ_- in (3) turns out not to be a problem.

The Lagrangian for the gauge field itself is obtained by constructing the gauge chiral potentials [5],

$$g\varphi_+ = \tfrac{1}{2}(\bar{D}D)_-[(1+g\Phi)^{-1}D_+(1+g\Phi)]\,,$$

$$g\varphi_- = \tfrac{1}{2}(\bar{D}D)_+[(1+g\Phi)D_-(1+g\Phi)^{-1}]\,, \tag{4}$$

which transform simply,

$$\varphi_\pm \to \exp[ig\Lambda_\pm]\varphi_\pm\exp[-ig\Lambda_\pm]\,, \tag{5}$$

whereupon one can build the supergauge invariant

$$\begin{aligned}4\,\mathcal{L}_\Phi &= \int d^2\theta_+ \operatorname{Tr}[\varphi_+^2] + \int d^2\theta_- \operatorname{Tr}[\varphi_-^2] \\ &= \int d^2\theta_+ d^2\theta_- \operatorname{Tr}[\{(1+g\Phi)^{-1}\bar{D}_+\Phi\}\tfrac{1}{2}(\bar{D}D)_-\{(1+g\Phi)^{-1}D_+\Phi\} \\ &\quad + (+\leftrightarrow -)]\,.\end{aligned} \tag{6}$$

Because the free part of the Lagrangian (6),

$$-\tfrac{1}{8}\,\mathrm{Tr}[\Phi\bar{D}_+(\bar{D}D)_-D_+\Phi + \Phi\bar{D}_-(\bar{D}D)_+D_-\Phi]$$

involves the vector projector $[4\partial^2 + (\bar{D}D)^2]$, it possesses no inverse. Hence the necessity for fixing a gauge. The Wess-Zumino gauge [5] has gained wide currency because it eliminates the non-polynomial part of the Lagrangian and reduces the interactions to very familiar forms. However, like the axial gauge of ordinary theories, it destroys the manifest supersymmetry of the problem and in some ways complicates the evaluation of higher order corrections because every component field has to be separately accounted for. Therefore we shall follow more recent work [6] which respects the supercovariance, and will add the gauge fixing term

$$\mathcal{L}_\alpha = \tfrac{1}{4}\int \mathrm{d}^2\theta_+\mathrm{d}^2\theta_-\ \mathrm{Tr}[(\bar{D}D)_+\Phi(\bar{D}D)_-\Phi]/\alpha\,. \tag{7}$$

This must be made up by the inclusion of the superghost Lagrangian $\mathcal{L}_C$ to maintain the unitarity in the theory; $\mathcal{L}_C$ is found by making the infinitesimal transformation

$$\delta\Phi = i\{\Lambda_-(1+g\Phi) - (1+g\Phi)\Lambda_+\} \tag{8}$$

setting $\Lambda_\pm = C_\pm\omega$, where ω and C are anticommuting, and in the standard way constructing,

$$\mathcal{L}_C = \tfrac{1}{2}i\int \mathrm{d}^2\theta_+\mathrm{d}^2\theta_-\ \mathrm{Tr}[\{\bar{C}_-(\bar{D}D)_- + \bar{C}_+(\bar{D}D)_+\}\{C_-(1+g\Phi)$$

$$-(1+g\Phi)C_+\}]\ . \tag{9}$$

At this juncture it may be worth remarking that our choice of gauge breaking term (7) is perhaps superior to the more usual choice, based on Ψ, because it leads to a simple polynomial fictitious Lagrangian (9) with interactions which bear a close resemblance to those enjoyed by the matter field (3).

The sum of (6), (7) and (9) is invariant under the global generalized BRS transformations

$$\delta\Phi = i\{C_-(1+g\Phi) - (1+g\Phi)C_+\}\omega\,,$$

$$\delta\bar{C}_\pm = -(\bar{D}D)_\mp\Phi\omega/2\alpha\,,$$

$$\delta C_\pm = -\tfrac{1}{2}i[C_\pm, C_\pm]\omega\,, \tag{10}$$

as one may readily check. To obtain the BRS identities themselves one introduces sources for the gauge field Φ, the fictitious fields C and also the composite fields appearing on the right-hand side of (10) into the vacuum generating functional:

$$Z[J, I] \equiv \int (\mathrm{d}\Phi)(\mathrm{d}^4C) \exp i \int \mathrm{d}^4x\, \mathcal{L}[\Phi, C, I]\,,$$

where the total Lagrangian (neglecting matter) equals

$$\mathcal{L} = \mathcal{L}_\Phi + \mathcal{L}_\alpha + \mathcal{L}_C + \int \mathrm{d}^2\theta_+ \mathrm{d}^2\theta_-\ \mathrm{Tr}[J\Phi + \bar{C}_+ J_- + \bar{J}_+ C_- + \bar{C}_- J_+ + \bar{J}_- C_+]$$

$$+ \int \mathrm{d}^2\theta_+ \mathrm{d}^2\theta_-\ \mathrm{Tr}(iI\{C_-(1 + g\Phi) - (1 + g\Phi)C_+\} - \tfrac{1}{2} igI_-[C_+, C_+]$$

$$- \tfrac{1}{2} igI_+[C_-, C_-])\,, \tag{11}$$

and the sources $J_\pm, \bar{J}_\pm, I(J, I_\pm)$ are anticommuting (commuting). Invariance of Z, and hence of $W = i \ln Z$, under the BRS variations yields the identity

$$\int \mathrm{d}^4x \mathrm{d}^4\theta\ \mathrm{Tr}\left[J \frac{\delta W}{\delta I} - \frac{1}{2\alpha}(\bar{D}D)_+ \frac{\delta W}{\delta J} J_+ - \frac{1}{2\alpha}(\bar{D}D)_- \frac{\delta W}{\delta J} J_- + \bar{J}_+ \frac{\delta W}{\delta I_+}\right.$$

$$\left. + \bar{J}_- \frac{\delta W}{\delta I_-}\right] = 0\,, \tag{12}$$

since

$$\delta[\delta\Phi] = \delta[C_\pm, C_\pm] = 0\,.$$

In practice one is interested in renormalization counter terms and thus the BRS identities for proper vertices. Hence one passes to the one-particle irreducible generating functional

$$\Gamma[\Phi, C, I] = W[J, I] - \int \mathrm{d}^4x \mathrm{d}^4\theta\ \mathrm{Tr}[J\Phi + \bar{C}_+ J_- + \bar{J}_+ C_- + \bar{C}_- J_+ + \bar{J}_- C_+]\,,$$

in terms of which the gauge invariance is expressed as

$$\int \mathrm{d}^4x \mathrm{d}^4\theta\ \mathrm{Tr}\left[\frac{\delta\Gamma}{\delta\Phi}\frac{\delta\Gamma}{\delta I} - \frac{(\bar{D}D)_+\Phi}{2\alpha}\frac{\delta\Gamma}{\delta\bar{C}_+} - \frac{(\bar{D}D)_-\Phi}{2\alpha}\frac{\delta\Gamma}{\delta\bar{C}_-} - \frac{\delta\Gamma}{\delta C_+}\frac{\delta\Gamma}{\delta I_-}\right.$$

$$\left. - \frac{\delta\Gamma}{\delta C_-}\frac{\delta\Gamma}{\delta I_+}\right] = 0\,. \tag{13}$$

The relevant Ward identities are obtained by functionally differentiating (13) so as to leave zero ghost number N, where one knows the assignments

$$N(\Phi) = N(J) = 0\,, \qquad N(I_\pm) = -2\,,$$

$$N(C_\pm) = N(J_\pm) = -N(\bar{C}_\pm) = -N(\bar{J}_\pm) = -N(I) = 1\,,$$

by inspection of (11). We shall examine the identities for the two-point functions in detail here. These are derived from the equation of motion for the fictitious fields

$$\tfrac{1}{2}(\bar{D}D)_{\pm}\delta\Gamma/\delta I = \delta\Gamma/\delta\bar{C}_{\pm}\ , \tag{14}$$

as well as (13); thus we consider the set

$$\tfrac{1}{2}(\bar{D}D)_{+}\delta^2\Gamma/\delta I\delta C_{\pm} = \delta^2\Gamma/\delta\bar{C}_{+}\delta C_{\pm}{}',$$

$$\tfrac{1}{2}(\bar{D}D)_{-}\delta^2\Gamma/\delta I\delta C_{\pm} = \delta^2\Gamma/\delta\bar{C}_{-}\delta C_{\pm}\ , \tag{15}$$

and

$$\int(\mathrm{d}1)\frac{\delta^2\Gamma}{\delta\Phi(1)\delta\Phi(2)}\frac{\delta^2\Gamma}{\delta I(1)\delta C_{\pm}(3)} = \frac{(\bar{D}D)_{-}(2)}{2\alpha}\frac{\delta^2\Gamma}{\delta\bar{C}_{+}(2)\delta C_{\pm}(3)}$$
$$+\frac{(\bar{D}D)_{+}(2)}{2\alpha}\frac{\delta^2\Gamma}{\delta\bar{C}_{-}(2)\delta C_{\pm}(3)}\ .$$

In the tree approximation these identities are trivially satisfied (note that the combination $\bar{C}_{\pm}C_{\mp}$ gives zero in this classical limit), but to verify their validity in one or higher loops we are obliged to define the expressions by a supersymmetric regularization. This leads us directly to the dimensional continuation of the entire formalism since it is the only practicable technique known to date for treating non-Abelian gauge theories.

3. Supersymmetric theories in arbitrary dimensions

In an earlier paper we described [9] the extension of supersymmetry to arbitrary dimensions $2l$. Here we shall summarise the relevant points:

(i) Spinors have $2\nu = 2^l$ components. Thus the integration measure of superspace is $\int \mathrm{d}^{2l}x\ \mathrm{d}^{2\nu}\theta$.

(ii) Majorana constraints demand that l take on the sequence of values $l = 2, 6, 10, \ldots$ or $l = 1, 5, 9, \ldots$. Hence to obtain the four-dimensional limit one sets $l = 2(2k+1)$ and makes a continuation in k down to $k = 0$.

(iii) The analogue of the γ_5 matrix which is $\Gamma_{-1} = -i^l\Gamma_0\Gamma_1\Gamma_2 \ldots \Gamma_{2l-1}$ can be used as in four dimensions to construct chiral projectors, $\tfrac{1}{2}(1 \pm i\Gamma_{-1})$ and covariant derivatives $D_{\pm}$. Note that now we have $(\theta_{\pm})^{\nu+1} = (D_{\pm})^{\nu+1} = 0$ since chiral spinors have ν independent anticommuting components.

(iv) Lagrangians which reduce to the four-dimensional counterparts can be constructed in two 'natural' ways. The first way typically takes for the kinetic Lagrangian

$$\mathcal{L}_1 \propto (\bar{D}D)^2 [\chi_+\chi_-] \ .$$

The second method instead adopts

$$\mathcal{L}_2 \propto (\bar{D}D)^\nu [\chi_+\chi_-] = \int \mathrm{d}^\nu\theta_+ \mathrm{d}^\nu\theta_- [\chi_+\chi_-] \ .$$

Because the first method leads to anomalous looking terms in Ward identities for the supercurrent [9] (the Lagrangian $\mathcal{L}_1$ gives an invariant action only in four dimensions) we shall, in this paper, abandon it in favour of the second method which has an invariant action in arbitrary dimensions. However, it must be pointed out that the number of derivatives occurring in the kinetic energy $(\partial)^\nu$ changes with the number of dimensions and this is unlike anything we have experienced before with dimensional regularization! Nevertheless it is still a satisfactory way of continuing the integrals because the index of the propagator $\nu = 2^{l-1}$ is different from that of the phase space $2l$. Indeed the odd thing is that as l increases the convergence properties of Feynman diagrams improve.

(v) The irreducible chiral superfields $\chi_\pm$ appearing above have 2^ν independent components and are picked out from the full scalar superfield χ, with its $2^{\nu+1}$ components, by applying the conditions $D_\pm \chi_\mp = 0$ exactly as in four dimensions.

We now wish to apply this technique to supergauge theory. Introduce the abbreviation

$$D^\nu = \epsilon^{A_1 \dots A_\nu} D_{A_1} \dots D_{A_\nu} / \nu! \ , \tag{17}$$

so that in four dimensions one has $D^2_\pm = \frac{1}{2}(\bar{D}D)_\pm$. When D^ν_- acts on a general superfield combination we are guaranteed a positive chiral field. Turning to the gauge potentials, we observe that for large l there exist a host of chiral gauge multispinors that can be invented:

$$g\varphi_{+A_1 \dots A_m} = D^\nu_- [(1+g\Phi)^{-1} D_{+A_1} \dots D_{+A_m} (1+g\Phi)] \tag{18}$$

or for short,

$$\varphi^m_+ = D^\nu_- [(1+g\Phi)^{-1} D^m_+ \Phi] \ , \qquad 0 < m < \nu \ .$$

All of these transform nicely as in (5). Hence we have the freedom to write a supergauge invariant kinetic energy,

$$\mathcal{L}_\Phi = \int \mathrm{d}^\nu\theta_+ \, \mathrm{Tr} \sum_{m+n=\nu} a_{mn} \varphi^m_+ \varphi^n_+ + (+ \leftrightarrow -) \ . \tag{19}$$

For $l = 2$ there is only one term in (19), namely a_{11}; but for higher l there are $\frac{1}{2}\nu$

terms (due to the symmetry of a) which can be exploited.

In any case (19) is singular and requires gauge fixing. We therefore add to (19) the term

$$\mathcal{L}_\alpha = \int \mathrm{d}^\nu\theta_+ \mathrm{d}^\nu\theta_- \ \mathrm{Tr}[D_+^\nu \Phi D_-^\nu \Phi]/\alpha\,, \tag{20}$$

exactly as in (7), to arrive at the equation of motion

$$\{\sum_m a_m(-1)^m [D_+^m D_-^\nu D_+^{\nu-m} + D_-^m D_+^\nu D_-^{\nu-m}] + (-1)^\nu [D_+^\nu D_-^\nu + D_-^\nu D_+^\nu]/\alpha\}\Phi = 0\,. \tag{21}$$

In four dimensions (21) reduces simply to

$$\{\bar{D}_+(\bar{D}D)_-D_+ + \bar{D}_-(\bar{D}D)_+D_- - [(\bar{D}D)_+(\bar{D}D)_- + (\bar{D}D)_-(\bar{D}D)_+]/\alpha\}\Phi = 0\,,$$

and, in the Fermi gauge ($\alpha = 1$) almost trivially yields $p^2\Phi = 0$. In the appendix we prove that in the general case the coefficients a for any l can be chosen such that

$$\sum_m a_m(-1)^{m-\nu} [D_+^m D_-^\nu D_+^{\nu-m} + D_-^m D_+^\nu D_-^{\nu-m}] = (p^2)^{\nu/2} - D_+^\nu D_-^\nu - D_-^\nu D_+^\nu\,. \tag{22}$$

Therefore also in arbitrary dimensions the Fermi gauge produces an equation free of covariant derivatives,

$$(p^2)^{\nu/2}\Phi = 0\,. \tag{23}$$

The rest of the analogy with four dimensions is much more direct. The fictitious superfield Lagrangian

$$\mathcal{L}_C = i\int \mathrm{d}^\nu\theta_+ \mathrm{d}^\nu\theta_- \ \mathrm{Tr}[(\bar{C}_-D_-^\nu + \bar{C}_+D_+^\nu)\{C_-(1+g\Phi) - (1+g\Phi)C_+\}]\,, \tag{24}$$

added to (19) and (20), is invariant under the BRS transformations (10), providing we substitute $D_\pm^\nu$ for $\frac{1}{2}(\bar{D}D)_\pm$. Likewise in the Ward identities (15) one should make the same substitution; otherwise there is no change.

4. Verification of the identities

We want now to check the self-energy identities

$$D_+^\nu(1)\delta^2\Gamma/\delta I(1)\delta C_\pm(2) = \delta^2\Gamma/\delta\bar{C}_+(1)\delta C_\pm(2)\,,$$

$$D_-^\nu(1)\delta^2\Gamma/\delta I(1)\delta C_\pm(2) = \delta^2\Gamma/\delta\bar{C}_-(1)\delta C_\pm(2)\,, \tag{25}$$

$$\int (\mathrm{d}1) \frac{\delta^2\Gamma}{\delta\Phi(1)\delta\Phi(2)} \frac{\delta^2\Gamma}{\delta I(1)\delta C_\pm(3)} = \frac{D_-^\nu(2)}{\alpha} \frac{\delta^2\Gamma}{\delta\bar{C}_+(2)\delta C_\pm(3)}$$

$$+ \frac{D_+^\nu(2)}{\alpha} \frac{\delta^2\Gamma}{\delta\bar{C}_-(2)\delta C_\pm(3)} , \tag{26}$$

since we possess an adequate regularization. Before doing so let us formally establish the transversality of the gauge field self-energy. Operate on (26) with $D_+^\nu(2)$, substitute the relations (25) and use the basic property

$$D_\pm^\nu D_\mp^\nu D_\pm^\nu = (-\partial^2)^{\nu/2} D_\pm^\nu . \tag{27}$$

Then one finds

$$\int (\mathrm{d}1) D_+^\nu(2) \frac{\delta^2\Gamma}{\delta\Phi(1)\delta\Phi(2)} \frac{\delta^2\Gamma}{\delta I(1)\delta C_\pm(3)} = \frac{(-\partial^2)^{\nu/2}}{\alpha} D_+^\nu(2) \frac{\delta^2\Gamma}{\delta I(2)\delta C_\pm(3)} ,$$

so, in the Fermi gauge and in momentum space,

$$\int (\mathrm{d}1) D_+^\nu(2) [\delta(1,2)(p^2)^{\nu/2} + \Pi(1,2)] \frac{\delta^2\Gamma}{\delta I(1)\delta C(3)}$$

$$= (p^2)^{\nu/2} D_+^\nu(2) \frac{\delta^2\Gamma}{\delta I(2)\delta C(3)} .$$

Therefore $D_+^\nu(2)\Pi(1,2) = 0$, and similarly $D_-^\nu(2)\Pi(1,2) = 0$. These conditions imply that the gauge field self energy part $\Pi(1,2)$ is transverse and may thereby be expressed in the form

$$\Pi(\theta, p, \theta') = [(p^2)^{\nu/2} - D_+^\nu D_-^\nu - D_-^\nu D_+^\nu] F(\theta, p, \theta') . \tag{28}$$

We will check this in a moment by explicit calculation. However for the present let us return to the verification of (25) and (26).

For a start we shall require the Feynman rules, up to order g since we are stopping at the one-loop level. (The gauge tadpole graphs are of order g^2 but vanish in this method of dimensional continuation too.) From (19) and (20) we deduce the Fermi gauge field propagator

$$\langle 0 | \mathrm{T}[\Phi\Phi'] | 0 \rangle = i\Delta = i\delta^{2\nu}(\theta - \theta')\Delta_c(x - x') , \tag{29}$$

which has the great virtue that $\Delta^n = 0$ for $n > 1$. For many purposes therefore the interaction $\chi_-^\dagger (1 + g\Phi)^{-1} \chi_-$ can be replaced by $\chi_-^\dagger (1 - g\Phi)\chi_-$, rectifying the imbal-

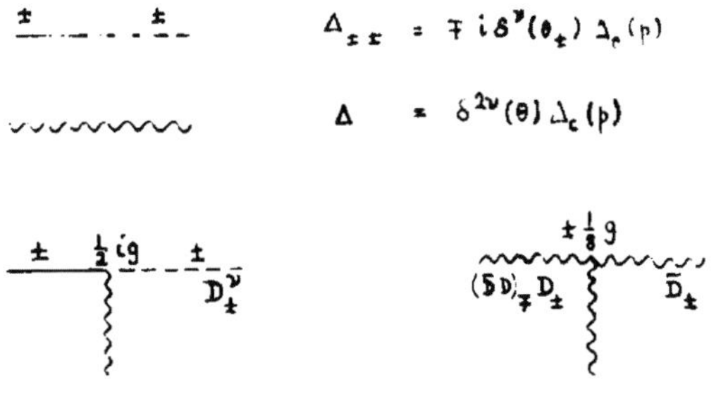

Fig. 1. The supercovariant Feynman rules to order g.

ance within (3). From (24) one obtains the fictitious field propagators

$$\langle 0|\mathrm{T}[C_\pm \bar{C}'_\pm]|0\rangle = i\Delta_{\pm\pm} = \mp D^\nu_\mp \delta^{2\nu}(\theta - \theta')\Delta_c(x - x')$$

$$= \mp\delta^\nu(\theta_\pm - \theta'_\pm)\Delta_c(x - x') \,, \tag{30}$$

$$\langle 0|\mathrm{T}[C_\pm \bar{C}'_\mp]|0\rangle = 0 \,.$$

Both in (29) and (30) the regularized causal propagator is

$$\Delta_c(x) = \int \mathrm{d}^{2l}x\; \mathrm{e}^{ip\cdot x}(p^2 + i0)^{-\nu/2} \,. \tag{31}$$

With the propagators in fig. 1 are also illustrated the rules for the vertices that couple the gauge and fictitious fields. These follow from the interaction Lagrangian

$$\tfrac{1}{2}ig\,\mathrm{Tr}\{\Phi[C_-, D^\nu_- \bar{C}_-] + \Phi[C_+, D^\nu_+ \bar{C}_+]\} \,. \tag{32}$$

For the gauge field itself we have only cited the rules in 4 dimensions, arising from

$$\tfrac{1}{8}g\,\mathrm{Tr}\{\Phi[\bar{D}_+\Phi, (\bar{D}D)_- D_+\Phi] - \Phi[\bar{D}_-\Phi, (\bar{D}D)_+ D_-\Phi]\} \,, \tag{33}$$

as they are too complicated to draw for arbitrary l.

In checking (25) it should be recognized that I is coupled to irreducible chiral fields $C_\pm$. Hence the generalization of (11) to $2l$ dimensions can be made to include the chiral projectors $D^\nu_\pm D^\nu_\mp/(-\partial^2)^{\nu/2}$ so as to give the interaction

$$i\,\mathrm{Tr}\left\{(1 + g\Phi)I \frac{D^\nu_+ D^\nu_-}{(-\partial^2)^{\nu/2}} C_- - I(1 + g\Phi)\frac{D^\nu_- D^\nu_+}{(-\partial^2)^{\nu/2}} C_+\right\} . \tag{34}$$

There is then very little to the actual verification of (25) since the diagrams that contribute to each side of the identity are in fact the same; it just suffices to show that, after regularization, the ghost self energy Σ is finite. The graphs for Σ in order g^2

Fig. 2. The fictitious field self-energy.

are depicted in fig. 2; quantitatively they give

$$\Sigma^{(2)}_{\pm\pm}(x,\theta) = \frac{\delta^2\Gamma^{(2)}}{\delta C_\pm \delta \bar{C}_\pm} = -g^2 D^\nu_\pm(\Delta D^\nu_\pm \Delta_{\pm\pm}) = \pm g^2 D^\nu_\pm(\delta^{2\nu}(\theta)\Delta^2_c(x))$$

$$= \pm g^2 \delta^\nu(\theta_\mp)\Delta^2_c(x)\,. \tag{35}$$

Including the tree approximation and transforming to momentum space,

$$\delta^2\Gamma/\delta C_\pm \delta\bar{C}_\pm = \mp i\delta^\nu(\theta_\mp)[1 + g^2 I(p)]\,, \tag{36}$$

where

$$I(p) = -i\int e^{ip\cdot x}\Delta^2_c(x)\mathrm{d}^{2l}x = -i\int \mathrm{d}^{2l}k\,[k^2]^{-\nu/2}[(k+p)^2]^{-\nu/2}/(2\pi)^{2l}$$

$$= \Gamma(\nu - l)\Gamma^2(\tfrac{1}{2}\nu - l)/(4\pi)^l\Gamma^2(\tfrac{1}{2}\nu)\Gamma(\nu - 2l)(p^2)^{\nu - l}\,, \tag{37}$$

is well defined for non-integer l but develops a pole at $l = \nu = 2$. The chirality changing parts $\delta^2\Gamma/\delta C_\pm \delta\bar{C}_\mp$ vanish in order g^2 (and to all orders) for the same reason that they disappear in tree approximation: namely they have the form of a $2l$ divergence $\int \mathrm{d}^{2\nu}\theta(\bar{C}_\pm D^\nu_\pm C_\mp)$ which cannot contribute to the action †. This implies that there is no fictitious mass counter term, only a kinetic energy one corresponding to wave function renormalization. At any rate the lesson that we learn from the example of Σ is that the covariant derivative combinatorics take care of themselves for any l as well as for four dimensions, leaving the convolution $I(p)$ of two regularized massless propagators. Therefore in evaluating the gauge field self-energy Π it is enough to carry out the D combinatorics for $l = 2$ and to show that Π is four dimensionally transverse; then to continue the result ⋆ to $2l$ dimensions because we have already proved the transversality for any l.

For determining Π we may first consider the fictitious contributions Π^C drawn

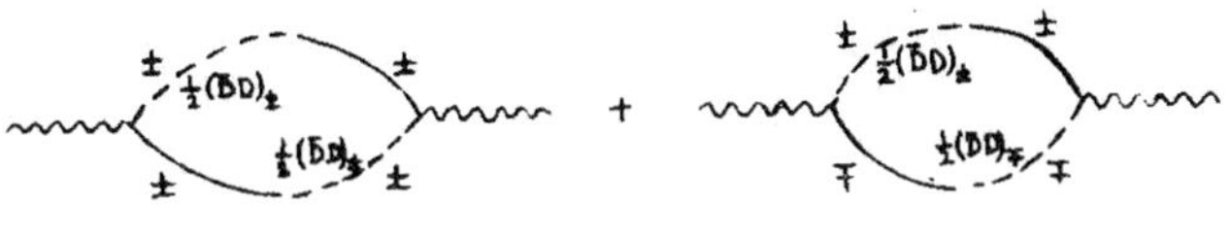

Fig. 3. Superghost contribution to the gauge field self-energy.

† The only other possible term one can construct, $\int \mathrm{d}^{2\nu}\theta(\bar{C}_\pm C_\mp)$, has the wrong dimension and never arises from the quantum loops.

⋆ Judging from the trouble we have experienced even to carry out the four-dimensional calculation we believe that the full $2l$-dimensional computation is probably unmanageable.

in fig. 3. These add up to

$$\Pi^C = -ig^2 [\tfrac{1}{2}(\bar{D}D)_+\Delta_{++} - \tfrac{1}{2}(\bar{D}D)_-\Delta_{--}]^2$$

$$= ig^2 [\{\tfrac{1}{4}(\bar{D}D)_+(\bar{D}D)_- + \tfrac{1}{4}(\bar{D}D)_-(\bar{D}D)_+\}\delta^{2\nu}(\theta)\Delta_c]^2$$

$$= ig^2 [\{\exp(\tfrac{1}{4}\bar{\theta}\partial\cdot\gamma\gamma_5\theta) + \exp(-\tfrac{1}{4}\bar{\theta}\partial\cdot\gamma\gamma_5\theta)\}\Delta_c]^2 = 4ig^2\Delta_c^2$$

(we have dropped a $\Delta_c\partial^2\Delta_c$ tadpole term above, consistently with the dimensional technique). The gauge field contributions Π^Φ are much more difficult to compute since they involve sewing together the numerous gauge field vertices of fig. 1 in their various permutations. There are a whole series of graphs to contend with but fortunately many of them cancel and one remains with the set drawn in fig. 4. To deter-

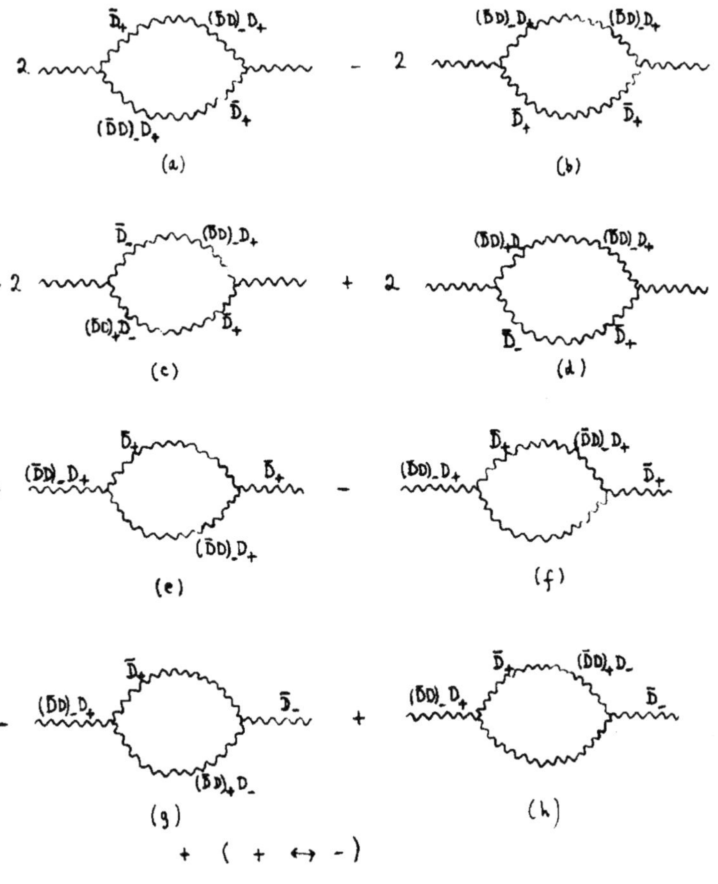

Fig. 4. Gauge field contributions to the gauge field self-energy.

mine these we have to use the property that on

$$\langle\varphi(x_1\theta_1)\varphi(x_2\theta_2)\rangle = f(x_1 - x_2 - \tfrac{1}{2}i\bar{\theta}_2\gamma\theta_1, \theta_1 - \theta_2)\,,$$

$$D_1 \to i(\partial/\partial\bar{\theta} - \tfrac{1}{2}i\partial\cdot\gamma\theta)\,, \qquad D_2 \to -i(\partial/\partial\bar{\theta} + \tfrac{1}{2}i\partial\cdot\gamma\theta)\,,$$

where $\theta = \theta_1 - \theta_2$ and $x = x_1 - x_2$. We then discover that diagrams 4b, 4d, 4g and 4h are identically zero upon adding the chiral conjugates. Without quoting the tedious details, the remaining diagrams sum up to

$$\Pi^{\Phi} = -ig^2\,[4 + \{\partial^2 + \tfrac{1}{4}(\bar{D}D)_+(\bar{D}D)_- + \tfrac{1}{4}(\bar{D}D)_-(\bar{D}D)_+\}(\bar{\theta}\theta)^2/16]\,\Delta_c^2\,.$$

Hence the total self-energy does indeed turn out to be transverse,

$$\Pi = \Pi^{C} + \Pi^{\Phi} = -\tfrac{1}{2}ig^2\,\{\partial^2 + \tfrac{1}{4}(\bar{D}D)_+(\bar{D}D)_- + \tfrac{1}{4}(\bar{D}D)_-(\bar{D}D)_+\}\delta^4(\theta)\Delta_c^2(x)\,,$$

this answer necessarily being the four-dimensional limit of

$$\Pi(p,\theta) = \tfrac{1}{2}g^2\,\{\partial^2 + D_+^{\nu}D_-^{\nu} + D_-^{\nu}D_+^{\nu}\}\delta^{2\nu}(\theta)I(p)\,. \tag{38}$$

A fortiori the identity (26) will be satisfied. Perhaps we should point out that the result (38) is the first supercovariant calculation in a *non-Abelian* gauge theory that has been carried out to date. When we pick out the vector projection from (38), the answer is familiar *,

$$\Pi^{ab}_{\mu\nu}(p) = \delta^{ab}C_2(G)g^2(p^2\eta_{\nu\mu} - p_\mu p_\nu)I(p)\,,$$

because it is the same as choosing the Fermi gauge and Wess-Zumino [5] supergauge (wherein the gauge vector, gauge ghost and gauge spinors contribute). Nevertheless, the point is that by preserving the sypersymmetry throughout one has essentially the same anomalous dimension $\gamma = -g^2C_2(G)/16\pi^2$ for all the components of the gauge superfield.

5. Renormalization counter terms

We will assume for the purposes of this paper that the supergauge theories are renormalizable. (The statement is patently obvious in the Wess-Zumino supergauge, but to prove it in a supersymmetric context [6] one has to demonstrate that the non-polynomiality in the theory is not dangerous.) We shall look for the repercus-

* We should not be dismayed by the curious residue $(1-2\ln 2)^{-1}$ in $I(p)$ at $l = 2$. It simply means that the anomalous dimension is determined by the formula $\lim_{l\to 2}(\nu - 1)g\partial(\ln Z)/\partial g$, rather than by $\lim_{l\to 2}(2-l)g\partial(\ln Z)/\partial g$, when working in a supersymmetric framework.

sions *vis a vis* the renormalization counter terms that need to be introduced. Let us therefore write down the most general supergauge invariant Lagrangian including possible counter terms:

$$\mathcal{L}(x,\theta) = Z\sum_m a_m[(1+gZ_gZ^{-1}\Phi)^{-1}D_+^m\Phi]D_-^\nu[(1+gZ_gZ^{-1}\Phi)^{-1}D_+^{\nu-m}\Phi]$$

$$+(+\leftrightarrow -)+(D_+^\nu\Phi D_-^\nu\Phi)/\alpha + i(\bar{C}_-D_-^\nu+\bar{C}_+D_+^\nu)\{\tilde{Z}(C_- - C_+)$$

$$+\tilde{Z}_g g(C_- - C_+)\} + i\hat{Z}I(C_- - C_+) + ig\hat{Z}_gI(C_- - C_+)$$

$$-\tfrac{1}{2}igZ_g'(I_-[C_+,C_+]+I_+[C_-,C_-])\,, \tag{40}$$

to be traced out and integrated over $d^{2l}x$ and $d^{2\nu}\theta$. Note that since the vector corrections are purely transverse,

$$\alpha_0 = Z\alpha\,. \tag{41}$$

Let us now apply the BRS identities to relate the various terms in (40). From (25) we see immediately that

$$\hat{Z} = \tilde{Z}\,. \tag{42}$$

Differentiating (25) further with respect to Φ,

$$D^\nu\delta^3\Gamma/\delta\Phi\delta I\delta C = \delta^3\Gamma/\delta\Phi\delta C\delta\bar{C}\,,$$

gives

$$\hat{Z}_g = \tilde{Z}_g\,. \tag{43}$$

To find the last few relations define [3]

$$\hat{\Gamma} = \Gamma - \int d^{2\nu}\theta\ \mathrm{Tr}(D_+^\nu\Phi D_-^\nu\Phi)/\alpha\,,$$

in order to get rid of the gauge fixing pieces in the primitive identity (13) as well as in (26). By taking a Φ derivative in (26),

$$\frac{\delta^3\hat{\Gamma}}{\delta\Phi^3}\frac{\delta^2\hat{\Gamma}}{\delta I\delta C} + \frac{\delta^2\hat{\Gamma}}{\delta\Phi^2}\frac{\delta^3\hat{\Gamma}}{\delta\Phi\delta I\delta C} = 0\,, \tag{44}$$

from which one deduces the Slavnov-Taylor connection between the wave function and coupling renormalizations for gauge and fictitious fields

$$Z_g\tilde{Z} - Z\tilde{Z}_g = 0 \qquad \text{or} \qquad Z_g/Z = \tilde{Z}_g/\tilde{Z}\,. \tag{45}$$

Similarly by taking ghost field derivatives,

$$\frac{\delta^3\hat{\Gamma}}{\delta C_+\delta C_+\delta I_-}\frac{\delta^2\hat{\Gamma}}{\delta C_+\delta \bar{C}_+}+\frac{\delta^3\hat{\Gamma}}{\delta\Phi\delta C_+\delta\bar{C}_+}\frac{\delta^2\hat{\Gamma}}{\delta I\delta C_+}=0\,,$$

and one finds that

$$Z_g'\tilde{Z}=\tilde{Z}_g\hat{Z} \qquad \text{or} \qquad Z_g'=\tilde{Z}_g\,.$$

Hence the BRS transformations themselves (10) undergo a pure multiplicative renormalization from their bare values.

After all is said and done there survive three independent constants, say Z, $\tilde{Z}$ and Z_g, in terms of which all the others are given. Contrary to usual Yang-Mills theory, in the supersymmetric version one has no need to treat separately the renormalization Z_4 of the four-point gauge field vertex (Φ^4) since this is automatically contained in (40) *via* the expansion in powers of Φ. As confirmation we can satisfy ourselves that the supergauge identity

$$\frac{\delta^4\hat{\Gamma}}{\delta\Phi^4}\frac{\delta^2\hat{\Gamma}}{\delta I\delta C}+\frac{\delta^3\hat{\Gamma}}{\delta\Phi^3}\frac{\delta^3\Gamma}{\delta\Phi\delta I\delta C}=0\,,$$

yields exactly what is required.

$$Z_4\tilde{Z}=Z_g\tilde{Z}_g \qquad \text{or} \qquad Z_4=Z_g^2/Z\,.$$

Similarly the higher point functions are supercovariantly renormalized.

We are extremely grateful to Professor A. Salam for constantly urging us to adopt the supersymmetric regularization and for his encouragement throughout the investigation. We are also very indebted to Dr. J. Strathdee for making available to us some of his unpublished notes on aspects of the dimensional continuation.

Appendix

We wish here to give the derivation of relation (22) of the text, given the freedom to adjust our $\frac{1}{2}\nu$ coefficients a. Beginning with the anticommutator $\{D,\bar{D}\}=p$, one can easily derive the relation

$$\sum_1^{\nu/2} a_m\,[D_+^m D_-^\nu D_+^{\nu-m}+D_-^m D_+^\nu D_-^{\nu-m}]=\sum_1^{\nu/2} a_m'\,[p^m\{D_-^{\nu-m},D_+^{\nu-m}\}$$

$$-p^{m-1}\{D_-^{\nu-m},D_+^{\nu-m}\}]=\sum_1^{\nu/2} A_m p^m\{D_-^{\nu-m},D_+^{\nu-m}\}\,. \tag{A.1}$$

But for even N one can also prove that

$$\{D_+^{N+1}, D_-^{N+1}\} = \sum_1^{\nu/2} B_m p^{2m+1} \{D_+^{N-2m}, D_-^{N-2m}\} .$$

Hence (A.1) can be re-expressed in the form

$$\sum_1^{\nu/2} b_m p^{2m} \{D_+^{\nu-2m}, D_-^{\nu-2m}\} . \tag{A.2}$$

Adding this to the gauge fixing term, we may arrange for all the $b_1, ..., b_{\nu/2}$ to vanish and that the final term $b_{\nu/2} = 1$. This then establishes (22).

References

[1] A.A. Slavnov, Theor. Math. Phys. 10 (1972) 99;
J.C. Taylor, Nucl. Phys. B33 (1971) 436.
[2] C. Becchi, A. Rouet and R. Stora, Comm. Math. Phys. 42 (1975) 127.
[3] H. Kluberg-Stern and J.B. Zuber, Phys. Rev. 12D (1975) 467;
B.W. Lee, Phys. Letters 46B (1974) 214;
G. Costa and M. Tonin, Nuovo Cimento Letters 14 (1975) 171;
J. Zinn-Justin, Lectures at Bonn Summer Institute, 1974 (Springer-Verlag).
[4] G. 't Hooft and M. Veltman, Nucl. Phys. 44B (1972) 189;
J. Ashmore, Nuovo Cimento Letters 4 (1972) 289;
C.G. Bollini and J. Giambiagi, Cimento 12B (1972) 20.
[5] J. Wess and B. Zumino, Nucl. Phys. B78 (1974) 1;
A. Salam and J. Strathdee, Phys. Letters 51B (1974) 353;
S. Ferrara and B. Zumino, Nucl. Phys. B79 (1974) 99;
R. Delbourgo, A. Salam and J. Strathdee, Phys. Letters 51B (1974) 475.
[6] S. Ferrara and O. Piguet, Nucl. Phys. B93 (1975) 261;
A.A. Slavnov, Nucl. Phys. B97 (1975) 155;
B. de Wit and D. Freedman, Phys. Rev. 12D (1975) 2286;
O. Piguet and A. Rouet, Nucl. Phys. B99 (1975) 458.
[7] S. Ferrara and O. Piguet, ref. [6].
F. Krause, M. Scheunert, J. Honerkamp and M. Schlindwein, Nucl. Phys. B95 (1975) 397;
R. Delbourgo, J. Phys. G1 (1975) 800.
[8] D.M. Capper, Nuovo Cimento 25A (1975) 259;
R. Delbourgo, Nuovo Cimento 25A (1975) 646;
J. Honerkamp, F. Krause, M. Scheunert and M. Schlindwein, Phys. Letters 53B (1974) 60;
K. Fujikawa and W. Lang, Nucl. Phys. B88 (1975) 61.
[9] R. Delbourgo and V.B. Prasad, J. Phys. G1 (1975) 380.

Part G
BRST QUANTIZATION

G.0. Commentaries G

With the advent of QCD as the gauge theory of the strong interaction and the proof of its renormalisability by 't Hooft (Nucl. Phys. B35, 167 (1971)), the quantisation of non-Abelian gauge groups became a prominent topic of research. The procedure was recognised as being more complicated than Abelian QED, where the Gupta-Bleuler procedure sufficed to cancel the contributions of the longitudinal photons by the time-like ones. Veltman (Nucl. Phys. B44, 189 (1972)) knew that for gluons this required the introduction of ghost scalar particles in intermediate states by examining the perturbation series; this would ensure unitarity at all levels. The concept was systematised by Fadeev & Popov (Phys. Lett. B25, 29 (1967)) by looking at the gauge trajectories of fields via the functional formalism whereby a ghost scalar and its dual make an entrance; despite their apparent conflict with the spin-statistics theorem these fictitious scalar fields became an accepted feature of non-Abelian gauge theories, the crucial point being that the ghosts were only needed at an intermediate stage and were nonexistent at the asymptotic, physical level.

Because the canonical quantisation process requires gauge fixing, the gauge symmetry must be broken in a formal sense. Becchi, Rouet & Stora (Comm. Math. Phys. 43, 127 (1975)) and Tyutin (arxiv: 0812.0580 (1975)) investigated the nature of the surviving symmetry when ghosts are included and discovered the famous BRST transformation connecting the gauge fields and ghosts; it features anticommuting coordinate parameters and simplifies enormously the treatment of renormalisation (J.C. Taylor, Nucl. Phys. B33, 436 (1971)) of QCD, as was demonstrated in detail by Kugo and Ojima (Prog. Theor. Phys. Supp. 66, 1 (1979)). Indeed one might characterise BRST symmetry as 'the symmetry of quantised QCD'.

Soon afterwards, Curci and Ferrari (Phys. Lett. B63, 91 (1976)) discovered a dual symmetry to BRST wherein the ghosts and duals are effectively interchanged. By introducing an auxiliary scalar field obeying ordinary statistics, Lautrup (Mat. Fys. Medd. Dan. Vid. Selsk 35, 11 (1967)) and Nakanishi (Prog. Theor. Phys. 37, 618 (1967)) tidied up the full set of transformations between gauge vectors and both scalar ghosts. The stage was set for treating the quantisation of gravitation.

G.1 Delbourgo R. and Ramón-Medrano M., Becchi-Rouet-Stora gauge identities for gravity. Nucl. Phys. B110, 467-472 (1976)

G.2 Delbourgo R. and Jarvis P.D., Extended BRS invariance and OSp(4/2) supersymmetry. J. Phys. A15, 611-625 (1982)

G.3 Delbourgo R., Jarvis P.D. and Thompson G., Local OSp(4/2) supersymmetry and extended BRS transformations for gravity. Phys. Lett. B109, 25-27 (1982)

In paper [G.1] we determined, for the first time, the analogue of the BRST transformations for gravitation. Thus quantisation again requires us to break the general covariance and settle upon the gauge of the metric. Feynman (Acta Phys. Pol. 24, 697 (1963)) anticipated that this would require the introduction of vector ghosts in order to preserve the unitarity of the theory. And the analysis in [G.1] confirmed this conjecture, supported by Stelle's treatment (Phys. Rev. D16, 953 (1977)) of higher derivative gravity. Paper [G.2] put the treatment of ghosts and duals in QCD on a nice group theoretic basis associated with the super Lie algebra OSp(4/2). Finally article [G.3] achieved the same thing for gravity. Of course there were parallel contributions by many others and the following papers ought to be read in parallel:

- K. Nishijima and M. Okawa, Prog. Theor. Phys. 60, 272 (1978)
- L. Bonora and M. Tonin, Phys. Lett. B98, 48 (1981)
- F.R. Ore and P. van Nieuwenhuizen, Nucl. Phys. B204, 317 (1982)
- L. Baulieu and J. Thierry-Mieg, Nucl. Phys. B197, 477 (1982)
- K. Nishijima, Nucl. Phys. B238, 601 (1984)
- N. Nakanishi and I. Ojima, Covariant Operator Formalism of Gauge Theories and Quantum Gravity, Chapter 4 (1990)
- M. Pasti and M. Tonin, Nuovo Cim. B69, 97 (1982)

The development of BRST transformations has not ceased. The papers by Malik and coworkers are taking the idea to its limits. For example see J. Phys. A43, 375403 (2010). The review by Baulieu (Phys. Repts. 129, 1 (1985)) is comprehensive and is worth reading too.

Nuclear Physics B110 (1976) 467–472
© North-Holland Publishing Company

BECCHI-ROUET-STORA GAUGE IDENTITIES FOR GRAVITY

R. DELBOURGO and M. Ramon MEDRANO *
Blackett Laboratory, Imperial College, Prince Consort Road, London SW7 2BZ

Received 7 April 1976

The Becchi-Rouet-Stora gauge transformations for the quantum theory of gravitation are deduced. The gauge identities for the graviton and vector ghost field are verified to one-loop order for the two-point functions

1. Introduction

The Slavnov-Taylor identities [1] for Yang-Mills theory and gravitation [2] are by now well-known, and have been explicitly verified in some instances. Although undoubtedly correct there are two difficulties about using them:

(i) they have a non-local appearance;

(ii) because they apply directly to the Green functions it takes a rather lengthy chain of argument [3] to get to the proper vertices and thereby determine the renormalization counter terms as well as the relations between them.

A significant advance was made last year by Becchi, Rouet and Stora (BRS) [4] who, by introducing a global anticommuting scalar parameter, were able to extend the gauge transformations to include the scalar ghost fields themselves. The beauty of this new approach is that the Ward identities are manifestly local and immediately applicable to the proper vertices. This was recently put to good use [5].

In this paper we present the BRS identities for gravity **. In sect. 2 we deal with the formalism, and the general expression for the gauge identities is deduced. In sect. 3 the gauge identities corresponding to the two-point functions, involving the graviton and vector ghost fields, are verified up to one-loop order; dimensional regularization techniques are used.

Where possible we try to draw the analogy with the corresponding Yang-Mills case

* Work supported by GIFT (Grupo Inter universitario de Fisica Teorica) Spain.
** Although we have independently rederived them, the credit for finding them should go to Dixon. They are mentioned in a short paragraph at the end of his thesis. We understand that Professor Deser (private communication) has also discovered them.

2. The BRS transformations for gravity

Begin with the Lagrangian for gravity plus fictitious vector fields in arbitrary dimensions [2] $2l$ (since we shall later regularize dimensionally in checking the resulting identities) in a covariant gauge specified by a parameter α,

$$\mathcal{L} = [\tilde{g}^{\rho\sigma}\tilde{g}_{\lambda\mu}\tilde{g}_{\kappa\nu} - 2\delta^{\sigma}_{\kappa}\delta^{\rho}_{\lambda}\tilde{g}_{\mu\nu} - \tilde{g}^{\rho\sigma}\tilde{g}_{\mu\kappa}\tilde{g}_{\lambda\nu}/(2l-2)]\,\partial_{\rho}\tilde{g}^{\mu\kappa}\partial_{\sigma}\tilde{g}^{\nu\lambda}/2k^2$$

$$- \partial_{\mu}\tilde{g}^{\mu\nu}\partial_{\kappa}\tilde{g}^{\kappa\lambda}\eta_{\nu\lambda}/2k^2\alpha + \bar{C}^{\nu}\partial^{\mu}D_{\mu\nu\lambda}C^{\lambda}\,, \tag{1}$$

where $\tilde{g}^{\mu\nu} = g^{\mu\nu}\sqrt{-g} = \eta^{\mu\nu} + k\phi^{\mu\nu}$ defines our graviton field and the differential operator D is given by

$$D_{\mu\nu\lambda} = -\partial_{\lambda}\phi_{\mu\nu} - \phi_{\mu\nu}\partial_{\lambda} + \eta_{\mu\lambda}\phi_{\kappa\nu}\partial^{\kappa} + \eta_{\nu\lambda}\phi_{\kappa\mu}\partial^{\kappa}$$

$$+ k^{-1}(\eta_{\mu\lambda}\partial_{\nu} + \eta_{\nu\lambda}\partial_{\mu} - \eta_{\mu\nu}\partial_{\lambda})\,. \tag{2}$$

The invariance of $\mathcal{L}$ under the gauge transformations $\delta\phi_{\mu\nu} = D_{\mu\nu\lambda}\zeta^{\lambda}$ is familiar and leads to the Slavnov-Taylor identities for gravity. However if one follows BRS it is natural to put $\zeta^{\lambda} = C^{\lambda}\omega$, where ω is the global anticommuting scalar parameter which is also used in Yang-Mills theory. Then the gauge transformations, generalised to include the vector ghosts C and $\bar{C}$, read

$$\delta\phi_{\mu\nu} = D_{\mu\nu\lambda}C^{\lambda}\omega\,, \qquad \delta\bar{C}^{\nu} = -\partial_{\mu}\phi^{\mu\nu}\omega/\alpha\,, \qquad \delta C_{\nu} = C^{\mu}\partial_{\mu}C_{\nu}\omega\,. \tag{3}$$

It is straightforward, if a bit tedious, to verify that (1) is invariant under the changes (3). This is entirely parallel to Yang-Mills theory where the transformations for the vector field ϕ_{μ} and the scalar ghost C read

$$\delta\phi_{\mu} = D_{\mu}C\omega\,, \qquad \delta\bar{C} = -\partial^{\mu}\phi_{\mu}\omega/\alpha\,, \qquad \delta C = -\tfrac{1}{2}g(C\times C)\omega\,. \tag{4}$$

Also for gravity we have $\delta(D_{\mu\nu\lambda}C^{\lambda}) = \delta(C^{\mu}\partial_{\mu}C_{\nu}) = 0$ on par with $\delta(D_{\mu}C) = \delta(C\times C) = 0$ for Yang-Mills.

Next, we follow the usual procedure of including (invariant) composite source fields $I^{\mu\nu}$, I^{μ} and external sources $J^{\mu\nu}$, J^{μ} and $\bar{J}^{\mu}$ in the action (notice the variation of $\bar{C}^{\nu}$ is linear in the fields due to the specific linear gauge condition chosen; therefore the introduction of a new source is not needed):

$$\mathcal{L}_J = J^{\mu\nu}\tilde{g}_{\mu\nu}/k + \bar{J}^{\mu}C_{\mu} + \bar{C}^{\mu}J_{\mu} + I^{\mu\nu}D_{\mu\nu\lambda}C^{\lambda} + I^{\mu}C_{\nu}\partial^{\nu}C_{\mu}\,. \tag{5}$$

Then under the transformations (3) we obtain the generalized BRS identity for the vacuum functional

$$\left[J^{\mu\nu}D_{\mu\nu\lambda}\frac{\delta}{\delta\bar{J}_{\lambda}} + \bar{J}_{\mu}\frac{\delta}{\delta I_{\mu}} + \frac{1}{\alpha}J_{\nu}\partial_{\mu}\frac{\delta}{\delta J_{\mu\nu}}\right]Z = 0\,. \tag{6}$$

The machinery for passing to the connected functional $W = -i \ln Z$ and the proper vertex functional $\Gamma = W - \int \phi J$ can also be taken over to this case and we obtain

$$\int dx \left[\frac{\delta\Gamma}{\delta\phi_{\mu\nu}} \frac{\delta\Gamma}{\delta I^{\mu\nu}} - \frac{\delta\Gamma}{\delta C_\mu} \frac{\delta\Gamma}{\delta I^\mu} + \frac{1}{\alpha} \partial^\mu \phi_{\mu\nu} \frac{\delta\Gamma}{\delta \bar{C}_\nu} \right] = 0 , \tag{7}$$

plus the useful equation of motion

$$\partial_\mu \frac{\delta\Gamma}{\delta I_{\mu\nu}} = \frac{\delta\Gamma}{\delta \bar{C}_\nu} . \tag{8}$$

The generalized BRS proper vertex identities then come by functionally differentiating (7) and (8) say to leave zero ghost number.

3. The BRS identities for the two-point functions

We shall check (in momentum space)

$$p_\mu \delta^2\Gamma/\delta I_{\mu\nu}\delta C_\rho = \delta^2\Gamma/\delta\bar{C}_\nu \delta C_\rho , \tag{9}$$

found directly from (8), and

$$\frac{\delta^2\Gamma}{\delta\phi_{\kappa\lambda}\delta\phi_{\mu\nu}} \frac{\delta^2\Gamma}{\delta I^{\mu\nu}\delta C_\rho} = p_\kappa \frac{\delta^2\Gamma}{\delta\bar{C}_\lambda \delta C_\rho} + p_\lambda \frac{\delta^2\Gamma}{\delta\bar{C}_\kappa \delta C_\rho} , \tag{10}$$

obtained by operating on (7) with $\delta^2/\delta\phi_{\kappa\lambda}\delta C_\rho$ and fixing on the Fermi gauge $\alpha = 1$. By Lorentz covariance,

$$\begin{aligned}
&\delta^2\Gamma/\delta\bar{C}_\lambda\delta C_\rho = d_{\lambda\rho}A + e_{\lambda\rho}B , \\
&\delta^2\Gamma/\delta I_{\mu\nu}\delta C_\rho = p_\rho d_{\mu\nu}F + p_\rho e_{\mu\nu}G + [d_{\mu\rho}p_\nu + d_{\nu\rho}p_\mu]H , \\
&\delta^2\Gamma/\delta\phi_{\kappa\lambda}\delta\phi_{\mu\nu} = [d_{\kappa\mu}d_{\lambda\nu} + d_{\lambda\mu}d_{\kappa\nu} - 2d_{\kappa\lambda}d_{\mu\nu}/(2l-1)]E_1 \\
&\quad + [d_{\kappa\mu}e_{\lambda\nu} + d_{\lambda\mu}e_{\kappa\nu} + d_{\kappa\nu}e_{\lambda\mu} + d_{\lambda\nu}e_{\kappa\mu}]E_2 + e_{\kappa\lambda}e_{\mu\nu}E_3 \\
&\quad + d_{\kappa\lambda}d_{\mu\nu}E_4 + [e_{\kappa\lambda}d_{\mu\nu} + d_{\kappa\lambda}e_{\mu\nu}]E_5 ,
\end{aligned} \tag{11}$$

where $d_{\mu\nu} \equiv \eta_{\mu\nu} - p_\mu p_\nu/p^2 \equiv \eta_{\mu\nu} - e_{\mu\nu}$. The identities (10) stipulate that

$$\begin{aligned}
&A - p^2 H = B - p^2 G = 0 , \\
&A - 2E_1 H = (2l-1)E_4 F + E_5 G = (2l-1)E_5 F + E_3 G - 2B = 0 .
\end{aligned} \tag{12}$$

Fig. 1.

Graphically the one-loop correction to $\delta^2\Gamma/\delta\bar{C}_\lambda\delta C_\rho$ is given by fig. 1. The Feynman rules can be read directly from (1), and the graviton propagator in the Fermi gauge is

$$D_{\rho\sigma,\lambda\mu}(p^2) = \frac{1}{2(p^2 + i\epsilon)} [\eta_{\rho\lambda}\eta_{\mu\sigma} + \eta_{\rho\mu}\eta_{\lambda\sigma} - \eta_{\rho\sigma}\eta_{\lambda\mu}] \; .$$

The values A and B referred to in eq. (11), up to one-loop order, are [6]

$$A/p^2 = 1 + I/8(2l - 1) \, , \qquad B/p^2 = 1 + I/8 \, ,$$

where

$$I \equiv \frac{k^2p^2}{(2\pi)^{2l}} \int \mathrm{d}^{2l}q/q^2(p-q)^2 \; .$$

The one-loop contribution to $\delta^2\Gamma/\delta I_{\mu\nu}\delta C_\rho$ corresponds to the graph (fig. 2) and the values F, G and H, occurring in (11) up to one-loop order, are [2]

$$F = 1 - 5I/8 \, , \qquad G = 1 + I/8 \, , \qquad H = 1 + I/8(2l - 1) \; .$$

It is worth noticing that the graph of fig. 2 without the current $I_{\mu\nu}$ attached already appeared in the Slavnov-Ward identities of gravity [2], nevertheless, there it had a somewhat "peculiar" origin, in so far that it did not appear obviously in the Lagrangian. However we see that in the BRS context this vertex appears naturally thanks to the introduction of composite fields and their corresponding sources.

Finally the one-loop contribution to $\delta^2\Gamma/\delta\phi_{\kappa\lambda}\delta\phi_{\mu\nu}$ corresponds to the graphs

Fig. 2.

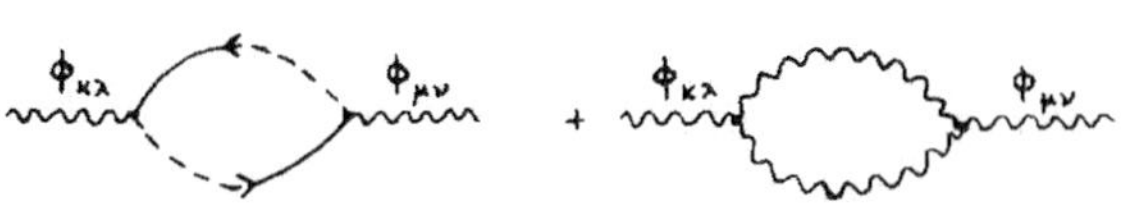

Fig. 3.

(fig. 3) and the values of $E_1 \ldots E_5$ up to one-loop order are [6] *

$$2E_2/p^2 = 1 + 0I\,,$$

$$[(2l-1)E_4 - E_5]/p^2 = 0 - I/4(l-1)\,,$$

$$[E_5 - E_3/(2l-1)]/p^2 = -2/(2l-1) - I/4(l-1)\,,$$

$$(2l-1)E_4/p^2 = -1/2(l-1) + (4l^5 - 28l^4 + 9l^3 + 62l^2 - 7l - 16)$$

$$\times\ I/32(l-1)^2(4l^2-1)\,.$$

One visibly finds that (12) are indeed satisfied to order k^2. It is interesting to note that the required relations look very different from the Slavnov-Taylor ones [2], and could not *a priori* have been guessed.

Evidently by differentiating (10) further with respect to ϕ one can get relations between multigraviton vertices and those involving the vector ghosts. We have every reason to believe that they will be obeyed since the dimensional regularization respects the gauge symmetry. Hence one will obtain relations between various possible counter terms.

4. Conclusion

We have shown that for gravity, as for its more simple non-Abelian gauge theory companion the Yang-Mills fields, one can find a global anticommuting parameter transformation (BRS) such that Ward identities for the proper vertices are easily deduced.

Although at present gravity still seems to be non-renormalisable, nevertheless we think that any future new attempt ro undertake the problem of renormalisability should be guided by and take advantage of the Becchi-Rouet-Stora identities exposed in the above sections.

References

[1] A.A. Slavnov, Theor. Math. Phys. 10 (1972) 99;
J.C. Taylor, Nucl. Phys. B33 (1971) 436.

* We have re-expressed the covariants of ref. [6] in terms of our projector covariants. Specifically the invariants T_i are given in terms of our E_i by the relations

$$E_1 = T_3\,, \quad E_2 = T_3 + T_5\,, \quad E_3 = T_1 + T_2 + 2T_3 + 2T_4 + 4T_5\,,$$

$$E_4 = T_2 - 2T_3/(2l-1)\,, \quad E_5 = T_2 + T_4\,.$$

[2] D.M. Capper and M. Ramón Medrano, Phys. Rev. D9 (1974) 1641.
[3] B.W. Lee, Phys. Letters 46B (1974) 214;
G. Costa and M. Tonin, Nuovo Cimento Letters 14 (1975) 171.
[4] C. Becchi, A. Rouet and R. Stora, Comm. Math. Phys. 42 (1975) 127;
J. Dixon, Oxford Thesis (1975); Nucl. Phys. B99 (1975) 420.
[5] H. Kluberg-Stern and J.B. Zuber, Phys. Rev. D12 (1975) 467.
[6] D.M. Capper, G. Leibbrandt and M. Ramón Medrano, Phys. Rev. D8 (1973) 4320.

J. Phys. A: Math. Gen. **15** (1982) 611–625. Printed in Great Britain

Extended BRS invariance and OSp(4/2) supersymmetry

R Delbourgo and P D Jarvis

Department of Physics, University of Tasmania, Hobart, Tasmania 7001

Received 9 June 1981

Abstract. A superfield action is proposed within an OSp(4/2) framework whose component form reproduces the covariant ξ-gauge Yang–Mills action, but with modified ghost-compensating terms. (The case $\xi = 0$ reduces to the usual Landau gauge.) 'Supertranslations' give rise to extended BRS transformations, and lead to constraints amongst the renormalisation constants. In addition, the system admits 'super-Lorentz' transformations, which mix vector and ghost fields. For other field representations, the ghost structure suggested by the space–time supersymmetry OSp(4/*G*) is also exhibited. This simplifies the rules for counting ghosts and their own ghosts.

1. Introduction and main results

It has long been recognised that the so-called BRS symmetry (Becchi *et al* 1975, 1976) of the gauge-fixing plus ghost-compensating Lagrangian in Yang–Mills theories (Feynman 1963, De Witt 1965, Faddeev and Popov 1967) has powerful implications for their quantisation and renormalisation. Subsequent investigations have revealed that an 'extended' BRS set can be constructed (Curci and Ferrari 1976, Ojima 1980), involving a two-parameter 'BRS group' where the roles of 'ghost' and 'antighost' can essentially be interchanged. Following earlier work on the unextended case (Ferrara *et al* 1977, Fujikawa 1978), Bonora and Tonin (1981) in an important recent paper have presented a concise derivation of the extended BRS transformations from a manifest superfield formalism, in which the BRS group consists of supertranslations in the a-number superspace coordinates.

Several arguments can be adduced leading to the possible existence of a supersymmetric BRS formalism. Firstly, the fact that the gauge potential $A_\mu^a(x)$ is accompanied by a-number 'ghost' fields, denoted here by $\omega^a(x)$ and $\bar{\omega}^a(x)$, all transforming in the adjoint representation of the gauge group, is highly reminiscent of Fermi–Bose supersymmetry, where gauge supermultiplets, including vector bosons and Majorana fermions, necessarily fall into the adjoint representation (for a review, see Fayet and Ferrara (1977)). Secondly, there is a natural 'auxiliary field' formalism for gauge fixing, whereby the usual covariant term $-(\partial^\mu A_\mu)^2/2\xi$ is replaced by

$$(\partial^\mu A_\mu)B + \tfrac{1}{2}\xi B^2,$$

where $B(x)$ is an auxiliary field of dimension two. $B(x)$ here plays a role similar to that of the auxiliary field in Fermi–Bose sypersymmetry. Finally, it is to be expected that a consistent supersymmetric derivation would provide a natural rationale for the existence of the two-parameter extended BRS group, in which the ghost and antighost fields ω and $\bar{\omega}$ are placed on an equal footing.

0305-4470/82/020611 + 15\$02.00 © 1982 The Institute of Physics

It is found (Bonora and Tonin 1981) that the appropriate supersymmetry involves a six-dimensional superspace with coordinates $(x_\mu, \theta, \bar{\theta})$. Making superfield expansions on the *a*-number coordinates θ and $\bar{\theta}$, Bonora and Tonin (1981) impose constraints on the 'supercurvature' and obtain sufficient conditions that supertranslations on θ and $\bar{\theta}$ correspond to extended BRS transformations, and the most general supersymmetric Lagrangian density is a two-gauge parameter generalisation of the usual Yang–Mills ghost Lagrangian. In a related paper, Bonora *et al* (1980) have justified their derivation in terms of a geometrical construction, enabling an interpretation to be given to terms such as 'connection' and 'curvature', which were undefined in the original work.

It is the purpose of the present paper to present an alternative formulation of BRS supersymmetry. We go beyond the work of Bonora and Tonin (1981) and Bonora *et al* (1980) in the sense that their group of 'supertranslations' in $(\theta, \bar{\theta})$ space is here enlarged to include transformations mixing x_μ and $(\theta, \bar{\theta})$. The appropriate supersymmetry is a real form of the inhomogeneous OSp(4/2) supergroup (Dondi and Jarvis 1979), consisting of supertranslations, Lorentz transformations, symplectic transformations in $(\theta, \bar{\theta})$ space, as well as 'supertranslations' and 'super-Lorentz' transformations (§ 2). With the help of this space–time supersymmetry, the definitions and transformation properties of the gauge potential and field strength follow naturally (§ 3), without further geometrical constructions.

The formalism of a gauge theory over six-dimensional superspace having been introduced, attention is restricted to a special class of potentials (beyond what gauge freedom alone would allow). An action for pure Yang–Mills theory, with gauge-fixing and ghost-compensating terms, is derived in § 3 as the component form of an appropriate superfield action, wherein the gauge potential belongs to the special class. The Lagrangian, equation (15), is a particular case of that of Bonora and Tonin (1981), wherein the two independent gauge parameters are made equal. There is a quartic ghost self-coupling, and the vector–ghost coupling attains a symmetrical form reminiscent of charged scalar electrodynamics. These additional couplings fade out as $\xi \to 0$, and in this limit with $\partial^\mu A_\mu = 0$, the model reduces to the conventional Landau gauge.

Supertranslations give rise to extended BRS transformations amongst the component fields (as in Bonora and Tonin 1981). These are exposed in § 4, and used to derive generating functional equations. The resulting BRS identities (for the effective action) imply renormalisability of the model. Relations between the renormalisation constants are exhibited, and verified to one-loop order. The only difference from the usual Yang–Mills case is in the fact that the longitudinal part of the vector boson propagator (and hence the gauge parameter ξ) receives a separate renormalisation. The transverse part continues to obey the usual Slavnov–Taylor identity. Moreover, the symmetry of the model ensures that all counterterms already have counterparts in the bare Lagrangian. In particular, renormalisation respects the symmetrical form of the vector–ghost interaction.

The space–time supersymmetry we propose also admits super-Lorentz transformations, which mix the coordinates x_μ and $(\theta, \bar{\theta})$. In contrast to supertranslations, the class of gauge potentials of interest is not super-Lorentz invariant. Nonetheless one can implement a corresponding set of vector–ghost mixing transformations (§ 5). The resulting identities for the effective action are also presented and checked.

Concluding remarks are made in § 6, together with a comparison with other approaches. Finally, the ghost structure suggested by the OSp(4/2) supersymmetry is given for other field representations, including the totally symmetrical and mixed-symmetry rank-three gauge fields. This represents a tidy and effective way of counting ghosts and superghosts for any chosen field representation.

2. Space–time supersymmetry

The space–time supersymmetry which we impose is a real form of the six-dimensional inhomogeneous orthosymplectic supergroup OSp(4/2) $\wedge$ $T_{4/2}$, the group of all super-linear transformations preserving the distance (Dondi and Jarvis 1979)

$$(X-Y)^2 \equiv (X-Y)_u g^{uv} (X-Y)_v$$

between points in superspace. Taking $X_u = (x_\mu, \theta_\alpha)$, where $\mu = 0, 1, 2, 3$ and $\alpha = 1, 2$, we have

$$X_u g^{uv} X_v = x^\mu x_\mu + \theta^\alpha \theta_\alpha. \tag{1}$$

Here the orthosymplectic metric is

$$g^{uv} = \begin{pmatrix} \eta^{\mu\nu} & 0 \\ 0 & \varepsilon^{\alpha\beta} \end{pmatrix}$$

where $\eta^{\mu\nu}$ is the usual diagonal Lorentz metric, and $\varepsilon^{\alpha\beta}$ is the 2×2 antisymmetric matrix,

$$\varepsilon^{\alpha\beta} = \begin{pmatrix} 0 & 1 \\ -1 & 0 \end{pmatrix}.$$

The real form is determined by that of the underlying Lie group (Parker 1980), which we take to be simply $O(3, 1)\times Sp(2, R)$, namely the Lorentz group together with 2×2 real symplectic transformations. The group $Sp(2, R)$ is more familiar as $SL(2, R)$, locally isomorphic to the three-dimensional Lorentz group $SO(2, 1)$. The θ_α thus transform as a spinor representation of this group.

Since the group matrices are real, one can in principle take θ_α to be real. However, to ensure the reality of the bilinear form (1), with the usual properties of complex conjugation for *a*-numbers, a different assignment must be made. The choice adopted in the usual BRS formalism (Bonora and Tonin 1981, Kugo and Ojima 1979),

$$\theta_1^* = \theta_1, \qquad \theta_2^* = -\theta_2,$$

is one possibility. However, the $Sp(2, R)$ symmetry ensures that for the present case, the choice $\theta_2 = \theta_1^*$ is an equally feasible alternative. This choice is adopted below, where we write $(\theta_1, \theta_2) = (\theta, \bar{\theta})$ (Dondi and Jarvis 1979). Most of the formalism goes through for either choice; in the former case, it is more usual to write $\theta_2 = i\bar{\sigma}$, $\theta_1 = \sigma$, and for the ghost fields $\bar{\omega} = i\bar{c}(x)$, $\omega(x) = c(x)$, with σ, $\bar{\sigma}$, $c(x)$ and $\bar{c}(x)$ real (Bonora and Tonin 1981, Kugo and Ojima 1979).

In addition to the usual translations and Lorentz transformations of the Poincaré group, the space–time supergroup includes symplectic rotations on θ_α,

$$(x_\mu, \theta_\alpha) \to (x_\mu, \tau_\alpha^\beta \theta_\beta), \tag{2}$$

where $\tau_\alpha^\gamma \varepsilon^{\alpha\beta} \tau_\beta^\delta = \varepsilon^{\gamma\delta}$. There are also the supertranslations

$$(x_\mu, \theta_\alpha) \to (x_\mu, \theta_\alpha + \varepsilon_\alpha), \tag{3}$$

and the super-Lorentz transformations,

$$(x_\mu, \theta_\alpha) \to (x_\mu + \lambda^\beta_\mu \theta_\beta, \theta_\alpha - \lambda^\nu_\alpha x_\nu). \tag{4}$$

Clearly, for compatibility with the chosen reality conditions of the θ_α, we must have

$$\varepsilon_2 = \varepsilon^*_2, \qquad \lambda^\mu_2 = \lambda^{\mu *}_1, \qquad \tau^2_2 = \tau^{1*}_1, \qquad \tau^1_2 = \tau^{2*}_1.$$

The superfield actions we shall construct require the group-invariant measure

$$\mathrm{d}^6 X = \mathrm{d}^4 x \, \mathrm{d}\theta \, \mathrm{d}\bar{\theta}.$$

That this is indeed invariant can be readily checked for supertranslations and symplectic rotations. For the super-Lorentz transformations, the same applies after use of the exponential formula to evaluate a superdeterminant.

3. Superfield formalism

In formulating a local gauge theory over superspace (Salam and Strathdee 1974), one encounters superfields

$$\Phi(x, \theta) = A(x) + \theta^\alpha \psi_\alpha(x) + \theta^\beta \theta_\beta B(x),$$

whose components in the Taylor expansion (a quadratic polynomial in θ) are ordinary fields. In particular, the gauge potential will be a superfield

$$\Phi_u(x, \theta) = (A_\mu(x), A_\alpha(x)) + (\text{higher-order terms}),$$

with $A_\mu(x)$ a c-number field, and $A_\alpha(x)$ an a-number field, transforming as the six-dimensional vector representation of OSp(4/2), and taking values in the Lie algebra of the gauge group (taken to be compact):

$$\Phi_u(x, \theta) = \Phi^a_u(x, \theta) T^a$$

where $[T^a, T^b] = \mathrm{i} f^{abc} T^c$, and f^{abc} are the totally antisymmetrical structure constants.

The gauge field strength $\Phi_{uv}(x, \theta)$ is a superfield transforming in the 17-dimensional graded-antisymmetrical tensor representation of OSp(4/2) (Dondi and Jarvis 1980, 1981). If we define the signature factor $[uv] = \pm 1$ such that $[\alpha\beta] = -1$ and $[\mu\nu] = [\mu\alpha] = +1$, we have, following the usual constructions (see, for example, Abers and Lee 1973),

$$\Phi_{uv} = \partial_u \Phi_v - [uv] \partial_v \Phi_u - \mathrm{i} e [\Phi_u, \Phi_v], \tag{5}$$

where e is the gauge coupling constant. Thus

$$\Phi_{\mu\nu} = \partial_\mu \Phi_\nu - \partial_\nu \Phi_\mu - \mathrm{i} e [\Phi_\mu, \Phi_\nu],$$

$$\Phi_{\mu\alpha} = \partial_\mu \Phi_\alpha - \partial_\alpha \Phi_\mu - \mathrm{i} e [\Phi_\mu, \Phi_\alpha],$$

$$\Phi_{\alpha\beta} = \partial_\alpha \Phi_\beta + \partial_\beta \Phi_\alpha - \mathrm{i} e [\Phi_\alpha, \Phi_\beta]_+,$$

with

$$\Phi_{\mu\nu} = -\Phi_{\nu\mu}, \qquad \Phi_{\mu\alpha} = -\Phi_{\alpha\mu}, \qquad \Phi_{\alpha\beta} = \Phi_{\beta\alpha}.$$

Gauge transformations of Φ_u and Φ_{uv} under (x, θ)-dependent elements $U(x, \theta)$ of the gauge group are given as usual by

$$\Phi'_u(x, \theta) = U^{-1} \Phi_u(x, \theta) U - (\mathrm{i}/e)(\partial_u U^{-1}) U, \qquad \Phi'_{uv}(x, \theta) = U^{-1} \Phi_{uv}(x, \theta) U. \tag{6}$$

Any gauge transformation can be uniquely decomposed (Bonora *et al* 1980) as a product of a purely x-dependent piece $U_0(x)$ and an x- and θ-dependent piece $U_1(x, \theta)$:

$$U(x, \theta) = \{\exp[-ie\Lambda(x)]\}\{\exp[-ie(\theta^\alpha \omega_\alpha(x) - \theta^\beta \theta_\beta B(x))]\} \equiv U_0 U_1. \quad (7)$$

At this stage we introduce the special class of gauge potential superfields $\Phi_u(x, \theta)$ which will be required in deriving the Yang–Mills action (Bonora and Tonin 1981). Namely, we restrict attention henceforth to those potentials which are related by a gauge transformation to the special form in which the only non-vanishing component field is the ordinary four-vector potential $A_\mu(x)$. Without loss of generality, we may take the gauge transformation in question to be of the $U_1(x, \theta)$ type (since $U_0(x)$ does not mix components). For the special class we have

$$\Phi_u(x, \theta) = U_1^{-1} \begin{pmatrix} A_\mu(x) \\ 0 \end{pmatrix} U_1 - \frac{i}{e}(\partial_u U_1^{-1}) U_1 \quad (8)$$

and

$$\Phi_{uv}(x, \theta) = U_1^{-1} \begin{pmatrix} F_{\mu\nu}(x) \\ 0 \\ 0 \end{pmatrix} U_1, \quad (9)$$

where $F_{\mu\nu}(x)$ is the usual Yang–Mills field strength.

It should be emphasised that this restriction is to be regarded as an additional assumption, beyond what is allowed from gauge freedom alone. For the present superfield formalism it corresponds in some sense to the usual procedure of gauge fixing and ghost compensation: this viewpoint is of course borne out by the final results.

The six-dimensional action is taken to be the sum of a gauge-independent piece and a gauge-dependent piece:

$$W(\Phi) = W_0(\Phi) + W_1(\Phi).$$

Observing that, if Φ_u belongs to the restricted class of potentials,

$$\Phi^{auv} \Phi^a_{vu} = (U_1^{-1} F_{\mu\nu} U_1)^a (U_1^{-1} F_{\nu\mu} U_1)^a = -F^{a\mu\nu} F^a_{\mu\nu}, \quad (10)$$

so that the usual Yang–Mills Lagrangian is formally invariant. The corresponding six-dimensional action is simply chosen to reflect this:

$$W_0 = \int X^2 \, d^6X \cdot \tfrac{1}{4} \Phi^{auv} \Phi^a_{vu}. \quad (11)$$

Here the action of $d\theta \, d\bar{\theta}$ is to pick out the coefficient of $\theta^\alpha \theta_\alpha$, namely $-\frac{1}{4} F^{a\mu\nu} F^a_{\mu\nu}$. Thus the supertranslation invariance of W_0 is also assured.

The choice of gauge-dependent action W_1 is not unique, and can only be guided by considerations of dimension, supersymmetry, and ultimately with a view to the required final form. The choice we adopt is

$$W_1 = \int d^6X \cdot 2\Phi^u \Phi_u / \xi, \quad (12)$$

where Φ_u is of the restricted form, and ξ is a real constant.

It remains to give $\Phi_u(x, \theta)$ and the component form of the action for the restricted class of potentials defined by (8) and (9). Expanding the exponential in (7), the finite group element $U_1(x, \theta)$ can be written

$$U_1(x, \theta) = 1 - ie\theta^\beta\omega_\beta + \tfrac{1}{2}ie\theta^\gamma\theta_\gamma(B - \tfrac{1}{2}ie\omega^\delta\omega_\delta).$$

From (8), we have

$$\Phi_\mu = A_\mu + \theta^\beta D_\mu\omega_\beta - \tfrac{1}{2}\theta^\gamma\theta_\gamma[D_\mu B + \tfrac{1}{2}e(D_\mu\omega^\delta)\times\omega_\delta],$$

$$\Phi_\alpha = \omega_\alpha + \theta^\beta(B\varepsilon_{\beta\alpha} - \tfrac{1}{2}e\omega_\beta\times\omega_\alpha) - \tfrac{1}{2}\theta^\gamma\theta_\gamma[-eB\times\omega_\alpha + \tfrac{1}{6}e^2(\omega_\alpha\times\omega^\delta)\times\omega_\delta], \tag{13}$$

where D_μ is the covariant derivative,

$$D_\mu B = \partial_\mu B + eA_\mu \times B,$$

etc. Taking the $\theta^\alpha\theta_\alpha$ coefficient of $2\Phi^u\Phi_u/\xi$ gives with (10) the Minkowski-space form of the action†:

$$W = \int d^4x\Big[-\tfrac{1}{4}F^{\mu\nu}\cdot F_{\mu\nu} + \left(\frac{2}{\xi}\right)(\partial^\mu A_\mu\cdot B + B^2) - \left(\frac{1}{\xi}\right)\partial^\mu\omega^\alpha\cdot D_\mu\omega_\alpha - \left(\frac{e^2}{12\xi}\right)\omega^\alpha\times\omega^\beta\cdot\omega_\alpha\times\omega_\beta\Big]. \tag{14}$$

In (13) and (14), the dot and cross products are generalisations of the usual vector notation:

$$\partial^\mu A_\mu\cdot B = \partial^\mu A^a_\mu B^a, \qquad (\omega_\alpha\times\omega_\beta)^c = f^{abc}\omega^a_\alpha\omega^b_\beta, \qquad \text{etc.}$$

With appropriate rescalings of B and ω_α, and expanding in terms of the spinor components ω and $\bar\omega$ the total Lagrangian becomes

$$\mathcal{L} = -\tfrac{1}{4}F^{\mu\nu}\cdot F_{\mu\nu} + (\partial^\mu A_\mu\cdot B + \tfrac{1}{2}\xi B^2) - \partial^\mu\bar\omega\cdot\partial_\mu\omega - \tfrac{1}{2}eA^\mu\cdot\bar\omega\times\overleftrightarrow{\partial}_\mu\omega + \tfrac{1}{8}e^2\xi(\bar\omega\times\omega)^2. \tag{15}$$

Elimination of the auxiliary field B leads to the usual covariant gauge-fixing term $-(\partial^\mu A_\mu)^2/2\xi$. The present Lagrangian *differs from the conventional one* in the form of the vector–ghost coupling, and the quartic ghost self-coupling. It is a particular case of that of Bonora and Tonin (1981), in which the two independent gauge parameters they introduce are held equal. The fields

$$B_\pm = B \pm \tfrac{1}{2}e\bar\omega\times\omega \tag{16}$$

are the same as B, $\bar B$ in their notation, and the identification is completed by adopting the standard reality assignment $\bar\omega = i\bar c$, $\omega = c$, with c, $\bar c$ real (see, for example, Kugo and Ojima (1979), and references therein). However, as emphasised above, the convention $\bar\omega = \omega^*$ is also feasible in the present formulation. The additional symmetry between ω and $\bar\omega$ forbids terms which would be formally non-Hermitian if $\bar\omega$ and ω are complex conjugates, and in particular leads the vector–ghost coupling to attain a form reminiscent of charged scalar electrodynamics. In the limit $\xi \to 0$, the additional terms fade out, and one is left with the conventional Landau gauge. The Feynman rules from (15) are the conventional ones for the vector field (see, for example, Abers and Lee 1973). Those for the ghost field are given in figure 1.

† It should be noted from (11) and (12) that the action is formally scale invariant at the classical level.

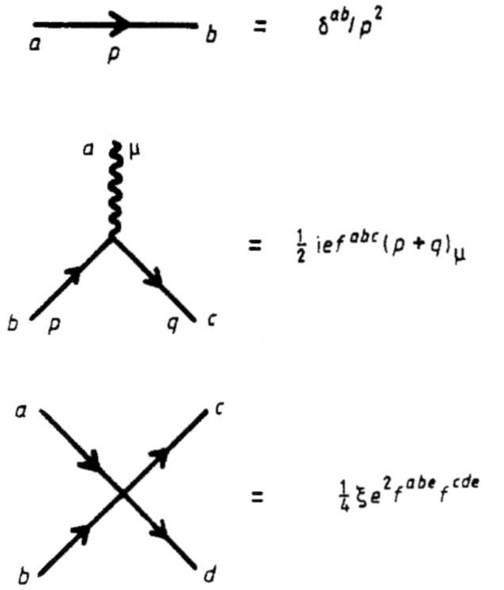

Figure 1. Feynman rules for ghost couplings.

4. Supertranslations and the effective action

The action (11) plus (12) is formally invariant under supertranslations

$$\delta\Phi_u(x,\theta)=\Phi_u(x,\theta+\varepsilon)-\Phi_u(x,\theta).$$

Moreover, it can be verified that the supertranslations respect the condition (8) defining the restricted class of gauge potentials. The component form of the supertranslations follows from (13). In terms of the spinor components ε and $\bar{\varepsilon}$, we have with (16)

$$\delta A_\mu=\bar{\varepsilon}D_\mu\omega-\varepsilon D_\mu\bar{\omega},\qquad \delta\omega ff\!-\!\tfrac{1}{2}e\bar{\varepsilon}\omega\times\omega-\varepsilon B_-,\qquad \delta\bar{\omega}=-\bar{\varepsilon}B_+ +\tfrac{1}{2}e\varepsilon\bar{\omega}\times\bar{\omega},$$

$$\delta B_+=0-e\varepsilon B_+\times\bar{\omega},\qquad \delta B_-=e\bar{\varepsilon}B_-\times\omega+0. \tag{17}$$

These transformations thus provide an invariance of the total Lagrangian, as also follows explicitly from (15). This is the so-called extended BRS invariance. The symmetrical form with respect to $\bar{\varepsilon}$ and ε is in keeping with the present treatment of ω and $\bar{\omega}$ (in fact, the dual δ_ε transformations can be obtained from the $\delta_{\bar{\varepsilon}}$ transformations by Hermitian conjugation, if the convention $\bar{\omega}=\omega^*$ is adopted). The fact that the supertranslations anticommute is reflected in the so-called nilpotency of the BRS transformations, namely from (17)

$$\delta_{\bar{\varepsilon}}B_+=0=\delta_\varepsilon B_+\times\bar{\omega},$$

and so on.

In following the implications of the BRS invariance for the quantised model and its renormalisation, it is convenient to eliminate B for $-(\partial^\mu A_\mu)/\xi$ (which is consistent with (17)). Also, we restrict attention only to the $\delta_{\bar{\varepsilon}}$-type BRS transformations. Finally, the nilpotency conditions allow the introduction of composite source terms in the Lagrangian:

$$\mathcal{L}_S=j^\mu\cdot A_\mu+\bar{J}\cdot\omega+\bar{\omega}\cdot J+\bar{I}^\mu\cdot D_\mu\omega-\tfrac{1}{2}e\bar{I}\cdot\omega\times\omega-K\cdot B_+-e\bar{K}\cdot B_-\times\omega. \tag{18}$$

The equations of motion following from $\mathcal{L}-\mathcal{L}_S$ read

$$\partial^\mu D_\mu\omega+\tfrac{1}{2}e\xi(B_-\times\omega)=J+\tfrac{1}{2}eK\times\omega+\tfrac{1}{4}e^2\bar{K}\times(\omega\times\omega), \tag{19}$$

plus a similar (but more complicated) equation for $\bar{\omega}$, plus

$$D^\mu F_{\mu\nu} + \frac{1}{\xi}\partial^\mu(\partial^\nu A_\nu) - \tfrac{1}{2}e\omega \times \overleftrightarrow{\partial}_\nu \omega = j_\nu - \bar{I}_\nu \times \omega - \frac{1}{\xi}\partial_\nu K + \frac{e}{\xi}\partial_\nu(\bar{K} \times \omega). \tag{20}$$

In (18) and (19), we have

$$B_\pm = -(\partial^\mu A_\mu)/\xi \pm \tfrac{1}{2}e\bar{\omega} \times \omega \tag{21}$$

in place of (16).

Equations (19) and (20) impose conditions on the vacuum functional

$$Z(j^\mu, J, \bar{J}, \bar{I}^\mu, \bar{I}, K, \bar{K}) = N\int \mathrm{d}[A, \omega, \bar{\omega}] \exp\left(\mathrm{i}\int \mathrm{d}x\, (\mathscr{L} - \mathscr{L}_\mathrm{S})\right);$$

for example

$$\partial^\mu \mathrm{i}\frac{\delta Z}{\delta \bar{I}^{\mu a}} = J + \frac{\xi \mathrm{i}}{2}\frac{\delta Z}{\delta \bar{K}^a} - \frac{ef^{abc}K^b \mathrm{i}}{2}\frac{\delta Z}{\delta \bar{J}^c} - \frac{ef^{abc}\bar{K}^b \mathrm{i}}{2}\frac{\delta Z}{\delta \bar{I}^c}. \tag{22}$$

Similarly from the supertranslations (17) one finds that

$$\int \mathrm{d}x\left(j_x^{\mu a}\mathrm{i}\frac{\delta Z}{\delta \bar{I}_x^{\mu a}} - \bar{J}_x^a \mathrm{i}\frac{\delta Z}{\delta \bar{I}_x^a} - \mathrm{i}\frac{\delta Z}{\delta K_x^a}J_x^a\right) = 0 \tag{23}$$

from the variation of $\mathscr{L}_\mathrm{S}$. It is usual to pass to the connected vacuum functional W via $Z = \mathrm{e}^{\mathrm{i}W}$ and thence to the effective action

$$\Gamma(A, \omega, \bar{\omega}, \bar{I}^\mu, \bar{I}, K, \bar{K}) = W(j, J, \bar{J}; \bar{I}^\mu, \bar{I}, K, \bar{K}) + \int \mathrm{d}x\, (j^\mu A_\mu + \bar{J}\omega + \bar{\omega}J)$$

with the correspondences

$$\delta\Gamma/\delta A^\mu = j_\mu, \qquad \delta W/\delta j_\mu = -A^\mu,$$

and so on. (22) and (23) become, for example,

$$\partial^\mu \frac{\delta\Gamma}{\delta \bar{I}^{\mu a}} = -\frac{\delta\Gamma}{\delta\bar{\omega}^a} + \frac{\xi}{2}\frac{\delta\Gamma}{\delta\bar{K}^a} + \tfrac{1}{2}ef^{abc}K^b\omega^c - \frac{ef^{abc}\bar{K}^b}{2}\frac{\delta\Gamma}{\delta\bar{I}^c} \tag{24}$$

and

$$\int \mathrm{d}x\left(\frac{\delta\Gamma}{\delta A_{\mu x}^a}\frac{\delta\Gamma}{\delta \bar{I}_x^{\mu a}} + \frac{\delta\Gamma}{\delta\omega_x^a}\frac{\delta\Gamma}{\delta\bar{I}_x^a} + \frac{\delta\Gamma}{\delta\bar{\omega}_x^a}\frac{\delta\Gamma}{\delta K_x^a}\right) = 0. \tag{25}$$

After differentiating (24) and (25) with respect to S, ω and $\bar{\omega}$ so as to leave zero ghost number, the equations translate into identities amongst the strongly connected Green functions, as follows. On the ω equation, apply

$$\frac{\delta}{\delta\omega^b}, \qquad \frac{\delta^2}{\delta\omega^b\,\delta A^{\nu c}}, \qquad \frac{\delta^3}{\delta\omega^b\,\delta A^{\nu c}\,\delta A^{\lambda d}}, \qquad \frac{\delta^3}{\delta\omega^b\,\delta\bar{\omega}^c\,\delta\omega^d}.$$

On the BRS identity, apply

$$\frac{\delta^2}{\delta\omega^b\,\delta A^{\nu c}}, \qquad \frac{\delta^3}{\delta\omega^b\,\delta A^{\nu c}\,\delta A^{\lambda d}}, \qquad \frac{\delta^4}{\delta\omega^b\,\delta A^{\nu c}\,\delta A^{\lambda d}\,\delta A^{\kappa e}},$$

$$\frac{\delta^3}{\delta\omega^b\,\delta\bar{\omega}^c\,\delta\omega^d}, \qquad \frac{\delta^4}{\delta\omega^b\,\delta\bar{\omega}^c\,\delta\omega^d\,\delta A^{\nu e}}.$$

Thus we obtain (integration over x is implied in (27), but not over other spatial arguments):

$$\partial^\mu_x \frac{\delta^2\Gamma}{\delta\bar I^{\mu a}\,\delta\omega^b} = -\frac{\delta^2\Gamma}{\delta\bar\omega^a\,\delta\omega^b}+\frac{\xi}{2}\frac{\delta^2\Gamma}{\delta\bar K^a_x\,\delta\omega^b}, \tag{26a}$$

$$\partial^\mu_x \frac{\delta^3\Gamma}{\delta\bar I^{\mu a}_x\,\delta\omega^b\,\delta A^{\nu c}} = -\frac{\delta^3\Gamma}{\delta\bar\omega^a_x\,\delta\omega^b\,\delta A^{\nu c}}+\frac{\xi}{2}\frac{\delta^3\Gamma}{\delta\bar K^a_x\,\delta\omega^b\,\delta A^{\nu c}}, \tag{26b}$$

$$\partial^\mu_x \frac{\delta^4\Gamma}{\delta\bar I^{\mu a}_x\,\delta\omega^b\,\delta A^{\nu c}\,\delta A^{\lambda d}} = -\frac{\delta^4\Gamma}{\delta\bar\omega^a_x\,\delta\omega^b\,\delta A^{\nu c}\,\delta A^{\lambda d}}+\frac{\xi}{2}\frac{\delta^4\Gamma}{\delta\bar K^a\,\delta\omega^b\,\delta A^{\nu c}\,\delta A^{\lambda d}}, \tag{26c}$$

$$\partial^\mu_x \frac{\delta^4\Gamma}{\delta\bar I^{\mu a}_x\,\delta\omega^b\,\delta\bar\omega^c\,\delta\omega^d} = -\frac{\delta^4\Gamma}{\delta\bar\omega^a_x\,\delta\omega^b\,\delta\bar\omega^c\,\delta\omega^d}+\frac{\xi}{2}\frac{\delta^4\Gamma}{\delta\bar K^a_x\,\delta\omega^b\,\delta\bar\omega^c\,\delta\omega^d}; \tag{26d}$$

$$\frac{\delta^2\Gamma}{\delta A^{\mu a}_x\,\delta A^{\nu c}}\cdot\frac{\delta^2\Gamma}{\delta\bar I^{\mu a}_x\,\delta\omega^b}+\frac{\delta^2\Gamma}{\delta\bar\omega^a_x\,\delta\omega^b}\cdot\frac{\delta^2\Gamma}{\delta K^a_x\,\delta A^{\nu c}}=0, \tag{27a}$$

$$\begin{aligned}&\frac{\delta^3\Gamma}{\delta A^{\mu a}_x\,\delta A^{\nu c}\,\delta A^{\lambda d}}\cdot\frac{\delta^2\Gamma}{\delta\bar I^{\mu a}_x\,\delta\omega^b}\;\frac{\delta^2\Gamma}{\delta\bar\omega^a_x\,\delta\omega^b}\cdot\frac{\delta^3\Gamma}{\delta K^a_x\,\delta A^{\nu c}\,\delta A^{\lambda d}}\\ &\quad+\left(\frac{\delta^2\Gamma}{\delta A^{\mu a}_x\,\delta A^{\nu c}}\cdot\frac{\delta^3\Gamma}{\delta\bar I^{\mu a}_x\,\delta\omega^b\,\delta A^{\lambda d}}+\frac{\delta^3\Gamma}{\delta\bar\omega^a\,\delta\omega^b\,\delta A^{\nu c}}\cdot\frac{\delta^2\Gamma}{\delta K^a_x\,\delta A^{\lambda d}}+(cd)\right)=0,\end{aligned} \tag{27b}$$

$$\begin{aligned}&\frac{\delta^4\Gamma}{\delta A^{\mu a}_x\,\delta A^{\nu c}\,\delta A^{\lambda d}\,\delta A^{\kappa e}}\cdot\frac{\delta^2\Gamma}{\delta\bar I^{\mu a}_x\,\delta\omega^b}+\frac{\delta^2\Gamma}{\delta\bar\omega^a_x\,\delta\omega^b}\cdot\frac{\delta^4\Gamma}{\delta K^a_x\,\delta A^{\nu c}\,\delta A^{\lambda d}\,\delta A^{\kappa e}}\\ &\quad+\left(\frac{\delta^2\Gamma}{\delta A^{\mu a}_x\,\delta A^{\nu c}}\cdot\frac{\delta^4\Gamma}{\delta\bar I^{\mu a}_x\,\delta\omega^c\,\delta A^{\lambda d}\,\delta A^{\kappa e}}+\frac{\delta^3\Gamma}{\delta A^{\mu a}_x\,\delta A^{\nu c}\,\delta A^{\lambda d}}\cdot\frac{\delta^3\Gamma}{\delta\bar I^{\mu a}_x\,\delta\omega^b\,\delta A^{\kappa e}}\right.\\ &\quad+\frac{\delta^3\Gamma}{\delta\bar\omega^a_x\,\delta\omega^b\,\delta A^{\kappa e}}\cdot\frac{\delta^3\Gamma}{\delta K^a_x\,\delta A^{\nu c}\,\delta A^{\lambda d}}\\ &\quad\left.+\frac{\delta^4\Gamma}{\delta\bar\omega^a_x\,\delta\omega^b\,\delta A^{\lambda d}\,\delta A^{\kappa e}}\cdot\frac{\delta^2\Gamma}{\delta K^a_x\,\delta A^{\nu c}}+(cde)\right)=0,\end{aligned} \tag{27c}$$

$$\begin{aligned}&\frac{\delta^2\Gamma}{\delta\bar\omega^c\,\delta\omega^a_x}\cdot\frac{\delta^3\Gamma}{\delta\bar I^a_x\,\delta\omega^b\,\delta\omega^d}+\left(\frac{\delta^3\Gamma}{\delta A^{\mu a}_x\,\delta\omega^b\,\delta\bar\omega^c}\cdot\frac{\delta^2\Gamma}{\delta\bar I^a_{\mu x}\,\delta\omega^d}-(bd)\right)\\ &\quad+\left(\frac{\delta^2\Gamma}{\delta\bar\omega^a_x\,\delta\omega^b}\cdot\frac{\delta^3\Gamma}{\delta K^a_x\,\delta\bar\omega^c\,\delta\omega^d}-(bd)\right)=0,\end{aligned} \tag{27d}$$

$$\begin{aligned}&\frac{\delta^2\Gamma}{\delta A^{\mu a}_x\,\delta A^{\nu e}}\cdot\frac{\delta^4\Gamma}{\delta\bar I^{\mu a}_x\,\delta\omega^b\,\delta\bar\omega^c\,\delta\omega^d}+\left(\frac{\delta^4\Gamma}{\delta A^{\mu a}_x\,\delta A^{\nu e}\,\delta\bar\omega^c\,\delta\omega^d}\cdot\frac{\delta^2\Gamma}{\delta\bar I^a_{\mu x}\,\delta\omega^b}-(bd)\right)\\ &\quad+\frac{\delta^3\Gamma}{\delta A^{\nu e}\,\delta\bar\omega^c\,\delta\omega^a_x}\cdot\frac{\delta^3\Gamma}{\delta\bar I^a_x\,\delta\omega^b\,\delta\omega^d}+\frac{\delta^4\Gamma}{\delta\bar\omega^a_x\,\delta\omega^b\,\delta\bar\omega^c\,\delta\omega^d}\cdot\frac{\delta^2\Gamma}{\delta K^a_x\,\delta A^{\nu e}}\\ &\quad+\left(\frac{\delta^3\Gamma}{\delta A^{\mu a}_x\,\delta\bar\omega^c\,\delta\omega^d}\cdot\frac{\delta^3\Gamma}{\delta\bar I^a_{\mu x}\,\delta\omega^b\,\delta A^{\nu e}}-(bd)\right)\\ &\quad+\left(\frac{\delta^2\Gamma}{\delta\bar\omega^a_x\,\delta\omega^b}\cdot\frac{\delta^4\Gamma}{\delta K^a_x\,\delta A^{\nu e}\,\delta\bar\omega^c\,\delta\omega^d}-(bd)\right)+\frac{\delta^2\Gamma}{\delta\omega^a_x\,\delta\bar\omega^c}\cdot\frac{\delta^4\Gamma}{\delta\bar I^a_x\,\delta\omega^b\,\delta\omega^d\,\delta A^{\nu e}}\end{aligned}$$

$$+\left(\frac{\delta^3\Gamma}{\delta\bar{\omega}^a_x\,\delta\omega^b\,\delta A^{\nu e}}\cdot\frac{\delta^3\Gamma}{\delta K^a_x\,\delta\bar{\omega}^c\,\delta\omega^d}-(bd)\right)=0. \tag{27e}$$

These identities can be verified at tree level using the Feynman rules (figure 1: those for the sources follow trivially from (18)). For example, (27*a*) becomes (figure 2)

$$(-\eta_{\mu\nu}k^2+k_\mu k_\nu(1-\xi^{-1})(-ik^\mu\delta^{bc})+(-i\xi^{-1}k_\nu)(k^2\delta^{bc})=0.$$

At one-loop order the identity is given in figure 3, and is verified by an explicit calculation.

Figure 2. A tree-level BRS identity.

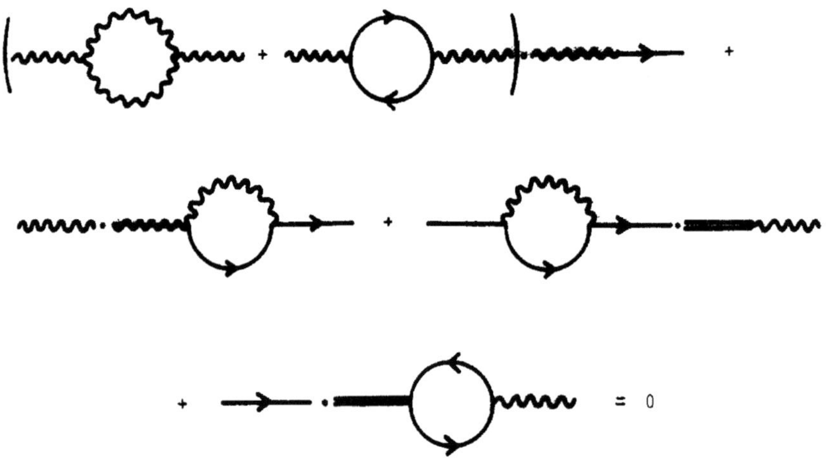

Figure 3. BRS identity at one-loop level.

Renormalisation is effected by introducing appropriate $(Z-1)\mathscr{L}$-type counterterms in order to cancel the quantum correction infinities of the effective action. The superficially divergent graphs correspond to the original terms in but *may* also include†

$$\frac{\delta^4\Gamma}{\delta A\,\delta A\,\delta\bar{\omega}\,\delta\omega}\ \text{(as in scalar electrodynamics)},\qquad \text{and}\ \frac{\delta^4\Gamma}{\delta\bar{K}\,\delta\omega\,\delta A\,\delta A}.$$

These functions do not arise at tree level but enter the general identities (26) and (27). Because they are not included in $\mathscr{L}$, we know that any potential divergences must cancel by symmetry requirements. An explicit check at one-loop level (taking zero external momentum for simplicity and ignoring infrared problems there) bears this out.

In a similar vein, it is the Sp(2) symmetry which dictates the renormalisation of the three-point ghost vertex $\delta^3\Gamma/\delta A\,\delta\bar{\omega}\,\delta\omega$. In spinor notation, the only possible coupling is

$$A^\mu\cdot\omega^\alpha\times\partial_\mu\omega_\alpha\propto A^\mu\cdot\bar{\omega}\times\overleftrightarrow{\partial}_\mu\omega,$$

† Note that $\delta^3\Gamma/\delta K\,\delta A\,\delta A$ vanishes identically.

since $\partial^\mu A_\mu \cdot \omega^\alpha \times \omega_\alpha \equiv 0$. The only slight complications occur in the source terms: additional contributions to $\delta^2\Gamma/\delta\bar{K}\,\delta\omega$ and $\delta^2\Gamma/\delta\bar{K}\,\delta A\,\delta\omega$ are handled by extra counterterms. Finally, then, the renormalised Lagrangian reads

$$\begin{aligned}\mathcal{L} = &-\tfrac{1}{4}ZF^a_{\mu\nu}F^{\mu\nu a} - (1/2\xi)Z''(\partial\cdot A^a)^2 - \tilde{Z}\partial^\mu\bar{\omega}^a\partial_\mu\omega_a - \tfrac{1}{2}e\tilde{Z}_e f^{abc}A^{\mu a}\bar{\omega}^b\overleftrightarrow{\partial}_\mu\omega^c\\ &+\tfrac{1}{8}e^2\xi\tilde{Z}_4(f^{abc}\bar{\omega}^b\omega^c)^2 + \tfrac{1}{4}e^4 Z_4(f^{abc}A^b_\mu A^c_\nu)^2 - j^{\mu a}A^a_\mu - \bar{J}^a\omega^a - \bar{\omega}^a J^a\\ &+\tfrac{1}{2}e\bar{\bar{Z}}_e f^{abc}\bar{\bar{I}}^a\omega^b\omega^c - \bar{I}^{\mu a}(\bar{Z}\partial_\mu\omega^a + e\bar{Z}_e f^{abc}A^b_\mu\omega^c)\\ &-K^a(\hat{Z}\partial\cdot A^a/\xi - \tfrac{1}{2}ef^{abc}\hat{Z}_e\bar{\omega}^b\omega^c)\\ &-ef^{abc}\bar{K}^a(\hat{\bar{Z}}_e\partial\cdot A^b/\xi + \tfrac{1}{2}e\hat{\bar{Z}}_4 f^{bde}\bar{\omega}^d\omega^e)\omega^c\\ &-e(\delta\hat{\bar{Z}}_e/\xi)f^{abc}\bar{K}^a A^{\mu b}\partial_\mu\omega^c - \delta\tilde{Z}\partial_\mu\bar{K}^a\partial^\mu\omega^a,\end{aligned}$$

with

$$F^a_{\mu\nu} \equiv \partial_\mu A^a - \partial_\nu A^a_\mu + eZ_eZ^{-1}f^{abc}A^b_\mu A^c_\nu. \tag{28}$$

By equating infinite parts†, the equations of motion and BRS identities translate into relationships amongst the different Z's. For example, ($27a$) yields

$$(p_\nu p^2/\xi)[(1-Z'')+(1-\bar{Z})-(1-\tilde{Z})-(1-\hat{Z})]=0.$$

The complete set of relations derivable from (26) and (27) in this manner is

$$Z/Z_e = \tilde{Z}/\tilde{Z}_e, \qquad Z_4 = 0, \qquad \tilde{Z}_4 = \tilde{Z}_e^2/Z'' \tag{29}$$

for field renormalisations, and

$$\begin{aligned}&\hat{Z} = Z''\bar{Z}/\tilde{Z}, &&\bar{Z}/\bar{Z}_e = Z/Z_e, &&\hat{\bar{Z}}_4 = \tilde{Z}_4,\\ &\bar{Z} = \tilde{Z} + \delta\tilde{Z}, &&\bar{Z}_e = \hat{Z}_e = \bar{\bar{Z}}_e = \tilde{Z}_e + \tfrac{1}{2}\delta\hat{\bar{Z}}_e = \tfrac{1}{2}(\tilde{Z}_e + \hat{\bar{Z}}_e),\end{aligned} \tag{30}$$

for source renormalisations. These identities may be verified to one-loop order using their calculated values given in table 1.

Table 1. Renormalisation constants in one-loop order. C is the adjoint representation Casimir invariant and $L = (e^2/16\pi^2)\log(\Lambda^2/\mu^2)$.

Z	$1+\dfrac{13-3\xi}{6}CL$	$\bar{Z}$	$1+\frac{3}{4}\xi CL$
		$\bar{Z}_e$	$1-\frac{1}{4}\xi CL$
Z_e	$1+\dfrac{17-9\xi}{12}CL$	$\hat{Z}$	1
		$\hat{Z}_e$	$1-\frac{1}{4}\xi CL$
$\tilde{Z}$	$1+\frac{1}{4}(3-\xi)CL$	$\hat{\bar{Z}}_e$	1
$\tilde{Z}_e$	$1-\frac{1}{2}\xi CL$	$\hat{\bar{Z}}_4$	$1-\frac{3}{4}\xi CL$
Z''	$1-\frac{1}{4}\xi CL$	$\bar{\bar{Z}}_e$	$1-\frac{1}{4}\xi CL$
Z_4	0	$\delta\tilde{Z}$	$\frac{1}{4}\xi CL$
$\tilde{Z}_4$	$1-\frac{3}{4}\xi CL$	$\delta\hat{\bar{Z}}_e$	$\frac{1}{2}\xi CL$

† Strictly speaking, one could introduce additional sources at the bare level for the extra terms involving $\delta\tilde{Z}$ and $\delta\hat{\bar{Z}}_e$. However, the results are unaffected at the one-loop level. Further, we have omitted a quantum-loop induced δZKK counterterm, which does not contribute to the equation of motion.

Equations (29) and (30) mean that infinities can be precisely associated with *multiplicative* renormalisations of the fields and coupling in the original bare Lagrangian (subscript o) wherever they appear:

$$A_o = Z^{1/2}A, \qquad \omega_o = \tilde{Z}^{1/2}\omega, \qquad e_o = eZ_e/Z^{3/2}, \qquad \xi_o = \xi Z/Z''. \tag{31}$$

In particular, (29) includes the Slavnov–Taylor identity (Taylor 1971, Slavnov 1972) $Z/Z_e = \tilde{Z}/\tilde{Z}_e$. By the same token the sources, and indeed the BRS transformations themselves, undergo multiplicative renormalisation.

5. Super-Lorentz transformations

The action (11) plus (12) is formally invariant under the super-Lorentz transformations

$$\delta\Phi_\mu(x,\theta) = \Lambda_\mu^{\ \nu}\Phi_\nu(\Lambda^{-1}(x,\theta)) - \Phi_\nu(x,\theta), \tag{32}$$

where $\Lambda(x,\theta)$ is given by (4). However, these transformations do *not* respect the condition (8) defining the restricted class of gauge potentials. In contrast to the case of supertranslations, the variations implied by (32) are incompatible with the parametrisation of the components in (13) in terms of just *four* fields $A_\mu(x)$, $\omega_\alpha(x)$ and $B(x)$.

A set of transformations can still be obtained from (32) by extracting the variations of $A_\mu(x)$, $\omega_\alpha(x)$ and $B(x)$ from the lowest-order components in (13), as if the remaining variations were consistent. In the case of $\delta A_\mu(x)$, (32) and (13) conspire to give a ghost-dependent gauge transformation given by

$$\delta A_\mu = D_\mu(\bar{\lambda}_\nu x^\nu\omega) - D_\mu(\lambda_\nu x^\nu\bar{\omega}) \tag{33}$$

in terms of the infinitesimal spinor components $\bar{\lambda}_\nu$ and λ_ν. Comparing (33) with (17*a*), a formal similarity with an x-dependent BRS transformation is seen. The complete set of transformations, in terms of the appropriately scaled variables, is

$$\begin{aligned} &\delta A = D_\mu(\bar{\lambda}_\nu x^\nu\omega), \qquad \delta\omega = \bar{\lambda}_\nu x^\nu\omega\times\omega, \qquad \delta\bar{\omega} = (2/\xi)\bar{\lambda}^\mu A_\mu - \bar{\lambda}_\nu x^\nu B_+, \\ &\qquad \delta B_+ = -(2/\xi)\bar{\lambda}^\mu D_\mu\omega, \qquad \delta B_- = -\xi\bar{\lambda}^\mu\partial_\mu\omega + \bar{\lambda}_\nu x^\nu B_-\times\omega, \end{aligned} \tag{34}$$

plus a similar set in terms of λ_ν, derived by Hermitian conjugation (cf (17)).

The basic set (34) further imply

$$\delta D_\mu\omega = \tfrac{1}{2}e\bar{\lambda}_\mu\omega\times\omega, \qquad \delta\omega\times\omega = 0, \qquad \delta B_-\times\omega = -(1/\xi)\bar{\lambda}^\mu\partial_\mu(\omega\times\omega),$$

from which

$$\delta\mathscr{L} = -(2/\xi)\bar{\lambda}^\mu F^a_{\mu\nu}\partial^\nu\omega^a. \tag{35}$$

Furthermore, it can be verified that the variations (34) are consistent with the equations of motion (19), in the absence of sources. Thus the auxiliary field B can be eliminated, as in the supertranslation case. The identity for the generating functional resulting from (34) and (35) is

$$\int dx\Big[j_\mu\cdot i\frac{\delta}{\delta\bar{J}} - \bar{I}_\mu\cdot i\frac{\delta}{\delta\bar{I}} + x_\mu\Big(j^\nu\cdot i\frac{\delta}{\delta\bar{I}^\nu} + J\cdot i\frac{\delta}{\delta K} - \bar{J}\cdot i\frac{\delta}{\delta\bar{I}}\Big)$$
$$-\frac{2}{\xi}\Big(J\cdot i\frac{\delta}{\delta j^\mu} - F_{\mu\nu}\Big(i\frac{\delta}{\delta j}\Big)\cdot\partial^\nu i\frac{\delta}{\delta\bar{J}} + K\cdot i\frac{\delta}{\delta\bar{I}^\mu} + \bar{K}\cdot i\frac{\delta}{\delta\bar{I}}\Big)\Big]Z = 0.$$

Some simplification is possible with the use of the equation of motion, (19). In fact, in the Landau gauge $\xi \to 0$, (34) and (35) are equivalent to (19), in the absence of composite sources. Passing to the effective action, one finds in general that

$$\begin{aligned}\int \mathrm{d}x \Big[& A_\mu \cdot \frac{\delta\Gamma}{\delta\bar{K}} - \omega \cdot \frac{\delta\Gamma}{\delta A^\mu} + \bar{I}^\mu \cdot \frac{\delta\Gamma}{\delta\bar{\bar{I}}} - \mathrm{i}\Delta_{\mu\lambda} \cdot \frac{\delta^2\Gamma}{\delta A^\lambda \delta\bar{K}} \\ & + x_\mu \left(\frac{\delta\Gamma}{\delta A^\lambda} \cdot \frac{\delta\Gamma}{\delta\bar{\bar{I}}_\lambda} + \frac{\delta\Gamma}{\delta\omega} \cdot \frac{\delta\Gamma}{\delta\bar{\bar{I}}} + \frac{\delta\Gamma}{\delta\bar{\omega}} \cdot \frac{\delta\Gamma}{\delta K} \right) \\ & + \frac{1}{\xi} \Big(2\partial^\lambda A_\lambda \cdot \frac{\delta\Gamma}{\delta\bar{\bar{I}}^\mu} - \partial_\mu\omega \cdot K + K \cdot \frac{\delta\Gamma}{\delta\bar{\bar{I}}^\mu} - e\bar{K} \cdot A_\mu \times \frac{\delta\Gamma}{\delta\bar{\bar{I}}} \\ & + 2\bar{K} \cdot \partial_\mu \frac{\delta\Gamma}{\delta\bar{\bar{I}}} - 2\mathrm{i}\partial^\kappa \Delta_{\kappa\lambda} \frac{\delta^2\Gamma}{\delta A^\lambda \delta\bar{\bar{I}}^\mu} + \mathrm{i}e\bar{K} \cdot \Delta_{\mu\lambda} \frac{\delta^2\Gamma}{\delta A^\lambda \times \delta\bar{\bar{I}}} \Big) \Big] = 0 \end{aligned} \qquad (36)$$

where

$$\Delta_{\mu\lambda} = \delta^2 W / \delta_i^{\ \mu} \delta j^\lambda,$$

and an additional integral over the spatial coordinate is understood in the appropriate $\Delta_{\mu\lambda}$ terms.

6. Conclusions

We have seen above that an alternative formulation of gauge fixing in pure Yang–Mills theory, based upon inhomogeneous OSp(4/2) space–time supersymmetry, leads naturally to the model Lagrangian (15). Here a covariant gauge-fixing term for the vector potential is accompanied by non-standard ghost couplings, including a four-point coupling (cf Das 1980). As a result of additional symmetry between ghost and anti-ghost fields, the Lagrangian is formally real if these are complex conjugates (in contrast to the usual case). The model is renormalisable in standard fashion, and the BRS transformations lead to the Slavnov–Taylor identity for the renormalisation of the transverse part of the vector propagator. The renormalisation constants are given to one-loop order in table 1.

Bonora *et al* (1980) have justified their superfield formalism in terms of a fibre bundle construction. In this and other geometrical approaches (Thierry-Mieg and Ne'eman 1979, Quirós *et al* 1980), no discussion has been given of a possible enlarged space–time supersymmetry. With this at hand, however, the model can be regarded as being obtained by dimensional reduction from the six-dimensional theory. Indeed, the condition (9) can be interpreted as the usual one of 'triviality in higher dimensions'. Naturally this breaks the fullOSp(4/2) supergroup, but respects supertranslations, and Sp(2) transformations.

The present formulation of gauge fixing and ghosts can be applied straightforwardly to other models. For an antisymmetrical rank-two tensor gauge field, the appropriate OSp(4/2) representation is the rank-two graded anti-symmetrical tensor representation (the 17-dimensional adjoint representation, as in (5)). This contains six gauge fields $A_{[\mu\nu]}$, eight ghosts $A_{\mu\alpha}$, and three scalar fields $A_{(\alpha\beta)}$, with graded dimensions $6-8+3=1$, as is appropriate for a scalar field (Siegel 1980, Marchetti and Tonin 1981). For gravity, the irreducible rank-two traceless graded-symmetrical tensor representation of OSp(4/2) is 18-dimensional (note that $17+(18+1)=36$), in accord with the usual assignment (Delbourgo and Medrano 1976) of two vector ghosts $g_{\mu\alpha}$ accompanying the gravition field $g_{(\mu\nu)}$, and graded dimension $10-8=2$. Including the OSp(4/2) trace, the reducible 19-dimensional representation may correspond to a scalar-tensor theory (see also Namazie and Storey 1979).

More generally, one can consider 'extended' ghost supersymmetries OSp$(4/G)$ in G dimensions, where G is even. For example, the rank-three graded-antisymmetrical tensor representation of OSp$(4/G)$ (including the gauge field $A_{[\lambda\mu\nu]}$) has graded dimension $\frac{1}{6}(4-G)(3-G)(2-G)$, suggesting that for $G=2$ or $G=4$, the theory has zero physical degrees of freedom. Similarly, consider the rank-three gauge field of mixed symmetry proposed as a representation of a massless spin-1 field (Curtright 1980). The appropriate OSp$(4/G)$ representation would appear to be the reducible rank-three tensor, of graded mixed symmetry type (with non-zero graded trace). The graded dimension (the same as that in SU$(4/G)$) is $\frac{1}{3}(4-G)(5-G)(3-G)$ (cf Bars and Balantekin 1981). The count of degrees of freedom is thus correct for $G=2$, while for $G=4$ the theory appears to be null. In the former case, the ghost assignments, from $O(4)\times Sp(2)$ reduction (the same as the $SU(4)\times SU(2)$ case: Dondi and Jarvis (1981) are: 20 gauge fields $A_{[\mu\nu]\lambda}$; 32 ghost fields $A_{\mu\nu\alpha}$; 16 gauge fields $A_{\mu\alpha\beta}$, and 2 ghost fields $A_{[\alpha\beta]\gamma}$. As a final example, the graded dimension of the irreducible totally graded-symmetrical graded-traceless rank-three tensor representation of OSp$(4/G)$ is $\frac{1}{6}(8-G)(3-G)(4-G)$, suggesting that the structure of a symmetric rank-three tensor gauge theory is for $G=2$: 20 gauge fields $A_{(\lambda\mu\nu)}$, 20 ghost fields $A_{(\lambda\mu)\alpha}$, 4 gauge fields $A_{\lambda[\alpha\beta]}$, supplemented by $4-2=2$ trace conditions, leaving two degrees of freedom appropriate to a massless (spin-3) gauge field. Because it is so concise and simple, we would recommend this method of counting the families of ghosts in preference to others.

Acknowledgments

PDJ acknowledges the support of a Queen Elizabeth Fellowship during this research. G Thompson is thanked for informative discussions.

Note added in proof. After this work was completed and submitted for publication we received a preprint (Columbia University No 196) by Baulieu and Thierry-Mieg (*Nucl. Phys.* to be published) which is rather similar in content. Our Lagrangian (15) corresponds to their Hamiltonian case. However, they do not develop the consequent Sp(2) symmetry nor the OSp(4/2) supersymmetric enlargement.

References

Abers E S and Lee B W 1973 *Phys. Rep.* **9**
Bars I and Balantekin A B 1981 *J. Math. Phys.* **22** 1149
Becchi C, Rouet A and Stora A 1975 *Commun. Math. Phys.* **42** 127
—— 1976 *Ann. Phys., NY* **98** 287
Bonora L, Pasti P and Tonin M 1980 *Padua Preprint IFPD* 24/80
Bonora L and Tonin M 1981 *Phys. Lett.* **98B** 48
Curci G and Ferrari R 1976 *Phys. Lett.* **63B** 91
Curtright T 1980 *Chicago Preprint EFI* 80/04
Das A 1980 *Maryland Preprint* 81–089
Delbourgo R and Medrano, R M 1976 *Nucl. Phys.* B **110** 467
De Witt B S 1965 *Dynamical Theory of Groups and Fields* (New York: Gordon and Breach)

Dondi P H and Jarvis P D 1979 *Phys. Lett.* **84B** 75; *Erratum* **87B** 403
——1980 *Z. Phys.* C **4** 201
——1981 *J. Phys. A: Math. Gen.* **14** 547
Faddeev L D and Popov V N 1967 *Phys. Lett.* **25B** 29
Fayet P and Ferrara S 1977 *Phys. Rep.* **32** 69
Ferrara S, Piguet O and Schweda M 1977 *Nucl. Phys.* B **119** 493
Feynman R P 1963 *Acta Phys. Polon.* **26** 697
Kugo T and Ojima I 1979 *Prog. Theor. Phys. Suppl.* **66**
Fujikawa K 1978 *Prog. Theor. Phys.* **59** 2045
Marchetti P A and Tonin M 1981 *Padua Preprint IFPD* 32/81
Namazie M A and Storey D 1979 *Nucl. Phys.* B **157** 170
Ojima I 1980 *Prog. Theor. Phys.* **64** 625
Parker M 1980 *J. Math. Phys.* **21** 689
Quirós M, de Urríes F J, Hoyos J, Mazón M L and Rodriguez E 1980 *Madrid Preprint IEM*
Salam A and Strathdee J 1974 *Nucl. Phys.* B **76** 477
Siegel W 1980 *Phys. Lett.* **93B** 170
Slavnov A A 1972 *Sov. Phys.-JETP* **10** 99
Taylor J C 1971 *Nucl. Phys.* B **33** 436
Thierry-Mieg J and Ne'eman Y 1979 *Ann. Phys., NY* **123** 247

Volume 109B, number 1,2 PHYSICS LETTERS 11 February 1982

LOCAL OSp(4/2) SUPERSYMMETRY AND EXTENDED BRS TRANSFORMATIONS FOR GRAVITY

R. DELBOURGO, P.D. JARVIS [1] and G. THOMPSON
Department of Physics, University of Tasmania, Hobart, Tasmania 7001, Australia

Received 20 August 1981

When the gravitational field and its fictitious partners are grouped into an OSp(4/2) supermultiplet, an extended BRS invariance emerges. The action and BRS transformations (ordinary and dual) are written in supersymmetric form, and the extended gauge identities are deduced, in parallel with recent work on Yang–Mills theory.

In a recent letter [1], we showed that the standard gauge-fixing procedure for gravity, leading to the well-known BRS invariance [2,3], does not allow a corresponding invariance under any simple set of dual BRS transformations, in contrast to the nonabelian vector case [4,5]. It was suggested that a supersymmetric framework [6,7] could shed light on the difficulty. In this letter we wish to report on a successful resolution of the problem using superfields, appropriate to OSp(4/2) supersymmetry [7], thereby demonstrating the existence of extended BRS transformations for gravity. The formulation entails the graviton field $g^{\mu\nu}$, the fictitious partners ω^μ and $\bar{\omega}^\nu$, and an auxiliary field B^μ. The derivation of the BRS transformations, and invariant actions, is perfectly straightforward, although as anticipated the resulting lagrangian may involve additional couplings. We will also demonstrate the direct analogy which exists between nonabelian vector theory and gravity, especially at the superfield level. Indeed we begin with a brief resumé of the formalism for Yang–Mills theory [7], in order to make the generalization to gravity more obvious.

Within an OSp(4/2) framework, the supervector field $\phi_u(X) \equiv T^a \phi_u^a(x,\theta)$ is taken as a gauge transform of a special potential:

$$\phi_u(X) = \begin{pmatrix} \phi_\mu(x,\theta) \\ \phi_\alpha(x,\theta) \end{pmatrix} = U_1 \begin{pmatrix} A_\mu(x) \\ 0 \end{pmatrix} U_1^{-1} + \frac{\mathrm{i}}{e} U_1^{-1} \partial_u U_1 \,, \tag{1}$$

where

$$U_1 = \exp\{-\mathrm{ie}[\theta^\alpha \omega_\alpha(x) - \tfrac{1}{2}\theta^\beta\theta_\beta B(x)]\}\,, \tag{2}$$

and $\mu = 0, 1, 2, 3; \alpha = 1, 2$. From (1) and (2) there follows the field strength

$$\phi_{uv} = U_1^{-1} \begin{pmatrix} F_{\mu\nu}(x) & \\ 0 & \\ \vdots & \\ 0 & \end{pmatrix} U_1 \,. \tag{3}$$

The superaction, including covariant gauge-fixing terms, then reads

$$W = \int \mathrm{d}^6 X \{-\tfrac{1}{4} X^2 F^{uva}[uv] F^a_{uv} + (2/\xi)\phi^{ua}\phi^a_u\}\,, \tag{4}$$

[1] Queen Elizabeth II Research Fellow.

Volume 109B, number 1,2 PHYSICS LETTERS 11 February 1982

where the sign factor $[uv]$ is defined so that $[\mu\nu] = [\mu\alpha] = [\alpha\mu] = +1$ and $[\alpha\beta] = -1$, and $d^6X = d^4x\, d^2\theta$. In component fields, this becomes

$$\mathcal{L} = -\tfrac{1}{4}F^{\mu\nu}F_{\mu\nu} + [(\tfrac{1}{2}B^2 + \partial A\cdot B) - \partial\omega^\alpha\cdot D\omega_\alpha - \tfrac{1}{12}e^2(\omega^\alpha\times\omega^\beta)\cdot(\omega_\alpha\times\omega_\beta)]/\xi\,. \tag{5}$$

The reality properties permit us to take $\omega_1 = C$, $\omega_2 = \mathrm{i}\bar{C}$ and we may redefine $B \to \xi B$; after these substitutions are made (5) assumes a more familiar-looking form. The more symmetrical (Sp(2) invariant) treatment given to the ghosts means that the action differs slightly from the standard one; indeed taking $\omega_2 = \bar{\omega}_1$ is now also permitted.

The generalization to gravity now suggests itself. One begins by assuming that the gravitational superfield $G^{uv} = [uv]G^{vu}$ is the gauge transform of a special superfield G_0^{uv}, namely

$$G^{uv}(X) = \begin{pmatrix} G^{\mu\nu}(x,\theta) \\ G^{\mu\alpha}(x,\theta) \\ G^{\alpha\beta}(x,\theta) \end{pmatrix} = \frac{\partial X^u}{\partial X_0^x}\frac{\partial X^v}{\partial X_0^y}\, G_0^{xy}\,[xv]\,[xy]\,, \tag{1'}$$

where G_0^{uv} is given by

$$G_0^{uv} = \begin{pmatrix} g^{\mu\nu}(x_0) & \\ & \epsilon^{\alpha\beta} \end{pmatrix}.$$

The coordinate transform analogous to (2) is parametrized by

$$x^\mu = x_0^\mu + \theta_0^\alpha\omega_\alpha^\mu(x_0) + \tfrac{1}{2}\theta_0^\beta\theta_{0\beta}B^\mu(x_0)\,, \qquad \theta^\alpha = \theta_0^\alpha\,. \tag{2'}$$

Written out explicitly, the components of G^{uv} (with argument x, not x_0) are

$$\begin{aligned} G^{\mu\nu} = {} & \{g^{\mu\nu} + \omega_\alpha^\mu\omega^{\alpha\nu}\} + \theta^\alpha[(g^{\lambda\nu}\partial_\lambda - B^\nu)\omega_\alpha^\mu + (g^{\lambda\mu}\partial_\lambda - B^\mu)\omega_\alpha^\nu - \omega_\alpha^\lambda\partial_\lambda(g^{\mu\nu} + \omega_\beta^\mu\omega^{\nu\beta})] \\ & + \tfrac{1}{2}\theta^\beta\theta_\beta[B^\lambda\partial_\lambda(g^{\mu\nu} + \omega_\alpha^\mu\omega^{\nu\alpha}) + (B^\mu\partial_\lambda g^{\lambda\nu} + B^\nu\partial_\lambda g^{\lambda\mu}) + 2B^\mu B^\nu + (B^\mu\omega_\alpha^\nu + B^\nu\omega_\alpha^\mu)\partial_\lambda\omega^{\lambda\alpha} + \partial_\kappa\omega^{\mu\alpha}g^{\kappa\lambda}\partial_\lambda\omega_\alpha^\nu \\ & + \partial_\kappa\omega^{\kappa\alpha}(g^{\lambda\mu}\partial_\lambda\omega_\alpha^\nu + g^{\lambda\nu}\partial_\lambda\omega_\alpha^\mu) + \omega^{\kappa\alpha}(\tfrac{1}{2}\omega_\alpha^\lambda\partial_\kappa + \partial_\kappa\omega_\alpha^\lambda)\partial_\lambda(g^{\mu\nu} + \omega_\beta^\mu\omega^{\nu\beta})]\,, \qquad G^{\mu\alpha} \neq 0\,, \quad G^{\alpha\beta} = \epsilon^{\alpha\beta} \end{aligned} \tag{6}$$

The supercoordinate changes (2′) ensure that the curvature vanishes in all the θ-directions:

$$R^{\mu\nu}(x,\theta) \neq 0\,, \quad R^{\mu\alpha} \equiv 0\,, \quad R^{\alpha\beta} \equiv 0\,. \tag{3'}$$

Again this is similar to (3).

The obvious choice of gravitational superaction from these considerations is

$$W = \int d^6X \left(\frac{X^2}{2K^2}\sqrt{-G}\,R^{uv}G_{vu} + \frac{1}{\xi K^2}\,G^{uv}I_{vu}\right), \tag{4'}$$

where $I_{uv} \equiv \begin{pmatrix}\eta_{\mu\nu} & 0 \\ 0 & \epsilon_{\alpha\beta}\end{pmatrix}$ is introduced [‡1] so as to break the general coordinate invariance and thereby fix the gauge. The breaking term is actually similar to that appearing in (4) when it is remembered that

$$G^{uv}I_{vu} = I^{ab}E^u{}_b E_a{}^v I_{vu}\,,$$

where $E^u{}_a$ is the sechsbein field appropriate to local OSp(4/2) symmetry [‡2].

The invariance of the measure $\sqrt{-G}\, d^6X$ ensures that the first part of the action reduces to the usual Einstein

[‡1] There is more freedom of manoeuvre in that $\sqrt{-G}$ factors can be incorporated into the breaking term. These issues, plus consideration of axial gauges etc., will be discussed in subsequent publications.

[‡2] The sechsbein field will be given greater prominence in future work since it plays a crucial role for fermionic matter.

Volume 109B, number 1,2 PHYSICS LETTERS 11 February 1982

form; the second part is obtained from the $\frac{1}{2}\theta^2$ end of (6). Hence in component fields,

$$\mathcal{L} = (1/2K^2)\sqrt{-g}R + (\eta_{\mu\nu}/\xi K^2)\{2B^\mu B^\nu + 2B^\mu(\partial_\lambda g^{\lambda\nu} + \omega^\nu_\alpha \partial_\lambda \omega^{\alpha\lambda}) + B^\lambda \partial_\lambda(g^{\mu\nu} + \omega^\mu_\alpha \omega^{\nu\alpha}) + \partial_\kappa \omega^{\mu\alpha} g^{\kappa\lambda} \partial_\lambda \omega^\nu_\alpha$$
$$+ \partial_\kappa \omega^{\kappa\alpha}[2g^{\lambda\mu}\partial_\lambda \omega^\nu_\alpha - \tfrac{1}{2}\omega^\lambda_\alpha \partial_\lambda(g^{\mu\nu} + \omega^\mu_\beta \omega^{\nu\beta})] + \partial_\lambda \omega^{\kappa\alpha}\partial_\kappa \omega^\lambda_\alpha (g^{\mu\nu} + \omega^\mu_\beta \omega^{\nu\beta})\} . \quad (5')$$

This is very close in spirit to (5). The analogue of the vector field A^μ is

$$g^{\mu\nu} + \tfrac{1}{2}\omega^\mu_\alpha \omega^{\nu\alpha} - \tfrac{1}{2}\eta^{\mu\nu}(g^{\kappa\lambda} + \tfrac{1}{2}\omega^\kappa_\alpha \omega^{\lambda\alpha})\eta_{\kappa\lambda} ,$$

for gravity; we even see the analogue of $(\omega \times \omega)^2$ in the last terms of (5′).

It remains to give the BRS and dual BRS transformations for gravity. These are obtained via the supertranslation

$$\delta G^{\mu\nu}(x, \theta) = G^{\mu\nu}(x, \theta + \epsilon) - G^{\mu\nu}(x, \theta) ,$$

which formally leaves (4′) invariant. Re-labelling $(\omega^\mu_1, \omega^\mu_2) = (\omega^\mu, \bar{\omega}^\mu)$ and $(\epsilon_1, \epsilon_2) = (\epsilon, \bar{\epsilon})$, one arrives at the extended BRS transformations

$$\delta g^{\mu\nu} = \bar{\epsilon}(\partial_\lambda \omega^\mu g^{\lambda\nu} + \partial_\lambda \omega^\nu g^{\nu\lambda} - \omega^\lambda \partial_\lambda g^{\mu\nu}) - \epsilon(\partial_\lambda \bar{\omega}^\mu g^{\lambda\nu} + \partial_\lambda \bar{\omega}^\nu g^{\mu\lambda} - \bar{\omega}^\lambda \partial_\lambda g^{\mu\nu}) ,$$
$$\delta \omega^\mu = -\bar{\epsilon}(\omega^\lambda \partial_\lambda \omega^\mu) + \epsilon(B^\mu + \bar{\omega}^\lambda \partial_\lambda \omega^\mu) , \quad \delta \bar{\omega}^\mu = \bar{\epsilon}(B^\mu - \omega^\lambda \partial_\lambda \bar{\omega}^\mu) + \epsilon(\bar{\omega}^\lambda \partial_\lambda \bar{\omega}^\mu) ,$$
$$\delta B^\mu = -\bar{\epsilon}\omega^\lambda \partial_\lambda B^\mu + \epsilon \bar{\omega}^\lambda \partial_\lambda B^\mu . \quad (7)$$

The BRS transformations correspond to $\epsilon \to 0$, while $\bar{\epsilon} \to 0$ provides the dual ones.

To study the consequences for the Green functions we add to the lagrangian appropriate source terms, including composite sources corresponding to the first variations of the natural combinations (cf. ref. [7])

$$B^\mu_+ = B^\mu + \bar{\omega}^\lambda \partial_\lambda \omega^\mu , \quad B^\mu_- = B^\mu - \omega^\lambda \partial_\lambda \bar{\omega}^\mu ,$$

second variations are guaranteed zero by nil-potency. Thus for ordinary BRS transformations we add (dual ones can be treated similarly):

$$-J_{\mu\nu}g^{\mu\nu} - \bar{J}_\mu \omega^\mu - J_\mu \bar{\omega}^\mu - L_\mu B^\mu_- - I_{\mu\nu}\delta g^{\mu\nu}/\delta\bar{\epsilon} - \bar{I}_\mu \delta\omega^\mu/\delta\bar{\epsilon} - I_\mu \delta\bar{\omega}^\mu/\delta\bar{\epsilon} - K_\mu \delta B^\mu_-/\delta\bar{\epsilon} .$$

This leads to

$$\int(-J_{\mu\nu}\delta/\delta I_{\mu\nu} + J_\mu \delta/\delta I_\mu + \bar{J}^\mu \delta/\delta\bar{I}^\mu - K_\mu \delta/\delta L_\mu)Z = 0 ,$$

for the generating functional and

$$\int[-(\delta\Gamma/\delta g^{\mu\nu})\delta\Gamma/\delta I_{\mu\nu} + (\delta\Gamma/\delta\bar{\omega}^\mu)\delta\Gamma/\delta I_\mu + (\delta\Gamma/\delta\omega_\mu)\delta\Gamma/\delta\bar{\omega}^\mu - (\delta\Gamma/\delta B^\mu_-)\delta\Gamma/\delta L_\mu] = 0 , \quad (8)$$

for the effective action.

The details of this work and a check on the validity of the identities (at the one-loop level anyway) will be presented in a fuller paper. There we shall amplify upon the geometrical restrictions imposed by (2′) and we will make much greater use of the sechsbein formalism.

References

[1] R. Delbourgo and G. Thompson, J. Phys. G7 (1981) L133.
[2] R. Delbourgo and M.R. Medrano, Nucl. Phys. B110 (1976) 467.
[3] K. Stelle, Phys. Rev. D16 (1977) 953.
[4] G. Curci and R. Ferrari, Phys. Lett. 63B (1976) 91.
[5] I. Ojima, Prog. Theor. Phys. 64 (1980) 625.
[6] L. Bonora and M. Tonin, Phys. Lett. 98B (1981) 48.
[7] R. Delbourgo and P.D. Jarvis, Extended BRS invariance and OSp(4/2) supersymmetry, Hobart preprint (April 1981), to be published in J. Phys. A.

Part H
MULTIQUARKS

H.0. Commentaries H

The isospin picture of the proton and neutron, as being two aspects of the nucleon, plays a central role in nuclear theory. Wigner enlarged the SU(2) isospin group to SU(4) by also combining the two fermionic spin states and showed that the resulting representations characterised the interactions and bound states of nucleons. The extension of that idea to particle physics was promoted in the early 1960's by Gursey & Radicati (Phys. Rev. Lett. 13, 173 (1964)), Pais (Phys. Rev. Lett. 13, 175 (1964)) and Sakita (Phys. Rev. 136, B1756 (1964)), offering an effective non-relativistic picture; it naturally led to the notion of an SU(6) non-relativistic classification group, comprising the u, d and s quarks, the only quarks known at the time. Thus baryons comprising spin $1/2^+$ and $3/2^+$ resonances were fitted into 56-fold multiplet whilst the mesons, comprising pseudoscalar 0^- and vector 1^- mesons, found their place in a 35 multiplet. The puzzle as to how the baryons could be held in a symmetric representation, contrary to Fermi-Dirac statistics, was only resolved when colour became firmly established, with the requirement that physical particles are devoid of colour — again only understood by the nature of chromodynamic forces. The generalisation to include extra quark flavours (c, b and t) then becomes obvious, although the variation in heavy quark masses casts a shadow on the idea; yet the enlarged group does do a good classification job even if it offers no explanation of the wide disparity in bound state masses.

H.1 Salam A., Delbourgo R. and Strathdee J., The covariant theory of strong interaction symmetries. Proc. Roy. Soc. A284, 146-158 (1965)

H.2 Delbourgo R., Higher meson resonances and the 4212 multiplet of SU(12). Phys. Lett. 15, 347-350 (1965)

H.3 Delbourgo R. and Salam A., Reggeization in supermultiplet theories. Phys. Rev. 186, 1516-1528 (1969)

H.4 Delbourgo R. and Rotelli P., A Veneziano model for scattering of meson and baryon supermultiplets. Nuovo Cim. 69, 412-418 (1970)

Finding the Lorentz covariant description became important after these supermultiplet states came to the fore. We did so in article [H.1] and in the process discovered that the proton to neutron magnetic moment ratio was 3/2 which is quite close to experiment; our treatment relied on the formulation of multispinors subject to Bargmann-Wigner equations, which broke down the U(12) symmetry to U(6)⊗U(6). Sakita & Wali (Phys. Rev. 139, B1355 (1965)) almost simultaneously with us adopted this approach. However, even ignoring the fact that quark masses are unequal, Beg & Pais (Phys. Rev. 137, B1514 (1965)) and others showed that the maintenance of U(12) symmetry for all strong interactions was impossible as it would violate unitarity — in keeping with the Coleman-Mandula theorem. In spite of this setback, the idea lives on as a neat way of organising the families of baryons and mesons and it has been revived in the context of 'heavy quark effective theory' by Isgur & Wise (Phys. Lett. B232, 13 (1989)).

When categorising excitations of the fundamental supermultiplets, people naturally imagined a picture of bound quarks; thus they regarded them as orbital or radial resonances in analogy to atomic states. By contrast, article [H.2] was the first paper to regard such resonances as multiquark or molecular excitations of the ground state multiplets, which seemed to complicate the description. However, the recent discoveries of tetraquark mesons and pentaquark baryons combinations shows that the idea is not so far-fetched. The article [H.3] took the multiquark description to extremes and was an attempt to fit the notion within a Regge trajectory picture (very fashionable at the time), while [H.4] tied the idea to the Veneziano dual resonance picture, via duality diagrams. For a more comprehensive and more varied discussion of these approaches, see the following reviews:

- S. Ishida, M. Ishida and T. Maeda. Prog. Theor. Phys. 104, 785 (2000)
- M. Neubert. Phys. Rep. 245, 259 (1994)

The covariant theory of strong interaction symmetries

By A. Salam,† F.R.S.
Imperial College, London

and R. Delbourgo and J. Strathdee
International Centre for Theoretical Physics, Trieste, Italy

(*Received* 18 *January* 1965)

A classification of particles is suggested based on a $\tilde{U}(12)$ symmetry scheme. This is a relativistic generalization of the $U(6)$ symmetry. The spin $\frac{1}{2}$ and $\frac{3}{2}$ baryons are each described by 20-component spinors which satisfy Bargmann–Wigner equations and belong to the $\underline{364}$ representation of the $\tilde{U}(12)$ group while the vector and p.s. mesons belong to the representation $\underline{143}$. The procedure for writing fully relativistic form factors is worked out in detail for baryon–meson and meson–meson cases.

The new results are the following:

(1) $\dfrac{F^C(q^2)}{F^M(q^2)} \propto 1 + \dfrac{q^2}{\langle 2\mu\rangle m}$, where F^C and F^M are (Sachs) electromagnetic form factors.

(2) $\mu_p = 1 + 2m/\langle\mu\rangle$, where $\langle\mu\rangle$ is the mean mass of the 1^- multiplet and m the nucleon mass.

(3) $\mu_{\rho, K^*} = 3$.

The conventional $U(6)$ results can be recovered by projecting to the positive energy subspace in the rest system for each particle. To any irreducible representation of the $U(6)$ there corresponds one irreducible representation of $\tilde{U}(12)$ and vice versa.

1. Introduction

The problem of finding a relativistic generalization of the $U(6)$ group structure has engaged considerable attention recently (see references A). In an earlier paper (Delbourgo, Salam & Strathdee 1965*b*, to be referred to as I) it was suggested that one way to write relativistic S-matrix elements is to embed $U(6)$ in a $\tilde{U}(12)$ group-structure. The present paper gives the detailed formalism for writing relativistic S-matrix elements in this theory. In particular we compute the two basic baryon–meson and meson–meson form factors. The generalization of the symmetry from $U(6)$ to $\tilde{U}(12)$ gives the following new results:

(1) There is essentially just one *relativistic* form factor in strong interaction physics of the octet baryons and the (1^-) and (0^-) mesons. The relevance of this result to the conventional nucleon electric and magnetic form factors is discussed in §6.

(2) In Bohr magnetons the magnetic moments of the proton and the neutron are

$$\mu_p = 1 + 2x, \quad \mu_n = -\tfrac{2}{3}(1+2x),$$

where $x = m_N/\langle\mu\rangle$ and $\langle\mu\rangle$ is the mean mass of the (1^-) multiplet. The experimental magnetic moment values are well reproduced if $\langle\mu\rangle \approx 1000\,\mathrm{MeV}$. This mass value is not far from the mean of m_ρ, m_ω, m_ϕ, etc.

† On leave of absence at the International Centre for Theoretical Physics, Trieste, Italy.

The formalism describes both the spin $\frac{1}{2}$ and $\frac{3}{2}$ baryons as 20-component† composite entities made from the basic 4-component (Dirac) quark (Gell-Mann 1964; Zweig 1964). In §2 we describe the $\tilde{U}(12)$ algebra; in §3 are computed the form factors incorporating full $\tilde{U}(12)$ symmetry. In §4 we specialize to the reduced symmetry group $U(3) \times \mathscr{L}_4$ (where $\mathscr{L}_4$ refers to the homogeneous Lorentz group), the inhomogeneous Lorentz group ($I\mathscr{L}_4$) being considered in §5. The reduction of the $\tilde{U}(12)$ symmetry to $U(3) \times (I\mathscr{L}_4)$ in §§4 and 5 correctly reproduces the final physical symmetry situation, and gives the *relativistic* expressions for the baryon form factors in a fundamentally broken $\tilde{U}(12)$ symmetry scheme. The exact $U(6)$ symmetry can be recovered from our expressions in the limit of zero momenta.

2. $\tilde{U}(12)$ AND ITS SUBGROUPS

We assume that the fundamental entity for strong interactions is a 12-component (Dirac) quark. The group structure $\tilde{U}(12)$ is defined by the algebra of the 144 matrices $F^{Ri} = \gamma^R T^i$, $R = 1, \ldots, 16$; $i = 0, \ldots 8$. Here

$$\gamma^R = 1, \quad \gamma_\mu, \quad \sigma_{\mu\nu} = \tfrac{1}{2}\mathrm{i}[\gamma_\mu, \gamma_\nu], \quad \mathrm{i}\gamma_\mu\gamma_5, \quad \gamma_5,$$

with γ_0 hermitian and $\boldsymbol{\gamma}$ antihermitian and the metric $(1, -1, -1, -1)$. The general $\tilde{U}(12)$ transformation on the quark field $\psi_A = \psi_{p\alpha}$ $(p = 1, 2, 3;\ \alpha = 1, 2, 3, 4)$ will be assumed to be

$$\delta\psi_{p\alpha} = \mathrm{i}(\epsilon^j + \epsilon_5^j\gamma_5 + \epsilon_\mu^j\gamma_\mu + \mathrm{i}\epsilon_{\mu5}^j\gamma_\mu\gamma_5 + \tfrac{1}{2}\epsilon_{\mu\nu}^j\sigma_{\mu\nu})_\alpha^\beta (T^j)_p^q \psi_{q\beta}, \tag{2·1}$$

where all 144 ϵ's are real and this property leaves $\bar{\psi}\psi = \psi^+\gamma_0\psi$ invariant.

For higher representations of $\tilde{U}(12)$ (made up compositively from quarks) the transformation (2·1) will take the form

$$\delta\Psi = \mathrm{i}(\epsilon^j F^j + \epsilon_5^j F_5^j + \epsilon_\mu^j F_\mu^j + \epsilon_{\mu5}^j F_{\mu5}^j + \tfrac{1}{2}\epsilon_{\mu\nu}^j F_{\mu\nu}^j)\Psi. \tag{2·2}$$

The general commutation rules of the generators F are listed in the appendix. The quadratic Casimir operator is

$$F^jF^j - F_5^jF_5^j + \tfrac{1}{2}F_{\mu\nu}^jF_{\mu\nu}^j + F_\mu^jF_\mu^j - F_{\mu5}^jF_{\mu5}^{j'}.$$

Inspection of the commutators for $\tilde{U}(12)$ reveals that a 72-component subalgebra is generated by the operators F^j, F_5^j, $F_{\mu\nu}^j$. This is the subgroup $W(6)$ (see references A) and in the fundamental representation has the generators T^j, $\gamma_5 T^j$, and $\sigma_{\mu\nu}T^j$. The expressions

$$F_\mu^jF_\mu^j - F_{\mu5}^jF_{\mu5}^j \quad \text{and} \quad F^jF^j - F_5^jF_5^j + \tfrac{1}{2}F_{\mu\nu}^jF_{\mu\nu}^j$$

are now separately invariant under $W(6)$. Note that $W(6)$ possesses the important 36-parameter subgroup $U(6)$ $(T^i\sigma_{ab};\ T^i;\ a, b = 1, 2, 3)$. This will later be identified with the $U(6)$ of Gürsey, Radicati & Sakita.

† Of all the approaches to the relativistic $U(6)$ theory listed in references A, the formulation nearest in spirit to our own is that of P. Roman & J. J. Agasshi. The mesons are described similarly in both approaches but the baryon states (and their underlying symmetries) are different. Thus the well-known eightfold of baryons appears in the Agasshi–Roman theory with four components contrasted with the 20 components for our case. These essential differences come about because we use Bargmann–Wigner equations of §5 to make sure we are dealing with particles having a definite mass as well as definite spin.

3. Some representations of $\tilde{U}(12)$ and their decomposition

A. The fundamental representation of $\tilde{U}(12)$ is the 12-component quark discussed above. Following the usual procedure we assign to this quark the baryon number $B = \frac{1}{3}$. The baryons are then to be constructed from three quark states and the mesons from quark–antiquark states.

These states decompose under $\tilde{U}(12)$ in the following way:

$$\left.\begin{aligned} \underline{12} \otimes \underline{12}^* &= \underline{1} + \underline{143}, \\ \underline{12} \otimes \underline{12} \otimes \underline{12} &= \underline{220} + \underline{364} + \underline{572} + \underline{572}. \end{aligned}\right\} \tag{3·1}$$

The $\underline{220}$ is completely antisymmetrical, the $\underline{364}$ completely symmetrical and the $\underline{572}$ is of the mixed symmetry type [2, 1].

Under the subgroup $W(6)$ these states reduce according to

$$\left.\begin{aligned} \underline{143} &= (35, 1) + (6, 6^*) + (6^*, 6) + (1, 35) + (1, 1), \\ \underline{220} &= (20, 1) + (15, 6) + (6, 15) + (1, 20), \\ \underline{364} &= (56, 1) + (21, 6) + (6, 21) + (1, 56), \\ \underline{572} &= (70, 1) + (21, 6) + (6, 21) + (1, 70). \end{aligned}\right\} \tag{3·2}$$

Under the subgroup $U(3) \otimes \tilde{U}(4)$, where $U(3)$ refers to the space of unitary spin matrices T^j, and $\tilde{U}(4)$ to that of the Dirac matrices γ^R, the contents are given by

$$\left.\begin{aligned} \underline{143} &= (8, 15) + (1, 15) + (8, 1), \\ \underline{220} &= (8, 20) + (10, 4) + (1, 20), \\ \underline{364} &= (10, 20') + (8, 20'') + (1, 4), \\ \underline{572} &= (10, 20'') + (8, 20) + (8, 20') + (1, 20'') + (8, 4), \end{aligned}\right\} \tag{3·3}$$

where in each bracket the first number denotes the $U(3)$ representation and the second the $\tilde{U}(4)$ representation. The three $\tilde{U}(4)$ representations denoted here by 20, 20′ and 20″ are of symmetry types $[1^3]$, [3] and [2, 1] respectively.

We note also the reduction of the products

$$\left.\begin{aligned} \underline{143} \otimes \underline{143} &= 1 + \underline{143}_F + \underline{143}_D + \underline{4212} + \underline{5005} + \underline{5005}^* + \underline{5940}, \\ \underline{364} \otimes \underline{364}^* &= \underline{1} + \underline{143} + \underline{5940} + \underline{126412}, \\ \underline{143} \otimes \underline{364} &= \underline{364} + \underline{572} + \underline{16016} + \underline{35100}. \end{aligned}\right\} \tag{3·4}$$

B. Since it is our intention to assign the baryon to the $\underline{364}$, we compute the expectation values of the 144-vector $(\gamma^R T^j)_A^B$ between 364 states, namely

$$J^{Rj} = \bar{\Psi}^{ABC} (\gamma^R T^j)_A^{A'} \Psi_{A'BC}, \tag{3·5}$$

where Ψ_{ABC} is fully symmetric and has the $U(3) \times \tilde{U}(4)$ decomposition

$$\Psi_{\alpha p \beta q \gamma r} = \frac{\sqrt{3}}{2\sqrt{2}} D_{\alpha\beta\gamma, pqr} + \epsilon_{pqr} V_{[\alpha\beta\gamma]} + \frac{1}{2\sqrt{6}} (\epsilon_{pqs} N_{[\alpha\beta]\gamma, r}{}^{s} + \epsilon_{qrs} N_{[\beta\gamma]\alpha, p}{}^{s} + \epsilon_{rps} N_{[\gamma\alpha]\beta, q}{}^{s}). \tag{3·6}$$

Here (α, β, γ) take the values 1, 2, 3, 4 and (p, q, r) the values, 1, 2, 3. $V_{[\alpha\beta\gamma]}$ is completely antisymmetric, $D_{\alpha\beta\gamma, pqr}$ is completely symmetric in both $\alpha\beta\gamma$ and pqr and $N_{[\alpha\beta]\gamma,}{}^{s}_{r}$ is of the symmetry type [2, 1], i.e.

$$\left.\begin{aligned} N_{[\alpha\beta]\gamma} + N_{[\beta\alpha]\gamma} &= 0, \\ N_{[\alpha\beta]\gamma} + N_{[\beta\gamma]\alpha} + N_{[\gamma\alpha]\beta} &= 0. \end{aligned}\right\} \tag{3·7}$$

After some algebra we find (specializing to $i \neq 0$ so that $(T^i)^p_p = 0$),

$$\begin{aligned} J^{Ri} = {} & \tfrac{3}{8}\bar{D}^{\alpha\beta\gamma,pqr}(\gamma^R)^{\alpha'}_{\alpha}(T^i)^{p'}_{p}\,D_{\alpha'\beta\gamma,p'qr} \\ & + \tfrac{1}{4}[\bar{D}^{\alpha\beta\gamma}(\gamma^R)^{\alpha'}_{\alpha}(T^i)^{p'}_{p}\,\epsilon_{p'qs}N_{[\alpha'\beta]\gamma,}{}^{s}_{r} + \bar{N}^{[\alpha\beta]\gamma,}{}^{r}_{s}\,\epsilon^{pqs}(\gamma^R)^{\alpha'}_{\alpha}(T^i)^{p'}_{p}\,D_{\alpha'\beta\gamma,p'qr}] \\ & + [\bar{V}^{[\alpha\beta\gamma]}(\gamma^R)^{\alpha'}_{\alpha}\,(T^i)^{p'}_{p}\,N_{[\beta\gamma]\alpha',}{}^{p}_{p'} + \bar{N}^{[\beta\gamma]\alpha,}{}^{p}_{p'}(T^i)^{p'}_{p}(\gamma^R)^{\alpha'}_{\alpha}\,V_{[\alpha'\beta\gamma]}] \\ & - \tfrac{1}{24}(\bar{N}^{[\beta\alpha]\gamma}(\gamma^R)^{\alpha'}_{\alpha}\,N_{[\alpha'\beta]\gamma})^{i}_{3D+5F} \\ & + \tfrac{1}{12}(\bar{N}^{[\beta\alpha]\gamma}(\gamma^R)^{\alpha'}_{\alpha}\,N_{[\alpha'\gamma]\beta})^{i}_{3D+2F}, \end{aligned} \tag{3·8}$$

where

$$\left.\begin{aligned} (\bar{N}N)^i_F &= \bar{N}^p_r(T^i)^q_p\,N^r_q - \bar{N}^p_r\,N^q_p(T^i)^r_q, \\ (\bar{N}N)^i_D &= \bar{N}^p_r(T^i)^q_p\,N^r_q + \bar{N}^p_r\,N^q_p(T^i)^r_q. \end{aligned}\right\} \tag{3·9}$$

Notice the appearance at this stage of the characteristic combinations $3D+2F$ and $3D+5F$ for the form factors involving the eightfold baryons.

4. Reduction of $\tilde{U}(4)$ to the homogeneous Lorentz group $\mathscr{L}_4$

A. We now specifically consider the space-time symmetries. So far it has been assumed that in its space-time behaviour the fundamental 4-component ($\tilde{U}4$) entity ψ_α transforms as

$$\psi_\alpha \to S^\beta_\alpha\,\psi_\beta,$$

where

$$S^\beta_\alpha = 1 + \mathrm{i}(\epsilon_5\gamma_5 + \epsilon_\mu\gamma_\mu + \mathrm{i}\epsilon_{\mu5}\gamma_\mu\gamma_5 + \tfrac{1}{2}\epsilon_{\mu\nu}\sigma_{\mu\nu})^\beta_\alpha\,\psi_\beta, \tag{4·1}$$

with $\bar{\psi}$ transforming as $\bar{\psi}^\alpha \to \bar{\psi}^\beta[S^{-1}]^\alpha_\beta$. With ϵ's real, these transformations preserve the invariance of $\bar{\psi}^\alpha\psi_\alpha$. The higher representations $\Psi^{\gamma\delta\ldots}_{\alpha\beta\ldots}$ of $\tilde{U}(4)$ transform as

$$\Psi^{\gamma\delta\ldots}_{\alpha\beta\ldots} \to S^{\alpha'}_{\alpha}S^{\beta'}_{\beta}\ldots(S^{-1})^{\gamma}_{\gamma'}(S^{-1})^{\delta}_{\delta'}\ldots. \tag{4·2}$$

The symmetry represented by (4·1), however, is too general. In space-time terms it corresponds, as is well known, to the full conformal group symmetry C_4. To make contact, however, with physical space-time symmetries of (at this stage) the *homogeneous Lorentz group* we must descend from $\tilde{U}(4)$ to $\mathscr{L}_4$. There are a number of ways of doing this which are not all necessarily equivalent so far as the underlying physics is concerned as will be discussed in §5.

Disregarding the problems connected with unitary spin, clearly the most direct symmetry reduction is achieved by taking

$$\epsilon_5 = \epsilon_\mu = \epsilon_{\mu5} = 0.$$

Now as is well known for this case (*though not for* $\tilde{U}(4)$) one can define an antisymmetric matrix $(C^{-1})^{\alpha\beta}$ (within the Dirac algebra) with the defining property that $C^{-1}\psi^T$ transforms similarly to $\bar{\psi}$. (Here ψ^T is the transpose of ψ.) In particular, $C^{-1}\psi^T\psi$ just like $\bar{\psi}\psi$ is an invariant. Clearly from the definition above, the antisymmetric matrix $C^{-1} = (C^{-1})^{\alpha\beta} = -(C^{-1})^{\beta\alpha}$ plays the role for $\mathscr{L}_4$ of the metric tensor. We may regard $(C^{-1})^{\alpha\beta}$ as a contravariant quantity (with two upper indices) and its inverse $C_{\alpha\beta}$ as the corresponding covariant $(C_{\alpha\beta}(C^{-1})^{\beta\gamma} = \delta^{\gamma}_{\alpha})$.

It is easy to show (Jauch & Rohrlich 1955) that the matrix C with the defining property above† can be realized by finding a matrix satisfying

$$(\gamma_\mu C)_{\alpha\beta} = (\gamma_\mu C)_{\beta\alpha},$$

where

$$(\gamma_\mu C)_{\alpha\beta} = (\gamma_\mu)^{\gamma}_{\alpha}\, C_{\gamma\beta}.$$

It is also easy to show that the 16 Dirac matrices $(\gamma^R C)_{\alpha\beta}$ fall into two distinct classes; the matrices $(\gamma_\mu C)_{\alpha\beta}$ and $(\sigma_{\mu\nu} C)_{\alpha\beta}$ are symmetric, and $C_{\alpha\beta}, (\gamma_5 C)_{\alpha\beta}, (\mathrm{i}\gamma_\mu\gamma_5 C)_{\alpha\beta}$ are antisymmetric. For writing symmetric and antisymmetric higher-rank 'spinors' in $\mathscr{L}_4$ these are the primary quantities one needs.

To illustrate consider the following examples:

(i) *Multi-spinor of rank* 2

A second-rank symmetric spinor must have the form

$$\Phi_{\{\alpha\beta\}} = [(\gamma_\mu C)\,\phi_\mu + \tfrac{1}{2}(\sigma_{\mu\nu} C)\,\phi_{\mu\nu}]_{\alpha\beta}. \tag{4·3}$$

Likewise the general antisymmetric spinor has the form

$$\Phi_{[\alpha\beta]} = [C\phi + (\gamma_5 C)\,\phi_5 + \mathrm{i}(\gamma_\mu\gamma_5 C)\,\phi_{\mu 5}]_{\alpha\beta}. \tag{4·4}$$

(ii) *Fully symmetric spinor of rank* 3

Consider $\Psi_{\alpha\beta\gamma}$ with full symmetry in α, β, γ. From symmetry in γ and β, one may write Ψ in the form

$$\Psi_{\alpha\beta\gamma} = \psi_{\alpha\mu}(\gamma_\mu C)_{\beta\gamma} + \tfrac{1}{2}\psi_{\alpha\mu\nu}(\sigma_{\mu\nu} C)_{\beta\gamma}. \tag{4·5}$$

We now show that full symmetry in α, β, γ is realized provided

$$\gamma_\mu \psi_\mu = 0, \tag{4·6}$$

$$\gamma_\mu \psi_{\mu\nu} + \mathrm{i}\psi_\nu = 0. \tag{4·7}$$

For, the three antisymmetric tensors $(C^{-1})^{\gamma\alpha}$, $(C^{-1}\gamma_5)^{\gamma\alpha}$ and $(\mathrm{i}C^{-1}\gamma_\mu\gamma_5)^{\gamma\alpha}$ must annihilate $\Psi_{\alpha\beta\gamma}$.

This gives

$$\psi_{\alpha\mu}(\gamma_\mu)^{\alpha}_{\beta} + \tfrac{1}{2}\psi_{\alpha\mu\nu}(\sigma_{\mu\nu})^{\alpha}_{\beta} = 0,$$

$$\psi_{\alpha\mu}(\gamma_\mu\gamma_5)^{\alpha}_{\beta} + \tfrac{1}{2}\psi_{\alpha\mu\nu}(\sigma_{\mu\nu}\gamma_5)^{\alpha}_{\beta} = 0,$$

$$\psi_{\alpha\mu}(\gamma_\mu\gamma_\lambda\gamma_5)^{\alpha}_{\beta} + \tfrac{1}{2}\psi_{\alpha\mu\nu}(\sigma_{\mu\nu}\gamma_\lambda\gamma_5)^{\alpha}_{\beta} = 0.$$

If we suppress Dirac indices, the first two equations give (4·6) and $\sigma_{\mu\nu}\psi_{\mu\nu} = 0$, and the last is equivalent to (4·7). Note that as a result of (4·6) and (4·7) the 40-component entity on the right side of (4·5) contains only 20 independent components.

† One could have made the transition $\tilde{U}(4) \to \mathscr{L}_4$ via the intermediate symmetry stage $\widetilde{Sp}(4)$ $(\epsilon_5 = \epsilon_{\mu 5} = 0)$. The $\widetilde{Sp}(4)$ transformation with 10 parameters $(\epsilon_\mu, \epsilon_{\mu\nu})$ also admits of the existence of this matrix C. All statements of §4 apply equally to theories with $\widetilde{Sp}(4)$ symmetry.

(iii) *Mixed spinor of rank* 3

The 20-component tensor $\Psi_{[\alpha\beta]\gamma}$ which satisfies the 'trace' condition†

$$\Psi_{[\alpha\beta]\gamma}+\Psi_{[\gamma\alpha]\beta}+\Psi_{[\beta\gamma]\alpha}=0 \tag{4·8}$$

can necessarily be written in the form

$$\Psi_{[\alpha\beta]\gamma}=(\gamma_5 C)_{\alpha\beta}\psi_\gamma+\mathrm{i}(\gamma_\mu\gamma_5 C)_{\alpha\beta}\psi_\gamma+C_{\alpha\beta}K_\gamma. \tag{4·9}$$

The mixed symmetry character of Ψ yields the constraint

$$\gamma_5\psi+\mathrm{i}\gamma_\mu\gamma_5\psi_\mu-K=0. \tag{4·10}$$

This follows on multiplying (4·8) by $(C^{-1})^{\alpha\beta}$.

B. To return to $\tilde{U}(12)$, we now decompose all irreducible $\tilde{U}(12)$ higher representations relative to the representations of $U(3)\otimes\mathscr{L}_4$, maintaining the over-all symmetry. Thus for the $(8\times 20'')$ part of the $\underline{364}$ representation, we write

$$N_{[\alpha\beta]\gamma,p}{}^{q}=C_{\alpha\beta}K_{\gamma,p}{}^{q}+(\gamma_5 C)_{\alpha\beta}N_{\gamma,p}{}^{q}+(\mathrm{i}\gamma_\mu\gamma_5 C)_{\alpha\beta}N_{\gamma\mu,p}{}^{q}.$$

The contribution of $N_{[\alpha\beta]\gamma,p}{}^{q}$ to the currents (3·8), namely

$$J^{Ri}(N)=-\tfrac{1}{24}(\overline{N}^{[\beta\alpha]\gamma}(\gamma^R)_\alpha^{\alpha'}N_{[\alpha'\beta]\gamma})_{3D+5F}+\tfrac{1}{12}(\overline{N}^{[\beta\alpha]\gamma}(\gamma^R)_\alpha^{\alpha'}N_{[\alpha'\gamma]\beta})_{3D+2F}$$

may be now written out in terms of N and N_μ. Thus

$$J^i(N)=\tfrac{1}{2}[2\overline{N}N+\mathrm{i}(\overline{N}_\mu\gamma_\mu N-\overline{N}\gamma_\mu N_\mu)+\overline{N}_\lambda\gamma_\lambda\gamma_\mu N_\mu+\overline{N}_\mu N_\mu]_F^i,$$

$$\begin{aligned}J^i_\mu(N)=\;&\tfrac{1}{6}\mathrm{i}(\overline{N}_\mu N-\overline{N}N_\mu)^i_{F-3D}\\&-\tfrac{1}{6}[\mathrm{i}(\overline{N}_\lambda\gamma_\lambda\gamma_\mu N-\overline{N}\gamma_\mu\gamma_\lambda N_\lambda)-\overline{N}_\lambda\gamma_\mu N_\lambda+\overline{N}_\lambda\gamma_\lambda\gamma_\mu\gamma_\nu N_\nu]^i_{3D+2F},\end{aligned}$$

$$\begin{aligned}J^i_{\mu\nu}(N)=\;&\tfrac{1}{6}\mathrm{i}(\overline{N}_\mu N_\nu-\overline{N}_\nu N_\mu)^i_{F-3D}\\&+\tfrac{1}{6}[2\overline{N}\sigma_{\mu\nu}N+\mathrm{i}(\overline{N}_\lambda\gamma_\lambda\sigma_{\mu\nu}N-\overline{N}\sigma_{\mu\nu}\gamma_\lambda N_\lambda)+\overline{N}_\lambda\sigma_{\mu\nu}N_\lambda+\overline{N}_\lambda\gamma_\lambda\sigma_{\mu\nu}\gamma_\kappa N_\kappa]^i_{3D+2F},\end{aligned}$$

$$\begin{aligned}J_{\mu5}{}^i(N)=\;&-\tfrac{1}{6}(\overline{N}\gamma_5 N_\mu+\overline{N}_\mu\gamma_5 N+\mathrm{i}\overline{N}_\lambda\gamma_\lambda\gamma_5 N_\mu-\mathrm{i}\overline{N}_\mu\gamma_5\gamma_\lambda N_\lambda)^i_{3D+5F}\\&-\tfrac{1}{6}[2(\overline{N}_\mu\gamma_5 N+\overline{N}\gamma_5 N_\mu)+(\overline{N}_\lambda\gamma_\lambda\gamma_5\gamma_\mu N+\overline{N}\gamma_\mu\gamma_5\gamma_\lambda N_\lambda)\\&\quad-2\mathrm{i}(\overline{N}_\mu\gamma_5\gamma_\lambda N_\lambda-\overline{N}_\lambda\gamma_\lambda\gamma_5 N_\mu)-\mathrm{i}\overline{N}_\lambda\gamma_\mu\gamma_5 N_\lambda+\mathrm{i}\overline{N}_\lambda\gamma_\lambda\gamma_\mu\gamma_5\gamma_\nu N_\nu]^i_{3D+2F},\end{aligned}$$

$$\begin{aligned}J_5{}^i(N)=\;&-\tfrac{1}{6}(2\overline{N}\gamma_5 N+\mathrm{i}\overline{N}_\mu\gamma_\mu\gamma_5 N-\mathrm{i}\overline{N}\gamma_5\gamma_\mu N_\mu)^i_{6D+7F}\\&+\tfrac{1}{6}(\overline{N}_\lambda\gamma_5 N_\lambda+\overline{N}_\lambda\gamma_\lambda\gamma_5\gamma_\mu N_\mu)^i_{3D+2F}.\end{aligned}$$

Note now that already the sacrosanct combinations $3D+2F$ and $3D+5F$ of $\tilde{U}(12)$ have disappeared, being replaced by their various linear combinations.

5. The inhomogeneous Lorentz group and the final expressions for the form factors

A. The work so far has been concerned with purely static considerations. We have computed the expectation values of matrices γ^R between (the homogeneous Lorentz group) multispinors of various symmetry properties. The formalism can have no physical content till these spinors are made to represent physical particles,

† $\Psi_{[\alpha\beta]\gamma}$ has 24 components; equation (4·8) states that the fully antisymmetric part of $\Psi_{[\alpha\beta]\gamma}$ vanishes. We are thus left with 20 independent components if equation (4·8) is satisfied.

i.e. till the formalism assures that they correspond to the representations of the inhomogeneous Poincaré group. One needs therefore, at this stage, some equations of motion which the spinors $\Psi_{\alpha\beta\gamma\ldots}$ must satisfy. This essentially is the point of departure of our work in comparison with other approaches to the problem and was stressed strongly in I.

Among the variety of higher spin equations available, we shall choose the simplest, and the least restrictive (though in many ways the most profound) set of equations: we generalize the Bargmann–Wigner (1948)† approach to the representations of the inhomogeneous Lorentz group. The approach works with the equations

$$\left.\begin{aligned}(\gamma p)_\alpha^{\alpha'}\Psi_{\alpha'\beta\gamma\ldots}(p) &= m\Psi_{\alpha\beta\gamma\ldots}(p),\\ (\gamma p)_\beta^{\beta'}\Psi_{\alpha\beta'\gamma\ldots}(p) &= m\Psi_{\alpha\beta\gamma\ldots}(p).\\ \cdots\cdots\cdots&\cdots\cdots\cdots\end{aligned}\right\} \tag{5·1}$$

These describe particles of (*a*) one definite mass m, (*b*) one definite spin (provided $\Psi_{\alpha\beta\gamma\ldots}$ is a spinor of a definite symmetry type), (*c*) the solutions of the equations pose no problems of negative energies or indefinite metrics and (*d*) the higher spinors transform 'visibly' as direct products of the fundamental quark.

We now examine the implications of applying these equations to the spinors of rank 2 and 3 considered previously:

(i) *Spinor of rank* 2

Write
$$\Phi_\alpha^\beta = [\phi+\gamma_5\phi_5+i\gamma_\mu\gamma_5\phi_{\mu5}+\gamma_\mu\phi_\mu+\tfrac{1}{2}\sigma_{\mu\nu}\phi_{\mu\nu}]_\alpha^\beta. \tag{5·2}$$

From
$$(\gamma p)_\alpha^{\alpha'}\Phi_{\alpha'}^\beta = m\Phi_\alpha^\beta,\quad (\gamma p)_{\beta'}^\beta\Phi_\alpha^{\beta'} = m\Phi_\alpha^\beta$$

we deduce
$$\left.\begin{aligned}\phi &= 0,\\ p_\mu\phi_5 = im\phi_{\mu5},&\quad p_\mu\phi_{\mu5} = -im\phi_5,\\ p_\mu\phi_\nu-p_\nu\phi_\mu = im\phi_{\mu\nu},&\quad p_\nu\phi_{\nu\mu} = -im\phi_\mu.\end{aligned}\right\} \tag{5·3}$$

Thus $(\phi_5, \phi_{\mu5})$ together describe (Kemmer 1939) a 0^- particle and $(\phi_\mu, \phi_{\mu\nu})$ describe a 1^- particle. The relation of the second-rank spinor with what is essentially the Kemmer theory of spin zero and spin one particles was first pointed out by Belinfante (1939; the spinor is called an undor in his terminology). Note that the assumption that the tensor $\Phi_\alpha^\beta(p)$ transforms as a quark–antiquark composite $\psi_\alpha(p)\,\bar{\psi}^\beta(p)$ (with no relative momentum) fixes the parities of the mesons unambiguously.‡

(ii) *Fully symmetric spinor of rank* 3

On account of full symmetry the three equations can be collapsed into a single equation
$$(\gamma p)_\alpha^{\alpha'}\Psi_{\alpha'\beta\gamma} = m\Psi_{\alpha\beta\gamma}.$$

Substituting the expression (4·5) into this equation and contracting it with $(C^{-1}\gamma_\mu)^{\beta\gamma}$, $(C^{-1}\gamma_\mu)^{\alpha\beta}$ and $(C^{-1}\sigma_{\mu\nu})^{\alpha\beta}$ we find

$$(\gamma p-m)\,\psi_\mu = 0, \tag{5·4}$$

$$p_\nu\psi_{\nu\mu} = -im\psi_\mu,\quad p_\mu\psi_\nu-p_\nu\psi_\mu = im\psi_{\mu\nu}. \tag{5·5}$$

† These authors state the equations for the case of full symmetry in $\alpha, \beta, \gamma, \ldots$. The extension to other irreducible spinors of mixed symmetry (defined in the manner of §4) presents no difficulties. It is this type of extension, however, which allows us to describe a spin $\frac{1}{2}$ particle, for example, by a 20-component spinor.

‡ Hereafter we assume that for the fundamental quark $P\psi P^{-1} = \gamma_0\psi$.

It is simple to show, moreover, that $\psi_\mu, \psi_{\mu\nu}$ satisfying (4·6) and (4·7) give a fully symmetrical $\Psi_{\alpha\beta\gamma}$. Thus the system is entirely equivalent to the Rarita–Schwinger formalism (Rarita & Schwinger 1941) for a particle of spin $\frac{3}{2}^+$, except that the components $\psi_{\mu\nu}$ now appear 'on par' with ψ_μ.

(iii) *Mixed spinor of rank* 3

Applied to equation (4·9), the first equation of motion

$$(\gamma p)_\gamma^{\gamma'}\Psi_{\alpha\beta\gamma'} = m\Psi_{\alpha\beta\gamma}$$

gives simply

$$(\gamma p - m)\,\psi_\mu = (\gamma p - m)\,\psi = (\gamma p - m)\,K = 0, \tag{5·6}$$

while the other pair of equations

$$(\gamma p)_\alpha^{\alpha'}\Psi_{\alpha'\beta\gamma} = m\Psi_{\alpha\beta\gamma}$$

give the relations

$$\left.\begin{aligned} K = 0, \quad p_\mu\psi_\nu - p_\nu\psi_\mu = 0, \\ p_\mu\psi = \mathrm{i}m\psi_\mu, \quad p_\mu\psi_\mu = -\mathrm{i}m\psi. \end{aligned}\right\} \tag{5·7}$$

Taken together the system clearly describes a particle of spin $\frac{1}{2}^+$.

B. With the Bargmann–Wigner equations our identification with physical particles of the $\tilde{U}(12)$ quantities Φ_B^A and Ψ_{ABC} is complete.

We here summarize the results:

(i) The regular representation $12 \times 12^*$ decomposes as

$$\Phi_B^A = (\phi^j + \gamma_5\phi_5^j + \mathrm{i}\gamma_\mu\gamma_5\phi_{\mu5}^j + \gamma_\mu\phi_\mu^j + \tfrac{1}{2}\sigma_{\mu\nu}\phi_{\mu\nu}^j)_\beta^\alpha\,(T^j)_q^p.$$

If the mesons are free with mass m, the implication of the equations of motion are

$$p_\mu\phi_5^j = \mathrm{i}m\phi_{\mu5}^j, \quad p_\mu\phi_{\mu5}^j = -\mathrm{i}m\phi_5^j \quad \text{for } (0^-) \text{ particles},$$

$$p_\mu\phi_\nu^j - p_\nu\phi_\mu^j = \mathrm{i}m\phi_{\mu\nu}^j, \quad p_\nu\phi_{\nu\mu}^j = -\mathrm{i}m\phi_\mu^j \quad \text{for } (1^-) \text{ particles},$$

$$\phi^j \equiv 0 \quad \text{for } (0^+).$$

Thus spin zero particles are represented by 5-component entities; spin one by ten components and

$$144 = 143 + 1 = 9 \times 10 + 9 \times 5 + 9.$$

The last nine are the so-called trivial components.

(ii) *Rank* 3

The 364 components of the fully symmetric $\tilde{U}(12)$ tensor Ψ_{ABC} decompose as

$$\underline{364} = (10, 20) + (8, 20) + (1, 4).$$

In detail:

$$\Psi_{\alpha p\beta q\gamma r} = D_{\alpha\beta\gamma,\,pqr} + \epsilon_{pqr}V_{[\alpha\beta\gamma]} + \frac{1}{2\sqrt{6}}\left(\epsilon_{pqs}N_{[\alpha\beta]\gamma,\,r}^{\;s} + \epsilon_{qrs}N_{[\beta\gamma]\alpha,\,p}^{\;s} + \epsilon_{rps}N_{[\gamma\alpha]\beta,\,q}^{\;s}\right),$$

where D is completely symmetric both in its spinor and unitary spin indices, $V_{[\alpha\beta\gamma]}$ is completely antisymmetric and $N_{[\alpha\beta]\gamma,\,r}^{\;s}$ has mixed symmetry in spinor indices and is traceless in unitary spin indices.

The equations of motion ensure that D (with is 20 components) describes a particle of spin $\frac{3}{2}$, N (with its 20 components) a particle of spin $\frac{1}{2}$ and V vanishes identically because of complete antisymmetry. The relative parities of the decimet and the octet are the same (and for the quark ψ transforming as $P\psi P^{-1} = \gamma_0 \psi$ the same as the quark). The detailed consequences of the equations are to allow us to write D and N in the forms:

$$D_{\alpha\beta\gamma,pqr}(p) = (\gamma_\mu C)_{\alpha\beta} D_{\gamma\mu,pqr}(p) + \frac{\mathrm{i}}{2m}(\sigma_{\mu\nu} C)_{\alpha\beta}[p_\mu D_{\gamma\nu,pqr}(p) - p_\nu D_{\gamma\mu,pqr}(p)],$$

with $$(\gamma p - m) D_{\mu,pqr}(p) = 0, \quad \gamma_\mu D_{\mu,pqr}(p) = 0, \tag{5·8}$$

and for the eightfold baryon N:

$$mN_{[\alpha\beta]\gamma,r}^{s}(p) = [(\gamma p + m)\gamma_5 C]_{\alpha\beta} N_{\gamma,r}^{s}(p),$$
$$(\gamma p - m) N_r^s(p) = 0. \tag{5·9}$$

It is important to remember that the $\tilde{U}(12)$ multiplets (which were decomposed relative to $U(3) \times \mathscr{L}_4$ before the Bargmann–Wigner equations were applied) now no longer possess the full symmetry.† The exact $U(6)$ limit, however, can still be recovered, by going for every particle to its rest frame ($p = 0$) and projecting out to the positive energy subspace.‡ The demands of 'relativistic completion' are incompatible with exact symmetry at any (but zero) momenta.

One point is worth emphasizing again at this stage. With the Bargmann–Wigner equations the $\underline{364}$ multiplet of $\tilde{U}(12)$ has exactly the content of the $\underline{56}$ of $U(6)$; likewise for the $\underline{143}$ of $\tilde{U}(12)$ which corresponds with the $\underline{35}$ of $U(6)$; no more and no less. This will happen for all multiplets. In a future paper we shall treat the $\underline{572}$ multiplet and show how its algebra provides the relativistic completion of the $\underline{70}$ of $U(6)$ in a one-one manner.

C. With (5·8) and (5·9) inserted into (3·8) and p and p' denoting the incoming and outgoing baryon momenta, we can now give the final expressions for the form factors. These are

$$J^i = \tfrac{1}{4}m^{-2}P^2[(\bar{N}N)_F + 3\bar{D}_\lambda D_\lambda] + \tfrac{3}{2}m^{-2}q_\lambda \bar{D}_\lambda q_\kappa D_\kappa, \tag{5·10}$$

$$J_\mu^i = \tfrac{1}{2}m^{-1}P_\mu(\bar{N}N)_F + \tfrac{1}{4}m^{-2}(\bar{N}r_\mu N)_{D+\frac{2}{3}F} + m^{-2}(\epsilon_{\mu\nu\kappa\lambda}P_\kappa q_\lambda \bar{D}_\nu N + \text{h.c.})$$
$$+ \tfrac{3}{4}m^{-2}P^2\bar{D}_\lambda\gamma_\mu D_\lambda + \tfrac{3}{2}m^{-2}q_\lambda\bar{D}_\lambda\gamma_\mu q_\kappa D_\kappa, \tag{5·11}$$

$$J_{\mu\nu}^i = \tfrac{1}{4}m^{-2}P^2[(\bar{N}\sigma_{\mu\nu}N)_{D+\frac{2}{3}F} + 3\bar{D}_\lambda\sigma_{\mu\nu}D_\lambda] + \tfrac{3}{2}m^{-2}q_\lambda\bar{D}_\lambda\gamma_\mu q_\kappa D_\kappa$$
$$+ \tfrac{1}{4}\mathrm{i}m^{-2}(P_\mu q_\nu - P_\nu q_\mu)(\bar{N}N)_{\frac{1}{3}F-D} + \mathrm{i}m^{-1}(\epsilon_{\mu\nu\kappa\lambda}P_\lambda\bar{D}_\kappa N + \text{h.c.}), \tag{5·12}$$

$$J_{\mu5}^i = \tfrac{1}{4}m^{-2}P^2(\mathrm{i}\bar{N}\gamma_\mu\gamma_5 N)_{D+\frac{2}{3}F}P + \tfrac{3}{4}\mathrm{i}m^{-2}P^2\bar{D}_\lambda\gamma_\mu\gamma_5 D_\lambda + \tfrac{3}{2}\mathrm{i}m^{-2}q_\lambda\bar{D}_\lambda\gamma_\mu\gamma_5 q_\nu D_\nu$$
$$- \tfrac{1}{2}\mathrm{i}m^{-2}P^2(\bar{D}_\mu N + \text{h.c.}) + \mathrm{i}m^{-2}(p'_\mu p_\lambda\bar{D}_\lambda N + \text{h.c.}), \tag{5·13}$$

† The residual symmetry will always depend on the type of equation applied to the spinor $\Psi_{\alpha\beta\gamma\ldots}$. For example if in place of equations (5·1), one had used equation (8) of paper I $(T\{\gamma P + \mathrm{i}\gamma^5\gamma W\}\psi = M\psi)$ the $W(6)$ symmetry would have formally survived.

‡ We shall here prove that the positive energy projection of all irreducible representations $\Psi_{\alpha\beta\gamma\ldots}(p)$ of $\tilde{U}(12)$ remain irreducible representations of $U(6)$ even after Bargmann–Wigner equations are applied provided $\mathbf{p} = 0$. Here by $U(6)$ we mean the group generated by the matrices T^i, $T^i\boldsymbol{\sigma}$ ($\boldsymbol{\sigma}$ are 2×2 Pauli matrices). The proof is elementary. For $\mathbf{p} = 0$ the positive energy projection of γp equals $m\mathbf{1}$, and $\alpha, \beta, \gamma, \ldots$ run over 1, 2 rather than 1, ..., 4. Thus the symmetry character of any Ψ is unaltered by the application of the set of equations and carries itself from $\tilde{U}(12)$ to $U(6)$.

$$J_5^i = \tfrac{1}{4}m^{-2}P^2(\bar{N}\gamma_5 N)_{D+\frac{2}{3}F} + (3)^{-\frac{1}{2}}m^{-1}(q_\lambda \bar{D}_\lambda N + \text{h.c.}) + \tfrac{3}{4}m^{-2}P^2\bar{D}_\lambda\gamma_5 D_\lambda + \tfrac{3}{2}m^{-2}q_\lambda\bar{D}_\lambda\gamma_5 q_\kappa D_\kappa, \quad (5{\cdot}14)$$

where $P = p+p'$, $q = p-p'$; $r_\mu \equiv \epsilon_{\mu\nu\kappa\lambda}P_\nu q_\kappa \gamma_\lambda\gamma_5$. Note that the coefficient multiplying $P_\mu/2m$ and $r_\mu/4m^2$ are the Sachs (1962) form factors F_C and F_M respectively. To recover the conventional $U(6)$ results for the form factors (Sakita 1964; Bég, Lee & Pais 1964) take $\mathbf{p} = \mathbf{p}' = 0$.

If we further make the identifications

$$i\mu\phi_{\kappa\lambda}^i = p_\kappa\phi_\lambda^i - p_\lambda\phi_\kappa^i,$$

$$i\mu\phi_{\lambda 5}^i = p_\lambda\phi_5^i, \quad \mu = \text{meson mass},$$

for the meson field interaction $\Phi_A^B J_B^A$, we arrive at the following predictions for the pseudoscalar and vector currents:

$$J_5 = \left(1+\frac{2m}{\mu}\right)\frac{P^2}{4m^2}[(\bar{N}\gamma_5 N)_{D+\frac{2}{3}F} + \bar{D}_\lambda\gamma_5 D_\lambda] + \frac{3}{2m^2}\left(1+\frac{2m}{\mu}\right)q_\lambda\bar{D}_\lambda\gamma_5 q_\kappa D_\kappa + \frac{1}{m}\left(1+\frac{2m}{\mu}\right)[q_\lambda\bar{D}_\lambda N + \text{h.c.}], \quad (5{\cdot}15)$$

$$J_\mu = \frac{P_\mu}{2m}\left(1+\frac{q^2}{2\mu m}\right)(\bar{N}N)_F + \left(1+\frac{2m}{\mu}\right)\left(\bar{N}\frac{r_\mu}{4m^2}N\right)_{D+\frac{2}{3}F} + \frac{3P^2}{4m^2}\bar{D}_\lambda\left[\left(1+\frac{2m}{\mu}\right)\gamma_\mu - \frac{P_\mu}{\mu}\right]D_\lambda + \frac{3}{2m^2}q_\lambda\bar{D}_\lambda\left[\left(1+\frac{2m}{\mu}\right)\gamma_\mu - \frac{P_\mu}{\mu}\right]q_\kappa D_\kappa - \frac{1}{m^2}\left(1+\frac{2m}{\mu}\right)[\epsilon_{\mu\nu\kappa\lambda}P_\nu q_\kappa\bar{D}_\lambda N + \text{h.c.}]. \quad (5{\cdot}16)$$

6. Results and conclusions

As shown in § 5C the baryon–meson (N^+NM) vertex contains just one form factor. Its relation with the electric and the magnetic form factors introduced by Sachs (1962) is obvious. Assuming that the photon ($F_{\mu\nu}$) couples with the composite structure of the baryon† through an effective gauge-invariant 'interaction' $f(q^2)\,F_{\mu\nu}M_{\mu\nu}^u$, where $M_{\mu\nu}^u$ is the U-scalar meson combination ($M^u = M^3 + (3)^{-\frac{1}{2}}M^8$), one would get for the electromagnetic form factors the expressions

$$J_\mu^C = \frac{P_\mu}{2m}\left(1+\frac{q^2}{2\mu m}\right)(\bar{N}N)_F\,F(q^2), \quad (6{\cdot}1)$$

$$J_\mu^M = \left(1+\frac{2m}{\mu}\right)\left(\bar{N}\frac{r_\mu}{4m^2}N\right)_{D+\frac{2}{3}F}, \quad (6{\cdot}2)$$

† To write a minimal photon interaction ($\gamma p \to \gamma p - e\gamma A$) seems highly inappropriate for a structure as complex as a baryon. To take an analogy, no one would normally contemplate using the 'minimal ansatz' for a helium 3 or tritium nucleus.

where $F(q^2) \propto f(q^2)/(q^2-\mu^2)$. As stated in the introduction we note that

$$\text{(i)} \qquad \left.\begin{aligned} \mu_p &= (1+2m/\mu) \quad \text{(Bohr magnetons)}, \\ \mu_n &= -\tfrac{2}{3}(1+2m/\mu). \end{aligned}\right\} \tag{6·3}$$

$$\text{(ii)} \qquad \frac{J^C}{J^M} = \frac{1+q^2/2\mu m}{1+2m/\mu}. \tag{6·4}$$

For small q^2 essentially there is therefore just one electromagnetic (Sachs) form factor (Barnes 1962).

(iii) The U_6 limit of $\tilde{U}(12)$ taken at the stage of equations (5·10) to (5·14) results in losing the 'anomalous' magnetic moment of the proton. Stated otherwise, the extension of the ($U(6)$) 35-fold of mesons to a ($\tilde{U}(12)$) 143-fold is essential to obtain the results (6·3).

(iv) For weak interactions (5·11) and (5·13) give the vector and axial-vector form factors. These reproduce the well-known $U(6)$ result $g_A/g_v = -5/3$ at zero momentum transfer (Kawarabayashi 1964; Rosen & Parksava 1964; Bég & Pais 1965; Altarelli, Buccella & Gatto 1964; Babu 1964).

(v) *Meson–meson vertex*

In general $\underline{143} \times \underline{143}$ contains $\underline{143}$ twice but for the special case of identical $\underline{143}$'s, there is just one coupling of the type

$$\mathscr{L} = \Phi_A^B \Phi_B^C \Phi_C^A.$$

Written out in terms of ϕ_5^i, ϕ_μ^i, etc., introduced in equation (5·2), $\mathscr{L}$ equals

$$f^{ijk}[3\phi_\mu^i \phi_\nu^j \phi_{\mu\nu}^k + 3\phi_{\mu5}^i \phi_{\mu\nu}^j \phi_{\nu5}^k + 6\phi_5^i \phi_\mu^j \phi_{\mu5}^k + \phi_{\mu\nu}^i \phi_{\nu\lambda}^j \phi_{\lambda\mu}^k]$$
$$+\tfrac{3}{2} d^{ijk} \epsilon_{\kappa\lambda\mu\nu} [\phi_k^i \phi_{\lambda\mu}^j \phi_{\nu5}^k + \tfrac{1}{2}\phi_{\kappa\lambda}^i \phi_{\mu\nu}^j \phi_5^k + (\text{terms in } \phi^i)]. \tag{6·5}$$

Introducing equations (5·3), we get for the effective vector current

$$J_\mu^i(p,p') = f^{ijk}\Big[-P_\mu\Big\{\Big(1+\frac{q^2}{2\mu^2}\Big)\phi_\nu^j(p')\,\phi_\nu^k(-p) - \Big(1-\frac{q^2}{\mu^2}\Big)\phi_5^j(p')\,\phi_5^k(-p)\Big\}$$
$$+3\{q_\nu \phi_\nu^j(p')\,\phi_\mu^k(-p) + q_\nu \phi_\nu^j(-p)\,\phi_\mu^k(p')\} - P_\mu q_\nu \phi_\nu^j(p')\, q_\lambda \phi_\lambda^k(-p)/\mu^2\Big]$$
$$-\frac{l}{\mu} d^{ijk} \epsilon_{\kappa\lambda\mu\nu} P_\lambda q_\nu \{\phi_k^j(p')\,\phi_5^k(-p) + \phi_5^j(p')\,\phi_\kappa^k(-p)\}. \tag{6·6}$$

Introducing the gauge-invariant electromagnetic coupling $f(q^2)\,F_{\mu\nu}\phi_{\mu\nu}$, we obtain a total magnetic moment of 3 Bohr magnetons and a quadrupole moment of -4 (in units of e/μ^2) for positively charged 1^- particles.

(vi) *The outlook*

With the effective baryon–meson and meson–meson vertices available it is a trivial step to write pole approximations for the strong interaction four-particle process. With this approximation as the starting-point all S-matrix techniques (like Mandelstam representation, Reggeization, analytic continuation now both in angular momentum and unitary spin) are available for determining the complete ($\tilde{U}(12)$) relativistic S-matrix theory. This is so because as a rule all that the S-matrix theory requires are 'Born approximations' as the 'input'. The higher 'fundamental' 4-particle vertices involving 4-particles (like $\bar{N}NMM$) which are analogous to 4-field Lagrangians (and knowledge of which is necessary for peripheral interactions) are no harder to write down, using the methods and techniques of this paper. Every one of these 'fundamental' interactions will now be momentum dependent—this dependence coming in a determinate manner from $\tilde{U}(12)$ symmetry. The conventional field theory Lagrangians are but local approximations of these 'fundamental' interactions.

Note that one of the ways in which the inhomogeneous Lorentz group breaks the full $\tilde{U}(12)$ symmetry is to force certain components of Φ_B^A to become 'redundant'. In a future paper we wish to exploit these 'redundant' components and use them as symmetry-breaking 'spurions'.

It would seem that the $\tilde{U}(12)$ ideas fulfil completely the dream of the $U(6)$ theorist, i.e. to write S-matrix elements whose momentum dependence is dictated by the internal symmetry group of SU_3.

Our thanks are due to Dr M. A. Rashid for helping with the content of the $\tilde{U}(12)$ representations. We are grateful to Drs J. Charap, P. T. Matthews, and J. C. Ward for numerous fruitful discussions.

Appendix

Our convention for the unitary spin matrices is the following: $T^i = \frac{1}{2}\lambda^i$, with the λ^i defined by Gell-Mann (1962). Thus $\mathrm{Tr}(T^iT^j) = \frac{1}{2}\delta^{ij}, f^{0jk} = 0$, and $d^{0jk} = \delta^{jk}(2/3)^{\frac{1}{2}}$.

$$[T^i, T^j] = \mathrm{i}f^{ijk}T^k, \quad \{T^i, T^j\} = d^{ijk}T^k.$$

From the fundamental representation, $F^{Ri} \equiv \gamma^R T^i$ we deduce the following commutators:

$$[F^i, F^j] = \mathrm{i}f^{ijk}F^k,$$

$$[F^i, F_5^j] = \mathrm{i}f^{ijk}F_5^k,$$

$$[F_5^i, F_5^j] = -\mathrm{i}f^{ijk}F^k,$$

$$[F^i, F_{\mu\nu}^j] = \mathrm{i}f^{ijk}F_{\mu\nu}^k,$$

$$[F_5^i, F_{\mu\nu}^j] = \tfrac{1}{2}\mathrm{i}f^{ijk}\epsilon_{\mu\nu\kappa\lambda}F_{\kappa\lambda}^k,$$

$$[F_{\kappa\lambda}^i, F_{\mu\nu}^j] = \mathrm{i}d^{ijk}(g_{\kappa\nu}F_{\lambda\mu}^k + g_{\lambda\mu}F_{\kappa\nu}^k - g_{\kappa\mu}F_{\lambda\nu}^k - g_{\lambda\nu}F_{\kappa\nu}^k)$$
$$+\mathrm{i}f^{ijk}\{(g_{\kappa\mu}g_{\lambda\nu} - g_{\lambda\mu}g_{\kappa\nu})F^k - \epsilon_{\kappa\lambda\mu\nu}F_5^k),$$

$$[F_\mu^i, F_\nu^j] = \mathrm{i}f^{ijk}g_{\mu\nu}F^k - \mathrm{i}d^{ijk}F_{\mu\nu}^k,$$

$$[F_\mu^i, F_{\nu 5}^j] = \mathrm{i}d^{ijk}g_{\mu\nu}F_5^k + \tfrac{1}{2}\mathrm{i}f^{ijk}\epsilon_{\mu\nu\kappa\lambda}F_{\kappa\lambda}^k,$$

$$[F_{\mu 5}^i, F_{\nu 5}^j] = -\mathrm{i}f^{ijk}g_{\mu\nu}F^k - \mathrm{i}d^{ijk}F_{\mu\nu}^k,$$

158 A. Salam, R. Delbourgo and J. Strathdee

and
$$[F^i, F^j_\mu] = \mathrm{i}f^{ijk}F^k_\mu,$$
$$[F^i, F^j_{\mu 5}] = \mathrm{i}f^{ijk}F^k_{\mu 5},$$
$$[F^i_5, F^j_\mu] = \mathrm{i}d^{ijk}F^k_{\mu 5},$$
$$[F^i_5, F^j_{\mu 5}] = \mathrm{i}d^{ijk}F^k_\mu,$$
$$[F^i_\lambda, F^j_{\mu\nu}] = \mathrm{i}d^{ijk}(g_{\lambda\mu}F^k_\nu - g_{\lambda\nu}F^k_\mu) - \mathrm{i}f^{ijk}\epsilon_{\lambda\mu\nu\kappa}F^k_{\kappa 5},$$
$$[F^i_{\lambda 5}, F^j_{\mu\nu}] = \mathrm{i}d^{ijk}(g_{\lambda\mu}F^k_{\nu 5} - g_{\lambda\nu}F^k_{\mu 5}) + \mathrm{i}f^{ijk}\epsilon_{\lambda\mu\nu\kappa}F^k_\kappa.$$

References

References A

Salam, A. 1965 *Phys. Lett.* (in the Press).
Delbourgo, R., Salam, A. & Strathdee, J. 1964*a* ICTP 64/7; 1964*b* ICTP 64/11 (referred to as I) (to be published).
Marshak, R. E. & Okubo, S. 1964 *Phys. Rev. Lett.* **13**, 818, and preprint (Rochester).
Feynmann, R. P., Gell-Mann, M. & Zweig, G. 1964 *Phys. Rev. Lett.* **13**, 678.
Bardacki, K., Cornwall, J. M., Freund, P. G. O. & Lee, B. W. 1964, 1965 *Phys. Rev. Lett.* **13**, 698; **14**, 48.
Fulton, T. & Wess, J. 1964 Preprints (Vienna).
Agasshi, J. J. & Roman, P. 1964 Preprint (Boston).
Bég, M. A. B. & Pais, A. 1965 *Phys. Rev.* (to be published).
Mahanthrappa, K. T. & Sudarshan, E. C. G. 1965 Preprint (Syracuse).
Wyld, H. W. 1965 Preprint (Illinois).

Other references

Altarelli, G., Buccella, F. & Gatto, R. 1964 Preprints (Florence).
Babu, P. 1964 Preprint (Bombay).
Bargmann, V. & Wigner, E. 1948 *Proc. Nat. Acad. Sci., Wash.*, **34**, 211.
Barnes, K. J. 1962 *Phys. Rev. Lett.* **1**, 166.
Bég, M. A. B., Lee, B. W. & Pais, A. 1964 *Phys. Rev. Lett.* **13**, 514.
Bég, M. A. B. & Pais, A. 1965 *Phys. Rev. Lett.* **14**, 51.
Belinfante, F. J. 1939 *Physica*, **6**, 870.
Gell-Mann, M. 1962 *Phys. Rev.* **125**, 1067.
Gell-Mann, M. 1964 *Phys. Lett.* **8**, 216.
Jauch, J. M. & Rohrlich, F. 1955 *The theory of photons and electrons.* New York: Addison-Wesley.
Kawarabayashi, K. 1964 ICTP 64/5.
Kemmer, N. 1939 *Proc. Roy. Soc.* A, **173**, 91.
Rarita, W. & Schwinger, J. 1941 *Phys. Rev.* **60**, 61.
Rosen, S. P. & Paksava, S. 1964 *Phys. Rev. Lett.* **13**, 773.
Sachs, R. G. 1962 *Phys. Rev.* **126**, 2256.
Sakita, B. 1964 *Phys. Rev. Lett.* **13**, 643.
Zweig, G. 1964 C.E.R.N. Rep. (unpublished).

Volume 15, number 4 PHYSICS LETTERS 15 April 1965

HIGHER MESON RESONANCES AND THE $\underline{4212}^{+}$ MULTIPLET OF $S\tilde{U}(12)$

R. DELBOURGO
International Centre for Theoretical Physics, Trieste

Received 8 March 1965

In its lowest meson multiplet of $\underline{35}$, the SU(6) symmetry scheme accommodates almost every one of the known 0^- and 1^- particles. The counterpart representation in $S\tilde{U}(12)$ is $\underline{143}^-$ leaving room for an extra 0^- singlet, which can be identified as the $\overline{X}^0$ 960 MeV resonance. Now the next meson multiplet occurring in $S\tilde{U}(12)$ in the $\underline{4212}^+$ and with the increasing number of recently reported parity (+) resonances * is our purpose in this note to attempt to fill up this representation and to determine the implications of $S\tilde{U}(12)$ symmetry [2] on the various decay modes. A very similar representation to the $\underline{4212}^+$ is the $\underline{5940}^+$ but just for the sake of simplicity it will be neglected - as it is, there remain a host of gaps in the $\underline{4212}^+$ even after we have fitted in all of the fairly well-established particles. Thus the $\underline{4212}^+$ comprises, in addition to the $\underline{189}$ of SU(6) another $\underline{35}+\underline{1}$ states **. Its physical content and our assignments are given in the accompanying table.

Following the notation of ref. 2 we briefly recapitulate for future comparison the results concerning the vertex function of three $\underline{143}^-$ multiplets. We let m and μ stand for the masses of the vector and scalar mesons $^+$; then the $\tilde{U}(12)$ invariant interaction is given by $^{++}$

$$\mathscr{L}=\frac{g^2}{16\mu^2m^2}\Big[im^2(m+2\mu)f^{ijk}(r-q)_\lambda\varphi^i_\lambda(p)\varphi^j_5(q)\varphi^k_5(r)+\text{cyclic}+2\mu(2m+\mu)d^{ijk}\epsilon_{\lambda\mu\kappa\nu}p_\kappa q_\nu\varphi^i_\lambda(p)\varphi^j_\mu(q)\varphi^k_5(r)+$$
$$+\text{cyclic}+i\mu^2f^{ijk}\varphi^i_\lambda(p)\varphi^j_\mu(q)\varphi^k_\nu(r)\{3m[(q-r)_\lambda g_{\mu\nu}+(r-p)_\mu q_{\nu\lambda}+(p-q)_\nu g_{\lambda\mu}]+4q_\lambda r_\mu p_\nu/m\}\Big] \tag{1}$$

and from the ρ decay width we deduce $^{+++}$

$$g_{\rho\pi\pi}=g(m+2\mu)/16\mu^2\approx 2.8\ .$$

* We base our work on [1].

** The reason why *more* physical states are accommodated is that $S\tilde{U}(12)$ tracelessness does not imply SU(6) tracelessness. This question will be discussed more fully in a future paper.

$^+$ An ad hoc mass splitting between 0^- and 1^- states has been introduced since they remain disjoint even after application of the equations of motion. This was not done in ref. 2. We make similar mass splitting later for the SU(3) multiplets within the $\underline{4212}^+$.

$^{++}$ As noted earlier (ref. 2a) the $1^-0^-0^-$ and $1^-1^-1^-$ couplings are pure f whereas the $1^-0^-1^-$ coupling is pure d. Note also that $\varphi\rightarrow\rho\pi$ is forbidden by Lipkin's assignment of the physical φ. Eq. (1) also gives

$$g_{\omega\rho\pi}/g_{\rho\pi\pi}=2\mu(2m+\mu)/m^2(m+2\mu)\approx 0.51/\mu$$

if we use mean masses of multiplets

$$m\approx 900\ \text{MeV}\ ,\qquad \mu\approx 400\ \text{MeV}\ .$$

This is in close agreement with the Gell-Mann, Sharp and Wagner estimate of $0.47/\mu_\pi$ from ω decay. (An error in the Lagrangian of ref. 2b, eq. (3.1), has been corrected to derive this result; a factor of 2 should multiply the term $\varphi^i_\kappa\varphi^i_{\lambda\mu}\varphi^k_{\nu5}$.)

$^{+++}$ If the $\rho\pi\pi$ coupling is evaluated at *zero* momentum transfer it gets depressed by a factor $2\mu(m+2\mu)^{-1}$ down to about 1.35. (This can be seen from eq. (6.6), ref. 2a. There is actually an error in the expression for the $\pi\pi$ contribution to the vector current. In place of the factor $(1-q^2/\mu^2)$ one should read $(1+q^2/2\mu^2)\rightarrow(1+q^2/2\mu m)$ with our present mass splitting of 0^- and 1^- nonets.) In ref. 2a we found $g_{\rho NN}=3\mu g_{\pi NN}/5(\mu+2m_N)\approx 1.3$ giving remarkably close agreement from the universality point of view.

Volume 15, number 4 PHYSICS LETTERS 15 April 1965

Table 1
Assignments of the parity (+) resonances within the $\underline{4212}^+$ multiplet of $S\tilde{U}(12)$. The data are taken from the tables of Rosenfeld et al. (Question marks denote less secure assignments or not well-established experimental values.)

0^+				1^+				2^+			
SU(3) Repn.	Particle	Mass (MeV)	Width (MeV)	SU(3) Repn.	Particle	Mass (MeV)	Width (MeV)	SU(3) Repn.	Particle	Mass (MeV)	Width (MeV)
$\underline{1}$	ABC ?	315	15 ?	$\underline{1}$	X_0 ?	960	10 ?	$\underline{1}$	K_1K_1 ?	1020	?
$\underline{1}$	σ	400	80 ?	$\underline{8}$	B	1215	120	$\underline{8}$	f_0	1250	100
$\underline{8}$	κ	725	10 ?		K_ρ	1215	60		A2 ?	1310	80
	φ'	520	70		KK* ?	1410	60		-	-	-
	-	-	-	$\underline{8}$	A1 ?	1090	125				
$\underline{8}$	-	-	-		-	-	-				
					-	-	-				
$\underline{27}$	-	-	-	$\underline{8}$	-	-	-				
				$\underline{10}$	-	-	-				
				$\underline{10}$*	-	-	-				

The decomposition of the $\underline{4212}^+$ (represented by $\Phi^{[AB]}_{[CD]}$) has already been given* . The SU(12) invariant interaction

$$\mathscr{L} = g\Phi^{[AB]}_{[CD]}(p)\,\Phi^{C}_{A}(q)\,\Phi^{D}_{B}(r) \tag{2}$$

reduces after some calculation to

$$\begin{aligned}
g^{-1}\mathscr{L} = &-(2\mu+M)\varphi^{[rs]}_{[pq]}(p)\,\varphi^{p}_{5r}(q)\,\varphi^{q}_{5s}(r)/12\mu + (M+m+\mu)[(M+m-\mu)\epsilon_{\kappa\lambda\nu\mu}(\mu p_\mu - Mq_\mu) - 2\epsilon_{\kappa\lambda\tau\mu}p_\mu q_\tau p_\nu]\times \\
&\times A_{\kappa\lambda}{}^{[rs]}_{[pq]}(p)\varphi^{p}_{5r}(q)\,\varphi^{q}_{\nu s}(r)/4\mu mM^2 + (M+2m)^2[-g_{\mu\nu}M^2 + 2p_\mu p_\nu]\varphi^{[rs]}_{[pq]}(p)\,\varphi^{p}_{\mu r}(q)\,\varphi^{q}_{\nu s}(r)/24m^2M^2 + \\
&+[2M^2(M+2m)^2 g_{\mu\kappa}g_{\nu\lambda} - p_\mu p_\nu(q-r)_\nu(q-r)_\lambda + 4M(M+2m)g_{\nu\lambda}p_\mu(q-r)_\kappa]S_{\kappa\lambda}{}^{[rs]}_{[pq]}(p)\varphi^{p}_{\mu r}(q)\varphi^{q}_{\nu s}(r)/8m^2M^2 + \\
&+(M+2\mu)^2\,\eta^{\{rs\}}_{\{pq\}}(p)\,\varphi^{p}_{5r}(q)\,\varphi^{q}_{5s}(r)/16\mu^2 + (M+2m)^2[-g_{\mu\nu}M^2+2p_\mu p_\nu]\,\eta^{\{rs\}}_{\{pq\}}(p)\varphi^{p}_{\mu r}(q)\varphi^{q}_{\nu s}(r)/16m^2M^2 + \\
&+[(M+\mu)^2 - m^2][-g_{\mu\lambda}\{(M+m)^2-\mu^2\} + p_\mu(r-q)_\lambda]U_\lambda{}^{\{rs\}}_{[pq]}(p)\,\varphi^{p}_{\mu r}(q)\,\varphi^{q}_{5s}(r)/8\sqrt{2}\,\mu mM^2 + \\
&+(M+2m)\times[M\epsilon_{\lambda\nu\kappa\mu}(mp_\kappa - Mr_\kappa) - 2\epsilon_{\lambda\nu\kappa\tau}p_\kappa r_\tau p_\nu]U_\lambda{}^{\{rs\}}_{[pq]}(p)\,\varphi^{p}_{\mu r}(q)\,\varphi^{q}_{\nu s}(r)/4\sqrt{2}\,m^2M^2 + \\
&\qquad\text{similar terms with } q \rightleftharpoons r, \qquad U^{\{\,\}}_{[\,]} \to U^{[\,]}_{\{\,\}}
\end{aligned} \tag{3}$$

We now turn to the decay properties predicted by (3):

2^+ Resonances

$S\tilde{U}(12)$ inhibits all but the interaction of the 2^+ with two 1^- mesons, a decay which is not physically realizable; consequently for the other available decay modes $2^+ \to 0^- + 0^-$, $2^+ \to 0^- + 1^-$ we must establish the *smallness* of the coupling constant (in relation to $g_{\rho\pi\pi}$ say) to indicate little symmetry breaking. The large decay widths (~ 100 MeV) would appear to belie this statement, but if due account is taken of the large phase space available and the *D-wave* nature of the decay into pseudoscalar mesons

* See especially footnote 6, ref. 2b.

Volume 15, number 4 PHYSICS LETTERS 15 April 1965

we find that $g_{f\pi\pi}$, $g_{f\omega\pi}$, $g_{A2\kappa\kappa}$, $g_{A2\pi\rho}$ are indeed small*. Thus from phenomenological couplings**

$$\mathscr{L}_{\binom{f\pi\pi}{A\kappa\kappa}} = g_{\binom{f\pi\pi}{A\kappa\kappa}} (q-r)_{\kappa} (q-r)_{\lambda} S_{\kappa\lambda}(p)\, \varphi_5(q)\, \varphi_5(r)/\mu$$

$$\mathscr{L}_{\binom{f\pi\omega}{A\pi\rho}} = g_{\binom{f\pi\omega}{A\pi\rho}} \epsilon_{\tau\mu\nu\kappa}(q-r)_{\tau} p_{\mu} \varphi_{\nu}(r)\, S_{\kappa\lambda}(p)(q-r)_{\lambda} \varphi_5(q)/\mu M \tag{4}$$

we obtain the decay widths

$$\Gamma_{\binom{f\pi\pi}{A\kappa\kappa}} = 8 g^2_{\binom{f\pi\pi}{A\kappa\kappa}} \left| q_{\binom{f\pi\pi}{A\kappa\kappa}} \right|^5 / 3\pi\mu^2 M^2_{\binom{f}{A}}$$

$$\Gamma_{\binom{f\omega\pi}{A\rho\pi}} = g^2_{\binom{f\omega\pi}{A\rho\pi}} \left| q_{\binom{f\omega\pi}{A\rho\pi}} \right|^5 / \pi\mu^2 M^2_{\binom{f}{A}} \tag{5}$$

to be compared with

$$\Gamma_{\rho\pi\pi} = g^2_{\rho\pi\pi} \left| q_{\rho\pi\pi} \right|^3 / 2\pi M^2_{\rho} \ . \tag{6}$$

Hence

$$g_{\rho\pi\pi} : g_{f\pi\pi} : g_{f\omega\pi} : g_{A\kappa\kappa} : g_{A\rho\pi} \approx 1 : \tfrac{1}{20} : \tfrac{1}{15} : \tfrac{1}{20} : \tfrac{1}{8} \ ,$$

consistent with approximate $S\tilde{U}(12)$ invariance.

<u>1^+ Resonances</u>

Placing the B, $K^*\bar{K}$ and $K\rho$ resonances in a U-octet of eq. (3) (in the A-octet the decay $B \to \omega + \pi$ would be forbidden by isospin) we obtain the decay widths

$$\Gamma_{\binom{B\omega\pi}{K^*K}} = 3 g^2_{\binom{B\omega\pi}{K^*K}} \left| q_{\binom{B\omega\pi}{K^*K}} \right| 8\pi M^2_{\binom{B}{K^*K}} \ . \tag{7}$$

From experimental values we get $g_{B\omega\pi}/g_{K^*K} \approx 0.75$ to be compared with the theoretical estimate of $\sqrt{6}/3 \approx 0.82$. No numerical statement is possible for the $K\rho$ resonance owing to the Q-value of -30 MeV! If we place the A1 into the second U-octet, the decay width is the same as in (7) so the experimental data give $g_{A1\rho\pi}/g_{B\omega\pi} \approx 0.6$, which is not in violent disagreement with the theoretical prediction of unity***.

<u>0^+ Resonances</u>

For simplicity we must assume that the singlet states are filled first; then if we place the ABC in the φ-singlet and σ in the η-singlet, $S\tilde{U}(12)$ gives the theoretical ratios

$$g_{\sigma\pi\pi}/g_{ABC\pi\pi} = 3(M_{\sigma} + 2\mu)^2 / 8\sqrt{2}\ \mu(M_{ABC} + 2\mu) \approx 0.9$$

$$g_{\sigma\pi\pi}/g_{B\omega\pi} = \frac{(M_{\sigma} + 2\mu)^2}{32\mu^2\sqrt{6}} \frac{\{\mathscr{L}(M+\mu)^2 - m^2\}\{(M+m)^2 - \mu^2\}}{24\sqrt{2}\ \mu m M^2} \approx 0.3 \ .$$

The experimental ratios of 1.6 and 0.7 respectively are at least of the same order of magnitude. As for the other SU(3) representations, it is conceivable that the ephemeral resonances φ' and κ fit partly into an octet, but in view of their rather dubious nature we will not attempt to extract any quantitative information about them.

* As a general rule we use mean masses of multiplets when making a theoretical comparison of coupling constants. However when calculating coupling constants from the experimental decay widths we use *physical* masses to obtain a realistic appraisal of phase space effects.

** To define dimensionless coupling constants we use for our mass scale the pseudoscalar meson; this is a questionable point about our procedure.

*** Placing the A1 into an A-octet would lead to a serious discrepancy for the coupling constant ratios.

Volume 15, number 4 PHYSICS LETTERS 15 April 1965

In summary these results never show big disagreement with the demands of $S\tilde{U}(12)$ invariance, and in the case of the better-established particles are sometimes numerically quite good. Nevertheless it would be premature to scrutinize or read too much into them in view of the tentative assignments of the few available (and still not experimentally well analyzed) resonances, mixing problems and parallel occurrence of $\underline{5940}^+$ states only serving to muddy the issue. However, the large multiplicity of these higher $S\tilde{U}(12)$ meson resonances seem to call for more exhaustive experimental research.

I am grateful to Professor Abdus Salam and Dr. M. A. Rashid and J. Strathdee for helpful comments. My thanks are also due to the IAEA for their hospitality at the International Centre for Theoretical Physics, Trieste.

References

1. A.H.Rosenfeld, A.Barbaro-Galtieri, W.H.Barkas, P.L.Bastien, J.Kriz and M.Ross, Rev.Mod.Phys. 35 (1964) 977.
2. R.Delbourgo, A.Salam and J.Strathdee, Proc.Roy.Soc. 284 (1965) 146.
   R.Delbourgo, M.A.Rashid, A.Salam and J.Strathdee, Proc.Roy.Soc., to be published.
   A.Salam, J.Strathdee, J.C.Charap and P.T.Matthews, Physics Letters 15 (1965) 184.
   See also B.Sakita and K.C.Wali, Phys.Rev.Letters, to be published.

* * * * *

PHYSICAL REVIEW VOLUME 186, NUMBER 5 25 OCTOBER 1969

Reggeization in Supermultiplet Theories

R. DELBOURGO* AND ABDUS SALAM†
International Atomic Energy Agency, International Centre for Theoretical Physics, Miramare-Trieste, Italy
(Received 27 January 1969)

Methods for incorporating higher-symmetry ideas into the phenomenology of Regge poles are reviewed. These methods, which were developed originally for treating models of the "quark-excitation" type, are extended to cover also the "orbital excitation" models with particular reference to the oscillator model.

I. INTRODUCTION

ONE of the rather surprising features of the present scene in particle physics is the increasing evidence of the relevance of $SU(6)$-like symmetry ideas[1] in describing hadron spectra on the one hand and, on the other hand, the comparative disregard of such symmetries and the strong correlations they may be expected to provide among residue functions—even as a crude guiding principle—by those working in Regge phenomenology.[2]

One possible reason for this disregard could be that detailed experimental confirmation (from decay data) of the validity of higher symmetries for coupling parameters and residues exists for the low-lying $SU(6)$ states only. A second and more practical reason is perhaps the nonavailablity of a simple, consistent, and detailed formalism embodying the marriage of Regge ideas to higher symmetries.[2] A beginning was made in this direction in a series of recent papers. Unfortunately, although the discussion was general, the details of the formalism were given for one specific model of Reggeized higher symmetries, specifically, the model based on a quark-excitation picture for higher resonances, where along a trajectory the total *quark* content for physical states (half the number of quarks plus antiquarks) increases by integer steps in the form N, $N+1$, $N+2$, $\cdots$, and it is the quark number N which is Reggeized. It is our purpose in this paper to consider in detail the rival models based on an orbital excitation picture of two- and three-quark composites [group-theoretically, models of the type $SU(6)\otimes O_L(3)$, with a Reggeization of the orbital quantum number L].

As is well known, the quark-excitation models predict "exotic" resonances with high values of strangeness and isotopic charges, while in the orbital models only the **1**'s, **8**'s, and **10**'s of $SU(3)$ make their appearance. The physical hadron spectrum may, in the end, prove to possess features of both models; the present evidence, however, seems to favor orbital models of lesser or greater complexity with the known baryon resonances apparently grouping themselves in multiplets of $(\mathbf{56},0^+)$, $(\mathbf{56},2^+)$, $(\mathbf{70},1^-)$, $\cdots$ and meson resonances in $(\mathbf{35},0^+)$,

---

* Present address: Imperial College, London, England.
† On leave of absence from Imperial College, London, England.
[1] For a recent review, see H. Harari, in *Proceedings of the Fourteenth International Conference on High-Energy Physics, Vienna, 1968* (CERN, Geneva, 1968), p. 195.
[2] The only systematic attempts in this direction that we know of are by P. G. O. Freund, Phys. Rev. **157**, 1412 (1967); R. Arnold, *ibid.* **162**, 1334 (1967); Y. Ne'eman, L. Horwitz, and N. Cabibbo, Phys. Letters **22**, 336 (1967).

(**35**,1⁻), ··· of $SU(6)\otimes O_L(3)$. We wish to stress that the great virtue of using symmetry ideas is that we do not require the physical existence of quarks; we obtain quark-model results, with their correct *relativistic* kinematics, without actually believing that such objects exist.

The plan of the paper is as follows. In Sec. II, we review the basis of the Reggeization procedures, given any rest symmetry group for particle multiplets. The most important concept here is the notion of *generalized helicity*. This is introduced and we then define the appropriate rotation functions needed for Reggeization. These functions are a generalization of the familiar $d_{\lambda\lambda'}{}^J(\theta)$ rotation functions of the group $O_J(3)$. In Sec. III, the equivalent M-function formalism for writing amplitudes using multispinors is introduced in terms of which actual calculations are made. We wish to stress with the greatest possible emphasis that this multispinor formalism, using Bargmann-Wigner equations to describe supermultiplets, is not just a luxury. Insofar as it embodies the correct kinematics[3] and (most important) provides a natural formalism into which *symmetry-breaking effects* (due to mass splittings within a supermultiplet) can be incorporated, the multispinor M-function formalism for scattering amplitudes is an *important ingredient* of the Reggeization scheme. One wishes one could stress this enough so that the unfortunate prejudice against learning what is basically a very simple and yet extraordinarily powerful technique could be overcome. Section IV deals with the detailed description of the oscillator excitation model and its application to Reggeized meson-baryon (MB) and baryon-baryon (BB) scattering. In Sec. V, we discuss briefly the kinematic-singularity problem and the question of Tollerization versus Reggeization of physical amplitudes as a means to cope with such singularities. Section VI discusses the situation where the singularities are removed by Gribov doubling of the meson multiplets. In a separate note, the formalism of this paper is applied to the problem of charge-exchange meson-baryon scattering to see if the Reggeization of $SU(6)\otimes O(3)$-like theories with the drastic decrease in the number of residues they provide gives a reasonable fit to the data.[4] There we show that, in fact, one can correlate all known processes with a one-parameter formula.

II. REGGEIZATION SCHEME FOR HIGHER SYMMETRIES: GENERAL CONSIDERATIONS

As stated in the Introduction, there are two distinct types of models of Reggeized higher symmetries.

A. Orbital Excitation Models

Here, higher symmetries combine *intrinsic* spin and unitary spin as in the original $SU_{F,S}(6)$ proposals emanating from Wigner's $SU(4)$ (F is the unitary spin index and S is intrinsic spin), while Reggeization proceeds for orbital momentum **L**. (Here $\mathbf{J}=\mathbf{L}+\mathbf{S}$.) The models we shall consider correspond to the following symmetry groups:

A. $SU(6)\otimes O_L(3)$,

A(i). $U(6)\otimes U(6)|_{F,S}\otimes O_N(4)$,

A(ii). $U(6)\otimes U(6)|_{F,S}\otimes SU_N(3)$.

(i) *O(4) orbital models.* The $U(6)\times U(6)$ intrinsic-spin–unitary-spin symmetry treats quark and antiquark spins as distinct and independent so that the intrinsic spin group contained[5] in $U(6)\otimes U(6)$ is the subgroup $SU_{Sq}(2)\otimes SU_{S\bar{q}}(2)$. As is well known, this group has the same structure as $O_S(4)$. From this point of view, a natural, though by no means essential, generalization of *orbital* angular momentum also is to consider four-dimensional orbital momenta, thereby enlarging $O_L(3)$ to $O_N(4)$, where N stands for the quantum number appearing in the eigenvalue $N(N+2)$ of the $O(4)$ Casimir operator.[6,7]

(ii) *U(3) orbital model.* A remarkable feature of the baryon spectrum known at present appears to be that all known particles belong to (**56**,L^{even}) and (**70**,L^{odd}). The fact that there appear to be two (**56**,0⁺) multiplets and no (**56**,1⁺) bears out the need for a radial quantum number N for classification purposes. A suggestion has been made that possibly the extra orbital degrees of freedom are associated with a harmonic-oscillator-like potential and the Reggeized quantum number is one of the Casimir operators of the three-dimensional harmonic-oscillator group $SU(3)$ rather than $O(4)$. This oscillator group $SU(3)$ is the same group familiar from

---

[3] The point to be reiterated is that in theories of $SU(6)$ variety the "Clebsch-Gordan" coupling coefficients contain mass-dependent kinematic factors. If the symmetry were exact, it would not matter how these things were computed. The symmetry, however, is not exact and the multispinor formalism has the advantage of explicitly stating this mass (and the kinematic) dependence. It is therefore relatively easy to take account of symmetry-breaking effects, for example, by introducing physical masses in place of mean multiplet masses in the kinematic factors.

[4] R. Delbourgo and Abdus Salam, Phys. Letters **28B**, 497 (1969).

[5] Another freedom which has not been exploited lies in that groups like $SU(6)$ and $U(6)\otimes U(6)$ admit of pseudoquark representations in addition to quark representations (pseudoquarks, like the antiquarks, carry opposite intrinsic parity to quarks). In fact, we shall see later that a Tollerization of physical amplitudes (in contrast to Reggeization) more or less forces one to take pseudoquarks seriously, and with them hadrons of unnatural intrinsic parity.

[6] The important physical example where rest states are appropriately classified in terms of a four-dimensional angular momentum group $O_N(4)$ is the case of the hydrogen atom, the four-dimensional character of the group being a reflection of the extra symmetries possessed by the Coulomb $1/r$ potential. As is well known, hydrogen atom energy levels fall on Regge trajectories in an N versus E^2 plot, where the principal quantum number N determines $N(N+2)$, the Casimir operator of the orbital $O_N(4)$. [$N=\frac{1}{2}(k_1+k_2)$, where k_1 and k_2 are the 2 three-dimensional angular momenta associated with the two independent subgroups $O_{k_1}(3)$ and $O_{k_2}(3)$ which make up $O_N(4)$: $O_N(4)\approx O_{k_1}(3)\times O_{k_2}(3)$. For the hydrogen atom, only states with $k_1=k_2$ are realized and $N=0, 1, 2, \cdots$.] The possible existence of such an orbital $O_N(4)$ in hadron spectroscopy has been speculated by Barut and Kleinert (see Ref. 7). The fact that two (**56**,0⁺) multiplets appear to be known seems to bear out the need for a radial quantum number like N for classification purposes (see further under the "oscillator model").

[7] A. O. Barut and H. Kleinert, Phys. Rev. **161**, 1464 (1967).

nuclear physics shell-model spectroscopy. It is discussed in detail further on.

B. Quark-Excitation Model

In a different category and contrasting with the spin-orbit coupling models considered above is the quark-excitation model which was treated in detail in the earlier papers.[8] Here one starts the rest symmetry $U(6)\otimes U(6)$ and Reggeizes one or more of the Casimir operators of this group. One of the simplest cases was the Reggeization of *total quark number* N, the physical particles lying along two master trajectories in N versus $(\text{mass})^2$ plot and $2N$ taking the values 3, 5, 7, $\cdots$ for baryons and 2, 4, 6, $\cdots$ for mesons.

Reggeization procedure. Now, even though the physical ideas behind the two types of models A and B are different, the techniques for applying Regge ideas to the high-energy behavior of scattering amplitudes are very similar. So we shall state these in generality for any particle classification group G.

(1) Neglecting small deviations from a mean mass, assume that *all hadron* states (at rest) can be classified as representations of a (rest) symmetry group G.

$G=SU(6)\otimes O(3)$	for orbital models of type A
$=[U(6)\otimes U(6)]\otimes O(4)$	for orbital models of type A(i)
$=[U(6)\otimes U(6)]\otimes U(3)$	for orbital models of type A(ii)
$=U(6)\otimes U(6)$	for quark-excitation models of type B.

(2) A significant empirical feature of the spectroscopy is that only some rather simple representations of these groups appear to be realized in nature—in general, these are representations characterized by just one quantum number N (Casimir invariant of G) besides baryon number.

(3) For every rest symmetry group G, there exists a generalized helicity subgroup which we shall denote as G_W—the generalized helicity[9] being denoted by W.

(4) The importance of the generalized helicity subgroup lies in that, if the symmetry were exact for three-point vertices, W spin must be conserved. Labeling physical states with N and W [in analogy with J and λ for $G=SU_J(2)$], we thus have, for the three-point function,

$$\langle W|T(E)|W_1W_2\rangle=\sum_{\zeta}\langle\zeta W|W_1W_2\rangle T_{W_1W_2}(E).\quad(1)$$

Here $\langle\zeta W|W_1W_2\rangle$ denotes the Clebsch-Gordan coefficient which in G_W couples $D^{W_1}\otimes D^{W_2}$ to D^W. (In general, there may be more than one independent coupling, so we have included a parameter ζ to distinguish among them.)

(5) W spin is also conserved for collinear scattering processes (forward scattering). Thus,

$$\langle W_3W_4|T(E)|W_1W_2\rangle = \sum_{\zeta\zeta' W}\langle W_3W_4|\zeta' W\rangle T_{\zeta'\zeta}(E)\langle\zeta W|W_1W_2\rangle.\quad(2)$$

(6) The noncollinear four-point functions exhibit conservation of coplanar symmetry which for models A, A(i), A(ii) is $SU(3)$, $[U(3)\otimes U(3)]\otimes O(2)$,

$$[U(3)\otimes U(3)]\otimes U(1)\,,$$

and for model B is $U(3)\otimes U(3)$.

(7) If we assume only that the subgroup symmetries (1)–(6) hold as empirical facts (at least to a fair approximation), there is the mathematical theorem[10] that we may express a nonforward scattering amplitude in terms of a *complete* set of suitably defined functions $d_{WW'}{}^N(\theta)$ as follows:

$$\langle W_3W_4|T(E,\theta)|W_1W_2\rangle = \sum_{N\zeta W\zeta' W'}\langle W_3W_4|\zeta' W'\rangle d_{WW'}{}^N(-\theta) \times T_{\zeta' W',\zeta W}{}^N(E)\langle\zeta W|W_1W_2\rangle.\quad(3)$$

Here,

$$d_{W'W}{}^N(\theta)=\langle NW'|e^{-i\theta J_2}|NW\rangle\quad(4)$$

are the generalized rotation functions—the matrix elements of the space rotation operator $e^{i\theta J_2}$—for the group G. The expansion theorem used above relies on the completeness notion which requires that we sum over a *one-parameter* family D^N of representations of G, since we are dealing with a function $T(\theta)$ of just one variable θ.

[8] R. Delbourgo, Abdus Salam, M. A. Rashid, and J. Strathdee, Phys. Rev. **170**, 1477 (1968); R. Delbourgo, A. Salam, and J. Strathdee, *ibid.* **172**, 1727 (1968); R. Delbourgo and H. A. Rashid, *ibid.* **176**, 2074 (1968).

[9] All rest symmetry groups G must contain the subgroups $SU_J(2)\otimes SU_F(3)$ $[SU_J(2)=SU_L(2)\oplus SU_S(2)]$. Embedding $SU_J(2)$ into the Lorentz structure $SL(2,C)$, one identifies the conventional helicity subgroup as that subgroup of $SU_J(2)$ whose elements commute with the Lorentz boosting operator J_{03}. To define generalized helicity, one may likewise imbed the respective rest symmetry groups G into the appropriate relativistic structures:

A	$SL_S(6,C)\otimes O_L(3,1)$,
A(i)	$\tilde{U}_S(12)\otimes O_N(4,1)$ or $\tilde{U}_S(12)\otimes O_N(4,2)\approx\tilde{U}_S(12)\otimes U_N(2,2)$,
A(ii)	$\tilde{U}_S(12)\otimes U_N(3,1)$,
B	$U_N(6,6)$.

The corresponding generalized helicity subgroups are

A	$U(3)\times U(3)\|_W\otimes O(2)$,
A(i)	$U_W(6)\otimes O(3)$,
A(ii)	$U_W(6)\otimes U(2)$,
B	$U_W(6)$.

Note that we are making the important distinction between $\tilde{U}_S(12)$ and $U(6,6)$ In its original formulation, $\tilde{U}(12)$ was the symmetry group combining *intrinsic spin* (S) and unitary spin (F), while we regard $U(6,6)$ as a noncompact rest symmetry which combines *total spin* (J) and (F). The two $U_W(6)$ groups in models A(i), A(ii), and model B are therefore different groups; they refer, respectively, to intrinsic spin S (which has to be combined with the orbital angular momentum to give J) and to J itself.

[10] For a fuller discussion, see Delbourgo, Salam, and Strathdee, Ref. 8.

FIG. 1. Trajectories from $O(4)_N$ excitations. (○ = normal parity; × = abnormal parity.)

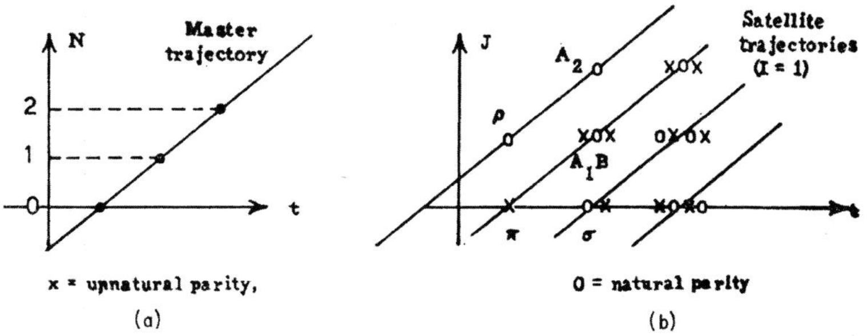

(8) We connect the expansion (7) with assumption (1) if we now assume that $T^N(E)$ exhibits poles in the complex N Casimir plane, corresponding to supermultiplets of group G; this reduces Eq. (3) to the form

$$\langle W_3W_4|T(E,\theta)|W_1W_2\rangle = \sum_{N\zeta W\zeta' W'} \langle W_3W_4|\zeta' W'\rangle \times g_{W_3W_4\zeta' W'}{}^N[d_{WW'}{}^N(-\theta)/(E^2-m_N^2)] \times g_{\zeta W W_1W_2}{}^N\langle \zeta W|W_1W_2\rangle. \quad (5)$$

(9) We can now pass to a Regge amplitude by making a Sommerfeld-Watson transformation:

$$\lim_{W\to\infty} \langle W_3W_4|T(E,\theta)|W_1W_2\rangle \sim \sum_{\zeta W\zeta' W'} g_{\zeta W' W_3W_4}{}^\alpha \times[d_{W'W}{}^\alpha(\theta)/\sin\pi\alpha(E)] g_{\zeta W W_1W_2}{}^\alpha\langle \zeta W|W_1W_2\rangle, \quad (6)$$

where $\alpha(m_N^2)=N$ is the supermultiplet trajectory function.

(10) The points on a "master" trajectory $\alpha(m_N^2)$ represent particles of differing spin values which the supermultiplet groups together. In Refs. 8 and 11, the mathematical reduction problem of expressing the general rotation functions $d^N(\theta)$ in terms of the Legendre polynomials $P^L(\theta)$ or $P^J(\theta)$ and their derivatives $[d^N(\theta)=\sum_{J,K} a_{J,K}P^J(\theta)]$ was discussed in detail. Physically this means that one master trajectory gives rise to a number of equally spaced satellite trajectories labeled with the parameter κ (in the conventional Regge ReJ-m^2 plane), all parallel to the master trajectory in the exact symmetry limit. (These satellites are not to be confused with the daughter trajectories considered by Freedman and Wang and by Toller.) In the $O(4)$ orbital scheme for mesons, for example, the following schematic picture may hold (see Fig. 1):

$N=0$	$S^P=1^-,0^-$;	$L^P=0^+$;	$J^P=0^-,1^-$				
$N=1$	$S^P=1^-,0^-$;	$L=1^-,0^+$;	$J^P=2^+,1^+,0^+$;	$1^+,1^-,0^-$			
$N=2$	$S^P=1^-,0^-$;	$L=2^+1^-0^+$;	$J^P=3^-,2^-,1^-,0^-$;	$2^+,1^+,0^+$;	$1^-,0^-$;	$2^-,1^+,0^-$.	

Note the rather obvious but extremely important circumstance that the leading satellite trajectory with the 0^- particles on it is automatically shifted downwards by one unit of J from the leading vector-tensor trajectory. The very high-energy behavior, naturally, is always dominated by the leading trajectory if the selection rules allow it to be exchanged.

(11) To take account of trajectory shifts due to symmetry breaking, we need mass formulas which, in general, may have the form (with L in place of J for the orbital models)

$$M^2=M^2(N,J,F) = M_0^2(N)+M_1^2(F)+JM_2^2(F), \quad (7)$$

where $N=J+K$, and F denotes the $SU(3)$ labels (including I and Y). To incorporate trajectory shifts due to symmetry breaking in the formalism, one may go back to the formula

$$T\approx\int \frac{dN}{\sin\pi N}\frac{b^Nd^N(-\theta)}{t-M^2(N)}, \quad (8)$$

write $d^N(-\theta)=\sum a_{KJ}P^J(-\theta)$, and, as an ansatz, replace $M^2(N)$ by $M^2(N,J,F)$, obtaining

$$T\sim\sum_K \int \frac{dJ}{\sin\pi J}\frac{b^{J+K}a_{KJ}P_J(-\theta)}{t-M^2(K,J,F)}. \quad (9)$$

The satellite trajectory functions $\alpha(K,J,F)$ are given as solutions of $t=M^2(K,J,F)$.

Unfortunately, no completely reliable theoretical method exists for computing these trajectory shifts. We

[11] R. Delbourgo, K. Koller, and R. Williams, J. Math. Phys. **10**, 957 (1969).

must therefore, at present, introduce the precise trajectory functions as part of empirical input. The utility of the supermultiplet Reggeization schemes is thus impaired, except for the hope that the residues are not so strongly affected by symmetry breaking as the trajectories. This appears to be the case for meson-baryon scattering (see Ref. 4).

(12) The rotation functions were computed in a previous publication[11] for a number of groups for simple classes of representations. As a general rule, a rotation function $d_{WW'}{}^{N}(\theta)$ is a sum of derivatives of the *basic* function $d_{[1][1]}{}^{N}$ (the function which appears in superscalar scattering with the exchange of a multiplet labeled with the quantum number N). This is analogous to the statement that the $d_{\lambda\lambda'}{}^{J}(\theta)$ in three dimensions can be expressed as sums of derivatives of $P_J(\theta)$. We list in the adjoining table these basic rotation functions for symmetry groups and representations of interest. G is the multiplet symmetry at rest, G_W the generalized helicity subgroup, $\mathcal{G}$ the embedding covariant group, and N labels the (one-parameter) class of representations (more precisely, we indicate the Young tableaux to which N refers).

(a) $G=U(\nu)\otimes U(\nu),\quad G_W=U(\nu),\quad \mathcal{G}=U(\nu,\nu).$

For representations (W_N,W_N) corresponding to Young tableaux $(N,0,0,\cdots,0;\, N,N,N,\cdots,N)$,

$$d^N(\theta)\propto C_N{}^{\frac{1}{2}\nu}(\cos\theta). \tag{10a}$$

(b) $G=U(2\nu),\quad G_W=U(\nu)\otimes U(\nu),\quad \mathcal{G}=SL(2\nu,c).$

For representations (W_N) described by tableaux $(N+1,\, N,\, \cdots,\, N)$,

$$d^N(\theta)\propto C_N{}^{\nu-\frac{1}{2}}(\cos\theta). \tag{10b}$$

(c) $G=U(\nu),\quad G_W=U(\nu-1),\quad \mathcal{G}=U(\nu,1).$

For representations (N) described by tableaux $(N,0,0,\cdots,0)$,

$$d^N(\theta)=(\cos\theta)^N. \tag{10c}$$

(d) $G=O(\nu),\quad G_W=O(\nu-1),\quad \mathcal{G}=O(\nu,1).$

For representations (N) described by $(N,0,0,\cdots,0)$,

$$d^N(\theta)\propto C_N{}^{\frac{1}{2}\nu-1}(\cos\theta). \tag{10d}$$

Proofs of statements (a), (b), and (d) are already in print[11]; a proof of (c) is given further on. The Reggeized components of models A, A(i), A(ii), and B are the cases (d) with $\nu=3$, (d) with $\nu=4$, (c) with $\nu=3$, and (a) with $\nu=6$, respectively.

III. COVARIANT FORMALISM FOR SCATTERING AMPLITUDES AND M FUNCTIONS

So far we have worked with the helicity formalism. In principle, all we need now are general expressions for rotation functions $d_{WW'}{}^{N}(\theta)$ in terms of derivatives of $d_{[1][1]}{}^{N}$ of Sec. II and formulas for the general Clebsch-Gordan coefficients $\langle W|W_1W_2\rangle$, etc. One can proceed perfectly well by listing these things with the use of sophisticated group-theory methods, including the spin-orbit coupling coefficients needed in models A, A(i), and A(ii). It so happens that one of the simplest ways of making these computations is to work *ab initio* in terms of an M-function approach using a multispinor formalism. Since this has the additional merit of exhibiting manifest covariance, of allowing crossing to be performed with ease, of automatically incorporating the threshold and other mass-dependent kinematic factors,[12] from now on we shall abandon the helicity framework and work consistently with the M functions.

A. Wave Functions of Particle Multiplets

Consider first the wave functions of the particle multiplets for the various models:

Model A: *Orbital excitations.* Represent $O(3)$ multiplets of $L=0,\,1,\,2,\,\cdots$ by *symmetric traceless* tensors

$$\phi(p),\quad \phi_\mu(p),\quad \phi_{(\mu_1\mu_2)}(p),\ \cdots,$$

with the restrictions

$$\begin{aligned} p_\mu\phi_\mu&=0,\\ p_{\mu_1}\phi_{(\mu_1\mu_2\cdots\mu_L)}&=0,\\ p^2\Phi_{(\mu_1\cdots\mu_L)}&=m_L{}^2\Phi_{(\mu_1\cdots\mu_L)}. \end{aligned} \tag{11}$$

A(i): *Hydrogen-like excitations.* Represent $SU(2)\otimes SU(2)\approx O(4)$ multiplets belonging to the representation $(\frac{1}{2}N,\frac{1}{2}N)$, $N=0,\,1,\,2,\,\cdots$, by the multispinors

$$\Phi,\ \Phi_\alpha{}^\beta,\ \Phi_{(\alpha_1\alpha_2)}{}^{(\beta_1\beta_2)},\ \cdots,$$

symmetric in α's and β's separately, satisfying $\alpha,\beta=1,\,2,\,3,\,4$, Bargman-Wigner equations:

$$\begin{aligned} &(p-m)_{\gamma_i}{}^{\alpha_i}\Phi_{(\alpha_1\cdots\alpha_i\cdots\alpha_N)}{}^{(\beta_1\cdots\beta_N)}(p)\\ &\quad=\Phi_{(\alpha_1\cdots\alpha_N)}{}^{(\beta_1\cdots\beta_i\cdots\beta_N)}(p+m)_{\beta_i}{}^{\gamma_i}=0. \end{aligned} \tag{12}$$

A(ii). Represent $SU(3)$ multiplets belonging to the representation[13] $[N]$, $N=0,\,1,\,2,\,\cdots$, by the fields

$$\phi(p),\quad \phi_\mu(p),\quad \phi_{\mu_1\mu_2}(p),\ \cdots,$$

which are symmetric but *not traceless* in their indices. The equations they satisfy are the same as for the $O(3)$ case:

$$\begin{aligned} p_\mu\phi_\mu&=0,\\ p_{\mu_1}\phi_{(\mu_1\mu_2\cdots\mu_N)}&=0,\\ p^2\phi_{(\mu_1\cdots\mu_N)}&=m_N{}^2\phi_{(\mu_1\cdots\mu_N)}. \end{aligned} \tag{13}$$

So much for the orbital part of the representations in the models A, A(i), and A(ii). For the intrinsic-spin-

[12] As stated in the Introduction, the explicit appearance of the mass factors permits one to devise mass-breaking prescriptions.

[13] See P. G. O. Freund, Nuovo Cimento 58, 519 (1968) The $O(3)$ decompositions of these tensors is as follows:

	Dimensionality N_{C_3}
$[0]\to J=0$	1
$[1]\to J=1$	3
$[2]\to J=2,\,0$	6
$[3]\to J=1,\,3$	10, and so on.

unitary-spin part, we employ the $\tilde{U}(12)$ formalism for models A(i) and A(ii) using the multispinors

$$\Phi_B{}^A(p),\quad A, B=1, \cdots, 12$$

for mesons $(6,\bar{6})$ of rest symmetry $U_S(6)\otimes U_S(6)$ and

$$\Psi_{(ABC)}(p),\quad A, B, C=1, \cdots, 12$$

(symmetric A, B, C) for baryons (56,1) and $\Psi_{[AB]C}$ antisymmetric in $[A,B]$ and satisfying the cyclic condition

$$\Psi_{[AB]C}+\Psi_{[BC]A}+\Psi_{[CA]B}=0$$

for the (70,1).

The Bargmann-Wigner equations[14] are

$$\begin{aligned}(p-m)_C{}^A\Phi_A{}^B=\Phi_A{}^B(p+m)_B{}^C=0,\\ (p-m)_D{}^A\Psi_{(ABC)}=(p-m)_D{}^A\Psi_{[AB]C}=0,\ \text{etc.}\end{aligned}\qquad(14)$$

Combining the spin–unitary-spin *and* the orbital degrees of freedom, the tensors have the final forms:

Model A(i):

$$\begin{aligned}&\Phi_B{}^A{}_{(\alpha_1\cdots\alpha_N)}{}^{(\beta_1\cdots\beta_N)} && \text{for } (6,\bar{6},N),\\ &\Psi_{(ABC)(\alpha_1\cdots\alpha_N)}{}^{(\beta_1\cdots\beta_N)} && \text{for } (56,1,N),\\ &\Psi_{[AB]C(\alpha_1\cdots\alpha_N)}{}^{(\beta_1\cdots\beta_N)} && \text{for } (70,1,N),\end{aligned}$$

with the equations of motion (12) and (14) stated earlier, and

Model A(ii):

$$\begin{aligned}&\Phi_B{}^A{}_{(\mu_1\cdots\mu_N)} && \text{for } (6,\bar{6},N),\\ &\Psi_{(ABC)(\mu_1\cdots\mu_N)} && \text{for } (56,1,N),\\ &\Psi_{[AB]C(\mu_1\cdots\mu_N)} && \text{for } (70,1,N),\end{aligned}$$

satisfying Eqs. (13) and (14). We shall be concentrating on model A(ii) in what follows.

Model B: *The quark excitation model* was dealt with in detail in Ref. 8. To complete the discussion for the case of the group $U_J(6)\times U_J(6)$, the appropriate multispinors belonging to the fully symmetrical representation (the so-called Feynman representation) are

$$\begin{aligned}&\Phi_A{}^B, \Phi_{(A_1A_2)}{}^{(B_1B_2)}, \Phi_{(A_1A_2A_3)}{}^{(B_1B_2B_3)}, \cdots && \text{for mesons},\\ &\Psi_{(A_1A_2A_3)}\Psi_{(A_1A_2A_3A_4)}{}^B, \cdots && \text{for baryons},\end{aligned}$$

satisfying Eqs. (14).

B. Three-Point Couplings

Given three multiplets of G for any of these models, one can easily write down the G_W invariant couplings in M-function form by noting that the three-particle momenta transform as G_W scalars. The rule then is to saturate all indices among themselves and with momentum tensors; the number of different ways of doing this giving the different types of couplings one can construct. To illustrate, consider the simple case of a quark $(6,1;0)_{-\frac{1}{2}p+q}$ and an antiquark $(1,\bar{6};0)_{\frac{1}{2}p+q}$ coupling with the $(6,\bar{6};N)_p$ meson multiplets.

$$\begin{aligned}\text{A(i)}\quad \mathcal{L}_{\text{eff}}&=\bar{v}^A(\tfrac{1}{2}p+q)u_B(-\tfrac{1}{2}p+q)\\ &\quad\times[G_0\delta_A{}^Bq_D{}^C+G_1\delta_A{}^C\delta_D{}^B]q_{\beta_1}{}^{\alpha_1}\\ &\quad\cdots q_{\beta_N}{}^{\alpha_N}\Phi_C{}^D{}_{(\alpha_1\cdots\alpha_N)}{}^{(\beta_1\cdots\beta_N)}(p);\quad(15\text{a})\\ \text{A(ii)}\quad &=\bar{v}^A(\tfrac{1}{2}p+q)u_B(-\tfrac{1}{2}p+q)[G_0\delta_A{}^Bq_D{}^C+G_1\delta_A{}^C\delta_D{}^B]\\ &\quad\times q_{\mu_1}\cdots q_{\mu_N}\Phi_C{}^D{}_{(\mu_1\cdots\mu_N)}(p);\quad(15\text{b})\\ \text{B}\quad \mathcal{L}_{\text{eff}}&=\bar{v}^A(\tfrac{1}{2}p+q)u_B(-\tfrac{1}{2}p+q)\\ &\quad\times[G_0\delta_A{}^Bq_{B_1}{}^{A_1}+G_1\delta_A{}^{A_1}\delta_{B_1}{}^B]q_{B_2}{}^{A_2}\\ &\quad\cdots q_{B_N}{}^{A_N}\Phi_{(A_1\cdots A_N)}{}^{(B_1\cdots B_N)}(p).\quad(15\text{c})\end{aligned}$$

We can cast all models in the differential form

$$\mathcal{L}_{\text{eff}}=\bar{v}^Au_B[g_0\delta_A{}^B+mg_1(\partial/\partial q_B{}^A)]\Phi_{(N)}(p,q),\quad(16)$$

where

$$\begin{aligned}\Phi_{(N)}(p,q)&\\ &=\mu^{-(N+1)}q_B{}^Aq_{\beta_1}{}^{\alpha_1}\cdots q_{\beta_N}{}^{\alpha_N}\\ &\qquad\times\Phi_A{}^B{}_{(\alpha_1\cdots\alpha_N)}{}^{(\beta_1\cdots\beta_N)}(p)\quad(17\text{a})\\ &=\mu^{-(N+1)}q_B{}^Aq_{\mu_1}\cdots q_{\mu_N}\Phi_A{}^B{}_{(\mu_1\cdots\mu_N)}(p)\quad(17\text{b})\\ &=\mu^{-N}q_{B_1}{}^{A_1}q_{B_2}{}^{A_2}\cdots q_{B_N}{}^{A_N}\Phi_{(A_1\cdots A_N)}{}^{(B_1\cdots B_N)}(p)\quad(17\text{c})\end{aligned}$$

is the coupling corresponding to scattering of supersinglets for each of these models. Note the distinction between $q_B{}^A$ and $q_\beta{}^\alpha$. For the $\tilde{U}(12)$, case $q_B{}^A=q_\beta{}^\alpha\delta_b{}^a$ where a, b refer to $SU(3)$ indices and α, β to $\tilde{U}(4)$ indices.

C. Supermultiplet Exchange Contributions to Four-Point Couplings and Computation of Rotation Functions

Just as the Legendre functions $d_{\lambda'\lambda}{}^J(\theta)$ can be "calculated" by considering the exchange of a spin-J particle in an M-function framework, so can the generalized rotation functions $d_{W'W}{}^N(\theta)$ for each of the models by exchanging a supermultiplet of quantum number N using the covariant couplings (15) written earlier. The procedures were illustrated in great detail in Ref. 8 for model B; here we quickly go over the methods again for models A(i) and A(ii). The basic functions $d_{[1][1]}{}^N(\theta)$ arise from scattering of supersinglets. Combining the three-point couplings with the $[N]$ propagators

$$\begin{aligned}\text{A(i)}\quad &\langle\Phi_A{}^B{}_{(\alpha_1\cdots\alpha_N)}{}^{(\beta_1\cdots\beta_N)}(p)\Phi_{B'}{}^{A'}{}_{(\beta_1'\cdots\beta_{N'})}{}^{(\alpha_1'\cdots\alpha_{N'})}(-p)\rangle(p^2-M^2)\\ &=(p+M)_A{}^{A'}(p-M)_{B'}{}^B\sum_{\alpha,\beta}(p+M)_{\alpha_1}{}^{\alpha_1'}\cdots(p+M)_{\alpha_N}{}^{\alpha_N'}(p-M)_{\beta_1}{}^{\beta_1}\cdots(p-M)_{\beta_{N'}}{}^{\beta_N},\quad(18\text{a})\\ \text{A(ii)}\quad &\langle\Phi_A{}^B{}_{(\mu_1\cdots\mu_N)}(p)\Phi_{B'}{}^{A'}{}_{(\mu_1'\cdots\mu_{N'})}(-p)\rangle(p^2-M^2)\\ &=(p+M)_A{}^{A'}(p-M)_{B'}{}^B\sum_\mu(-g_{\mu_1\mu_1'}+p_{\mu_1}p_{\mu_1'}/M^2)\cdots(-g_{\mu_N\mu_{N'}}+p_{\mu_N}p_{\mu_{N'}}/M^2),\quad(18\text{b})\end{aligned}$$

[14] A complete and detailed discussion of the $\tilde{U}(12)$ multispinors appears in *Proceedings of the International Seminar in High-Energy Physics and Elementary Particles, Trieste, 1965* (International Atomic Energy Agency, Vienna, 1965).

we get

$$(p^2-M^2)\langle\Phi_{(N)}(p,q)\Phi_{(N)}(-p,q')\rangle$$
$$=(|\mathbf{q}||\mathbf{q}'|/\mu^2)^{N+1}\times\cos\theta\, C_N{}^1(\cos\theta) \quad \mathrm{A(i)}$$
$$=(|\mathbf{q}||\mathbf{q}'|/\mu^2)^{N+1}\times\cos\theta(\cos\theta)^N \quad \mathrm{A(ii)}, \quad (19)$$

where

$$|\mathbf{q}||\mathbf{q}'|\cos\theta=-q\cdot q'+q\cdot pq'\cdot p/M^2, \quad (20)$$

showing that the excitation functions are $C_N{}^1(\cos\theta)$ and $(\cos\theta)^N$ for $O(4)$ and $U(3)$, respectively; both multiplied into a $U(6)\otimes U(6)$ spin factor $(\cos\theta)=d_{[1][1]}{}^{(6,\bar{6})}\times(\cos\theta)$ coming from the $(6,\bar{6})$ piece.[15]

More complicated functions can be discovered by differentiation; for example, in quark-antiquark scattering the derived d^N can be recovered from

$$[g_0\delta_A{}^B+mg_1(\partial/\partial q_B{}^A)]$$
$$\times[g_0\delta_{B'}{}^{A'}+mg_1(\partial/\partial q'_{A'}{}^{B'})]d_{[1][1]}{}^N(\theta) \quad (21)$$

by contraction over external wave functions. We thus have a complete M-function substitute for quark-antiquark scattering of the helicity formulation of the earlier section. More complicated cases of physical interest are described in Sec. IV for model A(ii).

IV. DETAILS OF MESON-BARYON AND BARYON-BARYON SCATTERING IN THE OSCILLATOR MODEL

In this section, we wish to discuss the harmonic-oscillator model in detail. From the slender evidence available, it appears that baryons group themselves as

$$\left.\begin{matrix}(56,0)\\(56,2)\end{matrix}\right\}\quad(70,1)$$

with (56,1) apparently missing. It would thus seem that we are realizing the quantum numbers $N=0$, 2, 4, of an $SU(3)$ group, corresponding to representations

$$L=0\,,$$
$$L=0,\,2\,,$$
$$L=0,\,2,\,4,\ \text{etc.}$$

Also, we know that $d_{[1][1]}{}^N\approx(\cos\theta)^N$ for all models in the asymptotic limit. So, we can in any case regard the oscillator model as *representative* of all models in the high-energy limit. Of course, to lower orders the models will differ from one another.

A. Wave Functions

Hereafter, we work entirely in the covariant framework provided by the auxiliary group $\tilde{U}(12)\otimes U(3,1)$. For M-function purposes, we adopt the fields

$$\phi_A{}^B{}_{(\mu_1\cdots\mu_N)},\quad \psi_{(ABC)(\mu_1\cdots\mu_{2N})},\quad \text{and}\ \psi_{[AB]C,(\mu_1\cdots\mu_{2N+1})}.$$

If we are interested in the Lorentz group components within the supermultiplets, we make, in the first step, the decompositions

$$\Phi_A{}^B{}_{(\mu_1\cdots\mu_N)}(p)$$
$$=(1/2\sqrt{2}M)[(\gamma\cdot p+M)$$
$$\times(\gamma_5P_{(\mu_1\cdots\mu_N)}-\gamma_\mu V_{\mu(\mu_1\cdots\mu_N)})]_A{}^B, \quad (22)$$

$$u_{ABC(\mu_1\cdots\mu_N)}(p)$$
$$=(1/2\sqrt{2}M)[(\gamma\cdot p+M)\gamma_\mu C]_{\alpha\beta}D_{(abc)\gamma\mu(\mu_1\cdots\mu_N)}$$
$$+(1/6\sqrt{2}M)\{[(\gamma\cdot p+M)\gamma_5C]_{\alpha\beta}$$
$$\times\epsilon_{abd}N_c{}^d{}_{\gamma(\mu_1\cdots\mu_N)}+\text{perms}\}, \quad (23)$$

$$u_{[AB]C(\mu_1\cdots\mu_N)}(p)$$
$$=(1/4M)[(\gamma\cdot p+M)\gamma_\mu C]_{\alpha\beta}\epsilon_{abd}\mathfrak{N}_c{}^d{}_{\gamma\mu(\mu_1\cdots\mu_N)}$$
$$+(1/2\sqrt{2})M[(\gamma\cdot p+M)\gamma_5C]_{\alpha\beta}\mathfrak{D}_{(abc)\gamma(\mu_1\cdots\mu_N)}$$
$$+[1/2(\sqrt{6})M]\{[(\gamma\cdot p+M)\gamma_5C]_{\beta\gamma}\epsilon_{bcd}\mathfrak{N}_a{}^d{}_{\alpha(\mu_1\cdots\mu_N)}$$
$$-[(\gamma\cdot p+M)\gamma_5C]_{\alpha\gamma}\epsilon_{acd}\mathfrak{N}_b{}^d{}_{\beta(\mu_1\cdots\mu_N)}\}$$
$$+(\epsilon_{abc}/12M)\{[(\gamma\cdot p+M)\gamma_5C]_{\alpha\gamma}\mathcal{Y}_{\beta(\mu_1\cdots\mu_N)}$$
$$+[(\gamma\cdot p+M)\gamma_5C]_{\beta\gamma}\mathcal{Y}_{\alpha(\mu_1\cdots\mu_N)}\}, \quad (24)$$

wherein the (N) excitations of the basic (quark) spin fields $P(0^-)$, $V(1^-)$, $N(\frac{1}{2}^+)$, $D(\frac{3}{2}^+)$, $\cdots$ is explicitly exhibited. In the second step, the excitations are reduced under $O(3)_L$ according to the further decomposition

$$\phi_{(\mu_1\cdots\mu_N)}(p)$$
$$=\hat{\phi}_{(\mu_1\cdots\mu_N)}{}^{(L=N)}(p)+\left[\frac{2}{N(N-1)(2N-1)}\right]^{1/2}$$
$$\times\sum_{kl}d_{\mu_k\mu_l}\hat{\phi}_{(\mu_1\cdots\bar{k}\bar{l}\cdots\mu_N)}{}^{(L=N-2)}(p)$$
$$+\text{lower orbital momenta}; \quad (25)$$

$d_{\mu\nu}\equiv-g_{\mu\nu}+p_\mu p_\nu/M^2$ and $\hat{\phi}$ are traceless in their indices. ($\bar{k}$ indicates that μ_k is absent, etc.) In the third step, the product of each of these $SU(2)_L$ irreducible components with the $SU(2)_S$ components is reduced out into total $SU(2)_J$ components. If we take a given orbital excitation L, the decomposition reads as follows: For fields of (quark) spin 1, $\frac{1}{2}$, and $\frac{3}{2}$ (the only cases of physical interest),

$$\phi_{\mu(\mu_1\cdots\mu_L)}=\phi_{(\mu\mu_1\cdots\mu_L)}{}^{(L+1)}+\frac{1}{[L(L+1)]^{1/2}}\sum_k\frac{p_\lambda}{M}\epsilon_{\lambda\mu\mu_k\nu}\phi_{(\mu_1\cdots\bar{k}\cdots\mu_L\nu)}{}^{(L)}$$
$$+\frac{1}{L}\left(\frac{2L-1}{2L+1}\right)^{1/2}\left[\sum_k d_{\mu\mu_k}\phi_{(\mu_1\cdots\bar{k}\cdots\mu_L)}{}^{(L-1)}-\frac{2}{2L-1}\sum_{kl}d_{\mu_k\mu_l}\phi_{(\mu\mu_1\cdots\bar{k}\bar{l}\cdots\mu_L)}{}^{(L-1)}\right], \quad (26)$$

[15] Actually, if Bose statistics is taken into account, only odd N values for A(i) and A(ii) and even N values for B are permitted.

$$\psi_{(\mu_1\cdots\mu_L)}=\psi_{(\mu_1\cdots\mu_L)}{}^{(L+\frac{1}{2})}+\frac{1}{[L(4L-1)]^{1/2}}\sum_k w_{\mu_k}\psi_{(\mu_1\cdots\bar{k}\cdots\mu_L)}{}^{(L-\frac{1}{2})}, \tag{27}$$

$$\begin{aligned}\psi_{\mu(\mu_1\cdots\mu_L)}&=\psi_{(\mu\mu_1\cdots\mu_L)}{}^{(L+\frac{3}{2})}\\
&+\frac{1}{[L(L+3)]^{1/2}}\Big[\sum_k\frac{p_\lambda}{M}\epsilon_{\lambda\mu\mu_k\nu}\psi_{(\mu_1\cdots\bar{k}\cdots\mu_L\nu)}{}^{(L+\frac{1}{2})}+\sum_k w_{\mu_k}\psi_{(\mu\mu_1\cdots\bar{k}\cdots\mu_L)}{}^{(L+\frac{1}{2})}-\tfrac{1}{3}Lw_\mu\psi_{(\mu_1\cdots\mu_L)}{}^{(L+\frac{1}{2})}\Big]\\
&+\frac{1}{L}\Big(\frac{2L-1}{2L}\Big)^{1/2}\Big[\tfrac{2}{3}\sum_k d_{\mu\mu_k}\psi_{(\mu_1\cdots\bar{k}\cdots\mu_L)}{}^{(L-\frac{1}{2})}-\frac{2}{2L-1}\sum_{kl}d_{\mu_k\mu_l}\psi_{(\mu\mu_1\cdots\bar{k}\bar{l}\cdots\mu_L)}{}^{(L-\frac{1}{2})}-\tfrac{1}{3}\sum_k\frac{p_\lambda}{M}\epsilon_{\lambda\mu\mu_k\nu}w_\nu\psi_{(\mu_1\cdots\bar{k}\cdots\mu_L)}{}^{(L-\frac{1}{2})}\Big]\\
&+\Big(\frac{6}{(L+3)!}\Big)^{1/2}\Big[\sum_{kl}d_{\mu\mu_k}w_{\mu_l}\psi_{(\mu_1\cdots\bar{k}\bar{l}\cdots\mu_L)}-\frac{2}{2L-1}w_\mu\sum_{kl}d_{\mu_k\mu_l}\psi_{(\mu_1\cdots\bar{k}\bar{l}\cdots\mu_L)}-\frac{2}{2L-1}\sum_{klm}d_{\mu_k\mu_l}w_{\mu_m}\psi_{(\mu\mu_1\cdots\bar{k}\bar{l}\bar{m}\cdots\mu_L)}\Big],\end{aligned} \tag{28}$$

where $\psi_{(\mu_1\cdots\mu_N)}{}^{(N+\frac{1}{2})}$ are Rarita-Schwinger fields of spin $N+\frac{1}{2}$ and we have used above the abbreviation for the relativistic spin operator $w_\mu\equiv p_\lambda\sigma_{\lambda\mu}\gamma_5/M$. In the exact-symmetry limit, we can summarize these reductions by sets of J trajectories, as depicted in Figs. 2–5, where we have assumed exchange degeneracy for ease of drawing.

B. Three-Point Couplings

The great importance of reductions (22)–(28) is apparent when one wishes to compare matrix elements of particular spin components, primarily in the three-point couplings which are relevant for correlating various decay parameters or Regge residues. As we mentioned earlier, these three-point vertices are constructed in a G_W-invariant manner by forming index invariants between fields and momenta.[16] For hadrons like $(6,\bar{6};0)$, $(6,\bar{6};1)$, $(56,1;0)$, $(70,1;1)$ and a meson supermultiplet of excitation N which we later Reggeize, we list below the effective Lagrangians of these hadrons, writing our couplings in a differential notation:

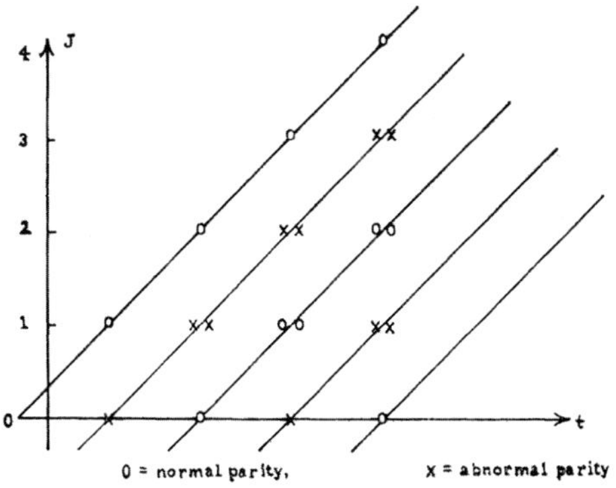

FIG. 2. Reggeized $SU(3)$ nonets of mesons from oscillator model $(6,\bar{6};N)$. (O = normal parity, × = abnormal parity.)

[16] The over-all coupling constants depend intrinsically on the excitation numbers of the three interacting particles (N_1,N_2,N_3), and we may even suspect that when one excitation number becomes large there is a corresponding falloff in the residue; but we do not attempt to investigate this aspect.

$(6,\bar{6};0)_{\frac{1}{2}p+q}-(6,\bar{6};0)_{\frac{1}{2}p-q}-(6,\bar{6};N)_p$:

$$\begin{aligned}\mathcal{L}=\mu^{1-N}\Phi_A{}^B(\tfrac{1}{2}p+q)\Phi_C{}^D(\tfrac{1}{2}p-q)&\Big[h_0{}^{(-)}\delta_B{}^C\delta_D{}^A+h_0'{}^{(-)}\mu^{-2}q_B{}^Aq_D{}^C\\
&+\mu h_1{}^{(+)}\Big(\delta_B{}^C\frac{\partial}{\partial q_A{}^D}+\delta_D{}^A\frac{\partial}{\partial q_C{}^B}\Big)\\
&+\mu h_1{}^{(-)}\Big(\delta_B{}^C\frac{\partial}{\partial q_A{}^D}-\delta_D{}^A\frac{\partial}{\partial q_C{}^B}\Big)\Big]\Phi_{(N)}(p,q);\end{aligned} \tag{29}$$

$(6,\bar{6};1)_{\frac{1}{2}p+q}-(6,\bar{6};0)_{\frac{1}{2}p-q}-(6,\bar{6};N)_p$:

$$\begin{aligned}\mathcal{L}=\mu^{1-N}\Phi_A{}^B{}_\mu(\tfrac{1}{2}p+q)\Phi_C{}^D(\tfrac{1}{2}p-q)&\Big[(h_{00}\delta_B{}^C\delta_D{}^A+\mu^{-2}h_{00}'q_B{}^Aq_D{}^C)q_\mu\\
&+(\mu^2h_{01}\delta_B{}^C\delta_D{}^A+h_{01}'q_B{}^Aq_D{}^C)\frac{\partial}{\partial q_\mu}\\
&+\mu q_\mu\Big(h_{10}\delta_B{}^C\frac{\partial}{\partial q_A{}^D}+h_{10}'\delta_D{}^A\frac{\partial}{\partial q_C{}^B}\Big)\\
&+\mu^3\frac{\partial}{\partial q_\mu}\Big(h_{11}\delta_B{}^C\frac{\partial}{\partial q_A{}^D}+h_{11}'\delta_D{}^A\frac{\partial}{\partial q_C{}^B}\Big)\Big]\Phi_{(N)}(p,q);\end{aligned} \tag{30}$$

$(56,1;0)_{\frac{1}{2}p-q}-(\overline{56},1;0)_{\frac{1}{2}p+q}-(6,\bar{6};N)_p$:

$$\begin{aligned}\mathcal{L}=m^{-N}\bar{u}^{(ACD)}(\tfrac{1}{2}p+q)u_{(BCD)}(-\tfrac{1}{2}p+q)\\
\times[g_0\delta_A{}^B+mg_1(\partial/\partial q_B{}^A)]\Phi_{(N)}(p,q);\end{aligned} \tag{31}$$

$(56,1;0)_{\frac{1}{2}p-q}-(\overline{70},1;1)_{\frac{1}{2}p+q}-(6,\bar{6};N)$:

$$\begin{aligned}\mathcal{L}=m^{-N}\bar{u}^{[AC]D}{}_\mu(\tfrac{1}{2}p+q)u_{(BCD)}(-\tfrac{1}{2}p+q)\\
\times[g_0q_\mu+m^2g_1(\partial/\partial q_\mu)](\partial/\partial q_B{}^A)\Phi_{(N)}(p,q).\end{aligned} \tag{32}$$

This set of formulas are some of the most crucial ones in this paper.

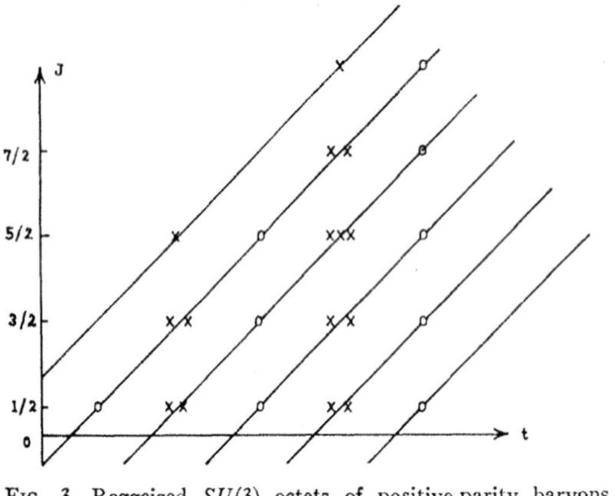

FIG. 3. Reggeized $SU(3)$ octets of positive-parity baryons (56,1; $2N$) and negative-parity baryons (70,1; $2N+1$) from oscillator model.

As before, we have used here the abbreviation

$$\Phi_{(N)}(p,q) \equiv q_{\mu_1}\cdots q_{\mu_N} q_B{}^A \Phi_A{}^B{}_{(\mu_1\cdots\mu_N)}(p) \qquad (33)$$

for the fully contracted meson field of excitation N. The superscripts ($\pm$) on the couplings h for meson coupling (with no baryons involved) refer to the even and odd N values of the exchange mesons. Bose statistics tells us that $h^+=0$ when N is odd and $h^-=0$ when N is even.[17]

C. Meson-Baryon (MB) and Baryon-Baryon (BB) Scattering

The two most important cases concern meson-baryon (MB) and baryon-baryon (BB) scattering. We investigate the pure symmetry limit (with all masses degenerate).

If we apply the coupling rules embodied by formulas (29)–(32), it is always possible to express the covariant M functions in the form

$$T = D_q D_{q'} \Delta_{(N)}(p;\, q,q')\,, \qquad (34)$$

where D stand for various differential operators whose order is *governed by the external excitation numbers*, and where $\Delta_{(N)}$ is the fully contracted propagator which occurs in the scattering of supersinglets:

$$\Delta_{(N)} = (|\mathbf{q}|\,|\mathbf{q}'|)^{N+1}\left[(\cos\theta_t)^{N+1}/(p^2-M_N{}^2)\right], \qquad (35)$$

$$|\mathbf{q}|\,|\mathbf{q}'|\,\cos\theta_t \equiv -q\cdot q' + q\cdot p\, q'\cdot p/M_N{}^2\,. \qquad (20)$$

In fact, $(\cos\theta)(\cos\theta)^N$ is the direct product of the basic representation functions $d_{[1][1]}{}^{(6,\bar{6})}(\theta)\, d_{[1][1]}{}^N(\theta)$ for $[U(6)\otimes U(6)]\otimes U(3)$ from which the general

$$d_{[W][W']}{}^{(6,\bar{6})}(\theta)\, d_{[W][W']}{}^N(\theta)$$

follow by the differentiations. Let us list the first few derivatives for later use:

$$\frac{\partial}{\partial q_A{}^B}\Delta_{(N)} = \frac{(\mathbf{q}\cdot\mathbf{q}')^N}{p^2-M^2}\,\frac{[(M+p)q'(M-p)]_B{}^A}{4M^2}\,, \qquad (36a)$$

$$\frac{\partial}{\partial q_\mu}\Delta_{(N)} = \frac{N(\mathbf{q}\cdot\mathbf{q}')^N|\mathbf{q}'|}{p^2-M^2}\left(-q_\mu{}'+\frac{p\cdot q'}{M^2}p_\mu\right), \qquad (36b)$$

$$\frac{\partial^2}{\partial q_\mu \partial q_A{}^B}\Delta_{(N)} = \frac{N(\mathbf{q}\cdot\mathbf{q}')^{N-1}|\mathbf{q}'|^2}{p^2-M^2}\left(-q_\mu+\frac{p\cdot q'}{M^2}p_\mu\right) \times\frac{[(M+p)q'(M-p)]_B{}^A}{4M^2}\,, \qquad (36c)$$

$$\frac{\partial^2}{\partial q_A{}^B \partial q'_{B'}{}^{A'}}\Delta_{(N)} = \frac{(\mathbf{q}\cdot\mathbf{q}')^N}{p^2-M^2}(M+p)_B{}^{B'}(M-p)_{A'}{}^A\,. \qquad (36d)$$

These Reggeize through the replacement

$$\frac{(\mathbf{q}\cdot\mathbf{q}')^N}{p^2-M^2} \to \frac{(\mathbf{q}\cdot\mathbf{q}')^{\alpha-1}}{\sin\pi(\alpha-1)} \quad (N\to\alpha-1)\,. \qquad (37)$$

It is well known that (37) gives poles at nonsense N

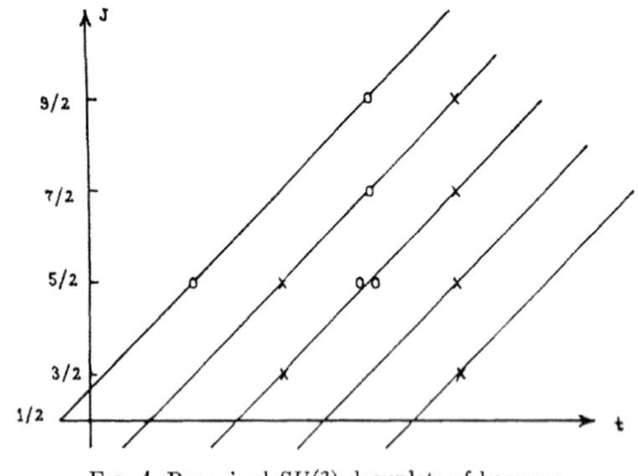

FIG. 4. Reggeized $SU(3)$ decuplets of baryons from oscillator model.

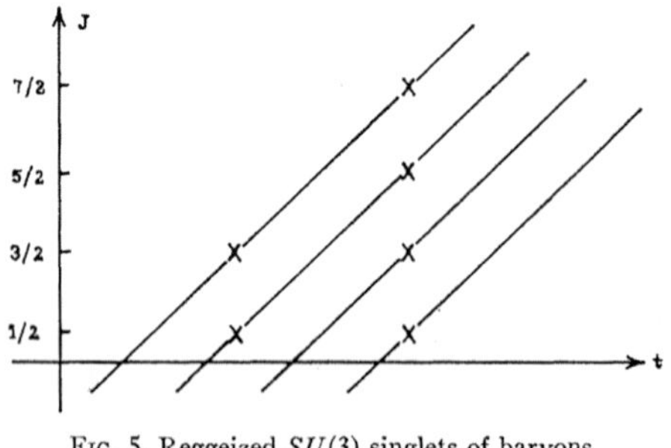

FIG. 5. Reggeized $SU(3)$ singlets of baryons from oscillator model.

[17] The subscripts on the dimensionless coupling constants refer to excitation numbers of the $U(6)\otimes U(2)_W$ subgroup representations; for instance, h_{11} in (30) corresponds to the (**35**;**3**) component of the exchanged meson, the "0" referring to singlets of the group, "1" to the first multiplet, "2" to the second, and so on. Note that, in contrast to model B, models A(i) and A(ii) can have no **405** component of $U(6)_W$ in couplings to the exchanged mesons. Also, observe the very small number of constants that appear, particularly for the baryons. This is the most powerful predictive feature of supermultiplet theory.

values ($\alpha=0, -1, -2, \cdots$). A recent suggestion to avoid these poles has been to replace $1/\sin\pi(\alpha-1)$ by $\Gamma(1-\alpha)$. This corresponds to introducing the Gell-Mann ghost-eliminating mechanism. It may be possible to devise other mechanisms for eliminating nonsense J values, but they would not fit into the orbital excitation picture so simply.

The above is all the apparatus one needs for studying the physical amplitudes of interest. We now summarize the formulas for BB and MB scattering:

$$T_{BB}=\bar{u}^{ACD}(\tfrac{1}{2}p+q')u_{BCD}(-\tfrac{1}{2}p+q)\bar{u}^{B'C'D'}(-\tfrac{1}{2}p+q')u_{A'C'D'}(\tfrac{1}{2}p+q')$$
$$\times m^{-2(N+1)}[g_0\delta_A{}^B+mg_1(\partial/\partial q_B{}^A)][g_0\delta_{B'}{}^{A'}+mg_1(\partial/\partial q_{A'}{}^{B'})]\Delta_{(N)}, \quad (38)$$

$$T_{MB}=\bar{u}^{ACD}(\tfrac{1}{2}p+q)u_{BCD}(-\tfrac{1}{2}p+q)\Phi_{A'}{}^{B'}(\tfrac{1}{2}p+q')\Phi_{C'}{}^{D'}(\tfrac{1}{2}p-q')(m\mu)^{-(N+1)}$$
$$\times\left[g_0\delta_B{}^A+mg_1\frac{\partial}{\partial q_A{}^B}\right]\left[h_0{}^{(-)}\delta_{B'}{}^{C'}\delta_{D'}{}^{A'}+\mu^{-2}h_0'{}^{(-)}q'_{B'}{}^{A'}q'_{D'}{}^{C'}\right.$$
$$\left.+\mu h_1{}^{(+)}\left(\delta_{B'}{}^{C'}\frac{\partial}{\partial q'_{A'}{}^{D'}}+\delta_{D'}{}^{A'}\frac{\partial}{\partial q'_{C'}{}^{B'}}\right)+\mu h_1{}^{(-)}\left(\delta_{B'}{}^{C'}\frac{\partial}{\partial q'_{A'}{}^{D'}}-\delta_{D'}{}^{A'}\frac{\partial}{\partial q'_{C'}{}^{B'}}\right)\right]\Delta_{(N)}. \quad (39)$$

These are the master formulas of this paper. For definitions of derivatives and wave functions $\Phi_B{}^A, \cdots$, see (22)–(28).

To illustrate the use of these master formulas, consider the charge-exchange MB processes. For those amplitudes, the couplings g_0 and h_0 are not relevant since they only describe elastic processes. Thus,

$$T_{\text{c.e.}}=(m\mu)^{-N-1}g_1h_1{}^{(+)}\bar{u}^{ACD}u_{BCD}\{\Phi,\Phi\}_{B'}{}^{A'}\frac{\partial}{\partial q_B{}^A}\frac{\partial}{\partial q'_{A'}{}^{B'}}\Delta_{(N)}$$
$$+(m\mu)^{-N-1}g_1h_1{}^{(-)}\bar{u}^{ACD}u_{BCD}[\Phi,\Phi]_{B'}{}^{A'}\frac{\partial}{\partial q_B{}^A}\frac{\partial}{\partial q'_{A'}{}^{B'}}\Delta_{(N)}. \quad (40)$$

Recall that ($\pm$) superscripts refer to even and odd N. As is well known, after Reggeization a signature factor needs to be introduced into the formalism (the simple argument of Reggeization in Sec. II does not automatically produce this). In what follows, whenever we write $g^{\pm}$, we shall assume that N-signature projections $\frac{1}{2}(1\pm e^{i\pi N})$ are to be included. Performing the differentiation and simplifying,

$$T_{\text{c.e.}}=[g_1h_1{}^{(+)}/(t-M^2)]\bar{u}\Gamma_+\{\Phi,\Phi\}\Gamma_-u(\mathbf{q}\cdot\mathbf{q}')^N+[g_1h_1{}^{(-)}/(t-M^2)]\bar{u}\Gamma_+[\Phi,\Phi]\Gamma_-u(\mathbf{q}\cdot\mathbf{q}')^N \quad (41)$$

with

$$(\Gamma_\pm)_A{}^B=(1/2M)(M\pm p)_A{}^B \quad (42)$$

and $t\equiv p^2=M^2$ at the pole. Let us focus our attention on only those reactions where the incoming meson is pseudoscalar and the target is a nucleon, while allowing the final meson and nucleon to be any other member of the $(6,\bar{6})$ and $(56,1)$ multiplets. Breaking up the outgoing mesons in formula (41) into vector and pseudoscalar parts and separating out $SU(3)$ components, the Reggeized amplitude reads

$$T_{\text{Regge}}=[1+M/2\mu]\bar{u}^{ACD}(\Gamma_+q'\Gamma_-/\mu)_\alpha{}^\beta u_{BCD}$$
$$\times[\tfrac{1}{2}\beta_-(1-e^{i\pi\alpha_-})\Gamma(1-\alpha_-)(P,P)_F(\mathbf{q}\cdot\mathbf{q}'/\mu m)^{\alpha_--1}+\tfrac{1}{2}\beta_+(1+e^{i\pi\alpha_+})\Gamma(1-\alpha_+)(P,P)_D(\mathbf{q}\cdot\mathbf{q}'/\mu m)^{\alpha_+-1}]_a{}^b$$
$$+[1+M/2\mu]\bar{u}^{ACD}(\Gamma_+[q',\gamma_\lambda]\gamma_5\Gamma_-/2\mu)_\alpha{}^\beta u_{BCD}[\tfrac{1}{2}\beta_-(1-e^{i\pi\alpha_-})\Gamma(1-\alpha_-)(P,V_\lambda)_D(\mathbf{q}\cdot\mathbf{q}'/\mu m)^{\alpha_--1}$$
$$+\tfrac{1}{2}\beta_+(1+e^{i\pi\alpha_+})\Gamma(1-\alpha_+)(P,V_\lambda)_F(\mathbf{q}\cdot\mathbf{q}'/\mu m)^{\alpha_+-1}]_a{}^b+[1+M/2\mu]\bar{u}^{ACD}(\Gamma_-\gamma_\lambda\gamma_5\Gamma_+)_\alpha{}^\beta u_{BCD}$$
$$\times[\tfrac{1}{2}\beta_-(1-e^{i\pi\alpha_-})\Gamma(1-\alpha_-)(P,V_\lambda)_F(\mathbf{q}\cdot\mathbf{q}'/\mu m)^{\alpha_--1}+\tfrac{1}{2}\beta_+(1+e^{i\pi\alpha_+})\Gamma(1-\alpha_+)(P,V_\lambda)_D(\mathbf{q}\cdot\mathbf{q}'/\mu m)^{\alpha_+-1}]_a{}^b. \quad (43)$$

We have distinguished between the two possible trajectories α_- and α_+ associated with the two signatures. The $(\mathbf{q}\cdot\mathbf{q}')^{\alpha_+-1}$ or $(\mathbf{q}\cdot\mathbf{q}')^{\alpha_--1}$ oscillator factors are characteristic orbital effects. Making a further decomposition into octet (N) and decuplet (D) pieces of the **56**, one gets

$$T_{PN\to PN}=\{a_F(\bar{N}N)_F+a_D(\bar{N}N)_D+\mu^{-1}[b_F(\bar{N}q'N)_F+b_D(\bar{N}q'N)_D]\}$$
$$\times[\tfrac{1}{2}\beta_-\Gamma(1-\alpha_-)(1-e^{i\pi\alpha_-})(P,P)_F(\mathbf{q}\cdot\mathbf{q}'/\mu m)^{\alpha_--1}+\tfrac{1}{2}\beta_+\Gamma(1-\alpha_+)(1+e^{i\pi\alpha_+})(P,P)_D(\mathbf{q}\cdot\mathbf{q}'/\mu m)^{\alpha_+-1}] \quad (44)$$

with $M^2=t$,

$$a_D=\left(1+\frac{M}{2\mu}\right)\left(1+\frac{2m}{M}\right)\frac{\mathbf{q}\cdot\mathbf{q}'}{m\mu}, \quad a_F=\left(\frac{2}{3}-\frac{M}{2m}\right)a_D, \quad (45a)$$

$$b_D=\left(1+\frac{M}{2\mu}\right)\left(1+\frac{2m}{M}\right)\left(1-\frac{M^2}{4m^2}\right),\quad b_F=\tfrac{2}{3}b_D. \tag{45b}$$

D, F having the usual $SU(3)$ connotations, as previously,

$$T_{PN\to PD}=\left(1+\frac{M}{2\mu}\right)\left(1+\frac{2m}{M}\right)\frac{1}{m^2\mu}\epsilon_{\mu\nu k\lambda}q_\mu' q_\nu p_k\bar{D}_\lambda N[\tfrac{1}{2}\beta_-(1-e^{i\pi\alpha_-})\Gamma(1-\alpha_-)(P,P)_F(\mathbf{q}\cdot\mathbf{q}'/\mu m)^{\alpha_--1}$$
$$+\tfrac{1}{2}\beta_+(1+e^{i\pi\alpha_+})\Gamma(1-\alpha_+)(P,P)_D(\mathbf{q}\cdot\mathbf{q}'/\mu m)^{\alpha_+-1}], \tag{46}$$

$$T_{PN\to VN}=\left(1+\frac{M}{2\mu}\right)\bar{u}\left\{\frac{\mathbf{q}\cdot\mathbf{q}'}{M^2\mu}p_\lambda-\frac{i\sigma_{k\lambda}q_k'}{\mu}+\frac{\{p,[q',\gamma_\lambda]\}}{4M\mu}\right\}\gamma_5 u[\tfrac{1}{2}\beta_-(1-e^{i\pi\alpha_-})\Gamma(1-\alpha_-)(P,P_\lambda)_D(\mathbf{q}\cdot\mathbf{q}'/\mu m)^{\alpha_--1}$$
$$+\tfrac{1}{2}\beta_+(1+e^{i\pi\alpha_+})\Gamma(1-\alpha_+)(P,V_\lambda)_F(\mathbf{q}\cdot\mathbf{q}'/\mu m)^{\alpha_+-1}]+\left(1+\frac{M}{2\mu}\right)\left(1+\frac{2m}{M}\right)\bar{u}\gamma_5 u\cdot\frac{p_\lambda}{M}$$
$$\times[\tfrac{1}{2}\beta_-(1-e^{i\pi\alpha_-})\Gamma(1-\alpha_-)(P,V_\lambda)_F(\mathbf{q}\cdot\mathbf{q}'/\mu m)^{\alpha_--1}+\tfrac{1}{2}\beta_+(1+e^{i\pi\alpha_+})\Gamma(1-\alpha_+)(P,V_\lambda)_D(\mathbf{q}\cdot\mathbf{q}'/\mu m)^{\alpha_+-1}]. \tag{47}$$

These are the amplitudes in the exact-symmetry limit with all masses degenerate. The following significant features may be noted:

(a) Barring differences due to signature, there is a common residue and trajectory function occurring in the characteristic combination

$$\left(1+\frac{2m}{M}\right)\left(1+\frac{M}{2\mu}\right)\beta\left(\frac{\mathbf{q}\cdot\mathbf{q}'}{m\mu}\right)^{\alpha-1}\Gamma(1-\alpha). \tag{48}$$

(b) The signature factors $(1\pm e^{i\pi\alpha_\pm})$ multiplied into $\Gamma(1-\alpha_\pm)$ mean that the even α-signature amplitudes vanish for $\alpha=-1,\ -3,\ \cdots$ and odd α-signature amplitudes vanish for $\alpha=0,\ -2,\ \cdots$. Hence, the vector $(-)$ trajectory gives amplitude zeros for $\alpha=0,\ -2,\ \cdots$ and the tensor $(+)$ trajectory for $\alpha=-1,\ -3,\ \cdots$; this applies to all the Regge amplitudes and not just the spin-flip components. We may therefore expect *dips* in the cross sections at these places. This is indeed borne out experimentally. Such dips are well known for π-N charge-exchange processes; the important remark from our point of view is that $\pi N\to\eta N$ also appears to show such dips at even-signature positions.

(c) In the forward direction, we expect to reproduce the predictions of Carter *et al.*,[18] since $SU(6)_W$ is conserved in this limit. Thus, $PN\to PD$ vanish owing to the M_1 nature of the coupling to DN of the *vector* and also the *tensor* trajectory.

V. TOLLERIZATION OF AMPLITUDES AND KINEMATIC SINGULARITIES

Before we compare our results with experiment, we must consider the $t\to 0$ limit where kinematic singularities make their appearance whenever spins are involved. This is shown quite clearly in the characteristic product

$$\left(1+\frac{2m}{\sqrt{t}}\right)\left(1+\frac{\sqrt{t}}{2\mu}\right)\beta(t),$$

which is not an analytic function of t, near $t=0$. As is well known, two types of mechanism have been proposed to remove these singularities: the first evasion—the statement that $\beta(t)$ must have a compensating zero; the second is the addition of conspiring trajectories. The most elegant formulation of conspiracies is the one proposed by Toller where one expands scattering amplitudes not in terms of a complete set of rotation functions of the rest group G, but in terms of the covariant embedding group $\mathcal{G}$ [in Toller's case, $G=O(3)$ and $\mathcal{G}=O(3,1)$]. A Tollerization procedure then replaces Reggeization; this, as is well known, leads to the parent and daughter phenomena.

For the quark-excitation model, Tollerization was studied in detail in Ref. 8. The rotation functions of the embedding groups for models studied here are the following:

		Embedding non-compact group	Rotation functions	
A	$\Rightarrow$	$O(3,1)$	$C_{\mathfrak{N}}{}^1(\cosh\zeta)$	
A(i)	$\Rightarrow$	$U(2,2)$	$C_{\mathfrak{N}}{}^{3/2}(\cosh\zeta)$	(49)
A(ii)	$\Rightarrow$	$U(3,1)$	$(\cosh\zeta)^N$	
B	$\Rightarrow$	$U(6,6)$	$C_{\mathfrak{N}}{}^{11/2}(\cosh\zeta)$.	

An alternative solution—and one not as general as Toller's—is to introduce conspiring trajectories following Gribov.[19] This is the solution most suited to the multispinor formalism for hadrons. It arises from the natural possibility of doubling afforded by quarks and pseudoquarks within a multispinor framework (this is the doubling first introduced by Gribov).

[18] J. Carter, J. Coyne, S. Meshkov, D. Horn, M. Kugler, and H. J. Lipkin, Phys. Rev. Letters **15**, 373 (1965).

[19] V. N. Gribov, L. Okun, and I. Pomeranchuk, Zh. Eksperim. i Teor. Fiz. **45**, 1114 (1963) [English transl.: Soviet Phys.—JETP **18**, 769 (1964)].

For mesons, for example, one is led to consider two trajectories coinciding at $M=t=0$ with identical residues but otherwise distinct, which correspond to $(6,\bar{6})$ and $(\bar{6},6)'$, the primes indicating the pseudoquark composites.

In accordance with these ideas, we must therefore add to the amplitudes (43) extra terms with the sign of M reversed, corresponding to $(6,\bar{6})\to(\bar{6},6)'$. For the MB scattering amplitudes, then, we typically meet the combination

$$\frac{1}{2}\left(1+\frac{2m}{M}\right)\left(1+\frac{M}{2\mu}\right)\beta(M)\Gamma(1-\alpha)s^{\alpha-1}$$
$$+\frac{1}{2}\left(1-\frac{2m}{M}\right)\left(1-\frac{M}{2\mu}\right)\beta'(M)\Gamma(1-\alpha')s^{\alpha'-1}. \tag{50}$$

Taking $\alpha(0)=\alpha'(0)$, $\beta(0)=\beta'(0)$, the $M=\sqrt{t}$ singularity disappears. In fact, if we suppose that for moderately small spacelike t, $\alpha'(t)=\alpha(t)$ and $\beta\approx\beta'\approx$const, the combination of terms[20] could sum to[21]

$$\left(1+\frac{m}{\mu}\right)\beta\Gamma(1-\alpha)s^{\alpha-1}.$$

Before applying our formalism to elastic processes, one has to make up one's mind about the Pomeranchukon. In the absence of any fundamental understanding of vacuum exchange, it is probably fair to regard the Pomeranchukon as a [fixed, $SU(3)$ scalar?] pole with $\alpha(0)=1$ which occurs in elastic processes to describe the background effects of inelastic channels via unitarity.

The remaining problem is symmetry breaking. We know it to be very important as far as trajectory shifts, which govern high-energy behaviors, are concerned. Failing a reliable theory of mass splitting between members of a supermultiplet, the only course open at present is to take the positions of the trajectories as empirical input (most significantly for the pion). With regard to the residues, one may hope that they do not change drastically, and this is what seems to be borne out by our preliminary analysis of data (Ref. 4). If this had not worked, our first prescription would have been to use physical masses in place of mean masses in kinematic factors.

VI. REGGEIZED MESON-BARYON SCATTERING AND COMPARISON WITH EXPERIMENT

We now list the final formulas and their main features for charge-exchange scattering processes that are dominated by the vector ($C=-1$) and tensor ($C=+1$) leading trajectories; namely, those in which both the initial and final meson are pseudoscalar. We shall, here, make an assumption outside the supermultiplet schemes proper—that of exchange degeneracy, i.e., $\alpha_+=\alpha_-$, $\beta_+=\beta_-$. The final expressions are as follows:

$T_{PN\to PN}$: Flip amplitude is

$$B=\left(1+\frac{m}{\mu}\right)\left(1-\frac{t}{4m^2}\right)$$
$$\times(\bar{N}q'N)_{D+(2/3)F}\frac{\beta}{\mu}\Gamma(1-\alpha)\left(\frac{s}{2m\mu}\right)^{\alpha-1}$$
$$\times[\tfrac{1}{2}(1-e^{i\pi\alpha})(PP)_F+\tfrac{1}{2}(1+e^{i\pi\alpha})(PP)_D] \tag{51a}$$

and the nonflip amplitude is

$$A'\approx A+\frac{s}{2m\mu}\left(1-\frac{t}{4m^2}\right)^{-1}B$$
$$=\left(1+\frac{t}{4m\mu}\right)(\bar{N}N)_F\beta\Gamma(1-\alpha)\left(\frac{s}{2m\mu}\right)^{\alpha}$$
$$\times[\tfrac{1}{2}(1-e^{i\pi\alpha})(PP)_F+\tfrac{1}{2}(1+e^{i\pi\alpha})(PP)_D], \tag{51b}$$

which enter in the differential cross section as

$$\frac{d\sigma}{dt}=\frac{1}{16\pi s^2}\left[\left(1-\frac{t}{4m^2}\right)|a'|^2\right.$$
$$\left.-\frac{t}{4m^2}\left(1-\frac{t}{4m^2}\right)^{-1}\left|\frac{sb}{2m\mu}\right|^2\right] \tag{52}$$

and give

$$\left(\frac{d\sigma}{dt}\right)_{PN\to PN}=\frac{|\beta\Gamma(1-\alpha)|^2}{16\pi s^2}\left(1-\frac{t}{4m^2}\right)\left(\frac{s}{2m\mu}\right)^{2\alpha}$$
$$\times\left\{\left(1+\frac{t}{4m\mu}\right)^2-\frac{t}{4m^2}\left[\frac{g_{D+(2/3)F}}{g_F}\left(1+\frac{m}{\mu}\right)\right]^2\right\}$$
$$\times|\tfrac{1}{2}(1-e^{i\pi\alpha})h_F+\tfrac{1}{2}(1+e^{i\pi\alpha})h_D|^2, \tag{53}$$

the F and D suffixes carrying the normal $SU(3)$ meanings.

$T_{PN\to PD}$: The baryon vertex is of the expected M_1 *type* all along the trajectory, as may be seen in the amplitude

$$C=\left(1+\frac{\bar{m}}{\mu}\right)\frac{1}{\bar{m}^2\mu}\epsilon_{\mu\nu\kappa\lambda}q_\mu'q_\nu p_\kappa\bar{D}_\lambda N\beta\Gamma(1-\alpha)\left(\frac{s}{2m\mu}\right)^{\alpha-1}$$
$$\times[\tfrac{1}{2}(1-e^{i\pi\alpha})(PP)_F+\tfrac{1}{2}(1+e^{i\pi\alpha})(PP)_D]. \tag{54}$$

Here, $\bar{m}$ is the mean of octet (N) and decuplet (D) masses. When this is substituted into the cross section

$$\frac{d\sigma}{dt}=\frac{1}{6\pi s^2}\left(1-\frac{t}{4\bar{m}^2}\right)\frac{s^2t}{\bar{m}^4\mu^2}|c|^2,$$

[20] There is the familiar problem of existence of particles on the Gribov-doubled trajectories. We have no new ideas on this except the conventional one that possibly the relevant residues vanish.

[21] The difference between Toller and Gribov conspiracies lies in that Toller would demand, in addition to Gribov doubling, the trajectories corresponding to $(36,1)^+$ and $(1,36)^+$.

one gets

$$\left(\frac{d\sigma}{dt}\right)_{PN\to PD} \approx -\frac{t|\beta\Gamma(1-\alpha)|^2}{6\pi s^2\bar{m}^2}\left(1-\frac{t}{4\bar{m}^2}\right)\left(\frac{s}{2\bar{m}\mu}\right)^{2\alpha}\left(1+\frac{\bar{m}}{\mu}\right)^2 \times|\tfrac{1}{2}(1-e^{i\pi\alpha})h_F+\tfrac{1}{2}(1+e^{i\pi\alpha})h_D|^2.$$

To make comparison with experiment for the $Y=0$ charge-exchange processes

$$\pi^-p\to\pi^0n,\quad K^-p\to\bar{K}^0n,\quad \pi^-p\to\eta n,$$
$$\pi^+p\to\pi^0N^{*++},\quad K^+p\to K^0N^{*++},\quad \pi^+p\to\eta N^{*++},$$

for which considerable data exist, we have used mean masses $\langle\mu\rangle\approx0.45$ GeV/c, $\langle m\rangle\approx1.15$ GeV/c, and $\langle\bar{m}\rangle\approx1.3$ GeV/c in the kinematic factors and the exchange-degeneracy approximation $\alpha_+\approx\alpha_-\approx\alpha\approx0.5+t$, in units of $(\mathrm{GeV}/c)^2$; we also took a constant residue β. Considering these very rough approximations, the agreement with the data is quite encouraging (Ref. 4).

In summary, the model has the following characteristics:

(1) It is based on the excitation scheme

$$[U(6)\otimes U(6)]_S\otimes U(3)_N,$$

though for Regge exchange of meson trajectories only the leading orbital excitation $L=N=\alpha-1$ is significant.

(2) The relativistic kinematics is provded by the embedding group $\tilde{U}(12)_S\otimes U(3,1)_N$ and produces significant kinematic factors in the Clebsch-Gordan coefficients.

(3) From the quark-antiquark nature of the mesons, $1/\sqrt{t}$ kinematic factors are encountered. A Gribov doubling of the exchanged trajectory $(6,\bar{6})$ and $(\bar{6},6')$ has been employed to eliminate these $1/\sqrt{t}$ singularities.

(4) At nonsense J values we have used the Gell-Mann mechanism for eliminating ghosts. Perhaps other mechanisms could be constructed, but they do not appear to fit so naturally into the excitation picture. Our model predicts zeros in the π charge-exchange reactions at $\alpha=0, 2, \cdots$, zeros in $\pi^\pm\to\eta$ reactions at $\alpha=-1, -3, \cdots$, and no significant dips in the K charge-exchange reactions. This last fact is due to cancellations of the signature factors in the Gell-Mann mechanism.

(5) The assumed exchange degeneracy can be relaxed by using separate α_+, β_+ and α_-, β_- for the even- and odd-signature pieces in place of the assumed common $\alpha=0.5+t$ and β.

(6) Inasmuch as we have neglected any possible t dependence of β, our simplified model admits no parameters except the one constant β. When β is fixed by the high-energy data and the Regge formula is extrapolated to the vector-meson mass, one finds $(g_{\rho\pi\pi})_{\mathrm{expt}}=2(g_{\rho\pi\pi})_{\mathrm{theoret}}$. This is a failure of the model, and we have no explanation for it.

(7) A turnover effect at small t is predicted by the model because of the largeness of the spin flip relative to the flipless amplitude: $(b/a)(s/2m\mu)=-(5/3)(1+m/\mu)\approx-6$. Moreover, the opposite relative sign of the two and the fact that a' is a polynomial in t means that an extension of the model to elastic scattering cross-section differences will give rise to a crossover effect. However, the positions of turnover and crossover points are incorrectly given by the model (which uses mean masses) at $t\approx0.05$ and 0.5 GeV/c. It is possible that mass shifts play an important role in altering these points.

(8) Density-matrix calculations, which have not been discussed, constitute important tests of the model. Observe, too, that in its present form the simple pole picture used in this section is unable to explain polarization effects as it deals with a *single* Regge amplitude.

(9) Reactions dominated by pion exchange, e.g., $T_{PN\to VN}$, $T_{NN\to NN}$, which will form the subject of a separate note, show interesting new features owing to the singular character of the pion residue and its nearness to the scattering region. For instance, in nucleon-nucleon charge-exchange scattering, the π and ρ contributions together give terms of the form

$$d\sigma/dt\approx[s^{\alpha_\pi}\Gamma(-\alpha_\pi)]^2+[s^{\alpha_\rho}\Gamma(1-\alpha_\rho)]^2 \qquad (55)$$

so that for small t and moderately large s the pion dominates and *gives the observed sharp forward peak*, but at larger values of t the ρ takes over. The reason why the pion gives a finite nonzero cross section even in the forward limit $t\to0$ is that the amplitude $\bar{u}\not{p}\gamma_5u\bar{u}'\not{p}\gamma_5u/p^2=(4m^2/t)\bar{u}\gamma_5u\bar{u}'\gamma_5u$ carries a desirable singular residue $1/t$, even though there is Gribov doubling.[22] The actual detailed analysis of pion-exchange reactions will be treated elsewhere.

[22] This $1/t$ singularity is of course cancelled by the t factor occurring in the nucleon $\gamma_5\otimes\gamma_5$ piece.

IL NUOVO CIMENTO VOL. LXIX A, N. 3 1° Ottobre 1970

Veneziano Model for the Scattering of Meson and Baryon Supermultiplets (*) (**).

R. DELBOURGO and P. ROTELLI

Physics Department, Imperial College - London

(ricevuto il 20 Aprile 1970)

Summary. — An adaptation of the multispinor formalism for evaluating duality diagrams is applied to the scattering of mesons $(6, \overline{6})$ from baryons (56, 1). The prescription automatically provides the « correct » D/F ratios and relative magnitudes of helicity flip/no flip amplitudes at the particle poles, and reproduces Reggeized supermultiplet predictions in the asymptotic region. However, the resonance structure is highly unphysical and consequently the low-energy predictions are very poor.

A proposal (1) was recently made for combining supermultiplet schemes and Veneziano models, since both ideas are based on pole dominance and are subject to « unitary corrections ». This proposal consisted of the following simple rule. Draw the legal duality diagram (2), contract SU_3 and $SU_{2,2}$ spinor indices as specified by the diagram, and multiply by a suitable Veneziano « form factor ». The details were presented for meson-meson scattering and although the results (which differed from earlier treatments) were valuable in that they synthesized such disparate processes as $\pi\pi \to \pi\pi$, $\pi\omega$, $\rho\rho$ etc. in a stringent way, they had limited application. We wish here to report on the details for

(*) To speed up publication, the authors of this paper have agreed to not receive the proofs for correction.

(**) The research reported in this document has been sponsored in part by the Air Force Office of Scientific Research OAR through the European Office of Aerospace Research, United States Air Force.

(1) R. DELBOURGO and P. ROTELLI: *Phys. Lett.*, to be published.

(2) H. HARARI: *Phys. Rev. Lett.*, **22**, 562 (1969); J. L. ROSNER: *Phys. Rev. Lett.*, **22**, 689 (1969).

meson-baryon scattering where the predictions (³) are more amenable to experimental comparison. The important point to appreciate is that supermultiplet techniques dictate the ratios of amplitudes with varying degrees of spin flip (as in the meson-meson case) and the *correct D/F ratios* of the meson-baryon couplings without having to invoke these separately (⁴). The Veneziano ansatz, as we shall see, provides a very rich and unphysical resonance structure. However, our model is expected to apply as a reasonable leading approximation to *inelastic* amplitudes (⁵) for all values of s, t, u in circumstances where the endemic mass degeneracies and parity doublings are deemed not to be especially significant. The symmetry breaking and mass splitting thus rest upon unitarizing the model.

There are only three duality diagrams for the process

$$(6, \overline{6})_{p_1} + (56, 1)_{p_2} \to (6, \overline{6})_{-p_3} + (56, 1)_{-p_4}$$

which, according to the advocated rule, give the total amplitude

$$\overline{\Psi}^{ABC}(-p_4)\Psi_{ABD}(p_2)[\Phi^E_C(p_3)\,\Phi^D_E(p_1)\,V_{\psi\varphi}(s,t) + (1\leftrightarrow 3,\, s\leftrightarrow u)] + \\ + \overline{\Psi}^{ABC}(-p_4)\Psi_{ADE}(p_2)\,\Phi^D_B(p_3)\,\Phi^E_C(p_1)\,V_{\psi\psi}(s,u)$$

with

$$V_{\psi\varphi}(s,t) = \beta_{\psi\varphi}\frac{\Gamma(\frac{3}{2}-\alpha_\psi(s))\,\Gamma(1-\alpha_\varphi(t))}{\Gamma(\frac{5}{2}-\alpha_\psi(s)-\alpha_\varphi(t))}$$

and

$$V_{\psi\psi}(s,u) = \beta_{\psi\psi}\frac{\Gamma(\frac{3}{2}-\alpha_\psi(s))\,\Gamma(\frac{3}{2}-\alpha_\psi(u))}{\Gamma(3-\alpha_\psi(s)-\alpha_\psi(u))}\,.$$

Such an amplitude will describe many reactions including production of $\frac{3}{2}^+$ and 1^- resonances, but from hereon in we shall concentrate on the scattering of 0^- and $\frac{1}{2}^+$ octets.

(³) For more conventional Veneziano models of $\pi\mathcal{N}$ scattering see K. IGI: *Phys. Lett.*, **28** B, 330 (1968); S. K. BOSE and K. C. GUPTA: University of Notre Dame preprint; J. NAMYSLOWSKI and M. SWIECKI: Polish Institute of Nuclear Research preprint; C. LOVELACE: CERN preprint TH. 1047 (25th June 1969); A. W. HENDRY, S. T. JONES and H. W. WYLD jr.: University of Illinois preprint.

(⁴) In the case of meson-meson scattering, exchange degeneracy and charge conjugation invariance are alone sufficient to determine the SU_3 structure except for the octet/singlet ratio.

(⁵) The Pomeron contribution to elastic amplitudes is, as always, outside the scope of duality but must necessarily appear in any reasonable unitarization procedure.

If we insert the standard wave functions

$$\Phi_A^B(p) = [(\gamma\cdot p + \mu)\gamma_5]_\alpha^\beta \varphi_a^b ,$$
$$\overline{\Psi}_{ABC}(p) = [(\gamma\cdot p + m)\gamma_5 c]_{\alpha\beta}\varepsilon_{abx}\psi^x_{\gamma,c}(p) + \text{permutations}$$

and let $\langle\ \rangle$ stand for SU_3 traces, we can compactly express the conventional A' and B' amplitudes as follows:

$$A' = [\langle 2\,\bar{4}\,3\,1\rangle - \langle \bar{4}\,2\,3\,1\rangle + \langle \bar{4}\,2\rangle\langle 3\,1\rangle]\mathscr{A}'_{\psi\varphi}(s,t) + [3\langle 2\,\bar{4}\,3\,1\rangle - 2\langle 4\,2\,3\,1\rangle +$$
$$+ 2\langle \bar{4}\,1\,2\,3\rangle + \langle \bar{4}\,2\rangle\langle 3\,1\rangle - 3\langle \bar{4}\,1\rangle\langle 3\,2\rangle]\mathscr{A}'_{\psi\psi}(s,u) + (1\leftrightarrow 3,\ s\leftrightarrow u)\,,$$

$$B = \tfrac{1}{3}[5\langle 2\,\bar{4}\,3\,1\rangle + \langle \bar{4}\,2\,3\,1\rangle - \langle \bar{4}\,2\rangle\langle 3\,1\rangle]\mathscr{B}_{\psi\varphi}(s,t) +$$
$$+ \tfrac{3}{2}[\langle \bar{4}\,3\,1\,2\rangle + \langle \bar{4}\,1\rangle\langle 3\,2\rangle]\mathscr{B}_{\psi\psi}(s,u) - (1\leftrightarrow 3,\ s\leftrightarrow u)\,,$$

where we have defined spin kinematic factors

$$\mathscr{A}'(s,t) \quad = [(s-u)(4m\mu + t) - (4\mu^2 - t)(4m^2 - t)]\,V_{\psi\varphi}(s,t)\,,$$
$$\mathscr{A}'_{\psi\psi}(s,u) = [(m^2-\mu^2)^2 + t(m+\mu)^2 - us)\,V_{\psi\psi}(s,u)\,,$$
$$\mathscr{B}_{\psi\varphi}(s,t) \quad = 4\mu(m+\mu)(4m^2 - t)\,V_{\psi\varphi}(s,t)\,,$$
$$\mathscr{B}_{\psi\psi}(s,u) \quad = 4\mu(m+\mu)(4m^2 - t)\,V_{\psi\psi}(s,u)\,.$$

In particular, for $\pi\mathscr{N}$ scattering the results read

$$A'^{(+)} = 9[\mathscr{A}'_{\psi\varphi}(s,t) + \mathscr{A}'_{\psi\varphi}(u,t)] + 10\mathscr{A}'_{\psi\psi}(s,u)\,,$$
$$A'^{(-)} = 3[\mathscr{A}'_{\psi\varphi}(s,t) - \mathscr{A}'_{\psi\varphi}(u,t)]\,,$$
$$B^{(+)} \ = 3[\mathscr{B}_{\psi\varphi}(s,t) - \mathscr{B}_{\psi\varphi}(u,t)]\,,$$
$$B^{(-)} \ = 5[\mathscr{B}_{\psi\varphi}(s,t) + \mathscr{B}_{\psi\varphi}(u,t)] + 3\mathscr{B}_{\psi\psi}(s,u)\,.$$

The most important theoretical formulae follow by going to

i) *Asymptotic s values.* At fixed t,

$$A'^{(+)} \sim 18s(4m\mu + t)\beta_{\psi\varphi}[R_\varphi(s) - R_\varphi(-s)]\,,$$
$$A'^{(-)} \sim 6s(4m\mu + t)\beta_{\psi\varphi}[R_\varphi(s) + R_\varphi(-s)]\,,$$
$$B^{(+)} \sim 12\mu(m+\mu)(4m^2 - t)\beta_{\psi\varphi}[R_\varphi(s) - R_\varphi(-s)]\,,$$
$$B^{(-)} \sim 20\mu(m+\mu)(4m^2 - t)\beta_{\psi\varphi}[R_\varphi(s) + R_\varphi(-s)]\,,$$

where

$$R_\varphi(s) = (-\alpha' s)^{\alpha_\varphi(t)-1}\Gamma(1-\alpha_\varphi(t)) ,$$

while, for fixed u,

$$A'^{(+)} \sim s[u + (m+\mu)^2][18\beta_{\psi\varphi} R_\psi(-s) - 10\beta_{\psi\psi} R_\psi(s)] ,$$
$$A'^{(-)} \sim -6s[u + (m+\mu)^2]\beta_{\psi\varphi} R_\psi(-s) ,$$
$$B^{(+)} \sim -12su(m+\mu)\beta_{\psi\varphi} R_\psi(-s) ,$$
$$B^{(-)} \sim 4su(m+\mu)[5\beta_{\psi\varphi} R_\psi(-s) + 3\beta_{\psi\psi} R_\psi(s)] ,$$

with

$$R_\psi(s) = (-\alpha' s)^{\alpha_\psi(u)-\frac{3}{2}}\Gamma(\tfrac{3}{2}-\alpha_\psi(u)) ,$$

where we have implied a universal slope α' for all trajectories.

ii) *Threshold values.* At $t = 4m^2$ ($\mathcal{N}\overline{\mathcal{N}}$ annihilation at rest) *all amplitudes vanish identically* (6). At the $\pi\mathcal{N}$ threshold $s = (m+\mu)^2$ the only relevant amplitudes are

$$A'^{(+)} = -3A'^{(-)} = -288m^2\mu^2 V_{\varphi\varphi}((m-\mu)^2, 0) .$$

iii) *Particle poles.*

a) To study the meson trajectories we shall first remark that the flipless amplitude A' is F-type at the baryon vertex while the flip amplitude B is $D + \frac{2}{3}F$ type. Taking the leading $(\cos\theta_s) \sim (t-u)$ behaviour of V at $\alpha_N(t) = N \geqslant 1$ yields as residues

$$a'^{(+)} = 9[(s-u)(4m\mu + t_N)\xi_+ - (4m^2 - t_N)(4\mu^2 - t_N)\xi_-]z^{N-1} ,$$
$$a'^{(-)} = 3[(s-u)(4m\mu + t_N)\xi_- - (4m^2 - t_N)(4\mu^2 - t_N)\xi_+]z^{N-1} ,$$
$$b^{(+)} = 12\mu(m+\mu)(4m^2 - t_N)\xi_+ z^{N-1} ,$$
$$b^{(-)} = 20\mu(m+\mu)(4m^2 - t_N)\xi_- z^{N-1} ,$$

where

$$z^{N-1} = -\beta_{\psi\varphi}(\alpha' N!)^{-1}[\tfrac{1}{2}\alpha'(s-u)]^{N-1}$$

and

$$\xi_\pm = (1 \pm e^{i\pi N}) .$$

---

(6) This is a well-known difficulty for supermultiplet schemes but can be circumvented by using momentum spurions, see, for example, F. HUSSAIN and P. ROTELLI: *Phys. Rev.*, **142**, 1013 (1966).

Thus we have the leading ρ-f trajectory accompained by lower (*e.g.* σ-ρ′) trajectories as in ππ scattering. At the ρ pole ($N=1$) the ratio $b^{(-)}/a'^{(-)}$ yields according to the vector-dominance model, a proton magnetic moment $4m(m+\mu)(4m\mu+m_\rho^2)^{-1}$.

b) For the baryon trajectory we note that the first pole at $\alpha_\psi(s)=\frac{3}{2}$, say, contains *eight types* of mass-degenerate particles; four spin-$\frac{3}{2}$ resonances lie on the leading trajectory and identify the presence of the Regge trajectories Δ_δ, Δ_γ, $\mathcal{N}_\delta$, $\mathcal{N}_\gamma$; four spin-$\frac{1}{2}$ resonances including the nucleon lie on the first daughter trajectory. Since we are working only with the leading Veneziano term, we shall ignore the daughter trajectories. (The leading trajectory also contains four more trajectories characterized by Δ_α, Δ_β, $\mathcal{N}_\alpha$, $\mathcal{N}_\beta$ with poles commencing at $\alpha_\psi=\frac{5}{2}$.)

For the spin-$\frac{3}{2}$ resonances we calculate their contributions to $\pi\mathcal{N}$ scattering via Feynman rules (7) and compare to leading order in $\cos\theta_s$ with appropriate isospin amplitudes in our model

$$M^{(\frac{3}{2})}=M^{(+)}-M^{(-)}\,,\qquad M^{(\frac{1}{2})}=M^{(+)}+2M^{(-)}\,,$$

to yield

$$g_\delta^2(\Delta)\;=\frac{\Delta}{m\alpha'}\big[(\Delta^2+(m+\mu)^2)(6\beta_{\psi\varphi}+5\beta_{\psi\psi})+2\Delta(m+\mu)(2\beta_{\psi\varphi}+3\beta_{\psi\psi})\big]\,,$$

$$g_\gamma^2(\Delta)\;=\frac{\Delta}{m\alpha'}\big[(\Delta^2+(m+\mu)^2)(-6\beta_{\psi\varphi}-5\beta_{\psi\psi})+2\Delta(m+\mu)(2\beta_{\psi\varphi}+3\beta_{\psi\psi})\big]\,,$$

$$g_\delta^2(\mathcal{N})=\frac{\Delta}{m\alpha'}\big[(\Delta^2+(m+\mu)^2)\,5(3\beta_{\psi\varphi}+\beta_{\psi\psi})-2\Delta(m+\mu)(13\beta_{\psi\varphi}+6\beta_{\psi\psi})\big]\,,$$

$$g_\gamma^2(\mathcal{N})=\frac{\Delta}{m\alpha'}\big[-5(\Delta^2+(m+\mu)^2)(3\beta_{\psi\varphi}+\beta_{\psi\psi})-2\Delta(m+\mu)(13\beta_{\psi\varphi}+6\beta_{\psi\psi})\big]\,.$$

It is natural to expect that the low-energy data are sensitive to the structure at the first Veneziano pole and in the case of scattering lengths to lower-order Veneziano terms. We shall see how this is borne out by our comparisons with experiment. Let us use mean masses $\Delta=m=1.15\ \text{GeV}/c^2$ $\mu=0.45\ \text{GeV}/c^2$ as required by a unique nucleon mass for both external and internal particles

---

(7) The vertices used both for isospin $\frac{3}{2}(\Delta)$ and $\frac{1}{2}(\mathcal{N})$ are

$$g_{\delta\ \text{or}\ \gamma}\frac{(p_2)_\lambda}{M}\overline{\mathcal{N}}_\lambda(q)(1\ \text{or}\ \gamma_5)\mathcal{N}(p_2)\qquad\qquad \text{for } J^P=\tfrac{3}{2}^+\ \text{or}\ \tfrac{3}{2}^-,$$

where M is the mass of the spin-$\frac{3}{2}$ particle, and projection operator

$$\left[-g_{\mu\nu}+\frac{1}{3}\gamma_\mu\gamma_\nu+\frac{1}{3M}(q_\mu\gamma_\nu-\gamma_\mu q_\nu)+\frac{2}{3M^2}q_\mu q_\nu\right](\gamma\cdot q+M)\,.$$

as well as by the symmetry limit of the supermultiplet schemes (8). Taking $\alpha' = 1(\text{GeV}/c)^{-2}$, $\alpha_\psi(0) = 0.18$, $\alpha_\varphi(0) = 0.5$ and hence $m_\rho^2 = 0.5$ ($\neq \mu$). This immediately predicts $(G_M/G_E)_{t=m_\rho^2} \simeq 3$. By comparing with the experimental high-energy $\pi\mathcal{N}$ charge-exchange data in the forward (or backward) direction we find that $\beta_{\psi\varphi} \simeq -0.5\ (\text{GeV}/c^2)^{-4}$. Whence the universal ρ coupling constant is given as

$$g_\rho^2 = -6(4m\mu + m_\rho^2)\beta_{\psi\varphi}/\alpha' \simeq 7.5\,,$$

a factor of four smaller than the experimental value $g^2_{\rho\,\text{exp}} \simeq 30$. It is interesting to note also that $\beta_{\psi\varphi}$ is the only parameter entering the expression for the scattering lengths $a_{\text{SL}}^{\frac{1}{2},\frac{3}{2}}$. Theoretically

$$a_{\text{SL}}^{\frac{3}{2}} = 4a_{\text{SL}}^{\frac{1}{2}} = -0.38\,\mu^{-1}\,,$$

in poor agreement with $a^{\frac{3}{2}}_{\text{SL (exp)}} \simeq -\frac{1}{2}a^{\frac{1}{2}}_{\text{SL (exp)}} = -0.087\,\mu^{-1}$. To determine $\beta_{\psi\psi}$ we go to residue of the experimentally dominant $\Delta_\delta(1236)$ pole. Its width fixes $g_\delta^2 \simeq 1600$ and hence $\beta_{\psi\psi} \simeq 50\ (\text{GeV}/c^2)^{-4}$. This value for $\beta_{\psi\psi}$ predicts a cross-section which from π-p backward scattering is an order of magnitude too big. It also means that the other three degenerate spin-$\frac{3}{2}$ resonances are ghosts.

Summarizing, we find that although our model contains the good features of the Reggeized supermultiplet schemes (9) and, in addition, incorporates duality, it is quantitatively a very poor first approximation to the low-energy structure of meson-baryon scattering. Other authors who have used much weaker Veneziano parametrizations (3) have by and large been forced to similar conclusions. Thus until such time as a convincing crossing-symmetric unitarization procedure is devised we must view the starting Veneziano amplitudes with some degree of suspicion.

* * *

We are most grateful to Dr. H. F. JONES for several helpful discussions on fermion trajectories, and to Prof. A. SALAM for some valuable comments.

---

(8) By using «physical» masses (in GeV/c²) $m = 0.94$, $\mu = 0.14$, $\Delta = 1.24$, $m_\rho = 0.76$ and $\alpha_\psi(0) = 0.18$, $\alpha_\varphi(0) = 0.5$, $\alpha' = 0.9\ (\text{GeV}/c)^{-2}$ we find, proceeding as in the text, $(G_M/G_E)_{t=m_\rho^2} \simeq 32/3$, $\beta_{\psi\varphi} = -2.2\ (\text{GeV}/c^2)^{-4}$, $g_\rho^2 = 14.0$, $a_{\text{SL}}^{\frac{3}{2}} = 4a_{\text{SL}}^{\frac{1}{2}} = -0.044\,\mu^{-1}$. $\beta_{\psi\psi} \simeq 30.0\ (\text{GeV}/c^2)^{-4}$ again producing $\mathcal{N}_\gamma\Delta_\gamma\mathcal{N}_\delta$ ghosts $g_\gamma^2(\Delta) \simeq -54$, $g_\delta^2(\mathcal{N}) \simeq -266$ and $g_\gamma^2(\mathcal{N}) \simeq -2.660$.

(9) R. DELBOURGO and A. SALAM: *Phys. Lett.*, **28** B, 497 (1968) and *Phys. Rev.* (to be published), which is in the nature of a review paper. The original idea of supermultiplet Reggeization was conceived by A. SALAM and J. STRATHDEE: *Phys. Rev. Lett.*, **19**, 339 (1967).

RIASSUNTO (*)

Si applica un adattamento del formalismo multispinoriale per la valutazione dei diagrammi di dualità allo scattering di mesoni $(6, \overline{6})$ su barioni (56, 1). Le ipotesi forniscono automaticamente i « corretti » rapporti D/F e i moduli relativi delle ampiezze con o senza flip di elicità nei poli delle particelle, e riproducono le previsioni di supermultipletto reggeizzato nella regione asintotica. D'altra parte la struttura della risonanza è altamente non fisica e di conseguenza le previsioni a bassa energia sono scarse.

(*) *Traduzione a cura della Redazione.*

Модель Венециано для рассеяния мезонных и барионных супермультиплетов.

Резюме (*). — При рассеянии мезонов $(6, \overline{6})$ барионами (56, 1) используется адаптация формализма высших спиноров для вычисления дуальных диаграмм. Это описание автоматически обеспечивает « правильные » D/F отношения и относительные величины для амплитуд с переворачиванием четности к амплитудам без переворачивания четности в полюсах частиц, и воспроизводит реджеизованные супермультиплетные предсказания в асимптотической области. Однако, резонансная структура является очень нефизичной и, следовательно, предсказания при низких энергиях являются очень плохими.

(*) *Переведено редакцией.*

Part I
QUARK LEVEL LINEAR SIGMA MODEL

I.0. Commentaries I

During the late 1960's, initiated by Gell-Mann (Physics 1, 63 (1964)), there was considerable activity in the area of current algebras and their application to weak and strong interaction scattering processes by Adler and others. Weinberg (Phys. Rev. D8, 605 (1973)) advanced our understanding of the subject to a large extent by considering the chiral properties of low energy processes, leading to momentum expansions of scattering amplitudes involving pseudo-scalar mesons — a procedure which now goes by the name of 'chiral perturbation theory'. In that description the chiral flavour symmetry is nonlinearly realised and the pions, kaons, etc. are regarded as Goldstone bosons. The article by Callan, Coleman, Wess and Zumino (Phys. Rev. 177, 2247 (1969)) neatly explains the process. During roughly the same period Nambu & Jona-Lasinio (Phys. Rev. 122, 345 (1961) and 124, 246 (1961)) developed a model involving fundamental fermions whereby the pseudo-scalars are dynamically generated as bound states. In their model the chiral symmetry is explicitly broken only by the small 'current' quark masses, which are supposed to produce small corrections to the results of chiral perturbation theory when moving to the mass shell from zero momentum.

I.1 Delbourgo R. and Scadron M.D., Proof of the Nambu-Goldstone realisation for vector-gluon-quark theories. J. Phys. G5, 1621-1631 (1979)
I.2 Delbourgo R. and Scadron M.D., Dynamical symmetry breaking and the sigma-meson mass in quantum chromodynamics. Phys. Rev. Lett. 48, 379-382 (1982)
I.3 Delbourgo R. and Scadron M.D., Dynamical generation of linear σ model SU(3) Lagrangian and meson nonet mixing. Int. J. Mod. Phys. A13, 657-676 (1990)

It was Schwinger (Ann. Phys. 2, 407 (1957)) who advocated a simple σ model for realising linearly an O(4) symmetry group — an approach which was unpopular at the time as it required the existence of a low-energy scalar particle. Nonetheless Gell-Mann and Levy (Nuovo Cim. 16, 705 (1960)) pursued this approach and applied it with some success to an isodoublet of fermions with the σ meson being responsible for the spontaneous breaking of the chiral symmetry.

In paper [I.1], with the arrival of QCD, it became important to prove that the Nambu-Goldstone realisation did in fact apply. In paper [I.2] we showed that with quarks the resultant σ bound state mass was approximately twice the mass of the associated quarks (whose dynamical masses were also spontaneously generated). In article [I.3] we extended this work to the three quarks (u, d, s) leading to a scalar meson nonet. These 0^+ states are now well established. Readers should also consult the review article by Tornquist (Z. Phys. C68, 647-660 (1995)) who was one of the first persons to take the linear realisation of chiral groups seriously. See also the review article in Part M whereby the mesons are regarded as bound states of the fermions and thus possess zero wave function renormalisation constants, with the only mass scale being determined by the pion decay constant.

J. Phys. G: Nucl. Phys., Vol. 5, No. 12, 1979. Printed in Great Britain

Proof of the Nambu–Goldstone realisation for vector-gluon–quark theories†

R Delbourgo and M D Scadron‡
Department of Physics, University of Tasmania, Hobart, Tasmania, Australia

Received 31 May 1979, in final form 10 July 1979

Abstract. We give a proof of the theorem that, in chiral-invariant vector-gluon theories, the spontaneous generation of fermion mass is associated with zero-mass pseudoscalar bosons. The proof relies heavily on the proper determination of Bethe–Salpeter kernels which respect the Ward identities correctly, and is exact to the extent that the propagator is self-consistently truncated. Given this result, we survey the implications of PCAC for the quark–gluon model.

1. Introduction

Owing to the ever-mounting evidence for the quark model in a vector-gluon (more specifically QFD and QCD) context, it is important to fortify its foundations in as firm a manner as possible. One of the cornerstones of the theory is 'current algebra—PCAC' and in particular the chiral nature of the light pseudoscalar mesons of which the pion is a member. It is now almost lore that zero-mass pseudoscalars can arise in a formally chiral-invariant theory of fermions like QCD via the Nambu–Goldstone mechanism (Nambu 1960, Goldstone 1961) of spontaneous breakdown of the vacuum chiral symmetry. Accompanying these massless bosons are massive fermions, and together they maintain the conservation law for the axial-vector current. This intriguing idea has a basis in a four-fermion model where Nambu and Jona-Lasinio (1961, hereafter referred to as NJL) showed explicitly that there is a formal equivalence between the mass-gap equation and the homogeneous Bethe–Salpeter equation for a pseudoscalar fermion–antifermion composite, at least in Hartree approximation. While there is ample and learned discussion (Baker *et al* 1964, Pagels 1973, Jackiw and Johnson 1973, Cornwall and Norton 1973, Eichten and Feinberg 1974) in the literature on this subject for related field theories, to our knowledge no proof for the theory of quarks *and vector gluons* exists which is quite as transparent as the NJL demonstration. Indeed, all of the discussions which we have tracked down are lacking in some respect, and for this reason we shall address ourselves once again to this problem and try to give a correct formulation of the theorem.

In §2 we will describe a number of quasi-proofs of the theorem and indicate their failings. For a correct proof it is essential that the equations which are invoked to describe the pseudoscalar bound state should respect the Ward identities. We demonstrate this in §3 and present some generalisations of the method in the following section. Finally, §5 deals with the implications of this theorem for the quark model and PCAC.

† Supported in part by the US/Australia Cooperative Science Programme, sponsored here by NSF.
‡ On leave from the University of Arizona, Tucson, Arizona 85721, USA.

0305-4616/79/121621 + 11$01.00 © 1979 The Institute of Physics

2. False proofs

All proofs of the Nambu–Goldstone conjecture are based on demonstrating the equivalence between the self-energy equation and the bound-state equation for a zero-mass pseudoscalar once the fermion is granted a non-zero mass. Thus the spontaneous mass generation for the spinor is tied to the masslessness of the pseudoscalar. We exhibit below various previous proofs of the conjecture in vector-gluon theories (which are chiral-symmetric for massless bare fermions) at various levels of sophistication and discuss why they are slightly deficient in one way or another.

The first indication that there is some truth in the conjecture comes by imitating the NJL programme, but this time with vector gluons. To lowest-order perturbation theory and introducing intermediate massive fermions, the self-energy reads

$$\Sigma(\gamma.p) \equiv C(p) + (D(p) - 1)\gamma.p = \mathrm{i}g^2 \int \gamma^\mu \frac{1}{\gamma.k - m}\gamma_\mu \frac{\bar{\mathrm{d}}^4k}{(k-p)^2} \tag{1}$$

in the Fermi gauge. Now it is tempting to consider the Bethe–Salpeter equation for a pseudoscalar composite (see figures 1 and 2) in a comparable approximation:

$$\Gamma_5(p;q) = -\mathrm{i}g^2 \int \gamma^\mu \frac{1}{\gamma.(k+\frac{1}{2}q) - m}\Gamma_5(k;q)\frac{1}{\gamma.(k-\frac{1}{2}q) - m}\gamma_\mu \frac{\bar{\mathrm{d}}^4k}{(k-p)^2}. \tag{2}$$

We find that in the limit $q \to 0$, with a trial form factor $\gamma_5 P$, that

$$P(p) = \mathrm{i}g^2 \int \gamma^\mu \frac{1}{k^2 - m^2}\gamma_\mu P(k)\frac{\bar{\mathrm{d}}^4k}{(k-p)^2}. \tag{3}$$

This bears an uncanny resemblance to the gap equation (with $m_0 = 0$)

$$\Sigma(m) = -m$$

if we simply forget about the $\gamma.p$ term in (1), if we also neglect the dependence of P on its argument and if we put the fermions on mass shell in (3). The resulting condition

$$1 = 4\mathrm{i}g^2 \int \frac{1}{(k^2 - m^2)}\frac{\bar{\mathrm{d}}^4k}{(k-p)^2}\bigg|_{p^2=m^2} \tag{4}$$

can then be interpreted as a relation between the coupling constant and the ratio (Λ/m), where Λ is a needed cut-off, a relation expected to hold in some more general way at more comprehensive levels of approximation for Σ for vector gluons.

Although these steps are permissible in the NJL model in Hartree approximation, unfortunately we cannot forget about p dependence in Σ and the Bethe–Salpeter kernel, nor can we replace P by a constant in (3), as this would conflict with the whole notion of

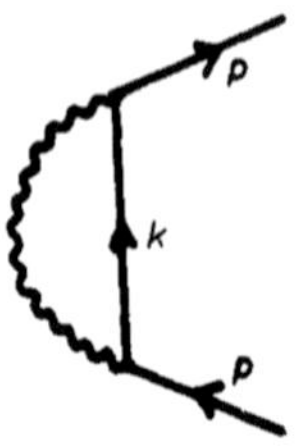

Figure 1. Lowest-order self-energy for a quark via gluon exchange.

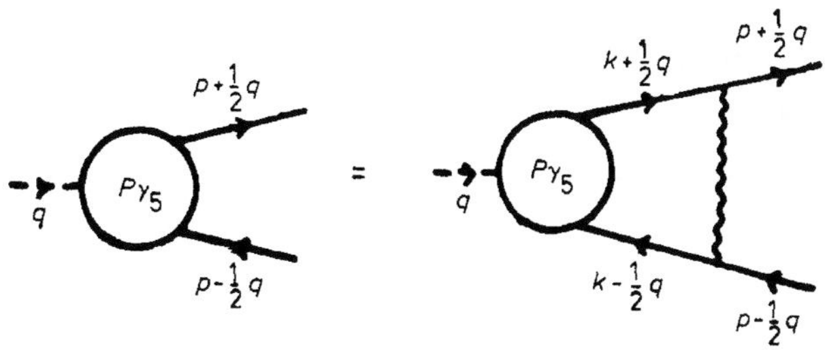

Figure 2. Typical Bethe–Salpeter equation for a bound quark–antiquark pseudoscalar state.

form factors for composite states. This difficulty was appreciated by Baker *et al* (1964) and by Pagels (1973), who suggested instead working with the more fully dressed equations†

$$\Sigma(p) = \mathrm{i}g^2 \int \gamma^\mu S(k)\gamma_\mu \frac{\mathrm{d}^4k}{(k-p)^2} \tag{5}$$

and

$$\Gamma_5(p;q) = -\mathrm{i}g^2 \int \gamma^\mu S(k+\tfrac{1}{2}q)\Gamma_5(k;q)S(k-\tfrac{1}{2}q)\gamma_\mu \frac{\mathrm{d}^4k}{(k-p)^2} \tag{6}$$

in which‡

$$S^{-1}(p) = \gamma . p + \Sigma(p) = C(p) + \gamma . pD(p) \tag{7}$$

is the unrenormalised inverse propagator (see figures 3 and 4). These authors then tried to establish the formal equivalence between (5) and (6). Their discussion is indeed much closer to the truth; however, if carefully followed through, their proof also breaks down, but in a less obvious way than the NJL analogue treatment.

Suppose one substitutes (7) into (5). Then one obtains the pair of coupled equations

$$C(p) = 4\mathrm{i}g^2 \int \frac{C(k)}{C^2(k) - k^2D^2(k)} \frac{\mathrm{d}^4k}{(k-p)^2} \tag{5a}$$

$$D(p) - 1 = 2\mathrm{i}g^2 \int \frac{D(k)}{C^2(k) - k^2D^2(k)} \frac{k.p}{p^2} \frac{\mathrm{d}^4k}{(k-p)^2}. \tag{5b}$$

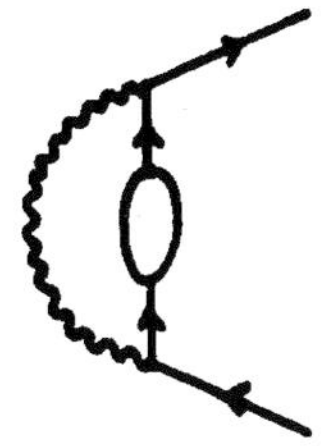

Figure 3. A self-consistent determination of quark self-energy.

† Actually they worked in the Landau gauge rather than the Fermi gauge, but that is irrelevant for the purpose of the theorem except in so far that the cut-off is not needed.

‡ In (7) C and D are actually functions of p^2. This is understood throughout the paper for scalar functions. We also use the notation $F'(p) \equiv \mathrm{d}F(p^2)/\mathrm{d}p^2$ for ease of writing.

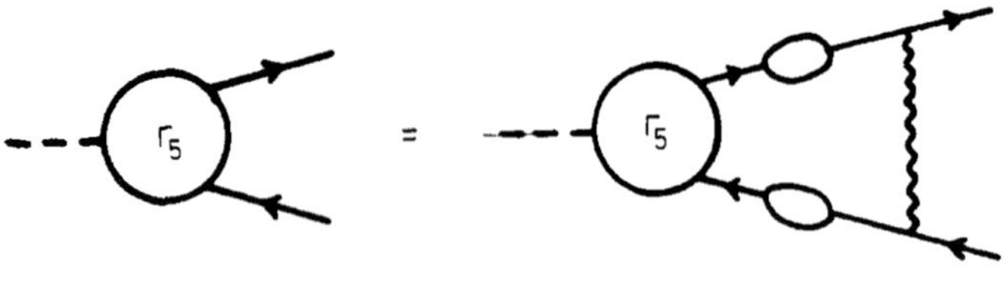

Figure 4. A (pseudoscalar) Bethe–Salpeter equation with dressed quarks.

Likewise in (6) one should, by analogy with (7), make the general expansion

$$\Gamma_5(p;q) = (P + \gamma . qA + p . q\gamma . pQ + [\gamma . p, \gamma . q]R)\gamma_5 \tag{8}$$

in which P, A, Q and R are functions of p^2, q^2 and $(p . q)^2$ —assuming that charge-conjugation invariance is not violated by the spontaneous breaking. One finds that R vanishes identically, a consequence of the chiral symmetry, and up to first order in q, the remaining functions satisfy three coupled integral equations:

$$P(p) = 4\mathrm{i}g^2 \int \frac{P(k)}{C^2(k) - k^2D^2(k)} \frac{\bar{\mathrm{d}}^4k}{(k-p)^2} \tag{6a}$$

$$A(p) = 2\mathrm{i}g^2 \int \left[\frac{C(k)D(k)P(k) - (C^2(k) + k^2D^2(k))A(k)}{(C^2(k) - k^2D^2(k))^2} + \tfrac{1}{3}\left(k^2 - \frac{(p . k)^2}{p^2}\right) \right.$$
$$\left. \times \frac{2D^2(k)A(k) + 2(C(k)D'(k) - C'(k)D(k))P(k) - (C^2(k) - k^2D^2(k))Q(k)}{(C^2(k) - k^2D^2(k))^2} \right] \frac{\bar{\mathrm{d}}^4k}{(k-p)^2} \tag{6b}$$

$$p^2Q(p) = 2\mathrm{i}g^2 \int \tfrac{1}{3}\left(\frac{4(p . k)^2}{p^2} - k^2\right) \frac{\bar{\mathrm{d}}^4k}{(k-p)^2}$$
$$\times \frac{2D^2(k)A(k) + 2(C(k)D'(k) - C'(k)D(k))P(k) - (C^2(k) - k^2D^2(k))Q(k)}{(C^2(k) - k^2D^2(k))^2}. \tag{6c}$$

While it is immediately obvious that $P \propto C$ will reduce (6*a*) to (5*a*), it is totally unclear how any combination of (6*b*) and (6*c*) can reduce to (5*b*) in order that we could claim have a complete equivalence between (5) and (6). In fact, we will shortly show that (6) is *not* the appropriate equation to study and it is not quite the right analogue of (5). Nevertheless, the manner in which (5*a*) comes out to be identical with (6*a*) shows that one is nearly on the right track. Pagels (1973), although appreciating the importance of having some p dependence in C, tended to force the proof by simply setting $D = 1$ on the right of (5)–(7) and by making various other approximations in the denominators of the integrands. We shall refrain from so doing, for if there is a theorem to be *proved* it should be derived *exactly*.

3. A correct proof based on the Ward identities

Let us suppose that we resort to some self-consistent approximation scheme, such as that of Baker *et al* (1963, 1967) which determines Σ according to (5). In setting up Bethe–Salpeter equations of whatever variety, it is essential to maintain the Ward–Takahashi identities at the composite level, since the properties we are seeking (masslessness of pseudoscalars for one) will otherwise be lost. In effect, this gauge constraint dictates the correct structure of the form factor equations analogous to (6). Thus it tells us that the right equation for the vector vertex part is

$$\Gamma_\lambda(p;q) = \gamma_\lambda - \mathrm{i}g^2 \int \gamma^\mu S(k + \tfrac{1}{2}q)\Gamma_\lambda(k;q)S(k - \tfrac{1}{2}q)\gamma_\mu \frac{\bar{\mathrm{d}}^4 k}{(k-p)^2}. \tag{9}$$

Multiplying (9) by q^λ is entirely consistent with (5) and the Ward–Takahashi identity

$$q^\lambda\Gamma_\lambda(k;q) = S^{-1}(k + \tfrac{1}{2}q) - S^{-1}(k - \tfrac{1}{2}q). \tag{10}$$

Likewise for the axial-vector vertex, the chiral symmetry forces us to write the inhomogeneous equation

$$\Gamma_{\lambda 5}(p;q) = \mathrm{i}\gamma_\lambda\gamma_5 - \mathrm{i}g^2 \int \gamma^\mu S(k + \tfrac{1}{2}q)\Gamma_{\lambda 5}(k;q)S(k - \tfrac{1}{2}q)\gamma_\mu \frac{\bar{\mathrm{d}}^4 k}{(k-p)^2} \tag{11}$$

as the only one consistent with the axial Ward identity

$$-\mathrm{i}q^\lambda\Gamma_{\lambda 5}(k;q) = S^{-1}(k + \tfrac{1}{2}q)\gamma_5 + \gamma_5 S^{-1}(k - \tfrac{1}{2}q) \tag{12}$$

and with the self-energy equation (5). Without guidance from these gauge identities we would have no clue about the validity of various approximations to the kernels of Bethe–Salpeter equations. For example, we have no means of knowing if the approximation on the pseudoscalar kernel made in (6) is at all consistent with (5). Having said that, we shall study (11) in place of (6); the pseudoscalar will be induced in the singular ($q^2 \to 0$) part of the axial vertex. To find it we first solve the identity (12), namely from

$$\begin{aligned} -\mathrm{i}q^\lambda\Gamma_{\lambda 5}(k;q) = [&\gamma.(k + \tfrac{1}{2}q)D(k + \tfrac{1}{2}q) - \gamma.(k - \tfrac{1}{2}q)D(k - \tfrac{1}{2}q) + C(k + \tfrac{1}{2}q) \\ &+ C(k - \tfrac{1}{2}q)]\gamma_5 \end{aligned}$$

we derive

$$\begin{aligned} \Gamma_{\lambda 5}(k;q) = {}& \mathrm{i}q^{-2}q_\lambda\gamma_5(C(k + \tfrac{1}{2}q) + C(k - \tfrac{1}{2}q)) + \tfrac{1}{2}(D(k + \tfrac{1}{2}q) + D(k - \tfrac{1}{2}q))\mathrm{i}\gamma_\lambda\gamma_5 \\ &+ \mathrm{i}\gamma.kk_\lambda\gamma_5(D(k + \tfrac{1}{2}q) - D(k - \tfrac{1}{2}q))/q.k + \Gamma^{\mathrm{T}}_{\lambda 5}(k;q) \end{aligned}$$

where the transverse axial components $\Gamma^{\mathrm{T}}_{\lambda 5}$ are divergenceless, $q^\lambda\Gamma^{\mathrm{T}}_{\lambda 5} \equiv 0$, and are of order q by analyticity arguments. Hence to order q, $\Gamma_{\lambda 5}$ is entirely determined by the axial Ward identity:

$$\Gamma_{\lambda 5}(k;q) \underset{q\to 0}{\to} 2\mathrm{i}q^{-2}q_\lambda\gamma_5 C(k) + \mathrm{i}\gamma_\lambda\gamma_5 D(k) + 2\mathrm{i}k_\lambda\gamma.k\gamma_5 D'(k) + \mathrm{O}(q). \tag{13}$$

Precisely because $C \neq 0$ (in order that the fermion acquire mass), $\Gamma_{\lambda 5}$ is singular as $q^2 \to 0$, and this is well known (Nambu 1960, Nambu and Jona-Lasinio 1961) to signal the appearance of the pseudoscalar. The residue is the function C and thus, in this limit, equation (11) reduces to the homogeneous equation (5*a*) without further ado. It only remains to show that the terms of order q^0 in (13) when inserted into (11) reproduce (5*b*) and we are finished with the proof. This is readily accomplished since, not surprisingly, one finds after a little work† that these pieces of the vertex equation are just derivatives of (5*b*). With hindsight we can also understand why the use of (6) and (8) is incorrect, for the

† The intermediate steps involve a p_ν differentiation of

$$p_\mu(D(p) - 1) = 2\mathrm{i}g^2 \int \frac{k_\mu D(k)}{C^2(k) - k^2 D^2(k)} \frac{\bar{\mathrm{d}}^4 k}{(k-p)^2}$$

and contracting separately with $p_\mu p_\nu$ and $\eta_{\mu\nu}$.

divergence of (11) shows that we are missing an inhomogeneous term (unity) on the right of (6*b*). That is, $A(p)$ should be replaced by $A(p) - 1$ on the left of (6*b*). This is where we differ from previous approaches. Furthermore we can anticipate the solutions of (6*a*), (6*b*) and (6*c*) through the divergence of (13) compared with (8), namely

$$P(p) = 2C(p) \qquad A(p) = D(p) \qquad Q(p) = 2D'(p). \tag{14}$$

The whole structure is closely knit and free of dubious steps: the proof is complete to the extent that the self-energy equation is approximated by (5). While (14) is true for all p^2 one might wish to set a scale on these functions and this can be done by imposing the normalisation gap condition $\Sigma(m) = -m$.

4. Generalisations

In order to improve on the self-consistent equation (5) it is necessary to include further quantum corrections to the meson propagator and/or the vertex function. In QED it is perfectly straightforward to dress up the photon; one simply replaces (5) with

$$\Sigma(p) = -\mathrm{i}g^2 \int \gamma^\mu S(k)\gamma^\nu D_{\mu\nu}(k-p)\bar{\mathrm{d}}^4k \tag{15}$$

where $D_{\mu\nu}$ is the corrected photon propagator in any chosen gauge. The counterparts of the Bethe–Salpeter equations (9) and (11) are

$$\Gamma_\lambda(p;q) = \gamma_\lambda + \mathrm{i}g^2 \int \gamma^\mu S(k+\tfrac{1}{2}q)\Gamma_\lambda(k;q)S(k-\tfrac{1}{2}q)\gamma^\nu D_{\mu\nu}(k-p)\bar{\mathrm{d}}^4k \tag{16}$$

$$\Gamma_{\lambda 5}(p;q) = \mathrm{i}\gamma_\lambda\gamma_5 + \mathrm{i}g^2 \int \gamma^\mu S(k+\tfrac{1}{2}q)\Gamma_{\lambda 5}(k;q)S(k-\tfrac{1}{2}q)\gamma^\nu D_{\mu\nu}(k-p)\bar{\mathrm{d}}^4k \tag{17}$$

and it is trivial to show, using (7), that the gauge identities (10) and (12) are respected (see figures 5, 6 and 7.) One can therefore repeat the entire proof of §3, the only difference being that the propagator $(k-p)^{-2}$ is replaced by something more complicated, but the steps are otherwise the same†.

At a more sophisticated level one may like to introduce vertex corrections and contemplate using the complete unrenormalised self-energy

$$\Sigma(p) = \mathrm{i}g^2 \int \gamma^\mu S(k)\Gamma^\nu(k,p)D_{\mu\nu}(k-p)\bar{\mathrm{d}}^4k. \tag{18}$$

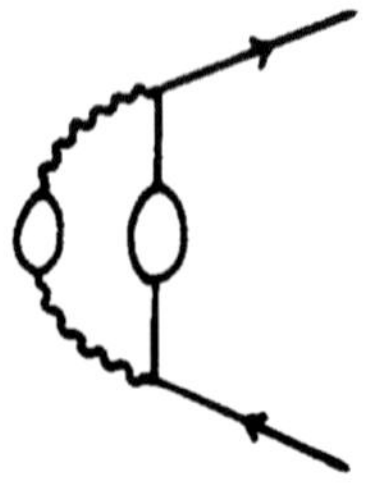

Figure 5. A self-consistent determination of quark self-energy with dressed gluons.

† One might worry that in QED there does not seem to be a Goldstone pseudoscalar. However, QED has finite electron mass and we are not certain that the bare electron is really massless; this point bears further investigation.

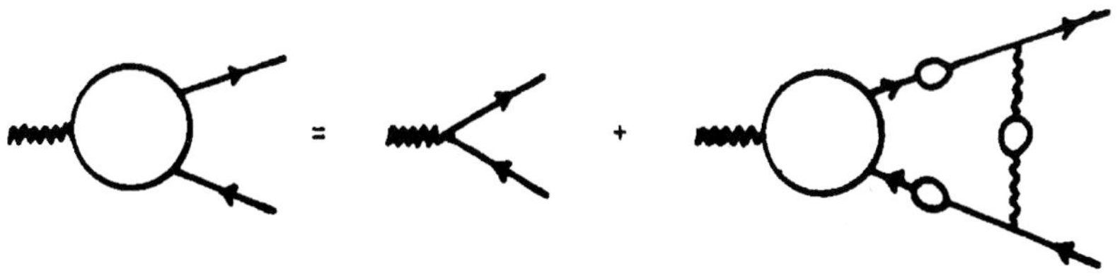

Figure 6. The vector or axial-vector Bethe–Salpeter equation with dressed quarks and gluons.

In that case the analogues of (16) and (17) are no longer simple. For instance, the one-sided equation

$$\Gamma_\lambda(p;q) = \gamma_\lambda + \mathrm{i}g^2 \int \gamma^\mu S(k+\tfrac{1}{2}q)\Gamma_\lambda(k;q)S(k-\tfrac{1}{2}q)\Gamma^\nu(k-\tfrac{1}{2}q, p-\tfrac{1}{2}q)D_{\mu\nu}(k-p)\bar{\mathrm{d}}^4k,$$

which represents a reasonable first guess, becomes inconsistent with (10). If one is prepared to be less ambitious and just introduce longitudinal vertex corrections to the vertex then one can apply the gauge technique (Delbourgo and West 1977, Delbourgo 1979) to rewrite (18) in the renormalised form ($m_0 \equiv 0$)

$$\int \frac{\rho(W)(W+\tilde{\Sigma}(\gamma.p, W))}{\gamma.p - W}\mathrm{d}W = 0 \tag{19}$$

where the renormalised $\tilde{\Sigma}$ is

$$\tilde{\Sigma}(p, W) = -m + \frac{(p-W)}{\pi}\int \frac{\mathrm{Im}\,\tilde{\Sigma}(W', W)\,\mathrm{d}W'}{(W'-p)(W'-W)}$$

and $\mathrm{Im}\,\tilde{\Sigma}$ is the absorptive part of the self-energy to order g^2 for a fermion of mass W.

One automatically has the representations

$$S(p+\tfrac{1}{2}q)\Gamma_\mu(p;q)S(p-\tfrac{1}{2}q) = \int \frac{1}{\gamma.(p+\frac{1}{2}q)-W}\gamma_\mu\frac{1}{\gamma.(p-\frac{1}{2}q)-W}\rho(W)\,\mathrm{d}W \tag{20}$$

$$S(p+\tfrac{1}{2}q)\Gamma_{\mu5}(p;q)S(p-k)$$

$$= \mathrm{i}\int \frac{1}{\gamma.(p+\frac{1}{2}q)-W}\left(\gamma_\mu - \frac{2Wq_\mu}{q^2}\right)\gamma_5\frac{1}{\gamma.(p-\frac{1}{2}q)-W}\rho(W)\,\mathrm{d}W \tag{21}$$

up to transverse vertex corrections of order q. The theorem here just amounts to the observation that the axial vertex perforce contains an induced pseudoscalar pole with residue determined by $\rho(W)$, i.e. by the self-energy equation (19) in gauge approximation. This residue is the analogue of the C form factor which arose in previous sections. The analogue of the D comes by expanding (21) to the next order.

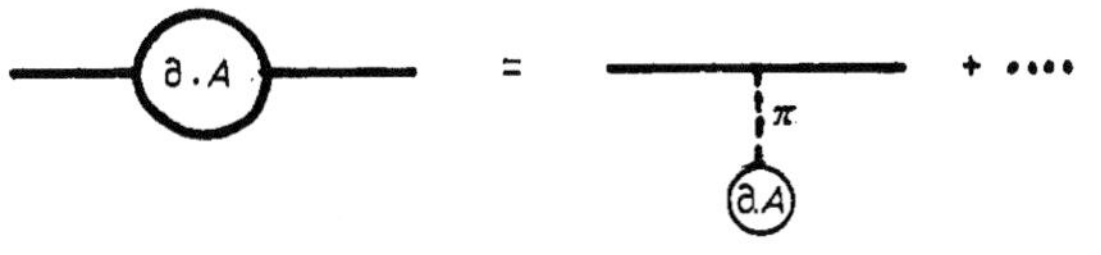

Figure 7. Nucleon or quark tadpole diagram.

The generalisation to a non-Abelian theory like QFD or QCD at the level of (15) is fairly clear. One introduces the spinor group representation matrices T^a and rewrites

$$\Sigma(p) = -\mathrm{i}g^2 \int T^a\gamma^\mu S(k)\gamma^\nu T^b D^{ab}_{\mu\nu}(k-p)\,\mathrm{d}^4k. \tag{22}$$

It is imperative now to work in the axial gauge $n^\mu A^a_\mu = 0$ if one wants to retain the naive analogues of (10) and (12),

$$q^\lambda\Gamma^a_\lambda(k;q) = S^{-1}(k+\tfrac{1}{2}q)T^a - T^aS^{-1}(k-\tfrac{1}{2}q) \tag{23}$$

$$-\mathrm{i}q^\lambda\Gamma^a_{\lambda 5}(k;q) = S^{-1}(k+\tfrac{1}{2}q)T^a\gamma_5 + \gamma_5 T^aS^{-1}(k-\tfrac{1}{2}q). \tag{24}$$

To be consistent with these Ward identities we have to include the triple meson vertex (VVV and AVV) contributions to the identities as follows:

$$\Gamma^a_\lambda(p;q) = \gamma_\lambda T^a + \mathrm{i}g^2 \int T^b\gamma^\mu S(k+\tfrac{1}{2}q)\Gamma^a_\lambda(k;q)S(k-\tfrac{1}{2}q)\gamma^\nu T^c D^{bc}_{\mu\nu}(k-p)\mathrm{d}^4k$$

$$+\tfrac{1}{2}\mathrm{i}g^2 \int f^{abc}\Gamma_\lambda{}^{\mu\nu}(k;q)D^{bb'}_{\mu\mu'}(k-\tfrac{1}{2}q)D^{cc'}_{\nu\nu'}(k+\tfrac{1}{2}q)\gamma^{\nu'}T^{c'}S(k-p)\gamma^{\mu'}T^{b'}\mathrm{d}^4k \tag{25}$$

$$\Gamma^a_{\lambda 5}(p;q) = \mathrm{i}\gamma_\lambda\gamma_5 T^a + \mathrm{i}g^2 \int T^b\gamma^\mu S(k+\tfrac{1}{2}q)\Gamma^a_{\lambda 5}(k;q)S(k-\tfrac{1}{2}q)\gamma^\nu T^c D^{bc}_{\mu\nu}(k-p)\mathrm{d}^4k$$

$$+\tfrac{1}{2}\mathrm{i}g^2 \int f^{abc}\Gamma_{\lambda 5}{}^{\mu\nu}(k;q)D^{bb'}_{\mu\mu'}(k-\tfrac{1}{2}q)D^{cc'}_{\nu\nu'}(k+\tfrac{1}{2}q)\gamma^{\nu'}T^{c'}S(k-p)\gamma^{\mu'}T^{b'}\mathrm{d}^4k \tag{26}$$

and use the Ward–Takahashi identities for the vector mesons among themselves (Delbourgo *et al* 1974, Kummer 1975). The divergences of (25) and (26) then reproduce (23) and (24) if we recall the group-theory combinatoric identity

$$T^bT^aT^b + \tfrac{1}{2}f^{abc}f^{bcd}T^d = (T^bT^b)T^a \tag{27}$$

which follows from $[T^a, T^b] = \mathrm{i}f^{abc}\,T^c$. Taking the soft limit of (26) along the lines of (13) we are guaranteed that the occurrence of a pseudoscalar bound state is consistent and equivalent to the initial self-consistency equations (22) and (24).

Finally, if we wish to incorporate some measure of vertex corrections for QCD by application of the gauge technique (Salam 1963, Delbourgo 1979) we again have to solve a self-consistency equation like (19) except that $\tilde{\Sigma}$ will be modified to incorporate the gluon dressing (which conventional wisdom gives as $D(q) \sim q^{-4}$). But in any case the important thing is that a non-trivial $\rho(W)$ will exist and, via the solution (21), a pseudoscalar bound state as well whose residue is fixed in terms of ρ. Of course the detailed value for the residue awaits a reliable computation of $\rho(W)$ for QCD.

5. PCAC in the quark–gluon model

It is often stated that the Goldberger–Treiman relation (GTR) and PCAC are properties of the hadronic σ model (where the pion is elementary rather than a composite) while the local commutation relations of current algebra are idealised from the quark model. The validity of the Nambu–Goldstone realisation for quark–gluon theories as just demonstrated, however, suggests that the GTR and PCAC also apply to the quark model. In this section we try to construct in a logical sequence the transition from the spontaneously broken chiral-invariant world with conserved axial-vector currents (CAC) and massless bare quarks $m_0 = 0$ to the chiral-broken world of PCAC and non-vanishing bare (i.e. current) quark masses $m_0 \neq 0$.

First consider the SU(2) × SU(2)-invariant world with $m_0 = 0$ and *constituent* (i.e. renormalised) non-strange quarks having masses $m_u = m_d = \hat{m}$ roughly determined from the aligned quark configurations:

$$\hat{m} \sim \tfrac{1}{2}m_\rho \sim \tfrac{1}{3}m_\Delta \sim 390\,\text{MeV}. \tag{28}$$

(For a more detailed analysis based on static magnetic moments, see Greenberg (1964) and de Rújula *et al* (1975).) Following Nambu (1960) we construct quark and nucleon matrix elements of the CAC current as $(p' - p = q)$

$$\langle q'|A^i_\mu(0)|q\rangle = \bar{q}'\tfrac{1}{2}\tau^i\left(\gamma_\mu\gamma_5 - \frac{2\hat{m}}{q^2}q_\mu\gamma_5\right)q \tag{29a}$$

$$\langle N'|A^i_\mu(0)|N\rangle = \bar{N}'\tfrac{1}{2}\tau^i g_A(q^2)\left(\gamma_\mu\gamma_5 - \frac{2m_N}{q^2}q_\mu\gamma_5\right)N. \tag{29b}$$

Then the (bound-state) massless pion transition to the vacuum

$$\langle 0|A^i_\mu(0)|\pi^j_0(q)\rangle = \mathrm{i}f^0_\pi q_\mu \delta^{ij} \tag{30}$$

generates poles in the currents (29) as depicted in figure 7, leading to the GTR for both quarks and hadrons:

$$f^0_\pi g_{\pi\bar{q}q} = \hat{m} \tag{31a}$$

$$f^0_\pi g_{\pi NN}(0) = m_N g_A(0). \tag{31b}$$

Since we expect $f^0_\pi \sim f_\pi \simeq 93$ MeV as determined by $\pi \to \mu\nu$, not only is (31*b*) reasonably well satisfied, but (31*a*) implies

$$g^2_{\pi\bar{q}q}/4\pi \sim 1, \tag{32}$$

a palatable result.

Now even after SU(3) breaking occurs (to be discussed shortly) the Goldstone nature of the kaon means that it too satisfies approximate relations analogous to (31) with *constituent* strange-quark mass m_s determined in similar manner to (28):

$$m_s \sim \tfrac{1}{2}m_\phi \sim \tfrac{1}{3}m_\Omega \sim 540\,\text{MeV}, \tag{33}$$

leading to the accepted ratio of roughly $m_s/\hat{m} \sim 1{\cdot}4$. Then the ratio of the kaon to pion GTR's for quarks gives

$$f_K/f_\pi \simeq \tfrac{1}{2}(m_s + \hat{m})/\hat{m} \simeq 1{\cdot}2, \tag{34}$$

in excellent agreement with experiment. Thus the GTR's and f_K/f_π have a quark-model interpretation—based primarily upon the Nambu–Goldstone realisation as verified in §§3 and 4.

Next we investigate the isoscalar 0^- mesons η and η' in the chiral-invariant world with $m_0 = 0$. The Nambu–Goldstone realisation is slightly modified in this case. That is, although the gluon vertex or, equivalently, the self-energy graph of figure 3 does *not* contribute to the η or η' masses, the *triangle anomaly* diagram of figure 8 (quark annihilation diagram in free-quark language) breaks the CAC condition.

Hence the octet η_8 still remains massless in this limit so that

$$\partial . A_\pi = \partial . A_K = \partial . A_{\eta_8} = 0, \tag{35}$$

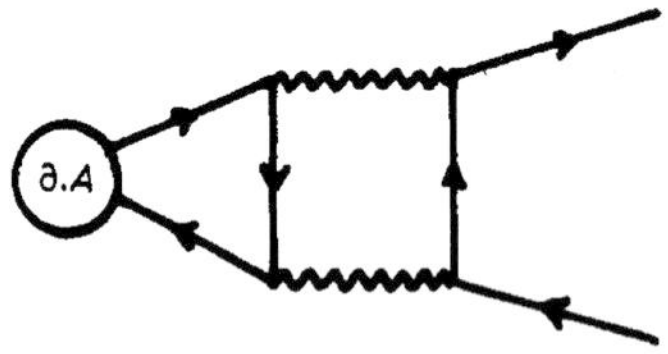

Figure 8. Triangle anomaly for singlet axial current at the quark–gluon level.

but the U(1) singlet η_0 could acquire a mass from the anomaly via the gluon field (see Adler 1969)

$$\partial . A_{\eta_0} = \frac{\sqrt{6}g^2}{64\pi^2}\varepsilon_{\mu\nu\alpha\beta}G_r^{\mu\nu}G_r^{\alpha\beta} \tag{36a}$$

$$\langle 0|\partial . A_{\eta_0}|\eta_0\rangle = f^0_{\eta_0}m^2_{\eta_0}. \tag{36b}$$

Now we may consider the SU(3) world of chiral symmetry *breaking* with Hamiltonian density $H = H_0 + H'$ and

$$[Q_i^5, H_0] = 0 \qquad [Q_i^5, H'] \neq 0. \tag{37}$$

In the quark–gluon model, the chiral-breaking Hamiltonian density is determined by the *current* quark mass matrix

$$H' = \bar{q}m_0 q = \hat{m}_0(\bar{u}u + \bar{d}d) + m_{os}\bar{s}s. \tag{38}$$

That is, $\hat{m}_0$ and m_{os} are the *bare* masses in the quark Lagrangian (no longer is $m_0 = 0$ as in §§2–4). These masses are supposedly much smaller than the constituent masses (28) and (33) and measure the departures of the axial current from the CAC limit via the quark-model commutation relations

$$\partial . A_\pi = -\mathrm{i}[Q_\pi^5, H'] = -\hat{m}_0\bar{q}\lambda_\pi\gamma_5 q \tag{39a}$$

$$\partial . A_K = -\mathrm{i}[Q_K^5, H'] = -\tfrac{1}{2}(m_{os} + \hat{m}_0)\bar{q}\lambda_K\gamma_5 q \tag{39b}$$

$$\partial . A_{\eta_{ns}} = -\mathrm{i}[Q_{ns}^5, H'] = -\hat{m}_0\bar{q}\lambda_{ns}\gamma_5 q \tag{39c}$$

$$\partial . A_{\eta_s} = -\mathrm{i}[Q_s^5, H'] = -m_{os}\bar{q}\lambda_s\gamma_5 q. \tag{39d}$$

The π and K masses are completely generated by the current quark masses as

$$\langle 0|\partial . A_\pi|\pi\rangle = f_\pi m_\pi^2 = \hat{m}_0\langle 0|\bar{q}\lambda_\pi\gamma_5 q|\pi\rangle \tag{40a}$$

$$\langle 0|\partial . A_K|K\rangle = f_K m_K^2 = -\tfrac{1}{2}(m_{os} + \hat{m}_0)\langle 0|\bar{q}\lambda_K\gamma_5 q|K\rangle \tag{40b}$$

with $\langle 0|\partial . A_{\eta,\eta'}|\eta,\eta'\rangle$ receiving contributions from both the current quark masses and from the triangle anomaly (36). In the latter case one must rediagonalise the η_{ns}–η_s states to obtain the diagonal η–η' states. A consistent diagonalisation procedure stated in Hamiltonian language (Jones and Scadron 1979) finds m_η^2 ($m_{\eta'}^2$) is 26% (71%) gluon anomaly with the remaining mass coming from the massive current quarks.

As to the actual values of the current quark masses, specific schemes depend upon light-plane dynamics. The conventional wisdom finds $m_{os}/\hat{m}_0 \simeq 25$ and $\hat{m}_0 \simeq 5$ MeV (see, e.g., a recent review by Weinberg 1977), while a more recent analysis deduces $m_{os}/\hat{m}_0 \simeq 5$ and $\hat{m}_0 \simeq 62$ MeV (Fuchs and Scadron 1978). But in any case, scaled to the constituent quark masses (28) and (33), either scheme leads to similar CAC-breaking effects in (40).

Moreover, the latter scheme is consistent with the observed πN σ term and GTR *discrepancy*, $\Delta = 1 - m_N g_A/f_\pi g_{\pi NN}$ (see Scadron and Jones 1974, Jones and Scadron 1975).

To summarise, the quark-model version of PCAC (39) is even more predictive than the σ-model version $\partial . A_\pi = f_\pi m_\pi^2 \phi_\pi$, etc, because the former not only reproduces the GTR's (31) and the meson transitions (40), but it also correctly predicts the ratio f_K/f_π in the constituent quark basis, the η–η' mixing, πN σ term and GTR discrepancy in the current quark basis. All of these results depend critically upon the underlying Nambu–Goldstone structure of the bound-state quark–antiquark pseudoscalar mesons, as we have generalised from the NJL analysis in §§3 and 4.

References

Adler S L 1969 *Phys. Rev.* **177** 2426
Baker M, Johnson K and Lee B W 1964 *Phys. Rev.* **133** B209
Baker M, Johnson K and Willey R 1963 *Phys. Rev. Lett.* **11** 526
—— 1967 *Phys. Rev.* **163** 1699
Cornwall J M and Norton R E 1973 *Phys. Rev. D* **8** 3338
Delbourgo R 1979 *Nuovo Cim.* A **49** 484
Delbourgo R, Salam A and Strathdee J 1974 *Nuovo Cim.* A **23** 23
Delbourgo R and West P 1977 *J. Phys. A: Math. Gen.* **10** 1049
Eichten E J and Feinberg F L 1974 *Phys. Rev.* D **10** 3254
Fuchs N H and Scadron M D 1978 *Purdue Preprint*
Goldstone J 1961 *Nuovo Cim.* **19** 154
Greenberg O W 1964 *Phys. Rev. Lett.* **13** 598
Jackiw R and Johnson K 1973 *Phys. Rev.* D **8** 2386
Jones R F and Scadron M D 1975 *Phys. Rev.* D **11** 174
—— 1979 *Imperial College Preprint*
Kummer W 1975 *Acta Phys. Aust.* **41** 315
Nambu Y 1960 *Phys. Rev.* **117** 648
Nambu Y and Jona-Lasinio G 1961 *Phys. Rev.* **122** 345
Pagels H 1973 *Phys. Rev.* D **7** 3689
de Rújula A, Georgi H and Glashow S 1975 *Phys. Rev.* D **12** 147
Salam A 1963 *Phys. Rev.* **130** 1287
Scadron M D and Jones H F 1974 *Phys. Rev.* D **10** 967
Weinberg S 1977 *Harvard Preprint*

VOLUME 48, NUMBER 6 PHYSICAL REVIEW LETTERS 8 FEBRUARY 1982

Dynamical Symmetry Breaking and the Sigma-Meson Mass in Quantum Chromodynamics

R. Delbourgo
Physics Department, University of Tasmania, Hobart 7001, Australia

and

M. D. Scadron
Physics Department, University of Arizona, Tucson, Arizona 85721
(Received 20 October 1981)

The spontaneous breakdown of chiral symmetry is analyzed dynamically via bound-state Bethe-Salpeter equations. While in general spontaneous mass generation is linked to a massless 0^- pion and no specific constraint on a massive 0^+ meson, for the particular theory of asymptotically free QCD it is shown that a 0^+ σ meson should exist with mass $m_\sigma \approx 600$ to 700 MeV.

PACS numbers: 11.30.Qc, 11.30.Rd, 12.35.Cn, 14.40.Cs

We are now accustomed to viewing the spontaneous breakdown of chiral symmetry from two radically different theoretical frameworks:

(i) *Lagrangian symmetry breaking.*—In the Gell-Mann–Lévy σ model,[1] the 0^- pion and 0^+ σ meson are assumed to be elementary particles which couple in a chiral-invariant manner to elementary quarks (or nucleons) in the fundamental Lagrangian. Spontaneous breakdown of chiral symmetry then means that the Lagrangian remains chirally invariant, but the σ field develops a nonzero vacuum expectation value, $\langle\sigma\rangle_0 = f_\pi \neq 0$, such that $f_\pi g_{\pi qq} = m_{qk} \neq 0$ with the pion remaining massless. There is, however, *no constraint on the value of* m_σ in the σ model.

(ii) *Dynamical symmetry breaking.*—In the Nambu–Jona-Lasinio (NJL) four-fermion (Hartree-Fock) approach,[2] which can be extended to vector-gluon theories[3] including QCD, gluons and quarks are assumed as elementary. The quark mass is then dynamically generated along with a bound-state massless pion such that the equations (DE) which dress the quark, giving it all of its mass, are *precisely* the same equations ($\mathrm{PBE}|_{q\to 0}$) which bind the quark and antiquark together in a pseudoscalar s wave at zero momentum transfer to form the massless pion,

$$\mathrm{DE} = \mathrm{PBE}|_{q\to 0}. \tag{1}$$

With an $m_{qk}\bar{q}q$ term then occurring in the renormalized Lagrangian, *appearing* to break the chiral symmetry, the Nambu-Goldstone pion arises as a massless pole in the axial current. But the unanswered question is as follows: Must the 0^+ σ meson also necessarily exist as a dynamical bound state in order to restore an apparent chiral symmetry to the renormalized Lagrangian? If so, *then what is the dynamical value of* m_σ in the quark model?

In this paper we shall work within the above dynamical-symmetry-breaking picture *specifically for the quark vector-gluon non-Abelian theory of QCD* and demonstrate that in the chiral limit, just as (1) is valid, then the asymptotic freedom property also requires

$$\mathrm{PBE}|_{q\to 0} = \mathrm{SBE}|_{q^2 = 4m_{qk}^2}, \tag{2}$$

where SBE denotes the 0^+ scalar p-wave binding equation evaluated at $q^2 = 4m_{qk}^2$. Thus, given (1) and (2), spontaneous breakdown of chiral symmetry for QCD means that when the quark acquires all of its mass $m_{qk} \neq 0$ via the gluon dressing equation, quark and antiquark automatically bind together to form in the chiral limit (a) a 0^- s-wave pion with $m_\pi = 0$, and (b) a 0^+ p-wave σ meson with $m_\sigma = 2m_{qk}$.

Indeed, in the original four-fermion model of NJL, the σ meson result (2) was discovered but not significantly exploited because no mass scale was then associated with m_σ or m_{qk}. Furthermore the NJL $(\bar{\psi}\psi)^2$ theory is nonrenormalizable and it was not clear to what extent (2) followed for "realistic" model field theories. Here we wish to stress that QCD is one such renormalizable theory for which (2) is valid and for which $m_\sigma = 2m_{qk}$ make good phenomenological sense.

To begin with, we remind the reader of the subtleties in proving the first spontaneous-breakdown condition (1) in the context of chiral-invariant quark-gluon field theories. In particular, the bound-state (homogeneous) Bethe-Salpeter equation depicted in Fig. 1 at $q \to 0$ only relates the 0^- spinless wave function $P(p^2)$ to the spinless component $C(p^2)$ of the dressing equation with inverse quark propagator $S^{-1}(p) = C(p^2) + \not{p}D(p^2)$. The spin-one nature of the gluon complicates the spontane-

© 1982 The American Physical Society

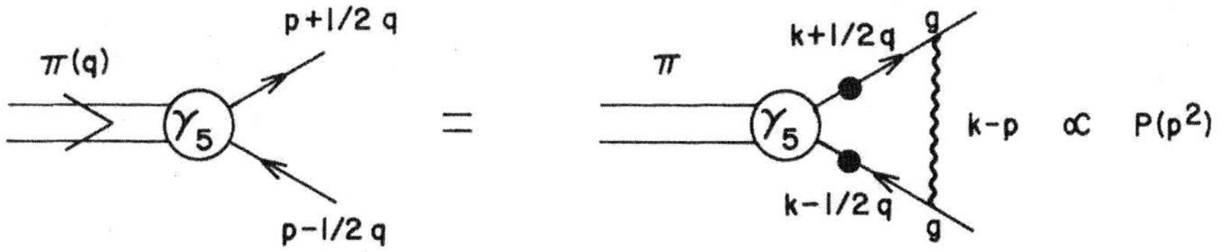

FIG. 1. Spinless component of the 0^- pseudoscalar Bethe-Salpeter bound-state equation for $P(p^2)$.

ous-breakdown condition (1) via the form factor $D(p^2)$. But it is the (spin-one) axial-vector Ward identity and an inhomogeneous $\gamma_\mu\gamma_5$ addition to Fig. 1 which disentangles the relations between the form factors, thus preserving (1) but only providing[3] $P(p^2)=2C(p^2)$.

In order to establish the σ-meson relation (2), it is necessary to demonstrate that the 0^+ wave function $U(p^2)$ satisfies the same bound-state Bethe-Salpeter equation corresponding to Fig. 2 as the Bethe-Salpeter equation for the 0^- wave function $P(p^2)$ of Fig. 1. However, to verify this relation in a general way, we must consider wave functions P and U which depend upon *both* invariants p^2 and q^2: $P=P(p^2,q^2)$, $U=U(p^2,q^2)$.

Then the most transparent way to decouple the q^2 from the p^2 dependence on the right-hand side of Figs. 1 and 2 is to employ dispersion relations at fixed q^2. Suppressing the non-Abelian character of the gluon, ignoring the momentum variation of the gluon coupling, and working in the Feynman gauge for simplicity, we find

$$\mathrm{Im}P(p^2,q^2)=2g^2\int d\rho_2(k)\,(k-p)^{-2}(-q^2)P(k^2,q^2=4m^2-4k^2), \tag{3a}$$

$$\mathrm{Im}U(p^2,q^2)=2g^2\int d\rho_2(k)\,(k-p)^{-2}(4m^2-q^2)U(k^2,q^2=4m^2-4k^2), \tag{3b}$$

where $d\rho_2$ represents two-body phase space and $(k\pm\frac{1}{2}q)^2=m^2$ requires $k\cdot q=0$ and $q^2=4m^2-4k^2$. Next we construct (unsubtracted) dispersion relations for P and U, respectively evaluated at $q^2=0$ and $q^2=4m^2$, so that the $-q^2$ and $4m^2-q^2$ factors in (3a) and (3b) are eliminated:

$$P(p^2,q^2=0)=\frac{1}{\pi}\int\frac{\mathrm{Im}P(p^2,q'^2)dq'^2}{q'^2-q^2}\bigg|_{q^2=0}=-\frac{2g^2}{\pi}\int dq^2\int\frac{d\rho_2(k)}{(k-p)^2}P(k^2,q^2=4m^2-4k^2), \tag{4a}$$

$$U(p^2,q^2=4m^2)=\frac{1}{\pi}\int\frac{\mathrm{Im}U(p^2,q'^2)dq'^2}{q'^2-q^2}\bigg|_{q^2=4m^2}=-\frac{2g^2}{\pi}\int dq^2\int\frac{d\rho_2(k)}{(k-p)^2}U(k^2,q^2=4m^2-4k^2). \tag{4b}$$

As an aside we note that the (Wick-rotated) O(4) invariant $p\cdot q$ does not appear in (4) because P and U are spin-zero scalar wave functions (with four-dimensional angular momentum $N=0$). Alternatively, the dispersive nature of (3) and (4) insures that the $k\cdot q=0$ constraint on the right-hand side of (3) does not translate to a $p\cdot q$ dependence on the left-hand side of (4).

At this point it is clear that *if* both P and U are in fact independent of the meson momentum-transfer invariant q^2, then (4a) and (4b) have identical structures, implying that

$$P(p^2)\propto U(p^2). \tag{5}$$

This result is obviously valid for the NJL $(\bar\psi\psi)^2$ model where all such momenta invariants, p^2 or q^2, are suppressed. But we maintain that (5) is

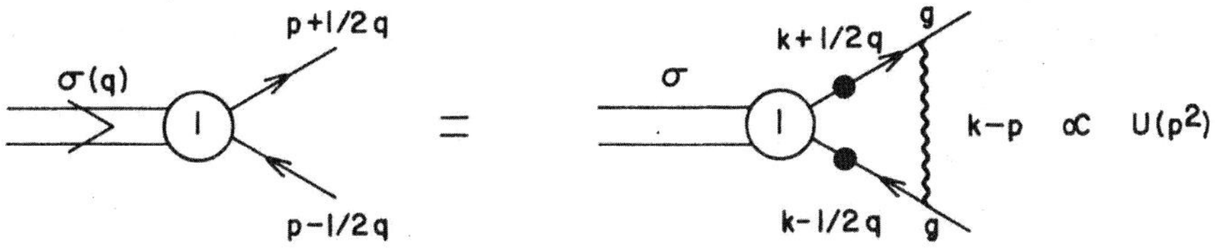

FIG. 2. The 0^+ scalar Bethe-Salpeter bound-state equation for $U(p^2)$.

VOLUME 48, NUMBER 6 PHYSICAL REVIEW LETTERS 8 FEBRUARY 1982

also true for the asymptotically free theory of QCD with running coupling constant $\alpha_s(p^2) = \pi d/\ln(p^2/\Lambda^2)$ and (low) energy scale of[4]

$$\Lambda \approx 150 \pm 50 \text{ MeV}. \tag{6}$$

Since the latter observation is our key point, we provide further elaboration. Asymptotic freedom requires not only a logarithmic falloff of the quark-gluon coupling for *large* p^2/Λ^2, but also that such couplings $g(p^2, q^2, p\cdot q)$ depend upon only *one* momentum invariant, i.e., p^2 in this case. Thus the desired q^2 suppression in the pseudoscalar and scalar wave functions of (4) follows for QCD when $p^2 \gg \Lambda^2$. However, the spontaneous generation of quark mass occurs for $m_{\text{dyn}}(p^2) = -C(p^2)/D(p^2) \neq 0$, where m_{dyn} is the dynamically generated quark mass appearing in the quark propagator and in the induced (Nambu-Goldstone) pseudoscalar component of the axial current

$$J_{\mu 5} \propto \gamma_\mu \gamma_5 - 2m_{\text{dyn}}(p^2)\gamma_5 q_\mu/q^2, \tag{7}$$

such that[5-7]

$$m_{\text{dyn}}(p^2 = m_{\text{dyn}}{}^2) \equiv m_{\text{dyn}} \approx 300\text{–}320 \text{ MeV}. \tag{8}$$

The latter mass scale is consistent with $m_{\text{dyn}} \sim \frac{1}{3} m_N$ being the chiral-limiting nonstrange constituent quark mass, and it is also slightly less than the presently accepted chiral-broken nonstrange constituent quark mass, $\hat{m}_{\text{con}} \approx 340$ MeV, as it should be. Then the asymptotic freedom condition $p^2 \gg \Lambda^2$ follows from (6) and (8) even for *low* p^2 *in the spontaneous-breakdown region.*

If instead Λ turns out to be larger, say $\Lambda \sim 300$–$500 \text{ MeV} \sim m_{\text{dyn}}$, the absence of q^2 dependence in P and U hence the validity of (5) still follow for QCD because of the coupling-constant freezeout[8] for $p^2 \lesssim 1 \text{ GeV}^2$. Thus the main feature in QCD which leads to (5) is the ultraviolet (or deep Euclidean) structure of the QCD coupling or equivalently the ultraviolet behavior of the dynamically generated quark mass, which behaves for large p^2 as[9]

$$m_{\text{dyn}}(p^2) = \frac{M^2}{p^2} m_{\text{dyn}}(M^2)\left(\frac{\ln(M^2/\Lambda^2)}{\ln(p^2/\Lambda^2)}\right)^{1-d}, \tag{9}$$

with d the anomalous dimension $d = 12(33-2n_f)^{-1}$. Indeed, when one combines (9) with the flavor-dependent current quark masses, the large-p^2 behavior of all such masses properly extrapolates[7] down to the spontaneous-breakdown region of $p^2 \sim (300 \text{ MeV})^2$.

To summarize in a slightly different manner, the asymptotic freedom property of QCD allows us to ignore the q^2 dependence of P and U so that (4) leads to (5): $P(p^2) \propto U(p^2)$. At the same time the axial Ward identity or equivalently (7) requires $P(p^2) = 2C(p^2)$ (i.e., $m_{\text{dyn}} = -C/D$) and all must be nonvanishing in order that massless pions exist. Note that there is no definite relation between P and U as there is between P and C. [This is because the σ meson does not appear as a pole in the vector current as the pion occurs in the axial-vector current (7).] Thus (5) connects Figs. 1 and 2 via the binding-equation relation (2) for m in (4) identified as m_{qk}. The latter quark mass in the quark loops of Figs. 1 and 2 must then correspond to $m_{\text{dyn}} \approx 300$–$320$ MeV in the chiral limit. Finally, therefore, we may make the identification for QCD

$$q^2 = 4m^2 \text{ in (4b)} \to m_\sigma = 2m_{\text{dyn}} \approx 600\text{–}640 \text{ MeV}. \tag{10}$$

Chiral-breaking corrections to (10) could increase m_σ to at most 700 MeV.

With regard to this numerical estimate for m_σ, nuclear theorists have long discovered the need for a 0^+ 2π-exchange isobar of mass 500–600 MeV in order to explain the $^3S_1$ NN nuclear force.[10] Moreover, low-energy (and subthreshold) πN scattering data lead to a πN σ term of magnitude[11] $\sigma_{\pi N} \approx 65$ MeV. Then in the context of the unrenormalized σ model, we deduce that[12]

$$\sigma_{\pi N} = (m_\pi{}^2/m_\sigma{}^2)m_N \approx 65 \text{ MeV}, \quad m_\sigma \approx 530 \text{ MeV}. \tag{11}$$

This result is slightly modified by factors of $g_A \approx 1.25$ when the σ model is renormalized so that (11) corresponds to a renormalized mass $m_\sigma \sim 600$–700 MeV. As a last piece of evidence, simultaneous fits to $\pi\pi \to \pi\pi, K\bar{K}$ data[13] require an $I = 0$ resonance in the 700–800 MeV region (the σ?). The associated large decay width $\Gamma_\sigma = 800 \pm 400$ MeV is not incompatible with the theoretical σ-model value of $\Gamma_\sigma \sim 400$ MeV. Nevertheless, analyses of $\sigma \to \pi^0\pi^0$ (which is not contaminated by the ρ background as is $\sigma \to \pi^+\pi^-$) via four-photon detection devices[14] should be carried out to further constrain the σ mass and width.

If in fact the σ meson is finally confirmed in the 600–700 MeV region, then our above analysis suggests the following theoretical conclusions:

(a) Because the dynamical binding-equation relation (2) is a model-dependent result, the presently accepted theory of asymptotically free QCD with a low energy Λ scale would then be indirectly verified in that it, and perhaps few other re-

VOLUME 48, NUMBER 6 PHYSICAL REVIEW LETTERS 8 FEBRUARY 1982

normalizable field theories, requires that a 0^+ σ meson should exist with $m_\sigma \sim 600-700$ MeV.

(b) Such a massive scalar $\bar{q}q$ meson is the chiral-symmetry-spontaneous-breakdown analog of the Higgs meson (minus the Schwinger mechanism giving mass to the vector bosons) now sought after to verify the $SU(2)\otimes U(1)$ spontaneously broken gauge theory.[15] If indeed the latter Higgs meson does exist, then perhaps its mass can be dynamically determined[16] as the σ mass is by QCD. This may be done in conjuction with the gauge technique.[17]

One of us (M.D.S.) is grateful to M. Halpern, R. Jacob, and A. Patrascioiu for informative comments. This work was supported in part by the U. S. Department of Energy Contract No. DE-AC02-80ER10663.

[1]M. Gell-Mann and M. Lévy, Nuovo Cimento 16, 705 (1960).

[2]Y. Nambu and G. Jona-Lasinio, Phys. Rev. 122, 345 (1961).

[3]R. Delbourgo and M. D. Scadron, J. Phys. G 5, 1621 (1979).

[4]S. J. Eidelman, L. M. Kurdadze, and A. I. Vainshtein, Phys. Lett. 82B, 278 (1979); R. K. Ellis, CERN Report No. TH3090, 1981 (to be published); S. Drell, in Proceedings of the Lepton-Photon Conference, Bonn, 1981 (to be published); S. Brodsky and P. Lepage, in Proceedings of the Summer Institute on Particle Physics, Stanford Linear Accelerator Center, 1981 (to be published).

[5]H. Pagels and S. Stokar, Phys. Rev. D 20, 2947 (1979).

[6]J. M. Cornwall, Phys. Rev. 22, 1452 (1980).

[7]M. D. Scadron, Rep. Prog. Phys. 44, 213 (1981); N. G. Fuchs and M. D. Scadron, to be published.

[8]M. Creutz, Phys. Rev. D 21, 2308 (1980).

[9]H. D. Politzer, Nucl. Phys. B117, 397 (1976).

[10]See, e.g., K. Erkelenz, Phys. Rep. 13C, 190 (1974).

[11]M. M. Nagels *et al.*, Nucl. Phys. B109, 1 (1979).

[12]See, e.g., H. F. Jones and M. D. Scadron, Phys. Rev. D 11, 174 (1975).

[13]See, e.g., P. Estabrooks, Phys. Rev. D 19, 2678 (1979).

[14]K. W. Lai, private communication.

[15]S. Weinberg, Phys. Rev. Lett. 19, 1264 (1967); A. Salam, in *Elementary Particle Theory*, edited by N. Svartholm (Almqvist & Wiksell, Stockholm, 1968).

[16]R. Jackiw and K. Johnson, Phys. Rev. D 8, 2386 (1973); J. M. Cornwall and R. E. Norton, Phys. Rev. D 8, 3338 (1973).

[17]R. Delbourgo and P. West, Phys. Lett. 22B, 96 (1977); R. Delbourgo and B. G. Kenny, J. Phys. G 7, 417 (1981).

International Journal of Modern Physics A, Vol. 13, No. 4 (1998) 657–676
© World Scientific Publishing Company

DYNAMICAL GENERATION OF LINEAR σ MODEL SU(3) LAGRANGIAN AND MESON NONET MIXING

R. DELBOURGO and M. D. SCADRON*

Physics Department, University of Tasmania, Hobart, Australia 7001

Received 21 April 1997

This paper is the SU(3) extension of the dynamically generated SU(2) linear σ model Lagrangian worked out previously using dimensional regularization. After discussing the quark-level Goldberger–Treiman relations for SU(3) and the related gap equations, we dynamically generate the meson cubic and quartic couplings. This also constrains the meson–quark coupling constant to $g = 2\pi/\sqrt{3}$ and determines the SU(3) scalar meson masses in a Nambu–Jona-Lasinio fashion. Finally we dynamically induce the U(3) pseudoscalar and scalar mixing angles in a manner compatible with data.

1. Introduction

In a recent paper,[1] we have extended the original spontaneously broken SU(2) linear σ model (LσM) to the quark-level dynamically generated LσM. The latter LσM is close in spirit to the four-fermion theory of NJL; only the tight-binding bound states with chiral-limiting mass $m_\pi = 0$ and $m_\sigma = 2m_q$ in the NJL approach become elementary particle states in the LσM scheme. Dimensional regularization and $Z = 0$ compositeness conditions are the key ingredients making the SU(2) theory extremely predictive. In this paper we generalize the dynamically generated LσM to SU(3) symmetry and also discuss meson nonet mixing.

Experimental signatures[2–4] of the elusive nonstrange isoscalar and strange isospinor scalar resonances σ(600–700) and κ(800–900) combined with recent theoretical observations on scalar mesons[5,6] make the original SU(2) and SU(3) linear sigma model (LσM) field theories[7,8] of interest once again. Specifically, a broad nonstrange scalar σ(400–900) was extracted in the last reference in Ref. 2 and supported in the 1996 PDG tables[3] (with an upper limit mass scale 300 MeV higher).

Such a scalar σ(400–900) has a mean value of $m_\sigma = 650$ MeV, which is in agreement with the prediction of the dynamically generated LσM.[1] This SU(2) LσM computed in one-loop order reproduces[1,9] many satisfying chiral-limiting results: $m_\pi = 0$, $m_\sigma = 2m_q$ (the latter two of course are true in the four-fermion NJL model[10]), vector-meson dominance (VMD) universality[11] $g_{\rho\pi\pi} = g_\rho$, the dynamically generated scale[1] $g_{\rho\pi\pi} = 2\pi$, the KSRF[12] rho mass $m_\rho = \sqrt{2}g_\rho f_\pi$, and Weinberg's[5]

*Permanent address: Physics Department, University of Arizona, Tucson AZ 85721, USA.

mended chiral symmetry decay width relation $\Gamma_\sigma = (9/2)\Gamma_\rho$. Moreover the semileptonic $\pi \to e\nu\gamma$ empirical[3] structure-dependent form factors are approximately recovered[13] from the SU(2) LσM quark and meson loops. Finally the observed[14] $a_0(984) \to \gamma\gamma$ radiative decay width has been obtained[15] using SU(3) quark and meson loops in the LσM.

Very recently, the SU(2) LσM Lagrangian density (shifted around the stable vacuum with $\langle\sigma\rangle = \langle\pi\rangle = 0$ and quarks now with mass m_q) having the interacting part for elementary quarks and π and σ mesons,

$$\mathcal{L}_{\rm int} = g\bar{\psi}(\sigma + i\gamma_5\boldsymbol{\tau}\cdot\boldsymbol{\pi})\psi + g'\sigma(\sigma^2 + \boldsymbol{\pi}^2) - \lambda(\sigma^2 + \boldsymbol{\pi}^2)^2/4 \tag{1a}$$

with chiral-limiting meson–quark and meson–meson couplings[7]

$$g = m_q/f_\pi\,, \qquad g' = m_\sigma^2/2f_\pi = \lambda f_\pi \tag{1b}$$

(for $f_\pi \approx 90$ MeV), has been *dynamically generated.*[1] Such dynamical generation is driven by the meson–quark interaction $g\bar{\psi}(\sigma + i\gamma_5\boldsymbol{\tau}\cdot\boldsymbol{\pi})\psi$ alone. This leads to the chiral-limiting meson masses $m_\pi = 0$, $m_\sigma = 2m_q$, meson–meson cubic and quartic couplings g', $\lambda = g'/f_\pi$ and also constrains the fundamental meson–quark coupling to $g = 2\pi/\sqrt{N_c}$. The latter coupling together with the NJL scalar σ mass follow from dimensional regularization considerations. However, these results are regularization independent as shown in the second reference in Ref. 1.

Since the analogous (but much more complex) SU(3) LσM Lagrangian[8] has only been considered in its (unshifted) spontaneously generated form (but also giving rise to interesting physics[16,17]), in this paper we try to dynamically generate the SU(3) LσM Lagrangian, the U(3) meson masses, couplings, and in addition, dynamically induce the empirical meson mixing angles.

In Sec. 2 we focus on the quark-level SU(3) Goldberger–Treiman (GT) relations and corresponding SU(3) "gap equations." Then in Sec. 3 we dynamically generate the nonstrange (NS) σ meson–$\pi\pi$ and $K\bar{K}$ couplings $g'_{\sigma_{\rm NS}\pi\pi}$, $g'_{\sigma_{\rm NS}KK}$ obtained from the vanishing chiral-limiting pseudoscalar meson masses $m_\pi = 0$, $m_K = 0$. The latter also gives rise to the strange (S) σ meson–$K\bar{K}$ coupling $g'_{\sigma_{\rm S}KK}$. Next in Sec. 4 we dynamically generate the SU(3)-broken scalar meson masses (but with $m_\pi = m_K = 0$)

$$m_{\sigma_{\rm NS}} = 2\hat{m}\,, \qquad m_\kappa = 2\sqrt{m_s\hat{m}}\,, \qquad m_{\sigma_{\rm S}} = 2m_s\,. \tag{2}$$

Here the nonstrange, kappa and strange scalar meson masses are $m_{\sigma_{\rm NS}}$, m_κ, $m_{\sigma_{\rm S}}$ and the nonstrange and strange constituent quark masses are $\hat{m}$, m_s, respectively.

In Sec. 5 we comment on "bootstrapping" the cubic and quartic meson–meson couplings from one-loop order to tree order based on the gap equations discussed in Sec. 2. Finally, in Sec. 6 we dynamically induce the U(3) quark–annihilation graphs in the SU(3) LσM. They simulate (but do not double count) the effects of nonperturbative QCD by predicting $\eta - \eta'$ and $\sigma - f_0$ mixing angles that in fact are compatible with data. The latter mixing approach, while fitted self-consistently, bypasses a direct nonperturbative calculation of the singlet U(3) meson masses. We summarize our dynamically generated findings in Sec. 7 and list in the Appendix the needed nonstrange and strange (quark basis) U(3) structure constants.

2. Quark Level GT Relations and Gap Equations

Using only constituent quark masses already induced through vacuum expectation values of scalar fields along with the meson–quark (chiral) coupling $g\bar{\psi}[(\sigma_{\rm NS} + i\gamma_5\boldsymbol{\lambda}\cdot\boldsymbol{\pi}) + \lambda^a(\kappa^a + i\gamma_5 K^a) + \cdots]\psi$, the quark loop pion and kaon decay constants depicted in Figs. 1 are in the chiral limit ($q_\pi \to 0$, $q_K \to 0$ but $m_s \neq \hat{m}$) with $đ^4p = d^4p(2\pi)^{-4}$,

$$if_\pi = 4N_c g \int \frac{đ^4p\,\hat{m}}{(p^2-\hat{m}^2)^2}\,, \qquad if_K = 4N_c g \int \frac{đ^4p\,\frac{1}{2}(m_s+\hat{m})}{(p^2-\hat{m}^2)(p^2-m_s^2)}\,. \tag{3a}$$

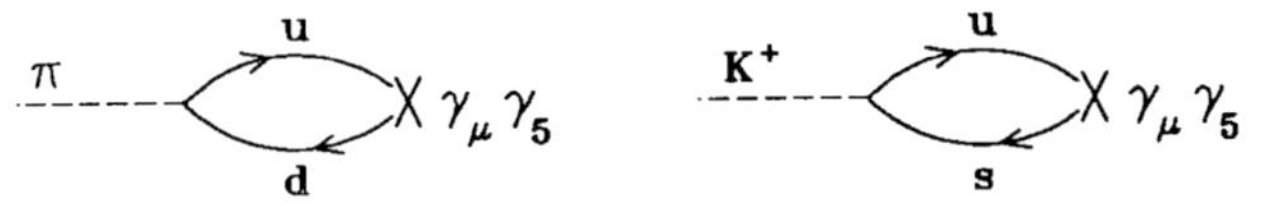

Fig. 1. Pion (a) and kaon (b) decay constants generated by quark loops.

Then invoking the quark-level pion GT relation in (1b) and its natural kaon generalization,[18]

$$f_\pi g = \hat{m}\,, \qquad f_K g = \frac{1}{2}(m_s + \hat{m})\,, \tag{3b}$$

Eqs. (3a) lead to the (log-divergent) CL gap equations

$$1 = -i4N_c g^2 \int^{\Lambda^2} \frac{đ^4p}{(p^2-\hat{m}^2)^2}\,, \tag{4a}$$

$$1 = -i4N_c g^2 \int^{\Lambda'^2} \frac{đ^4p}{(p^2-m_s^2)(p^2-\hat{m}^2)}\,. \tag{4b}$$

Since in the SU(3) LσM, the meson–quark coupling constant g is the *same* for pions and for kaons (in or away from the CL), the knowledge of g in (4a) and in (4b) fixes the log-divergent scales Λ and Λ'. In fact it has been shown[1] in the dynamically generated SU(2) LσM that for $N_c = 3$,

$$g = 2\pi/\sqrt{N_c} \approx 3.6276\,, \tag{5}$$

compatible with the nonstrange GT estimate in (3b) away from the CL $g = \hat{m}/f_\pi \approx$ 340 MeV/93 MeV $\approx$ 3.66. (We shall return to the derivation of (5) in Sec. 4.) Accordingly, the gap equations in (4) give the Euclidean integrals

$$1 = \int_0^{\Lambda^2/\hat{m}^2} \frac{(dq^2/\hat{m}^2)(q^2/\hat{m}^2)}{(1+q^2/\hat{m}^2)^2}\,, \qquad 1 \approx \int_0^{\Lambda'^2/m_s\hat{m}} \frac{(dq^2/m_s\hat{m})(q^2/m_s\hat{m})}{(1+q^2/m_s\hat{m})^2}\,. \tag{6}$$

The denominator in the second integral of (6) is the geometric average of the exact product $(1 + q^2/\hat{m}^2)(1 + q^2/m_s^2)$. Then both integrals in (6) being unity in turn leads to

$$\Lambda^2/\hat{m}^2 \approx \Lambda'^2/m_s\hat{m} \approx 5.3\,. \tag{7}$$

To verify that the dimensionless cutoffs in (7) make physical sense, we must first introduce a dimensionful scale. Returning to the CL nonstrange GT relation in (3b), we invoke the CL pion decay constant scale[19] $f_\pi^{\rm CL} \approx 90$ MeV, so that the CL nonstrange quark mass is (with divergenceless axial current $\partial A^\pi = 0$ generating the GT relation for pions)

$$f_\pi^{\rm CL} g = \hat{m}^{\rm CL} \approx (90 \text{ MeV})(3.6276) \approx 326 \text{ MeV}\,. \tag{8}$$

Away from the CL the ratio of the two GT relations in (3b) is fixed to the empirical value[12]

$$\frac{f_K}{f_\pi} = \frac{1}{2}\left(\frac{m_s}{\hat{m}} + 1\right) \approx 1.22 \quad \text{or} \quad \frac{m_s}{\hat{m}} \approx 1.44\,. \tag{9}$$

In fact this constituent quark mass ratio of about 1.4 is also known to hold for baryon magnetic dipole moments,[20] meson charge radii[21] and $K^* \to K\gamma$ decays.[22] However in the chiral limit we might expect $m_s/\hat{m} \leq 1.44$, say 4/3. Finally then the cutoff scales in (7) become

$$\Lambda \sim \sqrt{5.3}\hat{m}^{\rm CL} \sim 750 \text{ MeV}\,, \qquad \Lambda' \sim \sqrt{5.3 m_s^{\rm CL}\hat{m}^{\rm CL}} \sim 860 \text{ MeV}\,. \tag{10}$$

The above nonstrange cutoff scale of 750 MeV separates SU(2) LσM elementary particles (the u, d quarks, π and σ mesons, the latter taken as σ (650) as justified from Refs. 2, 3 and the discussion in the introduction) from the $\bar{q}q$ bound states with mass > 750 MeV ($\rho(770)$, $\omega(783)$, $A_1(1260)$, $f_2(1275)$, $A_2(1320)$). Likewise the qualitative isospinor cutoff scale in (10) of 860 MeV separates elementary $K(495)$, $\kappa(820)$ mesons and $m_s \approx 480$ MeV quarks (as we shall see later) from bound state $K^*(895)$, $K^{**}(1350)$ mesons in the SU(3) LσM. In field theory language, the merging of the elementary particle and bound state cutoff scales inferred from the gap equations (4) correspond to $Z = 0$ compositeness conditions,[23] whereby the scalar mesons $\sigma(650)$ and $\kappa(820)$ can be consistently treated either as elementary (in the LσM) or as bound states (in the NJL picture).

3. Dynamical Generation of the Cubic Meson Couplings

Consider now the Nambu–Goldstone massless pion and kaon in the chiral limit (CL). Starting only with the quark–meson coupling used in Sec. 2, the quark one-loop order pion self-energies are depicted in Figs. 2. In the CL the "vacuum polarization (VP)"-bubble-type amplitude of Fig. 2(a) is (displaying the two quark propagators in the denominator to parallel Eqs. (3) and (4))

$$M_{\rm VP}^\pi = -i4N_c 2g^2 \int \frac{đ^4p(p^2 - \hat{m}^2)}{(p^2 - \hat{m}^2)^2}\,, \tag{11a}$$

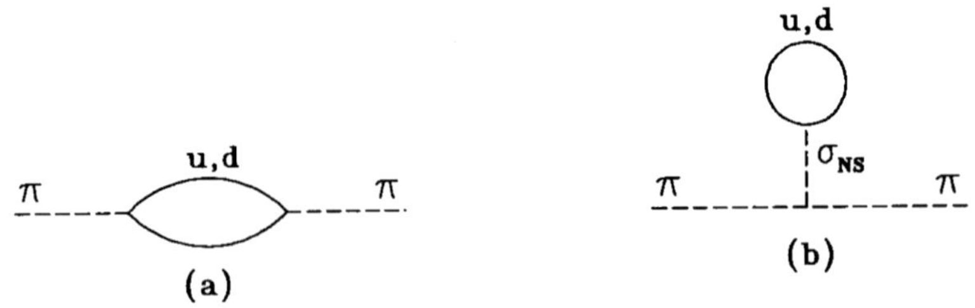

Fig. 2. Pion bubble (a) and quark tadpole (b) graphs.

where the factor of 2 in (11a) arises from u and d quark loops for $\pi^\circ qq$ couplings, or $(\sqrt{2})^2$ for $\pi^+ u\bar{d}$ couplings. Likewise the "quark tadpole" amplitude of Fig. 2(b) is in the CL.

$$M^{\pi}_{\text{qktad}} = \frac{i4N_c 2g2g'}{m^2_{\sigma_{\text{NS}}}} \int \frac{\bar{d}^4 p\, \hat{m}}{p^2 - \hat{m}^2} \,. \tag{11b}$$

The "new" $\sigma\pi\pi$ coupling $2g'$ (the factor 2 is from Bose symmetry) in Fig. 2(b) is dynamically generated so that the CL pion mass remains zero in one-loop order in the CL:

$$m^2_\pi = M^\pi_{\text{VP}} + M^\pi_{\text{qktad}} = 0\,, \tag{12a}$$

$$\left(g - \frac{2g'}{m^2_{\sigma_{\text{NS}}}}\hat{m}\right) \int \frac{\bar{d}^4 p}{(p^2 - \hat{m}^2)} = 0\,, \tag{12b}$$

$$g'_{\sigma_{\text{NS}}\pi\pi} = m^2_{\sigma_{\text{NS}}}/2f_\pi \,. \tag{12c}$$

Note that regardless of the two quadratic divergent integrals in Eqs. (11), the dynamically generated meson–meson tree-level coupling g' in (12c) "conspires" to keep the CL pion massless in (12a) and (12b).[1,9]

This SU(2) LσM result (12c) can be extended to SU(3) by considering the bubble and quark tadpole graphs in Figs. 3 which contribute to the kaon mass. The bubble amplitude for Fig. 3(a) is in the CL ($p_K \to 0$, $m_s \neq \hat{m}$),

$$M^K_{\text{VP}} = -i4N_c 2g^2 \int \frac{\bar{d}^4 p (p^2 - m_s \hat{m})}{(p^2 - m_s^2)(p^2 - \hat{m}^2)}\,, \tag{13a}$$

while the two nonstrange (NS) and strange (S) quark tadpole graphs of Fig. 3(b) and Fig. 3(c) have the respective CL amplitudes

$$M^K_{\text{qktadNS}} = i\frac{4N_c g g'_{\text{NS}}}{m^2_{\sigma_{\text{NS}}}} \int \frac{\bar{d}^4 p\, 2\hat{m}}{(p^2 - \hat{m}^2)}\,, \tag{13b}$$

$$M^K_{\text{qktadS}} = i\frac{4N_c\sqrt{2}g g'_S}{m^2_{\sigma_S}} \int \frac{\bar{d}^4 p\, m_s}{(p^2 - m_s^2)}\,. \tag{13c}$$

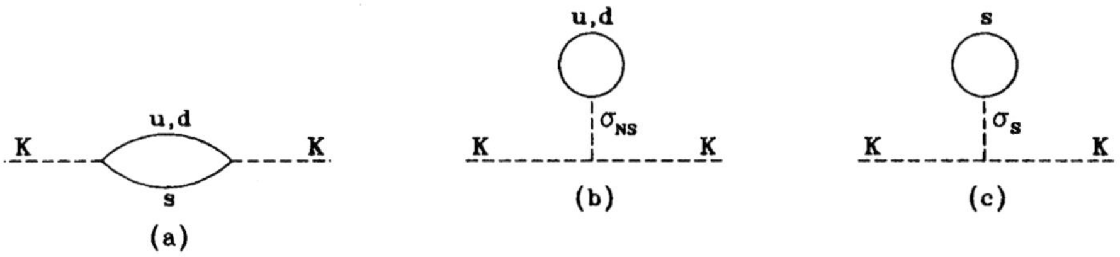

Fig. 3. Kaon bubble (a) and quark tadpole (b), (c) graphs.

Note that the factor of $2g^2$ in (13a) is due to the two $K^\circ sd$ vertices each with coupling $\sqrt{2}g$, while the factor of $2\hat{m}$ in (13b) counts 2 nonstrange quarks in the NS tadpole loop of Fig. 3(b). Finally the factor of $\sqrt{2}g$ in (13c) is due to the $\sigma_{\rm S}$SS coupling in Fig. 3(c). In order to combine Eqs. (13a)–(13c) so that the CL kaon mass remains zero,

$$m_K^2 = M_{\rm VP}^K + M_{\rm qktadNS}^K + M_{\rm qktadS}^K = 0\,, \tag{14a}$$

we invoke the partial fraction identity

$$\frac{(m_s+\hat{m})(p^2-m_s\hat{m})}{(p^2-m_s^2)(p^2-\hat{m}^2)} = \frac{\hat{m}}{p^2-\hat{m}^2} + \frac{m_s}{p^2-m_s^2}\,. \tag{14b}$$

Replacing the integrand of the kaon bubble amplitude in (13a) by the right-hand-side (RHS) of (14b), we see that the vanishing of m_K^2 in (14a) requires the two coefficients of the nonstrange loop integral to cancel and also the two coefficients of the strange loop integral to cancel. Thus we have dynamically generated *two* more tree-level meson cubic couplings in the CL:

$$g'_{\sigma_{\rm NS}KK} = m_{\sigma_{\rm NS}}^2/2f_K \tag{15a}$$

$$g'_{\sigma_{\rm S}KK} = m_{\sigma_{\rm S}}^2/\sqrt{2}f_K\,. \tag{15b}$$

The respective Clebsch–Gordan coefficients of 1, 1/2, $1/\sqrt{2}$ in (12c), (15a) and (15b) correspond to the SU(3) structure constants $d_{\rm NS\,33} = 1$, $d_{{\rm NS}\,KK} = 1/2$, $d_{{\rm S}\,KK} = 1/\sqrt{2}$, derived in the Appendix. Thus the dynamically generated cubic meson–meson couplings (12c), (15a) and (15b) indeed follow an SU(3) LσM pattern.

4. Scalar Cubic Couplings and Scalar Meson Masses

By chiral symmetry we expect the respective scalar–scalar–scalar meson couplings to be identical to the analog scalar–pseudoscalar–pseudoscalar couplings. For the SU(2) case, the two γ_5 vertices in the $\sigma_{\rm NS}\pi\pi$ loop of Fig. 4(a) reduce the divergence to

$$g'_{\sigma_{\rm NS}\pi\pi} = 2g\hat{m}\left[-i4N_cg^2\int\frac{d^4p(p^2-\hat{m}^2)}{(p^2-\hat{m}^2)^3}\right] = 2g\hat{m}\,, \tag{16a}$$

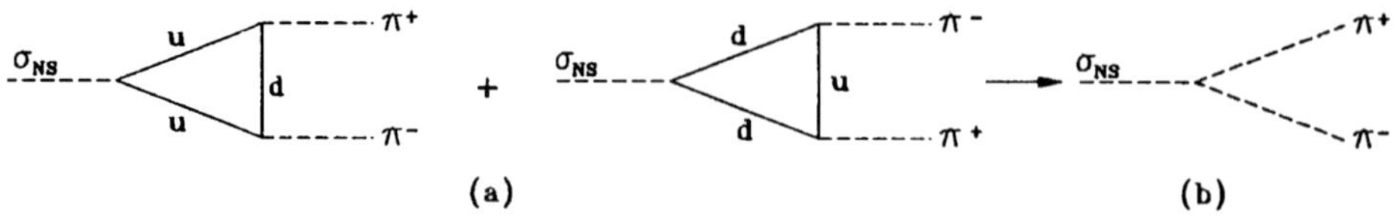

Fig. 4. Bootstrap of $g_{\sigma_{\rm NS}\pi\pi}$ quark triangle (a) to g' tree (b) graph.

by virtue of the gap equation (4a). But the two factors of unity replacing the γ_5's for the analogue $\sigma_{\rm NS}\sigma_{\rm NS}\sigma_{\rm NS}$ loop of Fig. 4(a) mean expanding out the trace in the CL gives[1]

$$g'_{\sigma_{\rm NS}\sigma_{\rm NS}\sigma_{\rm NS}} = 2g\left[-iN_c g^2 \int \frac{đ^4p\ {\rm Tr}(\not p + \hat m)^3}{(p^2-\hat m^2)^3}\right] = 6g\hat m\,, \tag{16b}$$

where we keep only the log divergent piece in the integral of (16b) since it dominates the coupling constant $g'_{\sigma_{\rm NS}\sigma_{\rm NS}\sigma_{\rm NS}}$. Only then is the tree-order chiral symmetry $g'_{\sigma_{\rm NS}\pi\pi} = g'_{\sigma_{\rm NS}\sigma_{\rm NS}\sigma_{\rm NS}}$ recovered in one-loop order.

Moreover both cubic loop couplings in (16) "bootstrap" the g' tree coupling $g' = m^2_{\sigma_{\rm NS}}/2f_\pi$ in the CL provided[1,9]

$$m_{\sigma_{\rm NS}} = 2\hat m\,, \tag{17a}$$

found by setting $g' = 2g\hat m = m^2_\sigma/2f_\pi$ and using the GTR $f_\pi g = \hat m$. This "shrinking" of quark loops to points in Eqs. (16) again corresponds to a $Z = 0$ compositeness condition.[23]

To dynamically generate (17a) we consider instead Figs. 5 representing $m^2_{\sigma_{\rm NS}}$ in the CL $p \to 0$. Using dimensional regularization, these graphs sum to Ref. 1

$$m^2_{\sigma_{\rm NS}} = 16iN_c g^2 \int đ^4p\left[\frac{\hat m^2}{(p^2-\hat m^2)^2} - \frac{1}{p^2-\hat m^2}\right] = \frac{N_c g^2 \hat m^2}{\pi^2}\,, \tag{17b}$$

using a Γ function identity $\Gamma(2-l)+\Gamma(1-l) \to -1$ in $2l = 4$ dimensions. Combining (17a) and (17b) one obtains the meson–quark coupling $g = 2\pi/\sqrt{N_c}$. The latter and (17a) can also be dynamically generated via the quark mass gap tadpole combined with the above dimensional regularization identity again.[1]

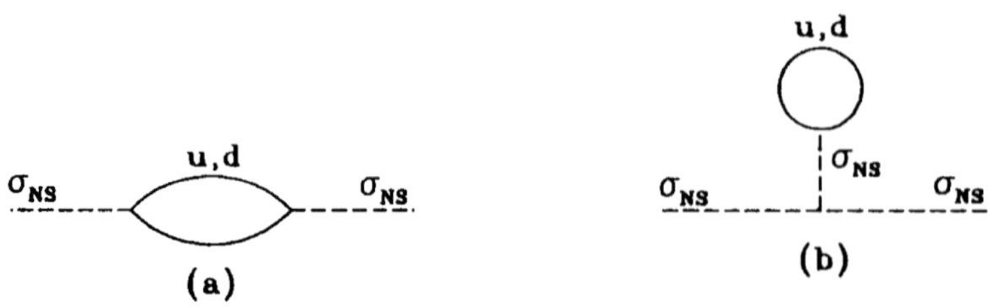

Fig. 5. Scalar $\sigma_{\rm NS}$ bubble (a) and quark tadpole (b) graphs.

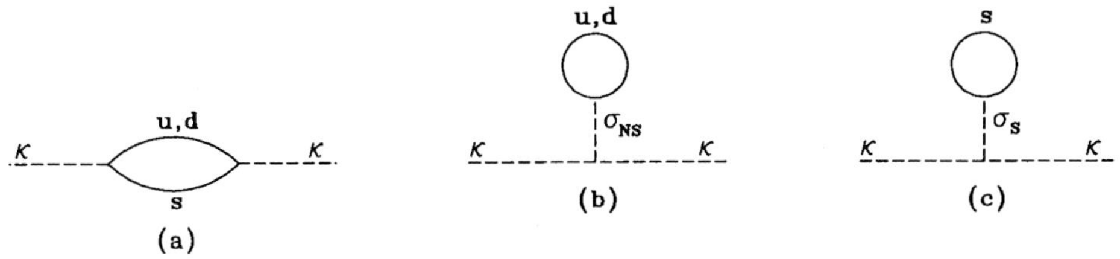

Fig. 6. Scalar kappa bubble (a) and quark tadpole (b), (c) graphs.

In a similar fashion, the scalar kappa meson self-energies of Figs. 6 have the bubble (VP) amplitude in the CL

$$M^{\kappa}_{\rm VP} = -i4N_c(\sqrt{2}g)^2 \int \frac{d^4p(p^2 + m_s\hat{m})}{(p^2 - m_s^2)(p^2 - \hat{m}^2)}\,. \tag{18a}$$

Again adding and subtracting $m_s\hat{m}$ terms to the numerator of (18a), cancelling the quadratic divergent VP part against the tadpole graphs of Fig. 6(b), 6(c) and using the gap equation (4b), we find in the CL

$$m_\kappa^2 = 0 + 2\cdot 2m_s\hat{m} \quad \text{or} \quad m_\kappa = 2\sqrt{m_s\hat{m}}\,. \tag{18b}$$

Finally for the purely strange meson self-energy graphs of Figs. 7, by analogy with the nonstrange scalar mass equation (17b), the graphs sum via the above dimensionless regularization identity to

$$m^2_{\sigma_{\rm S}} = 8iN_c g_{\rm S}^2 \int d^4p \left[\frac{m_s^2}{(p^2 - m_s^2)^2} - \frac{1}{p^2 - m_s^2}\right] = \frac{N_c g_{\rm S}^2 m_s^2}{2\pi^2}\,. \tag{19a}$$

Invoking the U(3) coupling $g_{\rm S} = \sqrt{2}g$ (which can be generated via the strange quark mass gap), Eq. (19a) generates the strange scalar meson mass

$$m_{\sigma_{\rm S}} = 2m_s\,. \tag{19b}$$

Thus we have dynamically generated the chiral-limiting (NJL-like) scalar masses (17a), (18b) and (19b) in the SU(3) LσM as indicated in Eq. (2).

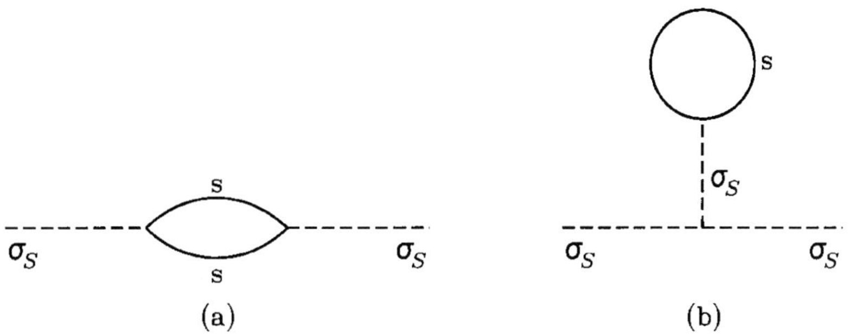

Fig. 7. Scalar $\sigma_{\rm S}$ bubble (a) and quark tadpole (b) graphs.

There are in fact two independent ways of extending these CL results away from the chiral limit so as to obtain the "physical" quark and scalar meson masses. More specifically with $m_\pi \neq 0$, the nonstrange CL relation (16) becomes

$$m_{\sigma_{\rm NS}}^2 - m_\pi^2 = (2\hat{m}^{\rm CL})^2 = (653\ {\rm MeV})^2 \quad \text{or} \quad m_{\sigma_{\rm NS}} \approx 668\ {\rm MeV}\,. \tag{20}$$

Alternatively with $f_\pi \approx 93$ MeV away from the chiral limit, the quark-level GT relation (8) becomes (still with $g = 2\pi/\sqrt{3}$),

$$\hat{m} = f_\pi g \approx (93\ {\rm MeV})(3.6276) \approx 337\ {\rm MeV}\,, \tag{21a}$$

in close agreement with the u constituent quark mass found from magnetic dipole moments.[20] Then a NJL-type estimate of the chiral-broken nonstrange scalar σ mass is

$$m_{\sigma_{\rm NS}} = 2\hat{m} \approx 674\ {\rm MeV}\,. \tag{21b}$$

Henceforth we will take $m_{\sigma_{\rm NS}} \approx 670$ MeV as the average between (20) and (21b).

The $I = 1/2$ scalar kappa meson with mass $m_\kappa \neq 0$ follows from an "equal-splitting-law"[24] compared to (20),

$$m_\kappa^2 - m_K^2 = m_{\sigma_{\rm NS}}^2 - m_\pi^2 \approx 0.43\ {\rm GeV}^2 \quad \text{or} \quad m_\kappa \approx 820\ {\rm MeV}\,. \tag{22}$$

On the other hand, the chiral-broken strange constituent quark mass found from (21a) and the ratio $m_s/\hat{m} \approx 1.4$ from (9) is

$$m_s = \hat{m}(m_s/\hat{m}) \approx 475\ {\rm MeV}\,, \tag{23a}$$

also in reasonable agreement with the magnetic moment determination.[20] Then the NJL-type estimate of the chiral-broken kappa mass in Eq. (2) scaled to (21a) above is

$$m_\kappa = 2\sqrt{m_s\hat{m}} \approx 2 \cdot 337\sqrt{1.4}\ {\rm MeV} \approx 805\ {\rm MeV}\,, \tag{23b}$$

with an average $m_\kappa \approx 810$ MeV, midway between (22) and (23b).

In the early 1970's, the particle data group (PDG) suggested the ground state kappa mass is in the 800–900 MeV region. Since 1974, however, this κ has been replaced by the $\kappa(1450)$. But scaled to the $\sigma(670)$ mass of (20) or (21), a κ in the 800–900 MeV region as in (22) and (23) is unavoidable. Nevertheless it is worth commenting on why a (peripherally produced) $\kappa(810)$ has not been observed. We suggest it is because of the soft pion theorem[25] (SPT) suppressing the $A_1 \to \pi(\pi\pi)_{\rm sw}$ decay rate due to the interfering $A_1 \to \sigma\pi$ amplitude. Likewise, a similar SPT suppresses the $\kappa(810) \to K\pi$ decay amplitude, explaining why the PDG tables no longer list the $\kappa(810)$. Specifically, the latter peripherally produced $\kappa(810)$ in $K^-p \to K^-\pi^+n$ is suppressed by the quark (box plus triangle) SPT chiral cancellation as in Ref. 25.

Henceforth we will take the ground state kappa mass at $m_\kappa \approx 810$ MeV as the average between (22) and (23b). It is satisfying that these elusive scalar σ and

κ masses were recently seen in polarization measurements (Svec *et al.*, Ref. 4) at 750 MeV and 887 MeV respectively, which are unaffected by the above soft pion theorem of Ref. 25.

We can also estimate the pure strange $\bar{s}s$ scalar meson mass in two ways. The equal-splitting-law analogue of (20) and (22), is

$$m_{\sigma_S}^2 - m_\kappa^2 = m_\kappa^2 - m_{\sigma_{NS}}^2 \quad \text{or} \quad m_{\sigma_S} \approx 930 \text{ MeV}, \tag{24}$$

while the NJL-like strange scalar mass from (2) using $m_s \approx 475$ MeV from (23a) is

$$m_{\sigma_S} = 2m_s \approx 950 \text{ MeV}, \tag{25}$$

with average mass $m_{\sigma_S} \approx 940$ MeV. Tornqvist and Roos in Ref. 2 claim the f_0 (980) is mostly an $\bar{s}s$ scalar meson. Accounting for the observed scalar mixing angle of 20° (scalar mixing is discussed in Eqs. (37)–(39) of Sec. 6), this observed f_0 (980) is compatible with the above predicted σ_S (930–950). The average "physical" chiral-broken scalar meson masses which we shall henceforth use in our dynamically generated SU(3) LσM are then

$$m_{\sigma_{NS}} \approx 670 \text{ MeV}, \qquad m_\kappa \approx 810 \text{ MeV}, \qquad m_{\sigma_S} \approx 940 \text{ MeV}. \tag{26}$$

5. Bootstrapping the Quartic Meson Lagrangian

Once the $Z = 0$ compositeness conditions[23] (or the quark mass gap and meson mass equations) are known, via the SU(3) gap equations (4) and SU(3) NJL equations (2), to shrink quark loops to LσM tree graphs, one should study how to induce the SU(3) quartic Lagrangian density

$$\mathcal{L}_{\text{quartic}}^{\text{L}\sigma\text{M}} = -\lambda[\sigma_{\text{NS}}^2 + \boldsymbol{\pi}^2 + \kappa^2 + K^2 + \sigma_{\text{S}}^2 + \eta_{\text{S}}^2]^2/4. \tag{27}$$

The U(2) nonstrange sector of (27) was investigated in Ref. 1 via the u, d quark box of Fig. 8(a), leading to the chiral limiting (CL) $\pi^0\pi^0 \to \pi^0\pi^0$ amplitude

$$T = -i8N_c g^4 \int đ^4p(p^2 - \hat{m}^2)^{-2} = 2g^2, \tag{28a}$$

by virtue of the log-divergent gap equation (4a). Similarly the $\pi^+\bar{K}^0 \to \pi^+\bar{K}^0$ quark box of Fig. 8(b) has CL amplitude

$$T = -i(\sqrt{2})^2 4N_c g^4 \int đ^4p(p^2 - \hat{m}^2)^{-1}(p^2 - m_s^2)^{-1}[p^2 - m_s\hat{m}] = 2g^2, \tag{28b}$$

by virtue of the partial fraction identity (14b) and gap equation (4). Likewise the $\eta_S\eta_S \to \eta_S\eta_S$ strange quark box of Fig. 8(c) has CL amplitude

$$T = -i4N_c g_S^4 \int đ^4p(p^2 - m_s^2)^{-2} = g_S^2 = 2g^2, \tag{28c}$$

due to the strange quark gap equation analogous to (4a) together with $g_S = \sqrt{2}g$.

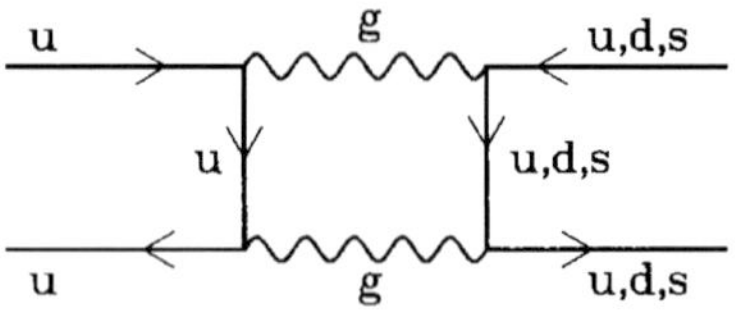

Fig. 8. Isoscalar gluon-mediated quark annihilation diagrams for intermediate QCD states.

Thus all three quark box graphs of Figs. 8 and Eqs. (28) have effective quartic (box) couplings in the chiral limit

$$\lambda_{\text{quartic box}} \to 2g^2\,, \tag{29a}$$

whereas the SU(3) LσM quartic Lagrangian (27) has LσM tree strength

$$\lambda = g'/f_\pi = (2\hat{m}g)/f_\pi = 2g^2\,, \tag{29b}$$

by virtue of the quark-level GT relation $\hat{m}/f_\pi = g$. The fact that $\lambda = 2g^2$ means that, starting from the meson–quark interaction, the SU(3) LσM quark box graphs of Figs. 8 and Eqs. (28) bootstrap back to the LσM quartic Lagrangian (27). These are all further examples of the $Z = 0$ compositeness condition helping to dynamically generate the entire SU(3) LσM Lagrangian.

6. Dynamically Inducing Mixing of Pseudoscalar and Scalar Meson States

Thus far, starting from the fundamental SU(3) meson–quark chiral interaction $g\bar{\psi}\lambda^i[S^i + i\gamma_5 P^i]\psi$, we have dynamically generated the LσM cubic and quartic meson–meson couplings, the chiral-limiting pseudoscalar and scalar SU(3) meson masses, and even the meson–quark couplings g and g_S. Taken together this forms the interacting part of the SU(3) linear sigma model (LσM) Lagrangian density

$$\mathcal{L}_{\text{int}}^{\text{L}\sigma\text{M}} = \mathcal{L}_{\text{meson–qk}} + \mathcal{L}_{\text{meson–meson}}\,. \tag{30}$$

It is then natural to study the additional U(3) mixing Lagrangian $\mathcal{L}_{\text{mixing}}$. In the spontaneously generated LσM scheme of Refs. 8, such an "input" $\mathcal{L}$ mixing term introduces extra mixing parameters in (30) which are to be determined by experiment. Alternatively in our dynamically generated approach to the SU(3) LσM, the predicted parameters in (30) already match observation without introducing new (arbitrary) parameters. That is, $\mathcal{L}_{\text{L}\sigma\text{M}}$ in (30) is an "output" Lagrangian rather than an input, and there is no additional $\mathcal{L}$ mixing Lagrangian.

In this dynamically generated LσM theory, the chiral-broken seven pseudoscalar (Nambu–Goldstone) meson masses m_π^2 and m_K^2 are inserted in the theory by hand and then the six chiral-broken scalar (NJL-like) masses of Eq. (26) will in turn dynamically generate (fit) the observed η and η' pseudoscalar along with the σ and f_0 scalar meson masses.

More specifically, for the case of pseudoscalar (P) meson states, the U(3) meson $\eta - \eta'$ mixing is generated by the quark annihilation amplitude β_P which turns a $\bar{u}u$ or $\bar{d}d$ meson P into a $\bar{u}u$, $\bar{d}d$ or $\bar{s}s$ meson P' state. This dynamical breaking of the OZI rule can be characterized in the language of QCD,[20,26] in the model-independent mixing approach of Ref. 27, or in terms of the SU(3) LσM.

To reach the same mixing conclusions (28)–(30) in the context of QCD, one observes that a singlet η_0 is twice formed via the "pinched" quark annihilation graph in Fig. 8. But such quark triangle graphs "shrink" to points in the LσM by the log-divergent gap equations (4), i.e. via $Z = 0$ conditions. Then $I = 0$ mesons take the place of QCD gluons in Fig. 8 so that one must consider the LσM meson loop graphs of Fig. 9 as simulating β_P for $\eta' - \eta$ mixing. In all of the above cases, one classifies the $I = 0$ nonstrange meson mass matrix as

$$M_P^2 = \begin{pmatrix} m_\pi^2 + 2\beta_P & \sqrt{2}\beta_P \\ \sqrt{2}\beta_P & 2m_K^2 - m_\pi^2 + \beta_P \end{pmatrix} \to \begin{pmatrix} m_\eta^2 & 0 \\ 0 & m_{\eta'}^2 \end{pmatrix}, \tag{31}$$

where the arrow indicates rediagonalization to the observed η and $\eta' I = 0$ states. Here nonstrange (NS) and strange (S) $I = 0$ meson states contribute to a unitary singlet state according to $|0\rangle = \sqrt{2/3}|\mathrm{NS}\rangle + \sqrt{1/3}|\mathrm{S}\rangle$, where $|\mathrm{NS}\rangle = |\bar{u}u + \bar{d}d\rangle/\sqrt{2}$ and $|\mathrm{S}\rangle = |\bar{s}s\rangle$.

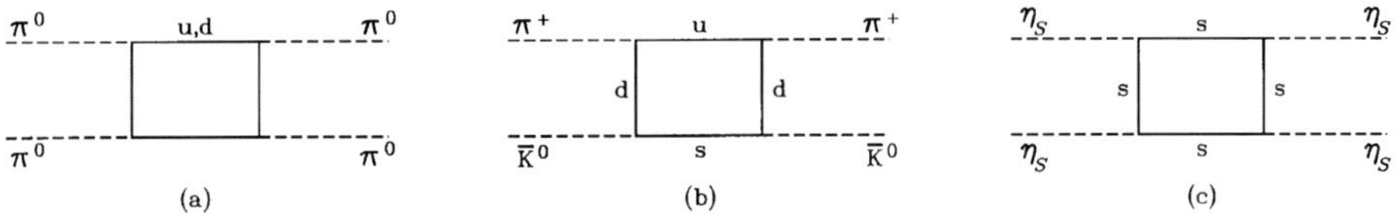

Fig. 9. Quark box graphs for $\pi^0\pi^0$ (a), $\pi^+\bar{\kappa}^0$ (b), $\eta_S\eta_S$ (c) scattering.

Note that the *one* parameter β_P on the LHS of (31) determines the *two* measured masses m_η^2 and $m_{\eta'}^2$ on the RHS of (31). Specifically, the trace of (31) requires

$$2m_K^2 + 3\beta_P = m_\eta^2 + m_{\eta'}^2 \quad \text{or} \quad \beta_P \sim 0.24\ \mathrm{GeV}^2\,, \tag{32a}$$

while the determinant of (31) gives

$$m_\pi^2(2m_K^2 - m_\pi^2) + (4m_K^2 - m_\pi^2)\beta_P = m_\eta^2 m_{\eta'}^2 \quad \text{or} \quad \beta_P \sim 0.28\ \mathrm{GeV}^2\,. \tag{32b}$$

Rather than work with isospin LσM intermediate states and dynamically generate β_P, it will be more straightforward to consider intermediate QCD glue (which is automatically flavor blind and $I = 0$) and "dynamically fit" β_P in (31) to the scale of (32) even in the context of the LσM.[28] There is then no need to introduce an additional SU(3) Lagrangian simulating this mixing; it is already built into the above quantum-mechanical picture.

Further fine tuning of (32) comes from accounting for the SU(3)-breaking ratio of the nonstrange and strange quark propagators in Fig. 9 via the constituent quark mass ratio $X = \hat{m}/m_s \approx 0.7$ from Eq. (9). In the latter case, keeping X free and weighting the off-diagonal NS–S β_P in (31) by X and the S–Sβ_P by X^2, the two parameters β_P and X on the LHS of this modified matrix (31) are uniquely determined by the η and η' masses to be[26]

$$\beta_P = \frac{(m_{\eta'}^2 - m_\pi^2)(m_\eta^2 - m_\pi^2)}{4(m_K^2 - m_\pi^2)} \approx 0.28 \text{ GeV}^2\,, \tag{33a}$$

$$X \approx 0.78\,. \tag{33b}$$

These latter two fitted parameters are compatible with (32) and with $X \approx 0.7$ from (9). Thus there is only the one nonperturbative parameter $\beta_P \approx 0.28$ GeV^2 to be explained in our dynamically fitted scheme. In fact this β_P can be partially understood from a perturbative two-gluon anomaly-type of graph.[29,30] Since the SU(3) nonstrange and strange pseudoscalar and scalar meson masses have already been dynamically generated in (26), our dynamically fitted U(3) extension in (33) cannot alter the masses in (26), but instead rotates the states via an $\eta' - \eta$ mixing angle. In effect, $\beta_P = m_{\eta_0}^2/3$ in (33) for $m_{\eta_0} \approx 915$ MeV regardless of the mixing scheme: QCD in Fig. 9 or the LσM induced by Figs. 8.

To this end one can recast the nonperturbative fitted pseudoscalar mixing scale of β_P in (33) in terms of the quark nonstrange (NS)–strange (S) basis pseudoscalar mixing angle ϕ_P with physical η and η' states defined by

$$|\eta\rangle = \cos\phi_P|\eta_{\rm NS}\rangle - \sin\phi_P|\eta_{\rm S}\rangle\,, \qquad |\eta'\rangle = \sin\phi_P|\eta_{\rm NS}\rangle + \cos\phi_P|\eta_{\rm S}\rangle\,, \tag{34a}$$

or equivalently in terms of the more familiar singlet-octet mixing angle

$$\theta_P = \phi_P - \tan^{-1}\sqrt{2}\,. \tag{34b}$$

Given the dynamically generated structure of the mixing mass matrix (31), the pseudoscalar mixing angle in (34) is predicted (fitted) to be[26]

$$\phi_P = \tan^{-1}\left[\frac{(m_{\eta'}^2 - 2m_K^2 + m_\pi^2)(m_\eta^2 - m_\pi^2)}{(2m_K^2 - m_\pi^2 - m_\eta^2)(m_{\eta'}^2 - m_\pi^2)}\right]^{1/2} \approx 42^\circ\,, \tag{34c}$$

or $\theta_P \approx -13^\circ$ since $\tan^{-1}\sqrt{2} \approx 55^\circ$ in (34b). We believe it significant that the "world data" in 1990 pointed to Ref. 27 $\theta_P = -14^\circ \pm 2^\circ$ or $\phi_P = 41^\circ \pm 2^\circ$, in good agreement with (34c).

In order to reconfirm this $\eta - \eta'$ mixing angle prediction (34c) specifically in the context of the LσM, we consider the radiative decays $\pi^\circ \to \gamma\gamma$, $\eta \to \gamma\gamma$, $\eta' \to \gamma\gamma$. In the former case, the usual u and d constituent quark triangle graphs lead to the $\pi^\circ\gamma\gamma$ amplitude $F\epsilon^{\alpha\beta\gamma\delta}k'_\alpha k_\beta \epsilon_\gamma^{*\prime}\epsilon_\delta^*$ with $F = \alpha/\pi f_\pi$. This of course is the ABJ[30] anomaly amplitude or the Steinberger[31] fermion loop result with $g_A = 1$ and $N_c = 3$ at the quark level. Moreover this $\pi^\circ\gamma\gamma$ amplitude is also the LσM prediction since there can be no three-pion (loop) correction. The resulting (LσM) decay rate is then

$$\Gamma(\pi^\circ\gamma\gamma) = m_\pi^3(\alpha/\pi f_\pi)^2/64\pi \approx 7.6 \text{ eV}, \tag{35a}$$

which is very close to experiment[12] 7.74 ± 0.55 eV.

For η, $\eta' \to \gamma\gamma$ decays, however, an additional constituent strange quark loop must be folded into the $\pi^\circ\gamma\gamma$ (LσM) rate prediction (35a). This leads to the decay rate ratios[26,27] constrained to recent data in Refs. 3 and 32

$$\frac{\Gamma(\eta\gamma\gamma)}{\Gamma(\pi^\circ\gamma\gamma)} = \left(\frac{m_\eta}{m_\pi}\right)^3 \left(\frac{5}{3}\right)^2 \cos^2\phi_P \left(1 - \frac{\sqrt{2}}{5}\frac{\hat{m}}{m_s}\tan\phi_P\right)^2 = 60 \pm 7, \tag{35b}$$

$$\frac{\Gamma(\eta'\gamma\gamma)}{\Gamma(\pi^\circ\gamma\gamma)} = \left(\frac{m'_\eta}{m_\pi}\right)^3 \left(\frac{5}{3}\right)^2 \sin^2\phi_P \left(1 + \frac{\sqrt{2}}{5}\frac{\hat{m}}{m_s}\cot\phi_P\right)^2 = 550 \pm 68. \tag{35c}$$

Using the constituent quark mass ratio $m_s/\hat{m} \approx 1.4$ from (9) or from Refs. 21 and 22, the LσM rate ratios in (35b)and (35c) respectively predict $\phi_P = 45^\circ \pm 2^\circ$ and $\phi_P = 38^\circ \pm 4^\circ$, which average to the pseudoscalar $\eta' - \eta$ mixing angle extracted from $\eta, \eta' \to \gamma\gamma$ observations,

$$\phi_P = 40.5^\circ \pm 3^\circ \quad \text{or} \quad \theta_P = -14^\circ \pm 3^\circ. \tag{36}$$

Again we believe it is significant that the phenomenological pseudoscalar mixing angle in (36) is compatible with the world average $-14^\circ \pm 2^\circ$ in Ref. 27 and with the dynamically fitted value $\phi_P \approx 42^\circ$ in (34c) obtained from quark-annihilation graphs for intermediate QCD states.

As regards the chiral analog scalar mixing angle ϕ_S, the parallel to the $\eta - \eta'$ angle ϕ_P in (34a) in the nonstrange–strange quark basis is defined via the physical $\sigma - f_0$ states

$$|\sigma\rangle = \cos\phi_S|\sigma_{NS}\rangle - \sin\phi_S|\sigma_S\rangle, \qquad |f_0\rangle = \sin\phi_S|\sigma_{NS}\rangle + \cos\phi_S|\sigma_S\rangle. \tag{37}$$

Instead of the dynamical fitted approach to ϕ_S obtained through the quark-annihilation amplitude β_S of Ref. 26 (yielding $\phi_S \sim 17^\circ$), given the already determined σ_{NS} and σ_S LσM scalar masses in (26), we can find ϕ_S via $\langle\sigma|f_0\rangle = 0$ and (37):

$$m^2_{\sigma_{NS}} = m^2_\sigma \cos^2\phi_S + m^2_{f_0}\sin^2\phi_S, \tag{38a}$$

$$m^2_{\sigma_S} = m^2_\sigma \sin^2\phi_S + m^2_{f_0}\cos^2\phi_S. \tag{38b}$$

These two equations (38) and the physical mass $m_{f_0} \approx 980$ MeV constrain m_σ and ϕ_S to the fitted values (with $m_{\sigma_{NS}} \approx 670$ MeV, $m_{\sigma_S} \approx 940$ MeV)

$$m_\sigma = \left[m^2_{\sigma_{NS}} + m^2_{\sigma_S} - m^2_{f_0}\right]^{1/2} \approx 610 \text{ MeV}, \tag{39a}$$

$$\phi_S = \sin^{-1}\left[\frac{m^2_{f_0} - m^2_{\sigma_S}}{m^2_{f_0} - m^2_\sigma}\right]^{1/2} \approx 20^\circ. \tag{39b}$$

For the pseudoscalar U(3) nonet (π, K, η, η'), we have used the dynamically generated SU(3) LσM and have self-consistently computed (dynamically fitted) the $\eta - \eta'$ mixing angle $\phi_P \approx 42°$ via the Nambu–Goldstone π and K masses leading to the nonperturbative mass matrix (31) and Eqs. (33) or mixing angles in (34). For the scalar U(3) nonet, however, we started with the SU(2) dynamically generated NJL–LσM NS and S scalar masses (26) and used equal-splitting laws to fit the $I = 1/2$ kappa mass (squared) half-way between as in the averages (26). Together with the observed $f_0(980)$ this led to the fitted scalar mixing angle $\phi_S \approx 20°$ and a slightly mixed $I = 0$ $\sigma(610)$ mass in (39). Note that the nearness of the observed $f_0(980)$ to our dynamically generated pure $\bar{s}s$ scalar mass $\sigma_S(940)$ is what is forcing ϕ_S in (39b) to be small. This parallels the $\bar{q}q$ vector case where the (nearby) $\phi(1020)$ vector meson is known to be almost purely $\bar{s}s$ strange.

All that remains undetermined in the latter nonet is the $I = 1$ scalar a_0 mass. Again it is the chiral equal-splitting laws [ESL] that require[24,33] in analogy with (22),

$$m_{a_0}^2 - m_{\eta_{NS}}^2 = m_{\sigma_{NS}}^2 - m_\pi^2 = m_\kappa^2 - m_K^2 \approx 0.43 \text{ GeV}^2 . \tag{40a}$$

Then with $\phi_P \approx 42°$ so that $m_{\eta_{NS}} \approx 760$ MeV by analogy with (38), Eq. (40a) predicts the fitted $I = 1$ a_0 mass to be for $m_{\sigma_{NS}} \approx 670$ MeV from (20) and (21),

$$m_{a_0} = [m_{\sigma_{NS}}^2 - m_\pi^2 + m_{\eta_{NS}}^2]^{1/2} \approx 1.00 \text{ GeV} . \tag{40b}$$

Of course this latter predicted mass is presumably the observed[12] $a_0(984)$. The fact that this $I = 1$ $a_0(984)$ is near the $I = 0$ $f_0(980)$ does *not* necessarily signal that both the a_0 and f_0 have the same (nonstrange) flavor quarks (as do the ρ and ω and the a_0 and f_0 in the alternative $\bar{q}q\bar{q}q$ scheme[34]). Rather, our $\bar{q}q$ SU(3) LσM picture of a mostly strange $f_0(980)$ and nonstrange $a_0(984)$ is based on the (standard) mixing equations of (38) and (39) and the (infinite momentum frame) ESLs of Eqs. (40). To support this latter $\bar{q}q$ LσM picture is the known almost purely strange vector meson $\phi(1020)$ being near this mostly strange $f_0(980)$. Moreover, Figs. 13 in the DM2 Collab. in Refs. 2 also suggests that the $f_0(980)$ is composed mostly of $\bar{s}s$ quarks.

In the context of this same LσM, the SU(2) meson–meson coupling $g'_{\sigma_{NS}\pi\pi} = (m_{\sigma_{NS}}^2 - m_\pi^2)/2g_\pi$ can be replaced by the SU(3) ESL coupling $g'_{\delta\eta_{NS}\pi} = (m_{a_0}^2 - m_{\eta_{NS}}^2)/2f_\pi$. Continuing to invoke SU(3) symmetry, one can then compute the scalar decay rate ratio as

$$\frac{\Gamma(f_0(980)\pi\pi)}{\Gamma(a_0(984)\eta\pi)} = \frac{3p_{f_0}}{2p_{a_0}} \left(\frac{\sin\phi_S}{\cos\phi_P}\right)^2 = \frac{37 \pm 7 \text{ MeV}}{57 \pm 11 \text{ MeV}} , \tag{41a}$$

where the observed rates are taken from the 1992 PDG tables.[32] For $\phi_P \approx 42°$ from (32c), the above (39a) requires (with momentum $p_{f_0} = 467$ MeV, $p_{a_0} = 319$ MeV)

$$\phi_S = 23° \pm 3° , \tag{41b}$$

compatible with (39b). The 1992 PDG tables[32] take the $a_0 \to \eta\pi$ rate as 57 ± 11 MeV, but the high-statistics $a_0 \to \eta\pi$ rate measured by Armstrong *et al.*,[35] of 95 ± 14 MeV predicts $\phi_S \approx 18^\circ \pm 3^\circ$ from (41a), more in line with (39b).

The above ground state $\bar{q}q$ scalar LSM nonet ($\sigma(610), \kappa(810), f_0(980), a_0(984)$) with dynamically fitted mixing angle $\phi_S \approx 20^\circ$ is qualitatively different from a $\bar{q}q\bar{q}q$ four-quark[34] or $\bar{K}K$ molecule[36] scheme. However the latter objections to a $\bar{q}q$ picture[37] based on the recent Crystal Ball[14] radiative decays $a_0 \to \gamma\gamma$ and $f_0 \to \gamma\gamma$ can also be understood in the SU(3) LσM (but not in a pure $\bar{q}q$ quark model). These narrow scalar decays have $\bar{q}q$ quark loops which interfere *destructively*[15] with SU(3) LσM meson loops and lead to rates ~ 0.5 keV or smaller as measured.[14,32]

7. Conclusion

In summary, we started in Sec. 2 only with the fundamental SU(3) meson–quark (chiral quark model) interaction

$$\mathcal{L}_{\text{meson-qk}} = g\bar{\psi}\lambda^i[S^i + i\gamma_5 P^i]\psi\,, \tag{42}$$

(where $i = 0, \ldots, 8$ and S_i, P_i are scalar and pseudoscalar elementary meson fields), with quark level SU(3) Goldberger–Treiman relations

$$f_\pi g = \hat{m}\,, \qquad f_K g = \frac{1}{2}(m_s + \hat{m})\,, \tag{3b}$$

ensuring $\partial A^j = 0$ for $j = 1 \cdots 8$. Then we dynamically generated the log-divergent chiral-limiting gap equations

$$1 = -i4N_c g^2 \int^\Lambda \frac{d^4p}{(p^2 - \hat{m}^2)^2}\,, \qquad 1 = -i4N_c g^2 \int^{\Lambda'} \frac{d^4p}{(p^2 - \hat{m}^2)(p^2 - m_s^2)}\,, \tag{4}$$

which self-consistently fixed the cutoffs to

$$\Lambda^2/\hat{m}^2 \approx \Lambda'^2/m_s\hat{m} \approx 5.3\,, \tag{7}$$

so that $\Lambda \sim 750$ MeV, or $\Lambda' \sim 860$ MeV in (10). This required all $I = 0, 1$ or $I = 1/2$ masses less than 750 MeV, 860 MeV to be elementary, such as $\hat{m}, m_s$ for quarks and m_π, m_K, $m_{\sigma_{NS}} \sim 670$ MeV, $m_\kappa \sim 810$ MeV for pseudoscalar and scalar mesons, respectively.

Next in Sec. 3 we dynamically generated the cubic meson SU(3) couplings in the CL,

$$g'_{\sigma_{NS}\pi\pi} = m^2_{\sigma_{NS}}/2f_\pi\,, \quad g'_{\sigma_{NS}KK} = m^2_{\sigma_{NS}}/2f_K\,, \quad g'_{\sigma_S KK} = m^2_{\sigma_S}/\sqrt{2}f_K\,, \tag{43}$$

and analogously for $\sigma\sigma\sigma$-like scalar couplings in Sec. 4 with the additional meson–quark SU(3)-limiting coupling constraint for $N_c = 3$

$$g = 2\pi/\sqrt{N_c} \approx 3.6276\,. \tag{5}$$

Recall that the cubic part of the standard[8] spontaneously broken SU(3) LσM Lagrangian density has the SU(3)-limiting form

$$\mathcal{L}^{\mathrm{L}\sigma\mathrm{M}}_{\mathrm{cubic}} = g' d^{ijk} S^i (S^j S^k + P^j P^k) . \tag{44a}$$

On the other hand, our dynamically generated SU(3) LσM Lagrangian also has the SU(3)-limiting structure of (44a), but away from the SU(3) limit it becomes

$$\begin{aligned}\mathcal{L}^{\mathrm{L}\sigma\mathrm{M}}_{\mathrm{cubic}} &= g'_{\pi\sigma_{\mathrm{NS}}\pi} \boldsymbol{\pi}\cdot\boldsymbol{\pi}\sigma_{\mathrm{NS}} + g'_{\pi\delta\eta_{\mathrm{NS}}} \boldsymbol{\delta}\cdot\boldsymbol{\pi}\eta_{\mathrm{NS}} + g'_{\pi\kappa K} \bar{K}\boldsymbol{\tau}\cdot\boldsymbol{\pi}\kappa \\ &\quad + g'_{K\delta K} \bar{K}\boldsymbol{\tau}\cdot\boldsymbol{\delta}K + g'_{K\sigma_{\mathrm{S}}K} \bar{K}K\sigma_{\mathrm{S}} + g'_{K\kappa\eta_{\mathrm{S}}} \bar{K}\kappa\eta_{\mathrm{S}} \\ &\quad + g'_{K\sigma_{\mathrm{NS}}K} \bar{K}K\sigma_{\mathrm{NS}} + g'_{K\kappa\eta_{\mathrm{NS}}} \bar{K}\kappa\eta_{\mathrm{NS}} .\end{aligned} \tag{44b}$$

Here we use the SU(3) partially-broken LσM meson–meson couplings with massive pseudoscalars[38]

$$g'_{\pi\sigma_{\mathrm{NS}}\pi} = (m^2_{\sigma_{\mathrm{NS}}} - m^2_\pi)/2f_\pi \approx 2.3\ \mathrm{GeV} , \tag{45a}$$

$$g'_{\pi\delta\eta_{\mathrm{NS}}} = (m^2_\delta - m^2_{\eta_{\mathrm{NS}}})/2f_\pi \approx 2.1\ \mathrm{GeV} , \tag{45b}$$

$$g'_{\pi\kappa K} = (m^2_\kappa - m^2_K)/2f_\pi \approx 2.2\ \mathrm{GeV} , \tag{45c}$$

$$g'_{K\delta K} = (m^2_\delta - m^2_K)/2f_K \approx 3.2\ \mathrm{GeV} , \tag{45d}$$

$$g'_{K\sigma_{\mathrm{S}}K} = (m^2_{\sigma_{\mathrm{S}}} - m^2_K)/2\sqrt{2}f_K \approx 2.1\ \mathrm{GeV} , \tag{45e}$$

and the more severely broken SU(3) LσM couplings

$$g'_{K\kappa\eta_{\mathrm{S}}} = (m^2_\kappa - m^2_{\eta_{\mathrm{S}}})/\sqrt{2}f_K \approx 0.10\ \mathrm{GeV} , \tag{46a}$$

$$g'_{K\sigma_{\mathrm{NS}}K} = (m^2_{\sigma\mathrm{NS}} - m^2_K)/4f_K \approx 0.45\ \mathrm{GeV} , \tag{46b}$$

$$g'_{K\kappa\eta_{\mathrm{NS}}} = (m^2_\kappa - m^2_{\eta_{\mathrm{NS}}})/4f_K \approx 0.20\ \mathrm{GeV} . \tag{46c}$$

In Sec. 4 we dynamically generated the NJL–LσM chiral-broken average scalar meson masses appearing in (44) and (45) as

$$\begin{aligned} m_{\sigma_{\mathrm{NS}}} &= 2\hat{m} \sim 670\ \mathrm{MeV} , \\ m_\kappa &= 2\sqrt{\hat{m}m_s} \sim 810\ \mathrm{MeV} , \\ m_{\sigma_{\mathrm{S}}} &= 2m_s \sim 940\ \mathrm{MeV} . \end{aligned} \tag{47}$$

Also in Sec. 5 the bootstrapping of the SU(3) quartic meson couplings in the Lagrangian were discussed, giving the usual value $\lambda = g'/f_\pi \approx 26$ in the CL.

Finally in Sec. 6 we focused on U(3) $\eta - \eta'$ and $\sigma - f_0$ particle mixing. In the original spontaneous broken SU(3) LσM,[8] (undetermined) particle mixing parameters were introduced in the extended version of the LσM Lagrangian (44a). However in our dynamically generated version of the SU(3) LσM, *no* additional mixing parameters enter the Lagrangian (44b). Rather, OZI violations generate quantum-mechanical particle mixing via the diagonalization of the mass matrix (31), but the resulting mixing parameters do not enter the dynamically generated Lagrangian (44b). Instead one dynamically fits the latter $\eta - \eta'$ and $\sigma - f_0$ mixing angles, in agreement with empirical data. This gives

$$\phi_P \approx 42^\circ\,, \qquad \phi_S \sim 20^\circ\,, \tag{48}$$

respectively for the U(3) nonets $(\pi(138), K(495), \eta(547), \eta'(958))_P$ and $(\sigma(610), \kappa(810), f_0(980), a_0(984))_S$. There is much recent data supporting this above SU(3) LσM nonet picture.[39]

In short, the theoretical dynamically generated and dynamically fitted SU(2) and SU(3) linear sigma model Lagrangians of Ref. 14 and here appear to give a good description of low energy strong interaction physics. Moreover the above LσM picture is a natural generalization of the four-quark Nambu–Jona-Lasinio dynamically generated scheme and is also compatible with low-energy QCD.[40]

Acknowledgments

One of the authors (MDS) appreciates discussions with A. Bramon, G. Clement, V. Elias, H. F. Jones, and R. Tarrach and is grateful for partial support from the U.S. Department of Energy. This research was also partially supported by the Australian Research Council.

Appendix

Here we translate standard symmetric SU(3) Cartesian structure constants $d_{oij} = \sqrt{2/3}\delta_{ij}$, $d_{338} = -d_{888} = 1/\sqrt{3}$, $d_{344} = d_{355} = 1/2$, $d_{366} = d_{377} = -1/2$, $d_{448} = d_{558} = d_{668} = d_{778} = -1/2\sqrt{3}$, to the strange (S)–nonstrange (NS) quark basis with

$$|\mathrm{NS}\rangle = \left|\frac{\bar{u}u + \bar{d}d}{\sqrt{2}}\right\rangle = \sqrt{\frac{2}{3}}|0\rangle + \sqrt{\frac{1}{3}}|8\rangle\,, \tag{A.1}$$

$$|\mathrm{S}\rangle = |\bar{s}s\rangle = \sqrt{\frac{1}{3}}|0\rangle - \sqrt{\frac{2}{3}}|8\rangle\,. \tag{A.2}$$

Then one finds

$$\begin{aligned} &d_{3\mathrm{NS,S}} = d_{3\mathrm{SS}} = d_{33\mathrm{S}} = 0\,, && d_{33\mathrm{NS}} = d_{\mathrm{NS,NS,NS}} = 1\,, \\ &d_{\mathrm{SSS}} = \sqrt{2}\,, && d_{0\mathrm{NS,NS}} = d_{0\mathrm{SS}} = \sqrt{\frac{2}{3}}\,, \\ &d_{8\mathrm{NS,NS}} = \frac{1}{\sqrt{3}}\,, && d_{8\mathrm{SS}} = -\frac{2}{\sqrt{3}}\,, \\ &d_{\mathrm{NS}KK} = \frac{1}{2}\,, && d_{\mathrm{S}KK} = \frac{1}{\sqrt{2}}\,. \end{aligned} \tag{A.3}$$

References

1. R. Delbourgo and M. D. Scadron, *Mod. Phys. Lett.* **A10**, 251 (1995). Here dimensional regularization leads to $g = 2\pi/\sqrt{3}$ and $m_\sigma = 2m_q$, but in fact these results are regularization independent. See R. Delbourgo, M. D. Scadron and A. A. Rawlinson, "Regularizing the quark-level linear σ model," submitted to *Int. J. Math. Phys. A*.
2. See, e.g., P. Estabrooks, *Phys. Rev.* **D19**, 2678 (1979); N. Biswas *et al.*, *Phys. Rev. Lett.* **47**, 1378 (1981); T. Akesson *et al.*, *Phys. Lett.* **B133**, 241 (1983); N. Cason *et al.*, *Phys. Rev.* **D28**, 1586 (1983); A. Courau *et al.*, *Nucl. Phys.* **B271**, 1 (1986); DM2 Collaboration (J. Augustin *et al.*), *Nucl. Phys.* **B320**, 1 (1989); M. Svec, *Phys. Rev.* **D53**, 2343 (1996); N. A. Tornquist and M. Roos, *Phys. Rev. Lett.* **76**, 1575 (1996).
3. Particle Data Group (R. M. Barnett *et al.*), *Phys. Rev.* **D54** Part I, 1 (1996).
4. M. Svec *et al.*, *Phys. Rev.* **D45**, 55, 1518 (1992); **D53**, 2343 (1996).
5. S. Weinberg, *Phys. Rev. Lett.* **65**, 1177 (1990).
6. D. Morgan and M. R. Pennington, *Z. Phys.* **C48**, 623 (1990); *Phys. Lett.* **B258**, 444 (1991); *Phys. Rev.* **D48**, 1185 (1993).
7. M. Gell-Mann and M. Lévy, *Nuovo Cimento* **16**, 705 (1960); also see V. DeAlfaro, S. Fubini, G. Furlan and C. Rossetti, *Currents in Hadron Physics* (North-Holland, Amsterdam, 1973) Chap. 5; more recently see P. Ko and S. Rudaz, *Phys. Rev.* **D50**, 6877 (1994).
8. M. Lévy, *Nuovo Cimento* **A52**, 23 (1967); S. Gasiorowicz and D. Geffen, *Rev. Mod. Phys.* **41**, 531 (1969).
9. T. Hakioglu and M. D. Scadron, *Phys. Rev.* **D42**, 941 (1990); **D43**, 2439 (1991); M. D. Scadron, *Mod. Phys. Lett.* **A7**, 497 (1992).
10. Y. Nambu and G. Jona-Lasinio, *Phys. Rev.* **122**, 345 (1961) (NJL).
11. J. J. Sakurai, *Ann. Phys.* **11**, 1 (1960).
12. K. Kawarabayashi and M. Suzuki, *Phys. Rev. Lett.* **16**, 255 (1966); Riazuddin and Fayyazuddin, *Phys. Rev.* **147**, 1071 (1966) (KSRF).
13. P. Pascual and R. Tarrach, *Nucl. Phys.* **146**, 509 (1978); A. Bramon and M. D. Scadron, *Europhys. Lett.* **19**, 663 (1992).
14. Crystal Ball Collaboration (H. Mariske *et al.*), *Phys. Rev.* **D41**, 3324 (1990).
15. S. Deakin, V. Elias, D. McKeon, M. D. Scadron and A. Bramon, *Mod. Phys. Lett.* **A9**, 995 (1994).
16. G. Clement and J. Stern, *Phys. Lett.* **B231**, 471 (1989).
17. J. A. McGovern and M. C. Birse, *Nucl. Phys.* **A506**, 367 (1990).
18. S. Gerosimov, *Sov. J. Nucl. Phys.* **29**, 259 (1979); R. Delbourgo and M. D. Scadron, *J. Phys.* **G5**, 1621 (1979).
19. M. D. Scadron, *Repts. Prog. Phys.* **44**, 213 (1981); S. A. Coon and M. D. Scadron, *Phys. Rev.* **C23**, 1150 (1981) show that $1 - f_\pi^{\rm CL}/f_\pi = m_\pi^2/8\pi^2 f_\pi^2 \approx 0.03$ for $f_\pi \approx 93$ MeV.
20. A. De Rújula, H. Georgi and S. Glashow, *Phys. Rev.* **D12**, 147 (1975); R. Delbourgo and D. Liu, *Phys. Rev.* **D53**, 6576 (1996).
21. C. Ayala and A. Bramon, *Euro. Phys. Lett.* **4**, 777 (1987) and references therein.
22. A. Bramon and M. D. Scadron, *Phys. Rev.* **D40**, 3779 (1989).
23. A. Salam, *Nuovo Cimento* **25**, 224 (1962); S. Weinberg, *Phys. Rev.* **130**, 776 (1962).
24. M. D. Scadron, *Mod. Phys. Lett.* **A7**, 669 (1992).
25. A. Ivanov, M. Nagy and M. D. Scadron, *Phys. Lett.* **B273**, 137 (1991).
26. H. F. Jones and M. D. Scadron, *Nucl. Phys.* **B155**, 409 (1979); M. D. Scadron, *Phys. Rev.* **D29**, 2079 (1984).
27. A. Bramon and M. D. Scadron, *Phys. Lett.* **B234**, 346 (1990); see also N. Isgur, *Phys. Rev.* **D12**, 3770 (1975).

28. Since the diagonalization of the 2×2 mass matrix in (31) is really a quantum-mechanical (and not a field-theoretical) procedure, we suggest that a "dynamical fit" in terms of external LσM states and internal QCD states is a consistent approach.
29. R. Delbourgo and M. D. Scadron, *Phys. Rev.* **D28**, 2345 (1983); S. R. Choudhury and M. D. Scadron, *Mod. Phys. Lett.* **A1**, 535 (1986).
30. S. L. Adler, *Phys. Rev.* **177**, 2426 (1969); J. S. Bell and R. Jackiw, *Nuovo Cimento* **60**, 47 (1969) (ABJ).
31. J. Steinberger, *Phys. Rev.* **76**, 1180 (1949).
32. Particle Data Group (L. Montanet *et al.*), *Phys. Rev.* **D50** Part I, 1173 (1994).
33. M. D. Scadron, *Phys. Rev.* **D26**, 239 (1982).
34. R. L. Jaffe, *Phys. Rev.* **D15**, 267, 281 (1977).
35. T. A. Armstrong *et al.*, *Z. Phys.* **C52**, 389 (1991).
36. J. Weinstein and N. Isgur, *Phys. Rev. Lett.* **48**, 659 (1982).
37. N. N. Achasov and G. N. Shestakov, *Z. Phys.* **C41**, 309 (1988).
38. See e.g. J. Schechter and Y. Ueda, *Phys. Rev.* **D3**, 2874 (1971).
39. See e.g. P. Estabrooks, Ref. 2, N. Cason *et al.*, Ref. 2; DM2 Collaboration Ref. 2 and M. Svec *et al.*, Ref. 4.
40. L. Baboukhadia, V. Elias and M. D. Scadron, "Linkage between QCD and the linear σ model," submitted for publication (1996).

Part J
QUANTUM EFFECTS IN GRAVITY

J.0. Commentaries J

When the quantum nature of gravity was fully investigated it became clear at the outset that the Einstein-Hilbert Lagrangian, identified as the spacetime curvature scalar R, could lead to unrenormalisable infinities; these problems can all be traced back to the dimensional scale M^{-2} of Newton's gravitational constant G_N. 't Hooft & Veltman (Ann. Inst. H. Poincare Phys. Theor. A20, 69 (1974)) tackled the problem of pure gravity to the first order and showed that this could lead to a rescaling of the constant, but Goroff & Sagnotti (Nucl. Phys. B266, 709 (1986)) went on to declare that unrenormalisability would persist in higher orders, as suspected from the start. This remains an outstanding problem today and various cures have been suggested; the main thrust of these proposals is to add higher order in curvature R terms to the classical Lagrangian. However with quadratic corrections a problem arises in as much as it can lead to negative norm degrees of propagation; this is quite separate from the vector ghosts needed to quantise the gravitational field. Arguments have been given that asymptotic behaviour will cure this difficulty and this remains to be proved; this area of research is quite active today and the case of an $F(R)$ Lagrangian is being doggedly pursued.

J.1 Delbourgo R., Salam A. and Strathdee J., Suppression of infinities in Einstein's gravitational theory. Nuovo Cim. Lett. 2, 354-359 (1969)

J.2 Delbourgo R. and Phocas-Cosmetatos P., Radiative corrections to the electron-graviton vertex. Nuovo Cim. Lett. 5, 420-422 (1972)

J.3 Delbourgo R. and Phocas-Cosmetatos P., Radiative corrections to the photon-graviton vertex. Phys. Lett. B41, 533-535 (1972)

Because perturbation theory leads to ever worse infinities accompanying loop expansions with powers of the gravitational constant, paper [J.1] was an attempt to bypass this problem by dropping perturbation expansions altogether and make use the full nonlinearity of gravity. We concentrated on the exponential representation of the graviton through the metric. Technical difficulties (see Part E) unfortunately precluded any significant development of this idea. On the other hand one might consider the quantum effects of electromagnetism on the classical (lowest order) gravitational vertices as they are of order of the fine structure constant α and might be amenable to experimental verification in future experiments, much like the anomalous magnetic moments of the leptons. That was the purpose of papers [J.2] and [J.3]. Though Behrends and Gastmans (Ann. Phys. 98, 225 (1976)) pointed out that we oversimplified the Feynman rules of the photon-graviton vertex, the conclusions of article [J.3] survive, namely that these corrections mostly affect high energy photons and require extraordinary precision data to be observed experimentally.

These sorts of effects have been carefully studied by Donoghue and coworkers; the reader is urged for instance to consult the article JHEP 1611, 117 (2016) and similar papers for the latest developments.

LETTERE AL NUOVO CIMENTO VOL. II, N. 8 11 Settembre 1969

Suppression of Infinities in Einstein's Gravitational Theory.

R. DELBOURGO (*), A. SALAM (**) and J. STRATHDEE

International Atomic Energy Agency
International Centre for Theoretical Physics - Miramare (Trieste)

(ricevuto il 4 Agosto 1969)

1. – It is an attractive conjecture that the universal gravitational coupling of all matter may provide a natural damping (1) of ultraviolet infinities such as those of self-charge and self-mass encountered in quantum field theories of matter. Unhappily, the plausibility arguments so far available in support of this conjecture rest on applications of functional techniques for the description of gravitational effects. These techniques, as is well known, do not offer a concrete computational procedure for a transcendental Lagrangian as complicated as that of Einstein. The only known practical technique which has been used in the past linearizes this Lagrangian; far from exhibiting any sign of the suppression of ultraviolet infinities, this method introduces in each order more virulent new infinities. It is the purpose of this note to draw attention to a computational procedure (2) developed recently for nonpolynomial Lagrangians which shows, firstly, that the gravitational Lagrangian is « renormalizable » and, secondly, exhibits explicitly the conjectured suppression of at least some of the ultraviolet infinities of other fields. We also indicate how a further sharpening of our methods may result in a suppression of all infinities in a field theory of matter. Although the magnitude of the effective cut-off introduced by gravitation is too high to have any practical bearing on calculations of self-mass and self-charge of particles, the mere fact that a natural damping mechanism may exist at all is of some significance.

2. – The Einstein Lagrangian

$$L = \frac{1}{\varkappa^2}\sqrt{-g}\, g^{\mu\nu}(\Gamma^{\lambda}_{\mu\varrho}\Gamma^{\varrho}_{\nu\lambda} - \Gamma^{\lambda}_{\mu\nu}\Gamma^{\varrho}_{\lambda\varrho}) \tag{1}$$

(*) Imperial College, London.

(**) On leave of absence from Imperial College, London.

(1) The conjecture that the gravitational interaction may suppress infinities in conventional field theory has been made before. Two references which we know of and which consider this aspect of the quantum theory of gravitation are: S. DESER: *Rev. Mod. Phys.*, **29**, 417 (1957); and B. S. DE WITT: *Phys. Rev.*, **162**, 1195 (1967).

(2) G. V. EFIMOV: *Sov. Phys. JETP*, **17**, 1417 (1963); *Nuovo Cimento*, **32**, 1046 (1964); *Nucl. Phys.*, **74**, 657 (1965); E. S. FRADKIN: *Nucl. Phys.*, **49**, 624 (1963); **76**, 588 (1966); R. DELBOURGO, A. SALAM and J. STRATHDEE: ICTP, Trieste, preprint IC/69/17 (to be published in *Phys. Rev.*).

is a nonpolynomial function when expressed in terms of the contravariant components, $g^{\mu\nu}$, of the metric tensor. The covariant components, $g_{\mu\nu}$, which enter the expression for $g = \det g_{\alpha\beta}$ and the Christoffel symbol

$$\Gamma^{\lambda}_{\mu\nu} = \tfrac{1}{2} g^{\lambda\varrho}(\partial_\mu g_{\nu\varrho} + \partial_\nu g_{\mu\varrho} - \partial_\varrho g_{\mu\nu}) \tag{2}$$

can be given as a ratio of two polynomials in $g^{\mu\nu}$.

Conventionally, the Lagrangian (1) is linearized by defining a new field variable, $h^{\mu\nu}$, by

$$g^{\mu\nu} = \eta^{\mu\nu} + \varkappa h^{\mu\nu}\,, \tag{3}$$

where $\varkappa$ denotes the gravitational constant and $\eta_{\mu\nu}$ the Minkowski metric. Expanding in powers of $\varkappa$ one finds

$$L = L_0 + L_{\text{int}}(\varkappa, h)\,, \tag{4}$$

where (3)

$$L_0 = -\tfrac{1}{4}(\partial_\mu h^{\lambda\varrho}\,\partial_\mu h^{\nu\varrho} - 2\partial_\varrho h^{\lambda\mu}\,\partial_\lambda h^{\varrho\mu} - \partial_\lambda h^{\mu\mu}\,\partial_\lambda h^{\varrho\varrho} + 2\partial_\mu h^{\mu\lambda}\,\partial_\lambda h^{\varrho\varrho}) \tag{5}$$

and

$$L_1 = \varkappa h^3 + \varkappa^2 h^4 + \dots\,. \tag{6}$$

Here we have collected all the bilinear terms, which are independent of $\varkappa$, in the free Lagrangian L_0 while, for L_{int} (which is a function of the h's and their derivatives) an expansion in powers of $\varkappa$ is implied. Perturbation calculations can be made in the usual way with vertices determined by (6) and a free propagator obtained from (5) together with a gauge condition. In a gauge of the Landau type (4), for example, one might use the propagator

$$G^{\varkappa\lambda,\mu\nu}(p) = \frac{1}{p^2 + i\varepsilon}\left\{\frac{1}{2}(d_{\varkappa\mu}d_{\lambda\nu} + d_{\varkappa\nu}d_{\lambda\mu}) - \frac{2}{3}d_{\varkappa\lambda}d_{\mu\nu} - d_{\varkappa\lambda}e_{\mu\nu} - e_{\varkappa\lambda}d_{\mu\nu} - 3e_{\varkappa\lambda}e_{\mu\nu}\right\}, \tag{7}$$

where $e_{\mu\nu} = p_\mu p_\nu / p^2 = \eta_{\mu\nu} - d_{\mu\nu}$. In succeeding orders, however, one finds integrals which diverge more and more virulently and calculations become pointless.

3. – To see how the new method works consider the simpler example of scalar « gravitation » described by the Lagrangian of Freund and Nambu (5),

$$\mathscr{L} = \frac{1}{2}\frac{(\partial\varphi)^2}{1 + 2\varkappa\varphi} = \frac{1}{2}(\partial\varphi)^2 - \frac{\varkappa\varphi(\partial\varphi)^2}{1 + 2\varkappa\varphi}\,. \tag{8}$$

---

(3) For brevity we shall implicitly contract indices with the Minkowskian metric $\eta_{\mu\nu}$. Thus, for example, $h_{\mu\mu} = \eta_{\mu\nu}h^{\mu\nu}$ and $\partial_\mu\partial_\mu = \eta^{\mu\nu}\partial_\mu\partial_\nu$, where $\eta_{\mu\nu} = \operatorname{diag}(+ - - -)$.

(4) This gauge is specified by the subsidiary condition $\partial_\mu h_{\mu\nu} - \frac{1}{2}\partial_\nu h_{\mu\mu} = 0$ which is the infinitesimal form of the Fock-de Donder condition for harmonic co-ordinates $\partial_\mu(\sqrt{-g}g^{\mu\nu}) = 0$. Harmonic co-ordinates are well adapted to the Minkowskian nature of the metric at infinity.

(5) P. G. O. Freund and Y. Nambu: *Phys. Rev.*, **174**, 1741 (1968).

From among the second-order terms in the nonpolynomial

$$\mathscr{L}_{\text{int}} = \frac{-\varkappa\varphi(\partial\varphi)^2}{1-2\varkappa\varphi} \tag{9}$$

one might single out for consideration the especially simple one

$$\varkappa^2\left\langle T\left(\frac{\varphi(x_1)}{1-2\varkappa\varphi(x_1)}\,\frac{\varphi(x_2)}{1-2\varkappa\varphi(x_2)}\right)\right\rangle : (\partial\varphi(x_1))^2(\partial\varphi(x_2))^2 : \tag{10}$$

which *contributes to graviton-graviton scattering.* The vacuum expectation value of the T-product in (10) may be evaluated as in the past by expanding $\mathscr{L}_{\text{int}}$ in powers of φ; the result is an infinite sum of highly singular terms $(\Delta(x))^n$ where $\Delta = \langle T(\varphi(x_1)\varphi(x_2))\rangle = -1/4\pi^2(x_1-x_2)^2$. Instead of this, one may use the new method which for the Symanzik region of external momenta gives the Green functions as Fourier transforms of configuration space amplitudes which are themselves given by definite integrals which can be looked upon as Borel sums of the infinite series in $(\Delta(x))^n$. For the expression in (10) the configuration-space integral is given by

$$F(\Delta) = \varkappa^2\left\langle T\left(\frac{\varphi(x_1)}{1-2\varkappa\varphi(x_1)}\,\frac{\varphi(x_2)}{1-2\varkappa\varphi(x_2)}\right)\right\rangle = \tag{11}$$

$$= \frac{\varkappa^2}{\pi}\int_{-\infty}^{\infty} dt_1\,dt_2 \exp[-t_1^2-t_2^2]\left[\frac{\sqrt{\Delta}(t_1+it_2)}{1+2\varkappa\sqrt{\Delta}(t_1+it_2)}\right]^2 = \frac{\varkappa^2}{\pi}\int_{-\infty}^{\infty} d\xi \exp[-\xi]\frac{\xi\Delta}{1-4\varkappa^2\Delta}$$

for spacelike separation $(x_1-x_2)^2<0$. For such separations Δ is real and positive. It is an important feature of the method that only spacelike separations are involved in the Fourier integrals for amplitudes in the Symanzik region. Values in the physical region must be obtained by analytic continuation. One can show ([6]) that the correct analyticity properties of the momentum-space amplitude—at least in the present case—are guaranteed if the configuration-space amplitude $F(\Delta)$ is represented by the principal value of the integral (11).

In this note we are not concerned with the problems of analyticity; our concern is with ultraviolet infinities or with the $x^2\to 0$ behaviour of (11). Clearly, $F(\Delta)\sim 1$ and is nonsingular although each term of its perturbation expansion in powers of $\varkappa$ is highly singular.

4. – This example illustrates a general feature of the new method. To determine the singularity behaviour of an S-matrix element, associate with each field φ a factor $M(\lim_{x\to 0}(T\varphi\varphi) = 1/x^2 \approx \lim_{M\to\infty} M^2)$ and with each derivative an additional power of M. Consider the limiting behaviour of $\mathscr{L}_{\text{int}}$ as $M\to\infty$. If $\mathscr{L}_{\text{int}}\sim M^4$, the theory possesses the conventional infinities which resemble those of self-mass, self-charge and boson-boson scattering irrespective of whether $\mathscr{L}_{\text{int}}$ is a polynomial in field variables of a nonpolynomial. We shall call such theories normal. If $\mathscr{L}_{\text{int}}\sim M^n$, $n>4$, the

([6]) B. W. LEE and B. ZUMINO: CERN preprint Th. 1053 (1969); R. FUKUDA: Tokyo University preprint (1969); E. KAPUSCIK, A. SALAM and J. STRATHDEE: ICTP, Trieste, preprint (in preparation).

infinities are likely to proliferate in the manner of conventionally unrenormalizable theories. These we call *abnormal*. If on the other hand $\mathscr{L}_{int} \sim M^n$ with $n < 4$ then we have a supernormal theory with fewer infinities than in a normal theory. If $n < 2$ there are likely to be no ultraviolet infinities. In the example considered $\varphi/(1 + 2\varkappa\varphi) \sim 1$ so that $n = 0$. We do not expect any ultraviolet infinities in any iteration of this operator.

5. – Let us consider the Einstein Lagrangian (1) from this point of view. Clearly if $g^{\mu\nu} \sim M$ then the covariant components behave like $g_{\mu\nu} \sim 1/M$ while $\sqrt{-g} \sim 1/M^2$ and, therefore, $L \sim M$. This is beautiful and we might have hoped that the theory would be supernormal. Unfortunately, to make the new calculational technique work we must separate off the free Lagrangian from L. Since $L_0 \sim M^4$ it follows that $L_{int} \sim M^4$ also. This shows that in the new treatment Einstein's gravitational theory may be expected to be no more than normal.

Consider now the interaction of gravitation with matter. The convergent behaviour of $g_{\mu\nu}$ and of $\sqrt{-g}$ gives rise to the damping effects mentioned at the beginning of this note. The coupling of gravitation to a self-interacting real scalar field, ϕ, for example, is represented by the Lagrangian

$$\tfrac{1}{2}\sqrt{-g}\,(g^{\mu\nu}\partial_\mu\phi\partial_\nu\phi - m^2\phi^2 + \lambda\phi^4)\,,$$

the kinematic part of which goes like M^3 rather than M^4 as it would in the absence of gravitation. Once again the separation of the free part of the meson Lagrangian $\tfrac{1}{2}\eta^{\mu\nu}\partial_\mu\phi\partial_\nu\phi$, necessary to make any calculations whatever, leaves behind $\mathscr{L}_{int}$ (matter) which behaves like M^4 though the infinities arising from the $\lambda\sqrt{-g}\,\phi^4 \approx M^3$ part of the interaction are likely to be suppressed. (We have not yet been able to prove it but it is our conjecture that whenever a separation of this type between $\mathscr{L}_f$ and $\mathscr{L}_{int}$ is made, which appears to enhance the singularity behaviour of the Lagrangian, there is a possibility of a still further summation of graphs which restores and reproduces in the final result the singularity behaviour to be expected from the total $\mathscr{L}$.)

We now consider the modifications to the gravity theory from consideration of gravi ton-fermion interaction. But before we do this there is a special complication in forming perturbation series with the massless graviton propagator (7) which must be mentioned. It was first shown by FEYNMAN (7) and later generalized by DE WITT (8), FADDEEV and POPOV (9) and MANDELSTAM (10) that in order to preserve S-matrix unitarity one must supplement the purely graviton graphs with closed loops of a (fictitious) zero-mass vector fermion. These can be accounted for by adding to the Lagrangian (1) an effective term

$$\tfrac{1}{2}\sqrt{-g}\,g^{\mu\nu}\partial_\mu A^*_\alpha\,\partial_\nu A_\alpha \tag{12}$$

with the stipulation that no external A-lines are admitted. From (12) we can separate a free Lagrangian for the A-field from which is obtained the propagator

$$G_{\mu\nu} = \frac{\eta_{\mu\nu}}{p^2 + i\varepsilon}\,. \tag{13}$$

---

(7) R. P. FEYNMAN: *Acta Phys. Polon.*, **24**, 697 (1963).

(8) B. S. DE WITT: *Phys. Rev.*, **162**, 1195, 1239 (1967).

(9) L. D. FADDEEV and V. N. POPOV: Kiev preprint ITF-67-36 and *Phys. Lett.*, **25** B, 29 (1967). These authors stress the importance of Fock-de Donder condition of ref. (4).

(10) S. MANDELSTAM: *Phys. Rev.*, **175**, 1604 (1968).

Accordingly we can associate the asymptotic factor M with A_μ and observe that the effective Lagrangian (12) goes like M^3 and so does not disturb the normality of the gravitational theory.

6. – Consider now the problem of coupling gravitation to fermions. As is well known, this necessitates the introduction of vierbein components $\lambda^{\mu a}(x)$. These fields are related to $g^{\mu\nu}$ by

$$g^{\mu\nu} = \lambda^{\mu a}\lambda^{\nu b}\eta_{ab}\,. \tag{14}$$

They transform under general co-ordinate transformations according to

$$\lambda^{\mu a} \to \frac{\partial \bar{x}^\mu}{\partial x^\nu}\lambda^{\nu a}$$

and under the group of local Lorentz transformations, Λ, according to

$$\lambda^{\mu a} \to \Lambda^a_b\,\lambda^{\nu b}\,.$$

In the classical treatment of Weyl ([11]) the local Lorentz transformations constitute a group of gauge transformations of the second kind and are quite distinct from and independent of the general co-ordinate transformations. It is amusing to remark, however, that by imposing the symmetry condition

$$\lambda^{\mu a} = \lambda^{a\mu} \tag{15}$$

one can associate a particular Lorentz transformation

$$\Lambda = \Lambda(\partial\bar{x}/\partial x, \lambda)$$

with each co-ordinate transformation. That is to say that the vierbein components can be reduced to ten independent ones which transform according to a nonlinear realization of the group of co-ordinate transformations. They constitute in fact the « Goldstone » finds ([12]) (analogous to the pion fields of chiral dynamics) and can be used to define the class of nonlinear realizations of the group of general co-ordinate transformations which become linear with respect to the Lorentz subgroup. From this point of view we can look upon the Dirac spinor transforming according to

$$\psi \to D(\Lambda)\,\psi\,, \qquad \Lambda \subset O_{3,1}\,,$$

as a nonlinear realization of the group of general relativity. Pursuing this analogy we can define the covariant derivative of ψ by

$$\psi_{;\mu} = \psi_{,\mu} - \tfrac{1}{4}B_{\mu ab}\,\sigma_{ab}\,\psi\,. \tag{16}$$

([11]) H. Weyl: *Zeits. Phys.*, **56**, 330 (1929); *Phys. Rev.*, **77**, 699 (1950).

([12]) A. Salam and J. Strathdee: ICTP, Trieste, preprint IC/68/105 (to appear in *Phys. Rev.*). The application of the methods developed in this reference to the gravitational field will be the subject of a forthcoming preprint.

where B_μ is given in terms of λ and its derivatives by

$$B_{\mu ab} = \tfrac{1}{2}\lambda_{\mu c}(\lambda^c_{\nu,\lambda} - \lambda^c_{\lambda,\nu})\lambda^\lambda_a \lambda^\nu_b - \tfrac{1}{2}(\lambda^\nu_a \lambda_{\nu b,\mu} - \lambda^\nu_b \lambda_{\nu a,\mu}) + \tfrac{1}{2}(\lambda^\nu_a \lambda_{\mu b,\nu} - \lambda^\nu_b \lambda_{\mu a,\nu})\,. \tag{17}$$

(It is understood that Greek indices are raised and lowered with $g^{\mu\nu}$ and $g_{\mu\nu}$ while Latin indices employ for this purpose the Minkowskian $\eta^{ab} = \eta_{ab}$.) With these definitions we can give the Lagrangian which describes the coupling of gravitation to the Dirac field [13]:

$$\sqrt{-g}\,\bar{\psi}\{i\gamma_a \lambda^{a\mu}(\psi_{,\mu} - \tfrac{1}{4}B_{\mu ab}\sigma_{ab}\psi) - m\psi\}\,, \tag{18}$$

where $\sqrt{-g} = \det(\lambda_{\mu a})$. In fact, we can everywhere express $g^{\mu\nu}$ in terms of $\lambda^{\mu\nu}$ and regard the latter as the basic field. We can retain the Einstein Lagrangian (1) with $g^{\mu\nu}$ expressed in terms of $\lambda^{\mu a}$ by formula (14).

The asymptotic power counting is now modified if we regard $\lambda^{\mu a}$ as the basic field and assume $\lambda^{\mu a} \sim M$. This implies $g^{\mu\nu} \sim M^2$ and not $\sim M$ as was the case with the assumption that $g^{\mu\nu}$ is the basic field. Correspondingly we find

$$\mathscr{L}_{\text{gravity}} \sim M^0\,,$$
$$\mathscr{L}_{\text{meson}} \sim M^2\,,$$
$$\mathscr{L}_{\text{fermion}} \sim M$$

and $\mathscr{L}_{\text{total}}$ appears supernormal. Once again the splitting-off of free Lagrangians which in each case go like M^4, makes all the resulting interactions behave like M^4. We believe that this is the point where future techniques must improve on the present one.

It is to be stressed that what we have called singularity behaviour of amplitudes is not the same as high-energy behaviour of on-shell S-matrix elements. The singularity behaviour which determines the ultraviolet infinities depends on the calculational techniques employed; this was in fact the intention of this note to demonstrate. From this point of view, the fact that the choice of $\lambda^{\mu a}$ as basic field gives $\mathscr{L}_{\text{gravity}} \sim M^0$, while the choice of $g^{\mu\nu}$ gives $\mathscr{L}_{\text{gravity}} \sim M$ poses no contradiction since the available calculational technique after splitting $\mathscr{L}_{\text{free}}$ from both Lagrangians leaves them with the same singularity behaviour $\mathscr{L}_{\text{int}} \approx M^4$.

* * *

Our thanks are due to an anonymous correspondent who sent in a number of suggestions after the first mimeographed draft of this note was circulated and which have led to a clarification of a number of points.

---

[13] There is some ambiguity in the definition of generally covariant Lagrangians. This question is discussed by T. W. B. KIBBLE: *Journ. Math. Phys.*, **2**, 212 (1961); and J. A. WHEELER and D. R. BRILL: *Rev. Mod. Phys.*, **29**, 465 (1957).

LETTERE AL NUOVO CIMENTO VOL. 5, N. 5 30 Settembre 1972

Radiative Corrections to the Electron-Graviton Vertex.

R. DELBOURGO and P. PHOCAS-COSMETATOS
Physics Department, Imperial College - London

(ricevuto il 21 Agosto 1972)

In an earlier publication (1) we set down the lowest-order electromagnetic corrections to the photon-graviton vertex. Owing to a cancellation between wave-function and vertex renormalizations, they were perfectly finite and unambiguous, though unfortunately not directly susceptible to experimental test. In this note we would like to pursue the same problem for the electron-graviton vertex following the same field-theoretic procedure, in contrast to the dispersive approach taken by PAGELS (2). Taken in conjunction with the $\gamma\gamma g$ vertex, the eeg vertex rounds off the radiative corrections of the classical stress tensor elements in electrodynamics.

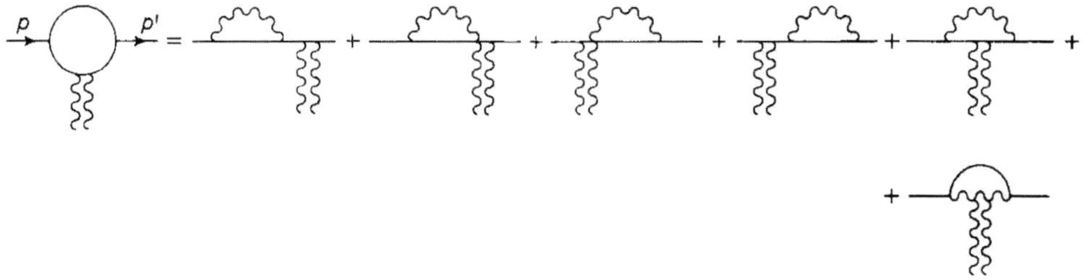

Fig. 1. – Feynman graphs to order α for eeg.

When evaluating these quantum-loop effects it is essential to consider all the graphs to the given order α as in Fig. 1; the inadvertent omission of any one of the diagrams otherwise spoils the remarkable cancellation of infinities. The Feynman rules for calculating the graphs have been stated previously and will not be repeated (1-3). The only point in the computation where the eeg vertex differs from the $\gamma\gamma g$ vertex is in the need for the electron mass renormalization (which is there in the absence of gravitational interactions anyhow). The counter term $m\bar{\psi}\psi|-g|^{-\frac{1}{2}}$ means that we will have

(1) R. DELBOURGO and P. PHOCAS-COSMETATOS: ICTP/71/35.
(2) H. PAGELS: *Phys. Rev.*, **144**, 1250 (1966).
(3) R. DELBOURGO and A. SALAM: PCAC *anomalies and gravitation*, IC/72/86.

to consider the supplementary diagrams of Fig. 2. When all the Feynman integrals are added there is an exact cancellation of infinities, corresponding to equality of wave-function and vertex renormalizations, as found for $\gamma\gamma g$.

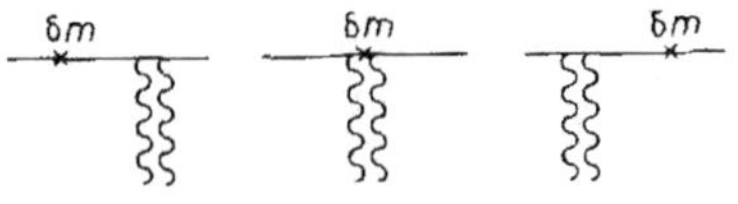

Fig. 2. – Self-mass vertex corrections to eeg.

To simplify the kinematics one can place the electrons on their mass shells so that only three form factors need to be computed:

$$\langle p'|\theta_{\mu\nu}(0)|p\rangle = \bar{u}(p')\big[\tfrac{1}{4}\{\gamma_\mu(p+p')_\nu + \gamma_\nu(p+p')_\mu\}G_1 + (p+p')_\mu(p+p')_\nu G_2 + \\ + (q_\mu q_\nu - q^2\eta_{\mu\nu})G_3\big]u(p)\,.$$

We stress once again that the $G_i(q^2)$ are obtained as finite quantities (*); specifically we find $G_1(0)=1$ and $G_2(q^2)=0$, so the interaction continues to be « minimal ». This is in agreement with Pagel's results (2) for the imaginary parts and the further assumptions he makes about the subtractions of the dispersion integrals, assumptions which are not necessary in our analysis. $G_3(q^2)$ is singular as $q^2\to 0$ as expected for a two-photon branch cut. It is perhaps worth-while to record the detailed expressions for the form factors;

$$G_1(q^2)-1 = \frac{\alpha}{2\pi}\int_0^1 dx\,dy\,dz\,\delta(1-x-y-z)\big[q^2z[1+xy(1-2z)(1-z)^{-2}]\cdot \\ \cdot[m^2(1-z)^2-q^2xy]^{-1} - z\ln[1-q^2xy/m^2(1-z)^2] - 2(1+z)\ln[1-q^2xy/m^2z^2]\big]\,,$$

$$G_2(q^2) = 0\,,$$

$$G_3(q^2) = -\frac{\alpha}{4\pi}\int_0^1 dx\,dy\,dz\,\delta(1-x-y-z)\,\big[m^2(2-z)(x-y)^2[m^2(1-z)^2-q^2xy]^{-1} + \\ + m^2[(1+z)(x-y)^2+(z-1)][m^2z^2-q^2xy]^{-1}\big]\,.$$

In conclusion we should like to point out that there are *no anomalies* (4) in the identities for the trace of the canonical stress tensor, providing one is careful to include the

(*) Barring the expected infra-red effects which find their resolution in the usual way.

(4) S. COLEMAN and R. JACKIW: *Ann. of Phys.*, **67**, 552 (1971). M. S. CHANOWITZ and J. ELLIS (*Phys. Lett.*, **40** B, 397 (1972)) find such anomalies. In our view this is because they neglect the sea-gull graphs coming from the γgee contact term. This contact term is necessarily present when the minimal substitution principle is applied to the lowest-order stress tensor element, irrespective of general covariance considerations.

sea-gull graphs in the analysis. For instance, at zero graviton momentum one gets for $\gamma\gamma\theta$

$$\langle J_\alpha(k)\,\theta_{\mu\mu}J_\beta(-k)\rangle = -m(\partial/\partial m)\langle J_\alpha(k)J_\beta(-k)\rangle = (k\cdot\partial/\partial k - 2)\,D^{-1}_{\alpha\beta}(k)$$

in agreement with notions of conformal symmetry (5). Likewise for eeθ we find

$$\langle j(p)\,\theta_{\mu\mu}\bar{j}(-p)\rangle = (-3 - m\partial/\partial m)\langle j(p)\bar{j}(-p)\rangle = (p\cdot\partial/\partial p - 4)\,S^{-1}(p)\,.$$

Thus if due care is exercised no inconsistencies arise.

---

(5) K. J. WILSON: *Phys. Rev. D*, **2**, 1473 (1970); C. G. CALLAN, S. COLEMAN and R. JACKIW: *Ann. of Phys.*, **59**, 42 (1970); D. G. BOULWARE, L. S. BROWN and R. PECCEI: *Phys. Rev.*, **3**, 1750 (1971).

Volume 41B, number 4 PHYSICS LETTERS 16 October 1972

RADIATIVE CORRECTIONS TO THE PHOTON-GRAVITON VERTEX

R. DELBOURGO and P. PHOCAS-COSMETATOS
Physics Department, Imperial College, London SW7 2BZ, UK

Received 1 August 1972

The lowest order electromagnetic corrections to the $\gamma\gamma g$ vertex due to a charged boson or fermion loop are evaluated and correspond to the effective quadrupole interaction $\frac{1}{4}\alpha(-1)^{2J}(2J+1)g^{\mu\nu}F_{\kappa\lambda}\overleftrightarrow{\partial}_\mu\overleftrightarrow{\partial}_\nu F_{\kappa\lambda}/720\pi M^2$ where J and M are the spin and mass of the particle in the quantum loop. Unfortunately the chances of subjecting this result to experimental test, either at present accelerator energies or by astronomical observation, are remote.

Because of gauge invariance, there is a strong resemblance between the kinematic structure of reaction amplitudes in electrodynamics and in gravitation [1]. This note is a result of an investigation into the nature of the quantum corrections to the classical gravitational vertices. The electromagnetic corrections to the electron-graviton vertex (effectively the electron stress tensor elements) have already been examined by Pagels [2], and here we study the counterpart problem for the photon which, at first glance, holds greater promise of being experimentally verifiable. We find that the order α correction to the photon-graviton vertex due to a massive charged particle loop (mass M, spin J) has the momentum space structure

$$(-1)^{2J}(2J+1)\,e^2 f k'\cdot\epsilon(k)\,k\cdot\epsilon'(k')(k+k')_\mu(k+k')_\nu\epsilon_{\mu\nu}(k-k')/1440\pi^2M^2 \tag{1}$$

which can be reexpressed in terms of the total effective Lagrangian,

$$= -\tfrac{1}{4}g^{\mu\nu}[g^{\kappa\lambda}F_{\mu\kappa}F_{\nu\lambda} + \alpha(-1)^{2J}(2J+1)F_{\kappa\lambda}\overleftrightarrow{\partial}_\mu\overleftrightarrow{\partial}_\nu F_{\kappa\lambda}/720\pi M^2 + \dots] \tag{2}$$

to this order in the gravitational ($f^2 = 8\pi G$) and electromagnetic couplings ($e^2 = 4\pi\alpha$). We also discover, as expected, that the infinite kinetic wavefunction and vertex renormalizations cancel *exactly*, so that (1) is a completely clean result, on a par with the calculation of the anomalous magnetic moment of the electron. In that sense it is rather a pity that we have been unable to find any way of confronting (1) against experiment, either through astronomical observations or in the laboratory – the basic reason is that, for the correction to stand out, one needs momentum transfers of the same order as $10^5\,M^2$ and these are simply not available with known sources of photons.

All the electromagnetic and gravitational interactions of the lightest particles with the lowest spin, the pion and electron, follow from the Lagrangian [3],

$$|-g|^{1/2}L(\varphi,\psi,g,A) = -\tfrac{1}{4}g^{\kappa\lambda}g^{\mu\nu}F_{\kappa\mu}F_{\lambda\nu} + \tfrac{1}{2}f^{-2}R(g) + g^{\kappa\lambda}(\partial_\kappa - ieA_\kappa)\varphi(\partial_\lambda + ieA_\lambda)\varphi^+ - \mu^2\varphi\varphi^+$$
$$+ L^{\mu n}\bar\psi\gamma_n(\tfrac{1}{2}i\overleftrightarrow{\partial}_\mu + eA_\mu)\psi - \bar\psi(m + i\epsilon^{klmn}B_{klm}\gamma_n\gamma_5)\psi$$

where we choose the electron and pion fields to have canonical weight 0. As usual, $g^{\mu\nu} = L^\mu_m L^\nu_n \eta^{mn}$ where L is the symmetric tetrad field which defines the graviton field to first order via $L = 1 + fh + \dots$ We can tabulate the Feynman rules provided by L to order f with on-mass shell gravitons:

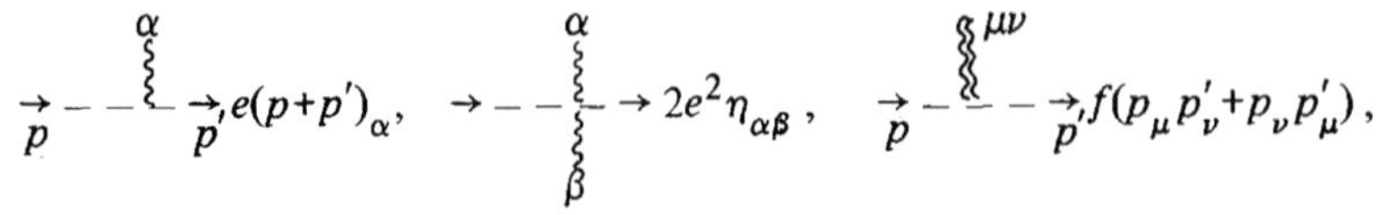

$e(p+p')_\alpha$, $2e^2\eta_{\alpha\beta}$, $f(p_\mu p'_\nu + p_\nu p'_\mu)$,

Volume 41B, number 4 PHYSICS LETTERS 16 October 1972

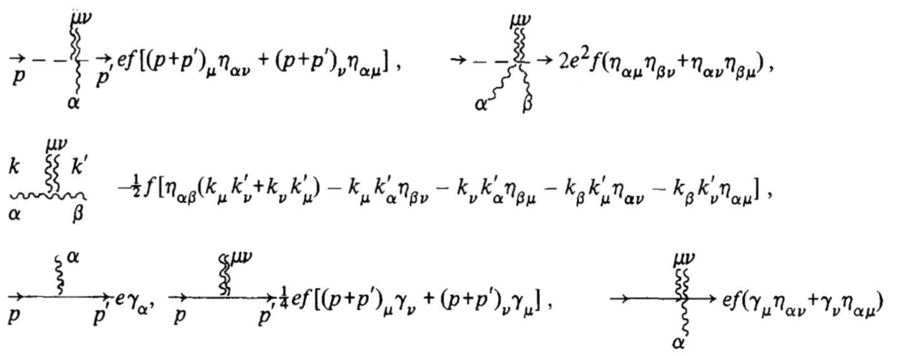

$$ef[(p+p')_\mu \eta_{\alpha\nu} + (p+p')_\nu \eta_{\alpha\mu}]\,, \qquad 2e^2 f(\eta_{\alpha\mu}\eta_{\beta\nu} + \eta_{\alpha\nu}\eta_{\beta\mu})\,,$$

$$-\tfrac{1}{4} f[\eta_{\alpha\beta}(k_\mu k'_\nu + k_\nu k'_\mu) - k_\mu k'_\alpha \eta_{\beta\nu} - k_\nu k'_\alpha \eta_{\beta\mu} - k_\beta k'_\mu \eta_{\alpha\nu} - k_\beta k'_\nu \eta_{\alpha\mu}]\,,$$

$$e\gamma_\alpha, \qquad \tfrac{1}{4} ef[(p+p')_\mu \gamma_\nu + (p+p')_\nu \gamma_\mu]\,, \qquad ef(\gamma_\mu \eta_{\alpha\nu} + \gamma_\nu \eta_{\alpha\mu})$$

and by sewing the vertices together we are able to construct the entire set of diagrams for the $\gamma\gamma g$ vertex to order α. For the π^+ loop they are drawn in fig. 1. For the e^- loop, the last three diagrams are absent.

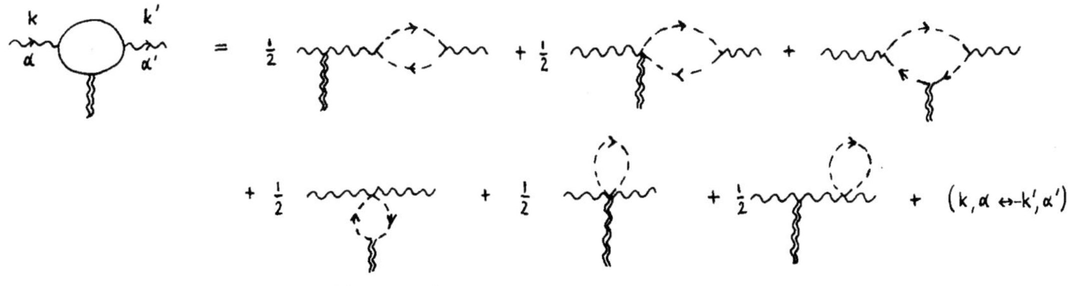

Fig. 1. Pion loop contributions to the $\gamma\gamma g$ vertex.

There is no point in giving the details of the calculation which are straightforward enough. The significant feature is the exact cancellation of all the corrections to the kinetic part of the bare vertex $\langle\gamma|\theta_{\mu\nu}|\gamma\rangle$, when the graphs are gauge-invariantly regularized. In particular, the logarithmic infinities associated with wave-function and vertex renormalizations are equal and opposite as prescribed by gauge invariance. We are left with a finite quadrupole-type correction, and for the pion and electron loops this can be expressed in the forms (1) or (2). In fact we expect the result to be true for any spin (not just $J = 0, M = m_\pi$ and $J = \frac{1}{2}, M = m_e$) because $(-1)^{2J}$ is the spin-statistics factor and $(2J+1)$ is simply the multiplicity of helicity-conserving gravitational couplings to the intermediate particle- anti-particle pair *. It is quite fair to draw a parallel between $\gamma\gamma g$ and $ee\gamma$ since in the latter case there is also cancellation of vertex and wave-function renormalizations, leaving one with the finite anomalous magnetic moment correction.

Obviously it is of interest to see if the frequency dependent effects implied by (1) can be tested experimentally. The first place which comes to mind is the bending of light from the sun, with the interaction governed by the *electron* loop. In order to arrive at a classical particle trajectory from (2), we compare the classical and quantum-mechanical (Born) cross-section formulae,

$$d\sigma/d\Omega = |T/8\pi M_\odot|^2 = -\tfrac{1}{2}\, db^2/d(\cos\theta)$$

which provides a differential relation between the impact parameter b and the scattering angle θ that carries all the classical dynamical information. At small θ the amplitude T is predominantly no flip and

*It is amusing to note that (1) becomes singular when $M = 0$, but that, on the other hand, neutrinos are uncharged. Weak interaction neutral currents which can couple to neutrinos would be very undesirable from our point of view.

Volume 41B, number 4 PHYSICS LETTERS 16 October 1972

$$\left|\frac{T^{++}}{8\pi M_\odot}\right| \approx \frac{4GM}{\theta^2}\left(1-\frac{\theta^2}{24}-\frac{4\alpha\omega^2\theta^2}{90\pi m^2}\right), \qquad \theta_\odot = \frac{4GM}{R_\odot}\left[1-\frac{1}{24}\left(1+\frac{16\alpha\omega^2}{15\pi m^2}\right)\left(\frac{4GM}{R_\odot}\right)^2 \log\left(\frac{R_\odot}{4GM}\right)+\ldots\right]. \tag{3}$$

The correction term being of order 0.1 × (Einstein value)$^2 \approx 10^{-11}$, we see that there is not the very faintest hope of verifying the (ω) frequency dependence even with the best X-ray sources! A heuristic argument due to Salam confirms this conclusion: since (1) has the form $\alpha t/180\pi m^2$ in the helicity conserving amplitude, it effectively corresponds to a correction $\alpha/180\pi m^2 r^3$ to the Newtonian potential. So even if it were possible to disentangle this from inhomogeneities of the sun, we would get an effective force on the photon of $2GM_\odot ER_\odot^{-2} \times (1+\alpha/180\pi m^2 R_\odot^2)$ and the resulting deflection carries many of the marks of (3). Presumably other cosmological tests of (1) also lie in the realms of fantasy.

Instead, therefore, let us try to look for repercussions of (1) in high-energy physics. The only possibility for making contact with elementary particles is to liken the 2^+ mesons to gravity [4]. Here too, however, we have drawn a blank:

(i) The minuscule decay rate $\Gamma_{f\to 2\gamma}$ is hardly likely to be measured in the forseeable future.

(ii) The $\gamma\gamma f$ coupling cannot be obtained from 2π photoproduction in the same way that $\gamma\gamma\pi^0$ is found from the Primakoff effect. The reason is that the background amplitude which interferes with the photon-exchange graph is not precisely known even at small momentum transfers.

(iii) If the Pomeron trajectory, or more loosely diffraction scattering, has similar characteristics [5] as graviton or f-exchange (one unit down in angular momentum) then we can look for possible effects of s-channel helicity nonconservation caused by (1). It is a simple matter to obtain the helicity amplitudes for high energy Compton scattering, $T^{++} \to \beta s^{\alpha_P(t)}[1-\alpha(-1)^{2J}(2J+1)t/180\pi M^2]$, $T^{+-} \to \beta s^{\alpha_P(t)}[-\alpha(-1)^{2J}(2J+1)t/180\pi M^2]$. The polarization effects will be maximized if we take the lightest hadron loop that couples to the Pomeron, i.e. the *pion*. Even so, they are tiny, being of order $0.0005t$ in GeV/c units, and giving only 1% effects in cross-sections even when $t = 10\,(\mathrm{GeV}/c)^2$.

We are forced to conclude that in spite of its simple nature, the radiative correction is just a nice academic exercise, because it evades any kind of experimental test. We hope we are wrong.

We thank Professor Abdus Salam for many pertinent remarks.

References

[1] S. Weinberg, Phys. Rev. 138 (1965) B988, and 135 (1964) B1049;
R. Delbourgo and A. Salam, Phys. Lett. 40B (1972) 381.
[2] H. Pagels, Phys. Rev. 144 (1966) 1250.
[3] R. Delbourgo, A. Salam and J. Strathdee, Nuovo Cim. Lett. 2 (1969) 354;
C.J. Isham, A. Salam and J. Strathdee, Phys. Rev. 3 (1971) D1805.
[4] R. Delbourgo, A. Salam and J. Strathdee, Nuovo Cim. 49 (1967) 593;
P.G.O. Freund, Phys. Rev. Lett. 16 (1966) 291, 424.
[5] H.F. Jones and A. Salam, Phys. Lett. 34B (1971) 149.

Part K
FEYNMAN DIAGRAMMATICS

K.0. Commentaries K

The calculation of Feynman diagrams in higher orders of perturbation theory has remained an active area of research for many years for a number of reasons: these endeavours affect the high precision of QED and QCD results and experiments, they determine the behaviour of the beta functions associated with low and high energy behaviour, and they provide a check on non-perturbative characteristics of Green functions. As a consequence numerous techniques have been devised for easier computations and specifically tailored for application through algebraic computer programs.

K.1 Broadhurst D.J., Delbourgo R. and Kreimer D., Unknotting the polarized vacuum of quenched QED. Phys. Lett. B366, 421-428 (1996)
K.2 Delbourgo R., Kalloniatis A. and Thompson G., Dimensional renormalisation: Ladders and rainbows. Phys. Rev. D54, 5373-5376 (1996)
K.3 Delbourgo R., Elliott D. and McAnally D., Dimensional renormalisation in ϕ^3 theory: Ladders and rainbows. Phys. Rev. D55, 5230-5233 (1997)
K.4 Davydychev A.I. and Delbourgo R., A geometrical angle on Feynman integrals, J. Math. Phys. 39, 4299-4334 (1998)
K.5 Davydychev A.I. and Delbourgo R., Explicitly symmetrical treatment of three-body phase space. J. Phys. A37, 4871-4886 (2004)
K.6 Bashir A., Concha-Sánchez Y. and Delbourgo R., 3-point off-shell vertex in scalar QED in arbitrary gauge and dimension. Phys. Rev. D76, 065009 (2007)

Paper [K.1], in which I played a minor role, was the determination of the beta function to fourth order in the fine structure constant for scalar electrodynamics; I merely suggested the use of the Duffin-Kemmer-Petiau for reducing the number of diagrams which needed to be considered in that calculation and left the experts (Broadhurst and Kreimer) in that subject to use their computer expertise in determining $\beta(\alpha)$. The ensuing result was new at the time and enlarged the evaluations made by Vermaseren and collaborators (Phys. Lett. B400, 379 (1997)) in that context.

Articles [K.2] and [K.3] concern the evaluation of renormalisation constants for certain classes of Feynman diagrams to all orders of perturbation theory, first for fermion-boson theories (in the limit as $D \to 4$) and secondly in ϕ^3 theory (as $D \to 6$). This is best done by examining the massless limit and converting the problems into differential equations. We showed there that the correct anomalous dimensions of the fields emerge by extracting the singular part in $1/(D-4)$ and $1/(D-6)$ respectively.

Article [K.4] was novel in that it proved that Feynman graphs can viewed geometrically as hypervolumes which in turn leads to new methods of working them out. Davydychev has greatly developed this geometrical approach into an art form and produced many interesting results which are useful to practitioners in this field. With three-body phase space the usual method of working out the answer produces an expression in which the symmetry of the result is implicit. Paper [K.5] tackles this problem in a manner that makes the symmetry explicit. Finally [K.6] is the first evaluation of the off-shell vertex function for scalar electrodynamics, in arbitrary gauge and dimension; the problem is a nice simplification of the spinor electrodynamics case.

ELSEVIER

11 January 1996

PHYSICS LETTERS B

Physics Letters B 366 (1996) 421–428

Unknotting the polarized vacuum of quenched QED ★

D.J. Broadhurst [a,1], R. Delbourgo [b,2], D. Kreimer [c,3]

[a] *Physics Department, Open University, Milton Keynes MK7 6AA, United Kingdom*

[b] *Department of Physics, University of Tasmania, GPO Box 252C, Hobart, Tasmania 7001, Australia*

[c] *Institut für Physik, Johannes Gutenberg-Universität, Postfach 3980, D-55099 Mainz, Germany*

Received 15 September 1995

Editor: P.V. Landshoff

Abstract

A knot-theoretic explanation is given for the rationality of the quenched QED beta function. At the link level, the Ward identity entails cancellation of subdivergences generated by one term of the skein relation, which in turn implies cancellation of knots generated by the other term. In consequence, each bare three-loop diagram has a rational Laurent expansion in the Landau gauge, as is verified by explicit computation. Comparable simplification is found to occur in scalar electrodynamics, when computed in the Duffin–Kemmer–Petiau formalism.

1. Introduction

The surprising rationality of the three- [1] and four-loop [2] quenched (i.e. single-electron-loop) terms in the QED beta function is an outstanding puzzle, which we here elucidate by giving the Ward identity $Z_1 = Z_2$ an interpretation in terms of the skeining relation that is the basis of the recent association of knots [3–6] with transcendental counterterms.

In Section 2 we study the intricate cancellations between transcendentals in all 6 of the methods (known to us) for calculating $\beta(a) \equiv \mathrm{d}a/\mathrm{d}\ln\mu^2 = \sum_n \beta_n a^{n+1}$, with a QED coupling $a \equiv \alpha/4\pi$. Our focus is the rationality of $\beta_3^{[1]} = -2$ [1] and $\beta_4^{[1]} = -46$ [2]. (Subscripts denote the number of loops; where necessary, superscripts denote the number of electron loops.) In Section 3 we compound the puzzle by exposing even more intricate cancellations of ζ_3 in three-loop scalar electrodynamics [7]. The argument from knot theory is given in Section 4, leading to specific predictions, confirmed by detailed calculations, performed using the techniques of [8–10]. Conclusions, regarding higher orders [2], unquenched (i.e., multi-electron-loop) contributions [11–13], and non-abelian gauge theories [14,15], are presented in Section 5.

★ Work supported in part by grant CHRX-CT94-0579, from HUCAM, and grant A69231484, from the Australian Research Council.

[1] E-mail: D.Broadhurst@open.ac.uk.

[2] E-mail: Delbourgo@physvax.phys.utas.edu.au.

[3] E-mail: Kreimer@dipmza.physik.uni-mainz.de.

2. Six calculations in search of an argument

There are (at least) 6 ways of obtaining the beta function of quenched QED. For each, we expose the delicate cancellation of transcendentals between diagrams, which cries out for explanation.

Method 1. Dyson–Schwinger skeleton expansion [1,16]. The Dyson–Schwinger equations give

0370-2693/96/$12.00 © 1996 Elsevier Science B.V. All rights reserved

SSDI 0370-2693(95)01343-1

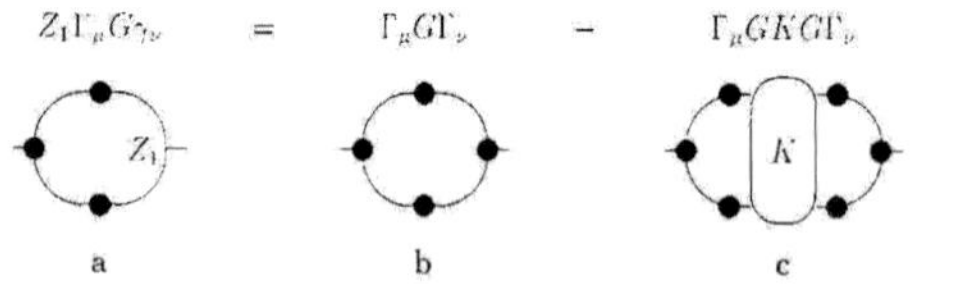

Fig. 1. Illustration of terms in the Dyson–Schwinger equation (1).

the photon self-energy, schematically, as [17]

$$\Pi_{\mu\nu} = Z_1\Gamma_\mu G\gamma_\nu = \Gamma_\mu G(1-KG)\Gamma_\nu\,, \tag{1}$$

where Γ_μ is the dressed vertex and G stands for the pair of dressed propagators in Fig. 1a. To obtain the second form, illustrated in Figs. 1b, 1c, one uses $\Gamma_\nu = Z_1\gamma_\nu + KG\Gamma_\nu$, where K is the kernel for e^+e^- scattering and loop integrations and spin sums are to be understood in the products. Now we expand each vertex to first order in the external momentum q: $\Gamma_\mu = \Gamma^0_\mu + q\cdot d\Gamma_\mu + O(q^2)$ where $d_\alpha \equiv \partial/\partial q_\alpha$ and the Ward identity gives $\Gamma^0_\mu \equiv \Gamma_\mu(p,p) = (\partial/\partial p_\mu)S^{-1}(p)$ in term of the inverse propagator. We use the low-momentum expansion

$$(\Gamma_\mu - \Gamma^0_\mu)G(1-KG)(\Gamma_\nu - \Gamma^0_\nu)$$
$$= (q\cdot d\Gamma_\mu)G(1-KG)(q\cdot d\Gamma_\nu) + O(q^3)\,, \tag{2}$$

differentiate twice, and make liberal use of $(1-KG)d_\alpha\Gamma_\nu = d_\alpha(KG)\Gamma^0_\nu + O(q)$, to obtain

$$\begin{aligned} &d_\alpha d_\beta \Pi_{\mu\nu} \\ &= \Gamma^0_\mu[\tfrac{1}{2}d_\alpha d_\beta G + (d_\alpha G)K(d_\beta G) \\ &+ 2(d_\alpha G)(d_\beta K)G + \tfrac{1}{2}G(d_\alpha d_\beta K)G \\ &+ (d_\alpha(GK))G(1-KG)^{-1}(d_\beta(KG)) \\ &+ (\alpha \leftrightarrow \beta)]\Gamma^0_\nu + O(q)\,, \end{aligned} \tag{3}$$

which entails only K and S. There is a very simple statement of this result: between the dressed zero-momentum vertices occur all and only the terms in $d_\alpha d_\beta(G(1-KG)^{-1})$ that give no subdivergences. In other words, the Dyson–Schwinger method kills maximal forests of subdivergences both on the left and on the right; only the overall divergence survives. Moreover, after cutting at a line on the left (or right), we may set both the mass and the external momentum to zero, with no danger of infrared divergence, and hence obtain the L-loop quenched beta function from finite massless two-point skeleton diagrams, with up to $(L-1)$ loops.

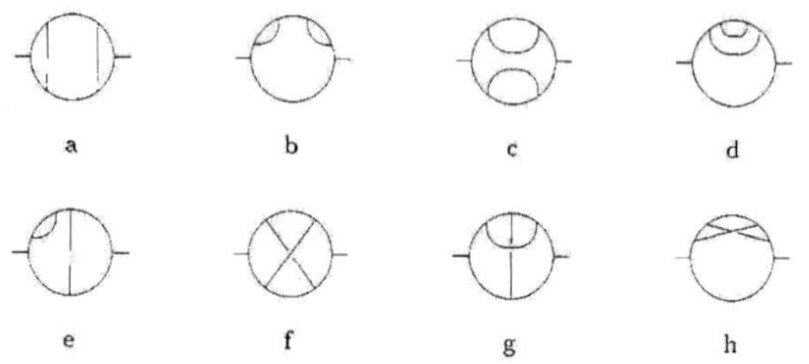

Fig. 2. Diagrams contributing to $\beta_3^{[1]}$, with photon lines drawn inside the electron loop.

We route half of the external momentum through the electron line and half through the positron line and find that each of the 5 terms in (3) yields a contraction independent contribution to $\beta_3^{[1]}$. To all orders, the first contribution to $\beta^{[1]}(a)$ is $\frac{4}{3}a^2(1-\frac{3}{2}\gamma^2+\gamma^3)$, where $\gamma \equiv \gamma_2^{[0]} = \xi a - \frac{3}{2}a^2 + \frac{3}{2}a^3 + O(a^4)$, is the quenched electron-field anomalous dimension [18] with a photon propagator $g_{\mu\nu}/k^2 + (\xi-1)k_\mu k_\nu/k^4$. The second term gives $\beta_2 = 4$. At 3 loops, all 5 terms contribute, giving a remarkable cancellation of ζ_3:

$$\begin{aligned} \beta_3^{[1]} &= -2\xi^2 \\ &+ \left[-\tfrac{2}{3}(7\xi-3)(\xi-3) + 2(\xi^2-6\xi-3)\zeta_3\right] \\ &+ (\xi-1)(\xi-5)\left[\tfrac{16}{3} - 4\zeta_3\right] \\ &+ \left[\tfrac{4}{3}(\xi^2+12\xi-23) + 2(\xi^2-6\xi+13)\zeta_3\right] + 8 \\ &= -2\,. \end{aligned} \tag{4}$$

In the Landau gauge, $\xi = 0$, we reproduce Rosner's ζ_3-cancellation: $0 = -6-20+26$ [1]. In the Feynman gauge, $\xi = 1$, the third term vanishes; the second and fourth are still transcendental and exactly cancel the fifth.

Method 2. Integration by parts of massive bubble diagrams [8]. We may study separately each of the 3-loop quenched diagrams of Fig. 2, using the techniques of [8], which reduce to algebra the calculation of 3-loop massive bubble diagrams, in $d \equiv 4-2\epsilon$ dimensions, thereby also yielding the finite parts of on-shell charge renormalization, used in [10] to establish the connection between 4-loop on-shell and minimally subtracted (unquenched) beta functions. The coefficient, d_n, of a^n/ϵ, in the sum of all n-loop, single-electron-loop, bare diagrams contributing to $1/Z_3$, differs from $\beta_n^{[1]}/n$; one must also take account of the quenched anomalous mass dimension,

$\gamma_{\rm m}^{[0]} = \sum_n \gamma_n a^n = 3a + \frac{3}{2}a^2 + \frac{129}{2}a^3 + O(a^4)$ [18]. To 4 loops, the effects of mass renormalization are also rational, giving agreement with 3-loop results [8,10]:

$$d_1 = \beta_1 = \tfrac{4}{3}, \quad d_2 = \tfrac{1}{2}\beta_2 - 2\beta_1\gamma_1 = -6,$$
$$d_3 = \tfrac{1}{3}\beta_3^{[1]} - 2\beta_2\gamma_1 - \beta_1\gamma_2 + 6\beta_1\gamma_1^2 = \tfrac{136}{3}, \qquad (5)$$

and generating a 4-digit prime in the quenched 4-loop coefficient:

$$\begin{aligned} d_4 &= \tfrac{1}{4}\beta_4^{[1]} - 2\beta_3^{[1]}\gamma_1 - \beta_2\gamma_2 + 8\beta_2\gamma_1^2 \\ &\quad + 8\beta_1\gamma_2\gamma_1 - \tfrac{2}{3}\beta_1\gamma_3 - \tfrac{64}{3}\beta_1\gamma_1^3 \\ &= -\tfrac{2969}{6}, \end{aligned} \qquad (6)$$

which string-inspired techniques [19] may eventually reproduce. In the simplest case [8] of contracting $d_\alpha d_\beta \Pi_{\mu\nu}|_{q=0}$ with $g_{\mu\nu}g_{\alpha\beta}$ in the Feynman gauge, we find cancelling coefficients of ζ_3/ϵ in the contributions of Figs. 2f and 2g, corresponding to ζ_3-cancellation between the fourth and second terms, respectively, of (4) at $\xi = 1$.

Method 3. Integration by parts of massless two-point diagrams [20,21]. Next we study the behaviour of the bare diagrams of Fig. 2 at large q^2 [21], where we may set $m = 0$ and have no need of mass renormalization, or differentiation. The integration-by-parts method of Chetyrkin and Tkachov [20] now suffices. It has been implemented in the program MINCER, whose test suite [22] evaluates all the diagrams of Fig. 2 in the Feynman gauge. Table 1 of [22] reveals that, in addition to the crossed-photon diagrams of Figs. 2f and 2g, the uncrossed-photon diagrams of Figs. 2a and 2e also entail ζ_3/ϵ. Moreover, the cancellation does not occur in the crossed and uncrossed sectors separately; the relative weights of ζ_3/ϵ in the contributions of the bare diagrams 2a, 2e, 2f, 2g are $3 : -6 : -1 : 4$.

Method 4. Infrared rearrangement of massless diagrams [14,23,24]. Infrared rearrangement is a technique for reducing the calculation of L-loop counterterms to that of $(L-1)$-loop massless two-point diagrams, by the subtraction of subdivergences in the MS scheme, followed by nullification of external momenta and appropriate cutting of massless bubble diagrams. As shown by Method 1, the Dyson–Schwinger equations make this unnecessary in QED. However, the technique prospered at the 3-loop level in QCD [14,23] to such an extent as to encourage the calculation of the 4-loop beta function of QED, with the rational quenched result $\beta_4^{[1]} = -46$ [2]. A measure of the seemingly miraculous cancellation of transcendentals in this method is afforded by the complicated combination of ζ_3 and ζ_5 that occurs in the *non*-abelian 4-loop QCD corrections to $R(e^+e^- \to \text{hadrons})$ [25].

Method 5. Propagation in a background field [26,27]. As observed in [26], the derivative $(d/da)(\beta^{[1]}(a)/a^2) = \sum_{n>1}(n-1)\beta_n^{[1]}a^{n-2}$ may be obtained from the coefficient of $aF_{\mu\nu}^2$ in the single-electron-loop contributions to the large-q^2 photon propagator in a background field $F_{\mu\nu}$, thereby providing a further alternative to infrared rearrangement. Two-loop massless background-field calculations have been performed in QCD [27,28], with results recently confirmed by a full analysis of the massive case [29]. From the two-loop corrections, $1 + (2C_A - C_F)\alpha_s/4\pi$, to the coefficient of $\langle \alpha_s G_{\mu\nu}^2\rangle$ in the correlator of the light-quark vector current of QCD, one immediately obtains $2\beta_3^{[1]}/\beta_2 = -1$, by setting $C_A = 0$, $C_F = 1$, which confirms the correctness of this method at 3 loops. A measure of the complexity of the ζ_3 cancellations is afforded by studying the diagram-by-diagram analysis of the appendix of [28].

Method 6. Crewther connection to deep-inelastic processes [30,31]. Finally, a useful check of $\beta_4^{[1]} = -46$ was obtained in [30] by taking the reciprocal of the 3-loop [31] radiative corrections to the Gross–Llewellyn-Smith sum rule, in the quenched abelian case:

$$\begin{aligned} \beta^{[1]}(a) &= \frac{\frac{4}{3}a^2}{1 - 3a + \frac{21}{2}a^2 - \frac{3}{2}a^3 + O(a^4)} \\ &= \tfrac{4}{3}a^2 + 4a^3 - 2a^4 - 46a^5 + O(a^6), \end{aligned} \qquad (7)$$

in precise agreement with [2]. This is an example of a Crewther connection [32] which is unmodified by renormalization in quenched QED. A measure of the complexity of the cancellations of ζ_3 and ζ_5 is afforded by studying the *non*-abelian terms in 3-loop deep-inelastic radiative corrections [31], which are replete with both transcendentals.

3. Cancellation of ζ_3 in scalar electrodynamics

Corresponding calculations in scalar electrodynamics (SED) are most conveniently performed using the Duffin-Kemmer-Petiau [33] (DKP) spinorial formalism for the charged scalar field, in which the Feynman rules are identical to those of fermionic QED, with a bare vertex $e_0\gamma_\mu$ and a bare propagator $S_0(p) = 1/(\not{p} - m_0)$. The only difference resides in the γ-matrices: in the DKP formalism $\not{p}$ is not invertible; instead one uses the fact that $\not{p}^3 = p^2\not{p}$ to obtain $S_0(p) = (\not{p}(\not{p} + m_0)/(p^2 - m_0^2) - 1)/m_0$. The trace of the unit matrix is $d + 1$, the trace of an odd number of γ-matrices vanishes, and the trace of an even number, γ_{μ_1} to $\gamma_{\mu_{2n}}$, is the sum of two terms [34]: $g_{\mu_1\mu_2}\cdots g_{\mu_{2n-1}\mu_{2n}}$ and the cyclic permutation $g_{\mu_2\mu_3}\cdots g_{\mu_{2n}\mu_1}$. The effect is to make all traces regular as $m_0 \to 0$, while automatically generating the many seagull terms of conventional scalar methods [17]. This is a great simplification, eliminating the need to include 21 seagull diagrams at 3 loops, which would lead to many more terms in $\mathrm{d}_\alpha \mathrm{d}_\beta \Pi_{\mu\nu}|_{q=0}$.

We compute the diagrams of Fig. 2 using the DKP formalism. At $q = 0$ (Method 2) we use the REDUCE [35] program RECURSOR [8], for 3-loop massive bubble diagrams; at $m = 0$, we use the REDUCE program SLICER [10], devised to check the results of [21] for the large-q^2 photon propagator in the $\overline{\mathrm{MS}}$ scheme. As in QED, we perform both calculations in an arbitrary gauge and contract $\mathrm{d}_\alpha \mathrm{d}_\beta \Pi_{\mu\nu}$ with $g_{\mu\nu}g_{\alpha\beta} + \lambda(g_{\mu\nu}g_{\alpha\beta} + g_{\mu\alpha}g_{\nu\beta} + g_{\nu\alpha}g_{\mu\beta})$, where λ is an arbitrary parameter, which affects the contributions of individual diagrams, but not the total result for the transverse self energy $\Pi_{\mu\nu}(q) = (q_\mu q_\nu - q^2 g_{\mu\nu})\Pi(q^2)$. At $m = 0$, the corresponding freedom is to contract with the tensor $g_{\mu\nu}q^2 + \lambda(d+2)q_\mu q_\nu$. On-shell SED mass renormalization is performed as in [8,9]: there are only two quenched self-energy diagrams at two loops; each is projected on-shell by taking the trace with $\not{p}(\not{p} + m)$ at $p^2 = m^2$, with a pole mass m. Unlike the QED case, the relation between bare and pole masses is infrared-singular in SED, though that causes no problem for the dimensionally regularized calculation of $Z_3 = 1/(1 + \Pi_0(0))$, where infrared singularities in the bare diagrams for $\Pi_0(0)$ are cancelled by those in $Z_{\mathrm{m}} \equiv m_0/m$. The on-shell methods of [8] then yield the quenched contributions

Table 1
Three-loop on-shell and MS coefficients of QED and SED beta functions

	$\beta_3^{[1]} + \beta_3^{[2]}$	
	QED	SED
on-shell	$-2 - \frac{224}{9} = -\frac{242}{9}$ [a]	$\frac{29}{2} - \frac{55}{18} = \frac{103}{9}$
MS	$-2 - \frac{44}{9} = -\frac{62}{9}$ [b]	$\frac{29}{2} - \frac{49}{18} = \frac{106}{9}$ [c]

[a]Ref. [11]. [b]Ref. [14]. [c]Ref. [7].

$$\begin{aligned}
&\frac{1}{Z_3^{[1]}} - 1 \\
&= \frac{4}{3\epsilon}a_{\mathrm{m}} + \frac{4(1 + 7\epsilon - 4\epsilon^3)}{\epsilon(2-\epsilon)(1-4\epsilon^2)}a_{\mathrm{m}}^2 \\
&+ \left(-\frac{2}{3\epsilon} + 16\zeta_2(5 - 8\ln 2) + \frac{1}{3}\zeta_3 + \frac{77}{9}\right)a_{\mathrm{m}}^3, \\
&= \frac{1}{3\epsilon}a_{\mathrm{m}} + \frac{4}{\epsilon(2-\epsilon)(1-4\epsilon^2)}a_{\mathrm{m}}^2 \\
&+ \left(+\frac{29}{6\epsilon} + 8\zeta_2(3 - 4\ln 2) + \frac{35}{12}\zeta_3 + \frac{136}{9}\right)a_{\mathrm{m}}^3,
\end{aligned} \tag{8}$$

for QED [8,10] and SED, respectively, where $a_{\mathrm{m}} = \Gamma(1+\epsilon)e_0^2/(4\pi)^{d/2}m^{2\epsilon}$ is a dimensionless coupling, with a pole mass m, and terms of order $a_{\mathrm{m}}^3\epsilon$ and a_{m}^4 are neglected. (There is no need to renormalize the bare charge, e_0, when dealing with the quenched contributions.) From the singular terms in (8) we read off the 3-loop beta function of quenched SED: $\tilde{\beta}^{[1]}(a) = \frac{1}{3}a^2 + 4a^3 + \frac{29}{2}a^4 + O(a^5)$. We have calculated the double-bubble term $\tilde{\beta}_3^{[2]}$ in the on-shell and MS schemes, obtaining agreement with [7] in the latter, as shown in Table 1.

For the 3-loop quenched $\overline{\mathrm{MS}}$ contributions to $\Pi(q^2)$, at large q^2, we obtain

$$\begin{aligned}
&\overline{\Pi}_3^{[1]}(q^2) \\
&= [-2\ln(\mu^2/q^2) - \tfrac{286}{9} - \tfrac{296}{3}\zeta_3 + 160\zeta_5]\bar{a}^3, \\
&= [+\tfrac{29}{2}\ln(\mu^2/q^2) + \tfrac{502}{9} - \tfrac{160}{3}\zeta_3 + 40\zeta_5]\bar{a}^3,
\end{aligned} \tag{9}$$

with $\bar{a} = \Gamma(1+\epsilon)\Gamma^2(1-\epsilon)e_0^2/\Gamma(1-2\epsilon)(4\pi)^{d/2}\mu^{2\epsilon}$, which suppresses ζ_2 in Laurent expansions. The QED result confirms [21]. The SED cancellations are even subtler: with $\xi = \lambda = 0$ the relative weights of ζ_3/ϵ from bare diagrams 2a, 2e, 2f, 2g, 2h are $18 : -36 : -5 : 22 : 1$.

4. The argument from knot theory

From the point of view of knot theory, as proposed in [3–6], the presence of transcendentals in counterterms can be traced to the knots that are obtained by skeining the link diagrams that encode the intertwining of loop momenta in Feynman diagrams. The absence of transcendentals in the quenched beta function does not, therefore, correspond to the absence of knots in the Feynman graphs, since the crossed-photon graphs of Figs. 2f, 2g, 2h all realize the link diagram whose skeining contains the trefoil knot [4]. Accordingly we expect to find ζ_3/ϵ in their divergent parts. To explain the cancellation of transcendentals, we must study the interplay between knot-theoretic arguments and the gauge structure of QED.

It was found in [3,4] that ladder topologies are free of transcendentals when the appropriate counterterms are added: after minimal subtraction of subdivergences, ladder graphs, such as in Figs. 2a, 2e, give rational terms in the Laurent expansion in powers of $1/\epsilon$. In [5,6], on the other hand, transcendentals corresponding to positive knots, with up to 11 crossings, were successfully associated with subdivergence-free graphs, up to 7 loops. Thus the skein relation played two distinct roles in previous applications: in [3] the so-called A part of the skein operation determined the subdivergences, while the B part gave no non-trivial knots in ladder topologies; in [5] there were no subdivergences associated with the A operation, while the knots from the B operation faithfully revealed the transcendentals resulting from nested subintegrations. In Figs. 2f, 2g, 2h we are now confronted with Feynman diagrams whose link diagrams generate the trefoil knot (via B) and also have subdivergences (corresponding to A).

We thus propose to associate the cancellation of transcendentals with the cancellation of subdivergences in the quenched beta function of QED, which is an immediate consequence of the Ward identity, $Z_1 = Z_2$.

To see the key role of the Ward identity, consider Fig. 2g. There is an internal vertex correction, which is rendered local by adding the appropriate counterterm graph. Due to the Ward identity, this counterterm graph is the same as that which compensates for the self-energy correction in Fig. 2e. In [3,4] it was shown that the latter counterterm could be interpreted as the A

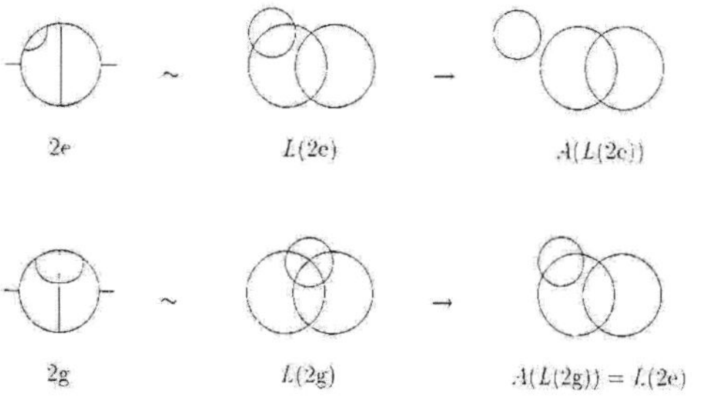

Fig. 3. The A part of the skein operation on the link diagrams for Figs. 2e, 2g.

part of the skein operation on the link diagram $L(2\mathrm{e})$ of Fig. 3, associated with the Feynman graph 2e. We assume that this is a generic feature of the relationship between skeining and renormalization and associate the corresponding counterterm for Feynman graph 2g with the term obtained from applying A twice to the link diagram $L(2\mathrm{g})$, which requires two skeinings to generate the same counterterm, along with the trefoil knot from the B term. The Ward identity thus becomes a relation between crossed and uncrossed diagrams, after skeining:

$$A(A(L(2\mathrm{g}))) = A(L(2\mathrm{e})) \Rightarrow A(L(2\mathrm{g})) = L(2\mathrm{e})\,. \tag{10}$$

The trefoil knot results, in this language, from $B(B(L(2\mathrm{g})))$, which generates, in general, a ζ_3/ϵ term from Fig. 2g, even after the subtraction of subdivergences. We now use the Ward identity (10) at the link level to obtain

$$B(B(L(2\mathrm{g}))) = B(B(A^{-1}(L(2\mathrm{e})))), \tag{11}$$

from which we see that it relates the transcendental counterterm from a torus-knot [4] topology, in Fig. 2g, to a knot-free [3] ladder topology, in Fig. 2e.

We conclude that ζ_3/ϵ should be absent from the bare diagram of Fig. 2g when it is calculated in the gauge where the bare diagram of Fig. 2e is free of ζ_3/ϵ, i.e., in the Landau gauge, where the latter is free of subdivergences. In Fig. 3 we summarize the argument. By a similar argument, we also conclude that the other graphs with the trefoil topology, namely Figs. 2f, 2h, should be free of ζ_3/ϵ in the Landau gauge.

Note that vanishing of the bare ladder diagram of Fig. 2e, in the Landau gauge, does not imply the van-

Table 2
QED contributions to $\Pi_3^{[1]}$, at $m = 0$ and $q = 0$, with $\xi = \lambda = 0$

	Laurent term			
	$\bar{a}^3/\epsilon^2$ $(m=0)$	$\bar{a}^3/\epsilon$ $(m=0)$	a_m^3/ϵ^2 $(q=0)$	a_m^3/ϵ $(q=0)$
diagram 2a	0	8/3	0	8/3
diagram 2b	0	0	−12/5	152/5
diagram 2c	0	0	72/5	8/5
diagram 2d	0	0	2/3	31/3
diagram 2e	0	0	0	−24
diagram 2f	−2/3	7/3	−2/3	17/3
diagram 2g	0	−28/3	0	−28/3
diagram 2h	2/3	11/3	−12	28
uncrossed	0	8/3	38/3	21
crossed	0	−10/3	−38/3	73/3
mass renorm	0	0	0	−46
total	0	−2/3	0	−2/3

ishing of $B(B(A^{-1}(L(2e))))$; it merely implies its rationality. The action of the A and B operators is non-trivial on the diagram of Fig. 2e; in the language of [3,4], skeining involves a change of writhe number in the subdivergence, and only terms with writhe number zero vanish in the Landau gauge, while the double application of the B operator in $B(B(A^{-1}(L(2e))))$ generates a non-vanishing writhe for the subdivergence.

To test these ideas, we evaluate the separate contributions of the QED diagrams of Fig. 2 to (8), (9) in the Landau gauge and find, indeed, that *each* bare diagram has a rational $1/\epsilon$ term in its Laurent expansion, at $\xi = 0$, for any value of the contraction parameter, λ. In any other gauge, the bare massless diagrams of Figs. 2a, 2e, 2f, 2g give non-zero, mutually cancelling, coefficients of ζ_3/ϵ. The $m = 0$ contributions to the Laurent expansion of $\Pi_3^{[1]}$ are given in Table 2, in the case $\xi = \lambda = 0$. The same rational behaviour is observed in the Landau gauge at $q = 0$ (again for any value of λ) with contributions also recorded in Table 2 (with $\lambda = 0$), along with that from mass renormalization of lower-loop diagrams.

To determine the corresponding gauge for SED, we study both of the two-loop DKP diagrams and find that each gives $1/\epsilon^2$ terms, proportional to $2\xi - 3$. Thus the gauge $\xi = \frac{3}{2}$ is the closest DKP analogy to the Landau gauge of QED. In any gauge, however, diagram 2h gives a ζ_3/ϵ term, proportional to $\lambda - 2$. (The absence of such a term in QED is attributable to the fact that the finite part of the two-loop fermion propagator does not involve ζ_3 at large momentum.) Accordingly we evaluate all $m = 0$ and $q = 0$ SED diagrams with $\xi = \frac{3}{2}$ and $\lambda = 2$ and find that every one is indeed rational, with a Laurent expansion recorded in Table 3. Again, we regard this as a significant success of skeining arguments, since fixing 2 parameters removes ζ_3/ϵ from each of the 5 diagrams 2a, 2e, 2f, 2g, 2h.

5. Conclusions

Complete cancellation of transcendentals from the beta function, at every order, is to be expected only in quenched QED and quenched SED, where subdivergences cancel between bare diagrams.

Contributions with more than one charged loop require scheme-dependent coupling-constant renormalization. Knot theory guarantees that double-bubble contributions are rational at the three-loop level, as found in [7,11] and shown in Table 1. But at four loops [2,10,12], and beyond [13], such multi-electron-loop contributions entail non-trivial knots, associated with zeta functions [4] and more exotic [5,6] transcendentals, whose cancellation is *not* underwritten by the Ward identity.

Likewise, we do not expect the beta functions of *non*-abelian gauge theories to be rational beyond the presently computed [14,15] three-loop level since coupling-constant renormalization is required to remove subdivergences generated by gluon- and ghost-loops in the gluon self-energy, which cannot be quenched without violating the Slavnov–Taylor identities. Indeed, it is still somewhat obscure how the rationality of β_3 comes about in QCD: as shown in [15], the renormalization constants $Z_{1,2,3}$, for the quark-gluon vertex, quark field, and gluon field, all involve ζ_3/ϵ at three-loops; only in $Z_\alpha = Z_1^2 Z_2^{-2} Z_3^{-1}$ does the cancellation occur. It may be possible to understand this in a three-loop background-field calculation, but we do not expect the intrinsically scheme-dependent non-abelian beta function to remain rational at higher orders in the MS scheme.

In fact, there are only two further quantities that we

Table 3
SED contributions to $\Pi_3^{[1]}$, at $m = 0$ and $q = 0$, with $\xi = \frac{3}{2}$ and $\lambda = 2$

	Laurent term					
	$\bar{a}^3/\epsilon^3$ $(m=0)$	$\bar{a}^3/\epsilon^2$ $(m=0)$	$\bar{a}^3/\epsilon$ $(m=0)$	$a_{\rm m}^3/\epsilon^3$ $(q=0)$	$a_{\rm m}^3/\epsilon^2$ $(q=0)$	$a_{\rm m}^3/\epsilon$ $(q=0)$
diagram 2a	1/16	55/96	4151/576	1/16	7/96	2501/576
diagram 2b	−13/8	−409/24	−16243/144	−13/8	−233/120	−8/45
diagram 2c	3/2	31/2	3703/36	3/2	43/20	203/45
diagram 2d	1/16	103/96	4387/576	1/16	−65/96	−131/576
diagram 2e	0	−5/48	41/32	0	−5/48	−1141/96
diagram 2f	−1/16	−9/32	911/192	−1/16	7/32	1013/192
diagram 2g	0	3/16	−917/288	0	3/16	−1349/288
diagram 2h	1/16	3/32	−1667/576	1/16	−29/32	−809/576
uncrossed	0	0	37/6	0	−1/2	−55/16
crossed	0	0	−4/3	0	−1/2	−13/16
mass renorm	0	0	0	0	1	109/12
total	0	0	29/6	0	0	29/6

expect to be rational beyond three loops: the quenched anomalous field and mass dimensions, $\gamma_2^{[0]}$ and $\gamma_{\rm m}^{[0]}$, of QED, whose rationality at the three-loop level is established in [15,18]. These are scheme-independent, since subdivergences cancel in the electron propagator, by virtue of the Ward identity, which here relates proper self-energy diagrams to one-particle-reducible diagrams.

One is tempted to seek a calculational method, based on the skein relation, to compute the rational anomalous dimensions of quenched QED, in terms of diagrams, with non-zero writhe numbers, that have been 'ladderized' by Ward identities, such as (11), which may eventually yield a rational calculus like that in [4]. The complexity of the rational contributions of Tables 2 and 3 indicates that such a calculus would be rather non-trivial.

In conclusion: all-order rationality of counterterms is specific to quenched abelian theories, where the cancellation of knots, from the B term [5,6] in the skein relation, matches the cancellation of subdivergences, from the A term [3,4]. This is exemplified by the rationality of all quenched three-loop bare QED diagrams, in the Landau gauge.

Acknowledgements

D.J.B. and D.K. thank the organizers and participants of the Pisa and Aspen multi-loop workshops for their interest in and comments on this work. Detailed discussions with Kostja Chetyrkin, John Gracey, Sergei Larin, Christian Schubert and Volodya Smirnov were most helpful. D.J.B. thanks Andrei Kataev and Eduardo de Rafael, for long-standing dialogues on the rationality of quenched QED, and the organizers of the UK HEP Institute in Swansea, where the computations were completed.

References

[1] J.L. Rosner, Phys. Rev. Lett. 17 (1966) 1190; Ann. Phys. 44 (1967) 11.

[2] S.G. Gorishny, A.L. Kataev, S.A. Larin and L.R. Surguladze, Phys. Lett. B 256 (1991) 81.

[3] D. Kreimer, Renormalization and Knot Theory, UTAS-PHYS-94-25, in: J. Knot Theor. Ramificat., to appear.

[4] D. Kreimer, Phys. Lett. B 354 (1995) 117.

[5] D.J. Broadhurst and D. Kreimer, Knots and numbers in ϕ^4 theory to 7 loops and beyond, OUT-4102-57, hep-ph/9504352, in: New Computing Techniques in Physics Research IV (World Scientific, Singapore), in press.

[6] D.J. Broadhurst and D. Kreimer, Systematic identification of transcendental counterterms with positive knots, OUT-4102-61, in preparation.

[7] K.G. Chetyrkin, S.G. Gorishny, A.L. Kataev, S.A. Larin and F.V. Tkachov, Phys. Lett. B 116 (1982) 455.

[8] D.J. Broadhurst, Z. Phys. C 54 (1992) 599.

[9] N. Gray, D.J. Broadhurst, W. Grafe and K. Schilcher, Z. Phys. C 48 (1990) 673;
D.J. Broadhurst, N. Gray and K. Schilcher, Z. Phys. C 52 (1991) 111.

[10] D.J. Broadhurst, A.L. Kataev and O.V. Tarasov, Phys. Lett. B 298 (1993) 445.

[11] E. de Rafael and J.L. Rosner, Ann. Phys. 82 (1974) 369.

[12] T. Kinoshita, H. Kawai and Y. Okamoto, Phys. Lett. B 254 (1991) 235;
H. Kawai, T. Kinoshita and Y. Okamoto, Phys. Lett. B 260 (1991) 193;
A.L. Kataev, Phys. Lett. B 284 (1992) 401.

[13] A. Palanques-Mestre and P. Pascual, Commun. Math. Phys. 95 (1984) 277;
J.A. Gracey, Mod. Phys. Lett. A 7 (1992) 1945;
D.J. Broadhurst, Z. Phys. C 58 (1993) 339.

[14] O.V. Tarasov, A.A. Vladimirov and A.Y. Zharkov, Phys. Lett. B 93 (1980) 429.

[15] S.A. Larin and J.A.M. Vermaseren, Phys. Lett. B 303 (1993) 334.

[16] K. Johnson, R. Willey and M. Baker, Phys. Rev. 163 (1967) 1699.

[17] J.D. Bjorken and S.D. Drell, Relativistic Quantum Fields (McGraw-Hill, New York, 1965).

[18] O.V. Tarasov, JINR P2-82-900 (1982);
S.A. Larin, NIKHEF-H/92-18, hep-ph/9302240, in: Proc. Intern. Baksan School on Particles and Cosmology, eds. E.N. Alexeev, V.A. Matveev, Kh.S. Nirov and V.A. Rubakov (World Scientific, Singapore, 1994).

[19] M.G. Schmidt and C. Schubert, DESY 94-189; hep-th/9410100.

[20] K.G. Chetyrkin and F.V. Tkachov, Nucl. Phys. B 192 (1981) 159;
F.V. Tkachov, Phys. Lett. B 100 (1981) 65.

[21] S.G. Gorishny, A.L. Kataev and S.A. Larin, Phys. Lett. B 273 (1991) 141; B 275 (1992) 512 (E); B 341 (1995) 448 (E).

[22] S.G. Gorishny, S.A. Larin, L.R. Surguladze and F.V. Tkachov, Comput. Phys. Commun. 55 (1989) 381.

[23] K.G. Chetyrkin, A.L. Kataev and F.V. Tkachov, Phys. Lett. B 85 (1979) 277; Nucl. Phys. B 174 (1980) 345.

[24] K.G. Chetyrkin and F.V. Tkachov, Phys. Lett. B 114 (1982) 340.

[25] S.G. Gorishny, A.L. Kataev and S.A. Larin, Phys. Lett. B 259 (1991) 144; Pisma Zh. Eksp. Teor. Fiz. 53 (1991) 121.

[26] K. Johnson and M. Baker, Phys. Rev. D 8 (1973) 1110.

[27] K.G. Chetyrkin, S.G. Gorishny and V.P. Spiridonov, Phys. Lett. B 160 (1985) 149;
G.T. Loladze, L.R. Surguladze and F.V. Tkachov, Phys. Lett. B 162 (1985) 363.

[28] L.R. Surguladze and F.V. Tkachov, Nucl. Phys. B 331 (199 35.

[29] D.J. Broadhurst, P.A. Baikov, V.A. Ilyin, J. Fleischer, O.V. Tarasov and V.A. Smirnov, Phys. Lett. B 329 (1994) 103;
D.J. Broadhurst, J. Fleischer and O.V. Tarasov, Z. Phys. C 60 (1993) 287.

[30] D.J. Broadhurst and A.L. Kataev, Phys. Lett. B 315 (1993) 179.

[31] S.A. Larin and J.A.M. Vermaseren, Phys. Lett. B 259 (1991) 345.

[32] R.J. Crewther, Phys. Rev. Lett. 28 (1972) 1421.

[33] R.J. Duffin, Phys. Rev. 54 (1938) 1114;
N. Kemmer, Proc. R. Soc. 166 (1938) 127; A 173 (1939) 91;
G. Petiau, Acad. R. Belg. Cl. Sci. Mem. Collect. 16 (2) (1936);
Z. Tokuoka and H. Tanaka, Prog. Theor. Phys. 8 (1952) 599;
I. Fujiwara, Prog. Theor. Phys. 10 (1953) 589.

[34] E. Fischbach, M.M. Nieto and C.K. Scott, J. Math. Phys. 14 (1973) 1760.

[35] A.C. Hearn, REDUCE user's manual, version 3.5, Rand publication CP78 (1993).

PHYSICAL REVIEW D VOLUME 54, NUMBER 8 15 OCTOBER 1996

BRIEF REPORTS

Brief Reports are accounts of completed research which do not warrant regular articles or the priority handling given to Rapid Communications; however, the same standards of scientifi quality apply. (Addenda are included in Brief Reports.) A Brief Report may be no longer than four printed pages and must be accompanied by an abstract.

Dimensional renormalization: Ladders and rainbows

R. Delbourgo*
Physics Department, University of Tasmania, GPO Box 252C, Hobart, Australia 7001

A. C. Kalloniatis†
Institut fur Theoretische Physik, University of Erlangen-Nuremberg, Staudtstrasse 7, Erlangen, D-91058 Germany

G. Thompson‡
International Centre for Theoretical Physics, Grignano-Miramare, Trieste, 34100 Italy
(Received 16 May 1996)

Renormalization factors are most easily extracted by going to the massless limit of the quantum fiel theory and retaining only a single momentum scale. We derive the factors and renormalized Green's functions to *all* orders in perturbation theory for rainbow graphs and vertex (or scattering) diagrams at zero momentum transfer, in the context of dimensional regularization, and we prove that the correct anomalous dimensions for those processes emerge in the limit $D \to 4$. [S0556-2821(96)00320-7]

PACS number(s): 11.10.Gh, 11.10.Jj, 11.10.Kk

I. INTRODUCTION

The connection between knot theory and renormalization theory [1] is one of the more exciting developments of fiel theory in recent years because it relates apparently different Feynman diagrams through the common topology of the associated knots. Thus it serves to explain why transcendental numbers for the renormalization constants Z occur in some diagrams [2] and not in others, thereby allowing the Feynman graphs to be grouped into equivalence classes. Sometimes, in gauge theories the Z factors within a particular class may cancel because of the existence of Ward identities, leaving a non-transcendental result for Z; this happens in electrodynamics of scalar and spinor particles to fourth order in the quenched limit and in chromodynamics to third order [3].

The class of graphs which correspond to ladders and rainbows are especially simple in this connection because they possess trivial knot topologies. Thus one may anticipate that Z factors for them are particularly easy to evaluate. Kreimer [1] has provided rules for extracting them within the framework of dimensional regularization, through the standard expedient of findin the simple $1/\epsilon = 2/(4-D)$ pole term arising in products of functions, after removing lower-order pole terms connected with subdivergences. Thus vertex diagrams bring in function factors of the type

$$_j\Delta(\epsilon) \equiv (p^2)^{\epsilon(j+1)} \int \frac{d^{4-2\epsilon}k}{(k^2)^{1+j\epsilon}(k+p)^2},$$

while rainbow graphs lead to products of

$$_j\Omega(\epsilon) \equiv (p^2)^{3+j\epsilon-D/2} \int [d^Dk/(k^2)^{2+j\epsilon}(k+p)^2].$$

It has to be said that, although the procedure is straightforward, extracting the $1/\epsilon$ term in nth order requires considerable graft. Kreimer has proven that the simple pole in ϵ is free of Riemann zeta functions.

In this paper we shall show that the problem can be solved to *all* orders in perturbation theory for ladders and rainbows [4], in the context of renormalization in dimensional regularization because of two fortuitous circumstances: (i) the Green's function satisfie a differential equation and (ii) this equation is actually soluble in terms of Bessel functions. The limit as $D \to 4$ may then be taken at the end and, as a useful check, the anomalous dimension properly emerges. (It is a rather delicate limit, requiring a saddle-point analysis of the integral representation of the Bessel function, since it looks quite singular.) We have successfully carried out this program for meson-fermion theories, both for vertex functions and rainbow diagrams; however we have not succeeded in solving the problem near $D=6$ for ϕ^3 theory because the differential equation is of fourth order and cannot be expressed in terms of standard functions; nevertheless we can obtain the answer in the limit $x \to 0$ or $p \to \infty$ for $D=6$.

In the next section we treat the vertex diagrams for scalar mesons, while the following section contains the analysis of the rainbow diagrams. The Appendix contains details of the vector meson case, which are rather more complicated.

II. VERTEX DIAGRAMS

We shall consider a theory of massless fermions ψ and mesons ϕ in D dimensions since the purpose of our work is to investigate the behavior of the Green's function as D

*Electronic address: bob.delbourgo@phys.utas.edu.au
†Electronic address: ack@theorie3.physik.uni-erlangen.de
‡Electronic address: thompson@ictp.trieste.it

 © 1996 The American Physical Society

tends to 4. Let $\gamma_{[r]}$ signify the product of $r\gamma$ matrices (of size $2^{D/2}\times 2^{D/2}$) normalized to unity, namely $\gamma_{[\mu_1\mu_2\cdots\mu_r]}$ so that we can write the meson-fermion interaction in the form

$$\mathcal{L}_{\rm int}=g\bar{\psi}\gamma_{[r]}\psi\phi^{[r]},$$

where $\phi^{[r]}$ is the corresponding tensor meson field The equation for the renormalized tensor vertex function $\Gamma_{[s]}$ at zero meson momentum, taking out the factor g, is

$$\Gamma_{[s]}(p)=Z\gamma_{[r]}\delta_s^r-ig^2\int \bar{d}^Dq\,\gamma_{[r]}\frac{1}{\gamma\cdot q}\times\Gamma_{[s]}(q)\frac{1}{\gamma\cdot q}\gamma_{[r']}\Delta^{rr'}(p-q), \qquad (1)$$

where $\bar{d}^Dq\equiv d^Dq/(2\pi)^D$. We shall assume that the massless meson propagator above, $\Delta^{rr'}(p-q)$, can be chosen in a Fermi-Feynman gauge so

$$\Delta^{rr'}=(-1)^r\eta^{rr'}/(p-q)^2,$$

where η stands for the diagonal Minkowskian metric pertaining to the tensor structure, specificall $\eta^{[\mu_1\nu_1}\cdots\eta^{\mu_r]\nu_r}$.

To make further progress we utilize the nonamputated Green's function,

$$G_{[r]}(p)=(1/\gamma\cdot p)\Gamma_{[r]}(p)(1/\gamma\cdot p),$$

to remain with the "simpler" linear integral equation

$$\gamma\cdot pG_{[s]}(p)\gamma\cdot p=Z\delta_s^r\gamma_{[r]}+i(-)^rc_r^sg^2\int \bar{d}^Dq\frac{G_{[s]}(q)}{(p-q)^2}, \qquad (2)$$

The nature of the couplings in massless theories means that the Green's function always stays proportional to $\gamma_{[s]}$ and can be decomposed into just two pieces [5,6]:

$$G_{[s]}(p)=\gamma_{[s]}A(p^2)+\gamma\cdot p\,\gamma_{[s]}\gamma\cdot pB(p^2). \qquad (3)$$

On the right-hand side of Eq. (2), $c_r^s\gamma_{[s]}=\gamma_{[r]}\gamma_{[s]}\gamma^{[r]}$, is essentially an element of the Fierz transformation matrix for any D, given by [5]

$$c_r^s=(-1)^{rs}\sum_q(-1)^q\binom{D-r}{s-q}\binom{r}{q}.$$

We shall convert the integral equation (2) into a differential equation by taking the Fourier transform. (In fact we could almost have done this from the start by writing the equation for the full Green's function in coordinate space.) This maneuver produces

$$\gamma\cdot\partial G_{[s]}(x)\gamma\cdot\overleftarrow{\partial}=Z\delta_s^r\gamma_{[r]}\delta^D(x)+ic_r^s(-)^rg^2\Delta_c(x)G_{[s]}(x), \qquad (4)$$

where the massless meson propagator is $i\Delta_c(x)=\Gamma(D/2-1)(-x^2+i\epsilon)^{1-D/2}/4\pi^{D/2}$. Because the coupling constant is a dimensionful quantity, we can defin a dimensionless strength $a=$ fin structure constant $/4\pi$ via

$$(-)^rc_r^sg^2\Gamma(D/2-1)\equiv 16\pi^{D/2}\mu^{4-D}a$$

upon introducing a mass scale μ. This simplifie the resulting expressions, as we can see in the purely scalar case, where there is but a single term and equation:

$$[\partial^2-(4a/x^2)(-\mu^2x^2)^{2-D/2}]G(x)=-Z\delta^D(x). \qquad (5)$$

The equation is readily solved in dimension $D=4$ yielding $G\propto(-x^2+i\epsilon)^{(-1-\sqrt{1+4a})/2}$. For $D\neq 4$ we can make progress by passing to a Euclidean metric ($r^2\equiv -x^2$):

$$\left[\frac{d^2}{dr^2}+\frac{D-1}{r}\frac{d}{dr}-\frac{4a}{r^2}(\mu r)^{4-D}\right]G(r)=Z\delta^D(r).$$

The whole point of the manipulation is that one is fortunately able to solve this equation (for $r\neq 0$ at firs) in terms of known functions, namely Bessel functions. The correct choice of solution, up to an overall factor, is

$$G(r)\propto r^{\epsilon-1}J_{1-1/\epsilon}[\sqrt{-4a}(\mu r)^\epsilon/\epsilon]; \quad D\equiv 4-2\epsilon$$

because in the limit as $a\to 0$ we recover the free-fiel solution r^{2D}. For dimensional reasons, let us carry out our renormalization so that $G(1/\mu)=\mu^{2-2\epsilon}$. With that convention, the vertex function reduces to

$$G(r)=(r/\mu)^{\epsilon-1}J_{1-1/\epsilon}[\sqrt{-4a}(\mu r)^\epsilon/\epsilon]/J_{1-1/\epsilon}(\sqrt{-4a}/\epsilon). \qquad (6)$$

Furthermore the correct singularity for the time-ordered function $\delta(x)$ emerges if we reinterpret $r^2=-x^2+i\epsilon$ above. We shall not worry at this stage whether a is positive or negative, the sign can vary with the model anyway, since we can easily continue the function from J to I as needed.

The problem presents itself: how does the four-dimensional result, with its anomalous scale $\gamma=\sqrt{1+4a}-1$, emerge from Eq. (6) as $\epsilon\to 0^-$, say? This is clearly a delicate limit because both the index and the argument of the Bessel function become infinitel large. Before answering this question, let us note that, in a perturbative expansion of Eq. (4), viz., a small argument expansion of J,

$$G(r)=r^{2\epsilon-2}\frac{[1+a(\mu r)^{2\epsilon}/\epsilon(2\epsilon-1)+a^2(\mu r)^{4\epsilon}/2\epsilon^2(2\epsilon-1)(3\epsilon-1)+\cdots]}{[1+a/\epsilon(2\epsilon-1)+a^2/2\epsilon^2(2\epsilon-1)(3\epsilon-1)+\cdots]},$$

the poles in ϵ cancel out to any particular order in a. For instance up to order a^2 we obtain, as $\epsilon \to 0$,

$$G(r)=r^{-2}[1+2(-a+a^2)\ln\mu r+2a^2(\ln\mu r)^2+\cdots]$$

which agrees precisely with the expansion of the anomalous dimension in the logarithmic terms. Returning to the limit of small ϵ, we will make use of the saddle-point method of obtaining asymptotic expansions of integrals. Suppose that $f(t)$ has a minimum at $t=\tau$ in the integral representation:

$$F=(-i/2\pi)\int_C \exp[f(t)]dt.$$

Then the saddle point method gives

$$F=\exp f(\tau)\frac{1}{\sqrt{2\pi f''(\tau)}}\left[1+\frac{f''''(\tau)}{8[f''(\tau)]^2}-\frac{5[f'''(\tau)]^2}{24[f''(\tau)]^3}+\frac{35[f'''(\tau)]^2}{384[f''(\tau)]^4}+\cdots\right].$$

As confirmatio of the correctness of the terms above we can verify that the Debye expansion [9] of the Bessel function is properly reproduced:

$$J_\nu(\nu/\cosh\tau)=\frac{-i}{2\pi}\int_{-i\pi+\infty}^{i\pi+\infty} dt e^{\nu(\sinh t/\cosh\tau - t)}$$

$$=\frac{e^{\nu(\tanh\tau-\tau)}}{\sqrt{2\pi\nu\tanh\tau}}\times\left[1+\frac{\coth\tau}{8\nu}\left(1-\frac{5}{3}(\coth\tau)^2\right)+\cdots\right],$$

because the integrand minimum occurs at $t=\tau$. Our case (6) is a variant of this. Working only to firs order in ϵ, and taking a negative initially, the integrand exponent is

$$(\sqrt{-4a}\sinh t+t)/\epsilon+\sqrt{-4a}\ln\mu r\sinh t-t$$

and is stationary at the complex value $t=\tau$, where

$$\cosh\tau=[(\epsilon-1)/\sqrt{-4a}(1+\epsilon\ln\mu r)].$$

Following through the mathematical steps, and omitting straightforward details, to order ϵ we end up with

$$G(r)=r^{2\epsilon-2}(\mu r)^{1-\epsilon-\sqrt{1+4a}}\left[1-\frac{2a\epsilon}{1+4a}\ln(\mu r)+\cdots\right]. \quad (7)$$

It is satisfying that this produces the all-orders (in coupling, a) result at four dimensions when $\epsilon\to 0$, with the correct anomalous scale.

The problem can be treated in much the same way for a pseudoscalar meson field The only possible difference is a change in sign of a, because of "γ_5" matrix anticommutation. As for the vector case ($r=s=1$), the procedure produces a pair of coupled equations for the two scalar components A and B of the Green's function, $G_\mu(p)=\gamma_\mu A(p^2)+\gamma\cdot p\gamma_\mu\gamma\cdot pB(p^2)$. A discussion of this case is given in the Appendix, where it is shown that the only easy limit is $D=4$; one find after Euclidean rotation that

$$A=ar^\beta,\quad B=br^{\beta+2},\quad a/b=c(\beta+2)/(\beta-2),$$

where

$$\beta=-1+\sqrt{5+\sqrt{16+4c+c^2}}\quad\text{and}\quad c=g^2/2\pi^2,$$

in four dimensions; we have chosen the root which reduces to the free fiel solution when $g=0$ although one can contemplate strictly nonperturbative solutions [10].

The difficult is symptomatic of what happens in ϕ^3 theory near six dimensions; in that case the Green's function, $G(p)=\Gamma(p)/p^4$ obeys the Fourier transformed equation [7],

$$(\partial^4-4a[(\mu r)^{6-D}/r^4])G(x)=Z\delta^D(r). \quad (8)$$

This is a differential equation of fourth order in r and its solution cannot readily be expressed in terms of familiar transcendental functions. However in the limit as $D\to 6$, it is quite simple to fin the (power law) solution:

$$G(r)\propto r^\beta,\quad \beta=-1-\sqrt{5-2\sqrt{4+a}},$$

which correctly reduces to the free fiel solution $\beta=-2$ when the coupling vanishes.

III. RAINBOW DIAGRAMS

We wish to treat the corrections to the fermion propagator in a similar manner, by considering the rainbow corrections. In such an approximation the rainbow graphs give rise to a self-energy, which is self-consistently determined according to

$$\Sigma_R(p)=-ig^2\int\frac{\overline{d^Dq}}{(p-q)^2}\left[\frac{1}{\gamma\cdot q}-\frac{1}{\gamma\cdot q}\Sigma_R(q)\frac{1}{\gamma\cdot q}\right],$$

with the unrenormalized propagator determined by

$$S_R(p)=1/\gamma\cdot p-(1/\gamma\cdot p)\Sigma_R(p)(1/\gamma\cdot p)$$

at this level [8]. This leads to the renormalized rainbow corrected propagator equation:

$$\gamma\cdot pS_R(p)\gamma\cdot p=Z_\psi\gamma\cdot p+ig^2\int\frac{\overline{d^Dq}}{(p-q)^2}S(q). \quad (9)$$

The nature of the massless problem is that one can always write $S_R(p)=\gamma\cdot p\sigma(p)$, and by Fourier transformation, convert Eq. (9) from an integral equation to a differential equation:

$$-i\gamma\cdot\partial\partial^2\sigma(x)=i\gamma\cdot\partial Z_\psi\delta^D(x)+ig^2\Delta_c(x)i\gamma\cdot\partial\sigma(x)$$

or

$$[\partial^2+ig^2\Delta_c(x)]\partial_\mu\sigma(x)=-Z_\psi\partial_\mu\delta^D(x). \quad (10)$$

Now for any function $f(\sqrt{x^2})$, using the two lemmas,

$$\partial_\mu f=x_\mu f'/\sqrt{x^2}, \quad (11a)$$

$$\partial_\mu\partial_\nu f=\left(\eta_{\mu\nu}-\frac{x_\mu x_\nu}{x^2}\right)\frac{f'}{\sqrt{x^2}}+\frac{x_\mu x_\nu}{x^2}f'', \quad (11b)$$

we can carry out an Euclidean rotation in order to arrive at the differential equation for the scalar function $\mathcal{S}\equiv d\sigma/dr$:

$$\left[\frac{d}{dr}\left(\frac{D-1}{r}+\frac{d}{dr}\right)+\frac{4a(\mu r)^{4-D}}{r^2}\right]\mathcal{S}(r)=-Z_\psi\delta^D(r). \quad (12)$$

In 4D this has the simple solution $\mathcal{S}(r)\propto r^{-1-2\sqrt{1+a}}$, in turn implying $S(p)\propto\gamma\cdot pp^{-4+2\sqrt{1+a}}$. However it is in fact possible to solve Eq. (12) for any D. The proper solution, normalized to $\mathcal{S}(1/\mu)=\mu^{3-2\epsilon}$ is

$$S(r)=r(r/\mu)^{\epsilon-2}\frac{J_{1-2/\epsilon}[\sqrt{-4a}(\mu r)^{\epsilon}/\epsilon]}{J_{1-2/\epsilon}(\sqrt{-4a}/\epsilon)}. \quad (13)$$

To get the rainbow propagator, we must firs integrate, $\sigma(r)=\int^r S(r)dr \propto \Sigma_{m=0}^{\infty} J_{2-2\epsilon+2m}(\sqrt{-4a}(\mu r)^{\epsilon}/\epsilon)$, and then Fourier transform to obtain $S_R(p)=\gamma\cdot p\sigma(p)$.

If the mesons are neither scalar nor pseudoscalar, but tensor, the coupling constant is multiplied by the factor c_1^r; that is all.

IV. CONCLUSIONS

We have demonstrated that it is possible to work out the all-orders solution of Green's functions for ladder and rainbow diagrams for any dimension D and that, in the limit as D approaches the physical dimension, the correct scaling dimension is obtained. We have exhibited fully how this happens for scalar theories, but have succeeded only to a limited extent in vector theories, because the equations are coupled and end up as fourth-order ones, with no transparent expression in terms of standard functions of mathematical physics. In any event, it is clear from the form of the Green's function that there are no transcendental constants in sight, even when we expand the answers perturbatively in terms of $\ln(\mu r)$, so that the renormalization constants are free of them. This confirm the findin of Kreimer for arbitrary ladder or rainbow order [2] and does not come as a surprise.

One can extend the ideas here to scattering processes which contain a single momentum scale, such as fermion-fermion scattering (again ladder graphs) for any D. It is a simple matter of taking the Fourier transform in particular channels and converting the momentum integral equations to differential ones in coordinate space. We shall not labor the issue in this paper since the steps are fairly obvious and can easily be fille in by the reader. What we have not solved for any D is the case of crossed ladders, when the kernel will presumably lead to transcendental Z constants; that is a task for the future.

ACKNOWLEDGMENTS

We would like to thank the Australian Research Council for providing financia support in the form of a small grant during 1995—when the majority of this work was carried out.

APPENDIX: THE VECTOR CASE

The vector vertex function

$$G_\mu(p)=A(p^2)\gamma_\mu+\gamma\cdot p\gamma_\mu\gamma\cdot pB(p^2),$$

upon Fourier transformation and tracing with γ_ν, produces the coordinate space equation,

$$(\partial^2\eta_{\mu\nu}-2\partial_\mu\partial_\nu)A+\partial^4\eta_{\mu\nu}B$$
$$=Z\eta_{\mu\nu}\delta(x)+ig^2(D-2)\Delta_c(x)$$
$$\times[\eta_{\mu\nu}A+(\partial^2\eta_{\mu\nu}-2\partial_\mu\partial_\nu)B].$$

Using lemmas (11), and identifying the terms multiplying $\eta_{\mu\nu}$ and $x_\mu x_\nu$, we arrive at the pair of coupled equations

$$\mathcal{O}[-A+\mathcal{O}B]+\frac{2}{2}\frac{dA}{dr}=ig^2(D-2)\Delta_c\left[A+\frac{2}{r}\frac{dB}{dr}-\mathcal{O}B\right],$$

$$\mathcal{Q}A=ig^2(D-2)\Delta_c\mathcal{Q}B,$$

where $\mathcal{O}\equiv[d^2/dr^2+(D-1)/r(d/dr)]$ and $\mathcal{Q}\equiv[d^2/dr^2-(1/r)(d/dr)]$. We have not suceeded in solving these equations in terms of familiar functions for $D\neq 4$. However in 4D, one can make considerable progress by looking for a power-law solution of the type $A(r)=ar^\beta$, $B(r)=br^{\beta+2}$. Simple calculation reveals that a solution exists provided that

$$a/b=c(\beta+2)/(\beta-2),$$

$$c^2+4c-\beta(\beta-2)(\beta+2)(\beta+4)=0,$$

$$c\equiv g^2/2\pi^2.$$

The quartic in the power exponent β is fortunately simple to solve in terms of the coupling (or c), the answer being

$$\beta=-1+\sqrt{5+\sqrt{16+4c+c^2}}=2+c/3-c^2/54\cdots,$$

so that $a/b\simeq 12+5c/3+\cdots$.

[1] D. Kreimer, Phys. Lett. B **354**, 111 (1995); "Renormalization and Knot Theory," Report No. UTAS-PHYS-94-25 and hep-th/9412045 (unpublished).

[2] D. Broadhurst and D. Kreimer, Int. J. Mod. Phys. C **6**, 51 (1995).

[3] D. Broadhurst, R. Delbourgo, and D. Kreimer, Phys. Lett. B **366**, 421 (1996).

[4] S. F. Edwards, Phys. Rev. **90**, 284 (1953). In particular see the appendix, which discusses the case of massive mesons (with massless particle interchange); the vertex turns out to be a hypergeometric function in that case.

[5] K. M. Case, Phys. Rev. **97**, 810 (1955); R. Delbourgo and V. Prasad, Nuovo Cimento A **21**, 32 (1974).

[6] Equation (3) is merely the most convenient way of writing out the decomposition; it can be transformed into the form $A\gamma_{[\mu_1\cdots\mu_r]}+C\Sigma_{j=1}^{r}p_{\mu_j}p^\lambda\gamma_{[\mu_1\ldots\mu_{j-1}\lambda\mu_{j+1}\cdots\mu_r]}$, by using the γ-matrix commutation relations. Here the tensor structure is more explicit. Note that C or B terms do not have an independent existence when $r=0$ (scalar) or when $r=D$ (pseudoscalar).

[7] Remember that for an interaction $g\phi^3$ in D dimensions, g has mass dimension $3-D/2$, while the propagator behaves as $1/r^4$. That is why the strength a is dimensionless in Eq. (8).

[8] Chains of self-energies are *not* being summed here. Were one to attempt that, it would be necessary to tackle the full Dyson-Schwinger equation $S^{-1}(p)=\gamma\cdot p+\Sigma(p)$, which would lead to a nonlinear equation for the propagator.

[9] M. Abramowitz and I. A. Stegun, *Handbook of Mathematical Functions* (Dover, New York, 1965). See especially Sec. 9.3.7.

[10] B. A. Arbuzov and A. T. Fillipov, Nuovo Cimento **38**, 284 (1965).

PHYSICAL REVIEW D VOLUME 55, NUMBER 8 15 APRIL 1997

Dimensional renormalization in ϕ^3 theory: Ladders and rainbows

R. Delbourgo* and D. Elliott†
University of Tasmania, GPO Box 252-21, Hobart, Tasmania 7001, Australia

D. S. McAnally‡
University of Queensland, St Lucia, Brisbane, Queensland 4067, Australia
(Received 18 November 1996)

The sum of all the ladder and rainbow diagrams in ϕ^3 theory near six dimensions leads to self-consistent higher order differential equations in coordinate space which are not particularly simple for arbitrary dimension D. We have now succeeded in solving these equations, expressing the results in terms of generalized hypergeometric functions; the expansion and representation of these functions can then be used to prove the absence of renormalization factors which are transcendental for this theory and this topology to all orders in perturbation theory. The correct anomalous scaling dimensions of the Green functions are also obtained in the six-dimensional limit. [S0556-2821(97)01208-3]

PACS number(s): 11.10.Gh, 11.10.Jj, 11.10.Kk

I. INTRODUCTION

In a recent paper [1] we managed to derive closed forms for ladder corrections to self-energy graphs (rainbows) and vertices, in the context of dimensional renormalization. We only succeeded in carrying out this program for Yukawa couplings near four dimensions, although we did obtain the differential equations pertaining to the ϕ^3 theory as well; but we were not able to solve the latter in simple terms. We have now managed to obtain closed expressions for ϕ^3 theory as well and wish to report the results here. The answers are indeed nontrivial and take the form of ${}_0F_3$ functions, which perhaps explains why they had eluded us so far. Interestingly, the closed form results for Yukawa-type models lead to Bessel functions with curious indices and arguments; but as these can also be written as ${}_0F_1$ functions, the analogy with ϕ^3 is close after all.

Given the exact form of the results, both for rainbows and ladders, we are able to test out Kreimer's [2] hypothesis about the connection between knot theory and renormalization theory with confidence, fully verifying that the renormalization factors for such topologies are indeed nontranscendental. At the same time we are able to determine the Z factors to any given order in perturbation theory and show that in the $D\to 6$ limit, the correct anomalous dimensions of the Green functions do emerge, which is rather satisfying. The various Z factors come out as poles in $1/(D-6)$ when the Green functions are expanded in the normal way as powers of the coupling constant, but the complete result produces the renormalized Green function to all orders in coupling for any dimension D.

In the next section we treat the vertex diagrams, converting the differential equation for ladders into hypergeometric form. Upon picking the correct solution we are able to do two things: (i) establish that in the $D\to 6$ limit one arrives at the correct anomalous scaling factor for the vertex function, and (ii) obtain the Z factors through a perturbative expansion of argument of the hypergeometric function. The case (i) is a bit tricky; it requires an asymptotic analysis, because the indices of the hypergeometric function as well as the argument diverge in the six-dimensional limit. The next section contains the analysis of the rainbow graphs; the equations are similar to the vertex case, but different solutions must be selected, resulting in a different anomalous dimension. It is nevertheless true that the self-energy renormalization constant remains nontranscendental. A brief concluding section ends the paper.

II. LADDER VERTEX DIAGRAMS

We will only treat the massless case, since this is sufficient to specify the Z factors once an external momentum scale is introduced. To further simplify the problem we shall consider the case where the vertex is at zero-momentum transfer, leaving just one external momentum p. The equation for the one-particle irreducible vertex Γ, in the ladder approximation, thereby reduces to

$$\Gamma(p)=Z+ig^2\int\frac{1}{q^2}\Gamma(q)\frac{1}{q^2}\frac{d^Dq/(2\pi)^D}{(p-q)^2}. \tag{1}$$

Letting $\Gamma(p)\equiv p^4G(p)$, the equation can be Fourier transformed into the coordinate space equation for G:

$$[\partial^4-ig^2\Delta_c(x)]G(x)=Z\delta^D(x), \tag{2}$$

where Δ_c is the causal Feynman propagator for arbitrary dimension D. Since the coupling g is dimensionful when $D\neq 6$, it is convenient to introduce a mass scale μ and define a dimensionless coupling parameter a via

$$\frac{g^2}{4\pi^{D/2}}\frac{\Gamma(D/2-1)}{(-x^2)^{1-D/2}}\equiv\frac{4a(\mu r)^{6-D}}{r^4}.$$

Then, rotating to Euclidean space ($r^2=-x^2$), the ladder vertex equation simplifies to

*Electronic address: bob.delbourgo@phys.utas.edu.au
†Electronic address: david.elliott@math.utas.edu.au
‡Electronic address: dsm@maths.uq.oz.au

 © 1997 The American Physical Society

$$\left[\left(\frac{d^2}{dr^2}+\frac{D-1}{r}\frac{d}{dr}\right)^2-\frac{4a(\mu r)^{6-D}}{r^4}\right]G(r)=Z\delta^D(r). \quad (3)$$

This is trivial to solve when $D=6$, since it becomes homogeneous for $r>0$ and the appropriate solution is

$$G(r)\propto r^b, \quad b=-1-\sqrt{5-2\sqrt{4+a}},$$

reducing to $G(r)\propto r^{-2}$ or $\Gamma(p)=1$ in the free field case ($a=0$); it represents a useful limit when analyzing the full equation (3), to which we now turn.

Let us define the scaling operator $\Theta_r=r(d/dr)$. This allows us to rewrite the square of the d'Alembertian operator as

$$\partial^4=\left[\frac{d^2}{dr^2}+\frac{D-1}{r}\frac{d}{dr}\right]^2$$
$$=r^{-4}(\Theta_r-2)\Theta_r(\Theta_r+D-4)(\Theta_r+D-2). \quad (4)$$

Hence for $r>0$ the original equation (3) reduces to the simpler form

$$[\Theta_\rho(\Theta_\rho-2)(\Theta_\rho+D-4)(\Theta_\rho+D-2)-4a\rho^{6-D}]G=0, \quad (5)$$

where $\rho=\mu r$ and Θ_ρ is the corresponding scaling operator. Next, rescaling the argument to $t=4a\nu^4\rho^{-1/\nu}$, with $\nu\equiv 1/(D-6)$, we obtain the hypergeometric equation

$$[\Theta_t(\Theta_t+2\nu)(\Theta_t-1-2\nu)(\Theta_t-1-4\nu)-t]G=0. \quad (6)$$

Being of fourth order, there are four linearly independent solutions:

$$_0F_3(b_1,b_2,b_3;t),$$
$$t^{1-b_1}{}_0F_3(2-b_1,b_2-b_1+1,b_3-b_1+1;t),$$
$$t^{1-b_2}{}_0F_3(2-b_2,b_3-b_2+1,b_1-b_2+1;t),$$
$$t^{1-b_3}{}_0F_3(2-b_3,b_1-b_3+1,b_2-b_3+1;t),$$

where $b_1\equiv 1+2\nu$, $b_2=-4\nu$, and $b_3=-2\nu$. The appropriate solution, which near $t=0$ behaves as r^{4-D} when $a=0$, is the last choice: namely,

$$G\propto t^{1+2\nu}{}_0F_3(2+2\nu,2+4\nu,1-2\nu;t).$$

Near $r=0$ this behaves like $r^{-2-1/\nu}$. Finally, renormalizing the Green function G to equal μ^{D-4} when $r=1/\mu$, the scale we introduced previously for the coupling constant, and restoring the original variables, we end up with the exact result

$$G(r)=r^{4-D}\frac{{}_0F_3\left(2-\frac{2}{6-D},2-\frac{4}{6-D},1+\frac{2}{6-D};\frac{4a(\mu r)^{6-D}}{(6-D)^4}\right)}{{}_0F_3\left(2-\frac{2}{6-D},2-\frac{4}{6-D},1+\frac{2}{6-D};\frac{4a}{(6-D)^4}\right)}. \quad (7)$$

To check that the poles in $(D-6)$ cancel out at any given order in perturbation theory, one simply expands the numerator and denominator in Eq. (7) to any particular power in the dimensionless coupling a and take the limit as $D\to 6$. For instance, to order a^3, with a little work one arrives at

$$r^2G(r)\to 1+\frac{a}{4}\ln(\mu r)+\frac{a^2}{64}\ln(\mu r)[1+2\ln(\mu r)]$$
$$+\frac{a^3}{1536}\ln(\mu r)[9+6\ln(\mu r)+4\ln^2(\mu r)]$$
$$+O(a^4). \quad (8)$$

It is most gratifying that this agrees perfectly with the expansion of the scaling index b obtained previously at $D=6$. The most significant point is that there is no sign of a transcendental constant in the singularities of the perturbation expansion for ϕ^3 theory near six dimensions, signifying that the renormalization constant Z is free of them, in agreement with the Kreimer hypothesis based on knot theory.

One last (rather difficult) check on our work is to see what happens directly to Eq. (7) as D approaches six, without having to invoke perturbation theory. For that an asymptotic analysis [3] based on the method of steepest descent (see for example, de Bruijn [4]) is needed. We start by making use of the Barnes integral representation [5,6] of the hypergeometric function:

$$_0F_3(b_1,b_2,b_3;t)$$
$$=\frac{1}{2i\pi}\int_{-i\infty}^{+i\infty}\frac{\Gamma(b_1)\Gamma(b_2)\Gamma(b_3)}{\Gamma(b_1+z)\Gamma(b_2+z)\Gamma(b_3+z)}\Gamma(-z)t^z dz.$$

In our case the b arguments lead us to evaluate the integral

$$I_\nu(r)\equiv\frac{1}{2i\pi}\int_{-i\infty}^{i\infty}\frac{\Gamma(2+4\nu)\Gamma(2+2\nu)\Gamma(1-2\nu)\Gamma(-z)}{\Gamma(2+4\nu+z)\Gamma(2+2\nu+z)\Gamma(1-2\nu+z)}$$
$$\times[4a\nu^4\rho^{-1/\nu}]^z dz \quad (9)$$

in the limit as $\nu\to\infty$. We shall show that as a function of ρ, I_ν behaves like $\rho^{1-\sqrt{5-2\sqrt{4+a}}}$. Remember that $G(r)\propto r^{-2-1/\nu}I_\nu(r)$.

For the method of steepest descents, suppose we write I_ν as

$$\frac{1}{2\pi i}\int_{-i\infty}^{i\infty}g_\nu(z)\exp[f_\nu(z)]dz,$$

where g_ν is a "slowly varying" function. We see from Eq. (9) that all the poles of the integrand lie on the positive real axis. If ζ is such that $\mathrm{Re}(\zeta)<0$ and $f'_\nu(\zeta)=0$, then an approximate evaluation of I_ν is given by

$$\alpha g_\nu(\zeta)\exp[f_\nu(\zeta)]/\sqrt{2\pi|f''_\nu(\zeta)|},$$

where

$$\alpha\equiv\exp[-i\arg(f''_\nu(\zeta))/2].$$

On applying the re ection formula for the gamma function [5] to both $\Gamma(1-2\nu)$ and $\Gamma(1-2\nu+z)$ appearing in Eq. (9), we find that we can write

$$g_\nu(z)=\rho^{-z/\nu}\frac{\sin\pi(2\nu-z)}{\sin(2\pi\nu)},$$

provided ν is not an integer, and

$$\exp[f_\nu(z)]=\frac{\Gamma(2+4\nu)\Gamma(2+2\nu)\Gamma(2\nu-z)\Gamma(-z)}{\Gamma(2+4\nu+z)\Gamma(2+2\nu+z)\Gamma(2\nu)}(4a\nu^4)^z.$$

Since

$$f'_\nu(z)=\ln(4a\nu^4)-[\psi(2\nu-z)+\psi(-z)+\psi(2+2\nu+z)+\psi(2+4\nu+z)],$$

where ψ denotes the psi (or digamma) function, we look for a zero at $z=-\xi\nu$ say, where $0<\xi<2$. Since we assume $\nu\gg1$ and since for $x\gg1$, $\psi(x)=\ln x+O(1/x)$, we find that ξ must satisfy the quartic

$$\xi(\xi+2)(\xi-2)(\xi-4)=4a.$$

The four solutions of this equation are

$$\xi=1\pm\sqrt{5\pm2\sqrt{4+a}}$$

and are all real if $0\le a<9/4$. In particular we shall choose the zero β say in (0,2) which is closest to the origin; that is

$$\beta=1-\sqrt{5-2\sqrt{4+a}}.$$

With this value of β we find

$$f''_\nu(-\beta\nu)\simeq(1-\beta)(4+2\beta-\beta^2)/(a\nu).$$

Since in fact $0<\beta<1$, we have that $\arg f''_\nu(-\beta\nu)=0$ so that $\alpha=1$. Again,

$$g_\nu(-\beta\nu)=\frac{\sin[(2+\beta)\pi\nu]}{\sin(2\pi\nu)}\rho^\beta$$

and, after some algebra,

$$\exp[f_\nu(-\beta\nu)]=\frac{4\sqrt{2\pi}\beta(\beta+2)}{a^{3/2}\nu^{1/2}}\left[\frac{16(2+\beta)}{(4-\beta)^2(2-\beta)}\right]^{2\nu}\times\exp(-4\beta\nu),$$

approximately. Consequently, for $\nu\gg1$ but not an integer, we find

$$_0F_3(2+2\nu,2+4\nu,1-2\nu;4a\nu^4\rho^{-1/\nu})\sim\frac{\rho^\beta\sin[(2+\beta)\pi\nu]}{a\sin(2\pi\nu)}\frac{4\beta(\beta+2)\exp(-4\beta\nu)}{(1-\beta)^{1/2}(4+2\beta-\beta^2)^{1/2}}\times\left[\frac{16(2+\beta)}{(4-\beta)^2(2-\beta)}\right]^{2\nu}.$$

Using this asymptotic expansion, we obtain simply from Eq. (7) that

$$G(r)=\mu^\beta r^{5-D-\sqrt{5-2\sqrt{4+a}}},\qquad(10)$$

which is just the scaling behavior at six dimensions which we were seeking. We have therefore fully verified the correctness of Eq. (7) in all the limits. The last step is to convert the answer to Minkowski space by making the familiar substitution $r^2\to-x^2+i\epsilon$.

III. RAINBOW DIAGRAMS

Let $\Delta_R(p)$ denote the renormalized ϕ propagator in rainbow approximation, so that $p^2\Delta_R(p)=1-\Sigma_R(p)/p^2$, where Σ_R is the rainbow self-energy. The propagator obeys the integral equation in momentum space

$$p^4\Delta_R(p)=Zp^2+ig^2\int\frac{d^Dk}{(2\pi)^D}\frac{\Delta_R(p-k)}{k^2},\qquad(11)$$

where Z now refers to the wave-function renormalization constant. As always we convert this into an x-space differential equation:

$$[\partial^4-ig^2\Delta_c(x)]\Delta_R(x)=-Z\partial^2\delta^D(x).\qquad(12)$$

Interestingly, this is exactly the same equation as Eq. (2), apart from the right-hand side, and it can therefore be converted into hypergeometric form by following the same steps as before. The only difference is that we should look for a different solution, because as $g\to0$, $\Delta_R(p)\to1/p^2$, or $\Delta_R(x)\sim(x^2)^{1-D/2}$.

A simple analysis shows the correct solution is

$$t^{1+4\nu}{}_0F_3(2+4\nu,2+6\nu,1+2\nu;t);$$

$$t=4a\nu^4(\mu r)^{-1/\nu},\quad \nu=1/(D-6),$$

because this reduces to r^{2-D} when $a=0$. Actually we can solve Eq. (12) directly at $D=6$ when $a\neq0$ because it is a simple homogeneous equation leading to

$$\Delta_R(x)\propto r^{-1-\sqrt{5+2\sqrt{4+a}}}$$

and thereby determine the anomalous dimension from the exponent of r. Anyhow, the exact solution of the rainbow sum for any D and renormalized at $r=1/\mu$ is here obtained to be

$$\Delta_R(r)=r^{2-D}\frac{{}_0F_3\left(2-\frac{4}{6-D},2-\frac{6}{6-D},1-\frac{2}{6-D};\frac{4a(\mu r)^{6-D}}{(6-D)^4}\right)}{{}_0F_3\left(2-\frac{4}{6-D},2-\frac{6}{6-D},1-\frac{2}{6-D};\frac{4a}{(6-D)^4}\right)}. \tag{13}$$

The numerator and denominator of Eq. (13), when expanded in powers of a will reproduce the (renormalized) perturbation series; to third order we find, in the limit as $D\to 6$, that all poles disappear and

$$r^4\Delta_R(r)=1-\frac{a}{12}\ln(\mu r)+\frac{a^2}{1728}\ln(\mu r)[11+6\ln(\mu r)]-\frac{a^3}{124416}\ln(\mu r)[103+66\ln(\mu r)+12\ln^2(\mu r)]+O(a^4).$$

This coincides perfectly with the expansion of the scaling exponent at $D=6$.

Lastly we need to show that the $D\to 6$ limit of Eq. (13) collapses to the scaling behavior found above, via an asymptotic analysis of the Barnes representation. We have indicated how this can be proven in the previous section and thus we skip the formal details to avoid boring the reader. The long and the short of the analysis is that no transcendentals enter into the above expressions for the self-energy (including their singularities, which are tied to the wave-function renormalization constant). These results confirm nicely the Kreimer [2] hypothesis that the Z factors will be simple rationals for such topologies.

IV. CONCLUSIONS

We have succeeded in evaluating an all-orders solution of Green functions for ladder and rainbow diagrams for any dimension D in ϕ^3 theory; the results are nontrivial, involving ${}_0F_3$ hypergeometric functions. We have demonstrated that, in the limit as $D\to 6$, the correct six-dimensional scaling behavior (which can be separately worked out) is reproduced. One can likewise determine the exact solutions for massless bubble ladder exchange in ϕ^4 theory, because the equations are very similar: they are also of fourth order and can be converted into hypergeometric form too [7].

More intriguing is the question of what happens when self-energy and ladder insertions are considered, so far as renormalization constants are concerned. A recent paper by Kreimer [8] has shown that such topologies with their disjoint divergences can produce transcendental Z in accordance with link diagrams that are of the $(2,q)$ torus knot variety, where the highest q is determined by the loop number. It would be interesting to show this result without resorting to perturbation theory by summing all those graphs exactly, as we have done in this paper. (Kreimer cautions that multiplicative renormalization may screen his new findings.) The generalization to massive propagators [9] does not seem beyond the realms of possibility either, although it has a marginal bearing on Z-factors.

ACKNOWLEDGMENTS

We thank the University of Tasmania Research Committee for providing a small grant which enabled this collaboration to take place.

[1] R. Delbourgo, A. Kalloniatis, and G. Thompson, Phys. Rev. D **54**, 5373 (1996).

[2] D. Kreimer, Phys. Lett. B **354**, 111 (1995); D. Kreimer, J. Knot Theory Ramif. (to be published).

[3] In Ref. [1], we deduced the Debye expansion of the Bessel function (and a variant of it) by this artifice, adopting a particularly helpful integral representation of Bessel function. In this paper we are obliged to use the Barnes representation as it is the only version which readily generalizes to ${}_pF_q$ hypergeometric functions; the stationary point analysis is completely different now and has been included for that reason.

[4] N. G. de Bruijn, *Asymptotic Methods in Analysis* (North-Holland, Amsterdam, 1958).

[5] M. Abramowitz and I. A. Stegun, *Handbook of Mathematical Functions* (Dover, New York, 1972).

[6] A. P. Prudnikov, Yu. A. Brychkov, and O. I. Marychev, *Integrals and Series (Additional Chapters)* (Nauka, Moscow, 1986); see p. 438, Eq. (12) in particular.

[7] In this connection, D. Broadhurst and D. Kreimer, Int. J. Mod. Phys. C **6**, 519 (1995), have investigated the nature of renormalization constants to seven-loop order for ϕ^4 theory and tied them to knot entanglements.

[8] D. Kreimer, "On knots in subdivergent diagrams," Mainz Report No. MZ-TH//96-31, 1996 (unpublished).

[9] S. F. Edwards, Phys. Rev. **90**, 284 (1953).

JOURNAL OF MATHEMATICAL PHYSICS VOLUME 39, NUMBER 9 SEPTEMBER 1998

A geometrical angle on Feynman integrals

A. I. Davydychev[a)]
Physics Department, University of Tasmania, GPO Box 252-21, Hobart, Tasmania, 7001 Australia and Institute for Nuclear Physics, Moscow State University, 119899, Moscow, Russia

R. Delbourgo[b)]
Physics Department, University of Tasmania, GPO Box 252-21, Hobart, Tasmania, 7001 Australia

(Received 22 December 1997; accepted for publication 29 April 1998)

A direct link between a one-loop N-point Feynman diagram and a geometrical representation based on the N-dimensional simplex is established by relating the Feynman parametric representations to the integrals over contents of $(N-1)$-dimensional simplices in non-Euclidean geometry of constant curvature. In particular, the four-point function in four dimensions is proportional to the volume of a three-dimensional spherical (or hyperbolic) tetrahedron which can be calculated by splitting into birectangular ones. It is also shown that the known formula of reduction of the N-point function in $(N-1)$ dimensions corresponds to splitting the related N-dimensional simplex into N rectangular ones. © *1998 American Institute of Physics*. [S0022-2488(98)00709-9]

I. INTRODUCTION

The development of techniques for an efficient calculation of one-loop N-point Feynman diagrams is very important for studying the leading and next-to-leading corrections to elementary particle processes within and beyond the Standard Model. Since any one-loop diagram can be reduced to a combination of scalar integrals,[1] in what follows we shall mainly deal with scalar Feynman integrals.

It is well known (see e.g., in Ref. 2) that results for the three- and four-point functions in four dimensions can be expressed in terms of dilogarithms (or related functions). The first explicit calculation of the general one-loop four-point function was given in Ref. 3 and later on, more compact results were presented in Refs. 2 and 4 (see also in Refs. 5 and 6). Moreover, it is known that (in four dimensions) the five-point function, the "pentagon," can be reduced to a linear combination of four-point functions[7–9] (see also in Refs. 10–13). Using linear dependence of the external momenta, a similar reduction procedure can be applied to the N-point functions with $N \geqslant 6$ (see in Refs. 14 and 8).

As a rule, explicit results for diagrams with several external legs possess a rather complicated analytical structure. For example, separate terms in a sum of dilogarithms may have singularities (cuts) which cancel in the whole sum. As the result, there are certain difficulties in describing analytic continuation to all regions (in external momenta and masses) of interest.

Another approach to the evaluation of one-loop integrals[15] makes it possible to represent them in terms of multiple hypergeometric functions. In Ref. 16, such results for the integrals with an arbitrary number of external legs were presented. As a rule, the corresponding hypergeometric series have a rather restricted region of convergence. In general, the problem of analytic continuation of occurring functions to all other regions of interest is very complicated.

The analytical structure of the results can be better understood if one employs a geometrical interpretation of kinematic invariants and other quantities. For example, the singularities of the general three-point function can be described pictorially through a tetrahedron constructed out of the external and internal momenta. This method can be used to derive Landau equations defining the positions of possible singularities[17] (see also in Refs. 18–20) and a similar approach can be applied to the four-point function[21,3] too.

[a)]Electronic mail: davyd@theory.npi.msu.su
[b)]Electronic mail: Bob.Delbourgo@utas.edu.au

 © 1998 American Institute of Physics

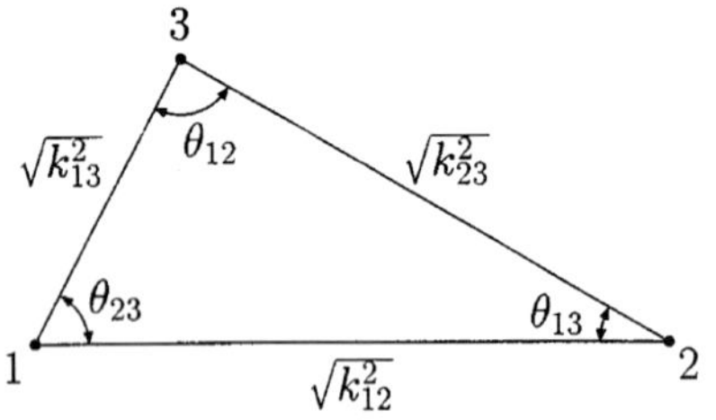

FIG. 1. The triangle associated with the massless three-point function.

Another well-known example of using geometrical ideas is the massless three-point function with arbitrary (off-shell) external momenta. Let us denote them as k_{12}, k_{23}, and k_{31}, so that $k_{12}+k_{23}+k_{31}=0$, and assume that all these momenta are time-like ($k_{jl}^2>0$). Using the standard notation for the "triangle" Källen function,

$$\lambda(x,y,z)=x^2+y^2+z^2-2xy-2yz-2zx, \tag{1.1}$$

and assuming that we are in the region $\lambda(k_{12}^2,k_{23}^2,k_{31}^2)\leqslant 0$, the result for the corresponding one-loop scalar Feynman integral (in four dimensions) can be neatly expressed as

$$\frac{2i\pi^2}{\sqrt{-\lambda(k_{12}^2,k_{23}^2,k_{31}^2)}}\{\mathrm{Cl}_2(2\theta_{12})+\mathrm{Cl}_2(2\theta_{23})+\mathrm{Cl}_2(2\theta_{31})\}, \tag{1.2}$$

where the angles θ_{12}, θ_{23}, and θ_{31} are nothing but those of the triangle with sides $\sqrt{k_{12}^2}$, $\sqrt{k_{23}^2}$, and $\sqrt{k_{31}^2}$, as shown in Fig. 1. Moreover, the denominator $\sqrt{-\lambda}$ is nothing but four times the area of this triangle. [Here we use the same normalization as in (2.1). The results for the massless three-point functions can be found, e.g., in Refs. 22, 2, 23. They are closely connected with the results for two-loop massive vacuum diagrams.[24–26] The functions related to the dilogarithm (including the Clausen function $\mathrm{Cl}_2(\theta)$) are defined in Appendix A (see also Ref. 27).]

In this paper, we discuss the geometrical interpretation of the kinematic variables related to the one-loop N-point functions. The generalization of the tetrahedron representation (which was used in the three-point case) is an N-dimensional simplex. Furthermore, we show that there is a direct transition from the Feynman parametric representation to the geometrical description connected with an N-dimensional simplex. We thereby arrive at a "geometrical" way of evaluating Feynman integrals.

The paper is organized as follows. In Sec. II we describe the connection of the N-point variables and the corresponding N-dimensional simplex. In Sec. III we show how one can "geometrize" the ordinary Feynman parametric representation. In Secs. IV, V, and VI we consider application of these geometrical ideas to the two-, three- and four-point functions, respectively. In Sec. VII we discuss how to use geometrical ideas for the reduction of N-point integrals in $(N-1)$ dimensions, and also for the calculation of the integrals with higher powers of denominators. Finally, in Sec. VIII, we summarize the main ideas and results.

II. BASIC SIMPLICES IN *N* DIMENSIONS

A. *N*-point function and the related simplex

Consider a one-loop scalar N-point "sun-type" diagram presented in Fig. 2. The corresponding Feynman integral is

$$J^{(N)}(n;\nu_1,\ldots,\nu_N)\equiv\int\frac{d^nq}{\Pi_{i=1}^N[(p_i+q)^2-m_i^2]^{\nu_i}}, \tag{2.1}$$

where n is the space–time dimension and ν_i are the powers of the propagators. (Here and below, the usual causal prescription for the propagators is understood, i.e., $1/[q^2-m^2]^\nu\leftrightarrow 1/[q^2-m^2+i0]^\nu$.) In general, it depends on $\frac{1}{2}N(N-1)$ momenta invariants k_{jl}^2 $(j<l)$, where

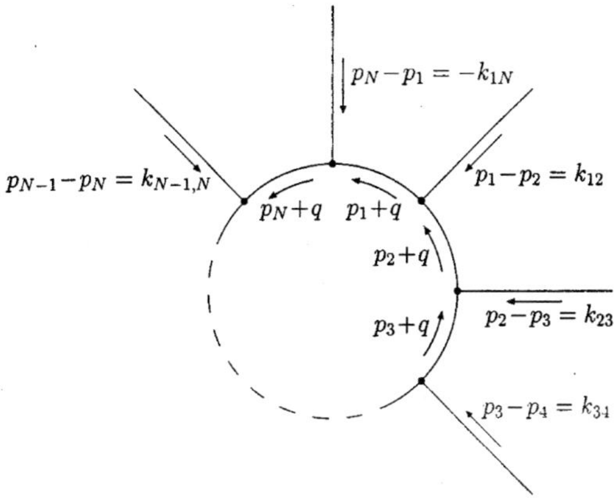

FIG. 2. The one-loop N-point diagram.

$$k_{jl}\equiv p_j-p_l, \tag{2.2}$$

and N masses m_i corresponding to the internal propagators. The momenta p_i are auxiliary in the sense that they all can be shifted by a constant vector without changing the momenta k_{jl} (2.2). We shall discuss this freedom below [see Eqs. (2.49)–(2.53)].

It is well known that, using the condition $\Sigma\alpha_i=1$, the standard quadratic form occurring in the denominator of the integrand of the Feynman parametric representation for these integrals [cf. Eqs. (3.1)–(3.2) below] can be rewritten in a homogeneous form,

$$\left[\sum_{j<l}\sum \alpha_j\alpha_l k_{jl}^2-\sum \alpha_i m_i^2\right]\Rightarrow-\left[\sum \alpha_i^2 m_i^2+2\sum_{j<l}\sum \alpha_j\alpha_l m_j m_l c_{jl}\right], \tag{2.3}$$

where (in some papers, the notation y_{jl} is used for $\pm c_{jl}$)

$$c_{jl}\equiv\frac{m_j^2+m_l^2-k_{jl}^2}{2m_jm_l}. \tag{2.4}$$

In the region between the corresponding two-particle pseudo-threshold, $k_{jl}^2=(m_j-m_l)^2$, and the threshold, $k_{jl}^2=(m_j+m_l)^2$, we have $|c_{jl}|<1$, and therefore in this region they can be understood as cosines of some angles τ_{jl},

$$c_{jl}=\cos\,\tau_{jl}=\begin{cases}1, & k_{jl}^2=(m_j-m_l)^2\\ -1, & k_{jl}^2=(m_j+m_l)^2\end{cases}. \tag{2.5}$$

The corresponding angles τ_{jl} are

$$\tau_{jl}=\arccos(c_{jl})=\arccos\left(\frac{m_j^2+m_l^2-k_{jl}^2}{2m_jm_l}\right)=\begin{cases}0, & k_{jl}^2=(m_j-m_l)^2\\ \pi, & k_{jl}^2=(m_j+m_l)^2\end{cases}. \tag{2.6}$$

The expressions in other regions should be understood in the sense of analytic continuation, using (when necessary) the causal prescription for the propagators. For example, when $k_{jl}^2<(m_j-m_l)^2(c_{jl}>1)$, i.e., below the pseudothreshold, we should interpret

$$\tau_{jl}=-i\ \mathrm{Arch}(c_{jl})=-\frac{i}{2}\ln\left(\frac{m_j^2+m_l^2-k_{jl}^2+\sqrt{\lambda(m_j^2,m_l^2,k_{jl}^2)}}{m_j^2+m_l^2-k_{jl}^2-\sqrt{\lambda(m_j^2,m_l^2,k_{jl}^2)}}\right), \tag{2.7}$$

where $\lambda(x,y,z)$ is the Källen function defined by Eq. (1.1). When $k_{jl}^2>(m_j+m_l)^2(c_{jl}<-1)$, i.e., above the threshold, we get

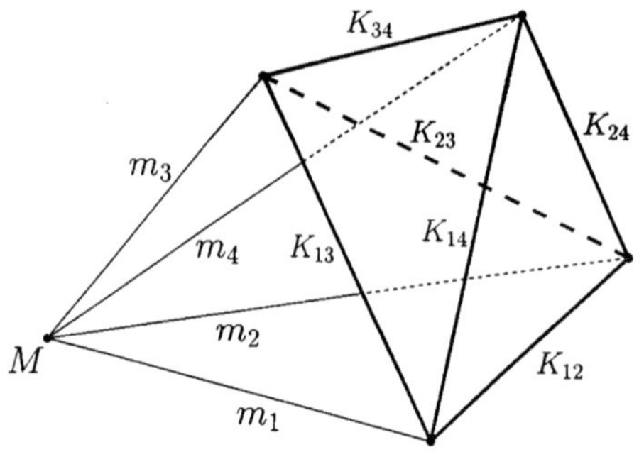

FIG. 3. The basic simplex for $N=4$.

$$\tau_{jl}=\pi+i\ \text{Arch}(-c_{jl})=\pi+\frac{i}{2}\ln\left(\frac{k_{jl}^2-m_j^2-m_l^2+\sqrt{\lambda(m_j^2,m_l^2,k_{jl}^2)}}{k_{jl}^2-m_j^2-m_l^2-\sqrt{\lambda(m_j^2,m_l^2,k_{jl}^2)}}\right). \tag{2.8}$$

Note that $\lambda(m_j^2,m_l^2,k_{jl}^2)$ is positive when $k_{jl}^2<(m_j-m_l)^2$ or $k_{jl}^2>(m_j+m_l)^2$.

When the angles τ_{jl} [defined by Eq. (2.6)] are real, the quadratic form on the right-hand side in Eq. (2.3) has a rather simple geometrical interpretation. Let us consider a set of Euclidean "mass" vectors whose lengths are m_i. Let them be directed so that the angle between the jth and the lth vectors is τ_{jl}. If we denote the corresponding unit vectors as a_i (so that the "mass" vectors are $m_i a_i$), we get

$$(a_j\cdot a_l)=\cos\ \tau_{jl}=c_{jl}. \tag{2.9}$$

This is also valid for the case $j=l$, since $\tau_{jj}=0$ and $c_{jj}=1$, cf. Eqs. (2.5)–(2.6). Now, we can represent Eq. (2.3) as

$$\sum \alpha_i^2 m_i^2+2\sum_{j<l}\sum \alpha_j\alpha_l m_j m_l c_{jl}=\left(\sum \alpha_i m_i a_i\right)^2. \tag{2.10}$$

In N dimensions, if we put all "mass" vectors together as emanating from a common origin, they, together with the sides connecting their ends, will define a *simplex* (or hypertetrahedron) which is the *basic* one for a given Feynman diagram. An example of such a simplex for $N=4$ is illustrated by Fig. 3. [Since the four-dimensional Euclidean space is understood, one should consider Fig. 3 just as an illustration (rather than precise picture).] It is easy to see that the length of the side connecting the ends of the jth and the lth mass vectors is $(m_j^2+m_l^2-2m_jm_lc_{jl})^{1/2}=(k_{jl}^2)^{1/2}$, so we shall call it a "momentum" side. In total, the *basic* N-dimensional simplex has $\frac{1}{2}N(N+1)$ sides, among them N mass sides (corresponding to the masses $m_1,\ldots,m_N$) and $\frac{1}{2}N(N-1)$ momentum sides (corresponding to the momenta k_{jl}, $j<l$), which meet at $(N+1)$ vertices. Each vertex is a "meeting point" for N sides. There is one vertex where all mass sides meet, which we shall call the *mass meeting point*, M. All other vertices are meeting points for $(N-1)$ momentum sides and one mass side. Furthermore, the number of $(N-1)$-dimensional hyperfaces is $(N+1)$. N of them (the *reduced* hyperfaces) can be obtained by excluding one mass side in turn, together with the corresponding vertex and the momentum sides meeting at that vertex. The reduced hyperface without the jth mass side is nothing but the $(N-1)$-dimensional simplex corresponding to an $(N-1)$-point function obtained from the N-point function by shrinking the jth propagator; this is equivalent to the power of propagator vanishing, $\nu_j=0$. The last hyperface, namely the *momentum* hyperface involves all momentum sides but does not include any mass sides. It is associated with the massless N-point function possessing the same external momenta as the basic one.

In general, in the N-point function only $(N-1)$ external momenta are independent, because they are related via the conservation law. We can choose, for example, the vectors $k_{1N},k_{2N},\ldots,k_{N-1,N}$. Considering all scalar products of these momenta, including their squares, we get $\frac{1}{2}N(N-1)$ invariants, as stated above. However, when n is integer and less than $(N-1)$, only n of the $(N-1)$ momenta are independent, e.g., $k_{1N}, k_{2N}, \ldots, k_{nN}$. Each of the remaining $(N-n-1)$ vectors k_{jN} $(j=n+1,\ldots,N-1)$ can be fixed by n scalar products $(k_{jN}\cdot k_{lN})$ $(l=1,\ldots,N-1)$. Hence the total number of independent momentum invariants for integer $n \leqslant N-1$ is

$$\tfrac{1}{2}n(n+1)+n(N-n-1)=\tfrac{1}{2}n(2N-n-1). \tag{2.11}$$

while for $n \geqslant N-1$ it remains equal to $\frac{1}{2}N(N-1)$; both expressions coincide for $n=N-1$. The linear dependence between the external momenta reveals itself in the degeneracy of the results associated with the vanishing of certain hypervolumes (see below).

We note that there is a difference between the "real" momenta of the N-point diagram, k_{jl}, and the vectors

$$K_{jl} \equiv m_j a_j - m_l a_l, \tag{2.12}$$

corresponding to the momentum sides of the simplex (the same applies to the momenta p_i and P_i discussed in Sec. II D). The momenta k_{jl} are defined in *pseudo-Euclidean* (Minkowski) space, and they are time-like ($k_{jl}^2>0$) in the region between the threshold and pseudo-threshold. The momentum sides of the simplex, K_{jl}, are vectors in *Euclidean* space, the (Euclidean) squares of their lengths (K_{jl}^2) being equal to k_{jl}^2. In what follows, we shall consider the K_{jl} as Euclidean analogs of the k_{jl} and, *vice versa*, the k_{jl} are pseudo-Euclidean analogs of the K_{jl}. Since the k_{jl}^2 $(j<l)$ form the complete basis of external invariants, all invariant scalars in Minkowski and Euclidean spaces are the same.

The matrix with the components (2.4),

$$\|c\| \equiv \|c_{jl}\| \equiv \begin{pmatrix} 1 & c_{12} & c_{13} & \cdots & c_{1N} \\ c_{12} & 1 & c_{23} & \cdots & c_{2N} \\ c_{13} & c_{23} & 1 & \cdots & c_{3N} \\ \cdots & \cdots & \cdots & \cdots & \cdots \\ c_{1N} & c_{2N} & c_{3N} & \cdots & 1 \end{pmatrix}, \tag{2.13}$$

is associated with many geometrical properties of the basic simplex. [The matrix (2.13) is nothing but the Gram matrix of the vectors $a_1, \ldots, a_N$.] In particular, we shall need its determinant,

$$D^{(N)} \equiv \det\|c_{jl}\|, \tag{2.14}$$

and the minors

$$D_{jl}^{(N-1)} \equiv \{\text{minor of (2.13) obtained by eliminating the } j\text{th row and the } l\text{th column}\}. \tag{2.15}$$

In particular, when $j=l$ the minor $D_{jj}^{(N-1)}$ is a *principal* one for the matrix (2.13).

The formulas for the *content* or hypervolume of the basic N-dimensional simplex and its hyperfaces, in terms of the determinant (2.14) and the minors (2.15), are well known in the N-dimensional Euclidean geometry (see for instance Refs. 28 and 29). Thus the content of the N-dimensional simplex is given by

$$V^{(N)} = \frac{1}{N!}\left(\prod_{i=1}^{N} m_i\right)\sqrt{D^{(N)}}. \tag{2.16}$$

The content of the jth $(N-1)$-dimensional reduced hyperface (which does not contain the jth mass side) is

$$\bar{V}_j^{(N-1)} = \frac{1}{(N-1)!}\left(\prod_{i\neq j} m_i\right)\sqrt{D_{jj}^{(N-1)}}, \tag{2.17}$$

whereas the content of the $(N-1)$-dimensional momentum hyperface is

$$\bar{V}_0^{(N-1)} = \frac{1}{(N-1)!} \sqrt{\Lambda^{(N)}}, \tag{2.18}$$

where $\Lambda^{(N)}$ is symmetric combination of the momenta defined in the same way as in Ref. 11,

$$\Lambda^{(N)} \equiv \det \| (k_{jN} \cdot k_{lN}) \| . \tag{2.19}$$

Note that the content of the basic simplex can also be presented as

$$V^{(N)} = \frac{1}{N} \bar{V}_0^{(N-1)} m_0 , \tag{2.20}$$

where m_0 is the distance between the mass meeting point and the momentum hyperface, i.e., the length of the vector of the height of the simplex, H_0. Therefore,

$$m_0 \equiv |H_0| = N \frac{V^{(N)}}{\bar{V}_0^{(N-1)}} = \left(\prod_{i=1}^{N} m_i \right) \sqrt{\frac{D^{(N)}}{\Lambda^{(N)}}}. \tag{2.21}$$

B. Normals, dihedral angles, and the dual matrix

The dihedral angles between the hyperfaces can be defined via the angles between their normals. Thus, we get $\frac{1}{2}N(N-1)$ dihedral angles between pairs of the reduced hyperfaces and N angles between the momentum hyperface and the reduced ones.

Again, let a_i be unit vectors directed along the mass sides of the basic N-dimensional simplex. The normals to the $(N-1)$-dimensional reduced hyperfaces can be defined as

$$n_{j\lambda} = \frac{\partial}{\partial a_j^{\lambda}} (\epsilon_{\nu_1 \ldots \nu_N} a_1^{\nu_1} \ldots a_N^{\nu_N}), \tag{2.22}$$

where $\epsilon_{\nu_1 \ldots \nu_N}$ is the completely antisymmetric tensor in N dimensions.

Using the well-known expression for the contraction of the product of two ϵ tensors in terms of the determinant with Kronecker δ symbols, it is easy to show that

$$(n_j \cdot n_l) = (-1)^{j+l} D_{jl}^{(N-1)} , \tag{2.23}$$

where $D_{jl}^{(N-1)}$ is defined in (2.15). In particular, n_j^2 is nothing but the principal minor $D_{jj}^{(N-1)}$ related to the $(N-1)$-point function associated with the corresponding reduced hyperface. The cosine of the angle $\tilde{\tau}_{jl}$ between the normals to the jth and the lth reduced hyperfaces is

$$\tilde{c}_{jl} \equiv \cos \tilde{\tau}_{jl} = \frac{(n_j \cdot n_l)}{\sqrt{n_j^2 n_l^2}} = \frac{(-1)^{j+l} D_{jl}^{(N-1)}}{\sqrt{D_{jj}^{(N-1)} D_{ll}^{(N-1)}}} . \tag{2.24}$$

In particular, this means that the matrix with the elements $\sqrt{D_{jj}^{(N-1)} D_{ll}^{(N-1)}} \tilde{c}_{jl} / D^{(N)}$ is the inverse of the matrix (2.13). Namely,

$$\|c_{jl}\|^{-1} = \frac{1}{D^{(N)}} \operatorname{diag}\left(\sqrt{D_{11}^{(N-1)}}, \ldots, \sqrt{D_{NN}^{(N-1)}} \right) \|\tilde{c}_{jl}\| \operatorname{diag}\left(\sqrt{D_{11}^{(N-1)}}, \ldots, \sqrt{D_{NN}^{(N-1)}} \right), \tag{2.25}$$

where $D^{(N)}$ is defined in (2.14). We shall call the matrix $\|\tilde{c}_{jl}\|$ the *dual* matrix (with respect to $\|c_{jl}\|$). [The matrix $\|\tilde{c}_{jl}\|$ is nothing but the Gram matrix of unit vectors directed along the normals n_i $(i=1,\ldots,N)$.] A trivial corollary of (2.25) is that

$$\tilde{D}^{(N)} \equiv \det \|\tilde{c}_{jl}\| = \frac{(D^{(N)})^{N-1}}{\prod_{i=1}^{N} D_{ii}^{(N-1)}} . \tag{2.26}$$

J. Math. Phys., Vol. 39, No. 9, September 1998 A. I. Davydychev and R. Delbourgo 4305

Note that the dihedral angles ψ_{jl} are related to the angles $\widetilde{\tau}_{jl}$ as

$$\psi_{jl}=\pi-\widetilde{\tau}_{jl}, \quad \cos\,\psi_{jl}=-\cos\,\widetilde{\tau}_{jl}. \tag{2.27}$$

The normal to the momentum hyperface can be defined as

$$n_{0\lambda}=-\epsilon_{\mu_1\ldots\mu_{N-1}\lambda}(K_{1N})^{\mu_1}\ldots(K_{N-1,N})^{\mu_{N-1}}. \tag{2.28}$$

If we represent $K_{jN}=(m_j a_j - m_N a_N)$ and use the fact that all terms involving two or more $m_N a_N$ disappear in (2.28) (due to the antisymmetry of the ϵ tensor), we get the following representation

$$n_{0\lambda}=-\left(\prod_{i=1}^{N} m_i\right)\sum_{j=1}^{N}\frac{n_{j\lambda}}{m_j}. \tag{2.29}$$

On one hand, the original definition (2.28) leads to

$$n_0^2=\det\|(k_{jN}\cdot k_{lN})\|\equiv\Lambda^{(N)}, \tag{2.30}$$

where $\Lambda^{(N)}$ is defined in (2.19). On the other hand, using (2.29) we obtain

$$n_0^2=\left(\prod_{i=1}^{N} m_i^2\right)\sum_{l=1}^{N}\frac{1}{m_l^2}F_l^{(N)}, \tag{2.31}$$

where [cf. Eq. (4) of Ref. 11]

$$F_l^{(N)}=\sum_{j=0}^{N}(-1)^{j+l}D_{jl}^{(N-1)}\frac{m_l}{m_j}=\frac{\partial}{\partial m_l^2}(m_l^2 D^{(N)}), \tag{2.32}$$

i.e., $F_l^{(N)}$ is the determinant obtained from $\det\|c_{jl}\|$ by substituting, instead of the lth column, the mass ratios m_l/m_j (where j is the line number). When taking the partial derivative in (2.32), it is implied that the set of independent variables involves m_i^2 and k_{jl}^2, rather than c_{jl}. In particular, the representation of $F_l^{(N)}$ in terms of determinants (2.32) (see also in Ref. 11) shows that they obey the following set of linear equations:

$$\sum_{l=1}^{N} c_{jl}F_l^{(N)}\frac{1}{m_l}=D^{(N)}\frac{1}{m_j}. \tag{2.33}$$

Another useful representation of n_0 is

$$n_0=-\frac{\Pi m_l}{\sqrt{D^{(N)}}}\sum_{i=1}^{N}\frac{1}{m_i}F_i^{(N)}a_i. \tag{2.34}$$

where, as usually, a_i are the unit vectors directed along the mass sides of the basic simplex. Using Eq. (2.33) (and remembering that $(a_j\cdot a_l)=c_{jl}$), one can easily see that the representation (2.34) is equivalent to Eqs. (2.28) and (2.29). Namely, n_0 (defined by (2.34)) is orthogonal to all $K_{jl}=m_j a_j - m_l a_l$ and n_0^2 is the same as in Eq. (2.31).

C. Splitting the basic simplex

The geometrical meaning of $F_l^{(N)}$ can be understood as follows. Using Eqs. (2.29) and (2.23), we get

$$(n_0\cdot n_l)=-\left(\prod_{i\neq l} m_i\right)F_l^{(N)}. \tag{2.35}$$

Therefore, the cosine of the angle $\widetilde{\tau}_{0l}$ between the normals to the momentum hyperface and the lth reduced hyperface is

$$\tilde{c}_{0l} \equiv \cos \tilde{\tau}_{0l} = \frac{(n_0 \cdot n_l)}{\sqrt{n_0^2 n_l^2}} = -\frac{(\Pi_{i \neq l} m_i) F_l^{(N)}}{\sqrt{\Lambda^{(N)} D_{ll}^{(N-1)}}}, \tag{2.36}$$

whereas the dihedral angle between these hyperfaces is given by

$$\psi_{0l} = \pi - \tilde{\tau}_{0l}, \quad \cos \psi_{0l} = -\cos \tilde{\tau}_{0l}. \tag{2.37}$$

The orthogonal projection of the ith reduced hyperface onto the momentum hyperface is

$$\bar{V}_i^{(N-1)} \cos \psi_{0i}. \tag{2.38}$$

The sum of all such projections should cover the whole momentum hyperface,

$$\sum_{i=1}^{N} \bar{V}_i^{(N-1)} \cos \psi_{0i} = \bar{V}_0^{(N-1)}. \tag{2.39}$$

Using Eqs. (2.36), (2.17), and (2.18), we see that condition (2.39) is equivalent to

$$\left(\prod_{j=1}^{N} m_j^2 \right) \sum_{l=1}^{N} \frac{F_l^{(N)}}{m_l^2} = \Lambda^{(N)}, \tag{2.40}$$

which is the case, cf. Eqs. (2.30) and (2.31). The condition (2.40) is equivalent to Eq. (38) of Ref. 11 (see also in Ref. 3). Equation (2.39) illustrates the geometrical meaning of (2.40). Moreover, we can see that, if we use the "height" H_0 for splitting the basic N-dimensional simplex into N rectangular ones whose "bases" correspond to the projections (2.38), the content of the ith rectangular simplex $V_i^{(N)}$ is proportional to $F_i^{(N)}$, namely,

$$V_i^{(N)} = \frac{1}{N} \bar{V}_i^{(N-1)} m_0 \cos \psi_{0i} = \frac{V^{(N)}}{\Lambda^{(N)}} \left(\prod_{l \neq i} m_l^2 \right) F_i^{(N)}. \tag{2.41}$$

In the same manner we can consider a projection onto one of the reduced hyperfaces. This gives

$$\sum_{l \neq i} \bar{V}_l^{(N-1)} \cos \psi_{il} + \bar{V}_0^{(N-1)} \cos \psi_{0i} = \bar{V}_i^{(N-1)} \tag{2.42}$$

or

$$\sum_{l=1}^{N} \bar{V}_l^{(N-1)} \tilde{c}_{il} + \bar{V}_0^{(N-1)} \tilde{c}_{0i} = 0. \tag{2.43}$$

In terms of $F_l^{(N)}$, Eq. (2.43) corresponds to definition (2.32).

The cosine of the angle τ_{0j} between the jth mass side and the height H_0 is

$$c_{0j} \equiv \cos \tau_{0j} = \frac{m_0}{m_j}. \tag{2.44}$$

Consider the matrix of cosines (similar to (2.13)) associated with the ith rectangular simplex. Its elements are

$$\begin{cases} c_{jl}, & \text{if } j \neq i \text{ and } l \neq i \\ c_{0l}, & \text{if } j = i \text{ and } l \neq i \\ c_{0j}, & \text{if } l = i \text{ and } j \neq i \\ 1, & \text{if } j = l = i. \end{cases} \tag{2.45}$$

Denoting

$$D_i^{(N)} \equiv \{\text{determinant of the matrix (2.45)}\}, \tag{2.46}$$

and using Eqs. (2.16) and (2.21), we get

$$V_i^{(N)} = \left(\prod_{l \neq i} m_l \right) \sqrt{\frac{D_i^{(N)}}{\Lambda^{(N)}}}. \tag{2.47}$$

Comparing Eqs. (2.41) and (2.47), we obtain the following result for the determinant (2.46):

$$D_i^{(N)} = \left(\prod_{l \neq i} m_l \right) \frac{(F_i^{(N)})^2}{\Lambda^{(N)}}. \tag{2.48}$$

D. Choice of the momenta p_i

Let us discuss possibilities for a choice of the momenta p_i in Eq. (2.1) and their Euclidean analogs P_i, such that $K_{jl} = P_j - P_l$. Due to Eq. (2.2), the p_i vectors emanating from an origin, together with the vectors k_{jl}, form an N-dimensional simplex in pseudo-Euclidean space. (For integer $n < N$, this simplex is degenerate, since one cannot use more than n dimensions for external vectors.) Analogously, in Euclidean space the vectors P_i together with the K_{jl} momentum hyperface (taken as a base), form an N-dimensional Euclidean simplex similar to the basic one. Using translational invariance of p_i, we can shift their origin as we like. In particular, we can make one of the p_i vectors zero, whereupon the origin of P_i coincides with one of the vertices belonging to the momentum hyperface. However, this way would break the symmetry of the problem. There are at least two reasonable, *symmetric* ways to fix the origin of p_i (and P_i):

(1) Choose the mass meeting point as the origin of P_i vectors, so that $P_i^2 = p_i^2 = m_i^2$ and the P_i vectors coincide with the mass sides, $P_i = m_i a_i$. In this case, the basic simplex becomes even "more basic," and the integral (2.1) can be expressed as

$$J^{(N)}(n; \nu_1, \dots, \nu_N) = \int \frac{d^n q}{\prod_{i=1}^N [q^2 + 2(p_i \cdot q)]^{\nu_i}} \Bigg|_{(p_j \cdot p_l) = m_j m_l c_{jl}}. \tag{2.49}$$

Such a choice of p_i simplifies study of singularities via Landau equations.[17,20] (In this case, the main Landau singularity is associated with $q \to 0$.) Note that, according to Eq. (2.34), in this case the vector of the height H_0 can be represented as

$$H_0 = \frac{\prod m_l^2}{\Lambda^{(N)}} \sum_{i=1}^N \frac{1}{m_i^2} F_i^{(N)} P_i. \tag{2.50}$$

However, this choice of p_i cannot be used for integer $n < N$.

(2) Pick, as the origin of P_i vectors, the point of intercept of the height H_0 and the momentum hyperface. This is nothing but a projection of the previous option onto the momentum hyperface. In this case (cf. Eq. (2.44)),

$$p_i^2 = P_i^2 = m_i^2 - m_0^2 = m_i^2 (1 - c_{0i}^2), \tag{2.51}$$

$$(p_j \cdot p_l) = (P_j \cdot P_l) = m_j m_l c_{jl} - m_0^2 = m_j m_l (c_{jl} - c_{0j} c_{0l}), \tag{2.52}$$

and the denominators in the integral (2.1) become

$$[(p_i + q)^2 - m_i^2] \Rightarrow [q^2 + 2(p_i \cdot q) - m_0^2]. \tag{2.53}$$

The projection of Eq. (2.34) onto the momentum hyperface shows that there exists the following relation which is valid for *this* choice of P_i,

$$\sum_{i=1}^N \frac{1}{m_i^2} F_i^{(N)} P_i = 0. \tag{2.54}$$

III. FROM FEYNMAN PARAMETERS TO NON-EUCLIDEAN GEOMETRY

A. Moving from linear to quadratic hypersurface

The standard Feynman parametric representation of a one-loop N-point integral in n dimensions (2.1), corresponding to the diagram presented in Fig. 2, reads

$$J^{(N)}(n;\nu_1,\ldots,\nu_N)=i^{1-n}\pi^{n/2}\,\frac{\Gamma\left(\Sigma\nu_i-\frac{n}{2}\right)}{\Pi\Gamma(\nu_i)}\int_0^1\cdots\int_0^1\frac{\Pi\alpha_i^{\nu_i-1}d\alpha_i\,\delta(\Sigma\alpha_i-1)}{[\sum\sum_{j<l}\alpha_j\alpha_l k_{jl}^2-\Sigma\alpha_i m_i^2]^{\Sigma\nu_i-n/2}}. \tag{3.1}$$

As we have already mentioned (2.3), the integrand can be written in a homogeneous form,

$$J^{(N)}(n;\nu_1,\ldots,\nu_N)=i^{1-2\Sigma\nu_i}\pi^{n/2}\,\frac{\Gamma\left(\Sigma\nu_i-\frac{n}{2}\right)}{\Pi\Gamma(\nu_i)}\int_0^1\cdots\int_0^1\frac{\Pi\alpha_i^{\nu_i-1}d\alpha_i\,\delta(\Sigma\alpha_i-1)}{[\Sigma\alpha_i^2m_i^2+2\sum\sum_{j<l}\alpha_j\alpha_l m_j m_l c_{jl}]^{\Sigma\nu_i-n/2}}, \tag{3.2}$$

where the "cosines" c_{jl} are defined in (2.4). Note that the limits of integration in Eqs. (3.1)–(3.2) can be extended from (0,1) to $(0,\infty)$, since the actual region of integration is defined by the δ function. It corresponds to a part of an $(N-1)$-dimensional hyperplane $\Sigma\alpha_i=1$ which is cut out by the conditions $\alpha_i\geqslant 0$ $(i=1,\ldots,N)$.

Now, let us consider the integral (3.2), and let us use a rescaling which is similar to one used in Refs. 21, 2, 4 (see also in Ref. 25, where the transformations of a general form have been discussed). Let us rescale $\alpha_i=m_i^{-1}\alpha_i'$, so that the δ function becomes

$$\delta\left(\Sigma\,\frac{\alpha_i'}{m_i}-1\right). \tag{3.3}$$

To restore the argument of the δ function in its original form, let us substitute

$$\alpha_i'=\mathscr{F}(\alpha_1'',\ldots,\alpha_N'')\alpha_i'', \quad \text{with } \mathscr{F}(\alpha_1,\ldots,\alpha_N)=\frac{\Sigma\alpha_i}{\Sigma\,\frac{\alpha_i}{m_i}}. \tag{3.4}$$

Note that the Jacobian of this substitution is $\mathscr{F}^N$ (cf. Appendix B).

Suppressing the primes, we arrive at the following representation:

$$J^{(N)}(n;\nu_1,\ldots,\nu_N)=i^{1-2\Sigma\nu_i}\pi^{n/2}\,\frac{\Gamma\left(\Sigma\nu_i-\frac{n}{2}\right)}{\Pi\Gamma(\nu_i)}\,\frac{1}{\Pi m_i^{\nu_i}}\int_0^1\cdots\int_0^1\frac{\Pi\alpha_i^{\nu_i-1}d\alpha_i\,\delta(\Sigma\alpha_i-1)}{\left(\Sigma\,\frac{\alpha_i}{m_i}\right)^{n-\Sigma\nu_i}(\alpha^T\|c\|\alpha)^{\Sigma\nu_i-n/2}}, \tag{3.5}$$

where we use matrix notation (2.13),

$$\alpha^T\|c\|\alpha\equiv\sum_{j=1}^{N}\sum_{l=1}^{N}c_{jl}\alpha_j\alpha_l=\sum\alpha_i^2+2\sum_{j<l}\sum\alpha_j\alpha_l c_{jl}. \tag{3.6}$$

Next, let us change the variables from α to α' via

$$\alpha_i=\mathscr{G}(\alpha_1',\ldots,\alpha_N')\alpha_i', \quad \text{with } \mathscr{G}(\alpha_1',\ldots,\alpha_N')=\frac{\Sigma(\alpha_i')^2+2\sum\sum_{j<l}\alpha_j'\alpha_l'c_{jl}}{\Sigma\alpha_i'}. \tag{3.7}$$

Here, the Jacobian of the substitution is $2\mathcal{G}^N$ (see Appendix B). Effectively, we put the quadratic form into the argument of the δ function. However, the linear denominator (coming from the $\mathcal{F}$ function) survives and we arrive at

$$J^{(N)}(n;\nu_1,...,\nu_N)=2i^{1-2\Sigma\nu_i}\pi^{n/2}\frac{\Gamma\left(\Sigma\nu_i-\frac{n}{2}\right)}{\Pi\Gamma(\nu_i)}\frac{1}{\Pi m_i^{\nu_i}}\int_0^\infty\cdots\int_0^\infty\frac{\Pi\alpha_i^{\nu_i-1}d\alpha_i}{\left(\Sigma\frac{\alpha_i}{m_i}\right)^{n-\Sigma\nu_i}}\delta(\alpha^T\|c\|\alpha-1), \tag{3.8}$$

where we continue to use the matrix notation (3.6).

B. The case $\Sigma\nu_i=n$

In the particular case $\Sigma\nu_i=n$ the linear denominator disappears and Eq. (3.5) gives

$$J^{(N)}(n;\nu_1,...,\nu_N)|_{\Sigma\nu_i=n}=i^{1-2\Sigma\nu_i}\pi^{n/2}\frac{\Gamma\left(\frac{n}{2}\right)}{\Pi\Gamma(\nu_i)}\frac{1}{\Pi m_i^{\nu_i}}\int_0^1\cdots\int_0^1\frac{\Pi\alpha_i^{\nu_i-1}d\alpha_i\delta(\Sigma\alpha_i-1)}{(\alpha^T\|c\|\alpha)^{\Sigma\nu_i-n/2}}. \tag{3.9}$$

Here all dependence on the external momenta and masses in the integral is through c_{jl}, Eq. (2.4).

Since the final integral (3.9) is the same as in the equal-mass case, we can formulate the following statement: Let $\Sigma\nu_i=n$, and let the result for the Feynman integral (3.1) with equal masses ($m_i=m$, $i=1,...,N$) be represented by a dimensionless function Ψ depending on the quantities $\bar{\bar{c}}_{jl}\equiv1-k_{jl}^2/(2m^2)$ (corresponding to Eq. (2.4) in the equal-mass case) as

$$J^{(n)}(n;\nu_1,...,\nu_N)|_{\Sigma\nu_i=n,\ m_i=m}=\frac{1}{m^{\Sigma\nu_i}}\Psi(\{\bar{\bar{c}}_{jl}\}). \tag{3.10}$$

Then the result for the corresponding integral with different masses m_i can be expressed in terms of *the same* function Ψ as

$$J^{(N)}(n;\nu_1,...,\nu_N)|_{\Sigma\nu_i=n}=\frac{1}{\Pi m_i^{\nu_i}}\Psi(\{c_{jl}\}), \tag{3.11}$$

where c_{jl} are defined in Eq. (2.4).

Simplification of the integrals $J^{(N)}(n;\nu_1,...,\nu_N)$ when $\Sigma\nu_i=n$ is not surprising; to some extent, this may be considered a generalization of the so-called "uniqueness" formula for massless triangle diagrams[30] to the case of massive N-point integrals.

Whenever $\Sigma\nu_i=n$ the denominator of the integrand of (3.8) disappears, and the result is

$$J^{(N)}(n;\nu_1,...,\nu_N)|_{\Sigma\nu_i=n}=2i^{1-2\Sigma\nu_i}\pi^{n/2}\frac{\Gamma\left(\frac{n}{2}\right)}{\Pi\Gamma(\nu_i)}\frac{1}{\Pi m_i^{\nu_i}}\int_0^\infty\cdots\int_0^\infty\prod\alpha_i^{\nu_i-1}d\alpha_i\delta(\alpha^T\|c\|\alpha-1). \tag{3.12}$$

In particular, when all $\nu_i=1$ $(n=N)$, our integrand is just a δ function,

$$J^{(N)}(N;1,...,1)=2i^{1-2N}\pi^{N/2}\frac{\Gamma\left(\frac{N}{2}\right)}{\Pi m_i}\int_0^\infty\cdots\int_0^\infty\prod d\alpha_i\delta(\alpha^T\|c\|\alpha-1). \tag{3.13}$$

Consider also a very special situation, when all nondiagonal c_{jl} vanish (i.e., $c_{jl}=\delta_{jl}$). Physically, this corresponds to the case where all external momenta squared take the mean values between the threshold and the pseudo-threshold, $k_{jl}^2=m_j^2+m_l^2$ (for $j<l$). In this case, all the angles τ_{jl} ($j\neq l$) are equal to $\frac{1}{2}\pi$. Employing the representation (3.12) we arrive at a very simple result for this special case,

$$J^{(N)}(n;\nu_1,\ldots,\nu_N)\big|_{\substack{\Sigma\nu_i=n\\ c_{jl}=\delta_{jl}}}=i^{1-2\Sigma\nu_i}\pi^{n/2}\frac{\Pi\Gamma\left(\frac{1}{2}\nu_i\right)}{2^{N-1}\Pi\Gamma(\nu_i)}\frac{1}{\Pi m_i^{\nu_i}}. \tag{3.14}$$

Imagine the N-dimensional Euclidean space of the parameters α_i. The quadratic argument of the δ function in (3.13) defines a $(N-1)$-dimensional hypersurface. Provided that all eigenvalues of this quadratic form are positive, this is an N-dimensional ellipsoid. Otherwise, this is an N-dimensional hyperboloid. So, the integral (3.13) is nothing but the measure (i.e., the content) of the part of this hypersurface which belongs to the region where all α's are non-negative. This region is a 2^Nth part of the whole N-dimensional space.

To calculate the content of this part of the hypersurface, it is reasonable to diagonalize the quadratic form. Suppose there is an N-dimensional rotation transforming the old coordinates α_i into the new ones, β_i, so that

$$\alpha^T\|c\|\alpha=\sum\alpha_i^2+2\sum_{j<l}\sum\alpha_j\alpha_l c_{jl}\Rightarrow\sum\lambda_i\beta_i^2. \tag{3.15}$$

Obviously, the product of λ's must be equal to the determinant $D^{(N)}$ of the matrix (2.13) corresponding to the quadratic form (3.6),

$$\lambda_1\ldots\lambda_N=D^{(N)}. \tag{3.16}$$

Note that this determinant $D^{(N)}$ is related to the content $V^{(N)}$ of the basic N-dimensional simplex via Eq. (2.16).

Now, let us assume that all eigenvalues λ_i are real and positive, i.e., the hypersurface defined by the quadratic form is an N-dimensional ellipsoid. After rotation from α's to β's, we get into the coordinate system corresponding to the principal axes of the ellipsoid. Then, we can rescale

$$\beta_i=\frac{\gamma_i}{\sqrt{\lambda_i}}, \tag{3.17}$$

the Jacobian of this transformation being $(\lambda_1\ldots\lambda_N)^{-1/2}=(D^{(N)})^{-1/2}$, according to (3.16). In this way, our N-dimensional ellipsoid is transformed into a hypersphere. Now all we need to calculate is the content of a part of this hypersphere which is cut out (in the space of γ_i) by the images of the hyperfaces restricting the region where all α_i are positive (in the space of α_i). This content, which we shall denote by $\Omega^{(N)}$, can be understood as the N-dimensional solid angle subtended by the above-mentioned hyperfaces. In terms of $\Omega^{(N)}$, the relevant integral (3.13) can be written as

$$J^{(N)}(N;1,\ldots,1)=i^{1-2N}\pi^{N/2}\frac{\Gamma\left(\frac{N}{2}\right)}{\Pi m_i}\frac{\Omega^{(N)}}{\sqrt{D^{(N)}}}. \tag{3.18}$$

An interesting fact is that $\Omega^{(N)}$ is also related to the basic simplex. To see this, let us note that effectively (in terms of α's), the transformation $\alpha_i\to\beta_i\to\gamma_i$ is equivalent to using the matrix $\|c\|$ as a new metric. Say, for a vector α the *new* length squared is

$$|\alpha|_c^2\equiv\alpha^T\|c\|\alpha. \tag{3.19}$$

For unit vectors $e_i^{(\alpha)}$, the *new* scalar product is

$$(e_j^{(\alpha)}\cdot e_l^{(\alpha)})_c=c_{jl}\equiv\cos\,\tau_{jl}. \tag{3.20}$$

Taking into account that $|e_i^{(\alpha)}|_c^2=1$, we see that in the space with the c-metric (i.e., in the space of γ_i) the angle between $e_j^{(\alpha)}$ and $e_l^{(\alpha)}$ is τ_{jl}. Therefore, we get nothing but the N-dimensional solid angle at the mass meeting point of the basic simplex. It defines the region of integration on the hypersurface of the unit hypersphere in the γ_i coordinate system. We also note that the cosines c_{jl} are invariant under simple rescaling of α's.

It is easy to see that the image of $e_i^{(\alpha)}$ in the γ_i-space is directed along the vector a_i. Moreover, according to Eq. (2.34), the image of the vector with the components $F_i^{(N)}/m_i$ is directed along the height H_0.

To summarize, the following statement is valid: The content of the N-dimensional solid angle $\Omega^{(N)}$ in the space of γ_i is equal to that at the vertex of the basic N-dimensional simplex where all mass sides meet. Moreover, the angles between the corresponding hyperfaces in the space of γ_i and those in the basic simplex are the same. Therefore, the result for the integral (3.13) can be expressed in terms of the content of the basic simplex and the content of its N-dimensional solid angle at the vertex which is common for all mass sides as

$$J^{(N)}(N;1,\ldots,1)=i^{1-2N}\pi^{N/2}\,\frac{\Gamma\left(\frac{N}{2}\right)}{N!}\,\frac{\Omega^{(N)}}{V^{(N)}}. \tag{3.21}$$

Looking at Eq. (3.21), we see that $\Omega^{(N)}$ is indeed the only thing which is to be calculated, since $V^{(N)}$ is known through Eq. (2.16). In the following sections, we shall consider the lowest examples to understand how to calculate $\Omega^{(N)}$.

Nevertheless, the above discussion leads us to the following significant statement: The content $\Omega^{(N)}$ is nothing but the content of a *non-Euclidean* $(N-1)$-dimensional simplex calculated in the spherical (or hyperbolic, depending on the signature of the eigenvalues λ_i) space of constant curvature. The sides of this non-Euclidean simplex are equal to the angles τ_{jl}. Therefore, the problem of calculating Feynman integrals is intimately connected with the problem of calculating the content of a simplex in non-Euclidean geometry.

C. The general case

To understand how to deal with the general case, when $\Sigma\nu_i\neq n$, we need some modification of the above transformations. First of all, we note that Eqs. (2.33) can be rephrased as

$$\sum_{l=1}^{N}\left(\sqrt{F_j^{(N)}}c_{jl}\sqrt{F_l^{(N)}}\right)\frac{\sqrt{F_l^{(N)}}}{m_l}=D^{(N)}\frac{\sqrt{F_j^{(N)}}}{m_j}. \tag{3.22}$$

In other words, the vector with components

$$f_i=\frac{\sqrt{F_i^{(N)}}}{m_i} \tag{3.23}$$

is an eigenvector of the matrix $\|C\|\equiv\|C_{jl}\|$ with the elements

$$C_{jl}=\left(\sqrt{F_j^{(N)}}c_{jl}\sqrt{F_l^{(N)}}\right), \tag{3.24}$$

and the corresponding eigenvalue is equal to $D^{(N)}$. Note that

$$f^2\equiv(f\cdot f)=\sum_{i=1}^{N}\frac{F_i^{(N)}}{m_i^2}=\frac{\Lambda^{(N)}}{\Pi_{i=1}^{N}m_i^2}, \tag{3.25}$$

$$\det\|C\|=D^{(N)}\prod_{j=1}^{N}F_j^{(N)}. \tag{3.26}$$

The idea is that it may be more convenient to use the quadratic form defined by the matrix $\|C_{jl}\|$, rather than $\|c_{jl}\|$, since we know one of its eigenvectors with the corresponding eigenvalue.

After using the masses m_i to rescale the Feynman parameters α_i and ''hiding'' the quadratic form into the argument of the delta function, we obtained the representation (3.8). Analogously, we can use $1/f_i=m_i/\sqrt{F_i^{(N)}}$, rather than m_i, to rescale the original α's. The result is

$$J^{(N)}(n;\nu_1,\ldots,\nu_N)=2i^{1-2\Sigma\nu_i}\pi^{n/2}\frac{\Gamma\left(\Sigma\nu_i-\frac{n}{2}\right)}{\Pi\Gamma(\nu_i)}\left(\prod f_i^{\nu_i}\right)\int_0^\infty\cdots\int_0^\infty\frac{\Pi\alpha_i^{\nu_i-1}d\alpha_i}{(\Sigma\alpha_i f_i)^{n-\Sigma\nu_i}}\,\delta(\alpha^T\|C\|\alpha-1), \tag{3.27}$$

where notation (3.6) has been used again. A nice feature of this representation is that in the denominator we have the contraction (scalar product) of α and the known eigenvector of the matrix $\|C\|$ defining the quadratic form (see Eqs. (3.22)–(3.24)).

If $\nu_1=\cdots=\nu_N=1$, we get

$$J^{(N)}(n;1,\ldots,1)=2i^{1-2N}\pi^{n/2}\Gamma\left(N-\frac{n}{2}\right)\left(\prod f_i\right)\int_0^\infty\cdots\int_0^\infty\frac{\Pi d\alpha_i}{(\Sigma\alpha_i f_i)^{n-N}}\,\delta(\alpha^T\|C\|\alpha-1). \tag{3.28}$$

Now, let us use the same transformations of variables as earlier. Namely, let us first rotate from α_i to the variables β_i such that the quadratic form becomes diagonal,

$$\alpha^T\|C\|\alpha=\sum_{i=1}^N\lambda_i\beta_i^2. \tag{3.29}$$

One of these β's, say β_N, is directed along the eigenvector f. Therefore, $\lambda_N=D^{(N)}$ and $\Pi_{i=1}^{N-1}\lambda_i=\Pi_{j=1}^N F_j^{(N)}$. Moreover, the denominator is also proportional to β_N. Finally, we rescale $\beta_i=\gamma_i/\sqrt{\lambda_i}$, and the denominator becomes

$$\left(\sum\alpha_i f_i\right)\Rightarrow\frac{1}{\Pi m_i}\sqrt{\frac{\Lambda^{(N)}}{D^{(N)}}}\gamma_N=\frac{1}{m_0}\gamma_N, \tag{3.30}$$

where m_0, defined in Eq. (2.21), is the distance between the mass meeting point and the momentum hyperface (i.e., the length of the height H_0).

Now, the transformation $\alpha_i\to\beta_i\to\gamma_i$ is equivalent to using the matrix $\|C\|$ as a new metric. For a vector α the *new* length squared is

$$|\alpha|_C^2\equiv\alpha^T\|C\|\alpha. \tag{3.31}$$

For unit vectors $e_i^{(\alpha)}$, the *new* scalar product is

$$(e_j^{(\alpha)}\cdot e_l^{(\alpha)})_C=C_{jl}, \tag{3.32}$$

and the corresponding cosine of the angle is

$$\frac{(e_j^{(\alpha)}\cdot e_l^{(\alpha)})_C}{|e_j^{(\alpha)}|_C|e_l^{(\alpha)}|_C}=\frac{C_{jl}}{\sqrt{C_{jj}C_{ll}}}=c_{jl}\equiv\cos\tau_{jl}. \tag{3.33}$$

[Note that in the γ_i-space with the C-metric the $e_i^{(\alpha)}$ vectors are not unit, $|e_i^{(\alpha)}|_C^2=C_{ii}=F_i^{(N)}$.] Again, as in the case of the c-metric, we get the same N-dimensional solid angle as that occurring at the mass meeting point of the basic simplex.

But we can go further. In particular, we can comprehend what the image of the vector f is. Consider the scalar products

$$(e_i^{(\alpha)}\cdot f)_C=(e_i^{(\alpha)}\cdot\|C\|f)=D^{(N)}(e_i^{(\alpha)}\cdot f)=D^{(N)}f_i, \tag{3.34}$$

$$(f\cdot f)_C=(f\cdot\|C\|f)=D^{(N)}(f\cdot f)=\frac{D^{(N)}\Lambda^{(N)}}{\Pi_{j=1}^{N}m_j^2}. \tag{3.35}$$

Hence, the cosine of the angle between the images of f and $e_i^{(\alpha)}$ is [cf. Eq. (2.44)]

$$\frac{(e_i^{(\alpha)}\cdot f)_C}{\sqrt{(e_i^{(\alpha)}\cdot e_i^{(\alpha)})_C(f\cdot f)_C}}=\sqrt{\frac{D^{(N)}}{\Lambda^{(N)}}}\prod_{j\neq i}m_j=\frac{m_0}{m_i}=\cos\,\tau_{0i}\equiv c_{0i}\,. \tag{3.36}$$

Therefore, in the basic simplex, with the mass sides m_i, etc., the image of the vector f is directed along the height H_0!

In terms of γ's, the integral becomes

$$J^{(N)}(n;1,...,1)=2i^{1-2N}\pi^{n/2}\Gamma\left(N-\frac{n}{2}\right)\frac{m_0^{n-N-1}}{\sqrt{\Lambda^{(N)}}}\int\limits_{\Omega^{(N)}}\cdots\int\frac{\Pi d\gamma_i}{\gamma_N^{n-N}}\,\delta\left(\sum\gamma_i^2-1\right), \tag{3.37}$$

where the integration goes over the interior of the N-dimensional solid angle $\Omega^{(N)}$ of the basic simplex. If we define the angle between the "running" unit vector and the Nth axis as θ, the above formula means that we should integrate over the hypersurface of the unit hypersphere with the "weight" $(\cos\,\theta)^{N-n}$, within the limits set by $\Omega^{(N)}$,

$$J^{(N)}(n;1,...,1)=i^{1-2N}\pi^{n/2}\Gamma\left(N-\frac{n}{2}\right)\frac{m_0^{n-N}}{\sqrt{D^{(N)}}\Pi m_i}\,\Omega^{(N;n)}, \tag{3.38}$$

with

$$\Omega^{(N;n)}\equiv\int\limits_{\Omega^{(N)}}\cdots\int\frac{d\Omega_N}{\cos^{n-N}\,\theta}. \tag{3.39}$$

Obviously, $\Omega^{(N;N)}=\Omega^{(N)}$. Using Eq. (2.16), the formula (3.38) can be exhibited in the following form:

$$J^{(N)}(n;1,...,1)=i^{1-2N}\pi^{n/2}\Gamma\left(N-\frac{n}{2}\right)\frac{m_0^{n-N}\Omega^{(N;n)}}{N!\,V^{(N)}}. \tag{3.40}$$

D. Splitting the *N*-dimensional solid angle

Now, let us again use the height H_0 to split the basic N-dimensional simplex into N rectangular ones, each time replacing one of the mass sides, m_i, by $m_0=|H_0|$. The content of the ith rectangular simplex $V_i^{(N)}$ is given by Eq. (2.41). Let us denote the corresponding N-dimensional solid angle as $\Omega_i^{(N)}$, so that

$$\Omega_i^{(N;n)}=\int\limits_{\Omega_i^{(N)}}\cdots\int\frac{d\Omega_N}{\cos^{n-N}\,\theta},\qquad \Omega^{(N;n)}=\sum_{i=1}^{N}\Omega_i^{(N;n)}. \tag{3.41}$$

Substituting (3.41) into Eq. (3.40), we get

$$J^{(N)}(n;1,...,1)=\sum_{i=1}^{N}\frac{V_i^{(N)}}{V^{(N)}}\,J_i^{(N)}(n;1,...,1), \tag{3.42}$$

where $J_i^{(N)}$ are the integrals corresponding to the rectangular simplices. Specifically, in the integral $J_i^{(N)}$ the internal masses are

$$m_1,\ldots,m_{i-1},m_0,m_{i+1},\ldots,m_N, \tag{3.43}$$

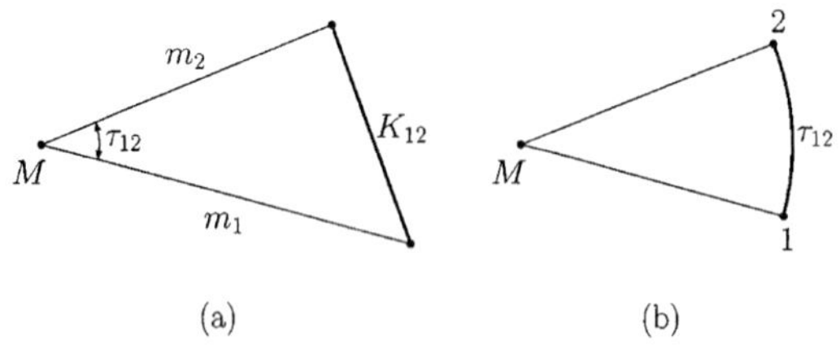

FIG. 4. Two-point case: (a) the basic triangle and (b) the arc τ_{12}.

while the momenta invariants are

$$\begin{cases} k_{jl}^2, & \text{if } j\neq i \text{ and } l\neq i \\ m_l^2-m_0^2, & \text{if } j=i \\ m_j^2-m_0^2, & \text{if } l=i \end{cases} . \tag{3.44}$$

Using (2.41), we arrive at

$$J^{(N)}(n;1,...,1)=\frac{1}{\Lambda^{(N)}}\left(\prod m_i^2\right)\sum_{i=1}^{N}\frac{1}{m_i^2}F_i^{(N)}J_i^{(N)}(n;1,...,1). \tag{3.45}$$

We are now ready to tackle some specific examples using the geometrical approach.

IV. TWO-POINT FUNCTION

Here, the two-dimensional basic simplex is a triangle and the angle τ_{12} lies between the two mass sides (see Fig. 4). We obtain

$$D^{(2)}=1-c_{12}^2=\sin^2\tau_{12}, \quad V^{(2)}=\tfrac{1}{2}m_1m_2\sin\tau_{12}, \quad \Omega^{(2)}=\tau_{12}, \tag{4.1}$$

$$m_0=m_1m_2\sqrt{\frac{D^{(2)}}{\Lambda^{(2)}}}, \quad \cos\tau_{0i}=\frac{m_0}{m_i}, \quad \tau_{01}+\tau_{02}=\tau_{12}, \quad \Lambda^{(2)}=k_{12}^2. \tag{4.2}$$

In two dimensions, from (3.21) we obtain the well-known result

$$J^{(2)}(2;1,1)=\frac{i\pi}{m_1m_2}\frac{\tau_{12}}{\sin\tau_{12}}, \tag{4.3}$$

while in three dimensions, Eqs. (3.40)–(3.41) yield

$$J^{(2)}(3;1,1)=i\pi^2\frac{1}{\sqrt{\Lambda^{(2)}}}\{\Omega_1^{(2;3)}+\Omega_2^{(2;3)}\}, \tag{4.4}$$

with

$$\Omega_i^{(2;3)}=\int_0^{\tau_{0i}}\frac{d\theta}{\cos\theta}=\ln\left(\frac{1+\sin\tau_{0i}}{1-\sin\tau_{0i}}\right). \tag{4.5}$$

Combining the logarithms, we get

$$J^{(2)}(3;1,1)=\frac{i\pi^2}{\sqrt{k_{12}^2}}\ln\left(\frac{m_1+m_2+\sqrt{k_{12}^2}}{m_1+m_2-\sqrt{k_{12}^2}}\right). \tag{4.6}$$

In the Euclidean region ($k_{12}^2<0$) the logarithm gives $\arctan(\sqrt{-k_{12}^2}/(m_1+m_2))$, and the result coincides with those presented in Refs. 11 and 31.

In four dimensions, the Γ function in front of the integral (3.38) becomes singular. Introducing dimensional regularization[32] to circumvent that difficulty, we get

$$J^{(2)}(4-2\varepsilon;1,1)=i\pi^{2-\varepsilon}\Gamma(\varepsilon)\,\frac{m_0^{1-2\varepsilon}}{\sqrt{\Lambda^{(2)}}}\,\{\Omega_1^{(2;4-2\varepsilon)}+\Omega_2^{(2;4-2\varepsilon)}\}, \tag{4.7}$$

with

$$\Omega_i^{(2;4-2\varepsilon)}=\int_0^{\tau_{0i}}\frac{d\theta}{\cos^{2-2\varepsilon}\theta}. \tag{4.8}$$

Expanding the integrand in ε and taking into account that

$$\int_0^{\tau}\frac{d\theta}{\cos^2\theta}\ln(\cos\theta)=\tan\tau\,\ln(\cos\tau)+\tan\tau-\tau, \tag{4.9}$$

we reach the well-known result (cf., e.g., in Refs. 2 and 33),

$$J^{(2)}(4-2\varepsilon;1,1)=i\pi^{2-\varepsilon}\Gamma(1+\varepsilon)\times\left\{\frac{1}{\varepsilon}+2-\ln m_1-\ln m_2+\frac{m_1^2-m_2^2}{k_{12}^2}\ln\frac{m_2}{m_1}-\frac{2m_1m_2}{k_{12}^2}\tau_{12}\sin\tau_{12}\right\}+\mathcal{O}(\varepsilon). \tag{4.10}$$

Furthermore, the representations (4.7)–(4.8) make it possible to construct next terms of the expansion in ε. For example, the ε term requires just

$$\int_0^{\tau}\frac{d\theta}{\cos^2\theta}\ln^2(\cos\theta)=\tan\tau(\ln^2(\cos\tau)+2\ln(\cos\tau)+2)-2\tau(1-\ln 2)-\mathrm{Cl}_2(\pi-2\tau) \tag{4.11}$$

(the corresponding result was obtained in Ref. 34; the Clausen function $\mathrm{Cl}_2(\theta)$ is defined in Appendix A). Moreover, the result for an arbitrary space–time dimension or ε can be obtained in terms of the Gauss hypergeometric function (cf. in Refs. 35 and 33), using

$$\Omega_i^{(2;4-2\varepsilon)}=\int_0^{\tau_{0i}}\frac{d\theta}{\cos^{2-2\varepsilon}\theta}=2\tan\tau_{0i}\;{}_2F_1\left(\left.\begin{matrix}1/2,\varepsilon\\3/2\end{matrix}\right|-\tan^2\tau_{0i}\right). \tag{4.12}$$

Using formulas of analytic continuation of the ${}_2F_1$ function, one can establish a connection with the result presented in Eq. (A7) of Ref. 33.

V. THREE-POINT FUNCTION

A. Geometrical picture and the three-dimensional case

Here the three-dimensional basic simplex is a tetrahedron with three mass sides (the angles between these mass sides are τ_{12}, τ_{13}, and τ_{23}) and three momentum sides. It is shown in Fig. 5(a). The volume of this tetrahedron is defined as

$$V^{(3)}=\tfrac{1}{6}m_1m_2m_3\sqrt{D^{(3)}}, \tag{5.1}$$

where (cf. Eqs. (2.13)–(2.14))

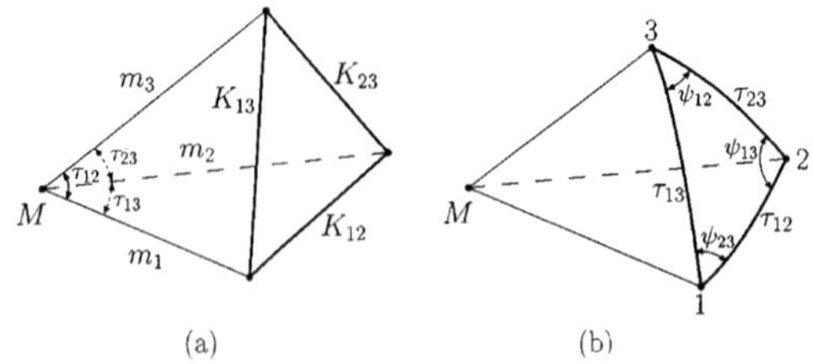

FIG. 5. Three-point case: (a) the basic tetrahedron and (b) the solid angle.

$$D^{(3)} \equiv \begin{vmatrix} 1 & c_{12} & c_{13} \\ c_{12} & 1 & c_{23} \\ c_{13} & c_{23} & 1 \end{vmatrix} = 1 - c_{12}^2 - c_{13}^2 - c_{23}^2 + 2c_{12}c_{13}c_{23}. \tag{5.2}$$

Furthermore, $\Omega^{(3)}$ is the usual solid angle at the vertex derived by the mass sides [cf. Fig. 5(b)]. Its value can be defined as the area of a part of the unit sphere cut out by the three planar faces adjacent to the vertex; in other words, this is the area of a spherical triangle corresponding to this section. The sides of this spherical triangle are obviously equal to the angles τ_{12}, τ_{13}, and τ_{23} while its angles, ψ_{12}, ψ_{13}, and ψ_{23}, are equal to those between the plane faces (we define ψ_{12} as the angle between the sides τ_{13} and τ_{23}, etc.). This definition is in agreement with Eq. (2.27). Using Eq. (2.24) or the well-known formulas of the spherical trigonometry, it is easy to show that

$$\cos \psi_{12} = \frac{\cos \tau_{12} - \cos \tau_{13} \cos \tau_{23}}{\sin \tau_{13} \sin \tau_{23}}, \quad \sin \psi_{12} = \frac{\sqrt{D^{(3)}}}{\sin \tau_{13} \sin \tau_{23}}, \tag{5.3}$$

and analogous expressions for ψ_{13} and ψ_{23}. The solid angle corresponding to spherical triangle can be obtained via

$$\Omega^{(3)} = \psi_{12} + \psi_{13} + \psi_{23} - \pi. \tag{5.4}$$

Using Eqs. (5.3) and (5.4), it is straightforward to show that

$$\begin{aligned} \Omega^{(3)} &= 2 \arccos\left(\frac{1 + c_{12} + c_{13} + c_{23}}{\sqrt{2(1+c_{12})(1+c_{13})(1+c_{23})}} \right) \\ &= 2 \arcsin\left(\sqrt{\frac{D^{(3)}}{2(1+c_{12})(1+c_{13})(1+c_{23})}} \right) \\ &= 2 \arctan\left(\frac{\sqrt{D^{(3)}}}{1 + c_{12} + c_{13} + c_{23}} \right). \end{aligned} \tag{5.5}$$

Finally, the result

$$J^{(3)}(3;1,1,1) = -\frac{i\pi^2}{2m_1 m_2 m_3} \frac{\Omega^{(3)}}{\sqrt{D^{(3)}}}, \tag{5.6}$$

with $\Omega^{(3)}$ defined by (5.5), corresponds to one obtained in Ref. 11 in a different way. Note that here we have derived it purely geometrically.

However, Eq. (5.4) cannot be easily generalized to the four-dimensional case. This is why it is instructive to reproduce the three-dimensional result (5.5) in a different manner, splitting the basic tetrahedron into rectangular ones (cf. Eq. (3.42)).

Consider a spherical triangle 123 on the unit sphere, cut out by the solid angle of the basic tetrahedron in the γ space, as illustrated in Fig. 6. The points 1,2,3 correspond to the intersections of the mass sides of the basic tetrahedron and the unit sphere. The sides of the triangle are equal

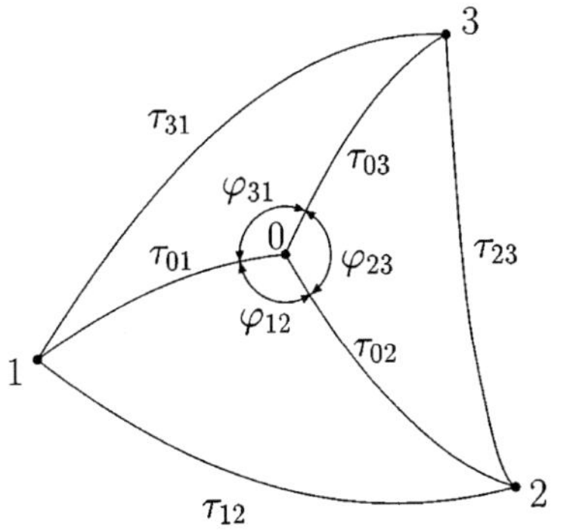

FIG. 6. The spherical triangle 123.

to the angles τ_{12}, τ_{23}, and τ_{31}. There is another point 0, also on the unit sphere, corresponding to the intersection of the height H_0 of the tetrahedron and the unit sphere. The point 0 is connected with each of the points 1,2,3. The corresponding sides 01, 02, and 03 are equal to τ_{01}, τ_{02}, and τ_{03}, respectively, so that

$$\cos\, \tau_{0i} = \frac{m_0}{m_i}, \quad m_0 = m_1 m_2 m_3 \sqrt{\frac{D^{(3)}}{\Lambda^{(3)}}}, \tag{5.7}$$

with (cf. Eq. (2.19))

$$\Lambda^{(3)} = -\tfrac{1}{4}[(k_{12}^2)^2 + (k_{13}^2)^2 + (k_{23}^2)^2 - 2k_{12}^2 k_{13}^2 - 2k_{12}^2 k_{23}^2 - 2k_{13}^2 k_{23}^2] = -\tfrac{1}{4}\lambda(k_{12}^2, k_{13}^2, k_{23}^2), \tag{5.8}$$

where $\lambda(x,y,z)$ is defined in Eq. (1.1) (cf. Fig. 1). The triangle 123 is thereby split into three triangles 012, 023, and 031. The angle between 01 and 02 is denoted as φ_{12}, etc. Obviously,

$$\varphi_{12} + \varphi_{23} + \varphi_{31} = 2\pi. \tag{5.9}$$

For the calculation of three-dimensional triangle integral, it does not really matter where the point 0 lies, though this becomes essential for the four-dimensional case. As we know, the $N=n$ integrals can be reduced to the integrals with equal masses. So, let us start with the equal mass case. In Fig. 6, this means that $\tau_{01} = \tau_{02} = \tau_{03} \equiv \tau_0$. Therefore, the spherical triangles 012, 023, and 031 are isosceles. Consider one of these triangles, say 012, as illustrated in Fig. 7. The sides 01 and 02 are equal to τ_0. The third side, 12, is equal to τ_{12}. The length of the perpendicular dropped from the point 0 to the side 12 is denoted as η_{12}. This perpendicular intersects the side 12 in a point T_{12} which in the equal-mass case divides the side 12 into two equal parts (each of them equals $\tau_{12}/2$). It also divides the angle φ_{12} into two equal parts. The angles at the vertices 1 and 2 (which are also equal in the equal-mass case) are denoted as κ_{12}. The lengths of two of the sides of a smaller isosceles triangle $01_\xi 2_\xi$, 01_ξ and 02_ξ, are $\tau_0 \xi$ $(0 \leq \xi \leq 1)$. The third side, $1_\xi 2_\xi$, is parallel (in the spherical sense) to the side 12 and its length is denoted as $\tau(\xi)$; obviously, $\tau(0) = 0$ and $\tau(1) = \tau_{12}$. The perpendicular $0T_{12}$ also splits it into two equal parts of the length $\tau(\xi)/2$. The length of the part of the perpendicular within the triangle $01_\xi 2_\xi$ is denoted as $\eta(\xi)$, whereas the angles at the vertices 1_ξ and 2_ξ are $\kappa(\xi)$; obviously, $\eta(0)=0$, $\eta(1)=\eta_{12}$, $\kappa(0)=(\pi - \varphi_{12})/2$, $\kappa(1)=\kappa_{12}$.

The main idea of introducing an auxiliary variable ξ is to simplify the limits of integration. We define the usual spherical integration angle θ in a way that it is the angle between the direction of H_0 and the "running" point. For the triangle 012 presented in Fig. 7, let us define that $\varphi=0$ corresponds to the direction of the perpendicular $0T_{12}$. In this way, the limits of the φ integration are $\pm\varphi_{12}/2$. The lower limit of the θ integration is zero, whereas the upper limit depends on φ and varies from τ_0 (at $\varphi = \pm\varphi_{12}/2$) to η_{12} (at $\varphi=0$). If we use, instead of θ, the variable ξ, its limits will be $0 \leq \xi \leq 1$, independently of φ. The result of an infinitesimal variation of ξ will be a thin strip based on the side $1_\xi 2_\xi$ (cf. Fig. 7). Moreover, all such strips from the triangles 012, 023 and

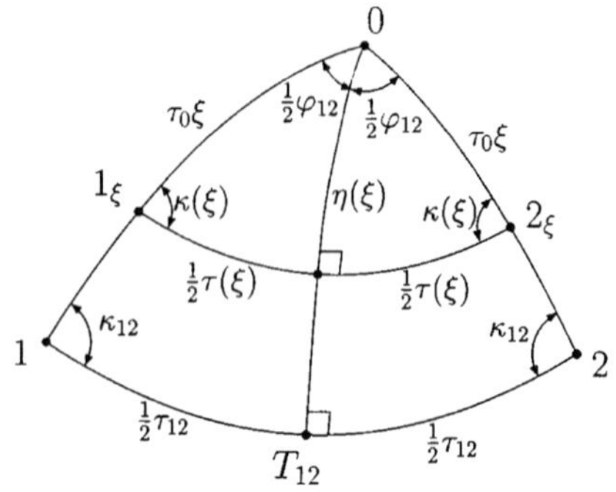

FIG. 7. The isosceles spherical triangle 012.

031 (cf. Fig. 6) can be ''glued'' together. So, the whole result of a variation in ξ is a thin closed spherical triangle. Alternatively, one can perform the azimuthal φ-integration first, and then evaluate the remaining ξ integral. As we shall see, both ways work in the three-point case.

Using formulas of spherical trigonometry (cf. Fig. 7), we get the following relations:

$$\sin(\tfrac{1}{2}\tau(\xi)) = \sin(\tfrac{1}{2}\varphi_{12})\sin(\tau_0\xi),$$

$$\sin\,\eta(\xi) = \sin\,\kappa(\xi)\sin(\tau_0\xi),$$

$$\cos\,\kappa(\xi) = \sin(\tfrac{1}{2}\varphi_{12})\cos\,\eta(\xi),$$

$$\tan\,\kappa(\xi) = (\cos(\tau_0\xi)\tan(\tfrac{1}{2}\varphi_{12}))^{-1},$$

$$\tan\,\eta(\xi) = \cos(\tfrac{1}{2}\varphi_{12})\tan(\tau_0\xi).$$

The situation for arbitrary φ is drawn in Fig. 8. This is a rectangular spherical triangle with an upper vertex 0 and one of the outgoing sides being the perpendicular of the length $\eta(\xi)$, as in Fig. 7. This side corresponds to $\varphi=0$. The angle at the vertex 0 is equal to φ and a running value of φ is understood. The side opposite to 0 is perpendicular to the one mentioned before, $\eta(\xi)$. This is a part of the corresponding side directed to the point 1_ξ in Fig. 7, i.e. the line of constant ξ. Its length is denoted as $\rho(\xi,\varphi)$. The third side or ''hypotenuse'' has length $\theta(\xi,\varphi)$. The vertex opposite to the $\eta(\xi)$ side is the running point with the coordinates θ, φ. The angle at this vertex is denoted as $\kappa(\xi,\varphi)$.

We want to express θ in terms of ξ and φ. Using the formulas of spherical trigonometry for the triangle in Fig. 8, analogous to ones written above, it is straightforward to show that

$$\tan\,\theta(\xi,\varphi) = \frac{\tan\,\eta(\xi)}{\cos\,\varphi} = \frac{\cos(\tfrac{1}{2}\varphi_{12})}{\cos\,\varphi}\tan(\tau_0\xi). \tag{5.10}$$

The area is obtained by integrating

$$\sin\,\theta\;d\theta\;d\varphi = -d(\cos\,\theta)d\varphi \Rightarrow -\frac{\partial(\cos\,\theta(\xi,\varphi))}{\partial\xi}\,d\xi\;d\varphi, \tag{5.11}$$

where, according to Eq. (5.10),

$$\cos\,\theta(\xi,\varphi) = \left(1 + \frac{\cos^2(\tfrac{1}{2}\varphi_{12})}{\cos^2\,\varphi}\tan^2(\tau_0\xi)\right)^{-1/2}. \tag{5.12}$$

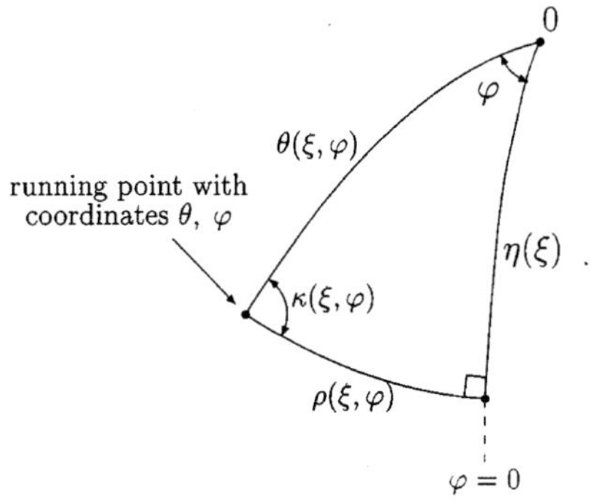

FIG. 8. Integration variables on the sphere.

If we first perform the φ-integral, this effectively corresponds to the area of the infinitesimal strip between two sides of constant ξ,

$$-2d\{\arctan(\cos(\tau_0\xi)\tan(\tfrac{1}{2}\varphi_{12}))\}. \tag{5.13}$$

Then, the remaining ξ integral gives just this arctan. Collecting the results for all triangles (cf. Fig. 6), we get the same as in Eq. (5.5). This can be seen if we observe that the expression (5.13) is nothing but $2d\kappa(\xi)$ and the spherical excess for the triangle $01_\xi 2_\xi$ is $(\varphi_{12}+2\kappa(\xi)-\pi)$. A nice thing about Eq. (5.13) is that in the ξ-integral we can glue all arctan functions into a symmetric answer at the *infinitesimal* level. The same final result can also be obtained when we first integrate over ξ, it being trivial since the integrand is a derivative with respect to ξ.

B. The four-dimensional case

If we consider the four-dimensional three-point function, the only (but very essential!) difference is that we should divide the integrand by $\cos\,\theta(\xi,\varphi)$, cf. Eq. (3.37). Therefore, the function in the integrand can be written as

$$-\frac{\partial}{\partial\xi}\ln(\cos\,\theta(\xi,\varphi)). \tag{5.14}$$

As in the previous case, one can integrate first over either φ or ξ; both ways lead to equivalent results. Let us integrate first over ξ, this gives

$$\ln\left(\frac{\cos\,\theta(0,\varphi)}{\cos\,\theta(1,\varphi)}\right)=\tfrac{1}{2}\ln\left(1+\frac{\tan^2\,\eta_{12}}{\cos^2\,\varphi}\right). \tag{5.15}$$

In the equal-mass case, the remaining φ-integral becomes

$$\int_0^{\varphi_{12}/2}d\varphi\,\ln\left(1+\frac{\tan^2\,\eta_{12}}{\cos^2\,\varphi}\right)=\frac{1}{2}\int_0^{\varphi_{12}}d\varphi'\left\{\ln(1+2\tan^2\,\eta_{12})+\ln\left(1+\frac{\cos\,\varphi'}{1+2\tan^2\,\eta_{12}}\right)\right.$$
$$\left.-\ln(1+\cos\,\varphi')\right\}, \tag{5.16}$$

where $\varphi'=2\varphi$. Using a result from Ref. 27 (p. 308, Eq. (38)), this integral can be expressed in terms of the Clausen function (see in Appendix A) as

$$\tfrac{1}{2}\tau_{12}\ln\left(\frac{\sin(\frac{1}{2}(\varphi_{12}+\tau_{12}))}{\sin(\frac{1}{2}(\varphi_{12}-\tau_{12}))}\right)+\tfrac{1}{2}\mathrm{Cl}_2(\varphi_{12}+\tau_{12})+\tfrac{1}{2}\mathrm{Cl}_2(\varphi_{12}-\tau_{12})-\mathrm{Cl}_2(\varphi_{12}). \tag{5.17}$$

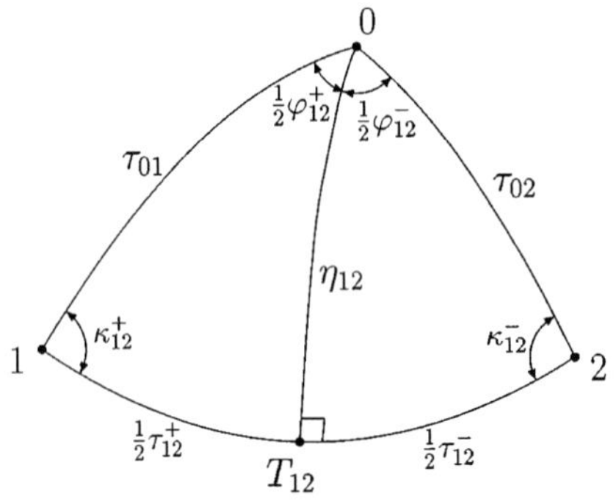

FIG. 9. An asymmetric spherical triangle 012.

The arguments of the Clausen functions have a very transparent geometric interpretation. We note that the logarithmic term can also be presented as

$$\tfrac{1}{2}\tau_{12} \ln\left(\frac{1+\sin\,\eta_{12}}{1-\sin\,\eta_{12}}\right). \tag{5.18}$$

The function (5.17) can also be expressed in terms of the generalized inverse tangent integral (A8) as

$$\mathrm{Ti}_2(\tan(\tfrac{1}{2}\tau_{12}),\tan(\tfrac{1}{2}\varphi_{12})) - \mathrm{Ti}_2(\tan(\tfrac{1}{2}\tau_{12}),-\tan(\tfrac{1}{2}\varphi_{12})). \tag{5.19}$$

The generalization to the case of unequal masses is easy. To understand why, let us consider Fig. 9. This is again a close-up of the triangle 012 from Fig. 6, but now it is asymmetric, because τ_{01} and τ_{02} are different. We denote the quantities in one of the triangles as $\varphi_{12}^{+}/2$, $\tau_{12}^{+}/2$, $\kappa_{12}^{+}/2$, and in another triangle as $\varphi_{12}^{-}/2$, $\tau_{12}^{-}/2$, $\kappa_{12}^{-}/2$. The height η_{12} is the same for both triangles. Obviously,

$$\tfrac{1}{2}(\varphi_{12}^{+}+\varphi_{12}^{-})=\varphi_{12} \ \text{and} \ \tfrac{1}{2}(\tau_{12}^{+}+\tau_{12}^{-})=\tau_{12}. \tag{5.20}$$

One can immediately see that the result for the integral over this asymmetric triangle is nothing but a half of the sum of the functions (5.17) labeled with plus and with minus. Namely,

$$\tfrac{1}{2}\tau_{12} \ln\left(\frac{1+\sin\,\eta_{12}}{1-\sin\,\eta_{12}}\right) + \tfrac{1}{4}\mathrm{Cl}_2(\varphi_{12}^{+}+\tau_{12}^{+}) + \tfrac{1}{4}\mathrm{Cl}_2(\varphi_{12}^{+}-\tau_{12}^{+}) - \tfrac{1}{2}\mathrm{Cl}_2(\varphi_{12}^{+})$$
$$+ \tfrac{1}{4}\mathrm{Cl}_2(\varphi_{12}^{-}+\tau_{12}^{-}) + \tfrac{1}{4}\mathrm{Cl}_2(\varphi_{12}^{-}-\tau_{12}^{-}) - \tfrac{1}{2}\mathrm{Cl}_2(\varphi_{12}^{-}). \tag{5.21}$$

Again, the geometric interpretation of the arguments of the Clausen functions is very transparent, thanks to Fig. 9.

Some useful relations are

$$\cos(\tfrac{1}{2}\tau_{12}^{+}) = \frac{\cos\,\tau_{01}}{\cos\,\eta_{12}}, \quad \cos(\tfrac{1}{2}\tau_{12}^{-}) = \frac{\cos\,\tau_{02}}{\cos\,\eta_{12}},$$

$$\sin(\tfrac{1}{2}\tau_{12}^{+}) = \sin\,\tau_{01}\,\sin(\tfrac{1}{2}\varphi_{12}^{+}), \quad \sin(\tfrac{1}{2}\tau_{12}^{-}) = \sin\,\tau_{02}\,\sin(\tfrac{1}{2}\varphi_{12}^{-}),$$

$$\tan(\tfrac{1}{2}\tau_{12}^{+}) = \sin\,\eta_{12}\,\tan(\tfrac{1}{2}\varphi_{12}^{+}), \quad \tan(\tfrac{1}{2}\tau_{12}^{-}) = \sin\,\eta_{12}\,\tan(\tfrac{1}{2}\varphi_{12}^{-}).$$

Worth noting is

$$\cos\,\eta_{12}=\frac{m_0\sqrt{k_{12}^2}}{m_1m_2\,\sin\,\tau_{12}}. \tag{5.22}$$

To get the complete result for the three-point integral in four dimensions, we should sum up the functions (5.21) for all three triangles 012, 023, and 031, and multiply by the overall factor (see Eq. (3.37)) which is

$$-i\pi^2\,\frac{1}{\sqrt{\Lambda^{(3)}}}. \tag{5.23}$$

Note that in this case we get $\sqrt{\Lambda^{(3)}}$ rather than $\sqrt{D^{(3)}}$ in the denominator.

VI. FOUR-POINT FUNCTION

A. The general case

In this case, the four-dimensional simplex has four mass sides and six momentum sides (see Fig. 3). It has five vertices and five three-dimensional hyperfaces. Four of these hyperfaces are the *reduced* ones, corresponding to three-point functions, whereas the fifth one is the momentum hyperface. In fact, this four-dimensional simplex is completely defined by its mass sides m_1, m_2, m_3, m_4 and six "planar" angles between them, τ_{12}, τ_{13}, τ_{14}, τ_{23}, τ_{24}, and τ_{34}. According to Eq. (2.16), the content (hypervolume) of this simplex equals

$$V^{(4)}=\frac{1}{24}\,m_1m_2m_3m_4\sqrt{D^{(4)}}, \tag{6.1}$$

with $D^{(4)}$ given by (2.13)–(2.14) at $N=4$,

$$D^{(4)}=\det\|c_{jl}\|,\quad \|c_{jl}\|=\begin{pmatrix} 1 & c_{12} & c_{13} & c_{14}\\ c_{12} & 1 & c_{23} & c_{24}\\ c_{13} & c_{23} & 1 & c_{34}\\ c_{14} & c_{24} & c_{34} & 1\end{pmatrix}. \tag{6.2}$$

The four-dimensional four-point function can be exhibited as [cf. Eqs. (3.18) and (3.21)]

$$J^{(4)}(4;1,1,1,1)=\frac{1}{12}\,i\pi^2\,\frac{\Omega^{(4)}}{V^{(4)}}=\frac{2i\pi^2}{m_1m_2m_3m_4}\,\frac{\Omega^{(4)}}{\sqrt{D^{(4)}}}. \tag{6.3}$$

So, the main problem is how to calculate $\Omega^{(4)}$.

In four dimensions, $\Omega^{(4)}$ is the value of the four-dimensional generalization of the solid angle at the vertex of the simplex where all four mass sides meet. In the spherical case, it can be defined as the volume of a part of the unit hypersphere which is cut out from it by the four three-dimensional reduced hyperfaces, each hyperface involving three mass sides of the simplex. This hypersection is a three-dimensional spherical tetrahedron whose six sides (edges) are equal to the angles τ_{jl}. [In the hyperbolic case, this is a hyperbolic tetrahedron whose volume can be obtained by analytic continuation (see below).]

It is illustrated in Fig. 10 where the ith vertex ($i=1,2,3,4$) corresponds to the intersection of the ith mass side of the basic simplex and the unit hypersphere. [It should be stressed that Fig. 10 is just an illustration (rather than a precise picture) since a realistic non-Euclidean tetrahedron should be understood as embedded into four-dimensional Euclidean space.] Furthermore, the dihedral angles of this *three*-dimensional spherical tetrahedron coincide with those of the *four*-dimensional basic simplex, ψ_{jl} (cf. Eqs. (2.24), (2.27)). The dual matrix $\|\widetilde{c}_{jl}\|$ as well as its determinant $\widetilde{D}^{(4)}$, cf. Eq. (2.26),

4322 J. Math. Phys., Vol. 39, No. 9, September 1998 A. I. Davydychev and R. Delbourgo

$$\|\tilde{c}_{jl}\| = \begin{pmatrix} 1 & \tilde{c}_{12} & \tilde{c}_{13} & \tilde{c}_{14} \\ \tilde{c}_{12} & 1 & \tilde{c}_{23} & \tilde{c}_{24} \\ \tilde{c}_{13} & \tilde{c}_{23} & 1 & \tilde{c}_{34} \\ \tilde{c}_{14} & \tilde{c}_{24} & \tilde{c}_{34} & 1 \end{pmatrix}, \quad \tilde{c}_{jl} = -\cos \psi_{jl}, \quad \tilde{D}^{(4)} = \det\|\tilde{c}_{jl}\|, \tag{6.4}$$

are usually referred to as the Gram matrix and determinant of the spherical or hyperbolic tetrahedron. For example, the dihedral angle at the edge 12 of the spherical tetrahedron in Fig. 10 is ψ_{34}, etc. (The digits labeling the edge and the dihedral angle at this edge should cover the whole set 1234.) We note that the dihedral angle is well defined in spherical and hyperbolic spaces, i.e. it remains unchanged along the given edge.

Unfortunately, there are no *simple* relations like (5.4) which might make it possible to express the volume of a spherical (or hyperbolic) tetrahedron in terms of its sides or dihedral angles. In fact, calculation of this volume in an elliptic or hyperbolic space is a well-known problem of non-Euclidean geometry (see e.g. Refs. 36–38). A standard way to solve this problem, say in spherical space, is to split an arbitrary tetrahedron into a set of birectangular ones. For example, the tetrahedron 1234 (shown in Fig. 10) is called *birectangular* (or *double-rectangular*) if (i) the edge 12 is perpendicular to the face 234 and (ii) the opposite edge, 34, is perpendicular to the face 123. It is easy to check that in this case three dihedral angles, ψ_{13}, ψ_{14}, and ψ_{24}, are right angles. The other three are usually denoted[39,40] as

$$\psi_{12} = \alpha, \quad \psi_{23} = \beta, \quad \psi_{34} = \gamma. \tag{6.5}$$

The volume of a birectangular tetrahedron is known (see below) and can be expressed in terms of Lobachevsky or Schläfli functions which can be related to dilogarithms (see Refs. 39 and 40).

The volume of the general tetrahedron must be a symmetrical function of the six edges τ_{jl}, or equivalently of the six dihedral angles ψ_{jl}. When we break it up into a sum of birectangular tetrahedra, this explicit symmetry may be lost, although of course it must be hidden in the properties of the sums of dilogarithmic functions.

The general way to split an arbitrary tetrahedron into a sum of birectangular ones is to fix a point 0, say inside the tetrahedron. [If this point (or the feet of some of the perpendiculars) happen to be outside the tetrahedron, this would mean that some of the volumes of the resulting birectangular tetrahedra should be taken with an opposite sign.] Then, connecting this point 0 with each of the vertices of the tetrahedron 1234, we split it into four smaller tetrahedra: 0123, 0124, 0134, and 0234. Consider one of these tetrahedra, say 0123. Dropping the perpendicular from 0 onto the face 123 and connecting the foot of this perpendicular F_{123} with the vertices 1, 2, and 3, we split 0123 into three *rectangular* tetrahedra, $012F_{123}$, $013F_{123}$, and $023F_{123}$. Then, each of these

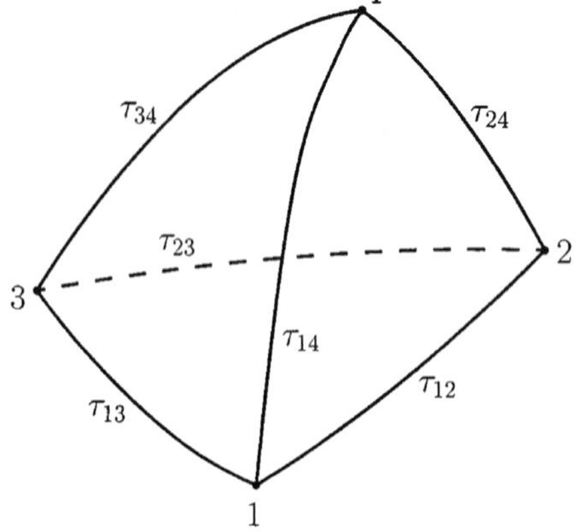

FIG. 10. The spherical tetrahedron 1234.

rectangular tetrahedra (say, $012F_{123}$) can be split into a sum of two *birectangular* tetrahedra by dropping a perpendicular from F_{123} onto the only side belonging to the original tetrahedron 1234 (the edge 12). In this way, we get $4\times3\times2=24$ birectangular tetrahedra.

Since we are free to choose the position of the point 0, there are (at least) two ways to reduce the number of terms involved in this volume splitting in the general case:

(i) The point 0 may coincide with one of the vertices of the original tetrahedron, say with the vertex 4 [cf. Fig. 11(a)]. In this case, we do not need to split into four smaller tetrahedra and can start just by dropping a perpendicular from the point 4 onto the face 123. In this way, we reduce the number of birectangular tetrahedra involved to six. However, the price we pay for this is the loss of explicit symmetry.

(ii) A symmetric choice of the point 0 can be related to the structure of the basic four-dimensional Euclidean simplex. In particular, we can take as the point 0 the intersection of the height H_0 and the unit hypersphere. Moreover, since the general four-point function can be reduced to the equal-mass case (see Eqs. (3.10)–(3.11)), it is enough to consider the case when the point 0 is equidistant from the vertices $1,2,3$, and 4. In this case, at the first stage of splitting we get four "*isosceles*" tetrahedra, signifying for each of them the three sides meeting at the point 0 are equal. Continuing the splitting, we see that due to "isoscelesness" the pairs of the tetrahedra produced at the last stage have equal values. Effectively, we get the sum of $4\times3=12$ volumes of birectangular tetrahedra. Although this is twice as much as in the case (i), the advantage is that we are able to preserve the explicit symmetry at all stages.

We note that the possibility to rescale the Feynman parameters in the case $n=N$ is closely connected with the freedom to choose the point 0 as we like. [See Eqs. (3.27)–(3.28). For $n=N$, the denominator disappears. In general, the f_i can take arbitrary positive values.] As we have seen, this rescaling preserves the angles τ_{jl} between the mass vectors, whereas the direction of the height H_0 varies. In particular, we can make it coincide with one of the mass vectors, which corresponds to the splitting (i) considered above. From this point of view, we lose the explicit symmetry since the corresponding rescaling is not symmetric.

Another possibility to reduce the number of birectangular volumes involved is to try to choose the point 0 in such a way that the volumes of the four tetrahedra (produced at the first stage of splitting) are equal (or proportional) to each other. In this case, the further splitting would give just $3\times2=6$ birectangular terms, i.e., the same number as in the case (i); but here we may preserve the explicit symmetry. However, it seems to be difficult (if possible at all) to define the position of the point 0 in this case, because one needs to solve a set of transcendental equations.

In some special examples described below we show that an additional symmetry of the diagram can also be employed.

B. Volume of birectangular tetrahedra in curved space

As we have seen (cf. Eq. (6.5)), the birectangular tetrahedron may be specified by three dihedral angles $\psi_{12}=\alpha$, $\psi_{23}=\beta$, $\psi_{34}=\gamma$, so that the corresponding Gram matrix $\|\tilde{c}_{jl}\|$ and its determinant (cf. Eq. (20) of Ref. 40) are

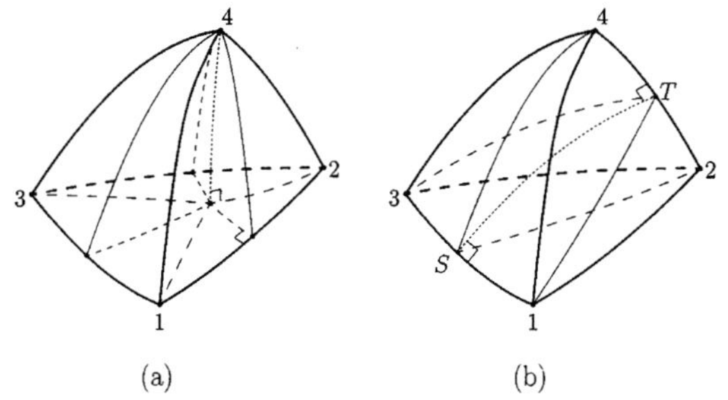

FIG. 11. Different ways of splitting the non-Euclidean tetrahedron.

$$\|\tilde{c}_{jl}\| = \begin{pmatrix} 1 & -\cos\alpha & 0 & 0 \\ -\cos\alpha & 1 & -\cos\beta & 0 \\ 0 & -\cos\beta & 1 & -\cos\gamma \\ 0 & 0 & -\cos\gamma & 1 \end{pmatrix}, \quad \tilde{D}^{(4)} = \sin^2\alpha\,\sin^2\gamma - \cos^2\beta. \tag{6.6}$$

The volume of this birectangular tetrahedron (i.e., its contribution to $\Omega^{(4)}$) can be presented as

$$\Omega^{(4)} = V(\alpha,\beta,\gamma) = \tfrac{1}{4}S(\pi/2-\alpha,\beta,\pi/2-\gamma). \tag{6.7}$$

The notation $S(\alpha,\beta,\gamma)$ corresponds to the Schläfli function as defined in Ref. 39, whereas the notation $V(\alpha,\beta,\gamma)$ was used in Ref. 40 and is proportional to the function originally used by Schläfli.[37] Note that function (6.7) is symmetric with respect to α and γ.

If we define (cf. Eq. (3.21) of Ref. 39)

$$X \equiv \frac{\cos\alpha\,\cos\gamma - \sqrt{\tilde{D}^{(4)}}}{\cos\alpha\,\cos\gamma + \sqrt{\tilde{D}^{(4)}}}, \tag{6.8}$$

the result for the volume can be expressed in terms of dilogarithms,[39]

$$\begin{aligned} 4V(\alpha,\beta,\gamma) &= \tfrac{1}{2}[\mathrm{Li}_2(Xe^{2i\alpha}) + \mathrm{Li}_2(Xe^{-2i\alpha})] + \tfrac{1}{2}[\mathrm{Li}_2(Xe^{2i\gamma}) + \mathrm{Li}_2(Xe^{-2i\gamma})] \\ &\quad - \tfrac{1}{2}[\mathrm{Li}_2(-Xe^{2i\beta}) + \mathrm{Li}_2(-Xe^{-2i\beta})] - \mathrm{Li}_2(-X) - \left(\frac{\pi}{2}-\alpha\right)^2 - \left(\frac{\pi}{2}-\gamma\right)^2 + \beta^2. \end{aligned} \tag{6.9}$$

When the condition $\tilde{D}^{(4)} \geqslant 0$ ($\sin\alpha\,\sin\gamma \leqslant \cos\beta$) is obeyed, the tetrahedron exists in the usual geometrical sense, and its volume is real. In this case, one can express $V(\alpha,\beta,\gamma)$ in terms of the real function $\mathrm{Li}_2(r,\theta) \equiv \mathrm{Re}[\mathrm{Li}_2(re^{i\theta})]$ (see Appendix A). However, one should distinguish between the cases $X \geqslant 0$ (or $\cos^2\alpha + \cos^2\beta + \cos^2\gamma \geqslant 1$) and $X \leqslant 0$ (or $\cos^2\alpha + \cos^2\beta + \cos^2\gamma \leqslant 1$). The corresponding expressions are

$$\begin{aligned} 4V(\alpha,\beta,\gamma)|_{X\geqslant 0} &= \mathrm{Li}_2(X,2\alpha) + \mathrm{Li}_2(X,2\gamma) - \mathrm{Li}_2(X,\pi-2\beta) - \mathrm{Li}_2(-X) \\ &\quad - \left(\frac{\pi}{2}-\alpha\right)^2 - \left(\frac{\pi}{2}-\gamma\right)^2 + \beta^2, \end{aligned} \tag{6.10}$$

$$\begin{aligned} 4V(\alpha,\beta,\gamma)|_{X\leqslant 0} &= \mathrm{Li}_2(-X,\pi-2\alpha) + \mathrm{Li}_2(-X,\pi-2\gamma) - \mathrm{Li}_2(-X,2\beta) - \mathrm{Li}_2(-X) \\ &\quad - \left(\frac{\pi}{2}-\alpha\right)^2 - \left(\frac{\pi}{2}-\gamma\right)^2 + \beta^2. \end{aligned} \tag{6.11}$$

In particular, when $X=-1$ we take into account that $\mathrm{Li}_2(1,\theta) = \frac{1}{4}(\pi-\theta)^2 - \frac{1}{12}\pi^2$ and get

$$V(\alpha,\beta,\gamma)|_{X=-1} = \tfrac{1}{4}\pi(\alpha+\beta+\gamma-\pi). \tag{6.12}$$

When $\alpha=\beta=\gamma=\frac{1}{2}\pi$, this gives $\frac{1}{8}\pi^2$, which agrees with Eq. (3.14) (at $N=4$, $\nu_i=1$).

When $\tilde{D}^{(4)} < 0$ ($\sin\alpha\,\sin\gamma < \cos\beta$), the tetrahedron does not exist in the usual geometrical sense (in the spherical case), but its volume (more precisely, the analytic continuation of the function corresponding to the volume) can be obtained from Eq. (6.9), and it is imaginary. In this case we define (according to Refs. 39, 40)

$$X = e^{-2i\delta}, \quad \text{so that} \quad \tan\delta = \frac{\sqrt{-\tilde{D}^{(4)}}}{\cos\alpha\,\cos\gamma}, \tag{6.13}$$

and we get

$$V(\alpha,\beta,\gamma)=(1/(8i))[\mathrm{Cl}_2(2\alpha+2\delta)-\mathrm{Cl}_2(2\alpha-2\delta)+\mathrm{Cl}_2(2\gamma+2\delta)$$
$$-\mathrm{Cl}_2(2\gamma-2\delta)-\mathrm{Cl}_2(\pi-2\beta+2\delta)+\mathrm{Cl}_2(\pi-2\beta-2\delta)+2\mathrm{Cl}_2(\pi-2\delta)]. \quad (6.14)$$

Using the connection between the Clausen function and the Lobachevsky function $L(\theta)$ (cf. Eq. (A10) in Appendix A), Eq. (6.14) can be rewritten in terms of the Lobachevsky function. [The function used in Ref. 40 which was denoted by a Cyrillic "L" can be expressed in terms of the Clausen function as $\frac{1}{2}\mathrm{Cl}_2(2\theta)$.] The same expression (6.14) gives the volume of a birectangular tetrahedron in the hyperbolic space (see Refs. 39 and 40).

C. Some cases with an additional symmetry

There are a few cases of particular practical interest because of their inherent higher symmetries. In particular, there is the most symmetrical case where all $c_{ij}=c$. Thus all dihedral angles of the full (regular) tetrahedron are equal, to ψ say. In that case the six birectangular tetrahedra corresponding to splitting (i) are all of equal size and $\Omega^{(4)}$ reduces to

$$\Omega^{(4)}=6V(\pi/3,\psi/2,\psi)=\tfrac{3}{2}S(\pi/6,\psi/2,\pi/2-\psi). \quad (6.15)$$

Coxeter gives the identities (Eqs. (4.31)–(4.32) of Ref. 39)

$$S(\pi/6,\psi/2,\pi/2-\psi)=4S(\pi/6,\pi/3,(\pi-\psi)/2)=\tfrac{2}{3}S((\pi-\psi)/2,\psi,(\pi-\psi)/2), \quad (6.16)$$

which he derived algebraically, and which can be used to convert $\Omega^{(4)}$ given by Eq. (6.7) into other forms. We will presently explain the geometrical meaning of these identities.

Another case of interest corresponds to equal mass scattering with equal internal masses; this means $c_{12}=c_{23}=c_{34}=c_{14}$ and in general are not equal to c_{13},c_{24}, which are associated with energy and scattering angle variables. This case has four of the tetrahedral sides equal and different from the other two (which are opposite to one another). Thus there are three distinct c's. We shall see that it is possible here to break up full tetrahedron into four birectangular ones by joining the midpoints of the unequal sides to render the problem more tractable and make the maximum use of what symmetry exists.

We shall carry out our analysis by relaxing the symmetry relations. First we consider the case where all six c_{ij} are equal to c and treat this in three different ways. Then we let $c_{13}\equiv c_s$ and $c_{24}\equiv c_t$ be unequal.

1. Completely symmetrical box diagram

Here all c_{jl} are equal and we are dealing with a completely regular tetrahedron. For example, in the equal-mass case this would mean that all k_{jl}^2 are equal (of course this represents an unphysical point in the Mandelstam diagram), to k^2 say, and all $c=1-k^2/(2m^2)$. We have

$$\|c_{jl}\|=\begin{pmatrix}1 & c & c & c\\ c & 1 & c & c\\ c & c & 1 & c\\ c & c & c & 1\end{pmatrix},\quad \|\tilde{c}_{jl}\|=\begin{pmatrix}1 & \tilde{c} & \tilde{c} & \tilde{c}\\ \tilde{c} & 1 & \tilde{c} & \tilde{c}\\ \tilde{c} & \tilde{c} & 1 & \tilde{c}\\ \tilde{c} & \tilde{c} & \tilde{c} & 1\end{pmatrix},\quad \tilde{c}=-\frac{c}{1+2c}. \quad (6.17)$$

This means that the dihedral angle ψ between each and every pair of hyperplanes equals $\arccos[c/(1+2c)]$. {The generalization of this totally symmetric case to an N-point function is quite easy and leads to $\cos\psi=c/[1+(N-2)c]$.}

We can evaluate the hypervolume in at least three ways (the first two correspond to the ways (i) and (ii) considered in the general case):

(i) By dropping a perpendicular from one corner onto the opposite hyperplane and another perpendicular onto one of the opposite sides, we subdivide the full tetrahedron into six equal birectangular tetrahedra [see Fig. 11(a)]. By inspection, each birectangular tetrahedron has dihedral angles ψ, $\psi/2$, and $\pi/3$. Therefore the total hypervolume equals

$$\Omega^{(4)}=6V(\psi,\psi/2,\pi/3)=\tfrac{3}{2}S(\pi/6,\psi/2,\pi/2-\psi). \tag{6.18}$$

(ii) We can alternatively consider the total volume as four times the volume contained in the tetrahedron generated by one base triangle and the "center of mass" point 0: $P_0=\frac{1}{4}(P_1+P_2+P_3+P_4)$ (here, we use the notation $P_i=m_i a_i$, cf. Eq. (2.49)). In this case the matrix associated with the tetrahedron 0123 is found to be

$$\|c_{jl}\|=\begin{pmatrix} 1 & c' & c' & c' \\ c' & 1 & c & c \\ c' & c & 1 & c \\ c' & c & c & 1 \end{pmatrix}, \quad c'=\frac{\sqrt{1+3c}}{2}. \tag{6.19}$$

leading to

$$\|\tilde{c}_{jl}\|=\begin{pmatrix} 1 & \tilde{c}' & \tilde{c}' & \tilde{c}' \\ \tilde{c}' & 1 & 1/2 & 1/2 \\ \tilde{c}' & 1/2 & 1 & 1/2 \\ \tilde{c}' & 1/2 & 1/2 & 1 \end{pmatrix}, \quad \tilde{c}'=-\sqrt{\frac{1+3c}{2(1+2c)}}. \tag{6.20}$$

This implies that there are two distinct dihedral angles,

$$\tfrac{2}{3}\pi \text{ and } \arccos\sqrt{\frac{1+3c}{2(1+2c)}}=\tfrac{1}{2}\arccos\left(\frac{c}{1+2c}\right)=\frac{\psi}{2} \tag{6.21}$$

in fact. Actually the result is hardly surprising since it can be deduced by simple inspection of the geometrical figure. Anyhow if we now divide 0123 into six identical birectangular tetrahedra by dropping a perpendicular from 0 onto the center of the triangle 123, these smaller tetrahedra possess dihedral angles $\psi/2$, $\pi/3$, $\pi/3$. Consequently we have an alternative geometrical expression for the full volume,

$$\Omega^{(4)}=24V(\pi/3,\pi/3,\psi/2)=6S(\pi/6,\pi/3,(\pi-\psi)/2). \tag{6.22}$$

Comparing this with the previous answer we have a geometrical explanation of one of the Coxeter identities (6.16). Furthermore, because (see Ref. 39, p. 18) there exists the differential relation

$$dS(\pi/6,\pi/3,\gamma)=-2\arccos\left(\frac{\cos\gamma}{\sqrt{4\cos^2\gamma-1}}\right)d\gamma, \tag{6.23}$$

we get the following integral representation of the four-dimensional solid angle (6.22):

$$\Omega^{(4)}=\pi^2-12\int_0^{(\pi-\psi)/2} d\gamma\,\arccos\left(\frac{\cos\gamma}{\sqrt{4\cos^2\gamma-1}}\right). \tag{6.24}$$

(iii) We take the two midpoints S and T of the opposite sides: $P_S=(P_1+P_3)/2$, $P_T=(P_2+P_4)/2$ [see Fig. 11(b)]. The full volume is then just four times that of the tetrahedron $12ST$, which is already birectangular as can be seen by inspection. In this case the matrix $\|c_{jl}\|$ associated with the tetrahedron $12ST$ has the following nondiagonal elements:

$$c_{12}=c, \quad c_{13}=c_{24}=\sqrt{\frac{1+c}{2}}, \quad c_{14}=c_{23}=c\sqrt{\frac{2}{1+c}}, \quad c_{34}=\frac{2c}{1+c}. \tag{6.25}$$

The corresponding Gram matrix of the tetrahedron $12ST$ is

$$\|\widetilde{c}_{jl}\| = \begin{pmatrix} 1 & 0 & \widetilde{c}' & 0 \\ 0 & 1 & 0 & \widetilde{c}' \\ \widetilde{c}' & 0 & 1 & \widetilde{c} \\ 0 & \widetilde{c}' & \widetilde{c} & 1 \end{pmatrix}, \tag{6.26}$$

with $\widetilde{c}$ and $\widetilde{c}'$ defined in Eqs. (6.17) and (6.20), respectively. From this we extract the three dihedral angles.

$$\alpha = \gamma = \tfrac{1}{2}\psi \quad \text{and} \quad \beta = \psi, \tag{6.27}$$

where ψ is nothing but the dihedral angle at the side 12, i.e., one of the dihedral angles of the full tetrahedron. This accords completely with the shape of $12ST$. Hence the total volume of the complete tetrahedron 1234 is given by

$$\Omega^{(4)} = 4V(\psi/2, \psi, \psi/2) = S((\pi-\psi)/2, \psi, (\pi-\psi)/2). \tag{6.28}$$

Comparing with the earlier expression, this provides a geometric meaning of the second Coxeter identity (6.16).

2. Partially symmetric box diagram

In this case we shall take four sides of the tetrahedron to be equal and two opposite sides to be unequal to them. In the equal-mass case, this would correspond to setting $k_{12}^2 = k_{23}^2 = k_{34}^2 = k_{14}^2 = k^2$, $k_{13}^2 = s$, $k_{24}^2 = t$, and we would get three different c's: $c = 1 - k^2/2m^2$, $c_s = 1 - s/2m^2$, $c_t = 1 - t/2m^2$ (s and t are just the Mandelstam variables). In this case much the simplest way of evaluating the volume of the tetrahedron is to break it up into four equal birectangular tetrahedra by taking the midpoints of the sides 13 and 24, as before; $P_S = (P_1 + P_3)/2$, $P_T = (P_2 + P_4)/2$. There is enough symmetry in the problem that the full tetrahedron is four times the birectangular tetrahedron $12ST$. The latter is described by the matrix $\|c_{jl}\|$ with the nondiagonal elements,

$$c_{12} = c, \quad c_{13} = \sqrt{\frac{1+c_s}{2}}, \quad c_{24} = \sqrt{\frac{1+c_t}{2}},$$

$$c_{23} = c\sqrt{\frac{2}{1+c_s}}, \quad c_{14} = c\sqrt{\frac{2}{1+c_t}}, \quad c_{34} = \frac{2c}{\sqrt{(1+c_s)(1+c_t)}}. \tag{6.29}$$

The nondiagonal elements of the "dual" matrix $\|\widetilde{c}_{jl}\|$ are

$$\widetilde{c}_{13} = -\sqrt{\frac{(1+c_s)(1+c_t)-4c^2}{2(1+c_t-2c^2)}}, \quad \widetilde{c}_{24} = -\sqrt{\frac{(1+c_s)(1+c_t)-4c^2}{2(1+c_s-2c^2)}},$$

$$\widetilde{c}_{34} = -c\sqrt{\frac{(1-c_s)(1-c_t)}{(1+c_s-2c^2)(1+c_t-2c^2)}}, \quad \widetilde{c}_{12} = \widetilde{c}_{14} = \widetilde{c}_{23} = 0. \tag{6.30}$$

We can now read off the three nontrivial dihedral angles,

$$\alpha = \arccos\left(\sqrt{\frac{(1+c_s)(1+c_t)-4c^2}{2(1+c_t-2c^2)}}\right), \quad \gamma = \arccos\left(\sqrt{\frac{(1+c_s)(1+c_t)-4c^2}{2(1+c_s-2c^2)}}\right),$$

$$\beta = \arccos\left(c\sqrt{\frac{(1-c_s)(1-c_t)}{(1+c_s-2c^2)(1+c_t-2c^2)}}\right). \tag{6.31}$$

Finally then the required volume of the full tetrahedron 1234 is obtained from

$$\Omega^{(4)}=4V(\alpha,\beta,\gamma)=S(\pi/2-\alpha,\beta,\pi/2-\gamma). \tag{6.32}$$

A well-known physical example of partially symmetric box diagram is related to the photon–photon scattering.[41] In this case, $k^2=0$, $c=1$, $\cos\beta=1$, $\beta=0$. Furthermore, $\widetilde{D}^{(4)}=\sin^2\alpha\sin^2\gamma-1$ and we should use formula (6.14) for $V(\alpha,0,\gamma)$.

D. The massless limit

In the case when all masses m_i vanish, the quantities c_{jl} become infinite (cf. Eq. (2.4)) and should be be considered as hyperbolic cosines. The determinant $D^{(4)}$, Eq. (6.2), has the following limit:

$$\begin{aligned}(m_1^2m_2^2m_3^2m_4^2D^{(4)})|_{m_i\to 0}\Rightarrow &\tfrac{1}{16}[(k_{12}^2k_{34}^2)^2+(k_{13}^2k_{24}^2)^2+(k_{14}^2k_{23}^2)^2\\ &-2k_{12}^2k_{34}^2k_{13}^2k_{24}^2-2k_{12}^2k_{34}^2k_{14}^2k_{23}^2-2k_{13}^2k_{24}^2k_{14}^2k_{23}^2]\\ &=\tfrac{1}{16}\lambda(k_{12}^2k_{34}^2,k_{13}^2k_{24}^2,k_{14}^2k_{23}^2)\end{aligned} \tag{6.33}$$

with $\lambda(x,y,z)$ defined by Eq. (1.1).

The elements of the dual matrix $\|\widetilde{c}_{jl}\|$ still can be interpreted as cosines of the dihedral angles, namely,

$$\widetilde{c}_{12}=\widetilde{c}_{34}=-\cos\psi_{34}=-\cos\psi_{12}=\frac{k_{13}^2k_{24}^2+k_{14}^2k_{23}^2-k_{12}^2k_{34}^2}{\sqrt{k_{13}^2k_{24}^2k_{14}^2k_{23}^2}}, \tag{6.34}$$

$$\widetilde{c}_{13}=\widetilde{c}_{24}=-\cos\psi_{24}=-\cos\psi_{13}=\frac{k_{14}^2k_{23}^2+k_{12}^2k_{34}^2-k_{13}^2k_{24}^2}{\sqrt{k_{14}^2k_{23}^2k_{12}^2k_{34}^2}}, \tag{6.35}$$

$$\widetilde{c}_{14}=\widetilde{c}_{23}=-\cos\psi_{23}=-\cos\psi_{14}=\frac{k_{12}^2k_{34}^2+k_{13}^2k_{24}^2-k_{14}^2k_{23}^2}{\sqrt{k_{12}^2k_{34}^2k_{13}^2k_{24}^2}}. \tag{6.36}$$

Therefore, in this situation we get nothing but an *ideal* hyperbolic tetrahedron (see pp. 39–40 of Ref. 40), i.e., the tetrahedron whose vertices are all at infinity. The pairs of opposite dihedral angles of the ideal tetrahedron are equal, $\psi_{12}=\psi_{34}$, $\psi_{13}=\psi_{24}$, $\psi_{14}=\psi_{23}$, whereas its volume is given by

$$\Omega^{(4)}=\frac{1}{2i}[\mathrm{Cl}_2(2\psi_{12})+\mathrm{Cl}_2(2\psi_{13})+\mathrm{Cl}_2(2\psi_{23})],\quad \psi_{12}+\psi_{13}+\psi_{23}=\pi, \tag{6.37}$$

according to Eq. (41) of Ref. 40 (see also Ref. 42).

First of all we see that Eqs. (6.33)–(6.36) depend only on the products $k_{12}^2k_{34}^2$, $k_{13}^2k_{24}^2$, and $k_{14}^2k_{23}^2$. This corresponds to the "glueing" of arguments in the massless case observed in Ref. 43. Moreover, the result (6.37) is of the same form as the massless three-point function in four dimensions (1.2). This accords completely with the reduction of the massless four-point function to the three-point function which has been proven in Ref. 43.

VII. REDUCTION OF INTEGRALS WITH $N>n$ AND RELATED PROBLEMS

A. Integrals with $N>n$

Consider Eq. (3.45) in the case when $n=N-1$,

$$J^{(N)}(N-1;1,...,1)=\frac{1}{\Lambda^{(N)}}\left(\prod m_i^2\right)\sum_{i=1}^{N}\frac{1}{m_i^2}F_i^{(N)}J_i^{(N)}(N-1;1,...,1). \tag{7.1}$$

The integrals on the right-hand side, $J_i^{(N)}(N-1;1,...,1)$, correspond to splitting the basic N-dimensional simplex into N rectangular ones. The ith rectangular simplex is obtained via substituting the ith mass side of the basic simplex by its height H_0 whose length is m_0. Therefore, the corresponding elements of the $\|c\|$ matrix (2.13) should be replaced by (cf. Eqs. (3.43)–(3.44))

$$c_{ji} \to \cos\,\tau_{0j} = \frac{m_0}{m_j}, \quad c_{il} \to \cos\,\tau_{0l} = \frac{m_0}{m_l}. \tag{7.2}$$

At this point let us consider what the representation (3.8) gives for $J_i^{(N)}$. The α-integrand becomes

$$\left(\sum_{l \neq i} \frac{\alpha_l}{m_l} + \frac{\alpha_i}{m_0} \right) \delta\left((\alpha^T \|c\| \alpha) \Big|_{\alpha_i=0} + 2 m_0 \alpha_i \sum_{l \neq i} \frac{\alpha_l}{m_l} + \alpha_i^2 - 1 \right). \tag{7.3}$$

One can see that the first factor in (7.3) is proportional to the derivative of the argument of the δ function with respect to α_i. Therefore, Eq. (7.3) can be expressed as

$$\frac{1}{2m_0} \frac{\partial}{\partial \alpha_i} \theta\left((\alpha^T \|c\| \alpha) \Big|_{\alpha_i=0} + 2 m_0 \alpha_i \sum_{l \neq i} \frac{\alpha_l}{m_l} + \alpha_i^2 - 1 \right). \tag{7.4}$$

Integrating over α_i, we get

$$J_i^{(N)}(N-1;1,...,1) = \frac{i^{1-2N} \pi^{(N-1)/2}}{m_0^2 \Pi_{l \neq i} m_l} \Gamma\left(\frac{N+1}{2} \right) \int_0^\infty \cdots \int_0^\infty \prod_{l \neq i} d\alpha_l \theta(1 - (\alpha^T \|c\| \alpha)|_{\alpha_i=0}). \tag{7.5}$$

Inserting

$$\frac{1}{N-1} \sum_{l \neq i} \frac{\partial \alpha_l}{\partial \alpha_l} = 1 \tag{7.6}$$

and integrating by parts, the integrand of (7.5) can be transformed as

$$\begin{aligned} \theta(1-(\alpha^T\|c\|\alpha)|_{\alpha_i=0}) &\Rightarrow -\frac{1}{N-1} \left(\sum_{l \neq i} \alpha_l \frac{\partial}{\partial \alpha_l} \right) \theta(1-(\alpha^T\|c\|\alpha)|_{\alpha_i=0}) \\ &\Rightarrow \frac{1}{N-1} \delta(1-(\alpha^T\|c\|\alpha)|_{\alpha_i=0}) \sum_{l \neq i} \alpha_l \frac{\partial}{\partial \alpha_l} (\alpha^T\|c\|\alpha)|_{\alpha_i=0} \\ &\Rightarrow \frac{2}{N-1} \delta((\alpha^T\|c\|\alpha)|_{\alpha_i=0} - 1). \end{aligned} \tag{7.7}$$

Using Eq. (3.8), we arrive at

$$J_i^{(N)}(N-1;1,...,1) = -\frac{1}{2m_0^2} J^{(N-1)}(N-1;1,...,1)|_{\text{without } i}. \tag{7.8}$$

Therefore, we get

$$J^{(N)}(N-1;1,...,1) = -\frac{1}{2D^{(N)}} \sum_{i=1}^{N} \frac{1}{m_i^2} F_i^{(N)} J^{(N-1)}(N-1;1,...,1)|_{\text{without } i}. \tag{7.9}$$

This corresponds to Eq. (47) of Ref. 11, when the number of external legs N is equal to the space–time dimension plus one. For $N=3,4,5$, Eq. (7.9) reproduces Eqs. (10), (8), (13) from Ref. 11, respectively. (For $N=3$, the explicit result was presented in Ref. 10.)

Similar reduction formulas have also been considered in Refs. 8–13. In Ref. 14 it was shown how the linear dependence of external momenta in the case $N \geq 6$ and for $n=4$ may be employed to reduce the N-point function to a set of $(N-1)$-point ones. A method for expressing the five-point function in four dimensions in terms of a sum of four-point functions was first shown in Ref. 7, whereas the reduction of the three-point function in two dimensions was studied in Ref. 10 (see also Ref. 18). These results were generalized in Refs. 8 and 9. For example, in Ref. 8 (see also Ref. 10) the representation for the coefficients corresponding to $F_l^{(N)}$ (Eq. (7) of Ref. 8) is given in terms of the propagators whose momenta are solutions to a certain set of equations corresponding to the other propagators. Moreover, it is possible to make a direct as opposed to step-by-step reduction to n-point functions for any $N>n$. More recently, some other approaches to this problem were suggested in Refs. 12 and 13.

It is also instructive to consider another derivation of Eq. (7.9). The representation (3.8) yields

$$J^{(N)}(N-1;1,...,1)=\frac{2i^{1-2N}\pi^{(N-1)/2}}{\Pi m_l}\Gamma\left(\frac{N+1}{2}\right)\int_0^\infty\cdots\int_0^\infty\Pi\, d\alpha_l\left(\sum_{i=1}^{N}\frac{\alpha_i}{m_i}\right)\delta(\alpha^T\|c\|\alpha-1). \tag{7.10}$$

Using Eq. (2.33) we get

$$\sum_{i=1}^{N}\frac{\alpha_i}{m_i}=\frac{1}{D^{(N)}}\sum_{i=1}^{N}\alpha_i\sum_{l=1}^{N}\frac{c_{il}}{m_l}F_l^{(N)}=\frac{1}{D^{(N)}}\sum_{i=1}^{N}\frac{F_i^{(N)}}{m_i}\sum_{l=1}^{N}c_{il}\alpha_l. \tag{7.11}$$

Repeating steps (7.3)–(7.7) for the resulting α-integrals, we arrive at the following result:

$$\frac{2i^{1-2N}\pi^{(N-1)/2}}{\Pi m_l}\Gamma\left(\frac{N+1}{2}\right)\int_0^\infty\cdots\int_0^\infty\Pi\, d\alpha_l\left(\sum_{l=1}^{N}c_{il}\alpha_l\right)\delta(\alpha^T\|c\|\alpha-1)$$
$$=-\frac{1}{2m_i}J^{(N-1)}(N-1;1,...,1)|_{\text{without } i}. \tag{7.12}$$

Substituting (7.11) and (7.12) into (7.10), we obtain once again Eq. (7.9).

A similar approach can also be used to obtain formulas connecting the integrals with different values of the space–time dimension.[13,44,45]

B. Integrals with higher powers of propagators

Equations like (7.12) can also be used for calculating integrals with higher powers of propagators. Such integrals can be associated with derivatives with respect to m_i^2, provided that m_i and k_{jl}^2 are considered as the set of independent variables. Specifically, consider the sum

$$c_{1i}m_1J^{(N)}(N+1;2,1,1,...,1)+c_{2i}m_2J^{(N)}(N+1;1,2,1,...,1)$$
$$+\cdots+c_{iN}m_NJ^{(N)}(N+1;1,...,1,2)\equiv\sum_{l=1}^{N}c_{il}m_lJ^{(N)}(N+1;\{1+\delta_{lj}\}). \tag{7.13}$$

(The notation for the set of arguments $\{\nu_j\}=\{1+\delta_{lj}\}$ means that all powers of denominators are equal to one, except for the ν_l which is equal to two. As usually, it is implied that $c_{ji}=c_{ij}$ and $c_{ii}=1$.) Employing the representation (3.8) for all integrals in (7.13), we obtain the same integral as in Eq. (7.12), multiplied by $(-\pi)$. Therefore, we get a system of equations which can be displayed in the matrix form,

$$\|c_{jl}\|\begin{pmatrix} m_1J^{(N)}(N+1;2,1,1,...,1)\\ m_2J^{(N)}(N+1;1,2,1,...,1)\\ \cdots\\ m_NJ^{(N)}(N+1;1,1,...,1,2)\end{pmatrix}=\frac{\pi}{2}\begin{pmatrix} m_1^{-1}J^{(N-1)}(N-1;1,...,1)|_{\text{without } 1}\\ m_2^{-1}J^{(N-1)}(N-1;1,...,1)|_{\text{without } 2}\\ \cdots\\ m_N^{-1}J^{(N-1)}(N-1;1,...,1)|_{\text{without } N}\end{pmatrix}. \tag{7.14}$$

The matrix inverse to $\|c_{jl}\|$ is given by Eq. (2.25). Applying $\|c_{jl}\|^{-1}$ to both sides, we get

$$\begin{pmatrix} (m_1/\sqrt{D_{11}^{(N-1)}})J^{(N)}(N+1;2,1,1,...,1) \\ (m_2/\sqrt{D_{22}^{(N-1)}})J^{(N)}(N+1;1,2,1,...,1) \\ \cdots \\ (m_N/\sqrt{D_{NN}^{(N-1)}})J^{(N)}(N+1;1,1,...,1,2) \end{pmatrix}$$

$$= \frac{\pi}{2D^{(N)}} \|\tilde{c}_{jl}\| \begin{pmatrix} m_1^{-1}\sqrt{D_{11}^{(N-1)}}J^{(N-1)}(N-1;1,...,1)|_{\text{without } 1} \\ m_2^{-1}\sqrt{D_{22}^{(N-1)}}J^{(N-1)}(N-1;1,...,1)|_{\text{without } 2} \\ \cdots \\ m_N^{-1}\sqrt{D_{NN}^{(N-1)}}J^{(N-1)}(N-1;1,...,1)|_{\text{without } N} \end{pmatrix}. \tag{7.15}$$

(Similar results can also be obtained using the integration-by-parts approach.[46,23])

Using the expression (2.24) for $\tilde{c}_{jN}$, we arrive at

$$J^{(N)}(N+1;\{1+\delta_{il}\}) = \frac{\pi}{2m_l D^{(N)}} \sum_{j=1}^{N} \frac{(-1)^{j+l}}{m_j} D_{jl}^{(N-1)} J^{(N-1)}(N-1;1,...,1)|_{\text{without } j}. \tag{7.16}$$

In particular, in the three-point four-dimensional case we end up with

$$J^{(3)}(4;1,1,2) = \frac{i\pi^2 \sin \tau_{12}}{2m_1 m_2 m_3^2 D^{(3)}} [\tau_{23} \cos \psi_{13} + \tau_{13} \cos \psi_{23} + \tau_{12}]. \tag{7.17}$$

The results for $J^{(3)}(4;1,2,1)$ and $J^{(3)}(4;2,1,1)$ can be obtained by permutation of indices. The sum of these three integrals is in agreement with the result obtained in Ref. 18.

It is easy to check the consistency of the results (7.16) and (7.9). Using the standard Feynman parametric representation (or the representation (3.8) involving the δ function of the quadratic form) it is straightforward to show that

$$\sum_{j=1}^{N} J^{(N)}(N+1;\{1+\delta_{ij}\}) = -\pi J^{(N)}(N-1;1,...,1). \tag{7.18}$$

[The generalization of this formula to the case of arbitrary n and the powers of the propagators ν_i can be found, e.g., in Ref. 47, Eq. (6).] Considering the sum of the integrals (7.16) and employing the relation (7.18), we arrive at Eq. (7.9).

VIII. CONCLUSION

Let us briefly summarize the main results of the present paper. We have proven that there is a direct link between Feynman parametric representation of a one-loop N-point function and the basic simplex in N-dimensional Euclidean space. To show this, we have changed the surface of integration in the Feynman parametric representation, from a linear one ($\Sigma \alpha_i = 1$) to a quadratic one ($\alpha^T \|c\| \alpha = 1$). The integral representation involving the δ function of this quadratic form is given by Eq. (3.8).

In the case $N=n$, the result for the Feynman integral turns out to be proportional to the ratio of an N-dimensional solid angle at the meeting point of the mass sides to the content of the N-dimensional basic simplex. (Moreover, in the case $N=n$ the result for the diagram with unequal masses can be expressed in terms of the function corresponding to the equal-mass case.) For example, to reproduce the result for the three-dimensional three-point function[11] we just require the expression for the area of a spherical triangle which can easily be calculated as the sum of the angles minus π (in general, an analytic continuation of the area function should be taken).

For the four-dimensional four-point function, the representation (3.8) provides a very interesting connection with the volume of the non-Euclidean (spherical or hyperbolic) tetrahedron. By splitting the non-Euclidean tetrahedron into birectangular ones, the latter can be expressed in terms of dilogarithms or related functions. (A compact integral representation for this volume can be found in Ref. 48.) Different ways of splitting correspond to different representations of the result for the four-point function. This ''freedom'' can be used to construct more compact representations (cf. Refs. 3, 2, 4).

In other cases, such as, for example, the four-dimensional three-point function, we get an extra factor in the integrand corresponding to a power of the cosine of the angle between the height of the basic simplex and the vector of integration. In general, the height of the basic simplex, H_0, plays an essential role in calculation of the integrals with $N \neq n$. It is used to split the basic Euclidean simplex into N rectangular simplices. When $N=n+1$, each integral corresponding to one of the resulting rectangular tetrahedra can be reduced to an $(N-1)$-point function. When $N<n$, this splitting also simplifies the calculation of separate integrals.

The derivations in this paper have been based on a purely geometrical approach to evaluation of N-point Feynman diagrams. In the resulting expressions, all arguments of functions arising possess a straightforward geometrical meaning in terms of the dihedral angles, etc. In particular, this is quite useful for choosing the most convenient kinematic variables to describe the N-point diagrams. We suggest that this approach can serve as a powerful tool for understanding the geometrical structure of loop integrals with several external legs, as well as the structure of the corresponding phase-space integrals (see, e.g., Ref. 49). More optimistically, it may shed light on analytical results for higher loops.

Note added in proof: After our manuscript was accepted for publication Ref. 50 appeared, which deals with nonrelativistic Feynman integrals. It cited Ref. 51 where the relativistic one-loop Feynman amplitudes (in Euclidean quantum field theory) were studied and expressed in terms of definite integrals over simplices. We also note that some representations for the Schläfli function (similar to those discussed in Sec. VI B) can be found in Ref. 52.

ACKNOWLEDGMENTS

One of the authors (A.I.D.) is grateful to the Australian Research Council for a grant which was used to support his visit to Hobart in 1996 (where an essential part of this work has been done), and to the Department of Physics, University of Tasmania for hospitality during the visit. A.I.D.'s research was partly supported by Grant Nos. INTAS–93-0744 and RFBR–96-01-00654.

APPENDIX A: DILOGARITHM AND RELATED FUNCTIONS

Here we present definitions and some representations of the dilogarithm and related functions. More detailed information can be found in Ref. 27.

The Euler's dilogarithm is defined by

$$\mathrm{Li}_2(z) \equiv -z \int_0^1 d\xi \, \frac{\ln \xi}{1-\xi z}. \tag{A1}$$

The Clausen function is related to the imaginary part of the dilogarithm,

$$\mathrm{Cl}_2(\theta) = \mathrm{Im}[\mathrm{Li}_2(e^{i\theta})] = -\int_0^\theta d\theta' \, \ln|2 \sin(\tfrac{1}{2}\theta')|, \tag{A2}$$

and it can also be represented as

$$\mathrm{Cl}_2(\theta) = -\sin\theta \int_0^1 \frac{d\xi \, \ln \xi}{1-2\xi \cos\theta + \xi^2}. \tag{A3}$$

Note that $\mathrm{Cl}_2(2\pi-\theta) = -\mathrm{Cl}_2(\theta)$.

If we consider dilogarithm of a general complex argument, $\mathrm{Li}_2(re^{i\theta})$, we get

$$\mathrm{Re}[\mathrm{Li}_2(re^{i\theta})] \equiv \mathrm{Li}_2(r,\theta), \tag{A4}$$

$$\mathrm{Im}[\mathrm{Li}_2(re^{i\theta})] = \omega \ln r + \tfrac{1}{2}[\mathrm{Cl}_2(2\theta) + \mathrm{Cl}_2(2\omega) + \mathrm{Cl}_2(2\chi)], \tag{A5}$$

with $\tan\omega = r \sin\theta/(1-r\cos\theta)$ and $\chi \equiv \pi - \theta - \omega$. We note that

$$\tfrac{1}{2}[\mathrm{Cl}_2(2\theta) + \mathrm{Cl}_2(2\omega) + \mathrm{Cl}_2(2\chi)] = -\sin\theta \int_0^r \frac{d\eta \, \ln \eta}{1-2\eta \cos\theta + \eta^2}. \tag{A6}$$

The imaginary part can also be rewritten as

$$\mathrm{Im}[\mathrm{Li}_2(re^{i\theta})]=\mathrm{Ti}_2(\tan\,\omega)-\mathrm{Ti}_2(\tan\,\omega,\tan\,\theta)=-r\,\sin\,\theta\int_0^1\frac{d\xi\,\ln\,\xi}{1-2r\xi\,\cos\,\theta+r^2\xi^2},\tag{A7}$$

where $\mathrm{Ti}_2(z)$ and $\mathrm{Ti}_2(z,a)$ are the ordinary and the generalized inverse tangent integrals, respectively,

$$\mathrm{Ti}_2(z)=\int_0^z\frac{dz'}{z'}\arctan\,z',\quad \mathrm{Ti}_2(z,a)=\int_0^z\frac{dz'}{z'+a}\arctan\,z'.\tag{A8}$$

In particular, $\mathrm{Ti}_2(z)=\mathrm{Ti}_2(z,0)$.

Finally the Lobachevsky function $L(\theta)$ is defined by

$$L(\theta)\equiv-\int_0^\theta d\theta'\,\ln(\cos\,\theta'),\quad (-\tfrac{1}{2}\pi\leqslant\theta\leqslant\tfrac{1}{2}\pi).\tag{A9}$$

It is related to Clausen's function by

$$\mathrm{Cl}_2(\theta)=-2L\left(\frac{\pi-\theta}{2}\right)+(\pi-\theta)\,\ln\,2,\quad L(\theta)=-\tfrac{1}{2}\mathrm{Cl}_2(\pi-2\theta)+\theta\,\ln\,2.\tag{A10}$$

APPENDIX B: ON THE RESCALING OF α'S

Consider the transformation

$$\alpha_i=\mathcal{G}(\alpha_1',\ldots,\alpha_N')\alpha_i',\quad i=1,\ldots,N,\tag{B1}$$

where the "scaling" function $\mathcal{G}$ is the same for all α's. The Jacobian of the transformation is

$$\text{Jacobian}=\left(\prod_{j=1}^N\alpha_j'\frac{\partial\mathcal{G}}{\partial\alpha_j'}\right)\begin{vmatrix}1+d_1 & 1 & \cdots & 1\\ 1 & 1+d_2 & \cdots & 1\\ \cdots & \cdots & \cdots & \cdots\\ 1 & 1 & \cdots & 1+d_N\end{vmatrix},\quad d_i\equiv\mathcal{G}\left(\alpha_i'\frac{\partial\mathcal{G}}{\partial\alpha_i'}\right)^{-1}.\tag{B2}$$

Since the determinant in (B2) is equal to

$$\left(\prod_{i=1}^N d_i\right)\left[1+\frac{1}{d_1}+\frac{1}{d_2}+\cdots+\frac{1}{d_N}\right],\tag{B3}$$

the expression for the Jacobian reduces to

$$\text{Jacobian}=\mathcal{G}^N\left(1+\frac{1}{\mathcal{G}}\sum_{j=1}^N\alpha_j'\frac{\partial\mathcal{G}}{\partial\alpha_j'}\right)=\mathcal{G}^N\left(1+\sum_{j=1}^N\alpha_j'\frac{\partial(\ln\,\mathcal{G})}{\partial\alpha_j'}\right).\tag{B4}$$

If $\mathcal{G}$ is a homogeneous function of α's of the order R, i.e.,

$$\mathcal{G}(Z\alpha_1',\ldots,Z\alpha_N')=Z^R\mathcal{G}(\alpha_1',\ldots,\alpha_N'),\tag{B5}$$

then

$$\sum_{j=1}^N\alpha_j'\frac{\partial\mathcal{G}}{\partial\alpha_j'}=R\mathcal{G}\ \text{and Jacobian}=[1+R]\mathcal{G}^N.\tag{B6}$$

In particular, if $\mathcal{G}=Q_r/Q'_{r'}$, where Q_r is a polynomial of degree r and $Q'_{r'}$ is a polynomial of degree r' in α', Jacobian$\rightarrow[1+(r-r')]\mathcal{G}^N$, i.e., $R=r-r'$. In the text we have encountered the situations when $r=r'=1$ $(R=0)$ and $r=2$, $r'=1$ $(R=1)$.

[1] L. M. Brown and R. P. Feynman, Phys. Rev. **85**, 231 (1952); G. Passarino and M. Veltman, Nucl. Phys. B **160**, 151 (1979).
[2] G. 'tHooft and M. Veltman, Nucl. Phys. B **153**, 365 (1979).
[3] A. C. T. Wu, Mat. Fys. Medd. K. Dan. Vidensk. Selsk. **33**(No. 3), 1 (1961).
[4] A. Denner, U. Nierste, and R. Scharf, Nucl. Phys. B **367**, 637 (1991).
[5] G. J. van Oldenborgh and J. A. M. Vermaseren, Z. Phys. C **46**, 425 (1990); G. J. van Oldenborgh, Phys. Lett. B **282**, 185 (1992).
[6] A. Denner, Fortschr. Phys. **41**, 307 (1993).
[7] F. R. Halpern, Phys. Rev. Lett. **10**, 310 (1963).
[8] B. Petersson, J. Math. Phys. **6**, 1955 (1965).
[9] D. B. Melrose, Nuovo Cimento **40A**, 181 (1965).
[10] G. Källén and J. Toll, J. Math. Phys. **6**, 299 (1965).
[11] B. G. Nickel, J. Math. Phys. **19**, 542 (1978).
[12] W. L. van Neerven and J. A. M. Vermaseren, Phys. Lett. B **137**, 241 (1984).
[13] Z. Bern, L. Dixon, and D. A. Kosower, Phys. Lett. B **302**, 299 (1993); Nucl. Phys. B **412**, 751 (1994).
[14] L. M. Brown, Nuovo Cimento **22**, 178 (1961).
[15] D. S. Kershaw, J. Math. Phys. **15**, 798 (1974); Phys. Rev. D **8**, 2708 (1973); E. E. Boos and A. I. Davydychev, Theor. Math. Phys. **89**, 1052 (1991); D. Kreimer, Z. Phys. C **54**, 667 (1992); Int. J. Mod. Phys. A **8**, 1797 (1993); L. Brücher, J. Franzkowski, and D. Kreimer, Mod. Phys. Lett. A **9**, 2335 (1994).
[16] A. I. Davydychev, J. Math. Phys. **32**, 1052 (1991); **33**, 358 (1992).
[17] L. D. Landau, Nucl. Phys. **13**, 181 (1959).
[18] G. Källen and A. Wightman, Mat. Fys. Skr. K. Dan. Vidensk. Selsk. **1**(No. 6), 1 (1958).
[19] S. Mandelstam, Phys. Rev. **115**, 1742 (1959); R. E. Cutkosky, J. Math. Phys. **1**, 429 (1960); J. C. Taylor, Phys. Rev. **117**, 261 (1960); Yu. A. Simonov, Sov. Phys. JETP **16**, 1599 (1963) [Zh. Eksp. Teor. Fiz. **43**, 2263 (1962)].
[20] R. J. Eden, P. V. Landshoff, D. I. Olive, and J. C. Polkinghorne, *The Analytic S-Matrix* (Cambridge University Press, Cambridge, 1966).
[21] R. Karplus, C. M. Sommerfield, and E. H. Wichmann, Phys. Rev. **114**, 376 (1959).
[22] J. S. Ball and T.-W. Chiu, Phys. Rev. D **22**, 2550 (1980); **23**, 3085(E) (1981); H. J. Lu and C. A. Perez, preprint SLAC-PUB-5809, 1992.
[23] A. I. Davydychev, J. Phys. A **25**, 5587 (1992).
[24] J. J. van der Bij and M. Veltman, Nucl. Phys. B **231**, 205 (1984); C. Ford, I. Jack, and D. R. T. Jones, *ibid.* **387**, 373 (1992); A. I. Davydychev and J. B. Tausk, *ibid.* **397**, 123 (1993).
[25] R. Scharf, doctoral thesis, Würzburg, 1994; R. Scharf and J. B. Tausk, Nucl. Phys. B **412**, 523 (1994).
[26] A. I. Davydychev and J. B. Tausk, Phys. Rev. D **53**, 7381 (1996).
[27] L. Lewin, *Polylogarithms and Associated Functions* (North–Holland, Amsterdam, 1981).
[28] D. M. Y. Sommerville, *An Introduction to the Geometry of N Dimensions* (Methuen, London, 1929).
[29] M. G. Kendall, *A Course in the Geometry of n Dimensions* (Charles Griffin, London, 1961).
[30] A. N. Vassiliev, Yu. M. Pis'mak, and Yu. R. Honkonen, Teor. Mat. Fiz. **47**, 291 (1981); N. I. Ussyukina, *ibid.* **54**, 124 (1983); D. I. Kazakov, Phys. Lett. **133B**, 406 (1983).
[31] A. K. Rajantie, Nucl. Phys. B **480**, 729 (1996).
[32] G. 'tHooft and M. Veltman, Nucl. Phys. B **44**, 189 (1972); C. G. Bollini and J. J. Giambiagi, Nuovo Cimento B **12**, 20 (1972).
[33] F. A. Berends, A. I. Davydychev, and V. A. Smirnov, Nucl. Phys. B **478**, 59 (1996).
[34] U. Nierste, D. Müller, and M. Böhm, Z. Phys. C **57**, 605 (1993).
[35] E. E. Boos and A. I. Davydychev, Teor. Mat. Fiz. **89**, 56 (1991).
[36] N. I. Lobatschefsky, *Imaginäre Geometrie, Kasaner Gelehrte Schriften, 1836* (Übersetzung mit Anmerkungen von H. Liebmann, Leipzig, 1904).
[37] L. Schläfli, Q. J. Math. **3**, 54 (1860); **3**, 97 (1860); *Gesammelte Matematische Abhandlungen* (Birkhäuser, Basel, 1953), Band II.
[38] J. L. Coolidge, *The Elements of Non-Euclidean Geometry* (Clarendon, Oxford, 1909); J. Böhm and E. Hertel, *Polyedergeometrie in n-Dimensionalen Räumen Konstanter Krümmung* (Birkhäuser, Basel, 1981).
[39] H. S. M. Coxeter, Q. J. Math. **6**, 13 (1935).
[40] E. B. Vinberg, Usp. Mat. Nauk **48**(No. 2), 17 (1993) [Russian Math. Surveys **48**(No. 2), 15 (1993)].
[41] R. Karplus and M. Neuman, Phys. Rev. **80**, 380 (1950); **83**, 776 (1951); B. De Tollis, Nuovo Cimento **32**, 757 (1964); **35**, 1182 (1965).
[42] J. Milnor, Bull. Am. Math. Soc. **6**, 9 (1982).
[43] N. I. Ussyukina and A. I. Davydychev, Phys. Lett. B **298**, 363 (1993); **305**, 136 (1993); D. J. Broadhurst, *ibid.* **307**, 132 (1993).
[44] Z. Bern and A. G. Morgan, Nucl. Phys. B **467**, 479 (1996).
[45] O. V. Tarasov, Phys. Rev. D **54**, 6479 (1996).
[46] F. V. Tkachov, Phys. Lett. **100B**, 65 (1981); K. G. Chetyrkin and F. V. Tkachov, Nucl. Phys. B **192**, 159 (1981).
[47] A. I. Davydychev, Phys. Lett. B **263**, 107 (1991).
[48] W.-Y. Hsiang, Q. J. Math. **39**, 463 (1988).
[49] B. Almgren, Ark. Fys. **38**(No. 7), 161 (1968); E. Byckling and K. Kajantie, *Particle Kinematics* (Wiley, New York, 1973).
[50] P. Wagner, J. Math. Phys. **39**, 2428 (1998).
[51] N. Ortner and P. Wagner, Ann. Inst. Henri Poincaré: Phys. Théor. **63**, 81 (1995).
[52] R. Kellerhals, Math. Ann. **285**, 541 (1989); in *Structural Properties of Polylogarithms*, edited by L. Lewin (AMS Math. Surveys and Monographs, Providence, RI, 1991), Vol. 37, p. 301.

INSTITUTE OF PHYSICS PUBLISHING JOURNAL OF PHYSICS A: MATHEMATICAL AND GENERAL

J. Phys. A: Math. Gen. **37** (2004) 4871–4886 PII: S0305-4470(04)73959-1

Explicitly symmetrical treatment of three-body phase space

A I Davydychev[1,2,3,4] and R Delbourgo[1]

[1] School of Mathematics and Physics, University of Tasmania, GPO Box 252-21, Hobart, Tasmania 7001, Australia
[2] Institute for Nuclear Physics, Moscow State University, 119992 Moscow, Russia
[3] Department of Physics, University of Mainz, Staudingerweg 7, D-55099 Mainz, Germany

E-mail: davyd@thep.physik.uni-mainz.de and bob.delbourgo@utas.edu.au

Received 17 December 2003, in final form 16 March 2004
Published 14 April 2004
Online at stacks.iop.org/JPhysA/37/4871 (DOI: 10.1088/0305-4470/37/17/016)

Abstract
We derive expressions for three-body phase space that are *explicitly* symmetrical in the masses of the three particles. We study geometrical properties of the variables involved in elliptic integrals and demonstrate that it is convenient to use the Jacobian zeta function to express the results in four and six dimensions.

PACS numbers: 11.80.Cr, 11.10.Kk, 12.30.Gp

1. Introduction

The subject of three-body (and, in general, N-body) relativistic phase space is as old as the hills and one might well think that all that there is to know is already known. In numerical and experimental terms this is indeed true: for a long time Dalitz plots [1, 2] have been routinely used in picturing data and they prove extremely helpful for picking out resonant intermediate states of particular spin by their preferential population of the plots. In the absence of any amplitude modulation by resonances or otherwise, the plots are at their blandest as they just represent three-body phase space.

In any multibody production such as $A + B \to 1 + 2 + \cdots + N$, the probability of the process is largely governed by the total momentum $p = p_A + p_B$, the masses of the final particles $m_1, m_2, \ldots, m_N$ relative to $\sqrt{p^2}$ and an overall coupling constant. Surely there is also the dynamics of production which modulates the coupling magnitude by intermediate state contributions, but the overall rate is mainly influenced by the unmodulated phase-space integral as written below. The case of $N = 3$ phase-space integrals and the manifest symmetry of the result upon the three masses of the product particles is the subject of this paper.

[4] Present address: Schlumberger, SPC, 155 Industrial Dr, Sugar Land, TX 77478, USA.

 © 2004 IOP Publishing Ltd Printed in the UK

One of the first comprehensive references on this subject is the paper by Almgren [3]. In his normalization, the integral over the N-particle phase space is defined as

$$I_N^{(D)}(p, m_1, \ldots, m_N) = \int \cdots \int \left\{ \prod_{i=1}^{N} \mathrm{d}^D p_i \delta(p_i^2 - m_i^2) \theta(p_i^0) \right\} \delta\left(\sum_{i=1}^{N} p_i - p \right) \quad (1.1)$$

where p is the total momentum. From now on, we will frequently use the notation $p = \sqrt{p^2}$, since usually it is easy to distinguish it from the cases when the four-dimensional vector p is meant. As a rule, we will also omit the arguments of $I_N^{(D)}$. In four dimensions ($D = 4$), we will denote $I_N \equiv I_N^{(4)}$ (this is the original Almgren's notation). More details about integrals in other dimensions can be found in [4] and in the rest of this paper. It is worth mentioning that $I_N^{(D)}$ is easy to work out for odd values of D, whereas considering even values of D brings in elliptic functions and is more difficult.

For kinematical reasons, it is clear that the results for the integrals (1.1) have no physical meaning if the absolute value of the momentum p is less than the sum of the masses. Therefore, in what follows we will imply that all results for $I_N^{(D)}$ are accompanied by $\theta\{p^2 - (m_1 + \cdots + m_N)^2\}$, without writing this theta function explicitly. In [5, 3], integral recurrence relations for I_N (at $D = 4$) were discussed. For an arbitrary dimension D, the generating relation can be presented as

$$I_N^{(D)}(p, m_1, \ldots, m_N) = \int \mathrm{d}s\, I_{R+1}^{(D)}(p, \sqrt{s}, m_{N-R+1}, \ldots, m_N) I_{N-R}^{(D)}(\sqrt{s}, m_1, \ldots, m_{N-R}). \quad (1.2)$$

Taking into account the theta functions associated with $I_{N-R}^{(D)}$ and $I_{R+1}^{(D)}$, one can see that the actual limits of the integration variable s in equation (1.2) extend from $\left(\sum_{i=1}^{N-R} m_i\right)^2$ to $\left(p - \sum_{i=N-R+1}^{N} m_i\right)^2$. Once we fix the subsets of masses in the arguments of the integrals on the r.h.s. of (1.2), the explicit symmetry gets lost. It is clear, however, that equation (1.2) still contains that symmetry, since one can split the masses $m_1, \ldots, m_N$ into these two subsets in any possible way. Another type of integral recurrence relations for $I_N^{(D)}$, with respect to the value of D, was considered in [6].

The simplest example is the two-particle phase space, $N = 2$. In this case, the phase-space integral (1.1) in four dimensions can be easily evaluated as

$$I_2 = \frac{\pi}{2p^2} \sqrt{\lambda(p^2, m_1^2, m_2^2)} \quad (1.3)$$

where

$$\lambda(x, y, z) \equiv x^2 + y^2 + z^2 - 2xy - 2yz - 2zx \quad (1.4)$$

is nothing but the well-known Källen function [7].

Using equations (1.2) and (1.3) (for the case $D = 4$, $R = 1$), one can obtain the following integral representation [3, 8] for the three-particle ($N = 3$) phase space:

$$I_3 = \frac{\pi^2}{4p^2} \int_{s_2}^{s_3} \frac{\mathrm{d}s}{s} \sqrt{(s - s_1)(s - s_2)(s_3 - s)(s_4 - s)} \quad (1.5)$$

with

$$s_1 = (m_1 - m_2)^2 \qquad s_2 = (m_1 + m_2)^2 \qquad s_3 = (p - m_3)^2 \qquad s_4 = (p + m_3)^2 \quad (1.6)$$

so that $s_1 \leqslant s_2 \leqslant s_3 \leqslant s_4$. The result of the calculation of the integral (1.5) can be expressed in terms of the elliptic integrals [3, 8] (for convenience, we collect the definitions and relevant properties of elliptic integrals in the appendix),

$$I_3 = \frac{\pi^2}{4p^2\sqrt{Q_+}}\left\{\frac{1}{2}Q_+\left(m_1^2+m_2^2+m_3^2+p^2\right)E(k)\right.$$
$$+4m_1m_2[(p-m_3)^2-(m_1-m_2)^2][(p+m_3)^2-m_3p+m_1m_2]K(k)$$
$$+8m_1m_2\left[\left(m_1^2+m_2^2\right)\left(p^2+m_3^2\right)-2m_1^2m_2^2-2m_3^2p^2\right]\Pi\left(\alpha_1^2,k\right)$$
$$\left.-8m_1m_2\left(p^2-m_3^2\right)^2\Pi\left(\alpha_2^2,k\right)\right\} \tag{1.7}$$

where we use the following notations:

$$Q_+ \equiv (p+m_1+m_2+m_3)(p+m_1-m_2-m_3)(p-m_1+m_2-m_3)(p-m_1-m_2+m_3)$$
$$Q_- \equiv (p-m_1-m_2-m_3)(p-m_1+m_2+m_3)(p+m_1-m_2+m_3)(p+m_1+m_2-m_3) \tag{1.8}$$

$$k \equiv \sqrt{\frac{Q_-}{Q_+}} \qquad \alpha_1^2 = \frac{(p-m_3)^2-(m_1+m_2)^2}{(p-m_3)^2-(m_1-m_2)^2} \qquad \alpha_2^2 = \frac{(m_1-m_2)^2}{(m_1+m_2)^2}\alpha_1^2. \tag{1.9}$$

We note that in [8] the notation $q_{\pm\pm} \equiv (p\pm m_3)^2-(m_1\pm m_2)^2$ was used. In particular, we have $Q_+ = q_{++}q_{--}$, $Q_- = q_{+-}q_{-+}$, $k^2 = q_{+-}q_{-+}/(q_{++}q_{--})$, $\alpha_1^2 = q_{-+}/q_{--}$. Note that $Q_\pm$ differ by the sign of p only.

It is clear from definition (1.1) that I_3 should be a symmetrical function of the three masses m_1, m_2, m_3. The representation (1.7) in terms of elliptic integrals is however *not* explicitly symmetrical in the masses, although it must be implicitly so. One may, of course, generate a symmetrical form by averaging the unsymmetrical-looking expressions over the three possible permutations of m_i, but this would be 'cheating' since each of them should be symmetrical by itself, although this is hardly transparent.

Note that the quantities Q_+ and Q_- (and, therefore, the argument k) are totally symmetric in m_1, m_2, m_3. (In fact, they are symmetric in all four arguments p, m_1, m_2, m_3.) Therefore, the term containing $E(k)$ in equation (1.7) is also symmetric. The function $K(k)$ itself is also symmetric, but its coefficient is not symmetric. We also note that the product of Q_+ and Q_- produces the quantity

$$D_{123} \equiv Q_+Q_- = [p^2-(m_1+m_2+m_3)^2][p^2-(-m_1+m_2+m_3)^2]$$
$$\times[p^2-(m_1-m_2+m_3)^2][p^2-(m_1+m_2-m_3)^2] \tag{1.10}$$

that occurs in recurrence relations for the sunset diagram (see, e.g., in [9, 10]). It should be noted that the imaginary part of the sunset diagram is proportional to the three-particle phase-space integral. For instance, in the notation of [8], $\mathrm{Im}(T_{123}) = -4\pi^{-1}I_3$. We also note that ρ_N^D considered in [4, 6] are related to $I_N^{(D)}$ as $\rho_N^D = (2\pi)^{N+D-ND}I_N^{(D)}$.

For equal masses, $m_1 = m_2 = m_3 \equiv m$, equation (1.7) yields

$$\frac{\pi^2}{4p^2}\sqrt{(p-m)(p+3m)}\left\{\frac{1}{2}(p-m)(p^2+3m^2)E(k_{\mathrm{eq}})-4m^2pK(k_{\mathrm{eq}})\right\} \tag{1.11}$$

with

$$k_{\mathrm{eq}} = \sqrt{\frac{(p+m)^3(p-3m)}{(p-m)^3(p+3m)}}. \tag{1.12}$$

Some other special cases of equation (1.7) are described in [8, 11].

This paper is devoted to a new way of exhibiting the results in an *explicitly* symmetrical manner. To do this, we will employ another integral representation for I_3, in terms of

Mandelstam variables s, t, u [12] and the Kibble cubic $\Phi(s, t, u)$ [13]. In particular, we will show that it is convenient to present the result (1.7) in terms of Jacobi Z function whose definition is given in the appendix.

2. Phase-space integrals

As an illustration, let us demonstrate how the connection with the Dalitz figure can be derived directly from definition (1.1). The D-dimensional vector p can be presented as $(p^0, \mathbf{p})$, where $\mathbf{p}$ is the $(D-1)$-dimensional Euclidean vector of space components. Without loss of generality, we can work in the centre-of-mass frame, $p = (p^0, \mathbf{0})$. Using the integral representation

$$\delta\left(\sum_{i=1}^{N} p_i - p\right) = \frac{1}{(2\pi)^D} \int \mathrm{d}^D x \, \exp\left\{ \mathrm{i} \sum_{i=1}^{N} (p_i x) - \mathrm{i}(px) \right\} \tag{2.1}$$

with $(px) = p^0 x^0$, we get

$$I_N^{(D)} = \frac{1}{(2\pi)^D} \int \mathrm{d}^D x \, \mathrm{e}^{-\mathrm{i}p^0 x^0} \left\{ \prod_{i=1}^{N} \int \mathrm{d}^D p_i \delta\left(p_i^2 - m_i^2\right) \theta\left(p_i^0\right) \mathrm{e}^{\mathrm{i}(p_i x)} \right\}. \tag{2.2}$$

(Similar method was used in [14].) Integrating over $(D-1)$-dimensional angles of $\mathbf{p}_i$ we get

$$\int \mathrm{d}^D p_i \delta\left(p_i^2 - m_i^2\right) \theta\left(p_i^0\right) \mathrm{e}^{\mathrm{i}(p_i x)} = \frac{(2\pi)^{(D-1)/2}}{2\xi^{(D-3)/2}} \int_0^\infty \frac{\rho_i^{(D-1)/2} \mathrm{d}\rho_i}{\sqrt{\rho_i^2 + m_i^2}} J_{(D-3)/2}(\rho_i \xi) \, \mathrm{e}^{\mathrm{i}x^0 \sqrt{\rho_i^2 + m_i^2}} \tag{2.3}$$

with $\rho_i \equiv |\mathbf{p}_i|$ and $\xi \equiv |\mathbf{x}|$. In the four-dimensional case the Bessel function reduces to an elementary function, $J_{1/2}(\rho_i \xi) = [2/(\pi \rho_i \xi)]^{1/2} \sin(\rho_i \xi)$. We note an analogy with the calculation of Feynman integrals in the coordinate space [15, 22], when each massive propagator yields a (modified) Bessel function.

Let us consider, for example, the two-particle phase space. Then, the integration over ξ gives us

$$\int_0^\infty \xi \mathrm{d}\xi \, J_\nu(\rho_1 \xi) J_\nu(\rho_2 \xi) = 2\delta\left(\rho_1^2 - \rho_2^2\right)$$

with $\nu = (D-3)/2$, so that we can put $\rho_1 = \rho_2 \equiv \rho$, whereas the integration over x^0 yields another delta function, $\delta\left(p - \sqrt{\rho^2 + m_1^2} - \sqrt{\rho^2 + m_2^2}\right)$ in the centre-of-mass frame. The resulting integral

$$I_2^{(D)} = \frac{\pi^{(D-1)/2}}{2\Gamma\left(\frac{D-1}{2}\right)} \int_0^\infty \frac{\rho^{D-2} \mathrm{d}\rho}{\sqrt{\rho^2 + m_1^2}\sqrt{\rho^2 + m_2^2}} \, \delta\left(p - \sqrt{\rho^2 + m_1^2} - \sqrt{\rho^2 + m_2^2}\right) \tag{2.4}$$

can be easily evaluated, yielding (see, e.g., in [6])

$$I_2^{(D)} = \frac{\pi^{(D-1)/2}}{(2p)^{D-2}\Gamma\left(\frac{D-1}{2}\right)} \left[\lambda\left(p^2, m_1^2, m_2^2\right)\right]^{(D-3)/2} \tag{2.5}$$

where λ is defined in equation (1.4). For $D = 4$, equation (2.5) reduces to the well-known answer (1.3).

For the three-particle phase-space integral we get

$$I_3^{(D)} = \frac{2^{(D-7)/2} \pi^{D-2}}{\Gamma\left(\frac{D-1}{2}\right)} \int_0^\infty \frac{\mathrm{d}\xi}{\xi^{(D-5)/2}} \int_{-\infty}^{\infty} \mathrm{d}x^0 \, \mathrm{e}^{-\mathrm{i}p^0 x^0} \times \prod_{i=1}^{3} \int_0^\infty \frac{\rho_i^{(D-1)/2} \mathrm{d}\rho_i}{\sqrt{\rho_i^2 + m_i^2}} J_{(D-3)/2}(\rho_i \xi) \, \mathrm{e}^{\mathrm{i}x^0 \sqrt{\rho_i^2 + m_i^2}}. \tag{2.6}$$

Here we can integrate over ξ, using (see [16])

$$\int_0^\infty \frac{\mathrm{d}\xi}{\xi^{\nu-1}} J_\nu(\rho_1\xi) J_\nu(\rho_2\xi) J_\nu(\rho_3\xi) = \frac{2\theta\{-\lambda(\rho_1^2,\rho_2^2,\rho_3^2)\}[-\lambda(\rho_1^2,\rho_2^2,\rho_3^2)]^{\nu-1/2}}{\pi^{1/2}\Gamma\left(\nu+\frac{1}{2}\right)(8\rho_1\rho_2\rho_3)^\nu} \tag{2.7}$$

(with $\nu = (D-3)/2$), where λ is the Källen function (1.4). In fact, in our case, when all $\rho_i \geqslant 0$,

$$\theta\{-\lambda(\rho_1^2,\rho_2^2,\rho_3^2)\} = \theta(\rho_1+\rho_2-\rho_3)\theta(\rho_2+\rho_3-\rho_1)\theta(\rho_3+\rho_1-\rho_2), \tag{2.8}$$

i.e. it equals 1 when one can compose a triangle with sides ρ_1, ρ_2, ρ_3, and gives 0 otherwise (cf equation (11) of [17]).

Introducing notation $\sigma_i = \sqrt{\rho_i^2 + m_i^2}$ and integrating over x^0 (getting a δ function) we arrive at

$$I_3^{(D)} = \frac{\pi^{D-2}}{\Gamma(D-2)} \int_{m_1}^\infty \int_{m_2}^\infty \int_{m_3}^\infty \mathrm{d}\sigma_1\,\mathrm{d}\sigma_2\,\mathrm{d}\sigma_3 \delta(p-\sigma_1-\sigma_2-\sigma_3)$$
$$\times \left[-\lambda\left(\sigma_1^2-m_1^2,\sigma_2^2-m_2^2,\sigma_3^2-m_3^2\right)\right]^{(D-4)/2} \theta\left\{-\lambda\left(\sigma_1^2-m_1^2,\sigma_2^2-m_2^2,\sigma_3^2-m_3^2\right)\right\}. \tag{2.9}$$

In four dimensions the factor $[-\lambda]^{(D-4)/2}$ disappears and, geometrically, we need to calculate a closed area on the plane $\sigma_1 + \sigma_2 + \sigma_3 = p^0 \equiv p$, with the boundary of the figure described by

$$\lambda\left(\sigma_1^2-m_1^2,\sigma_2^2-m_2^2,\sigma_3^2-m_3^2\right) = 0 \qquad \sigma_1+\sigma_2+\sigma_3 = p. \tag{2.10}$$

Furthermore, introducing Mandelstam-type variables

$$s = p^2 + m_3^2 - 2p\sigma_3 \qquad t = p^2 + m_1^2 - 2p\sigma_1 \qquad u = p^2 + m_2^2 - 2p\sigma_2 \tag{2.11}$$

satisfying

$$s + t + u = m_1^2 + m_2^2 + m_3^2 + p^2 \equiv w_0 \tag{2.12}$$

one arrives at another integral representation (the limits of integration are discussed below),

$$I_3^{(D)} = \frac{\pi^{D-2}}{4p^{D-2}\Gamma(D-2)} \iiint \mathrm{d}s\,\mathrm{d}t\,\mathrm{d}u\,\delta(s+t+u-w_0)\,[\Phi(s,t,u)]^{(D-4)/2}\,\theta\{\Phi(s,t,u)\} \tag{2.13}$$

where

$$\Phi(s,t,u) = -\frac{1}{16p^2}\lambda\left\{\lambda\left(s,m_3^2,p^2\right),\lambda\left(t,m_1^2,p^2\right),\lambda\left(u,m_2^2,p^2\right)\right\} \tag{2.14}$$

can also be written in a more familiar Kibble cubic form [13],

$$\Phi(s,t,u) = stu - s\left(m_1^2m_2^2 + p^2m_3^2\right) - t\left(m_2^2m_3^2 + p^2m_1^2\right) - u\left(m_3^2m_1^2 + p^2m_2^2\right)$$
$$+2\left(m_1^2m_2^2m_3^2 + p^2m_1^2m_2^2 + p^2m_2^2m_3^2 + p^2m_3^2m_1^2\right) \tag{2.15}$$

provided that the condition (2.12) is satisfied. In particular, in four dimensions we have

$$I_3 = \frac{\pi^2}{4p^2} \iiint \mathrm{d}s\,\mathrm{d}t\,\mathrm{d}u\,\delta(s+t+u-w_0)\theta\{\Phi(s,t,u)\}. \tag{2.16}$$

According to definition (2.11) in terms of σ_i, one can see that the maximal values of s, t and u (corresponding to the upper limits of integration in equations (2.13) and (2.16)) are $s_{\max} = (p-m_3)^2$, $t_{\max} = (p-m_1)^2$ and $u_{\max} = (p-m_2)^2$. To define the minimal values of s, t and u, the familiar Dalitz–Kibble plot given in figure 1 is useful. Due to

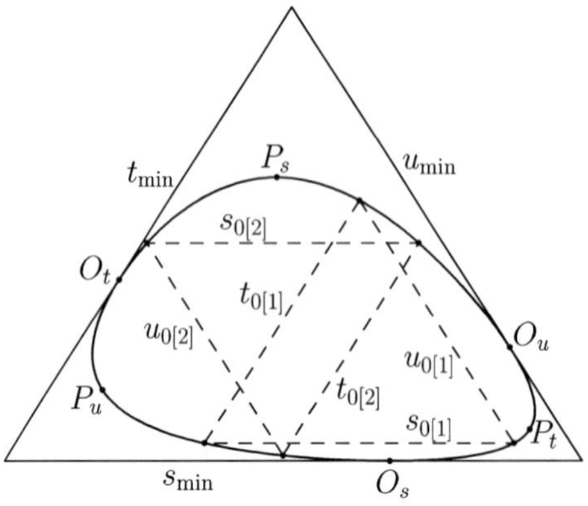

Figure 1. The Dalitz–Kibble integration area.

the condition (2.12), the region of integration is restricted by a triangle, with $s \geqslant s_{\min} = (m_1+m_2)^2$, $t \geqslant t_{\min} = (m_2+m_3)^2$ and $u \geqslant u_{\min} = (m_1+m_3)^2$. Moreover, due to the theta function $\theta\,\{\Phi(s,t,u)\}$ the region of integration is in fact restricted by the interior of the cubic curve $\Phi(s,t,u) = 0$, see figure 1. Within that area, the Mandelstam variables s, t and u take their minimal values in points O_s, O_t and O_u, respectively, whereas their maximal values correspond to the points P_s, P_t and P_u. (The dashed triangles will be discussed in section 4.)

The function $\Phi(s,t,u)$ has a maximum within the region of integration. For equal masses, the maximal value $\Phi_{\max} = \frac{1}{27}p^2(p^2-9m^2)^2$ occurs at $s = t = u = \frac{1}{3}(p^2+3m^2)$. For the general unequal masses, one needs to solve a fourth-order algebraic equation to find the position of the maximum.

We note that the representation (2.14) can be extracted from equation (5.39) of [18], using symmetry properties. Our $\Phi(s,t,u)$ corresponds to $-G\big(s,t,p^2,m_2^2,m_1^2,m_3^2\big)$, in the notations of [18]. The G-function is symmetric with respect to the permutations of three pairs of arguments, (s,t), $\big(p^2,m_2^2\big)$ and $\big(m_1^2,m_3^2\big)$. Although the authors presume from their equation (5.39) that 'from a practical point of view this identity is not very useful', we found that its symmetric form is certainly helpful in understanding the structural properties of phase-space integrals.

3. Geometrical interpretation

Let us introduce

$$c_{12} = \frac{s-m_1^2-m_2^2}{2m_1m_2} \qquad c_{23} = \frac{t-m_2^2-m_3^2}{2m_2m_3} \qquad c_{13} = \frac{u-m_1^2-m_3^2}{2m_1m_3}. \tag{3.1}$$

Then, the function $\Phi(s,t,u)$ can be presented as a Gram determinant,

$$\Phi(s,t,u) = 4m_1^2m_2^2m_3^2 \begin{vmatrix} 1 & c_{12} & c_{13} \\ c_{12} & 1 & c_{23} \\ c_{13} & c_{23} & 1 \end{vmatrix} \tag{3.2}$$

whereas the δ function becomes

$$\delta\big(s+t+u-m_1^2-m_2^2-m_3^2-p^2\big)$$
$$\Rightarrow \delta\left(m_1^2+m_2^2+m_3^2+2m_1m_2c_{12}+2m_2m_3c_{23}+2m_1m_3c_{13}-p^2\right). \tag{3.3}$$

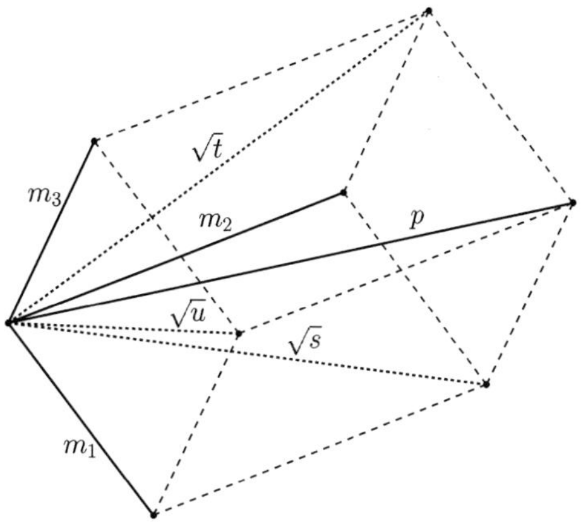

Figure 2. The parallelepiped interpretation.

In this way, we get

$$I_3 = \frac{2\pi^2}{p^2} m_1^2 m_2^2 m_3^2 \iiint \mathrm{d}c_{12}\,\mathrm{d}c_{13}\,\mathrm{d}c_{23}\theta\left(\begin{vmatrix} 1 & c_{12} & c_{13} \\ c_{12} & 1 & c_{23} \\ c_{13} & c_{23} & 1 \end{vmatrix}\right)$$
$$\times\, \delta\left(m_1^2 + m_2^2 + m_3^2 + 2m_1m_2c_{12} + 2m_2m_3c_{23} + 2m_1m_3c_{13} - p^2\right) \qquad (3.4)$$

where the integration extends over $c_{jl} \geqslant 1$.

If one were to interpret c_{jl} as the cosines of the angles between the m_j and m_l sides of a vertex of a parallelepiped (formed by m_1, m_2 and m_3, see figure 2), then all these quantities would have a straightforward geometrical interpretation. Namely, $\Phi(s, t, u)$ would be 4{volume of parallelepiped}^{2}, whereas the δ function would tell us that the 'principal' diagonal of this parallelepiped should be equal to p. In this case, the quantities $\sqrt{s}$, $\sqrt{t}$ and $\sqrt{u}$ could be identified as the diagonals of the faces of the parallelepiped, see figure 2. Moreover,

$$\frac{p^2 + m_1^2 - t}{2pm_1} \qquad \frac{p^2 + m_2^2 - u}{2pm_2} \qquad \text{and} \qquad \frac{p^2 + m_3^2 - s}{2pm_3} \qquad (3.5)$$

could be understood as cosines of the angles between the diagonal p and the m_i sides of the parallelepiped. In other words, these are the angles between p and m_i in triangles with sides $(p, m_1, \sqrt{t})$, $(p, m_2, \sqrt{u})$ and $(p, m_3, \sqrt{s})$, respectively.

Using this geometrical figure, we can mention a rather interesting geometrical meaning of equation (2.14). Namely, it tells us that the volume of the parallelepiped is $(8/p)$ times the area of triangle whose sides are given by the areas of triangles formed out of the principal diagonal p, one of the face diagonals ($\sqrt{s}$, $\sqrt{t}$ or $\sqrt{u}$), and the appropriate m_3, m_1 or m_2 side.

However, when we are above the threshold, $p^2 > (m_1+m_2+m_3)^2$, the quantities c_{jl} exceed one and therefore the expressions should be understood in the sense of analytic continuation, i.e. as hyperbolic cosines. The same is valid for the triangles $(p, m_3, \sqrt{s})$, etc: they should also be understood in the sense of analytic continuation, since $p \geqslant m_3 + \sqrt{s}$, etc. Therefore, the quantities σ_i/m_i should also be understood as hyperbolic cosines, whereas $\sqrt{\sigma_i^2 - m_i^2}/m_i$ are hyperbolic sines.

Nevertheless, in the region below the threshold (which we need, for instance, to describe the real part of the sunset diagram), this geometrical figure can have direct meaning, generalizing the figure we had for the one-loop two-point function [19].

4. Kibble cubic characteristics

Suppose

$$(s_0, t_0, w_0 - s_0 - t_0) \qquad (s_0, w_0 - s_0 - u_0, u_0) \qquad (w_0 - t_0 - u_0, t_0, u_0) \tag{4.1}$$

all are the roots of the equation $\Phi(s,t,u) = 0$. Then, we can present $\Phi(s,t,u)$ as

$$\Phi(s,t,u) = stu - st_0u_0 - s_0tu_0 - s_0t_0u + 2s_0t_0u_0. \tag{4.2}$$

Furthermore, if we shift the Mandelstam variables as

$$s = s_0 + s' \qquad t = t_0 + t' \qquad u = u_0 + u' \tag{4.3}$$

subject to the condition

$$s' + t' + u' = w_0 - s_0 - t_0 - u_0 \equiv w_0' \tag{4.4}$$

then

$$\Phi(s,t,u) \Rightarrow s't'u' + s_0t'u' + s't_0u' + s't'u_0. \tag{4.5}$$

Using equation (4.2) and defining

$$c_{tu} \equiv \sqrt{\frac{t_0u_0}{tu}} \qquad c_{st} \equiv \sqrt{\frac{s_0t_0}{st}} \qquad c_{su} \equiv \sqrt{\frac{s_0u_0}{su}} \tag{4.6}$$

we arrive at another Gram determinant representation for $\Phi(s,t,u)$ (cf equation (3.2)),

$$\Phi(s,t,u) = stu \begin{vmatrix} 1 & c_{tu} & c_{st} \\ c_{tu} & 1 & c_{su} \\ c_{st} & c_{su} & 1 \end{vmatrix}. \tag{4.7}$$

There are (at least) two sets of solutions (4.1) that can be described as

$$s_0 = \frac{A_1A_2}{A_3} \qquad t_0 = \frac{A_2A_3}{A_1} \qquad u_0 = \frac{A_1A_3}{A_2} \tag{4.8}$$

so that equation (4.2) yields

$$\Phi(s,t,u) = stu - A_1^2t - A_2^2u - A_3^2s + 2A_1A_2A_3. \tag{4.9}$$

The first set of solutions corresponds to

$$A_1 \equiv pm_1 + m_2m_3 \qquad A_2 \equiv pm_2 + m_3m_1 \qquad A_3 \equiv pm_3 + m_1m_2. \tag{4.10}$$

For this set, we have

$$w_0' = w_0 - s_0 - t_0 - u_0 = \frac{m_1m_2m_3pQ_-}{A_1A_2A_3} \tag{4.11}$$

$$c_{tu} = \frac{pm_3 + m_1m_2}{\sqrt{tu}} \qquad c_{st} = \frac{pm_2 + m_1m_3}{\sqrt{st}} \qquad c_{su} = \frac{pm_1 + m_2m_3}{\sqrt{su}}. \tag{4.12}$$

Note that if we change $p \to -p$ in equation (4.10), this would also be a solution, which would correspond to a 'non-physical' branch of the Kibble cubic.

The second set of solutions corresponds to

$$\begin{aligned} A_1 &\equiv \tfrac{1}{2}(p^2 + m_1^2 - m_2^2 - m_3^2) \qquad A_2 \equiv \tfrac{1}{2}(p^2 - m_1^2 + m_2^2 - m_3^2) \\ A_3 &\equiv \tfrac{1}{2}(p^2 - m_1^2 - m_2^2 + m_3^2). \end{aligned} \tag{4.13}$$

For this set, we get

$$w_0' = w_0 - s_0 - t_0 - u_0 = -\frac{Q_+Q_-}{16A_1A_2A_3} = -\frac{D_{123}}{16A_1A_2A_3} \tag{4.14}$$

$$c_{tu} = \frac{p^2 - m_1^2 - m_2^2 + m_3^2}{2\sqrt{tu}} \qquad c_{st} = \frac{p^2 - m_1^2 + m_2^2 - m_3^2}{2\sqrt{st}} \qquad c_{su} = \frac{p^2 + m_1^2 - m_2^2 - m_3^2}{2\sqrt{su}}. \tag{4.15}$$

It should be noted that the value of s_0 corresponding to the second set satisfies

$$\lambda(s_0, m_1^2, m_2^2) = \lambda(s_0, p^2, m_3^2) \tag{4.16}$$

i.e. the areas (or their analytical continuations) of triangles with sides $(\sqrt{s_0}, m_1, m_2)$ and $(\sqrt{s_0}, p, m_3)$ are equal. Analogously,

$$\lambda(t_0, m_2^2, m_3^2) = \lambda(t_0, p^2, m_1^2) \qquad \lambda(u_0, m_1^2, m_3^2) = \lambda(u_0, p^2, m_2^2). \tag{4.17}$$

Moreover, one can get the direct geometrical interpretation of the quantities (4.15) through the familiar parallelepiped shown in figure 2. Namely, c_{tu} is nothing but the cosine between the face diagonals $\sqrt{t}$ and $\sqrt{u}$. Accordingly, c_{su} is the cosine of the angle between $\sqrt{s}$ and $\sqrt{u}$ diagonals, whilst c_{st} is the cosine of the angle between $\sqrt{s}$ and $\sqrt{t}$ diagonals. If we construct a tetrahedron using the $\sqrt{s}$, $\sqrt{t}$ and $\sqrt{u}$ diagonals then, according to equation (4.7), $\Phi(s, t, u)$ would represent 36 times its volume squared.

In the Dalitz–Kibble plot shown in figure 1 we connect the points (4.1) for each of the two sets by dashed lines, introducing subscripts [1] and [2] for the first and the second set, respectively. The two resulting 'dashed' triangles indicate that the two sets of solutions are complementary to each other. Namely, the boundary of the Dalitz plot confines the products tu, st and su as follows:

$$\begin{aligned}
(pm_3 + m_1 m_2)^2 &\leqslant tu \leqslant \tfrac{1}{4}\left(p^2 - m_1^2 - m_2^2 + m_3^2\right)^2 \\
(pm_2 + m_1 m_3)^2 &\leqslant st \leqslant \tfrac{1}{4}\left(p^2 - m_1^2 + m_2^2 - m_3^2\right)^2 \\
(pm_1 + m_2 m_3)^2 &\leqslant su \leqslant \tfrac{1}{4}\left(p^2 + m_1^2 - m_2^2 - m_3^2\right)^2
\end{aligned} \tag{4.18}$$

or, equivalently,

$$\begin{aligned}
&(t_0 u_0)_{[1]} \leqslant tu \leqslant (t_0 u_0)_{[2]} \qquad (s_0 t_0)_{[1]} \leqslant st \leqslant (s_0 t_0)_{[2]} \\
&(s_0 u_0)_{[1]} \leqslant su \leqslant (s_0 u_0)_{[2]}.
\end{aligned} \tag{4.19}$$

In other words, the first and the second sets yield, respectively, the minimal and the maximal values of tu, st and su.

Let us consider the corresponding values of the 'cosines' c_{su}, c_{st} and c_{tu}. For the first set, c_{su}, c_{st} and c_{tu} would vary between 1 and $\cos\varphi_i$ $(i = 1, 2, 3)$, respectively, where

$$\begin{aligned}
&\cos\varphi_1 = \frac{2(pm_1 - m_2 m_3)}{p^2 + m_1^2 - m_2^2 - m_3^2} \qquad \cos\varphi_2 = \frac{2(pm_2 - m_3 m_1)}{p^2 - m_1^2 + m_2^2 - m_3^2} \\
&\cos\varphi_3 = \frac{2(pm_3 - m_1 m_2)}{p^2 - m_1^2 - m_2^2 + m_3^2}.
\end{aligned} \tag{4.20}$$

For the second set, c_{su}, c_{st} and c_{tu} would vary between 1 and $1/\cos\varphi_i$. This means that we need to understand them in the sense of analytic continuation.

The angles φ_i will be very important below. Their sines can be presented as

$$\begin{aligned}
&\sin\varphi_1 = \frac{\sqrt{Q_+}}{p^2 + m_1^2 - m_2^2 - m_3^2} \qquad \sin\varphi_2 = \frac{\sqrt{Q_+}}{p^2 - m_1^2 + m_2^2 - m_3^2} \\
&\sin\varphi_3 = \frac{\sqrt{Q_+}}{p^2 - m_1^2 - m_2^2 + m_3^2}.
\end{aligned} \tag{4.21}$$

It is interesting that the corresponding Gram determinant can be factorized as

$$\begin{vmatrix} 1 & -\cos\varphi_3 & -\cos\varphi_2 \\ -\cos\varphi_3 & 1 & -\cos\varphi_1 \\ -\cos\varphi_2 & -\cos\varphi_1 & 1 \end{vmatrix} = \frac{1}{k^2}\sin^2\varphi_1 \sin^2\varphi_2 \sin^2\varphi_3. \tag{4.22}$$

Equation (4.22) can be used to express k in terms of φ_i. We also note that

$$\tan\frac{\varphi_3}{2} = \sqrt{\frac{(p-m_3)^2-(m_1-m_2)^2}{(p+m_3)^2-(m_1+m_2)^2}} \tag{4.23}$$

and similarly for φ_1 and φ_2. In particular, one can see that at the threshold, $p = m_1 + m_2 + m_3$, the angles φ_i are related to the angles θ_i from equation (20) of [20] (see also [10]) as $\varphi_i = \pi - 2\theta_i$, and

$$(\varphi_1+\varphi_2+\varphi_3)|_{p=m_1+m_2+m_3} = \pi. \tag{4.24}$$

We can also consider associated angles ψ_i, such that

$$\sin\psi_i = k\sin\varphi_i \qquad \cos\psi_i = \sqrt{1-k^2\sin^2\varphi_i}. \tag{4.25}$$

Explicitly, we get

$$\sin\psi_3 = \frac{\sqrt{Q_-}}{p^2-m_1^2-m_2^2+m_3^2} \qquad \cos\psi_3 = \frac{2(pm_3+m_1m_2)}{p^2-m_1^2-m_2^2+m_3^2} \tag{4.26}$$

etc. For these angles, we get

$$\begin{vmatrix} 1 & \cos\psi_3 & \cos\psi_2 \\ \cos\psi_3 & 1 & \cos\psi_1 \\ \cos\psi_2 & \cos\psi_1 & 1 \end{vmatrix} = \frac{1}{k^2}\sin^2\psi_1 \sin^2\psi_2 \sin^2\psi_3. \tag{4.27}$$

5. A naturally symmetric representation

Using the representation (4.2) for $\Phi(s,t,u)$, in terms of s_0, t_0 and u_0, the three-body phase-space integral can be written as

$$I_3 = \frac{\pi^2}{4p^2}\iiint \mathrm{d}s\,\mathrm{d}t\,\mathrm{d}u\,\delta(s+t+u-w_0)\theta(stu - st_0u_0 - s_0tu_0 - s_0t_0u + 2s_0t_0u_0) \tag{5.1}$$

with $w_0 = p^2+m_1^2+m_2^2+m_3^2$. Integrating over u yields

$$I_3 = \frac{\pi^2}{4p^2}\iint \mathrm{d}s\,\mathrm{d}t\,\theta\{(st-s_0t_0)(w_0-s-t) - st_0u_0 - s_0tu_0 + 2s_0t_0u_0\}. \tag{5.2}$$

Then, integrating over t, we basically obtain the difference between the roots of the quadratic argument of the θ function, which is

$$\frac{1}{s}\sqrt{s^4 - 2w_0s^3 + \left(w_0^2+2s_0t_0+2s_0u_0-4t_0u_0\right)s^2 - 2(w_0t_0+w_0u_0-4u_0t_0)s_0s + s_0^2(t_0-u_0)^2}.$$

It is easy to check that for both sets of (s_0, t_0, u_0) the square root takes the familiar form (1.5), which yields the non-symmetric result (1.7) in terms of elliptic integrals.

Starting from the representation (2.13), one can easily generalize the result (1.5) to the D-dimensional case as

$$I_3^{(D)} = \frac{\pi^{D-1}}{(4p)^{D-2}\Gamma^2\left(\frac{D-1}{2}\right)}\int_{s_2}^{s_3}\frac{\mathrm{d}s}{s^{D/2-1}}[(s-s_1)(s-s_2)(s_3-s)(s_4-s)]^{(D-3)/2} \tag{5.3}$$

with s_i given in equation (1.6). Another way to derive the representation (5.3) is to use the recurrence relation (1.2),

$$I_3^{(D)} = \int_{s_2}^{s_3} \mathrm{d}s\, I_2^{(D)}(p, \sqrt{s}, m_3) I_2^{(D)}(\sqrt{s}, m_1, m_2) \tag{5.4}$$

and substitute the result (2.5) for $I_2^{(D)}$. The result (5.3) corresponds to equation (9) of [4]. (We note that the overall factor on the r.h.s. of equation (9) of [4] should be corrected: $(32\pi)^{2-2\ell}$ should be changed into $\frac{1}{2}(16\pi)^{2-2\ell}$, with $\ell = D/2$).

Using representation (5.3), it is easy to see (just substituting $s = x^2$) that all *odd*-dimensional phase-space integrals can be expressed in terms of polynomial functions (see, e.g., in [21–23]),

$$I_3^{(3)} = \frac{\pi^2}{2p}(p - m_1 - m_2 - m_3) \tag{5.5}$$

$$\begin{aligned} I_3^{(5)} = \frac{\pi^4}{60p^3}(p - m_1 - m_2 - m_3)^3 \bigg[& \frac{1}{7}(p - m_1 - m_2 - m_3)^4 + (m_1 + m_2 + m_3)p^3 \\ & - 2\left(m_1^2 + m_2^2 + m_3^2\right)p^2 + \left(m_1^3 + m_2^3 + m_3^3\right)p + 12m_1m_2m_3p \\ & - (m_1 + m_2 + m_3)(m_1 + m_2)(m_2 + m_3)(m_3 + m_1) \\ & + 4m_1m_2m_3(m_1 + m_2 + m_3) \bigg] \end{aligned} \tag{5.6}$$

etc, which are explicitly symmetric in the masses m_i. However, the results in *even* dimensions appear to be less trivial.

It is instructive to consider the two-dimensional case, $D = 2$. Then, the integral (5.3) yields just the elliptic integral $K(k)$,

$$I_3^{(2)} = \int_{s_2}^{s_3} \frac{\mathrm{d}s}{\sqrt{(s - s_1)(s - s_2)(s_3 - s)(s_4 - s)}} = \frac{2}{\sqrt{Q_+}} K(k). \tag{5.7}$$

This is of course explicitly symmetric in the masses without further ado. On the other hand, using the δ function in equation (5.1), we can insert $1 = (s + t + u)/w_0$ in the integrand, and then consider the three resulting terms (with s, t and u) separately. In this way, we arrive at an alternative expression,

$$\begin{aligned} I_3^{(2)} = \frac{1}{w_0} \Bigg\{ & \int_{s_2}^{s_3} \frac{s\, \mathrm{d}s}{\sqrt{(s - s_1)(s - s_2)(s_3 - s)(s_4 - s)}} + \int_{t_2}^{t_3} \frac{t\, \mathrm{d}t}{\sqrt{(t - t_1)(t - t_2)(t_3 - t)(t_4 - t)}} \\ & + \int_{u_2}^{u_3} \frac{u\, \mathrm{d}u}{\sqrt{(u - u_1)(u - u_2)(u_3 - u)(u_4 - u)}} \Bigg\} \end{aligned} \tag{5.8}$$

where the roots t_i and u_i can be obtained from s_i given in equation (1.6) by proper permutation of the masses m_i. Each of the integrals involved in equation (5.8) can be expressed in terms of Jacobian Z function (see the appendix). For example,

$$\int_{s_2}^{s_3} \frac{s\, \mathrm{d}s}{\sqrt{(s - s_1)(s - s_2)(s_3 - s)(s_4 - s)}} = \frac{\sin\varphi_3}{\sin\varphi_1 \sin\varphi_2} K(k) - K(k) Z(\varphi_3, k) \tag{5.9}$$

where φ_i are nothing but the three angles defined in equations (4.20) and (4.21). Comparing the resulting expression with the original result (5.7), we obtain a very useful relation between the three $Z(\varphi_i, k)$ functions,

$$Z(\varphi_1, k) + Z(\varphi_2, k) + Z(\varphi_3, k) = k^2 \sin\varphi_1 \sin\varphi_2 \sin\varphi_3. \tag{5.10}$$

Let us now consider the four-dimensional integral $I_3^{(4)} \equiv I_3$, namely, its representation (1.5). A useful observation is that the result would be simpler if we managed to get rid of s in the denominator. In particular, it would contain just one elliptic integral Π, rather than two. How are we to eliminate s in the denominator? Again, using the δ function in equation (5.1), we can insert $1 = (s+t+u)/w_0$ in the integrand, and then consider the three resulting terms (with s, t and u) separately. For the term with s, we perform the t and u integrations and arrive at the same integral as in equation (1.5), but without s in the denominator. In two other integrals, we just integrate in a different order, leaving as the last one the t or u integration, respectively. In this way, we obtain for the integral (5.1)

$$\frac{\pi^2}{4p^2 w_0}\int\int\int \mathrm{d}s\,\mathrm{d}t\,\mathrm{d}u(s+t+u)\delta(s+t+u-w_0)\theta(stu - st_0u_0 - s_0tu_0 - s_0t_0u + 2s_0t_0u_0)$$

$$= \frac{\pi^2}{4p^2 w_0}\left\{\int_{s_2}^{s_3} \mathrm{d}s\sqrt{(s-s_1)(s-s_2)(s_3-s)(s_4-s)}\right.$$

$$+\int_{t_2}^{t_3} \mathrm{d}t\sqrt{(t-t_1)(t-t_2)(t_3-t)(t_4-t)}$$

$$\left.+\int_{u_2}^{u_3} \mathrm{d}u\sqrt{(u-u_1)(u-u_2)(u_3-u)(u_4-u)}\right\} \qquad (5.11)$$

where, as before, the roots t_i and u_i can be obtained from s_i given in equation (1.6) by permutation of the masses.

Using the formulae given in [24] along with equations (A.8) and (A.11), the s-integral in equation (5.11) can be calculated in terms of a Jacobian Z function (see the appendix),

$$\int_{s_2}^{s_3} \mathrm{d}s\sqrt{(s-s_1)(s-s_2)(s_3-s)(s_4-s)}$$

$$= \sqrt{Q_+}\left\{2\left(p^2m_3^2 - m_1^2m_2^2\right)K(k)\frac{Z(\varphi_3,k)}{\sin\varphi_3} + \frac{1}{6}Q_-K(k)\right.$$

$$\left.+\frac{1}{6}\left[\left(p^2-m_1^2-m_2^2+m_3^2\right)^2 + 8\left(p^2m_3^2+m_1^2m_2^2\right)\right][E(k)-K(k)]\right\} \qquad (5.12)$$

where φ_3 is one of the three angles defined in equations (4.20) and (4.21).

Collecting the results for all three integrals and using the relation (5.10), we arrive at the symmetric result

$$I_3 = \frac{\pi^2}{8p^2}\left\{\sqrt{Q_+}\left(p^2+m_1^2+m_2^2+m_3^2\right)[E(k)-K(k)]\right.$$

$$\left.+Q_+K(k)\left[\frac{Z(\varphi_1,k)}{\sin^2\varphi_1}+\frac{Z(\varphi_2,k)}{\sin^2\varphi_2}+\frac{Z(\varphi_3,k)}{\sin^2\varphi_3}\right]\right\}. \qquad (5.13)$$

This symmetric result can also be presented in terms of the elliptic integrals Π, using (see in [24])

$$K(k)Z(\varphi_i,k) = \cot\varphi_i\sqrt{1-k^2\sin^2\varphi_i}\,[\Pi(k^2\sin^2\varphi_i,k) - K(k)]. \qquad (5.14)$$

In principle, one can also derive the result (5.13) directly from the non-symmetric representation (1.7) (see [25]), in a tedious way relying on the use of several relations collected in the appendix, including the addition formula (A.9) for Jacobi Z functions.

It is worth noting that in a similar way one can obtain results for higher even dimensions D. For instance, in six dimensions we get

$$
\begin{aligned}
I_3^{(6)} = \frac{\pi^4}{144p^4}\Bigg\{ & \frac{Q_+^{1/2}}{20}[E(k)-K(k)]\Big[192\big(p^8+m_1^8+m_2^8+m_3^8\big) - 112\big(p^4+m_1^4+m_2^4+m_3^4\big)^2 \\
& - 6\big(p^2+m_1^2+m_2^2+m_3^2\big)^4 - 156\big(p^6+m_1^6+m_2^6+m_3^6\big)\big(p^2+m_1^2+m_2^2+m_3^2\big) \\
& + 83\big(p^4+m_1^4+m_2^4+m_3^4\big)\big(p^2+m_1^2+m_2^2+m_3^2\big)^2\Big] \\
& + \frac{1}{40}Q_- Q_+^{1/2} K(k)\big[3\big(p^2+m_1^2+m_2^2+m_3^2\big)^2 - 16\big(p^4+m_1^4+m_2^4+m_3^4\big)\big] \\
& + \frac{3}{4}\frac{Q_+^{5/2}K(k)}{\sin\varphi_1\sin\varphi_2\sin\varphi_3}\left[\frac{Z(\varphi_1,k)}{\sin^2\varphi_1} + \frac{Z(\varphi_2,k)}{\sin^2\varphi_2} + \frac{Z(\varphi_3,k)}{\sin^2\varphi_3}\right] \\
& - \frac{3}{8}Q_+^2\big(p^2+m_1^2+m_2^2+m_3^2\big)K(k)\left[\frac{Z(\varphi_1,k)}{\sin^4\varphi_1} + \frac{Z(\varphi_2,k)}{\sin^4\varphi_2} + \frac{Z(\varphi_3,k)}{\sin^4\varphi_3}\right]\Bigg\}.
\end{aligned}
\tag{5.15}
$$

As an alternative way to obtain results for higher values of D, the approach of the paper [9] may be used.

In the equal-mass case,

$$
\varphi_1 = \varphi_2 = \varphi_3 \equiv \varphi_{\mathrm{eq}} \qquad \sin\varphi_{\mathrm{eq}} = \frac{\sqrt{(p-m)(p+3m)}}{p+m} \qquad \cos\varphi_{\mathrm{eq}} = \frac{2m}{p+m}. \tag{5.16}
$$

Here, using equation (5.10) we get

$$
Z(\varphi_{\mathrm{eq}}, k_{\mathrm{eq}}) = \tfrac{1}{3}k_{\mathrm{eq}}^2 \sin^3\varphi_{\mathrm{eq}} \tag{5.17}
$$

with k_{eq} defined in equation (1.12). In this way, we reproduce equation (1.11), whereas for $D = 6$ equation (5.15) yields

$$
\begin{aligned}
\frac{\pi^4}{2880p^4}\sqrt{(p-m)^3(p+3m)}\big\{ & (p^4-9m^4)(p^4-42p^2m^2+9m^4)[E(k_{\mathrm{eq}})-K(k_{\mathrm{eq}})] \\
& + (p+m)^3(p-3m)(p^4-36p^2m^2+27m^4)K(k_{\mathrm{eq}})\big\}.
\end{aligned}
\tag{5.18}
$$

We note that equation (5.17) yields a reduction formula of $Z(\varphi, k)$, for a special case when

$$
k = \frac{\sqrt{1-2\cos\varphi}}{\sin\varphi(1-\cos\varphi)}.
$$

Another interesting limit corresponds to the case when one of the masses vanishes (for example, $m_3 \to 0$). This corresponds to the case $k \to 1$, when $E(k)$ is finite ($E(1) = 1$) whereas $K(k)$ develops logarithmic singularity. At $m_3 = 0$, $\cos\varphi_3 < 0$ and $\varphi_3 > \pi/2$, so that we need to use equation (A.7). Using equations listed in [24], we get

$$
\lim_{k\to 1}\{K(k)[\pm Z(\varphi,k) - \sin\varphi]\} = -\frac{1}{2}\ln\left(\frac{1+\sin\varphi}{1-\sin\varphi}\right) \tag{5.19}
$$

where plus or minus should be used for $\varphi < \pi/2$ or $\varphi > \pi/2$, respectively. Let us consider equation (5.13). Using equations (5.19) and (4.21) we see that singular terms containing $K(k)$

cancel, and we arrive at the following result:

$$\lim_{m_3\to 0} I_3 = \frac{\pi^2}{8p^2}\left\{\sqrt{Q_+}(p^2+m_1^2+m_2^2) + \frac{1}{2}(p^2-m_1^2-m_2^2)^2 \ln\left(\frac{p^2-m_1^2-m_2^2+\sqrt{Q_+}}{p^2-m_1^2-m_2^2-\sqrt{Q_+}}\right)\right.$$
$$-\frac{1}{2}(p^2+m_1^2-m_2^2)^2 \ln\left(\frac{p^2+m_1^2-m_2^2+\sqrt{Q_+}}{p^2+m_1^2-m_2^2-\sqrt{Q_+}}\right)$$
$$\left.-\frac{1}{2}(p^2-m_1^2+m_2^2)^2 \ln\left(\frac{p^2-m_1^2+m_2^2+\sqrt{Q_+}}{p^2-m_1^2+m_2^2-\sqrt{Q_+}}\right)\right\} \tag{5.20}$$

where $Q_+ = \lambda(p^2, m_1^2, m_2^2)$ in this limit. It is easy to check that this expression is equivalent to known results (see, e.g., [3, 8]). The advantage of our approach is that the symmetry with respect to any of the remaining masses is always explicit, whereas non-symmetric expressions such as equation (1.7) lead to the answers which are not explicitly symmetric (cf equation (57) of [8]).

6. Conclusion

We have considered several representations for the three-particle phase space, exploring their symmetry properties and geometrical meaning. It was shown that the angles φ_i defined in equations (4.20) and (4.21) are convenient to describe the results for the three-particle phase-space integral I_3. In terms of the Jacobian Z function (related to the elliptic integral Π through equation (5.14)), the result for I_3 in four dimensions is given in equation (5.13). It is very compact and explicitly symmetric with respect to all masses m_i. Note that the three zeta functions $Z(\varphi_i, k)$ are connected through the relation (5.10). This relation can be obtained by comparing the representation (5.7) for two-dimensional integral $I_3^{(2)}$ with another representation obtained by using the delta function properties.

In this way, we have shown how to transcribe the unsymmetric evaluation (1.2) of the phase-space integral into a form which is manifestly symmetric in the masses of the three decay products. Of course, the practical importance of this exercise is rather restricted, since (1.2) can be worked out numerically anyhow. Nevertheless, our result has an elegant structure and theoretical significance as it bears upon properties of elliptic functions which arise from elimination of variables in equations (2.13) and (2.16).

We have also considered the six-dimensional case. The result for $I_3^{(6)}$ is given in equation (5.15), also expressed in terms of $Z(\varphi_i, k)$.

Acknowledgments

We are pleased to acknowledge financial support from the Australian Research Council under grant number A00000780. Partial support from the Deutsche Forschungsgemeinschaft (AD) is also acknowledged.

Appendix. Elliptic integrals

The normal elliptic integrals of the first and second kind are defined as

$$F(\varphi, k) = \int_0^{\sin\varphi} \frac{dt}{\sqrt{(1-t^2)(1-k^2t^2)}} = \int_0^{\varphi} \frac{d\psi}{\sqrt{1-k^2\sin^2\psi}} \tag{A.1}$$

$$E(\varphi, k) = \int_0^{\sin\varphi} \mathrm{d}t \sqrt{\frac{1-k^2t^2}{1-t^2}} = \int_0^{\varphi} \mathrm{d}\psi \sqrt{1-k^2\sin^2\psi}. \tag{A.2}$$

At $\varphi = \pi/2$ we get the complete elliptic integrals,

$$K(k) = F\left(\tfrac{\pi}{2}, k\right) = \int_0^1 \frac{\mathrm{d}t}{\sqrt{(1-t^2)(1-k^2t^2)}} = \frac{\pi}{2}\, {}_2F_1\left(\left.\begin{matrix}-\frac{1}{2}, \frac{1}{2}\\ 1\end{matrix}\right| k^2\right) \tag{A.3}$$

$$E(k) = E\left(\tfrac{\pi}{2}, k\right) = \int_0^1 \mathrm{d}t \sqrt{\frac{1-k^2t^2}{1-t^2}} = \frac{\pi}{2}\, {}_2F_1\left(\left.\begin{matrix}\frac{1}{2}, \frac{1}{2}\\ 1\end{matrix}\right| k^2\right) \tag{A.4}$$

$$\Pi(c, k) = \int_0^1 \frac{\mathrm{d}t}{(1-ct^2)\sqrt{(1-t^2)(1-k^2t^2)}} = \frac{\pi}{2} F_1\left(\tfrac{1}{2}; 1, \tfrac{1}{2}; 1 | c, k^2\right) \tag{A.5}$$

where F_1 is the Appell hypergeometric function of two arguments.

The Jacobian zeta function, $Z(\beta, k)$, is defined through

$$K(k)Z(\beta, k) = K(k)E(\beta, k) - E(k)F(\beta, k). \tag{A.6}$$

We will assume that $0 \leqslant k < 1$. (In the limit $k \to 1$, $F(\beta, k)$ and $K(k)$ are singular.) From the definition (A.6) it is obvious that $Z\left(\frac{\pi}{2}, k\right) = 0$. Moreover, using symmetry properties of $E(\beta, k)$ and $F(\beta, k)$ (see in [24]),

$$E\left(\tfrac{\pi}{2} + \delta, k\right) = 2E(k) - E\left(\tfrac{\pi}{2} - \delta, k\right) \qquad F\left(\tfrac{\pi}{2} + \delta, k\right) = 2K(k) - F\left(\tfrac{\pi}{2} - \delta, k\right)$$

we get

$$Z\left(\tfrac{\pi}{2} + \delta, k\right) = -Z\left(\tfrac{\pi}{2} - \delta, k\right). \tag{A.7}$$

To represent the elliptic functions $\Pi(\alpha_i, k)$ occurring in equation (1.7) in terms of Z functions, we can use

$$\Pi\left(\alpha_i^2, k\right) = K(k) + \frac{\alpha_i K(k) Z(\beta_i, k)}{\sqrt{\left(1-\alpha_i^2\right)\left(k^2-\alpha_i^2\right)}} \tag{A.8}$$

with $\beta_i = \arcsin(\alpha_i/k)$. Equation (A.8) corresponds to case III on p 229 of [24], when $0 \leqslant \alpha_i^2 < k^2$.

The following addition formulae from [24] (p 34, equation (142.01)) are needed:

$$Z(\beta_1, k) \pm Z(\beta_2, k) = Z(\varphi_\pm, k) \pm k^2 \sin\beta_1 \sin\beta_2 \sin\varphi_\pm \tag{A.9}$$

where the angles

$$\varphi_\pm = 2\arctan\left[\frac{\sin\beta_1\sqrt{1-k^2\sin^2\beta_2} \pm \sin\beta_2\sqrt{1-k^2\sin^2\beta_1}}{\cos\beta_1 + \cos\beta_2}\right]. \tag{A.10}$$

In fact, the angles φ_- and φ_+ correspond to the angles φ_1 and φ_2 (see equations (4.20) and (4.21)), respectively. Moreover, using the same addition formula (A.9) with $\beta_{1,2}$ substituted by $\varphi_{1,2}$, we get the symmetric connection (5.10) between $Z(\varphi_i, k)$ $(i = 1, 2, 3)$.

To derive the result given in equation (5.12), one can use the integral tables of [24], along with the following relation:

$$2Z(\beta_1, k) = -Z(\varphi_3, k) + \frac{2k^2\sin^3\beta_1\cos\beta_1\sqrt{1-k^2\sin^2\beta_1}}{1-k^2\sin^4\beta_1}. \tag{A.11}$$

It corresponds to the last two lines of equation (141.01) on p 33 of [24], where $\varphi \leftrightarrow \pi - \varphi_3$.

References

[1] Dalitz R H 1953 *Phil. Mag.* **44** 1068
[2] Fabri E 1954 *Nuovo Cimento* **11** 479
[3] Almgren B 1968 *Ark. Fys.* **38** 161
[4] Bashir A, Delbourgo R and Roberts M L 2001 *J. Math. Phys.* **42** 5553
[5] Hagedorn R 1962 *Nuovo Cimento* **25** 1017
[6] Delbourgo R and Roberts M L 2003 *J. Phys. A: Math. Gen.* **36** 1719
[7] Källen G 1964 *Elementary Particle Physics* (Reading, MA: Addison-Wesley)
[8] Bauberger S, Berends F A, Böhm M and Buza M 1995 *Nucl. Phys.* B **434** 383
[9] Tarasov O V 1997 *Nucl. Phys.* B **502** 455
[10] Davydychev A I and Smirnov V A 1999 *Nucl. Phys.* B **554** 391
[11] Aste A and Trautmann D 2003 *Can. J. Phys.* **81** 1433
[12] Mandelstam S 1955 *Proc. R. Soc. Lond.* A **233** 248
Mandelstam S 1959 *Phys. Rev.* **115** 1741
[13] Kibble T W B 1960 *Phys. Rev.* **117** 1159
[14] Arbuzov B A, Boos E E, Kurennoy S S and Turashvili K Sh 1986 *Yad. Fiz.* **44** 1565
Arbuzov B A, Boos E E, Kurennoy S S and Turashvili K Sh 1986 *Sov. J. Nucl. Phys.* **44** 1017 (Engl. Transl.)
[15] Mendels E 1978 *Nuovo Cimento* A **45** 87
Mendels E 2002 *J. Math. Phys.* **43** 3011
[16] Prudnikov A P, Brychkov Yu A and Marichev O I 1990 *Integrals and Series* vol 2 (New York: Gordon and Breach)
[17] Davydychev A I 2000 *Phys. Rev.* D **61** 087701
[18] Byckling E and Kajantie K 1973 *Particle Kinematics* (New York: Wiley)
[19] Davydychev A I and Delbourgo R 1998 *J. Math. Phys.* **39** 4299
[20] Berends F A, Davydychev A I and Ussyukina N I 1998 *Phys. Lett.* B **426** 95
[21] Rajantie A K 1996 *Nucl. Phys.* B **480** 729
Rajantie A K 1998 *Nucl. Phys.* B **513** 761 (erratum)
[22] Groote S, Körner J G and Pivovarov A A 1999 *Nucl. Phys.* B **542** 515
Groote S and Pivovarov A A 2000 *Nucl. Phys.* B **580** 459
[23] Roberts M L 2003 *PhD Thesis* University of Tasmania, Hobart
[24] Byrd P F and Friedman M D 1954 *Handbook of Elliptic Integrals for Engineers and Physicists* (Berlin: Springer)
[25] Davydychev A I and Delbourgo R 2002 *Proc. 15th Biennial Congress of the Australian Institute of Physics (Sydney, Australia, July 2002)* (Adelaide: Causal Productions) p 144 (*Preprint* hep-th/0209233)

PHYSICAL REVIEW D **76,** 065009 (2007)

3-point off-shell vertex in scalar QED in arbitrary gauge and dimension

A. Bashir,[1] Y. Concha-Sánchez,[1] and R. Delbourgo[2]

[1]*Instituto de Física y Matemáticas, Universidad Michoacana de San Nicolás de Hidalgo, Apartado Postal 2-82, Morelia, Michoacán 58040, Mexico*

[2]*School of Mathematics and Physics, University of Tasmania, Locked Bag 37 GPO, Hobart 7001, Australia*

(Received 5 May 2007; published 18 September 2007)

We calculate the complete one-loop off-shell three-point scalar-photon vertex in arbitrary gauge and dimension for scalar quantum electrodynamics. Explicit results are presented for the particular cases of dimensions 3 and 4 both for massive and massless scalars. We then propose nonperturbative forms of this vertex that coincide with the perturbative answer to order e^2.

DOI: 10.1103/PhysRevD.76.065009 PACS numbers: 11.15.Tk, 11.30.Rd, 12.20.−m

I. INTRODUCTION

The nonperturbative structure of Green functions in gauge field theories has turned out to be a challenging problem. Aside from the complicated non-Abelian scenario of quantum chromodynamics (QCD), even simpler examples such as quantum electrodynamics (QED) have proved a hard nut to crack in the nonperturbative regime. Nevertheless, gauge covariance relations, such as the Ward-Fradkin-Green-Takahashi identity (WFGTI) [1], and the Landau-Khalatnikov-Fradkin transformations (LKFT) [2] contain vital clues about the Green functions. Guided by such relations, extensive work has been carried out to construct nonperturbative Green functions [3–5]. As well, perturbation theory is a reliable guide when constraining acceptable structures in the weak coupling limit [6–11].

In the context of perturbation theory, a systematic study of spinor QED was initiated by Ball and Chiu [6]. They decomposed the vertex into a "longitudinal" part which ensures that the WFGTI is satisfied, and a "transverse" part. In a basis where kinematic singularities are avoided, they gave off-shell results for the one-loop transverse vertex in 4 dimensions in the Fermi-Feynman gauge. Later on, Kızılersü, Reenders, and Pennington extended this result to an arbitrary covariant gauge [7]. Results for massless and massive QED3 were obtained afterwards [9–12]. These results were then generalized to arbitrary dimension by Davydychev, Osland, and Saks [13] in the realm of QCD (from which all QED results can be inferred). Whereas the bare fermion-boson vertex in a minimal coupling gauge theory is merely γ^μ, in general the vertex can be expanded out in terms of 12 spin amplitudes constructed from γ^μ and two independent four-momenta [14]. The WFGTI fixes four coefficients of the 12 spin amplitudes in terms of the fermion functions comprising the longitudinal component. The transverse part thus involves eight vectors with eight unconstrained scalar coefficients that depend on the gauge parameter ξ, the space-time dimension $d = 2\ell$, fermion masses, and three kinematical invariants (k^2, p^2, q^2); so this is a complicated problem even at one-loop order.

One might hope that, in the absence of spinorial matrices, scalar quantum electrodynamics (SQED) can offer a simpler platform to study nonperturbative solutions [15]. In this scenario, the 3-point vertex can be written in terms of just two independent four-momenta. The WFGTI fixes the coefficient of one of these. Therefore, there is only one unconstrained function which defines the transverse vertex—representing an 8-fold simplification of spinor QED/QCD. The trade-off is that additional four-point interactions occur in SQED. Thus the 1-loop scalar-photon vertex involves two additional Feynman diagrams. Ball and Chiu [6] carried out this calculation for massive scalars in the Fermi-Feynman gauge ($\xi = 1$) for $d = 4$. In this article, we extend their work to arbitrary dimension d and gauge ξ involving the one-loop scalar propagator along the way.

There are several reasons why this calculation is helpful: (i) it keeps track of the correct gauge covariance properties of the Green functions; (ii) one can take on-shell limits to check the gauge invariance of physical observables—this is not possible [13] if one only has results near four dimensions; (iii) SQED anyway has interest in lower dimensions, for example, nonperturbative SQED in $2 + 1$ and $0 + 1$ dimensions has been examined by [16,17], respectively; (iv) three dimensional field theories contain several features of corresponding four dimensional field theories at high temperatures [18].

We have organized the article as follows. In Sec. II we introduce the notation to calculate the three-point vertex, discuss its decomposition in the light of WFGTI, and give the expressions for one-loop scalar propagator and the longitudinal component of the three-point vertex. In Sec. III we evaluate the complete one-loop vertex in arbitrary gauge and dimensions and hence deduce an expression for its transverse component. We suggest three simple and natural constructions for its nonperturbative counterpart in Sec. IV and finish by discussing the so-called transverse Takahashi identities in Sec. V. An appendix serves to summarize many useful expressions arising from the Feynman integrals.

 © 2007 The American Physical Society

II. PRELIMINARIES

We shall start by setting out the notation, discussing the WFGTI and decomposing the 3-point vertex into longitudinal and transverse components. We then make use of the scalar propagator to present the longitudinal part fully at 1-loop order. Constant reference is made here and in the next section to various (mainly hypergeometric) functions which are listed in an appendix.

A. Notation

We define the bare quantities in the usual form: the scalar propagator $S^0(p) = 1/(p^2 - m^2)$, the photon propagator $\Delta^0_{\mu\nu} = -[g_{\mu\nu}p^2 - (1-\xi)p_\mu p_\nu]/p^4$, the 3-point vertex $\Gamma^0_\mu = (k+p)_\mu$, and the 4-point double photon vertex $e^2\Gamma^0_{\mu\nu} = e^2 g_{\mu\nu}$, where ξ is the general covariant gauge parameter (such that $\xi = 0$ corresponds to Landau gauge) and e is the usual QED coupling constant. The 3-point vertex up to one-loop order is diagrammatically represented in Fig. 1; it can be written in terms of two vectors alone, namely, k^μ and p^μ or, if preferred, $P^\mu \equiv (p+k)^\mu$ and $q^\mu \equiv (k-p)^\mu$. Because of the presence of the four-point vertex, there are two additional diagrams to be calculated in addition to the usual one required for the spinor QED.

The full 3-point vertex satisfies the usual WFGTI:

$$q_\mu \Gamma^\mu(k,p) = S^{-1}(k) - S^{-1}(p) \tag{1}$$

and has the nonsingular limit

$$\Gamma^\mu(p,p) = \partial S^{-1}(p)/\partial p_\mu \tag{2}$$

when $k \to p$. We can use (1) to construct the "longitudinal" part of the vertex :

$$\Gamma^\mu_L(k,p) = \frac{S^{-1}(k) - S^{-1}(p)}{k^2 - p^2}(k+p)^\mu. \tag{3}$$

The full vertex can then be written as

$$\Gamma^\mu(k,p) = \Gamma^\mu_L(k,p) + \Gamma^\mu_T(k,p), \tag{4}$$

where the "transverse" part satisfies

$$q_\mu \Gamma^\mu_T(k,p) = 0, \qquad \Gamma^\mu_T(p,p) = 0, \tag{5}$$

and can be expanded out only in terms of the basis vector

$$\begin{aligned} T^\mu(k,p) &= k\cdot q p^\mu - p\cdot q k^\mu \\ &= [q^\mu(k^2 - p^2) - (k+p)^\mu q^2]/2. \end{aligned} \tag{6}$$

Thus the full vertex is

$$\begin{aligned} \Gamma^\mu(k,p) &= \frac{S^{-1}(k) - S^{-1}(p)}{k^2 - p^2}(p+k)^\mu \\ &\quad + \tau(k^2, p^2, q^2) T^\mu(k,p). \end{aligned} \tag{7}$$

The coefficient τ is a Lorentz scalar function of k and p, and can be expressed in terms of the 3 invariants, k^2, p^2, and q^2. Thus knowing only one unknown function τ is sufficient to fix the full 3-point vertex completely in SQED, the rest being tied to the scalar propagator S.

B. Longitudinal vertex

At one-loop the scalar propagator is given by two diagrams but because massless tadpole type diagrams are zero in dimensional regularization (which we are adopting), only the first diagram contributes, see Fig. 2.

In arbitrary dimensions $d = 2\ell$ and gauge ξ, the inverse propagator at one-loop is given by

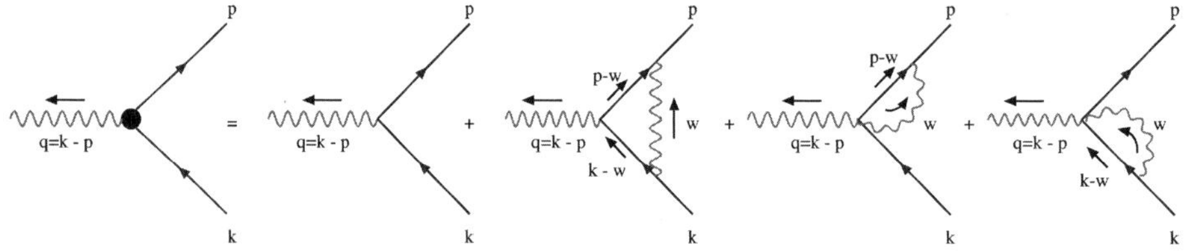

FIG. 1 (color online). One-loop 3-point vertex in SQED.

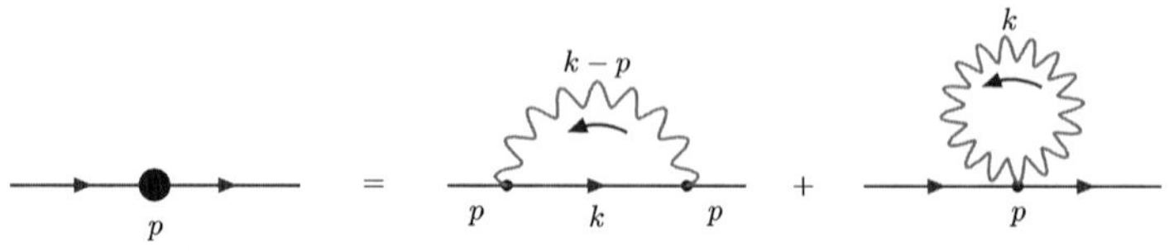

FIG. 2 (color online). One-loop scalar propagator.

$$S^{-1} = \frac{-e^2}{m^2}\left(\frac{m^2}{4\pi}\right)^\ell \Gamma(1-\ell) \times \left\{1 - 2\frac{(m^2+p^2)}{m^2}{}_2F_1\left(2-\ell,1;\ell;\frac{p^2}{m^2}\right) + (1-\xi)\frac{(m^2-p^2)^2}{m^4}{}_2F_1\left(3-\ell,2;\ell;\frac{p^2}{m^2}\right)\right\} \quad (8)$$

and readily yields the longitudinal part of the 3-point vertex at one loop:

$$\Gamma_L^\mu(k,p) = \frac{e^2\pi^2(k+p)^\mu}{(2\pi)^{2\ell}(k^2-p^2)}\{2Q_1(k)(m^2+k^2) - 2Q_1(p)(m^2+p^2) + (1-\xi)[(m^2-p^2)^2Q_3(p) - (m^2-k^2)^2Q_3(k)]\}, \quad (9)$$

where the functions $Q_i(p)$ are tabulated in the appendix.

III. ONE-LOOP VERTEX

In this section, we shall evaluate the complete off-shell one-loop vertex for all ξ and ℓ. Subtracting the longitudinal part from that produces the remaining transverse part.

A. The full vertex

The complete one-loop correction to the vertex is the sum of the three contributions that correspond to the last three graphs of Fig. 1:

$$\Lambda^\mu = \Lambda_1^\mu(k,p) + \Lambda_2^\mu(p) + \Lambda_2^\mu(k). \quad (10)$$

The first contribution involving only 3-point vertices is given by:

$$\Lambda_\mu^1 = \frac{-ie^2}{(2\pi)^{2\ell}}\{4(k\cdot p)(k+p)_\mu J^{(0)} + [-8(k\cdot p)g_\mu{}^\nu - 2(k+p)_\mu(k+p)^\nu]J_\nu^{(1)} + 4(k+p)^\nu J_{\mu\nu}^{(2)} + (k+p)_\mu K^{(0)} - 2K_\mu^{(1)} + (\xi-1)[(k+p)_\mu K^{(0)} + 4(k+p)_\mu p^\alpha k^\beta I_{\alpha\beta}^{(2)} - 8p^\alpha k^\beta I_{\mu\alpha\beta}^{(3)} - 2(k+p)_\mu \times (k+p)^\alpha J_\alpha^{(1)} + 4(k+p)^\alpha J_{\mu\alpha}^{(2)} - 2K_\mu^{(1)}]\}, \quad (11)$$

where

$$K^{(0)} = \int d^dw \frac{1}{[(p-w)^2-m^2][(k-w)^2-m^2]}, \qquad K_\mu^{(1)} = \int d^dw \frac{w_\mu}{[(p-w)^2-m^2][(k-w)^2-m^2]},$$
$$J^{(0)} = \int d^dw \frac{1}{w^2[(p-w)^2-m^2][(k-w)^2-m^2]}, \qquad J_\mu^{(1)} = \int d^dw \frac{w_\mu}{w^2[(p-w)^2-m^2][(k-w)^2-m^2]},$$
$$J_{\mu\nu}^{(2)} = \int d^dw \frac{w_\mu w_\nu}{w^2[(p-w)^2-m^2][(k-w)^2-m^2]}, \qquad I^{(0)} = \int d^dw \frac{1}{w^4[(p-w)^2-m^2][(k-w)^2-m^2]}, \quad (12)$$
$$I_\mu^{(1)} = \int d^dw \frac{w_\mu}{w^4[(p-w)^2-m^2][(k-w)^2-m^2]}, \qquad I_{\alpha\beta}^{(2)} = \int d^dw \frac{w_\alpha w_\beta}{w^4[(p-w)^2-m^2][(k-w)^2-m^2]},$$
$$I_{\mu\alpha\beta}^{(3)} = \int d^dw \frac{w_\mu w_\alpha w_\beta}{w^4[(p-w)^2-m^2][(k-w)^2-m^2]}.$$

The results of computing these integrals are provided in detail in the appendix where we also compare them with other calculations in the literature for some particular cases of d. The two Λ_2^μ contributions contain the 4-point vertices. They are relatively simple to evaluate as they contain only propagator type loops. Thus we only quote the final result:

$$\Lambda_2^\mu(p) = \frac{e^2\pi^2 p^\mu}{(2\pi)^{2\ell}}\left\{\left[3+\frac{m^2}{p^2}\right]Q_1(p) - \frac{\pi^{\ell-2}}{p^2}\Gamma(1-\ell)(m^2)^{\ell-1} + (\xi-1)\frac{(p^2-m^2)}{p^2} \times [Q_1(p) + (p^2-m^2)Q_3(p)]\right\}. \quad (13)$$

Equations (10)–(13) form the complete one-loop scalar-photon vertex for any ξ and ℓ at the one-loop level. This is a generalization to arbitrary dimension and gauge of the work of Ball and Chiu [6] who only examined the case $\xi = 1$, $\ell = 2$, using cutoff regularization. The explicit answers for the integrals (12) are stated in the appendix; as a general abbreviation we will write $X^{(0)} = i\pi^2 X_0/2$ in what follows and use $\{d, m\}$ as a superscript (or subscript) to signify dimension and mass, as and when needed.

B. The transverse vertex

The transverse vertex is obtained by subtracting the longitudinal vertex, Eq. (9), from the full one, Eqs. (10)–(13), at one loop. Carrying out this exercise, we arrive at the following coefficient τ of the transverse vector T^μ for massive scalars:

$$\tau_{d,m}(k^2,p^2,q^2) = \frac{e^2\pi^2}{2(2\pi)^d\Delta^2}\Big\{(k^2-2m^2+p^2-4k\cdot p)[-K_0+(m^2+k\cdot p)J_0] + \frac{2Q_1(p)}{k^2-p^2}[p^2(p^2-3k\cdot p) + k^2(k\cdot p-3p^2)-2m^2(p^2+k\cdot p)] - \frac{2Q_1(k)}{k^2-p^2}[k^2(k^2-3k\cdot p)+p^2(k\cdot p-3k^2)-2m^2(k^2+k\cdot p)] + (\xi-1)(m^2-k^2)(m^2-p^2)\Big[J_0-(k\cdot p+m^2)I_0 - \frac{2Q_3(p)}{k^2-p^2}(k\cdot p+p^2)+\frac{2Q_3(k)}{k^2-p^2}(k\cdot p+k^2)\Big]\Big\}, \quad (14)$$

where I_0, J_0, K_0, Q_{1-6} are explicitly stated in the appendix. $\Delta^2 \equiv (k\cdot p)^2 - k^2p^2 = (k\cdot q)^2 - k^2q^2$. Thus $4\Delta^2 = \lambda(k^2,p^2,q^2) = [p^4+k^4+q^4-2p^2k^2-2p^2q^2-2k^2q^2]$, the Källén function, and is related to the $(2\times\text{area})^2$ of a triangle with sides $\sqrt{k^2}$, $\sqrt{p^2}$, $\sqrt{q^2}$. In the massless limit, the above expression reduces to

$$\tau_{d,0} = \frac{e^2\pi^2}{2(2\pi)^d\Delta^2}\Big\{(k^2+p^2-4k\cdot p)[(k\cdot p)J_0^{d,0}-K_0^{d,0}] + \frac{2Q_1^{d,0}(p)}{k^2-p^2}[p^4-3(k^2+k\cdot p)p^2+k^2k\cdot p] - \frac{2Q_1^{d,0}(k)}{k^2-p^2}[k^4-3(p^2+k\cdot p)k^2+p^2k\cdot p] + (\xi-1)k^2p^2\Big[J_0^{d,0}-k\cdot pI_0^{d,0} - \frac{2Q_3^{d,0}(p)}{k^2-p^2}(k\cdot p+p^2) + \frac{2Q_3^{d,0}(k)}{k^2-p^2}(k\cdot p+k^2)\Big]\Big\}. \quad (15)$$

In the massive case for small $\epsilon = 2-\ell$ one gets

$$\tau_{4-2\epsilon,m} = \frac{\alpha}{8\pi\Delta^2}\Big\{(k^2-2m^2+p^2-4k\cdot p)[(m^2+k\cdot p)J_0^{4-2\epsilon,m}+2\mathcal{S}] + \frac{2L(p)}{k^2-p^2}(p^2(p^2-3k\cdot p)+k^2(k\cdot p-3p^2) - 2m^2(p^2+k\cdot p)) - \frac{2L(k)}{k^2-p^2}(k^2(k^2-3k\cdot p)+p^2(k\cdot p-3k^2)-2m^2(k^2+k\cdot p)) + (\xi-1)(m^2-k^2) \times(m^2-p^2)\Big[J_0^{4-2\epsilon,m} - \frac{2}{\chi}(k\cdot p+m^2)\Big(-q^2\mathcal{S}+\frac{p^2[(p^2-m^2)q^2+2m^2(k^2-p^2)]L(p)}{(p^2-m^2)^2} + \frac{k^2[(k^2-m^2)q^2-2m^2(k^2-p^2)]L(k)}{(k^2-m^2)^2}\Big) + 2\frac{(k\cdot p+p^2)(p^2+m^2)}{(k^2-p^2)(p^2-m^2)^2}L(p) - 2\frac{(k\cdot p+k^2)(k^2+m^2)}{(k^2-p^2)(k^2-m^2)^2}L(k)\Big]\Big\}. \quad (16)$$

The quantities $\mathcal{S}$, L can be found early on in the appendix. In the massless limit for small ϵ one reads off

$$\tau_{4-2\epsilon,0} = \frac{\alpha}{8\pi\Delta^2}\Big\{(k^2+p^2-4k\cdot p)\Big(k\cdot pJ_0^{4-2\epsilon,0}+\ln\frac{q^4}{p^2k^2}\Big) + \frac{(k^2+p^2)q^2-8p^2k^2}{p^2-k^2}\ln\frac{k^2}{p^2} + (\xi-1)k^2p^2\Big[J_0^{4-2\epsilon,0}+2\frac{p^2(k^2+k\cdot p)\ln(-p^2)}{k^2-p^2} - 2\frac{k^2(p^2+k\cdot p)}{k^2-p^2}\ln(-k^2)+2k\cdot p\ln(-q^2)\Big]\Big\}. \quad (17)$$

Note that for $\xi = 1$, it agrees with Eq. (2.9) of [6]. In the massive case with $d = 3$ we end up with

$$\tau_{3,m} = \frac{e^2}{16\pi\Delta^2}\Big\{(k^2-2m^2+p^2-4k\cdot p)\Big[(m^2+k\cdot p)J_0^{3,m}-2I\Big(\frac{q^2}{4}\Big)\Big] + 4(k^2(k^2-3k\cdot p)-2(k^2+k\cdot p)m^2 + (k\cdot p-3k^2)p^2)\frac{I(k^2)}{(k^2-p^2)} - 4(k^2(k\cdot p-3p^2)+p^2(p^2-3k\cdot p)-2m^2(k\cdot p+p^2))\frac{I(p^2)}{(k^2-p^2)} + (\xi-1)(m^2-k^2)(m^2-p^2)\Big[J_0^{3,m} - \frac{4m(k^2+k\cdot p)}{(k^2-m^2)^2(k^2-p^2)} + \frac{4m(p^2+k\cdot p)}{(m^2-p^2)^2(k^2-p^2)} - \frac{1}{\chi}\Big[2(m^2+k\cdot p)\Big(\frac{J_0}{2}(m^2+k\cdot p)q^2+m\Big(\frac{(k^2-m^2)q^2-(k^2+m^2)(k^2-p^2)}{(k^2-m^2)^2} - \frac{k^2-p^2+q^2}{m^2-p^2}\Big)\Big)\Big]\Big]\Big\}, \quad (18)$$

where χ is a geometrical quantity listed in the appendix. The result (18) simplifies remarkably in the massless limit:

$$\tau_{3,0} = \frac{e^2}{2\mathcal{K}\mathcal{P}\mathcal{Q}}\Big[\frac{\mathcal{K}^2+2\mathcal{Q}\mathcal{K}+\mathcal{P}^2+2\mathcal{P}\mathcal{Q}}{(\mathcal{K}+\mathcal{P})(\mathcal{K}+\mathcal{P}+\mathcal{Q})} + (\xi-1)\frac{1}{4}\Big], \quad (19)$$

where we have adopted the Euclidian notation $\sqrt{-k^2} = \mathcal{K}$, $\sqrt{-p^2} = \mathcal{P}$, $\sqrt{-q^2} = \mathcal{Q}$.

IV. ON THE NONPERTURBATIVE VERTEX

The one-loop expression for $\tau(k^2, p^2, q^2)$ provides a guide as to its possible form in the strong coupling regime. Any nonperturbative *ansatz* for the transverse vertex should reduce to the perturbative result evaluated above. Equations (8) and (14) suggest what τ might resemble for general e^2:

$$\begin{aligned}\tau_{d,m}(k^2,p^2,q^2) = {}& \frac{1}{4\Delta^2}\frac{[S^{-1}(k,\xi=1) - S^{-1}(p,\xi=1)]}{[(m^2+k^2)Q_1(k) - (m^2+p^2)Q_1(p)]} \times \Big\{(k^2 - 2m^2 + p^2 - 4k\cdot p)[-K_0 + (m^2 + k\cdot p)J_0] \\ & + \frac{2Q_1(p)}{k^2-p^2}[p^2(p^2 - 3k\cdot p) + k^2(k\cdot p - 3p^2) - 2m^2(p^2 + k\cdot p)] - \frac{2Q_1(k)}{k^2-p^2}[k^2(k^2 - 3k\cdot p) \\ & + p^2(k\cdot p - 3k^2) - 2m^2(k^2 + k\cdot p)]\Big\} + \frac{1}{2\Delta^2}\frac{[S^{-1}(k,\xi-1) - S^{-1}(p,\xi-1)]}{[(m^2-k^2)^2 Q_3(k) - (m^2-p^2)^2 Q_3(p)]}(m^2-k^2)(m^2-p^2) \\ & \times \Big\{J_0 - (k\cdot p + m^2)I_0 - \frac{2Q_3(p)}{k^2-p^2}(k\cdot p + p^2) + \frac{2Q_3(k)}{k^2-p^2}(k\cdot p + k^2)\Big\}. \end{aligned} \quad (20)$$

The notation here means that $S(p, \xi = 1)$ is the scalar propagator in the Fermi-Feynman gauge, whereas, $S(p, \xi - 1)$ is the coefficient of the scalar propagator proportional to $(\xi - 1)$. By construction, this expression reproduces the one-loop transverse vertex in the weak coupling regime. In specific dimensions and for the massless case, it simplifies. Of course its form is not unique but it is perhaps the simplest nonperturbative extension of our earlier results for any ξ and d. We expect that an identical two-loop calculation will help us pin down the exact structure better. Because of the lack of Dirac matrices, this two-loop calculation is not as formidable a task as for spinor QED or QCD. We are currently in the process of carrying it out.

Another approach is to tie in the asymptotic behavior with the anomalous dimension of the scalar field in 4d. To see how this is done, return to the one-loop self energy as obtained previously in Eq. (8); by using contiguity relations of hypergeometric functions, this can be cast in the simpler form

$$\begin{aligned}\Sigma(p) = {}& \frac{-e^2}{m^2}\left(\frac{m^2}{4\pi}\right)^{\ell}\Gamma(1-\ell)\Big[1 + 2(\ell-1)(1-\xi) \\ & + \{(1-2\ell) - \xi(3-2\ell)\}\left(1 + \frac{p^2}{m^2}\right) \\ & \times {}_2F_1\left(2-\ell, 1; \ell; \frac{p^2}{m^2}\right)\Big].\end{aligned}$$

According to the procedure for self-consistent regularization by higher-order corrections [19] we then make the substitution $\ell \to 2 + \gamma$ in Eq. (21) and renormalize by ensuring that the propagator behaves as $(-p^2)^{1+\gamma}$ as $p^2 \to \infty$. This gives the self-consistent asymptotic equation

$$\begin{aligned}\left(-\frac{p^2}{m^2}\right)^{1+\gamma} \sim {}& \frac{e^2}{16\pi^2}\frac{\Gamma^2(1+\gamma)}{\Gamma(2\gamma+2)}[(\xi-3)\Gamma(-\gamma) \\ & + 2(1-\xi)\Gamma(1-\gamma)]\left(1 + \frac{p^2}{m^2}\right)\left(1 - \frac{p^2}{m^2}\right)^{\gamma},\end{aligned} \quad (21)$$

which fixes $\gamma = (3 - \xi)e^2/16\pi^2 + O(e^4)$. But anyway it produces a nonperturbative form of the 4d propagator

$$\begin{aligned}S^{-1}(p) \simeq {}& \frac{e^2}{16(\gamma+1)\pi^2}[(\xi-3)\Gamma(-\gamma) + 2(1-\xi)\Gamma(1-\gamma)] \\ & \times\left(1 + \frac{p^2}{m^2}\right){}_2F_1(-\gamma, 1; \gamma+2; p^2/m^2).\end{aligned} \quad (22)$$

[This nonperturbative method succeeds in the ultraviolet but fails in the infrared limit $p^2 \to m^2$, when the propagator $S \sim 1/(p^2 - m^2)^{1+(3-\xi)e^2/8\pi^2}$. For infrared exponentiation it is much easier to resort to the gauge technique [3].] Exactly the same procedure can be applied to the transverse vertex. If τ is expressed in Feynman parametric form,

$$(p^2 - k^2)\tau(p,k,q) = \frac{4e^2\Gamma(2-\ell)}{(4\pi)^\ell}\int_0^1 d\sigma(1-\sigma)^{\ell-2}[(m^2-p^2\sigma)^{\ell-2} - (p\to k)] + \frac{e^2(2m^2+k^2+p^2-2q^2)\Gamma(3-\ell)}{(4\pi)^\ell}\int_0^1 d\sigma(1-\sigma)^{\ell-1}\int_{-1}^1 du u\mathcal{D}^{\ell-3} + \frac{e^2(\xi-1)(k^2-m^2)(p^2-m^2)\Gamma(4-\ell)}{(4\pi)^\ell}\int_0^1 d\sigma(1-\sigma)^{\ell-2}\int_{-1}^1 du u\mathcal{D}^{\ell-4}, \tag{23}$$

with $\mathcal{D} \equiv m^2 - q^2(1-\sigma)(1-u^2)/4 - p^2\sigma(1+u)/2 - k^2\sigma(1-u)$, then with the above substitution, an *ansatz* for the nonperturbative transverse vertex in 4d emerges:

$$(p^2 - k^2)\tau(p,k,q) = \frac{4e^2\Gamma(-\gamma)}{(4\pi)^2}\int_0^1 d\sigma(1-\sigma)^{\gamma}[(m^2-p^2\sigma)^{\gamma} - (p\to k)] + \frac{e^2(2m^2+k^2+p^2-2q^2)\Gamma(1-\gamma)}{(4\pi)^2}\int_0^1 d\sigma(1-\sigma)^{1+\gamma}\int_{-1}^1 du u\mathcal{D}^{\gamma-1} + \frac{e^2(\xi-1)(k^2-m^2)(p^2-m^2)\Gamma(2-\gamma)}{(4\pi)^2}\int_0^1 d\sigma(1-\sigma)^{\gamma}\int_{-1}^1 du u\mathcal{D}^{\gamma-2}, \tag{24}$$

That the anomalous dimension of the scalar field makes an appearance should come as no surprise: the WFGTI is at work. What is rather interesting about the form (24) is that for $q^2 = 0$ it takes the form $[F(p^2) - F(k^2)]/(p^2 - k^2)$ even though it is associated with the transverse piece; but for $q^2 \neq 0$ this particular structure disappears as one can see from the form of $\mathcal{D}$. Note anyway that (23) and (24) both have intrinsic dependences on all three variables p^2, k^2, and q^2 (or $p \cdot k$).

A third way of going nonperturbative relies upon dispersion relations; while these are well established for the two-point function, in the form of the Lehmann-Källén representation, they are trickier for the vertex function but can nevertheless be found as follows for graphs with triangular topology. Make the change of variable $\sigma \to m^2/W^2$ in Eq. (23)—so that W^2 runs from m^2 to ∞—in the denominator $\mathcal{D}$. This means we can generally write

$$S^{-1}(p) = \int_{m^2}^\infty dW^2 \frac{\rho(W^2)}{[p^2 - W^2 + i\epsilon]}, \quad \tau(p,k,q) = \int_{m^2}^\infty dW^2 \int_{-1}^1 du \frac{\mathcal{P}(W^2,u)}{[p^2(1+u)/2 + k^2(1-u)/2 + q^2(1-u^2)(W^2-m^2)/4 - W^2 + i\epsilon]}. \tag{25}$$

The idea is then to determine ρ and $\mathcal{P}$ self-consistently through the Schwinger-Dyson equations for the propagator and the vertex; the latter inevitably brings in the 4-point function, but we can use its own WGFTI to approximate it by connected 3-point graphs. While we have not solved this problem for $\mathcal{P}$, the idea has been taken to fruition [20] for ρ in SQED and QED, giving results that coincide with perturbation theory up to order e^4 for the charged field propagators. There is much more work involved in obtaining the spectral function $\mathcal{P}$ accurately to order e^4 and higher and this has not yet been done. These *ansätze* are unlikely to be the whole story. However, one may ask how close these are to the real vertex and how these can be compared to each other. By construction, our *ansätze* agree with perturbtion theory at the one-loop order.

(i) *Ansatz* (20) agrees with perturbation theory to $\mathcal{O}(e^2)$ in all momentum regimes, dimensions, and gauges. To see whether this relation between the vertex and the propagator survives at $\mathcal{O}(e^4)$, one needs to know these Green functions at that order.

(ii) *Ansatz* (24) would be in accord with the real vertex in 4 dimensions for large k^2 and p^2 but would fail for k^2 and $p^2 \approx m^2$. As one knows γ to $\mathcal{O}(e^2)$, it amounts to knowing the vertex to all orders in the leading logarithm approximation for asymptotically large values of momenta.

(iii) *Ansatz* (25) would agree with perturbation theory order by order depending upon the exact knowledge of the ρ and $\mathcal{P}$ functions to a given order. In principle, one could evaluate these functions nonperturbatively through SDEs. However, this exercise for $\mathcal{P}$ is a hard nut to crack.

A full two-loop calculation of the vertex should narrow down possible forms of any *ansatz*. Techniques for the two-loop vertex calculation, have been developed in [21–23]. All the master integrals for massless two-loop vertex diagram with three off-shell legs have been calculated in [24]. These advances indicate that the calculation of the two-loop transverse vertex should not be too difficult, at least for the massless case. This work is under progress.

V. ON THE TRANSVERSE TAKAHASHI IDENTITIES

Equation (20) is effectively a Ward-identity type relation linking the transverse vertex to the scalar propagator. There have been attempts to look for formal relations of this kind. Takahashi [25] discovered what are called transverse identities whose implications for the vertex have been examined for spinor QED [26–29]. In the case of SQED, as there is just one unknown which remains undetermined by the conventional WFGTI, it is tempting to look for a transverse Takahashi identity, hoping one might be able to determine the three-point vertex more realistically.

It should be noted that the general form of the vertex (7) shows that the transverse coefficient contributes to both basis vectors $P_\mu = (p+k)_\mu$ and $q_\mu = (k-p)_\mu$. The curl of the vertex $q_\mu\Gamma_\nu - q_\nu\Gamma_\mu$ will eliminate the component of Γ proportional to q, leaving us with the kinematic mixture $q_\mu P_\nu - q_\nu P_\mu$, multiplying the coefficient

$$\frac{S^{-1)}(p) - S^{-1}(k)}{p^2 - k^2} + q^2\tau(p^2, k^2, q^2)/2,$$

which is unwieldy. However the same kinematic combination can also be obtained by forming the "modified curl" $P_\mu\Gamma_\nu - P_\nu\Gamma_\mu$; this has the advantage of killing off the longitudinal part of Γ and *only bringing in* τ. We suggest that new identities involving the modified curl are more appropriate and will prove more promising for SQED.

ACKNOWLEDGMENTS

We thank A. Raya and M. E. Tejeda for valuable discussions. We acknowledge CIC, CONACyT, and COECYT for their financial support under grants No. 4.10, No. 46614-I, and No. C8070218-4, respectively.

APPENDIX

In this appendix we summarize the results for various integrals involved in the calculation of the 3-point vertex for quick reference. We write down the results for arbitrary d as well as small $\epsilon = 2 - \ell$ and $d = 3$. This way we aim to present all the integrals including the ones which have not been considered in the articles [6,7,9–11]. Wherever possible, we compare the results of specific cases with the ones in the above-mentioned articles.

1. Subsidiary quantities

There are a number of quantities which arise in various integrations that constantly appear later on, so we shall summarize them first and invoke them as they turn up. Unless specified we work in general dimension 2ℓ:

$$\begin{aligned}
\mathcal{S} &= \sqrt{1 - \frac{4m^2}{q^2}}\ln\left(\frac{\sqrt{1-\frac{4m^2}{q^2}}+1}{\sqrt{1-\frac{4m^2}{q^2}}-1}\right)\\
&= 2\sqrt{4m^2/q^2 - 1}\arctan(1/\sqrt{4m^2/q^2-1}),\\
I(p^2) &\equiv (1/\sqrt{-p^2})\arctan\sqrt{-p^2/m^2},\\
L(p^2) &\equiv (1 - m^2/p^2)\ln(1 - p^2/m^2),\\
Q_1(k) &\equiv (\pi m^2)^{\ell-2}\Gamma(1-\ell)\,{}_2F_1(2-\ell, 1; \ell; k^2/m^2),\\
\ell Q_2(k) &\equiv (\pi m^2)^{\ell-2}\Gamma(2-\ell)\,{}_2F_1(2-\ell, 1; \ell+1, k^2/m^2),\\
m^2 Q_3(k) &\equiv (\pi m^2)^{\ell-2}\Gamma(1-\ell)\,{}_2F_1(3-\ell, 2; \ell, k^2/m^2),\\
m^2 Q_4(k) &\equiv (\pi m^2)^{\ell-2}(2-\ell)\Gamma(-\ell)\\
&\quad\times {}_2F_1(3-\ell, 2; \ell+1; k^2/m^2),\\
Q_5(k) &\equiv -2i\pi^2(\pi m^2)^{\ell-2}\Gamma(2-\ell)\\
&\quad\times {}_2F_1(1/2, 1; \ell-1; 4m^2/q^2),\\
Q_6(p) &\equiv i\pi^\ell(\ell-3)(m^2)^{\ell-3}\Gamma(1-\ell)\\
&\quad\times {}_2F_1(1, 4-\ell; \ell; p^2/m^2),\\
\chi &\equiv m^2(k^2-p^2)^2 + (m^2-k^2)(m^2-p^2)q^2.
\end{aligned}$$

χ represents $(12 \times \text{volume})^2$ of a tetrahedron constructed with base triangle lengths $\sqrt{-k^2}$, $\sqrt{-p^2}$, $\sqrt{-q^2}$ and lateral sides to the apex of lengths $m, m, 0$; thus it has geometrical significance. It is worth noting the zero mass limits of the Q_i as we will consider such situations later:

$$\begin{aligned}
Q_1^{d,0}(k) &= -(-\pi k^2)^{\ell-2}\Gamma^2(\ell-1)\Gamma(2-\ell)/\Gamma(2\ell-2)\\
&= -Q_2^{d,0}(k),\\
2\pi Q_3^{d,0}(k) &= (-\pi k^2)^{\ell-3}\Gamma^2(\ell-2)\Gamma(3-\ell)/\Gamma(2\ell-4)\\
&= \pi Q_4^{d,0}(k)\\
(m^2)^{2-\ell}Q_5^{d,0}(k) &\to -2i\pi^\ell\Gamma(2-\ell),\\
Q_6^{d,0}(k)/m^2 &\to i\pi^\ell(-k^2)^{\ell-4}\Gamma(\ell)\Gamma(\ell-2)/\Gamma(2\ell-4)\\
\chi^{d,0} &= k^2p^2q^2 \text{ very simply.}
\end{aligned}$$

2. The K integrals

$K^{(0)}$ in the list (12) for arbitrary $d = 2\ell$ equals

$$\begin{aligned}
i\pi^2 K_0/2 &= K^{(0)}\\
&= i\pi^\ell\Gamma(2-\ell)(m^2)^{\ell-2}{}_2F_1\left(1, 2-\ell; \frac{3}{2}; \frac{q^2}{4m^2}\right),
\end{aligned} \tag{A1}$$

$$K_{d,0}^{(0)} = i\pi^\ell(-q^2)^{\ell-2}\frac{\Gamma^2(\ell-1)\Gamma(2-\ell)}{\Gamma(2\ell-2)}. \tag{A2}$$

Corresponding massive and massless expressions in the neighborhood of 4d ($\epsilon = 2-\ell$) are:

$$K^{(0)}_{4-2\epsilon,m} = i\pi^2[C - \mathcal{S}],$$
$$K^{(0)}_{4-2\epsilon,0} = i\pi^2[C - \ln(-q^2/m^2)], \quad \text{(A3)}$$

where

$$C = \frac{1}{\epsilon} - \gamma - \ln(\pi m^2) + 2. \quad \text{(A4)}$$

The first of the results (A3) agrees with Eqs. (44, 46–48) of [7]. When $d = 3$, this integral simplifies even more:

$$K^{(0)}_{3,m} = i\pi^2 I(q^2/4), \qquad K^{(0)}_{3,0} = i\pi^3/\sqrt{-q^2}. \quad \text{(A5)}$$

Expressions (A5) coincide with (A1) of [10] and (A3) of [11], respectively. For the $K^{(1)}_\mu$ integral in (12), it is easy to show that $K^{(1)}_\mu = (p + k)_\mu K^{(0)}/2$.

3. The J integrals

The $J^{(0)}$ integral can be found in various sources [7,8,10,11,30,31] the most general case (massive scalars and any d) has been discussed in [31]; we shall simply cite the known answers. J_0 is probably the most difficult one to work out as it brings in dilogarithmic or Spence (Sp) functions when ℓ is integer. For any ℓ the massless case has been given a completely elegant representation by Davydychev [30]:

$$J^{(0)} = 2i(-i\pi)^\ell (k^2, p^2, q^2)^{\ell-2} \frac{\Gamma^2(\ell-1)\Gamma(2-\ell)}{\Gamma(2\ell-2)} \times \left[\frac{(p^2q^2)^{2-\ell}}{p^2+q^2-k^2} {}_2F_1\left(1, 1/2; \ell - 1/2, -\frac{\Delta}{2(p^2+q^2-k^2)^2}\right) + \text{two perms} - \pi \frac{\Gamma(2\ell-2)}{\Gamma^2(\ell-1)} (2\Delta)^{\ell-3/2}\Theta \right] \quad \text{(A6)}$$

For massive scalars, like we have, the result $J_0^{4,m}$ in 4d is too lengthy (and uninformative) to quote. It is given in Eq. (16) of Ref. [7] and involves Spence functions of complicated arguments. In 3d the result is [11] easy to state:

$$J_0^{3,m} = \eta(k,p) I(\eta^2(k,p)\chi/4) + \eta(p,k) I(\eta^2(p,k)\chi/4);$$
$$\eta(k,p) = \frac{m^2(k^2-p^2)(2m^2-k^2-p^2)+\chi}{\chi(m^2-k^2)}. \quad \text{(A7)}$$

In the massless limit one obtains [8,10] from all of these forms,

$$J_0^{4,0} = \frac{2}{\Delta}\left[Sp\left(\frac{p\cdot q - \Delta}{-p^2}\right) - Sp\left(\frac{p\cdot q + \Delta}{-p^2}\right) + \frac{1}{2}\left(\frac{p^2 + p\cdot q - \Delta}{p^2 + p\cdot q + \Delta}\right) \ln\left(\frac{q^2}{p^2}\right)\right], \quad \text{(A8)}$$

$$J_0^{3,0} = \pi(\mathcal{K}^2 + \mathcal{P}^2)/\mathcal{K}\mathcal{P}\mathcal{Q}; \qquad \mathcal{K} \equiv \sqrt{-k^2},$$
$$\mathcal{P} \equiv \sqrt{-p^2}, \qquad \mathcal{Q} \equiv \sqrt{-q^2}. \quad \text{(A9)}$$

4. The J_μ integral

In its most general form, this can be written as

$$J^{(1)}_\mu = \frac{i\pi^2}{2}[k_\mu J_A(k,p) + p_\mu J_B(k,p)];$$
$$J_A(k,p) = J_B(p,k).$$

We find

$$J_A(k,p) = -\frac{1}{2\Delta^2}\{[p^2 - k\cdot p]K_0 + [p^2(k^2-m^2) - k\cdot p(p^2-m^2)]J_0 + 2p^2Q_1(p) - 2k\cdot pQ_1(k)\}. \quad \text{(A10)}$$

In the massless case,

$$J_A^{d,0}(k,p) = -\frac{1}{2\Delta^2}\{[p^2 - k\cdot p]K_0^{d,0} + p^2[k^2 - k\cdot p]J_0^{d,0} + 2p^2Q_1^{d,0}(p) - 2k\cdot pQ_1^{d,0}(k)\}. \quad \text{(A11)}$$

For small $\epsilon = 2 - \ell$, Eq. (A10) reduces to the following expressions (see (A16, A17, 3.9(a,b,c)) of [6] and (39–44) of [7])

$$J_A^{4-2\epsilon,m} = \frac{1}{2\Delta^2}\{-(m^2 p\cdot q + p^2 k\cdot q)J_0^{4-2\epsilon,m} + 2k\cdot pL(k) - 2p^2L(p) - 2p\cdot q\mathcal{S}\},$$
$$J_A^{4-2\epsilon,0} = \frac{1}{2\Delta^2}\{p^2(k\cdot p - k^2)J_0^{4-2\epsilon,0} + 2(p^2 - k\cdot p) \times \ln(-q^2) - 2p^2\ln(-p^2) + 2k\cdot p\ln(-k^2)\}, \quad \text{(A12)}$$

Similar results for $d = 3$ are:

$$J_A^{3,m} = \frac{1}{2\Delta^2}\{[p^2(k\cdot p - k^2) + m^2(p^2 - k\cdot p)]J_0^{3,m} + 2(k\cdot p - p^2)I(q^2/4) + 4p^2I(p^2) - 4k\cdot pI(k^2)\},$$
$$J_A^{3,0} = \frac{1}{2\Delta^2}\left\{[p^2(k\cdot p - k^2)]J_0^{3,0} - \frac{2\pi}{\sqrt{-q^2}}(p^2 - k\cdot p) + \frac{2\pi p^2}{\sqrt{-p^2}} - \frac{2\pi k\cdot p}{\sqrt{-k^2}}\right\}. \quad \text{(A13)}$$

The first of these expressions coincides with Eq. (A5) of [11] and the second with Eq. (A3) of [10] after the appropriate change of notation.

5. The $J^{(2)}_{\mu\nu}$ integral

From symmetry considerations, the integral $J^{(2)}_{\mu\nu}$ of the list (12) can be expanded out as follows:

$$J^{(2)}_{\mu\nu} = \frac{i\pi^2}{2}\left[\frac{g_{\mu\nu}}{2\ell}K_0 + \left(k_\mu k_\nu - g_{\mu\nu}\frac{k^2}{2\ell}\right)J_C + \left(p_\mu k_\nu + k_\mu p_\nu - g_{\mu\nu}\frac{(k\cdot p)}{\ell}\right)J_D + \left(p_\mu p_\nu - g_{\mu\nu}\frac{p^2}{2\ell}\right)J_E\right].$$

The coefficients J_C, J_D, and J_E in the above expressions are:

$$\begin{aligned}
J_C(k,p) &= \frac{1}{4(\ell-1)\Delta^2}\{[(2\ell-2)(p^2-m^2)k\cdot p - (2\ell-1)(k^2-m^2)p^2]J_A - (p^2-m^2)p^2 J_B \\
&\quad + [(2-\ell)p^2 + (\ell-1)k\cdot p]K_0 - 4(\ell-1)k\cdot pQ_2(k)\}, \\
J_E(k,p) &= J_C(p,k), \\
J_D(k,p) &= \frac{1}{8(\ell-1)\Delta^2}\{[2\ell k\cdot p(k^2-m^2) + (2-2\ell)k^2(p^2-m^2)]J_A + [2\ell(p^2-m^2)k\cdot p + (2-2\ell)p^2(k^2-m^2)]J_B \\
&\quad - [(\ell-1)q^2 + 2k\cdot p]K_0 + 4(\ell-1)[k^2Q_2(k) + p^2Q_2(p)]\}.
\end{aligned} \tag{A14}$$

In the massless case we obtain the following expressions:

$$\begin{aligned}
J_C^{d,0}(k,p) &= \frac{1}{4(\ell-1)\Delta^2}\{[(2\ell-2)p^2k\cdot p - (2\ell-1)k^2p^2]J_A^{d,0} - p^4J_B^{d,0} + [(2-\ell)p^2 + (\ell-1)k\cdot p]K_0^{d,0} \\
&\quad - 4(\ell-1)k\cdot pQ_2^{d,0}(k)\}, \\
J_D^{d,0}(k,p) &= \frac{1}{8(\ell-1)\Delta^2}\{[2\ell k\cdot pk^2 + (2-2\ell)k^2p^2]J_A^{d,0} + [2\ell p^2k\cdot p + (2-2\ell)p^2k^2]J_B^{d,0} \\
&\quad - [(\ell-1)q^2 + 2k\cdot p]K_0^{d,0} + 4(\ell-1)[k^2Q_2^{d,0}(k) + p^2Q_2^{d,0}(p)]\}.
\end{aligned} \tag{A15}$$

Then for small $\epsilon = 2-\ell$, we arrive at (compare Eq. (A16) with Eqs. (A18–A20) of [6] and Eqs. (49–51) of [7])

$$\begin{aligned}
J_C^{4-2\epsilon,m} &= \frac{1}{4\Delta^2}\{2p^2 + 2m^2k\cdot p/k^2 - 2k\cdot p\mathcal{S} + 2(k\cdot p)(1-m^2/k^2)L(k) + [2k\cdot p(p^2-m^2) + 3(m^2-k^2)p^2]J_A^{4-2\epsilon,m} \\
&\quad + p^2(m^2-p^2)J_B^{4-2\epsilon,m}\}, \\
J_D^{4-2\epsilon,m} &= \frac{1}{4\Delta^2}\{2k\cdot p[(k^2-m^2)J_A^{4-2\epsilon,m} + (p^2-m^2)J_B^{4-2\epsilon,m} - 1] - [m^2 - k^2\mathcal{S} + (k^2-m^2)L(k) \\
&\quad + k^2(p^2-m^2)J_A^{4-2\epsilon,m}] - [m^2 - p^2\mathcal{S} + (p^2-m^2)L(p) + p^2(k^2-m^2)J_B^{4-2\epsilon,m}]\}.
\end{aligned} \tag{A16}$$

$$\begin{aligned}
J_C^{4-2\epsilon,0} &= \frac{1}{4\Delta^2}\{p^2[2k\cdot p - 3k^2]J_A^{4-2\epsilon,0} - p^4J_B^{4-2\epsilon,0} + 2p^2 + 2k\cdot p\ln(k^2/q^2)\}, \\
J_D^{4-2\epsilon,0} &= \frac{1}{4\Delta^2}\{[k^2(2k\cdot p - p^2]J_A^{4-2\epsilon,0} + [p^2(2k\cdot p - k^2]J_B^{4-2\epsilon,0} + k^2\ln(q^2/k^2) + p^2\ln(q^2/p^2) - 2k\cdot p\}.
\end{aligned} \tag{A17}$$

Also for $d = 3$,

$$\begin{aligned}
J_C^{3,m} &= \frac{1}{2\Delta^2}\left\{[(p^2-m^2)k\cdot p - 2p^2(k^2-m^2)]J_A^{3,m} - p^2(p^2-m^2)J_B^{3,m} + \frac{2k\cdot p}{k^2}(m^2-k^2)I(k^2)\right. \\
&\quad \left. + (k\cdot p + p^2)I(q^2/4) - 2m\frac{k\cdot p}{k^2}\right\}, \\
J_D^{3,m} &= \frac{1}{4\Delta^2}\{[k^2(3k\cdot p - p^2) - m^2(3k\cdot p - k^2)]J_A^{3,m} + [p^2(3k\cdot p - p^2) - m^2(3k\cdot p - p^2)]J_B^{3,m} \\
&\quad - 2(m^2-k^2)I(k^2) - 2(m^2-p^2)I(p^2) - (k+p)^2I(q^2/4) + 4m\},
\end{aligned} \tag{A18}$$

$$\begin{aligned}
J_C^{3,0} &= \frac{1}{2\Delta^2}\{p^2(k\cdot p - 2k^2)J_A^{3,0} - p^4J_B^{3,0} - \pi k\cdot p/\sqrt{-k^2} + \pi p\cdot(p+k)/\sqrt{-q^2}\}, \\
J_D^{3,0} &= \frac{1}{4\Delta^2}\{k^2(3k\cdot p - p^2)J_A^{3,0} + p^2(3k\cdot p - p^2)J_B^{3,0} + \pi k^2/\sqrt{-k^2} + \pi p^2/\sqrt{-p^2} - \pi(k+p)^2/\sqrt{-q^2}\}.
\end{aligned} \tag{A19}$$

Equation (A18) is in agreement with (A6, A7) of [11] and Eq. (A19) with (A4) of [10].

6. The $I^{(0)}$ integral

The massive I_0 integral in arbitrary dimensions is given by

$$I^{(0)} = \frac{1}{2\chi}\{4(2-\ell)q^2(m^2+k\cdot p)J^{(0)} + (4m^2-q^2)Q_5(q) + [(q^2-2m^2)(p^2-m^2)+2m^2(k^2-m^2)]Q_6(p) + [(q^2-2m^2)(k^2-m^2)+2m^2(p^2-m^2)]Q_6(k)\}. \tag{A20}$$

Hence the massless case reduces to

$$I^{(0)}_{d,0} = \frac{i\pi^2}{k^2p^2}[(2-\ell)k\cdot pJ_0^{d,0} - k^2Q_3^{d,0}(k) - p^2Q_3^{d,0}(p) + q^2Q_3^{d,0}(q)]. \tag{A21}$$

When $\epsilon = 2-\ell$ is small, we have

$$I^{(0)}_{4-2\epsilon,m} = i\pi^2\left\{\frac{1}{\chi}\left[-q^2\mathcal{S} + p^2\frac{[(p^2-m^2)q^2+2m^2(k^2-p^2)]}{(p^2-m^2)^2}L(p) + k^2\frac{[(k^2-m^2)q^2-2m^2(k^2-p^2)]}{(k^2-m^2)^2}L(k)\right] - \frac{C-2}{(p^2-m^2)(k^2-m^2)}\right\},$$

$$I^{(0)}_{4-2\epsilon,0} = \frac{i\pi^2}{k^2p^2}\left[2-C+\ln\left(\frac{k^2p^2}{q^2m^2}\right)\right], \tag{A22}$$

whereas for $d=3$,

$$I^{(0)}_{3,m} = \frac{1}{\chi}\left\{q^2(m^2+k\cdot p)J^{(0)}_{3,m} + i\pi^2 m\left[\frac{q^2(k^2-m^2)-(k^2-p^2)(k^2+m^2)}{(k^2-m^2)^2} + \frac{q^2(p^2-m^2)+(k^2-p^2)(p^2+m^2)}{(p^2-m^2)^2}\right]\right\},$$

$$I^{(0)}_{3,0} = \frac{k\cdot p}{k^2p^2}J^{(0)}_{3,m}. \tag{A23}$$

The first of Eq. (A23) agrees with (A8) of [11].

7. The $I^{(1)}_\mu$ integral

In analogy with the integral $J^{(1)}_\mu$, $I^{(1)}_\mu$ can be expanded out as

$$I^{(1)}_\mu = \frac{i\pi^2}{2}[k_\mu I_A(k,p) + p_\mu I_B(k,p)];$$

$$I_B(k,p) = I_A(p,k), \tag{A24}$$

where

$$I_A(k,p) = \frac{1}{2\Delta^2}\{[k\cdot p(p^2-m^2) - p^2(k^2-m^2)]I_0 + [k\cdot p - p^2]J_0 + 2k\cdot pQ_3(k) - 2p^2Q_3(p)\}.$$

In the massless case

$$I_A^{d,0}(k,p) = \frac{1}{2\Delta^2}\{p^2[k\cdot p - k^2]I_0^{d,m} + [k\cdot p - p^2]J_0^{d,m} + 2k\cdot pQ_3^{d,0}(k) - 2p^2Q_3^{d,0}(p)\}. \tag{A25}$$

Near 4 dimensions

$$I_A^{4-2\epsilon,m} = \frac{1}{2\Delta^2}\left\{-k\cdot qJ_0^{4-2\epsilon,m} - 2q^2(m^2-k^2)\times(k^2-k\cdot p)\mathcal{S}/\chi + \frac{2L(p)}{(m^2-p^2)}[p^2-k\cdot p + p^2q^2(k^2-m^2)(m^2+k\cdot p)/\chi] + 2k^2q^2(m^2+k\cdot p)L(k)/\chi\right\}, \tag{A26}$$

$$I_A^{4-2\epsilon,0} = \frac{1}{2\Delta^2}\left\{[k\cdot p - p^2]J_0^{4-2\epsilon,0} - 2[k\cdot p - k^2]\times\frac{\ln(-q^2)}{k^2} + 2\frac{k\cdot p}{k^2}\ln(-p^2) - 2\ln(-k^2)\right\}. \tag{A27}$$

Equation (A26) is in agreement with the expression (53) of [7]. Similar answers for $d=3$ are

$$I_A^{3,m} = \frac{2}{\Delta^2}\left\{[k\cdot p(p^2-m^2) - p^2(k^2-m^2)]\frac{I_0^{3,m}}{4} + [k\cdot p - p^2]\frac{J_0^{3,m}}{4} + \frac{mp^2}{(m^2-p^2)^2} - \frac{mk\cdot p}{(m^2-k^2)^2}\right\}, \tag{A28}$$

$$I_A^{3,0} = -\frac{\pi}{k^2\sqrt{-q^2k^2p^2}}. \tag{A29}$$

Equation (A28) agrees with (A11) of [11].

8. The $I^{(2)}_{\mu\nu}$ integral

The integral $I^{(2)}_{\mu\nu}$ of the list (12) may be decomposed as follows:

$$I^{(2)}_{\mu\nu} = \frac{i\pi^2}{2}\left[\frac{g_{\mu\nu}}{2\ell}J_0 + \left(k_\mu k_\nu - g_{\mu\nu}\frac{k^2}{2\ell}\right)I_C + \left(p_\mu k_\nu + k_\mu p_\nu - g_{\mu\nu}\frac{k\cdot p}{\ell}\right)I_D + \left(p_\mu p_\nu - g_{\mu\nu}\frac{p^2}{2\ell}\right)I_E\right], \tag{A30}$$

$$I_E(k,p) = I_C(p,k), \tag{A31}$$

where, in arbitrary dimensions,

$$I_C(k,p) = \frac{1}{4(\ell-1)\Delta^2}\{[(2\ell-2)(p^2-m^2)k\cdot p - (2\ell-1)(k^2-m^2)p^2]I_A - (p^2-m^2)p^2 I_B + [(2\ell-2)k\cdot p - (2\ell-1)p^2]J_A - p^2 J_B + 2p^2 J_0 + 4(\ell-1)k\cdot p Q_4(k)\}, \tag{A32}$$

$$I_D(k,p) = \frac{1}{8(\ell-1)\Delta^2}\{[2\ell(k^2-m^2)k\cdot p - (2\ell-2) \times (p^2-m^2)k^2]I_A + [2\ell(p^2-m^2)k\cdot p - (2\ell-2)(k^2-m^2)p^2]I_B + [2\ell k\cdot p - (2\ell-2)k^2]J_A - 4k\cdot pJ_0 + [2\ell k\cdot p - (2\ell-2)p^2]J_B - 4(\ell-1) \times [p^2 Q_4(p) + k^2 Q_4(k)]\}. \tag{A33}$$

For the massless case, we arrive at the following simplified results

$$\begin{aligned} I_C^{d,0} &= \frac{1}{4(\ell-1)\Delta^2}\{p^2[(2\ell-2)k\cdot p - (2\ell-1)k^2]I_A^{d,0} - p^4 I_B^{d,0} + [(2\ell-2)k\cdot p - (2\ell-1)p^2]J_A^{d,0} \\ &\quad - p^2 J_B^{d,0} + 2p^2 J_0^{d,0} + 4(\ell-1)k\cdot p Q_4^{d,0}(k)\}, \\ I_D^{d,0} &= \frac{1}{4(\ell-1)\Delta^2}\{[\ell k\cdot p - (\ell-1)p^2](J_B^{d,0} + k^2 I_A^{d,0}) + [\ell k\cdot p - (\ell-1)k^2](J_A^{d,0} + p^2 I_B^{d,0}) \\ &\quad - 2k\cdot p J_0^{d,0} - 2(\ell-1)[p^2 Q_4^{d,0}(p) + k^2 Q_4^{d,0}(k)]\}. \end{aligned} \tag{A34}$$

Near 4 dimensions, these expressions yield

$$\begin{aligned} I_C^{4-2\epsilon,m} &= \frac{1}{4\Delta^2}\left\{2p^2 J_0^{4-2\epsilon,m} - \frac{4k\cdot p}{k^2}\left(1 + \frac{m^2 L(k)}{(k^2-m^2)}\right) + \{2k\cdot p - 3p^2\}J_A^{4-2\epsilon,m} - p^2 J_B^{4-2\epsilon,m}\right. \\ &\quad \left. + [-2k\cdot p(m^2-p^2) + 3p^2(m^2-k^2)]I_A^{4-2\epsilon,m} + p^2(m^2-p^2)I_B^{4-2\epsilon,m}\right\}, \\ I_D^{4-2\epsilon,m} &= \frac{1}{4\Delta^2}\left\{-2(k\cdot p)J_0^{4-2\epsilon,m} + 2\left(1 + \frac{m^2 L(k)}{(k^2-m^2)}\right) + 2\left(1 + \frac{m^2 L(p)}{(k^2-m^2)}\right) + (2k\cdot p - k^2)J_A^{4-2\epsilon,m}\right. \\ &\quad + (2k\cdot p - p^2)J_B^{4-2\epsilon,m} + [k^2(m^2-p^2) - 2k\cdot p(m^2-k^2)]I_A^{4-2\epsilon,m} \\ &\quad \left. + [p^2(m^2-k^2) - 2k\cdot p(m^2-p^2)]I_B^{4-2\epsilon,m}\right\}, \end{aligned} \tag{A35}$$

$$\begin{aligned} I_C^{4-2\epsilon,0} &= \frac{1}{4\Delta^2}\left\{2p^2 J_0^{4-2\epsilon,0} - 4\frac{k\cdot p}{k^2} + (2k\cdot p - 3p^2)J_A^{4-2\epsilon,0} - p^2 J_B^{4-2\epsilon,0} - p^4 I_B^{4-2\epsilon,0} + p^2(2k\cdot p - 3k^2)I_A^{4-2\epsilon,0}\right\}, \\ I_D^{4-2\epsilon,0} &= \frac{1}{4\Delta^2}\{-2k\cdot p J_0^{4-2\epsilon,0} + 4 + (2k\cdot p - k^2)J_A^{4-2\epsilon,0} + (2k\cdot p - p^2)J_B^{4-2\epsilon,0} - k^2(p^2 - 2k\cdot p)I_A^{4-2\epsilon,0} \\ &\quad - p^2(k^2 - 2k\cdot p)I_B^{4-2\epsilon,0}\}. \end{aligned} \tag{A36}$$

Finally for $d = 3$, we have

$$I_C^{3,m} = \frac{1}{2\Delta^2}\bigg\{2p^2 J_0^{3,m} + [p^2(k\cdot p - 2k^2) - m^2(k\cdot p - 2p^2)]I_A^{3,m} - p^2(p^2 - m^2)I_B^{3,m} + (k\cdot p - 2p^2)J_A^{3,m} - p^2 J_B^{3,m} + \frac{2mk\cdot p}{k^2(m^2-k^2)} - \frac{2k\cdot p}{k^2}I(k^2)\bigg\},$$

$$I_D^{3,m} = \frac{1}{4\Delta^2}\bigg\{-4k\cdot pJ_0^{3,m} + [k^2(3k\cdot p - p^2) - m^2(3k\cdot p - k^2)]I_A^{3,m} + [p^2(3k\cdot p - k^2) - m^2(3k\cdot p - p^2)]I_B^{3,m} + (3k\cdot p - k^2)J_A^{3,m} + (3k\cdot p - p^2)J_B^{3,m} - \frac{2m}{m^2-k^2} - \frac{2m}{m^2-p^2} + 2I(k^2) + 2I(p^2)\bigg\},$$

$$I_C^{3,0} = \frac{1}{2\Delta^2}\bigg\{p^2(k\cdot p - 2k^2)I_A^{3,0} - p^4 I_B^{3,0} + (k\cdot p - 2p^2)J_A^{3,0} - p^2 J_B^{3,0} - \frac{\pi k\cdot p}{k^2\sqrt{-k^2}} - \frac{4\pi p^2}{\sqrt{-k^2p^2q^2}}\bigg\},$$

$$I_D^{3,0} = \frac{1}{4\Delta^2}\bigg\{k^2(3k\cdot p - p^2)I_A^{3,0} + p^2(3k\cdot p - k^2)I_B^{3,0} + (3k\cdot p - k^2)J_A^{3,0} + (3k\cdot p - p^2)J_B^{3,0} + \frac{\pi}{\sqrt{-k^2}} + \frac{\pi}{\sqrt{-p^2}} + \frac{8k\cdot p}{\sqrt{-k^2p^2q^2}}\bigg\}. \tag{A37}$$

9. The $I^{(3)}_{\mu\alpha\beta}$ integral

The integral $I^{(3)}_{\mu\alpha\beta}$ comes contracted with vectors p^α and k^β so it is straightforward to show that

$$-8p^\alpha k^\beta I^{(3)}_{\mu\alpha\beta} = -2p^\alpha(k^2 - m^2)I^{(2)}_{\mu\alpha} - 2p^\alpha J^{(2)}_{\mu\alpha} + i\pi^2 p_\mu[Q_2(p) - (p^2 - m^2)Q_4(p)] - 2k^\beta(p^2 - m^2)I^{(2)}_{\mu\beta} - 2k^\beta J^{(2)}_{\mu\beta} + i\pi^2 k_\mu[Q_2(k) - (k^2 - m^2)Q_4(k)]. \tag{A38}$$

Therefore, we have $p^\alpha k^\beta I^{(3)}_{\mu\alpha\beta}$ in terms of integrals we already know.

[1] J. C. Ward, Phys. Rev. **78**, 182 (1950); E. S. Fradkin, Zh. Eksp. Teor. Fiz. **29**, 258 (1955) [Sov. Phys. JETP **2**, 361 (1956)]; H. S. Green, Proc. Phys. Soc. London Sect. A **66**, 873 (1953); Y. Takahashi, Nuovo Cimento **6**, 371 (1957).

[2] L. D. Landau and I. M. Khalatnikov, Zh. Eksp. Teor. Fiz. **29**, 89 (1956) [Sov. Phys. JETP **2**, 69 (1956)]; E. S. Fradkin, Sov. Phys. JETP **2**, 361 (1956); K. Johnson and B. Zumino, Phys. Rev. Lett. **3**, 351 (1959); B. Zumino, J. Math. Phys. (N.Y.) **1**, 1 (1960); S. Okubo, Nuovo Cimento **15**, 949 (1960); I. Bialynicki-Birula, Nuovo Cimento **17**, 951 (1960).

[3] A. Salam, Phys. Rev. **130**, 1287 (1963); J. Strathdee, Phys. Rev. **135**, B1428 (1964); R. Delbourgo and P. West, Phys. Lett. B **72**, 96 (1977); R. Delbourgo, Nuovo Cimento Soc. Ital. Fis. A **49**, 484 (1979); R. Delbourgo and B. W. Keck, J. Phys. G **6**, 275 (1980); R. Delbourgo, Aust. J. Phys. **52**, 681 (1999).

[4] D. C. Curtis and M. R. Pennington, Phys. Rev. D **42**, 4165 (1990); **44**, 536 (1991); **48**, 4933 (1993); A. Bashir and M. R. Pennington, Phys. Rev. D **50**, 7679 (1994); **53**, 4694 (1996).

[5] A. Bashir, Phys. Lett. B **491**, 280 (2000); A. Bashir and A. Raya, Phys. Rev. D **66**, 105005 (2002); A. Bashir and R. Delbourgo, J. Phys. A **37**, 6587 (2004); A. Bashir and A. Raya, in Proceedings of the 2004 International Workshop on Dynamical Symmetry Breaking, Nagoya, Japan, 2004, p. 257; Nucl. Phys. **B709**, 307 (2005); Nucl. Phys. B, Proc. Suppl. **141**, 259 (2005); arXiv:hep-ph/0511291; AIP Conf. Proc. **892**, 245 (2007).

[6] J. S. Ball and T-W. Chiu, Phys. Rev. D **22**, 2542 (1980).

[7] A. Kızılersü, M. Reenders, and M. R. Pennington, Phys. Rev. D **52**, 1242 (1995).

[8] A. Bashir, Kızılersü, and M. R. Pennington, Phys. Rev. D **57**, 1242 (1998).

[9] A. Bashir, A. Kızılersü, and M. R. Pennington, arXiv:hep-ph/9907418.

[10] A. Bashir, A. Kızılersü, and M. R. Pennington, Phys. Rev. D **62**, 085002 (2000).

[11] A. Bashir and A. Raya, Phys. Rev. D **64**, 105001 (2001).

[12] G. S. Adkins, M. Lymberopoulos, and D. D. Velkov, Phys. Rev. D **50**, 4194 (1994).

[13] A. I. Davydychev, P. Osland, and L. Saks, Phys. Rev. D **63**, 014022 (2000).

[14] J. Bernstein, *Elementary Particles and Their Currents* (Freeman, San Francisco, 1968).

[15] A. Salam and R. Delbourgo, Phys. Rev. **135**, B1398 (1964); R. Delbourgo and P. West, J. Phys. A **10**, 1049 (1977); R. Delbourgo, J. Phys. A **10**, 1369 (1977).

[16] A. B. Waites, Ph.D. thesis, University of Tasmania, Hobart, Australia, 1994.

[17] C. Savkli, F. Gross, and J. Tjon, Yad. Fiz. **68**, 874 (2005) [Phys. At. Nucl. **68**, 842 (2005)].

[18] D. J. Gross, R. D. Pisarski, and L. G. Yaffe, Rev. Mod. Phys. **53**, 43 (1981).

[19] R. Delbourgo, J. Phys. A **36**, 11697 (2003); D. Kreimer, Nucl. Phys. B, Proc. Suppl. **135**, 238 (2004).

[20] C.N. Parker, J. Phys. A **17**, 2873 (1984); R. Delbourgo and R.B. Zhang, J. Phys. A **17**, 3593 (1984).

[21] O.V. Tarasov, Phys. Rev. D **54**, 6479 (1996); J. Fleischer, F. Jegerlehner, and O.V. Tarasov, Nucl. Phys. **B566**, 423 (2000).

[22] N.I. Ussyukina and A.I. Davydychev, Phys. Lett. B **298**, 363 (1993); **332**, 159 (1994); **348**, 503 (1995).

[23] R. Bonciani, P. Mastrolia, and E. Remiddi, Nucl. Phys. **B661**, 289 (2003); **B676**, 399 (2004); **B702**, 359 (2004); P. Mastrolia and E. Remiddi, Nucl. Phys. **B664**, 341 (2003).

[24] T.G. Birthwright, E.W.N. Glover, and P. Marquard, J. High Energy Phys. 09 (2004) 042.

[25] Y. Takahashi, in *Quantum Field Theory* (Elsevier, New York, 1986), p. 19.

[26] K.-I. Kondo, Int. J. Mod. Phys. A **12**, 5651 (1997).

[27] H.-X He, F.C. Khanna, and Y. Takahashi, Phys. Lett. B **480**, 222 (2000).

[28] M.R. Pennington and R. Williams, J. Phys. G **32**, 2219 (2006).

[29] H.-X He and F.C. Khanna, Int. J. Mod. Phys. A **21**, 2541 (2006).

[30] A.I. Davydychev, Phys. Rev. D **61**, 087701 (2000).

[31] A.I. Davydychev, J. Math. Phys. (N.Y.) **33**, 358 (1992).

Part L
PROPERTY COORDINATES

L.0. Commentaries L

While the 'standard model' of particle physics reigns supreme for now, there have been numerous recent attempts at truly unifying the strong and electroweak forces of nature with gravity. They hail from Einstein's attempts at combining electromagnetism with gravity geometrically which consumed the latter years of his life. He was never completely convinced of the Kaluza-Klein approach of invoking higher dimensional spaces as they led to infinite towers of heavy states; nonetheless that is the basis and inspiration for many of the present attempts at unification, with the added incentive of tackling all the known forces. The inclusion of spinor degrees of freedom, as per supersymmetry, has complicated the higher-dimensional picture even more by twinning the known bosonic states with fermionic states, and vice versa. Unfortunately the experimental evidence for superpartners from the highest energy colliders and other sources is nonexistent; thus the unification hope remains as elusive as ever.

The works described below take a different tack. They enlarge the description of events by incorporating mathematically *scalar* anticommuting variables which act as coordinates for properties; they serve to characterise the 'what' of an event in addition to 'where-when'. Effectively they add fermionic variables to the four-dimensional bosonic backdrop. (Choosing these extra Grassmann coordinates to be scalar has to be done with care as the spin-statistics theorem should not be violated of course.) Because they are Lorentz scalar, expansions in property coordinates terminate rapidly and do not lead to spin profusion as some extensions to supersymmetry would have. Their melanges characterise the various 'generations'. Also each fermionic coordinate addition corresponds to reducing the overall effective dimension by unity; so it is our hope that in the final theory the universe will contain zero effective dimensions.

L.1 Delbourgo R. and Zhang R., Fermionic dimensions and Kaluza-Klein theory. Phys. Lett. B202, 296-300 (1989)

L.2 Delbourgo R., Jarvis P.D. and Warner R.C., Models for fermion generations based on five fermionic coordinates. Aust. J. Phys. 44, 135-147 (1991)

L.3 Delbourgo R., Jarvis P.D. and Warner R.C., Grassmann coordinates and Lie algebras for unified models. J. Math. Phys. 34, 3616-3641 (1993)

L.4 Delbourgo R. and Stack P.D., Where-when-what: The general relativity of space-time-property, Int. J. Mod. Phys. A29, 1450023 (2014)

L.5 Delbourgo R. and Stack P.D., General relativity for N properties, Mod. Phys. Lett. A31, 1650019 (2016)

L.6 Delbourgo R., The force and gravity of events, Mod. Phys. Lett. A31, 1630015 (2016)

In paper [L.1] we were basically flexing our muscles by simply appending two real fermionic coordinates to spacetime. We were encouraged to pursue this idea further by discovering that the curvature superscalar could lead to a unification of gravity and electromagnetism. There we also showed that doubling the coordinates, corresponding to an SU(2) algebra, could lead to a Yang-Mills Lagrangian; there were also indications that one could marry gravitational with electroweak interactions by distinguishing between left and right properties.

Papers [L.2] and [L.3] focussed on the number of properties needed to incorporate the known particles in their several generations and introduced the notion of self-duality for cutting down the number of independent multiplets that can arise in property expansions; we also listed the conditions needed for an anomaly-free theory. In [L.4] we emphasised that these fermionic coordinates were somehow fundamental for fully describing the nature of an event and we demonstrated that with one electricity property one could mimic the conclusions of [L.1] by unifying gravity and electromagnetism, with the inclusion of a cosmological term. (This was achieved by an algebraic computer programme developed by Paul Stack.) By the time [L.5] was written we had become fully conversant with the computational techniques and were able to determine the complete Lagrangian arising in such a scheme for any number of properties. Finally [L.6] is in the nature of a review article of this method for unifying the forces of nature, which is quite distinct from other approaches, and which comprises the standard model. It spells out the prediction of another quark having charge $-4/3$ and another possibly sterile neutrino.

In respect of anticommuting variables (perceived as annihilation/creation operators) the following papers by Casalbuoni and colleagues deserve special mention:

- R. Casalbuoni, Nuovo Cim A33, 115 (1976)
- A. Barducci, R. Casalbuoni, and L. Lusanna, Nuovo Cim. Lett. 19, 581 (1977)
- R. Casalbuoni and R. Gatto, Phys. Lett. B90, 1 (1980)
- R. Casalbuoni, S. De Curtis, D. Dominici, and R. Gatto, Phys. Lett. B155, 95 (1985)

as they are quite close in spirit with our own researches.

Volume 202, number 3 PHYSICS LETTERS B 10 March 1988

FERMIONIC DIMENSIONS AND KALUZA–KLEIN THEORY

R. DELBOURGO and R.B. ZHANG
Physics Department, University of Tasmania, Hobart, Tasmania, Australia 7005

Received 7 December 1987

Instead of appending extra bosonic dimensions to spacetime and needing to exorcise the higher modes, it is possible to construct Kaluza–Klein models in which the additional coordinates are fermionic and the higher modes do not arise. We erect a unified gravity/Yang–Mills theory on such a grassmannian framework and then discuss possible generalisations to other internal groups.

One of the major nuisances in ordinary Kaluza–Klein (KK) models is the persistence of higher modes associated with the extra bosonic dimensions [1]. The standard apology is to assume that the extra coordinates, corresponding to the compact(?) internal group, "curl up on themselves" into a very small size, thereby raising the excitations to Planck mass energies – although it must be admitted that the evidence and the mechanism for this contraction is not compelling [2]; in any event the higher modes always enter the quantised version and they do not ameliorate the renormalisability problem of current models [3]. For this and other reasons [4], KK models have, for the second time in their history, fallen out of favour and people have turned to string models for unification of forces. In this paper we wish to suggest that if the extra KK coordinates are instead taken as anticommuting, the difficulty with the higher modes essentially disappears and the KK construction may again become viable. (We will shortly present a unified gravity/Yang–Mills theory to substantiate this claim.) The inclusion of additional fermionic dimensions is anyhow a good way of visualising spin [5] and the trick can also be used to great effect in order to quantise gauge models in a BRST invariant manner [6]. In fact we are inclined to assert that the "existence" of extra Grassmann coordinates is more than just a computational device and should be taken quite seriously – it may well be the reason why we "cannot directly see" the extra dimensions.

Given that superspace leads one to the concept of extra Fermi coordinates [7], it is rather surprising that, apart from implementing BRST-invariant quantisation, and of course pursuing the enlargement of multiplets to supermultiplets, nobody has (as far as we know) thought of appending these dimensions in the true spirit of KK. To show that there are no obstacles to constructing a KK model with extra Grassmann dimensions, we shall append two anticommuting coordinates ξ^1 and ξ^2 to the commuting spacetime coordinates x^μ and with the same mass dimension. We need a minimum of two real ξ to carry any sort of internal structure, even a U(1) group. Further ξ are necessary if the internal symmetry is more elaborate.

With conjugation consistently defined as the operation [5]

$$(\xi^m)^* = -\xi_m = \xi^n \eta_{nm}\,, \quad \eta_{12} = 1\,,$$

the Sp(2) "rotation" operators

$$J_{mn} = \mathrm{i}(\xi_m \partial/\partial\xi^n + \xi_n \partial/\partial\xi^m)\,,$$

can be put into correspondence with SU(2) generators as follows, and the ξ then comprise a two-spinor representation:

$$J_{12} \to \mathrm{i}\sigma_3\,, \quad J_{11} \to \sigma_1 + \mathrm{i}\sigma_2\,, \quad J_{22} \to \sigma_1 - \mathrm{i}\sigma_2\,.$$

Volume 202, number 3 PHYSICS LETTERS B 10 March 1988

A particular subgroup, which we would like to associate with electromagnetism, is generated by U(1) rotations

$$\xi \to \exp[ie\Lambda\sigma_3]\xi ,$$

leaving $(\xi)^2 \equiv 2\xi^1\xi^2$ invariant. Given these rules and the basic "raising and lowering" index rules,

$$\xi^m = \eta^{mn}\xi_n , \quad \eta^{lm}\eta_{mn} = \delta^l{}_n , \quad \text{etc.,}$$

we are in a position to propose a correct ansatz for combined gravity–electromagnetism. First of all one has to recognise that the supermetric will be bosonic in the Bose–Bose and Fermi–Fermi sectors, but grassmannian in the Bose–Fermi sector. One must also pay heed to the fact that one is dealing with a twofold integration measure which will force us to extract two powers of ξ in any lagrangian by the Berezin integration rules. (With more complicated internal groups higher powers of ξ will be needed.) And last but not least one does not wish to introduce at the classical level, ghost fields with the wrong spin statistics. These restrictions lead us to the metric ansatz [#1]

$$G_{MN}(X) = \begin{pmatrix} g_{\mu\nu}(x)(1 - e^2\xi^2/2\kappa^2) - e^2\xi^2 A^2 & -e(A_\mu\xi)_n \\ -e(A_\nu\xi)_m & \eta_{mn} \end{pmatrix} \tag{1}$$

having the inverse

$$G^{MN}(X) = \begin{pmatrix} g^{\mu\nu}(x)(1 + e^2\xi^2/2\kappa^2) & e(\xi A^\mu)^n \\ e(\xi A^\nu)^m & \eta^{mn}(1 + \frac{1}{2}\xi^2 e^2 A^2) \end{pmatrix} . \tag{1'}$$

In (1), A stands for the matrix-valued field $iA\sigma_3$, e is the unit of charge and κ^2 is proportional to the newtonian gravitational constant. Note the similarity with ordinary five-dimensional KK theory. In our case the differential dx^5 is, roughly speaking, replaced by $\xi\, d\xi$ which is a nilpotent to third order: this is the key to the removal of unwanted higher modes.

It is helpful to write the line element in the form

$$ds^2 = dx^\mu g_{\mu\nu} dx^\nu (1 - e^2\xi^2/2\kappa^2) + [d\xi^m + (e\xi A_\nu)^m dx^\nu]\eta_{mn}[d\xi^n + (e\xi A_\mu)^n dx^\mu] .$$

Then one can easily see that under a "phase transformation" of the kind

$$\xi^m \to \xi^n[\exp(ie\Lambda\sigma_3)]_n^m$$

the field A must transform in the correct manner,

$$A_\mu \to A_\mu - \partial_\mu \Lambda .$$

This also leads one to the sechsbein $E_M{}^A$ and its inverse $(E_M{}^A E_A{}^N = \delta_M{}^N)$,

$$E_M{}^A = \begin{pmatrix} e_\mu{}^\alpha(1 - e^2\xi^2/4\kappa^2) & e(\xi A_\mu)^a \\ 0 & \delta_m{}^a \end{pmatrix}, \quad E_A{}^M = \begin{pmatrix} e_\alpha{}^\mu(1 + e^2\xi^2/4\kappa^2) & -e(\xi A_\alpha)^m \\ 0 & \delta_a{}^m \end{pmatrix}, \tag{2}$$

as well as the connection

$$\Phi_A{}^B = \begin{pmatrix} \Phi_\alpha{}^\beta & \Phi_\alpha{}^b \\ \Phi_a{}^\beta & \Phi_a{}^b \end{pmatrix},$$

[#1] Our notation is to reserve upper case Latin letters for supercoordinates and indices: lower case Greek letters are for spacetime indices, while lower case Latin letters are for Grassmann indices. Early letters correspond to tangent space, later letters to world coordinates. The flat metric is taken to be

$$I^{AB} = \begin{pmatrix} \eta^{\alpha\beta} & 0 \\ 0 & \eta^{ab} \end{pmatrix}.$$

Volume 202, number 3 PHYSICS LETTERS B 10 March 1988

which takes its values in the OSp(4/2) algebra and has the symmetry property (see footnote 1)

$$\Phi^{AB} = -[AB]\Phi^{BA} \quad \text{with} \quad \Phi^{AB} \equiv I^{AC}\Phi_C{}^B .$$

It is possible to solve for the connection by ensuring that

$$\mathrm{D}E^A = \mathrm{d}E^A + E^B\Phi_B{}^A = \left(-\epsilon^\alpha \frac{\mathrm{d}\xi^m \xi_m}{2\kappa^2} e^2 \quad 0 \right), \quad \epsilon^\alpha \equiv e_\mu{}^\alpha \, \mathrm{d}x^\mu ,$$

transforms covariantly. After some work we discover that

$$\Phi_A{}^B = \begin{pmatrix} \Phi^{(g)}{}_\alpha{}^\beta - (\xi\mathscr{A} + \mathrm{d}\xi)^c (e\kappa\xi F_\alpha{}^\beta)_c & -e(\xi F_{\mu\alpha})^b \mathrm{d}x^\mu \\ -e(F_\mu{}^\beta\xi)_a \mathrm{d}x^\mu & e\mathscr{A}_a{}^b \end{pmatrix},$$

where $\Phi^{(g)}$ is the normal gravitational connection and we have used the differential geometric abbreviation $\mathscr{A} = A_\mu \mathrm{d}x^\mu$, etc.

All this is with the express aim of finding the curvature tensor

$$R_A{}^B = \mathrm{d}\Phi_A{}^B + \Phi_A{}^C\Phi_C{}^B = \tfrac{1}{2}\mathrm{d}x^N \mathrm{d}x^M (R_{MN})_A{}^B ,$$

with the components expressed in terms of the connection as follows:

$$(R_{MN})_A{}^B = \partial_M(\Phi_N)_A{}^B - [MN]\partial_N(\Phi_M)_A{}^B + [NA][NC](\Phi_M)_A{}^C(\Phi_N)_C{}^B - [MN][MA][MC](\Phi_N)_A{}^C(\Phi_M)_C{}^B .$$

After some computation, we arrive at the scalar curvature

$$R = [B]E_B{}^N E_A{}^M [AN] I^{AC} (R_{MN})_C{}^B = \left(1 + \frac{\xi^2 e^2}{2\kappa^2}\right) R^{(g)} - e^2 (\xi F_{\alpha\beta})^a (F^{\beta\alpha})_a .$$

Consequently, when we proceed to construct an action for the gauge fields, in the normal way,

$$S = \int \mathrm{d}^4x \, \mathrm{d}^2\xi \, \sqrt{G} \, R/8e^2 ,$$

we inevitably finish up with the unified gravity/electromagnetic lagrangian

$$\mathscr{L} = \sqrt{-g}\left(\frac{R^{(g)}}{4\kappa^2} - \tfrac{1}{4} F_{\mu\nu}F^{\mu\nu} \right).$$

Furthermore, it is relatively easy to see how to incorporate matter fields within this formalism. Take a charged scalar field first. It must be described by a complex *fermionic* superfield,

$$\Phi(X) \equiv \xi^a \varphi_a(x) , \quad \Phi^\dagger(X) = \xi_a \dot{\varphi}^{\dagger a}(x)$$

in order (a) to carry a U(1) representation and (b) to embody the correct ξ factor when performing the subsequent Grassmann integration. One straightforwardly establishes that the KK action yields the "right" answer

$$S_\varphi = \int \mathrm{d}^6X \sqrt{G} \, G^{MN} \partial_M \Phi^\dagger \partial_N \Phi = \int \mathrm{d}^4x \sqrt{-g} \left(g^{\mu\nu}(\partial_\mu + \mathrm{i}eA_\mu)\varphi(\partial_\nu - \mathrm{i}eA_\nu)\varphi^\dagger + \frac{2e^2}{\kappa^2}\varphi^\dagger\varphi \right).$$

Notice the presence of a mass term for φ arising from the grassmannian part of the metric in combination with det(G). (Actually, it is possible to modify the value of the mass in the non-abelian generalisation by including additional factors of the type $1-\xi^2$.) A similar artifice works for the Dirac field. We define the complex *Bose* superfield

$$\Psi(X) = \xi^a \psi_a(x) ,$$

and identify the Grassmann gamma-matrix Γ^a with $(M\xi+\partial/M)^a \mathrm{i}\gamma_5$, introducing a mass factor M [#2]. The generalised Dirac action then properly reduces to

$$S_\psi = \int \mathrm{d}^6 X\, E\bar{\Psi} E_A{}^M \Gamma^A D_M \Psi = \int d^4x\,(\det e)\{\bar{\psi}[e_\alpha{}^\mu\gamma^\alpha(D_\mu+\mathrm{i}eA_\mu)-\mathrm{i}M\gamma_5]\psi+\bar{\psi}\, e\sigma^{\alpha\beta}F_{\alpha\beta}\mathrm{i}\gamma_5\psi/M\}\,.$$

Notice the occurrence of a Pauli interaction, just as it appears in the conventional theory. This spells trouble for renormalisation, but that was a lost cause anyway for R-gravity.

A more interesting development is the generalisation to $\mathrm{U}(n)$ symmetries say. Here we extend the coordinate ξ^1 to an n-fold spinor of the internal group, namely $(\xi^1,\xi^3,\xi^5,\ldots)$ and similarly deal with the conjugate ξ^2. Take the SU(2) case for definiteness where we must cope with a $2\oplus2$ spinor ξ^m where the metric is

$$I_{mn}=\begin{pmatrix} 0 & 0 & 1 & 0\\ 0 & 0 & 0 & 1\\ -1 & 0 & 0 & 0\\ 0 & -1 & 0 & 0\end{pmatrix},\quad \xi\equiv\begin{pmatrix}\xi^1\\ \xi^3\\ \xi^2\\ \xi^4\end{pmatrix}.$$

The bottom two components transform conjugately to the upper two components and for the electromagnetic matrix $A^\mu\sigma_3$ must be substituted the full $\mathrm{SU}(2)\otimes\mathrm{U}(1)$ matrix $A^{\mu i}\tau_i\otimes\sigma_3$: similarly for its curl, F. Here the coordinate invariance of the line element [#3]

$$\mathrm{d}s^2=\mathrm{d}x^\mu g_{\mu\nu}(x)\,\mathrm{d}x^\nu[1-2(e^2\xi^2/\kappa^2)^2]+\,[\mathrm{d}\xi^m+e(\xi A_\nu)^m\mathrm{d}x^\nu]\eta_{mn}[\mathrm{d}\xi^n+e(\xi A_\mu)^n\mathrm{d}x^\mu](1-2e^2\xi^2/\kappa^2)\,,$$

under SU(2) rotations of the fermionic coordinates

$$\xi\to\exp(\mathrm{i}\boldsymbol{\tau}\cdot\Lambda\sigma_3)\,\xi\equiv U\xi$$

immediately gives us the normal transformation rule of the gauge field,

$$A\to U(A-\mathrm{i}\partial/e)\,U^{-1}\,.$$

Also the field F, entering the superconnection, drops out quite simply:

$$F=\mathrm{d}A+eA\wedge A\,.$$

The classical action too is acceptable and, after integrating over the ξ, the lagrangian including fermion sources reads,

$$\mathscr{L}/\sqrt{-g}=R^{(g)}/16\kappa^2-\tfrac{1}{4}\boldsymbol{F}_{\mu\nu}\cdot\boldsymbol{F}_{\mu\nu}+\bar{\psi}(e_\alpha{}^\mu\mathrm{i}\gamma^\alpha\nabla_\mu-\mathrm{i}M\gamma_5+e\sigma^{\alpha\beta}F_{\alpha\beta}\mathrm{i}\gamma_5/M)\psi\,.$$

Undoubtedly one can append an extra two (BRST) Grassmann variables in order to quantise the model and bring out the BRST symmetry most transparently, as has already been done for conventional KK models [8].

Thus it appears that we have a KK scheme which can accommodate the gauge and matter fields without bringing in higher modes associated with extra bosonic dimensions. Unification of chromodynamics and gravity is immediate if the fermionic coordinates (and their conjugates) form a six-dimensional entity and no heed is paid to the spacetime nature of the fields, beyond summing appropriately over the spinor indices. However if one is envisaging a unification of gravity and electroweak theory then more careful attention must be paid to the spin properties of the ξ. Success seems within reach if one uses two left-handed ξ plus one right-handed ξ (both doubled for spin); and if one assigns the matter fields to a supermultiplet

$$\Psi(X)=\xi_{\mathrm{R}}{}^r\psi_{r\mathrm{R}}(x)+\xi_{\mathrm{L}}{}^l\psi_{l\mathrm{L}}(x)\,,\quad r=1,2\,;\quad l=1,2,3,4\,,$$

[#2] The need for a single power of ξ^a in Γ^a is clear if the term $\Gamma^a\partial_a$ is to reproduce a mass term. The $\mathrm{i}\gamma_5$ is needed to ensure $\{\Gamma^a,\Gamma^a\}=0$, and for the rest $[\Gamma^a,\Gamma^b]=2\eta^{ab}(2+2\xi\partial_\xi+\xi^2M^2+\partial_\xi^2/M^2)$ reduces to the scaling operator when acting on wavefunctions that are linear in ξ. However there is more flexibility in identifying Γ^a than we have suggested and the last word on this has surely not been spoken.

[#3] Note the additional factors of ξ which enter the metric so as to guarantee that the Berezin integral will reduce the action to the right four-dimensional form.

Volume 202, number 3 PHYSICS LETTERS B 10 March 1988

and the Higgs boson to the supermultiplet

$$\Phi(X)=\xi_{\mathrm{R}r}\xi^{\mathrm{L}l}\varphi_l^{\,r}(x)\ .$$

We believe that this new approach to KK shows considerable promise and we intend to tackle specific models in a future publication, where details of the scheme, which have inevitably been omitted in this compressed letter, will be spelled out.

This work was supported by an ARGS grant.

References

[1] A. Salam and J. Strathdee, Ann. Phys. 141 (1982) 316.
[2] T. Applequist and A. Chodos, Phys. Rev. D 28 (1983) 772;
A. Chodos and E. Myers, Ann. Phys. 156 (1984) 412;
M.A. Rubin and B.D. Roth, Nucl. Phys. 226 (1983) 444.
[3] R. Delbourgo and R.O. Weber, Nuovo Cimento A 92 (1986) 347.
[4] E. Witten, Shelter Island Conf. Proc. (MIT Press, Cambridge, 1985).
[5] R. Delbourgo, Grassmann Wavefunctions and Intrinsic Spin, Int. J. Mod. Phys., to be published.
[6] L. Bonora, P. Pasti and M. Tonin, Nuovo Cimento 64 A (1981) 307;
R. Delbourgo and P.D. Jarvis, J. Phys. A 15 (1982) 611;
L. Baulieu and J. Thierry-Mieg, Nucl. Phys. B 197 (1982) 477;
F. Ore and P. van Nieuwenhuizen, Nucl. Phys. B 204 (1982) 317;
R. Delbourgo, P. Jarvis and G. Thompson, Phys. Rev. D 26 (1982) 775.
[7] A. Salam and J. Strathdee, Nucl. Phys. B 76 (1974) 477;
S. Ferrara, J. Wess and B. Zumino, Phys. Lett. B 51 (1974) 239.
[8] R. Delbourgo, P. Jarvis and G. Thompson, J. Phys. A 15 (1982) 2813;
Y. Ohkuwa, Phys. Lett. B 114 (1982) 315.

Aust. J. Phys., 1991, **44**, 135–47

Models for Fermion Generations based on Five Fermionic Coordinates*

R. Delbourgo, P. D. Jarvis[A] *and R. C. Warner*

Department of Physics, University of Tasmania,
G.P.O. Box 252C, Hobart, Tas. 7001, Australia.
[A] Alexander von Humboldt Fellow.

Abstract

We show that a limited range of options for fermion families may be neatly encompassed in a spacetime augmented by five Grassmann internal coordinates if we require that the superfields are self-dual in an SU(5) sense. Amongst the possibilities is a family of just three standard model generations. We consider the nature of Higgs fields in this formalism and the form of possible gauge symmetries.

1. Introduction

The idea that internal space is founded not on bosonic but on fermionic coordinates has many attractions. Perhaps the strongest is the fact that internal multiplet representations are strongly circumscribed by the terminating and antisymmetric character of series expansions in such coordinates. The concept has been advanced in earlier papers by Dondi and Jarvis (1980), Casalbuoni and Gatto (1979, 1980), Delbourgo, *et al.* (1988) and Krolikowski (1989) but elements of freedom in the construction mean that the various formulations have differed to a considerable extent. Nevertheless, each model is subject to strong constraints and has to confront the recently established fact that only three light generations of standard particle families occur in nature. (We refer to the LEP collider experiments counting neutrino species via the Z-decay width. For a review see Denegri *et al.* 1990.)

A model based on five internal complex coordinates θ (Delbourgo 1989) has been advocated as an economical way of ensuring the correct multiplet structure and at the same time being capable of containing at least three families. A scheme was offered there in which superfields were expanded in either even *or* odd powers of θ, with the quarks and lepton fields appearing as coefficients in 5's and 10's, according to the standard SU(5) assignments once the quantum numbers of the coordinates are specified. In the most recent attempt at this unification (Delbourgo and White 1990), a predilection towards bosonic superfields was shown: particle multiplets were tied to *odd* powers in θ, and generations were connected with contracted $\bar{\theta}\theta$ factors, since this does not affect the SU(5) character of the states. As a result three families of quarks and leptons arose; but in addition a neutral singlet plus two

* Dedicated to Professor Ian McCarthy on the occasion of his sixtieth birthday.

0004-9506/91/020135$05.00

additional 5's were entrained, the latter causing anomaly discomfiture. The alternative version in which superfields are fermionic was excluded because those expansions in *even* powers in θ, augmented by $\bar{\theta}\theta$, will not admit three generations whatever other attractions they may possess.

In this paper we wish to explore a broader version of this Grassmann coordinate model, including the possibility of utilising both Bose and Fermi superfields to describe fermions, by returning to the basic idea of θ expansion to *all* orders. The additional concept we shall incorporate (to keep the resulting proliferation of fermion fields under control) is the notion of Grassmann self-duality. At present some detailed features of the dual operation remain to be refined, so that we restrict ourselves to a catalogue of the small number of possibilities that we see with our present understanding of duality. It is hoped that the few scenarios we outline may be of interest, and may even strike a chord of familiarity with experienced model-builders. In the near future we hope to narrow the options further, at least with regard to a more refined duality rule. For the present, the consequence is a classification scheme in which we find that it appears possible to construct models with two, three or four standard model generations (with the occasional right-handed neutrino), and the possibility of including either of a pair of nonstandard but anomaly-free fermion families, which contain both normal and exotic SU(3)×SU(2)×U(1) fermion representations. We shall describe in some detail how the Grassmannian SU(5) duality idea can serve to restore the balance between 5's and 10's in the previous models, and how when combined with anomaly considerations it leads us to the other models indicated. We then discuss briefly some aspects of the possible gauging of symmetries in the Grassmann picture, and make some remarks about Higgs fields and interaction Lagrangians.

Table 1. Full set of SU(5) multiplets contained in the coordinate expansion of a superfield Ψ

The column and row numbers s and r correspond to the term $(\bar{\theta})^s(\theta)^r$ in the Taylor series for Ψ

$r\backslash s$	0	1	2	3	4	5
0	1	5	10	$\overline{10}$	$\overline{5}$	1
1	$\overline{5}$	1+24	5+$\overline{45}$	10+$\overline{40}$	$\overline{10}$+$\overline{15}$	$\overline{5}$
2	$\overline{10}$	$\overline{5}$+45	1+24+75	5+$\overline{45}$+$\overline{50}$	10+$\overline{40}$	$\overline{10}$
3	10	$\overline{10}$+40	$\overline{5}$+45+50	1+24+75	5+$\overline{45}$	10
4	5	10+15	$\overline{10}$+40	$\overline{5}$+45	1+24	5
5	1	5	10	$\overline{10}$	$\overline{5}$	1

2. Superfield Structure, Fermions and Duality

Our internal space is based on five complex fermionic θ coordinates and their conjugates $\bar{\theta}$. Superfields are functions of these variables and their components emerge through the antisymmetric θ products of the Taylor expansion. In order to clarify the significance of the duality and hermiticity constraints that we have in mind, let us begin by listing the full set of SU(5) multiplets, standard and exotic, contained in such a superfield, before any conditions are imposed. The set is summarised in Table 1, where the column s refers to the power of $\bar{\theta}$ and the row r refers to the power of θ: a totality of 1024 states. We first discuss the question of available (and desirable) fermion representations which has been the major focus of our research to date, postponing to later sections the matter of Higgs fields and gauge symmetries.

We hope to extract our fermion degrees of freedom from this array. It should be noted that an alternative (more simplistic?) way to partially combat the problem of proliferating representations is to disregard mixed tensor representations of SU(5) in the set of monomials, i.e. to discard all the multiplets that are not 1's, 5's and 10's. This is the essential content of the models mentioned in the Introduction where two possibilities of three generations with unwelcome additional 5's, or of two generations plus two additional 10's, could be produced from considering bosonic or fermionic superfields (odd or even numbers of Grassmann powers) respectively. Those representations can be easily seen running diagonally down the table, the additional powers of θ and $\bar{\theta}$ for each diagonal step down the table being contracted so as to leave the SU(5) nature unchanged.

One example of the possibilities of Grassmann duality will be to suggest that we can use it to combine these two unsatisfactory models into one—not with seven generations of 5's and $\overline{10}$'s as might be feared, but with precisely three.

For our first set of models we shall demand that all 5's and $\overline{10}$'s have a fixed chirality (right-handed with our assignments of the θ). Further, we shall insist that the $\bar{5}$'s and 10's have the opposite chirality in order to identify them as the conjugates of the previous multiplets; in other words we shall require that the superfield Ψ has some kind of hermiticity property. This essentially produces a halving of the states, with an alternation of chiralities throughout the upper right triangle (Jarvis and White 1990), and the lower left triangle which represents its conjugate. This consideration is implicit in the previous models since we do not wish at present to generate SU(5) mirror representation fermion models. Because the monomials residing on the main diagonal are self-conjugate, the question of their behaviour under hermiticity will be a point of some subtlety. One may regard these terms as representing fields which would have insufficient degrees of freedom to represent fermions. It would be possible to assume that the superfield is antisymmetric in a block diagonal sense such that components of type $(\bar{\theta})^r\psi(r,r)(\theta)^r$ vanish. Thus all $r=s$ components, like $\bar{\theta}^I\phi_I^J\theta_J$, along the main diagonal, i.e. those 1's, 24's and 75's, would be eliminated.

Clearly, even with some type of hermiticity condition there are too many multiplets for comfort and we need to find some method of reducing them. The previous models carried out this reduction by restricting to fermionic or bosonic superfields, but suffered from the fact that this produced sets of representations which were unacceptable from an anomaly standpoint. If we were to simply take all the 5's and $\overline{10}$'s in Table 1 we would have an excessive seven generation (but anomaly-free) model. Within the scheme of taking all the unmixed tensor products of θ's there would also be a single right-handed neutrino from the θ^5 monomial.

An appropriate cutting tool here, and in models based on the full set of monomials, is Grassmann duality, a concept which was introduced previously (Delbourgo and White 1990) in an effort to construct invariant Lagrangians as Berezinian integrals; here we want to make more extensive use of the idea.

The key point is that taking the dual of a superfield (denoted by a breve sign) does not disturb the SU(5) properties of the superfield components. It is implemented by the following rule: if a superfield is expanded in the form

$$\Phi(\theta) = \sum_{r,s=0}^{5} (\bar{\theta})^s(\theta)^r \phi(r,s),$$

the dual superfield employs the self-same component fields, but assigned to duals of the θ monomials:

$$\breve{\Phi} = \sum_{r,s=0}^{5} (\bar{\theta})^{5-r}(\theta)^{5-s} \phi(r,s),$$

where, to be concrete, for the present work we interpret the dual of the typical multicoordinate term $\bar{\theta}^I \theta_P \theta_Q \theta_R$ to be $\epsilon_{PQRST}\bar{\theta}^S\bar{\theta}^T\epsilon^{IJKLM}\theta_J\theta_K\theta_L\theta_M/2!4!$, as an example of the general rule

$$(\bar{\theta})^s(\theta)^r \rightarrow (\bar{\theta})^{5-r}(\theta)^{5-s}.$$

In particular, we note that polynomials of degree 5 in the Grassmann coordinates $(\theta,\ \bar{\theta})$ are self-dual in the above sense up to possible phase factors. That is to say a superfield of the type $\Phi = (\bar{\theta})^5 N$ is invariant under the duality transformation, i.e. $\Phi = \breve{\Phi}$; whereas the dual of 1 is $(\bar{\theta})^5(\theta)^5 = (\bar{\theta}\theta)^5/5!$, and so on. In particular, while phase factors associated with duality are likely to be harmless in terms of reducing numbers of degrees of freedom for most entries in the table, they are critical (when combined with the rule for identification with the dual component) for the fate of the self-dual monomials on the cross-diagonal. These phases also interact with the hermiticity conditions, especially on the leading diagonal. They can also be sufficently constraining to eliminate an entire class (e.g. even powers) of monomials.

In the context of Table 1, hermitian conjugation corresponds to a reflection about the main diagonal whereas duality, with its substitution rule $r,s \rightarrow (5-s),(5-r)$ corresponds to reflection about the cross diagonal. It can be shown that for a set of monomials of given order, $(\bar{\theta})^s(\theta)^r$, the sign factor associated with taking the dual using the rule above alternates ($\pm$) with the number of $(\bar{\theta}\theta)$ contractions in the monomial.

To obtain the halving of states in the upper right triangle we shall impose a duality condition on the superfield Ψ. We are still refining the exact form of the dual transformation, in search of a compelling group-theoretic or super-geometrical formulation that will prescribe phases. For the present work we will use the dual transformation rule above with its ordering choices etc., and characterise some possible detailed phase assignments by including them in the imposition of the duality condition. When these phases are constructed using operators, such as those counting numbers of θ's etc., in monomials, they can lead to complicated restrictions on degrees of freedom.

This brings us to our first set of models. Taking only the unmixed tensor representations of SU(5) we reconsider the set of fermions contained in the bosonic superfield, i.e. odd monomials mentioned previously (Delbourgo and White 1990), but we impose the duality condition

$$\Psi = (-1)\breve{\Psi}.$$

With the dual rule above this produces a theory with two generations of 5's and $\overline{10}$'s, since it identifies the representations about the cross diagonal and eliminates the 5 and the singlet, but not the $\overline{10}$, on the cross diagonal. Of course by invoking self-duality instead of anti-self-duality we could have produced a model with three 5's and one $\overline{10}$ and one right-handed weak singlet neutrino.

As a third option we could excise all the representations on the cross-diagonal leaving two 5's and one $\overline{10}$. This would be a matter of selecting a slightly different dual identification, such as

$$\Psi = -(-1)^{\nu}\breve{\Psi},$$

where the operator ν counts the number of $\bar{\theta}\theta$ contractions in each monomial of the Ψ superfield expansion.

The point of this last case is that when taken together with a fermionic superfield (again over just the completely antisymmetric representations) with some type of dual folding about the cross-diagonal, which will therefore contain one 5 and two $\overline{10}$'s, we have three generations of fermion representations appropriate to the standard model. The exact nature of the dual relation used for the fermionic superfield should be less important as there are no self-dual monomials involved. The option of keeping all three cross-diagonal representations (5, $\overline{10}$, 1) would lead to a fourth generation distinguished by having a right-handed neutrino. It should be noted that the duality relation used above for the bosonic superfield is unsuitable for the fermionic one as the phases introduced by the operator $(-1)^{\nu}$ lead to the field components all being set to zero. Note that using the explicit basic dual rule above on all the monomials in Table 1 directly leads via (anti)-self-duality to various combinations of 5's and $\overline{10}$'s, but none of them is suitable from the standpoint of anomaly cancellation.

It might be thought unaesthetic to use both bosonic and fermionic superfields from the point of view of statistics, or it might be regarded as artificial to restrict consideration to the unmixed tensors of SU(5). Accordingly we now turn to an examination of the possibilities that open up if we extend our analysis to a more democratic acceptance of all the monomials of Table 1, rather than just the maximally contracted set. This gives us more options and as we are still developing our detailed idea of duality we choose, for the present, to be guided by searches for acceptable sets of representations.

To see what may be appropriate, we examine the requirement of anomaly cancellation. In Table 2 we list the contribution to the anomaly (King 1981) for each of the representations in question.

Table 2. Anomaly coefficients for SU(5) representations of fixed chirality

SU(5) representation	1, 24, 75	5, 10	15	40	45	50
Anomaly coefficient	0	1	9	16	6	15

Let us count the left-handed multiplets in the upper triangle remembering that $\bar{N}_R$ is the conjugate of N_L and possesses precisely the same anomaly coefficient. There are 6 singlets, 4×24's and 2×75's along the main diagonal, which do not contribute. This leaves $7{\times}\bar{5}$'s, $7{\times}10$'s, $3{\times}45$'s, $1{\times}15$'s, $1{\times}50$'s, $2{\times}\overline{40}$'s and one singlet off the diagonal ($s > r$). We consider first the possibility of taking both odd and even monomials as we did above. The anomaly coefficients along the middle of the second row cancel out as they consist of a standard SU(5) family plus the exotic combination (King 1981), $\overline{10}_R+\overline{15}_R+40_R+\overline{45}_R$. Hence if we are going to invoke some sort of duality condition to halve the upper triangle in this extended model, we must seek a mechanism whereby, along the cross diagonal, the $(\bar{\theta})^3(\theta)^2$ and the singlet $(\bar{\theta})^5$ are eliminated, without also destroying the $(\bar{\theta})^4(\theta)$ term which provides the 10 and 15; otherwise the anomaly cancellation will be endangered.

Once again the subtle aspects of duality are only really needed for the bosonic part of the superfield (odd monomials). Therefore in order to get an anomaly-free theory we take, for the odd monomials,

$$\Psi = (-1)^{\nu+(|\bar{n}-n|+1)/2}\,\breve{\Psi}$$

as the duality condition on Ψ, where the operators n, $\bar{n}$ and ν count the numbers of θ's, $\bar{\theta}$'s and $\bar{\theta}\theta$ contractions in the terms of the superfield expansion. This has the desired effect on the cross-diagonal (by construction). This dual relation cannot be extended to the even monomials because in combining it with either hermiticity or antihermiticity relations

$$\Psi = \pm\bar{\Psi}$$

the conflict of the two conditions eliminates them all. The situation for the even monomials is as before—they are simply identified across the cross-diagonal.

Table 3. Set of SU(5) multiplets contained in the coordinate expansion of bosonic and fermionic superfields after imposition of the duality condition

Asterisks signify Taylor components related to those explicitly listed by conjugation and/or duality while zeros denote vanishing components. The R and L subscripts signify right- and left-handed chiralities

$r\backslash s$	0	1	2	3	4	5
0	0	5_R	10_L	$\overline{10}_R$	$\bar{5}_L$	0
1	*	0	$5_R+\overline{45}_R$	$10_L+\overline{40}_L$	$\overline{10}_R+\overline{15}_R$	*
2	*	*	0	0	*	*
3	*	*	0	0	*	*
4	*	0	*	*	0	*
5	0	*	*	*	*	0

The only families that survive both these constraints are 3 standard generations of 5's and $\overline{10}$'s, and one exotic family consisting of the SU(5) right-handed multiplets $\overline{10}+\overline{15}+40+\overline{45}$, as shown in Table 3.

The inclusion of this last set appears a small price to pay for an anomaly-free grand supermultiplet, considering how many states we started from! It could of course also be regarded as an extravagant way to use up a surplus $\overline{10}$! For all we know, SU(5) theory (see O'Raifeartaigh 1986 and Ross 1984 for reviews) may need these fields when the accelerators push on to higher energies and new physics opens up. They should do no harm provided one can arrange that they become sufficiently heavy. It should be pointed out that the exotic family above itself (just) leads to a failure of SU(5) asymptotic freedom so that some of its members at least will need to be given masses at the GUT scale, or alternatively the GUT symmetry group to be gauged needs to be larger than SU(5). In this regard it is perhaps worth recalling (see King 1981 for details) that this combination is part of a 144 dimensional vector-spinor representation of SO(10), so that if in some modification of the present scenario it were possible to retain three singlet fermions (e.g. from the leading diagonal) to complete the standard generations as 16's of SO(10), as might be possible with a more subtle dual rule, then there would be a possibility of embedding the present collection of fermions into an asymptotically free set of SO(10) representations as discussed by King (1981).

A third class of models emerges if we persevere in attempting to obtain the fermions only from a bosonic superfield by extending the approach of Delbourgo and White (1990) to the larger set of representations in Table 1. In this case we are only considering the odd monomials and there is another possible assignment of chiralities which involves opposite chirality for alternate rows.

Table 4. Set of SU(5) multiplets contained in the coordinate expansion of fermionic superfields after imposition of the simple duality condition for the alternative chirality assignment

Asterisks signify Taylor components related to those explicitly listed by conjugation and/or duality while zeros denote vanishing components. The *R* and *L* subscripts signify right- and left- handed chiralities

$r\backslash s$	0	1	2	3	4	5
0	0	5_R	0	$\overline{10}_R$	0	1_R
1	*	0	$5_L+\overline{45}_L$	0	$\overline{15}_L$	0
2	0	*	0	$5_R+\overline{50}_R$	0	*
3	*	0	*	0	*	0
4	0	*	0	*	0	*
5	*	0	*	0	*	0

Taking this set of representations and using the simple dual rule of this paper we find that the condition

$$\Psi = \breve{\Psi}$$

leads to the elimination of the cross-diagonal representations $\overline{45}$ and $\overline{10}$, leaving one standard $5_R+\overline{10}_R$ generation, one singlet, one 5_L, one 5_R, and the anomaly-free combination $\overline{15}_L+\overline{45}_L+\overline{50}_R$ as shown in Table 4. The latter set of representations contains quarks and leptons with conventional quantum numbers as well as exotic particles. Again this set of representations is not asymptotically free in SU(5) and so we will need to see that some of the fermions (preferably the colour octets and sextets) become massive at the unification scale.

While the SU(2) representations of the normal fields are not all simple replicas of the conventional standard model generations, it is interesting that the lepton sector of this model system contains exactly three sets of left- and right-handed neutrinos and charged partners, with the only surplus being a doubly charged lepton and its antiparticle. In the colour sector there are three charge 2/3 and four charge −1/3 triplets, together with one set each of charge 4/3 and 5/3 triplets. In addition there are also sextets and octets.

Amusingly, this model has just enough leptons for three generations of left–right symmetric type and just enough charge 2/3 quarks for three generations, while there is a surplus of 'down' quarks. As indicated above, the SU(2) assignments of the fermions are not all of the standard type and it may be necessary to invoke some mixing before the physical particles emerge. It should be remarked that the 5_L and 5_R are associated with different θ monomials so that they cannot have bare Lagrangian mass terms. One can regard the order of the monomial (or the number ν of $\bar{\theta}\theta$ contractions) as a distinguishing quantum number.

If the third generation of leptons had not been experimentally observed it would have been possible to entertain the idea of a variant of the bosonic superfield model which retained only the self-dual monomials of the cross-diagonal. This would have involved a further change in the chirality assignment compared to that in Table 4, requiring that while the basic alternation should be retained, it would be supplemented by an alternation for successive contractions within each (s,r) set. The resulting model would have particle content similiar to the preceding system except for the absence of the 5_L and 5_R, and the fact that the 5_R, $\overline{45}_L$ and $\overline{10}_R$ would be associated with different θ monomials to the case above. In such a model two essentially conventional left–right symmetric generations could have been found (with the additional doubly charged lepton) together with a third generation of unaccompanied quarks, while the imbalance in numbers of 'up' and 'down' type quarks would also have been avoided. Exotic coloured particles and higher charge quarks and leptons would of course remain.

To summarise this section—we find that within the Grassmann coordinate framework for describing how fermion representations and generations emerge, there are only a small number of models that are acceptable from an anomaly point of view, and that could confront experiment with any chance of success, once duality concepts are incorporated. The simplest choice would appear to be exactly three standard model generations. Except for the possible addition of a fourth generation containing a right-handed neutrino, the other alternatives involve a very non-standard presentation of the candidates for known particles, as well as an assortment of exotic ones.

The issues of mass generation, the representations of Higgs fields and their Lagrangians, mixing of the fermions in the Grassmann model, as well as the question of the exact form of the gauge group for the symmetries of each model (which may well depend on the nature of the decimation of the set of monomials in Table 1) will clearly be of great importance in determining whether any of the present models can be promoted to a full unified scheme, rather than just a motivating tool for selecting representations. The natural SU(5) symmetry that is introduced by the use of the five complex Grassmann variables may be reduced by the identifications made by duality conditions, although it might be part of a larger symmetry group, as we shall discuss in a later section. Another feature of this scheme is that the requirement to produce action terms from Berezin integrations over products of superfields provides a major restriction on the nature of possible interactions.

3. Superfields for Higgs Scalars

The question of how to incorporate the Higgs scalars is far from obvious, whichever approach is adopted for the fermions. It is not clear for example if both Bose and Fermi superfields should be utilised. In any case the fields must be a subset of the monomials in Table 1, but now with opposite statistics fields attached. It is still true that we are faced once again with too many SU(5) multiplets and that we need to pare them down by some means or another, even if we tried to limit ourselves to just the unmixed tensor products of θ's. However, it should be noted that to break SU(5) gauge symmetry, we are obliged to include a Higgs 24, so that terms from the main diagonal of Table 1 are also needed, suggesting that we should consider a hermitian Higgs bosonic superfield Φ at the very least; beyond that it may be self-dual or anti-self-dual—we have no means of telling, since there is as yet no direct evidence for any Higgs bosons! The question of also including a fermionic Higgs superfield (odd monomials) could be regarded as depending on whether the fermions were obtained from both odd and even power monomials or not.

The restriction to a hermitian bosonic superfield leads, under simple duality or anti-duality conditions, to three SU(5) singlets, two 24's, two 10's, one $\overline{40}$, one $\overline{5}$ and one 75. The 24's are welcome while the presence of only a single $\overline{5}$ seems a strong constraint, but most of the remainder appear to be of little immediate utility. If we also include a fermionic superfield once again the details of the duality conditions become important and in the absence of anomaly cancellation constraints the options are quite open.

The most reductive step in a duality condition approach would be to eliminate the entire cross-diagonal; this would provide then two more 5's, one $\overline{10}$, and one $\overline{45}$. As we shall see below the self-dual monomials are in any case inadequate as standard Higgs fields. The $\overline{45}$ should be a bonus for splitting the leptons from the down quarks if the interaction terms resemble usual GUT models, and the extra 5's may be welcome (or not), but from both superfields the rest are a distinct embarrassment of riches. At present we can see no natural way to exorcise them.

One appealing idea, from the point of view of economy (although somewhat outside our main use of duality), would be to require that the fermionic Higgs superfield consists only of the self-dual monomials of θ, an impressive collection of $1+10+15+5+45+50$, but unfortunately this leads to a dead end in terms of conventional Higgs fields, since only the quadratic power of such a superfield (five powers of Grassmann coordinates) can survive Berezin integration—with no possibility of a quartic self-coupling and no classical vacuum expectation values for the scalar fields! However we should point out that such a superfield can couple in Yukawa fashion to the fermions. If some fermions condense out, or if masses are to be produced radiatively for some generations, then this formulation might possess some virtues. It might also have some uses in conjunction with the SU(5) scalar associated with the zeroth power monomial of the bosonic superfield.

A duality choice for the fermionic superfield which might be appropriate to the first of our exotic representation models is to adopt a duality condition for the hermitian Higgs fields,

$$\Phi = (-1)^{\nu+(|\bar{n}-n|-1)/2}\breve{\Phi}$$

complementary to that used for the fermions. Compared with the model outlined above where the entire cross-diagonal was eliminated we would now have an extra 5, $\overline{45}$ and $\overline{10}$.

Turning to the simple models where we discard all the representations beyond 1, 5 and 10 [which means abandoning the standard way of breaking SU(5) via 24's] we have the following possible Higgs superfield terms.

A self-dual bosonic superfield choice

$$\Phi = \breve{\Phi}$$

yields three SU(5) scalars,

$$[1+(\bar{\theta}\theta)^5/5!]\aleph(1), \qquad [(\bar{\theta}\theta)+(\bar{\theta}\theta)^4/4!]\aleph(2), \qquad [(\bar{\theta}\theta)^2/2+(\bar{\theta}\theta)^3/3!]\aleph(3),$$

one $\overline{5}$,

$$[1+(\bar{\theta}\theta)]\theta^4\bar{\Delta}(1),$$

and two 10's,

$$[1+(\bar{\theta}\theta)^3/3!]\theta^2\bar{\Upsilon}(1), \qquad [(\bar{\theta}\theta)+(\bar{\theta}\theta)^2/2]\theta^2\bar{\Upsilon}(2).$$

If we also include a fermionic Higgs superfield then, taking once more the opposite duality relation to the fermion model, i.e.

$$\Xi = (-1)^\nu \breve{\Xi},$$

we obtain three additional 5's, two $\overline{10}$ and a singlet. These consist of the superfield terms

$$[1+(\bar{\theta}\theta)^4/4!]\bar{\theta}\Delta(2), \qquad [(\bar{\theta}\theta)+(\bar{\theta}\theta)^3/3!]\bar{\theta}\Delta(3), \qquad [1+(\bar{\theta}\theta)^2/2!]\bar{\theta}^3\Upsilon(3)$$

and the surviving self-dual monomials which we group as

$$\bar{\theta}^5\aleph(4), \qquad (\bar{\theta}\theta)\bar{\theta}^3\bar{\Upsilon}(4), \qquad (\bar{\theta}\theta)^2\bar{\theta}\Delta(4),$$

and their conjugates. The free Lagrangian for these fields, taken together, arises painlessly through the superintegral,

$$\int d^5\bar{\theta}\, d^5\theta\, \partial\Phi^\dagger . \partial\Phi = \sum_{r=1}^{4} [\partial\bar{\aleph}(r).\partial\aleph(r) + \partial\bar{\Delta}(r).\partial\Delta(r) + \partial\bar{\Upsilon}(r).\partial\Upsilon(r)].$$

It is now quite feasible to construct Φ^4 self-interactions. We come across a fair number of terms although there are restrictions arising from the requirements of Berezin integration. Up to hermitian conjugation, we have listed below all the Φ^4 terms which involve the 5's, Δ's and singlets, $\aleph$'s (neglecting the 10's, Y because they are all charged and thus have zero vacuum expectation values):

$$\aleph^4(1),\quad \aleph^2(1)\aleph^2(2),\quad \aleph^2(1)\aleph^2(3),\quad \aleph^2(1)\aleph^2(4),$$
$$\aleph(1)\aleph(2)\aleph^2(3),\quad \aleph(1)\aleph^2(2)\aleph(3),\quad \aleph^3(2)\aleph(3),$$
$$\bar{\Delta}(2)\Delta(2)\aleph^2(1),\quad \bar{\Delta}(2)\Delta(2)\aleph(1)\aleph(2),\quad \bar{\Delta}(2)\Delta(2)\aleph(2)\aleph(3),\quad \bar{\Delta}(2)\Delta(2)\aleph^2(3),$$
$$\bar{\Delta}(2)\Delta(3)\aleph(1)\aleph(2),\quad \bar{\Delta}(2)\Delta(3)\aleph(1)\aleph(3),\quad \bar{\Delta}(2)\Delta(3)\aleph(2)\aleph(3),$$
$$\bar{\Delta}(3)\Delta(3)\aleph^2(1),\quad \bar{\Delta}(3)\Delta(3)\aleph^2(2),\quad \bar{\Delta}(3)\Delta(3)\aleph(1)\aleph(3),$$
$$\bar{\Delta}(1)\Delta(1)\aleph^2(1),\quad \bar{\Delta}(4)\Delta(4)\aleph^2(1),\quad \bar{\Delta}(1)\Delta(1)\aleph(1)\aleph(2),$$
$$\bar{\Delta}(2)\Delta(2)\bar{\Delta}(2)\Delta(1),\quad \bar{\Delta}(3)\Delta(3)\bar{\Delta}(3)\Delta(2),\quad \bar{\Delta}(1)\Delta(2)\bar{\Delta}(2)\Delta(1).$$

The constraints on couplings of these sets of Higgs fields due to the θ structure of superfields shows that they are far from being trivial clones. It should also be pointed out that some of the terms above appear dangerous, as they involve cubic factors of some fields which are untamed by corresponding quartics. One should note that combining the various Higgs fields into superfields for each SU(5) representation, *or* packing the representations into one or two superfields for the fermions (e.g. sorted by odd or even powers of θ's) and similiarly for all the Higgs fields, would enforce strict relationships between the coefficients of the above self-interactions and also between Yukawa terms.

4. Gauge Symmetries

We now turn to the gauge fields associated with these Grassmann models. The question of what to gauge involves consideration of the symmetries of the matter field sector. By virtue of the formulation with five complex anticommuting coordinates there is an obvious action of SU(5) on the monomials that comprise the superfields, with generators

$$F_I^J \equiv \bar{\theta}^J\partial/\partial\bar{\theta}^I - \theta_I\partial/\partial\theta_J.$$

We have accordingly decomposed the monomials and fields into SU(5) representations throughout our paper. We have also used the SU(5) anomaly conditions, but that can be taken simply as a compact check that an anomaly-free SU(3)×SU(2)×U(1) theory emerges at low energy. Standard SU(5) GUT theories are increasingly under threat in the light of results from proton lifetime experiments, and perhaps one could gauge just the standard model symmetries, regarding the number of θ's as a key to the number of generations, rather than as an invocation of SU(5). The presence of the exotic particle representations in some of our schemes suggests that a larger gauge group than SU(5) might be desirable, or that some of the fields should become massive at the unification scale, or both.

The full set of monomials represented by Table 1 does have further symmetries. There is a standard Clifford type representation of SO(10) that involves the extra generators

$$F_{KL} \equiv \bar{\theta}^K\bar{\theta}^L + \partial^2/\partial\theta_K\partial\theta_L$$

and their Hermitian conjugates together with the trace part

$$\bar{\theta}^J \partial/\partial\bar{\theta}^J - \theta_I \partial/\partial\theta_I .$$

In addition, the full array of superfield monomials, being constructed from products of Grassmann variables, forms a Grassmann algebra and this has a set of continuous automorphisms, generated by even derivations of the form

$$F_J^p \equiv (\bar{\theta})^{p-i}(\theta)^i \partial/\partial\bar{\theta}^J, \quad p = 1,3,5$$

and their conjugates. These are candidate symmetries as they map the degrees of freedom into themselves. There are further transformations on the monomials which we may schematically represent as

$$F_q^p \equiv (\bar{\theta})^{p-i}(\theta)^i \ \partial^q/(\partial\bar{\theta})^{q-j}(\partial\theta)^j .$$

We are presently carrying out a detailed study of the algebra of all these generators. Recently a related treatment of higher derivative operators which uses ordering in polynomials of powers and differentiations in such a way as to ensure that each operator generates interchanges of only one pair of monomials has been presented by Eyal (1990). The question of how many of these transformations should be regarded as symmetries awaits resolution.

It remains at this stage to see whether any of these extra symmetries can be maintained in the face of the decimation of the set of Grassmann monomials. SU(5) naturally acts *within* each monomial and some other symmetries, e.g. some transformations between the eventually sorted out generations, may survive. With regard to SU(5) gauge theories it should be remembered that the Grassmann models have greater constraints on possible Lagrangian terms than standard schemes. Such terms must not only be SU(5) scalars but must also contain the correct number of θ's to survive the Berezin integration. Complete phenomenological models using the fermion and Higgs representations of the current paper have yet to be constructed. Once that is achieved the role of these restrictions, acting like extra quantum numbers, needs further examination before SU(5) Grassmann unified models can be declared untenable. One feature of standard GUT theories which may have links with the Grassmann schemes is the introduction of discrete symmetries into the theory; this may be connected with the identifications made between monomials by the duality conditions. Such issues will be relevant to detailed discussions of asymptotic freedom, radiative effects, etc.

5. For the Future

In order to determine the mass matrices for the sources, one first needs to construct a semiclassical renormalisable potential for the various Higgs generations and the vacuum expectation values for each one of the uncharged fields. A superfield version of this is indicated but it may be necessary to distinguish between superfields with opposite statistics, before coupling them as $\lambda\Phi^4$. Next we must study all the Yukawa interactions between Higgs and fermions (exotica as well), and derive the mass terms and mixings, including the many terms encountered previously for 1's, 5's and 10's, again from an appropriate superfield interaction.

The full nature of the symmetry algebra and possible corresponding gauge fields is being mapped out and the concept of generations has also to be elucidated. Clearly there is a great deal of work confronting us before we can come to any definite conclusions. Still, we feel that the method holds considerable promise, because of the elegant way in which the generations emerge and the drastic diminution of states enforced by SU(5) Grassmann duality.

Acknowledgments

This research was supported by an ARC Grant (number A68931751). PDJ acknowledges contracts from the ARC and the University of Tasmania.

References

Casalbuoni, R., and Gatto, R. (1979, 1980). *Phys. Lett.* B **88**, 306; B **90**, 81.
Delbourgo, R. (1989). *Mod. Phys. Lett.* A **4**, 1381.
Delbourgo, R., Twisk, S. E., and Zhang, R. B. (1988). *Mod. Phys. Lett.* A **3**, 1073.
Delbourgo, R., and White, M. (1990). *Mod. Phys. Lett.* A **5**, 355.
Denegri, D., Sadoulet, B., and Spiro, M. (1990). *Rev. Mod. Phys.* **62**, 1.
Dondi, P. H., and Jarvis, P. D. (1980). *Z. Phys.* C **4**, 201.
Eyal, O. (1990). Techniques of using Grassmann variables for realizing some algebras, Karlsruhe (preprint).
Jarvis, P. D., and White, M. (1990). Fermion masses from supersymmetric dynamics in proper time, University of Tasmania preprint, submitted to the XXV International Conference on HEP.
King, R. C. (1981). *Nucl. Phys.* B **185**, 133.
Krolikowski, W. (1989). *Acta Phys. Polon.* B **19**, 599.
O'Raifeartaigh, L. (1986) 'Group Structure of Gauge Theories' (Cambridge Univ. Press).
Ross, G. G. (1984). 'Grand Unified Theories' (Frontiers in Physics Series, Benjamin/Cummings Publishing: Menlo Park).

Manuscript received 8 August, accepted 2 November 1990

Grassmann coordinates and Lie algebras for unified models

R. Delbourgo, P. D. Jarvis,[a] and Roland C. Warner[b]
Physics Department, University of Tasmania, GPO Box 252C, Hobart, Australia 7001

(Received 15 January 1993; accepted for publication 15 January 1993)

A variety of Lie algebras and certain classes of representations can be constructed using Grassmann variables regarded as Lorentz scalar coordinates belonging to an internal space. The generators are realized as combinations of multilinear products of the coordinates and derivative operators, while the representations emerge as antisymmetric polynomials in the variables and are thus severely restricted. The nature of these realizations and the interconnections between various subalgebras, for N independent complex anticommuting coordinates, is explored. The addition of such Grassmann coordinates to the usual spacetime manifold provides a natural superfield setting for a unified theory of symmetries of elementary particles. The particle content can be further restricted by imposing discrete symmetries (Lie algebra automorphisms). For the case $N=5$ some anomaly free choices of multiplets are derived through the imposition of specific superfield duality conditions.

I. INTRODUCTION

Over many years a vast amount of research has been carried out on unified models of elementary particles and their interaction forces, based on extensions to Minkowski space-time. These may involve the introduction of extra coordinates which are commuting variables, as in Kaluza–Klein theories[1] and bosonic string theories,[2] or anticommuting coordinate variables which are usually endowed with spinorial Lorentz labels in the superspace[3] approach to supersymmetry[4] and supergravity.[5] Both extensions have been frequently employed in recent years in Kaluza–Klein supergravity theories[6,7] and superstring theories.[8,9] By contrast unified models with extensions to the coordinate manifold founded on Lorentz scalar anticommuting coordinates have not been as extensively explored.[10–12] Yet in many ways, not the least of which are the highly constrained nature of the resulting representations and the character of the symmetry generators, they are a more promising line of research for unified descriptions of the internal symmetry quantum numbers of the matter fields of contemporary elementary particle physics. In particular a candidate unified model[11,12] characterized by five complex Grassmann variables offers an unusual method of handling the fermion generation problem with perhaps a relatively modest excess of as yet unseen particles.

Lorentz scalar Grassmann variables are also used in treatments of the ghost degrees of freedom in the BRST approach to the quantization of gauge theories and systems with constraints.[13] They also arise in models where the internal symmetry is graded in some way,[14,15] while Lorentz vector labels for Grassmann coordinates occur in the "old" formalism for the superstring.[16]

In this paper the transformations and possible symmetry algebras associated with unified models based on complex Lorentz scalar Grassmann coordinates are considered. We discuss the various Lie algebras and superalgebras that have a natural association with the introduced Grassmann coordinates, their interrelations and the decomposition of their (antisymmetric)

---

[a] Alexander von Humboldt Fellow.
[b] roland.warner@phys.utas.edu.au

0022-2488/93/34(8)/3616/26/$6.00
© 1993 American Institute of Physics

representations under branchings to several interesting subalgebras, and the role of Lie algebra automorphisms in truncating the space of representations and in reducing the symmetry algebra.

N complex Grassmann coordinates are introduced. Polynomials in these coordinates will be identified with the particle states or fields, while the generators of transformations are represented by polynomials in the coordinates and coordinate derivatives. Monomials representing the terms of a superfield or Taylor series expansion form a natural basis for the fields. A range of candidate symmetry algebras is presented. Choices are further constrained by the interplay between phenomenologically desirable restrictions on the already limited representations and the extent to which subalgebras of the transformation algebra can be simultaneously retained as symmetries. While it is of interest to demonstrate some rather unconventional realizations of a variety of Lie algebras, our main purpose is to exploit these results for the construction of unified models of elementary particles and as an example in this area we focus on the case $N=5$ which provides the most direct link to present phenomenology.

The paper is organized in the following manner. In the next section we consider the Lie algebra which acts on the Grassmann coordinate monomials and some useful subalgebras. The role of $sp(2N)$ and $so(2N)$ in reshufflings of the monomials in the general superfield is studied in Sec. III. Then the role of discrete symmetries (Lie algebra automorphisms) is examined in the context of imposing constraints on superfields so as to reduce the particle content. In Sec. V we apply these ideas to $su(5)$ as a case of possible physical relevance. We exhibit two anomaly-free models[12] which emerge naturally and discuss their symmetries. The paper concludes with a brief discussion of prospects for gauging the algebras, and brief consideration of more involved symmetry transformations from within the present scheme.

II. LIE ALGEBRA GENERATORS AND REPRESENTATIONS

A. States as monomials of Grassmann variables

We follow convention and label our complex Grassmann (anticommuting) coordinates θ_i $(i=1,...,N)$ and denoting their conjugates $\bar{\theta}^i$. One can also view these as $2N$ anticommuting real variables ϑ_I, where $I=1,2,...,2N$, and $\theta_i=\vartheta_i+i\vartheta_{i+N}$. The particle states or space-time dependent fields are expressible as components of a superfield expanded in a terminating power series in the Grassmann coordinate variables, a typical term being

$$\bar{\theta}^{j_1}\bar{\theta}^{j_2}\cdots\bar{\theta}^{j_q}\theta_{i_1}\theta_{i_2}\cdots\theta_{i_p};\quad p,q\leqslant N. \tag{1}$$

Since no Grassmann index can appear more than once, the above monomial comprises $N!N!/p!(N-p)!q!(N-q)!$ independent antisymmetric components. Ranging over p and q, we end up with a total of 2^{2N} different terms, as expected. Obviously the full set of monomials spans the product of the Grassmann algebras generated by the θ_i and $\bar{\theta}^i$ with a grading of the monomials into odd and even families. Each term, or more generally some combination of them, belongs to a multiplet of a Lie algebra.

B. Algebra of transformations of Grassmann monomials

As the most general transformations are reshufflings of these monomials, it is clear that the maximal algebra which operates on them is the graded algebra $gl(2^{2N-1}/2^{2N-1})$. It is from this superalgebra and its subalgebras that the transformations and possible symmetries relevant to our unified models can be sought. The full set of possible states — all the monomials of the Grassmann coordinates — fall naturally into the 2^{2N} dimensional fundamental representation of this superalgebra, and the decomposition of that representation into representations of the subalgebras is of direct importance for the description of the organized variety of matter multiplets. The task of merely enumerating all the subalgebras of $gl(2^{2N-1}/2^{2N-1})$ would be an essentially mechanical exercise. Our purpose in this paper is to display some realizations of Lie algebras using Grassmann variables which may have some unusual character, and to discuss those decompositions of $gl(2^{2N-1}/2^{2N-1})$ that have a natural association with our complex Grassmann coordinates.

To construct generators of rearrangements of the monomials we also need the canonically conjugate operators—the derivatives with respect to the coordinates: $\partial^i \equiv \partial/\partial\theta_i$, and $\bar{\partial}_i \equiv \partial/\partial\bar{\theta}^i$. The anticommutation relations between coordinates and derivatives are identical to those between fermionic annihilation and creation operators,

$$\{\theta_i,\partial^j\} = \{\bar{\theta}^j,\bar{\partial}_i\} = \delta_i^j, \tag{2}$$

revealing a Clifford algebra, so that one can also equivalently cast all of the following in Fock space formalism.

Operators which mix the monomials consist of various powers of coordinates and derivatives. A convenient ordering is given by the set,

$$F^{\bar{i}_1 \ldots \bar{i}_m k_1 \ldots k_p}_{j_1 \ldots j_n \bar{l}_1 \ldots \bar{l}_q} \equiv \bar{\theta}^{i_1} \ldots \bar{\theta}^{i_m} \theta_{j_1} \cdot \cdots \theta_{j_n} \partial^{k_1} \cdots \partial^{k_p} \bar{\partial}_{l_1} \cdots \bar{\partial}_{l_q}. \tag{3}$$

A suitable bracketing abbreviation for families of operators of similar degree, that suppresses index labels is

$$(F)^{\bar{m}p}_{n\bar{q}} \equiv (\bar{\theta})^m (\theta)_n (\partial)^p (\bar{\partial})_q. \tag{4}$$

It is also possible to approach things in a more synthetic fashion; at least at the level where all the Grassmann coordinates are treated without distinction, which has an obvious underlying $u(N)$ structure. The most obvious even $(F)^{\bar{m}p}_{n\bar{q}}$ operators, the $(F)^1_1$ and $(F)^{\bar{1}}_{\bar{1}}$, combine to generate a $u(N)$ which acts on the monomials and which preserves the 'scalar' $\Sigma_k \bar{\theta}^k \theta_k$ (henceforth we take summation over repeated indices as implied). This algebra, denoted $u(N)_{\mathscr{F}}$, can be realized with the operators $\mathscr{F}^i{}_j = \bar{\theta}^i \bar{\partial}_j - \theta_j \partial^i$, for $i,j = 1,2,\ldots,N$. All the remaining $(F)^{\bar{m}p}_{n\bar{q}}$'s can be regarded as tensor operators transforming under this algebra and all the monomials similarly belong to tensor product representations of $u(N)_{\mathscr{F}}$ antisymmetric in the θ's and the $\bar{\theta}$'s which transform as $\bar{N}$ and N representations, respectively.

There are clearly many $u(N)_{\mathscr{F}}$ representations in the decomposition of the fundamental and adjoint representations of $gl(2^{2N-1}/2^{2N-1})$. For the fundamental representation this is part of the attraction of the Grassmann coordinate approach to unified models—there are only a finite number of states and hence representations, and furthermore, various $u(N)$ representations will typically appear more than once, leading to a natural framework for the repetitive generations of matter fields. An important issue is whether the symmetry algebra can be larger than the $u(N)_{\mathscr{F}}$ just described, given that the algebra of transformations between the states is clearly larger. The 'decomposition' approach that we follow for most of this paper offers advantages in determining the various subalgebras which operate on the monomials, while the 'build-up' viewpoint is useful for checking the realizations of subalgebras as we see in Sec. II F.

While the $(F)^{\bar{m}p}_{n\bar{q}}$ forms are convenient for verifying specific (anti)commutation relations between transformation generators, a significant insight is obtained using the $\mathscr{M} = 2N$ real Grassmann variables ϑ_I. The operators corresponding to the $(F)^{\bar{m}p}_{n\bar{q}}$'s in this coordinate scheme may be defined by

$$G_{J_1 \ldots J_q}{}^{I_1 \ldots I_p} \equiv \vartheta_{J_1} \ldots \vartheta_{J_q} \partial^{I_1} \ldots \partial^{I_p}, \tag{5}$$

where the p- and q- dimensional sets of labels are selected from $1,2,\ldots,\mathscr{M}$, and an analogous $(G)^p_q$ shorthand notation will be adopted.

We define the operator $\mathscr{E} = \partial^1\partial^2\cdots\partial^{\mathscr{M}}\vartheta_{\mathscr{M}}\cdots\vartheta_2\vartheta_1$ which projects the ϑ-independent term of any expression. $\mathscr{E}$ has the normal-ordered expansion

$$\mathscr{E}=1-\sum_I \vartheta_I\partial^I+\sum_{I<J}\vartheta_I\vartheta_J\partial^J\partial^I+\cdots+\vartheta_1\vartheta_2\ldots\vartheta_{\mathscr{M}}\partial^{\mathscr{M}}\ldots\partial^2\partial^1$$

$$=1-\sum_I G_I{}^I+\sum_{I<J}G_{IJ}{}^{JI}-\sum_{I<J<K}G_{IJK}{}^{KJI}+\cdots+G_{12\ldots\mathscr{M}}{}^{\mathscr{M}\ldots21}. \tag{6}$$

As shown by Eyal[17] this projection operator can be used to construct a realization of the elementary matrices of the superalgebra $gl(2^{2N-1}/2^{2N-1})$. Specifically, the operator

$$E_{K_1K_2\ldots K_n}{}^{L_1L_2\ldots L_m}=\vartheta_{K_1}\vartheta_{K_2}\cdots\vartheta_{K_n}\mathscr{E}\partial^{L_1}\partial^{L_2}\cdots\partial^{L_m} \tag{7}$$

transforms $\vartheta_{L_1}\vartheta_{L_2}\cdots\vartheta_{L_m}$ into $\vartheta_{K_1}\vartheta_{K_2}\cdots\vartheta_{K_n}$, leaving everything else unchanged. It is clear that these operators can be expanded using Eq. (6) into a sum of the $(G)_q^p$ family of operators, showing that a particular class of $E_{K_1\ldots K_n}{}^{L_1\ldots L_m}$, which we may call $(E)_n^m$ for short, contains $(\vartheta)_q(\partial)^p$ operator monomials from $(\vartheta)_n(\partial)^m$ to $(\vartheta)_{\mathscr{M}}(\partial)^{\mathscr{M}}$, although some $(G)_q^p$'s will appear with nonzero coefficients only for exceptional choices of the labels $\{K_i,L_j\}$.

These expressions allow a simple view of the even part of the $gl(2^{2N-1}/2^{2N-1})$ superalgebra. The two mutually commuting sets of even-to-even and odd-to-odd $(E)_n^m$'s generate the $gl(2^{2N-1})\oplus gl(2^{2N-1})$ even subalgebra.

We now present those subalgebras which are most relevant to the complex Grassmann coordinate viewpoint. We begin with the subalgebra $so(4N)\oplus so(4N)$. It is helpful to note that the 2^{2N} states of the fundamental representation of $gl(2^{2N-1}/2^{2N-1})$, the elements of the Grassmann algebra generated by the ϑ_I's, fall into the two distinct 2^{2N-1} dimensional definite 'chiral' spinor representations, Δ_+ and Δ_- of $so(4N)$, corresponding to the odd and even monomials, respectively.

C. $gl(2^{2N-1}/2^{2N-1})\downarrow so(4N)\oplus so(4N)$

The even part of $gl(2^{2N-1}/2^{2N-1})$ contains an $so(4N)\oplus so(4N)$ subalgebra that can be realized by combining the standard procedure for generating $so(2\mathscr{M})$ from a Clifford algebra of $\mathscr{M}$ ϑ_I's and $\mathscr{M}$ $\partial/\partial\vartheta_I$'s, with the projection operators $\mathscr{P}_{\text{odd}}$ and $\mathscr{P}_{\text{even}}$ which select the odd or even families of ϑ_I monomials. These operators are most transparently expressed as sums of projection operators for monomials of each degree,

$$\mathscr{P}_{\text{even}}=\mathscr{E}+\sum_{I<J}\vartheta_I\vartheta_J\mathscr{E}\partial^J\partial^I+\cdots+\vartheta_1\vartheta_2\ldots\vartheta_{\mathscr{M}}\mathscr{E}\partial^{\mathscr{M}}\ldots\partial^2\partial^1, \tag{8}$$

$$\mathscr{P}_{\text{odd}}=\sum_I\vartheta_I\mathscr{E}\partial^I+\sum_{I<J<K}\vartheta_I\vartheta_J\vartheta_K\mathscr{E}\partial^K\partial^J\partial^I$$

$$+\sum_{I_1<I_2<\cdots<I_{\mathscr{M}-1}}\vartheta_{I_1}\vartheta_{I_2}\ldots\vartheta_{I_{\mathscr{M}-1}}\mathscr{E}\partial^{I_{\mathscr{M}-1}}\ldots\partial^{I_2}\partial^{I_1}. \tag{9}$$

The $so(4N)\oplus so(4N)$ algebra is generated by

$$\mathscr{U}_o{}^{KL}=\mathscr{P}_{\text{odd}}\partial^K\partial^L,\quad \mathscr{L}_{oIJ}=\vartheta_I\vartheta_J\mathscr{P}_{\text{odd}},\quad \mathscr{A}_{oJ}{}^I=\vartheta_J\mathscr{P}_{\text{even}}\partial^I, \tag{10}$$

and

$$\mathscr{U}_e{}^{KL}=\mathscr{P}_{\text{even}}\partial^K\partial^L,\quad \mathscr{L}_{eIJ}=\vartheta_I\vartheta_J\mathscr{P}_{\text{even}},\quad \mathscr{A}_{eJ}{}^{I}=\vartheta_J\mathscr{P}_{\text{odd}}\partial^I. \tag{11}$$

The projectors can also be expanded out using Eq.(6) as

$$\begin{aligned}\mathscr{P}_{\text{even}}=1-\sum_K G_K{}^K+2\sum_{K<L} G_{KL}{}^{LK}-4\sum_{K<L<M} G_{KLM}{}^{MLK}+\cdots\\ -2^{\mathscr{M}-2}\sum_{K_1<K_2<\cdots<K_{\mathscr{M}-1}} G_{K_1K_2\cdots K_{\mathscr{M}-1}}{}^{K_{\mathscr{M}-1}\ldots K_2K_1}\\ +2^{\mathscr{M}-1}G_{12\ldots\mathscr{M}}{}^{\mathscr{M}\ldots 21},\end{aligned} \tag{12}$$

$$\begin{aligned}\mathscr{P}_{\text{odd}}=\sum_K G_K{}^K-2\sum_{K<L} G_{KL}{}^{LK}+4\sum_{K<L<M} G_{KLM}{}^{MLK}+\cdots\\ +2^{\mathscr{M}-2}\sum_{K_1<K_2<\cdots<K_{\mathscr{M}-1}} G_{K_1K_2\cdots K_{\mathscr{M}-1}}{}^{K_{\mathscr{M}-1}\ldots K_2K_1}\\ -2^{\mathscr{M}-1}G_{12\ldots\mathscr{M}}{}^{\mathscr{M}\ldots 21}.\end{aligned} \tag{13}$$

Completeness implies $\mathscr{P}_{\text{odd}}+\mathscr{P}_{\text{even}}=1$. We can now display the $so(4N)\oplus so(4N)$ generators directly in terms of the $(G)^{p}_{q}$'s as

$$\begin{aligned}\mathscr{U}_o{}^{IJ}=\sum_K G_K{}^{KIJ}-2\sum_{K<L} G_{KL}{}^{LKIJ}+4\sum_{K<L<M} G_{KLM}{}^{MLKIJ}\\ +\cdots+2^{\mathscr{M}-3}\sum_{K_1<K_2<\cdots<K_{\mathscr{M}-2}} G_{K_1K_2\ldots K_{\mathscr{M}-2}}{}^{K_{\mathscr{M}-2}\cdots K_2K_1IJ},\end{aligned} \tag{14}$$

$$\begin{aligned}\mathscr{L}_{oIJ}=\sum_K G_{IJK}{}^{K}-2\sum_{K<L} G_{IJKL}{}^{LK}\\ +\cdots+2^{\mathscr{M}-3}\sum_{K_1<K_2<\cdots<K_{\mathscr{M}-2}} G_{IJK_1K_2\cdots K_{\mathscr{M}-2}}{}^{K_{\mathscr{M}-2}\cdots K_2K_1},\end{aligned} \tag{15}$$

$$\mathscr{A}_{oI}{}^{J}=G_I{}^J-\sum_K G_{IK}{}^{KJ}+2\sum_{K<L} G_{IKL}{}^{LKJ}+\cdots-2^{\mathscr{M}-2}\delta_I{}^J G_{12\ldots\mathscr{M}}{}^{\mathscr{M}\ldots 21}, \tag{16}$$

$$\begin{aligned}\mathscr{U}_e{}^{IJ}=G^{IJ}-\sum_K G_K{}^{KIJ}+2\sum_{K<L} G_{KL}{}^{LKIJ}-4\sum_{K<L<M} G_{KLM}{}^{MLKIJ}\\ +\cdots-2^{\mathscr{M}-3}\sum_{K_1<K_2<\cdots<K_{\mathscr{M}-2}} G_{K_1K_2\cdots K_{\mathscr{M}-2}}{}^{K_{\mathscr{M}-2}\cdots K_2K_1IJ},\end{aligned} \tag{17}$$

$$\begin{aligned}\mathscr{L}_{eIJ}=G_{IJ}-\sum_K G_{IJK}{}^{K}+2\sum_{K<L} G_{IJKL}{}^{LK}\\ +\cdots-2^{\mathscr{M}-3}\sum_{K_1<K_2<\cdots<K_{\mathscr{M}-2}} G_{IJK_1K_2\cdots K_{\mathscr{M}-2}}{}^{K_{\mathscr{M}-2}\cdots K_2K_1},\end{aligned} \tag{18}$$

$$\mathscr{A}_{eI}{}^{J} = \sum_{K} G_{IK}{}^{KJ} - 2 \sum_{K<L} G_{IKL}{}^{LKJ} + \cdots + 2^{\mathscr{M}-2} \delta_I{}^J G_{12\ldots\mathscr{M}}{}^{\mathscr{M}\cdots 21}. \tag{19}$$

A few remarks about the decomposition of the fundamental representation of $gl(2^{2N-1}/2^{2N-1})$ are appropriate. It is straightforward to verify that the reducible spinor representation $\Delta \equiv \Delta_+ \oplus \Delta_-$ of $so(4N)$ decomposes under the branching $so(4N) \downarrow so(2N) \oplus so(2)$ into the direct sum of all the antisymmetric tensor powers of the fundamental vector representation of $so(2N)$, i.e., the Grassmann algebra generated by the ϑ_I's which we assembled into the fundamental representation of $gl(2^{2N-1}/2^{2N-1})$ also forms the basic spinor of $so(4N)$. One can also verify that the antisymmetric tensor representations of $so(2N)$ of odd and even rank belong to the decompositions of the Δ_+ and Δ_- chiral spinors of $so(4N)$, respectively, under the same branching. From the way the two $so(4N)$ algebras have been constructed above one deduces that the decomposition of the fundamental representation of $gl(2^{2N-1}/2^{2N-1})$ under the branching to $so(4N) \oplus so(4N)$ is to $1 \otimes \Delta_+ \oplus \Delta_- \otimes 1$, so the two $so(4N)$'s each act on just one of the chiral subspaces.

The generators of the diagonal $so(4N)$ subalgebra act on both chiral spinors but do not mix them, and it is simple to see that the operators $\mathscr{P}_{\text{odd}}$ and $\mathscr{P}_{\text{even}}$ act like the chiral projection operators $(1+\Gamma)/2$ and $(1-\Gamma)/2$, respectively, yielding the "chirality" operator

$$\begin{aligned}\Gamma &= \mathscr{P}_{\text{odd}} - \mathscr{P}_{\text{even}} \\ &= -1 + 2\sum_{K} G_K{}^K - 4\sum_{K<L} G_{KL}{}^{LK} + 8\sum_{K<L<M} G_{KLM}{}^{MLK} \\ &\quad + \cdots + 2^{\mathscr{M}-1} \sum_{K_1<K_2<\cdots<K_{\mathscr{M}-1}} G_{K_1K_2\cdots K_{\mathscr{M}-1}}{}^{K_{\mathscr{M}-1}\ldots K_2K_1} - 2^{\mathscr{M}} G_{12\ldots\mathscr{M}}{}^{\mathscr{M}\ldots 21}.\end{aligned} \tag{20}$$

There is thus an elegant way in which the elements of the Grassmann algebra of monomials falls into these spinor representations of $so(4N)$. Although the realization of $so(4N)$ using a Clifford algebra of $4N$ elements, or equivalently $2N$ pairs of fermionic creation and annihilation operators is standard,[18] this elaboration which essentially involves the realization of the irreducible chiral spinor projections via the grading of the monomials and the links with $gl(2^{2N-1}/2^{2N-1})$, and the identification of Γ is perhaps less well known.

D. Subalgebras of $so(4N) \oplus so(4N)$

We present a few relevant examples of these:

- $so(4N) \oplus so(4N) \downarrow u(2N) \oplus u(2N)$ generated by $\mathscr{A}_{oI}{}^{J}$ and $\mathscr{A}_{eI}{}^{J}$.
- $so(4N) \oplus so(4N) \downarrow so(2N) \oplus so(2N) \oplus so(2N) \oplus so(2N)$ generated by $\mathscr{U}_o^{IJ} \pm \mathscr{L}_{oIJ}$, $\mathscr{A}_{oI}{}^{J} - \mathscr{A}_{oJ}{}^{I}$ and $\mathscr{U}_e^{IJ} \pm \mathscr{L}_{eIJ}$, $\mathscr{A}_{eI}{}^{J} - \mathscr{A}_{eJ}{}^{I}$.
- $so(4N) \oplus so(4N) \downarrow u(2N) \oplus u(2N) \downarrow sp(2N) \oplus sp(2N)$ generated by symmetrization of the indices on the $u(2N)$ operators, $\mathscr{A}_{oI}{}^{J} + \mathscr{A}_{oJ}{}^{I}$ and $\mathscr{A}_{eI}{}^{J} + \mathscr{A}_{eJ}{}^{I}$.
- $so(4N) \oplus so(4N) \downarrow u(2N) \oplus u(2N) \downarrow so(2N) \oplus so(2N)$ generated by antisymmetrization of the indices on those operators, $\mathscr{A}_{oI}{}^{J} - \mathscr{A}_{oJ}{}^{I}$ and $\mathscr{A}_{eI}{}^{J} - \mathscr{A}_{eJ}{}^{I}$. This realization is the direct sum of the $so(2N)$ diagonal subalgebras in the separate odd ('o') and even ('e') sectors of the four-fold $so(2N)$ above. An alternative realization would be to take the direct sum of all the operators there with respect to combining odd and even operators.
- Finally, obvious branchings from $so(2N) \oplus so(2N)$ and $sp(2N) \oplus sp(2N)$ to $u(N) \oplus u(N)$ can be discerned.

TABLE I. The main branchings of the algebra $so(4N)$ of interest as the symmetries of the Grassmann variable monomials and as candidate symmetry algebras for the duality-constrained theory.

	$so(4N)$	
$sp(2N) \oplus sp(2)$	$su(2N) \oplus u(1)$ $sp(2N) \oplus u(1)$ $so(2N)_A \oplus u(1)$	$so(2N) \oplus so(2N)$ $so(2N)_{B,C} \oplus u(1)$ $u(N) \oplus u(N)$
	$su(N) \oplus u(1) \oplus u(1)$	

At any stage in this progression one can also form the corresponding diagonal subalgebra within each direct sum. The most significant of these subalgebras is the diagonal subalgebra $so(4N) \oplus so(4N) \downarrow so(4N)$.

E. *so*(4*N*) and its subalgebras

A vast simplification occurs in forming the $so(4N)$ diagonal subalgebra. The generators simplify to the standard Clifford algebra construction of $so(4N)$ involving all the bilinears that can be formed from the $\mathscr{M} = 2N$ coordinates and their derivatives, namely, G_{IJ}, G^{KL}, and $G_I{}^J$'s. These may be conveniently translated into the complex Grassmann coordinate basis, namely, all the $(F)_2$,$(F)_{\bar{2}}$,$(F)_{1\bar{1}}$,$(F)^2$,$(F)^{\bar{2}}$,$(F)^{\bar{1}1}$,$(F)_1^{\bar{1}}$,$(F)_{\bar{1}}^1$,$(F)_1^1$, and $(F)_{\bar{1}}^{\bar{1}}$ generators.

While the question of the branchings of $so(4N)$ is naturally realization independent and a piece of standard group theory, our interest is in determining the action of the operators on superfields and the resulting behavior of their components. We commit ourselves to a particular realization of the states; different realizations of subalgebras, even though they may be related by Lie algebra automorphisms, are of specific interest for their action on the basis of monomials. In particular it is convenient to have the $u(N)$ subalgebra in the form $\mathscr{F}^i{}_j \equiv \bar{\theta}^i \bar{\partial}_j - \theta_j \partial^i$.

In Table I the relevant branching chains of subalgebras of $so(4N)$ down to $su(N)$ are displayed. Discussion of the general $so(2N) \downarrow so(2N-K) \oplus so(K)$ branchings is omitted.

We now discuss various realizations of these subalgebras that find later application as candidate symmetries.

1. so(4*N*)↓ *u*(2*N*)

The largest relevant subalgebra is $u(2N)$ and there are four realizations of interest. These incorporate between them all the (F) operators of the realization of $so(4N)$ listed above. The first, $u(2N)_\alpha$, generated by $\mathscr{A}_{oI}{}^J + \mathscr{A}_{eI}{}^J$, is the diagonal subalgebra of the $u(2N) \oplus u(2N)$ discussed in the previous subsection. The $u(2N)_\alpha$ is also contained within the $so(4N)$ realization of interest, and is generated by all those (F)'s containing one coordinate and one derivative, the $(F)_{\bar{1}}^{\bar{1}} - (F)_1^1$,$(F)_{\bar{1}}^1 + (F)_1^1$ $(F)^{\bar{1}1}$ and $(F)_{1\bar{1}}$ operators.

A second realization, $u(2N)_\beta$, involves the generators $(F)_{\bar{1}}^{\bar{1}} - (F)_1^1$ and $(F)_{\bar{1}}^{\bar{1}} + (F)_1^1$ again, but this time the algebra is completed by the $(F)_1^{\bar{1}}$ and $(F)_{\bar{1}}^1$ classes.

The last relevant pair of realizations, $u(2N)_\gamma$ and $u(2N)_\delta$, have in common the generators[19] $(F)_{\bar{1}}^{\bar{1}} - (F)_1^1$, $(F)_{(1\bar{1})}$ and $(F)^{(\bar{1}1)}$, and are completed by $(F)_2 + (F)_{\bar{2}}$, $(F)^{\bar{2}} + (F)^2$, $(F)_{\bar{1}}^1 - (F)_1^{\bar{1}}$, or $(F)_2 - (F)_{\bar{2}}$,$(F)^{\bar{2}} - (F)^2$, $(F)_{\bar{1}}^1 + (F)_1^{\bar{1}}$, respectively.

Such realizations can be more transparently presented at greater length in terms of the coordinates and derivatives. For example, the realization $u(2N)_\gamma$ consists of

$$\partial^i \bar{\partial}_j - \bar{\theta}^i \theta_j, \quad \theta_i \bar{\partial}_j + \theta_j \bar{\partial}_i, \quad \bar{\theta}^i \partial^j + \bar{\theta}^j \partial^i, \quad \theta_i \theta_j + \bar{\partial}_i \bar{\partial}_j, \quad \bar{\theta}^i \bar{\theta}^j + \partial^i \partial^j, \quad \text{and} \quad \mathscr{F}^i{}_j. \tag{21}$$

The realizations of the decompositions of the $u(2N)$ under some of the branchings in Table I will also be required.

- $so(4N)\downarrow u(2N)\equiv su(2N)\oplus u(1)\downarrow so(2N)\oplus u(1)$. Under this branching $u(2N)_\alpha$ decomposes to $so(2N)_\alpha\oplus u(1)_\alpha$, where $so(2N)_\alpha$ is generated by the $(F)^{\bar 1}_{\bar 1}-(F)^1_1$, $(F)^{[\bar 1 1]}$ and $(F)_{[1\bar 1]}$ operators, with $\mathscr{S}=\mathrm{Tr}((F)^{\bar 1}_{\bar 1}+(F)^1_1)-N$ for $u(1)_\alpha$.

The realization of the corresponding decomposition of $u(2N)_\beta\downarrow so(2N)_\beta\oplus u(1)_\beta$ is most naturally generated by

$$\bar\theta^i\bar\partial_j+\theta_i\partial^j,\quad \bar\theta^i\theta_j-\bar\theta^j\theta_i,\quad \partial^i\bar\partial_j-\partial^j\bar\partial_i,\quad \text{and}\quad \mathscr{D}=\bar\theta^k\bar\partial_k-\theta_k\partial^k.$$

The $u(2N)_\gamma$ and $u(2N)_\delta$ realizations also have natural decompositions, to $so(2N)_\gamma\oplus u(1)_\gamma$ generated by $(F)^{\bar 1}_{\bar 1}-(F)^1_1,(F)_2+(F)_{\bar 2}$, $(F)^2+(F)^{\bar 2}$ with $u(1)_\gamma$ generator $\mathscr{T}=-i\mathrm{Tr}((F)^{\bar 1}_{\bar 1}-(F)^{1}_{1})$, and $so(2N)_\delta\oplus u(1)_\delta$ generated by $(F)^{\bar 1}_{\bar 1}-(F)^1_1,(F)_2-(F)_{\bar 2}$, $(F)^2-(F)^{\bar 2}$ with $u(1)_\delta$ generated by $\mathscr{R}=\mathrm{Tr}((F)^{\bar 1}_1+(F)^1_{\bar 1})$, respectively.

- $so(4N)\downarrow u(2N)\equiv su(2N)\oplus u(1)\downarrow sp(2N)\oplus u(1)$. This branching leads to four realizations of $sp(2N)\oplus u(1)$. For three cases— $u(2N)_\alpha$, $u(2N)_\gamma$, and $u(2N)_\delta$—the $sp(2N)$ subalgebra, which we denote $sp(2N)_{\alpha\gamma\delta}$, is realized by $(F)^{\bar 1}_{\bar 1}-(F)^1_1$, $(F)^{(\bar 1 1)}$, and $(F)_{(\bar 1 1)}$, while the same realizations $\mathscr{S}$, $\mathscr{T}$, and $\mathscr{R}$ appear as above for their respective $u(1)$'s.

  The most straightforward realization of the decomposition $u(2N)_\beta\equiv su(2N)_\beta\oplus u(1)_\beta\downarrow sp_\beta(2N)\oplus u(1)_\beta$ is generated by

$$\bar\theta^i\bar\partial_j+\theta_i\partial^j,\quad \bar\theta^i\theta_j+\bar\theta^j\theta_i,\quad \partial^i\bar\partial_j+\partial^j\bar\partial_i,\quad \text{and}\quad \mathscr{D}.$$

- $so(4N)\downarrow u(2N)\equiv su(2N)\oplus u(1)\downarrow u(N)\oplus u(N)$. Various interesting realizations of this branching occur for our four $u(2N)$'s. $u(2N)_\alpha$ and $u(2N)_\beta$ have a subalgebra obtained by simply separating the $\bar\theta$ and θ sectors, which we denote $u(N)_{\bar\theta}\oplus u(N)_\theta$, realized by $(F)^{\bar 1}_{\bar 1}$ and $(F)^1_1$ operators.

  $u(2N)_\alpha$ also has a realization of the $u(N)\oplus u(N)$ subalgebra, generated by $\bar\theta^i\bar\partial_j+\theta_i\partial^j$ and $\bar\theta^i\partial^j+\theta_i\bar\partial_j$, which emerges naturally from the viewpoint of the real Grassmann variables ϑ_I, when the (I,J) indices in the realization of $u(2N)_\alpha$ by $\mathscr{A}^{\,J}_{oI}+\mathscr{A}^{\,J}_{eI}$ are restricted to the (i,j) and $(i+N,j+N)$ sectors.

  $u(2N)_\gamma$ and $U(2N)_\delta$ have branchings to $u(N)\oplus u(N)$ realized by $(F)^{\bar 1}_{\bar 1}-(F)^1_1$ together with $(F)^{\bar 1}_1+(F)^1_{\bar 1}$ and $(F)^{\bar 1}_1-(F)^1_{\bar 1}$, respectively, and both these realizations also occur in the decomposition of $u(2N)_\beta$.

2. so(4N)↓so(2N) ⊕ so(2N)

First we have the obvious case of the direct sum subalgebra generated by $[\mathscr{U}^{IJ}_o\pm\mathscr{L}_{oIJ}]+[\mathscr{U}^{IJ}_e\pm\mathscr{L}_{eIJ}]$, and $[\mathscr{A}_{oI}^{\ J}-\mathscr{A}_{oJ}^{\ I}]+[\mathscr{A}_{eI}^{\ J}-\mathscr{A}_{eJ}^{\ I}]$. In a second realization the sets of mutually commuting sectors in $so(2N)\oplus so(2N)$ are generated by making the separation of θ's and $\bar\theta$'s among the $so(4N)$ operators, i.e., selecting $(F)_2,(F)^2,(F)^1_1$, and $(F)_{\bar 2},(F)^{\bar 2},(F)^{\bar 1}_{\bar 1}$. This realization has a diagonal subalgebra $so(2N)_\gamma$ formed by taking the sum of the corresponding generators. There is also an obvious realization of the further branching $so(2N)\oplus so(2N)\downarrow u(N)\oplus u(N)$, leading to the $u(N)_{\bar\theta}\oplus u(N)_\theta$ discussed above.

3. so(4N)↓sp(2N) ⊕ sp(2)

There is a single realization of the symplectic decomposition with the $sp(2N)$ realized as before by $(F)^{\bar 1}_{\bar 1}-(F)^1_1$, $(F)_{(1\bar 1)}$, $(F)^{(1\bar 1)}$, and the $sp(2)$ being generated by combinations of the three operators that provide the distinct $u(1)$'s discussed above, i.e., $\mathscr{S}^+\equiv(\mathscr{R}+i\mathscr{T})/2=\mathrm{Tr}((F)^{\bar 1}_1)$, $\mathscr{S}^-\equiv(\mathscr{R}-i\mathscr{T})/2=\mathrm{Tr}((F)^1_{\bar 1})$, and $\mathscr{S}\equiv\mathrm{Tr}((F)^1_1+(F)^{\bar 1}_{\bar 1})-N$.

Note that the $u(N)_{\mathscr{F}}$ generators $\mathscr{F}^{i}_{\ j}$ in the class $(F)^{\bar{1}}_{\bar{1}}-(F)^{1}_{1}$, are common to most of the above algebras. While this subalgebra also appears in $u(2N)_{\beta}$ there do not appear to be obvious realizations of $so(2N)\oplus u(1)$ or $sp(2N)\oplus u(1)$ branchings for that realization. This difficulty arises, from the $u(N)_{\mathscr{F}}$ viewpoint, because the various $(F)^{\bar{m}p}_{n\bar{q}}$ operators comprising the coset $u(2N)_{\beta}/u(N)_{\mathscr{F}}$ realization do not easily decompose into symmetric and antisymmetric tensor representations of $u(N)_{\mathscr{F}}$.

F. Structure coefficients in the $(F)^{\bar{m}p}_{n\bar{q}}$ basis

The commutation relations between the various $(F)^{\bar{m}p}_{n\bar{q}}$ generators can be computed directly. We display some examples. The subalgebra $u(N)_{\bar{\theta}}\oplus u(N)_{\theta}$ of $so(4N)$ above, has the simple expected form:

$$[F^{p}_{q},F^{r}_{s}]=\delta^{p}_{s}F^{r}_{q}-\delta^{r}_{q}F^{p}_{s},\quad [F^{\bar{p}}_{\bar{q}},F^{\bar{r}}_{\bar{s}}]=\delta^{r}_{q}F^{\bar{p}}_{\bar{s}}-\delta^{p}_{s}F^{\bar{r}}_{\bar{q}},\quad [F^{\bar{p}}_{\bar{q}},F^{r}_{s}]=0. \tag{22}$$

These are augmented by other $(F)^{\bar{m}p}_{n\bar{q}}$'s to produce the subalgebras mentioned above.

In the $so(2N)_{\gamma}\oplus u(1)_{\gamma}$ and $so(2N)_{\delta}\oplus u(1)_{\delta}$ cases, which only differ in some sign factors in some combinations of $(F)^{\bar{m}p}_{n\bar{q}}$'s involved, we encounter as well as the $u(N)_{\mathscr{F}}$ generators, the $N(N-1)$ generators $F_{ij}\pm F_{\bar{i}\bar{j}}$, $F^{ij}\pm F^{\bar{i}\bar{j}}$ which complete these $so(2N)$ realizations, and the $u(1)$ generators $F^{\bar{k}}_{k}\mp F^{k}_{\bar{k}}$, with the typical commutators

$$[F_{ij}\pm F_{\bar{i}\bar{j}},F^{kl}\pm F^{\bar{k}\bar{l}}]=\delta^{[k}_{p}\delta^{l]}_{[j}\delta^{q}_{i]}(F^{\bar{p}}_{\bar{q}}-F^{p}_{q}), \tag{23a}$$

$$[F_{ij}\pm F_{\bar{i}\bar{j}},F^{k}_{l}-F^{\bar{k}}_{\bar{l}}]=\delta^{k}_{[i}\delta^{p}_{j]}(F_{pl}\pm F_{\bar{p}\bar{l}}), \tag{23b}$$

$$[F_{ij}\pm F_{\bar{i}\bar{j}},F^{\bar{k}}_{k}\mp F^{k}_{\bar{k}}]=[F^{kl}\pm F^{\bar{k}\bar{l}},F^{\bar{k}}_{k}\mp F^{k}_{\bar{k}}]=0, \tag{23c}$$

$$[F^{\bar{k}}_{k}\mp F^{k}_{\bar{k}},F^{m}_{l}-F^{\bar{m}}_{\bar{l}}]=0. \tag{23d}$$

The $so(2N)_{\alpha}\oplus u(1)_{\alpha}$ involves the $N(N-1)$ generators $F^{\bar{i}j}-F^{\bar{j}i}$, $F_{i\bar{j}}-F_{j\bar{i}}$ to complete the $so(2N)_{\alpha}$ and the $u(1)_{\alpha}$ generator $F^{\bar{k}}_{\bar{k}}+F^{k}_{k}$, with the commutation relations

$$[F^{\bar{i}j}-F^{\bar{j}i},F_{k\bar{l}}-F_{l\bar{k}}]=\delta^{[i}_{p}\delta^{j]}_{[k}\delta^{q}_{l]}(F^{\bar{p}}_{\bar{q}}-F^{p}_{q}), \tag{24a}$$

$$[F^{\bar{i}j}-F^{\bar{j}i},F^{k}_{l}-F^{\bar{k}}_{\bar{l}}]=\delta^{[i}_{p}\delta^{j]}_{l}(F^{\bar{p}k}-F^{\bar{k}p}), \tag{24b}$$

$$[F_{i\bar{j}}-F_{j\bar{i}},F^{k}_{l}-F^{\bar{k}}_{\bar{l}}]=\delta^{p}_{[j}\delta^{k}_{i]}(F_{p\bar{l}}-F_{l\bar{p}}). \tag{24c}$$

For $sp(2N)_{\alpha\gamma\delta}\oplus sp(2)$, the extra $N(N+1)+3$ generators to the diagonal $u(N)_{\mathscr{F}}$ are $F_{i\bar{j}}+F_{j\bar{i}}$, $F^{\bar{i}j}+F^{\bar{j}i}$ completing the $sp(2N)$ and the $sp(2)$ generators $F^{k}_{\bar{k}}$, $F^{\bar{k}}_{k}$, $F^{k}_{k}+F^{\bar{k}}_{\bar{k}}$, with commutation relations,

$$[F_{i\bar{j}}+F_{j\bar{i}},F^{\bar{l}k}+F^{\bar{k}l}]=\delta^{(l}_{q}\delta^{k)}_{(j}\delta^{p}_{i)}(F^{q}_{p}-F^{\bar{q}}_{\bar{p}}), \tag{25a}$$

$$[F^{k}_{\bar{k}},F^{\bar{l}}_{l}]=N-(F^{m}_{m}+F^{\bar{m}}_{\bar{m}}),\quad [F^{\bar{k}}_{k},F^{l}_{l}+F^{\bar{l}}_{\bar{l}}]=-2F^{\bar{m}}_{m},\quad [F^{k}_{\bar{k}},F^{l}_{l}+F^{\bar{l}}_{\bar{l}}]=2F^{m}_{\bar{m}}, \tag{25b}$$

in addition to the commutators with the $u(N)_{\mathscr{F}}$ generators that simply express the tensorial nature of the various $(F)^{\bar{m}p}_{n\bar{q}}$.

In addition to the cases above the full $so(4N)$ contains the $F^{k}_{\bar{j}}+F_{j\bar{i}}$ operators that form part of the $u(2N)_{\delta}$ realization completing the set of $2N(4N-1)$ generators.

In the next section we turn to the study of the relations between the various monomials under the assortment of transformation algebras that we have presented.

III. ACTION OF THE SUBALGEBRAS ON MONOMIALS

It is helpful to arrange the monomials in a tabular arrangement where the rows denote the number of θ's and the columns denote the number of $\bar{\theta}$'s.

One can regard this as a generalization of a block matrix in the sense that the tensor indices provide further labels inside each row and column location, but it must be noted that the number of monomial terms, i.e., the dimension of the various $u(N)$ tensors, varies widely from site to site. We can also consider the corresponding array of component fields, potentially also carrying Lorentz properties, in a superfield model of elementary particles where the role of the Grassmann coordinates is to give internal symmetry attributes to the various fields.

In this array form it is easy to discuss the action of the various transformation generators. Each multiplication by $\bar{\theta}$ or θ moves to the right by one column or down by one row, respectively. Conversely, every differentiation with respect to $\bar{\theta}$ moves one column to the left while those with respect to θ move up one row. Since the $u(N)_{\bar{\theta}} \oplus u(N)_{\theta}$ generators all involve one multiplication and one corresponding differentiation they leave us in the same location in the array. These generators transform the θ and $\bar{\theta}$ factors in the monomials separately. The diagonal $u(N)_{\mathscr{F}}$ subalgebra generated by the $\mathscr{F}^{i}_{\ j}$'s transforms the monomial labels as N and $\bar{N}$ representations as appropriate. As $u(N)_{\mathscr{F}}$ preserves the $\bar{\theta}^k\theta_k$ contraction it can be convenient to organize the monomials of given degree $(\bar{\theta})^p(\theta)^q$ into irreducible $su(N)_{\mathscr{F}}$ representations by extracting all the trace factors, e.g., by decomposing the $(\bar{\theta})^1(\theta)^2$ terms as $\bar{\theta}^i\theta_j\theta_m - \delta^i_{[j}\theta_{m]}(\bar{\theta}^k\theta_k)/(N-1)$, and $(\bar{\theta}^k\theta_k)\theta_m$. Note that the orthogonal combination of generators—the $F^{\bar{i}}_{\bar{j}} + F^{i}_{\ j}$ in the coset $(u(N)_{\bar{\theta}} \oplus u(N)_{\theta})/u(N)_{\mathscr{F}}$—mix these representations.

Two $u(1)$ generators also leave the positions in the array and the individual diagonal $su(N)_{\mathscr{F}}$ representions unchanged. One, $\mathscr{D} = \bar{\theta}^k\bar{\partial}_k - \theta_k\partial^k$, the trace of $su(N)_{\mathscr{F}}$, counts the difference between the number of $\bar{\theta}$'s and the number of θ's in a monomial, while the other, $\mathscr{S} = \theta_k\partial^k + \bar{\theta}^k\bar{\partial}_k - N$ counts the total number of Grassmann coordinates in the monomial. The eigenvalues of $\mathscr{S}$ and $\mathscr{D}$ can be used to provide an alternative (rotated) set of labels or coordinates for the array of monomials in place of the row and column numbers, while the remaining generators shuffle the monomials around.

The generators of the coset $u(2N)_\beta/(u(N)_{\bar{\theta}} \oplus u(N)_\theta)$ move parallel to the leading diagonal preserving the $\mathscr{D}$ eigenvalues, while those of the coset $u(2N)_\alpha/(u(N)_{\bar{\theta}} \oplus u(N)_\theta)$ move parallel to the cross-diagonal changing $\mathscr{D}$ eigenvalues while preserving those of $\mathscr{S}$. The actions of the $u(2N)_\gamma$ and $u(2N)_\delta$ realizations are more complicated as they involve combinations of generators that move in all directions on the grid. Here we concentrate on the generators of the realizations of various subalgebras of the $u(2N)$'s. The $sp(2)$ generators, $\mathscr{S}^+ = \bar{\theta}^k\theta_k$ and $\mathscr{S}^- = \partial^k\bar{\partial}_k$ alter the eigenvalues of $\mathscr{S}$ by adding or removing factors of the scalar $\bar{\theta}^k\theta_k$, i.e., they move diagonally back and forth across the grid, while other operators which commute with $\mathscr{S}$, in particular the $\bar{\theta}^i\partial^j + \epsilon\bar{\theta}^j\partial^i$ and $\theta_i\bar{\partial}_j + \epsilon\theta_j\bar{\partial}_i$ generators in $sp(2N)_{\alpha\gamma\delta}/u(N)_{\mathscr{F}}$ (for $\epsilon = +1$) or in the realization of $so(2N)_\alpha/u(N)_{\mathscr{F}}$ (for $\epsilon = -1$), shift along the cross-diagonals. The coset generators in $so(2N)_\gamma/u(N)_{\mathscr{F}}$ and $so(2N)_\delta/u(N)_{\mathscr{F}}$ have a more complicated action on the monomials. They transform a single monomial into combinations of pairs of monomials consisting of those two places right and those two places up from the starting point for $(F)^{\bar{2}} + (F)^2$, or those two places down and those two places left in the array for $(F)_2 + (F)_{\bar{2}}$.

At this point it is should be recalled that as we are considering here only the action of the bilinear and hence even generators, the odd and even monomials remain uncoupled by the transformations. Thus the array can be regarded as simply holding in a checkerboard fashion the two inequivalent spinor representations Δ_+ and Δ_- of $so(4N)$ and so we can connect each one of the even monomials to all the other even monomials using the $so(4N)$ generators, and similarly for the odd sector. It is appropriate to commence at one of the simple monomials, such as the singlets 1, $\theta_1\theta_2...\theta_N$, $\bar{\theta}^1\bar{\theta}^2...\bar{\theta}^N$ or $\bar{\theta}^1\bar{\theta}^2...\bar{\theta}^N\theta_N...\theta_2\theta_1 = (\bar{\theta}^k\theta_k)^N/N!$. The action of the various generators described above gives some insight into the arrangement of the monomials as collections of $u(N)_{\mathcal{F}}$ representations into representations of the corresponding subalgebras of $so(4N)$. For example, starting at $\bar{\theta}^1\bar{\theta}^2...\bar{\theta}^N$, the application of one of the generators of $sp(2N)_{\alpha\gamma\delta}/u(N)_{\mathcal{F}}$, i.e., the $F_{i\bar{j}} + F_{j\bar{i}}$, generates an element of the $N(N+1)/2$-dimensional $u(N)_{\mathcal{F}}$ representation within the $(\theta)^1(\bar{\theta})^{N-1}$ collection of monomials. Successive applications of these operators generate an $sp(2N)$ representation along the leading cross-diagonal. In a similar way the action of the $so(2N)_\alpha/u(N)_{\mathcal{F}}$ coset operators $F_{i\bar{j}} - F_{i\bar{j}}$ produces from that same singlet state the $N(N-1)/2$-dimensional $u(N)_{\mathcal{F}}$ representation which contains the remaining $(\theta)^1(\bar{\theta})^{N-1}$ monomials, and subsequent action of these operators leads to a collection of the appropriate $u(N)_{\mathcal{F}}$ representations to form an $so(2N)$ multiplet on the same cross-diagonal.

These "ladders" of $u(N)_{\mathcal{F}}$ representations can be ascended using the corresponding conjugate operators $F^{\bar{i}j} \pm F^{\bar{j}i}$, and of course the same $so(2N)$ or $sp(2N)$ representation can be produced by applying these operators to the singlet $\theta_1\theta_2...\theta_N$ at the other end of the cross-diagonal.

An alternative treatment of the states of the cross-diagonal as $u(N)_{\mathcal{F}}$ representations assembled into a $u(2N)$ representation uses the generators of $u(2N)_\alpha/u(N)_{\mathcal{F}}$, starting on the corner singlet states $(\bar{\theta})^N/N!$ or $(\theta)^N/N!$.

The $sp(2)$ generators $\mathcal{S}^+$ and $\mathcal{S}^-$ may be used to step perpendicularly away from these states on the leading cross-diagonal to the other monomials with the same $u(N)_{\mathcal{F}}$ content, but of degree $N \pm 2j, 1 \leqslant j \leqslant [N/2]$. When acting on the $sp(2N)$ representation along the cross-diagonal this procedure will generate the remainder of the $sp(2N) \oplus sp(2)$ representation.

An alternative way to move around on half of the checkerboard of monomials would be to start at the other (invariably even) singlets 1 or $\bar{\theta}^1\bar{\theta}^2...\bar{\theta}^N\theta_N...\theta_2\theta_1$ and act with the $sp(2)$ generators to move to one of the singlets in the leading diagonal, i.e., increase the $\mathcal{S}$ eigenvalue, and then branch out cross-diagonally. The $so(2N)_\alpha/u(N)_{\mathcal{F}}$ operators can be used directly, but the $sp(2N)_{\alpha\gamma\delta}/u(N)_{\mathcal{F}}$ generators annihilate $\bar{\theta}^k\theta_k$, so $(u(N)_{\bar{\theta}} \oplus u(N)_\theta)/u(N)_{\mathcal{F}}$ coset generators must be applied to the diagonal scalars first.

At this point an important difference between odd and even N stands out. For even N the four singlet states at the corners of the array can all be linked by the even generators and the discussions above lead to all the even monomials starting from any singlet. For the case when N is odd the singlets $\theta_1\theta_2...\theta_N$ and $\bar{\theta}^1\bar{\theta}^2...\bar{\theta}^N$ belong to the odd monomials while the singlets 1 and $\bar{\theta}^1\bar{\theta}^2...\bar{\theta}^N\theta_N...\theta_2\theta_1$ are in the even monomial sector and so the procedures above generate both sectors.

Now we return to the action of the realizations $so(2N)_\gamma$ and $so(2N)_\delta$. Again the situation is different for odd and even N. For odd N the $so(2N)$ representations of odd monomials thus constructed from each of the odd singlets, $\theta_1\theta_2...\theta_N$ and $\bar{\theta}^1\bar{\theta}^2...\bar{\theta}^N$, are distinct, being conjugates from the point of view of their $u(N)_{\mathcal{F}}$ tensor content. There are also distinct representations of even monomials generated from the 1 and $\bar{\theta}^1\bar{\theta}^2...\bar{\theta}^N\theta_N...\theta_2\theta_1$ singlets. For even N the even monomial representations generated from any singlet with these operators are the same.

It is not essential to commence constructing representations from singlet states, and of course for even N a nontrivial starting point like θ_i is needed to generate the odd monomials at all. Of necessity we would require fermionic (odd) generators from the enveloping superalgebra to move between monomials of odd and even degree and these are contained in the odd components of $(F)^{\bar{m}p}_{n\bar{q}}$; while this would lead us in the direction of internal supersymmetry and would involve the interesting task of studying the supersubalgebras of $gl(2^{2N-1}/2^{2N-1})$, we shall ignore them for the moment. This is not simply a matter of restraining the scope of our present study; for our intended application of Grassmann coordinate ideas to unified models of internal symmetries it is clear that to describe a family of particles by the components of a particular superfield carrying a given spin and statistics, e.g., a bosonic spinor superfield, we will only wish to retain either odd or even monomial terms in the field expansion, if there are not to be problems with the spin-statistics theorem. Accordingly it appears that internal supersymmetry will be a subtle matter to include in our schemes.

IV. DISCRETE SYMMETRIES AND SUPERFIELD CONSTRAINTS

Our program is to formulate unified models of elementary particles using the Grassmann coordinates in a superfield framework. Thus it will be necessary to construct field theoretical action functionals giving rise to appropriate dynamics for the individual superfield components, regarded as conventional fields over the space-time submanifold. The strategy is then to build actions using functions of the superfields themselves (together with suitable derivatives of them), and then to integrate over *all* of the Grassmann and space-time coordinates.[11]

The present paper focuses on the *algebraic* aspects of the scheme as a foundation for a unified gauge theory; however in order to make the connection to physical applications along the above lines some further properties of the superfields need to be established. Superfields must carry spin and statistics properties, while the component fields carry further Grassmann labels in keeping with the overall superfield assignment. This means that superfields of definite spin and statistics will contain either only odd or only even monomials.

The Grassmann schemes, while having the agreeable property of specifying in advance the nature of the matter representations through superfield expansion, generally lead to a proliferation of such multiplets because of the exponential dependence on N. In order to restrict the number of particles various superfield constraints related to Lie algebra automorphisms of finite order are considered. We call these *duality constraints* because of their similarity to dual transformations in differential geometry. In Sec. V the formalism is applied to the particular case of $N=5$.

A. Inner products and superfield duality

To discuss inner products of superfields, the standard tool of Berezin integration over the Grassmann coordinates is introduced.[20] Taking superfields Φ, Ψ having expansions over $\mathscr{M} = 2N$ Grassmann coordinates $\vartheta_K, 1 \leqslant K \leqslant \mathscr{M}$, of the form

$$\Phi(x,\vartheta)=\sum_{p=0}^{\mathscr{M}} \vartheta_{K_1}\vartheta_{K_2}\cdots\vartheta_{K_p}\phi^{K_1K_2\cdots K_p}/p!,$$

the natural inner product induced by the Berezin integral is

$$\langle\Phi,\Psi\rangle=\int d\vartheta_{\mathscr{M}}\cdots d\vartheta_2 d\vartheta_1 \Phi^*\Psi$$

$$=\sum_{p=0}^{\mathscr{M}} \epsilon_{K_1K_2\cdots K_{\mathscr{M}}}\phi^{*K_p\cdots K_2K_1}\psi^{K_{p+1}K_{p+2}\cdots K_{\mathscr{M}}}/p!(\mathscr{M}-p)!, \tag{26}$$

where $\epsilon_{K_1K_2\cdots K_{\mathscr{M}}}$ is the $\mathscr{M}$-dimensional alternating symbol, while $*$ denotes complex conjugation and reversal of the ordering of Grassmann variables in monomials. Note that in this example the ϑ's are real Grassmann coordinates, and with respect to this inner product,[21] the ϑ_K regarded as operators on monomials are Hermitian. In the complex basis, θ_i and $\bar{\theta}^i$ are conjugates and both must be integrated over. For various field theoretic purposes, such as constructing bilinear (kinetic) action terms for component fields, a diagonal inner product of the form

$$(\Phi,\Psi)\sim\sum_{p=0}^{\mathscr{M}} (\phi^{K_1K_2\cdots K_p})(\psi^{K_1K_2\cdots K_p})/p!$$

is more desirable and so clearly an operation on the superfields analogous to the differential geometric Hodge dual must be introduced to undo the effect of the $\epsilon_{K_1K_2\cdots K_{\mathscr{M}}}$ tensor in the Berezin integral inner product of Eq. (26).

The standard definition of the Hodge dual $\mathscr{M}-p$ form to a p-form[22] requires not only an orientable manifold but also a suitable metric η^{KL}. Similarly in the Grassmann case we require an invariant tensor η to raise indices and so we shall be led to consider a variety of η-dependent *dual* superfields ${}^{\eta}\Phi$,

$$({}^{\eta}\Phi)^{K_p\ldots K_2K_1}\equiv(-1)^{[\mathscr{M}/2]}\eta^{K_1L_1}\cdots\eta^{K_pL_p}\epsilon_{L_1L_2\ldots L_{\mathscr{M}}}\phi^{L_{p+1}\ldots L_{\mathscr{M}}}/(\mathscr{M}-p)! \tag{27}$$

with corresponding inner product

$$(\Phi,\Psi)_{\eta}\equiv\langle\Phi,{}^{\eta}\Psi\rangle=\sum_{p}\phi^{*K_1\cdots K_p}\eta_{K_1L_1}\cdots\eta_{K_pL_p}\psi^{L_1\cdots L_p}/p! \tag{28}$$

Each of the order p terms in the sum in this inner product is separately invariant under the transformations associated with the symmetries of the η tensor, so that a variety of p-dependent phase factors could be introduced into the definition of the η-dual of a component field Eq. (27), without disrupting the invariance. Conversely it is also apparent that the inner product could be associated with a larger invariance algebra if the relative signs were correctly chosen to correlate different terms so that monomial-degree changing operators from the transformation algebra also left the inner product unchanged.

We also note that the invariant tensor η may be skew-symmetric, as in one example below, and in such cases it is necessary to define index raising and lowering rules with greater precision than in the case where η is a symmetric tensor. The 'left-above to right-below' summation rule should be taken to apply in the equations above, i.e., $U^I=\eta^{IJ}U_J$ and $V_J=V^I\eta_{IJ}$, and for consistency of $\eta_{IJ}=\eta^{NM}\eta_{NI}\eta_{MJ}$ one takes $\eta^{IJ}\eta_{JK}=-\delta^I_J$ as defining the inverse.

In standard manifold theory here is a further situation in which a dual can be defined, which again requires an additional geometrical input, namely, when the manifold possesses a complex structure.[22] In terms of the notation of Eq. (27) the η-tensor is simply reinterpreted as a tensor with square -1 satisfying appropriate additional conditions.

We find it more useful to continue our discussion of duality in the complex basis. In this framework it is clear that the role of the Berezin integral in an inner product is to ensure that the superfield bilinear integrand contains each upper and each lower index exactly once. It is necessary to expand in terms of traceless component fields, to deal with irreducible representations of $su(N)_{\mathscr{F}}$, i.e., $\phi^{\beta_1\beta_2\cdots\beta_q}_{\alpha_1\alpha_2\cdots\alpha_p}\delta^{\alpha_p}_{\beta_q}=0$, etc., and so we express superfields as

$$\Phi(x,\theta,\bar\theta)=\sum_{p,q=0}^{N}\sum_{\nu>0}\frac{\bar\theta^{\alpha_1}\bar\theta^{\alpha_2}\cdots\bar\theta^{\alpha_p}}{p!}\frac{(\bar\theta^k\theta_k)^{\nu}}{\nu!}\frac{\theta_{\beta_1}\theta_{\beta_2}\cdots\theta_{\beta_q}}{q!}\phi^{(\nu)\beta_1\beta_2\cdots\beta_q}_{\alpha_1\alpha_2\cdots\alpha_p}. \tag{29}$$

This allows a complete labeling of the superfield degrees of freedom using the $su(N)_{\mathscr{F}}$ representation carried by the component field tensor indices, the ν label and $(p-q)$, as the latter two labels discriminate between the various recurrences of the representations of $su(N)_{\mathscr{F}}$. Note that $p-q$ is the eigenvalue of the $u(1)$ operator $\mathscr{D}$ on the corresponding monomials for all ν. In this basis we can envisage an inner product of the form

$$(\Phi,\Psi)\equiv\langle\Phi,\check\Psi\rangle\equiv\int\left(\prod d\theta d\bar\theta\right)\Phi^*\check\Psi \tag{30}$$

associated with what we term[11] the *Grassmann dual* $\check\Phi$, with components

$$(\breve{\Phi})^{(\nu)\beta_1\beta_2\cdots\beta_q}_{\alpha_1\alpha_2\cdots\alpha_p}=\Lambda_{p,q,\nu}\left(\frac{N!}{\nu!(N-\nu-p-q)!}\right)\phi^{(N-\nu-p-q)\beta_1\beta_2\cdots\beta_q}_{\phi\alpha_1\alpha_2\cdots\alpha_p} \tag{31}$$

which gives rise to a manifestly $su(N)_{\mathscr{G}}$ invariant inner product of the form

$$(\Phi,\Psi)\sim\sum_{p,q=0}^{N}\sum_{\nu}(\phi^{(\nu)\beta_1\cdots\beta_q}_{\alpha_1\cdots\alpha_p})^*\psi^{(\nu)\beta_1\cdots\beta_q}_{\alpha_1\cdots\alpha_p}/p!q! \tag{32}$$

and with scope for a larger invariance algebra depending once again on the choices of the phase factors $\Lambda_{p,q,\nu}$ in the dual operation. In fact by careful sign choices one can expect to construct an invariant potentially leading to a $gl(2^{2N-1})$ invariant bilinear, diagonal in all the component fields for superfields composed of odd or even sector monomials. That is not necessarily desirable given that the nature of the statistics carried by the component fields has yet to be included, but conveys the general idea—once again there are clearly many possible sign choices. In our present approach this $su(N)_{\mathscr{G}}$ invariant dual is a useful starting point but it is of interest to contemplate duals with larger invariances. While the original motivation for dual transformations was in terms of constructing actions and inner products, our present focus is on the use of self-duality or anti-self-duality constraints to restrict the number of matter representations in unified models of elementary particles, and to see the generation replication of families of representations. Accordingly in what follows we shall eventually move from the concrete basis of monomials in complex Grassmann coordinates to consider polynomials instead, labeled abstractly by the representation labels of their quantum numbers under for example $su(N)_{\mathscr{G}}\oplus u(1)_{\mathscr{D}}\oplus u(1)_{\mathscr{T}}$. This scheme allows us to study the sort of restrictions that can be placed on superfields by an interplay of different compatible duality transformations without excessive tensor analysis.

B. Lie algebra automorphisms

The classical study of automorphisms of semisimple Lie algebras is intimately connected with the problem of categorizing embeddings of subalgebras.[18,23] In the case of involutive automorphisms the subalgebras emerging are the maximal symmetric ones, and the results provide a classification of the real forms of the complex simple Lie algebras. Physically this structure has been applied to a discussion of charge conjugation operators in grand unified models;[24] we shall be more concerned with the use of these discrete symmetries in defining generalized internal parity operators and restricting the physical particle spectrum.

We follow the notation of Gilmore.[18] Generically an involutive automorphism $\sigma:\sigma^2=1$, effects a decomposition of the Lie algebra $\mathbf{g}$ of a Lie group G into a direct sum $\mathbf{g}=\mathbf{k}\oplus\mathbf{t}$, such that $\sigma(\mathbf{k})=+\mathbf{k}$ and $\sigma(\mathbf{t})=-\mathbf{t}$, with $[\mathbf{k},\mathbf{k}]\subset\mathbf{k}$, $[\mathbf{t},\mathbf{t}]\subset\mathbf{k}$, and $[\mathbf{k},\mathbf{t}]\subset\mathbf{t}$. The available reflection operators σ are classified;[18,23] the question of interest is the explicit form of a given σ acting on an arbitrary irreducible representation of $\mathbf{g}$.

By definition $\mathbf{k}$ commutes with the action of the desired σ and so the latter is diagonal (with eigenvalues ±1) on multiplets of $\mathbf{k}$ in any irreducible representation of $\mathbf{g}$. Moreover $\mathbf{k}$-multiplets connected by successive applications of the coset generators $\mathbf{t}$ must carry alternating signs. Thus the procedure is to assign eigenvalues relative to a suitable starting point, for example the multiplet containing the highest weight vector. In practice it is always possible to find an integer grading of $\mathbf{g}$ relative to an appropriate $u(1)$ subalgebra for which $\mathbf{k}=\mathbf{g}_0$ and $\mathbf{t}=\mathbf{g}_{+1}+\mathbf{g}_{-1}$, so that the alternation of sign is automatic. This procedure is illustrated below for the case of the automorphisms corresponding to the $u(2N)_\gamma$ subalgebra of $so(4N)$, and its $sp(2N)_\gamma$ and $so(2N)_\gamma$ subalgebras, making use of the explicit form of the generators and their action on $(\bar{\theta})^n(\theta)^m$ monomials.

A slight elaboration of the previous discussion obtains when there are two automorphisms σ and τ which have invariant subalgebras $\mathbf{k}$ and $\mathbf{h}$ such that $\mathbf{g}\supset\mathbf{k}\supset\mathbf{h}$. In this case both $\mathbf{k}$ and

t split into $\pm$ 1 eigenspaces of τ, and σ and τ manifestly commute. This structure is implicit in further discussion of the $so(4N) \supset u(2N) \supset sp(2N)$ (or $so(2N)$) $\supset u(N)$ branching chains below.

The $u(2N)_\gamma$ generators have been identified above in the $u(N)_{\mathscr{S}} \approx su(N)_{\mathscr{S}} \oplus u(1)_{\mathscr{D}}$ basis. Of more interest for the corresponding automorphism are the $2N(2N-1)$ generators of the $so(4N)/u(2N)_\gamma$ coset:–

$$\begin{gathered}\bar{\theta}^i\bar{\partial}_j+\theta_j\partial^i,\quad \theta_i\theta_j-\bar{\partial}_i\bar{\partial}_j,\quad \bar{\theta}^i\bar{\theta}^j-\partial^i\partial^j,\\ \theta_i\bar{\partial}_j-\theta_j\bar{\partial}_i,\quad \bar{\theta}^i\partial^j-\bar{\theta}^j\partial^i,\quad \partial^i\bar{\partial}_j+\bar{\theta}^i\theta_j,\end{gathered} \tag{33}$$

which can be arranged into pairs of ± 2 eigenstates of the $u(1)$ charge $\mathscr{T} \equiv -i(\mathscr{S}^+ - \mathscr{S}^-)$ which commutes with the $u(2N)_\gamma$ generators, since $u(2N)_\gamma \approx su(2N)\gamma \oplus u(1)_{\mathscr{T}}$. The trace-free parts of these coset operators can act as raising and lowering operators for $\mathscr{T}$, and they can also change the $u(N)_{\mathscr{S}}$ representation labels. For the spinor representations of $so(4N)$ carried by the θ, $\bar{\theta}$ monomials, the eigen-polynomials of σ can be constructed by acting with these coset generators on one of the $u(N)_{\mathscr{S}}$ singlets identified above. For example starting from the right hand corner of the $N \times N$ grid $(\bar{\theta}^1\bar{\theta}^2...\bar{\theta}^N)$, the $u(2N)_\gamma$ operators and the coset generators mentioned above will generate linear combinations of the monomials at all of those neighboring squares in the checkerboard which have the same "color" or grading (total number of Grassmann coordinates modulo 2).

Alternatively and more simply, the $N \times N$ grid of monomials presently labeled using the rotated coordinates $\mathscr{D}$ and $\mathscr{S}$ can be *replaced* by another superfield basis using polynomials of varying degree but retaining the groupings into $su(N)_{\mathscr{S}}$ representations, and using the $\mathscr{D}$ and $\mathscr{T}$ eigenvalues as the new rotated grid coordinates in place of those of $\mathscr{D}$ and $\mathscr{S}$. In terms of the original array of monomials this involves forming linear combinations along diagonals. The sites of the grid can be treated as before regarding $u(N)_{\mathscr{S}}$ properties but now represent polynomials with the difference between numbers of $\bar{\theta}$'s and θ's (the eigenvalue of $\mathscr{D}$) constant for a given site. The weight diagram is identical, since the spectrum of $\mathscr{T}$ is the same as that of $\mathscr{S}$ in $sp(2)$, but now it is clear that the $u(2N)_\gamma$ multiplets (antisymmetric tensors of rank up to N) run cross-diagonally and that the signs with respect to σ alternate with $|\mathscr{T}|$ away from the leading cross-diagonal which contains the corner singlet. In a similar way the actions of the automorphisms corresponding to the $so(2N)_\gamma$ and $sp(2N)_\gamma$ subalgebras can be identified.

For the $so(2N)_\gamma$ case the antisymmetric tensors of $u(2N)$ remain irreducible, so there is no alternation of sign down the cross diagonals in the new $N \times N$ grid with the $\mathscr{D}$, $\mathscr{T}$ rotated coordinates, except for the rank N case where the rank-N antisymmetric tensor decomposes into self-dual and anti-self dual parts with respect to $so(2N)_\gamma$. It can be seen that for the $so(4N)/u(2N)_\gamma$ coset generators, which transform as two rank-two antisymmetric tensors of $so(2N)_\gamma$ distinguished by having $\mathscr{T}$-eigenvalues ± 2, the action of the $so(2N)_\gamma$-preserving automorphism is respected by assigning the common τ eigenvalue $+1$ *or* -1. In terms of the formal treatment given, there are always two natural choices τ and $\sigma\tau$ which differ in sign on **t**. This overall sign choice in the action of τ is in any case determined by the assignment for the starting singlet.

In the $sp(2N)$ case, the $so(4N)/u(2N)_\gamma$ coset generators Eq. (33) transform under $sp(2N)_\gamma$ as two second rank antisymmetric tensors and two singlets, but now the tensors and the trace terms (singlets) must belong to different eigenspaces of τ so that $sp(2N)_\gamma$ multiplets which share cross diagonal positions have differing signs. Thus for example the $u(2N)_\gamma$ multiplet along the leading cross diagonal decomposes into antisymmetric tensors of rank $N, N-2, ..., 1$ or 0 with alternating signs. This is also consistent with the fact that the $N(2N-1)$ generators of the $u(2N)_\gamma/sp(2N)_\gamma$ coset,

$$(F)_2+(F)_{\bar{2}},(F)^2+(F)^{\bar{2}},(F)_1^{\bar{1}}-(F)_{\bar{1}}^{1}, \tag{34}$$

will be associated with the odd eigenspace of τ and will enforce the sign alternation *within* $u(2N)$ multiplets. The last link in our chains of involutive automorphisms is associated with the $sp(2N) \supset u(N)_{\mathcal{T}}$ and $so(2N) \supset u(N)_{\mathcal{T}}$ branchings.

C. Superfield constraints

The final stage is to establish the link with the superfield description of the $so(4N)$ spinor representations. The basis of our approach is that restrictions on particle content are to be achieved by means of superfield projections on to definite 'internal parity' multiplets. Given that the procedure should be covariant with respect to the physically identified $su(N)_{\mathcal{T}}$ algebra, it is natural to assume that the parities in question are involutive automorphisms of the $so(4N)$ associated with $su(N)_{\mathcal{T}}$ via the Clifford algebra construction.

In order to give the explicit construction of superfield constraints based on the various automorphisms identified we return to the superfield duals $\check{\Phi}$ and ${}^{\eta}\Phi$ of the previous subsection. As they are involutive by construction a decomposition into self-dual and anti-self-dual components such as

$$\Phi=\tfrac{1}{2}(\Phi+{}^{\eta}\Phi)+\tfrac{1}{2}(\Phi-{}^{\eta}\Phi)$$

can be made. A similar decomposition holds in terms of $\check{\Phi}$. Furthermore, one can ask for the invariance group of such superfield duals; namely, the set of transformations on the superfield which commute with the dual operation. The idea is to associate a $u(2N)_{\gamma}$ invariance with the Grassmann dual $\check{\Phi}$ and the appropriate subalgebra of $u(2N)_{\gamma}$ with the choice of invariant tensor η_{KL} for the case of ${}^{\eta}\Phi$. As we have seen in Sec. III, the natural choices for η_{KL} are those corresponding to the subalgebras $so(2N)$ (for symmetric η), and $sp(2N)$ (for antisymmetric η). In the former case

$$\eta_{KL}=\delta_{KL} \tag{35}$$

in the real basis $\vartheta_i=\frac{1}{2}(\theta_i+\bar{\theta}^i)$, and $\vartheta_{i+N}=(1/2i)(\theta_i-\bar{\theta}^i)$, and in the latter case the nonzero components are

$$\eta_{ij+N}=\frac{i}{2}\,\delta_{ij}=-\eta_{i+Nj}, \quad \text{for } 1\leqslant i\leqslant N. \tag{36}$$

To complete this discussion we note some further branching patterns. First, for the $su(N)$ invariant subalgebra within $so(2N)$ or $sp(2N)$ (wherein sign factors are acquired by successive application of the appropriate coset operators) the automorphism is generated by $\tilde{\Phi}(\theta,\bar{\theta})=\Phi(\theta,-\bar{\theta})$ and so $su(N)_{\mathcal{T}}$ multiplets acquire signs related to $(\mathcal{S}-\mathcal{D})/2$ (which commutes with $\mathcal{T}$).

With these remarks it is clear that the $\pm$ 1 eigenspaces of the generalized internal 'parity' operators can be projected out from the superfield by means of constraints such as $\Phi=\pm\check{\Phi}$, $\Phi=\pm{}^{\eta}\Phi$, $\tilde{\Phi}=\pm\Phi$, or more complicated combinations, and the abstract method outlined above serves to identify the surviving multiplets after projection.

We have not tried to classify all the available involutive automorphisms for which superfield constraints can be implemented. Discussion has centered around automorphisms associated in a natural way with just some of the subalgebras branching to the physical $su(N)_{\mathcal{T}}$ algebra. Duals and automorphisms may also be sought for others such as $su(N)\oplus su(N)$, or the 'direct sum' subalgebras like $so(2N)\oplus so(2N)$ and the diagonal $so(2N)$.

V. APPLICATION TO $N=5$ AND SPECIFIC MODELS

We have seen that the state space realized by polynomials in N complex Grassmann coordinates can be divided up into $su(N)$ representations, both for convenience and because it

TABLE II. The full set of $su(5)$ multiplets contained in the coordinate expansion of a superfield Ψ, p and q, the column and row numbers, correspond to the term $(\bar{\theta})^p(\theta)^q$ in the Taylor series for Ψ.

$q\backslash p$	0	1	2	3	4	5
0	1	5	$10]$	$\overline{10}$	$\bar{5}$	1
1	$\bar{5}$	$1+24$	$5+\overline{45}$	$10+\overline{40}$	$\overline{10}+\overline{15}$	$\bar{5}$
2	$\overline{10}$	$\bar{5}+45$	$1+24+75$	$5+\overline{45}+\overline{50}$	$10+\overline{40}$	$\overline{10}$
3	10	$\overline{10}+40$	$\bar{5}+45+50$	$1+24+75$	$\bar{5}+\overline{45}$	10
4	5	$10+15$	$\overline{10}+40$	$\bar{5}+45$	$1+24$	5
5	1	5	10	$\overline{10}$	$\bar{5}$	1

is expected that at least an $su(N)$ symmetry can be realized, providing a suitable framework for classifying the matter fields of a unified model of elementary particles. Previous studies[12] have shown that the observed three generations of fermions can be accommodated within a model with five complex Grassmann coordinates, or $N=5$ in the present terminology. Although this system provides a packaging of fermions in $su(5)$ representations both familiar and unfamiliar from conventional $su(5)$ models,[25] it also provides more copies and types of representations than are required. For this reason, and because the representations that are present will conflict with the anomaly if we assume that there will be a gauging of the $su(5)$ symmetry, we employed further "Grassmann duality" constraints to restrict the number of independent fermion degrees of freedom. Here we reanalyze the constraints and rederive those models,[12] but using more refined definitions of dual operators, as an application of the results of the previous section on discrete symmetries related to Lie algebra automorphisms, and the superfield duality transformations which implement them. Our present understanding of the way in which chirality is assigned to superfield components is incomplete, but we have managed to find some possible anomaly free assignments containing three fermion generations.

It should be mentioned that *conventional* $su(5)$ models are widely regarded as ruled out by the renormalization group extrapolations that can be made from recent precision experiments at LEP,[26] and also the question of proton lifetime, which requires a higher unification scale such as that associated with supersymmetric $su(5)$models. The details of the action in a nominal Grassmann coordinate model will face additional constraints because all the action terms are to be constructed as Berezin integrals over superfields. Thus there will be severe limits on the sorts of interaction vertices that can appear. This could prevent the occurrence of the conventionally 'dangerous' operators that lead to proton decay. With regard to the unification of couplings in renormalization group extrapolations of precision data from the weak interactions, there are many complications in the Grassmann models in terms of the extra fields present, their couplings and the question of symmetry breaking that require further exploration. In any case, the $N=5$ case is at least realistic enough to be a good proving ground for the Grassmann approach, and the same techniques can equally be applied in other cases.

A. Preliminaries

As a prelude we consider the relevant $su(5)$ representations themselves, the way they occur in the checkerboard pattern of θ and $\bar{\theta}$ monomials in this case, and how they are assembled into multiplets of the algebras in the $so(20)$ chain. Table II summarizes the situation. Within that grid rows increase with the power of θ and columns with the power of $\bar{\theta}$. As expected there are $2^{10}=1024$ states altogether representing the inequivalent spinor representations Δ_+ and Δ_- of $so(20)$. We reiterate that this grid can also be labeled by the eigenvalues of $\mathscr{D}$ and $\mathscr{S}$ with notional axes aligned with the diagonal and cross-diagonal in the Table.

It will be more convenient in what follows to have the 5×5 grid rewritten in the polynomial basis with the same $su(5)$ representations as above, but with polynomials for which $\mathscr{D}$

TABLE III. The spinor representations $\Delta_\pm$ of $so(20)$ decomposed into $su(5)$ reps., plotted by their eigenvalues of $\mathscr{D}$ (vertically) and $\mathscr{T}$ (horizontally), which run from -5 to 5. Branchings into antisymmetric irreducible tensors of $u(10)$, $sp(10)$, and $so(10)$ can be inferred by inspecting the columns at constant $\mathscr{T}$. The phases of the relevant representations under the various automorphisms as deduced from the properties of the adjoint representation appear in the underbraces.

					$\underbrace{1}_{+++}$					
				$\bar{5}$		$\bar{5}$				
			$\underbrace{10}_{+++}$		$\underbrace{\overline{10}}_{+++}$ $\underbrace{\overline{15}}_{++-}$		$\underbrace{\overline{10}}_{-++}$			
		10		10 $\overline{40}$		10 $\overline{40}$		10		
	$\underbrace{5}_{+++}$		$\underbrace{5}_{---}$ $\underbrace{\overline{45}}_{-+-}$		$\underbrace{5}_{+++}$ $\underbrace{\overline{45}}_{+-+}$ $\underbrace{\overline{50}}_{+++}$		$\underbrace{5}_{---}$ $\underbrace{\overline{45}}_{-+-}$		$\underbrace{5}_{+++}$	
1		1 24		1 24 75		1 24 75		1 24		1
	$\underbrace{\bar{5}}_{++-}$		$\underbrace{\bar{5}}_{--+}$ $\underbrace{45}_{-++}$		$\underbrace{\bar{5}}_{++-}$ $\underbrace{45}_{+--}$ $\underbrace{50}_{++-}$		$\underbrace{\bar{5}}_{--+}$ $\underbrace{45}_{-++}$		$\underbrace{\bar{5}}_{++-}$	
		$\overline{10}$		$\overline{10}$ 40		$\overline{10}$ 40		$\overline{10}$		
			$\underbrace{10}_{-+-}$		$\underbrace{10}_{+-+}$ $\underbrace{15}_{+++}$		$\underbrace{10}_{-+-}$			
				5		5				
					$\underbrace{1}_{++-}$					

and $\mathscr{T}$ are good quantum numbers, in place of the monomials labeled by $\mathscr{D}$ and $\mathscr{S}$.

The $su(5)$ multiplets of Table II are retained in this polynomial basis but are reoriented by 90° and displayed in Table III to emphasize the $\mathscr{D}$ and $\mathscr{T}$ coordinates. The utility of this arrangement is that it is $\mathscr{T}$ which commutes with the diagonal $u(10)_\gamma \supset sp(10)_\gamma \supset u(5)_{\mathscr{T}}$ and $u(10)_\gamma \supset so(10)_\gamma \supset u(5)_{\mathscr{T}}$ chains. In fact $u(10)_\gamma \approx su(10)_\gamma \oplus u(1)_{\mathscr{T}}$, while the $\mathscr{D}$ label comes from the final $u(5)_{\mathscr{T}}$ branching $u(5) \approx su(5)_{\mathscr{T}} \oplus u(1)_{\mathscr{D}}$.

Table III enables the decomposition of the odd and even spinors into antisymmetric tensors of $su(10)_\gamma$ to be read off directly, accompanied by their $u(1)_{\mathscr{T}}$ charges. In turn, these can be branched to $sp(10)_\gamma$ and $so(10)_\gamma$. In all three cases the irreducible multiplets can be identified by their content with respect to $su(5)_{\mathscr{T}} \oplus u(1)_{\mathscr{D}}$. The (horizontal) action of the commuting $sp(2)$ is obvious, and the $sp(10)_\gamma \oplus sp(2)$ decomposition can also be inferred.

Finally, the same data, which are provided for the spinor representations of $so(20)$ ($\theta,\bar{\theta}$ polynomials) by Table III, are also needed for the adjoint representation. The actual realization in terms of $\theta,\bar{\theta}$ differential operators has been extensively discussed in earlier sections; here we merely require the weight diagram and the $su(10)_\gamma$, $sp(10)_\gamma$, and $so(10)_\gamma$ content, labeled by $u(1)_{\mathscr{T}}$ and identified by their $su(5)_{\mathscr{T}} \oplus u(1)_{\mathscr{D}}$ decompositions. This information is provided in Table IV.

B. Discrete symmetries and superfield constraints

We now consider the discrete automorphism symmetries and the superfield duality constraints which implement them. Naturally we shall be concerned with the automorphisms in the chains $so(20) \supset u(10) \supset sp(10) \supset u(5)$ and $so(20) \supset u(10) \supset so(10) \supset u(5)$. These branchings (to maximal symmetric subalgebras at each stage[18,23]) correspond in each case to three successive commuting automorphisms, generically σ_1, σ_2, σ_3, which must be simultaneously diagonal in the $su(5) \oplus u(1)_{\mathscr{T}} \oplus u(1)_{\mathscr{D}}$ basis. In turn, the σ_i eigenvalues within

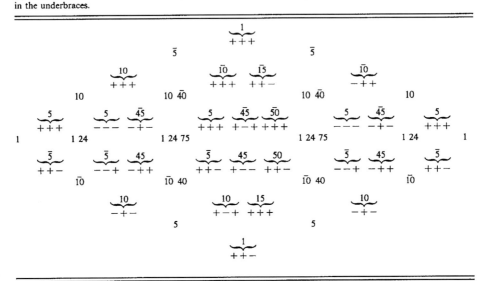

TABLE IV. The adjoint representation of $so(20)$ decomposed into $su(5)$ reps., plotted by their eigenvalues of $\mathscr{D}$ and $\mathscr{T}$, which run from -2 to 2. The phases associated with the gradings under the various automorphisms are displayed.

$\overline{10}$ ($-++$)	$\overline{10}$ ($+--$), $\overline{15}$ ($++-$)	$\overline{10}$ ($-++$)
24 ($-+-$), 1 ($---$)	24 ($+-+$), 1 ($+++$), 1 ($+++$), 24 ($+++$)	1 ($---$), 24 ($-+-$)
10 ($-++$)	10 ($+--$), 15 ($+--$)	10 ($-++$)

irreducible representations, within the spinors $\Delta_\pm$ in this case, follow once those in any particular submultiplet, such as that containing the highest weight, are assigned, provided the parity of the shift operators amongst the generators is known. Thus the $\mathbf{g}\supset(\mathbf{k}\oplus\mathbf{t})\supset(\mathbf{h}\oplus\cdots)$ decomposition of the adjoint representation is also vital and has already been depicted in Table IV.

In the decomposition of the original odd eigenspace $\mathbf{t}$ of σ_1 with respect to $\mathbf{h}$ there is a natural choice of two possible assignments of ± 1 eigenvalues of σ_2 as mentioned earlier; including the σ_3 eigenspaces there are thus four possibilities in all; however not every one will give rise to distinct projections in practice. For the chain $so(20)\supset u(10)\supset sp(10)\supset u(5)$, one favorable choice is presented in Table IV for the adjoint representation of $so(20)$ and in Table III for the spinors $\Delta_\pm$, where the notation is that the eigenvalues of $(\sigma_1,\sigma_2,\sigma_3)$ are listed for the $su(5)\oplus u(1)_{\mathscr{T}}\oplus u(1)_{\mathscr{D}}$ multiplets which appear in the identical location on the corresponding weight diagram.

C. *su*(5) models

As already mentioned, some possible $su(5)$ models emerged from previous work[12] using a more naive definition of the dual transformation and requiring the introduction of complicated phase operators. Here we rederive them in the present Grassmann framework, using the machinery of discrete symmetries and superfield duality. Although the discussion of charge conjugation is incomplete (complex conjugation is an involutive automorphism and should perhaps be related to branching chains and invariant subalgebras), two viable anomaly-free assemblies of $su(5)$ fermion representations that were found[12] have been recovered. In each case the first question to be addressed is the space-time spinor structure of the component fields arising from the superfield expansions.

A variety of alternatives for associating chirality with the components come readily to mind. The simplest assumption would be to choose both the odd and even superfields simply as Dirac spinors, without making any chiral projections. This would clearly lead to left–right symmetric models rather distant from the observed situation. Even if we chose the polynomials to correspond to the standard $su(5)$ chiral representations each putative fermion generation $(10_R+\bar{5}_R$, for example), would be accompanied by a left-handed counterpart $(5_L+\overline{10}_L)$ in the conjugate position in the $\mathscr{D},\mathscr{T}$ weight diagram. In order to avoid such duplication, it is natural to adopt a Hermiticity condition and assume that these pairs are related by charge conjugation, viz. $\overline{10}_L=(\overline{10}_R)^c$, etc., as spinor fields. This approach leaves the question of chirality assignments for the additional $su(5)$ representations unanswered.

Alternatively, a definite chirality for an entire superfield could be assumed, and Hermiticity of action functionals built in explicitly, or one could reasonably suppose that the chirality is fixed within given representations of the largest global symmetry, such as $\Delta_\pm$ for $so(20)$, but that it might vary between such representations, or correlations between internal symmetries and chirality could be envisaged. The following models exhibit some of these possibilities.

TABLE V. Surviving multiplets of Model 1. The states eliminated by imposing the duality conditions, $\tilde{\Psi}_o = -{}^{\eta}\check{\Psi}_o$ and $\check{\Psi}_e = \Psi_e$, are denoted by ∘ and • respectively, while those eliminated by the final differential constraint are marked by *'s.

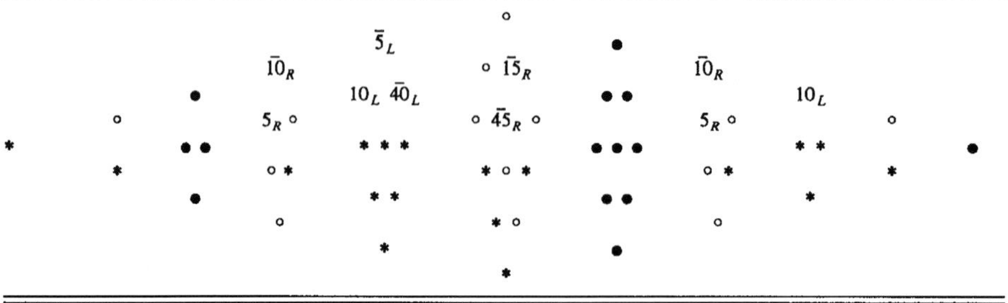

1. *Model 1: Three standard generations plus exotic particles*

In the first model we take the odd polynomials to be right-handed, and the even ones to be left-handed chiral spinors. This corresponds to linking space-time chirality to the $so(20)$ chirality of the $\Delta_\pm$ representations. Superfield duality constraints can now be applied to restrict the representation content. Recall that the various dual transformations $\check{\Phi}$, ${}^{\eta}\Phi$, $\tilde{\Phi}$ of Sec. IV C with the η tensor of Eq. (36), implement the automorphisms in the $so(20) \supset u(10) \supset sp(10) \supset u(5)$ chain. We impose the constraint $\tilde{\Psi}_o = -{}^{\eta}\check{\Psi}_o$ for the odd superfield Ψ_o, and the constraint $\check{\Psi}_e = \Psi_e$ for the even superfield Ψ_e.

The surviving component fields are readily identified by referring to Table III. With the sign conventions adopted in Sec. IV C, those $su(5)_{\mathscr{F}} \oplus u(1)_{\mathscr{D}} \oplus u(1)_{\mathscr{T}}$ multiplets for which the product of the three internal parities is odd are retained in Ψ_o, while for Ψ_e the multiplets retained are those whose $u(10)$ parity is even. At this stage real representations of $su(5)_{\mathscr{F}} \oplus u(1)_{\mathscr{D}}$ are involved for Ψ_e. Similarly for Ψ_o, some of the $su(5) \oplus u(1)_{\mathscr{D}}$ multiplets which would otherwise provide welcome material for the generations of fermionic matter are accompanied by conjugate partners. Such states—the majority of the components—would typically acquire unification scale masses in a generic GUT model with a broken $u(1)_{\mathscr{T}}$ gauge symmetry, although a global $u(1)_{\mathscr{T}}$ might protect them. Neither of these scenarios is attractive from the low energy viewpoint. The number of states is either insufficient or excessive and there are also problems with anomalies.

The most straightforward resolution of these difficulties is the application of a further superfield constraint, namely, a folding along the $\mathscr{T}$ axis or the leading diagonal of the polynomial array. Although there is an initial invariance $su(5)_{\mathscr{F}} \oplus u(1)_{\mathscr{D}} \oplus sp(2)$ at the Lie algebra level which might be exploited by other automorphisms, we shall take the restriction here simply as a differential constraint on the superfield itself which constrains the spectrum of $\mathscr{D}$ to an appropriate band. A natural way to achieve this is via a constraint of the form $\mathscr{D}(\mathscr{D}-1)(\mathscr{D}-2)...(\mathscr{D}-5)\Psi = 0$ which eliminates those components of both odd and even superfields which lie below the $\mathscr{T}$ axis. A similar constraint, $(\mathscr{D}-1)(\mathscr{D}-2)...(\mathscr{D}-4)\Psi = 0$, would allow only the middle rows of the upper $\mathscr{D}$ plane to remain. In the absence of more geometrical or phenomenological justification, the exact choice is not critical; the self-conjugate $su(5)$ multiplets 1, 24, 75 which may be involved are anomaly free by themselves and can in any case easily acquire unification scale masses.

The final set of surviving states is displayed in Table V . Consultation of Table VI confirms that this set is anomaly free with respect to gauging $su(5)_{\mathscr{F}}$ and has enough material for three standard fermion generations $5_R + \bar{10}_R$. There also appears the exotic but anomaly-free[27] combination $(40+\bar{10}+\bar{15}+\bar{45})_R$, raising the interesting possibility that once $su(5)_{\mathscr{F}}$ is broken the mixing of one of the $\bar{10}$'s with a 40 from the exotic set may give a hint as to the different nature

TABLE VI. Anomaly coefficients for $su(5)$ representations of fixed chirality.

$su(5)$ representation	1,24,75	5,10	15	40	45	50
Anomaly coefficient	0	1	9	16	6	15

of the t quark relative to the first two generations. For aspects of the interpretation of the weak interaction sector see Ref. 28.

2. Model 2: Incomplete generations from one $so(20)$ superfield

The second model compactly uses only one irreducible $so(20)$ superfield (the spinor carried by the odd monomials), but is unusual in that space-time chirality is required to alternate within the weight diagram. If $so(20)$ is still to be regarded as a global symmetry in this case, then to certain internal symmetry generators there should be adjoined a discrete space-time parity transformation.

In this case the dual transformations $\check{\Phi}$, ${}^{\eta}\Phi$, $\tilde{\Phi}$ of Sec. IV C, with the invariant tensor η of Eq.(35), implement automorphisms with the invariant subalgebra $so(10)$ in the $so(20) \supset u(10) \supset so(10) \supset u(5)$ chain, for which the branching rules for the odd spinor superfield can be inferred from Table III. In this model each multiplet of $su(5)_{\mathscr{F}} \oplus u(1)_{\mathscr{D}}$ is identified with the charge conjugate multiplet (of opposite chirality) in the appropriate reflected position in the weight diagram. The superfield Ψ_o is constrained to be even with respect to the $u(10)_\gamma$ Grassmann dual, $\Psi_o = \check{\Psi}_o$ which removes the states with $\mathscr{T} = \pm 3$. The surviving multiplets are displayed in Table VII.

Although there may not be enough material to provide correct strong and electroweak physics for the known fermion structure in the standard $su(3) \oplus su(2) \oplus u(1)$ model, even aside from a 'missing' t quark, this model is technically attractive in its Grassmann formulation in that it arises from a single superfield, it corresponds to a known[27] but unconventional anomaly free set $(2 \times 5 + \overline{50} + \overline{10})_R + (5 + \overline{15} + \overline{45})_L$, and there is an interesting interplay between space-time parity and $so(10)$ parity, which dictates the alternation of chirality within the selected $u(10)$ multiplets, so that for example this alternates between the self-dual and anti-self-dual rank 5 antisymmetric tensors $[1^5]_\pm$ and between the rank 3 and rank 7 antisymmetric tensors, $[1^3]$ and $[1^7]$. An extended candidate gauge symmetry $su(5)$

TABLE VII. Surviving multiplets in Model 2. The $so(10)$ multiplets 126+10 occur separately in columns. In this model all the even superfield components are absent: — denotes even locations for cross-reference with Table III. The o's denote entries which have been eliminated by the $u(10)_\gamma$ Grassmann duality constraint, $\Psi_o = \check{\Psi}_o$ imposed in this model. States related by conjugation are to be identified, and not taken as independent degrees of freedom.

					1_R					
				—		—				
			o		$\overline{10}_R\overline{15}_L$		o			
		—		—		—		—		
	5_R		o		$5_R\overline{45}_L\overline{50}_R$		o		5_L	
—		—		—		—		—		—
	$\overline{5}_R$		o		$\overline{5}_L45_R50_L$		o		$\overline{5}_L$	
		—		—		—		—		
			o		10_L15_R		o			
				—		—				
					1_L					

$\oplus\ u_{\mathscr{Q}}(1)$ emerges in this case because the $\mathscr{D}$ assignments are anomaly free. The outlying 5's of opposite chirality and $\mathscr{T}$ quantum numbers (Table II) mean that the latter symmetry, if gauged, would not be anomaly free in this scenario.

VI. CONCLUDING REMARKS

Grassmann coordinate schemes offer a natural framework for explaining the occurrence of particle generations and the introduction of superfield duality constraints in Ref. 12 provided the first prospects for reducing the overabundance of states. In the present paper we have put those early concepts on a firm footing using the tools of Lie algebra automorphisms. This has been possible because we now have elucidated the nature of the Lie algebras that transform the superfield components and we have seen how the duality conditions reduce the size of the residual symmetry algebra as well as cutting away the excess fermion content. In these concluding remarks we wish to outline the options for the further developments of the Grassmann coordinate scheme.

In the present paper we have also applied our general formalism to the case of five complex Grassmann coordinates, and demonstrated how the fermion matter superfield can be reduced so as to obtain three-generation type models. In addition to observing the reduction in size of the candidate symmetry algebra from the originally extensive transformation algebra, in parallel with the restriction in fermion matter representations, we observe that in these particular $N=5$ models there is no immediate impediment to elevating the residual global symmetry algebra to a local gauge symmetry, as the resulting collections of fermion multiplets appear in anomaly-free combinations.

While the questions of global symmetries in Grassmann coordinate schemes have been clarified by the present paper, there remain numerous further aspects to be investigated before a complete picture emerges. If we adhere to a standard view of the natural framework for unified theories, it is clear that two major sectors remain to be developed: gauge fields and symmetry breaking. The question of gauging the residual symmetry algebra has yet to be studied at the level of coupling matter superfields to gauge fields and constructing superfield actions for gauge field degrees of freedom. If we conservatively follow the Higgs mechanism approach to symmetry breaking then the nature of Higgs scalar matter representations arranged in a superfield and suitable kinetic actions and symmetry-breaking potentials will also require attention.

Let us first briefly consider the Higgs sector. We encounter here an embarrassment of riches similar to or worse than that for fermionic matter, and the nature of the duality conditions that might be applied are unclear at this point, beyond the requirement, in $su(5)$ models, that there be room for a standard 5 and 24: this leaves plenty of scope as can be seen from a glance at the variety of $su(5)$ representations in Table II. The replications of multiplets could be more difficult to come to grips with for Higgs fields than in the case of fermion generations. It may well be that in schemes like Model 2, we should associate the Higgs supermultiplet with part of the Grassmann-even multiplet Δ_-, as this is the easiest way of ensuring that spin-statistics is obeyed without resorting to special arguments. Unfortunately we have little guide from experiment about the Higgs sector, since there is yet no sign of even a single elementary Higgs boson. All one can do at this stage is lean on theoretical aesthetics and economy. Another aspect motivating a desire for a restriction in the number of Higgs representations is the issue of asymptotic freedom. This might appear a lost cause with large numbers of fermion and scalar multiplets of the candidate gauge groups, but may not really be a problem if the "massless" states below the unification scale are sufficiently few that the standard model strong and electroweak couplings can run as expected. We previously found[12] that, because the various Higgs field components are incorporated into superfields, there are complicated constraints on their couplings in a superfield action and generations of Higgs fields would be far from trivial clones. In fact the construction of a Higgs potential would appear to be a substantial challenge.

Given the surplus of fermions (with some rather exotic quantum numbers) that seem to occur in some models, it might also prove attractive to consider dynamical symmetry breaking scenarios as alternatives to the introduction of a potentially large number of elementary scalars.

Turning to the issue of gauge symmetries, we remark that our approach to the question of candidate gauge symmetry algebras has been conservative in the present work and we have focussed on the $so(4N)$ subalgebra of the $gl(2^{2N-1}/2^{2N-1})$ superalgebra of transformations and on subalgebras of $so(4N)$. This provided a natural framework for the use of involutive automorphisms associated with the branchings to symmetric subalgebras, and it is promising that in the models which emerged for the $N=5$ case, with the residual global symmetry algebras $su(5)$ or $su(5) \oplus u(1)$, the fermion representations were free of anomalies and so these are natural candidates for gauge symmetries.

There are also some more exotic aspects to the transformation superalgebra and its subalgebras. We present some indications of promising avenues.

The $su(N)_{\mathcal{G}}$ generators do not change the degree of Grassmann coordinate monomials and can be regarded as generation-conserving operators. If we are looking for generators of generation-changing processes it appears we must go beyond this set and include other operators from the transformation algebra of $(G)^{p}_{q}$'s or their complex basis equivalents the $(F)^{\bar{m}p}_{n\bar{q}}$'s. Indeed the existence of such additional generators is one of the most interesting features of the Grassmann formalism. Taking the complete set of even operators would mean starting with $gl(2^{2N-1})$ which seems enormous, and even taking a holomorphic/antiholomorphic splitting and treating just the even sectors of segregated $(F)^{\bar{m}0}_{0\bar{q}}$ and $(F)^{\bar{0}p}_{n\bar{0}}$ classes of operators would still be extensive.

Two facets of the transformation superalgebra are particularly deserving of further study. First of all it should be noted that the transformation superalgebra contains as a subsuperalgebra the Lie superalgebra $W(2N)$ of Cartan type in the classification of Kac,[29] also called a hyperclassical superalgebra by Ramond.[30] The $W(2N)$ superalgebra is generated by the operators $(G)^{p}_{1}$, which can also be considered as Grassmann vector fields in view of their single derivative operator. This highlights their similiarity to generators of general coordinate transformations in relativity, and is another caution against making assumptions that all generators should be associated with physical gauge fields. Making the further restriction to $p = 1,3,...,2N-1$ yields the even derivations on the Grassmann algebra composed of the monomials. These even derivations generate one parameter continuous automorphisms of the Grassmann algebra of monomials,[29] suggesting an intriguing possible alternative scenario for symmetries. The $su(N)_{\mathcal{G}}$ symmetry can be associated with the concept of equality between the various Grassmann coordinates, while the $W(2N)$-algebra would be associated with symmetries of the algebra of monomials itself. The connections of subsuperalgebras of the W-superalgebras with various super-geometrical invariants also merit further exploration.

Other possible extended symmetry algebras emerge from the fact that there is a finer grading for the $(G)^{p}_{1}$ operators, namely, by the degree (p) of the monomial they carry, schematically

$$[(G)^{p}_{1},(G)^{k}_{1}] \sim (G)^{p+k-1}_{1}. \tag{37}$$

This suggests that a hierarchy of subsuperalgebras containing the classes $(G)^{1}_{1}$ and $\{(G)^{p}_{1}, \forall p{:}p' \leqslant p \leqslant 2N\}$ or more precisely the even parts of these superalgebras, could be considered. Similar subalgebras can be naturally constructed in the complex basis.

It should be noted from Eq. (37) that these classes are nilpotent under the commutation operation for $p > N+1$. This suggests some weird possibilities for gauge algebras. There are in addition some puzzles to resolve here if one proposes using these operators as gauge symmetry generators, as naive use of the Hermiticity operation[11] in the complex basis would seem to require, for example, the inclusion of the conjugate generators $(F)^{\bar{1}0}_{0\bar{m}}$ to partner the $(F)^{\bar{m}0}_{01}$'s and this takes us out of the $W(2N)$ subalgebra of the transformation algebra.

This general viewpoint of looking beyond the quadratic operators of the $so(4N)$ algebra can also show us a variety of subalgebras of the transformation superalgebra which can be regarded as embodying the duality concept. The class of operators $(G)^{\mathscr{M}-1}_{\mathscr{M}-1}$ provides another realization of $u(2N)$, but one which only acts on the $\mathscr{M}$-plet of monomials $\epsilon_{IK_1K_2\cdots K_{\mathscr{M}-1}}\vartheta^{K_1}\vartheta^{K_2}\cdots\vartheta^{K_{\mathscr{M}-1}}$. There are corresponding operators in the complex basis in the classes $(F)^{\overline{N-1}N}_{N\overline{N-1}}$ and $(F)^{\bar{N}N-1}_{N-1\bar{N}}$ which can be seen to bear a dual relation to $(F)^{\bar{0}1}_{1\bar{0}}$ and $(F)^{\bar{1}0}_{0\bar{1}}$, respectively. An explicit representative of the first class would be

$$\epsilon_{ik_1k_2\cdots k_{N-1}}\epsilon^{jl_1l_2\cdots l_{N-1}}\bar{\theta}^{k_1}\bar{\theta}^{k_2}\cdots\bar{\theta}^{k_{N-1}}(\theta_p\partial^p)^N\bar{\partial}_{l_1}\bar{\partial}_{l_2}\cdots\bar{\partial}_{l_{N-1}}/N!((N-1)!)^2. \quad (38)$$

In the complex basis it is also clear that one can construct 'dual' operators which maintain segregation of the θ's and their conjugates, and another partner to the $u(N)_{\mathscr{F}}$ algebra is one generated by

$$\tilde{\mathscr{F}}_j{}^i=\epsilon_{jk_1k_2\cdots k_{N-1}}\epsilon^{il_1l_2\cdots l_{N-1}}(\bar{\theta}^{k_1}\bar{\theta}^{k_2}\cdots\bar{\theta}^{k_{N-1}}\bar{\partial}_{l_1}\bar{\partial}_{l_2}\cdots\bar{\partial}_{l_{N-1}}$$
$$-\theta_{l_1}\theta_{l_2}\cdots\theta_{l_{N-1}}\partial^{k_1}\partial^{k_2}\cdots\partial^{k_{N-1}})/((N-1)!)^2. \quad (39)$$

Thus we have an $su(N)\oplus su(N)$ algebra with generators $\mathscr{F}^i_j-\tilde{\mathscr{F}}^i_j$ and $\tilde{\mathscr{F}}^i_j$. The greater the number of gauge fields, the better the prospects for asymptotic freedom become, since the effect of the additional exotic matter multiplets are counteracted, although as indicated above some of these additional symmetry generators act only on a subset of the monomials.

Even after settling upon a gauge group, either from within $so(4N)$ or from one of these extended symmetry possibilities we need to construct the actions for the complete set of duality constrained matter superfields coupled to gauge fields. While earlier work[11] may provide a general guide, there will be some important differences in the present framework. The physical component fields are now associated with superfields as self-dual polynomials in θ and $\bar{\theta}$, not plain monomials, which as we showed in preliminary studies[12] leads to potentially more complicated interactions and mixings, and it should no longer be necessary to invoke the duality operator explicitly for its original role of forming action terms, as well as calling on the Hermitian conjugate superfields, since the superfields themselves are now constrained to be self-dual. Once again we point out that the Berezin rules of integration over the Grassmann coordinates may serve to eliminate some interactions which would otherwise appear permitted by simply considering the unified gauge symmetry, just as in conventional supersymmetry actions they can prevent the emergence of certain Lagrangian terms.

To view the present framework as a rather involved machine, whose output is simply the identification of some interesting collections of fermions with nonanomalous potential gauge symmetries, would be to misconstrue the Grassmann unified model program. Unlike conventional model building which consists of taking particular symmetry groups and then selecting representations, here the possible symmetry structures and multiplets are simultaneously delineated by the number of Grassmann coordinates chosen. Thus $\Delta_\pm$ is not just an elegant choice of representation for packaging various generations, flavors, colors, etc. of fermions in an $so(4N)$ theory, but rather constitutes the totality of possible superfield components for N complex Grassmann coordinates. From these constrictions the inescapable feature of *limited* repetitions of a *limited* variety of representations emerges very naturally. That is why we believe that future developments of these ideas could be quite exciting.

ACKNOWLEDGMENTS

The authors thank Ioannis Tsochantjis for his involvement and interest, and for many cogent remarks about Lie algebra automorphisms. Discussions with Tim Baker, Ming Yung, and David McAnally are also acknowledged. R.C.W. thanks Michael Peskin for a useful conversation about the $(G)^p_q$ algebra, the SLAC Theory seminar for their comments, and Pierre Ramond for discussions. He also thanks the SLAC Theory Group and the Institute for Fundamental Theory at the University of Florida for hospitality. P.D.J. thanks Paul Sorba and Nino Sciarrino for discussions on internal supersymmetry. Finally, this work would not have been possible without the guidance provided by group theoretical software packages such as SCHUR[31] and SimpLie.[32]

This research project and R.C.W. are supported by funds from an Australian Research Council (ARC) grant. P.D.J. acknowledges the ARC and the inspiration and support of the Alexander von Humboldt Foundation for part of this work.

[1] Th. Kaluza, Sitz. Preuss. Akad. Wiss. Berlin. Math. Phys., K1, 966 (1921); O. Klein, Z. Phys. **37**, 895 (1926). The developments from these papers through to the extension to supersymmetric theories are briefly reviewed with an extensive bibliography in Chap. 5 of Ref. 7 below.

[2] The bosonic string is comprehensively treated in the reviews of Superstring theory (Refs. 8 and 9) below.

[3] See, e.g., S. J. Gates, M. T. Grisaru, M. Rocek, and W. Siegel, *Superspace or One Thousand and One Lessons in Supersymmetry* (Benjamin, New York, 1983).

[4] P. Fayet and S. Ferrara, Phys. Rep. C **32**, 249 (1977); H. P. Nilles, *ibid.* **110**, 1 (1984); see also the reprint volumes by S. Ferrara, *Supersymmetry* (North-Holland, Amsterdam, World Scientific, Singapore, 1987).

[5] P. van Nieuwenhuizen, Phys. Rep. C **68** (1981); J. Wess and J. Bagger, *Supersymmetry and Supergravity* (Princeton University, Princeton, 1983); P. C. West, *Introduction to Supersymmetry and Supergravity* (World Scientific, Singapore, 1986); and the reprint volume by M. Jacob, *Supersymmetry and Supergravity* (North-Holland, Amsterdam, World Scientific, Singapore, 1986).

[6] M. J. Duff, B. E. W. Nilsson, and C. N. Pope, Phys. Rep. C **130**, 1 (1986); M. J. Duff, in *Physics in Higher Dimensions*, edited by T. Piran and S. Weinberg (World Scientific, Singapore, 1986).

[7] See the reprint volumes of A. Salam and E. Sezgin, *Supergravities in Diverse Dimensions* (North-Holland, Amsterdam, World Scientific, Singapore, 1989).

[8] M. Green, J. H. Schwarz, and E. Witten, *Superstring Theory* (Cambridge University, Cambridge, 1986).

[9] See the reprint volumes by J. H. Schwarz, *Superstrings, the First 15 Years* (World Scientific, Singapore, 1985).

[10] R. Casalbuoni and R. Gatto, Phys. Lett. B **88**, 306 (1979); **90**, 81 (1980); P. H. Dondi and P. D. Jarvis, Z. Phys. C **4**, 201 (1980); R. Delbourgo and R. B. Zhang, Phys. Rev. D **38**, 2490 (1988); P. Ellicott and D. J. Toms, Class. Quantum Gravit. **6**, 1033 (1989).

[11] R. Delbourgo, Mod. Phys. Lett. A **4**, 1381 (1989); R. Delbourgo and M. White, *ibid.* **5**, 355 (1990); R. Delbourgo, L. M. Jones, and M. White, Phys. Rev. D **41**, 679 (1990).

[12] R. Delbourgo, P. D. Jarvis, and R. C. Warner, Aust. J. Phys. **44**, 135 (1991).

[13] L. Bonora and M. Tonin, Phys. Lett. B **98**, 48 (1981); R. Delbourgo and P. D. Jarvis, J. Phys. A **15**, 611 (1982).

[14] Y. Ne'eman, Phys. Lett. B **81**, 190 (1979); D. B. Fairlie, *ibid.* **82**, 97 (1979); J. G. Taylor, *ibid.* **83**, 331 (1979); P. H. Dondi and P. D. Jarvis, *ibid.* **84**, 75 (1979).

[15] P. D. Jarvis, J. Math. Phys. **31**, 1783 (1990); Y. Ne'eman and S. Sternberg, Proc. Natl. Acad. Sci. USA, **87**, 7875 (1990); A. Connes and J. Lott,"Particle Models and Noncommutative Geometry," preprint IHES/P/90/23 (March 1990); F. Hussain and G. Thompson, Phys. Lett. B **260**, 359 (1991); R. Coquereaux, *ibid.* **261**, 449 (1991); B. S. Balakrishna, F. Gursey, and K. C. Wali, Phys. Rev. D **44**, 3313 (1991).

[16] This is sometimes referred to as the spinning string: See, for example, Sec. 6 of J. H. Schwarz, in *Physics in Higher Dimensions*, edited by T. Piran and S. Weinberg (World Scientific, Singapore, 1986) and the discussion in Chap. 4 of Ref. 8 above.

[17] O. Eyal, Techniques of Using Grassmann Variables for Realizing Some Algebras, Karlsruhe preprint KA-THEP-1990-3 (1990).

[18] R. Gilmore, *Lie Groups, Lie Algebras, and some of their Applications* (Wiley, New York, 1974).

[19] Parentheses around subscripts denote symmetrized indices $A_{(ij)} \equiv A_{ij} + A_{ji}$, while square brackets denote antisymmetrization.

[20] F. A. Berezin, *The Method of Second Quantization* (Academic, New York, 1966); B. DeWitt, *Supermanifolds* (Cambridge University, Cambridge, 1984).

[21] J. W. van Holten, Nucl. Phys. B **339**, 158 (1990).

[22] Y. Choquet-Bruhat and C. DeWitt-Morette with M. Dillard-Bleick, *Analysis, Manifolds, and Physics* (North-Holland, Amsterdam, 1982); T. Eguchi, P. B. Gilkey, and A. J. Hanson, Phys. Rep. C **66**, 213 (1980).

[23] J. F. Cornwell, *Group Theory in Physics* (Academic, New York, 1984), Vol 2, Chap. 14; F. R. Gantmacher, *Theory*

of Matrices (Chelsea, New York, 1959), Vol. 1; V. Kac, *Infinite Dimensional Algebras* (Cambridge University, Cambridge, 1990), Chap. 8.

[24] R. Slansky, in *First Workshop on Grand Unification*, edited by P. H. Frampton, S. L. Glashow, and A. Yildiz (Math Sci, Brookline, 1980).

[25] G. G. Ross, *Grand Unified Theories* (Benjamin-Cummings, Menlo Park, 1984); *Unification of the Fundamental Particle Interactions*, edited by S. Ferrara, J. Ellis, and P. van Nieuwenhuizen (Plenum, New York, 1980).

[26] J. Ellis, S. Kelley, and D. V. Nanopoulos, Phys. Lett. B **249**, 441 (1990); U. Amaldi, W. de Boer, and H. Fürstenau, Phys. Lett. B **260**, 447 (1991).

[27] R. C. King, Nucl. Phys. B **185**, 133 (1981).

[28] R. Delbourgo and R. Roleda, Int. J. Mod. Phys. A **7**, 5213 (1992).

[29] V. G. Kac, Ad. Math. **26**, 8 (1977).

[30] P. Ramond, in *Supersymmetry in Physics*, edited by V. A. Kostelecký and D. K. Campbell (North-Holland, Amsterdam, 1985).

[31] SCHUR 5.0. © Schur Software Associates (1990).

[32] SimpLie™ W. McKay, J. Patera, and D. Rand (CRM, Montreal, 1990).

International Journal of Modern Physics A
Vol. 29, Nos. 3 & 4 (2014) 1450023 (20 pages)
© World Scientific Publishing Company
DOI: 10.1142/S0217751X14500237

WHERE-WHEN-WHAT: THE GENERAL RELATIVITY OF SPACE-TIME–PROPERTY

ROBERT DELBOURGO* and PAUL D. STACK†

School of Mathematics and Physics, University of Tasmania,
Locked Bag 37 GPO, Hobart, Tasmania 7001, Australia
**bob.delbourgo@utas.edu.au*
†pdstack@utas.edu.au

Received 29 November 2013
Accepted 6 January 2014
Published 29 January 2014

We develop the general relativity of extended spacetime–property for describing events including their properties. The anticommuting nature of property coordinates, augmenting spacetime $(\mathbf{x}, t)$, allows for the natural emergence of generations and for the simple incorporation of gauge fields in the spacetime–property sector. With one electric property, this results in a geometrical unification of gravity and electromagnetism, leading to a Maxwell–Einstein Lagrangian plus a cosmological term. Addition of one neutrinic and three chromic properties should lead to unification of gravity with electroweak and strong interactions.

Keywords: Anticommuting; property; geometrical; unification; electromagnetism; gravity.

PACS numbers: 11.10.−z, 11.10.Kk, 12.10.−g

1. A Full Description of events

In a static universe, nothing happens. All systems continue inertially, never change and never interact; that sort of universe is the ultimate nonevent. Even speaking of an "observer" is a contradiction in terms, because every observer would be incommunicado and unaware of anything and everything. So, more to the point, a static universe is purely hypothetical and logically inconceivable. On the contrary, the real universe is in a state of flux. It is punctuated by a succession of events from which we gain the notion of space and time, as the scenario unfolds. Historically, the spacetime arena has held center stage ever since the ideas of relativity took hold about 100 years ago and we are accustomed to characterizing an event by its spacetime location, even to the extent of describing events geometrically. Indeed, the geometry of curved spacetime has revolutionized our ideas about gravity ever since Einstein's development of general relativity.

When an event occurs, it amounts to a change in circumstances, whereby an object alters its motion and possibly character as it engages with others. (It is the impact of these changes in spacetime, which underpins the process of observation and the Heisenberg uncertainty principle.) Thus, when a photon is emitted and reabsorbed by two charged objects, we interpret this as a succession of two events resulting in an electric interaction between the charged objects. And since quantum field theory came into being, we recognize this through a trilinear interaction between the charged object and a photon with the conservation of total energy–momentum at the event vertex, which is ensured by taking an integral over all spacetime of the interacting fields. So much is well understood. However, saying that an event has happened *there* and *then* does not fully specify the event; we must in addition specify *what* exactly has happened: what sort of transaction has taken place and what properties may have been exchanged. For instance, when a proton emits a neutron and charged pion, virtually, we have to add that information and then the event becomes fully explicated. This is normally done by introducing quantum fields with particular labels and interacting in a manner that usually conserves some quantum number, such as isospin for the nuclear example.

As physicists, we are well aware that particle properties do proliferate, but they can be systematized using group theory of some "internal group," resulting in certain types of representations of various Lie algebras. These can be constructed from some fundamental representation as a result of some basic dynamics connected with primitive constituents like quarks and leptons. As new particles are discovered, it occasionally becomes necessary to enlarge the group to accommodate features or properties that do not fall comfortably within the conventional picture; this is how progress in quantum field theory has been achieved and it has led to its ultimate incarnation as the standard model, wherein $\mathrm{U}(1) \times \mathrm{SU}(2) \times \mathrm{SU}(3)$ reigns supreme. Nevertheless some issues remain unresolved, such as the generation problem, where repetitions of particle families lead to further concepts about "horizontal groups" for systematizing them — although what the correct group and representations are is still unsettled. Most worrying of all is the number of parameters needed to characterize the interactions and masses of the (light) three generations in the standard model, though it must be admitted that those parameters are sufficient to describe a *vast* number of experimental facts. This has stimulated research into looking for some kind of grand unification for reducing the number of parameters and it has spawned a large number of interesting new concepts. Supersymmetry and superstrings, based on enlarged dimensions, are the foremost amongst these as they automatically solve the fine-tuning problem and naturally incorporate gravity, but we should not discount other ideas such as technicolor, preons, noncommuting spacetime variables, deformation groups, and so on.

Underlying all these extensions, the question arises: how to enumerate properties and characteristics of events at a fundamental particulate level. Traditionally, one specifies the associated quantum fields by attaching as many labels to them as needed to describe the properties and by ensuring that the events due to their mutual interactions conserve whatever properties remain intact overall; in other words, adapting to what the experiments dictate. This approach is predictive to the extent that if there are any missing components of the group representations, then they must eventually show up experimentally and conform to the overarching symmetry. And if the interactions do not preserve the expected symmetries, they are nowadays thought to be due to symmetry breaking effects coming from a yet unknown cause, possibly spontaneously generated.

In an effort to characterize the nature of an event, we have, over a number of years, suggested that it may be possible to specify attributes or "properties" of particles by connecting them with a few anticommuting Lorentz *scalar* coordinates (and their conjugates or opposites). An event can thereby be fully described by a conglomerate there-then-what label; by analogy to momentum conservation, the conservation of overall property, such as charge, is guaranteed by integrating appropriately over all property space. To that end we have suggested that the spacetime manifold with coordinates x, should be enlarged to a supermanifold[1–5] by attaching a few complex anticommuting ζ coordinates, with fields being regarded as functions of $X = (x, \zeta, \bar{\zeta})$. The anticommuting character of fermionic quantum fields is nothing new and the use of Grassmann variables has featured in the Becchi–Rouet–Stora–Tyutin (BRST) quantization of gauge theories, and more especially in the invention of supersymmetry — although the latter makes them Lorentz spinor rather than scalar. Once a property is ascribed, it cannot be doubly ascribed, corresponding to the fact the square of a given Grassmann variable vanishes. But because the conjugate property is available, one can build up property scalars that may be multiplied with other properties, and so on. This is how one might conceive families of systems with similar net attributes.[6,7] Furthermore, the anticommuting character of a fixed number of properties means that we can never encounter an infinite number of states, unlike bosonic extra dimensional spaces modeled along Kaluza–Klein lines. The mathematics of graded spaces is quite well understood now and we will take full advantage of this in what follows.

Previous investigations have shown that with a minimum of five independent ζ, one can readily accommodate all the known particles and their *generations*. We might tentatively call these properties: charge or electricity, neutrinicity and chromicity. The use of five is no accident, being based solidly on economical grand unified groups such as SU(5) and SO(10); what is more interesting is that the correct representations of those algebras come out automatically. (In our case, Sp(10) is the overarching internal group.) Since addition of Grassmann variables has the mathematical effect of reducing the *net* "dimension" of the space, there is the tantalizing prospect of a universe with zero total dimension. We have elaborated on some consequences of such a scheme,[8–10] including the appearance of a few new particles and the possibility of reducing the number of parameters appearing in the standard model, because we have just nine Higgs fields but only one coupling constant attached to all known fermions.[11]

The purpose of this paper is to describe the general relativity of spacetime–property along the lines originally devised by Einstein. Our aim is quite modest: we wish to see if one can unite electromagnetism with general relativity geometrically in a way which differs from Klein–Kaluza and (spinorial) supergravity in as much as the extended coordinates are *scalar* but anticommuting. Having established that, it means one may contemplate generalizations which include other forces, without producing infinite towers of excited states. This paper should be seen as a first step in that direction; further avenues for research are mentioned in the conclusions and include Higgs fields and possibly ghost fields associated with gauge-fixing at the quantized level. The extended metric will consist a x–x sector, a ζ–x sector and a ζ–$\bar{\zeta}$ sector; the space–space sector involves the gravitational field as well as some extra $\bar{\zeta}\zeta$ terms, the space–property sector brings in the gauge fields connected with property propagation, and the curved property–property sector can be the source of the cosmological term, as it happens; it is no surprise that the communicators of property or gauge fields should reside where they are, but the association of property curvature with a cosmological term is perhaps more intriguing. It is important to state that fermion fields, such as the gravitino, have no place in the extended metric as they would cause a conflict with Lorentz invariance. Because the fermionic superfield Ψ_α carries spinor indices, it must be studied separately and in that regard our approach differs radically from conventional spinorial supersymmetry; we give a preliminary treatment of fermions towards the end of the article.

As we are dealing with a graded manifold or supermanifold, it is *essential* to make sure that the supercoordinate transformations carry the correct commutation factors. This is carefully explained in Secs. 2–5, where our conventions are established and summarized to make the paper self-contained. A good notation is of course vital[12,13] and the ordinary general relativity convention with its placement of indices conflicts with conventional differentiation as a left operation (as a rule for the average reader); we have had to compromise on that as some traditional ideas are not easily overthrown. Asorey and Lavrov[14] have written a nice exposition of these ideas but they have instead chosen to take right derivatives, which makes for marginally simpler formulae but does not conform with traditional ideas about left differentiation, which we have religiously adhered to. In Sec. 3, we delineate transformation properties of supertensors and the supermetric. Then, we pay attention to the definition of covariant derivatives with particular application to the super-Riemann tensor $\mathcal{R}$; its supersymmetry properties are obtained and the super-Ricci and superscalar curvature are derived. The Bianchi identities follow in Sec. 5. For the remainder of the paper, we consider the case of just one property, such as charge, and in Sec. 6, we write down the most general metric, including the electromagnetic field. Unsurprisingly, we show that gauge transformations can be construed as supercoordinate transformations associated with phase changes on ζ. We then find the superdeterminant of this metric in Sec. 7, as well as the Palatini form of the superscalar curvature. In Sec. 8, we go on to evaluate the super-Riemann tensor components and determine the super-Ricci tensor and full superscalar curvature. This leads to the field equations and it is rather pleasing that the Einstein–Maxwell Lagrangian emerges very naturally, together with a cosmological contribution. Further, the electromagnetic stress tensor presents itself as a purely geometrical addition to the extended Einstein tensor. All such calculations are greatly assisted by an algebraic computer program for handling anticommuting variables as well as ordinary ungraded ones, which has been developed by one of us (PDS) using *Mathematica.* A penultimate Sec. 9 details the inclusion of matter fields, but our treatment of fermions there is to be regarded as preliminary at this stage. The conclusions close the paper and an appendix collates a list of generalized Christoffel symbols, curvature components and super-vielbeins that are needed at intermediate steps in the main text.

2. Extended Transformations and Notation

The addition of extra anticommutative coordinates to spacetime results in a graded manifold, where the standard spacetime is even and the property sector is odd. The notation used in this paper will be to define uppercase italicized indices (M, N, L, etc.) to run over all the dimensions of spacetime–property and hence have mixed grading. Lower case italicized indices (m, n, l, etc.) will correspond to even graded spacetime, and Greek characters (μ, ν, λ, etc.) will correspond to the odd graded property sector. The grading of an index given by $[M]$,[a] is $[m] = 0$ and $[\mu] = 1$. Later on, we reserve early letters of the alphabet (a, α, etc.) to signify flat or tangent space.

Our starting point is the transformation properties of contravariant and covariant vectors; from these, we can build up how a general tensor should transform. We will make use of Einstein summation convention, but it has to be done carefully. We pick a convention of always summing a contravariant index followed immediately by a covariant index (up then down). This results in contravariant and covariant vectors transforming as follows:

$$V'^M = V^N \frac{\partial X'^M}{\partial X^N}, \quad V'_M = \frac{\partial X^N}{\partial X'^M} V_N . \tag{1}$$

The scalar $V^M V_M$ then correctly transforms into itself:

$$V'^M V'_M = V^N \frac{\partial X'^M}{\partial X^N} \frac{\partial X^L}{\partial X'^M} V_L = V^N \delta_N{}^L V_L = V^N V_N , \tag{2}$$

since the (left) chain rule given by

$$\frac{\partial X'^M}{\partial X^N} \frac{\partial X^L}{\partial X'^M} = \delta_N{}^L . \tag{3}$$

From (1), one can build up the transformation properties of any tensor, by taking it to behave like a corresponding product of vectors. For example, a rank two covariant

[a] Asorey and Lavrov use the notation ϵ_M instead of $[M]$.

tensor T_{MN} has to transform like $V_M V_N$

$$V'_M V'_N = \frac{\partial X^R}{\partial X'^M} V_R \frac{\partial X^S}{\partial X'^N} V_S$$

$$= (-1)^{[R]([S]+[N])} \frac{\partial X^R}{\partial X'^M} \frac{\partial X^S}{\partial X'^N} V_R V_S \,. \tag{4}$$

In these manipulations, we have adhered to the traditional convention of writing derivatives on the left, so the sign factor arising in (4) is due to permuting V_R through the partial derivative. In this way, we find how T_{MN} transforms:

$$T'_{MN} = (-1)^{[R]([S]+[N])} \frac{\partial X^R}{\partial X'^M} \frac{\partial X^S}{\partial X'^N} T_{RS} \,. \tag{5}$$

Thus in (5), we do not have an immediate, direct up–down summation; the sign factor is introduced to compensate for this. In this manner, it is not hard to derive sign factors for any sort of tensor.

3. On Metrics and Supertensors

The metric supertensor G_{MN} is chosen to be graded symmetric, $G_{MN} = (-1)^{[M][N]} G_{NM}$ because it is associated with the generalized spacetime–property separation $ds^2 = dX^N\, dX^M\, G_{MN}$, which is overall bosonic. As with standard general relativity, the metric can be used to raise and lower indices; however the direct up–down summation rule must be strictly obeyed. This means that for vectors:[b]

$$V^M G_{MN} = V_N \,, \quad G^{MN} V_N = V^M \,. \tag{6}$$

If this order is not followed, then the resulting vector (or tensor) will not transform correctly according to the rules in Sec. 2. When raising or lowering indices of a supertensor, an adjoining up–down summation is sometimes impossible; in that event, a sign factor like in (4) must again be included to compensate for this, using the same argument.

For illustration, consider a tensor T_{MN} whose second index N we wish to raise to get $T_M{}^N$. To work out the sign factor required, look at a product of vectors instead, say $T_{MN} = U_M V_N$; then

$$T_M{}^N = U_M V^N = U_M G^{NL} V_L = (-1)^{[M]([N]+[L])} G^{NL} T_{ML} \,. \tag{7}$$

The sign factor ensures that $(-1)^{[M]([N]+[L])} G^{NL} T_{ML}$ behaves like $T_M{}^N$. This procedure extends to any tensor.

The inverse metric multiplies the covariant metric as follows:

$$G^{MN} G_{NL} = \delta^M{}_L = (-1)^{[M]} \delta_L{}^M \,, \tag{8}$$

$$(-1)^{[N]} G_{MN} G^{NL} = \delta_M{}^L \,. \tag{9}$$

[b]Note that in Ref. 7, the G with raised indices differs from the present G by a factor $(-1)^{[N]}$.

These equations are consistent with each other (the metric and its inverse being graded symmetric) and with the transformation properties of the singlet $\delta_M{}^L$. They are also consistent with Asorey and Lavrov.[14] In particular notice from (8) that the trace operation introduces a negative sign where the fermionic sector is concerned, as is well known.

4. Covariant Derivatives and the Riemann Supertensor

The connection coefficients of standard general relativity in the case of zero torsion are defined to be:

$$\Gamma_{mn}{}^k = (g_{lm,n} + g_{ln,m} - g_{mn,l})g^{lk}/2\,. \tag{10}$$

We take this as the starting point for our covariant derivative, extending it to a graded manifold by allowing for sign factors.

$$\Gamma_{MN}{}^K = ((-1)^{X_{LMN}}G_{LM,N} + (-1)^{Y_{LNM}}G_{LN,M} - (-1)^{Z_{MNL}}G_{MN,L})G^{LK}/2\,, \tag{11}$$

where X_{LMN}, Y_{LNM} and Z_{MNL} need to be determined so as to guarantee that the covariant derivative of a covariant vector transforms correctly as a rank 2 covariant tensor. We write the covariant derivative in semicolon notation as

$$A_{M;N} = (-1)^{W_{MN}}A_{M,N} - \Gamma_{MN}{}^K A_K\,, \tag{12}$$

where again W_{MN} is a sign factor to be found. Expanding this out and finding the conditions on the signs so that all second derivatives cancel and the remaining terms transform as a rank 2 covariant tensor, we arrive at

$$A_{M;N} = (-1)^{[M][N]}A_{M,N} - \Gamma_{MN}{}^K A_K\,, \tag{13}$$

with

$$\begin{aligned}\Gamma_{MN}{}^K = \big[&(-1)^{[M][N]+[L]}G_{ML,N} + (-1)^{[L]}G_{NL,M}\\ &- (-1)^{[L][M]+[L][N]+[L]}G_{MN,L}\big]G^{LK}/2\,.\end{aligned} \tag{14}$$

In a similar manner, one may establish that

$$A^M{}_{;N} = (-1)^{[M][N]}(A^M{}_{,N} + A^L\Gamma_{LN}{}^M)\,, \tag{15}$$

$$\begin{aligned}T_{LM;N} = (-1)^{[N]([L]+[M])}\big[&T_{LM,N} - \Gamma_{NL}{}^K T_{KM}\\ &- (-1)^{[L]([M]+[K])}\Gamma_{NM}{}^K T_{LK}\big]\,.\end{aligned} \tag{16}$$

The curious factors of $(-1)^{[M][N]}$, etc. arise from the mismatch between left derivatives clashing with the convention of placing subscript such as $_{,M}$ on the right and we are stuck with this inappropriateness. Anyhow, with these constructions of the covariant derivative, it is pleasing to check that $G_{MN;L}$ vanishes, as expected. And from equations such as (13), (15) and (16), one may deduce the rules for covariant derivatives of any supertensor.[c]

The Riemann curvature tensor arises in the normal way (with suitable sign factors):

$$(-1)^{[J]} A_J \mathcal{R}^J{}_{KLM} = A_{K;L;M} - (-1)^{[L][M]} A_{K;M;L} \,. \tag{17}$$

Carrying out the algebraic manipulations, we obtain

$$\begin{aligned} \mathcal{R}^J{}_{KLM} = (-1)^{[J]([K]+[L]+[M])} \big[& (-1)^{[K][L]} \Gamma_{KM}{}^J{}_{,L} \\ & - (-1)^{[K][M]+[L][M]} \Gamma_{KL}{}^J{}_{,M} \\ & + (-1)^{[L][M]} \Gamma_{KM}{}^R \Gamma_{RL}{}^J - \Gamma_{KL}{}^R \Gamma_{RM}{}^J \big] \,. \end{aligned} \tag{18}$$

This is the graded version of the standard Riemann curvature tensor.

It only remains to work out the fully covariant Riemann curvature tensor, if only to check its graded symmetry properties. Thus, we lower with the metric,

$$\mathcal{R}_{JKLM} = (-1)^{([I]+[J])([K]+[L]+[M])} \mathcal{R}^I{}_{KLM} G_{IJ} \,, \tag{19}$$

resulting in

$$\begin{aligned} \mathcal{R}_{JKLM} = (-1)^{[J]([K]+[L]+[M])} \big[& (-1)^{[K][L]} \Gamma_{KM}{}^I{}_{,L} \\ & - (-1)^{[K][M]+[L][M]} \Gamma_{KL}{}^I{}_{,M} \\ & + (-1)^{[L][M]} \Gamma_{KM}{}^R \Gamma_{RL}{}^I - \Gamma_{KL}{}^R \Gamma_{RM}{}^I \big] G_{IJ} \,. \end{aligned} \tag{20}$$

It is then not too hard to discover the expected graded symmetry relations,

$$\mathcal{R}_{KJLM} = -(-1)^{[J][K]} \mathcal{R}_{JKLM} \,, \tag{21}$$

$$\mathcal{R}_{JKML} = -(-1)^{[L][M]} \mathcal{R}_{JKLM} \,, \tag{22}$$

$$\mathcal{R}_{LMJK} = (-1)^{([J]+[K])([L]+[M])} \mathcal{R}_{JKLM} \,. \tag{23}$$

5. Bianchi Identities

The Riemann curvature tensor also satisfies the Bianchi identities. The first cyclic identity is readily established from (20)–(23):

$$(-1)^{[K][M]} \mathcal{R}_{JKLM} + (-1)^{[M][L]} \mathcal{R}_{JMKL} + (-1)^{[L][K]} \mathcal{R}_{JLMK} = 0 \,. \tag{24}$$

The second (differential) Bianchi identity, involving the covariant derivative of the curvature tensor, is most easily uncovered by proceeding to a "local frame" wherein

[c]In this context, the rule for covariant differentiation of a product is:

$$\begin{aligned} (A^K B_L C^M \cdots)_{;N} = A^K{}_{;N} (-1)^{[N]([L]+[M]+\cdots)} B_L C^M \cdots + A^K (-1)^{[N]([M]+\cdots)} B_{L;N} C^M \cdots \\ + A^K B_L (-1)^{[N](\cdots)} C^M{}_{;N} \cdots + \cdots \,, \text{ etc.} \end{aligned}$$

the Christoffel symbol (but not its derivative) vanishes; in that case, the tensor reduces to $\mathcal{R}_{JKLM;N} = (-1)^{[N]([J]+[K]+[L]+[M])}\mathcal{R}_{JKLM,N}$. With this simplification, there emerges the identity

$$(-1)^{[L][N]}\mathcal{R}_{JKLM;N} + (-1)^{[N][M]}\mathcal{R}_{JKNL;M} + (-1)^{[M][L]}\mathcal{R}_{JKMN;L} = 0\,. \quad (25)$$

To get the contracted version of the second Bianchi identity, involving the Ricci tensor, we look at

$$G^{LJ}[(-1)^{[L][N]}\mathcal{R}_{JKLM;N} + (-1)^{[N][M]}\mathcal{R}_{JKNL;M} + (-1)^{[M][L]}\mathcal{R}_{JKMN;L}] = 0\,. \quad (26)$$

This results in

$$\mathcal{R}_{KM;N} - (-1)^{[N][M]}\mathcal{R}_{KN;M} + (-1)^{[M][L]+[L][N]+[K][L]}G^{LJ}\mathcal{R}_{JKMN;L} = 0\,, \quad (27)$$

wherein the Ricci tensor has the graded symmetry, $\mathcal{R}_{KM} = (-1)^{[M][K]}\mathcal{R}_{MK}$. One last contraction with G^{MK} gives

$$\mathcal{R}_{;N} = 2(-1)^{[M][N]}\mathcal{R}^M{}_{N;M}\,, \quad (28)$$

which can be written in the form $\mathcal{G}^M{}_{N;M} = 0$, where

$$\mathcal{G}^M{}_N = \mathcal{R}^M{}_N - \delta^M{}_N\mathcal{R}/2\,, \quad (29)$$

is the graded version of the Einstein tensor. Having established all the necessary equations with the requisite sign factors, we are in a position to tackle a simple but important case, featuring one property, namely charge, and ensuing electromagnetism.

6. Gauge Changes as Property Transformations

To begin tackling the case of one property, an ansatz for the metric has to be made which incorporates the property coordinates. With everything flat, the metric distance in the manifold $X = (x, \zeta, \bar{\zeta})$ is given by

$$ds^2 = dX^A\,dX^B\,\eta_{BA} = dx^a\,dx^b\,\eta_{ba} + d\zeta\,d\bar{\zeta}\,\eta_{\bar{\zeta}\zeta} + d\bar{\zeta}\,d\zeta\,\eta_{\zeta\bar{\zeta}}\,,$$

where $\eta_{\zeta\bar{\zeta}} = -\eta_{\bar{\zeta}\zeta} = \ell^2/2$ and η_{ba} is Minkowskian. Notice that, we are obliged to introduce a fundamental length ℓ so as to ensure that the separation has the correct physical dimensions of length^2 because the ζ are being taken as dimensionless.[d] *This should be construed as the tangent space.* We easily spot that it is invariant under Lorentz transformations and global phase transformations on ζ. However, it is not invariant under local x-dependent phase transformations and we are obliged to introduce the gauge field to correct for this, as we shall soon show.

[d]We have chosen to use the complex $\bar{\zeta}$, ζ description rather than the real coordinates ξ, η (where $\zeta = \xi + i\eta$) because it lends itself more easily to group analysis when one enlarges the number of property coordinates.

To proceed to curved space we follow the standard method of invoking the tetrad formalism, but generalized to a graded space. (The metric is of course, a product of appropriate frame vectors $\mathcal{E}$, which provide the curvature.) We have also been guided by the Kaluza–Klein metric: the standard general relativistic metric is to be contained in the spacetime–spacetime sector and gauge fields must reside in the spacetime–property sector, but we have allowed for U(1) invariant property curvature coefficients, denoted by c_i. The results below are not as *ad hoc* as they may seem; for now, we are just interested in seeing how far one may mimic Klein–Kaluza by using an anticommuting extension to spacetime rather than a commuting one. (The various terms which arise in the metric below do not allow for fermion contributions as they would carry a Lorentz spinor index and would conflict with Lorentz invariance.) We envisage that the c_i are expectation values of chargeless Higgs or dilaton fields, which ought to be considered in the most general situation, left for future research.

The frame vectors $\mathcal{E}_M{}^A$ that curve the space are stated in App. A.2 and provide the cure for local phase invariance. They generate the metric in the usual manner[15,16] via

$$G_{MN} = (-1)^{[B]+[B][N]} \mathcal{E}_M{}^A \eta_{AB} \mathcal{E}_N{}^B \,. \tag{30}$$

The entries are tightly constrained by the fact that G_{mn} and $G_{\zeta\bar{\zeta}}$ have to be bosonic, while $G_{m\zeta}$ has to be fermionic in a commutational sense. Further, they only admit expansions up to $\bar{\zeta}\zeta$; that is why the electromagnetic field A_m multiplied by ζ appears in $G_{m\zeta}$; in principle, one could also include in that sector an anticommuting ghost field C_m times $\bar{\zeta}\zeta$, as one encounters in quantum gravity, but at a semiclassical level, we are ignoring this aspect of the problem. Putting this all together results in the following metric

$$G_{MN} = \begin{pmatrix} G_{mn} & G_{m\zeta} & G_{m\bar{\zeta}} \\ G_{\zeta n} & 0 & G_{\zeta\bar{\zeta}} \\ G_{\bar{\zeta}n} & G_{\bar{\zeta}\zeta} & 0 \end{pmatrix}, \tag{31}$$

where

$$\begin{aligned} G_{mn} &= g_{mn}(1 + 2c_1\bar{\zeta}\zeta) + e^2\ell^2 A_m A_n \bar{\zeta}\zeta \,, \\ G_{m\zeta} = G_{\zeta m} &= -\frac{ie\ell^2 A_m \bar{\zeta}}{2} \,, \\ G_{m\bar{\zeta}} = G_{\bar{\zeta}m} &= -\frac{ie\ell^2 A_m \zeta}{2} \,, \\ G_{\zeta\bar{\zeta}} = -G_{\bar{\zeta}\zeta} &= \frac{\ell^2(1 + 2c_2\bar{\zeta}\zeta)}{2} \,. \end{aligned} \tag{32}$$

A couple of general observations: the charge coupling e accompanies the electromagnetic potential A and the constants c_i are allowed in the frame vectors to provide phase invariant property curvature rather like mass enters the Schwarzschild metric; the space–property metric is guaranteed to be anticommuting through the factor ζ.

This inverse metric stays graded symmetric, $G^{MN} = (-1)^{[M][N]}G^{NM}$, and transforms correctly as a rank 2 covariant tensor. It can be derived from (8) or (9). Its elements are

$$\begin{aligned} G^{mn} &= g^{mn}(1 - 2c_1\bar{\zeta}\zeta)\,, \\ G^{m\zeta} &= G^{\zeta m} = ieA^m\zeta\,, \\ G^{m\bar{\zeta}} &= G^{\bar{\zeta}m} = -ieA^m\bar{\zeta}\,, \\ G^{\zeta\bar{\zeta}} &= -G^{\bar{\zeta}\zeta} = 2(1 - 2c_2\bar{\zeta}\zeta)/\ell^2 - e^2A^mA_m\bar{\zeta}\zeta\,. \end{aligned} \tag{33}$$

Now, suppose that we make a spacetime dependent U(1) phase transformation in the property sector:

$$x' = x\,, \quad \zeta' = e^{i\theta(x)}\zeta\,, \quad \bar{\zeta}' = e^{-i\theta(x)}\bar{\zeta}\,. \tag{34}$$

Then, from the general transformation rules such as (5) and its contravariant counterpart, we readily find that

$$eA'_m = eA_m + \partial_m\theta\,, \tag{35}$$

which shows that the field A_m acts as a gauge field under variations in charge phase. This can be checked for all components of the metric G_{MN} from the transformation rule (5). On the other hand, G^{mn} remains unaffected and thus is gauge-invariant in the sense of (34) and (35). The same comments apply to $\mathcal{R}_{mn}$ and $\mathcal{R}^{mn}$; the former varies with gauge, but the latter does not.

7. Metric Superdeterminant and Palatini Form

To produce the field equations, we require the superdeterminant or Berezinian of the metric, which is given by Berezin and DeWitt[1] to be

$$s\det(X) = \det(A - BD^{-1}C)\det(D)^{-1}\,, \tag{36}$$

for a graded matrix of the form:

$$X = \begin{bmatrix} A & B \\ C & D \end{bmatrix}. \tag{37}$$

Given our metric (32), this turns out to be:

$$s\det(G_{MN}) = \frac{4}{\ell^4}\det(g_{mn})[1 + (8c_1 - 4c_2)\bar{\zeta}\zeta] \tag{38}$$

or for short,

$$\sqrt{-G_{\cdot\cdot}} = \frac{2}{\ell^2}\sqrt{-g_{\cdot\cdot}}\,[1 + (4c_1 - 2c_2)\bar{\zeta}\zeta]\,. \tag{39}$$

The absence of the gauge potential should be noted.

While on the subject of the superdeterminant, we note that in general, $(\sqrt{G..})_{,M} = \sqrt{G..}(-1)^{[N]}\Gamma_{MN}{}^{N}$, and not just for our particular G. As a direct consequence, $[\sqrt{G..}A^M]_{;M} = [\sqrt{G..}(-1)^{[M]}A^M]_{,M}$, which impacts on the graded Gauss' theorem. Further, using the derivative identity,

$$G^{LK}{}_{,M} = -(-1)^{[M][L]}G^{LN}\Gamma_{NM}{}^{K} - (-1)^{[K]([N]+[L])}G^{KN}\Gamma_{NM}{}^{L}\,, \tag{40}$$

we can establish a useful lemma:

$$\begin{aligned}\left[\sqrt{G..}\,G^{MK}\right]_{,L} &= (-1)^{[N]}\sqrt{G..}\,\Gamma_{LN}{}^{N}G^{MK} \\ &\quad - \sqrt{G..}\left[(-1)^{[L][M]}G^{MN}\Gamma_{NL}{}^{K}\right. \\ &\quad \left. + (-1)^{[K]([L]+[M])}G^{KN}\Gamma_{NL}{}^{M}\right],\end{aligned} \tag{41}$$

so that $\left[\sqrt{G..}\,G^{LK}\right]_{,L} = -\sqrt{G..}\,G^{LM}\Gamma_{ML}{}^{K}$ quite simply. Then because, under an integral sign, the total derivative terms $[(-1)^{[L]}\sqrt{G..}\,G^{MK}\Gamma_{KM}{}^{L}]_{,L}$, and $[(-1)^{[L]}\sqrt{G..}\,G^{MK}\Gamma_{KL}{}^{L}]_{,M}$ both effectively give zero, one can show that

$$\sqrt{G..}\,(-1)^{[L]}G^{MK}\left[(-1)^{[L]([M]+[K])}(\Gamma_{KM}{}^{L})_{,L} - (\Gamma_{KL}{}^{L})_{,M}\right] \tag{42}$$

$$= 2(-1)^{[L]}\sqrt{G..}\,G^{MK}\left[\Gamma_{KL}{}^{N}\Gamma_{NM}{}^{L} - \Gamma_{KM}{}^{N}\Gamma_{NL}{}^{L}\right]. \tag{43}$$

This means that sum of the first two (double) derivative terms in $\sqrt{G..}\,\mathcal{R}$ is exactly double the sum of the last two terms, apart from a sign change; in other words, the scalar curvature can be reduced to Palatini form, even in the graded case:

$$\sqrt{G..}\,\mathcal{R} \to (-1)^{[L]}\sqrt{G..}\,G^{MK}\left[(-1)^{[L][M]}\Gamma_{KL}{}^{N}\Gamma_{NM}{}^{L} - \Gamma_{KM}{}^{N}\Gamma_{NL}{}^{L}\right]. \tag{44}$$

This can help to simplify some of the calculations and it also endorses the correctness of all our graded sign factors.

8. The Ricci Tensor and Superscalar Curvature

From Eq. (14) and the metric given in (32), one may calculate the Christoffel symbols, $\Gamma_{MN}{}^{K}$. A list of these can be found in App. A.1. Using these connections in (20), one may determine the fully covariant Riemann curvature tensor, $\mathcal{R}_{JKLM}$. This can be a painful process and is where an algebraic computer program developed by one of us (PDS) comes in handy, for it minimizes the possibility of errors. Even after making use of it and the symmetry properties of $\mathcal{R}$ there are a large number of components. We have not bothered to list them as they are so numerous and not particularly enlightening. However, the contracted Ricci tensor,

$$\mathcal{R}_{KM} = (-1)^{[K][L]}G^{LJ}\mathcal{R}_{JKLM}\,, \tag{45}$$

has fewer entries, so we have provided a list of them[e] and their contravariant counterparts,

$$\mathcal{R}^{JL} = (-1)^{[M]} G^{JK} \mathcal{R}_{KM} G^{ML}\,, \tag{46}$$

in App. A.2; the latter are gauge invariant. Finally, the Ricci superscalar can be found by contraction with the metric,

$$\mathcal{R} = G^{MK} \mathcal{R}_{KM}\,. \tag{47}$$

In a frame that is locally flat in spacetime, the spacetime component of the contravariant Ricci tensor reduces to

$$\mathcal{R}^{mn} = 4g^{mn} c_1[1 + (2c_2 - 6c_1)\bar{\zeta}\zeta]/\ell^2 - e^2\ell^2 F^{ml} F^{n}{}_{l}\bar{\zeta}\zeta/2\,, \tag{48}$$

and the curvature superscalar collapses to

$$\mathcal{R} = 8[4c_1 - 3c_2 + c_1(8c_2 - 10c_1)\bar{\zeta}\zeta]/\ell^2 - e^2\ell^2 F^{nl} F_{nl}\bar{\zeta}\zeta/4\,. \tag{49}$$

Both expressions (48) and (49) are gauge independent. By making use of them and the superdeterminant (39), we may evaluate first, the total Lagrangian density for electromagnetic property,

$$\mathcal{L} = \int d\zeta\, d\bar{\zeta}\, \sqrt{-G_{..}}\, \mathcal{R} \propto -\frac{1}{4} F_{mn} F^{mn} + \frac{48(c_1 - c_2)^2}{e^2\ell^4}\,,$$

and second, the Einstein tensor in flat spacetime:

$$\int d\zeta\, d\bar{\zeta}\, \sqrt{-G_{..}}\, (\mathcal{R}^{km} - \mathcal{R}G^{km}/2)$$
$$\propto \left[48c_2(c_1 - c_2)g^{km}/e^2\ell^4 - (F^{kl}F^{m}{}_{l} - F_{ln}F^{ln}g^{km}/4)\right].$$

The familiar expression for the electromagnetic stress tensor, namely $T^{km} \equiv F^{k}{}_{l}F^{lm} + F_{nl}F^{nl}g^{km}/4$ emerges naturally and becomes part of the geometry. But, we also recognize a cosmological constant term that is largely determined by the magnitude $(c_2 - c_1)/\ell^4$. (As an aside, we have verified that (48) and (49), remain true in a general frame, not necessarily locally flat.)

Including gravity by curving spacetime means including the standard gravitational curvature R and will render (48) and (49) generally covariant. (One has to be careful here to track factors of $\bar{\zeta}\zeta$, as one will be integrating over property.) It is straightforward to see that the gravitational part of the superscalar $\mathcal{R}^{(g)}$ is $R(1-2c_1\bar{\zeta}\zeta)$, while the super-Ricci tensor $\mathcal{R}^{(g)km}$ contains $R^{km}(1-4c_1\bar{\zeta}\zeta)$. In consequence, we may evaluate the full gravitational-electromagnetic Lagrangian through

[e]We are reasonably certain that those expressions, though complicated, are correct because we have checked that the differential Bianchi identity (28) is obeyed and this is a highly nontrivial test.

the property integral:

$$\mathcal{L} = \int d\zeta\, d\bar{\zeta}\, \sqrt{-G_{..}}\, \mathcal{R}$$

$$= 2e^2\sqrt{-g_{..}}\left[\frac{2(c_1-c_2)R}{e^2\ell^2} - \frac{F_{mn}F^{mn}}{4} + \frac{48(c_1-c_2)^2}{e^2\ell^4}\right], \quad (50)$$

wherein, we recognize

$$16\pi G_N \equiv \kappa^2 = \frac{e^2\ell^2}{2(c_1-c_2)}, \qquad \Lambda = \frac{12(c_2-c_1)}{\ell^2}.$$

To verify that the entire setup is consistent and free of error, we may determine the gravitational variation δG_{MN} which equals $\delta g_{mn}(1+2c_1\bar{\zeta}\zeta)$. Hence, the gravitational field equation is obtained through

$$0 = \int d\zeta\, d\bar{\zeta}\, \sqrt{-G_{..}}\,(1+2c_1\bar{\zeta}\zeta)(\mathcal{R}^{km} - G^{km}\mathcal{R}/2)$$

$$= \sqrt{-g_{..}}\left[\frac{4(c_1-c_2)}{\ell^2}\left(R^{km} - \frac{1}{2}g^{km}R\right) - T^{km} - \frac{48(c_1-c_2)^2}{\ell^4}g^{km}\right]. \quad (51)$$

This is just what we would have obtained from (50). In any case, we see that the universal coupling of gravity to stress tensors T has a factor $8\pi G_N \equiv \kappa^2/2 = e^2\ell^2/4(c_1-c_2) > 0$. The result is to make the cosmological term go negative and, what is probably worse, it has a value which is inordinately larger than the tiny experimental value found by analyses of supernovae! (All cosmological terms derived from particle physics, except for exactly zero, share the same problem.) Numerically speaking, $\kappa \simeq 5.8 \times 10^{-19}$ (GeV)$^{-1}$ means $\ell \sim 10^{-18}$ (GeV)$^{-1}$ is Planckian in scale. Of course the magnitude of the miniscule cosmological constant $\Lambda \sim 4 \times 10^{-84}$ (GeV)^2 is at variance with Planckian expectations by the usual factor of 10^{-120}, which is probably the most mysterious natural ratio. So far as our scheme is concerned, we are disappointed but not particularly troubled by the wrong sign of Λ because it can readily be reversed by extra property curvature coefficients, when we enlarge the number of properties (as we have checked when enlarging the number of properties to at least two). The magnitude of Λ is quite another matter because it will require some extraordinary fine-tuning, even after fixing the sign.

9. Inclusion of Matter Fields

The conventional results which we obtained for electromagnetism plus gravity, through the property of electricity, merely confirm the fact that our scheme is perfectly viable and offers a novel perspective on nature. We anticipate that when one incorporates other properties, like chromicity and neutrinicity, then the usual picture of Quantum Chromodynamics (QCD) plus gravity plus electroweak theory will emerge. For now, we wish to exhibit some preliminary research concerning inclusion of matter fields, despite being limited to the single property of electric charge.

9.1. *Scalar field*

Adhering to the tenets of the spin-statistics connection, we begin by assuming that a superscalar field $\Phi(X)$ is overall bosonic and can be expanded into even powers of $\bar{\zeta}\zeta$; thus, it has the general form $\Phi(x,\zeta,\bar{\zeta}) = U(x) + V(x)\bar{\zeta}\zeta$. Note that, we could have included in Φ two anticommuting scalar ghost fields in the combination $\bar{\zeta}C + \bar{C}\zeta$; such ghost fields have a place in quantum theory but, with their incorrect spin-statistics, cannot be regarded as physical asymptotic states. The same comment applies to the spacetime–property sector, where we could have included vector ghost fields of the type C_m, $\bar{C}_m$ multiplying $(1 + c'\bar{\zeta}\zeta)$. We have ignored these extras as we are only dealing with semiclassical electromagnetism/gravity for the purpose of this investigation, but they are sure to come in their own when quantization of the scheme is undertaken. By imposing self-duality, $U(x) = V(x) = \varphi(x)$, Φ may be reduced to the form[6,11]

$$\Phi(X) = \varphi(x)(1 + \bar{\zeta}\zeta)/2\,. \tag{52}$$

Necessarily φ carries zero charge and is a far cry from a Higgs field. (In fact, to obtain the correct quantum numbers of the Higgs field, it is imperative to attach three chromic properties to charge.) As we will be coupling this field to the supermetric, which brings in the superdeterminant $2\sqrt{-G_{..}}/\ell^2$, we shall introduce an extra factor of ℓ^2 to eliminate this scale and we will also ignore c_i curvature in what follows except from what the mixed x–ζ sector produces.

A mass term in the Lagrangian of $\mu^2\varphi^2/2$ will arise through the property integral

$$\left(\frac{\ell^2}{2}\right)\int d\zeta\, d\bar{\zeta}\,\sqrt{-G_{..}}\,\mu^2\Phi^2 = \int d\zeta\, d\bar{\zeta}\,\sqrt{-g_{..}}\,(1 + 2\bar{\zeta}\zeta)\mu^2\varphi^2/4\,. \tag{53}$$

The kinetic term is more interesting, because it adds to the mass owing to the property components in Φ; specifically this can be attributed to the ζ, $\bar{\zeta}$ derivatives, which add a piece to the mass. Thus, we consider

$$\left(\frac{\ell^2}{2}\right)\int d\zeta\, d\bar{\zeta}\,\sqrt{-G_{..}}\,G^{MN}\partial_N\Phi\partial_M\Phi\,. \tag{54}$$

Upon inserting the metric from (32), we find that the contributions from the gauge field cancel out, *as they must*, and we are left with

$$\int d\zeta\, d\bar{\zeta}\,\sqrt{-g_{..}}\,[(1 + 2\bar{\zeta}\zeta)\,g^{mn}\partial_n\phi\partial_m\phi/4 + \bar{\zeta}\zeta\varphi^2/\ell^2]\,. \tag{55}$$

The only feasible way, then, to cancel off mass, in order to obtain a massless scalar field in this scheme is to match the φ^2/ℓ^2 from the property kinetic energy to the previously constructed mass term (53).

9.2. *Spinor field*

In seeking a generalization of the Dirac equation to incorporate the graded derivative, we need to bear in mind that the electromagnetic potential is embedded in the spacetime–property frame vector $E_A{}^M$; therefore, we first need to determine the inverse vielbein, obtained via the condition $\mathcal{E}_M{}^A E_A{}^N = \delta_M{}^N$. The components are listed in App. A.2, where it will be seen that the vector potential is held in the space–property sector via $E_a{}^\zeta$ and $E_a{}^{\bar\zeta}$. Since the Dirac operator has a natural extension from $i\gamma^a e_a{}^m \partial_m$ to $i\Gamma^A E_A{}^M \partial_M$, it is vital to include the graded derivative $\partial/\partial\zeta$ at the very least.

Just as Dirac was obliged to enlarge spinors from two to four components, in order to go from nonrelativistic electrons to relativistic ones, so too we are forced to extend the space in order to deal with the property derivatives. (There may be other ways to attain that goal.) Since the Dirac operator will act on a spinorial superfield, we have been led to consider an extended field $\Psi(X) = \theta\bar\zeta\psi(x)$ and the representation:

$$\Gamma^a = \gamma^a\,, \quad \ell\Gamma^\zeta = \frac{2i\partial}{\partial\theta}\,, \quad \ell\Gamma^{\bar\zeta} = 2i\theta\,, \tag{56}$$

wherein θ is *another* complex scalar a-number and we eventually have to integrate over θ and $\bar\theta$. (The Γ^ζ, $\Gamma^{\bar\zeta}$ act like fermionic annihilators.) The action of the extended Dirac operator then yields

$$\begin{aligned} i\Gamma^A E_A{}^M \partial_M \Psi &= \left[i\gamma^a e_a{}^m \partial_m + e\gamma^a A_a \bar\zeta \frac{\partial}{\partial\bar\zeta} + \frac{2}{\ell}(1 - f\bar\zeta\zeta)\frac{\partial^2}{d\theta\, d\bar\zeta}\right]\theta\bar\zeta\psi \\ &= \theta\bar\zeta\gamma^a e_a{}^m (i\partial_m + eA_m)\psi - \frac{2}{\ell}(1 - f\bar\zeta\zeta)\psi\,. \end{aligned} \tag{57}$$

This means that when we include the adjoint $\bar\Psi \equiv -\bar\psi\zeta\bar\theta$ and integrate over the subsidiary θ, ζ, we end up with the normal gauge invariant spinorial Lagrangian density:

$$\begin{aligned} \mathcal{L} &= \int (d\zeta\, d\bar\zeta)(d\theta\, d\bar\theta)\bar\Psi(X)\left[i\Gamma^A E_A{}^M \partial_M - \mathcal{M}\right]\Psi(X) \\ &= \bar\psi(x)\left[\gamma^a e_a{}^m (i\partial_m + eA_m) - \mathcal{M}\right]\psi(x)\,. \end{aligned} \tag{58}$$

Very likely, there exists a more elegant way of reaching (58) but however this is done, the coupling of the charged fermion to the spacetime–property vielbein, which contains A, is critical. The representation (56) will surely need revisiting in order to encompass chirality, if the basic fermions are taken as left-handed, especially if we attach charge conjugate left-handed pieces, in order to encompass all spin states.

10. Conclusions

The framework underlying our research was inspired by supersymmetry, but instead of using auxiliary *spinor* coordinates we have made them *scalar* and connected them with something *tangible*, namely property or attribute. This point is important because all physical events are described by changes in momentum and/or property. From this perspective, systematization of property, with the natural occurrence of generations, becomes a guiding principle. An obvious criticism of the approach is that the superstructure has only led to the standard Einstein–Maxwell Lagrangian, which is hardly an earth-shattering conclusion! True, but by geometrizing spacetime–property, we have succeeded in reinterpreting gauge fields as the messengers of property in a larger graded curved space, besides offering a new viewpoint on the nature of events; as a bonus, we see that curvature in property space can act as a source of a cosmological constant — this with just one property coordinate — even if its value is ridiculous. Furthermore addition of further ζ coordinates offers a natural path to group theory classification[f] without entraining infinite towers of states as one gets with bosonic extensions to spacetime.

We foresee no intrinsic difficulty in extending the work to QCD or to electroweak theory, though the algebraic manipulations will perforce be more intricate. Going all the way, we anticipate an extension of our calculations to five property coordinates (ζ^0 to ζ^4) and the distillation of the final group to that of the standard model will mean that property curvature coefficients $c_{\rm col}$, $c_{\rm ew}$ are to be associated with color and electroweak invariants, $\bar{\zeta}_i\zeta^i$ and $(\bar{\zeta}_0\zeta^0 + \bar{\zeta}_4\zeta^4)$ respectively — perhaps engendered by expectation values of chargeless Higgs or dilaton fields. For the future, these are just the most prominent issues that come to mind with correction of the Λ sign foremost among them and the quantization of the scheme in the present framework as the next step. There are surely several research avenues[g] to explore in the present picture, many more than we have envisaged.

In the end, this geometrical approach of uniting gravity with other natural forces through a larger graded spacetime–property manifold may turn out to be quite misguided. That would be disappointing as it is hard to imagine what other original way, one could unify the gravitational field with other fields. As a fallback position, we could be ultraprudent by abandoning the unification goal: just introduce gauge fields in the time-honored way, ensuring that differentiation is gauge covariant under property transformations, by replacing ordinary derivatives ∂ with $D = \partial + ieA\cdot T$, where the generators T are represented by property rotations acting on matter superfields; e.g. the charge operator by $(\bar{\zeta}\partial/\partial\bar{\zeta} - \zeta\partial/\partial\zeta)$. That would be a backward step and even then, we might be confronted by insurmountable obstacles. However, for now, the geometric scheme seems flexible enough and accords well with our general understanding of fundamental particle physics and its content as well as gravitation. Should our framework fall by the wayside, there is no other recourse than to persist with variants of grand theories which are currently on the market,

[f] To stress this point we mention that we have succeeded in obtaining the combined Yang–Mills–Gravity Lagrangian for two property coordinates. The details are much more intricate than the one ζ case considered in this paper and will be submitted for publication separately.

[g] Looking beyond the horizon of this paper, since the gravitational coupling to any stress tensor is universal and since we have married couplings with fields it means that these interaction couplings will need to be universal too. At low energies QCD and Quantum Electrodynamics (QED) coupling constants e and g are widely different so, unless the curvature coefficients c_e and c_w are taken to be different which is entirely possible, we envisage a scenario where these couplings are running and unified at a GUT-like scale ℓ with $1/\alpha \simeq 40$; they only look different when we run down from $e^2(\ell^2) = g^2(\ell^2)$ to electroweak/strong interaction scales.

in the hope that experiment will, at sufficiently high energy, substantiate one of them. Failing that, we trust that one day somebody will conceive a radically new description of events that will lead to new insights with *testable* predictions.

Acknowledgments

We would like to thank Dr. Peter Jarvis for much helpful advice and for his knowledgeable comments about supergroups, their representations and their dimensions.

Appendix A. Christoffel Symbols, Vielbeins and Curvature Components

A.1. *The graded Christoffel connections*

From definition (14) and the metric elements (32), one may derive the following components of the Christoffel symbols:

$$\Gamma_{mn}{}^{l} = \Gamma^{[g]}{}_{mn}{}^{l} + e^2\ell^2(A_nF_{mk} + A_mF_{nk})g^{kl}\bar{\zeta}\zeta/2\,,$$

$$\Gamma_{mn}{}^{\zeta} = \frac{\zeta}{2}\left[ie\big(2A^k\Gamma^{[g]}_{mnk} - A_{m,n} - A_{n,m}\big) - 2e^2A_mA_n - \frac{4c_1}{\ell^2}g_{mn}\right],$$

$$\Gamma_{mn}{}^{\bar{\zeta}} = \frac{\bar{\zeta}}{2}\left[ie\big(A_{m,n} + A_{n,m} - 2A^k\Gamma^{[g]}_{mnk}\big) - 2e^2A_mA_n - \frac{4c_1}{\ell^2}g_{mn}\right],$$

$$\Gamma_{\zeta n}{}^{l} = \Gamma_{n\zeta}{}^{l} = \bar{\zeta}\big[ie\ell^2F_{kn}g^{kl}/4 - c_1\delta_n{}^{l}\big]\,,$$

$$\Gamma_{\bar{\zeta}n}{}^{l} = \Gamma_{n\bar{\zeta}}{}^{l} = \zeta\big[ie\ell^2F_{kn}g^{kl}/4 + c_1\delta_n{}^{l}\big]\,,$$

$$\Gamma_{\zeta n}{}^{\zeta} = \Gamma_{n\zeta}{}^{\zeta} = -ieA_n - \left[\frac{e^2\ell^2}{4}A^kF_{kn} + ie(c_1 - 2c_2)A_n\right]\bar{\zeta}\zeta\,,$$

$$\Gamma_{\zeta n}{}^{\bar{\zeta}} = \Gamma_{n\zeta}{}^{\bar{\zeta}} = \Gamma_{\bar{\zeta}n}{}^{\zeta} = \Gamma_{n\bar{\zeta}}{}^{\zeta} = 0\,,$$

$$\Gamma_{\bar{\zeta}n}{}^{\bar{\zeta}} = \Gamma_{n\bar{\zeta}}{}^{\bar{\zeta}} = ieA_n - \left[\frac{e^2\ell^2}{4}A^kF_{kn} + ie(2c_2 - c_1)A_n\right]\bar{\zeta}\zeta\,,$$

$$\Gamma_{\zeta\bar{\zeta}}{}^{l} = \Gamma_{\bar{\zeta}\zeta}{}^{l} = 0\,,$$

$$\Gamma_{\zeta\bar{\zeta}}{}^{\zeta} = -\Gamma_{\bar{\zeta}\zeta}{}^{\zeta} = -2c_2\zeta\,,$$

$$\Gamma_{\zeta\bar{\zeta}}{}^{\bar{\zeta}} = -\Gamma_{\bar{\zeta}\zeta}{}^{\bar{\zeta}} = -2c_2\bar{\zeta}\,,$$

$$\Gamma_{\zeta\zeta}{}^{l} = \Gamma_{\zeta\zeta}{}^{\zeta} = \Gamma_{\zeta\zeta}{}^{\bar{\zeta}} = \Gamma_{\bar{\zeta}\bar{\zeta}}{}^{l} = \Gamma_{\bar{\zeta}\bar{\zeta}}{}^{\zeta} = \Gamma_{\bar{\zeta}\bar{\zeta}}{}^{\bar{\zeta}} = 0\,.$$

Above, $\Gamma^{(g)}$ signifies the purely gravitational connection and $F_{mn} \equiv A_{n,m} - A_{m,n}$ is the standard Maxwell tensor. These connections are essential in determining the full super-Riemann components.

A.2. *Ricci tensor components*

A concise list of the Ricci supertensor components (contravariant and covariant) is as follows, where we neglect spacetime curvature:

$$\begin{aligned}
\mathcal{R}^{km} &= 4g^{mk}c_1[1+(2c_2-6c_1)\bar\zeta\zeta]/\ell^2 - e^2\ell^2 F^k{}_l F^{ml}\bar\zeta\zeta/2\,,\\
\mathcal{R}^{k\zeta} &= 4iec_1A^k\zeta/\ell^2 - ieF^{kl}{}_{,l}\zeta/2\,,\\
\mathcal{R}^{k\bar\zeta} &= -4iec_1\bar\zeta A^k/\ell^2 + ie\bar\zeta F^{kl}{}_{,l}/2\,,\\
\mathcal{R}^{\zeta\bar\zeta} &= 8[3c_2(1-2c_2\bar\zeta\zeta) - 2c_1(1-c_1\bar\zeta\zeta)]/\ell^4\\
&\quad - e^2(4A^mA_mc_1/\ell^2 + F_{mn}F^{mn}/4 + A_mF^{nm}{}_{,n})\bar\zeta\zeta\,,\\
\mathcal{R}_{km} &= 4c_1g_{km}[1-2(c_1-c_2)\bar\zeta\zeta]/\ell^2 - 4e^2(2c_1-3c_2)A_kA_m\bar\zeta\zeta\\
&\quad + e^2\ell^2g^{nl}\big[A_{k,n}A_{m,l} - A_{n,m}A_{l,k}\big]\bar\zeta\zeta/2\,,\\
\mathcal{R}_{k\zeta} &= 2ie(2c_1-3c_2)\bar\zeta A_k + ie\ell^2\bar\zeta F^l{}_{k,l}/4\,,\\
\mathcal{R}_{k\bar\zeta} &= 2ie(2c_1-3c_2)A_k\zeta + ie\zeta\ell^2F^l{}_{k,l}/4\,,\\
\mathcal{R}_{\zeta\bar\zeta} &= [6c_2-4c_1+4(c_2-c_1)(3c_2-c_1)\bar\zeta\zeta] - \ell^4e^2F^{kl}F_{kl}\,\bar\zeta\zeta/16\,.
\end{aligned}$$

Other components are derivable from symmetry properties of $\mathcal{R}_{MN}$.

A.3. *Vielbeins*

From the frame vectors, namely $\mathcal{E}_M{}^A$, whose components are

$$\begin{aligned}
&\mathcal{E}_m{}^a = (1+c_1\bar\zeta\zeta)e_m{}^a\,, && \mathcal{E}_m{}^\zeta = -ie\bar\zeta A_m\,, && \mathcal{E}_m{}^{\bar\zeta} = -ieA_m\zeta\,,\\
&\mathcal{E}_\zeta{}^a = 0\,, && \mathcal{E}_\zeta{}^\zeta = 0\,, && \mathcal{E}_\zeta{}^{\bar\zeta} = (1+c_2\bar\zeta\zeta)\,,\\
&\mathcal{E}_{\bar\zeta}{}^a = 0\,, && \mathcal{E}_{\bar\zeta}{}^\zeta = -(1+c_2\bar\zeta\zeta)\,, && \mathcal{E}_{\bar\zeta}{}^{\bar\zeta} = 0\,,
\end{aligned}$$

we may derive the super-vielbeins $E_A{}^N$, obtained via $\mathcal{E}_M{}^AE_A{}^N = \delta_M{}^N$. In this way, we arrive at the set:

$$\begin{aligned}
&E_a{}^m = e_a{}^m(1-c_1\bar\zeta\zeta)\,, && E_a{}^\zeta = ieA_a\zeta\,, && E_a{}^{\bar\zeta} = -ie\bar\zeta A_a\,,\\
&E_\zeta{}^m = 0\,, && E_\zeta{}^\zeta = 0\,, && E_\zeta{}^{\bar\zeta} = -(1-c_2\bar\zeta\zeta)\,,\\
&E_{\bar\zeta}{}^m = 0\,, && E_{\bar\zeta}{}^\zeta = (1-c_2\bar\zeta\zeta)\,, && E_{\bar\zeta}{}^{\bar\zeta} = 0\,.
\end{aligned}$$

These expressions are required in Subsec. 9 B. As a useful check on their correctness, we may ascertain that $G^{MN} = (-1)^{[A][M]}\eta^{AB}E_B{}^ME_A{}^N$, emerges properly. For example, we directly arrive at

$$\begin{aligned}
G^{\zeta\bar\zeta} &= \eta^{ab}E_b{}^\zeta E_a{}^{\bar\zeta} - 2E_{\bar\zeta}{}^\zeta E_\zeta{}^{\bar\zeta}/\ell^2 + 2E_\zeta{}^\zeta E_{\bar\zeta}{}^{\bar\zeta}/\ell^2\\
&= \eta^{mn}(ieA_n\zeta)(-i\bar\zeta eA_m) + 2(1-2c_2\bar\zeta\zeta)/\ell^2\\
&= 2(1-2c_2\bar\zeta\zeta)/\ell^2 - e^2A^mA_m\bar\zeta\zeta\,.
\end{aligned}$$

References

1. F. A. Berezin, *The Method of Second Quantization* (Academic Press, Boston, 1966).
2. B. DeWitt, *Supermanifolds* (Cambridge University Press, Cambridge, 1984).
3. F. A. Berezin, *Introduction to Superanalysis* (D.Reidel, Dordrecht, 1987).
4. A. Rogers, *Supermanifolds: Theory and Applications* (World Scientific, Singapore, 2007).
5. E. Witten, Notes on supermanifolds and integration, arXiv:1209.2199v2.
6. R. Delbourgo, P. D. Jarvis and R. Warner, *J. Math. Phys.* **34**, 3616 (1993).
7. R. Delbourgo and R. B. Zhang, *Phys. Rev. D* **38**, 2490 (1988).
8. R. Delbourgo, *J. Phys. A* **39**, 5175 (2006).
9. R. Delbourgo, *J. Phys. A* **39**, 14735 (2006).
10. R. Delbourgo, *Int. J. Mod. Phys. A* **22**, 4911 (2007).
11. R. Delbourgo, arXiv:1202.4216.
12. R. Delbourgo, P. D. Jarvis and G. Thompson, *Phys. Lett. B* **109**, 25 (1982).
13. R. Delbourgo, P. D. Jarvis and G. Thompson, *Phys. Rev. D* **26**, 775 (1982).
14. M. Asorey and P. M. Lavrov, *J. Math. Phys.* **50**, 013530 (2009).
15. S. Carroll, *Spacetime and Geometry* (Addison Wesley, San Francisco, 2004).
16. S. Weinberg, *Gravitation and Cosmology* (J. Wiley & Sons, New York, 1972).

Modern Physics Letters A
Vol. 31, No. 3 (2016) 1650019 (7 pages)
© World Scientific Publishing Company
DOI: 10.1142/S021773231650019X

General relativity for N properties

Robert Delbourgo* and Paul D. Stack†

School of Physical Sciences, University of Tasmania, G. P. O. Box 37, Hobart, TAS 7005, Australia
**bob.delbourgo@utas.edu.au*
†pdstack@utas.edu.au

Received 7 November 2015
Accepted 16 November 2015
Published 5 January 2016

We determine the coefficients of the terms multiplying the gauge fields, gravitational field and cosmological term in a scheme whereby properties are characterized by N anticommuting scalar Grassmann variables. We do this for general N, using analytical methods; this obviates the need for our algebraic computing package which can become quite unwieldy as N is increased.

Keywords: Grassmann; property; relativity.

PACS Nos.: 11.10.Kk, 11.10.Nx

1. Structure of the Lagrangian Terms

Over the last few years, we have developed a scheme whereby the gravity is unified with other force fields by embracing them all in a supermetric which features spacetime augmented by Lorentz scalar anticommuting coordinates ζ^μ. These ζ specify the characteristics or properties of an event,[1–4] in addition to location and time x^m. This scheme[a] resulted in a Lagrangian which contained the Yang–Mills, gravitational and cosmological terms, consistent with general coordinate invariance, but which pointed to the need for coupling constant unification at an appropriately high length scale l at least when chirality was involved. We were able to make progress for the case of a small number of properties by using of an algebraic computation package like Mathematica; this provided ample confirmation for the gauge invariance of the result by explicit computation of the super-Ricci scalar. The final answer was dependent on a number of property curvature coefficients and the calculation

[a]One can always make any gauge field Lagrangian consistent with general relativity by ensuring invariance under spacetime coordinate changes through the gravitational metric or vierbein but consistency does not mean proper unification.

became progressively more difficult (and expensive in machine time) as the number N of properties got larger, in a factorial sense. In this paper, we shall describe a way of extracting the result for any N by an analytical method which significantly obviates the need for extensive computer calculation and is a major advance in tackling the practical case of $N = 5$ (or $N = 10$ if we distinguish chiralities) which comprises the standard model.

The trick which leads to such an advance is based upon the Palatini form of the Ricci superscalar $\mathcal{R}$:

$$\mathcal{R} = (-1)^{[L]} G^{MK} [(-1)^{[L][M]} \Gamma_{KL}{}^{N} \Gamma_{NM}{}^{L} - \Gamma_{KM}{}^{N} \Gamma_{NL}{}^{L}] , \tag{1}$$

where $M = m$ for spacetime or $M = \mu$ for graded property; the Christoffel symbol Γ is defined by

$$\begin{aligned} 2\Gamma_{MN}{}^{K} \equiv [(-1)^{[M][N]} G_{MR,N} + G_{NR,M} \\ - (-1)^{[R]([M]+[N])} G_{MN,R}](-1)^{[R]} G^{RK} , \end{aligned} \tag{2}$$

all derivatives are left-sided and the square brackets denote the grading (0 or 1) of the appropriate index. The significant remark is that $\mathcal{R}$ formally possesses the structure $G^{KL} G^{MN} G^{RS} (\partial_{.} G_{..} \partial_{.} G_{..})_{KLMNRS}$ with the index subscripts $KLMNRS$ distributed between the derivatives and metrics within the brackets ( ). It then becomes an exercise in picking out typical terms which can contribute to the gravitational curvature, the gauge fields and the cosmological constant.

2. The Supermetric

As explained in previous work, the metric (derived from triangular frame vectors) which contains gravity and the gauge fields (denoted by A and ignoring coupling constants) has the following components:

$$x - x \text{ sector}, \; G_{mn} = g_{mn} C + l^2 \bar{\zeta} (A_m A_n + A_n A_m) \zeta C'/2 , \tag{3}$$

$$x - \zeta \text{ sector}, \; G_{m\nu} = -il^2 (\bar{\zeta} A_m)^{\bar{\nu}} C'/2 , \tag{4}$$

$$\zeta - \bar{\zeta} \text{ sector}, \; G_{\mu\bar{\nu}} = l^2 \delta_{\mu}{}^{\nu} C'/2 . \tag{5}$$

In the most general situation, expressions C, C' represent polynomials of permitted *gauge-invariant* curvature terms:

$$C \equiv 1 + \sum_{r=1}^{N} c_r Z^r , \qquad C' \equiv 1 + \sum_{r=1}^{N} c'_r Z^r ; \qquad Z \equiv \bar{\zeta}\zeta . \tag{6}$$

The inverse metric components are readily found:

$$x - x \text{ sector}, \; G^{mn} = g^{mn} C^{-1} , \tag{7}$$

$$x - \zeta \text{ sector}, \; G^{m\nu} = i(A^m \zeta)^{\nu} C^{-1} , \tag{8}$$

$$\zeta - \bar{\zeta} \text{ sector}, \; G^{\mu\bar{\nu}} = [2\delta_{\nu}{}^{\mu}/l^2 - (\bar{\zeta} A^m)^{\bar{\nu}} (A_m \zeta)^{\mu}] C'^{-1} . \tag{9}$$

A general rotation of the property coordinates, $\zeta \to \exp[i\Theta(x)]\zeta$ then just corresponds to a gauge transformation of the force fields, so we anticipate that the Ricci superscalar $\mathcal{R}$ must turn out to be gauge invariant. Indeed it is, as verified multiple times through Mathematica evaluations.

To further the calculation of the dependence of the Lagrangian on the property curvature coefficients c_r and c'_r, we will require the Berezinian[5,6] of the metric. The latter is actually gauge invariant so we can set $A \to 0$ to evaluate it. A simple calculation produces

$$\sqrt{G..} = \sqrt{g..}C^2(l^2C'/2)^{-N} . \tag{10}$$

We will also need the Grassmann integrals,

$$\int (d^N\zeta\, d^N\bar{\zeta})Z^N = (-1)^{\langle N\rangle}N! , \tag{11}$$

$$\int (d^N\zeta\, d^N\bar{\zeta})Z^{N-1}(\bar{\zeta}H\zeta) = (-1)^{\langle N\rangle}(N-1)!\,\mathrm{Tr}\,H , \tag{12}$$

where $\langle M\rangle$ signifies int$[M/2]$ and H stands for a general U(N) matrix in property space.

Before continuing, we will find it useful to use alternative parametrizations to (6), since they help to simplify the ensuing analysis; namely write

$$C = \exp\left[-\sum_{r=1}^{N} a_r Z^r\right], \qquad C' = \exp\left[-\sum_{r=1}^{N} a'_r Z^r\right]; \qquad Z \equiv \bar{\zeta}\zeta . \tag{13}$$

Thus derivatives are easily found,

$$\begin{aligned} \frac{\partial C}{\partial \zeta} &= \sum_{r=1}^{N} \bar{\zeta} r a_r Z^{r-1} C \equiv \bar{\zeta} DC , \\ \frac{\partial C}{\partial \bar{\zeta}} &= -\sum_{r=1}^{N} r a_r Z^{r-1}\zeta C \equiv -D\zeta C , \quad \text{etc.} \end{aligned} \tag{14}$$

Of course there is a simple translation table between parametrizations (6) and (13):

$$a_1 = -c_1 , \qquad a_2 = {c_1}^2/2 - c_2 , \qquad a_3 = -{c_1}^3/3 + c_1c_2 - c_3 , \tag{15}$$

$$a_4 = {c_1}^4/4 - {c_1}^2c_2 + c_2/2 + c_1c_3 - c_4 , \tag{16}$$

$$a_5 = -{c_1}^5/5 + {c_1}^3c_2 - c_1{c_2}^2 - {c_1}^2c_3 + c_2c_3 - c_5 , \quad \text{etc.} \tag{17}$$

In this a-parametrization, $\sqrt{G..} = \sqrt{g..}\,(2/l^2)^N \exp\left[\sum_{r=1}^{N}(Na'_r - 2a_r)Z^r\right]$.

3. Determination of the Various Parts of the Lagrangian

We are after the integral of the superscalar curvature, $\int (d^N\zeta\, d^N\bar{\zeta})\sqrt{G..}\,\mathcal{R}$, which will produce three sorts of terms: the purely gravitational bit $R^{[g]}$, the gauge contribution proportional to Tr $F.F$, where F is the generalized curl of the gauge field

R. Delbourgo & P. D. Stack

and finally the constant, cosmological part. The tactic is to identify relevant bits of each by picking out appropriate pieces of $G^{KL}G^{MN}G^{RS}(\partial.G..\partial.G..)_{KLMNRS}$.

3.1. *The gravitational term*

The gravitational curvature arises from the structure $g^{kl}g^{mn}g^{rs}(\partial.g..\partial.g..)_{klmnrs}$, which itself comes from $G^{kl}G^{mn}G^{rs}(\partial.G..\partial.G..)_{klmnrs}$ and therefore carries the factor C^{-1} as can be ascertained from Eqs. (3) and (7). Including the Berezinian, we see that

$$\sqrt{G..}\mathcal{R} \supset \sqrt{g..}\, R^{[g]}\, (2/l^2)^N \exp\left[\sum_1^N (Na'_r - a_r)Z^r\right]. \tag{18}$$

Upon ζ integration, we deduce that

$$\int (d^N\zeta\, d^N\bar{\zeta})\sqrt{G..}\mathcal{R} \supset \sqrt{g..}\, R^{[g]}(-1)^{\langle N\rangle}\left(\frac{2}{l^2}\frac{d}{dZ}\right)^N \exp\left[\sum_1^N (Na'_r - a_r)Z^r\right]\Bigg|_{Z=0} \tag{19}$$

which we can later convert into c-form, if desired.

3.2. *The gauge field term*

The curl $F_{mn} = A_{n,m} - A_{m,n} + i[A_n, A_m]$, which we are sure arises in the Lagrangian, can be picked out by focussing on the first derivative of the gauge field and ignoring the other parts as these will come automatically. Now the gauge field occurs in the $x - \zeta$ component $G_{m\nu}$ of the metric and is attached to a factor of ζ as well as C'. Therefore, we need only examine terms of the type $G^{km}G^{ln}G^{\sigma\bar{\rho}}(\partial_k G_{l\bar{\rho}}\partial_m G_{n\sigma})$ and these engender a factor $(l^2C'/2)C^{-2}(\bar{\zeta}F.F\zeta)$ upon contraction over indices. It follows that, apart from a proportionality factor,

$$\sqrt{G..}\mathcal{R} \supset \sqrt{g..}(2/l^2)^{N-1}(\bar{\zeta}F.F\zeta)\exp\left[\sum_{r=1}^N (N-1)a'_r Z^r\right]. \tag{20}$$

Integration over property (see Eq. (12)) yields

$$\int (d^N\zeta\, d^N\bar{\zeta})\sqrt{G..}\mathcal{R} \propto \sqrt{g..}\,\mathrm{Tr}(F^{mn}F_{mn})(-1)^{\langle N\rangle}\left(\frac{2}{l^2}\frac{d}{dZ}\right)^{N-1}$$

$$\times \exp\left[\sum_{r=1}^N (N-1)a'_r Z^r\right]\Bigg|_{Z=0}. \tag{21}$$

One readily checks via the case $N = 1$ that the proportionality factor needed is just $-1/2$.

3.3. *The cosmological term*

This piece, which like $\sqrt{G}$ does not depend on A, is a bit more complicated because it can arise from three types of contribution:

$$(G^{kl}G^{mn} \text{ or } G^{km}G^{ln})G_{kl,\rho}G_{mn,\bar{\sigma}}G^{\bar{\sigma}\rho}\,,$$

$$(G^{kl}G_{kl,\bar{\mu}}G_{\rho\bar{\sigma},\nu})(G^{\nu\bar{\mu}}G^{\bar{\sigma}\rho} \text{ or } G^{\rho\bar{\mu}}G^{\bar{\sigma}\nu})\,,$$

$$(G^{\kappa\bar{\lambda}}G_{\bar{\lambda}\kappa,\bar{\mu}}G_{\rho\bar{\sigma},\nu})(G^{\nu\bar{\mu}}G^{\bar{\sigma}\rho} \text{ or } G^{\rho\bar{\mu}}G^{\bar{\sigma}\nu})\,.$$

Each of these has to be taken with multiplicative factors α_N, β_N and γ_N, respectively and may depend on N through the contraction $G^{\bar{\rho}\nu}G_{\nu\bar{\rho}} \propto N$. Thus, we can ascertain via comparison with the $N = 1, 2$ and 3 cases that $\alpha_N = 6$ is N-independent, β_N is linear in N, namely $\beta_N = -4(2N+1)$; and that $\gamma_N = (2N+1)(N+1)$ is N-quadratic. (Alternatively, α, β and γ can be painstakingly determined from first principles.) Now,

$$G^{kl}G^{mn}G_{kl,\rho}G_{mn,\bar{\sigma}}G^{\bar{\sigma}\rho} \text{ entrains } (2/l^2)C'^{-1}C^{-2}\left(Z\frac{dC}{dZ}\frac{dC}{dZ}\right),$$

$$G^{kl}G_{kl,\bar{\mu}}G_{\rho\bar{\sigma},\nu}G^{\nu\bar{\mu}}G^{\bar{\sigma}\rho} \text{ entrains } (2/l^2)C'^{-2}C^{-1}\left(Z\frac{dC}{dZ}\frac{dC'}{dZ}\right),$$

$$G^{\kappa\bar{\lambda}}G_{\bar{\lambda}\kappa,\bar{\mu}}G_{\rho\bar{\sigma},\nu} \text{ entrains } (2/l^2)\,C'^{-2}C^{-1}\left(Z\frac{dC'}{dZ}\frac{dC'}{dZ}\right).$$

Altogether we can conclude that the cosmological term arises from

$$\sqrt{G..}\mathcal{R} \supset \sqrt{g..}(2/l^2)^{N+1}Z \times \exp\left[\sum_{r=1}^{N}((N+1)a'_r - 2a_r)Z^r\right] \times [\alpha_N D^2 + \beta_N DD' + \gamma_N D'^2]\,, \tag{22}$$

with

$$\alpha_N = 6\,, \qquad \beta_N = -4(2N+1)\,, \qquad \gamma_N = (2N+1)(N+1)\,.$$

All that is left is to integrate (12) over property.

3.4. *The full result for any N*

Upon ζ integration we end up with the totality,

$$(-1)^{\langle N\rangle}\left(\frac{l^2}{2}\right)^N \int (d^N\zeta\, d^N\bar{\zeta})\sqrt{G..}\mathcal{R}$$

$$= \sqrt{g..}\left(\frac{d}{dZ}\right)^N \left(R^{[g]}\,e^{\sum(Na'_r - a_r)Z^r} - \frac{l^2}{4N}\operatorname{Tr} F.F\, Z\, e^{\sum(N-1)a'_r Z^r} + \frac{2Z}{l^2}e^{\sum((N+1)a'_r - 2a_r)Z^r}\{6D^2 - 4(2N+1)DD' + (2N+1)(N+1)D^2\}\right)\Bigg|_{Z=0}. \tag{23}$$

R. Delbourgo & P. D. Stack

Table 1. Coefficients of terms multiplying gravity, gauge and cosmological pieces (up to $N = 3$, $C = C'$ for $N = 3$).

N	$R^{[g]}$	Tr $F.F$	Cosmological constant
1	$(2/l^2)(c_1 - c'_1)$	$-1/2$	$(24/l^4)(c_1 - c'_1)^2$
2	$(8/l^4)(2c_1c'_1 - 3{c'_1}^2 - c_2 + 2c'_2)$	$-c'_1/l^2$	$-(16/l^6)(24c_1c_2 - 38{c_1}^2c'_1 - 40c_2c'_1 + 110c_1{c'_1}^2 - 75{c'_1}^3 - 40c_1c'_2 + 60c'_1c'_2)$
3	$(96/l^6)(2{c_1}^3 - 3c_1c_2 + c_3)$	$(4/l^4)(3{c_1}^2 - 2c_2)$	$-(1152/l^8)(5{c_1}^4 - 10{c_1}^2 + 2{c_2}^2 + 3c_1c_3)$

Converting from a_r to c_r via (15) to (17), the reader can verify that the results for $N = 1, 2, 3$ stated in previous papers emerge correctly. These are shown in Table 1.

In this way, the coefficients can be fully determined analytically for any N *without resorting to algebraic computer packages.*

4. Application to $N = 4$

We may apply the above technique to the case where charge and color (electricity and chromicity properties) are taken together. Since QED and QCD are parity invariant we need not concern ourselves with different properties for handedness, as one would need to for electroweak theory. Let indices on ζ of $i = 1, 2, 3$ refer to color (red, green, blue, carrying charge 1/3 as per $\bar{D}$ quarks) and 4 refer to electronic charge (-1). This means combining the coupling of the gluon fields B_m with the electromagnetic field A_m in the property-spacetime sector:

$$G_{m4} = il^2\zeta^{\bar{4}}\, eA_m/2\,; \qquad G_{mi} = il^2[-\zeta^{\bar{i}}\, eA_m/3 + \zeta^{\bar{j}} f {B_m}^{j\bar{i}}]/2\,. \tag{24}$$

The burning issue is whether we are forced to assume that the couplings e and f must merge at some high energy scale in order to ensure gravitational universality (as we needed to when discussing chirality) or whether there is sufficient freedom allowing them to differ from each other. In fact, we shall presently see that the latter holds.

To that end we will *simplify* the argument by adopting an overall set of curvatures $C = C'$ arising in (3) to (5). However, because we are dealing with a direct product $\mathrm{U}(1) \times \mathrm{SU}(3)$ gauge group, we have at our disposal two independent property invariants: $\zeta^{\bar{4}}\zeta^4$ and $\zeta^{\bar{i}}\zeta^i$. Of particular interest is the possibility of

$$C = 1 + \cdots + c_e(\zeta^{\bar{4}}\zeta^4)(\zeta^{\bar{i}}\zeta^i)^2 + c_f(\zeta^{\bar{i}}\zeta^i)^3 + \cdots ,$$

involving two curvature constants c_e and c_f. As we are interested in the gauge field contributions to the Lagrangians, we must focus on terms having the structure $G^{\mu\bar{\nu}}G^{kl}G^{mn}(\partial_m G_{k\bar{\nu}})(\partial_n G_{l\nu})$, which entrain an overall factor C^{-1} multiplying the flat field case. In particular, we find that

$$G^{4\bar{4}}G^{kl}G^{mn}(\partial_m G_{k\bar{4}})(\partial_n G_{l4}) \to g^{km}g^{ln}e^2\zeta^{\bar{4}}F_{kl}F_{mn}\zeta^4\,,$$
$$G^{i\bar{j}}G^{kl}G^{mn}(\partial_m G_{k\bar{j}})(\partial_n G_{li}) \to g^{km}g^{ln}[e^2\zeta^{\bar{4}}F_{kl}F_{mn}\zeta^4/3 + f^2\zeta^{\bar{i}}(E_{kl}E_{mn})^{i\bar{j}}\zeta^j]\,, \tag{25}$$

where $F_{mn} \equiv A_{n,m} - A_{m,n}$ and $E_{mn} \equiv B_{n,m} - B_{m,n} + if[B_n, B_m]$ are the standard "curls" of the electromagnetic and color fields, respectively.

Now remembering that for four properties $\sqrt{G..} = (2/l^2)^4\sqrt{g..}C^{-2}$ we deduce that the sum of the gauge field contributions will be held in the expression

$$\mathcal{R}\sqrt{G..} \supset \left[1 - 3c_e(\zeta^{\bar{4}}\zeta^4)(\zeta^{\bar{i}}\zeta^i)^2 - 3c_f(\zeta^{\bar{i}}\zeta^i)^3 + \cdots\right].$$
$$g^{km}g^{ln}[4e^2\zeta^{\bar{4}}F_{kl}F_{mn}\zeta^4/3 + f^2\zeta^{\bar{i}}(E_{kl}E_{mn})^{i\bar{j}}\zeta^j]. \tag{26}$$

It only remains to integrate over the four properties to discover the gauge field Lagrangian, (including appropriate factors of l^2)

$$\int (d^4\zeta\, d^4\bar{\zeta})\sqrt{G..}\mathcal{R} \supset -(12/l^2)[4c_f e^2 F.F + c_e f^2\,\mathrm{Tr}(E.E)]\,. \tag{27}$$

Clearly all one needs to do is to set $c_e f^2 = 4c_f e^2$ and we maintain a uniform gravitational constant in the ensuing work, *without forcing equality of the color and electromagnetic couplings.* Relaxing the assumptions $C = C'$ and the form of C makes it even easier to ensure uniformity of G_N.

5. Conclusions

We have described a "first-principles" way of determining the Ricci coefficients for spacetime curvature, property curvature (embodied in the cosmological constant) and gauge field Lagrangians, which arise from the super-Ricci scalar. The method represents a major advance as it unshackles us from relying on a computer algebra package, which struggles timewise as the number of properties rises. The final results just depend on the property curvature coefficients which enter the supermetric while maintaining gauge covariance and they have been listed in Table 1. The procedure puts us in a strong position for handing electroweak theory and the full standard model unification with gravity.

References

1. R. Delbourgo and P. D. Stack, *Int. J. Mod. Phys. A* **29**, 1450023 (2014).
2. P. D. Stack and R. Delbourgo, *Int. J. Mod. Phys. A* **30**, 1550005 (2015).
3. R. Delbourgo and P. D. Stack, *Int. J. Mod. Phys. A* **30**, 1550095 (2015).
4. P. D. Stack and R. Delbourgo, *Int. J. Mod. Phys. A* **30**, 1550211 (2015).
5. F. A. Berezin, *Commun. Math. Phys.* **40**, 153 (1975).
6. B. S. DeWitt, *Phys. Rep.* **19**, 295 (1975).

Modern Physics Letters A
Vol. 31, No. 16 (2016) 1630015 (14 pages)
© World Scientific Publishing Company
DOI: 10.1142/S0217732316300159

World Scientific
www.worldscientific.com

The force and gravity of events

Robert Delbourgo
School of Physical Sciences, University of Tasmania, Hobart, Tasmania 7001, Australia
bob.delbourgo@utas.edu.au

Published 17 May 2016

Local events are characterized by "where", "when" and "what". Just as (bosonic) spacetime forms the backdrop for location and time, (fermionic) property space can serve as the backdrop for the attributes of a system. With such a scenario I shall describe a scheme that is capable of unifying gravitation and the other forces of nature. The generalized metric contains the curvature of spacetime and property separately, with the gauge fields linking the bosonic and fermionic arenas. The super-Ricci scalar can then automatically yield the spacetime Lagrangian of gravitation and the Standard Model (plus a cosmological constant) upon integration over property coordinates.

Keywords: Properties; unification; gravity; forces.

1. An Algebraic Framework for Events

The material which I will present is sufficiently different from other attempts at unification of forces that I rather fancy, A. Salam might have given it a nod of approval. Two years ago, at the Dyson 90th anniversary conference, I outlined[1] how it is possible to unify gravity with the simplest of all forces, electromagnetism — Einstein's eldorado — simply by appending a single complex anti-commuting Lorentz *scalar* variable to spacetime, *not a spinor*; importantly no infinite KK modes arise. My partner in crime (Paul Stack) and I have made considerable progress since then and I will now try to summarize how to unify gravity with the other forces of nature through a relatively simple supermetric. Our attempts in this direction have been motivated by the present parlous state of particle physics and the snail's pace of progress in this area over the last 40 years. Here is a statement which may bring me some opprobium: namely, apart from the timely discovery of the Higgs boson, emergence of multiquark states and significant astrophysical advances, there is very little to celebrate in our attempts to unravel nature at the most basic level. This is in spite of determined, quasi-herculean efforts of theorists who have persistently espoused/promoted very clever ideas. So far, Nature stubbornly refuses to cooperate by providing us with unequivocal experimental signs of SUSY, strings/branes and other ingenious proposals. It seems that the simple Standard Model of particles and cosmology still rules. Nonetheless, its plethora of parameters have spurred theorists to search for generalizations of the Standard Model which may help to cut down the number of arbitrary constants and leave room for mysterious dark matter. Many schemes have been put forward. These usually add other gauge fields, sterile particles, invoke enlarged groups and introduce scalar fields, perhaps associated with cosmological inflation. My feeling is that these ideas are very much hit-or-miss and they do seem to lack a fundamental basis. I think Salam might have looked askance at them. Anyhow here goes ...

For many years we have become accustomed to the notion of spacetime events, with local fields (belonging to representations of some gauge group) interacting at a particular site and time. The $x = (t, \mathbf{x})$ spacetime continuum serves as the backdrop for the "when" and "where" of an event. But, until one specifies the fields involved in the interaction, the "what" of the action is left open, to be determined by experiment. Now we should realize that any event necessarily consists of a transaction or a change of property at a location. (The transaction is usually communicated by a gauge field.) It occurred to me that it might be possible to provide a mathematical backdrop for "properties" or "attributes" of the participating fields by invoking a property space with its own set of coordinates. As far as we can tell there seem to be a finite number of quantum numbers or properties in nature. So the basic idea is to put some mathematics into the "what" of the event by invoking anti-commuting (Lorentz scalar) coordinates ζ; these should serve to provide the setting for the gauge groups and particle attributes and fields should be functions of these ζ as well as spacetime location x. The *full* action is to be integrated over the properties ζ like one does for x. The reason why I have picked ζ as anti-commuting is because when an object is endowed by several such properties, the melange is necessarily finite; and since the square of a property vanishes it means that once a fundamental constituent possesses that attribute it cannot doubly have it. Of course, since we are dealing with quantum mechanics in the long run, these properties must be complex so the anti-attribute $\bar{\zeta}$ should be permitted. By combining properties with anti-properties one can build up "generations" of particles possessing the same *overall* attributes. In some sense, N-extended supersymmetry is based on the same idea but it suffers badly from spin state proliferation.

The question is how many property coordinates ζ are needed? There must be enough to describe the visible world. The pioneers of unified forces[2,3] have forged the way and provided the inspiration. Despite some criticisms to which these full gauge groups have been subjected, I have opted for SU(5) and SO(10) gauge models; these have many attractive features, so for now I will suppose that there are five independent[a] complex ζ. Later on we will be forced to subtly enlarge this number in order to reflect the incontrovertible fact that fermions of distinct chiralities — through their electroweak characteristics — behave quite differently at low energy; thus experiment obliges us to distinguish between left and right properties.

[a] I have found that four ζ are definitely insufficient to produce three generations at least.

2. Mathematical Description

By enlarging spacetime x with ζ we hope to encompass *all* possible fundamental events. Even though the term has been overused we will assume that there exist "superfields" $\Phi(X)$ and $\Psi(X)$ which are functions of the super-coordinate $X^M \equiv (x^m, \zeta^\mu, \zeta^{\bar{\mu}})$. The idea is that an integral over products of just one or two superfields can provide the entire action for every event. The calculus for handling the combination of bosonic x and fermionic ζ is well-established[b] and the graded character of X means that Berezinian integration is to be adopted for property integration, with super-determinants coming into play. By curving the superspace we will automatically be able to describe gravity and the other forces of nature, as we shall see.

But let us start with flat space and assume parity conservation; presently we shall improve on this by adding gauge fields and parity violation. With five ζ we are dealing with an overarching Sp(10) group. The supermetric distance for flat OSp(1, 3/10) is

$$ds^2 = dx^a\, dx^b\, \eta_{ba} + \ell^2 (d\zeta^{\bar{\alpha}}\, d\zeta^{\beta} \eta_{\beta\bar{\alpha}} + d\zeta^{\alpha}\, d\zeta^{\bar{\beta}}\, \eta_{\bar{\beta}\alpha})/2\,, \tag{1}$$

where η_{ba} is Minkowskian and $\eta_{\beta\bar{\alpha}} = -\eta_{\bar{\alpha}\beta} = \delta_\beta{}^\alpha$; also a fundamental length scale ℓ must be introduced because we are presuming that property ζ is dimensionless.

The Bose fields are to be associated with even powers of ζ and its conjugate, while the Fermi fields are connected with odd powers. Let us reserve the labels 1, 2, 3 for color property or "chromicity" and 0, 4 to neutrinicity, electricity. The quantum numbers which are ascribed to these, viz.

$$\text{Charge:}\ \ Q(\zeta^0, \zeta^{\bar{1}}, \zeta^{\bar{2}}, \zeta^{\bar{3}}, \zeta^4) = (0, 1/3, 1/3, 1/3, -1) \tag{2}$$

$$\text{Fermion Number:}\ \ F(\zeta^0, \zeta^{\bar{1}}, \zeta^{\bar{2}}, \zeta^{\bar{3}}, \zeta^4) = (1, -1/3, -1/3, -1/3, 1) \tag{3}$$

really only come to life when one introduces the gauge fields, as we soon will. Given the assignments (2), the lepton doublet generations are connected with (ζ^0, ζ^4), multiplied by powers of $\zeta^{\bar{\rho}}\zeta^{\rho}$; the quark generations arise more subtly.

Component fields ϕ and ψ emerge[4] when we expand Φ and Ψ as polynomials in ζ and $\bar{\zeta}$. Fermions are to be associated with odd powers and bosons with even powers of attributes. Charge conjugation of course corresponds to the "reflection" operation $\zeta \leftrightarrow \bar{\zeta}$ and we may define a duality operation (that does not affect the SU(5) representations) under which $(\zeta)^r(\bar{\zeta})^s \leftrightarrow (\zeta)^{5-s}(\bar{\zeta})^{5-r}$. By imposing self-duality *or* anti-self-duality on the superfields we can greatly reduce the number of independent component fields arising in the ζ-expansion. This is detailed in Ref. 4. In amongst the boson Φ states are nine color neutral uncharged mesons of which the combination $\zeta^4\zeta^{\bar{1}}\zeta^{\bar{2}}\zeta^{\bar{3}}$ is recognizable as the Standard Model Higgs.

[b]We developed this from scratch as we wanted to adhere to Einstein up notation for coordinates and traditional left operations like differentiation. Also, we wanted to settle the notation to our own satisfaction. Hereafter, Latin letters signify spacetime and Greek letters signify property. Early letters of the alphabet connote flatness while later letters imply curved space.

R. Delbourgo

However, the quark isomultiplets ψ which exist in Ψ are slightly different from the Standard Model! The up- and down-quarks come as two weak isodoublets/singlets and part of a weak *isotriplet/isodoublet/isosinglet* contained in SU(5) representations of dimension 45. Thus,

$$\begin{pmatrix} U^{[\bar\mu\bar\nu]} \sim \zeta^{\bar\mu}\zeta^{\bar\nu}\zeta^{0} \\ D^{[\bar\mu\bar\nu]} \sim \zeta^{\bar\mu}\zeta^{\bar\nu}\zeta^{4} \end{pmatrix}, \qquad \begin{pmatrix} U'^{[\bar\mu\bar\nu]} \sim \zeta^{\bar\mu}\zeta^{\bar\nu}\zeta^{0}\zeta^{\bar 4}\zeta^{4} \\ D'^{[\bar\mu\bar\nu]} \sim \zeta^{\bar\mu}\zeta^{\bar\nu}\zeta^{4}\zeta^{\bar 0}\zeta^{0} \end{pmatrix}, \qquad \begin{pmatrix} U''^{\lambda} \sim \zeta^{\lambda}\zeta^{\bar 4}\zeta^{0} \\ D''^{\lambda} \sim \zeta^{\lambda}(\zeta^{\bar 0}\zeta^{0}, \zeta^{\bar 4}\zeta^{4}) \\ X''^{\lambda} \sim \zeta^{\lambda}\zeta^{\bar 0}\zeta^{4} \end{pmatrix}$$

implies the existence of a brand new quark X'' (of charge $-4/3$) in a third generation. Though X'' may be more massive than even the top quark, the consequence at lower energy scales is that we do not expect the CKM matrix to be quite unitary.[c] Probably the best way to find X is via a high energy electron-positron collider? Other predictions of the scheme are that heavy leptons should be seen as well as unaccompanied (massive?) D-type quarks. If none of these signals eventuates then it is back to the drawing board and a re-examination to see if any of these ideas about property is salvageable or if the disease is terminal.

3. Force Fields

The most interesting feature of our scheme is the way that gauge fields enter and tie in with the quantum number assignments. We note that a flat metric in X is only invariant under global SU(N) unitary rotations of the N attributes. But as soon as we make them local or x-dependent, so that

$$\zeta^{\mu} \to \zeta'^{\mu} = [\exp(i\Theta(x))]^{\mu\bar\nu}\zeta^{\nu} \tag{4}$$

we find that there is an inconsistency in the transformation rules for the metric; we are forced to "curve" the space and introduce gauge fields to repair the fault. The way to do this is to write the generalized event (separation)^2 as

$$ds^2 = dX^M\, dX^N G_{NM}\,; \qquad G_{NM} = \mathcal{E}_N{}^B \mathcal{E}_M{}^A \eta_{AB}(-1)^{[B][M]}\,, \tag{5}$$

where the metric arises through frame vectors $\mathcal{E}$ and the grading is defined in the usual way: $[m] = 0$, $[\mu] = 1$. Thus the transformations rules for G,

$$G'_{SR}(X') = \left(\frac{\partial X^M}{\partial X'^R}\right)\left(\frac{\partial X^N}{\partial X'^S}\right) G_{NM}(X)(-1)^{[S]([R]+[M])}\,, \tag{6}$$

under the local rotations of ζ demand that we introduce components $G_{n\mu}$, $G_{n\bar\mu}$ which have a vectorial character; they should be overall fermionic and must somehow involve the gauge field V as this is the communicator of property across spacetime. A few moment's reflection (neglecting coupling constants for the present)

[c]This meshes in with the observation that an isotriplet couples more strongly with the charged W-boson than an isodoublet and therefore the known decay width of the top quark requires a correspondingly smaller V_{tb} coupling to W.

leads one to the identification $\mathcal{E}_m{}^\alpha = -iV_m{}^{\alpha\bar{\nu}}\zeta^\nu$, which is very similar to the way that the em field makes an appearance in the original Klein–Kaluza model; there is really very little room for manoeuvre and the appearance is indeed entirely natural: gauge fields transmit property from one place and time to the next so they ought to arise in the spacetime-attribute sector. The only liberty permitted to us is to multiply by polynomials in property scalars $Z \equiv \zeta^{\bar{\mu}}\zeta^\mu$, since these are gauge invariant and carry no quantum numbers. We might say that inclusion of these polynomials corresponds to "curving" property space.

Using that freedom, the only metric which is fully consistent with local SU(N) gauge transformations is

$$\begin{pmatrix} G_{mn} & G_{m\nu} & G_{m\bar{\nu}} \\ G_{\mu n} & G_{\mu\nu} & G_{\mu\bar{\nu}} \\ G_{\bar{\mu}n} & G_{\bar{\mu}\nu} & G_{\bar{\mu}\bar{\nu}} \end{pmatrix}$$

$$= \begin{pmatrix} g_{mn}C + \ell^2\bar{\zeta}\{V_m, V_n\}\zeta C'/2 & -i\ell^2(\bar{\zeta}V_m)^{\bar{\nu}}C'/2 & i\ell^2(V_m\zeta)^\nu C'/2 \\ -i\ell^2(\bar{\zeta}V_n)^{\bar{\mu}}C'/2 & 0 & \ell^2\delta_\mu{}^\nu C'/2 \\ i\ell^2(V_n\zeta)^\mu C'/2 & -\ell^2\delta_\nu{}^\mu C'/2 & 0 \end{pmatrix}. \quad (7)$$

Here $C(Z) = 1+\sum_{n=1}^N c_n Z^n$, $C'(Z) = 1+\sum_{n=1}^N c'_n Z^n$ are independent polynomials of order N in Z which are allowed without destroying the gauge symmetry. One then readily checks that the rule (6) just corresponds to the usual gauge transformation: $iV'_m(x') = \exp[i\Theta(x)](iV_m(x) + \partial_m)\exp[-i\Theta(x)]$. If we just demand subgroup gauge symmetry, we can relax the conditions on the Z polynomials and have them invariant under local subgroup rotations, so more property curvature coefficients c_n can be entertained. We will come back to this when considering QCD plus QED and electroweak theory in such a framework.

The procedure from hereon is pretty straightforward,[5] paying very particular attention to orders of terms and signs that are due to grading. One first constructs the super-Ricci scalar $\mathcal{R}$ from the Christoffel symbols

$$\begin{aligned} 2\Gamma_{MN}{}^K = {} & [(-1)^{[M][N]}G_{ML,N} + G_{NL,M} \\ & - (-1)^{[L]([M]+[N])}G_{MN,L}](-1)^{[L]}G^{LK} \end{aligned} \quad (8)$$

via the Palatini form

$$\mathcal{R} = G^{MK}\mathcal{R}_{KM} = (-1)^{[L]}G^{MK}[(-1)^{[L][M]}\Gamma_{KL}{}^N\Gamma_{NM}{}^L - \Gamma_{KM}{}^N\Gamma_{NL}{}^L]. \quad (9)$$

Secondly one integrates $\mathcal{R}$ over property. This leads to the gravitational and gauge field Lagrangian plus a cosmological term.[6] (As a bonus, the stress tensor T_{mn} of the gauge fields is automatically incorporated in $\mathcal{R}_{mn}$ when we extract the resulting "equations of motion".) The coefficients in front of these terms depend on the number of properties[7] and on the property curvature coefficients but they all have the generic form

$$\left(\frac{\ell^2}{2}\right)^{2(N-1)} \int d^N\zeta\, d^N\bar{\zeta}\,\sqrt{G..}\,\mathcal{R} = \frac{\mathcal{A}R^{[g]}}{\ell^2} + \mathcal{B}\,\mathrm{Tr}(F.F) + \frac{\mathcal{C}}{\ell^4}\,, \tag{10}$$

where $F_{mn} \equiv V_{n,m} - V_{m,n} + i[V_m, V_n]$ and the N-dependent coefficients $\mathcal{A}$, $\mathcal{B}$, $\mathcal{C}$ are listed in Ref. 7.

The matter fields and their Lagrangians are then introduced,

$$\mathcal{L}_\phi = \int d^N\zeta\, d^N\bar{\zeta}\,\sqrt{-G..}\,G^{MN}\partial_N\Phi\partial_M\Phi\,, \tag{11}$$

$$\mathcal{L}_\psi = \int d^N\zeta\, d^N\bar{\zeta}\,\sqrt{-G..}\,\bar{\Psi}i\Gamma^A E_A{}^M\partial_M\Psi\,. \tag{12}$$

The gauge and gravitational interactions of the component fields (ϕ, ψ) then just fall out, but these sometimes require wave function renormalizations due to influence of the property curvature coefficients c_n — coefficients which are absent in flat space. The key point is that the gauge fields couple correctly to the matter fields through the vielbein term

$$E_A{}^M\partial_M \supset e_a{}^m[\partial_m + i(V_m\zeta)^\mu\,\partial_\mu - i(\bar{\zeta}V_m)^{\bar{\mu}}\,\partial_{\bar{\mu}}]\,,$$

so the property derivative is compensated by a further property coordinate attached to the gauge field V; this is our version of covariant differentiation. Incidentally I ought to declare that such complicated calculations were originally carried using an algebraic computer package devised by Paul Stack and, after time-consuming computation, they always produced gauge- and coordinate-invariant results. Knowing this always happened, we have since been able to find a shorter analytic way of picking out the correct terms in (10)–(12) by a procedure which can be generalized to any number of attributes and dispense with Mathematica. Finally, to (11) and (12) we may add the renormalizable super-Yukawa self interactions $\Psi\Phi\Psi$ and $V(\Phi) \simeq \Phi^4$ in the usual manner, with the aim of generating a mass term through the expectation values held in the chargeless fields within $\langle\Phi\rangle$.

Before moving on, three comments about the fermion fields deserve particular mention. Firstly the adjoint field $\bar{\Psi}$ has to be carefully defined with appropriate signs[8] in property space to produce a series of terms $\bar{\psi}\psi$, after integrating over ζ. Secondly, $\bar{\zeta}\psi$ and their charge conjugates $\psi^{(c)}\zeta$ both appear in the full expansion of $\Psi(\zeta, \bar{\zeta})$ and they simply lead to a doubling of the eventual answers; thus we can simplify calculations by "halving" the expansion of Ψ to $\Psi \supset \bar{\zeta}\psi$ terms. Thirdly and intriguingly, we have to extend the concept of Dirac γ matrices to super Γ matrices, such that $(\Gamma^A P_A)^2 = \eta^{AB}P_B P_A$. In spacetime we get the standard $\Gamma^a = \gamma^a$ with $\{\gamma^a, \gamma^b\} = 2\eta^{ab}$, but in the property sector one needs to ensure that the "square-rooted" Γ^α are *fermionic* and obey

$$[\Gamma^\alpha, \Gamma^\beta] = [\Gamma^{\bar{\alpha}}, \Gamma^{\bar{\beta}}] = 0\,, \qquad [\Gamma^\alpha, \Gamma^{\bar{\beta}}] = 2\eta^{\alpha\bar{\beta}} = 2\delta_\beta{}^\alpha\,. \tag{13}$$

In the same way that Dirac introduced 4×4 matrices and made novel use of the Clifford algebra for spacetime, we must do something similar for property space. We can arrange for the commutators (13) to be satisfied by augmenting property space with auxiliary coordinates θ^α, setting $\Gamma^\alpha \equiv \sigma_+\theta^\alpha$, $\Gamma^{\bar\alpha} \equiv \sigma_-\partial/\partial\theta^\alpha$, and making sure that Ψ is multiplied by the projected singlet $\Theta \equiv (1+\sigma_3)\theta^1\theta^2\cdots\theta^N/2$, over which one eventually integrates.[d] There are probably less extravagant ways of doing this.

4. Electric and Chromic Relativity

To see how all this works out, consider QED and QCD which involve one attribute called electricity plus three "chromicity" properties (commonly termed red, green, blue). Thus, we confine ourselves to coordinates ζ^1 to ζ^4 and combine both chiralities in Dirac fields since those interactions are blind to parity. As we are confining ourselves to U(1) × SU(3) we are dealing with two sets of gauge fields within the fuller SU(4): the em field A and the gluon fields B, having coupling constants e and f respectively. One identifies the frame vectors $\mathcal{E}_m{}^\kappa = -i(fB_m - eA_m/3)^{\kappa\bar\iota}\zeta^\iota$, $\mathcal{E}_m{}^4 = ieA_m$, leading to the basic metric elements

$$G_{m4} = i\ell^2\zeta^{\bar 4}eA_mC'/2, \quad G_{m\iota} = i\ell^2[\zeta^{\bar\iota}eA_m/3 - \zeta^{\bar\kappa}fB_m{}^{\kappa\bar\iota}]C'/2\,, \tag{14}$$

which may be multiplied by polynomials in two *distinct* invariants $\zeta^{\bar\kappa}\zeta^\kappa$ and $\zeta^{\bar 4}\zeta^4$. I should point out that it is the interactions (14) which actually determine the charge and color assignments stated in (2) and (3). Also the coupling *must* accompany the gauge fields in order to produce the correct interactions with matter fields.

To simplify the subsequent argument about the resulting interactions, I will assume that the property curvature polynomials are common to spacetime and property space:

$$C = C' = 1 + \cdots + c_e(\zeta^{\bar 4}\zeta^4)(\zeta^{\bar\kappa}\zeta^\kappa)^2 + c_f(\zeta^{\bar\kappa}\zeta^\kappa)^3\,. \tag{15}$$

As we are dealing with four properties, we find that the Berezinian[9,10] is $\sqrt{-G..} = (2/\ell^2)^4\sqrt{-g..}C^{-2}$. A careful analytical calculation shows that the super-Ricci scalar contains the following gauge field combination:

$$\mathcal{R}\sqrt{-G..} \supset [1 - 3c_e(\zeta^{\bar 4}\zeta^4)(\zeta^{\bar\kappa}\zeta^\kappa)^2 - 3c_f(\zeta^{\bar\kappa}\zeta^\kappa)^3 + \cdots]$$
$$\sqrt{-g..}g^{km}g^{ln}[4e^2\zeta^{\bar 4}F_{kl}F_{mn}\zeta^4/3 + f^2\zeta^{\bar\kappa}(E_{kl}E_{mn})^{\kappa\bar\iota}\zeta^\iota]\,, \tag{16}$$

where $F_{mn} = A_{n,m} - A_{m,n}$ and $E_{mn} = B_{n,m} - B_{m,n} + if[B_n, B_m]$ are the standard "curls" of the electromagnetic and gluon fields. The last step is to integrate over the four properties. Including appropriate scaling factors of ℓ^2 one gets

$$\int (d^4\zeta\, d^4\bar\zeta)\mathcal{R}\sqrt{-G..} \supset (-12\sqrt{-g..}/\ell^2)[4c_fe^2F.F + c_ef^2\,\mathrm{Tr}(E.E)]\,. \tag{17}$$

Last but not least, we must ensure gravitational universality; so we have to set $c_ef^2 = 4c_fe^2$, which is perfectly feasible without demanding equality of the color

[d] Of course the adjoint $\bar\Psi$ contains the conjugate singlet $\bar\Theta = (1+\sigma_3)\theta^{\bar N}\cdots\theta^{\bar 2}\theta^{\bar 1}/2$.

and electromagnetic couplings. If we relax the assumption that $C = C'$, it is even easier to ensure universality of Newton's constant $G_{\rm N}$.

The color and electromagnetic interactions of the matter fields Ψ, Φ emerge from (11) and (12) exactly as expected. See Ref. 8. I shall not delve into that because the story is not quite complete and is therefore likely to be misleading: we have neglected neutrinicity (the fifth property ζ^0) so the ensuing generations are not the physical ones, as sketched in Sec. 3. To correct for this, we must turn to the leptons.

5. Electroweak Relativity

The application of our scheme to the original electroweak model[11–13] of leptons requires an interesting extension of previous work[14] and leads to an intriguing prediction about the weak mixing angle, not to mention the prediction of two leptonic generations. The fact that the weak isospin and hypercharge assignments of the leptons change with chirality obliges us to invoke distinct properties ζ_L and ζ_R for left- and right-handed leptons respectively to which the gauge fields latch on (through the frame vectors). The full SU(4) gauge field $V^{\mu\bar{\nu}}$, acting on the pair of doublets $(\zeta^0_L, \zeta^4_L, \zeta^0_R, \zeta^4_R)$ is not needed; only the restricted $\mathrm{SU}(2)_L \times \mathrm{U}(1)$ rotations demand attention. Thus, we re-interpret $V_m = L_m + R_m$, with

$$L_m = (g\mathbf{W}_m.\tau - g' B_m)/2\,, \qquad R_m = g' B_m(\tau_3 - 1)/2\,, \tag{18}$$

possessing the standard weak hypercharge assignments

$$Y(\zeta^0_L, \zeta^4_L, \zeta^0_R, \zeta^4_R) = (-1, -1, 0, -2)\,. \tag{19}$$

It must also be understood that L is to be associated with the left property derivative $\partial/\partial\zeta_L$ and R is to be associated with the right property derivative $\partial/\partial\zeta_R$; g and g' are the usual coupling constants tied to the weak triplet $\mathbf{W}$ and weak singlet hypercharge B respectively.

It is sufficiently general for our purposes to take the polynomial property curvatures C and C' to be equal and direct products of quadratic left- and right-handed polynomials:

$$\begin{gathered} C = C_R C_L = [1 + c_R Z_R + c_{RR} Z_R^2][1 + c_L + c_{LL} Z_L^2]\,; \\ Z_R \equiv \bar{\zeta}_R \zeta_R,\ Z_L \equiv \bar{\zeta}_L \zeta_L\,. \end{gathered} \tag{20}$$

These enter in the metric components:

$$G_{m\zeta_L} = -i\ell^2 \bar{\zeta}_L L_m C/2\,; \qquad G_{m\zeta_R} = -i\ell^2 \bar{\zeta}_R R_m C/2\,, \tag{21}$$

$$G_{\zeta_L\bar{\zeta}_L} = G_{\zeta_R\bar{\zeta}_R} = \ell^2 C/2\,, \qquad G_{\zeta_L\zeta_R} = G_{\bar{\zeta}_L\bar{\zeta}_R} = G_{\zeta_L\bar{\zeta}_R} = G_{\zeta_R\bar{\zeta}_L} = 0\,. \tag{22}$$

The remaining metric element reads

$$G_{mn} = C[g_{mn} + (\text{gauge field terms})]\,. \tag{23}$$

Factorizability of C simplifies the calculations enormously when we integrate over the whole eight properties: $\int d^2\zeta_R\, d^2\bar{\zeta}_R\, d^2\zeta_L\, d^2\bar{\zeta}_L$.

The various contributions to the super-Ricci scalar drop out as follows, bearing in mind that $\sqrt{-G..} = (2/\ell^2)^4\sqrt{-g..}(C_R C_L)^{-3}$. There are three terms:

$$\int d^2\zeta_R..d^2\bar{\zeta}_L\,\sqrt{-G..}\mathcal{R} \supset 36\sqrt{-g..}(2/\ell^2)^4 R^{[g]}(2c_R^2 - c_{RR})(2c_L^2 - c_{LL})\,, \quad (24)$$

$$\int d^2\zeta_R..d^2\bar{\zeta}_L\,\sqrt{G..}\mathcal{R} \supset -\frac{3}{2}\sqrt{g..}\left(\frac{2}{\ell^2}\right)^3 [c_L(3c_R^2 - 2c_{RR})(g^2\mathbf{W}_{mn}.\mathbf{W}^{mn} + g'^2 B_{mn}B^{mn}) + g'^2 2c_R(3c_L^2 - 2c_{LL})B_{mn}B^{mn}]\,, \quad (25)$$

$$\int d^2\zeta_R..d^2\bar{\zeta}_L\,\sqrt{G..}\mathcal{R} \supset 12\sqrt{g..}(2/\ell^2)^5[(4c_L c_{LL} - 5c_L^2)(2c_R^2 - c_{RR}) + (L \leftrightarrow R)]\,, \quad (26)$$

where $W_{mn} \equiv W_{n,m} - W_{m,n} + ig[W_n, W_m]$ and $B_{mn} \equiv B_{n,m} - B_{m,n}$. The full answer is the sum of (24)–(26).

Universality of gravity at the semiclassical level anyway (and the correct normalization of the gauge fields) is guaranteed when we set

$$c_L(3c_R^2 - 2c_{RR})(g^2 - g'^2) = 2c_R(3c_L^2 - 2c_{LL})g'^2\,,$$

which is readily arranged. But much more intriguing is the fact that if the property curvature is parity conserving so that $c_R = c_L = c$, $c_{RR} = c_{LL} = c_2$ and implying that all parity violation comes from the gauge fields in the frame vectors, then $g^2 = 3g'^2$. Thus the weak angle reduces to 30°. It makes good sense because the property curvature C polynomial accompanies the gravitational field and, as far as we know, unquantized gravity does not know the left from the right. So this restriction seems very natural and the value of the weak angle is a consequence of gravitational universality in this framework; it is not a result of invoking a higher group or anomaly cancellation, as some other analyses[15,16] would have.

Turning to the matter fields, we can reduce the number of components by invoking selfduality[e] (corresponding to symmetry about the cross-diagonal in the superfield expansions). Ignoring the charge conjugate terms, which simply double the results below, two fermion generations, ψ and ψ' arise from expanding Ψ. Using the shorthand symbols $Z_L \equiv \bar{\zeta}_L\zeta_L, Z_R \equiv \bar{\zeta}_R\zeta_R$ as in (20), we get

$$2\Psi = \bar{\zeta}_L[\psi_L(1 + Z_R^2/2) + \psi'_L Z_R](1 + Z_L) + (L \leftrightarrow R)\,, \quad (27)$$

$$2\bar{\Psi} = [\overline{\psi_L}(1 + Z_R^2/2) + \overline{\psi'_L} Z_R]\zeta_L(1 + Z_L) + (L \leftrightarrow R)\,. \quad (28)$$

[e]SU(2) duality, indicated by $\times$, stipulates that $1^\times = Z^2/2, (\zeta^\mu)^\times = (\zeta^\mu)Z, Z^\times = Z$, $(\eta_{\mu\nu}\zeta^\mu\zeta^\nu)^\times = -\eta_{\mu\nu}\zeta^\mu\zeta^\nu$, where $Z = \zeta^{\bar{\mu}}\zeta^\mu$. Vice versa, and likewise for the hermitian conjugate combinations. Thus, the self-dual combinations are $(1+Z^2/2)$, Z and $\zeta(1+Z)$ with $\eta_{\mu\nu}\zeta^\mu\zeta^\nu \to 0$. We apply this separately to left and right leptonic properties in the following equations, corresponding to the subgroup $\mathrm{SU}(2)_L \times \mathrm{SU}(2)_R$.

R. Delbourgo

Since chirality ensures that $\overline{\psi_L}\psi_L = \overline{\psi_R}\psi_R = 0$, we find that a mass term arising from the product $\bar{\Psi}\Psi$ has insufficient powers of ζ to give a nonzero answer; thus a *mass term vanishes identically* and this is a good thing because it indicates that we need to couple fermions to bosons before one can generate mass. The kinetic term is fine however; in flat space,

$$-\int d^2\zeta_R..d^2\bar{\zeta}_L\,\bar{\Psi}i\gamma.\partial\Psi = \overline{\psi_L}i\gamma.\partial\psi_L + \overline{\psi'_L}i\gamma.\partial\psi'_L + (L \leftrightarrow R)\,. \tag{29}$$

Regarding the bosons, we recall that the self-dual combinations are $(1 + Z^2/2)$ and Z with $\zeta\zeta \to 0$, separately for left- and right-handed properties. Hence the fully self-dual, hermitian superBose field Φ is

$$\begin{aligned}2\Phi &= \varphi(1+Z_L^2/2)(1+Z_R^2/2) + \varphi' Z_L Z_R + \Lambda Z_L(1+Z_R^2/2) + PZ_R(1+Z_L^2/2)\\ &\quad + [\bar{\zeta}_R\phi\zeta_L + \bar{\zeta}_L\phi^\dagger\zeta_R + \phi'\zeta_R\zeta_L + \bar{\zeta}_L\bar{\zeta}_R\phi'^\dagger](1+Z_L)(1+Z_R)\,.\end{aligned} \tag{30}$$

If we further restrict ourselves to fields of even parity under the operation $\zeta_R \leftrightarrow \zeta_L$, we find $\Lambda = P \equiv \chi/\sqrt{2}$, $\varphi' = 0, \phi = \phi'$, so the expansion (30) reduces to

$$\begin{aligned}2\Phi &= \varphi(! + Z_L^2/2)(1+Z_R^2/2) + \varphi' Z_L Z_R\\ &\quad + \chi[Z_L(1+Z_R^2/2) + Z_R(1+Z_L^2/2)]/\sqrt{2}\\ &\quad + [\bar{\zeta}_R\phi\zeta_L + \bar{\zeta}_L\phi^\dagger\zeta_R](1+Z_L)(1+Z_R)\,.\end{aligned} \tag{31}$$

The normalization factors have been concocted so that

$$-\int d^2\zeta_R..d^2\bar{\zeta}_L\,\Phi^2 = -\varphi^2 - \varphi'^2 - \chi^2 + \mathrm{Tr}(\phi^2)\,. \tag{32}$$

In (31) the quartet $\phi^{\mu\bar{\nu}} = (\phi_0 I + \phi.\tau)^{\mu\bar{\nu}}/\sqrt{2}$ consists of a singlet and a triplet. The quantum numbers I_{3L}, Y, $Q = I_{3L} + Y/2$ of the components read:

$$Y(\varphi,\varphi',\chi) = (0,0,0)\,, \qquad I_{3L}(\varphi,\varphi',\chi) = (0,0,0)\,, \qquad Q(\varphi,\varphi',\chi) = (0,0,0)\,;$$

$$Y(\phi^{0\bar{0}},\phi^{0\bar{4}},\phi^{4\bar{0}},\phi^{4\bar{4}}) = (1,1,-1,-1) \qquad 2I_{3L}(\phi^{0\bar{0}},\phi^{0\bar{4}},\phi^{4\bar{0}},\phi^{4\bar{4}}) = (-1,1,-1,1)\,;$$

$$Q(\phi^{0\bar{0}},\phi^{0\bar{4}},\phi^{4\bar{0}},\phi^{4\bar{4}}) = (0,1,-1,0)\,.$$

The Higgs boson will be associated with $\phi_0 + \phi_3$, as we will presently discover. For that identification we need to consider the super-Yukawa and gauge field interactions in flat spacetime, before we curve spacetime with gravity.

With L and R gauge fields defined in (18), the vielbeins which correspond to the metric elements (21)–(23) are:

$$\begin{pmatrix} E_a{}^m & E_a{}^\mu & E_a{}^{\bar{\mu}}\\ E_\alpha{}^m & E_\alpha{}^\mu & E_\alpha{}^{\bar{\mu}}\\ E_{\bar{\alpha}}{}^m & E_{\bar{\alpha}}{}^\mu & E_{\bar{\alpha}}{}^{\bar{\mu}}\end{pmatrix} = \frac{1}{\sqrt{C}}\begin{pmatrix} e_a{}^m & i[(L_a\zeta_L)+(R_a\zeta_R)]^\mu & -i[(\bar{\zeta}_L L_a)+(\bar{\zeta}_R R_a)]^{\bar{\mu}}\\ 0 & \delta_\alpha{}^\mu & 0\\ 0 & 0 & \delta_{\bar{\alpha}}{}^{\bar{\mu}}\end{pmatrix}. \tag{33}$$

Thus the fermion kinetic energy can be written in the form $\bar{\Psi}i\Gamma^A D_A\Psi$, where

$$D_A = E_A{}^M\partial_M = E_A{}^m\partial_m + E_A{}^\mu\partial_\mu + E_A{}^{\bar{\mu}}\partial_{\bar{\mu}}$$

acts like a covariant derivative. Let V serve as a generic gauge field; the action of $i\gamma^a D_a$ on $f(Z)(\bar{\zeta}\psi)$ is to give $f(Z)\bar{\zeta}\gamma.(i\partial_a + V_a)\psi$ and on $f(Z)(\bar{\psi}\zeta)$ is to give $f(Z)(i\partial_a + V_a)\bar{\psi}\gamma^a\zeta$. So when we integrate over property we end up precisely with the usual gauge field interaction $\bar{\psi}\gamma^a(i\partial_a + V_a)\psi$ for each of the two generations, which in the leptonic case translates into

$$\overline{\psi_L}\gamma^a(i\partial_a + L_a)\psi_L + \overline{\psi_R}\gamma^a(i\partial_a + R_a)\psi_R + (\psi \to \psi')\,.$$

This is unsurprising; interpreting $(\psi^0, \psi^4) = (\nu, l)$, one ends up with the standard

$$\begin{aligned}\mathcal{L}_\psi &= \bar{l}\gamma.(i\partial - eA)l + \bar{\nu}i\gamma.\partial\nu + \frac{e}{\sqrt{2}\sin\theta}[\overline{\nu_L}\gamma.W^+l_L + \overline{l_L}\gamma.W^-\nu_L] \\ &\quad + \frac{e}{\sin 2\theta}(\overline{\nu_L}\gamma.Z\nu_L) + e\tan\theta(\overline{l_R}\gamma.Zl_R) - e\cot 2\theta(\overline{l_L}\gamma.Zl_L) \\ &\quad + (l,\nu) \to (l',\nu')\,, \end{aligned} \tag{34}$$

where $\cos\theta = g/\sqrt{g^2+g'^2}$, $\sin\theta = g'/\sqrt{g^2+g'^2}$, $e = gg'/\sqrt{g^2+g'^2}$. Equation (34) simplifies to a considerable extent when $\theta = 30°$, as indicated by gravitational universality, because the Z field then interacts purely axially with the charged lepton, in contrast to the purely vectorial electromagnetic field.

But when we come to the bosons we discover something new. Acting with the covariant derivative on the Bose superfield,

$$D_a\Phi = [E_a{}^m\partial_m + E_a{}^\mu\partial_\mu + E_a{}^{\bar{\mu}}\partial_{\bar{\mu}}]\Phi = [\partial_a + i(V_a\zeta)^\mu\partial_\mu - i(\bar{\zeta}V_a)^{\bar{\mu}}\partial_{\bar{\mu}}]\Phi\,.$$

Referring to Eq. (31) we obtain

$$\begin{aligned}2D\Phi.D\Phi &= (1+2Z_L)(1+2Z_R)[\bar{\zeta}_R\{\partial\phi + i(\phi L - R\phi)\}\zeta_L\bar{\zeta}_L \\ &\quad \times \{\partial\phi + i(\phi R - L\phi)\zeta_R\}]\end{aligned} \tag{35}$$

plus terms which disappear when integrated over property. If we concentrate on the uncharged fields held in the quartet ϕ, viz. $\phi^{0\bar{0}}$ and $\phi^{4\bar{4}}$, that occur on the diagonal (or equivalently ϕ_0 and ϕ_3), we find that

$$\begin{aligned}2\,\mathrm{Tr}[(\phi R - L\phi)(\phi L - R\phi)] &\to \frac{1}{2}g^2W^+W^-(\phi_+^2 + \phi_-^2) + \frac{1}{4}\phi_+^2(gW_3 - g'B)^2 \\ &\quad + \frac{1}{4}\phi_-^2(gW_3 + g'B)^2 - g'^2\phi_-^2\,; \quad \phi_\pm \equiv \phi_0 \pm \phi_3\,.\end{aligned}$$

In order to recover the standard vector meson masses, we must therefore take

$$\langle\phi_-\rangle = 0 \quad \text{or} \quad \langle\phi_0\rangle = \langle\phi_3\rangle \quad \text{and} \quad \langle\phi_+\rangle = v\,,$$

for the expectation values, whereupon

$$\langle 2\,\mathrm{Tr}[(\phi R - L\phi)(\phi L - R\phi)]\rangle \to \frac{1}{2}v^2 g^2 W^+W^- + \frac{1}{4}v^2(g^2+g'^2)Z^2$$
$$= \frac{e^2v^2}{2\sin^2\theta}W^+W^- + \frac{e^2v^2}{\sin^2 2\theta}Z^2\,. \qquad (36)$$

All is as it should be and the em field A remains massless.

Given these expectation values, we turn to the Yukawa interaction of the super-Bose field Φ with the super-fermion field Ψ. Before launching into this we need to remind ourselves that in order to get masses for leptons *as well as neutrinos*, we have to consider the Higgs doublet H as well as its doublet counterpart $i\tau_2 H^*$. In our context it means that we have to consider ϕ as well as $\tau_2\phi^*\tau_2$. Since we will be integrating over the ζ and the fermion pieces involve $\bar\zeta_R\zeta_L$ or $\bar\zeta_L\zeta_R$, we need to pick out matching Bose pieces. Using the acceptable combination[f] $\hat\phi = c_l\tau_2\phi^*\tau_2 + s_l\phi$, in place of ϕ, we then find that

$$-8\sqrt{2}\,\bar\Psi\Phi\Psi \supset (\bar\zeta_R\hat\phi\zeta_L + \bar\zeta_L\hat\phi^\dagger\zeta_R)(1+2Z_L)(1+2Z_R)$$
$$\cdot\,[\bar\zeta_L\zeta_R(\overline{\psi_R} + Z_R\overline{\psi'_R})(\psi_L + Z_L\psi'_L) + (R \leftrightarrow L)]\,.$$

Consequently, integrating over property produces a mixture of the two generations:

$$-16\int d^2\zeta_R..d^2\bar\zeta_L\;\bar\Psi\Phi\Psi = (2\bar\psi + \bar\psi')\hat\phi(2\psi+\psi') \equiv 5\bar{\hat\psi}\hat\phi\hat\psi\,.$$

Taking expectation values of Φ to generate a fermionic mass term, and recalling that $\langle\phi_-\rangle = 0$, the Yukawa term (including a coupling constant $\mathfrak{g}$) reduces to

$$5\mathfrak{g}(\overline{\nu_l},\bar l)\begin{pmatrix} c_l\langle\phi_+\rangle & 0 \\ 0 & s_l\langle\phi_+\rangle\end{pmatrix}\begin{pmatrix}\nu_l \\ l\end{pmatrix} = 5v\mathfrak{g}(c_l\overline{\nu_l}\nu_l + s_l\bar l l)\,. \qquad (37)$$

The other mixture $\check\psi = (-\psi + 2\psi')/\sqrt{5}$ does not acquire a mass in this model. If we were to stretch credulity and pretend we have a decent model for leptons we would be inclined to associate $\hat\psi$ with the muonic doublet and $\check\psi$ with the electronic one; but all this is academic: we really need the three color properties to corall the known leptonic generations.

The last thing to consider is the effect of spacetime curvature (through $e_m{}^a$ or g_{mn}) and of property curvature $C(Z)$ on the above results. The effect of e is very simple: it just serves to make the interactions generally covariant and we have nothing more to add to that. The effect of C enters through the Berezinian

$$\sqrt{G_{..}} = \sqrt{g_{..}}(2/\ell^2)^4 C^{-2}$$
$$\propto (1 - 2c_R - 2c_{RR}Z_{RR} + 3c_R^2 Z_R^2)(1 - 2c_L - 2c_{LL}Z_{LL} + 3c_L^2 Z_L^2)\,.$$

[f] With such a combination, $\mathrm{Tr}\,\hat\phi^2 = (c_l^2 + s_l^2)(\phi_0^2 + \phi^2) + 2c_l s_l(\phi_0^2 - \phi^2)$. So taking expectation values, $\mathrm{Tr}\,\langle\hat\phi\rangle^2 = 2(c_l^2+s_l^2)v^2 \to 2v^2$ if we interpret $c_l \equiv \cos\theta_l$, $s_l \equiv \sin\theta_l$.

It is subtler and causes mixing as well as wave function renormalization. To see what happens, consider the kinetic term of the fermions and simplify the argument by assuming the property curvature is blind to parity as we did before to recover a weak mixing angle of 30°. In that case, using the expanded

$$\begin{aligned}\sqrt{G_{..}} = \sqrt{g_{..}}(2/\ell^2)^4[1 - 2c(Z_R + Z_L) + 4c^2 Z_R Z_L \\ + (3c^2 - 2c_2)\{Z_R^2(1 - 2cZ_L) + Z_L^2(1 - 2cZ_R)\} \\ + \{(3c^2 - 2c_2)Z_L Z_R\}^2],\end{aligned} \tag{38}$$

we obtain, after ζ integration, the kinetic term ($D \equiv \partial - iV$),

$$\begin{aligned}\sqrt{g_{..}}[(1-c)\{\bar{\psi}i\gamma.D\psi + \bar{\psi'}i\gamma.D\psi' - 2c(\bar{\psi}i\gamma.D\psi' + \bar{\psi'}i\gamma.D\psi)\} \\ + c(c^2 + 2c_2)\bar{\psi}i\gamma.D\psi]\end{aligned} \tag{39}$$

which reduces to (29) when $c \to 0$. Thus the curving of property engenders source field mixing and wave function renormalization, without affecting the coupling of the gauge field V configuration. Similar conclusions apply to the Bose sector.

6. Generalizations and Conclusions

I have outlined the main consequences of a mathematical scheme for handling the "when-where-what" of events by an enlarged coordinate backdrop, part being commuting (spacetime) and part anti-commuting (property). It automatically produces a finite number of generations of elementary particles and provides a framework that unifies gravity with the other forces of nature. We treated the case of strong and electromagnetic interactions SU(3) × U(1) corresponding to three chromicity and one electricity property, making for a total of four P-conserving properties. Then we considered augmenting these by neutrinicity to describe electroweak theory and there we found the need to distinguish between left and right leptons. Thus the minimal number of properties ζ for encompassing the known forces is 5 (or 7 if we double up for leptonic handedness). The full story requires the use of them all and I admit to not having properly tackled that yet. It is a daunting business as you have seen from the calculations presented earlier. We went on to show that if the property curvature coefficients respect parity — which befits semiclassical gravity at any rate — the weak mixing angle must equal 30° to guarantee gravitational universality. Also we proved that the simplest generalization of the standard electroweak model resulted in *two* lepton generations, one massive and one massless, and in addition was able to reproduce what we know about vector masses. We remain nonplussed as to how to constrain the coefficients c_n which curve property and we are still searching for a principle that will do the job.

To fully handle the complete SU(3) × SU(2)$_L$ × U(1) gauge group, rather than bits and pieces, will require more calculational acrobatics and is left for future research. Suffice it to say that we have come across obstacles and have so far circumvented them all. Whether we will be able to overcome looming problems is quite another matter: it may well turn out that the predictions which emerge will not be able to withstand experimental scrutiny. We have set our sights on reproducing the Standard Model, with the particle generations automatically catered for. If this succeeds, one can look farther afield, seeing as we have barely scratched the surface of the scheme. A left–right symmetric picture beckons; sterile states that do not interact with the basic constituents exist aplenty in the expansions of Ψ and Φ and, if we think fancifully, may have connections to dark matter; finally the quantization via BRST seems to find a natural place in our framework since it introduces anti-commuting scalar variables attached to the ghost fields, leading to an Sp(2) translation group. On a more cautionary note, the future may judge the entire approach as being completely misguided; after all, just one ugly fact can slay a beautiful hypothesis. The history of physics is littered with such failures. If so, the present scheme can be buried with lots of other valiant attempts in the graveyard of failed theories, but its ghost may linger awhile.

Acknowledgments

I wish to express my thanks to Dr. Paul Stack for his computational wizardry in Mathematica and his numerous accurate contributions to this topic. If there are any errors in this paper they are entirely my own. Also I am indebted to Dr. Peter Jarvis for his insights and encouragement over the years. Finally, I would like to record the generous support I have received from the organizers of this splendid meeting.

References

1. R. Delbourgo, *Int. J. Mod. Phys. A* **28**, 1330051 (2013).
2. H. Georgi and S. Glashow, *Phys. Rev. Lett. B* **32**, 438 (1974).
3. H. Fritzsch and P. Minkowski, *Ann. Phys.* **93**, 193 (1975).
4. R. Delbourgo, P. D. Jarvis and R. C. Warner, *Aust. J. Phys.* **44**, 135 (1991).
5. R. Delbourgo and P. D. Stack, *Int. J. Mod. Phys. A* **29**, 50023 (2014).
6. P. D. Stack and R. Delbourgo, *Int. J. Mod. Phys. A* **30**, 1550005 (2015).
7. R. Delbourgo and P. D. Stack, *Mod. Phys. Lett. A* **31**, 1650019 (2016).
8. P. D. Stack and R. Delbourgo, *Int. J. Mod. Phys. A* **30**, 1550211 (2015).
9. F. A. Berezin, *Commun. Math. Phys.* **40**, 153 (1975).
10. B. S. DeWitt, *Phys. Rep.* **19**, 295 (1975).
11. S. L. Glashow, *Nucl. Phys.* **22**, 579 (1961).
12. S. Weinberg, *Phys. Rev. Lett.* **19**, 1264 (1967).
13. A. Salam, *Eighth Nobel Symposium*, ed. N. Svartholm (Almquist Wiksell, 1968).
14. R. Delbourgo and P. D. Stack, *Int. J. Mod. Phys. A* **30**, 1550095 (2015).
15. S. Dimopoulos and D. E. Kaplan, *Phys. Lett. B* **531**, 127 (2002).
16. L. E. Ibanez, *Phys. Lett. B* **303**, 65 (1993).

Part M
REVIEWS

M.0. Commentaries M

There are just two reviews in this section. The first is a review, written in 1975, of various ways for handling the infinities arising in quantum field theory; little has changed since then except for the inclusion of zeta-function (Phys. Rev. D40, 436 (1989)) and regularisation by dimensional reduction (Nucl. Phys. B167, 479 (1980)) formalisms. The second article concerns the generalisation of the Schwinger-Gell-Mann-Levy sigma model from nucleons to quarks, with the extra condition that the scalar masses are dynamically generated as quark-antiquark bound states, in the spirit of the Nambu-Jona-Lasinio model; thus zero wave function renormalisation constants apply and serve to reduce the number of parameters needed to describe the system.

The reader is advised to consult the reputable review journals, such as Reviews of Modern Physics, Physics Reports, Fortschritte der Physik, etc. in order to investigate more recent developments.

M.1 Delbourgo R., How to deal with infinite integrals in quantum field theory. Rep. Prog. Phys. 39, 345-399 (1976)

M.2 Scadron M.D., Rupp G. and Delbourgo R., The quark-level linear σ model. Fortschritte der Phys. 61, 994-1027 (2013)

Article [M.1] deals with several treatments of infinities, both infrared and ultraviolet, and how they come about: analytic regularisation, dimensional continuation and Pauli-Villars methods are covered. The main issue is how to preserve vector gauge invariance when handling these infinite terms. Incidentally, a good review of the dimensional method is given by Leibbrandt (Rev. Mod. Phys. 47, 849 (1975)). Review [M.2] is a comprehensive treatment of scalar mesons in the context of the σ model and how this affects observable results; the only scales appearing in this highly constrained approach are the pion decay constant and the current quark masses.

How to deal with infinite integrals in quantum field theory

R DELBOURGO

Blackett Laboratory of Physics, Imperial College, Prince Consort Road, London SW7 2BZ

Abstract

Infinite integrals arising in perturbative expansions to quantum field theory have to be defined by means of a regularization procedure before they can be cancelled by a renormalization of the physical parameters in the theory. After a rapid survey of traditional regularization schemes we describe fairly recent developments relying on point-splitting, non-polynomial interactions, analytic regularization and dimensional continuation. Among various critical tests of the schemes we consider vacuum polarization, scattering in an external field and the axial anomaly.

This review was completed in March 1976.

Rep. Prog. Phys. 1976 **39** 345–399

Contents

1. Introduction

Soon after the initial formulation of quantum electrodynamics came the recognition that, like its classical counterpart, the theory was beset by infinities, and that these infinities had to be properly understood before any calculation could be considered reliable. Classical ideas about self-energy effects for point-charged particles pointed the way towards a consistent treatment of the infinities and sometime later a renormalization programme was developed whereby the infinities of electrodynamics are absorbed into a redefinition of the physical parameters (mass, charge) of the fully quantized theory. Central to the success of the programme is the fact that in electrodynamics the infinities at any given order in perturbation theory do not proliferate and are never worse than quadratic—the theory is 'renormalizable'. Subsequently, a large class of theories were discovered in which the infinities were renormalizable in that sense and were thus amenable to reliable computations in perturbation theory. More recently, this limitation to renormalizable quantum field theory has proved a powerful constraint in constructing acceptable models of interacting particles which unify electromagnetic and weak (and possibly strong) couplings. However, it should be pointed out that non-renormalizable theories are disregarded only because perturbation expansions in these cases are useless, and for no other reason.

In carrying out the renormalization programme it is crucial to isolate the infinities in a sensible, systematic way; that is, we have to make the infinite integrals finite. The objective is achieved by a 'regularization' procedure whereby one (or several) parameters M are introduced which by some means render the integrals finite; in the limit as $M^{-1}\to 0$ the infinities are recovered. The idea is not so very different from the way that improper functions like the Dirac δ function can be regarded as improper limits of ordinary functions, e.g.

$$\delta(x)=\lim_{M\to\infty}\int_{-M}^{M} e^{ipx}\,dp/2\pi=\lim_{M\to\infty}\frac{\sin Mx}{\pi x}.$$

One can devise any number of regularization procedures which isolate the infinities and it is obviously quite impossible to cover them all. Therefore we shall concentrate on the half-dozen or so procedures which have gained wide acceptance because they are relatively simple to implement; indeed, some of them even have physical meaning. As the title of the article suggests the discussion will tend to be somewhat mathematical and we will offer no apologies for what is essentially a technical exercise.

A few words about the organization of this review. For those unfamiliar with quantum field theory we have devoted §2 to explaining how the divergent Feynman integrals arise in the first place. Being essentially a summary of the theory of quantized fields it is perhaps somewhat terse in parts and we have not provided every step in deriving the various formulae. However, we have tried to make succeeding equations at least follow plausibly from one another. Also we have omitted references to what is essentially an old topic. However, our language has been couched in modern terms whereby functional derivatives with respect to external sources are taken to derive the Green's functions in the theories. This simplifies and rationalizes a lot of the more traditional formalism based on the interaction picture, etc which one finds in all but the latest textbooks. Of course, for the expert this chapter is superfluous—

a glance through it will be sufficient for understanding the notation. (Other conventions are given for completeness in an appendix.)

The subject proper begins in §3 where we pinpoint the character and origin of the 'ultraviolet' infinities as the small-distance or large-momentum behaviour of the integrands. (For 'infrared' infinities it is the small-momentum behaviour of the integrand which is instead relevant.) We introduce two traditional regularization methods, one of which operates more usefully in p space (kinetic modifications of the propagator) and the other in x space (point-splitting). In such methods the 'cutoff' parameter M^{-1} or ϵ plays a purely mathematical role and is removed at the end. However, there do exist certain models in which cutoffs appear naturally and thus play a physical role since they do *not* disappear at the end. We give two examples of this, finite-volume lattice field theory and non-polynomial interactions, each governed by its own intrinsic scale.

In the following section we outline the bare bones of renormalization theory—this tells us what to do with the regularized infinities—and calculate a number of constants which renormalize the physical parameters in a few simple models. The finite renormalized transition amplitudes are largely independent of the renormalization constants. However, in exceptional circumstances the perturbation calculations do leave behind finite anomalous corrections which survive the limit $M^{-1}\to 0$, the example of the axial anomaly being a case in point. Turning to infrared infinities, these normally cancel in electrodynamics when *all* soft-photon corrections are taken into account (internally and externally emitted). We demonstrate this cancellation in first order for scattering in an external field, using the traditional artifice of giving the photon a small mass (to isolate the infrared infinity) simply in order to compare the technique with later regularization schemes.

The last two sections contain the most recent developments, analytic and dimensional regularization techniques. Both methods are tested *vis-à-vis* vacuum polarization, the axial anomaly and the mechanism of infrared cancellation.

2. Genesis of the infinite integrals

As we have stated in the introduction the aim of this section is to provide a short self-contained synopsis of renormalizable quantum field theory, explaining the relevant steps which lead to the perturbation expansions of the transition elements or Green's functions for those readers to whom the subject is new. There exist many excellent textbooks on quantum field theory to which the reader can refer to as and when certain subtleties arise in the course of the development or when the transition from one formula to the next is perhaps abrupt.

2.1. Lagrangian field theory

Hamilton's action principle can be taken as a convenient starting point for classical particle mechanics. It stipulates that the action

$$W_{12}=\int_{t_1}^{t_2} L[q, \dot{q}; j]\, \mathrm{d}t \tag{2.1}$$

is an extremum under variations $\delta q(t)$ of the particle coordinates which vanish at the end points. $j(t)$ is the 'external source', responsible for explicit time dependence in the problem. To pass to classical field theory one replaces the dynamical coordinate

$q(t)$ by the field variable $\Phi(\boldsymbol{x}, t)$ and one writes the Lagrangian as an integral over (an appropriate region of) space of the Lagrange density:

$$L[\Phi, \dot{\Phi}; J] = \int \mathscr{L}[\Phi, \partial_\mu \Phi; J] \, \mathrm{d}^3\boldsymbol{x}.$$

Thus the action is an integral over space-time of a number of field variables Φ, their first derivatives, and external sources J:

$$W_{12} = \int_{t_1}^{t_2} \mathscr{L}[\Phi, \partial\Phi; J] \, \mathrm{d}^4x. \tag{2.2}$$

The stationary character of W leads us to a simple generalization of the usual Euler–Lagrange equations, viz

$$\delta\mathscr{L}/\delta\Phi - \partial_\mu[\delta\mathscr{L}/\delta(\partial_\mu\Phi)] = 0 \tag{2.3}$$

for each distinct field Φ. These are the 'equations of motion'. The physical content of the field theory is fully determined once the number of fields Φ and the functional form of $\mathscr{L}$ are specified. Usually $\mathscr{L}$ consists of a free part (a bilinear in the Φ) plus a number of interaction terms; it will prove flexible enough and very useful to introduce any explicit space–time dependence on the theory through the linear term $J(x)\,\Phi(x)$. Three examples below typify most of the respectable field theories.

(*a*) Suppose there is only one real scalar field ϕ. Since it is required to satisfy the Klein–Gordon equation in the absence of any interaction, we write

$$\mathscr{L} = \tfrac{1}{2}[(\partial\phi)^2 - \mu^2\phi^2] - V(\phi) - j\phi \tag{2.4}$$

where $V(\phi)$ is any function of ϕ. The resulting equation of motion is

$$(\partial^2 + \mu^2)\,\phi + V'(\phi) + j = 0. \tag{2.5}$$

A famous example is the $V = g\phi^4$ model.

(*b*) If the scalar field is complex it undergoes phase transformations and carries charge. In fact, the electrodynamics of mesons with charge e is obtained by the minimal substitution rule,

$$\begin{aligned} \partial\phi &\to \mathrm{D}\phi \equiv (\partial + \mathrm{i}eA)\,\phi \\ \partial\phi^\dagger &\to \mathrm{D}\phi^\dagger \equiv (\partial - \mathrm{i}eA)\,\phi^\dagger \end{aligned} \tag{2.6}$$

where A_μ is the (real) electromagnetic field potential. The Lagrangian

$$\mathscr{L} = \mathrm{D}\phi^\dagger.\mathrm{D}\phi - \mu^2\phi^\dagger\phi - \tfrac{1}{4}F_{\kappa\lambda}F^{\kappa\lambda} - j^\dagger\phi - \phi^\dagger j - j^\lambda A_\lambda \tag{2.7}$$

in which appears the Maxwell field

$$F_{\kappa\lambda} \equiv \partial_\kappa A_\lambda - \partial_\lambda A_\kappa \tag{2.8}$$

remains invariant (for $j = j^\dagger = j_\lambda = 0$) under the gauge transformations

$$\begin{aligned} &\phi \to \mathrm{e}^{\mathrm{i}\Lambda}\phi, && \phi^\dagger \to \mathrm{e}^{-\mathrm{i}\Lambda}\phi^\dagger, && A \to A - \partial\Lambda/e, \\ &\mathrm{D}\phi \to \mathrm{e}^{\mathrm{i}\Lambda}\mathrm{D}\phi, && \mathrm{D}\phi^\dagger \to \mathrm{e}^{-\mathrm{i}\Lambda}\mathrm{D}\phi^\dagger, && F \to F. \end{aligned} \tag{2.9}$$

Of the ensuing equations of motion,

$$\begin{aligned} (\mathrm{D}^\lambda\mathrm{D}_\lambda + \mu^2)\,\phi &= -j \\ (\partial^2\eta_{\mu\nu} - \partial_\mu\partial_\nu)\,A^\nu &= \mathrm{i}e\phi^\dagger\overleftrightarrow{\mathrm{D}}_\mu\phi + j_\mu \end{aligned} \tag{2.10}$$

the first is gauge covariant and the second is gauge invariant in the absence of external sources, since then the electromagnetic current $\mathrm{i}e\phi^\dagger\overleftrightarrow{\mathrm{D}}\phi$ is conserved. In that case $\partial.A$ is a free massless field.

(*c*) If we want to deal with spin $\frac{1}{2}$ particles carrying charge e (protons, positrons, ...) we have to substitute ϕ by a spinor field ψ and the Klein–Gordon equation by the Dirac equation. After applying the minimal substitution rules in the free Dirac equation, and including external sources, we obtain the complete spinor electrodynamics Lagrangian,

$$\mathscr{L}=\bar{\psi}[\gamma.(\tfrac{1}{2}\mathrm{i}\overleftrightarrow{\partial}-eA)-m]\,\psi-\tfrac{1}{4}F_{\kappa\lambda}F^{\kappa\lambda}-\bar{\eta}\psi-\bar{\psi}\eta-j^{\lambda}A_{\lambda}. \tag{2.11}$$

This leads to the spinor and photon field equations

$$(\mathrm{i}\gamma.\partial-m)\,\psi=e\gamma A\psi+\eta$$
$$(\partial^2\eta_{\mu\nu}-\partial_\mu\partial_\nu)\,A^\nu=e\bar{\psi}\gamma_\mu\psi+j_\mu. \tag{2.12}$$

Again, for $\eta=\bar{\eta}=j_\lambda=0$ the electromagnetic current $e\bar{\psi}\gamma\psi$ is conserved and $\partial.A$ becomes a free field.

2.2. *Quantization*

It is possible to pass from a Lagrangian to a Hamiltonian description by the conventional procedure of isolating all the dynamical fields in $\mathscr{L}$ (those that possess a time derivative) and replacing their rates of change by the canonical momenta $\Pi\equiv\delta\mathscr{L}/\delta\dot{\Phi}$ wherever they appear in the expression for $\mathscr{H}(\Phi, \boldsymbol{\partial}\Phi, \Pi; J)\equiv\Pi\dot{\Phi}-\mathscr{L}$. One may go on to express the classical Hamiltonian equations of motion in the form of Poisson brackets, e.g. $\dot{F}=(F, H)$ etc, in order to pass to the quantized theory by replacing the Poisson brackets with commutators. The fields have now to be interpreted as operators and one must take care to specify, where necessary, the order of non-commuting operators. The Lagrangian itself, like the action, is a Hermitian operator. After quantization the fundamental Poisson bracket assumes the form of an equal-time commutation relation

$$[\Phi(\boldsymbol{x}, t), \Pi(\boldsymbol{y}, t)]_{\mp}=\mathrm{i}\delta^3(\boldsymbol{x}-\boldsymbol{y}). \tag{2.13}$$

The $\mp$ index stipulates that we have to take commutators for boson fields and anticommutators for fermion fields in order to maintain the normal spin-statistics connection (otherwise local quantum field theory becomes acausal). At equal times the remaining commutators $[\Phi, \Phi]$ and $[\Pi, \Pi]$ must vanish. For the examples of §2.1 this canonical quantization works as follows.

(*a*) Here $\pi=\dot{\phi}$ and one simply gets

$$\mathscr{H}=\tfrac{1}{2}[\pi^2+(\boldsymbol{\partial}\phi)^2+\mu^2\phi^2]+j\phi+V(\phi) \tag{2.14}$$

with

$$[\phi(\boldsymbol{x}, t), \dot{\phi}(\boldsymbol{y}, t)]=\mathrm{i}\delta^3(\boldsymbol{x}-\boldsymbol{y})$$

and

$$[\phi, \phi]=[\dot{\phi}, \dot{\phi}]=0 \qquad \text{at equal times.} \tag{2.15}$$

(*b*) There is a problem in dealing with gauge vector bosons like the electromagnetic field, stemming from the fact that A_0 is not a true dynamical variable but can be eliminated from the equations of motion. Likewise, the spatial divergence $\partial.A$ is not physically relevant. In fact, only the two transverse spatial components of A are true quantum degrees of freedom to which the canonical quantization procedure applies. Nevertheless, rather than proceed non-covariantly one can make good use of the gauge invariance of the theory (for vanishing J) to pick the covariant Lorentz gauge whereby all physical matrix elements of $\partial.A$ vanish. This amounts to adding to (2.7) and (2.11) the gauge-fixing term

$$\mathscr{L}_{\text{gauge}}=\tfrac{1}{2}(\partial.A)^2 \tag{2.16}$$

and insisting on $\langle\Psi'|\partial.A|\Psi\rangle=0$ for all physical state vectors Ψ. Actually it suffices to impose $\partial.A^+|\Psi\rangle=0$ where A^+ is the positive frequency component of A (involving annihilation operators as we shall see). After including (2.16) the electromagnetic field equation (2.10) simplifies to

$$\partial^2 A_\mu=\mathrm{i}e\phi^\dagger\overleftrightarrow{\mathrm{D}}_\mu\phi+j_\mu. \tag{2.17}$$

Providing the electromagnetic current is conserved $\partial.A$ is a massless free field and the whole scheme is self-consistent.

Adopting this covariant approach, we meet a four-vector canonical momentum $\Pi^\nu=\delta\mathscr{L}/\delta\dot{A}_\nu=-\dot{A}_\nu$, as well as the (complex) momentum conjugate to the charged scalar field, $\pi=\delta\mathscr{L}/\delta\dot{\phi}=\mathrm{D}_0\phi^\dagger$, which enter into the Hamiltonian density

$$\mathscr{H}=-\tfrac{1}{2}[\Pi_\lambda\Pi^\lambda+\boldsymbol{\partial}A_\lambda.\boldsymbol{\partial}A^\lambda]+[\pi^\dagger\pi+\mathbf{D}\phi^\dagger.\mathbf{D}\phi+\mu^2\phi^\dagger\phi]+j^\dagger\phi+\phi^\dagger j+j^\lambda A_\lambda. \tag{2.18}$$

The quantization rules read

$$[\phi(\boldsymbol{x},t),\dot{\phi}^\dagger(\boldsymbol{y},t)]=\mathrm{i}\delta^3(\boldsymbol{x}-\boldsymbol{y})$$
$$[A_\mu(\boldsymbol{x},t),\dot{A}_\nu(\boldsymbol{y},t)]=-\mathrm{i}\eta_{\mu\nu}\delta^3(\boldsymbol{x}-\boldsymbol{y}) \tag{2.19}$$

with other equal-time commutators zero. The negative definite contributions to $\mathscr{H}$ coming from A_0 and Π_0 present no problem in the long run because they cancel against the longitudinal spatial components of $\boldsymbol{A}$ and $\boldsymbol{\Pi}$.

(*c*) Again, it is convenient to pick the Lorentz gauge $\partial.A=0$. There is the electromagnetic momentum field Π^μ as well as the spinor momentum

$$\pi=\delta\mathscr{L}/\delta\dot{\psi}=\mathrm{i}\bar{\psi}\gamma_0=\mathrm{i}\psi^\dagger.$$

We prefer to use $\bar{\psi}$ in place of π when working out

$$\mathscr{H}=\bar{\psi}(\tfrac{1}{2}\mathrm{i}\boldsymbol{\gamma}.\overleftrightarrow{\boldsymbol{\partial}}+m)\psi+e\bar{\psi}\gamma A\psi-\tfrac{1}{2}[\Pi_\lambda\Pi^\lambda+\boldsymbol{\partial}A_\lambda.\boldsymbol{\partial}A^\lambda]+\bar{\psi}\eta+\bar{\eta}\psi+j^\lambda A_\lambda. \tag{2.20}$$

The only non-vanishing canonical commutators are (2.19) and

$$\{\psi(\boldsymbol{x},t),\psi^\dagger(\boldsymbol{y},t)\}=\mathrm{i}\delta^3(\boldsymbol{x}-\boldsymbol{y}). \tag{2.21}$$

2.3. Free fields and propagators

When all external sources J and self sources (i.e. couplings) vanish, the fields Φ are free. If we make a Fourier decomposition of Φ, the momentum space components have to lie on the mass (m) shell and must therefore contain a factor $\delta(p^2-m^2)$. Therefore the expansions of the free scalar (ϕ), spinor (ψ) and vector (A_μ) fields read:

$$\phi(x)=\underset{p}{\mathsf{S}}\,[a(p)\,\mathrm{e}^{-\mathrm{i}p.x}+b^\dagger(p)\,\mathrm{e}^{\mathrm{i}p.x}]$$
$$\psi(x)=\underset{p}{\mathsf{S}}\sum_{\lambda=\pm\frac{1}{2}}[a(p,\lambda)\,u(p,\lambda)\,\mathrm{e}^{-\mathrm{i}p.x}+b^\dagger(p,\lambda)\,v(p,\lambda)\,\mathrm{e}^{\mathrm{i}p.x}] \tag{2.22}$$
$$A_\mu(x)=\underset{p}{\mathsf{S}}\sum_{\lambda=\pm1}[a(p,\lambda)\,\epsilon_\mu(p,\lambda)\,\mathrm{e}^{-\mathrm{i}p.x}+a^\dagger(p,\lambda)\,\epsilon_\mu^*(p,\lambda)\,\mathrm{e}^{\mathrm{i}p.x}]$$

where we introduce the notation

$$\mathop{\mathsf{S}}_p \equiv \int \mathrm{d}^4p\,\theta(p)\,\delta(p^2-m^2) \equiv \int \mathrm{d}^3\boldsymbol{p}/2(\boldsymbol{p}^2+m^2)^{1/2}(2\pi)^3$$
$$\delta^3(\boldsymbol{p},\boldsymbol{p}') = 2(\boldsymbol{p}^2+m^2)^{1/2}(2\pi)^3\,\delta^3(\boldsymbol{p}-\boldsymbol{p}') \qquad (2.23)$$

appropriate to continuously varying momenta and integrations over all space (box normalization is considered afterwards). In (2.22) u and v denote positive- and negative-energy solutions of the free Dirac equation, and ϵ are transverse polarization vectors for a massless gauge field ($\boldsymbol{p}.\boldsymbol{\epsilon}=p.\epsilon=0$); their properties are listed in the appendix. Note that for real ϕ, $a=b$.

The quantization rules (2.19) and (2.21) imply the momentum space commutation relations

$$[a(\boldsymbol{p}),\, b^\dagger(\boldsymbol{p}')] = \delta^3(\boldsymbol{p},\boldsymbol{p}'), \qquad \text{scalar}$$
$$\{a(\boldsymbol{p},\lambda),\, b^\dagger(\boldsymbol{p}',\lambda')\} = \delta^3(\boldsymbol{p},\boldsymbol{p}')\,\delta_{\lambda\lambda'}, \qquad \text{spinor} \qquad (2.24)$$
$$[a(\boldsymbol{p},\lambda),\, a^\dagger(\boldsymbol{p}',\lambda')] = \delta^3(\boldsymbol{p},\boldsymbol{p}')\,\delta_{\lambda\lambda'}, \qquad \text{real vector}$$

with other commutators vanishing. At this point the theory is said to be 'second quantized' since the field degrees of freedom, as well as the traditional observables like momentum etc, achieve the status of quantum operators quite visibly. Following the analogy with quantum mechanics for the simple harmonic oscillator where commutators similar to (2.24) were first encountered, we can set up a complete 'Fock-space' basis by acting on a 'vacuum state' with the free-field creation operators, $a^\dagger$ for particles and $b^\dagger$ for the antiparticles. For instance, with real ϕ the completeness relation of these particular basis vectors reads

$$1 = \sum_n \mathop{\mathsf{S}}_{p_1} \dots \mathop{\mathsf{S}}_{p_n} |p_1 \dots p_n\rangle \frac{1}{n!} \langle p_1 \dots p_n|$$
$$|p_1 \dots p_n\rangle \equiv a^\dagger(p_1) \dots a^\dagger(p_n)|0\rangle. \qquad (2.25)$$

The Bose or Fermi symmetry of the state is thus automatic.

Of fundamental importance to later work are the free-field vacuum expectation values,

$$\langle 0|\phi(x)\,\phi^\dagger(y)|0\rangle \equiv \mathrm{i}\Delta_+(x-y)$$
$$\langle 0|\phi^\dagger(y)\,\phi(x)|0\rangle \equiv -\mathrm{i}\Delta_-(x-y) \qquad (2.26)$$

and specially constructed combinations like

$$\langle 0|[\phi(x),\,\phi^\dagger(y)]|0\rangle = \mathrm{i}[\Delta_+(x-y)+\Delta_-(x-y)] \equiv \mathrm{i}\Delta(x-y)$$
$$\langle 0|T[\phi(x)\,\phi^\dagger(y)]|0\rangle \equiv \mathrm{i}[\theta(x-y)\,\Delta_+(x-y) - \theta(y-x)\,\Delta_-(x-y)] \qquad (2.27)$$
$$\equiv \mathrm{i}\Delta_c(x-y)$$
$$\langle 0|R[\phi(x)\,\phi^\dagger(y)]|0\rangle \equiv \mathrm{i}\theta(x-y)\,\Delta(x-y) \equiv \mathrm{i}\Delta_R(x-y).$$

These define the free-field propagators Δ_i. By their nature $\Delta_\pm$ and Δ satisfy the homogeneous equation $(\partial^2+m^2)\,\Delta_\pm=0$ whereas the time-ordered propagator Δ_c and the retarded function Δ_R are true Green's functions in that they obey the inhomogeneous equation

$$(\partial^2+m^2)\,\Delta_{c,R}(x) = -\delta^4(x).$$

We can evaluate $\Delta_\pm$ (and hence all other) functions from first principles by inserting a one-particle† state in (2.26) to get

† The multiparticle states give zero for free fields, not interacting fields.

$$\mathrm{i}\Delta_+(x-y) = \mathop{\mathsf{S}}_p \langle 0|\phi(x)|p\rangle\langle p|\phi^\dagger(y)|0\rangle = \mathop{\mathsf{S}}_p \mathrm{e}^{-\mathrm{i}p.(x-y)}$$
$$= \int \text{đ}^4p\ \delta(p^2-m^2)\ \theta(p)\ \mathrm{e}^{-\mathrm{i}p.(x-y)}. \tag{2.28}$$

The Fourier transforms and basic properties of the other Δ_i are detailed in the next section.

In terms of these propagation functions one can deduce the unequal-time free-field commutators,

$$\begin{aligned} [\phi(x), \phi^\dagger(y)] &= \mathrm{i}\Delta(x-y) \\ \{\psi(x), \bar\psi(y)\} &= \mathrm{i}(\mathrm{i}\gamma.\partial+m)\ \Delta(x-y) \equiv \mathrm{i}\ \mathsf{S}\,(x-y) \\ [A_\mu(x), A_\nu(y)] &= -\mathrm{i}\eta_{\mu\nu} \lim_{m\to 0} \Delta(x-y) \equiv -\mathrm{i}\eta_{\mu\nu}D(x-y). \end{aligned} \tag{2.29}$$

In keeping with the quantization rules,

$$\mathrm{i}\Delta(x-y) = \mathop{\mathsf{S}}_p [\mathrm{e}^{-\mathrm{i}p.(x-y)} - \mathrm{e}^{\mathrm{i}p.(x-y)}]$$

is zero for spacelike separations (equal times) and satisfies

$$\mathrm{i}\dot\Delta(\boldsymbol{x}-\boldsymbol{y}, 0) = \delta^3(\boldsymbol{x}-\boldsymbol{y}).$$

It should be pointed out that for photons

$$[\partial.A, A_\nu] = -\mathrm{i}\ \partial_\nu D \neq 0.$$

The implication is that one may not impose $\partial.A|\Psi\rangle = 0$ on physical states. In fact, the only consistent procedure is to invoke the subsidiary condition

$$0 = \mathrm{i}\ \partial.A^+|\Psi\rangle \equiv \mathop{\mathsf{S}}_k k.\epsilon(k,\lambda)\ \mathrm{e}^{-\mathrm{i}k.x}a^\dagger(k,\lambda)|\Psi\rangle$$

assuring transversality of ϵ. By hermiticity this is sufficient to make $\langle\Psi|\partial.A|\Psi'\rangle = 0$.

2.4. *External sources, vacuum-generating functional*

For the present let us set all coupling constants equal to zero and study the field theory in the presence of external sources. We concentrate on the real scalar field and then generalize to other cases. Thus we begin with

$$(\partial^2+\mu^2)\ \phi^j = -j \tag{2.30}$$

where j is a c number. A formal solution to (2.30) can be found by the well known Green's function technique,

$$\phi^j(x) = \phi(x) + \int_{-\infty}^{\infty} \Delta_{\mathrm{R}}(x-y)\ \mathrm{d}^4y j(y) = \phi(x) + \int_{-\infty}^{t} \Delta(x-y)\ \mathrm{d}^4y j(y). \tag{2.31}$$

Here $\phi^j(x)$ denotes the solution to the complete equation, a function of time through j, while $\phi(x)$ is a free field which coincides with $\phi^j(x)$ in the limit as $t\to-\infty$ when the effect of $j(x)$ fades out *by assumption*. Because (2.31) corresponds to a simple c number translation of the free field, the commutation relations are undisturbed. Indeed, we can effect this shift by means of a unitary transformation

$$\phi^j(x) = U^{-1}(t, -\infty)\ \phi(x)\ U(t, -\infty)$$
$$U(\tau_1, \tau_2) = T[\exp(-\mathrm{i}\textstyle\int_{\tau_2}^{\tau_1} \phi(y)\ \mathrm{d}^4y j(y))]$$
$$\equiv \sum_{n=0}^{\infty} \frac{(-\mathrm{i})^n}{n!} T[\textstyle\int_{\tau_2}^{\tau_1} \phi(y_1)\ \mathrm{d}^4y_1 j(y_1) \ldots \int_{\tau_2}^{\tau_1} \phi(y_n)\ \mathrm{d}^4y_n j(y_n)]. \tag{2.32}$$

It is important to insert the time-ordering factor in (2.32) to ensure the causal evolution of the system and guarantee the correctness of the formulae

$$U(\tau_1, \tau_3)\, U(\tau_3, \tau_2) = U(\tau_1, \tau_2) = U^{-1}(\tau_2, \tau_1)$$
$$\dot{U}(t, -\infty) = -\mathrm{i} \int \mathrm{d}^3\boldsymbol{x}\phi(x) j(x)\, U(t, -\infty).$$

All this means is that the eigenstates appropriate to the externally coupled field $\phi^j(x)$ evolve in time according to

$$|\Psi, t\rangle^j = U^{-1}(t, -\infty)|\Psi\rangle = T[\exp \mathrm{i} \textstyle\int_{-\infty}^{t} \phi(x)\, \mathrm{d}^4x j(x)]|\Psi\rangle \tag{2.33}$$

where Ψ are the constant eigenstates of the free-field theory. At any time t the $|\Psi, t\rangle^j$ of course form a complete set.

The time dependence of the theory is totally determined once all the 'S-matrix' transition elements are known:

$$\begin{aligned} {}^j\langle \Psi' t' | \Psi t\rangle^j &= \langle \Psi' | U(t', -\infty)\, U^{-1}(t, -\infty)|\Psi\rangle = \langle \Psi'|U(t', t)|\Psi\rangle \\ &= \langle \Psi' | T[\exp(-\mathrm{i} \textstyle\int_t^{t'} \phi(y)\, \mathrm{d}^4y j(y))]|\Psi\rangle. \end{aligned} \tag{2.34}$$

Since the basis vectors $|\Psi\rangle$ are themselves generated by acting with ϕ on the vacuum this is equivalent to knowing the free-field matrix elements

$$\langle 0|\phi(x_1') \ldots \phi(x_n')\, U(t', t)\, \phi(x_1) \ldots \phi(x_n)|0\rangle.$$

Better still, it suffices to know all the time-ordered Green's functions,

$$G_{t't}(x_1 \ldots x_n) = \langle 0|T[\phi(x_1) \ldots \phi(x_n) \exp(-\mathrm{i} \textstyle\int_t^{t'} \phi(y)\, \mathrm{d}^4y j(y))]|0\rangle. \tag{2.35}$$

The true S-matrix elements correspond to taking the limit $t' \to \infty$, $t \to -\infty$ and thereby giving the sources free rein. Hereafter we omit the labels t' and t on G in this particular limit.

In order to derive explicit closed forms for G it is useful to introduce the notion of a functional derivative. This is defined according to the formulae

$$\frac{\delta}{\delta j(x)} \int j(y)\, \mathrm{d}^4y \phi(y) = \phi(x)$$

or

$$\delta j(y)/\delta j(x) = \delta^4(x-y). \tag{2.36}$$

Also to save a lot of writing we use the abbreviation

$$\mathsf{S}\, j\phi \equiv \textstyle\int j(y)\, \mathrm{d}^4y \phi(y).$$

With these notational conveniences out of the way one gets from (2.35)

$$\begin{aligned} G(x_1 \ldots x_n) &= \mathrm{i}^n \delta^n \langle 0|T[\exp(-\mathrm{i}\, \mathsf{S}\, \phi j)]|0\rangle / \delta j(x_1) \ldots \delta j(x_n) \\ &= \mathrm{i}^n \delta^n Z[j] / \delta j(x_1) \ldots \delta j(x_n) \end{aligned} \tag{2.37}$$

where we have defined the 'vacuum-generating functional'

$$Z[j] \equiv \langle 0|T[\exp(-\mathrm{i}\, \mathsf{S}\, \phi j)|0\rangle = {}^j\langle 0\infty|0-\infty\rangle^j. \tag{2.38}$$

A closed form for Z itself is found by observing that

$$\begin{aligned} G(x_1 \ldots x_n) &= \langle 0|T[\phi(x_1) \ldots \phi(x_n)\, U(\infty, -\infty)]|0\rangle \\ &= \langle 0|T[U(\infty, t_1)\, \phi(x_1)\, U(t_1, t_2) \ldots \\ &\qquad \times U(t_{n-1}, t_n)\, \phi(x_n)\, U(t_n, -\infty)]|0\rangle \\ &= {}^j\langle 0\infty|T[\phi^j(x_1) \ldots \phi^j(x_n)]|0-\infty\rangle^j \end{aligned} \tag{2.39}$$

are nothing else but the time-ordered expectation values for the externally coupled fields. In particular,

$$G(x) = {}^{j}\langle 0\infty \,|\, \phi^{j}(x) \,|\, 0-\infty\rangle^{j} = \mathrm{i}\delta Z[j]/\delta j(x)$$

by virtue of (2.30), must obey the differential equation

$$(\partial^2 + \mu^2)\, G(x) = -j(x)\, Z[j].$$

Hence Z obeys the functional differential equation

$$(\partial^2 + \mu^2)\, \delta Z[j]/\delta j(x) = \mathrm{i}j(x)\, Z[j] \tag{2.40}$$

with the boundary value $Z(0) = 1$ from (2.38). The correct causal solution of (2.40) is

$$\begin{aligned} Z[j] &= \exp\,[-\tfrac{1}{2}\mathrm{i}\textstyle\int j(x)\, \mathrm{d}^4x \Delta_{\mathrm{c}}(x-y)\, \mathrm{d}^4y j(y)] \\ &\equiv \exp\,[-\tfrac{1}{2}\mathrm{i}\, \mathsf{S}\, j\Delta_{\mathrm{c}} j] \end{aligned} \tag{2.41}$$

from which we can obtain all G using (2.37). For instance, the two-point Green's function is given by

$$\begin{aligned} G(x_1, x_2) &= -\delta^2[\exp\,(-\tfrac{1}{2}\mathrm{i}\, \mathsf{S}\, j\Delta_{\mathrm{c}} j)]/\delta j(x_1)\, \delta j(x_2) \\ &= [\mathrm{i}\Delta_{\mathrm{c}}(x_1 - x_2) + \textstyle\int \Delta_{\mathrm{c}}(x_1 - y_1)\, \mathrm{d}^4y_1 j(y_1) \int \Delta_{\mathrm{c}}(x_2 - y_2)\, \mathrm{d}^4y_2 j(y_2)]\, Z[j] \end{aligned}$$

and can be pictorially represented as in figure 1. Similarly, the n-point Green's functions are obtained by connecting all points with causal propagator ($\mathrm{i}\Delta_{\mathrm{c}}$) lines, or by attaching them to the external sources j.

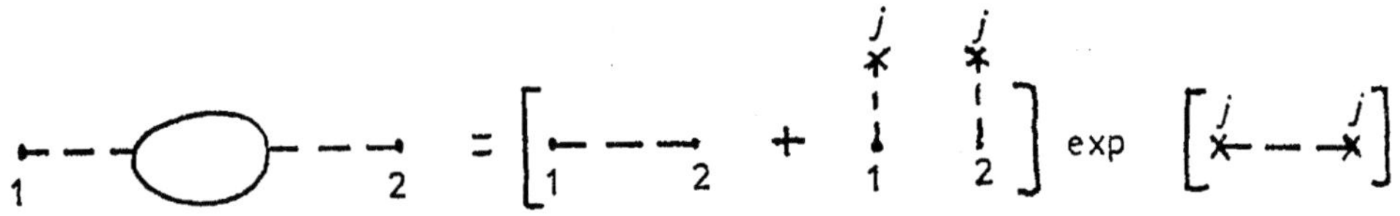

Figure 1. The two-point Green's function in the presence of an external source.

The entire formalism can be generalized to arbitrary numbers and types of field Φ by including a source $J(x)$ for each of them. To take an example, the vacuum functional in spinor electrodynamics is

$$Z[j_\mu, \bar{\eta}, \eta] = \exp \mathrm{i}\, \mathsf{S}\, (\tfrac{1}{2} j^\lambda D_{\mathrm{c}} j_\lambda - \bar{\eta} S_{\mathrm{c}} \eta) \tag{2.42}$$

and the coefficients of its functional series expansions in j_λ, η and $\bar{\eta}$ give the Green's functions of the theory. We should remark that for spinor fields the sources η, $\bar{\eta}$, or their functional derivatives, are anticommuting to guarantee the Fermi statistics, and one must therefore pay careful heed to orders of factors. Also, for photons the necessity of choosing a gauge is evident from the fact that the operator $(\partial^2\eta_{\mu\nu} - \partial_\mu\, \partial_\nu)$ on the left of (2.12) has no inverse; thus the addition of the gauge-fixing term in the Lagrangian (2.16) is vital.

2.5. Interacting fields and perturbative expansions

The quantum field theory becomes non-trivial when we begin to take account of self-interactions among the fields: spontaneous particle creation and destruction with or without the intermediary external sources. In trying to determine the Green's functions of the fully interacting theory we can mimic the arguments of §2.4 by supposing that, like J, the couplings are switched off in the infinite past causing the theory to become free in that limit. For simplicity we shall consider first the simple scalar field model (2.4). Following much the same steps as before, it is fairly plausible to quote the unitary operator which relates ϕ^j to its free-field limit ϕ in the infinite past, viz

$$U(t,t')=T[\exp(-\mathrm{i}\int_{t'}^{t}\{V[\phi(y)]+j(y)\,\phi(y)\}\,\mathrm{d}^4y] \qquad (2.43)$$

in the limit $t'\to-\infty$. The unitary S-matrix elements

$$S_{\Psi'\Psi}=\lim_{t,-t'\to\infty}\langle\Psi't'|\Psi t\rangle$$

are again completely determined from the time-ordered Green's functions (2.39) arising in the functional series expansion of the vacuum transition amplitude

$$Z[j]=\sum_{n=0}^{\infty}(-\mathrm{i})^n\int\mathrm{d}^4x_1\ldots\mathrm{d}^4x_n j(x_1)\ldots j(x_n)\,G(x_1\ldots x_n)/n! \qquad (2.44)$$

in accordance with (2.37). However, the new vacuum functional

$$Z[j]=\langle 0|T[\exp(-\mathrm{i}\,\mathsf{S}\{V+\phi j\})]|0\rangle \qquad (2.45)$$

involves the ***whole interaction Hamiltonian*** in the exponent. Unfortunately, it is now no longer possible to evaluate Z in closed form as in (2.41). ***Faute de mieux*** one is obliged to resort to perturbation methods which we shall now describe.

Within (2.35) $\phi(x)$ is obtained by functionally differentiating with $\delta/\delta j(x)$. Therefore it is fairly obvious that we can re-express (2.45) in the form

$$Z[j]=\exp[-\mathrm{i}\,\mathsf{S}\,V(\mathrm{i}\delta/\delta j)]\langle 0|T[\exp-\mathrm{i}\,\mathsf{S}\,\phi j]|0\rangle$$
$$\equiv\sum_N\frac{(-\mathrm{i})^N}{N!}\left[\mathsf{S}\,V\left(\mathrm{i}\frac{\delta}{\delta j}\right)\right]^N\exp[-\tfrac{1}{2}\mathrm{i}\,\mathsf{S}\,j\Delta_c j] \qquad (2.46)$$

an infinite series in the interaction. Since V normally contains a coupling constant we thereby get a power series expansion in the coupling. To understand the significance of (2.46) take the case $V(\phi)=g\phi^4$. Every time V operates as $g(\delta/\delta j)^4$ it joins together four j extremities occurring in the primitive vacuum functional (2.41). Thus we can associate a four-point 'vertex' with this particular V. Carrying this out to second order in g we obtain

$$Z[j]=1+3\mathrm{i}g\Delta_c^2(0)\int\mathrm{d}^4x-24g^2\int\mathrm{d}^4x\,\mathrm{d}^4y\,\Delta_c^4(x-y)$$
$$+\frac{1}{2!}(3\mathrm{i}g\Delta_c^2(0)\int\mathrm{d}^4x)^2-72g^2\Delta_c^2(0)\int\mathrm{d}^4x\,\mathrm{d}^4y\,\Delta_c^2(x-y)+\mathrm{O}(g^3,j^2). \quad (2.47)$$

This has been expressed diagrammatically in figure 2, where in each graph one has to integrate out the space–time position of each vertex. (Note than $Z(0)$ carries at least the infinite factor $\int\mathrm{d}^4x=\delta^4(0)$—the volume of space–time—and $\Delta_c(0)$, so an expression like (2.47) has only formal meaning.)

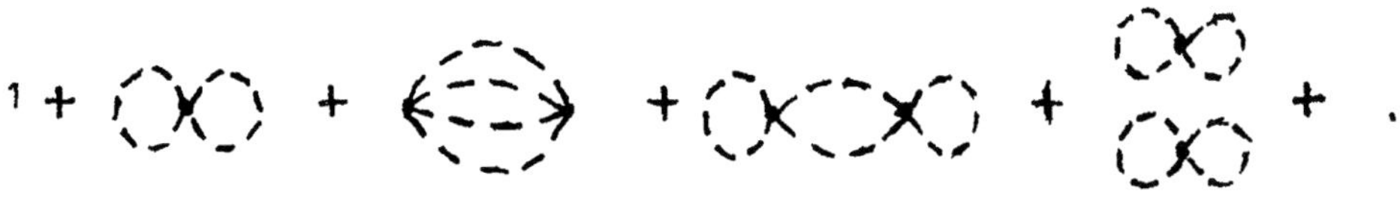

Figure 2. The vacuum-generating functional $g\phi^4$ theory.

The two-point G is derived from (2.46) by differentiating twice with respect to j. In essence, this corresponds to breaking a line in the series of diagrams of figure 2 in all possible ways, and then multiplying the answer into the vacuum graphs themselves. This is depicted in figure 3 where, again, one has to integrate out the position of each V vertex. The 'connected' Green's function $C(x, y)$ is obtained by simply dividing

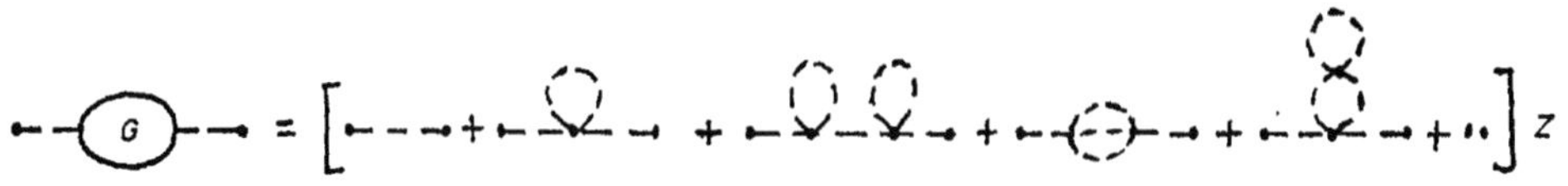

Figure 3. The two-point Green's function in $g\phi^4$ theory.

out the vacuum-to-vacuum amplitude. This factorizing out of Z can be done as well for the general connected Green's function $C(x_1 \ldots x_n)$. At a formal level it is automatically achieved if one uses

$$\mathrm{i}W[j] = \ln Z[j]$$
$$C(x_1 \ldots x_n) = \mathrm{i}^{n+1}\delta^n W[j]/\delta j(x_1) \ldots \delta j(x_n). \tag{2.48}$$

2.6. Feynman graphs in simple models, Ward identities

Let us continue by computing some typical Green's functions for the more physical example of electrodynamics. Again as our calculational tool we introduce external sources J. The J will disappear at the end if one wants the transition amplitudes of the purely self-interacting theory. In scalar electrodynamics

$$Z[j] = \langle 0|T[\exp -\mathrm{i} \textstyle\int (eA^\mu \phi^\dagger(\overleftrightarrow{\partial}_\mu + \mathrm{i}eA_\mu)\phi + j^\dagger\phi + \phi^\dagger j + j^\lambda A_\lambda)]|0\rangle$$
$$= \exp\left[-\mathrm{i}e \int \frac{\delta}{\delta j_\mu}\frac{\delta}{\delta j}\left(\overleftrightarrow{\partial}_\mu + \mathrm{i}e\frac{\delta}{\delta j_\mu}\right)\frac{\delta}{\delta j^\dagger}\right] \exp\left[\mathrm{i}\int (\tfrac{1}{2}j^\lambda D_c j_\lambda - j^\dagger \Delta_c j)\right] \tag{2.49}$$

gives the collection of Feynman graphs shown in figure 4. These are found by joining up the vertices $\mathrm{i}e\phi^\dagger \overleftrightarrow{\partial}\phi A$ (emanating three lines) and $e^2A^2\phi^\dagger\phi$ (emanating four lines) through contractions with photon propagators $\mathrm{i}D_c$ (wavy lines) or meson propagators

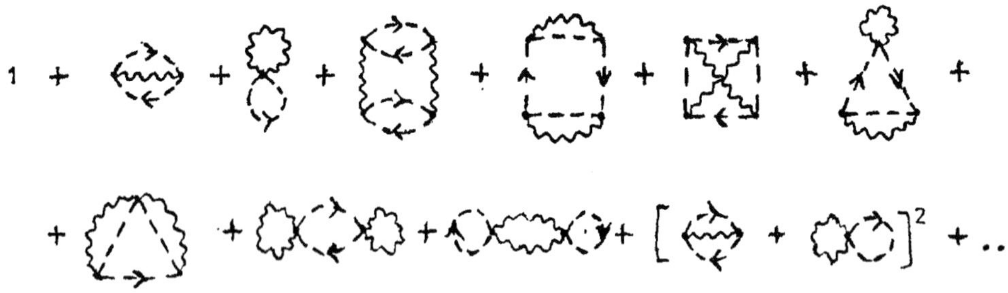

Figure 4. The vacuum functional in scalar electrodynamics.

$\mathrm{i}\Delta_c$ (broken lines). The arrow on the complex field propagators is to remind us of the sequence $\langle\phi\phi^\dagger\rangle$ associated with the charge carried by the line. Observe that Z or W is given as an infinite power series in the fine-structure constant $\alpha = e^2/4\pi$, rather than e.

The connected two-point Green's functions (the so-called 'complete or dressed propagators') for mesons and photons are found by applying $\mathrm{i}^2\delta^2/\delta j^\dagger(x)\,\delta j(y)$ and

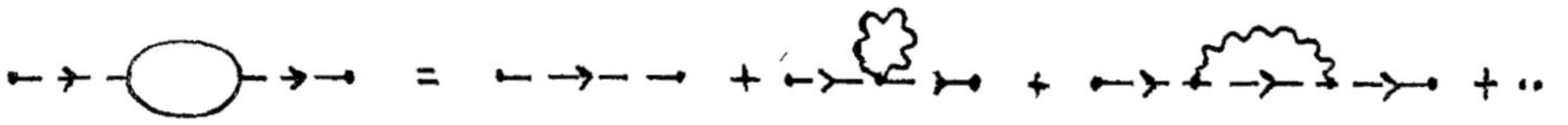

Figure 5. The meson propagator in scalar electrodynamics.

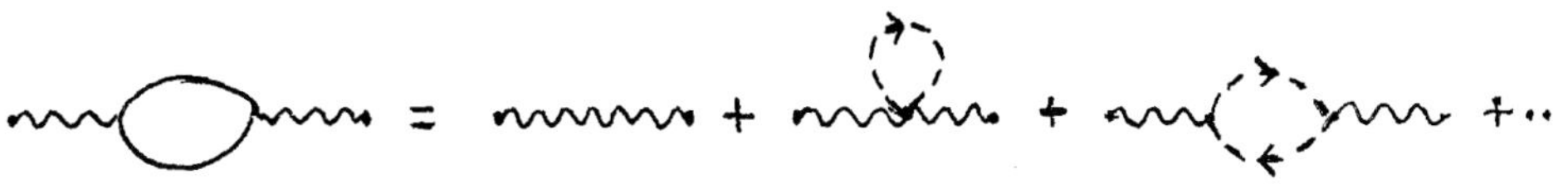

Figure 6. The photon propagator in scalar electrodynamics.

$i^2\delta^2/\delta j_\mu(x)\,\delta j_\nu(y)$ on (2.49) and discarding the disconnected graphs (vacuum amplitude factor). They can be obtained by cutting a meson or photon line in figure 4. To order α the results are summarized in figures 5 and 6, which quantitatively† read:

$$\begin{aligned}\Delta_c'(x_1-x_2)=&\,\Delta_c(x_1-x_2)+4ie^2\int\Delta_c(x_1-y)\,d^4y\,\Delta_c(y-x_2)\,D_c(0)\\&+ie^2\int\{\Delta_c(x_1-y_1)\overleftrightarrow{\partial}_{y_{1\mu}}\Delta_c(y_1-y_2)\overleftrightarrow{\partial}_{y_2}{}^{\mu}\Delta_c(y_2-x_2)\}\,d^4y_1\\&\times d^4y_2\,D_c(y_1-y_2)+\dots \qquad (2.50)\end{aligned}$$

$$\begin{aligned}D_{c\mu\nu}'(x_1-x_2)=&-\eta_{\mu\nu}D_c(x_1-x_2)-2ie^2\eta_{\mu\nu}\int D_c(x_1-y)\,d^4y\,D_c(y-x_2)\,\Delta_c(0)\\&-ie^2\int D_c(x_1-y_1)\,d^4y_1\{\Delta_c(y_1-y_2)\overleftrightarrow{\partial}_{y_{1\mu}}\Delta_c(y_2-y_1)\overleftrightarrow{\partial}_{y_{2\nu}}\}\\&\times d^4y_2\,D_c(y_2-x_2)+\dots. \qquad (2.51)\end{aligned}$$

The three-point meson–photon vertex $C_\mu(x_1, x, x_2)$ is found by a triple functional derivative $i^3\delta^3/\delta j^\dagger(x_1)\,\delta j_\mu(x)\,\delta j(x_2)$ and basically corresponds to attaching a photon line in all possible ways to the diagrams of figure 5. In the ensuing diagrams (figure 7) one notices that it is possible to factorize out complete propagators from C and write

$$C_\lambda(x_1, x, x_2)=i\int\Delta_c'(x_1-y_1)\,\Gamma_\lambda(y_1, y, y_2)\,\Delta_c'(y_2-x_2)\,D_c'(y-x)\,d^4y_1\,d^4y\,d^4y_2 \quad (2.52)$$

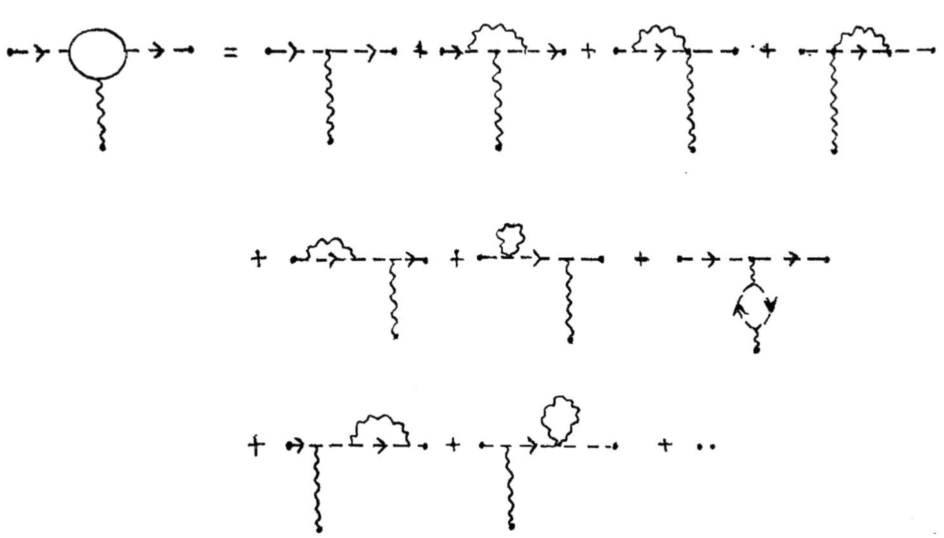

Figure 7. The photon–two-meson vertex in scalar electrodynamics.

† Note that the graphs $\langle 0|A_\mu|0\rangle$ vanish by Lorentz invariance.

with all the relevant physics being contained in the 'proper vertex' Γ_λ. This is a general result. We can always factor out complete propagators from the connected Green's functions $C(x_1 \ldots x_n)$ to leave us with the 'proper' vertex functions $\Gamma(y_1 \ldots y_n)$ whose Fourier transforms (on shell) yield directly the transition amplitudes $\langle \ldots p_n, \infty | p_1, \ldots, -\infty \rangle$, upon contraction with external wavefunctions. The translational invariance of the theory implies that, like C, Γ is a function of differences of space–time arguments and this automatically leads to the energy-momentum conserving factor $\delta^4(p_1 + \ldots - p_n)$ in the Fourier transforms.

With spinors instead of scalars the electrodynamics Feynman diagrams are fewer in number since there is no basic four-point vertex (joining two fermions with two photons) to worry about. Now the vacuum functional is simply

$$Z[j] = \exp\left[-ie \,\mathsf{S}\, \frac{\delta}{\delta\eta} \gamma_\mu \frac{\delta}{\delta j_\mu} \frac{\delta}{\delta\bar{\eta}}\right] \exp\left[i \,\mathsf{S}\, (\tfrac{1}{2} j^\lambda D_c j_\lambda - \bar{\eta} S_c \eta)\right]. \qquad (2.53)$$

The graphs are pretty much the same as figures 4, 5, 6 and 7 of scalar electrodynamics with the notable omission of 'tadpole' terms $\Delta_c(0)$ and $D_c(0)$ and 'seagull' diagrams. In place of (2.50) and (2.51) the complete fermion and photon propagators read

$$S_c'(x_1 - x_2) = S_c(x_1 - x_2) + ie^2 \int S_c(x_1 - y_1)\, \gamma_\mu S_c(y_1 - y_2)\, \gamma^\mu S_c(y_2 - x_2) \times D_c(y_1 - y_2)\, d^4y_1\, d^4y_2 + \ldots \qquad (2.54)$$

$$D_{c_{\mu\nu}}'(x_1 - x_2) = -\eta_{\mu\nu} D_c(x_1 - x_2) + ie^2 \int D_c(x_1 - y_1)\, d^4y_1 \times \mathrm{Tr}\, [S_c(y_1 - y_2)\, \gamma_\mu S_c(y_2 - y_1)\, \gamma_\nu]\, d^4y_2 D_c(y_2 - x_2) + \ldots \qquad (2.55)$$

to order α. The characteristic 'closed-loop' factor of -1 which occurs in (2.55) is due to the anticommutativity of η derivatives whenever a spinor trace needs to be taken within a Green's function.

We conclude this section with a brief discussion about Ward identities—these play an important part later in deciding upon the suitability of particular regularization schemes. Gauge theories, such as electrodynamics, are invariant under a series of local gauge transformations like (2.9), for zero external sources. Because the vacuum states carry no quantum numbers the vacuum functional Z is itself gauge invariant† even in the presence of sources, i.e. $\delta Z/\delta\Lambda(x) = 0$ for an infinitesimal gauge transformation $\delta\Lambda(x)$. Now under a gauge variation, only the source terms and the gauge-fixing term (2.16) are altered. Since

$$\delta\phi = i\delta\Lambda\phi, \qquad \delta\phi^\dagger = -i\delta\Lambda\phi^\dagger$$

and

$$\delta A_\mu = -\partial(\delta\Lambda)/e, \qquad \delta(\partial . A)^2 = -2(\partial . A)\, \partial^2(\delta\Lambda)/e$$

we obtain the relation

$$0 = \delta Z/\delta\Lambda = \langle 0 | T[(j^\dagger\phi - \phi^\dagger j - i\partial . j/e - i\, \partial^2\, \partial . A/e) \exp i\, \mathsf{S}] | 0 \rangle$$

which is interpreted as the functional differential equation

$$\left(iej^\dagger \frac{\delta}{\delta j^\dagger} - iej \frac{\delta}{\delta j} - i\, \partial^\mu j_\mu + \partial^2\, \partial_\mu \frac{\delta}{\delta j_\mu}\right) Z[j] = 0. \qquad (2.56)$$

† One can prove this statement in the functional integral formulation, $Z = \int (d\Phi) \exp(-i\int(\mathscr{H} + J\Phi))$, from the gauge invariance of the integration measure $(d\Phi)$.

By applying further $\delta/\delta J$ derivatives on (2.56) and setting $J=0$ at the end one obtains the Ward–Takahashi identities which express the original gauge invariance of the source-free theory, identities which are, in principle, held to be valid to all orders in perturbation theory, providing the regularization procedure which gives meaning to the formal expressions for the Green's functions respects the gauge symmetry. For example, if one operates on (2.56) with $\delta^2/\delta j(y)\,\delta j^\dagger(z)$ there results the identity

$$\partial_x{}^2\,\partial_x{}^\mu G_\mu(x;y,z)=e[\delta^4(x-z)\,G(x,y)-\delta^4(x-y)\,G(z,x)] \qquad (2.57)$$

that relates the three-point meson–photon vertex to the two-point meson propagator. Similarly, for higher-point meson Green's functions one can appreciate how the gauge identities relate the divergence of the connected $(n+1)$-point function to differences between connected n-point functions with one photon line deleted. Where there are no external charged particles the situation is even simpler; if we act on (2.56) with $\delta/\delta j_\nu$ there follows

$$\partial_x{}^2\,\partial_{x_\mu}\delta^2 Z/\delta j_\mu(x)\,\delta j_\nu(y)=\mathrm{i}\,\partial_x{}^\nu\delta^4(x-y)=-\mathrm{i}\,\partial_x{}^\nu\,\partial^2 D_\mathrm{c}(x-y)$$

or $\qquad (2.58)$

$$\partial_x{}^2\,\partial_x{}^\mu[D_{\mathrm{c}\mu\nu}{}'(x-y)-D_{\mathrm{c}\mu\nu}(x-y)]=0$$

i.e. the transversality of all quantum corrections to the photon propagator. Similarly, one may prove that purely photon† proper Green's functions are divergenceless at every photon leg, independently of the e^2 perturbation expansion.

2.7. Power counting and infinities

The perturbative expansions (2.47), (2.50), (2.55) etc are meaningless as they stand, for they are riddled with infinities. Prominent among these are $\delta^4(0)$ and the tadpole

$$\mathrm{i}\Delta_\mathrm{c}(0)=\lim\,[-1/4\pi^2(x^2-\mathrm{i}0)]=\mathrm{i}\int \bar{\mathrm{d}}^4p(p^2-m^2+\mathrm{i}0)^{-1} \qquad (2.59)$$

which is visibly infinite when $x\to 0$, or at the upper limit $p\to\infty$ of the momentum integral. Another typical infinity is more disguised: the products of propagators taken between the same space–time points, for example

$$\int \mathrm{d}^4x\,\mathrm{e}^{\mathrm{i}p\cdot x}[\Delta_\mathrm{c}{}^4(x)\quad\text{or}\quad\Delta_\mathrm{c}{}^3(x)\quad\text{or}\quad\Delta_\mathrm{c}{}^2(x)] \qquad \text{in } g\phi^4 \text{ theory}$$
$$\int \mathrm{d}^4x\,\mathrm{e}^{\mathrm{i}p\cdot x}[D_\mathrm{c}(x)\,\partial\,\partial\Delta_\mathrm{c}(x)\quad\text{or}\quad\partial\Delta_\mathrm{c}(x)\,\partial\Delta_\mathrm{c}(x)] \qquad \text{in scalar ED}$$
$$\int \mathrm{d}^4x\,\mathrm{e}^{\mathrm{i}p\cdot x}[D_\mathrm{c}(x)\,S_\mathrm{c}(x)\quad\text{or}\quad S_\mathrm{c}(x)\,S_\mathrm{c}(x)] \qquad \text{in spinor ED.}$$

The problem here is that for integer $n>1$, the small-distance behaviour $(x\to 0)$ of $\Delta_\mathrm{c}{}^n(x)\sim(1/x^2)^n$ is not $\int\mathrm{d}^4x$ integrable. We designate this troublesome short-distance behaviour of products of propagators (or for that matter field operators) as symptomatic of an 'ultraviolet' infinity, since in momentum space it translates into a non-integrable high-frequency dependence $(p\to\infty)$ of the Fourier transform of $\Delta_\mathrm{c}{}^n$.

The large-distance behaviour of the integrands can sometimes lead to another kind of divergence, the 'infrared infinity'. This happens in the mass shell limit of Green's functions due to the emission of low momentum massless particles (soft photons). A

† In usual electrodynamic theories where charge conjugation invariance is respected the number of photon lines is necessarily even—Furry's theorem.

rough indication of the existence of such a singularity is provided by the first-order spinor propagator correction due to virtual emission and re-absorption of a photon,

$$\int S_c(x)\, D_c(x)\, e^{ip.x}\, d^4x.$$

We are already aware of the ultraviolet difficulties as $x \to 0$. Consider in contrast $x \to \infty$. In this limit we may crudely set $S_c(x) \sim e^{-mr}$, $D_c(x) \sim 1/r^2$, $e^{ip.x} \sim e^{r\sqrt{p^2}}$ where $r^2 = -x^2$ and we neglect the angular part of $p.x$. Evidently for $p^2 = m^2$ the exponential damping is removed. This is the signal for an infrared infinity, and a more careful evaluation does indeed confirm its existence.

The aim of this review is to outline various regularization procedures which attempt to define the otherwise meaningless perturbation graphs. Any such procedure seeks to modify the Δ functions by introducing auxiliary regularization parameters, which may or may not possess physical significance. In the limit as the parameters disappear the infinities rear up again. However, in devising a regularization technique one tries as far as possible to respect the initial symmetries and invariances of the Lagrangian; if this can be done the basic properties of the classical theory are not destroyed by the quantum radiative corrections and the regularization may be deemed successful insofar as it defines the 'correct' infinities. In exceptional circumstances one can actually show that no regularization will respect the invariance in question, and in those cases the consequences of regularization become quite physical, leading to the so-called quantum 'anomalies'.

Naturally, not all perturbation graphs are superficially divergent; indeed, for renormalizable models like $g\phi^4$ and scalar and spinor electrodynamics the infinities prove to be the exception rather than the rule and for that reason can be treated systematically by a renormalization programme. When the Feynman graphs are anyway finite the regularization process becomes irrelevant because the integrals remain well defined as the regularization parameters disappear. Thus the only Green's functions which have to be treated carefully are the ones with potential infinities. It is here that we can make good use of Dyson's power-counting formula; it is tailor-made for determining the large-momentum behaviour of the G and thereby diagnosing any incipient infinities. The formula is found for $g\phi^4$ theory, say, by noting the topological relation

$$E + 2I = 4N$$

where N is the number of (four-point) vertices, equal to the order of g, I is the number of internal propagators connecting such vertices and E is the number of external legs of the connected Green's functions. Since ultimately we have to perform $(N-1)$ integrations—the final Nth integration just expresses energy-momentum conservation—over Δ products we have, crudely,

$$\prod_{i=1}^{N-1} \int d^4y_i\, e^{ik_i.y_i}[\Delta(y)]^I \sim \int (d^4y)^{N-1}\, e^{ik.y}(y^{-2})^I \sim k^{4-E}$$

as the large-k behaviour of the integral, by purely dimensional reasoning. Thus for $E > 4$ the k dependence damps out nicely. On the other hand, for $E \leqslant 4$ the small-distance behaviour of the integrand is dangerous. For obvious reasons the terminology

$$\int d^4y \approx \text{quartic infinity } [E=0]$$
$$\int d^4y/y^2 \approx \text{quadratic infinity } [E=2]$$
$$\int d^4y/y^4 \approx \text{logarithmic infinity } [E=4]$$

has come to be used. Superficially, only Green's functions with 0, 2 and 4 points must be remedied by regularization.

The analogous power-counting formulae for spinor electrodynamics are the topological relations

$$N=\tfrac{1}{2}E_\psi+I_\psi=E_A+2I_A$$

and the ultraviolet behaviour

$$\int(\mathrm{d}^4y)^{N-1}[\Delta(y)]^{I_A}[S(y)]^{I_\psi}\,\mathrm{e}^{\mathrm{i}k.y}\sim\int(\mathrm{d}^4y)^{N-1}\,y^{-2I_A}y^{-3I_\psi}\,\mathrm{e}^{\mathrm{i}k.y}\sim k^{4-E_A-3E_\psi/2}$$

since for small y the spinor propagator $S(y)\sim\gamma.\partial\Delta\sim\gamma.y/y^4$. Of the five primitive infinities (i) $E_A=E_\psi=0$ (vacuum graphs), (ii) $E_A=2$, $E_\psi=0$ (photon propagator), (iii) $E_A=0$, $E_\psi=2$ (spinor propagator), (iv) $E_A=1$, $E_\psi=2$ (spinor–photon vertex), (v) $E_A=4$, $E_\psi=0$ ($\gamma\gamma$ scattering), the first is immaterial to the physics because it is always divided out, the third and fourth are logarithmic because there are never any linear infinities by symmetrical integration, the second is only logarithmic and the fifth is zero by gauge invariance. Scalar electrodynamics is rather similar except for the additional infinities in photon–meson scattering (related to the photon–meson vertex by gauge invariance) and meson–meson scattering; the latter requires a primitive $g\phi^4$ vertex for consistency of renormalization.

We must emphasize that the power-counting formula as given above is only capable of characterizing the 'superficial divergences'. It is quite often true that a superficially convergent Green's function can contain infinities due to divergent sub-integrations. These sub-infinities have to be consistently treated by the renormalization scheme before one can believe any of the calculations of superficially finite diagrams.

3. Isolation of the infinities

In this section we shall examine more closely the origin of the infinities and in the process we will describe the more traditional methods of regularizing them, which rely on *ad hoc* modifications of the propagators. We shall also sketch the more sophisticated point-splitting techniques. The final section discusses briefly some physical cases in which the propagator modifications appear naturally owing to the existence of intrinsic scales.

3.1. The Δ functions

Because all the perturbative expansions involve integrals over sums and products of Δ_c, our examination of the infinities must begin with a careful study of these functions. Even in many non-perturbative schemes one generally makes certain assumptions about analyticity which sooner or later require knowledge of the functions (Schwinger 1949a, Rivier 1949)

$$\Delta_i(x)=\int_{C_i}\bar{\mathrm{d}}^4p\;\mathrm{e}^{-\mathrm{i}p.x}/(p^2-m^2)$$

appropriate to a spinless particle. The different Δ, labelled by i, are described by choices of integration contour C_i around the singularities $p_0=\pm(\mathbf{p}^2+m^2)^{1/2}$ in the p_0 plane. Since

$$(\partial^2+m^2)\,\Delta_i(x)=-\int_{C_i}\bar{\mathrm{d}}^4p\;\mathrm{e}^{-\mathrm{i}p.x}$$

it follows that when the path is located in a finite region of the p_0 plane we are dealing with a homogeneous function,

$$(\partial^2+m^2)\,\Delta_i(x)=0$$

whereas when the contour passes through the point at infinity, the function is an inhomogeneous (Green's) function

$$(\partial^2+m^2)\,\Delta_i(x)=-\delta^4(x).$$

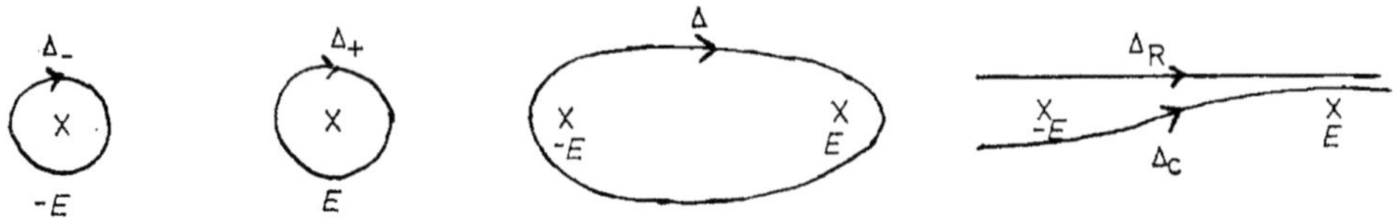

Figure 8. Contour integrals for some important Δ functions.

In figure 8 we draw the contours which define the two fundamental homogeneous functions Δ_+ and Δ_- as well as the paths for three other useful functions $\Delta(x)$, $\Delta_R(x)$ and $\Delta_c(x)$,

$$\Delta(x)=\Delta_+(x)+\Delta_-(x), \qquad \Delta_R(x)=\theta(x)\,\Delta(x)$$
$$\Delta_c(x)=\theta(x)\,\Delta_+(x)-\theta(-x)\,\Delta_-(x). \tag{3.1}$$

Using the step function in time $\theta(x)$ one may alternatively express all Δ_i in terms of the two basic functions Δ and Δ_c.

By definitions of the contours one has the Fourier representations

$$i\Delta_+(x)=\int\theta(p)\,\delta(p^2-m^2)\,e^{-ip.x}\,\bar{d}^4p=\mathop{\mathsf{S}}_p e^{-ip.x}$$
$$i\Delta_-(x)=-\int\theta(-p)\,\delta(p^2-m^2)\,e^{-ip.x}\,\bar{d}^4p=-\mathop{\mathsf{S}}_p e^{ip.x}$$
$$i\Delta(x)=\int\epsilon(p)\,\delta(p^2-m^2)\,e^{-ip.x}\,\bar{d}^4p=\mathop{\mathsf{S}}_p (e^{-ip.x}-e^{ip.x})$$
$$\Delta_R(x)=\int(p^2-m^2-i\epsilon(p)\,0)^{-1}\,e^{-ip.x}\,\bar{d}^4p$$
$$\Delta_c(x)=\int(p^2-m^2+i0)^{-1}\,e^{-ip.x}\,\bar{d}^4p. \tag{3.2}$$

All reality and reflexion properties follow from

$$\Delta_+(x)=-\Delta_-(-x)=\Delta_-^*(x)$$

for instance, $\Delta(x)=-\Delta(-x)$. And if one remembers that when $x^2<0$ a homogeneous Lorentz transformation from $x\to-x$ can be effected without altering the *invariant* Δ_i one also deduces

$$\theta(-x^2)\,\Delta_i(x)=\theta(-x^2)\,\Delta_i(-x)$$

whence $\theta(-x^2)\,\Delta(x)=0$, i.e. Δ vanishes for space-like arguments, in particular on a $t=0$ surface.

One may derive explicit expressions for the Δ in terms of Bessel functions. As we shall later be interested in the Δ functions for arbitrary dimensions $2l$, we will give the Bessel representations (Gelfand and Shilov 1964, Akyeampong and Delbourgo 1975a) for any integer $2l$ and get the four-dimensional answers by simply taking the limit $l\to2$ at the end. To work out Δ_+ we need the volume element for time-like p in arbitrary dimensions. After carrying out $(2l-2)$ angular integrations this reads

$$\int d^{2l}p=\int_0^\infty\int_0^\infty(p^2)^{l-1}\,dp^2(\sinh\zeta)^{2l-2}\,d\zeta\,\pi^{l-1/2}/\Gamma(l-\tfrac{1}{2}) \tag{3.3}$$

where $\zeta=\tanh^{-1}(|\mathbf{p}|/p_0)$ is the rapidity. For positive time-like x, directed along the time axis, we also have

$$p.x=(p^2x^2)^{1/2}\cosh\zeta.$$

Inserting these in (3.2) one readily establishes that

$$\theta(x^2)\ \Delta_+(x)=\theta(x^2)\frac{e^{-i\pi l}}{(2\pi)^l}\left(\frac{m}{\sqrt{x^2}}\right)^{l-1}\tfrac{1}{2}\pi H_{l-1}{}^{(2)}(m\sqrt{x^2})$$

providing $\mathrm{Re}\, l>\frac{1}{2}$. $H_{l-1}{}^{(2)}$ is the second-kind Hankel function. For negative time-like x one similarly finds that Δ_+ is given in terms of the complex conjugate Hankel function $H_{l-1}{}^{(1)}$; and finally for space-like x one meets the modified Bessel function K_{l-1}. Gathering the various parts,

$$\begin{aligned}i(2\pi)^l\ \Delta_+(x)=\theta(-x^2)(m/\sqrt{-x^2})^{l-1}\ K_{l-1}(m\sqrt{-x^2})-\tfrac{1}{2}i\pi\theta(x^2)(m/\sqrt{x^2})^{l-1}\\ \times\{\theta(x)\ e^{-i\pi(l-1)}H_{l-1}{}^{(2)}(m\sqrt{x^2})-\theta(-x)\ e^{i\pi(l-1)}H_{l-1}{}^{(1)}(m\sqrt{x^2})\}.\end{aligned}\tag{3.4}$$

We can gain a deeper understanding of this formula by looking at what happens when we pass through the light cone $x^2=0$. For $x^2<0$ the argument of K reveals a branch cut running from $x^2=0$ to ∞ when l is real. In fact, above the cut K has an imaginary argument and can be replaced by $H^{(2)}$; below the cut this changes to $H^{(1)}$. Therefore it is enough to consider the function $(m/\sqrt{-x^2})^{l-1}\ K_{l-1}(m\sqrt{-x^2})$. To get $\Delta_+(x)$ we assert that:

(i) when $x^2<0$, Δ_+ is the above function;
(ii) when $x^2>0$, $t>0$, Δ_+ is the function defined above the right-hand cut;
(iii) when $x^2>0$, $t<0$, Δ_+ is the function defined below the right-hand cut.

Similar arguments can be put forward for Δ_-, except that we must go to opposite sides of the cut for time-like x. We can summarize all of this by the formula

$$\begin{aligned}i(2\pi)^l\ \Delta_\pm(x)=\pm\ \theta(-x^2)(m/\sqrt{-x^2})^{l-1}\ K_{l-1}(m\sqrt{-x^2})\\ \pm\ \theta(x^2)(m/(-x^2\pm i\epsilon(x)\ 0)^{1/2})^{l-1}\ K_{l-1}(m(-x^2\mp i\epsilon(x)\ 0)^{1/2}).\end{aligned}\tag{3.5}$$

Combining terms we deduce from (3.1) that

$$(2\pi)^l\ \Delta(x)=-\pi\theta(x^2)(m/\sqrt{x^2})^{l-1}\ J_{1-l}(m\sqrt{x^2})\tag{3.6}$$
$$i(2\pi)^l\ \Delta_c(x)=(m/(-x^2+i0)^{1/2})^{l-1}\ K_{l-1}(m(-x^2+i0)^{1/2})$$

whose zero mass limit is readily found from the behaviour of Bessel functions for small arguments:

$$\begin{aligned}iD_\pm(x)&=\pm\frac{\Gamma(l-1)}{4\pi^l}\left(\frac{1}{-x^2\mp i\epsilon(x)\ 0}\right)^{l-1}\\ D(x)&=-\frac{\theta(x^2)\ \epsilon(x)}{2\pi^{l-1}\Gamma(2-l)}\left(\frac{1}{x^2}\right)^l\\ iD_c(x)&=\frac{\Gamma(l-1)}{4\pi^l}\left(\frac{1}{-x^2+i0}\right)^{l-1}.\end{aligned}\tag{3.7}$$

Since we shall be principally interested in the four-dimensional limit it is worthwhile recording the explicit expressions in this case:

$$\begin{aligned}\Delta(x)&=\epsilon(x)[\theta(x^2)\ J_1(m\sqrt{x^2})/4\pi m\sqrt{x^2}-\delta(x^2)/2\pi]\\ &\xrightarrow[x\to 0]{}\epsilon(x)[-\delta(x^2)/2\pi+m^2\theta(x^2)/8\pi+O(|x^2|)]\\ &\xrightarrow[x\to\infty]{}\epsilon(x)\ \theta(x^2)(2m/\pi x^3)^{1/2}\cos\phi(m\sqrt{x^2})/4\pi\end{aligned}\tag{3.8}$$

and

$$\begin{aligned}4\pi^2 i\Delta_c(x) &= \theta(x^2)\, i\pi m[J_1(m\sqrt{x^2}) - iY_1(m\sqrt{x^2})]/2\sqrt{x^2} \\ &\qquad + \theta(-x^2)\, mK_1(m\sqrt{-x^2})/\sqrt{(-x^2)} - i\pi\delta(x^2) \\ &\xrightarrow[x\to 0]{} (-x^2+i0)^{-1} + \tfrac{1}{4}m^2 \ln(\tfrac{1}{4}m^2|x^2|) + \tfrac{1}{4}i\pi m^2\theta(x^2) + O(x\ln x) \qquad (3.9) \\ &\xrightarrow[x\to\infty]{} (m\pi/2x^3)^{1/2}[\theta(x^2)\, i\, e^{-i\phi(m\sqrt{x^2})} + \theta(-x^2)\, e^{-m\sqrt{-x^2}}].\end{aligned}$$

where ϕ denotes the asymptotic phase of the Bessel function. Note too that the massless propagators are exactly the functions for small x.

3.2. *Sums and products of propagators*

When a set of fields Φ_r interact just via the linear combination $\Sigma_r \rho_r \Phi_r$, it is sensible to introduce the external source coupling $J(\Sigma_r \rho_r \Phi_r)$. Equation (2.46) then clearly generalizes to

$$Z[J] = \sum_N \frac{(-i)^N}{N!}\left[\mathsf{S}\, V\left(i\frac{\delta}{\delta J}\right)\right]^N \exp\left[-\tfrac{1}{2}i\, \mathsf{S}\, J \sum_r \rho_r^2 \Delta_{cr} J\right] \qquad (3.10)$$

and, as expected, one encounters the propagator sum $\Sigma_r \rho_r^2 \Delta_c(x|m_r)$ where m_r is the mass of the free (spinless) field Φ_r. Then in the vicinity of the light cone, the dominant behaviour is given by

$$\sum_r 4\pi^2 i\Delta_c(x|m_r)\,\rho_r^2 = (-x^2+i0)\sum_r \rho_r^2 + \tfrac{1}{4}\sum_r \rho_r^2 m_r^2 \ln(\tfrac{1}{4}m_r^2|x^2|) + \tfrac{1}{4}i\pi\sum_r \rho_r^2 m_r^2\theta(x^2) + \dots. \qquad (3.11)$$

This leads us straight away to one of the oldest regularization procedures: namely, to take linear combinations of free-field propagators such that the worst singularities in (3.11) are eliminated. However, in order to produce such smoothing for $x\to 0$ it is vital, for the cancellations to take place, that some of the ρ_r^2 be negative. For instance, by including three or more terms, we can arrange that

$$\Sigma\,\rho_r^2 = \Sigma\,\rho_r^2 m_r^2 = \Sigma\,\rho_r^2 m_r^2 \ln m_r = 0 \qquad (3.12)$$

providing some of the ρ_r are imaginary. But imaginary ρ_r means that the Lagrangian has become complex and leads to a non-unitary S matrix and unphysical effects in the shape of negative probabilities! Luckily such disasters may be avoided at any accessible finite momentum if we take the limit $m_r\to\infty$ for all unphysical fields possessing these negative norms. In this way the only region where Δ_r differs significantly from zero is right in the vicinity of the light cone, and correspondingly the ρ_r which make for the cancellations in (3.12) become vanishingly small. The procedure thus becomes a purely mathematical device for enabling us to handle these singular Δ, since it is devoid of physical content—the catastrophes only occur for infinitely great momenta.

The equivalent description in momentum space leads to further insights. The analogous statement here is that the sum,

$$\sum_r \rho_r^2 \Delta_c(p|m_r) = \sum_r \rho_r^2 (p^2 - m_r^2 + i0)^{-1}$$

for choices such as (3.12) becomes more strongly damped as $p\to\infty$ than the single mass propagator. For instance, with one additional term ($r=2$) the modification is precisely Feynman's (1949) regularization prescription

$$(k^2-m^2)^{-1}\to(k^2-m^2)^{-1}-(k^2-M^2)^{-1}=(m^2-M^2)/(k^2-m^2)(k^2-M^2)$$

or

$$-\int_{m^2}^{M^2} \mathrm{d}m^2(k^2-m^2)^{-2} \qquad (3.13)$$

with the limit $M\to\infty$ having to be taken in the end. The form of the regularized inverse propagator, $(k^2-m^2)(k^2-M^2)/(m^2-M^2)$, implies that the interacting field combination is effectively described by a free Lagrangian which involves higher derivatives of the field than the first, viz

$$\tfrac{1}{2}\phi(\partial^2+m^2)(\partial^2+M^2)\,\phi/(M^2-m^2)$$

or perhaps more correctly,

$$\tfrac{1}{2}[(\partial^2\phi)^2+(m^2+M^2)(\partial\phi)^2+m^2M^2\phi^2]/(M^2-m^2).$$

For obvious reasons this artifice is sometimes referred to as 'kinetic regularization'. Formula (3.13) does, at the same time, suggest a further regularization procedure, which is to take a series of mass derivatives (enough to render the momentum space convolutions finite) and then integrate back over mass, e.g.

$$(k^2-m^2)^{-1}\to\int_{m^2}^{M^2}\mathrm{d}m^2\int_{m^2}^{M^2}\mathrm{d}m^2\left(\frac{\partial}{\partial m^2}\right)^2(k^2-m^2)^{-1}$$
$$=(M^2-m^2)^2/(k^2-m^2)(k^2-M^2)^2 \qquad \text{etc.} \qquad (3.14)$$

Here the regularizing mass M, which is eventually taken to infinity, appears at the end point of the mass integration.

These are just one or two of the more traditional methods of regularization. It should be made clear that whatever technique is chosen, and we shall study the most popular ones presently, the consequences have to be followed through to the end before we make $M\to\infty$. This is not to say that any regularization is as good as any other; on the contrary, some are definitely superior to others in that they allow us to preserve the initial symmetries of the Lagrangian at every stage of the perturbation calculation.

In any case this much is clear: in perturbation expansions of quantum field theory we are continually having to deal with 'improper functions' and integrals over them. Even Δ itself is an improper function as we soon realize from the step and delta functions which enter into expressions (3.7)–(3.9). In fact, strictly speaking, Δ has no meaning on its own. Rather, it is a 'distribution' defined (in a linear fashion) by its overlap with a well behaved function $F(x)$ i.e.

$$\int[\Delta(x)]^N F(x)\,\mathrm{d}^4x<\infty, \qquad N\leqslant 4.$$

More generally we require of our 'test functions' F that the overlap with the Green's functions be all finite:

$$\int G(x_1\ldots x_n)\,F(x_1\ldots x_n)\,\mathrm{d}^4x_1\ldots\mathrm{d}^4x_n<\infty.$$

This places strong restrictions on the choice of F. We have to have F continuous in its several arguments as well as its partial derivatives (up to a certain specified order depending on N) with bounded terms in the Taylor series expansion near $x=0$;

also, of course, F must fall off sufficiently rapidly at infinity for the integral to exist in the infrared limit. Given these conditions it is true that the limit $M\to\infty$ in the regularized Green's functions can be taken with confidence:

$$\int G(x_1 \ldots x_n)\, F(x_1 \ldots x_n) \prod_i \mathrm{d}^4x_i = \lim_{M\to\infty} \int G(x_1 \ldots x_n|M)\, F(x_1 \ldots x_n) \prod_i \mathrm{d}^4x_i$$

and the regularization procedure is placed on a sound basis.

3.3. Small-distance behaviour, point-splitting

As we have repeatedly stressed, the source of our troubles is the highly singular nature of the products $(\Delta(x_i-x_j))^n$ which occur in the perturbation expansions of the connected Green's functions $C(x_1 \ldots x_n)$ when one or more pairs of coordinates coalesce, $x_i-x_j\to 0$. It will be instructive to study this problem in a particular example as this will suggest to us a new regularization method, one which is specifically tied to coordinate space rather than momentum space. Consider the lowest order correction to the photon propagator in spinor electrodynamics due to virtual emission and re-absorption of an electron–positron pair, and characterized by the 'vacuum-polarization' tensor,

$$\begin{aligned}\Pi_{\mu\nu}(x-y) &= \mathrm{i}e^2\langle 0|T[j_\mu(x)\, j_\nu(y)]|0\rangle \\ &= \mathrm{i}e^2\langle 0|T[\bar\psi(x)\,\gamma_\mu\psi(x)\,\bar\psi(y)\,\gamma_\nu\psi(y)]|0\rangle \\ &= \mathrm{i}e^2\,\mathrm{Tr}\,[\gamma_\mu S_\mathrm{c}(x-y)\,\gamma_\nu S_\mathrm{c}(y-x)]\end{aligned} \tag{3.15}$$

in first order. Near the light cone, since

$$S_\mathrm{c}(x)\to \mathrm{i}\gamma\,.\,\partial(\mathrm{i}/4\pi^2x^2) = \gamma\,.\,x/2\pi^2x^4$$

it follows that the ultraviolet behaviour of Π is given by

$$\begin{aligned}\Pi_{\mu\nu}(x) &\to -\mathrm{i}e^2\,\mathrm{Tr}\,[\gamma_\mu\gamma\,.\,x\gamma_\nu\gamma\,.\,x]/4\pi^4(x^2)^4 \\ &= \mathrm{i}e^2(2x_\mu x_\nu - x^2\eta_{\mu\nu})/\pi^4(x^2)^4.\end{aligned} \tag{3.16}$$

The Ward identity (2.58) stipulates that $\partial^\mu\Pi_{\mu\nu}(x)=0$. Certainly if we take the divergence of (3.16) we get zero for all non-zero x, and we might easily conclude that the answer is gauge invariant. Such an argument is much too swift! The difficulty is precisely at the light cone $x\to 0$ where (3.16) needs to be properly defined. To take an analogy, we know that $D_\mathrm{c}(x)\sim 1/x^2$, and we can easily show that $\partial^2 D_\mathrm{c}(x)=0$ for any non-zero x; yet we also know that at the light cone the answer is $\delta^4(x)$.

Just to show how easily one is led to a gauge-violating answer, consider the Fourier transform

$$\Pi_{\mu\nu}(k) = \int \mathrm{e}^{\mathrm{i}kx}\Pi_{\mu\nu}(x)\,\mathrm{d}^4x$$

in the low-energy limit $k\to 0$, where we can replace $x_\mu x_\nu$ by $\frac{1}{4}\eta_{\mu\nu}x^2$ within the integrand upon symmetrical integration:

$$\Pi_{\mu\nu}(0) = -\mathrm{i}e^2 \int \mathrm{d}^4x\,\eta_{\mu\nu}/\pi^4(x^2)^3.$$

Making a Euclidean rotation $x_0\to \mathrm{i}x_0\equiv x_4$ and defining $r^2=-x^2$ the integral reduces to

$$\Pi_{\mu\nu}(0) \sim \eta_{\mu\nu}e^2 \int_{0+}^{\infty} \mathrm{d}r^2/r^4$$

which is visibly divergent near $r=0$. Excluding the origin by introducing a minimum radius cutoff at $r^2=M^{-2}$ we obtain a leading quadratic infinity $\Pi_{\mu\nu}\sim e^2\eta_{\mu\nu}M^2/\pi^2$ by naively following these steps.

The lesson is that we have to be much more careful in our approach to the small-distance limit $x \to y$ in (3.15). One way of defining the singularity has been proposed by Johnson (1964), which is to indulge in further point-splitting (Schwinger 1959) when taking products of field operators at the same place. In particular, he splits the fermion bilinear $\bar{\psi}(x)\,\Gamma\psi(x)$ symmetrically into $\bar{\psi}(x+\frac{1}{2}\epsilon)\,\Gamma\psi(x-\frac{1}{2}\epsilon)$ and takes the limit $\epsilon \to 0$ at the very end of the calculation, performing a suitable average† over ϵ components in the process. Since such a split destroys the gauge invariance of the original bilinear, the symmetry is restored (Schwinger 1959, Mandelstam 1962) by including the line integral factor $\exp\left(-ie\int_{x-\frac{1}{2}\epsilon}^{x+\frac{1}{2}\epsilon} d\xi^\mu A_\mu(\xi)\right)$. One thereby arrives at the 'path-dependent', but gauge-invariant, currents

$$J(x|\epsilon) = \bar{\psi}(x+\tfrac{1}{2}\epsilon)\,\Gamma\psi(x-\tfrac{1}{2}\epsilon)\exp\left[-ie\int_{x-\frac{1}{2}\epsilon}^{x+\frac{1}{2}\epsilon} d\xi^\mu A_\mu(\xi)\right]. \tag{3.17}$$

Especially interesting is the vector current $\Gamma = \gamma_\mu$ (and later the axial and pseudoscalar currents). To work out its divergence we have to use

$$\mathrm{i}\,\partial_\mu \exp\left[-ie\int d\xi^\mu A_\mu\right] = e[A_\mu(x+\tfrac{1}{2}\epsilon) - A_\mu(x-\tfrac{1}{2}\epsilon)]\exp[\ldots]$$

and the equations of motion (2.12). It is then straightforward to prove that

$$\begin{aligned}\partial^\mu J_\mu(x|\epsilon) &= 2e\bar{\psi}(x+\tfrac{1}{2}\epsilon)\,\gamma\{A(x+\tfrac{1}{2}\epsilon) - A(x-\tfrac{1}{2}\epsilon)\}\,\psi(x-\tfrac{1}{2}\epsilon)\exp[\ldots] \\ &\underset{\epsilon\to 0}{\to} e\epsilon_\mu\bar{\psi}(x+\tfrac{1}{2}\epsilon)\,\gamma_\nu F^{\mu\nu}(x)\,\psi(x-\tfrac{1}{2}\epsilon)\exp[\ldots].\end{aligned} \tag{3.18}$$

Current conservation is proved only if one can show that the ϵ singularities of (3.18) are harmless insofar that all matrix elements $\langle\Psi|\partial^\mu J_\mu(x|\epsilon)|\Psi'\rangle$ equal zero when $\epsilon \to 0$. This can be proved for Lagrangian models without axial vector mesons (Hagen 1969b).

One can successfully show that point-splitting results in a transverse $\Pi_{\mu\nu}$ without any quadratic infinities. The crucial reason why the method works is because the line integral itself contributes to $\delta J_\mu/\delta A_\nu$. This causes (3.15) to be amended to

$$\begin{aligned}\Pi_{\mu\nu}(k|\epsilon) &= \mathrm{i}e^2\int d^4x\, e^{\mathrm{i}k.x} \\ &\quad\times[\langle 0|T[\bar{\psi}(x+\tfrac{1}{2}\epsilon)\,\gamma_\mu\psi(x-\tfrac{1}{2}\epsilon)\,\bar{\psi}(\tfrac{1}{2}\epsilon)\,\gamma_\nu\psi(-\tfrac{1}{2}\epsilon)]|0\rangle \\ &\qquad - \int_{x-\frac{1}{2}\epsilon}^{x+\frac{1}{2}\epsilon} d\xi_\nu\delta^4(\xi)\langle 0|T[\bar{\psi}(x+\tfrac{1}{2}\epsilon)\,\gamma_\mu\psi(x-\tfrac{1}{2}\epsilon)]|0\rangle] \\ &= \mathrm{i}e^2\int d^4x\, e^{\mathrm{i}k.x}\,\mathrm{Tr}\,[\gamma_\mu\{S_c(x-\epsilon)\,\gamma_\nu S_c(-\epsilon-x) + \mathrm{i}S_c(-\epsilon)\int_{x-\frac{1}{2}\epsilon}^{x+\frac{1}{2}\epsilon} d\xi_\nu\delta^4(\xi).\end{aligned}$$

Taking the p transforms of the ϵ-dependent terms and fixing upon a linear path,

$$\begin{aligned}\int_{x-\frac{1}{2}\epsilon}^{x+\frac{1}{2}\epsilon} d\xi_\nu\, e^{-\mathrm{i}q.\xi} &= \tfrac{1}{2}\epsilon_\nu\int_{-1}^{1} ds\,\exp[-\mathrm{i}q.(x+\tfrac{1}{2}s\xi)] = 2\epsilon\sin(\tfrac{1}{2}q.\epsilon)/q.\epsilon \\ &= \epsilon_\nu[1-(q.\epsilon)^2/24 + O(\epsilon^4)].\end{aligned}$$

Therefore

$$\begin{aligned}\Pi_{\mu\nu}(k|\epsilon) &= \mathrm{i}e^2\int \bar{d}^4p\,\mathrm{Tr}\,[\gamma_\mu\, e^{\mathrm{i}(2p-k).\epsilon}S_c(p+\tfrac{1}{2}k)\,\gamma_\nu S_c(p-\tfrac{1}{2}k) \\ &\qquad + \mathrm{i}\gamma_\mu\epsilon_\nu\, e^{\mathrm{i}p.\epsilon}S_c(p)\{1-(k.\epsilon)^2/24+\ldots\}] \\ &\underset{\epsilon\to 0}{\to} \mathrm{i}e^2\int \bar{d}^4p\,\mathrm{Tr}\,[\gamma_\mu S_c(p+\tfrac{1}{2}k)\,\gamma_\nu S_c(p-\tfrac{1}{2}k) \\ &\qquad + \gamma_\mu\{1+(k.\partial/\partial p)^2/24+\ldots\}\,\partial S_c(p)/\partial p_\nu].\end{aligned} \tag{3.19}$$

† It is here that ambiguities in the procedure reveal themselves. Sometimes ϵ is taken as a four-vector and $\overline{\epsilon_\mu\epsilon_\nu} = \frac{1}{4}\eta_{\mu\nu}\epsilon^2$ used; sometimes ϵ is a spatial three-vector with $\overline{\epsilon_i\epsilon_j} = \frac{1}{3}\delta_{ij}\epsilon^2$.

The quadratic divergence of order k^0 is thereby cancelled and one is left with a nicely transverse $\Pi_{\mu\nu}(k)$, albeit logarithmically infinite.

3.4. Feynman integrals in momentum space

The singularity problem in x space translates into p space as a lack of definition of the Fourier integrals occasioned by the bad behaviour in the integrand for $p \to \infty$ (ultraviolet) or $p \to 0$ (infrared). Most textbook discussions of regularization are couched in momentum space language. There are some advantages in going to Fourier transforms and in this section we shall describe how the p-space Feynman integrals are carried out. Very many of the techniques and formulae that are exposed below will be constantly referred to in later sections.

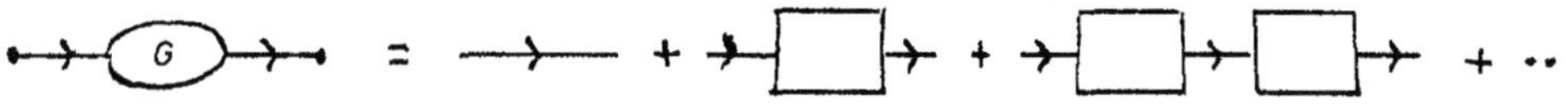

Figure 9. Self-energy iterates in the complete propagator.

The most obvious advantage of going into momentum space is that chains of diagrams (x-space convolutions), such as are drawn in figure 9, become a very simple geometric series (p-space products). For instance, the complete scalar propagator sums into the series

$$\Delta_c'(p) = \Delta_c(p) - \Delta_c(p)\,\Pi(p)\,\Delta_c(p) + \Delta_c(p)\,\Pi(p)\,\Delta_c(p)\,\Pi(p)\,\Delta_c(p) - \dots$$

or

$$\Delta_c'^{-1}(p) = \Delta_c^{-1}(p) + \Pi(p) \tag{3.20}$$

where $\Pi(p)$ denotes all 'proper self-energy' contributions. In $g\phi^4$ theory to order g^2, one has

$$\Pi(p) = -12ig\Delta_c(0) - 96g^2 \int e^{ip.x}\Delta_c^3(x)\,d^4x + 144g^2\Delta_c(0)\int d^4x\Delta_c^2(x) + \dots . \tag{3.21}$$

Similarly, in spinor electrodynamics, the dressed propagators can be organized into the formulae

$$S_c'^{-1}(p) = S_c^{-1}(p) + \Sigma\,(p) \tag{3.22}$$
$$D_{c_{\mu\nu}}'^{-1}(k) = D_{c_{\mu\nu}}^{-1}(k) + \Pi_{\mu\nu}(k)$$

where, according to (2.54) and (2.55),

$$\Sigma(p) = ie^2 \int \gamma_\mu S_c(p+k)\,\gamma^\mu D_c(k)\,\bar{d}^4k + O(e^4)$$
$$\Pi_{\mu\nu}(k) = ie^2 \int \mathrm{Tr}[\gamma_\mu S_c(p+\tfrac{1}{2}k)\,\gamma_\nu S_c(p-\tfrac{1}{2}k)]\,\bar{d}^4p + O(e^4). \tag{3.23}$$

The ultraviolet difficulty is the divergence of the momentum loop integrations at the upper end owing to $S(p) \sim 1/p$, $D(k) \sim 1/k^2$. In fact, this is not the only difficulty. There is a hidden infrared divergence as we go to the mass shell $\gamma.p \to m$ in the derivatives of Σ there for soft ($k \to 0$) photon momentum. The infrared infinity is logically different from the ultraviolet one and may be regularized differently. Of course, if the regularization scheme succeeds in defining infrared *and* ultraviolet infinities at one go, that is all to the good.

(3.21) and (3.23) are the simplest illustrations of the general Feynman integral

$$I(p_r) = \int \mathcal{N}(k, p_r)\,\mathcal{D}^{-1}(k, p_r)\,\bar{d}^{2l}k \tag{3.24}$$

taken over a loop momentum k, which is determined by a set of external momenta p_r. (To save repetition we shall work out the formulae in $2l$-dimensional space from the outset—the four-dimensional integrals are easily extracted by setting $l=2$ everywhere.) Typically the numerator $\mathcal{N}$ depends on the spins of the exchanged particles and may therefore be endowed with a number of spinor and tensor indices; the denominator is a product of factors,

$$\mathscr{D}=\prod_{r=1}^{N}[(k+p_r)^2-m_r^2+\mathrm{i}0]^{\nu_r}. \tag{3.25}$$

Although the ν_r are usually integers, we shall not assume this in the following because in analytic regularization one is specifically interested in complex ν values.

The product (3.25) can be neatly combined by means of the identity† (Feynman 1949)

$$\prod_{r=1}^{N}\frac{\Gamma(\nu_r)}{A_r^{\nu_r}}=\left(\prod_r\int_0^1\alpha_r^{\nu_r-1}\,\mathrm{d}\alpha_r\right)\delta(1-\alpha_1-\ldots-\alpha_N)$$
$$\times\frac{\Gamma(\nu_1+\ldots+\nu_N)}{[A_1\alpha_1+\ldots+A_N\alpha_N]^{\nu_1+\ldots+\nu_N}} \tag{3.26}$$

where the α_r are the 'Feynman parameters'. If the integral over k exists (and for non-integer l *or* non-integer ν one can check *a posteriori* that it does) then the k origin of integration can be shifted‡ to leave us with

$$I(p)=(\prod\textstyle\int_0^1\alpha^{\nu-1}\,\mathrm{d}\alpha)\,\delta(1-\sum\alpha)\int\mathcal{N}(k,p,\alpha)[k^2-M^2(p,m,\alpha)]^{-\Sigma\nu} \tag{3.27}$$

where M^2 is a Lorentz invariant scalar function of the four momenta, the α and the internal masses m. In the shifted numerator $\mathcal{N}$ we need only retain those terms that are even in k and simplify those as follows:

$$\int f(k^2)\,k_\mu k_\nu\,\bar{\mathrm{d}}^{2l}k=(2l)^{-1}\,\eta_{\mu\nu}\int k^2f(k^2)\,\bar{\mathrm{d}}^{2l}k$$
$$\int f(k^2)\,k_\kappa k_\lambda k_\mu k_\nu\,\bar{\mathrm{d}}^{2l}k=[2l(2l+2)]^{-1}(\eta_{\mu\nu}\eta_{\kappa\lambda}+\eta_{\mu\kappa}\eta_{\nu\lambda}+\eta_{\mu\lambda}\eta_{\nu\kappa})\int k^4f(k^2)\,\bar{\mathrm{d}}^{2l}k \tag{3.28}$$

by Lorentz covariance, as and when they arise. The loop integral is thus reduced to a sum of terms like

$$\int(k^2)^T(k^2-m^2)^{-\Sigma}\,\bar{\mathrm{d}}^{2l}k.$$

Providing that the external momenta p are space-like it is permissible to rotate k_0 to $\mathrm{i}k_0$ (i.e. there are no singularities in the k_0 plane one cannot avoid) and turn the integral into

$$\mathrm{i}\int(-q^2)^T(-q^2-m^2)^{-\Sigma}\,\bar{\mathrm{d}}^{2l}q$$

where $q^2=-k^2$ and the q metric is just the Euclidean $\delta_{\mu\nu}$. Now in a $2l$-dimensional Euclidean space the polar decomposition of q

$$q_\mu=(q,\theta_1,\theta_2,\ldots,\theta_{2l-1})$$

† This is itself proved by writing $\Gamma(\nu)\,A^{-\nu}=\int_0^\infty\mathrm{d}\rho\,\mathrm{e}^{-A\rho}\rho^{\nu-1}$, scaling to $\rho_r=\alpha_r\sum_s\rho_s$ and integrating over $\sum_s\rho_s$.

‡ For integer l and ν the integral does not exist when it fails to converge and one can pick up surface terms in shifting the origin.

gives the volume element

$$\mathrm{d}^{2l}q = q^{2l-1}\,\mathrm{d}q.\sin^{2l-2}\theta_1\,\mathrm{d}\theta_1.\sin^{2l-3}\theta_2\,\mathrm{d}\theta_2.\ldots.\sin\theta_{2l-2}\,\mathrm{d}\theta_{2l-2}.\mathrm{d}\theta_{2l-1}$$

with $0 \leqslant \theta_i < \pi$, except for the last $0 \leqslant \theta_{2l-1} < 2\pi$.

We can thereby integrate over all angles, using

$$\textstyle\int_0^\pi \sin^{2\mu}\theta\,\mathrm{d}\theta = \Gamma(\frac{1}{2})\,\Gamma(\mu+\frac{1}{2})/\Gamma(\mu+1)$$

to simplify the momentum integral:

$$\textstyle\int \mathrm{d}^{2l}q f(q^2) = \int \mathrm{d}q^2 (q^2)^{l-1} f(q^2)\,\pi^l/\Gamma(l). \tag{3.29}$$

Finally, therefore,

$$-\mathrm{i}\int \bar{\mathrm{d}}^{2l}k\,\frac{(k^2)^T}{(k^2-m^2)^\Sigma} = \frac{(-1)^{T-\Sigma}\,\pi^l}{(2\pi)^{2l}\,\Gamma(l)}\int_0^\infty \frac{(q^2)^{T+l-1}\,\mathrm{d}q^2}{(q^2+m^2)^\Sigma}$$
$$= \frac{(-1)^{T-\Sigma}\,\Gamma(l+T)\,\Gamma(\Sigma-l-T)}{(4\pi)^l\,\Gamma(l)\,\Gamma(\Sigma)(m^2)^{\Sigma-l-T}} \tag{3.30}$$

which has to be integrated over the Feynman parameters α. This is no problem for space-like p. If we continue p up to time-like values singularities migrate into the α range of integration and by deforming the α contour appropriately we can even obtain $I(p)$ for physical p.

Observe that (3.30) has singularities when $T+l-\Sigma=0, 1, \ldots.$ This signifies that the original integral was badly defined for l and ν integers due to an ultraviolet infinity. However, away from these points the integral has meaning. At this juncture we ought to consider defining the integrals $\int \mathrm{d}^4k$, $\int \mathrm{d}^4k(k^2-m^2)^{-1}$, etc which formally arise in four dimensions all the time. The old fashioned approach is to rotate contours to Euclidean space and cut them off at large $q^2=-k^2=M^2$. Such a prescription gives

$$-\mathrm{i}\int \bar{\mathrm{d}}^4k\,\frac{(k^2)^T}{(k^2-m^2)^\Sigma} \to \frac{(-1)^{T-\Sigma}}{16\pi^2}\int_0^{M^2} \frac{(q^2)^{T+1}\,\mathrm{d}q^2}{(q^2+m^2)^\Sigma}$$
$$= \frac{(-1)^{T-\Sigma}}{16\pi^2}(M^2)^{T-\Sigma+2}\int_0^1 \mathrm{d}x\,x^{T+1}(x+m^2/M^2)^{-\Sigma}. \tag{3.31}$$

So for large M,

$$-16\pi^2\mathrm{i}\textstyle\int \bar{\mathrm{d}}^4k \to \frac{1}{2}M^4$$
$$16\pi^2\mathrm{i}\textstyle\int \bar{\mathrm{d}}^4k(k^2-m^2)^{-1} \to M^2$$
$$-16\pi^2\mathrm{i}\textstyle\int \bar{\mathrm{d}}^4k(k^2-m^2)^{-2} \to \ln(M^2/m^2) \qquad \text{etc.}$$

We can be more sophisticated and regard the integrals as special integer limits of l, Σ and T within (3.30). For instance

$$-\mathrm{i}\int \bar{\mathrm{d}}^4k \overset{?}{=} \lim_{\Sigma\to 0} -\mathrm{i}\int \bar{\mathrm{d}}^4k(k^2-m^2)^{-\Sigma} = \lim_{\Sigma\to 0}\frac{(-m^2)^{2-\Sigma}}{16\pi^2(1-\Sigma)(2-\Sigma)} \tag{3.32}$$

? or

$$\lim_{l\to 2,\, m\to 0} -\mathrm{i}\int \bar{\mathrm{d}}^4k\left(\frac{k^2}{k^2-m^2}\right) = \lim_{l\to 2,\, m\to 0} -\Gamma(1-l)\left(\frac{m^2}{4\pi}\right)^l. \tag{3.33}$$

That there are inherent ambiguities in this approach is evident in both answers. In (3.32) the LHS ought not to depend on m, but the RHS does; in (3.33) the result appears to depend on the order of the limits. Similarly, the interpretation of

$$-\mathrm{i}\int k^{-2}\,\bar{\mathrm{d}}^4k \stackrel{?}{=} \lim_{\Sigma\to1,\,m\to0} -\mathrm{i}\int \bar{\mathrm{d}}^4k(k^2-m^2)^{-\Sigma} = \lim_{\Sigma\to1,\,m\to0} \frac{(-m^2)^{2-\Sigma}}{16\pi^2(1-\Sigma)(2-\Sigma)} \tag{3.34}$$

?
or

$$\lim_{l\to2,\,m\to0} -\mathrm{i}\int \bar{\mathrm{d}}^{2l}k(k^2-m^2)^{-1} = \lim_{l\to2,\,m\to0} -\frac{\Gamma(1-l)(m^2)^{l-1}}{(4\pi)^l} \tag{3.35}$$

has to be made with excessive care.

3.5. Natural cutoffs, non-polynomial interactions

In the next section we shall carry out a more thorough study of the infinities and what to do with them. For the present it is appropriate to ask if there exist any systems which are free of infinities and the mechanism that softens the singularities. Perhaps the simplest example of a system which naturally possesses cutoffs is a crystal; it has a finite size (volume L^3) setting the scale for large x behaviour and a small scale (volume a^3) determined by the cell size which comes into its own for small x. It is, of course, true that such a system is not Lorentz invariant; however, in the limit $L\to\infty$, $a\to0$, special relativity should be restored.

For the moment forget the small-distance behaviour (i.e. that the field disturbance is only required at lattice points $\boldsymbol{x}=n_r\boldsymbol{a}_r$ where the $\boldsymbol{a}_r$ are primitive lattice vectors) and concentrate on the large-scale properties of the crystal. We adopt periodic boundary conditions

$$\phi(\boldsymbol{x}+\boldsymbol{L})=\phi(\boldsymbol{x}) \qquad \text{where} \qquad \boldsymbol{L}=L(1,1,1).$$

This quantizes the momentum to the (infinite) discrete set of values

$$k_i=2\pi n_i/L, \qquad n_i \text{ integer.}$$

The continuum normalizations and integrals used previously in (2.23) must be amended as follows:

$$\mathop{\mathsf{S}}_{p} f(p)\to\sum_{n} L^{-3}F(\boldsymbol{n})/2(m^2+2\pi n^2/L^2)^{1/2}\equiv\mathop{\mathsf{S}}_{n} F(\boldsymbol{n})$$

$$\delta^3(\boldsymbol{p},\boldsymbol{p}')\to2L^3(m^2+2\pi n^2/L^2)\prod_i\delta_{n_in_i'}\equiv\delta^3(\boldsymbol{n},\boldsymbol{n}') \tag{3.36}$$

where $F(\boldsymbol{n})\equiv f(2\pi\boldsymbol{n}/L)$. The dependence on the volume of space (the size of the box) is now manifest. The transition probabilities are obtained from (2.34) by taking $t'=-t=\frac{1}{2}L$ large†.

The intensive quantity of interest is the *transition rate per unit volume*

$$\lim_{L\to\infty}\langle\Psi'\tfrac{1}{2}L|\Psi-\tfrac{1}{2}L\rangle/L^{-4}\equiv\Gamma_{\Psi'\Psi}. \tag{3.37}$$

Recalling the S-matrix element is eventually given as an integral over centre-of-mass variables,

$$\langle\Psi'|\Psi\rangle=\int \mathrm{d}^4x\,\mathrm{e}^{\mathrm{i}(p-p')\cdot x}\,M_{\Psi'\Psi}$$

† One could logically take the infinite time limit independently of the spatial limit, of course.

the space–time integration supplies the factor

$$\prod_{\mu} \frac{\sin \frac{1}{2}(p-p')_\mu L}{\frac{1}{2}(p-p')_\mu} \to \delta^4(p-p') \qquad \text{for} \qquad L \to \infty$$

and one finally arrives at

$$\Gamma_{\Psi'\Psi} = \lim_{L\to\infty} \frac{1}{L^4} \prod_{\mu} \left[\frac{\sin \frac{1}{2}(p-p')_\mu L}{\frac{1}{2}(p-p')_\mu L}\right] |M_{\Psi'\Psi}|^2$$
$$= \delta^4(p-p')|M_{\Psi'\Psi}|^2. \tag{3.38}$$

This discussion just serves to show that there are no difficulties in principle as we increase the size of the box, if we are concerned with intensive variables. Furthermore, it also indicates that if we encounter factors like $\delta^4(0)$ in the continuum limit, these are as likely as not the volume of space–time in which the physical process is evolving. The vacuum transition rate Γ_{00} contains just such a factor, although this happens not to be especially worrying because it is always divided out anyhow†.

The cutoff operating at small distances is much more interesting. Providing we are only investigating the x properties over scales of the order of the cell size, we can take the infinite crystal limit $L\to\infty$. Basically then, we have the converse situation to (3.36) viz $x_i = aN_i$ assume the infinite discrete values, while the momentum p varies continuously within the range $2\pi/a$. Here the appropriate replacements are

$$\int d^3x g(x) = \sum_N G(N), \qquad \int d^3p \to \prod_i \int_{-\pi/a}^{\pi/a} dp_i. \tag{3.39}$$

We can appreciate the cutoff role of a by working out the tadpole term,

$$i\Delta_c(0) = i \int \bar{d}p_0 \int \bar{d}^3p (p_0{}^2 - p^2 - m^2 + i0)^{-1}$$
$$= \int \bar{d}^3p/2(p^2+m^2)^{1/2} \sim \begin{cases} (2ma^3)^{-1}, & m \neq 0 \\ a^{-2}, & m = 0. \end{cases} \tag{3.40}$$

Similarly, integrals which were formerly divergent, like $\int \Delta^2(x)\, d^4x$, become finite and dependent on a.

Realizing that fundamental scales offer the possibility of damping out infinities let us examine some Lorentz covariant models in which this mechanism operates successfully: the damping afforded by non-polynomial interactions (Fradkin 1963, Efimov 1964, Salam 1970). If only to point the way, imagine a model in which the propagator is modified from $iD_c(x)$ to $(-iD_c{}^{-1}(x) - 4\pi^2a^2)^{-1} \propto (x^2+a^2)^{-1}$. We can find a Lagrangian which gives rise to such a two-point function by making the expansion

$$[-iD_c{}^{-1}(x) - 4\pi^2a^2]^{-1} = iD_c(x)[1 + 4\pi^2a^2 iD_c(x) + (4\pi^2a^2 iD_c(x))^2 + \ldots]$$

and recognizing that

$$\langle 0|T[\phi^n(x)\,\phi^n(0)]|0\rangle = [iD_c(x)]^n\, n! \tag{3.41}$$

for a massless scalar field ϕ. Hence the Lagrangian

$$\mathscr{L} = \tfrac{1}{2}(\partial\phi)^2 - \sum_{n=1}^{\infty} \phi^n(2\pi a)^{n-1}/n[(n-1)!]^{1/2}$$

† However, in certain cases where the kinetic energy Lagrangian takes the form $\frac{1}{2}(\partial\phi)^2\, U(\phi)$ one must add canonical quantization terms $\delta^4(0) \ln U(\phi)$ in which the space–time volume plays a more prominent role.

would be capable of yielding such a soft Green's function when evaluated to second order in the interaction. More generally, we can characterize a non-polynomial interaction by a potential which possesses an *infinite* series expansion in the field†,

$$V(\phi)=\sum_n v_n{:}\phi^n{:}/n!. \tag{3.42}$$

The coefficients v_n contain a scale a, to an appropriate power, ensuring that V has the right dimensions; it is this scale a which eventually acts as the cutoff. In second order we are concerned with the two-point function,

$$\langle 0|\,T[V(x)\,V(0)]\,|0\rangle=\sum_n v_n{}^2[\mathrm{i}\Delta_c(x-y)]^n/n!. \tag{3.43}$$

This series (3.43) is to be interpreted as an asymptotic expansion ($\Delta\to0$) of the actual matrix element. Going further, we may replace it by a Sommerfeld–Watson contour integral,

$$\sum_n(-1)^n f_n/n!\equiv\mathrm{i}\int\Gamma(-z)f(z)/2\pi$$

and then take the Fourier transform (Salam and Strathdee 1970); thus‡

$$\int \mathrm{d}^4x\;\mathrm{e}^{\mathrm{i}k.x}\langle 0|\,T[V(x)\,V(0)]\,|0\rangle=\frac{\mathrm{i}}{2\pi}\int_c \mathrm{d}z\Gamma(-z)\,\mathrm{e}^{-\mathrm{i}\pi z}v^2(z)\,\Delta_c(k,z) \tag{3.44}$$

where

$$\Delta_c(k,z)\equiv\int \mathrm{d}^4x\;\mathrm{e}^{\mathrm{i}k.x}[\mathrm{i}\Delta_c(x)]^z \tag{3.45}$$

is the so-called 'superpropagator'. A closed form for $\Delta_c(k,z)$ is only known in the massless limit,

$$D_c(k,z)=-\mathrm{i}(4\pi)^{2(1-z)}\,\Gamma(2-z)/\Gamma(z)(-k^2)^{2-z}. \tag{3.46}$$

(This has poles at integer $z\geqslant2$ and, less obviously, poles at $z=-n$ with residues proportional to $(\partial/\partial k)^{2(n-1)}\,\delta^4(k)$.) We can appreciate how infinities might be damped out in these non-polynomial models by supposing that the series (3.43) starts with $n=2$, i.e. *all* the terms in (3.42) are diseased with possible infinities. However, the contour c in (3.44) threads to the *left* of $z=2$ and the asymptotic behaviour (assuming that c can be straightened out) is governed by the first singularity on the left which is dictated by $D(k,z)$ and the properties of $v(z)$. Quite often there is a pole at $z=1$ and this then determines the asymptotic behaviour of (3.44) as $k\to\infty$ (or $x\to0$) in a calculable way governed by the cutoff scale a. The ultraviolet behaviour is thereby nicely damped.

All this may seem rather contrived. Nevertheless, there are several examples in physics where non-polynomial Lagrangians do arise naturally. We shall quote two.

(i) Derivatively coupled pseudoscalar theory, characterized by

$$\mathscr{L}=\bar\psi(\mathrm{i}\gamma.\partial-m)\,\psi+\tfrac12[(\partial\phi)^2-\mu^2\phi^2]+\mathrm{i}g\bar\psi\gamma\gamma_5\psi.\partial\phi/\mu. \tag{3.47}$$

When one re-defines the spinor field by means of a gauge rotation to $\chi=\exp\,(g\gamma_5\phi/\mu)\,\psi$, there results the equivalent theory

$$\mathscr{L}=\bar\chi[\mathrm{i}\gamma.\partial-m\exp\,(-2g\gamma_5\phi/\mu)]\,\chi+\tfrac12[(\partial\phi)^2-\mu^2\phi^2] \tag{3.48}$$

† The normal ordering symbol :: used in (3.42) signifies that all tadpole graphs have already been extracted, so that $\langle V\rangle=0$.

‡ Subject to certain ambiguities in continuations to non-integer z.

wherein the interaction is an exponential Lagrangian. It is best to treat it as a whole, and not to make a perturbation expansion in the scale g/μ, if one is interested in the $x\to 0$ behaviour. Thus, using (3.41),

$$\langle 0|T[e^{a\phi(x)}\, e^{a\phi(0)}]|0\rangle = \exp[ia^2\Delta_c(x)] \xrightarrow[x\to 0]{} \exp[-a^2/4\pi^2x^2] \qquad (3.49)$$

has an essential singularity at the light cone. For $a^2<0$ the zero in $\exp(-a^2/x^2)$ as $-x^2\to 0+$ is so strong that it can be multiplied by any fixed power $\Delta_c{}^n(x)$ to still give zero. Actually, $a^2>0$ and the exponential (3.49) goes complex for time-like x. The sensible prescription (Okubo 1954, Volkov 1968, Lehmann and Pohlmeyer 1971) is to define (3.49) as the average, $\frac{1}{2}(\lim_{a+i0} + \lim_{a-i0})$.

(ii) All fields are coupled to gravity. Since the gravitational field involves the Newtonian mass scale $G^{-1/2}$ it has been conjectured (Salam 1970) that gravity acts as a universal cutoff when momenta attain that scale (of course, when other interactions are included damping scales may set in at smaller momenta). To get an idea of this consider the gravitational interactions of a scalar meson,

$$\mathscr{L} = \mathscr{L}_{\text{Einstein}}(g) + \tfrac{1}{2}\sqrt{-g}(g^{\mu\nu}\,\partial_\mu\phi\,\partial_\nu\phi - \mu^2\phi^2). \qquad (3.50)$$

Take the parametrization $g^{\mu\nu} = (\eta^{\mu\nu} + 2\kappa h^{\mu\nu})^{1/2}$ to define the graviton field h where $16\pi\kappa^2 \equiv G$, the Newtonian constant. Also simplify the argument by adopting a scalar version of gravity wherein $h^{\mu\nu} = \eta^{\mu\nu}h$. Then

$$\sqrt{-g} = (1+2\kappa h)^{-1} = \sum_0^\infty (-2\kappa)^n\, h^n$$

and, making use of (3.49), we arrive at the two-point function

$$\langle 0|T[\{-g(x)\}^{1/2}\{-g(0)\}^{1/2}]|0\rangle = \sum_0^\infty [4\kappa^2 iD_c(x)]^n\, n!. \qquad (3.51)$$

This formally divergent series can be Borel summed, by means of the representation $n! = \int \zeta^n \exp(-\zeta)\,d\zeta$, to yield

$$\langle 0|T[\{-g(x)\}^{1/2}\{-g(0)\}^{1/2}]|0\rangle = \mathscr{P}\int_0^\infty e^{-\zeta}[1-4i\zeta\kappa^2 D_c(\kappa)]^{-1}\,d\zeta \qquad (3.52)$$

a principal value prescription (Delbourgo *et al* 1969, Lee and Zumino 1969, Fried 1967) having been taken to avoid imaginary parts in the multipropagator at space-like x. The function on the right of (3.52) is thus the principal branch of the exponential integral function and for small x behaves as $(\pi^2x^2/\kappa^2)\ln(x^2/\kappa^2)$, exhibiting the gravitational cutoff. We will return to regularization via non-polynomial interactions later in the context of analytic regularization.

In conclusion, it should be mentioned that any basic theory endowed with a fundamental scale is potentially capable of alleviating the infinity problem. Nonlinear realizations of chiral symmetry and quantum field theory in de Sitter space are two very different examples of this.

4. Regularization and renormalization

In renormalizable theories the number of superficial infinities (as given by the Dyson formula) is limited to just a few lower point Green's functions. It is this fact which permits us to treat them systematically by means of a renormalization programme whereby the infinities are absorbed into re-definition of the relevant physical parameters, the masses and coupling constants. Renormalization theory is a subject in its own right lying outside the scope of this review, so we shall only detail those aspects of the theory which directly concern the infinities themselves, namely the evaluation of 'renormalization constants'.

4.1. *Elements of renormalization*

Consider first $g\phi^4$ theory. Neglecting vacuum graphs the theory is afflicted by three infinities, a quadratic and logarithmic divergence in the two-point function $C(x_1, x_2)$, and a logarithmic divergence in the four-point function $C(x_1 \dots x_4)$. Insofar as it is always formally possible to make (momentum) Taylor series expansions of the C about certain specified points, once the integrals have been regularized, we can introduce counter-terms in the Lagrangian to fix the C at certain values for those specified momenta. These counter-terms must then carry the ultraviolet infinities since the remainder terms in the Taylor series are all finite as the regularization parameter disappears. In $g\phi^4$ one has to perform mass, wavefunction and coupling renormalizations in order that every Green's function is finite and finally expressible in terms of the physical parameters of the theory. The renormalized Lagrangian

$$\mathscr{L} = \tfrac{1}{2} Z_\phi[(\partial\phi)^2 - Z_\mu \mu^2 \phi^2] - Z_g g \phi^4 \tag{4.1}$$

is obtained by re-scaling,

$$\phi = Z_\phi^{-1/2}\phi_0, \qquad \mu^2 = Z_\mu^{-1}\mu_0^2, \qquad g = Z_g^{-1} Z_\phi^2 g_0$$

the field, mass and coupling (subscript 0) occurring originally in the Lagrangian. The object is to ensure that the complete propagator and proper scattering vertex,

$$\begin{aligned} \Delta'^{-1}(p) &= Z_\phi(p^2 - Z_\mu \mu^2) + \Pi(p^2) \\ \Gamma(p_1 \dots p_4) &= Z_g g + T_4(p_1 \dots p_4) \end{aligned} \tag{4.2}$$

are finite to every order in g. The Bogoliubov, Parasiuk, Hepp and Zimmermann (BPHZ) procedure (which we will mostly follow) is more specific since it actually fixes the values of Δ' and Γ_4 on the mass shell and the symmetry point:

$$\begin{aligned} &\Delta'^{-1}(p) = (p^2 - \mu^2) + \mathrm{O}(p^2 - \mu^2)^2 \\ &\Gamma_4(p_i) = g \qquad \text{for} \qquad p_i^2 = \mu^2,\ p_i \cdot p_j = -\tfrac{1}{3}\mu^2 \quad (i \neq j) \end{aligned} \tag{4.3}$$

the so-called 'normalization' conditions. This fixes the Z:

$$\begin{aligned} Z_\phi &= 1 - \mathrm{d}\Pi(p^2)/\mathrm{d}p^2 \big|_{p^2=\mu^2} \\ Z_\mu &= 1 - Z_\phi^{-1}\Pi(\mu^2)/\mu^2 \\ Z_g &= 1 - g^{-1}T(p_i)\big|_{\text{sym point}}. \end{aligned} \tag{4.4}$$

In practice, the conditions (4.4) are imposed iteratively via an expansion in g:

$$Z = 1 + \sum_{n=1}^{\infty} g^n Z^{(n)}.$$

The successive $Z^{(n)}$ in the series are construed to cancel the infinities of that order, g^n, in Π and T and are determined in part by the $Z^{(n)}$ of lower order. To first order since

$$\Pi(p^2) = 12ig\Delta_c(0) = 12ig \int \bar{\mathrm{d}}^4 p/(p^2 - \mu^2)$$

is p-independent and

$$T(p_i) = -288ig^2 \sum_{ij} \int d^4x\, e^{i(p_i+p_j)\cdot x}\, \Delta_c^2(x)$$
$$= -288ig^2 \sum_{ij} \int \bar{d}^4k/[k^2-\mu^2][(k+p_i+p_j)^2-\mu^2]$$
$$= -288ig^2 \sum_{ij} \iint \bar{d}^4k\, d\alpha[k^2+(p_i+p_j)^2\,\alpha(1-\alpha)-\mu^2]^{-2}$$
$$Z_\phi^{(1)}=0, \qquad Z_\mu^{(1)} = -12i \int \bar{d}^4k/\mu^2(k^2-\mu^2)$$
$$Z_g^{(1)} = 864i \iint \bar{d}^4k\, d\alpha[k^2-\mu^2+4\mu^2\alpha(1-\alpha)/3]^{-2} \tag{4.5}$$

are the BPHZ values of the renormalization constants. To next order the relevant diagrams and counter-terms are listed in figure 10. The sub-infinities (of order g) having already been subtracted off, it is the second loop integral infinity which must be cancelled by $Z^{(2)}$ to satisfy the normalization conditions (4.3).

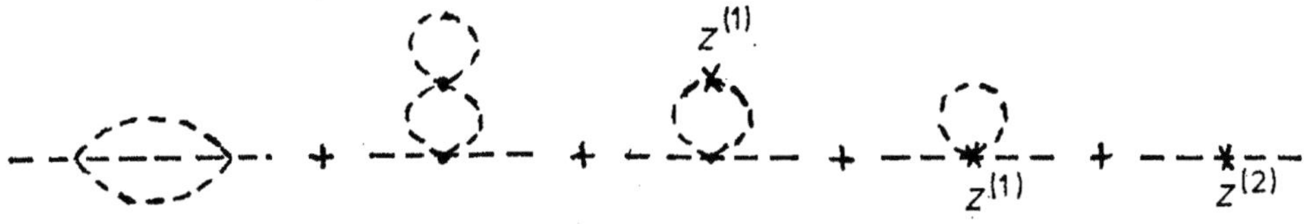

Figure 10. Renormalization counter-terms to order g^2 in $g\phi^4$ theory.

Scalar electrodynamics is only slightly more elaborate. Although it is true that no primary $g(\phi^\dagger\phi)^2$ term appears in the Lagrangian (2.7), it must be introduced as a basic subtraction term for renormalization purposes. Presupposing that a gauge invariant programme can be carried through, the theory, written in terms of re-normalized physical fields, has

$$\mathscr{L} = Z_\phi(\partial_\mu - ieZ_eZ_\phi^{-1}A_\mu)\,\phi^\dagger.(\partial^\mu + ieZ_eZ_\phi^{-1}A^\mu)\,\phi - \mu^2 Z_\mu Z_\phi \phi^\dagger\phi$$
$$-\tfrac{1}{4}Z_A(\partial_\mu A_\nu - \partial_\nu A_\mu)(\partial^\mu A^\nu - \partial^\nu A^\mu) + gZ_g(\phi^\dagger\phi)^2 \tag{4.6}$$

where the original bare quantities (subscript 0) are given by

$$\phi_0 = Z_\phi^{1/2}\phi, \qquad A_0 = Z_A^{1/2}A, \qquad \mu_0^2 = Z_\mu\mu^2$$
$$e_0 = Z_eZ_\phi^{-1}Z_A^{-1/2}e, \qquad g_0 = Z_gZ_\phi^{-2}g.$$

Our assumption that gauge invariance and regularization are compatible implies that in (3.22), $\Pi_{\mu\nu} = (k_\mu k_\nu - k^2\eta_{\mu\nu})\,\Pi_A$, whence

$$D_{c_{\mu\nu}}'^{-1}(k) = -k^2\eta_{\mu\nu}[Z_A + \Pi_A(k^2)] + k_\mu k_\nu \Pi_A(k^2). \tag{4.7}$$

The normalization condition for the photon thereby fixes Z_A to equal

$$Z_A \equiv 1 - \Pi_A(0). \tag{4.8}$$

As for the vertex renormalization constant Z_e, we note first that the proper meson–photon diagrams described by

$$e^{-1}\Gamma_\mu(p',p) = (p'+p)_\mu[Z_e + I(p'^2, p^2, (p'-p)^2)]$$
$$+(p'-p)_\mu(p'^2-p^2)\,H(p'^2, p^2, (p'-p)^2)$$

have intrinsic divergences solely in the I component, which are cancelled by Z_e,

$$Z_e \equiv 1 - I(\mu^2, \mu^2, 0). \tag{4.9}$$

Moreover, since the Ward–Takahashi identity (2.57) gives

$$(p'-p)^\mu \Gamma_\mu(p',p) = e[\Delta'^{-1}(p') - \Delta'^{-1}(p)]$$

or $\qquad$ (4.10)

$$\Gamma_\mu(p,p) = e\,\partial\Delta^{-1}(p)/\partial p_\mu$$

it follows that the wavefunction and vertex renormalizations must be equal, $Z_e = Z_\phi$, whatever gauge-invariant regularization method one selects. Hence the charge renormalization is only governed by the photon line, $e = Z_A{}^{1/2} e_0$. Finally, we must point out that the gauge symmetry determines the renormalization constant for photon–meson ($eZ_\phi A^2\phi^\dagger\phi$) scattering in terms of the others.

There is a little less to spinor electrodynamics, except for some Dirac algebra:

$$\mathscr{L} = Z_\psi\bar{\psi}(i\gamma.\partial - mZ_m)\,\psi - eZ_e\bar{\psi}\gamma.A\psi - \tfrac{1}{4}Z_A F_{\mu\nu}F^{\mu\nu}. \qquad (4.11)$$

The photon renormalization is no different to (4.7), but the spinor renormalization is something new. We have to ensure that

$$\begin{aligned} S'^{-1}(p) &= Z_\psi(\gamma.p - mZ_m) + \Sigma(p) \qquad (4.12)\\ &\equiv Z_\psi(\gamma.p - mZ_m) + \gamma.pK(p^2) - mJ(p^2) \end{aligned}$$

satisfies the normalization condition $S^{-1}(p) \to (\gamma.p - m)$ near the mass shell limit $\gamma.p \to m$. This determines the mass and wavefunction renormalization constants for the fermion ($\psi_0 = Z_\psi{}^{1/2}\psi$, $m_0 = Z_m m$) as follows:

$$\begin{aligned} Z_\psi &= 1 - K(m^2) + 2m^2\,\mathrm{d}(J-K)/\mathrm{d}p^2\,|_{p^2=m^2} \qquad (4.13)\\ Z_m &= 1 + Z_\psi{}^{-1}[K(m^2) - J(m^2)]. \end{aligned}$$

Last, but not least, the vertex renormalization $Z_e = e_0 Z_\psi Z_A{}^{1/2}/e$ is found from the general expression

$$\begin{aligned} e^{-1}\Gamma_\mu(p',p) = \gamma_\mu&[Z_e + L(p^2, p'^2, (p'-p)^2)]\\ &+ (p'-p)^\nu[\gamma_\nu, \gamma_\mu]\,G(p^2, p'^2(p-p')^2) + \ldots \qquad (4.14) \end{aligned}$$

by requiring that the γ_μ form factor† be unity on the mass shell:

$$Z_e \equiv 1 - L(m^2, m^2, 0). \qquad (4.15)$$

The Ward identity analogous to (4.10) then gives $Z_e = Z_\psi$.

4.2. Evaluation of renormalization constants

The renormalization constants Z are the only formally divergent quantities in these quantum field theories. Although the physical quantities are independent of their 'values' it is of some interest to see how the Z are 'calculated' in practice because this tells us something about the bare parameters of the theory e_0, m_0, g_0, ..., and in certain models the bare constants take on prescribed values. Moreover, the dependence of the Z on the cutoff mass M is of crucial significance in renormalization group equations. Let us then first list some formal integrals which define the Z before we introduce the M regularization.

The $Z^{(1)}$ of $g\phi^4$ have already been listed in (4.5). Next consider scalar electrodynamics‡. The meson self-energy is obtained from (2.50)

† It is only this form factor which requires an infinite subtraction.

‡ There is a bit more to it than spinor electrodynamics because we have to worry about tadpole graphs.

$$\Pi(p^2)=ie^2\int\frac{d^4k}{k^2}\left\{\frac{(2p+k)^2}{(p+k)^2-\mu^2}-4\right\}$$
$$=ie^2\int d^4k\left\{-\frac{4}{k^2}+\int_0^1 d\alpha\,\frac{[k^2+p^2(2-\alpha)^2]}{[k^2+p^2\alpha(1-\alpha)-\mu^2\alpha]^2}\right\}. \tag{4.16}$$

The tadpole diagram adds to the mass but is irrelevant to the computation of Z_ϕ from (4.4). With the photon self-energy,

$$\Pi_{\mu\nu}(k)=-ie^2\int\frac{d^4p}{(p^2-\mu^2)}\left\{\frac{(2p+k)_\mu(2p+k)_\nu}{(p+k)^2-\mu^2}-2\eta_{\mu\nu}\right\} \tag{4.17}$$

we remark that

$$k^\mu\Pi_{\mu\nu}(k)=-ie^2\int d^4p\left\{\frac{(2p-k)_\nu}{p^2-\mu^2}-\frac{(2p+k)_\nu}{(p+k)^2-\mu^2}\right\}$$

would vanish identically providing one could shift origin in the integrand. As for the photon–meson vertex to order e^3,

$$\Gamma_\nu=-ie^2\int\frac{d^4k}{k^2}\left\{\frac{(2p+k).(2p'+k)(p+p'+2k)_\nu}{[(p+k)^2-\mu^2][(p'+k)^2-\mu^2]}-\frac{2(2p'+k)_\nu}{(p'+k)^2-\mu^2}-\frac{2(2p+k)_\nu}{(p+k)^2-\mu^2}\right\} \tag{4.18}$$

a linear divergent, we can again verify the Ward identity (4.10) only in the most formal way. The same applies to the Ward identities for the proper meson–photon scattering graphs:

$$k^\mu\Gamma_{\nu\mu}(p'k',pk)=e[\Gamma_\nu(p',p'+k)-\Gamma_\nu(p-k',p)]$$
$$k'^\nu\Gamma_{\nu\mu}(p'k',pk)=e[\Gamma_\mu(p+k,p)-\Gamma_\mu(p',p'-k)]. \tag{4.19}$$

The regularization of $g\phi^4$ theory can be very simple-minded as there is no particular symmetry to retain. Thus one may simply cut off the momentum space integrals in (4.5) at large $k^2=-M^2$ or equivalently we can modify the meson propagator in the manner of (3.14). This gives regularized

$$Z_\mu^{(1)}=-12i\int d^4k(\mu^2-M^2)^2/(k^2-\mu^2)(k^2-M^2)^2\,\mu^2$$
$$=-24i\iint d^4k\,d\alpha(\mu^2-M^2)^2\,\alpha/[k^2-M^2\alpha-\mu^2(1-\alpha)]^3\,\mu^2$$
$$=-\frac{3}{4\pi^2}\int d\alpha\,\frac{(M^2-\mu^2)^2\,\alpha/\mu^2}{M^2\alpha+\mu^2(1-\alpha)}\sim-\frac{3M^2}{4\pi^2\mu^2} \tag{4.20}$$
$$Z_g^{(1)}=-1728i\iint d^4k\,d\alpha\int_{\mu^2}^{M^2}\frac{[1-4\alpha(1-\alpha)/3]\,d\mu^2}{[k^2-\mu^2+4\mu^2\alpha(1-\alpha)/3]^3}$$
$$=-54\int_0^1 d\alpha\int_{\mu^2}^{M^2}d\mu^2/\mu^2\sim-54\ln(M^2/\mu^2).$$

We cannot be so strict with electrodynamics for reasons of gauge invariance. Thus it is no good modifying the charge meson propagator as in (3.14) without also changing the vertices of the theory; otherwise, one is certain to violate the gauge identities, for instance

$$\frac{\partial}{\partial p_\mu}\frac{(p^2-\mu^2)(p^2-M^2)}{\mu^2-M^2}\neq 2p_\mu.$$

However, for the meson self-energy what one can do is to alter the photon propagator from k^{-2} to

$$\int_{\kappa^2}^{M^2}d\kappa^2\int_{\kappa^2}^{M^2}\frac{2\,d\kappa^2}{(k^2-\kappa^2)^3}=\frac{(\kappa^2-M^2)^2}{(k^2-\kappa^2)(k^2-M^2)^2}$$

to isolate both ultraviolet ($M\to\infty$) and infrared ($\kappa\to 0$) infinities. This does not spoil the gauge symmetry and yields in this order

$$\left(\frac{\partial}{\partial\kappa^2}\right)^2\Pi(p^2)=\frac{e^2}{16\pi^2}\int_0^1 d\alpha\left[-\frac{4}{\kappa^2}+\frac{(1-\alpha)^2\{p^2(\alpha^2+2\alpha-4)+2\mu^2\alpha+2\kappa^2(1-\alpha)\}}{[\mu^2+\kappa^2(1-\alpha)-p^2\alpha(1-\alpha)]^2}\right] \quad (4.21)$$

$$Z_\phi=1-\lim_{\kappa\to 0}\int_{\kappa^2}^{M^2}d\kappa^2\int_{\kappa^2}^{M^2}d\kappa^2\left(\frac{\partial}{\partial\kappa^2}\right)^2\frac{\partial\Pi(p^2)}{\partial p^2}\bigg|_{\mu^2}$$

$$=1-\frac{e^2}{16\pi^2}\lim_{\substack{\kappa\to 0\\ M\to\infty}}\int_0^1 d\alpha\left[(4-6\alpha+3\alpha^2)\ln\left(\frac{\mu^2\alpha^2+M^2(1-\alpha)}{\mu^2\alpha^2+\kappa^2(1-\alpha)}\right)\right.$$

$$\left.+\mu^2(2-\alpha)^2\left(\frac{\alpha(1-\alpha)}{\mu^2\alpha^2+\kappa^2(1-\alpha)}-\frac{\alpha(1-\alpha)}{\mu^2\alpha^2+M^2(1-\alpha)}\right)\right] \quad (4.22)$$

$$Z_\mu=1-\lim_{\kappa\to 0}\int_{\kappa^2}^{M^2}d\kappa^2\int_{\kappa^2}^{M^2}d\kappa^2\left(\frac{\partial}{\partial\kappa^2}\right)^2\Pi(\mu^2)/\mu^2$$

$$\sim 1-3e^2M^2/4\pi^2 \qquad \text{as} \qquad M\to\infty.$$

There was no difficulty in setting $\kappa=0$ for the mass renormalization. By contrast, Z_ϕ (and for that matter $\Pi(p^2)-\Pi(\mu^2)$) is infrared-divergent due to a logarithmic singularity at the lower end of the α integral when κ vanishes (this quite apart from an ultraviolet infinity $e^2\ln M^2/8\pi^2$).

The photon self-energy requires even greater care. One could resort to ϵ splitting, for instance, as we did in spinor ED, but we prefer to use the Pauli and Villars (1949) regularization technique (Stückelberg and Rivier 1948) at this stage. We know that we have to tamper with the charged meson lines in (4.17); we also know that modifying each line separately is no good. The correct way to maintain gauge invariance is to regularize the loop as a whole rather than treat individual lines. In this Pauli–Villars approach (4.17) is modified to

$$\Pi_{\mu\nu}(k)\to\int_{\mu^2}^{M^2}d\mu^2\int_{\mu^2}^{M^2}d\mu^2\left(\frac{\partial}{\partial\mu^2}\right)^2\Pi_{\mu\nu}(k)$$

$$=\frac{e^2}{16\pi^2}\int d\mu^2\int d\mu^2(k_\mu k_\nu-k^2\eta_{\mu\nu})\int_0^1 d\alpha(1-2\alpha)^2[\mu^2-k^2\alpha(1-\alpha)]^{-2}$$

or

$$\Pi_A(k^2)=\frac{e^2}{16\pi^2}\int_0^1 d\alpha(1-2\alpha)^2\left[\ln\left(\frac{M^2-k^2\alpha(1-\alpha)}{\mu^2-k^2\alpha(1-\alpha)}\right)+\frac{\mu^2-M^2}{M^2-k^2\alpha(1-\alpha)}\right]$$

whence the photon wavefunction renormalization constant,

$$Z_A=1+\frac{e^2}{16\pi^2}\int_0^1 d\alpha(1-2\alpha)^2[\ln(M^2/\mu^2)-1+\mu^2/M^2]$$

$$\sim 1-\frac{e^2}{24\pi^2}\ln\left(\frac{M}{\mu}\right). \quad (4.23)$$

The tactics for evaluating the vertex part Γ_μ of (4.18) are simpler: just regularize the photon line as for the meson self-energy. This will expose the infrared and ultraviolet infinities and will permit a check of the relation $Z_e=Z_\phi$. We shall defer a consideration of this to the end of the section.

4.3. Anomalies

From the few cases that we have examined we might easily conclude that in the final analysis (after the Z subtractions have been performed) the physical matrix elements are independent of the regularizing parameter such as M or ϵ. This is false: in exceptional circumstances the regularization leaves its imprint in the form of a finite calculable anomaly. A famous illustration is the axial anomaly (Adler 1969, Bell and Jackiw 1969) and we shall treat it in various ways: in this section we adopt point-splitting (Hagen 1969a, Jackiw and Johnson 1969). This anomaly concerns a fermion field interacting with a vector gauge field (the photon A_μ say) via a conserved current j_μ. The theory is regularized to maintain the gauge symmetry, or $\partial^\mu j_\mu = 0$, at every stage. Thus the vector Ward identities are true to every order (M^{-1} or ϵ) in the regularizing parameter. Next one considers the axial and pseudoscalar currents,

$$J_{\mu 5} = \mathrm{i}\bar{\psi}\gamma_\mu\gamma_5\psi \qquad \text{and} \qquad J_5 = \bar{\psi}\gamma_5\psi.$$

Naively one would deduce from the equations of motion (2.12) the operator identity

$$\partial^\mu J_{\mu 5} = -2mJ_5$$

and in the limit as $m \to 0$ that the axial current is conserved. Unfortunately, the answer is wrong because *any* photon-gauge-conserving choice of regularization destroys the axial identity.

For instance, suppose that one indulges in the ϵ-separation method, whereby currents are constructed as in (3.17). This wrecks the naive chiral properties of the currents; for treating the photon as a classical field,

$$\begin{aligned}
&\partial^\mu J_{\mu 5}(x|\epsilon) + 2mJ_5(x|\epsilon) \\
&\quad = -2e\bar{\psi}(x+\tfrac{1}{2}\epsilon)\,\gamma_\nu\gamma_5\{A^\nu(x+\tfrac{1}{2}\epsilon) - A^\nu(x-\tfrac{1}{2}\epsilon)\}\,\psi(x-\tfrac{1}{2}\epsilon)\exp\left[-\mathrm{i}e\textstyle\int A.\mathrm{d}\xi\right] \\
&\quad \underset{\epsilon\to 0}{\sim} -e\epsilon_\mu\bar{\psi}(x+\tfrac{1}{2}\epsilon)\,\gamma_\nu\gamma_5 F^{\mu\nu}\psi(x-\tfrac{1}{2}\epsilon)\exp\left[-\mathrm{i}e\textstyle\int A.\mathrm{d}\xi\right] \qquad (4.24)\\
&\quad = \mathrm{i}e\epsilon_\mu F^{\mu\nu}\,\mathrm{Tr}\,\{\gamma_\nu\gamma_5 S_\mathrm{c}'(x-\tfrac{1}{2}\epsilon, x+\tfrac{1}{2}\epsilon)\exp[\ldots]\}.
\end{aligned}$$

Were it not for the singularities in ϵ of the trace on the RHS of (4.24) we would obtain the naive axial identity. To reveal these singularities one makes a perturbation expansion of the fermion propagator in powers of the photon field,

$$[\mathrm{i}\gamma.\partial - e\gamma.A - m]\, S_\mathrm{c}'(x, x') = \delta^4(x-x')$$

$$S_\mathrm{c}'(x, x') = S_\mathrm{c}(x, x') + \int S_\mathrm{c}(x, y)\, e\gamma.A(y)\, S_\mathrm{c}(y, x')\,\mathrm{d}^4y + \ldots$$

keeps the first term and drops† the exponent in (4.24):

$$\begin{aligned}
&\partial^\mu J_{\mu 5}(x|\epsilon) + 2mJ_5(x|\epsilon) \\
&\quad = \mathrm{i}e^2\epsilon_\mu F^{\mu\nu}\,\mathrm{Tr}\,\{\gamma_\nu\gamma_5 \textstyle\int S_\mathrm{c}(x-\tfrac{1}{2}\epsilon, y)\,\gamma.A(y)\,S_\mathrm{c}(y, x+\tfrac{1}{2}\epsilon)\,\mathrm{d}^4y\} \\
&\quad = \mathrm{i}e^2\epsilon_\mu F^{\mu\nu}\textstyle\int \bar{\mathrm{d}}^4k A^\lambda(k)\,\mathrm{e}^{-\mathrm{i}k.x}\,\mathrm{Tr}\,\{\gamma_\nu\gamma_5\int \bar{\mathrm{d}}^4p\,\mathrm{e}^{-\mathrm{i}p.\epsilon}S_\mathrm{c}(p)\,\gamma_\lambda S_\mathrm{c}(p+k)\} \\
&\quad = 4\mathrm{i}e^2\epsilon_\mu F^{\mu\nu}\textstyle\int \bar{\mathrm{d}}^4k A^\lambda(k)\,\mathrm{e}^{-\mathrm{i}k.x}\int \bar{\mathrm{d}}^4p\,\mathrm{e}^{-\mathrm{i}p.\epsilon}\epsilon_{\rho\sigma\nu\lambda}p^\rho k^\sigma\Delta_\mathrm{c}(p)\,\Delta_\mathrm{c}(p+k).
\end{aligned}$$

Now

$$\int \bar{\mathrm{d}}^4p\,\mathrm{e}^{-\mathrm{i}p.\epsilon}\Delta_\mathrm{c}(p)\,\Delta_\mathrm{c}(p+k) \sim \mathrm{i}\ln(m^2\epsilon^2)/16\pi^2 \qquad \text{as} \qquad \epsilon\to 0$$

$$\partial^\mu J_{\mu 5}(x|\epsilon) + 2mJ_5(x|\epsilon) \to \frac{1}{8\pi^2}\,\epsilon_\mu F^{\mu\nu}F^{\sigma\lambda}\epsilon_{\rho\sigma\nu\lambda}\frac{\partial}{\partial\epsilon_\rho}(\ln\epsilon^2) = \frac{e^2}{4\pi^2}\frac{\epsilon_\mu\epsilon^\rho}{\epsilon^2}\,\epsilon_{\rho\sigma\nu\lambda}F^{\mu\nu}F^{\sigma\lambda}.$$

† All quantities which are dropped give finite contributions that vanish when $\epsilon \to 0$.

Averaging four-dimensionally over ϵ we arrive at the anomalous axial identity in the limit $\epsilon \to 0$

$$\partial^{\mu} J_{\mu 5} = -2mJ_5 + e^2 \epsilon_{\mu\nu\kappa\lambda} F^{\mu\nu} F^{\kappa\lambda}/16\pi^2. \tag{4.24a}$$

The Adler anomaly on the right of (4.24) corresponds to a finite correction in the two-photon matrix elements of the axial divergence which does not vanish even for massless fermions. One might think the result is an artifact of the method. However, Adler (1969) has shown the generality of the result by proving the impossibility of maintaining current conservation at every leg of the vector–vector–axial vertex arising from a fermion loop (figure 11).

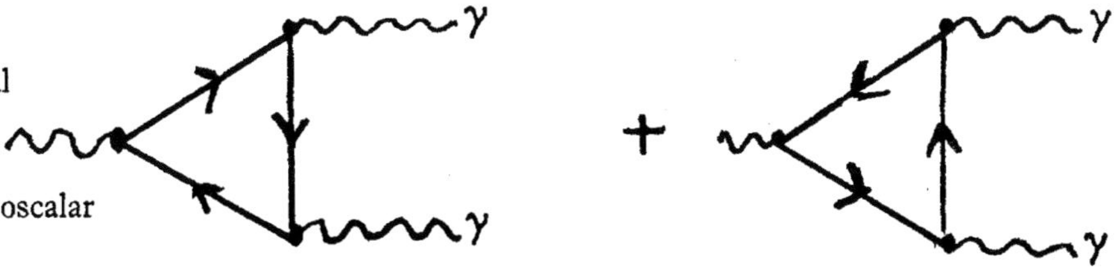

Figure 11. The triangle anomaly.

4.4. *Regulator fields*

We pointed out that the replacement of the free propagator $\Delta_c(x|m^2)$ by the weighted sum $\Sigma \rho_r^2 \Delta(x|m_r^2)$ could be understood as the introduction of auxiliary fields ϕ_r (mass m_r) which interact with the sources in much the same way as the original physical field ϕ. This approach to regularization leads us directly to the concept of regulator fields (Pais 1947, Gupta 1953) and affords us a better appreciation of the Pauli–Villars technique.

Recall that the proper way to maintain gauge invariance for $\Pi_{\mu\nu}$ is to regularize the entire charged loop line. We can understand the failure of kinetic regularization for each propagator because the regularized free Lagrangian of §3.2 clearly destroys the gauge symmetry. We would do better to use gauge-covariant derivatives in the Lagrangian:

$$[(\mathrm{D}^2\phi^\dagger)(\mathrm{D}^2\phi) + (\mu^2 + M^2)\,\mathrm{D}\phi^\dagger.\mathrm{D}\phi + \mu^2 M^2 \phi^\dagger \phi]/(M^2 - \mu^2)$$

but unfortunately this device enormously complicates the (multiphoton) vertices that have to be considered (Kroll 1966). Much better to treat the continuous charged line as a whole by adding new heavy 'regulator fields' Φ_r with EM interactions (Pais and Uhlenbeck 1950) induced by minimal substitution:

$$\mathrm{D}\Phi_r^\dagger.\mathrm{D}\Phi_r - \mu_r^2 \Phi_r^\dagger \Phi_r = (\partial + \mathrm{i}e_r A)\,\Phi_r^\dagger.(\partial - \mathrm{i}e_r A)\,\Phi_r - \mu_r^2 \Phi_r^\dagger \Phi_r.$$

That is, in the EM current, include

$$J_\lambda = \sum_r e_r(\Phi_r^\dagger \mathrm{i}\overleftrightarrow{\mathrm{D}}_\lambda \Phi_r + \bar{\Psi}_r \gamma_\lambda \Psi_r) \tag{4.25}$$

where e_r is the charge of the rth regulator (Bose Φ or Fermi Ψ). If we then sum over all charged-field contributions to vacuum polarization, (4.17) is altered to

$$\Pi_{\mu\nu} = -\mathrm{i} \sum_r e_r^2 \int \frac{\bar{\mathrm{d}}^4 p}{(p^2 - \mu_r^2)} \left\{ \frac{(2p+k)_\mu (2p+k)_\nu}{(p+k)^2 - \mu_r^2} - 2\eta_{\mu\nu} \right\} \tag{4.26}$$

a result which becomes finite if we choose our charges† such that

$$\sum e_r^2 = \sum e_r^2 \mu_r^2 = \sum e_r^2 \ln \mu_r = 0. \tag{4.27}$$

(One must take $\mu_r \to \infty$ and e_r imaginary, for $r > 1$, to avoid unphysical consequences at the end.) The final result is equivalent to the mass derivative procedure which yields (4.23).

Note that adding charged regulators does nothing to alleviate the ultraviolet problem of the original charged-particle self-energy. The only way to get the damping is to introduce heavy vector mesons $A_{\mu s}$ enjoying the same type of 'coupling' with the charged fields as the photon itself. This then gives the regulated meson self-energy

$$\Pi(p^2) = \mathrm{i} \sum_s e_s^2 \int \frac{\bar{\mathrm{d}}^4 k}{(k^2 - M_s^2)} \left\{ \frac{(2p+k)^2}{(p+k)^2 - \mu^2} - 4 \right\} \tag{4.28}$$

and effectively reproduces (4.21) and (4.22).

As a nice application of the regulator field method let us return to the axial current anomaly (4.24). When we introduce one spinor regulator field Ψ having mass M in addition to the physical spinor field ψ (mass m) the regularized axial divergence relation reads

$$\partial^\mu J_{\mu 5}(x|m) - \partial^\mu J_{\mu 5}(x|M) = -2mJ_5(x|m) + 2MJ_5(x|M) \tag{4.29}$$

in an obvious notation, where by construction the infinities in the two-photon-pseudovector matrix are cancelled. If all the matrix elements of $J_\mathrm{R}(x|M)$ were to behave as M^{-2} (or better) as $M \to \infty$ we could disregard the regular field contributions altogether in (4.29). But this is not the case. It is true that the vector–vector–axial matrix element, having been rendered vector-gauge-invariant and therefore carrying an external factor of third order in momentum‡, multiplies an integral of order m^{-2} and is thus safe as $m \to \infty$. However, the vector–vector–pseudoscalar element is given by

$$\begin{aligned} &\int \mathrm{d}^4x\, \mathrm{d}^4x' \exp[\mathrm{i}(k.x + k'.x')] \langle 0 | T[J_\mu(x)\, J_\nu(x')\, J_5(0)] | 0 \rangle \\ &= -2\mathrm{i}e^2 \,\mathrm{Tr} \int \bar{\mathrm{d}}^4 p [\gamma_\nu S_\mathrm{c}(p)\, \gamma_\mu S_\mathrm{c}(p+k)\, \gamma_5 S_\mathrm{c}(p'-k)] \\ &= \epsilon_{\mu\nu\kappa\lambda} k^\kappa k'^\lambda \frac{me^2}{2\pi^2} \iint \mathrm{d}\alpha\, \mathrm{d}\beta \frac{\theta(1-\alpha-\beta)}{[2k.k'\alpha\beta - m^2]} \end{aligned} \tag{4.30}$$

when $k^2 = k'^2 = 0$. Although it is finite one sees that the matrix element of MJ_5 in the limit $M \to \infty$, $-e^2 \epsilon_{\mu\nu\kappa\lambda} k^\kappa k'^\lambda / 4\pi^2$, is non-zero (Hagen 1969b). Multiplying it into polarization vectors $\epsilon^\mu(k)\, \epsilon^\nu(k')$ one recognizes this regulator field contribution as none other than the Adler anomaly (4.24). Theories in which axial mesons (and their currents) play a primary role are fraught with axial anomalies in some other matrix elements as well.

Nor is the idea of an anomalous Ward identity (Bardeen 1969, Brown *et al* 1969) confined to the axial divergence. Lagrangian models of massless particles interacting via dimensionless couplings appear, on the face of it, to be scale-invariant because they involve no dimensional parameters, and correspondingly one expects to meet a conserved scaling current. Unfortunately, one is obliged to introduce regulator fields to define the quantum corrections, and these have associated with them a heavy mass M

† In even more sophisticated versions one can even take continuous integrals over mass ensuring, nevertheless, that $\int e^2(\mu)\, \mathrm{d}\mu^2 = \int e^2(\mu)\, \mu^2\, \mathrm{d}\mu^2 = \int e^2(\mu) \ln \mu\, \mathrm{d}\mu^2 = 0$.

‡ It has the form $(k+k')_\rho\, \epsilon_{\mu\nu\kappa\lambda} k^\kappa k'^\lambda$ for physical photons ($k^2 = k'^2 = 0$).

scale. Taking $M \to \infty$ can be shown to produce an anomaly (Callan *et al* 1970) in the divergence of the scaling current, analogously to the axial anomaly. Again the regularization yields something physical (Chanowitz and Ellis 1972).

4.5. Infrared cancellations

So far we have mostly neglected the infrared divergences. The reason we have done so is because it is well known that in electrodynamics infrared divergences of mass-shell matrix elements are cancelled (Schwinger 1949a, b, Eriksson 1961, Frautschi *et al* 1961) in any given order of α by divergences in bremsstrahlung processes (soft-photon emission) to leave a finite answer wherein the sensitivity ΔE of charged-particle energy detection plays a prominent role. We want to reproduce the standard argument here for the simplest example we know (Rutherford scattering) because we shall afterwards examine unconventional ways of regularizing the infrared infinities.

The standard method of isolating the infrared divergences is to grant the photon a small mass κ. The relevant diagrams which contribute to scattering in a weak external field can be abstracted from figure 7 by coupling the electromagnetic current (photon line) to external field, with the proviso that we halve the meson self-energy contributions when we place the charged particles on shell†. Now (4.16) and (4.18) with photon mass κ give

$$\Pi(p^2) - \Pi(\mu^2) = ie^2 \int \frac{\bar{d}^4k}{k^2-\kappa^2}\left[\frac{(2p+k)^2}{(p+k)^2-\mu^2} - \frac{k^2+4p.k+\mu^2}{k^2+2p.k}\right]$$

$$\Gamma_\nu|_{p^2=p'^2=\mu^2} = -ie^3 \int \frac{\bar{d}^4k}{(k^2-\kappa^2)}\frac{1}{(k^2+2p.k)(k^2+2p'.k)} \times [2k_\nu(4p.p'-k^2)+(p'+p)_\nu(4p.p'-2p.k-2p'.k-3k^2)].$$

Since we are concerned with the infrared divergence as $\kappa \to 0$ only small values of the photon momentum k matter in the integrand. Therefore in the soft-k approximation,

$$\Pi(p^2) - \Pi(\mu^2) \approx ie^2(p^2-\mu^2) \int \bar{d}^4k\mu^2/(k^2-\kappa^2)(k.p)^2$$

$$\Gamma_\nu| \approx -ie^3(p'+p)_\nu \int \bar{d}^4kp.p'/(k^2-\kappa^2)\, p.kp'.k.$$

Hence, to order e^3 the scattering vertex is dominated by

$$\Gamma_\nu = e(p'+p)_\nu \left[1+\tfrac{1}{2}ie^2 \int \frac{\bar{d}^4k}{k^2-\kappa^2}\left\{\frac{\mu^2}{(k.p)^2}+\frac{\mu^2}{(k.p')^2}-\frac{2p.p'}{(k.p)(k.p')}\right\}\right. \tag{4.31}$$

in the infrared‡. On the other hand, the order e^2 photon emission diagrams of figure 12 provide the amplitude

$$M_{\mu\nu}\epsilon^{\nu*}(k) = e^2\epsilon^{\nu*}(k)[2\eta_{\mu\nu} - (2p'+k)_\mu\, \Delta_C(p'+k)(p'+p+k)_\nu - (2p-k)_\mu\, \Delta_C(p-k)(p'+p-k)_\nu]$$

$$\underset{k\to 0}{\to}\; e^2(p'+p)_\nu\left[\frac{p.\epsilon^*}{p.k}-\frac{p'.\epsilon^*}{p'.k}\right]. \tag{4.32}$$

† Owing to the square root factor in wavefunction renormalization.

‡ We may disregard the infrared finite photon self-energy part in this limit if we only want to exhibit the leading infrared cancellation.

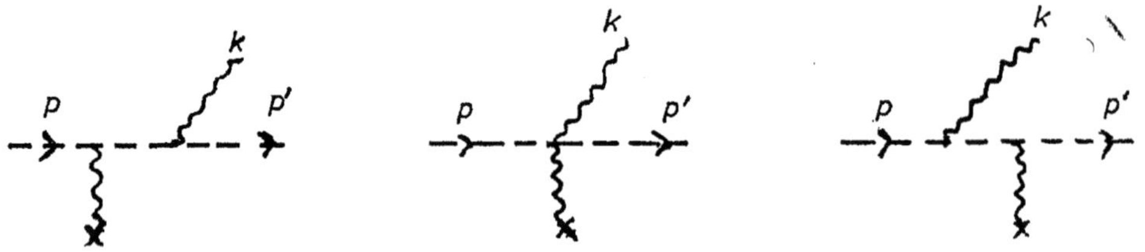

Figure 12. Bremsstrahlung diagrams for charged meson scattering in a weak external field.

If A stands for the external electromagnetic potential, the κ-regularized cross section is given by

$$|A.\Gamma|^2 = |(p'+p).A|^2 \left[1 - ie^2 \int \frac{đ^4k}{k^2-\kappa^2}\left\{\frac{\mu^2}{(k.p)^2} + \frac{k^2}{(k.p')^2} - \frac{2p.p'}{(k.p)(k.p')}\right\}\right]$$

plus the integral up to maximum detectable photon energy ΔE of

$$\underset{\gamma}{\mathrm{S}}\, |A.M.\epsilon^*|^2 = -|(p'+p).A|^2\, e^2 \int^{\Delta E} đ^4k\delta(k^2-\kappa^2)\,\theta(k)\left(\frac{p}{p.k} - \frac{p'}{p'.k}\right)^2.$$

Combining the terms and letting $d\sigma^{(n)}$ stand for the cross section to order e^{2n},

$$d\sigma^{(2)} = d\sigma^{(1)}\left[1 - e^2 \int đ^4k \left\{\frac{\mu^2}{(k.p)^2} + \frac{\mu^2}{(k.p')^2} - \frac{2p.p'}{(k.p)(k.p')}\right\}\right.$$
$$\left.\times\{\theta(k)\,\theta(\Delta E - k)\,\delta(k^2-\kappa^2) - i(k^2-\kappa^2+i0)^{-1}\}\right]. \quad (4.33)$$

But for physical p and p' the location of singularities in the k_0 plane shows that

$$\int đ^4k(p'.p)/(k.p'-i0)(k.p-i0)(k^2-\kappa^2+i0)$$
$$= -i\int đ^4k\theta(k)\,\delta(k^2-\kappa^2)(p'.p)/(k.p')(k.p).$$

Hence an infrared cancellation takes place in (4.33) between soft virtual photons and soft real photon emission to leave

$$\frac{d\sigma^{(2)}}{d\sigma^{(1)}} = 1 - e^2 \int đ^4k \left\{\frac{\mu^2}{(k.p)^2} + \frac{\mu^2}{(k.p')^2} - \frac{2p.p'}{(k.p)(k.p')}\right\}\theta(k-\Delta E)\,\delta(k^2-\kappa^2) \quad (4.34)$$

which is perfectly well behaved for non-zero ΔE even in the limit $\kappa \to 0$. The integral (4.34) has been evaluated in various limits of p for small ΔE; for instance, in the non-relativistic limit one finds (Schwinger 1949b)

$$\frac{d\sigma^{(2)}}{d\sigma^{(1)}} = \frac{e^2}{6\pi^2}\frac{(p-p')^2}{m^2}\left[\ln\left(\frac{2\Delta E}{\mu}\right) - \frac{11}{24}\right]. \quad (4.35)$$

5. Analytic regularization

Except for point-splitting and non-polynomial interactions, the regularization procedures we have described until now have always relied on introducing heavy unphysical particles in some form or another. For the remainder of this work we shall discuss fairly recent developments in techniques of regularization which are conceptually quite different.

5.1. Analytic mass regularization

This was the precursor of analytic propagator regularization and makes a good introduction to the subject. To understand its origins, we return to the expression for the lowest order vacuum polarization,

$$\left(\frac{\partial}{\partial\mu^2}\right)^2 \Pi_{\rho\sigma}(k) = \frac{e^2}{16\pi^2}(k_\rho k_\sigma - k^2\eta_{\rho\sigma})\int_0^1 \mathrm{d}\alpha \frac{(1-2\alpha)^2}{[\mu^2 - k^2\alpha(1-\alpha)]^2}.$$

Recognizing that the integrand is a particular value of

$$\left(\frac{\partial}{\partial\mu^2}\right)^z \ln[\mu^2 - k^2\alpha(1-\alpha)] = (-1)^{1-z}\,\Gamma(z)[\mu^2 - k^2\alpha(1-\alpha)]^{-z} \tag{5.1}$$

a formula which can be analytically continued to arbitrary values (Kallen 1951, Karlson 1954) of z, we see that a possible gauge-invariant definition of the vacuum polarization is

$$\begin{aligned}\Pi_{\rho\sigma}(k) &= \lim_{z\to1}\frac{-e^2}{16\pi^2}(k_\rho k_\sigma - k^2\eta_{\rho\sigma})\left(-\frac{\partial}{\partial\mu^2}\right)^{z-1}\int_0^1 \mathrm{d}\alpha(1-2\alpha)^2\ln[\mu^2 - k^2\alpha(1-\alpha)]\\ &= \lim_{z\to1}\frac{e^2}{16\pi^2}(k_\rho k_\sigma - k^2\eta_{\rho\sigma})\int_0^1 \mathrm{d}\alpha(1-2\alpha)^2\,\Gamma(z-1)[\mu^2 - k^2\alpha(1-\alpha)]^{1-z}.\end{aligned} \tag{5.2}$$

The previous $\log(M/\mu)$ infinity is transformed into a pole at $z=1$.

Very much the same idea can be used to evaluate the meson self-energy (4.21). In this case one generalizes the formula

$$(\partial/\partial\kappa^2)^2(\kappa^2\ln\kappa^2 - \kappa^2) = \kappa^{-2}$$

to

$$(-\partial/\partial\kappa^2)^z(\kappa^2\ln\kappa^2 - \kappa^2) = \Gamma(z-1)(\kappa^2)^{1-z}$$

and interprets, in an obvious manner,

$$\begin{aligned}\Pi(p^2) = \frac{e^2}{16\pi^2}\lim_{z\to1}\int_0^1 \mathrm{d}\alpha\Bigg[&-\frac{4\Gamma(z-2)}{(\kappa^2)^{z-2}} - \frac{2(\alpha-1)^{z-1}\,\Gamma(z-2)}{[\mu^2+\kappa^2(1-\alpha)-p^2\alpha(1-\alpha)]^{z-2}}\\ &-(\alpha-1)^{1+z}\frac{\{p^2(2-\alpha)^2+\mu^2(1-\alpha)\}\,\Gamma(z-1)}{[\mu^2+\kappa^2(1-\alpha)-p^2\alpha(1-\alpha)]^{z-1}}\Bigg].\end{aligned} \tag{5.3}$$

The ultraviolet (and infrared) infinities are again manifested as poles at the physical limit $z=1$. This method of taking mass derivatives to an arbitrary integer z and then continuing analytically (Riesz 1949) down to $z=1$ is acceptable when it is perfectly obvious with respect to which mass one should differentiate. It rapidly gets out of hand where there are several masses in the problem.

5.2. Analytic propagator regularization

In an effort to put the above problem into clearer perspective Speer (1968) suggested that the modification†

$$\begin{aligned}(k^2-m^2)^{-1} \to -\lim_{z\to1}\left(\frac{\partial}{\partial m^2}\right)^z \ln(k^2-m^2) &= \lim_{z\to1}\Gamma(z)(k^2-m^2)^{-z}\\ &= \lim_{z\to1}\textstyle\int_0^\infty \alpha^{z-1}\exp[-\alpha(k^2-m^2)]\,\mathrm{d}\alpha\end{aligned} \tag{5.4}$$

† Up to z-dependent factors which reduce to unity at $z=1$.

be made for all the internal propagators in a Feynman diagram. This regularizes all the Feynman integrals (Guttinger 1966, Bollini *et al* 1964, Caianiello *et al* 1969) such as (3.30) because Σ is kept away from the dangerous integer values till the end; the infinities are converted into singularities at $z=1$. For example, consider meson-meson scattering in $g\phi^4$ theory. Analytic regularization gives us

$$T_4(p_i) = -288ig^2 \lim_{z\to 1} \sum_{ij} \Gamma^2(z) \int d^4k[k^2-\mu^2)^{-z}[(k+p_i+p_j)^2-\mu^2]^{-z}$$

$$= \frac{18g^2}{\pi^2} \lim_{z\to 1} \sum_{ij} \int_0^1 d\alpha \frac{[\alpha(1-\alpha)]^{z-1}\,\Gamma(2z-2)}{[(p_i+p_j)^2\,\alpha(1-\alpha)-\mu^2]^{2z-2}} \tag{5.5}$$

and

$$Z_g = 1 - \frac{54g}{\pi^2} \lim_{z\to 1} \int_0^1 d\alpha[\alpha(1-\alpha)]^{z-1}\,\Gamma(2z-2)[4\mu^2\alpha(1-\alpha)/3-\mu^2]^{2-2z}. \tag{5.6}$$

After subtracting the infinite coupling renormalization the physical scattering amplitude is well behaved as $z\to 1$. This example is the lowest order manifestation of the singular series

$$Z(g, z) = \sum_{n=0}^{\infty} \frac{P_n(g)}{(z-1)^n} \tag{5.7}$$

which states that the renormalization constants are given by a Laurent expansion about $z=1$ with polynomial residues (of degree g^n). That there are no other types of singularity at $z=1$ arising in the perturbation series is, at first sight, a surprising theorem. However, when one realizes that these Z correspond to local Lagrangian counter-terms the result is essential for making a success of the renormalization programme.

We remark that the Laurent expansion (5.7) corresponds to extracting purely the pole terms from Green's functions like (5.5), e.g.

$$Z_g = 1 - T_4/g \sim 1 - 27g/\pi^2(z-1).$$

It differs from the BPHZ prescription (5.6) by a finite calculable part and corresponds to choosing different normalizations at the subtraction points for the superficially divergent C. This brings us to the question of ambiguities of analytic regularization, namely the multiplication of modified propagators (5.4) by arbitrary functions $f(z)$ such that $f(1)=1$. If we carry out such a change in (5.5) the effect is simply to alter the finite part of the integral,

$$\lim_{z\to 1} f(z)\, I(z)\, \Gamma(2z-2) = \frac{I(1)}{2(z-1)} - 2c + \tfrac{1}{2}[f'(1)+I'(1)] + O(z-1).$$

Therefore if we renormalize according to (5.7) the physical amplitude undergoes a calculable renormalization. (Of course, if we adopt the BPHZ procedure (5.6) is correspondingly affected but the physical amplitude stays the same.)

There is one snag with analytic regularization in its primitive form (5.4): it is not a gauge-invariant operation. Even to lowest order, Ward identities such as (4.10) are destroyed:

$$\partial(p^2-m^2)^z/\partial p_\mu = z(p^2-m^2)^{z-1}\, 2p_\mu \neq 2p_\mu.$$

We should not expect $\Pi_{\mu\nu}$ to be transverse if we tamper with the individual charged-particle lines without at the same time changing the vertices of the theory (Kroll 1966,

Slavnov 1972, Lee and Zinn Justin 1972). To restore the gauge symmetry one needs to express the analytically regularized Lagrangian in a gauge-invariant form. For scalar electrodynamics say

$$
\begin{aligned}
\mathscr{L}_z &= \phi^\dagger(D^2+m^2)^z\phi - \tfrac{1}{4}F_{\mu\nu}(\partial^2)^{z-1}F^{\mu\nu} \\
&= \phi^\dagger(\partial^2+m^2+\mathrm{i}e.A.\overleftrightarrow{\partial}+e^2A^2)^z\phi - \tfrac{1}{4}F_{\mu\nu}(\partial^2)^{z-1}F^{\mu\nu} \\
&= \phi^\dagger[(\partial^2+m^2)^z+z(\partial^2+m^2)^{z-1}(\mathrm{i}eA.\overleftrightarrow{\partial}+e^2A^2)-\tfrac{1}{2}z(z-1)(\partial^2+m^2)^{z-2} \\
&\quad\times e^2A_\mu A_\nu\overleftrightarrow{\partial}^\mu\overleftrightarrow{\partial}^\nu+\mathrm{O}(e^3)]\,\phi-\tfrac{1}{4}F_{\mu\nu}(\partial^2)^{z-1}F^{\mu\nu}
\end{aligned}
\tag{5.8}
$$

implies that we need to consider an infinite number of photon–meson vertices. Letting $\Delta_z^{-1}(p)\equiv(p^2-m^2)^z/\Gamma(z)$, we require to order e^2 the one- and two-photon vertices

$$\Gamma_\mu(p',p)=e(p'+p)_\mu\,[\Delta_z^{-1}(p')-\Delta_z^{-1}(p)]/(p'^2-p^2) \tag{5.9}$$

and

$$
\begin{aligned}
\Gamma_{\mu\nu}(p'k',pk)/e^2 &= 2\eta_{\mu\nu}[\Delta_z^{-1}(p+k)-\Delta_z^{-1}(p'-k)]/[(p+k)^2-(p'-k)^2] \\
&\quad -\frac{(p'+p+k)_\nu(p'+p-k')_\mu}{(p+k)^2-(p-k')^2}\left[\frac{\Delta_z^{-1}(p+k)-\Delta_z^{-1}(p')}{(p+k)^2-p'^2}\right. \\
&\qquad\qquad\left.-\frac{\Delta_z^{-1}(p'-k)-\Delta_z^{-1}(p')}{(p'-k)^2-p'^2}\right] \\
&\quad -\frac{(p'+p-k)_\nu(p'+p+k')_\mu}{(p+k)^2-(p'-k)^2}\left[\frac{\Delta_z^{-1}(p+k)-\Delta_z^{-1}(p)}{(p+k)^2-p^2}\right. \\
&\qquad\qquad\left.-\frac{\Delta_z^{-1}(p-k')-\Delta_z^{-1}(p)}{(p-k')^2-p^2}\right]
\end{aligned}
$$

which automatically satisfy (4.10) and (4.19). Then it is easily verified that

$$\Pi_{\mu\nu}(k)=-\mathrm{i}e^2\int \bar{\mathrm{d}}^4p\,\Delta_z(p)[\Gamma_\mu(p,p+k)\,\Gamma_\nu(p+k,p)\,\Delta_z(p+k)-\Gamma_{\mu\nu}(pk,pk)]$$

correctly obeys

$$k^\mu\Pi_{\mu\nu}(k)=-\mathrm{i}e^2\int \bar{\mathrm{d}}^4p[\Gamma_\nu(p+k,p)\,\Delta_z(p+k)-\Gamma_\nu(p,p-k)\,\Delta_z(p)]$$

because for $z\neq 1$ one can shift origin confidently.

The same method can be extended to spinors. The standard procedure is to modify the fermion propagator from $S_c(p)$ to

$$S_z(p)=\Gamma(z)(\gamma.p+m)/(p^2-m^2)^z. \tag{5.10}$$

Since

$$
\begin{aligned}
\partial S_z^{-1}(p)/\partial p_\mu &\propto \partial[(\gamma.p-m)(p^2-m^2)^{z-1}]/\partial p_\mu \\
&= \gamma_\mu(p^2-m^2)^{z-1}+2(z-1)(\gamma.p-m)(p^2-m^2)^{z-2}
\end{aligned}
$$

one can again see the necessity for changing the primary vertices to comply with gauge invariance. However, the approach is rather difficult to implement and makes heavy work of even the simplest fermion loop calculation. In an attempt to simplify all of this Breitenhohner and Mitter (1968, 1972) (see also Collop 1975) offer an alternative prescription for analytic regularization which is closer in spirit to (5.1). They suggest breaking up the initial fermion loop numerator in the integrand,

$$P(m,p_i)=\mathrm{Tr}\,[\Gamma_1(\gamma.p_1+m)\,\Gamma_2\ldots\Gamma_n(\gamma.p_n+m)]$$

in the form

$$P(m, p_i) = P_1(m^2, p_i) + mP_2(m^2, p_i).$$

Next, they define a continued polynomial

$$\begin{aligned} Q(\kappa^2) &= P_1(\kappa^2, p_i) + mP_2(\kappa^2, p_i) \\ &= [(m+\kappa)\, P(\kappa; p_i) - (m-\kappa)\, P(-\kappa, p_i)]/2\kappa \end{aligned} \tag{5.11}$$

and finally they advocate substituting for the original Feynman integral

$$I(p_i) = \int \mathrm{d}^4 p \int \prod_i \mathrm{d}\alpha_i \delta(1-\Sigma\alpha)\, \Gamma(n)\, P(m, p_i)[\Sigma\, \alpha_i p_i^2 - m^2 + \mathrm{i}0]^{-n}$$

the analytically regularized expression

$$I_z(p_i) = \int \mathrm{d}^4 p \int \prod_i \mathrm{d}\alpha_i \delta(1-\Sigma\,\alpha) \left[\left(-\frac{\partial}{\partial \kappa^2}\right)^{n-1} \frac{Q(\kappa^2)\,\Gamma(z)}{[\kappa^2 - m^2 + \mathrm{i}0]^z}\right]_{\kappa^2 = \Sigma \alpha p^2} \tag{5.12}$$

which has eventually to be taken in the limit $z \to 1$—it may require subtraction if infinite.

We can check that their prescription is gauge invariant by going back to vacuum polarization in spinor electrodynamics. Thus

$$\begin{aligned} \Pi_{\mu\nu}(k) &= \mathrm{i}e^2 \int \mathrm{d}^4 p\, \mathrm{Tr}\, [\gamma_\mu S_c(p+k)\, \gamma_\nu S_c(p)] \\ &= \mathrm{i}e^2 \int\int \mathrm{d}^4 p\, \mathrm{d}\alpha P_{\mu\nu}(m)/[p^2 + k^2\alpha(1-\alpha) - m^2]^2 \end{aligned}$$

where

$$P_{\mu\nu}(m) = 4[\eta_{\mu\nu}\{-\tfrac{1}{2}p^2 + k^2\alpha(1-\alpha) + m^2\} - 2\alpha(1-\alpha)\, k_\mu k_\nu]$$

is automatically of P_1 type. Consequently,

$$\begin{aligned} \Pi_z^{\mu\nu}(k) &= \mathrm{i}e^2 \int\int \mathrm{d}^4 p\, \mathrm{d}\alpha \left[\left(-\frac{\mathrm{d}}{\mathrm{d}\kappa^2}\right) \frac{P^{\mu\nu}(\kappa)}{(\kappa^2 - m^2)^z}\right]_{\kappa^2 = p^2 + k^2\alpha(1-\alpha)} \\ &= \frac{e^2}{2\pi^2}(k^\mu k^\nu - k^2\eta^{\mu\nu}) \int \mathrm{d}\alpha\, \alpha(1-\alpha)\, \Gamma(z-1)[k^2\alpha(1-\alpha) - m^2]^{1-z}. \end{aligned} \tag{5.13}$$

Essentially the same result is found using analytic mass regularization (see (5.2) for the counterpart scalar case).

One can go somewhat further and recover the axial anomaly by these methods. The crucial element in the new approach is the non-commutative nature of mass multiplication and regularization operations. Thus, since the pseudoscalar current element (4.30),

$$\Gamma_5 = \int \mathrm{d}\alpha\, \mathrm{d}\beta \theta(1-\alpha-\beta)\, m/(K^2 - m^2)$$

is of type P_2 (referring to (5.11)) the prescription gives

$$(\Gamma_5)_z = \int \mathrm{d}\alpha\, \mathrm{d}\beta \theta(1-\alpha-\beta)\, m/(\kappa^2 - m^2)^z \big|_{\kappa^2 = K^2}$$

to be contrasted with

$$(m\Gamma_5)_z = \int \mathrm{d}\alpha\, \mathrm{d}\beta \theta(1-\alpha-\beta)\, \kappa^2/(\kappa^2 - m^2)^z \big|_{\kappa^2 = K^2}.$$

One sees that

$$(m\Gamma_5)_z - m(\Gamma_5)_z = \int \mathrm{d}\alpha\, \mathrm{d}\beta \theta(1-\alpha-\beta)/(K^2 - m^2)^{z-1}$$

and in the limit $z \to 1$ provides a finite anomaly, which is nothing less than the Adler term.

5.3. Massless superpropagators

The analytically regularized massless propagator

$$D_z(k)=\Gamma(z)(k^2+\mathrm{i}0)^{-z} \tag{5.14}$$

is easily Fourier transformed (Bollini *et al* 1964). In $2l$ dimensions, ($r^2\equiv -x^2$),

$$\begin{aligned} D_z(x) &= \int \bar{\mathrm{d}}^{2l}k\, \mathrm{e}^{-\mathrm{i}k.x}\Gamma(z)(k^2+\mathrm{i}0)^{-z} \\ &= \mathrm{i}(2\pi)^{-l}\,\Gamma(z)\, r^{1-l} \int_0^\infty \mathrm{d}\kappa J_{l-1}(\kappa r)\, \mathrm{e}^{\mathrm{i}\pi z}\kappa^{l-2z} \\ &= \mathrm{i}\, \mathrm{e}^{\mathrm{i}\pi z}\Gamma(l-z)(-x^2+\mathrm{i}0)^{z-l}/2^{2z}\pi^l. \end{aligned} \tag{5.15}$$

Thus the four-dimensional limit of (5.15) is proportional to $(D_c(x))^{2-z}$, which is reminiscent of the superpropagator (3.45) and (3.46) found in the context of non-polynomial Lagrangians. An application to the electron self-mass problem reveals the similarities in technique but differences in interpretation of the two methods. We start with the formal expression for

$$\delta m=(m-m_0)=m(1-Z_m)=-Z_\psi^{-1}\Sigma(p)|_{\gamma.p=m}$$

obtained in lowest order from

$$\begin{aligned} \Sigma(p) &= \mathrm{i}e^2 \int \mathrm{d}^4x\, \mathrm{e}^{\mathrm{i}p.x}\gamma_\mu S_c(x)\,\gamma^\mu D_c(x) \\ &\approx -2e^2 \int \mathrm{d}^4x\, \mathrm{e}^{\mathrm{i}p.x}\mathrm{i}D_c(x)\,\gamma.\partial\mathrm{i}D_c(x) \end{aligned} \tag{5.16}$$

where we have neglected the intermediate electron mass for simplicity.

In analytic regularization (5.16) is re-interpreted as the $z\to 1$ limit of

$$\begin{aligned} \underset{z}{\Sigma}(p) &= -2e^2 \int \mathrm{d}^4x\, \mathrm{e}^{\mathrm{i}p.x}\mathrm{i}D_z(x)\,\gamma.\partial\mathrm{i}D_c(x) \\ &= -\mathrm{i}e^2\,\mathrm{e}^{\mathrm{i}\pi z}2^{-2-2z}\pi^{-4}\Gamma(2-z)\,\gamma.\partial/\partial p \int \mathrm{d}^4x\, \mathrm{e}^{\mathrm{i}p.x}(-x^2+\mathrm{i}0)^{z-4} \\ &= -e^2\,\mathrm{e}^{\mathrm{i}\pi z}\Gamma(z-3)\,\gamma.p(-p^2)^{1-z}/32\pi^2. \end{aligned} \tag{5.17}$$

Hence $\delta m=\lim_{z\to 1} e^2m/64\pi^2(z-1)$ will be a divergent quantity corresponding to *infinite* bare mass.

On the other hand, in a non-polynomial context (5.16) represents the first term of an infinite series

$$\begin{aligned} \Sigma(p) &= -2e^2 \int \mathrm{d}^4x\, \mathrm{e}^{\mathrm{i}p.x} \sum_{n=0}^{\infty} v_n{}^2[\mathrm{i}D_c(x)]^{n+1}\,\gamma.\partial\mathrm{i}D_c(x) \\ &= -\mathrm{i}e^2\pi^{-1}\int_{0-} \mathrm{d}z v^2(z)\,\Gamma(-z)\,\mathrm{e}^{-\mathrm{i}\pi z}\int \mathrm{d}^4x\, \mathrm{e}^{\mathrm{i}p.x}\gamma.\partial[\mathrm{i}D_c(x)]^{z+2}/(z+2) \\ &= \mathrm{i}e^2\pi^{-1}\int_{0-} \mathrm{d}z v^2(z)[\Gamma(-z)]^2(4\pi)^{-2-2z}\,\gamma.p(p^2)^z/\Gamma(z+3). \end{aligned} \tag{5.18}$$

Usually this result is finite. For instance, if the non-polynomality is exponential, $V(\phi)=\exp(\lambda\phi)$, $V(z)=\lambda^z$ and the contour integral is dominated by the first singularity on the right, a *double* pole at $z=0$ whose residue gives

$$\delta m\approx e^2m\ln(\lambda^2m^2/16\pi^2)/8\pi^2. \tag{5.19}$$

It has been speculated that gravity (Isham *et al* 1971) with its characteristic scale $\lambda^{-1}\approx 10^{17}$ GeV acts as a cutoff in electrodynamics and that the electron mass is purely electromagnetic in origin, i.e. $m=\delta m$ or $m_0=0$. Certainly (5.19) is of the correct order of magnitude in accordance with this view. The muon mass is, of course, a separate issue.

5.4. Infrared problems

We may round off the subject of analytic regularization by returning to the infrared cancellation mechanism. Providing we keep away from $z=1$ in (5.4) or (5.14) the Feynman integrals are perfectly well defined as distributions; indeed, analytic regularization treats both ultraviolet and infrared infinities on an equal footing. In particular there is no need to introduce a small photon mass. Since the infrared infinities must reappear when $z \to 1$ we may ask how they are cancelled by soft bremsstrahlung processes. The answer is found by looking back at the example of scattering in a weak field.

Since there are no relevant electron loops in this problem it is enough to regularize the photon line. This gives the corrected vertex

$$\Gamma_z^\nu = e(p'+p)^\nu \left[1+\tfrac{1}{2}ie^2 \int \frac{\bar{d}^4k\,\Gamma(z)}{(k^2+i0)^z}\left\{\frac{\mu^2}{(k.p)^2}+\frac{\mu^2}{(k.p')^2}-\frac{2p.p'}{(k.p)(k.p')}\right\}\right] \qquad (5.20)$$

instead of (4.31). The singularities in the k_0 plane of $(k^2+i0)^{-z}(k.p-i0)^{-1}(k.p'-i0)^{-1}$ consist of two poles and two branch points (at $k_0 = \pm |\mathbf{k}| \mp i0$). Closing the contour above reduces the integral to

$$\Gamma_z^\nu = e(p'+p)^\nu[1+\pi e^2 \int \bar{d}^4k \,\mathrm{Disc}_+ (k^2)^{-z}\{\ldots\}]. \qquad (5.21)$$

The only remaining question concerns soft-photon emission. However, once we realize that

$$i\,\mathrm{Disc}_+ (-k^2)^{-1} \equiv i\theta(k)[(-k^2+i0)^{-1}-(-k^2-i0)^{-1}] = \theta(k)\,\delta(k^2)$$

is nothing but the photon phase-space factor, it is clear that in the analytic context one must regard

$$i\,\mathrm{Disc}_+ (-k^2)^{-z} \equiv i\theta(k)[(-k^2+i0)^{-z}-(-k^2-i0)^{-z}] = -2\theta(k)\,\theta(k^2)\sin \pi z/|k^2|^z \qquad (5.22)$$

as the photon phase space†. The bremsstrahlung cross section is then obtained by integrating the square of (4.32) up to a maximum photon energy ΔE. When added to (5.21) one obtains the cross section

$$\frac{d\sigma^{(2)}}{d\sigma^{(1)}} = 1+2e^2 \int \bar{d}^4k \left\{\frac{\mu^2}{(k.p)^2}+\frac{\mu^2}{(k.p')^2}-\frac{2p.p'}{(k.p)(k.p')}\right\} \theta(k-\Delta E)\frac{\sin \pi z}{|k^2|^z}$$

which is perfectly well behaved for $\Delta E \neq 0$ even as $z \to 1$; it necessarily reproduces (4.35) in the non-relativistic limit.

It is a plausible hypothesis that the cancellation mechanism works in general, providing we are careful to use the phase-space formula (5.22) whenever photons are emitted. To our knowledge an explicit proof of this to *all* orders is lacking.

6. Dimensional regularization

This is the last of the regularization methods that we shall try to cover. For reasons that will become apparent as we proceed it has gained enormously in popularity and now supersedes most other techniques. It differs in principle from them in that

† In a distributional sense

$$\lim_{z\to n} \frac{\theta(\xi)\xi^{-z}}{\Gamma(1-z)} = \delta^{(n-1)}(\xi)$$

so we correctly recover the usual phase space as $z \to 1$ in (5.22).

the propagators of the theory are left alone; what is changed is the dimensionality of phase space (Ashmore 1972, Cicuta and Montaldi 1972, Bollini and Giambiagi 1972a, 'tHooft and Veltman 1972). The idea of dimensional continuation comes from observation that the product $\Delta_c^2(x)$, which is so badly defined in four dimensions, exists perfectly well in two or three dimensions. Thus its Fourier transform,

$$\int \mathrm{đ}^{2l}p/[p^2-m^2][(p+k)^2-m^2]$$

is finite for $l=1$ or $\frac{3}{2}$ but not for $l=2$, as one can see trivially by power counting.

6.1. *Dimensional continuation*

Feynman integrals are normally taken over four space–time degrees of freedom. After performing various angular integrations, with the help of Feynman parameters, the final integration is over the length of the four-vector; it is at this ultimate stage that infinities, if they occur, are exposed as end-point singularities. Now there is no difficulty in practice in doing the self-same calculation for arbitrary integer dimensions as we indicated in §3.4. All that happens is that the radial (and angular) phase-space factors are altered in a well defined manner (Bollini and Giambiagi 1972b). The final radial integration then depends on the dimensionality ($2l$) of space–time†. The idea of dimensional regularization is to define the integral by analytic continuation from l regions where it exists to other l values. Here the infinities show up as singularities in the l plane (poles in Γ functions actually) which will require renormalization. Of course, the Feynman integral may involve a number of vector or spinor labels and a possible source of worry is what meaning one should attach to these for non-integer dimensions. Fortunately, this difficulty never arises when we come to deal with physical transition rates: these are always expressed in the form of scalar products for which the l continuation is straightforward.

A simple example which shows the simplicity and power of dimensional regularization is the Fourier transform of $(\Delta_c(x))^N$. For ultraviolet infinities we can effectively use the massless propagator in arbitrary dimensions

$$\mathrm{i}D_c(x)=\Gamma(l-1)(-x^2+\mathrm{i}0)^{1-l}/4\pi^l.$$

Then with $\kappa^2\equiv -k^2$,

$$\begin{aligned}\mathrm{i}\int \mathrm{d}^{2l}x\, \mathrm{e}^{\mathrm{i}k.x}[\mathrm{i}D_c(x)]^N &= (2\pi)^l\,\kappa^{1-l}\int_0^\infty \mathrm{d}r\, r^l J_{l-1}(\kappa r)[\Gamma(l-1)\,r^{2-2l}/4\pi^l]^N \\ &= (4\pi)^{l-lN}[\Gamma(l-1)]^N(-k^2+\mathrm{i}0)^{lN-l-N}\,\Gamma(l+N-lN)/\Gamma(lN-N) \qquad (6.1)\end{aligned}$$

is well behaved except when $(N-1)\,l=N,\ N+1,\ \ldots,$ *and* $l=1,\ 0,\ -1,\ \ldots.$ In particular, the range $1<l<2$ is safe. At $l=2$ there is a pole which has to be subtracted away by a renormalization counter-term. Thus we envisage that renormalization constants at any given order in perturbation theory are given by a Laurent series (Ashmore 1973, Speer 1974, Butera *et al* 1974, de Vega and Schaposnik 1974)

$$Z=\sum_n P_n(g)/(l-2)^n \qquad (6.2)$$

where, if the theory is renormalizable, the residues $P_n(g)$ are polynomials ('tHooft 1973a, b, Collins and Macfarlane 1974, 'tHooft and Veltman 1972) of degree g^n in

† To be more precise we are supposing that space and not time is continued and that the metric becomes $\eta_{\mu\nu}=\mathrm{diag}\,(1,\ -1,\ -1,\ \ldots,\ -1)$.

the dimensionless couplings g and other parameters (mass ratios, possibly) in the theory.

To check the validity of (6.2) consider $g\phi^4$ theory. (For arbitrary $2l$, g has dimensions M^{4-2l} so the dimensionless variable is truly the coupling $g\mu^{2l-4}$.) Now equation (4.5) and above are replaced by

$$T_4(p_i) = -288\mathrm{i}g^2 \sum_{ij} \iint \bar{\mathrm{d}}^{2l}k \, \mathrm{d}\alpha[k^2+(p_i+p_j)^2\alpha(1-\alpha)-\mu^2]^{-2}$$
$$=\frac{288g^2}{(4\pi)^l}\sum_{ij}\Gamma(2-l)[\mu^2-\alpha(1-\alpha)(p_i+p_j)^2]^{l-2}$$
$$\sim 54g^2/\pi^2(2-l) \qquad \text{near} \qquad l=2.$$

In the minimal subtraction procedure (6.2) we would simply put $Z_g=1-54g/\pi^2(2-l)$ to this order. On the other hand, in the BPHZ procedure we would instead set

$$Z_g=1-54g\mu^{2l-4}\Gamma(2-l)\int \mathrm{d}\alpha[1-4\alpha(1-\alpha)/3]^{l-2}$$

which differs by a finite part from the pure pole term. This brings us to possible ambiguities in dimensional continuation due to the freedom to multiply all Feynman integrals like (3.30) by any function $f(l)$ having the boundary value $f(2)=1$. In complete analogy with the discussion in §5.2 for analytic regularization we discover that the only change is a finite renormalization of the physical amplitudes and that if we insist in normalizing the proper vertices according to (4.4) the net effect is to alter (Kang 1975) the bare parameters $\mu_0, g_0, \ldots$ of the theory owing to re-evaluation of the Z.

Aside from this, the integrals are well defined because the point $l=2$ is avoided until the very end. The prescription also stipulates the value of the tadpole graph:

$$\mathrm{i}\Delta_c(0)=\mathrm{i}\int \bar{\mathrm{d}}^{2l}k/(k^2-m^2)=\Gamma(1-l)(m^2)^{l-1}/(4\pi)^l. \qquad (6.3)$$

In particular, since for $1<l\neq 2$ (6.3) vanishes for $m=0$ it is consistent to set the massless tadpole $\mathrm{i}D_c(0)$ equal to *zero* in dimensional regularization (Capper and Leibbrandt 1974). By the same token we interpret $\int \mathrm{d}^{2l}k\equiv 0$.

6.2. Field theory in arbitrary dimensions

We can put the logic of dimensional continuation (Wilson 1970) on a firmer basis by regarding the physical theory as the $l=2$ limit of a bona fide theory in arbitrary integer $2l$ dimensions. It is well known how to continue vectors, spinors, γ matrices, etc to any l so this part of the analysis is quite straightforward, e.g. spinor electrodynamics is still given by

$$\mathscr{L}=\bar{\psi}(\mathrm{i}\gamma.\partial-m+e\gamma.A)\,\psi-\tfrac{1}{4}(\partial_\mu A_\nu-\partial_\nu A_\mu)(\partial^\mu A^\nu-\partial^\nu A^\mu)$$

with A_μ interpreted as a $2l$-component vector and ψ as a 2^l-component spinor†. The theory is still invariant under the vector gauge transformations (2.9) and correspondingly the photon current $j_\mu=\bar{\psi}\gamma_\mu\psi$ is still conserved in the absence of external sources. Most importantly, it follows that dimensional continuation will respect the vector gauge identities. Computationally this is rather obvious because the naive

† We restrict l to integer values if the charge conjugation matrix is to exist with the property $C\gamma_\mu C^{-1}=-\tilde{\gamma}_\mu$, giving the photon negative charge parity.

Feynman rules and the relation between proper vertices are untouched by the dimensional continuation.

Let us return to scalar electrodynamics for the last time to emphasize these points. The formal integrals (4.16) and (4.17) are hardly affected but can now be integrated over the loop momentum without any strictures about shift of origin:

$$\Pi(p^2) = \mathrm{i}e^2 \iint \bar{\mathrm{d}}^{2l}k\, \mathrm{d}\alpha [k^2 + p^2(2-\alpha)^2]/[k^2 + p^2\alpha(1-\alpha) - \mu^2\alpha]^2$$
$$= \frac{e^2}{(4\pi)^l}\Gamma(1-l)\int_0^1 \mathrm{d}\alpha\, \frac{\mu^2\alpha l - p^2\{(1-l)(2-\alpha)^2 + l\alpha(1-\alpha)\}}{[\mu^2\alpha - p^2\alpha(1-\alpha)]^{2-l}} \tag{6.4}$$

$$\Pi_{\mu\nu}(k) = \frac{e^2}{(4\pi)^l}\Gamma(2-l)(k_\mu k_\nu - k^2\eta_{\mu\nu})\int_0^1 \mathrm{d}\alpha\, \frac{(1-2\alpha)^2}{[\mu^2 - k^2\alpha(1-\alpha)]^{2-l}}. \tag{6.5}$$

In deriving these (gauge-invariant) answers it is crucial to use the formulae (3.28) appropriate to any l. The ensuing renormalization constants are particular examples of (6.2):

$$\begin{aligned} Z_\phi &\sim 1 + e^2/8\pi^2(2-l) \\ Z_\mu &\sim 1 + 3e^2/16\pi^2(2-l) \\ Z_A &\sim 1 - e^2/48\pi^2(2-l). \end{aligned} \tag{6.6}$$

After inclusion of the Z the Green's functions become finite. For instance

$$D'^{-1}{}_{\mu\nu}(k) = -k^2\eta_{\mu\nu}\left[1 - \frac{e^2}{16\pi^2}\int_0^1 \mathrm{d}\alpha(1-2\alpha)^2 \ln\{1 - k^2\alpha(1-\alpha)/\mu^2\} + \ldots\right]$$
$$+ k_\mu k_\nu \text{ terms.} \tag{6.7}$$

Dimensional regularization can also be applied to abnormal amplitudes† and for extracting anomalies in Ward identities (Akyeampong and Delbourgo 1974, 1975b). To find the axial anomaly we must first define the generalization of pseudoscalar and pseudovectors to arbitrary dimensions. The natural choices $\bar{\psi}\gamma_0\gamma_1 \ldots \gamma_{2l-1}\psi$ and $\bar{\psi}\gamma_\mu\gamma_0 \ldots \gamma_{2l-1}, \psi$ are not appropriate since they do not allow an e^2 transition element for $\pi^0 \to 2\gamma$ when $l \geqslant 3$. Instead, we have to identify the less obvious candidates

$$\bar{\psi}\gamma_{[\kappa}\gamma_\lambda\gamma_\mu\gamma_{\nu]}\psi \equiv J_{[\kappa\lambda\mu\nu]} \qquad \text{as the pseudoscalar current}$$
$$\mathrm{i}\bar{\psi}\gamma_{[\lambda}\gamma_\mu\gamma_{\nu]}\psi \equiv J_{[\lambda\mu\nu]} \qquad \text{as the axial vector current.}$$

The $2l$-dimensional analogue of the Ward identity (4.29) turns out to be

$$\partial_{[\kappa}J_{\lambda\mu\nu]} = -2mJ_{[\kappa\lambda\mu\nu]} + \bar{\psi}(\tfrac{1}{2}\mathrm{i}\overleftrightarrow{\partial} + eA)^\rho\{\gamma_\rho, \gamma_{[\kappa\lambda\mu\nu]}\}\,\psi \tag{6.8}$$

where the last term does not exist for $l=2$ because the anticommutator vanishes there. When we take the two-photon matrix elements of (6.8) the axial and pseudoscalar currents correctly yield the vector gauge-invariant amplitudes at $l=2$. The last 'evanescent' term produces the axial anomaly. Thus it has associated the Feynman integral

$$2\mathrm{i}e^2 \int \bar{\mathrm{d}}^{2l}p\, \mathrm{Tr}\,[\{2p.\gamma, \gamma_{[\kappa\lambda\mu\nu]}\}\, S_c(p-k')\,\gamma_\sigma S_c(p)\,\gamma_\rho S_c(p+k)]$$
$$= 2\mathrm{i}e^2 \int \bar{\mathrm{d}}^{2l}p\, \frac{\mathrm{Tr}\,[\{2p.\gamma, \gamma_{[\kappa\lambda\mu\nu]}\}\,\gamma.k'\gamma_\sigma\gamma.p\gamma_\rho\gamma.k]}{[(p-k')^2 - m^2][p^2 - m^2][(p+k)^2 - m^2]} \tag{6.9}$$

whose numerator‡ carries a factor $(l-2)$ from the trace while the loop integration

† Amplitudes which are overall pseudoscalar, pseudovector, ...

‡ Since the anticommutator vanishes in four dimensions.

gives a singularity $(l-2)^{-1}$ because of the divergence in the momentum integral. The product is precisely the axial anomaly.

The argument also suggests when anomalies may be expected in a dimensional context (Akyeampong and Delbourgo 1975b). A four-dimensional Ward identity carries a possible anomaly when the $2l$-dimensional counterpart identity contains an evanescent term which disappears for four dimensions. For graphs with no closed loops (the classical limit of the theory) the anomalous terms can be discarded, but for closed loops (the quantum corrections in the theory) they can lead to non-zero answers. The dilation current anomaly is just such an example.

6.3. *Infrared problems*

Providing we keep away from $l=2$ all infinities, infrared and ultraviolet, are regularized. When $l\to 2$, however, they turn up in different ways. Consider

$$Z_\phi = 1 - \frac{e^2\mu^{2l-4}}{(4\pi)^l}\,\Gamma(2-l)\int_0^1 \mathrm{d}\alpha\,\alpha^{2l-4}\,[\alpha(1-\alpha)\,l-(2-\alpha)^2 \\ -(1-\alpha)(2-\alpha)^2(2-l)/\alpha]. \qquad (6.10)$$

The Γ function has a pole at $l=2$—this is the ultraviolet infinity. The infrared infinity instead manifests itself as an end-point singularity in the parametric integral:

$$\textstyle\int_0^1 \mathrm{d}\alpha\,\alpha^{2l-5} \equiv (2l-4)^{-1}$$

which would be absent if the photon were accorded a small mass κ. We may, in fact, isolate the infrared infinities by picking out the $\ln(\kappa/\mu)$ dependences in the renormalized Green's functions.

That the operations $\kappa\to 0$ and $l\to 2$ do not commute (Marciano and Sirlin 1975, Girotti 1975) is irrelevant so far as infrared cancellations are concerned. For that reason it is quite sufficient to set $\kappa=0$ from the start and just deal with the dimensionally regularized expressions. We can see the cancellation taking place (Gastmans and Meuldermans 1973) for the example of meson scattering in a weak field. The vertex corrected by a virtual photon line is

$$\Gamma_\nu = e(p'+p)_\nu\left[1+\tfrac{1}{2}\mathrm{i}e^2\int\frac{\bar{\mathrm{d}}^{2l}k}{k^2}\left\{\frac{\mu^2}{(k.p)^2}+\frac{\mu^2}{(k.p')^2}-\frac{2p.p'}{(k.p)(k.p')}\right\}\right. \qquad (6.11)$$

in the infrared limit. On the other hand, soft-photon emission is still described by amplitude (4.32). This must, however, be integrated over $2l$-dimensional phase space in working out the transition rate. When combined with the square of (6.11) one gets

$$\frac{\mathrm{d}\sigma^{(2)}}{\mathrm{d}\sigma^{(1)}} = 1-e^2\int \bar{\mathrm{d}}^{2l}k\left\{\frac{\mu^2}{(k.p)^2}+\frac{\mu^2}{(k.p')^2}-\frac{2p.p'}{(k.p)(k.p')}\right\} \\ \times[\theta(k)\,\theta(\Delta E-k)\,\delta(k^2)-\mathrm{i}(k^2+\mathrm{i}0)^{-1}] \\ = 1-e^2\int\bar{\mathrm{d}}^{2l}k\{\ldots\}\,\theta(k-\Delta E)\,\delta(k^2) \qquad (6.12)$$

for a maximum detectable photon energy ΔE. With the cancellation visibly done in (6.12) one can pass immediately to $l=2$.

6.4. *Recent developments*

It would be fair to say that the majority of the physics community has been won over to dimensional regularization (Leibbrandt 1976) for three very good reasons.

(i) The utter simplicity of the calculations contrasts favourably with more elaborate old fashioned ways.

(ii) The complete respect for vector gauge symmetry (Abelian or not) leaves intact the formal Ward identities.

(iii) Anomalies emerge naturally when the symmetry (scale, chiral, . . .) is lost in the continuation away from four dimensions. Since non-Abelian gauge theories have been at the centre of some remarkable advances in electromagnetic and weak interaction theory, dimensional regularization has proved a most powerful tool in seeing through the renormalization programme for unified gauge models where other regularizations have proved deficient or else too complicated to pursue. Nor has progress been limited to Yang–Mills theory, for dimensional techniques have done valuable service in quantum gravity (Capper *et al* 1973, 'tHooft 1973a, b), with its associated gauge identities, despite non-renormalizability.

Physics at asymptotic momenta ($p\to\infty$ or $p\to 0$), embodied in the renormalization group equations, has also benefited greatly from dimensional regularization ('tHooft 1973)—the proof of asymptotic freedom of non-Abelian gauge theories was only established this way. Perhaps the only places where the method leaves something to be desired is in weak interactions in which the role of the chiral projectors becomes problematic (Delbourgo and Prasad 1974) and in supersymmetric theories where even for the simplest models dimensional regularization is more difficult to use than rudimentary kinetic regularization (Delbourgo and Prasad 1975). These problems are under intensive study at present and one may hope that they will be favourably resolved in the near future.

Acknowledgments

I would like to thank Professor M D Scadron for his hospitality at the University of Arizona where part of this work was carried out. Also I am very grateful to colleagues, too numerous to mention individually, for the benefit of their personal insights.

Appendix. Notations and conventions

We work in natural units whereby $c=\hbar=1$. All units have dimensions expressed in terms of mass (M) or length ($L\sim M^{-1}$). The action $\int\mathscr{L}\,\mathrm{d}^4x$ is dimensionless and implies that $(\mathscr{L})=(M^4)$. This then fixes the dimensions of the fields and couplings appearing in the Lagrangian. For instance in examples (*a*) to (*c*) (§2.1),

$$[\phi]=[A_\mu]=[M], \qquad [\psi]=[M^{3/2}], \qquad [g]=[e]=[M^0].$$

Later when we extend the integrals to $2l$ dimensions, $(\mathscr{L})=(M^{2l})$ whereupon

$$[\phi]=[A_\mu]=[M^{l-1}], \qquad [\psi]=[M^{l-1/2}], \qquad [g]=[e^2]=[M^{4-2l}].$$

Four vectors of position and momentum are written as $x_\mu=(t,\boldsymbol{x})$ and $p_\mu=(E,\boldsymbol{p})$, etc and we adopt a time-like Minkowski metric η whereby

$$x.p=x^\mu p_\mu=x_\mu\eta^{\mu\nu}p_\nu=tE-\boldsymbol{x}.\boldsymbol{p}.$$

The gradient operator is written $\partial^\mu = \partial/\partial x_\mu = (\partial/\partial t, \partial/\partial \mathbf{x})$. For Fourier transforms we repeatedly use the notation

$$\text{đ}k = dk/2\pi, \qquad \bar{\delta}(k) = 2\pi\delta(k).$$

Thus $\bar{\delta}^4(p) = \int d^4x \exp(ip.x)$.

The Poincaré transformations $x_\mu \to \Lambda_\mu{}^\nu x_\nu + a_\mu$ have associated the Hilbert space operators $J_{\mu\nu}$, P_λ obeying the algebra

$$[J_{\kappa\lambda}, J_{\mu\nu}] = i(\eta_{\lambda\mu} J_{\kappa\nu} + \eta_{\kappa\nu} J_{\lambda\mu} - \eta_{\kappa\mu} J_{\lambda\nu} - \eta_{\lambda\nu} J_{\kappa\mu})$$
$$[J_{\mu\nu}, P_\lambda] = i(\eta_{\nu\lambda} P_\mu - \eta_{\mu\lambda} P_\nu)$$

and are implemented by the unitary operator

$$U(a, \Lambda) = \exp[ia.P] \exp[-\tfrac{1}{2}\theta^{\mu\nu} J_{\mu\nu}].$$

For fields the conventions are as follows.

(i) Scalar fields

$$U(a, \Lambda)\, \phi(x)\, U^{-1}(a, \Lambda) = \phi(\Lambda x + a).$$

(ii) Spinor fields

Dirac matrices: $\{\gamma_\mu, \gamma_\nu\} = 2\eta_{\mu\nu}$ with

$$\gamma_\mu{}^\dagger = \gamma_0\gamma_\mu\gamma_0, \qquad \tilde{\gamma}_\mu = -C^{-1}\gamma_\mu C$$
$$\gamma_5 \equiv \gamma_0\gamma_1\gamma_2\gamma_3 = -\gamma_5{}^\dagger = C\tilde{\gamma}_5 C^{-1}$$

and

$$C = i\gamma_0\gamma_2.$$

The spinor generators $S_{\mu\nu} = \tfrac{1}{2}\sigma_{\mu\nu} = \tfrac{1}{4}i[\gamma_\mu, \gamma_\nu]$ enter in Poincaré transformations:

$$U(a, \Lambda)\, \psi(x)\, U^{-1}(a, \Lambda) = D^{1/2}(\Lambda^{-1})\, \psi(\Lambda x + a)$$

via the matrix $D^{1/2}(\Lambda) = \exp(-\tfrac{1}{4}i\theta^{\mu\nu}\sigma_{\mu\nu})$.

The adjoint spinor $\bar{\psi} = \psi^\dagger\gamma_0$ transforms oppositely, i.e.

$$U(a, \Lambda)\, \bar{\psi}(x)\, U^{-1}(a, \Lambda) = \bar{\psi}(\Lambda x + a)\, D^{1/2}(\Lambda).$$

The c-number solutions of the Dirac equation for positive and negative energies referred to in the text are defined by

$$(\gamma.p - m)\, u(p, \lambda) = 0, \qquad (\gamma.p + m)\, v(p, \lambda) = 0$$

and $v = C\tilde{\bar{u}}$. They satisfy the completeness relation

$$\bar{u}(p, \lambda')\, u(p, \lambda) = 2m\delta_{\lambda'\lambda}, \qquad \bar{v}(p, \lambda)\, v(p, \lambda') = -2m\delta_{\lambda'\lambda}$$
$$\sum_\lambda u(p, \lambda)\, \bar{u}(p, \lambda) = (\gamma.p + m), \qquad \sum_\lambda v(p, \lambda)\, \bar{v}(p, \lambda) = (\gamma.p - m).$$

(iii) Vector fields

Here we have the spin representation $S_{\mu\nu} \to \Sigma_{\mu\nu}$ where

$$(\Sigma_{\mu\nu})_{\kappa\lambda} = i(\eta_{\mu\kappa}\eta_{\nu\lambda} - \eta_{\mu\lambda}\eta_{\nu\kappa}).$$

The analogous transformation formula

$$U(a, \Lambda)\, A_\rho(x)\, U^{-1}(a, \Lambda) = D_{\rho\sigma}{}^1(\Lambda^{-1})\, A^\sigma(\Lambda x + a)$$

involves the matrix $D^1(\Lambda) = \exp(-\frac{1}{2}i\theta^{\mu\nu}\Sigma_{\mu\nu})$.

The free-field solutions (polarization vectors) of a spin-one field $\epsilon_\mu(k, \lambda)$ satisfy $k^\mu \epsilon_\mu(k, \lambda) = 0$. They are

$$\epsilon_\mu(k, 0) = (|\boldsymbol{k}|, E\hat{\boldsymbol{k}})/m, \qquad \epsilon_\mu(k, \pm 1) = (0, \boldsymbol{n}_1 \pm i\boldsymbol{n}_2)^{-1/2}$$

where $\boldsymbol{n}_i . \boldsymbol{k} = 0$, $\boldsymbol{n}_i . \boldsymbol{n}_j = \delta_{ij}$. They form an orthonormal set,

$$\epsilon_\mu{}^*(k, \lambda)\, \epsilon^\mu(k, \lambda') = -\delta_{\lambda\lambda'}.$$

The longitudinal state ($\lambda = 0$) is irrelevant for real photons.

References

Adler S 1969 *Phys. Rev.* **177** 2426
Akyeampong DA and Delbourgo R 1974 *Nuovo Cim.* A **17** 58
—— 1975a *Nuovo Cim.* A **19** 141
—— 1975b *Nuovo Cim.* A **19** 249
Ashmore J 1972 *Lett. Nuovo Cim.* **4** 289
—— 1973 *Comm. Math. Phys.* **29** 177
Bardeen WA 1969 *Phys. Rev.* **184** 1848
Bell JS and Jackiw R 1969 *Nuovo Cim.* A **60** 47
Bollini CG and Giambiagi J 1972a *Phys. Lett.* **40B** 5661
—— 1972b *Nuovo Cim.* B **12** 20
Bollini CG, Giambiagi JJ and Gonzales DA 1964 *Nuovo Cim.* **31** 550
Breitenhohner P and Mitter H 1968 *Nucl. Phys.* B **7** 443
—— 1972 *Nuovo Cim.* A **10** 655
Brown RW, Shih CF and Young BN 1969 *Phys. Rev.* **186** 1491
Butera P, Cicuta GM and Montaldi E 1974 *Nuovo Cim.* A **19** 513
Caianiello ER, Guerra F and Marinaro M 1969 *Nuovo Cim.* A **60** 713
Callan C, Coleman S and Jackiw R 1970 *Ann. Phys., NY* **59** 42
Capper DM and Leibbrandt G 1974 *J. Math. Phys.* **15** 795
Capper DM, Leibbrandt G and Ramon Medrano M 1973 *Phys. Rev.* D **8** 4370
Chanowitz M and Ellis S 1972 *Phys. Lett.* **40B** 317
Cicuta GM and Montaldi E 1972 *Lett. Nuovo Cim.* **4** 329
Collins JC and Macfarlane A 1974 *Phys. Rev.* D **10** 1201
Collop DJ 1975 *Nucl. Phys.* B **85** 415
Delbourgo R and Prasad VB 1974 *Nuovo Cim.* A **23** 257
—— 1975 *J. Phys. G: Nucl. Phys.* **1** 377
Delbourgo R, Salam A and Strathdee J 1969 *Phys. Rev.* **187** 1999
Efimov GV 1963 *Sov. Phys.-JETP* **17** 1417
—— 1964 *Nuovo Cim.* **32** 1046
Eriksson KE 1961 *Nuovo Cim.* **19** 1010
Feynman RP 1949 *Phys. Rev.* **76** 769
Fradkin ES 1963 *Nucl. Phys.* **49** 624
Frautschi S, Suura H and Yennie DR 1961 *Ann. Phys., NY* **13** 379
Fried HM 1967 *Nuovo Cim.* A **52** 1333
Gastmans R and Meuldermans R 1973 *Nucl. Phys.* B **63** 277
Gelfand IM and Shilov GE 1964 *Generalized Functions* vol 1 (London: Academic Press)
Girotti HO 1975 *University of Trieste Preprint* IC/75/62
Gupta SN 1953 *Proc. Phys. Soc.* A **66** 129
Guttinger W 1966 *Fortschr. Phys.* **14** 489
Hagen CR 1969a *Phys. Rev.* **177** 2622
—— 1969b *Phys. Rev.* **188** 2416

'tHooft G 1973a *Nucl. Phys.* B **61** 455
—— 1973b *Nucl. Phys.* B **62** 444
'tHooft G and Veltman M 1972 *Nucl. Phys.* B **44** 189
Isham OJ, Salam A and Strathdee J 1971 *Phys. Rev.* D **3** 1805
Jackiw R and Johnson K 1969 *Phys. Rev.* **182** 1459
Johnson K 1964 *Brandeis Lecture Notes* (Englewood Cliffs, New Jersey: Prentice Hall)
Kallen G 1951 *Ark. Phys.* **5** 130
Kang JS 1976 *Phys. Rev.* D **13** 851
Karlson E 1954 *Ark. Phys.* **7** 221
Kroll NM 1966 *Nuovo Cim.* A **45** 65
Lee BW and Zinn Justin J 1972 *Phys. Rev.* D **5** 3121
Lee BW and Zumino B 1969 *Nucl. Phys.* B **13** 671
Lehmann H and Pohlmeyer K 1971 *Comm. Math. Phys.* **20** 101
Leibbrandt G 1976 *Rev. Mod. Phys.* **47** 849
Mandelstam S 1962 *Ann. Phys., NY* **19** 1
Marciano WJ and Sirlin A 1975 *Nucl. Phys.* B **88** 86
Okubo S 1954 *Prog. Theor. Phys.* **11** 80
Pais A 1947 *Kon. Ned. Akad. Wet.-Verb.* D **1** 19
Pais A and Uhlenbeck GE 1950 *Phys. Rev.* **79** 145
Pauli W and Villars F 1949 *Rev. Mod. Phys.* **21** 434
Riesz M 1949 *Acta Math.* **81** 1
Rivier D 1949 *Helv. Phys. Acta* **22** 265
Salam A 1970 *Coral Gables Conference* (London: Gordon and Breach)
Salam A and Strathdee J 1970 *Phys. Rev.* D **1** 3296
Schwinger J 1949a *Phys. Rev.* **75** 651
—— 1949b *Phys. Rev.* **76** 790
—— 1959 *Phys. Rev. Lett.* **3** 296
Slavnov AA 1972 *Theor. Math. Phys.* **13** 1064
Speer E 1968 *J. Math. Phys.* **9** 1404
—— 1974 *J. Math. Phys.* **15** 1
Stückelberg ECG and Rivier D 1948 *Phys. Rev.* **74** 218
de Vega HJ and Schaposnik M 1974 *J. Math. Phys.* **15** 1998
Volkov MK 1968 *Comm. Math. Phys.* **7** 289
Wilson K 1970 *Phys. Rev.* D **2** 1473

Fortschr. Phys. **61**, No. 11, 994 – 1027 (2013) / **DOI** 10.1002/prop.201300022

The quark-level linear σ model

Michael D. Scadron[1,*], **George Rupp**[2,**], and **Robert Delbourgo**[3,***]

[1] Physics Department, University of Arizona, Tucson, AZ 85721, USA

[2] Centro de Física das Interacções Fundamentais, Instituto Superior Técnico, Universidade de Lisboa, 1049-001 Lisboa, Portugal

[3] School of Mathematics and Physics, University of Tasmania GPO Box 252-21, Hobart 7001, Australia

Received 21 August 2013, accepted 2 September 2013
Published online 10 October 2013

This review of the quark-level linear σ model (QLLσM) is based upon the dynamical realization of the pseudoscalar and scalar mesons as a linear representation of $SU(2) \times SU(2)$ chiral symmetry, with the symmetry weakly broken by current quark masses. In its simplest $SU(2)$ incarnation, with two non-strange quark flavors and three colors, this nonperturbative theory, which can be selfconsistently bootstrapped in loop order, is shown to accurately reproduce a host of low-energy observables with only one parameter, namely the pion decay constant f_π. Extending the scheme to $SU(3)$ by including the strange quark, equally good results are obtained for many strong, electromagnetic, and weak processes just with two extra constants, viz. f_K and $\langle \pi | H_{\text{weak}} | K \rangle$. Links are made with the vector-meson-dominance model, the BCS theory of superconductivity, and chiral-symmetry restoration at high temperature. Finally, these ideas are cautiously generalized to the electroweak sector, including the W, Z, and Higgs bosons, and also to CP violation.

© 2013 WILEY-VCH Verlag GmbH & Co. KGaA, Weinheim

1 Introduction

The magnitude of the strong interaction between the hadrons precludes the use of perturbation theory (PT) and this has been understood for a very long time. Only in the asymptotic high-energy regime, where the QCD coupling becomes logarithmically small, does it make any sense to use PT, and then only for the interactions involving gluons with their effective coupling. Thus it has been the goal of particle physicists to use nonperturbative schemes in order to tackle, with any semblance of reliability or conviction, the low-energy features of hadronic interactions. The sense of this approach is highlighted by the fact that the current quark masses are so much smaller than the constituent quark masses within the hadron, so that the extra mass is provided by the cloud of mesons and gluons which comprise the sum total. Foremost amongst these nonperturbative approaches has been the application of spontaneously broken chiral symmetry, accompanied by current algebra. For a good while the nonlinear realization of chiral symmetry at zero energy was used, together with an expansion in powers of momentum in order to get away from that particular limit. Unfortunately this has led to a plethora of expansion parameters and it blunts the use of the nonlinear theory predictions.

However it has also been found that the linear realization of chiral symmetry at the quark level is an alternative way of handling the low-energy properties of hadrons, without abandoning spontaneous breaking concepts introduced by Nambu [1]. In particular the Gell-Mann–Lévy model [2, 3], constrained by a vanishing of the renormalization constants of the mesons (which makes them composite states) is extremely predictive for a host of observed phenomena. Indeed it turns out that for pionic interactions essentially *every* low energy feature is determined completely just by one scale, the pion weak decay

* E-mail: scadron@physics.arizona.edu
** Corresponding author E-mail: george@ist.utl.pt, Phone: +351 218 419 103, Fax: +351 218 419 143
*** E-mail: Bob.Delbourgo@utas.edu.au

WILEY InterScience DISCOVER SOMETHING GREAT

© 2013 WILEY-VCH Verlag GmbH & Co. KGaA, Weinheim

constant $f_\pi \simeq 92.2$ MeV. In the chiral limit, where the pion mass vanishes, all the other constants are totally fixed. Thus with just three colors, one can determine that

- the pion-quark coupling is $g = 2\pi/\sqrt{3}$;
- the quartic pion-pion interaction is $\lambda = 2g^2 = 8\pi^2/3$;
- the constituent nonstrange quark mass is $\hat{m} = f_\pi g \simeq 335$ MeV;
- the sigma meson partner to the pion has a mass $m_\sigma = 2\hat{m} \simeq 670$ MeV.

It is even possible to extrapolate away from the chiral limit, by allowing for small current quark masses (which mar the chiral symmetry slightly) and thereby determine the deviations.

All this is explained in detail in Sects. 1, 2, and 3. There, as in succeeding sections, we compare the predictions of the quark-level linear σ model (QLLσM) with other methods, based on other premises. In Sect. 4 we revisit the compositeness condition ($Z = 0$), and show how this can be used to set a demarcation scale between scalar and vector mesons. The importance of chiral cancellations is reviewed in Sect. 5, as it explains the vanishing of certain amplitudes, which might otherwise be quite large. This also affects the π-N so-called sigma term associated with scattering lengths, treated in Sect. 6. Sect. 7 is devoted to the pion charge radius, which is again fully determined in terms of f_π as $r_\pi = \hbar c\sqrt{3}/2\pi f_\pi \simeq 0.61$ fm, and can be contrasted with values obtained through the vector dominance model, incidentally explaining the value of the ρ-π-π coupling constant as well. The breakdown of chiral symmetry at higher temperatures is considered next (Sect. 8), and this occurs at a critical temperature of $2f_\pi$; comparisons with the BCS (Bardeen, Cooper, Schrieffer [4]) and NJL (Nambu–Jona-Lasinio) [5]) models are described there, too.

In Sect. 9, we review the extension to $SU(3)$ by inclusion of the strange quark mass. Now the ratio f_K/f_π fixes the constituent strange quark mass to be about 470 MeV, but its couplings to the quarks remain the same as the pion's. Furthermore, the κ analogue of the σ meson is estimated to be about 797 MeV in mass. This is consonant with equal-mass splitting laws between scalar and pseudoscalar mesons, and in the process we review the mixing-angle parameters. Sect. 10 covers e.m. decay rates, as they constitute clean tests of all that has gone before and generally fit the data very well, including the isoscalar scalar $f_0(500)$ [6] meson. The other light scalar isoscalar $f_0(980)$, as well as its isovector partner $a_0(980)$, are dealt with in more detail in Sect. 11, in particular concerning their strong decays. Sects. 12 and 13 are devoted to weak decays, which are governed by one new scale: the $K, \bar{K}$ matrix element of the weak Hamiltonian H_w, or equivalently the transition element $\langle\pi|H_w|K\rangle$. Ramifications of these ideas to other weak decays are also treated. The e.m. form factors of mesons, their role in certain weak decays, as well as the estimation of meson polarizabilities form the subject of Sect. 14. In Sect. 15 we establish a link between the critical temperature, mentioned in Sect. 8, and the BCS theory of superconductivity with its characteristic energy gap.

Sect. 16 makes an analogy between the QLLσM and the standard electroweak model. In that picture the Higgs boson is regarded largely as a top-antitop scalar bound state with a mass of about 315 GeV, set by a weak decay constant of $f_w \simeq 246$ GeV. This picture is consonant with a weak KSRF (Kawarabayashi, Suzuki, Riazuddin, Fayyazuddin) [7, 8] relation and the observed masses of the weak vector bosons plus the weak mixing angle. Possible implications of recently observed [9, 10] Higgs-like signals at the large hadron collider (LHC) of CERN are discussed as well. We conclude this review in Sect. 17 by an analysis of CP violation, as supposed to arise from a nonstandard $WW\gamma$ vertex.

2 $SU(2)$ QLLσM

First we state the $SU(2)$ quark-level linear σ model (QLLσM) Lagrangian density, with interacting part

$$\mathcal{L}_{\text{int}}^{\text{L}\sigma\text{M}} = g\bar{\psi}(\sigma + i\gamma_5\vec{\tau}\cdot\vec{\pi})\psi + g'\sigma(\sigma^2 + \vec{\pi}^2) - \frac{\lambda}{4}(\sigma^2 + \vec{\pi}^2)^2 , \quad (1)$$

 © 2013 WILEY-VCH Verlag GmbH & Co. KGaA, Weinheim

with the chiral-limiting (CL) pion-quark and meson-meson couplings

$$g \;=\; \frac{m_q}{f_\pi} \quad , \quad g' \;=\; \frac{m_\sigma^2}{2f_\pi} = \lambda f_\pi \; . \tag{2}$$

This QLLσM is in the spirit of the original Gell-Mann–Lévy LσM [2,3], but for quarks [11,12] rather than for nucleon fermions, and also with Nambu–Goldstone [1,13,14] pseudoscalar pions, having vanishing mass in the chiral limit, i.e., $m_\pi^{\rm CL} = 0$. Note, however, that the nonstrange pion and sigma mesons in Eq. (1) are quantum fields which both vanish in the CL. Such a vanishing does not occur in [2,3] for spin-$1/2$ nucleons, in contrast with the present QLLσM scheme. Furthermore, in the chiral limit the nonstrange constituent quark mass $\hat{m}_{\rm con} = (m_u + m_d)/2$ is half the mass of the σ meson, i.e.,

$$m_\sigma^{\rm CL} \;=\; 2\hat{m}_{\rm con}^{\rm CL} \; , \tag{3}$$

which is valid for both the QLLσM of Eq. (1) and the nonlinear NJL [5] model. The Goldberger–Treiman relation (GTR) [15] for the QLLσM reads $\hat{m}_{\rm con} = f_\pi g$, which should be compared with the GTR for nucleons, viz. $g_A m_N = f_\pi g_{\pi NN}$, but now at the quark level, with g_A=1 for constituent quarks [16].

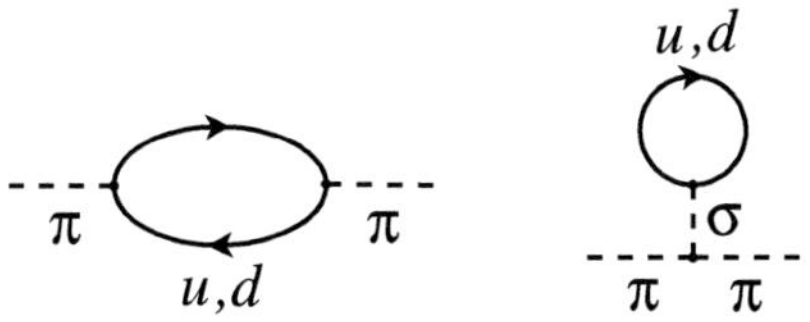

Fig. 1 Pion-selfenergy quark-loop graphs. Left: bubble; right: σ tadpole.

Next we follow [17,18], and compute the pseudoscalar pion mass, which vanishes in the CL, via the selfenergy graphs of Figs. 1 and 2. The quark loops (QL) of Fig. 1 give a pion mass squared

$$m_{\pi,{\rm QL}}^2 \;=\; -i8N_c g^2 \left(1 - \frac{2g' f_\pi}{m_\sigma^2}\right) \int \frac{\bar{d}^4 p}{p^2 - m_q^2} \;=\; 0 \; , \tag{4}$$

with $\bar{d}^4 p \equiv d^4 p/(2\pi)^4$. Now, $m_{\pi,{\rm QL}}^2$ vanishes identically in the CL, since then $g' = m_\sigma^2/2f_\pi$. As for the meson-loop (ML) graphs in Fig. 2, we first invoke the partial-fraction identity

$$\frac{g'^2}{(p^2 - m_\sigma^2)(p^2 - m_\pi^2)} \;=\; \frac{\lambda}{2}\left[\frac{1}{p^2 - m_\sigma^2} - \frac{1}{p^2 - m_\pi^2}\right] , \tag{5}$$

using $g'^2 = \lambda(m_\sigma^2 - m_\pi^2)/2$. Then, the contributions from the diagrams in Fig. 2 become

$$m_{\pi,{\rm ML}}^2 \;=\; i(-2\lambda + 5\lambda - 3\lambda) \int \frac{\bar{d}^4 p}{p^2 - m_\pi^2} \;+\; i(2\lambda + \lambda - 3\lambda) \int \frac{\bar{d}^4 p}{p^2 - m_\sigma^2} \;=\; 0 \; . \tag{6}$$

Note that the coefficients of the quadratically divergent graphs in Fig. 2 vanish *identically.* Combining Eqs. (4) and (6) generates the vanishing Nambu-Goldstone pion mass in the CL

$$m_\pi^2 \;=\; 0\,|_{\text{quark loops}} \;+\; 0\,|_{\pi\text{ loops}} \;+\; 0\,|_{\sigma\text{ loops}} \;=\; 0 \; . \tag{7}$$

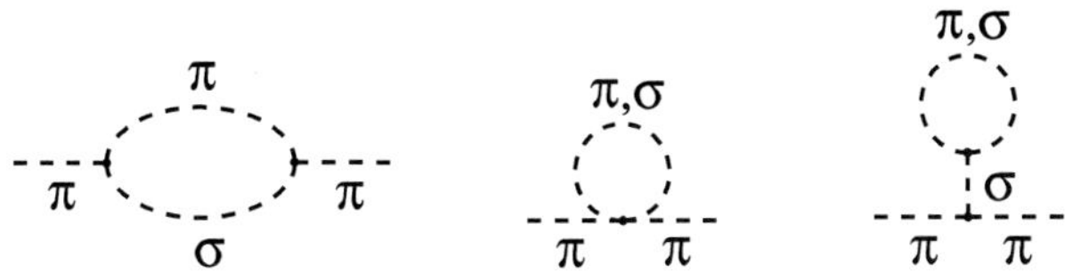

Fig. 2 Pion-selfenergy meson-loop graphs. Left: bubble; middle: "snail"; right: tadpole.

© 2013 WILEY-VCH Verlag GmbH & Co. KGaA, Weinheim

Such is the subtle beauty of chiral symmetry!

Now we determine the π^0 decay constant f_π. The latest data in the Particle Data Group (PDG) [6] tables give[1]

$$f_\pi = (92.21 \pm 0.15)\ \text{MeV}\ . \tag{8}$$

Also, the pion-quark coupling is [19, 20]

$$g_{\pi qq} = \frac{2\pi}{\sqrt{N_c}} \simeq 3.6276\ , \tag{9}$$

a value to be reconfirmed shortly, for $N_c = 3$ colors. Thus, the nonstrange constituent quark mass found via the GTR becomes

$$\hat{m}_{\text{con}} = f_\pi g_{\pi qq} = (92.21\ \text{MeV})\ \frac{2\pi}{\sqrt{3}} \simeq 334.5\ \text{MeV}\ . \tag{10}$$

This value is slightly less than a quick estimate resulting from combining the proton mass with its magnetic moment, i.e., for $\hat{m}_{\text{con}} = (m_u + m_d)_{\text{con}}/2$,

$$\hat{m}_{\text{con}} \simeq \frac{m_p}{\mu_p} \simeq \frac{938.27\ \text{MeV}}{2.7928} \simeq 336.0\ \text{MeV}\ . \tag{11}$$

A more accurate computation from the proton magnetic moment, which takes into account a 4 MeV mass difference between the down and the up quark, yields the prediction [21]

$$\hat{m}_{\text{con}} \simeq 337.5\ \text{MeV}\ , \tag{12}$$

which is just a little bit higher than in Eqs. (10,11).

Finally, we work in the CL to generate f_π^{CL} via a once-subtracted dispersion relation [22, 23] (involving no arbitrary parameters as in chiral perturbation theory):

$$f_\pi - f_\pi^{\text{CL}} = \frac{m_\pi^2}{\pi} \int_0^\infty \frac{\Im\text{m}\, f_\pi(q^2)\, dq^2}{q^2(q^2 - m_\pi^2)} = \frac{m_\pi^2}{8\pi^2 f_\pi}\ . \tag{13}$$

With $f_\pi = 92.21$ MeV and an average pion mass $m_\pi \simeq 137$ MeV, we thus find

$$1 - \frac{f_\pi^{\text{CL}}}{f_\pi} = \frac{m_\pi^2}{8\pi^2 f_\pi^2} \simeq 2.8\%\ , \tag{14}$$

which in turn predicts

$$f_\pi^{\text{CL}} = f_\pi(1 - 0.028) \simeq 89.63\ \text{MeV}\ . \tag{15}$$

Then, via the GTR, we get

$$\hat{m}_{\text{con}}^{\text{CL}} = f_\pi^{\text{CL}} g_{\pi qq} \simeq (89.63\ \text{MeV})\ \frac{2\pi}{\sqrt{3}} \simeq 325.1\ \text{MeV}\ . \tag{16}$$

[1] Note that the PDG quotes $f_{\pi^\pm}$, which by definition is larger by a factor $\sqrt{2}$.

 © 2013 WILEY-VCH Verlag GmbH & Co. KGaA, Weinheim

3 Dynamically generating the $SU(2)$ QLLσM

Following [17, 18], we first compute the nonstrange quark loop in Fig. 3, leading to the log-divergent gap equation (LDGE)

$$1 = -i4N_c g^2 \int \frac{d\!\!^-{}^4p}{(p^2 - m_q^2)^2} , \tag{17}$$

due to the quark-loop integral for the neutral pion decay constant combined with the quark-level GTR (10). This LDGE also holds for the nonlinear NJL scheme in [5], and leads to many low-energy theorems [24]. Furthermore, the quark tadpole graph of Fig. 4 generates a counter-term mass gap

$$m_q = -\frac{8iN_c g^2}{m_\sigma^2} \int \frac{d\!\!^-{}^4p}{p^2 - m_q^2} m_q . \tag{18}$$

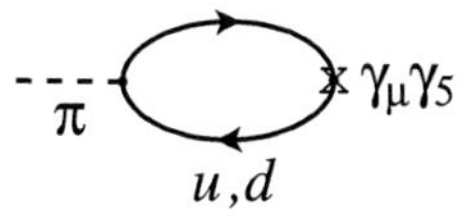

Fig. 3 Pion nonstrange quark loop for the LDGE in Eq. (17).

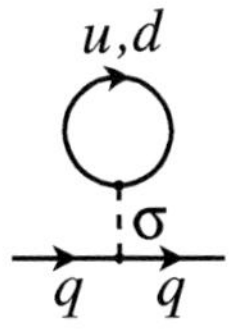

Fig. 4 Quark-selfenergy tadpole graph.

Then, canceling out the m_q scale gives

$$m_\sigma^2 = -8iN_c g^2 \int \frac{d\!\!^-{}^4p}{p^2 - m_q^2} . \tag{19}$$

Lastly, the σ bubble plus σ tadpole graphs of Fig. 5 in the CL generate the counter-term relation

$$m_{\sigma,\mathrm{CL}}^2 = 16iN_c g^2 \int d\!\!^-{}^4p \left[\frac{m_q^2}{(p^2 - m_q^2)^2} - \frac{1}{p^2 - m_q^2} \right]_{\mathrm{CL}} . \tag{20}$$

Substituting Eqs. (17,19) into Eq. (20) leads to the CL relation

$$m_{\sigma,\mathrm{CL}}^2 = -4m_{q,\mathrm{CL}}^2 + 2m_{\sigma,\mathrm{CL}}^2 \quad \Rightarrow \quad m_{\sigma,\mathrm{CL}} = 2m_q^{\mathrm{CL}} , \tag{21}$$

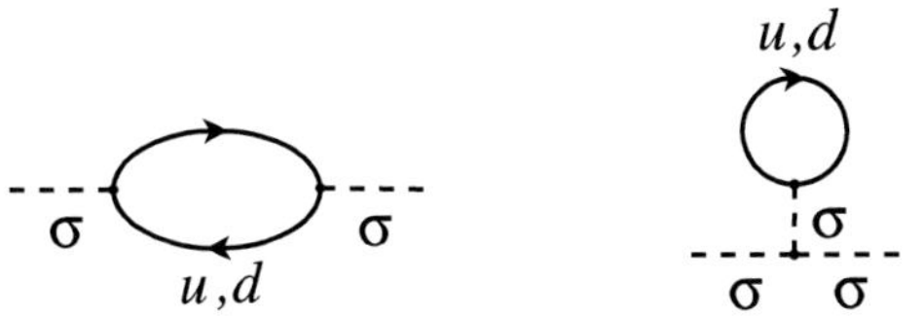

Fig. 5 Sigma-meson selfenergy graphs. Left: quark bubble; right: σ tadpole.

© 2013 WILEY-VCH Verlag GmbH & Co. KGaA, Weinheim

but now for the QLLσM rather than for the nonlinear NJL model. Note that Eq. (21), together with Eq. (16), predicts $m_\sigma^{\rm CL} \simeq 650.3$ MeV.

Moreover, the integral difference in Eq. (20) leads to the dimensional-regularization lemma (DRL) [17], via a Wick rotation:

$$\int đ^4p \left[\frac{m_q^2}{(p^2 - m_q^2)^2} - \frac{1}{p^2 - m_q^2} \right] = -\frac{i m_q^2}{16\pi^2} \,. \tag{22}$$

This result follows in many regularization schemes (dimensional, analytic, ζ-function, Pauli-Villars) and also in a scheme-independent manner [25,26]. Then, substituting Eq. (22) back into Eq. (20) leads to

$$m_\sigma^{\rm CL} = \frac{\sqrt{N_c}}{\pi}\, g\, m_q^{\rm CL} \,, \tag{23}$$

which predicts, also using Eq. (21) and $N_c = 3$, the crucial coupling

$$g = g_{\pi qq} = \frac{2\pi}{\sqrt{N_c}} \simeq 3.6276 \,. \tag{24}$$

In fact, the latter result also holds [19, 20] in infrared-QCD studies.

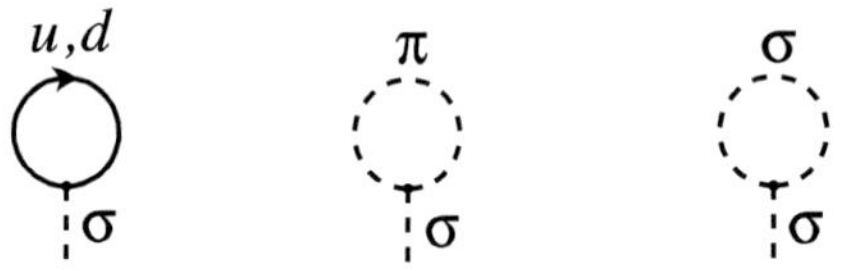

Fig. 6 Sigma tadpole graphs. Left: quark loop; middle: π loop; right: σ loop.

Moreover, B. W. Lee's null tadpole condition [27], resulting from the vanishing of the sum of the three tadpole graphs of Fig. 6 in the CL, reads

$$0 = \langle\sigma\rangle = -8iN_c\, g\, m_q \int \frac{đ^4p}{p^2 - m_q^2} + 3ig^2 \int \frac{đ^4p}{p^2} + 3ig' \int \frac{đ^4p}{p^2 - m_\sigma^2} \,, \tag{25}$$

where the factors 3 are due to combinatorics. Now, we drop the middle massless-tadpole term, due to the vanishing of the π mass in the CL, and scale the first and third quadratically divergent integrals to m_q^2 and m_σ^2, respectively. Using next the identity $g' = m_\sigma^2/2f_\pi$ along with the GTR $m_q = f_\pi g$, but not needing Eq. (24), the scale $1/f_\pi$ cancels out, which results in [17]

$$N_c (2 m_q^{\rm CL})^4 = 3 m_{\sigma,{\rm CL}}^4 \,. \tag{26}$$

Since $m_{\sigma,{\rm CL}} = 2m_q^{\rm CL}$ from Eq. (21), this implies $N_c = 3$. There are indeed many methods to find [17] $N_c = 3$.

The standard way to verify $N_c = 3$ is via the $\pi^0 \to 2\gamma$ quark-loop decay amplitude

$$\left|F_{\pi^0\to 2\gamma}\right| = \frac{\alpha N_c}{3\pi f_\pi} \simeq 0.02519\ {\rm GeV}^{-1}, \quad \text{for } N_c = 3 \,, \tag{27}$$

which predicts a decay rate

$$\Gamma_{\pi^0\to 2\gamma} = \frac{m_{\pi^0}^3 \left|F_{\pi^0\to 2\gamma}\right|^2}{64\pi} \simeq 7.76\ {\rm eV} \,. \tag{28}$$

 © 2013 WILEY-VCH Verlag GmbH & Co. KGaA, Weinheim

The latter is very near data [6], with $\tau_{\pi^0} \simeq 8.52 \times 10^{-17}$ s:

$$\Gamma^{\text{PDG}}_{\pi^0 \to 2\gamma} = 0.9882 \, \frac{\hbar}{\tau_{\pi^0}} \simeq 7.63 \text{ eV} , \tag{29}$$

where 0.9882 is the $\pi^0 \to 2\gamma$ branching fraction.

Since the B. W. Lee null tadpole condition (Eq. (25)) holds, the *true* vacuum corresponds to $\langle \sigma \rangle = \langle \pi \rangle = 0$, and not to the false vacuum needed in the Gell-Mann–Lévy [2, 3] nucleon-level LσM for spontaneous symmetry breaking. Moreover, with $g' = m_\sigma^2 / 2 f_\pi$ in the CL, the meson-type GTR $g' = \lambda f_\pi$ requires

$$\lambda = 2g^2 , \tag{30}$$

which is valid in both tree and one-loop order, the latter being also true via the LDGE in Eq. (17). This nonperturbative bootstrap scale along with $g = 2\pi/\sqrt{3}$ (Eq. (24)) requires

$$\lambda = 2g^2 = \frac{8\pi^2}{3} \simeq 26.3 , \tag{31}$$

which also holds at one-loop level, owing to the LDGE.

4 $Z = 0$ compositeness condition

Following [17], we return to the LDGE integral in Eq. (17), but now cut off in the ultraviolet (UV) region via a parameter Λ. Then, the equation becomes, for $X = \Lambda^2 / m_q^2$,

$$1 = -i4N_c g^2 \int_0^\Lambda \frac{đ^4 p}{(p^2 - m_q^2)^2} = \ln(X+1) - \frac{X}{X+1} . \tag{32}$$

This implies $X \simeq 5.3$, so that in the CL the UV cutoff scale becomes

$$\Lambda \simeq \sqrt{5.3}\, m_q^{\text{CL}} \simeq 2.302\,(325.1 \text{ MeV}) \simeq 749 \text{ MeV} . \tag{33}$$

Now, the 749 MeV UV scale separates the elementary particles $\pi(137)$ and $\sigma_{\text{CL}}(650)$ from the $q\bar{q}$ bound states $\rho(775)$, $\omega(783)$, $f_0(980)$, $a_0(980)$, $a_1(1260)$, and so forth. This is called a $Z = 0$ compositeness condition (CC) [28–30]. In the QLLσM, the condition follows from the renormalization constant being

$$Z = 1 - \frac{N_c g^2}{4\pi^2} , \tag{34}$$

which vanishes for $N_c = 3$, since $g = 2\pi/\sqrt{3}$. For more details, we refer to [30]. When meson loops are folded in, the UV cutoff equation changes to [18]

$$1 = \ln(X'+1) - \frac{X'}{X'+1} + \frac{1}{6} , \tag{35}$$

where the extra term amounts to $\lambda/16\pi^2$, with $\lambda = 8\pi^2/3$ (Eq. (31)). Given Eq. (35), we predict $X' \simeq 4.15$, leading to a reduced UV scale

$$\Lambda' \simeq \sqrt{4.15}\, m_q^{\text{CL}} \simeq 662 \text{ MeV} . \tag{36}$$

This value is quite near $m_{\sigma,\text{CL}} \simeq 650$ MeV, which mass even increases slightly away from the CL:

$$m_\sigma = \sqrt{m_{\sigma,\text{CL}}^2 + m_\pi^2} \simeq 664.1 \text{ MeV} . \tag{37}$$

© 2013 WILEY-VCH Verlag GmbH & Co. KGaA, Weinheim

Either $\Lambda' \simeq 662$ MeV or $m_\sigma \simeq 664$ MeV are about 85 MeV less than the usual $Z = 0$ CC cutoff at 749 MeV in Eq. (33). The very similar energy scales in Eqs. (36) and (37) indicate that the inclusion of meson loops leads to a "double counting" of $q\bar{q}$ states as partially elementary and partially bound states. This issue is addressed in more detail in [18].

Specifically, the nonstrange $q\bar{q}$ pion $\pi(137)$ is an elementary particle in the QLLσM, but the also nonstrange $q\bar{q}$ scalar resonance $\sigma(664)$ can be treated as either elementary or a bound state [30]. This may be one of the reasons why it has been so difficult to experimentally identify the $f_0(500)$, with a listed [6] mass range of 400–550 MeV. Nevertheless, the π and the σ can be treated in a current-algebra fashion as "chiral partners" [31].

5 Chiral shielding

The same kind of chiral cancellations as employed in the formulation of the QLLσM in Sects. 2 and 3 can be invoked to explain the smallness of certain decay or scattering amplitudes, or even their nonobservation. Here, we shall focus on two processes, viz. $a_1(1260) \to \pi(\pi\pi)_{S\text{-wave}}$ and $\gamma\gamma \to \pi\pi$.

For conserved axial currents ($\partial \cdot A = 0$), the leading quark-loop pion propagator can be shielded via the Dirac matrix *identity* [32,33]

$$\frac{1}{\not{p} - m} 2m\gamma_5 \frac{1}{\not{p} - m} = -\gamma_5 \frac{1}{\not{p} - m} - \frac{1}{\not{p} - m}\gamma_5 \,. \tag{38}$$

Then, as $p_\pi \to 0$, the $a_1(1260) \to \pi(\pi\pi)_{S\text{-wave}}$ box and triangle graphs of Fig. 7 sum up to *zero,* in the CL. That is, for $p_\pi \to 0$,

$$M^{\text{box}}_{a_1\to 3\pi} \longrightarrow -\frac{1}{f_\pi} M_{a_1\to\sigma\pi} \,, \tag{39}$$

$$M^{\text{triangle}}_{a_1\to 3\pi} \longrightarrow \frac{1}{f_\pi} M_{a_1\to\sigma\pi} \,. \tag{40}$$

So the total $a_1(1260) \to 3\pi$ soft-momentum amplitude is

$$M^{\text{total}}_{a_1\to 3\pi} = M^{\text{box}}_{a_1\to 3\pi} + M^{\text{triangle}}_{a_1\to 3\pi} \longrightarrow 0 \,, \tag{41}$$

which is in agreement with the old experimental decay rate

$$\Gamma_{a_1(1260)\to\pi(\pi\pi)_{S\text{-wave}}} \lesssim (1 \pm 1)\ \text{MeV} \tag{42}$$

reported in the 1990 PDG tables [34], on the basis of the analysis of [35]. On the other hand, the lone $M_{a_1\to\sigma\pi}$ amplitude, which corresponds to only the triangle graph in Fig. 7, is *not* small, as confirmed by the experimental [36] decay rate

$$\Gamma_{a_1(1260)\to\sigma\pi} \sim (130 \pm 40)\ \text{MeV}\,. \tag{43}$$

This is one of the cleanest checks of chiral cancellations in the QLLσM.

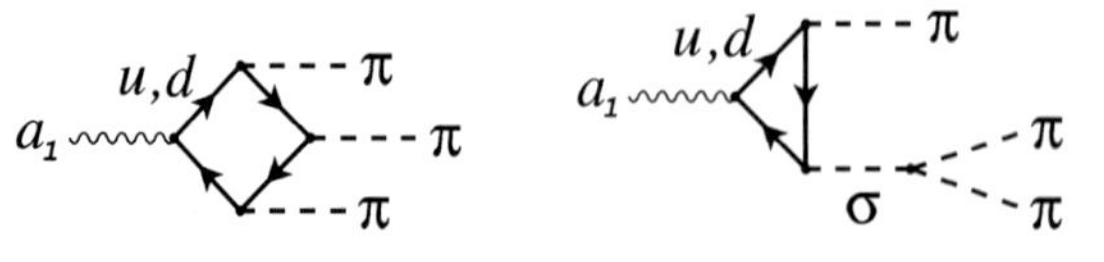

Fig. 7 Graphs for the decay [6] $a_1(1260) \to 3\pi$. Left: quark box; right: quark triangle.

 © 2013 WILEY-VCH Verlag GmbH & Co. KGaA, Weinheim

Another confirmation comes from the process $\gamma\gamma \to \pi\pi$, whose rate should vanish as $s \to m_\sigma^2$. Namely, just as in the above a_1 case, there is a quark-box and a quark-triangle contribution, as depicted in Fig. 8, leading to a total amplitude

$$\langle \pi\pi|\gamma\gamma\rangle \;=\; -\frac{\langle \pi\pi|\gamma\gamma\rangle_{\text{box}}}{f_\pi} + \frac{\langle \pi\pi \to \sigma|\gamma\gamma\rangle_{\text{triangle}}}{f_\pi} \;\to\; 0\,. \tag{44}$$

This result is compatible with Crystal Ball data [37], which revealed a tiny $\gamma\gamma \to \pi^0\pi^0$ cross section of the order of 10 nb at energies around the σ mass. Four years earlier, Kaloshin and Serebryakov [38] had predicted a $\gamma\gamma \to \pi^+\pi^-$ cross section exactly of this magnitude at the mass of a then hypothetical scalar $\epsilon(700)$ resonance. On the other hand, the $f_0(500) \to \gamma\gamma$ decay rate is quite large, viz. of the order of 3–4 keV [39–43]. Again, this is due to the fact that now only the triangle graph in Fig. 8 contributes.

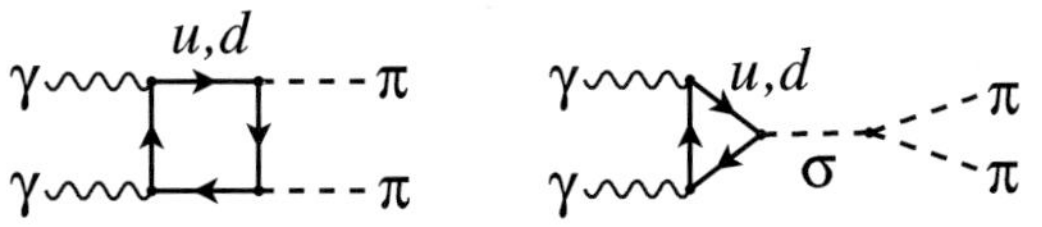

Fig. 8 Graphs for the process $\gamma\gamma \to \pi\pi$. Left: quark box; right: quark triangle.

As final examples, we should mention the processes $\pi^- p \to \pi^-\pi^+ n$ and $K^- p \to K^-\pi^+ n$ [32,33], in which the S-wave amplitudes for the $\pi^-\pi^+$ and $K^-\pi^+$ final states, respectively, are suppressed once again because of a cancellation between a box and a triangle diagram. The triangles correspond to the decays of the scalar resonances $f_0(500)$ (alias σ) and $K_0^*(800)$ (alias κ), respectively, which have been so hard to observe experimentally, exactly because of chiral shielding.

6 Linking the nonstrange current quark mass scale with the πN σ-term

The PDG [6] now lists the light current quark masses as $m_u = 2.3^{+0.7}_{-0.5}$ MeV and $m_d = 4.8^{+0.7}_{-0.3}$ MeV, estimated in a mass-independent subtraction scheme such as $\overline{\text{MS}}$, at a scale $\mu \simeq 2$ GeV [6]. These current quarks are generally believed to be dressed by gluons so as to acquire constituent masses of a few hundreds of MeV, like $\hat{m}_{\text{con}}$ in the QLLσM. However, this dressing is a highly nonperturbative and nonlinear process, which does not allow to write down simple relations between current and constituent masses. Nevertheless, chiral symmetry does allow to estimate an *effective* nonstrange current quark mass, as the difference between the constituent mass and the dynamical mass, i.e.,

$$\hat{m}_{\text{cur}} \;=\; \hat{m}_{\text{con}} - \hat{m}_{\text{dyn}}\;. \tag{45}$$

Then, in the limit $m_\pi \to 0$, we get $\hat{m}_{\text{cur}} \to 0$ and $\hat{m}_{\text{con}} \to \hat{m}_{\text{dyn}}$. Unfortunately, Eq. (45) is only a rough relation, because $\hat{m}_{\text{con}}$ and $\hat{m}_{\text{dyn}}$ are on nearby mass shells. To fine-tune Eq. (45) in the low-energy region, we invoke infrared QCD, stating that the dynamical quark mass should run as [44][2]

$$\hat{m}_{\text{dyn}}(p^2) \;=\; \frac{\hat{m}^3_{\text{dyn}}}{p^2}\,, \tag{46}$$

where for consistency

$$\hat{m}_{\text{dyn}} \;\equiv\; \hat{m}_{\text{dyn}}(p^2 = \hat{m}^2_{\text{dyn}})\;. \tag{47}$$

[2] Also see [45], which combined current quark masses with structure functions, finding $\hat{m}_{\text{cur}} \sim 62$ MeV and $(m_s/\hat{m})_{\text{cur}} \sim 5$, close to $\hat{m}_{\text{cur}} \simeq 68.3$ MeV and $(m_s/\hat{m})_{\text{cur}} \simeq 6.26$ above.

© 2013 WILEY-VCH Verlag GmbH & Co. KGaA, Weinheim

On the other hand, $\hat{m}_{\text{dyn}}$ can be estimated from the nucleon mass as

$$\hat{m}_{\text{dyn}} \simeq \frac{m_N}{3} \simeq 313 \text{ MeV} . \quad (48)$$

Using now the nonstrange constituent quark mass of $\hat{m}_{\text{con}} \simeq 337.5$ from Eq. (12) above, Eqs. (45–48) yield [44]

$$\hat{m}_{\text{cur}} = \left(337.5 - \frac{313^3}{337.5^2}\right) \text{MeV} \simeq 68.3 \text{ MeV} . \quad (49)$$

Note that this effective current quark mass away from the CL is remarkably close to half the average pion mass

$$\bar{m}_\pi = \frac{139.57 + 134.98}{2} \text{ MeV} \simeq 137 \text{ MeV} . \quad (50)$$

Next we study the nucleon, which is a nonstrange qqq state, and the πN σ term at the Cheng–Dashen (CD) point [46] $\bar{t} \equiv 2\bar{m}_\pi^2 = 2\mu^2$. Early estimates were

$$\sigma_{\pi N}(\bar{t}) = (66 \pm 9) \text{ MeV [47]} , \quad \sigma_{\pi N}(\bar{t}) = (65 \pm 6) \text{ MeV [48]} ,^3$$
$$\sigma_{\pi N}(\bar{t}) = (70 \pm 6) \text{ MeV [50]} , \quad \sigma_{\pi N}(\bar{t}) = (64 \pm 8) \text{ MeV [51]} .$$

More recently [52], a slightly larger value of (71 ± 9) MeV was found, but still perfectly compatible with the first few analyses. The average of these five numbers gives $\sigma_{\pi N}(2\mu^2) \simeq 67$ MeV, which is surprisingly close to the effective current quark mass $\hat{m}_{\text{cur}} \simeq 68.3$ MeV in Eq. (49). Note that both $\sigma_{\pi N}$ and $\hat{m}_{\text{cur}}$ are measures of $SU(2)$ chiral symmetry breaking.

An overview of the πN sigma term can be found in [53]. Starting point is the quenched-lattice prediction by the APE Collaboration [54, 55]

$$\sigma^{\text{APE}}_{\pi N,\,\text{quenched}} = (24.5 \pm 2.0) \text{ MeV} . \quad (51)$$

This result is near the Gell-Mann–Oakes–Renner (GMOR) [56] perturbative value

$$\sigma^{\text{GMOR}}_{\pi N} = \frac{m_\Xi + m_\Sigma - 2m_N}{2} \frac{m_\pi^2}{m_K^2 - m_\pi^2} \simeq 26 \text{ MeV} . \quad (52)$$

The nonperturbative, nonquenched (NQ) addition to $\sigma_{\pi N}$ stems from the σ-meson tadpole graphs [57–59], yielding

$$\sigma^{\text{NQ}}_{\pi N} = \frac{m_{\pi^0}^2}{m_\sigma^2} m_N = \frac{134.98^2}{664.1^2} 938.9 \text{ MeV} \simeq 38.8 \text{ MeV} .^4 \quad (53)$$

Then, the total σ term is predicted to be

$$\sigma_{\pi N} = \sigma^{\text{GMOR}}_{\pi N} + \sigma^{\text{NQ}}_{\pi N} \simeq (26 + 39) \text{ MeV} = 65 \text{ MeV} , \quad (54)$$

which is very close to the average value of 67 MeV from the five analyses above.

The theoretical estimate of 65 MeV is also near the infinite-momentum-frame (IMF) value [61]

$$\sigma_{\pi N} = \frac{m_\Xi^2 + m_\Sigma^2 - 2m_N^2}{2m_N} \frac{m_\pi^2}{m_K^2 - m_\pi^2} \simeq 63 \text{ MeV} . \quad (55)$$

[3] Also see [49].

[4] This value is remarkably close to the mass of a hypothetical new light boson, for which evidence has been found [60] in experimental data.

 © 2013 WILEY-VCH Verlag GmbH & Co. KGaA, Weinheim

Note that in the IMF tadpoles are suppressed.

For comparison, the revised chiral-perturbation-theory (ChPT) value now is 60 MeV [62],[5] which should follow from the positive and coherent sum of *four* terms, i.e.,

$$\sigma_{\pi N}(\bar{t}) = \sigma_{\pi N}^{\text{GMOR}} + \sigma_{\pi N}^{\text{HOChPT}} + \sigma_{\pi N}^{\bar{s}s} + \sigma_{\pi N}^{t\text{-dep.}} \tag{56}$$

$$\simeq (25 + 10 + 10 + 15)\ \text{MeV} = 60\ \text{MeV}\,. \tag{57}$$

Here, the second term on the right-hand side arises from higher-order ChPT, the third one from the strange-quark sea, and the fourth is a t-dependent contribution due to going from $t = 0$ to the CD point, where the πN background is minimal. Leutwyler [62] concluded: *"The three pieces happen to have the same sign."* Of course, for things to work out right, all *four* pieces must have the same sign, including the GMOR term. Note, however, that very recently an $N_f = 2$ ChPT analysis of πN data [65] managed to extract a value as large as $\sigma_{\pi N} = (59 \pm 7)$ MeV. The good news is that, besides $\Delta(1232)$ degrees of freedom, no contribution from the strange-quark sea was now needed, in agreement with data [66, 67] and [57–59], but in stark conflict with earlier ChPT analyses like in [62]. Clearly, the QLLσM amounts to a much simpler and more straightforward approach, reproducing the data without any problem.

Summarizing, we have shown is this section that the QLLσM effective current quark mass of 68.3 MeV is *very near* the πN σ term prediction, both via tapoles ($\simeq$ 65 MeV) and using the IMF ($\simeq$ 63 MeV), which is in turn fully compatible with the experimental analyses.

7 Pion charge radius for the LσM and VMD schemes

We now return to [18], noting that it took until 1979 [68, 69][6] before the QLLσM was employed to calculate the pion charge radius, in the CL, viz.

$$r_{\pi,\text{CL}}^{\text{QLL}\sigma\text{M}} = \frac{\hbar c\sqrt{N_c}}{2\pi f_\pi^{\text{CL}}} \simeq 0.61\ \text{fm}\,, \tag{58}$$

for $N_c = 3$, $\hbar c \simeq 197.3$ MeV fm, and where $f_{\pi,\text{CL}}$ is now $\simeq 89.63$ MeV (see Eq. (15)). Stated another way, taking $g = 2\pi/\sqrt{3}$ (as we consistently do) and invoking the GTR $\hat{m}_{\text{con}}^{\text{CL}} = f_\pi^{\text{CL}} g \simeq 325.1$ MeV, r_π can also be expressed as

$$r_{\pi,\text{CL}}^{\text{QLL}\sigma\text{M}} = \frac{\hbar c}{\hat{m}_{\text{con}}^{\text{CL}}} \simeq 0.61\ \text{fm}\,. \tag{59}$$

Recall that the original vector-meson-dominance (VMD) prediction was [71]

$$r_{\pi,\text{CL}}^{\text{VMD}} = \frac{\sqrt{6}\,\hbar c}{m_\rho} \simeq 0.62\ \text{fm}\,, \tag{60}$$

where we use the PDG [6] mass $m_\rho \simeq 775.5$ MeV.

As for measurements of r_π, the PDG tables [6] report an average value of (0.672 ± 0.008) fm. There is a tight link between the QLLσM and VMD predictions for r_π, as stressed in [18], viz.

$$\frac{r_{\pi,\text{CL}}^{\text{VMD}}}{\hbar c} = \frac{\sqrt{6}}{m_\rho} \simeq \frac{1}{\hat{m}_{\text{con}}^{\text{CL}}} = \frac{\sqrt{3}}{2\pi f_\pi^{\text{CL}}} = \frac{r_{\pi,\text{CL}}^{\text{QLL}\sigma\text{M}}}{\hbar c}\,. \tag{61}$$

[5] Also see [63, 64].

[6] Also see [70].

© 2013 WILEY-VCH Verlag GmbH & Co. KGaA, Weinheim

For quark loops (QL) alone, another (cf. Eq. (13)) once-subtracted dispersion relation, evaluated at $q^2 = 0$, gives in the CL [72]

$$r^2_{\pi,\mathrm{QL}} = \frac{6}{\pi}\int_0^\infty \frac{\Im\mathrm{m}\, F_\pi(q^2)\, dq^2}{(q^2)^2} = \frac{N_c(\hbar c)^2}{4\pi^2 f^2_{\pi,\mathrm{CL}}}, \tag{62}$$

with the form factor normalized to $F_\pi(q^2 = 0) = 1$. Note that Eq. (62) precisely amounts to the square of Eq. (58) above, which is presumably how [68, 69] arrived at the result.

Concerning the relation between the one-loop-order QLLσM with quark loops alone and the tree-level VMD model, the $Z = 0$ compositeness condition and the cutoff $\Lambda \simeq 749$ MeV $< m_\rho$ from Eq. (33) suggest the ρ meson is an external $q\bar{q}$ bound state. Then, the LDGE in Eq. (17) leads to [17, 24]

$$g_{\rho\pi\pi} = g_\rho \left[-i4N_c g^2 \int^\Lambda \frac{đ^4 p}{(p^2 - m_q^2)^2} \right] = g_\rho\,, \tag{63}$$

which is Sakurai's [71] VMD universality relation.

If meson loops (see Fig. 2) are added in Eq. (63), the QLLσM $g_{\rho\pi\pi}$ coupling becomes [18]

$$g_{\rho\pi\pi} = g_\rho + \frac{1}{6}\, g_{\rho\pi\pi} \quad\Longrightarrow\quad \frac{g_{\rho\pi\pi}}{g_\rho} = \frac{6}{5} = 1.2\,. \tag{64}$$

Here, $1/6 = \lambda/16\pi^2$ as in Eq. (35).

On the other hand, from data [6] we get, for $p = 363$ MeV and $\Gamma_{\rho\pi\pi} = 149.1$ MeV,

$$|g_{\rho\pi\pi}| = m_\rho \sqrt{\frac{6\pi\Gamma_{\rho\pi\pi}}{p^3}} \simeq 5.9444\,, \tag{65}$$

while the $\rho^0 \to e^+e^-$ decay rate $\Gamma_{\rho\bar{e}e} \simeq 7.04$ keV [6] requires

$$|g_\rho| = \alpha \sqrt{\frac{4\pi m_\rho}{3\Gamma_{\rho\bar{e}e}}} \simeq 4.9569\,. \tag{66}$$

So from data we obtain the ratio

$$\left| \frac{g_{\rho\pi\pi}}{g_\rho} \right| \simeq \frac{5.9444}{4.9569} \simeq 1.199\,. \tag{67}$$

The agreement with the theoretical QLLσM prediction in Eq. (64), which has further improved over the years [72], is simply stunning.

8 Chiral-symmetry-restoration temperature

Next we deal with strong interactions at nonzero temperature, with $\hat{m}_{\mathrm{dyn}} \to 0$ as $T \to T_c$, where T_c is the critical temperature. Now, in the CL $\hat{m}_{\mathrm{con}}$ and m_σ "melt" [73] to zero at T_c, according to [74]

$$\hat{m}^{\mathrm{CL}}_{\mathrm{con}}(T) = \hat{m}^{\mathrm{CL}}_{\mathrm{con}} - \frac{8N_c\, g^2\, \hat{m}^{\mathrm{CL}}_{\mathrm{con}}}{m^2_{\sigma,\mathrm{CL}}} \frac{T^2}{2\pi^2} \mathcal{J}_+(0)\,, \tag{68}$$

where

$$\mathcal{J}_+(0) = \int_0^\infty \frac{x}{e^x + 1}\, dx = \frac{\pi^2}{12}\,. \tag{69}$$

 © 2013 WILEY-VCH Verlag GmbH & Co. KGaA, Weinheim

So at $T = T_c$ the left-hand side of Eq. (68) vanishes, yielding

$$m^2_{\sigma,\mathrm{CL}} = \frac{N_c\, g^2\, T_c^2}{3} \,. \tag{70}$$

Using $m_{\sigma,\mathrm{CL}} = 2\hat{m}^{\mathrm{CL}}_{\mathrm{con}}$ and the GTR $\hat{m}^{\mathrm{CL}}_{\mathrm{con}} = f^{\mathrm{CL}}_{\pi} g$, we get [75], setting $N_c = 3$,

$$T_c = 2f^{\mathrm{CL}}_{\pi} \simeq 179.3\ \mathrm{MeV}\,. \tag{71}$$

Alternatively, we follow the BCS [4] procedure, by first determining the Debye cutoff k_D at $T = 0$ [73], i.e.,

$$k_D = \hat{m}_{\mathrm{dyn}} \sinh \frac{\pi}{2\alpha_s} \simeq 1252\ \mathrm{MeV}\,, \tag{72}$$

where we have taken $\hat{m}_{\mathrm{dyn}} \simeq 313$ MeV from Eq. (48), and $\alpha_s \simeq 0.75$ at the scale of m_σ [73]. So when $\hat{m}_{\mathrm{con}}(T_c) = 0$, or

$$1 = \frac{2\alpha_s}{\pi} \int_0^{\frac{k_D}{2T_c}} \frac{\tanh x}{x}\, dx\,, \tag{73}$$

the upper limit of this integral is found to be [73]

$$\frac{k_D}{2T_c} \simeq 3.58 \implies T_c \simeq \frac{1252}{7.16}\ \mathrm{MeV} \simeq 174.9\ \mathrm{MeV}\,. \tag{74}$$

Lastly, one can study the nonlinear NJL [5] model, with cutoff $\Lambda \simeq 749$ MeV (Eq. (33)), to derive [74]

$$T_c^2 \simeq 4f^2_{\pi,\mathrm{CL}} - \frac{9\hat{m}^4_{\mathrm{con,CL}}}{8\pi^2\Lambda^2} \simeq (172.8\ \mathrm{MeV})^2\,. \tag{75}$$

For comparison, let us just mention[7] the results of some lattice computations, which all give a T_c in the range 157–182 MeV, with error bars accounted for. So our predictions in Eqs. (71,74,75) are fully compatible with the lattice. Note that the energy scales in these equations can be converted to a Kelvin temperature scale through division by the Boltzmann constant k.

9 $SU(3)$ extension of the QLLσM

In order to extend the $SU(2)$ QLLσM to $SU(3)$ [82], we must first determine the strange constituent quark mass $m_{s,\mathrm{con}}$. The most straightforward way to do so, in the context of the QLLσM, is by defining a GTR for the kaon, viz. [21, 83]

$$\frac{m_{s,\mathrm{con}} + \hat{m}_{\mathrm{con}}}{2} = f_K\, g_{Kqq}\,, \quad \text{with} \quad g_{Kqq} = g_{\pi qq} = \frac{2\pi}{\sqrt{3}}\,, \tag{76}$$

where the left-hand side reflects the quark content of the kaon, and $f_K \simeq 1.197 f_\pi$ (see [6], p. 949). Dividing by $\hat{m}_{\mathrm{con}} = f_\pi\, g_{\pi qq}$ then yields

$$\left(\frac{m_{s,\mathrm{con}} + \hat{m}_{\mathrm{con}}}{2\hat{m}_{\mathrm{con}}}\right) = \frac{f_K}{f_\pi} \simeq 1.197\,, \tag{77}$$

[7] See [76–81].

© 2013 WILEY-VCH Verlag GmbH & Co. KGaA, Weinheim

from which we obtain

$$m_{s,\text{con}} = \left(2\frac{f_K}{f_\pi} - 1\right)\hat{m}_{\text{con}} \simeq 1.394 \times 337.5 \text{ MeV} \simeq 470.5 \text{ MeV} . \tag{78}$$

We may also estimate m_s roughly from the vector-meson masses m_ϕ and m_ρ, since the $\phi(1020)$ is (mostly) $s\bar{s}$ and the $\rho(770)$ is $n\bar{n}$ ($n = u, d$). Using the PDG [6] masses 1019.5 MeV and 775.5 MeV, respectively, we thus get

$$2(m_{s,\text{con}} - \hat{m}_{\text{con}}) \simeq m_\phi - m_\rho \simeq 244 \text{ MeV} , \tag{79}$$

which gives $m_{s,\text{con}} \simeq (337.5 + 122)$ MeV $\simeq 460$ MeV, in reasonable agreement with the GTR value of 470.5 MeV in Eq. (78).

Coming now to the effective strange current quark mass $m_{s,\text{cur}}$, we may estimate it from the kaon and pion masses away from the CL. In a similar fashion as we derived in Eqs. (49,50) that $\hat{m}_{\text{cur}}$ is about half the pion mass, we now write [44]

$$\bar{m}_K \simeq m_{s,\text{cur}} + \hat{m}_{\text{cur}} , \tag{80}$$

where $\bar{m}_K \simeq 495.7$ MeV is the average kaon mass [6]. So this predicts [44]

$$m_{s,\text{cur}} \simeq \bar{m}_K - \hat{m}_{\text{cur}} \simeq (495.7 - 68.3) \text{ MeV} = 427.4 \text{ MeV} . \tag{81}$$

This in turn gives the ratio [44]

$$\frac{m_{s,\text{cur}}}{\hat{m}_{\text{cur}}} \simeq \frac{427.4}{68.3} \simeq 6.26 . \tag{82}$$

Note that this ratio is much smaller than the value of 25 advocated in ChPT [84]. However, generalized ChPT [85] admits quark-mass ratios considerably smaller than 25. Moreover, also a light-plane approach [86] predicts a ratio between 6 and 7, compatible with Eq. (82).

Let us now use $m_{s,\text{con}}$ estimated above to calculate scalar and pseudoscalar masses besides m_σ and m_π. In Eqs. (16,21,37) we determined the isoscalar scalar mass to be $m_{\sigma,\text{CL}} = 2\hat{m}^{\text{CL}}_{\text{con}} \simeq 650.3$ MeV in the CL, and $m_\sigma = (m^2_{\sigma,\text{CL}} + m^2_\pi)^{1/2} \simeq 664.1$ MeV away from it. Then we predict for the isodoublet scalar κ (alias $K^*_0(800)$ [6]), using Eqs. (78,12), a mass of [44]

$$m_\kappa \simeq 2\sqrt{m_{s,\text{con}}\hat{m}_{\text{con}}} \simeq 797 \text{ MeV} , \tag{83}$$

which is very near the E791 data [87] at (797 ± 19) MeV, and also compatible with the average of the *experimental* masses reported in the PDG listings [6].[8] Moreover, a mass of 797 MeV is also in reasonable agreement with the κ pole positions of $(727 - i263)$ MeV, $(714 - i228)$ MeV, and $(745 - i316)$ MeV found in the coupled-channel quark-model calculations of [88–90], which all correspond to S-wave $K\pi$ resonances peaking at roughly 800 MeV.

Next we shall employ $SU(3)$ equal-mass-splitting laws (EMSLs) [82, 91, 92] to check the differences between squared scalar and pseudoscalar masses, i.e.,

$$\begin{aligned} m^2_\sigma - m^2_\pi &\simeq (0.6641^2 - 0.1366^2) \text{ GeV}^2 \simeq 0.42 \text{ GeV}^2 , \\ m^2_\kappa - m^2_K &\simeq (0.797^2 - 0.4957^2) \text{ GeV}^2 \simeq 0.39 \text{ GeV}^2 , \\ m^2_{a_0} - \bar{m}^2_\eta &\simeq (0.985^2 - 0.753^2) \text{ GeV}^2 \simeq 0.40 \text{ GeV}^2 , \end{aligned} \tag{84}$$

[8] Note that the quoted [6] *"our average"* $K^*_0(800)$ mass of (682 ± 29) MeV is strongly biased towards the low value found in a theoretical analysis, and does not represent the average of the experimental observations.

 © 2013 WILEY-VCH Verlag GmbH & Co. KGaA, Weinheim

where $\bar{m}_\eta$ is the average η, η' mass. So all three EMSLs have about the same $SU(3)$ chiral-symmetry-breaking scale.

Next we review η-η' mixing. In the flavor basis $(n\bar{n}, s\bar{s})$, η-η' mixing can be written as [82, 93–95]

$$|\eta_{n\bar{n}}\rangle = \quad |\eta\rangle \cos\phi_P + |\eta'\rangle \sin\phi_P \tag{85}$$

$$|\eta_{s\bar{s}}\rangle = -\,|\eta\rangle \sin\phi_P + |\eta'\rangle \cos\phi_P \ , \tag{86}$$

which requires squared masses

$$m^2_{\eta_{n\bar{n}}} = (m_\eta \cos\phi_P)^2 + (m_{\eta'} \sin\phi_P)^2 \tag{87}$$

$$m^2_{\eta_{s\bar{s}}} = (m_\eta \sin\phi_P)^2 + (m_{\eta'} \cos\phi_P)^2 \ , \tag{88}$$

with sum (for any angle)

$$m^2_{\eta_{n\bar{n}}} + m^2_{\eta_{s\bar{s}}} = m^2_\eta + m^2_{\eta'} \ . \tag{89}$$

From the structure of the pseudoscalar mass matrix, one can then derive [82, 93–95] for the mixing angle ϕ_P the expressions

$$\begin{aligned}\phi_P &= \arctan\sqrt{\frac{(m^2_{\eta'} - 2m^2_K + m^2_\pi)(m^2_\eta - m^2_\pi)}{(2m^2_K - m^2_\eta - m^2_\pi)(m^2_{\eta'} - m^2_\pi)}} \\ &\simeq 41.9^\circ\end{aligned} \tag{90}$$

or — equivalently —

$$\phi_P = \arctan\sqrt{\frac{m^2_{\eta_{n\bar{n}}} - m^2_\eta}{m^2_{\eta'} - m^2_{\eta_{n\bar{n}}}}} \simeq 41.9^\circ \ . \tag{91}$$

In Eq. (90) we have substituted $m_\eta = 547.85$ MeV and $m_{\eta'} = 957.78$ MeV [6], as well as the isospin-averaged kaon and pion masses, while in Eq. (91) the theoretical mass $m_{\eta_{n\bar{n}}} = 758.56$ MeV from Eq. (87) has been used (Cf. $m_{\eta_{s\bar{s}}} = 801.29$ MeV from Eq. (88)). This ϕ_P is not only well within the wide range $\simeq 35^\circ$–45° of experimentally [6] determined mixing angles, but also close to the value favored by a coupled-channel model study of the $a_0(980) \to \pi\eta$ line shape [90]. Moreover, a mixing angle of $\phi_P \simeq 42^\circ$ allows to reproduce several e.m. processes involving the η or the η', as we shall show in the next section. Finally, several other works [96–99] also arrived at a pseudoscalar mixing angle of about 42°.

To conclude this section, we look at mixing in the scalar-meson sector, viz. between the σ ($f_0(500)$) and the $f_0(980)$. Now, the mass of the $f_0(980)$ is known reasonably well, namely at (990 ± 20) MeV [6], but the PDG mass of the σ is listed in the wide interval 400–550 MeV, and moreover denoted as "Breit-Wigner mass" or "K-matrix pole" [6]. However, the σ is clearly a very broad ($\Gamma \sim$ 400–700 MeV) non-Breit-Wigner resonance, due to the nearby $\pi\pi$ threshold and the Adler zero [100, 101] beneath. So the effective σ mass will always be model dependent, and may very well come out above the mentioned range of 400–550 MeV. Then, we may use the scalar-meson equivalent of Eq. (89) to write

$$m_{\bar{\sigma}} = \sqrt{m^2_{\sigma_{n\bar{n}}} + m^2_{\sigma_{s\bar{s}}} - m^2_{f_0}} \ , \tag{92}$$

where f_0 is short for $f_0(980)$. With the QLLσM/NJL relations $m_{\sigma_{n\bar{n}}} = 2\hat{m}_{\text{con}} \simeq 675$ MeV and $m_{\sigma_{s\bar{s}}} = 2m_{s,\text{con}} \simeq 941$ MeV, we thus estimate $m_{\bar{\sigma}} \simeq 617$ MeV, not too far from $m_\sigma \simeq 664.1$ MeV in Eq. (37). In order to get the scalar mixing angle ϕ_S, we take the scalar versions of Eqs. (87,88) and subtract one from the other, which gives

$$\begin{aligned}m^2_{\sigma_{n\bar{n}}} - m^2_{\sigma_{s\bar{s}}} &= (m^2_{\bar{\sigma}} - m^2_{f_0})(1 - 2\sin^2\phi_S) = \\ &\quad (m^2_{\bar{\sigma}} - m^2_{f_0})\cos 2\phi_S \ ,\end{aligned} \tag{93}$$

© 2013 WILEY-VCH Verlag GmbH & Co. KGaA, Weinheim

and so

$$\phi_S = \frac{1}{2} \arccos \frac{m^2_{\sigma_{s\bar{s}}} - m^2_{\sigma_{n\bar{n}}}}{m^2_{f_0} - m^2_{\tilde{\sigma}}} \simeq 21.1^\circ \, . \tag{94}$$

Note that [102] already estimated $\phi_S \sim 20^\circ$ in a similar way.

10 Electromagnetic decays, quark loops, and meson loops

In this section, we shall compute various e.m. decay rates of pseudoscalar, vector, and scalar mesons using quark loops, and also meson loops when justified because of phase space.

In terms of a Levi–Civita amplitude $\mathcal{M}$, the rate for a pseudoscalar (P) meson decaying into two photons is [103]

$$\Gamma_{P\to 2\gamma} = \frac{m_P^3 |\mathcal{M}_{P\to 2\gamma}|^2}{64\pi} \, , \tag{95}$$

the rate for a vector (V) meson decaying into a P meson and a photon is [103]

$$\Gamma_{V\to P\gamma} = \frac{p_{P\gamma}^3 |\mathcal{M}_{V\to P\gamma}|^2}{12\pi} \, , \tag{96}$$

and the rate for a P meson decaying into a V meson and a photon is [103]

$$\Gamma_{P\to V\gamma} = \frac{p_{V\gamma}^3 |\mathcal{M}_{P\to V\gamma}|^2}{4\pi} \, . \tag{97}$$

In Eqs. (96,97), $p_{P\gamma}$ and $p_{V\gamma}$ are three-momenta in the decaying particle's rest frame.

Table 1 Experimental [6] and theoretical amplitudes from u/d quark loops, for several e.m. decays of light pseudoscalar and vector mesons. For details on amplitudes, see [103]. Note, however, that the present values of f_π, g_ρ, $\hat{m}_{\text{con}}$, and $m_{s,\text{con}}$ have been used.

Decay	Γ^{exp} (MeV)	$\lvert\mathcal{M}^{\text{exp}}\rvert$ (GeV^{-1})	$\lvert\mathcal{M}^{\text{th}}\rvert$ (GeV^{-1})
$\pi^0 \to \gamma\gamma$	$(7.74 \pm 0.55) \times 10^{-6}$	0.0252 ± 0.0009	0.0252
$\eta \to \gamma\gamma$	$(5.1 \pm 0.3) \times 10^{-4}$	0.025 ± 0.001	0.0255
$\eta' \to \gamma\gamma$	$(4.3 \pm 0.3) \times 10^{-3}$	0.032 ± 0.001	0.0344
$\eta' \to \rho\gamma$	0.060 ± 0.004	0.41 ± 0.02	0.413
$\eta' \to \omega\gamma$	$(6.2 \pm 0.5) \times 10^{-3}$	0.14 ± 0.01	0.152
$\rho^\pm \to \pi^\pm\gamma$	0.067 ± 0.007	0.22 ± 0.01	0.206
$\rho^0 \to \eta\gamma$	0.044 ± 0.003	0.48 ± 0.02	0.460
$\omega \to \pi^0\gamma$	0.70 ± 0.03	0.69 ± 0.02	0.617
$\omega \to \eta\gamma$	$(3.9 \pm 0.2) \times 10^{-3}$	0.14 ± 0.01	0.140
$\phi \to \pi^0\gamma$	$(5.4 \pm 0.3) \times 10^{-3}$	0.040 ± 0.002	0.041
$\phi \to \eta\gamma$	0.056 ± 0.002	0.21 ± 0.01	0.208
$\phi \to \eta'\gamma$	2.7 ± 0.1	0.22 ± 0.01	0.212

Now we are in a position to analyse several mesonic decays with one or two photons in the final state. Starting with the two-photon decays of P mesons, let us recall the famous quark-loop amplitude for $\pi^0 \to \gamma\gamma$, viz.

$$|\mathcal{M}_{\pi^0\to\gamma\gamma}| = \frac{e^2 N_c}{12\pi^2 f_\pi} = \frac{\alpha N_c}{3\pi f_\pi} \simeq 0.0252 \, \text{GeV}^{-1} \, , \tag{98}$$

 © 2013 WILEY-VCH Verlag GmbH & Co. KGaA, Weinheim

where we have substituted $N_c = 3$ and [6] $f_\pi = 92.21$ MeV. The theoretical amplitude is in perfect agreement with the amplitude extracted from the observed [6] rate, using Eq. (95),

$$|\mathcal{M}^{\text{exp}}_{\pi^0\to\gamma\gamma}| = \left(\frac{64\pi\Gamma_{\pi^0\to\gamma\gamma}}{m^3_{\pi^0}}\right)^{1/2} \simeq (0.0252 \pm 0.0009)\ \text{GeV}^{-1}\ . \tag{99}$$

where we have used the experimental [6] value $\Gamma_{\pi^0\to\gamma\gamma} = (7.74 \pm 0.55)$ eV. This result encourages us to estimate the pseudoscalar mixing angle ϕ_P from the observed two-photon widths of the η and the η'. The amplitudes for $\eta \to \gamma\gamma$ and $\eta' \to \gamma\gamma$ read [103]

$$|\mathcal{M}_{\eta\to\gamma\gamma}| = \frac{\alpha N_c}{9\pi f_\pi}\left(5\cos\phi_P - \sqrt{2}\,\frac{\hat{m}_{\text{con}}}{m_{s,\text{con}}}\sin\phi_P\right)\ , \tag{100}$$

$$|\mathcal{M}_{\eta'\to\gamma\gamma}| = \frac{\alpha N_c}{9\pi f_\pi}\left(5\sin\phi_P + \sqrt{2}\,\frac{\hat{m}_{\text{con}}}{m_{s,\text{con}}}\cos\phi_P\right)\ , \tag{101}$$

with $\hat{m}_{\text{con}} \simeq 337.5$ MeV and $m_{s,\text{con}} \simeq 470.5$ MeV from Eqs. (10,78), respectively. If we now take $\phi_P = 41.9^\circ$, the latter theoretical amplitudes become $|\mathcal{M}_{\eta\to\gamma\gamma}| \simeq 0.0255$ GeV^{-1} and $|\mathcal{M}_{\eta'\to\gamma\gamma}| \simeq 0.0344$ GeV^{-1}, to be compared with the extracted experimental [6] ones $|\mathcal{M}^{\text{exp}}_{\eta\to\gamma\gamma}| \simeq (0.025\pm0.001)$ GeV^{-1} and $|\mathcal{M}^{\text{exp}}_{\eta'\to\gamma\gamma}| \simeq (0.032 \pm 0.001)$ GeV^{-1}, respectively. In Table 1, we list the theoretical and experimental amplitudes of the π^0, η, and η' $P \to 2\gamma$ decays, as well as those of nine $P \to V\gamma$ and $V \to P\gamma$ processes, several of which involving an η or η' meson. For the precise form of the amplitudes concerning the $P \to V\gamma$ and $V \to P\gamma$ decays, see [103]. Let us just mention that, in the case of decays involving the ω or the ϕ, the small vector mixing angle ϕ_V, which expresses the deviation from ideal flavor mixing in this sector, plays an important role. For instance, the decay $\phi \to \pi^0\gamma$, which would vanish for ideal mixing, determines to a large extent the value of ϕ_V, optimized at 3.8° [103] and allowing to reproduce the other rates with an ω or ϕ as well. The overall agreement with data in Table 1 is spectacular, except for the decay $\omega \to \pi^0\gamma$, which is nevertheless only about 10% off. Finally, the quark-loop approach to e.m. decays of mesons, in the spirit of the QLLσM, also works quite well for several strange, charm, and even charmonium V states [103].

To conclude this section, we consider the two-photon decays of the light scalar mesons $f_0(500)$ (alias σ), $f_0(980)$, and $a_0(980)$, with rates given by

$$\Gamma_{S\to2\gamma} = \frac{m^3_S|\mathcal{M}_{S\to2\gamma}|^2}{64\pi}\ , \tag{102}$$

just as in the case of P mesons. Dealing first with the σ, in the NJL limit, i.e., for $m_\sigma = 2\hat{m}_{\text{con}}$, the amplitude takes the simple form [104]

$$|\mathcal{M}_{\sigma\to\gamma\gamma}| = \frac{5\alpha N_c}{9\pi f_\pi}\ , \tag{103}$$

where the factor $5/3$ with respect to the π^0 amplitude stems from the fact that, for an isoscalar, the contributions from the u and the d quark loops add up, in contrast with the π^0 case. Assuming that the σ is purely $n\bar{n} = (u\bar{u}+d\bar{d})/\sqrt{2}$, with mass $m^{\text{NJL}}_\sigma = 2\hat{m}_{\text{con}} = 675$ MeV, we obtain a quark-loop rate of 2.70 keV [43].[9] This rate would becomes 2.57 keV, if we used in Eq. (103) the value $m_\sigma = 664.1$ MeV from Eq. (37). But away from the NJL limit, the correct gauge-invariant quark-loop amplitude becomes [43]

$$\mathcal{M}^{n\bar{n}}_{\sigma\to\gamma\gamma} = \frac{5\alpha}{3\pi f_\pi}2\xi_n[2 + (1 - 4\xi_n)I(\xi_n)]\ , \tag{104}$$

[9] Also see [105].

© 2013 WILEY-VCH Verlag GmbH & Co. KGaA, Weinheim

where $\alpha = e^2/4\pi$, $\xi_j = m_j^2/m_\sigma^2$, and $I(\xi)$ is the triangle loop integral given by

$$I(\xi)\begin{cases} = \dfrac{\pi^2}{2} - 2\log^2\left[\sqrt{\dfrac{1}{4\xi}} + \sqrt{\dfrac{1}{4\xi} - 1}\right] + 2\pi i \log\left[\sqrt{\dfrac{1}{4\xi}} + \sqrt{\dfrac{1}{4\xi} - 1}\right] & (\xi \leq 0.25)\,, \\ = 2\arcsin^2\left[\sqrt{\dfrac{1}{4\xi}}\right] & (\xi \geq 0.25)\,. \end{cases} \tag{105}$$

Substitution of $m_\sigma = 664.1$ MeV then yields a rate of 2.39 keV, while allowing for an $s\bar{s}$ admixture with scalar mixing angle $\phi_S = 21.1°$ further reduces the rate to 1.84 keV. However, one now has to include meson loops as well here, which are not negligible at all, contrary to the π^0, η, and η' cases, because of phase space. A complete analysis of such contributions was carried out in [43], including pion, kaon, κ ($K_0^*(800)$), and $a_0(980)$ loops. The net effect of these loops is a very sizable increase of the rate, resulting now in a value of 3.39 keV, which should be compared with the recent analyses yielding 3.1–4.1 keV [39, 40] and 3.1–3.9 keV [41, 42]. [Note that the quark-loop-only two-photon rate of the σ is significantly smaller than the one reported in [43], due to the experimentally updated [6] value $f_K = 1.197 f_\pi$, leading to a constituent strange quark mass $m_{s,\text{con}} = 470.5$ MeV, but principally because of the scalar mixing angle $\phi_s = 21.1°$ used here, which is more realistic than the one employed in [43]. Nevertheless, the total $\sigma \to 2\gamma$ rate, including meson loops, is very close to the value of 3.5 keV found in the latter paper. This can be understood from the interference effects among the various quark-loop and meson-loop contributions.]

The case of the $f_0(980)$ is trickier, as its mass is slightly larger than twice the strange quark mass $m_s \simeq 470.5$ MeV, so that we are beyond the NJL limit. Assuming for the moment that this limit holds approximately, we can estimate the quark-loop amplitude as [104]

$$|\mathcal{M}_{f_0(980)\to\gamma\gamma}| = \frac{\alpha N_c g_{f_0}^{s\bar{s}}}{9\pi m_{s,\text{con}}} = \frac{\sqrt{2}\alpha N_c \hat{m}_{\text{con}}}{9\pi f_\pi m_{s,\text{con}}}\,, \tag{106}$$

which gives a two-photon width of about 0.33 keV, compatible with the average experimental [6] value $0.29^{+0.07}_{-0.06}$ keV. However, much more serious than the small violation of the NJL limit is the presence of an $n\bar{n}$ admixture in the $f_0(980)$, corresponding to a nonvanishing scalar mixing angle, as also suggested by the allowed [6] $f_0(980) \to \pi\pi$ decay mode. Namely, the effect of an $n\bar{n}$ component is enhanced by a factor of roughly 25 [106], since the electric charge of the u quark is twice that of the s quark, which makes the prediction of $\Gamma_{f_0(980)\to\gamma\gamma}$ highly unstable. The necessary inclusion of meson loops, too, will also add to the uncertainty, although a (partial) cancellation of $n\bar{n}$ and meson-loop contributions is a plausible possibility. Concretely, including the same meson loops as above for the σ, a not unreasonable scalar mixing angle of 18° is required to obtain an $f_0(980) \to 2\gamma$ rate of 0.29 keV. However, caution is recommended because of the very strong sensitivity of this result to the precise value of ϕ_S.

Finally, the two-photon width of the $a_0(980)$ is the most difficult one in the framework of the QLLσM, and probably in any effective description with quark degrees of freedom. The reason is that the $a_0(980)$ is way beyond the NJL limit, as $m_{a_0(980)} = (980 \pm 20)$ MeV [6] and $\hat{m}_{\text{con}} = 337.5$ MeV, so that dispersive effects will arise from the quark loops. If one simply discards the corresponding imaginary parts — because of quark confinement — and includes meson loops, the QLLσM prediction [104] may be compatible with the experimental [6] value $\Gamma_{a_0(980)\to\gamma\gamma} = (0.30 \pm 0.10)$ keV.

11 Scalar mesons $a_0(980)$ and $f_0(980)$

Next we revisit the $a_0(980)$ and $f_0(980)$ scalar mesons, and study their strong decays. The PDG tables [6] now list the isovector $a_0(980)$ and isoscalar $f_0(980)$ with central masses of 990 ± 20 MeV and $980 \pm$

 © 2013 WILEY-VCH Verlag GmbH & Co. KGaA, Weinheim

20 MeV, respectively. Henceforth, we shall refer to these scalars in any equations simply as a_0 and f_0. In the QLLσM, they are both bound states heavier than 749 MeV, separated from the elementary $q\bar{q}$ mesons $\sigma(664)$ and $\pi(137)$, as suggested by the $Z = 0$ compositeness condition in Sect. 4 above. Note that also the pseudoscalars $\eta_{n\bar{n}}$ and $\eta_{s\bar{s}}$, introduced in the previous section, are bound states, as $m_{\eta_{n\bar{n}}} \simeq 758.56$ MeV and $m_{\eta_{s\bar{s}}} \simeq 801.29$ MeV.

Now we estimate the strong-interaction decay rate for the process $a_0 \to \eta\pi$, which is approximately given by [104]

$$\Gamma_{a_0 \to \eta\pi} = \frac{p}{8\pi}\left[\frac{2g_{a_0\eta_{n\bar{n}}\pi}\cos\phi_P}{m_{a_0}}\right]^2 \simeq 135 \text{ MeV} , \tag{107}$$

using $p = 319$ MeV [6], $\phi_P \simeq 41.9°$, along with the bound-state CL coupling

$$g_{a_0\eta_{n\bar{n}}\pi} = \frac{m_{a_0}^2 - m_{\eta_{n\bar{n}}}^2}{2f_\pi^{\text{CL}}} \simeq 2.15 \text{ GeV} , \tag{108}$$

the latter being near the QLLσM coupling $\lambda f_\pi^{\text{CL}} = \left(m_\sigma^{\text{CL}}\right)^2/2f_\pi^{\text{CL}} = 2\hat{m}_{\text{con}}\, g_{\pi qq} \simeq 2.359$ GeV. Furthermore, the PDG tables report [6] the branching ratio $\Gamma_{a_0\to K\bar{K}}/\,\Gamma_{a_0\to\eta\pi} = 0.183 \pm 0.024$, as well as the two $a_0 \to K\bar{K}$ rates $\Gamma_{a_0\to K\bar{K}} \simeq 24$ MeV and 25 MeV. On the theoretical side, the N/D approach to unitarized ChPT [107] gives 24 MeV, while a much earlier analysis [108] yielded 25 MeV. Thus, we take the average rate $\Gamma_{a_0\to K\bar{K}} \simeq 24.5$ MeV to predict

$$\Gamma_{a_0\to\eta\pi} \simeq \frac{\Gamma_{a_0\to K\bar{K}}}{0.183} \simeq 134 \text{ MeV} , \tag{109}$$

which is very near Eq. (107) above.

Next we study the $I = 0$ scalar meson $f_0(980)$, and show that it is mostly an $s\bar{s}$ bound state. Data [6] finds the e.m. branching ratio

$$\frac{B(\phi(1020) \to f_0\gamma)}{B(\phi(1020) \to a_0\gamma)} = \frac{(3.22 \pm 0.19) \times 10^{-4}}{(7.6 \pm 0.6) \times 10^{-5}} = 4.24 \pm 0.42 . \tag{110}$$

Since we know that the vector $\phi(1020)$ is almost a pure $s\bar{s}$ state, Eq. (110) clearly suggests the isoscalar $f_0(980)$ is mostly an $s\bar{s}$ scalar bound state (the isovector $a_0(980)$ has no strange-quark content).

Additional information comes from data on strong decays involving the $f_0(980)$. First of all, there is the non-observation [6] of the decay $a_1(1260) \to f_0(980)\pi$, whereas $a_1(1260) \to \sigma\pi$ *has* been seen [6]. This again confirms that $f_0(980)$ is mainly $s\bar{s}$. Also, the observed [6, 109] rate ratio

$$\frac{\Gamma_{f_0\to\pi\pi}}{\Gamma_{f_0\to\pi\pi} + \Gamma_{f_0\to K\bar{K}}} = 0.84^{+0.02}_{-0.02} \quad\Rightarrow\quad \frac{\Gamma_{f_0\to K\bar{K}}}{\Gamma_{f_0\to\pi\pi}} \simeq 0.19 \tag{111}$$

is very near the observed branching ratio $\Gamma_{a_0\to K\bar{K}}/\Gamma_{a_0\to\eta\pi} = 0.183$ mentioned above.

Now we estimate the $f_0 \to \pi\pi$ partial width as [104]

$$\Gamma_{f_0\to\pi\pi} = \frac{p}{8\pi}\frac{3}{4}\left[\frac{2g_{\sigma\pi\pi}\sin\phi_S}{m_{f_0}}\right]^2 \simeq 42.2 \text{ MeV} , \tag{112}$$

where we have used $p = 471$ MeV [6], the QLLσM coupling $g_{\sigma\pi\pi} = \lambda f_\pi^{\text{CL}} = 2.357$ GeV, and $\phi_S \simeq 21.1°$ from Eq. (94). This partial width is compatible with the E791 [110] value of $(44 \pm 2 \pm 2)$ MeV, and so lends further support to a scalar mixing angle of roughly 21°.

Finally, also weak interactions can be used to show that the $f_0(980)$ is dominantly an $s\bar{s}$ state, by modeling [111] the decay $D_s^+ \to f_0(980)\pi^+$, with an observed [6] partial width of about 2×10^{-14} GeV, via a W^+-emission process.

© 2013 WILEY-VCH Verlag GmbH & Co. KGaA, Weinheim

Fortschr. Phys. **61**, No. 11 (2013) 1013

12 Nonleptonic weak decays $K \to 2\pi$

In the present and the next section, we apply our QLLσM approach to nonleptonic kaon decays and the $\Delta I = 1/2$ rule, by introducing a weak Hamiltonian. Nambu [1] tried to link the (chiral) GTR-conserving axial currents with semileptonic weak $\pi \to \mu\nu$ decay. Instead, we begin by extracting the nonleptonic $K \to 2\pi$ weak amplitudes from the recent observed data [6], viz. (in units of GeV)

$$|\mathcal{M}_{K_S \to \pi^+\pi^-}| = m_{K_S}\sqrt{\frac{8\pi\Gamma_{+-}}{q_a}} \simeq 39.204 \times 10^{-8} , \tag{113}$$

$$|\mathcal{M}_{K_S \to \pi^0\pi^0}| = m_{K_S}\sqrt{\frac{16\pi\Gamma_{00}}{q_b}} \simeq 36.657 \times 10^{-8}, \tag{114}$$

$$|\mathcal{M}_{K^+ \to \pi^+\pi^0}| = m_{K^+}\sqrt{\frac{8\pi\Gamma_{+0}}{q_c}} \simeq 1.8125 \times 10^{-8} . \tag{115}$$

Here, the center-of-mass (CM) momenta (in MeV) $q_a = 206$, $q_b = 209$, and $q_c = 205$, with decay rates (in 10^{-16} GeV) $\Gamma_{+-} = 50.825$, $\Gamma_{00} = 22.563$, and $\Gamma_{+0} = 0.10995$. The average $\Delta I = 1/2$ scale from Eqs. (113,114) is

$$|\mathcal{M}_{K_S \to 2\pi}|^{\text{avg.}}_{\Delta I = 1/2} \simeq \frac{39.204 + 36.657}{2} \times 10^{-8}\,\text{GeV} \simeq 37.93 \times 10^{-8}\,\text{GeV} , \tag{116}$$

about 21 times the much smaller $\Delta I = 3/2$ scale in Eq. (115).

The first-order-weak (FOW) and second-order-weak (SOW) quark-model-based scales originate from the SOW soft-kaon theorem [22], due to s-d single-quark-line (SQL) weak transitions, generated by the SOW quark loop of Fig. 9. This SOW scale reads

$$|\langle K^0 | H_{W,Z} | \bar{K}^0 \rangle| = 2\beta_w^2 m_{K^0}^2 = m_{K^0} \Delta m_{K_{LS}} , \tag{117}$$

so that the observed [6] $\Delta m_{K_{LS}} \simeq 3.484 \times 10^{-12}$ MeV, for [6] $m_{K^0} \simeq 497.614$ MeV, implies the dimensionless weak scale [112]

$$|\beta_w| = \sqrt{\frac{\Delta m_{K_{LS}}}{2m_{K^0}}} \simeq 5.917 \times 10^{-8} . \tag{118}$$

Note that the GIM scheme [113] estimated $|\beta_w| \sim 5.6 \times 10^{-8}$.

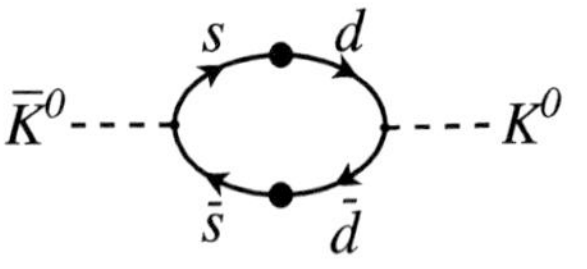

Fig. 9 Second-order-weak $\bar{K}^0 \leftrightarrow K^0$ transition.

Then, the FOW quark loop of Fig. 10 generates the $\Delta I = 1/2$ $K_L \to \pi^0$ transition [112, 114, 115]

$$|\langle \pi^0 | H_w | K_L \rangle|_{\Delta I = 1/2} = \frac{2|\beta_w| m_{K^0}^2 f_K}{f_\pi} \simeq 3.507 \times 10^{-8}\,\text{GeV}^2 , \tag{119}$$

with [6] $f_K/f_\pi \simeq 1.197$ and the β_w scale from Eq. (118), taking the final-state π^0 on the K_L mass shell via PCAC. The crucial FOW scale in Eq. (119) is compatible with the average of 11 different data sets [116], such as $K \to 2\pi$, $K \to 3\pi$, $K_S \to 2\gamma$, $K_L \to 2\gamma$, ...:

$$|\langle \pi^+ | H_w | K^+ \rangle| = |\langle \pi^0 | H_w | K_L \rangle| = (3.59 \pm 0.05) \times 10^{-8}\,\text{GeV}^2 . \tag{120}$$

 © 2013 WILEY-VCH Verlag GmbH & Co. KGaA, Weinheim

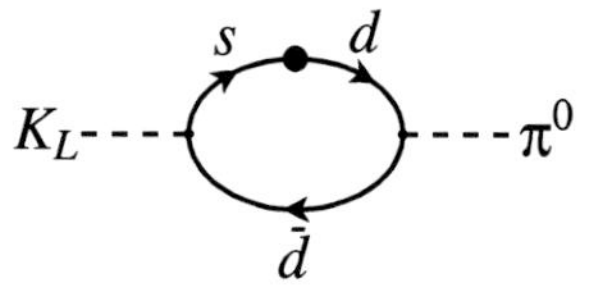

Fig. 10 First-order-weak $K_L \leftrightarrow \pi^0$ transition.

Note that the predicted $\Delta I = 1/2$ weak scale in Eq. (119) is only 2.4% below the central value in Eq. (120). Also, the latter equation confirms the scalar σ meson, alias $f_0(500)$ [6], as the "chiral partner" of π^0 [31].

Specifically, we can estimate the $\Delta I = 1/2$ $K \to 2\pi$ amplitude via the σ-pole graph of Fig. 11, predicting [115]

$$|\langle\pi\pi|H_w|K_S\rangle|_{\Delta I=1/2} \simeq \left|\frac{\langle\sigma|H_w|K_S\rangle\, m_\sigma^2}{(m_\sigma^2 - m_{K^0}^2 + i m_\sigma \Gamma_\sigma) f_\pi}\right| \tag{121}$$

$$\simeq \frac{(3.507\times 10^{-8})\,0.4409}{(0.6654^2)\,0.09221}\ \text{GeV} \simeq 37.87\times 10^{-8}\ \text{GeV}\,, \tag{122}$$

where we have used $m_\sigma \simeq 664.1$ MeV, $m_{K^0} = 497.6$ MeV, $\Gamma_\sigma \simeq 600$ MeV, and $f_\pi = 92.21$ MeV [6]. Note that this $\Delta I = 1/2$ estimate lies right between the values from data in Eqs. (113,114).

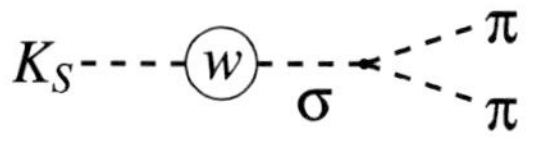

Fig. 11 $K_S \to 2\pi$ σ-pole graph.

Finally, note that pion PCAC (manifest in the nucleon LσM [2, 3]) requires, from the weak-interaction chiral commutator $[Q + Q_5, H_w] = 0$, that

$$|\langle\pi\pi|H_w|K_S\rangle| \simeq \frac{1}{f_\pi}\,|\langle\pi|[Q_5^\pi, H_w]|K_S\rangle| \simeq \frac{1}{f_\pi}\,|\langle\pi^0|H_w|K_L\rangle| \tag{123}$$

$$\simeq \frac{3.507\times 10^{-8}\ \text{GeV}^2}{0.09221\ \text{GeV}} \simeq 38.03\times 10^{-8}\ \text{GeV}\,. \tag{124}$$

This scale is again right in between the values in Eqs. (113,114), and extremely close to the $\Delta I = 1/2$ estimate in Eq. (122).

In passing, we confirm that the $\Delta I = 3/2$ scale in Eq. (115) is definitely small, as already mentioned above, by estimating the W-emission (WE) $K^+ \to \pi^+\pi^0$ amplitude in Fig. 12, i.e.,

$$\left|\langle\pi^+\pi^0|H_w|K^+\rangle\right|_{\text{WE}} = \left|\frac{G_F V_{ud} V_{us}}{2\sqrt{2}}\,(m_{K^+}^2 - m_{\pi^0}^2) f_\pi\right| \simeq 1.885\times 10^{-8}\ \text{GeV}\,, \tag{125}$$

for [6] $G_F = 11.6637\times 10^{-6}$ GeV^{-2}, $|V_{ud}| = 0.97419$, $|V_{us}| = 0.2257$, and $f_\pi = 92.21$ MeV. This WE estimate is indeed near the observed $\Delta I = 3/2$ amplitude in Eq. (115).

13 $K \to 2\pi$ weak tadpole scale and the $\Delta I = 1/2$ rule

As an alternative to estimating the $K_S \to \pi^0\pi^0$ rate via the σ-pole graph of Fig. 11, we could compute it through the tadpole graph of Fig. 13 [117–119]. This implies [115, 119] the amplitude magnitude, via

© 2013 WILEY-VCH Verlag GmbH & Co. KGaA, Weinheim

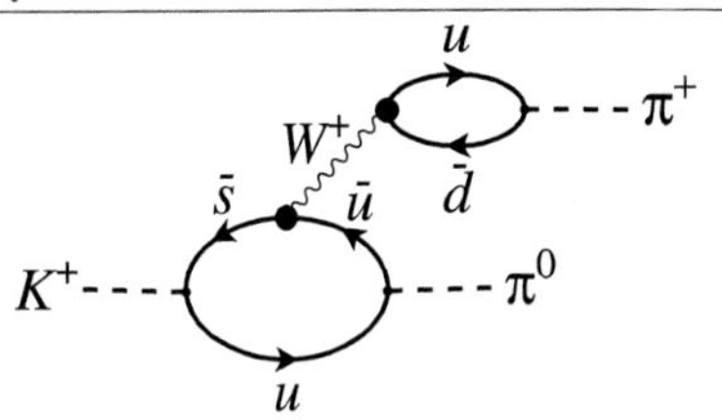

Fig. 12 W-emission $K^+ \to \pi^+\pi^0$ graph.

PCAC (for $f_\pi = 92.21$ MeV),

$$|\mathcal{M}_{K_S\to 2\pi^0}| \simeq \frac{1}{2f_\pi^2}\left(1-\frac{m_{\pi^0}^2}{m_{K_S}^2}\right)|\langle 0|H_w|K_S\rangle| \ , \tag{126}$$

or

$$|\langle 0|H_w|K_S\rangle| \simeq \frac{2f_\pi^2}{1-\dfrac{m_{\pi^0}^2}{m_{K_S}^2}}|\mathcal{M}_{K_S\to 2\pi^0}| \simeq 0.6737\times 10^{-8}\ \text{GeV}^3 \ , \tag{127}$$

using Eq. (114). This scale is not far from the pion PCAC FOW-SQL weak amplitude, in units of GeV^3,

$$|\langle 0|H_w|K_S\rangle| \simeq 2f_\pi\, 2|\beta_w|\, m_{K_S}^2\frac{f_K}{f_\pi} \simeq 0.6468\times 10^{-8} \ , \tag{128}$$

where we have used $\beta_w \simeq 5.9166\times 10^{-8}$ from Eq. (118), and [6] $f_K/f_\pi \simeq 1.197$.

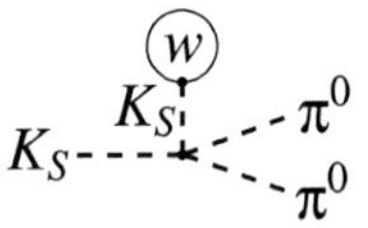

Fig. 13 $K_S \to \pi^0\pi^0$ tadpole graph.

Another check on $\langle 0|H_w|K_S\rangle$ is via the radiative decays $\pi^0 \to 2\gamma$ and $K_L \to 2\gamma$, the latter process also involving a weak transition. Given the very successful $\pi^0 \to 2\gamma$ scale in Eqs. (27,28), for $N_c = 3$, the analogue $K_L \to 2\gamma$ amplitude is

$$|F_{K_L\to 2\gamma}| = \sqrt{\frac{64\pi\Gamma_{K_L\to 2\gamma}}{m_{K_L}^3}} \simeq 0.3389\times 10^{-8}\ \text{GeV}^{-1} \ , \tag{129}$$

from the observed rate [6] $\Gamma_{K_L\to 2\gamma} \simeq 0.70376\times 10^{-20}$ GeV. Note that, using the FOW weak scale $\langle \pi^0|H_w|K_L\rangle \simeq 3.507\times 10^{-8}$ GeV^2 from Eq. (119), along with $|F_{\pi^0\to 2\gamma}| \simeq 0.02519$ from Eq. (27), the theoretical Levi–Civita $K_L \to 2\gamma$ amplitude obeys

$$|F_{K_L\to 2\gamma}|_{\text{th.}} \simeq \frac{|F_{\pi^0\to 2\gamma}|\,|\langle \pi^0|H_w|K_L\rangle|}{m_{K_L}^2 - m_{\pi^0}^2} \simeq 0.3851\times 10^{-8}\ \text{GeV}^{-1} \ . \tag{130}$$

Then the theoretical radiative tadpole scale is

$$|\langle 0|H_w|K_S\rangle|^{\text{rad.}}_{\text{th.}} = \left|\frac{F_{K_L\to 2\gamma}}{F_{\pi^0\to 2\gamma}}\right|(m_{K_L}^2 - m_{\pi^0}^2)\,2f_\pi \simeq 0.6468\times 10^{-8}\ \text{GeV}^3 \ , \tag{131}$$

© 2013 WILEY-VCH Verlag GmbH & Co. KGaA, Weinheim

where we have substituted the theoretical $K_L \to 2\gamma$ amplitude from Eq. (130). The value in Eq. (131) is essentially equal to the pion PCAC FOW-SQL amplitude in Eq. (128), which gives us confidence in our tadpole approach.

With hindsight, Weinberg [120, 121][10] showed that this "truly weak" kaon tadpole *cannot* be rotated away, as sometimes thought. The reasonable agreement among the different analyses of the kaon tadpole scale in Eqs. (127,128,131) confirms Weinberg's result [120, 121].

14 Meson form factors

Following closely [72, 123], we study how meson form factors, normalized as $F(q^2=0) = 1$, are compatible with the $SU(2)$ and $SU(3)$ QLLσM scheme. Specifically, the charged pion and kaon vector e.m. currents are defined as

$$\langle \pi^+(q')|V^\mu_{\text{e.m.}}|\pi^+(q)\rangle = F_\pi(q^2)(q'+q)^\mu, \tag{132}$$

$$\langle K^+(q')|V^\mu_{\text{e.m.}}|K^+(q)\rangle = F_K(q^2)(q'+q)^\mu. \tag{133}$$

Then the quark loops for the $SU(3)$ QLLσM theory predict, in the CL,

$$F^{\text{CL}}_{\pi,\text{QLL}\sigma\text{M}}(k^2) = -i4g^2N_c\int_0^1 dx \int đ^4p \left[p^2 - \hat{m}_q^2 + x(1-x)k^2\right]^{-2}, \tag{134}$$

$$F^{\text{CL}}_{K,\text{QLL}\sigma\text{M}}(k^2) = -i4g^2N_c\int_0^1 dx \int đ^4p \left[p^2 - \hat{m}_{sn}^2 + x(1-x)k^2\right]^{-2}, \tag{135}$$

where $\hat{m}_q = (m_u + m_d)/2$ and $m_{sn} = (m_s + \hat{m}_q)/2$. The logarithmic divergence in Eqs. (134,135) can be minimized via a rerouting procedure [124, 125], also using the LDGE in Eq. (17), which gives

$$F^{\text{CL}}_{\pi,\text{QLL}\sigma\text{M}}(0) = -i4g^2N_c\int đ^4p \left[p^2 - \hat{m}_q^2\right]^{-2} = 1, \tag{136}$$

$$F^{\text{CL}}_{K,\text{QLL}\sigma\text{M}}(0) = -i4g^2N_c\int đ^4p \left[p^2 - \hat{m}_{sn}^2\right]^{-2} = 1, \tag{137}$$

for $m_\pi \to 0$ and $m_K \to 0$ in the CL. Note the detailed comment in [124, 125] that rerouting one-half of the loop momenta in the opposite direction removes the apparent log divergence of the integrals in Eqs. (134,135), leading to the finite integrals in Eqs. (136,137), while also justifying the needed gauge invariance of the vector currents defined in Eqs. (132,133).

Next we study the $\pi^+ \to e^+\nu\gamma$ radiative decay form factors, namely the vector form factor F_V and the axial-vector one F_A. Now, the experimental status of the latter two observables has varied a lot over the years. For instance, in 1989 [126] reported the measured values $F_V = 0.023$ and $F_A = 0.021$, albeit with very large error bars. However, three years earlier the same collaboration measured [127] the ratio $\gamma = F_A/F_V \simeq 0.7$, though again with a huge error. Now, the observed [126] value $F_V = 0.023$ was in reasonable agreement with the old conserved-vector-current (CVC) prediction [128], at a zero value of the invariant $e^+\nu$ mass squared q^2,

$$F_V(0) = \frac{\sqrt{2}m_{\pi^+}}{8\pi^2 f_\pi} \simeq 0.027. \tag{138}$$

On the other hand, the measured [127] value $\gamma \simeq 0.7$ was compatible with the $SU(2)$ QLLσM prediction [129], from the sum of a nonstrange quark triangle plus a pion loop,

$$\gamma = \frac{F_A(0)}{F_V(0)} = 1 - \frac{1}{3} = \frac{2}{3}. \tag{139}$$

[10] Also see [122].

© 2013 WILEY-VCH Verlag GmbH & Co. KGaA, Weinheim

However, extension of vector-meson dominance to axial-vector dominance [129] reduces the latter prediction to $\gamma \simeq 0.5$. Much more recently, F_V and F_A have been measured [130] with improved accuracy and an only mild dependence on q^2, resulting in a ratio $\gamma = 0.46 \pm 0.07$, and so compatible with axial-vector dominance.

Next we consider charged-pion polarization for the process $\gamma\gamma \to \pi^+\pi^-$. The quantities effectively measured are combinations of the electric and magnetic polarizabilities α_{π^+} and β_{π^+}, respectively, viz. [31, 131]

$$(\alpha - \beta)_{\pi^+} \quad \text{and} \quad (\alpha + \beta)_{\pi^+} \,. \tag{140}$$

Now, general chiral symmetry requires the latter combination to vanish [131], which is compatible with the experimental result [132] $(\alpha + \beta)_{\pi^+} = (1.4 \pm 3.1_{\text{stat}} \pm 2.5_{\text{syst}}) \times 10^{-43}\,\text{cm}^3$ $(10^{-4}\,\text{fm}^3)$. So we shall focus on the electric polarizability α_{π^+} only. A simple QLLσM estimate predicts

$$\alpha_{\pi^+}^{\text{QLL}\sigma\text{M}} \simeq \frac{(\hbar c)^3 \alpha_{\text{QED}}\, \gamma_{\text{QLL}\sigma\text{M}}}{8\pi^2 m_\pi f_\pi^2} \simeq 3.99 \times 10^{-4}\,\text{fm}^3 \,, \tag{141}$$

with $\alpha_{\text{QED}} = e^2/4\pi$, $\gamma_{\text{QLL}\sigma\text{M}} \simeq 2/3$ from Eq. (139), and where we have used $\hbar c \simeq 197.3$ MeV fm, $f_\pi \simeq 92.21$ MeV. A more detailed calculation with quark and pion loops yields [131]

$$\alpha_{\pi^+}^{\text{QLL}\sigma\text{M}} \simeq \frac{2.17\,(\hbar c)^3 \alpha_{\text{QED}}}{16\pi^2 m_\pi f_\pi^2} \simeq 6.5 \times 10^{-4}\,\text{fm}^3 \,, \tag{142}$$

where the factor $2.17 \simeq 5/3 + 0.5$ stems from N_c times the sum of the squares of the u and d quark charges in the quark loops, i.e., $5/3 = 3\left((2/3)^2 + (-1/3)^2\right)$, plus a Feynman-integral contribution of about 0.5 from the pion loop. The value of $6.5 \times 10^{-4}\,\text{fm}^3$ in Eq. (142) agrees quite well with a recent analysis [133] based on several experiments, resulting in a value

$$(\alpha - \beta)_{\pi^+} = 13^{+2.6}_{-1.9} \times 10^{-4}\,\text{fm}^3 \,, \tag{143}$$

which should be divided by two in order to compare with α_{π^+}, if one indeed assumes that the sum $(\alpha + \beta)_{\pi^+}$ vanishes or is very small. These values are also in reasonable agreement with the NJL prediction [134] $(\alpha - \beta)_{\pi^+} = 9.39 \times 10^{-4}\,\text{fm}^3$, and moreover with dispersion sum rules [134]. However, [130] measured $\alpha_{\pi^+} = (2.78 \pm 0.10) \times 10^{-4}\,\text{fm}^3$, so that also for this observable some controversy persists.

For a very recent summary of experimental results on pion polarizabilites over the years (excluding [130]), see [135], and for a detailed discussion of nucleon polarizabilities and their relation to the two-photon width of the σ meson, see [136–138].

Lastly, we study the semileptonic weak $K^+ \to \pi^0 e^+ \nu$ ($K_{\ell 3}$) decay. The nonrenormalization theorem [139] for the QLLσM says [140]

$$f_+(0) = 1 - \frac{g^2}{8\pi^2}\left[\frac{m_{s,\text{con}}}{\hat{m}_{\text{con}}} - 1\right]^2 \simeq 0.974 \,, \tag{144}$$

where we have used that $m_{s,\text{con}}/\hat{m}_{\text{con}} \simeq 2f_K/f_\pi - 1 \simeq 1.394$ [6]. A prior estimate [3], based on $K_{\ell 2}$ and $K_{\ell 3}$ decays, but compatible with the QLLσM, found

$$f_+(0) = \frac{1}{1.23}\frac{f_K}{f_\pi} \simeq \frac{1.197}{1.23} \simeq 0.973 \,. \tag{145}$$

Both approaches are in agreement with the data (see [72] for details).

 © 2013 WILEY-VCH Verlag GmbH & Co. KGaA, Weinheim

15 Superconductivity and the $SU(2)$ Goldberger-Treiman relation

Now we shall try to make a link between the energy gap in superconductivity and the QLLσM, via the critical temperature described in Sect. 8.

The theory of superconductivity was first understood by Bardeen, Cooper, and Schrieffer (BCS) in [4]. In Eq. (3.30) of this paper, they found that the ratio of the energy gap 2Δ and the critical temperature T_c is given by

$$\frac{2\Delta}{k_B T_c} \simeq 3.50 , \tag{146}$$

where k_B is the Boltzmann constant, expressed as $k_B \simeq 8.62 \times 10^{-5}\,\mathrm{eV\,K^{-1}}$ [6]. Recall that, at the quark level, with g_A=1, and in the CL, the GTR gives, with $\hat{m}_{\text{dyn}} \simeq m_N/3 \simeq 313$ MeV and $f_\pi^{\text{CL}} \simeq 89.63$ MeV,

$$g_{\pi qq} \simeq \frac{\hat{m}_{\text{dyn}}}{f_\pi^{\text{CL}}} \simeq 3.49 . \tag{147}$$

Note that, with $\Delta \to \hat{m}_{\text{dyn}}$ and $k_B T_c \to 2 f_\pi^{\text{CL}}$, Eq. (146) converts into Eq. (147). This connection between condensed-matter and particle physics was stressed in [141]. In some sense, this is the spirit of Nambu's original work [1]. Actually, the BCS ratio in Eq. (146) can be written mathematically as [4]

$$\left(\frac{2\Delta}{k_B T_c}\right)_{\text{BCS}} = 2\pi e^{-\gamma_E} \simeq 3.528 , \tag{148}$$

where $\gamma_E \simeq 0.5772$ is the Euler constant. On the other hand, the QLLσM pion-quark coupling is selfconsistently bootstrapped to the value (cf. Eq. (24))

$$g_{\pi qq} = \frac{2\pi}{\sqrt{3}} \simeq 3.628 , \tag{149}$$

still remarkably close to BCS value in Eq. (148).

Looking directly at condensed-matter phenomenology, data for 2H–$NbSe_2$, with a critical temperature $T_c = 7.2$ K, finds [142, 143] an energy gap $2\Delta = 17.2\,\mathrm{cm^{-1}}$, which expressed in the inverse wave length $1/\lambda = \nu/c = h\nu/hc = E/2\pi\hbar c$ yields

$$\frac{2\Delta}{k_B T_c} \simeq \frac{2\pi\,(197.3\,\mathrm{MeV\,fm})\,(17.2 \times 10^{-13}\,\mathrm{fm^{-1}})}{(8.62 \times 10^{-5}\,\mathrm{eV\,K^{-1}})\,(7.2\,\mathrm{K})} \simeq 3.44 . \tag{150}$$

Another experiment [144], using an Rb_3C_{60} superconductor, reports a ratio $\Delta/k_B = 53$ K for $T_c = 29.4$ K, which gives

$$\frac{2\Delta}{k_B T_c} = \frac{106}{29.4} \simeq 3.61 . \tag{151}$$

So the average of these two experimental results is very close to the theoretical BCS predictions in Eqs. (146,148), but remarkably enough also to the QLLσM values in Eqs. (147,149). As a matter of fact, Nambu found [145], not long after the pioneering BCS paper [4], that gauge invariance is a valid concept for superconductivity, but with radiative photons replaced by acoustical phonons.

16 Dynamically generating the top-quark and scalar-Higgs masses

In this section we shall apply QLLσM ideas to the gauge bosons $W^\pm$ and Z, as well as the top quark and Higgs boson, by analogy with the low-energy sector.

© 2013 WILEY-VCH Verlag GmbH & Co. KGaA, Weinheim

The Higgs mass in the electroweak Standard Model (EWSM) is a free parameter, but experiment now indicates a direct lower-mass search limit of about 114 GeV, with 95% CL [6]. On the other hand, a global fit to precision electroweak data, gathered in the course of many years at LEP, Tevatron, and other accelerators yields a range of ($m_H = 94^{+29}_{-24}$) GeV ($m_H < 152$ GeV, 95% C.L.) [6]. However, these estimates are based on *perturbative* EWSM calculations, which leaves some room for alternative scenarios, also due to the triviality problem [146, 147].[11] Moreover, the analyses themselves may be less trustworthy than generally assumed [148–150]. Finally, clear Higgs-like signals have been seen very recently by the ATLAS [9] and CMS [10] Collaborations at LHC, i.e., at a mass of about 125 GeV. Further experiments will be needed, though, to confirm and determine the quantum numbers of the observed boson.

Here, we shall try to estimate the Higgs mass in a *nonperturbative* (NP) framework, in the spirit of the QLLσM.

The EWSM couplings are $g_W^2/8m_W^2 = G_F/\sqrt{2}$, for [6] $G_F = 11.6637 \times 10^{-6}$ GeV^{-2}, which gives, for $m_W = 80.385$ GeV, [6]

$$|g_w| \;=\; \sqrt{\frac{8G_F}{\sqrt{2}}}\, m_W \;\simeq\; 0.65295\,. \tag{152}$$

Furthermore, the NP vacuum expectation value (VEV) f_w is extracted from G_F as

$$f_w \;=\; \frac{1}{\sqrt{\sqrt{2}G_F}} \;\simeq\; 246.22\ \text{GeV}\,. \tag{153}$$

Then, quadratically divergent tadpole graphs can be made to cancel, in the spirit of B. W. Lee's NP null tadpole condition [27] (see Fig. 6), which allows to predict [21, 151–154] the scalar Higgs mass. Alternatively, and in analogy with the scalar σ meson (cf. Eqs. (3,21)), the EWSM Higgs boson is dominated by a scalar $t\bar{t}$ state, which in the CL would yield a mass $m_H^{\text{CL}} = 2m_t$. Beyond the CL this value is then reduced to

$$m_H \;=\; \sqrt{(2m_t)^2 - (2m_W^2 + m_Z^2)} \;\simeq\; 314.9\ \text{GeV}\,, \tag{154}$$

for the measured [6] heavy masses (in GeV)

$$m_t \;\simeq\; 173.5\,, \quad m_W \;\simeq\; 80.385\,, \quad m_Z \;\simeq\; 91.1876\,, \tag{155}$$

where we have dropped the (small) errors as well as the negligible contributions from the much lighter other quarks.

Alternatively, we can use the modified Kawarabayashi-Suzuki-Riazuddin-Fayyazudin (KSRF) [7] approach in the electroweak sector. As an illustration, for the ρ meson the KSRF relation predicts

$$m_\rho \;=\; \sqrt{2}\, g_{\rho\pi\pi} f_\pi \;\simeq\; 775.18\ \text{MeV}\,, \tag{156}$$

which is remarkably close to the PDG [6] value $m_\rho = 775.49$ MeV. The weak-interaction KSRF analogue is obtained by the substitutions [21]

$$m_\rho \to M_W \quad , \quad g_{\rho\pi\pi} \to \frac{g_w}{2} \quad , \quad \sqrt{2}\, f_\pi \to f_w\,, \tag{157}$$

where the weak coupling simulates $g_\rho \tau^+/2$, and the charged W requires a $\sqrt{2}$ in the weak (VEV) decay constant. Thus, the strong-interaction KSRF relation in Eq. (156) translates into the NP weak KSRF relation

$$m_W \;=\; \frac{1}{2}\, g_w\, f_w\,, \tag{158}$$

[11] Also see the references in [146] to earlier work on $\lambda\phi^4$ theory.

 © 2013 WILEY-VCH Verlag GmbH & Co. KGaA, Weinheim

which is precisely the famous EWSM relation from Eqs. (152,153) [113, 155, 156]. Moreover, using $\Gamma_{Z\to e^+e^-} = 83.91$ MeV [6], one can predict [21], via VMD,

$$|g_Z| = e\bar{e}\sqrt{\frac{m_Z}{12\pi\Gamma_{Z\to e^+e^-}}} \simeq 0.5076\,, \tag{159}$$

as $e^2/4\pi \simeq 1/137.036$ and $\bar{e}^2/4\pi \simeq 1/128.93$ (at the m_Z scale).

Now, the tree-level vector and axial-vector couplings of the Z to the electron get modified, from $g_V^e = -1/2 + 2\sin^2\theta_w$ and $g_A^e = -1/2$ to [6]

$$g_A^e = -0.5064(1) \quad , \quad g_V^e = -0.0398(3)\,, \tag{160}$$

by radiative corrections. Nevertheless, Z remains largely axial, since $\sin^2\theta_w = 0.23108(5)$ (see [6], review on electroweak model, Table 10.2, NOV scheme). The difference $V-A$ coupling becomes

$$g_{V\text{–}A}^e = 0.4666\,, \tag{161}$$

which is reasonably close to the EWSM value $2\sin^2\Theta_w = 0.462$. This is also supported by the ratio of Eq. (158) to the conventional EWSM rate for the Z, namely

$$\Gamma_{Ze^+e^-} = \left(\frac{g_w}{4}\right)^2 \frac{M_Z^3}{12\pi M_W^2} = 82.94\ \text{MeV}\,, \tag{162}$$

which is near the experimental value of 83.91 MeV. This yields the alternative NP expression

$$\sin^2\Theta_w = 1 - (g_w g_Z/4e\bar{e})^2 = 0.2316\,, \tag{163}$$

and so $2\sin^2\Theta_w = 0.4632$, again close to the value of $g_{V\text{–}A}^e$ in Eq. (161).

We may now estimate the very heavy top-quark mass m_t via a NP GTR, as we did for the lighter quarks. Here we have to be careful to take account of an EWSM factor of $2\sqrt{2}$ and the (V-A) VMD coupling $g_Z/2$. In this way we get [21]

$$m_t = 2\sqrt{2}\, f_w\, \frac{|g_Z|}{2} = \sqrt{2}\times 246.22\ \text{GeV}\times 0.5076 \simeq 176.8\ \text{GeV}\,, \tag{164}$$

not far away from the PDG [6] value of $(173.5 \pm 0.6 \pm 0.8)$ GeV. If we now use our theoretical prediction for m_t in Eq. (164) to estimate the Higgs mass via Eq. (154), we find about 322 GeV. It is curious to notice that, very recently, indications of a narrow 325 GeV scalar resonance were found [157] in unpublished, "exotic" CDF data [158].[12]

Lastly, we remark that the top-quark width is huge, viz. $2.0^{+0.7}_{-0.6}$ GeV [6], strongly dominated by the $t \to bW^+$ mode. Note that the latter process is purely weak, though not at all weak in the common sense. Therefore, $t\bar{t}$ physics, and in particular a scalar $t\bar{t}$ state, should be dealt with by NP methods. Thus, we think the Higgs might be selfconsistently described as such a state, being both elementary and composite, just like the σ meson in the QLLσM.

In the next section, we shall revisit the Higgs mass in the context of CP violation.

17 CP violation and Higgs mass

In this section, we deal with CP violation (CPV), but in an NP approach *within* the Standard Model, inspired by the QLLσM, and based on a careful analysis of the most recent CPV data. Moreover, we model CPV in a way that allows to make another estimate of the scalar Higgs mass.

[12] Also see [159].

© 2013 WILEY-VCH Verlag GmbH & Co. KGaA, Weinheim

In [115,119], it was noted that the presently observed [6] branching ratio (BR) for CP-conserving (CPC) $K_S \to 2\pi$ weak decays is extremely near the CPV $K_L \to 2\pi$ BR, i.e.,

$$B\left(K_S \to \frac{\pi^+\pi^-}{\pi^0\pi^0}\right)_{\text{CPC}} = \frac{(69.20 \pm 0.05)\,\%}{(30.69 \pm 0.05)\,\%} = 2.255 \pm 0.004\,, \tag{165}$$

$$B\left(K_L \to \frac{\pi^+\pi^-}{\pi^0\pi^0}\right)_{\text{CPV}} = \frac{(1.967 \pm 0.010) \times 10^{-3}}{(8.64 \pm 0.06) \times 10^{-4}} = 2.277 \pm 0.02\,. \tag{166}$$

Not only are the CPC and CPV scales in Eqs. (165,166) near each other, but the CPC scale in Eq. (165) is near [115] the $\Delta I = 1/2$ scale of 2, due to the σ pole and also the tadpole graph for $K_S \to 2\pi$ weak decays. Also, the present CPC radiative BR [6] is about

$$B\left(K_S \to \frac{\pi^+\pi^-\gamma}{\pi^+\pi^-}\right)_{\text{CPC}} \simeq \frac{1.79 \times 10^{-3}}{69.2 \times 10^{-2}} \simeq 2.59 \times 10^{-3}\,. \tag{167}$$

Moreover, the approximate CPV radiative BR [6] is

$$B\left(K_L \to \frac{\pi^+\pi^-\gamma}{\pi^+\pi^-}\right)_{\text{CPV}} \simeq \left|\frac{\eta_{+-\gamma}}{\eta_{+-}}\right|^2 B\left(K_S \to \frac{\pi^+\pi^-\gamma}{\pi^+\pi^-}\right)_{\text{CPC}} \simeq (2.87 \pm 0.16) \times 10^{-3}\,, \tag{168}$$

using [6] $|\eta_{+-\gamma}| = (2.35 \pm 0.07) \times 10^{-3}$ and $|\eta_{+-}| = (2.232 \pm 0.011) \times 10^{-3}$. Equation (168) is 11% higher than Eq. (167) [115, 119, 160], whereas the radiative e.m. scale is

$$\frac{\alpha}{\pi} \simeq \frac{1}{137.036\,\pi} \simeq 2.323 \times 10^{-3}\,, \tag{169}$$

which is 3.9% higher than $|\eta_{+-}|$ above. Now, it was remarked long ago [161, 162] that the scale of α/π could be the relevant measure for CPV $K \to 2\pi$ weak decays.

In the spirit of the present QLLσM review, we follow [115] and theoretically compute the radiatively (rad) corrected $\Delta I = 1/2$ SQL scale for $K^0 \to \pi^+\pi^-$ minus unity, i.e.,

$$\left|\frac{\eta_{+-}}{\eta_{00}}\right|_{\text{SQL}}^{\text{rad}} - 1 = 2\,\frac{\alpha}{\pi} = 4.646 \times 10^{-3}\,, \tag{170}$$

near CPV data minus unity, viz.

$$\left|\frac{\eta_{+-}}{\eta_{00}}\right|_{\text{data}}^{\text{CPV}} - 1 = \frac{(2.232 \pm 0.011) \times 10^{-3}}{(2.221 \pm 0.011) \times 10^{-3}} - 1 \simeq 4.95 \times 10^{-3}\,. \tag{171}$$

Note, too, that the indirect CPV scale [6]

$$|\varepsilon| = \frac{2\eta_{+-} + \eta_{00}}{3} = (2.228 \pm 0.011) \times 10^{-3} \tag{172}$$

is quite near the α/π scale in Eq. (169), and about one-half the SQL scales in Eqs. (170,171).

Other measures of CPV are the phase angles δ_L (also called A_L) and ϕ_{+-}, related by [119, 163]

$$\delta_L = 2|\varepsilon| \cos\phi_{+-} \simeq 3.232 \times 10^{-3} \tag{173}$$

via Eq. (172), and by the observed [6] angle $\phi_{+-} = (43.51 \pm 0.05)^\circ$. The latter value of ϕ_{+-} is near [164]

$$\phi_{+-} = \arctan\left[\frac{2\Delta m_{LS}}{\Gamma_{K_S}}\right] \simeq 43.47^\circ\,, \tag{174}$$

© 2013 WILEY-VCH Verlag GmbH & Co. KGaA, Weinheim

extracted from [6] $\Delta m_{LS} = 3.484 \times 10^{-12}$ MeV and $\Gamma_{K_S} = \hbar/\tau_{K_S} = 7.351 \times 10^{-12}$ MeV. Lastly, another measure of the CPV phase angle δ_L is the rate asymmetry in semileptonic weak decays [6], i.e.,

$$\delta_L = \frac{\Gamma_{K_L\to\pi^-\ell^+\nu} - \Gamma_{K_L\to\pi^+\ell^-\bar{\nu}}}{\Gamma_{K_L\to\pi^-\ell^+\nu} + \Gamma_{K_L\to\pi^+\ell^-\bar{\nu}}} = (3.32 \pm 0.06) \times 10^{-3} , \tag{175}$$

which is not far from the value in Eq. (173).

Now we look at a possible source of CPV in the context of the SM Cabibbo-Kobayashi-Maskawa (CKM) [165–168] matrix, which for convenience we write in the original parametrization due to Kobayashi and Maskawa [168] (see also the PDG [6] CKM review), viz. [115, 119]

$$V = \begin{pmatrix} V_{ud} & V_{us} & V_{ub} \\ V_{cd} & V_{cs} & V_{cb} \\ V_{td} & V_{ts} & V_{tb} \end{pmatrix} \to \begin{pmatrix} c_1 & -s_1 & 0 \\ s_1 & c_1 & 0 \\ 0 & 0 & -c_2c_3(1+i\delta) \end{pmatrix} . \tag{176}$$

Here, we have introduced a small CPV complex phase δ by writing the almost unity V_{tb} element as $-c_2c_3e^{i\delta} \simeq -c_2c_3(1+i\delta)$, in the limit of a real $SU(4)$ Cabibbo submatrix, with $\theta_1 = -\theta_C$ and $\theta_2, \theta_3 \to 0$ [115, 119]. An SM mechanism that may give rise to such a complex phase is a nonstandard $WW\gamma$ vertex of the form [115, 119, 169]

$$\langle \gamma_\mu(q) W_\beta | W_\alpha \rangle = ie\lambda_w \, \epsilon_{\alpha\beta\mu\sigma} \, q^\sigma , \tag{177}$$

contributing to the tree-level and loop-order $t \to bW^+$ mixing graphs of Fig. 14. Evaluation of these graphs yields [115, 119]

$$-V_{tb} \simeq 1 + i\lambda_w \frac{\alpha}{\pi} \ln\left(1 + \frac{\Lambda^2}{m_t^2}\right) , \tag{178}$$

where Λ is an ultraviolet (chiral) cutoff. If we assume that this cutoff is given by the Higgs mass, Eqs. (176, 178) lead to

$$\ln\left(1 + \frac{m_H^2}{m_t^2}\right) = c_2c_3 \frac{\pi}{\alpha} \delta \simeq 1.428 , \tag{179}$$

where we have used $c_2c_3 = 0.99913$ [6], and $\delta \simeq 3.32 \times 10^{-3}$ from Eq. (175). Substituting now the predicted $m_t \simeq 176.8$ GeV from Eq. (164), we obtain

$$m_H \simeq \sqrt{e^{1.428} - 1}\, m_t \simeq 314.8 \text{ GeV} , \tag{180}$$

which is very close to the EWSM value of 314.9 GeV in Eq. (154), and also reasonably near the KSRF value of 322 GeV, both predicted in Sect. 16 above (also see [21]). Moreover, all three NP Higgs scales are roughly compatible with the observed [6] $m_t = (173.5 \pm 0.6 \pm 0.8)$ GeV as well.

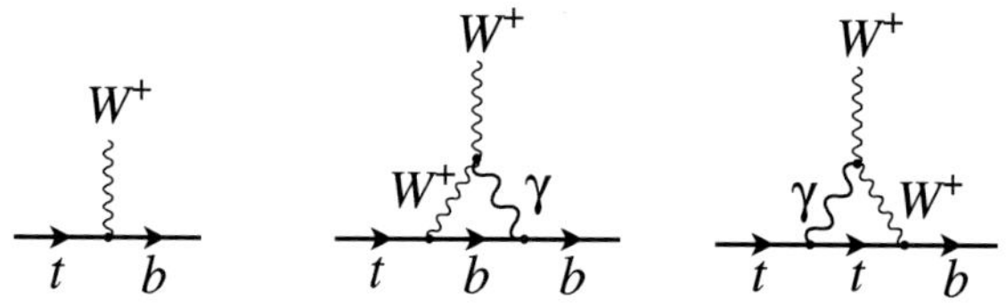

Fig. 14 $t \to bW^+$ transition. Left: tree graph; middle and right: nonstandard $W\gamma W$ loop graphs.

In conclusion, we should emphasize that our *nonperturbative Standard-Model* scenarios for CPV and the Higgs are in no way related to the usual *perturbative field-theory approach beyond the SM.* Another NP approach to CPV in the SM can be found in [170].

© 2013 WILEY-VCH Verlag GmbH & Co. KGaA, Weinheim

18 Concluding remarks

In this review article we have shown how a huge number of strong, electromagnetic, and weak processes fit very well within a scheme whereby chiral symmetry is spontaneously generated in a *linear* representation. The pion, sigma meson, and their $SU(3)$ partners arise dynamically through the quark interactions, and they govern a large body of data with very few parameters indeed. The only *small* corrections to the chiral limit are due to the current quark masses of the lightest quarks (u, d, s).

Sects. 1–15 above covered the many facets of this description. One might think of extrapolating the ideas to the top-quark sector as we did in Sects. 16 and 17, but this is more speculative and perhaps also more problematic because one is moving a fair way from the chiral (zero-mass) limit. An attempt along such lines is fraught with difficulties since the chiral-symmetry-breaking corrections are surely more substantial.

Our description of the Higgs boson as primarily a $t\bar{t}$ composite yielded a mass of about 315 GeV. As mentioned in Sect. 16, this is far removed from the preferred values resulting from fits to electroweak data, though these limits should be interpreted with some caution [148–150]. We are also aware of the recent Higgs-like signals at the LHC, observed by the ATLAS collaboration [9] at 126 GeV and by the CMS collaboration [10] at 125 GeV. However, insufficient statistics has not allowed so far to pin down the spin-parity of the found state, which could be a scalar, pseudoscalar, or a tensor boson, in view of the seen two-photon decay mode. Thus, one cannot exclude yet an interpretation of the data by e.g. a technipion, as predicted in certain (extended) technicolor models, which will have very similar decay modes [171–173], albeit with different angular distributions. Alternatively, the enhancement at 125 GeV might be just one of several threshold enhancements due to a possible substructure in the weak-interaction sector [174]. So we await with considerable interest the high-statistics measurements to be done after the LHC ugrade in a couple of years, which will hopefully allow to carry out the required partial-wave analyses of the observed boson's decay products. Nevertheless, irrespective of the possible confirmation of a Higgs-like scalar at about 125 GeV, the existence of another scalar with a mass of roughly 320 GeV is not out of the question.

Despite the power and simplicity of the QLLσM, other, more traditional approaches to the LσM have been appearing in the literature, even very recently. For instance, in [175] a LσM with only mesonic degrees of freedom and global chiral symmetry was formulated, whose Lagrangian contains elementary vector and axial-vector fields, besides the usual pseudoscalar and scalar ones. Note that this is a very complicated, perturbative model with several adjustable parameters, but in principle applicable to both the light scalar nonet and the scalars in the energy region 1.3–1.7 GeV. More relevant for the present NP theory is a formal selfconsistent generalization [176, 177] of the QLLσM by dynamical generation beyond one loop, including sunset-type diagrams. Such an approach, based on the imposed exact cancellation of quadratic divergences, may eventually allow an asymptotically free formulation of strong interactions, as an effective alternative [176] to QCD, also at higher energies. An asymptotically free QLLσM belongs [178] to a class of non-Hermitian yet PT-symmetric field theories, which have been actively pursued by especially C. Bender [179] and collaborators.

As a final remark, let us stress once again the importance of the NP nature of the QLLσM, as formulated in the present review. Due to the applied bootstrap principle, all couplings of the theory are selfconsistently interrelated via dynamical generation and loop shrinking [17], leaving no model parameters, save the — experimentally fixed [6] — pion weak decay constant f_π. In spite of this lack of freedom, or more likely thanks to it, a wealth of low-energy observables can be described, some even with amazing precision. No other effective theory of strong interactions comes even close in performance.

Acknowledgements One of us (MDS) is deeply indebted to his former co-authors A. Bramon, V. Elias, N. H. Fuchs, H. F. Jones, and N. Paver for their valuable contributions to the development of the QLLσM, and also to E. van Beveren, F. Kleefeld, and many others (see references below) for collaboration on several related applications. The figures in this review have been produced with the graphical package *SCRIBBLE* [180], and we thank P. Nogueira for helpful suggestions. This work was supported by the *Fundação para a Ciência e a Tecnologia* of the *Ministério da Ciência, Tecnologia e Ensino Superior* of Portugal, under contracts nos. CERN/FP/83502/2008 and CERN/FP/109307/2009.

 © 2013 WILEY-VCH Verlag GmbH & Co. KGaA, Weinheim

References

[1] Y. Nambu, Phys. Rev. Lett. **4**, 380–382 (1960).
[2] M. Gell-Mann and M. Levy, Nuovo Cim. **16**, 705 (1960).
[3] V. de Alfaro, S. Fubini, G. Furlan, and C. Rossetti, in: Currents in Hadron Physics (North-Holland Publ., Amsterdam, 1973), Chap. 5.
[4] J. Bardeen, L. Cooper, and J. Schrieffer, Phys. Rev. **106**, 162 (1957).
[5] Y. Nambu and G. Jona-Lasinio, Phys. Rev. **122**, 345–358 (1961).
[6] J. Beringer et al., Phys. Rev. D **86**, 010001 (2012).
[7] K. Kawarabayashi and M. Suzuki, Phys. Rev. Lett. **16**, 255 (1966).
[8] Riazuddin and Fayyazuddin, Phys. Rev. **147**, 1071–1073 (1966).
[9] G. Aad et al., Phys. Lett. B **716**, 1–29 (2012).
[10] S. Chatrchyan et al., Phys. Lett. B **716**, 30–61 (2012).
[11] M. Gell-Mann, Phys. Lett. **8**, 214–215 (1964).
[12] G. Zweig, "An SU_3 model for strong interaction symmetry and its breaking", in: Developments in the Quark Theory of Hadrons, edited by D. Lichtenberg and S. Rosen (Hadronic Press, Nonantum, MA, 1980), pp. 22–101, Also see CERN Reports TH-401 and TH-412 (1964).
[13] J. Goldstone, Nuovo Cim. **19**, 154–164 (1961).
[14] J. Goldstone, A. Salam, and S. Weinberg, Phys. Rev. **127**, 965–970 (1962).
[15] M. Goldberger and S. Treiman, Phys. Rev. **110**, 1478–1479 (1958).
[16] S. Weinberg, Phys. Rev. Lett. **65**, 1181–1183 (1990).
[17] R. Delbourgo and M. Scadron, Mod. Phys. Lett. A **10**, 251–266 (1995).
[18] A. Bramon, Riazuddin, and M. Scadron, J. Phys. G **24**, 1–12 (1998).
[19] V. Elias and M. Scadron, Phys. Rev. Lett. **53**, 1129 (1984).
[20] V. Elias and M. Scadron, Phys. Rev. D **30**, 647 (1984).
[21] M. Scadron, R. Delbourgo, and G. Rupp, J. Phys. G **32**, 735–745 (2006).
[22] M. Scadron, Rept.Prog.Phys. **44**, 213–292 (1981).
[23] S. Coon and M. Scadron, Phys. Rev. C **23**, 1150–1153 (1981).
[24] T. Hakioglu and M. Scadron, Phys. Rev. D **42**, 941–944 (1990).
[25] R. Delbourgo, M. Scadron, and A. Rawlinson, Mod. Phys. Lett. A **13**, 1893–1898 (1998).
[26] M. Scadron, Acta Phys. Polon. B **32**, 4093–4104 (2001).
[27] B. Lee, Chiral Dynamics (Gordon and Breach, New York, 1972).
[28] A. Salam, Nuovo Cim. **25**, 224–227 (1962).
[29] S. Weinberg, Phys. Rev. **130**, 776–783 (1963).
[30] M. Scadron, Phys. Rev. D **57**, 5307–5310 (1998).
[31] E. van Beveren, F. Kleefeld, G. Rupp, and M. D. Scadron, Mod. Phys. Lett. A **17**, 1673 (2002).
[32] L. R. Babukhadia, Y. Berdnikov, A. Ivanov, and M. Scadron, Phys. Rev. D **62**, 037901 (2000).
[33] A. Ivanov, M. Nagy, and M. Scadron, Phys. Lett. B **273**, 137–140 (1991).
[34] J. Hernandez et al., Phys. Lett. B **239**, 1–515 (1990).
[35] R. Longacre, Phys. Rev. D **26**, 82–90 (1982).
[36] D. Asner et al., Phys. Rev. D **61**, 012002 (2000).
[37] H. Marsiske et al., Phys. Rev. D **41**, 3324 (1990).
[38] A. Kaloshin and V. Serebryakov, Z. Phys. C **32**, 279–290 (1986).
[39] M. Pennington, Phys. Rev. Lett. **97**, 011601 (2006).
[40] M. Pennington, T. Mori, S. Uehara, and Y. Watanabe, Eur. Phys. J. C **56**, 1–16 (2008).
[41] G. Mennessier, S. Narison, and W. Ochs, Phys.Lett. B **665**, 205–211 (2008).
[42] G. Mennessier, S. Narison, and X. G. Wang, Phys.Lett. B **696**, 40–50 (2011).
[43] E. van Beveren, F. Kleefeld, G. Rupp, and M. D. Scadron, Phys. Rev. D **79**, 098501 (2009).
[44] M. Scadron, F. Kleefeld, and G. Rupp, Europhys. Lett. **80**, 51001 (2007).
[45] N. H. Fuchs and M. Scadron, Phys. Rev. D **20**, 2421 (1979).
[46] T. Cheng and R. F. Dashen, Phys. Rev. Lett. **26**, 594 (1971).
[47] H. Nielsen and G. Oades, Nucl. Phys. B **72**, 310–320 (1974).

© 2013 WILEY-VCH Verlag GmbH & Co. KGaA, Weinheim

[48] G. Hohler, F. Kaiser, R. Koch, and E. Pietarinen, Handbook Of Pion-Nucleon Scattering, Physics Data 12-1 (Fachinformationszentrum, Karlsruhe, 1979).
[49] G. Hohler, H. Jakob, and R. Strauss, Phys. Lett. B **35**, 445–449 (1971).
[50] M. Olsson and E. Osypowski, J. Phys. G **6**, 423 (1980).
[51] R. Koch, Z. Phys. C **15**, 161–168 (1982).
[52] M. Olsson, Phys. Lett. B **482**, 50–56 (2000).
[53] M. Scadron, PiN Newslett. **13**, 362–366 (1997).
[54] S. Cabasino et al., Phys. Lett. B **258**, 195–201 (1991).
[55] S. Cabasino et al., Phys. Lett. B **258**, 202–206 (1991).
[56] M. Gell-Mann, R. Oakes, and B. Renner, Phys. Rev. **175**, 2195–2199 (1968).
[57] G. Clement, M. D. Scadron, and J. Stern, Z. Phys. C **60**, 307–310 (1993).
[58] G. Clement, M. D. Scadron, and J. Stern, J. Phys. G **17**, 199–204 (1991).
[59] S. A. Coon and M. D. Scadron, J. Phys. G **18**, 1923–1932 (1992).
[60] E. van Beveren and G. Rupp, arXiv:1202.1739 [hep-ph].
[61] M. Scadron, Mod. Phys. Lett. A **7**, 669–676 (1992).
[62] H. Leutwyler, "Nonperturbative methods", in: 26th Intern. Conf. on High Energy Physics, (Dallas, 6–12 Aug. 1992), pp. 185–211, Rapporteur Talk.
[63] J. Gasser, H. Leutwyler, and M. Sainio, Phys. Lett. B **253**, 252–259 (1991).
[64] J. Gasser, H. Leutwyler, and M. Sainio, Phys. Lett. B **253**, 260–264 (1991).
[65] J. Alarcon, J. Martin Camalich, and J. Oller, Phys. Rev. D **85**, 051503 (2012).
[66] R. D. Young, J. Roche, R. D. Carlini, and A. W. Thomas, Phys. Rev. Lett. **97**, 102002 (2006).
[67] D. B. Leinweber, S. Boinepalli, A. W. Thomas, P. Wang, A. G. Williams et al., Phys. Rev. Lett. **97**, 022001 (2006).
[68] R. Tarrach, Z. Phys. C **2**, 221–223 (1979).
[69] S. Gerasimov, Yad.Fiz. **29**, 513–522 (1979), Erratum: ibid. (1980), Sov. J. Nucl. Phys. **32**, 156.
[70] V. Bernard, B. Hiller, and W. Weise, Phys. Lett. B **205**, 16 (1988).
[71] J. Sakurai, Annals Phys. **11**, 1–48 (1960).
[72] M. D. Scadron, F. Kleefeld, G. Rupp, and E. van Beveren, Nucl. Phys. A **724**, 391–409 (2003).
[73] D. Bailin, J. Cleymans, and M. Scadron, Phys. Rev. D **31**, 164 (1985).
[74] N. Bilic, J. Cleymans, and M. Scadron, Int. J. Mod. Phys. A **10**, 1169–1180 (1995).
[75] J. Cleymans, A. Kocic, and M. Scadron, Phys. Rev. D **39**, 323–328 (1989).
[76] F. Karsch, E. Laermann, and A. Peikert, Nucl. Phys. B **605**, 579–599 (2001).
[77] Z. Fodor and S. Katz, J. High Energy Phys. **0404**, 050 (2004).
[78] P. Petreczky, J. Phys. G **30**, S1259–S1262 (2004).
[79] C. Bernard et al., Phys. Rev. D **71**, 034504 (2005).
[80] A. Ali Khan et al., Phys. Rev. D **63**, 034502 (2001).
[81] Y. Nakamura, V. Bornyakov, M. Chernodub, Y. Mori, S. Morozov et al., AIP Conf. Proc. **756**, 242–244 (2005).
[82] R. Delbourgo and M. Scadron, Int. J. Mod. Phys. A **13**, 657 (1998).
[83] R. Delbourgo and M. Scadron, J. Phys. G **5**, 1621 (1979).
[84] J. Gasser and H. Leutwyler, Phys. Rept. **87**, 77–169 (1982).
[85] S. Descotes-Genon, L. Girlanda, and J. Stern, J. High Energy Phys. **0001**, 041 (2000).
[86] H. Sazdjian and J. Stern, Nucl. Phys. B **94**, 163 (1975).
[87] E. Aitala et al., Phys. Rev. Lett. **89**, 121801 (2002).
[88] E. van Beveren, T. Rijken, K. Metzger, C. Dullemond, G. Rupp et al., Z. Phys. C **30**, 615–620 (1986).
[89] E. van Beveren and G. Rupp, Eur. Phys. J. C **22**, 493–501 (2001).
[90] E. van Beveren, D. Bugg, F. Kleefeld, and G. Rupp, Phys. Lett. B **641**, 265–271 (2006).
[91] M. Scadron, Phys. Rev. D **26**, 239–247 (1982).
[92] R. Delbourgo and M. Scadron, Phys. Rev. Lett. **48**, 379–382 (1982).
[93] H. Jones and M. Scadron, Nucl. Phys. B **155**, 409 (1979).
[94] M. Scadron, Phys. Rev. D **29**, 2076 (1984).
[95] D. Klabucar, D. Kekez, and M. D. Scadron, J. Phys. G **27**, 1775–1784 (2001).
[96] A. Bramon and M. Scadron, Phys. Lett. B **234**, 346 (1990).

 © 2013 WILEY-VCH Verlag GmbH & Co. KGaA, Weinheim

[97] A. Bramon, R. Escribano, and M. Scadron, Eur. Phys. J. C **7**, 271–278 (1999).
[98] T. Feldmann, P. Kroll, and B. Stech, Phys. Rev. D **58**, 114006 (1998).
[99] T. Feldmann, Int. J. Mod. Phys. A **15**, 159–207 (2000).
[100] S. L. Adler, Phys. Rev. **137**, B1022–B1033 (1965).
[101] S. L. Adler, Phys. Rev. **139**, B1638–B1643 (1965).
[102] R. Delbourgo, D. S. Liu, and M. D. Scadron, Phys. Lett. B **446**, 332–335 (1999).
[103] R. Delbourgo, D. S. Liu, and M. Scadron, Int. J. Mod. Phys. A **14**, 4331–4346 (1999).
[104] M. D. Scadron, G. Rupp, F. Kleefeld, and E. van Beveren, Phys. Rev. D **69**, 014010 (2004), Erratum: ibid (2004), **69**, 059901.
[105] M. Schumacher and M. D. Scadron, Fortschr. Phys. **61**, 703 (2013).
[106] F. Kleefeld, E. van Beveren, G. Rupp, and M. D. Scadron, Phys. Rev. D **66**, 034007 (2002).
[107] J. Oller and E. Oset, Phys. Rev. D **60**, 074023 (1999).
[108] A. Astier, L. Montanet, M. Baubillier, and J. Duboc, Phys. Lett. B **25**, 294 (1967).
[109] V. Anisovich, V. Nikonov, and A. Sarantsev, Phys. Atom. Nucl. **65**, 1545–1552 (2002).
[110] E. Aitala et al., Phys. Rev. Lett. **86**, 765–769 (2001).
[111] E. van Beveren, G. Rupp, and M. D. Scadron, Phys. Lett. B **495**, 300–302 (2000), Erratum: ibid (2001), **509**, 365.
[112] R. Delbourgo and M. Scadron, Lett.Nuovo Cim. **44**, 193–198 (1985).
[113] S. Glashow, J. Iliopoulos, and L. Maiani, Phys. Rev. D **2**, 1285–1292 (1970).
[114] M. Scadron and V. Elias, Mod. Phys. Lett. A **10**, 1159–1168 (1995).
[115] M. D. Scadron, G. Rupp, and E. van Beveren, Mod. Phys. Lett. A **19**, 2267–2278 (2004).
[116] J. Lowe and M. Scadron, Mod. Phys. Lett. A **17**, 2497–2512 (2002).
[117] M. Scadron, Phys. Lett. B **95**, 123–127 (1980).
[118] R. Karlsen and M. Scadron, Phys. Rev. D **45**, 4108–4112 (1992).
[119] S. Choudhury and M. Scadron, Phys. Rev. D **53**, 2421–2429 (1996).
[120] S. Weinberg, Phys. Rev. D **8**, 605–625 (1973).
[121] S. Weinberg, Phys. Rev. D **8**, 4482–4498 (1973).
[122] B. McKellar and M. Scadron, Phys. Rev. D **27**, 157 (1983).
[123] M. D. Scadron, F. Kleefeld, G. Rupp, and E. van Beveren, AIP Conf. Proc. **660**, 311–324 (2003).
[124] N. Paver and M. Scadron, Nuovo Cim. A **78**, 159–171 (1983).
[125] L. Ametller, C. Ayala, and A. Bramon, Phys. Rev. D **29**, 916 (1984).
[126] S. Egli et al., Phys.Lett. B **222**, 533 (1989).
[127] S. Egli et al., Phys.Lett. B **175**, 97 (1986).
[128] V. Vaks and B. Ioffe, Nuovo Cim. **10**, 342 (1958).
[129] A. Bramon and M. Scadron, Europhys. Lett. **19**, 663–667 (1992).
[130] M. Bychkov, D. Pocanic, B. VanDevender, V. Baranov, W. H. Bertl et al., Phys.Rev.Lett. **103**, 051802 (2009).
[131] M. Scadron, Phys. Atom. Nucl. **56**, 1595–1603 (1993).
[132] Y. Antipov, V. Batarin, V. Bezzubov, N. Budanov, Y. Gorin et al., Z. Phys. C **26**, 495 (1985).
[133] L. Fil'kov and V. Kashevarov, Phys. Rev. C **73**, 035210 (2006).
[134] B. Hiller, W. Broniowski, A. A. Osipov, and A. H. Blin, Phys.Lett. B **681**, 147–150 (2009).
[135] L. Fil'kov and V. Kashevarov, PoS C D **09**, 036 (2009).
[136] M. Schumacher, Eur. Phys. J. A **30**, 413–422 (2006), Erratum: ibid (2007), **32**, 121.
[137] M. Schumacher, AIP Conf. Proc. **1030**, 129–134 (2008).
[138] M. Schumacher, Eur. Phys. J. C **67**, 283–293 (2010).
[139] M. Ademollo and R. Gatto, Phys. Rev. Lett. **13**, 264–265 (1964).
[140] N. Paver and M. Scadron, Phys. Rev. D **30**, 1988 (1984).
[141] B. Green and M. Scadron, Physica B **305**, 175 (2001).
[142] R. Sooryakumar and M. Klein, Phys. Rev. Lett. **45**, 660 (1980).
[143] P. Littlewood and C. Varma, Phys. Rev. Lett. **47**, 811 (1981).
[144] R. Kiefl, W. MacFarlane, K. Chow, S. Dunsiger, T. Duty, T. Johnston, J. i Schneider, J. Sonier, L. Brard, R. Strongin, J. Fischer, and A. SmithIII., Phys. Rev. Lett. **70**, 3987 (1993).
[145] Y. Nambu, Phys. Rev. **117**, 648–663 (1960).

© 2013 WILEY-VCH Verlag GmbH & Co. KGaA, Weinheim